Microbial Zoonoses

Series Editor
Subhash Chandra Parija, Sri Balaji Vidyapeeth, A Deemed-to-be Un
Pondicherry, India

The series on Microbial Zoonoses provides a holistic overview of various zoonotic pathogens including parasitic, fungal, bacterial, and viral. The individual volume provides conceptual knowledge and recent advancements on the various aspects of these pathogens, infectious disease caused by these organisms, control strategies, and their public health importance.

This book series discusses the pathogen life-cycle, disease spectrum and their impact on human health, modern diagnostic methods, treatment, host-pathogen interactions, and the importance of public health control measures including prophylactic measures in human and animal hosts.

Special emphasis on this series is on the immunological response to these pathogens, molecular epidemiology, and genomics, novel diagnostic techniques, and antimicrobial resistance in zoonotic pathogens.

Each book consists of two parts. Part A incorporates section which deals with chapters on the nomenclature of parasites and classification of zoonoses, life-cycle of the organism, pathogenesis, pathology, diagnosis, epidemiology and their clinical manifestations.

Part B of the book includes chapters that summarize the host-pathogen interactions, the immunological response of the host, new vaccines and novel treatment regimens against the pathogens. It also discusses strategies for control and prevention of microbial zoonoses.

Given its scope, the individual books in this book series is of interest to professionals, scientists, and postgraduate students in medicine, veterinary medicine, public health, and diagnostic laboratories.

Subhash Chandra Parija
Abhijit Poddar
Editors

Textbook of Bacterial Zoonoses

Springer

Editors
Subhash Chandra Parija
Formerly at Sri Balaji Vidyapeeth
(Deemed to be University)
Pondicherry, India

Abhijit Poddar
Centre for Bio-Policy Research (CBPR)
Mahatma Gandhi Medical Advanced
Research Institute, Sri Balaji Vidyapeeth
(Deemed to be University)
Pondicherry, India

ISSN 2662-5288 ISSN 2662-5296 (electronic)
Microbial Zoonoses
ISBN 978-981-95-6614-3 ISBN 978-981-95-6615-0 (eBook)
https://doi.org/10.1007/978-981-95-6615-0

This Springer imprint is published by the registered company Springer Nature Singapore Pte Ltd.
The registered company address is: 152 Beach Road, #21-01/04 Gateway East, Singapore 189721, Singapore

Lotus Feet of Divine, the Lord Jagannath, Lord Subhadra, and Lord Balabhadra
Supreme Power, who has willed and given us the strength to take up this work.
My Father
Late Managovinda Parija
Mother
Late Nishamani Parija
Wife Jyotirmayee Parija
and
My professional colleagues and mentors, for their constant guidance
Subhash Chandra Parija
My family, friends, and mentors
for their unwavering support and belief in me.
Abhijit Poddar

Preface

The enduring relationship between humans and the animal kingdom has considerably influenced our collective evolution. From our early interactions with animals as hunter-gatherers to the deliberate domestication of specific species, this profound bond has facilitated pathways for microbial transmission. Consequently, the transmission of pathogenic bacteria from animals to humans, resulting in zoonotic diseases, represents a phenomenon as ancient as human civilisation itself. Approximately 60% of newly emerging infectious diseases are zoonotic, with 70% of these arising from wildlife sources.

Currently, bacterial zoonoses pose a significant global public health challenge, affecting both human and animal populations on various levels. This field of bacterial zoonoses is both dynamic and constantly changing. From the well-known and widespread threats of salmonellosis, brucellosis, leptospirosis, and zoonotic tuberculosis to the emergence of pathogens such as *Coxiella burnetii* and the urgent concern of antimicrobial-resistant (AMR) strains across diverse bacterial groups, the situation is worsening. Additionally, several lesser-known bacterial zoonoses, including erysipeloid, dermatophilosis, and various Rickettsial infections such as scrub typhus and murine typhus, continue to impose significant burdens, especially on vulnerable groups that often suffer from insufficient surveillance and control efforts. These diseases are particularly common in developing regions, yet they remain a noticeable presence even in highly industrialised countries.

In the context of human health, bacterial zoonoses substantially contribute to morbidity and mortality, exerting considerable pressure on healthcare infrastructures and individual health outcomes. The range of illnesses includes mild, self-limiting conditions as well as severe, life-threatening infections, often requiring prolonged treatment and recovery periods. In animal populations, these diseases cause considerable suffering, decreased productivity in livestock such as reduced milk or meat yield and reproductive failures, and significant economic impacts on farmers and the agricultural sector. Additionally, they pose risks to the health and conservation of wildlife species.

The emergence and resurgence of bacterial zoonoses are increasingly associated with complex ecological, economic, and social transitions. Factors such as climate change, deforestation, intensification of agriculture, urban sprawl, and the expansion of global trade have collectively heightened interactions among humans, animals, and the environment. In this context, the 'One Health' approach—now officially endorsed by the WHO, FAO, WOAH, and UNEP—has emerged as an essential framework for understanding, preventing, and responding to zoonotic threats. The One Health concept emphasises the interconnectedness of human, animal, and environmental health and highlights the necessity for interdisciplinary collaboration in addressing such diseases.

Given this complex and changing landscape, a comprehensive resource that synthesises the latest scientific knowledge while balancing detailed information with accessible presentation is urgently needed. It requires a truly multidisciplinary approach that integrates biological, epidemiological, clinical, diagnostic, and control aspects. This *Textbook of Bacterial Zoonoses* aims to fill that crucial gap. We have developed it as an essential resource for individuals working across various disciplines who are interested in zoonotic bacterial diseases. Our goal is to provide a foundational understanding for both undergraduate and postgraduate students. Additionally, this book will serve as a key reference for faculty, public health practitioners, researchers, and policymakers within the fields of medicine, veterinary science, public health, and related health professions.

We extend our sincere appreciation to all contributing authors whose expertise and unwavering dedication were essential in bringing this ambitious project to fruition. Our publishing partner, Springer Nature, has consistently demonstrated exceptional support and innovation throughout the development of this text. We sincerely hope this volume will reach a wide international audience, offering invaluable insights to all who are dedicated to understanding and addressing bacterial zoonoses.

Pondicherry, India Subhash Chandra Parija
Pondicherry, India Abhijit Poddar

Acknowledgements

We are pleased to convey our sincere appreciation to all individuals whose contributions have rendered this textbook possible. The authors and co-authors of the chapters have demonstrated exceptional commitment and scholarly rigor throughout this endeavour. Despite their numerous professional commitments, they submitted high-quality chapters punctually, and we extend our heartfelt gratitude for their dedication. We wish to record our appreciation and gratitude to Professor Dr Abhijit Chaudhury, Tirupathi, in particular for his valuable guidance and advices.

Dr Subhash Chandra Parija expresses his sincere gratitude to his family members—late Kailash Chandra Parija, Er Kukumina Ray, Er Subhasis Ray, Er Rajkumar Parija, Smrithi Parija, Dr Madhuri Parija, Dr Ajay Halder, Er Mayuri Parija, Er Shailesh Nandan, and grandchildren Sri Harihar, Sri Ram, and Shyama, and nephew Sidhartha Das—for their continuous encouragement, patience, and unwavering support during the preparation of this book.

Dr Abhijit Poddar is deeply grateful to his entire family, friends, and mentor, Dr S.R. Rao, for their unwavering support and strength. Sanchita and Indira have been a constant source of inspiration throughout the development of this academic pursuit.

Finally, we sincerely thank our publishing partner, Springer Nature, whose professionalism and enthusiasm for this project helped guide the book through all production stages. Their ongoing dedication to promoting academic publishing is truly appreciated.

Pondicherry, India — Subhash Chandra Parija
Pondicherry, India — Abhijit Poddar

About the Book

Genesis

Bacterial infections continue to constitute a substantial portion of the global infectious disease burden, often transcending socio-economic and geographic boundaries. Although there is considerable international attention directed towards prominent bacterial diseases such as tuberculosis and cholera, numerous bacterial zoonoses remain insufficiently recognised, underreported, and frequently misunderstood. This lack of clarity not only distorts our comprehension of their actual prevalence but also impedes prompt and effective public health interventions.

Since 1940, over 60% of the approximately 400 emerging or re-emerging infectious diseases identified have been zoonotic in origin, with a considerable proportion attributable to bacteria. Bacterial zoonoses such as brucellosis, leptospirosis, anthrax, Q fever, and rickettsioses represent only a fraction of the broader issue. Numerous other diseases remain neglected despite their significant impact on human and animal health, particularly in regions characterised by intensive livestock farming, wet markets, close contact with wildlife, or widespread antimicrobial misuse. These infections frequently traverse the human-animal-environment interface, underscoring the importance of early detection, precise diagnosis, and comprehensive control strategies.

Unlike viral or parasitic diseases, numerous bacterial zoonoses remain susceptible to treatment with antibiotics; however, the escalating problem of antimicrobial resistance (AMR) has complicated this paradigm. Bacterial zoonotic pathogens are increasingly discovered to carry multi-drug resistance genes, which are occasionally shared across different species and national boundaries. This concerning trend endangers public health and food security, and necessitates immediate international collaboration.

Although bacteriology is well represented within the discipline of human microbiology, zoonotic bacterial infections often remain peripheral in medical education and policy discussions. In contrast, veterinary medicine continues to emphasise zoonotic bacteria given their dual importance for animal and public health. This disciplinary divide is increasingly seen as a limitation in effectively addressing shared health challenges.

This *Textbook of Bacterial Zoonoses* endeavours to bridge the existing gap by offering a comprehensive and integrated understanding of the diagnosis, epidemiology, treatment, prevention, and control of bacterial zoonotic infections. It is meticulously designed to serve as a foundational resource for undergraduate and postgraduate students in medicine, veterinary and allied sciences, public health professionals, policymakers, and all individuals engaged in the management of infectious diseases. This volume represents the culmination of a collective effort by over 90 dedicated contributors, each providing their expertise to address this multidisciplinary and urgent public health issue.

The Content Organisation

This textbook comprises three sections that present a logical flow of information on the strong foundation of bacterial zoonoses and discuss the specifics of various bacterial zoonotic infections. The content begins with fundamental bacteriological principles, moves through detailed disease-specific descriptions, and concludes with future-oriented perspectives relevant to emerging zoonotic threats and public health preparedness.

Part I: General Bacteriology provides the foundational framework necessary to understand bacterial zoonoses in depth. This section covers essential topics, including bacterial taxonomy, ecology, host–pathogen relationships, and bacterial immunology, which form the foundation for understanding the complex interactions involved in zoonotic infections. Recognising the increasing importance of technology in biological sciences, we have incorporated chapters on bacterial genomics and proteomics, which are rapidly transforming our understanding of pathogen biology, diagnostics, and surveillance. In addition, this part offers insights into the epidemiology of bacterial zoonoses, diagnostic approaches, including both conventional and molecular methods, and a comprehensive overview of antimicrobial resistance, which continues to complicate the treatment of many bacterial infections. A concluding chapter focuses on prevention, surveillance, and control strategies, reflecting the One Health approach that connects human, animal, and environmental health in the management of zoonotic diseases.

Part II: Zoonotic Bacterial Infections constitutes the core of the textbook and offers a systematic classification of zoonotic diseases based on bacterial taxonomy and clinical significance. The section is methodically divided into seven subsections, each addressing a group of pathogens that exhibit shared microbiological or clinical characteristics. The initial group encompasses zoonotic Gram-positive bacterial infections, including *Staphylococcus*, *Streptococcus*, *Bacillus anthracis*, *Listeria monocytogenes*, among others, that hold importance in both public and veterinary health. The subsequent group examines Gram-negative bacterial infections such as *Brucella*, *Salmonella*, *Escherichia coli* (EHEC), *Yersinia*, *Pasteurella*, and *Campylobacter*, which are frequently associated with food-borne illnesses and outbreaks. Mycobacterial infections, comprising *Mycobacterium tuberculosis* Complex and non-tuberculous mycobacteria, are addressed in a dedicated section owing to their distinctive diagnostic and epidemiological challenges.

The subsequent sections explore spirochaetal zoonoses, including leptospirosis, Lyme disease, and relapsing fever, followed by a comprehensive account of rickettsial infections, such as scrub typhus, spotted fevers, and typhus group infections, among others. Obligate intracellular bacterial infections, such as Q fever, psittacosis, bartonellosis, and ehrlichiosis, are presented separately to emphasise their complex life cycles and diagnostic challenges. Lastly, a segment on minor yet clinically significant zoonotic infections introduces less frequently encountered bacteria such as *Capnocytophaga*, *Trueperella*, *Arcobacter*, and *Corynebacterium*, which are increasingly recognised due to changes in animal exposure, diagnostics, and susceptibility of immunocompromised hosts. Each chapter within this section adheres to a uniform structure encompassing taxonomy, morphology, cultivation, genomics, pathogenesis, epidemiology, clinical manifestations in humans and animals, diagnostics, treatment, and control strategies.

Part III: Future Perspectives in Bacterial Zoonoses examines the broader implications and contemporary challenges associated with zoonotic bacterial infections. It commences with a chapter on the human and animal bacteriome, investigating the role of microbial communities in health, immunity, and disease aetiology. A dedicated chapter on biosafety practices underscores the significance of laboratory containment, personal protection, and outbreak preparedness, especially for professionals handling high-risk bacterial pathogens. This section also considers the influence of geoclimatic changes on disease emergence and the evolving distribution of bacterial zoonoses. The concluding chapter addresses the economic impact of bacterial zoonotic infections, encompassing both direct healthcare expenditures and indirect losses in agriculture, trade, and livelihood security, thereby linking the burden of infectious diseases with socio-economic development.

The Value Addition

Alongside the standard features of a textbook, we have introduced three key value additions throughout the chapters of this volume to enhance student engagement, clinical application, and research focus. These have been carefully integrated to support critical thinking, practical learning, and academic exploration in the field of bacterial zoonoses.

The first added value is a case study presented in each chapter. These case scenarios are inspired by real-life clinical or field situations and are included to encourage analytical thinking. They offer a platform for students to apply theoretical knowledge to practical settings, ranging from clinical diagnosis and treatment to outbreak management and field-level decision-making. Each case is followed by a discussion that guides learners through clinical reasoning, diagnostic strategies, and interpretation of laboratory results. This interactive approach aims to make learning more reflective, engaging, and practice-oriented.

The second value addition is a curated list of unanswered research questions related to each bacterial pathogen or disease. These research prompts aim to encourage curiosity and act as a starting point for future investigations by postgraduate students, young researchers, and early-career academics. The questions encompass a broad range—diagnostic limitations, pathogenesis, host-pathogen interactions, antimicrobial resistance mechanisms, vaccine development, and One Health–based intervention models.

The third value addition in this textbook is an in-depth focus on diagnostics. Given the crucial role of accurate and timely diagnosis in managing and controlling zoonotic bacterial infections, each chapter includes a dedicated and detailed section on diagnostic methods. This covers conventional bacteriological techniques (culture, staining, biochemical identification), serological assays, molecular tools (PCR, real-time PCR, isothermal amplification), and modern advances such as MALDI-TOF and point-of-care technologies.

The fourth and unique value addition is the inclusion of Part III: Future Perspectives in Bacterial Zoonoses. This dedicated section expands the scope of the textbook beyond disease description and management, encouraging readers to consider the broader ecological, technological, and socio-economic aspects of zoonotic bacterial diseases. It features chapters on the role of the bacteriome in health and disease, laboratory biosafety and preparedness, the effects of climate and geoclimatic change on disease emergence, and the economic impacts of bacterial zoonotic infections on health systems, agriculture, and global trade. These forward-looking themes aim to provide students and professionals with a comprehensive and anticipatory understanding of zoonotic threats in the twenty-first century.

In summary, the *Textbook of Bacterial Zoonoses* has been carefully curated to serve as a comprehensive, up-to-date, and interdisciplinary reference for undergraduate and postgraduate students, academic faculty, clinicians, veterinarians, microbiologists, epidemiologists, and public health policymakers. It reflects the urgent need for integrative thinking and coordinated action in managing zoonotic threats within the One Health paradigm, and it is our hope that it will support education, research, and effective disease control efforts both locally and globally.

Our sincere appreciation extends to the contributors, esteemed professionals in microbiology, infectious diseases, veterinary science, and public health, who have graciously shared their expertise to develop this thorough resource. Their commitment to the progress of bacterial zoonotic disease research is manifest in the depth and clarity of their contributions.

As editors, we hope this textbook will inspire readers to confront the challenges of bacterial zoonoses with sufficient knowledge, innovation, and resolve. We encourage you to engage critically with the material, to ask questions, and to contribute to the collective effort to fight these difficult infections.

Wish you the best for a good reading of the book!

Subhash Chandra Parija
Abhijit Poddar

Contents

Editors and Contributors

About the Editors

Subhash Chandra Parija, MBBS, MD, PhD, DSc, FRCPath, was the Vice-Chancellor of Sri Balaji Vidyapeeth, a deemed-to-be university, Pondicherry, India, and is the Former Director of the Jawaharlal Institute of Postgraduate Medical Education and Research (JIPMER), Pondicherry, India. With nearly 45 years of teaching and research experience in the field of medical microbiology, he is a Food and Agriculture Organization (FAO) expert and has been consulted to draft guidelines on preventing foodborne parasites. He was on the MD examination boards at Colombo University, Sri Lanka; Sultan Qaboos University, Oman; and the University of Malaya, Malaysia. He received a DSc from Madras University for his contributions to the field of medical parasitology. He holds 1 granted patent, 1 published patent, 23 copyrights, and 1 technology transfer. He has authored or edited 19 books, including the most popular *Textbook of Medical Microbiology & Immunology*, *Effective Medical Communication: The A, B, C, D, E of It*, *Textbook of Parasitic Zoonoses*, and *Textbook of Fungal Zoonoses and Sapronoses*, all published by Springer. He has published more than 450 papers in reputable national and international journals. He has been honoured with more than 29 awards including the Medical Council of India's Dr. B.C. Roy National Award and the National Academy of Medical Sciences' Dr. P.N. Chuttani Oration Award. He founded the Indian Academy of Tropical Parasitology (IATP), the only professional organisation of medical parasitologists in India.

Abhijit Poddar, PhD, is an experienced policy expert and biotechnologist with over 15 years of experience promoting regulatory reforms in India and fostering South-South and North-South collaboration in health security. He currently serves as an Associate Professor at the MGM Advanced Research Institute, Sri Balaji Vidyapeeth (Deemed to be University), Pondicherry, India. In his current academic role, he has demonstrated significant leadership in institution building. He played a key role in establishing the MGM Advanced Research Institute, a Centre for Bio-Policy Research, a Virus Research and Diagnostic Lab (VRDL), and a Zebrafish facility for preclinical studies. His research interests are diverse, bridging applied wet-lab science with public policy. They include microbial biotechnology, the use of Zebrafish as a preclinical model, and a strong focus on the 'One Health' framework, covering biosecurity, pandemic preparedness, and the governance of emerging technologies relevant to zoonotic disease control. He has supervised PhD and MSc students and played a key role in designing innovative, future-ready courses in Biosecurity, Bio-entrepreneurship, and One Health, pioneering pedagogical methods such as tabletop exercises. He has published extensively, with more than 30 research articles in peer-reviewed journals, including *IJSEM*, *Lancet Global Health*, *Nature Pediatric Research*, and *Health Security*; holds two Indian patents; and has contributed to numerous national policy documents.

Contributors

Kumar S. Abhishek Department of Microbiology, All India Institute of Medical Sciences, Jodhpur, Rajasthan, India

Jacob Abraham Institute of Animal Health and Veterinary Biologicals, Palode, Kerala, India

Jyotsna Agarwal Department of Microbiology, Dr. Ram Manohar Lohia Institute of Medical Sciences, Lucknow, Uttar Pradesh, India

Prasad Andhare Department of Life Sciences, School of Sciences, GSFC University, Vadodara, Gujarat, India

Shampa Anupurba Department of Microbiology, Institute of Medical Sciences, Banaras Hindu University, Varanasi, Uttar Pradesh, India

Mashila Arasan Mahatma Gandhi Medical Advanced Research Institute, Sri Balaji Vidyapeeth (Deemed-to-be-University), Pondicherry, India

Flábio Ribeiro de Araújo Embrapa Beef Cattle, Campo Grande, MS, Brazil

T. R. Arun ICMR-National Institute for One Health, Nagpur, Maharashtra, India

Kokila Banerjee Drs. Tribedi and Roy Diagnostic Laboratory, Kolkata, West Bengal, India

Tuhina Banerjee Department of Microbiology, Institute of Medical Sciences, Banaras Hindu University, Varanasi, Uttar Pradesh, India

S. B. Barbuddhe ICAR-National Meat Research Institute, Hyderabad, Telangana, India

Barnamoy Bhattacharjee Department of Microbiology, Silchar Medical College & Hospital, Silchar, Assam, India

Sanjay Bhattacharya Department of Microbiology, Tata Medical Center, Kolkata, West Bengal, India

Malabika Biswas Department of Microbiology, Tata Medical Center, Kolkata, West Bengal, India

Rakhee Biswas Department of Microbiology, Jawaharlal Institute of Postgraduate Medical Education & Research, Pondicherry, India

Fatih Buyuk Department of Microbiology, Faculty of Veterinary Medicine, Kafkas University, Kars, Turkey

Debadatta Dhar Chanda Department of Microbiology, Silchar Medical College & Hospital, Silchar, Assam, India

Abhijit Chaudhury Department of Microbiology, Sri Venkateswara Institute of Medical Sciences University, Tirupati, Andhra Pradesh, India

Sandhya Chouhan Department of Biomedical Engineering, Central University of Rajasthan, Ajmer, Rajasthan, India

School of Medical Science and Technology, Indian Institute of Technology, Kharagpur, West Bengal, India

Subarnarekha Chowdhury Department of Biophysics, All India Institute of Medical Sciences, New Delhi, India

Suprativ Das School of Bio Sciences & Technology, Vellore Institute of Technology, Vellore, Tamil Nadu, India

Surojit Das Department of Biomedical Laboratory Science and Management, Vidyasagar University, Midnapore, West Bengal, India

Sangeeta Deka Department of Microbiology, Nagaon Medical College & Hospital, Nagaon, Assam, India

Yitayew Demessie Department of Veterinary Biomedical Sciences, College of Veterinary Medicine and Animal Sciences, University of Gondar, Gondar, Ethiopia

Pankaj Dhaka Centre for One Health, College of Veterinary Science, Guru Angad Dev Veterinary and Animal Sciences University, Ludhiana, Punjab, India

Yaren Ersoy Department of Microbiology, Faculty of Veterinary Medicine, Kafkas University, Kars, Turkey

George C. Fthenakis Veterinary Faculty, University of Thessaly, Karditsa, Greece

Shradha R. Gaonkar Department of Microbiology, Royal Hospital, Margao, Goa, India

Emanuelle Baldo Gaspar Embrapa Dairy Cattle, Juiz de Fora, MG, Brazil

Tapashi Ghosh School of Tropical Medicine, Kolkata, West Bengal, India

Ujjala Ghoshal Department of Microbiology, All India Institute of Medical Sciences, Kalyani, West Bengal, India

Aditi Gupta Department of Microbiology, All India Institute of Medical Sciences, Jodhpur, Rajasthan, India

Akanksha Gupta Department of Microbiology, Dr. Ram Manohar Lohia Institute of Medical Sciences, Lucknow, Uttar Pradesh, India

Vangara Harika Department of Microbiology, Sri Venkateswara Institute of Medical Sciences University, Tirupati, Andhra Pradesh, India

Amita Jain All India Institute of Medical Sciences, Raebareli, Uttar Pradesh, India

Susanthika Jayachandiran Mahatma Gandhi Medical Advanced Research Institute (MGMARI), Sri Balaji Vidyapeeth (Deemed to be University), Pondicherry, India

Pradeep Jothimani Mahatma Gandhi Medical Advanced Research Institute, Sri Balaji Vidyapeeth (Deemed-to-be-University), Pondicherry, India

Deepjyoti Kalita Department of Microbiology, All India Institute of Medical Sciences, Guwahati, Assam, India

Derya Karatas Yeni Department of Microbiology, Faculty of Veterinary Medicine, Necmettin Erbakan University, Konya, Turkey

Kesavarajan Govindarajan Karthikeyan Department of Microbiology, Sri Lalithambigai Medical College and Hospital, Faculty of Medicine, Dr. MGR Educational and Research Institute, Maduravoyal, Adayalampattu, Chennai, Tamil Nadu, India

Angeliki I. Katsafadou Faculty of Public and One Health, University of Thessaly, Karditsa, Greece

Department of Environmental Health Sciences, Yale School of Public Health, New Haven, CT, USA

Archana Keche Department of Microbiology, All India Institute of Medical Sciences, Raipur, Chhattisgarh, India

Yogendra Keche Department of Pharmacology, All India Institute of Medical Sciences, Raipur, Chhattisgarh, India

Prathyusha Kokkayil Department of Microbiology, All India Institute of Medical Sciences, Patna, Bihar, India

Mukesh Kumar Department of Biophysics, All India Institute of Medical Sciences, New Delhi, India

School of Allied Health Sciences, Galgotias University, Greater Noida, USA

Mahalakshmi Kumaresan Department of Microbiology, Jawaharlal Institute of Postgraduate Medical Education & Research (JIPMER), Pondicherry, India

Namrata Kumari Department of Microbiology, Indira Gandhi Institute of Medical Sciences, Patna, Bihar, India

Aditya Kundu Department of Microbiology, All India Institute of Medical Sciences, Kalyani, West Bengal, India

Geetha Lakshmi Dr. M.G.R. Educational and Research Institute, Chennai, Tamil Nadu, India

Jagadish Mahanta ICMR-Regional Medical Research Centre for Northeast, Dibrugarh, Assam, India

Ashoka Mahapatra Department of Microbiology, All India Institute of Medical Sciences, Bhubaneswar, Odisha, India

Sourav Maiti Diagnostic Division, Consortium of Accredited Healthcare Organizations (CAHO), Kolkata, West Bengal, India

Diptanu Majumder Department of Microbiology, All India Institute of Medical Sciences, Kalyani, Nadia, West Bengal, India

Sudipto Mandal Department of Microbiology, Ramakrishna Mission Vidyamandira, Howrah, West Bengal, India

Sefinew Alemu Mekonnen Department of Veterinary Epidemiology and Public Health, College of Veterinary Medicine and Animal Sciences, University of Gondar, Gondar, Ethiopia

Shravani Mitra Formerly at Division of Bacteriology, ICMR-National Institute of Cholera and Enteric Diseases, Kolkata, West Bengal, India

Srujana Mohanty Department of Microbiology, All India Institute of Medical Sciences, Bhubaneswar, Odisha, India

Tribhuban Mohan Mohapatra Institute of Medical Sciences, Banaras Hindu University, Varanasi, Uttar Pradesh, India

Subham Mookerjee Department of Zoology, Ramakrishna Mission Vivekananda Centenary College, Khardaha, West Bengal, India

Anirban Mukherjee Department of Microbiology, Gurudas College, Calcutta University, Kolkata, West Bengal, India

Rajeshwar V. Naik Department of Internal Medicine, Royal Hospital, Margao, Goa, India

Manoj Nair Institute of Animal Health and Veterinary Biologicals, Palode, Kerala, India

P. Niveditha ICAR-National Meat Research Institute, Hyderabad, Telangana, India

Vignan Foundation for Science, Technology and Research, Guntur, Andhra Pradesh, India

Mahendra Pal Narayan Consultancy of Veterinary Public Health and Microbiology, Bharuch, Gujarat, India

Subhash Chandra Parija Former Vice Chancellor, Sri Balaji Vidyapeeth (Deemed to be University), Pondicherry, India

Girish Patil Department of Microbiology, All India Institute of Medical Sciences, Raipur, Chhattisgarh, India

Paula Adas Pereira Suniga IAGRO – State Agency for Animal and Plant Health Defence, Água Clara, MS, Brazil

Abhijit Poddar Mahatma Gandhi Medical Advanced Research Institute, Sri Balaji Vidyapeeth (Deemed to be University), Pondicherry, India

Sakhare Geetanjali Pramod Department of Microbiology, Tata Medical Center, Kolkata, West Bengal, India

Bharath Prithiviraj Reckitt Health US, Microbiome Science Platforms, Montvale, NJ, USA

Anil Jacob Purty Department of Community Medicine, Pondicherry Institute of Medical Sciences, Pondicherry, India

Nonika Rajkumari Department of Microbiology, All India Institute of Medical Sciences, Kalyani, Nadia, West Bengal, India

Vikrant Singh Rajput Department of Biomedical Engineering, Central University of Rajasthan, Ajmer, Rajasthan, India

Dhamodharan Ramasamy Mahatma Gandhi Medical Advanced Research Institute (MGMARI), Sri Balaji Vidyapeeth (Deemed to be University), Pondicherry, India

Deepak B. Rawool ICAR-National Meat Research Institute, Hyderabad, Telangana, India

Lenita Ramires dos Santos Embrapa Beef Cattle, Campo Grande, MS, Brazil

Mallika Sen Gupta Department of Microbiology, All India Institute of Medical Sciences, Kalyani, West Bengal, India

Ajanta Sharma Department of Microbiology, Gauhati Medical College & Hospital, Guwahati, Assam, India

Shashikala Department of Clinical Microbiology, Pondicherry Institute of Medical Sciences, Pondicherry, India

Shilpshri V. Shinde Department of Biotechnology, Nagpur Veterinary College, Nagpur, Maharashtra, India

Sumukh Shourya Department of General Medicine, Bharati Vidyapeeth Medical College, Pune, Maharashtra, India

Saurabh RamBihariLal Shrivastava Datta Meghe Medical College, Datta Meghe Institute of Higher Education and Research, Nagpur, Maharashtra, India

Mukesh Kumar Department of Biophysics, All India Institute of Medical Sciences, New Delhi, India

Rakesh Singh Department of Microbiology, Jawaharlal Institute of Postgraduate Medical Education & Research (JIPMER), Pondicherry, India

Kuntal Kumar Sinha Department of Microbiology, All India Institute of Medical Sciences-Jodhpur, Jodhpur, Rajasthan, India

Sumetha Suga Deiva Suga Department of Microbiology, A.C.S. Medical College and Hospital, Dr M.G.R Educational and Research Institute, Chennai, Tamil Nadu, India

Sukanya Sugumaran Department of Microbiology, ACS Medical College & Hospital, Dr M.G.R Educational and Research Institute, Chennai, Tamil Nadu, India

Roja Suresh Mahatma Gandhi Medical Advanced Research Institute (MGMARI), Sri Balaji Vidyapeeth (Deemed to be University), Pondicherry, India

Vibhor Tak Department of Microbiology, All India Institute of Medical Sciences-Jodhpur, Jodhpur, Rajasthan, India

Bhaskar Thakuria Department of Microbiology, All India Institute of Medical Sciences, Patna, Bihar, India

George Th. Tsangaris Proteomics Research Unit, Biomedical Research Foundation of Academy of Athens, Athens, Greece

Dhwani Upadhyay Department of Life Sciences, Parul Institute of Applied Sciences, Parul University, Vadodara, Gujarat, India

Natarajan Vaghesan Department of Microbiology, Sri Lalithambigai Medical College and Hospital, Faculty of Medicine, Dr. MGR Educational and Research Institute, Maduravoyal, Adayalampattu, Chennai, Tamil Nadu, India

Balamurali Venkatesan Department of Microbiology, Sri Lalithambigai Medical College and Hospital, Faculty of Medicine, Dr. M.G.R. Educational and Research Institute, Adayalampattu, Chennai, Tamil Nadu, India

Deepthi Vijay Centre for One Health, College of Veterinary Science, Guru Angad Dev Veterinary and Animal Sciences University, Ludhiana, Punjab, India

Kumar Vikram Department of Microbiology, ESIC Medical College and Hospital, Joka, Kolkata, West Bengal, India

Part I

General Bacteriology

1 Bacterial Taxonomy

Abhijit Chaudhury and Subhash Chandra Parija

Learning Objectives

1. To understand the detailed and updated classification of bacteria
2. To learn the fundamental principles underlying bacterial classification
3. To become familiar with modern taxonomic tools and their impact on the reclassification of certain bacteria

Introduction

The term "taxonomy" is derived from the Greek *taxis* ("arrangement") and *nomos* ("law"), and it is the science of the systematics of all living organisms. The aim of taxonomy is to group organisms based on common factors to construct a hierarchical structure. It can be defined as "the science that manages the orderly arrangement of living things. It follows a hierarchical system that attempts to account for evolutionary relationships. The main goal is to establish a natural system that accurately reflects the organization found in nature." The trinity of taxonomy comprises classification, nomenclature, and identification. *Classification* is the orderly arrangement of organisms based on similarity; *nomenclature* is the naming of the organisms as derived from the classification, and *identification* is the method to characterize the organism to know if an organism has been classified correctly. To these three aspects of taxonomy, a fourth facet of *phylogeny* has been added, which is the history of the evolution of a species or group.

Taxonomy Through the Ages

Western scientific taxonomy started in Greece more than 2000 years ago, but the foundation of modern taxonomy was laid by the Swedish botanist Carl Linnaeus (1707–1778). His *Species Plantarum* (1753) and *Systema Naturae* (1758) were two seminal books on the plants and animals known at that time, and he classified all living beings into two groups: Plantae and Animalia, and introduced the binomial nomenclature system using genus and species name. The Linnean system of two kingdoms existed for almost 100 years. With the invention of microscopes, newer microscopic forms of living beings were being discovered, which were not mentioned in the Linnean system. In 1866, Ernst H. Haeckel introduced the third kingdom, Protista, to include the bacterial and protozoan populations. Ferdinand Cohn is credited with the first documented description of bacterial classification in 1875 when he defined six bacterial genera based on

A. Chaudhury (✉)
Department of Microbiology, Sri Venkateswara Institute of Medical Sciences University, Tirupati, Andhra Pradesh, India

S. C. Parija
Formerly at Sri Balaji Vidyapeeth (Deemed to be University), Pondicherry, India

S. C. Parija, A. Poddar (eds.), *Textbook of Bacterial Zoonoses*, Microbial Zoonoses,
https://doi.org/10.1007/978-981-95-6615-0_1

morphological characteristics. Starting from the 1870s till the middle of the past century, bacterial identification and the resultant classification and nomenclature were purely phenotypic. It started with the morphological description of bacteria based on the size, shape, and structure, which was then extended to the macromolecular constituents of the bacterial cell composition. These included, among others, cellular fatty acids, respiratory quinones, polar lipids, and peptidoglycan types, which gave rise to the use of chemotaxonomy in bacterial taxonomy studies. Although chemotaxonomic characters play an important role in taxonomy, they suffer from a lack of reproducibility. The other parameters that have been extensively used for classification and are still the basis of identification of common pathogenic bacteria are the biochemical reactions exhibited by different families, genera, and species. These reactions are primarily based on the enzymatic properties of bacteria involved in carbohydrate, protein, and lipid metabolism.

Microorganisms had been predominantly classified under the Botanical Code (*International Code of Nomenclature for algae, fungi, and plants or Botanical Code*, 1867) till 1947 because bacteria had traditionally been considered fungi. A seminal change happened in 1930 at the First International Congress of Microbiology in Paris. For the first time, it was decided that bacteria and viruses should be separated with their own unique codes. This initiated the revised edition of the *International Code of Nomenclature of Bacteria and Viruses* in 1958, which is now known as the *International Code of Nomenclature of Prokaryotes (Prokaryotic Code)* and includes archaea but has removed viruses.

The distinction between prokaryotes and eukaryotes was the starting point for further revision in the taxonomy of organisms. In the year 1969, R. H. Whittaker, an American taxonomist, proposed a five-kingdom classification: Monera, Protista, Fungi, Plantae, and Animalia, where the kingdom Monera consisted of all prokaryotes while all others were eukaryotes. The kingdom Monera included eubacteria, cyanobacteria (blue-green algae), and archaea. In the 1960s, genetic methods were slowly being introduced, and two popular tools used in the classification of bacteria were G+C content in DNA and DNA–DNA hybridization (DDH) to find the relatedness of species. Revolutionary development happened in the 1970s with the introduction of (i) polymerase chain reaction; (ii) Sanger's sequencing methods, along with the description of 16S rRNA for prokaryotes; and (iii)18S rRNA for eukaryotes, which could be used as genetic and molecular markers to organize all organisms. The sequencing of these markers by Woese and Fox led to the reorganization of all living units into three domains: (i) eubacteria, comprising all typical bacteria; (ii) archaebacteria, containing methanogenic bacteria; and (iii) eukaryotes. According to the authors, the eubacteria group has three major subdivisions: the blue-green bacteria and chloroplasts, the "Gram-positive" bacteria, and the "Gram-negative" bacteria. Whole genome sequencing for bacteria, which started with *Haemophilus influenzae* in 1995, has helped calculate the overall genome-related indices and has replaced DNA–DNA hybridization in recent years. Whole genome sequences can be used to design a phylogenetic framework on which systematic classification can be based. It also has greater resolution power compared to the 16S rRNA gene for deriving both ancient and recent relationships between the strains. In 1998, Thomas Cavalier-Smith introduced the six-kingdom model. Bacteria were treated as a single kingdom, and eukaryotes were divided into five kingdoms: Protozoa, Animalia, Fungi, Plantae, and Chromista. He used the terminologies of superkingdom, subkingdom, branch, infrakingdom, superphylum, subphylum, and infraphylum to avoid excessive kingdoms and phyla in taxonomy. The genomic methods have been refined over the past two or three decades with the development of high-throughput next-generation sequencing. Genome metrics have been applied to arrive at the overall genome relatedness index (OGRI) for classifying bacteria. DNA–DNA hybridization is considered the gold standard for bacterial taxonomy even in the present era. However, OGRIs have started replacing the hybridization technique due to better reproducibility, improved genomic data, and being economical. The OGRIs include average nucleotide identity (ANI) and digital DNA–DNA hybridization (dDDH). In the ANI method, genome sequences of two strains are first compared through an alignment algorithm like BLAST. In the next step, the average identity of the hits is computed in silico by using specific software like BioEdit. If the ANI value is ≥95%, the two strains are taken to be of the same species. OrthoANI and dDDH are other alternative genomic metrics that have been created. ANI has been proven to be a powerful method to measure the evolutionary distance, which correlates with the DDH values.

Thus, at present, it is observed that there has been a gradual transition from a purely phenotypic classification method till the 1960s to a combination of phenotypic and genotypic methods in the present era, together with the addition of phylogeny in the modern-day taxonomic information. With the advancements in genome sequencing, there has been a better armamentarium currently, not only to define the species and qualify the microbial diversity but also to study the evolutionary history of an organism. Figure 1.1 provides the timeline of the changing pattern in bacterial taxonomy.

Bacterial Taxonomy: Present Status

Bergey's Manual of Systematic Bacteriology remains the Bible for bacterial classification and was first published in 1923 as *Manual of Determinative Bacteriology*. At present, it runs into five volumes, published successively from 2001 to 2012. These include *The Archaea and the Deeply Branching and Phototrophic Bacteria* (Volume (Vol) 1); *The Proteobacteria, Part A: Introductory Essays; Part B: The Gammaproteobacteria; Part C: Alpha-, beta-, delta- and epsilon- proteobacteria* (Vol 2); *The Firmicutes* (Vol 3), *The Bacteroidetes, Spirochaetes, Tenericutes (Mollicutes) & others* (Vol 4); and *The Actinobacteria* (Vol 5). With the rapidly changing scenario in bacterial taxonomy brought about by genomic studies, *Bergey's Manual of Systematics of Archaea and Bacteria* (BMSAB), the first edition, has been published online since 2015. It has replaced and provides an expansion upon the second edition of *Bergey's Manual of Systematic Bacteriology*, last published in 2012.

Zoonotic medically important bacteria primarily belong to the groups of Proteobacteria, Firmicutes, Actinobacteria, and other miscellaneous bacteria described in Vol 4 of *Bergey's Manual*. The bacteria are distributed among 23 phyla. These phyla are further subdivided into 28 classes. The list has widened subsequently, and more than 40 phyla have been recognized to date. Within the Gram-negatives, the phylum "Proteobacteria" has been subdivided into five classes: Alphaproteobacteria (*Rhodospirilla, Rhizobia*, and *Rickettsias*), Betaproteobacteria (*Neisseria*), Gammaproteobacteria (*Pseudomonads*), Deltaproteobacteria (*Desulfos* and *Myxos*), and Epsilonproteobacteria (*Campylobacters*). In addition to these, other medically important Gram-negative phyla include Chlamydiae, Spirochaetes, Fusobacteria, and Bacteroidetes. The Gram-positive bacteria have been divided into two phyla based on the G+C content: Firmicutes with a low and Actinobacteria with a high G+C percentage. Box 1.1 gives a representation of the taxonomic hierarchy commonly used in the literature.

Box 1.1 Taxonomic Hierarchy Used in Bacterial Nomenclature

Domain: Bacteria
Phylum: Pseudomonadota
Class: Gammaproteobacteria
Order: Enterobacterales
Family: Enterobacteriaceae
Genus: *Klebsiella*
Species: *Klebsiella oxytoca*

Proteobacteria *(Pseudomonadota)*

These bacteria were first described by Woese in 1987 as "purple bacteria and their relatives" and named Proteobacteria by Stackebrandt in 1988. In the domain of bacteria, this is the largest phylum with more than 500 genera and 2000 species, which includes diverse and heterogeneous bacte-

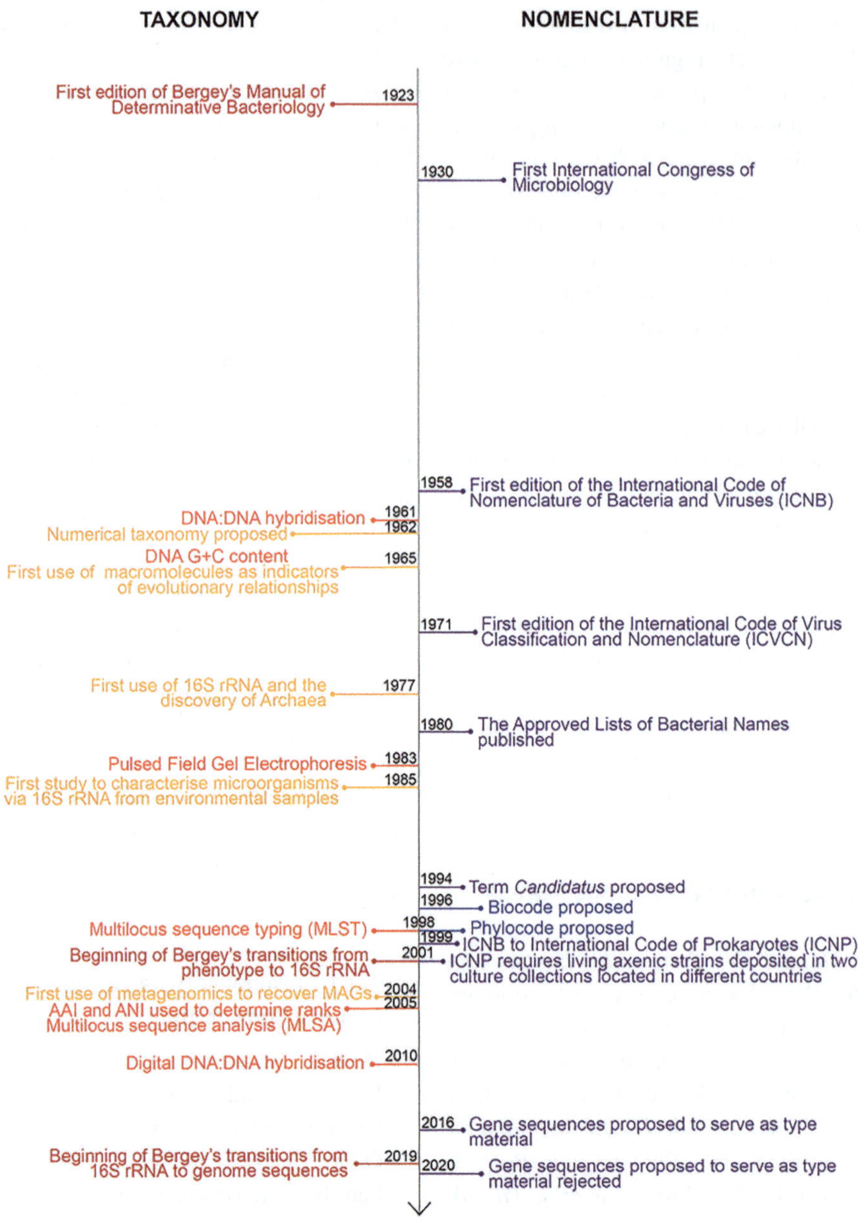

Fig. 1.1 Chronology of the prokaryote taxonomy in the past 100 years. (Image source: Reproduced from Hugenholtz et al. (2021), Creative Commons Attribution 4.0 International License)

ria (hence the name Proteo-, from *Proteus*, the Greek God of the sea, who could assume various shapes). The members of this phylum are frequent human and animal pathogens and are Gram-negative in nature, with the presence of lipopolysaccharides in their outer membranes. Of the five classes mentioned above, four contain pathogenic members.

(a) Alphaproteobacteria: Many of these bacteria are obligate intracellular in nature. It includes the following pathogenic genera: *Rickettsia, Bartonella, Brucella,* and *Ehrlichia.*
(b) Betaproteobacteria: It includes a diverse group of bacteria, of which the pathogenic genera are *Bordetella, Neisseria,* and *Burkholderia.*

(c) Gammaproteobacteria: It is the most heterogeneous class, which is divided into 15 established orders and 250 genera. Of them, about 20 genera and around 50 species are human pathogens. The important orders containing the pathogenic members are Aeromonadales, Cardiobacteriales, Enterobacterales, Legionellales, Pseudomonadales, Pasteurellales, and Vibrionales. The well-known family of Enterobacteriaceae underwent a significant revision in 2016, and a new order of Enterobacterales was created, consisting of more than 250 species in seven families: Enterobacteriaceae, Erwiniaceae, Hafniaceae, Morganellaceae, Pectobacteriaceae, Yersiniaceae, and Budviciaceae.

(d) Epsilonproteobacteria (*Campylobacterota*): Members of this class are microaerophilic and contain the important pathogens of the genera *Campylobacter* and *Helicobacter*.

Other Gram-Negative Phyla

Volume 4 of *Bergey's Manual of Systematic Bacteriology* includes bacteria that are mostly unrelated to each other. These include the following phyla, some of whose members are human and/or animal pathogens: Bacteroidota, Fusobacterota, Spirochaetota, Tenericutes (Mollicutes and Mycoplasmatota), and Chlamydiota.

Firmicutes (Bacillota)

These are low–G+C content (avg. 40%) Gram-positive prokaryotes with more than 1300 species and 235 genera. Important pathogens in this group belong to the taxa of *Bacillus*, *Clostridium*, *Enterococcus*, *Erysipelothrix*, *Listeria*, *Peptococcus*, *Staphylococcus*, *Streptococcus*, and *Veillonella*.

Actinobacteria (Actinomycetota)

These are also Gram-positive bacteria but have a high G+C content (avg. >60%). Of the various classes, only the class Actinomycetia contains human/animal pathogens, including the genus *Mycobacterium*.

Practical Considerations: The Way Forward

"The true taxonomist is a man with a mission; he often leads a cloistered life, protected from the vexations and frustrations of the everyday world, and he may well wear blinkers as opaque as any worn by a horse. He is more likely to have academic interests than to be an applied worker ... and an Englishman rather than an American. Living a life of seclusion, safe in his small laboratory, and surrounded by his books, his microscope (and perhaps his computer tape), he has an unconcern for the mundane application of his work. But science has a way of making itself useful, and the useful application of classification is identification." These words were used by ST Cowan, the renowned bacteriologist, while delivering a lecture at a symposium in the USA in 1964. This holds true when we come down from the elevated and complex world of taxonomy to the day-to-day management of a clinical bacteriology laboratory, be it medical or veterinary. The clinical microbiologist may not be aware of all the recent changes in nomenclature and classification of the limited number of bacteria encountered in daily operations. It would be too far-fetched to imagine that the clinician getting the report will be aware of the changing scenario in bacterial taxonomy. The last sentence in the above lines by Cowan sums up the aim of a clinical bacteriologist: *to identify the isolate to the correct genus and/or species level.*

Till the advent of the 16S rRNA and sequencing technology, bacterial identification had a phenotypic approach to identify the bacteria to the species level without going into the phylogenetic information. Edwards and Ewing are credited with the seminal work firmly laying the foundations of the phenotypic principles for the characterization, classification, and identification of bacteria, particularly of the Enterobacteriaceae family. Their classification system relied on the classification and identification on the overall morphologic and biochemical pattern of a species by using a large number of biochemical tests and bacterial strains. Sneath, in 1957, first used the theories of the French biologist Addanson (1727–1806) and showed that all phenotypic features had equal merits in the characterization of an organism, which led to the introduction of *Numerical Taxonomy* by Sneath and Sokal in 1962. It helped in improving the validity of phenotypic identification by using 100–200 tests and calculating coefficients of similarity between strains and species. Later on, chemotaxonomic characterization was introduced by the analysis of bacterial macromolecular constitution. Needless to mention that the vast majority of identification and nomenclature of common and uncommon bacteria were made during this pre-genotypic era, and the genus and species names of most of them remain unchanged and validated even to this time. Genetic studies during the past four or five decades have helped arrive at a polyphasic classification system incorporating both phenotypic and phylogenetic data. This has resulted in the reclassification and new nomenclature of a number of bacteria. This has resulted in some confusion among the bench workers in clinical microbiology laboratories and the clinicians, who are more familiar with the older names of the isolated bacteria. Renaming *Mycobacterium terrae* into *Mycolicibacter terrae* may not cause many problems among the above stakeholders, but the change in name from *Brucella anthropi* to *Ochrobactrum anthropi* or *Staphylococcus sciuri* to *Mammaliicoccus sciuri* is sure to cause confusion at multiple levels. Readers are advised to go through Prinzi and Moore's publication regarding the impact of nomenclature changes in clinical microbiology. Table 1.1 lists the nomenclature changes among the important bacteria causing human or animal infections. This may help properly understand the reports generated by a 16S rRNA analysis or the data generated by automated identification systems and the Matrix-assisted laser desorption/ionization-time of flight (MALDI-TOF).

The clinical microbiology laboratory heavily depends on the morphological, cultural, and biochemical characteristics of bacteria for identification. This phenotypic analysis is generally enough for genus and/or species delineation. This is because most pathogenic bacteria belong to well-established phenotypes, and this consistency helps in quick diagnostics, which is crucial for managing patients or sick animals. Sequencing and phylogeny are important when a novel strain is encountered whose identity cannot be resolved using common phenotypic features.

Case Study

The genus *Mycobacterium* (with more than 190 species) contains some prominent pathogens as well as many environmental strains. All these can cause opportunistic infections in immunocompromised hosts and also infections in immunocompetent subjects. Although considered a homogeneous genus, it underwent a major taxonomic change in 2018. When the phylogenetic tree was constructed using 150 *Mycobacterium* genomes based on 1941 core proteins, it was found that there were five distinct groups within the genus *Mycobacterium* at the highest level, which were designated as the "*Tuberculosis–Simiae*," "*Terrae*," "*Triviale*," "*Fortuitum–Vaccae*," and "*Abscessus–Chelonae*" clades. Based on the phylogenomic analyses and a number of molecular signatures, taxonomists have recently revised the classification of the genus *Mycobacterium* based on advanced molecular and phylogenetic studies. The original genus has now been divided into an emended genus *Mycobacterium*, which includes the *Tuberculosis–Simiae* clade and comprises all the major human pathogens. In addition, four new genera have been proposed: *Mycolicibacterium* gen. nov. (representing the *Fortuitum–Vaccae* clade), *Mycolicibacter* gen. nov. (the *Terrae* clade), *Mycolicibacillus* gen. nov. (the *Triviale* clade), and *Mycobacteroides* gen. nov. (the *Abscessus–Chelonae* clade).

This reclassification has led to the renaming of several well-known species. For instance,

Table 1.1 Recent changes in the names of common pathogenic bacteria

SN	New name	Old/conventional name
1	*Clostridioides difficile*	*Clostridium difficile*
2	*Mycobacteroides abscessus*	*Mycobacterium abscessus*
3	*Mycobacteroides chelonae*	*Mycobacterium chelonae*
4	*Mycolicibacterium fortuitum*	*Mycobacterium fortuitum*
5	*Mycolicibacillus triviale*	*Mycobacterium triviale*
6	*Mycolicibacter terrae*	*Mycobacterium terrae*
7	*Cutibacterium acnes*	*Propionibacterium acnes*
8	*Falseniella ignava*	*Facklamia ignava*
9	*Mammaliicoccus sciuri*	*Staphylococcus sciuri*
10	*Mammaliicoccus lentus*	*Staphylococcus lentus*
11	*Klebsiella aerogenes*	*Enterobacter aerogenes*
12	*Pseudescherichia vulneris*	*Escherichia vulneris*
13	*Brucella anthropi*	*Ochrobactrum anthropi*
14	*Capnocytophaga canis* sp. *nov.*	*Capnocytophaga canimorsus*
15	*Stutzerimonas nosocomialis*	*Pseudomonas nosocomialis*
16	*Klebsiella shigelloides*	*Plesiomonas shigelloides*
17	*Fusobacterium animalis* sp. nov.	*Fusobacterium nucleatum* subsp. *animalis*
18	*Fusobacterium nucleatum* subsp. *vincentii*	*Fusobacterium vincentii* sp. nov.
19	*Glaesserella parasuis* gen. nov.	*Haemophilus parasuis*
20	*Epilithonimonas hominis*	*Chryseobacterium hominis*

Mycobacterium abscessus is now referred to as *Mycobacteroides abscessus*, and *Mycobacterium fortuitum* has been reclassified as *Mycolicibacterium fortuitum.* These changes reflect the growing understanding of genetic diversity within the group and ensure a more accurate representation of their evolutionary relationships (Table 1.1).

This new classification has drawn a fair number of criticisms since it may affect patient care and create confusion among clinicians and microbiologists about their true significance. On the other hand, sound arguments have been put forward to justify educating ourselves on these recent changes.

1. How may the changes in nomenclature affect patient care?
2. How can the laboratory communicate the changes to clinicians and the antimicrobial stewardship team?
3. How can the clinical microbiology laboratory incorporate the changes and make them an integral part of their regular workflow, and is it necessary?

Research Questions

1. Horizontal gene transfer in bacteria between different species and even genera can alter the basic genetic makeup of a bacterium. Can this affect the species definition in taxonomic classification?
2. How can we correlate genome species with pathogenicity since other species beyond the established species produce the same clinical entities?
3. How can the addition of novel species to the growing list of organisms and nomenclature changes affect antibiotic susceptibility testing methods and standards in a clinical microbiology laboratory? This is important for patient care and can impact the antibiotic stewardship program.

Further Readings

Agarwal RK, Parija SC, Sanyal SC. Taxonomic studies on *Vibrio* and related genera. Indian J Med Res. 1980;71:340–53.

Carroll KC, Munson E, Butler-Wu SM, Patrick S. Point-counterpoint: what's in a name? Clinical microbiology laboratories should use nomenclature based on current taxonomy. J Clin Microbiol. 2023;61:e01732–22.

Cavalier-Smith T. A revised six-kingdom system of life. Biol Rev. 1998;73:203–66.

Cowan ST. Principles and practice of bacterial taxonomy-a forward look. J Gen Microbiol. 1965;39:148–58.

Edwards PR, Ewing WH. Identification of enterobacteriaceae. 2nd ed. Minneapolis: Burgess Publishing Co.; 1962.

Gupta RS, Lo B, Son J. Phylogenomics and comparative genomic studies robustly support division of the Genus Mycobacterium into an emended genus Mycobacterium and four novel genera. Front Microbiol. 2018;9:67.

Helene LCF, Klepa MS, Mariangela H. New insights into the taxonomy of bacteria in the genomic era and a case study with Rhizobia. Int J Microbiol. 2022;2022:4623713.

Hugenholtz P, Chuvochina M, Oren A, Parks DH, Soo RM. Prokaryotic taxonomy and nomenclature in the age of big sequence data. ISME J. 2021;15:1879–92.

Parks DH, Chuvochina M, Waite DW, Rinke C, Skarshewski A, Chaumeil PA, Hugenholtz P. A standardized bacterial taxonomy based on genome phylogeny substantially revises the tree of life. Nat Biotechnol. 2018;36:996–1004.

Prinzi AM, Moore NM. Change of plans: overview of bacterial taxonomy, recent changes of medical importance, and potential areas of impact. Open Forum Infect Dis. 2023;10(7):ofad269. https://doi.org/10.1093/ofid/ofad269.

Sneath PHA, Sokal RR. Numerical taxonomy. Nature. 1962;193:855–60.

Tortoli E, Brown-Elliott BA, Chalmers JD, et al. Same meat, different gravy: ignore the new names of mycobacteria. Eur Respir J. 2019;54:190079.

Woese CR, Fox GE. Phylogenetic structure of the prokaryotic domain: the primary kingdoms. Proc Natl Acad Sci. 1977;74:5088–90.

Bacterial Ecology

2

Suprativ Das and Anirban Mukherjee

Learning Objectives

1. To learn in detail about the classification of bacteria based on their energy, carbon, and electron sources, along with their habitat
2. To understand the role of bacteria in regulating various biogeochemical cycles
3. To learn about the application of modern research techniques in bacterial ecology

Introduction

Bacteria, ranging from 0.5 to 50 μm, are tiny, unicellular organisms on the Earth's surface. The diverse metabolic functions of bacteria enable them to survive in different environmental conditions. They are classified into autotrophs, heterotrophs, lithotrophs, and organotrophs based on their energy, carbon, and electron sources. Autotrophs are organisms capable of synthesising food using sunlight (photoautotrophs) or inorganic substances (chemoautotrophs). Heterotrophs are organisms that obtain their energy by the oxidation of organic substances. Lithotrophs and organotrophs, on the other hand, exhibit a vast biochemical diversity that enables them to survive in extremes of temperature. The different types of microbes show various ways of interacting while living in different environments to maintain ecological stability. Seven significant ecological relationships exist: mutualism, commensalism, parasitism, competition, antagonism, predation, and cooperation.

Research methods supporting bacterial ecological investigations include techniques with specific strengths and limitations during usage. Traditional culture-based methods allow researchers to demonstrate bacterial activities through direct examination. However, these procedures restrict their ability to study many bacteria because many are unculturable. Modern DNA approaches, starting with the DNA sequencing of 16S rRNA genes and metagenomics, have transformed the field of study to identify unculturable organisms and bacterial communities, along with their structure and function. Fluorescent *in situ* hybridisation (FISH) and stable isotope probing (SIP) have been used to visualise bacterial populations through fluorescence and monitor their metabolic functions. These analytic approaches encounter several difficulties, including data analysis bias, high expenses, and the requirement for specialised research equipment.

S. Das
School of Bio Sciences & Technology, Vellore Institute of Technology, Vellore, Tamil Nadu, India

A. Mukherjee (✉)
Department of Microbiology, Gurudas College, Calcutta University, Kolkata, West Bengal, India

S. C. Parija, A. Poddar (eds.), *Textbook of Bacterial Zoonoses*, Microbial Zoonoses,
https://doi.org/10.1007/978-981-95-6615-0_2

Metabolic Diversity in Bacteria

Bacteria exhibit a vast degree of metabolic diversity. This allows them to inhabit virtually every environment on Earth. This diversity stems from their ability to utilise a wide range of energy and carbon sources through various biochemical pathways.

Nutrient Requirement

Based on their mode of nutrition, bacteria are classified into several categories (Table 2.1):

Autotrophs

Microorganisms that obtain their own food by using inorganic substances, such as sunlight or chemical energy, are called autotrophs. Photoautotrophs and chemoautotrophs are two of its subtypes. Organisms that use sunlight as the sole energy source and CO_2 as the carbon source are photoautotrophs. They convert CO_2 into organic compounds through photosynthesis and release oxygen. Examples include green plants, cyanobacteria, and green sulphur bacteria. Organisms that obtain their energy by oxidising inorganic substances, including H_2S, NH_3, and Fe, are known as chemoautotrophs. They fix carbon in the environment. *Thiobacillus* species is an example of this group.

Heterotrophs

Organisms that satisfy carbon and energy requirements by oxidising organic compounds are heterotrophs. They can be further classified into chemoheterotrophs and photoheterotrophs. Microorganisms that obtain food and energy by metabolising organic compounds are known as chemoheterotrophs. An example is *Escherichia coli*, which lives in the intestines of humans and animals. Microorganisms that use sunlight as the sole energy source and organic compounds, such as carbohydrates and alcohol, as the carbon sources are known as photoheterotrophs. Three prominent examples of photoheterotrophic organisms include purple non-sulphur bacteria, green non-sulphur bacteria, and *Heliobacteria.*

Lithotrophs

Lithotrophs are microorganisms that utilise inorganic substances (mainly of mineral origin), such as iron, ammonia, and hydrogen sulphide, as electron donors to obtain energy for survival. They are divided into two major subtypes: photolithotrophs and chemolithotrophs. Photolithotrophs are microorganisms that derive energy from sunlight while using inorganic compounds as electron donors. Photolithography occurs widely in plants, but cyanobacteria demonstrate microbial photolithographic behaviour through oxygenic photosynthesis, which uses water as an electron source.

Table 2.1 Classification of bacteria based on their nutritional requirements

Category	Energy source	Carbon source	Electron donor	Examples
Photoautotroph	Sunlight	CO_2	Inorganic (H_2S and H_2O)	Cyanobacteria, green plants, algae, green sulphur bacteria
Photoheterotroph	Sunlight	Organic	Organic compounds	Green non-sulfur bacteria, heliobacteria
Photolithotroph	Sunlight	Can be organic or inorganic	Inorganic	Cyanobacteria
Photoorganotroph	Sunlight	Can be organic or inorganic	Organic compounds	Purple non-sulfur, green non-sulfur bacteria, heliobacteria
Chemoautotroph	Chemical energy	CO_2	CO_2	*Thiobacillus* species
Chemoheterotroph	Chemical energy	Organic	Organic or inorganic compounds	*Escherichia coli*
Chemolithotroph	Chemical energy	Can be organic or inorganic	Inorganic compounds	Sulfur-oxidising bacteria
Chemoorganotroph	Chemical energy	Can be organic or inorganic	Organic compounds	*Clostridium botulinum*

Chemolithotrophs derive energy by chemically oxidising inorganic substances without sunlight. All organisms of these groups belong to microorganisms, consisting of aerobic and anaerobic species. Nitrogen, sulphur, and iron biological cycles depend heavily on chemolithotrophic processes. For example, sulphur-oxidising bacteria like *Thiobacillus* spp. oxidise hydrogen sulphide to elemental sulphur to produce energy.

Organotrophs
Organotrophs are microorganisms that obtain their energy by oxidising organic molecules. They utilise organic molecules as electron donors for their metabolic processes. Organotrophs can be further divided into chemoorganotrophs and photoorganotrophs. Microorganisms that derive energy from the chemical breakdown of organic molecules, such as carbohydrates and lipids, are called chemoorganotrophs. They use oxygen as terminal electron acceptors and can perform aerobic and anaerobic respiration. *E. coli* is an example of a chemoorganotroph. Photoorganotrophs are microorganisms that use sunlight as the sole energy source and organic molecules as the electron source to carry out the metabolic processes. Examples include non-sulphur bacteria, purple non-sulphur bacteria, and heliobacteria.

Oxygen Requirements

Microorganisms require oxygen to carry out their metabolism. Based on the oxygen requirement, they are classified into obligate aerobes, facultative anaerobes, obligate anaerobes, aerotolerant anaerobes, and capnophiles.

Obligate Aerobes
These are microorganisms that grow strictly in the presence of oxygen, which is used as a terminal electron acceptor. Examples include *Pseudomonas aeruginosa*, *Mycobacterium tuberculosis,* and *Bacillus subtilis*.

Facultative Anaerobes
Microorganisms that grow both in the presence and absence of oxygen are termed facultative anaerobes. When oxygen is available, they respire aerobically but switch to anaerobic respiration without oxygen. Examples include *Staphylococcus aureus* and yeasts.

Obligate Anaerobes
Microorganisms that cannot grow in the presence of oxygen are referred to as obligate anaerobes. They lack the enzymes necessary to detoxify oxygen species (ROS) and are harmed by oxygen. Examples include *Clostridium botulinum* and *Bacteroides fragilis.*

Aerotolerant Anaerobes
Organisms that do not require oxygen for growth but can tolerate its presence are called aerotolerant anaerobes. Examples include *Lactobacillus* spp.

Microaerophiles
They are referred to as microorganisms that can survive at low oxygen concentrations but become sensitive to high oxygen levels. Examples include *Campylobacter jejuni* and *Helicobacter pylori.*

Capnophiles
Capnophiles are microorganisms that require a higher concentration of carbon dioxide for growth. Example includes *Neisseria gonorrhoeae* and *Campylobacter* spp.

Factors Affecting Ecology

The ecological success of bacterial communities is a direct consequence of the intricate relationship between biotic and abiotic factors. These shape the bacterial diversity, abundance, activity, and interactions within various environments.

Abiotic Factors

Temperature
Temperature controls metabolic rates by categorising bacteria into psychrophiles or mesophiles and thermophiles or hyperthermophiles within distinct temperature ranges. The thermo-

philic organism *Thermus aquaticus* survives in the extreme heat conditions of hot springs, which other organisms cannot tolerate.

pH

The pH level of an environment regulates bacterial activity through enzymes while determining which bacterial type will survive as acidophiles, neutrophiles, or alkaliphiles. Dissolved chemicals in the environment determine how effectively bacteria use their enzymes. The acidophile bacterium *Acidithiobacillus ferrooxidans* exists best within such acidic conditions as a pH of 1–5.

Oxygen Content

The availability of oxygen plays a critical role in determining the bacterial respiratory behaviour. Specifically, the oxygen concentration influences whether bacteria are classified as obligate aerobes, obligate anaerobes, or facultative anaerobes. Additionally, it determines whether certain bacteria can adapt to varying oxygen levels, such as aerotolerant anaerobes, which can survive in the presence of oxygen without utilising it, and microaerophiles, which require only low levels of oxygen for growth. Oxygen availability is essential for the regulation of bacterial metabolic processes. For instance, *Clostridium botulinum* is an obligate anaerobe, which means that it can only survive and grow in an oxygen-free environment.

Nutrient Availability

Growth depends on how essential nutrients, including carbon, nitrogen, phosphorus, and sulphur, are available for the process. Key components required for bacterial development also supply them with the necessary energy. *Rhizobium* spp., which occurs in terrestrial environments, grows best when exposed to conditions that enable it to contribute to plant–soil relationships through nitrogen gas fixation, leading to better plant development.

Salinity

Osmotic balance dependence and salinity determine bacterial characteristics as either halophiles or non-halophiles. Salinity impacts the osmotic balance of bacterial cells. *Halobacterium salinarum* experiences optimal growth in an excessively saline environment, while the growth of other organisms is inhibited.

Water Content

Water availability directly influences metabolic activity, while xerophiles demonstrate survival capabilities under dry conditions. At the same time, water controls bacterial metabolic processes. Organisms can survive severely dry conditions without hydration, like that of *Aspergillus penicillioides*.

Biotic Factors

Microorganisms

Microorganisms are present in diverse environments. They play a significant role in regulating the nutrient cycle by transforming complex organic matter into simpler forms. Decomposers are crucial in converting dead organic matter into simpler compounds, releasing nutrients into the environment.

Plants

Primary producers, such as plants, create the base of most ecosystems. Plants transform sunlight into chemical energy through photosynthesis while producing oxygen, along with organic substances that support the survival of other living beings. They manage the oxygen and carbon cycle of nature. They also maintain soil fertility.

Humans and Animals

Humans are responsible for agriculture, urbanisation, and deforestation, which affect the nutrient cycle by causing pollution, loss of habitat, and climate change. Animals, including pollinators (butterflies, bees, and bats), help in the dispersal of seeds and allow plants to colonise new areas, thus spreading genetic diversity.

Bacterial Habitat

The microbial kingdom of bacteria extends across numerous habitats on Earth since these tiny organisms adapt well to diverse ecological niches.

Terrestrial Bacteria

They are the microorganisms that dwell in soil, rocks, and other land-based environments. They play a pivotal role in regulating the nutrient cycle, decomposing organic matter, and maintaining soil's diversity—for example, *Rhizobium* spp. The root nodules exhibit a symbiotic relationship with the plants by fixing the atmospheric nitrogen to make it usable by plants. *Bacillus subtilis*, another terrestrial bacterium, plays a significant role in decomposing organic matter, thus improving soil fertility. Terrestrial bacteria demonstrate rapid growth on new substrates, enabling them to be utilised for bioremediation and space exploration.

Aquatic Bacteria

Aquatic microbes are present in freshwater, marine, and brackish water environments. They play a major role in the ecosystem by enhancing nutrient cycling, decomposition, and primary production. For example, *Cyanobacterium*, a photosynthetic bacterium, is a primary producer that captures sunlight to produce oxygen. *Pseudomonas aeruginosa*, which is present in aquatic environments, is often associated with the formation of biofilms. Recently, it has been used in bioremediation.

Extremophilic Bacteria

Extremophiles are microorganisms that thrive in extreme environmental conditions, usually inhospitable to most other life forms. They survive in extremes of temperature, pH, salinity, and pressure, including hot springs, deep-sea vents, and acid lakes. Extremophiles include (i) thermophiles surviving at high temperatures ranging from 60 to 80 °C (e.g. *Thermus aquaticus*); (ii) hyperthermophiles, which can grow above 80 °C (*Pyrolobus fumarii*); (iii) psychrophiles growing below 15 °C (*Psychrobacter* spp.); (iv) halophiles living in high salt concentrations (e.g. *Halobacterium salinarum*); (v) acidophiles, growing at, for example, a pH of 1–5 (*Acidithiobacillus ferrooxidans*); (vi) alkaliphiles, surviving at a pH of more than 9 (*Bacillus alcalophilus*); (vii) barophiles or piezophiles surviving at high pressure (e.g. *Thermococcus barophilus*); and (viii) xerophiles surviving at extremely dry conditions (e.g. *Aspergillus penicillioides*). Taq polymerase extracted from *Thermus aquaticus* has been used in the polymerase chain reaction (PCR) technology. Two exceptional extremophile species include *Deinococcus radiodurans*, which displays radiation resistance, and *Halobacterium salinarum*, which demonstrates salt tolerance in salt flats.

Bacterial Interactions

Bacteria form different kinds of relationships with other organisms. These relationships can be beneficial, neutral, or harmful to one or both organisms. These interactions are pivotal in nutrient cycling, disease, and symbiosis.

Symbiotic Relationship

Symbiosis refers to the close and long-term relationship of two different organisms. It can be mutualism, commensalism, parasitism, or amensalism.

Mutualism

It is defined as a symbiotic relationship between two organisms where both benefit. They are needed for the survival of the ecosystem. Nitrogen-fixing bacteria, such as *Rhizobium*, reside in the root nodules of leguminous plants and convert atmospheric nitrogen to ammonia, making it readily available to the plant for growth. In return, the plant helps by providing shelter and nutrients, such as carbohydrates, to the bacteria.

Commensalism

It is the type of interaction where one organism benefits from the interaction, but the other is neither benefited nor harmed. Bacteria living on the skin or in the mucous membranes of humans and animals come under his category. *Staphylococcus epidermidis,* present on human skin, obtains nutrients from the human body. Sometimes, it can protect the skin by outcompeting harmful pathogens.

Parasitism

Parasitic interactions are interactions in which one organism benefits at the expense of another. Bacteria that cause diseases in plants, animals, and humans are examples of parasitic interactions. *M. tuberculosis*, *E. coli*, and *Streptococcus pneumoniae* evade the host immune system, utilise its resources, and cause diseases such as tuberculosis, food poisoning, and pneumonia, respectively. *Wolbachia* spp. infects the insect hosts, manipulating their reproductive system to benefit their propagation.

Amensalism

It is the type of interaction where one organism is harmed, while the other one remains unaffected. For example, *Penicillium* spp. secretes penicillin, which inhibits the growth of nearby bacteria. *Penicillium* spp. is not harmful, but it gains no advantage due to the inhibition of others. *Lactobacillus* species produce lactic acid, which lowers the pH of the environment and inhibits the growth of pathogens like *Salmonella* or *E. coli*. However, despite producing lactic acid, *Lactobacillus* remains unaffected.

Competition and Antagonism

Competition and antagonism are two meaningful bacterial interactions influencing ecological balance and mould microbial communities. Bacteria compete for scarce resources like oxygen, nutrition, or space. When bacteria compete for the same resources, it is termed indirect competition. When one group of bacteria inhibits the growth of another group of bacteria, it is termed direct competition. For example, *B. subtilis* and *E. coli* often compete for glucose in an environment. The form of interaction where one organism inhibits the growth of others by producing toxins or antibiotics is referred to as antagonism. *Lactobacillus* produces bacteriocins that restrain the growth of other bacterial species.

Predation and Cooperation

Multiple complex relationships between bacteria include predation and cooperation as part of their diverse interaction network. A bacterium that acts as a predator kills a prey bacterium to obtain nutritional value. Through its life cycle, *Bdellovibrio bacteriovorus* attacks Gram-negative bacteria by entering their periplasmic space, consuming the cell materials until the prey is lysed for progeny release. *Myxococcus xanthus* operates as a social predator that hunts in groups by producing enzymes and antibiotics, which cause lysis in prey such as *E. coli*.

On the other hand, bacterial cooperation involves interactions where bacteria work together to achieve mutual benefits. Bacterial communication systems, namely, quorum sensing and biofilms, are two significant mechanisms that bacteria use to cooperate. *P. aeruginosa* construct biofilms, which allow cells to distribute nutrients while creating a shielding zone against environmental dangers.

Biogeochemical Cycles

Nitrogen Cycle

The nitrogen cycle is a biogeochemical pathway that transfers nitrogen between the atmosphere and three environments, including water, soil, and living creatures. Life depends on nitrogen because this element functions as the principal structural element in amino acids, proteins, and nucleic acids. Atmospheric nitrogen (N_2) exists in a form that most living organisms cannot utilise. Microbial processes change nitrogen from its non-utilisable form into a form that living creatures use. Figure 2.1 shows a diagrammatic representation of the cycle.

The steps of the nitrogen cycle are as follows:

Nitrogen Fixation

The conversion of atmospheric N_2 into ammonia (NH_3) occurs through nitrogen-fixing bacterial processes involving *Rhizobium* in legume root nodules and cyanobacteria. The Haber–Bosch

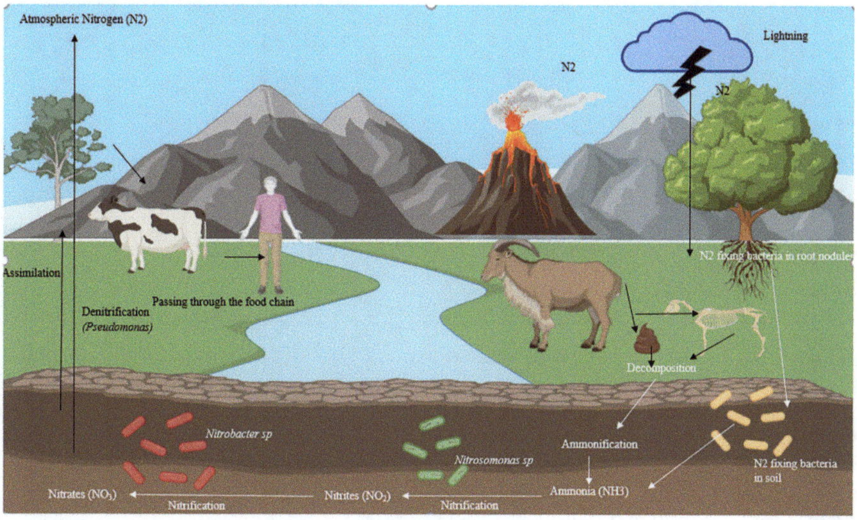

Fig. 2.1 Nitrogen cycle. (Image source: Authors' personal collection)

industrial process, along with natural lightning, functions alongside bacterial processes to convert atmospheric nitrogen into usable ammonia (NH_3).

Nitrification

The bacterium *Nitrosomonas* converts ammonia (NH_3) into nitrites (NO_2^-), which *Nitrobacter* then transforms into nitrates (NO_3^-).

$$2NH_3 + 3O_2 \rightarrow 2NO_2^- + 2H^+ + 2H_2O$$

$$2NO_2^- + O_2 \rightarrow 2NO_3^-$$

Assimilation

Through root intake, plants acquire nitrates (NO_3^-) and ammonium (NH_4^+), which they use to synthesise proteins and DNA. Animals' nitrogen life cycle works through their ingestion of plants and the consumption of other animals. This process transfers nitrogen through the food chain, supporting growth and reproduction.

Ammonification

Organic waste and dead organisms decompose under the actions of decomposing bacteria and fungi, leading to ammonium (NH_4^+) formation. Ammonification plays a significant role by transforming organic matter nitrogen into soil-replenishing ammonium (NH_4^+).

Denitrification

Nitrate-consuming anaerobic bacteria such as *Pseudomonas* and *Clostridium* transform NO_3^- into atmospheric N_2 gas through a microbial process. This step completes the cycle by returning nitrogen to the atmosphere, preventing over-accumulation in the environment.

$$2NO_3^- \rightarrow N_2 + 3O_2$$

Carbon Cycle

Life depends on carbon as an essential element because this element shapes the fundamental structures of all organic molecules, including carbohydrates, proteins, and lipids. Through this cycle, the Earth continues to provide carbon, sustaining life forms while maintaining a stable climate state.

Figure 2.2 shows a diagrammatic representation of the cycle. The steps are mentioned below.

Carbon Enters the Atmosphere

Natural processes that release carbon dioxide into the atmosphere include respiration, volcanic eruptions, decomposition, and human-caused processes of burning fossil fuels and deforestation.

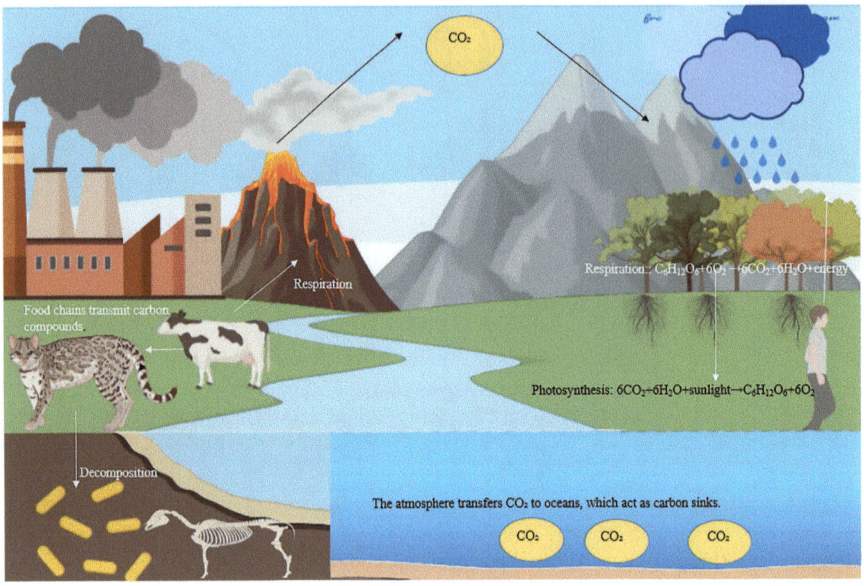

Fig. 2.2 Carbon cycle. (Image source: Authors' personal collection)

Photosynthesis

Through photosynthesis, plants and algae use atmospheric CO_2 to produce the chemical compound glucose ($C_6H_{12}O_6$) and simultaneously create oxygen (O_2). The creation of the food chain begins with this process ($6CO_2 + 6H_2O + \text{sunlight} \rightarrow C_6H_{12}O_6 + 6O_2$).

Carbon in Food Chains

Food chains transmit carbon compounds from plants to herbivores, who, in turn, pass them to carnivores. Animals use carbon dioxide to create tissue elements through this process.

Respiration

Cellular respiration causes plants and animals to emit CO_2 as they create energy for their bodily functions ($C_6H_{12}O_6 + 6O_2 \rightarrow 6CO_2 + 6H_2O + \text{energy}$).

Decomposition

After the death of plants and animals, bacteria break down the complex organic matter into simpler substances by releasing CO_2 and CH_4 into the soil.

Carbon Storage in Fossil Fuels

Through deep geological time, decomposed organic matter eventually changes to fossil fuels, including coal, river oil, river gas, and natural gas.

Combustion

When fossil fuels are burned through combustion processes, they release carbondioxide (CO_2) emissions that increase airborne carbon levels.

Oceanic Carbon Exchange

The atmosphere transfers CO_2 to oceans, which act as carbon sinks. Marine organisms use absorbed carbon to create calcium carbonate ($CaCO_3$) shells, leading to their remains becoming limestone after death. The ocean poses a double impact by absorbing CO_2 from the air while returning parts of this material to the atmosphere.

Sulphur Cycle

It is a biogeochemical cycle. In this cycle, sulphur is transformed from one form to another through the Earth's lithosphere, atmosphere, hydrosphere, and biosphere. Sulphur is a crucial element in life because it is present in vital biomolecules, such as amino acids (including methionine and cysteine) and vitamins. Figure 2.3 shows a diagrammatic representation of the cycle. The steps are discussed below:

Fig. 2.3 Sulphur cycle. (Image source: Authors' personal collection)

Weathering of Rocks and Precipitation

Weathering processes enable rocks and minerals to release stored sulphur deposits into the soil. The sulphur interacts with oxygen to create soluble sulphates (SO_4^{2-}) that plants and microorganisms can use. Volcanic eruptions release SO_2 into the atmosphere, which combines with water and precipitates on the Earth's surface as acid rain.

$$SO_2 + H_2O \rightarrow H_2SO_3$$

$$2H_2SO_3 + O_2 \rightarrow 2H_2SO_4 \left(\text{acid rain}\right)$$

Assimilation by Plants and Microbes

Plants acquire sulphates through soil, integrating them into natural organic compounds like amino acids. Microorganisms require sulphates for their metabolic activities, and they successfully take up these compounds when needed.

Transfer Through Food Chains

Living organisms acquire sulphur by eating plants and other animals. Organic compounds containing sulphur travel through the food chain via animal tissue cells.

Decomposition

After the death of plants and animals, bacteria break down the complex organic matter into simpler substances by releasing H_2S and elemental sulphur into the soil.

Oxidation of Hydrogen Sulphide

During decomposition, the photosynthetic bacteria, such as *Chlorobium,* along with chemolithotrophic bacteria, oxidise hydrogen sulphide to produce either elemental sulphur or sulphates.

Reduction of Sulphates

Sulphate-reducing bacteria, *Desulfovibrio desulfuricans*, reduce sulphates into hydrogen sulphide gas in environments lacking oxygen.

Release into the Atmosphere

The atmosphere receives sulphur gases, including SO_2 and H_2S, when volcanic eruptions, fossil fuel burning, and decomposition events occur. The cycle ends when sulphur-containing molecules fall back to Earth through precipitation events.

Phosphorus Cycle

The phosphorus cycle tracks the natural pathways of phosphorus, which move between the three major components of the Earth: the lithosphere, the hydrosphere, and the biosphere. All living organisms require phosphorus as an essential component, which is present in cell membranes, DNA, RNA, ATP, and phospholipid structures. Figure 2.4 shows a diagrammatic representation of the cycle. The steps are discussed below:

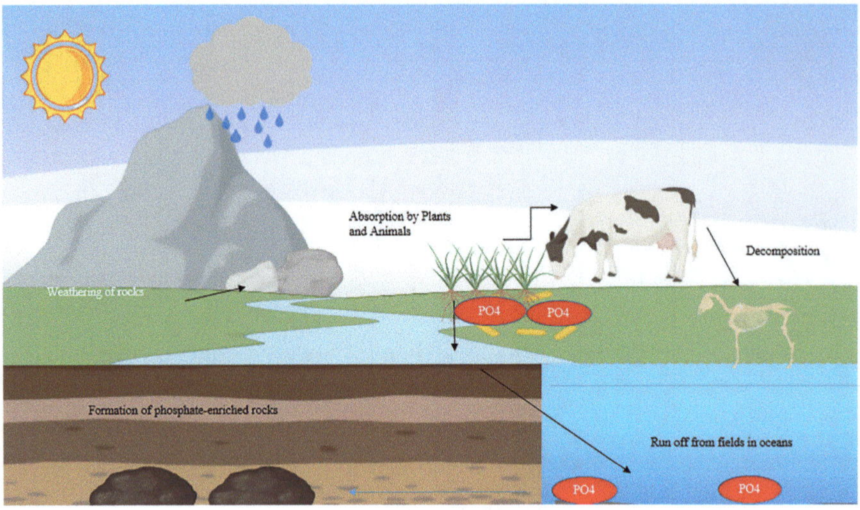

Fig. 2.4 Phosphorus cycle. (Image source: Authors' personal collection)

Weathering

Weathering processes, such as rain and erosion, break down these rocks, releasing phosphate salts into the soil and water. This process makes phosphorus available to other forms of life.

Absorption by Plants

Plants obtain dissolved phosphate salts from both soil and water for their biological needs to synthesise nucleic acids and ATP. Aquatic plants gather their inorganic phosphorus from the base of the water layers. The limited availability of phosphate compounds in the soil and water restricts plant growth.

Absorption by Animals

Animals derive phosphorus by ingesting plant matter and animal food sources. Phosphorus enters their body tissues, bones, and teeth through metabolic activity.

Decomposition

Following the death of plants and animals, decomposing microorganisms such as bacteria and fungi transform organic substances into inorganic phosphate, which the soil or water absorbs. Recovery of phosphorus within ecosystems occurs through its repetitive cycling.

Sedimentation

Excess phosphorus from aquatic environments ultimately becomes deposited as sediments at the bottom of these water environments. Sediments accumulate over extensive periods before tectonic activities, leading to the formation of new phosphate-enriched rocks.

Oxygen Cycle

The oxygen cycle is a biogeochemical process that tracks oxygen molecules between the Earth's atmosphere, biosphere, hydrosphere, and lithosphere. Earth receives ongoing life support from the oxygen cycle, which tightly interacts with the carbon cycle and other vital processes. Figure 2.5 shows a diagrammatic representation of the cycle. The steps are discussed below:

Photosynthesis

During photosynthesis, oxygen-producing organisms, such as plants and algae, alongside cyanobacteria, transform CO_2 and H_2O into $C_6H_{12}O_6$, along with O_2. The production of O_2 through photosynthesis generates most atmospheric oxygen.

$$6CO_2 + 6H_2O + \text{sunlight} \rightarrow C_6H_{12}O_6 + 6O_2$$

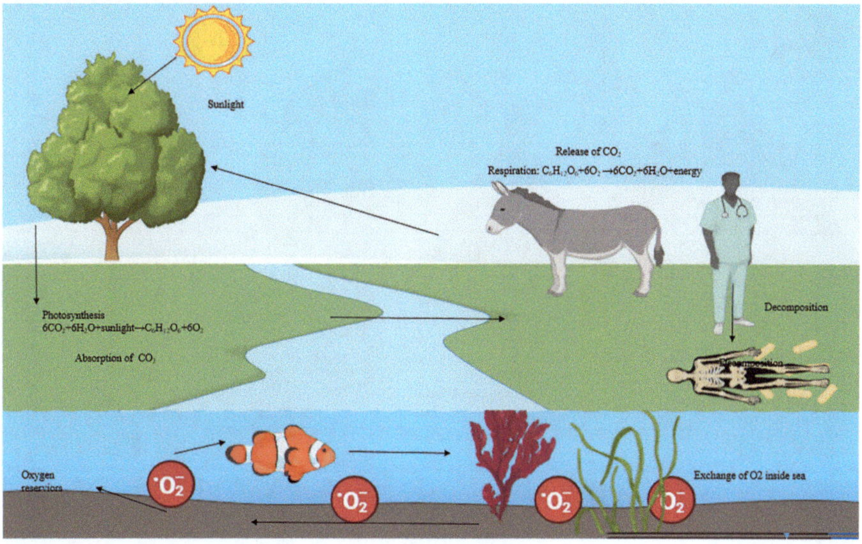

Fig. 2.5 Oxygen cycle. (Image source: Authors' personal collection)

Respiration

All organisms that require oxygen use it for cellular respiration when they break down glucose to generate energy. Such chemical reactions produce carbon dioxide as a secondary product of the process.

$$C_6H_{12}O_6 + 6O_2 \rightarrow 6CO_2 + 6H_2O + \text{energy}$$

Decomposition

After death, decomposers utilise organic materials through microbial and fungal activity while consuming atmospheric oxygen and releasing CO_2 into the air.

Oxygen Exchange in Oceans

The oxygen cycle depends heavily on ocean functioning. Ocean phytoplankton produce oxygen through photosynthesis, but marine organisms require this oxygen for their metabolic process of respiration. Oxygen molecules move freely between water and atmospheric gas compounds.

Formation of Oxygen Reservoirs

The transmission of oxygen-carrying calcium carbonates occurs over geological time frames as the detritus from living organisms is incorporated into sedimentary and rocky formations within the lithosphere. The reservoirs store oxygen in non-gaseous forms.

Iron Cycle

Earth's crust, oceans, living organisms, and the atmosphere constitute the four domains through which iron continuously circulates as a part of its biogeochemical cycles. Almost all life forms depend on iron as an essential micronutrient to function, as this mineral supports oxygen transport via haemoglobin and photosynthesis. All these facilitate their chlorophyll production and cellular respiration through electron transport. Figure 2.6 shows a diagrammatic representation of the cycle. The steps are discussed below:

Weathering of Rocks

Iron is released into the soil and water by the weathering of iron-rich rocks and minerals. Volcanic eruptions also contribute to the atmospheric iron deposition. The released iron exists as soluble ferrous iron or insoluble ferric iron.

Absorption by Plants

Plants absorb soluble ferrous iron from the soil through their roots. Iron is essential for producing chlorophyll, which enables photosynthesis.

Fig. 2.6 Iron cycle. (Image source: Authors' personal collection)

In aquatic ecosystems, phytoplankton use dissolved iron for similar purposes.

Transfer Through Food Chains

Animals acquire iron by consuming plants or other animals. Iron is a critical component of haemoglobin in vertebrates, facilitating oxygen transport in blood. A deficiency in iron leads to anaemia, impairing oxygen delivery.

Decomposition

After the death of plants and animals, bacteria break down the complex organic matter into simpler substances by releasing iron back into its elemental form.

Microbial Oxidation and Reduction

Microorganisms play a significant role in cycling iron between its ferrous forms under oxygenated conditions. In this, iron-oxidising bacteria convert Fe^{+2} to Fe^{+3} under oxygenated conditions. Examples are *Acidithiobacillus ferrooxidans, Leptothrix ochracea, Sphaerotilus natans, Gallionella ferruginea, Mariprofundus ferrooxydans, and Rhodobacter ferrooxidans.* In contrast, iron-reducing bacteria convert Fe^{+3} to Fe^{+2} under anaerobic conditions. Examples are *Geobacter metallireducens, Shewanella oneidensis, Deferribacter desulfuricans, Geothrix fermentans,* and *Thermoanaerobacter siderophilus.*

Ecology of Zoonotic Bacteria

Zoonotic bacteria, capable of transmission between animals and humans, exist within intricate ecological systems. Their survival, transmission, and emergence are shaped by a complex interplay of factors. These include siderophiles, their reservoirs in diverse host species and the environment, various modes of transmission, their ability to persist outside a host, and dynamic interactions with hosts. Moreover, this is influenced by external factors like human activities and environmental changes.

Reservoirs

Zoonotic bacteria are maintained in nature within reservoirs, which can be other animals (wild or domestic) or environmental niches. These reservoirs serve as long-term hosts or environments where the bacteria can survive and multiply without necessarily causing disease in the reservoir itself.

Animal Reservoirs

A vast array of animal species harbour zoonotic bacteria. Wild animals like rodents (e.g. *Borrelia burgdorferi,* causing Lyme disease, and hantaviruses), bats (implicated in various viral zoonoses, though bacterial examples exist), and birds are significant reservoirs. Domesticated animals, including livestock (cattle, pigs, and poultry—sources of *Salmonella*, *Campylobacter*, and *Brucella*) and companion animals (dogs and cats—can carry *Pasteurella* and *Bartonella*), also play a crucial role due to their close contact with humans. The health status of the reservoir population, its density, and movement patterns all influence the potential for bacterial spillover.

Environmental Reservoirs

Some zoonotic bacteria survive and even multiply in the environment outside of a host. Soil and water can serve as reservoirs for *Bacillus anthracis* (anthrax) and *Leptospira* species (leptospirosis). The persistence of bacteria in these environmental reservoirs is influenced by factors such as temperature, moisture, pH, and nutrients.

Transmission Routes

The movement of zoonotic bacteria from their reservoirs to human populations occurs through various transmission routes:

Direct Contact

This involves physical contact with an infected animal or its bodily fluids (saliva, blood, urine, and faeces). Examples include handling livestock, petting animals, or being bitten or scratched.

Indirect Contact

This occurs through contact with environments or objects contaminated by infected animals. This can include contact with contaminated soil, water, or surfaces in animal habitats or farms.

Vector-Borne Transmission

Arthropod vectors like ticks, mosquitoes, and fleas can transmit zoonotic bacteria between animals and humans. Lyme disease, for instance, is transmitted by ticks carrying *Borrelia burgdorferi.*

Food-Borne Transmission

Consuming contaminated food products of animal origin (undercooked meat, unpasteurised milk, and contaminated eggs) is a major route for *Salmonella*, *Campylobacter*, and certain strains of *E. coli*. Contamination can occur at various food production, processing, or preparation stages.

Waterborne Transmission

Ingesting water contaminated with animal faeces containing zoonotic bacteria can lead to infection. *Leptospira* species are a prime example of waterborne zoonotic bacteria.

Environmental Factors

Environmental factors play a significant role in zoonotic bacteria's survival, distribution, and transmission. Temperature, precipitation, humidity, and soil composition can affect the persistence of bacteria in the environment and the survival and activity of vectors and reservoir hosts. Changes in these factors, often linked to climate change, can alter the geographical range of reservoirs and vectors, leading to the emergence of zoonotic diseases in new areas.

The Impact of Human Activities

Human activities are increasingly recognised as major drivers of zoonotic disease emergence and spread. The intensification of livestock farming, for instance, can lead to increased animal density, facilitating the transmission of bacteria among animals and consequently increasing the risk of spillover to humans through direct contact or food contamination. Moreover, the use of antimicrobials in agriculture contributes to the development and spread of antimicrobial-resistant zoonotic bacteria. Furthermore, as human populations expand and encroach into natural habitats through urbanisation and land use change, contact between humans, livestock, and wildlife increases, creating new interfaces for pathogen exchange. Deforestation, habitat fragmentation, and urbanisation alter the distribution and behaviour of reservoir hosts and vectors, potentially increasing human exposure to zoonotic bacteria. Finally, increased global travel and trade of animals and animal products facilitate the rapid spread of zoonotic bacteria across geographical boundaries.

Bacterial Ecology Research Methods

Bacterial ecology research methods use techniques and approaches to study bacteria and their interaction with the environment and other organisms. These methods have evolved from traditional culture-based techniques to modern molecular and computation-based approaches. They aim to understand bacteria's complex relationships and roles in the ecosystem.

Culture-Based Methods

These methods involve culturing diverse microorganisms under laboratory conditions to study their morphological and physiochemical characteristics (Fig. 2.7). MacConkey agar plates facilitate the differentiation of bacteria that ferment lactose, aiding in the isolation of pure bacterial cultures. Mannitol salt agar is used to identify *Staphylococcus* species, allowing them to be isolated from specimens contaminated with other bacteria. Stab cultures are used for the short-term storage of bacterial isolates.

Culturomics, a recent technique, helps grow diverse organisms by providing different environmental conditions. These methods enable direct behavioural monitoring of bacteria, and they are affordable. The inability to culture environmental bacteria remains a significant drawback since most bacteria fail to grow in culture. The culture-based techniques confer crucial validation of molecular findings and examination of bacterial behaviour.

DNA-Based Methods

DNA-based methods help study bacterial communities, their interactions, and their roles in the ecosystem by analysing their DNA.

PCR-Based Methods

PCR enables the rapid amplification of specific DNA sequences (Fig. 2.8). This forms the basis for various applications, including (i) detection by identifying the presence of specific bacterial taxa or genes in a sample, (ii) determining the abundance of specific bacterial groups or genes within a community by quantitative PCR (qPCR), and (iii) determining sequencing-specific PCR-amplified gene regions (e.g. 16S rRNA and functional genes) by amplicon sequencing to assess diversity and relative abundance.

DNA–DNA Hybridisation (DDH)

This technique measures the overall genomic similarity between two bacterial isolates by assessing the degree of hybridisation between their denatured DNA strands (Fig. 2.9). Higher hybridisation percentages indicate closer genetic relatedness, providing insights into species boundaries and evolutionary relationships.

16S rRNA Gene Sequencing

By sequencing a specific region of the highly conserved 16S ribosomal RNA gene, bacterial taxa present in a sample can be identified and phylogenetic trees can be constructed. This method is a cornerstone of microbial ecology for assessing diversity and understanding evolutionary relationships.

Metagenomics (Shotgun Sequencing)

This approach involves directly sequencing the total DNA extracted from an environmental sample. It bypasses the need for culturing individual species. It provides a comprehensive snapshot of the genetic potential and taxonomic diversity of the entire microbial community. Analysis of the resulting large datasets can reveal the functional genes present and the metabolic capabilities of the community.

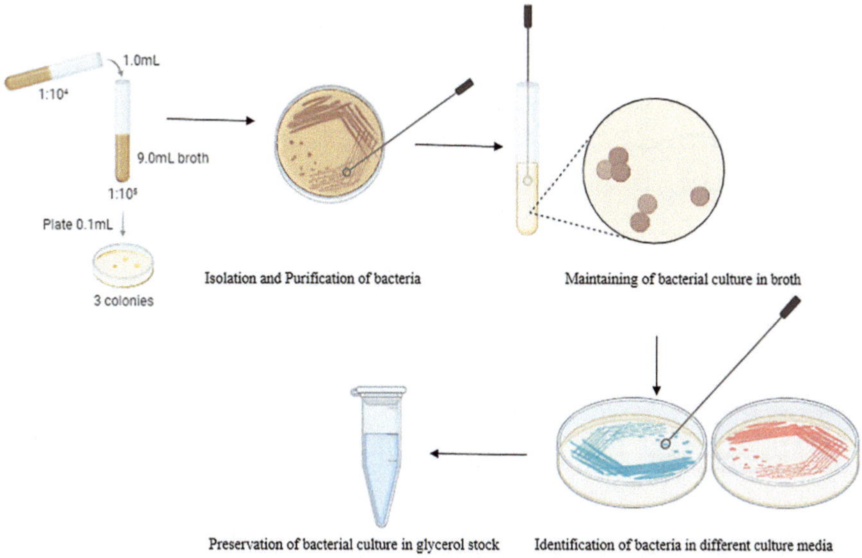

Fig. 2.7 Isolation and maintenance of bacterial culture in the lab. (Image source: Authors' personal collection)

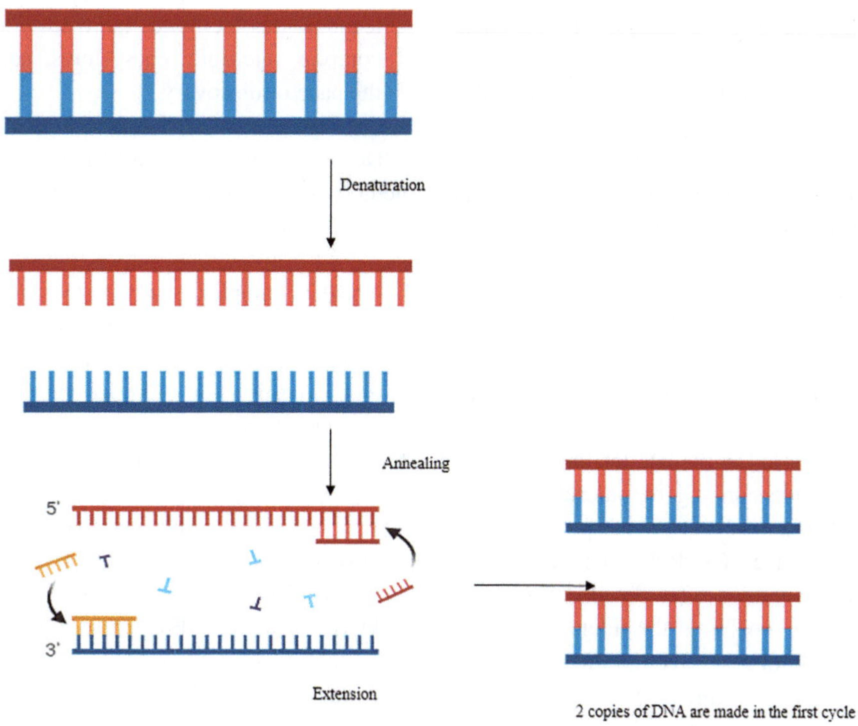

Fig. 2.8 Principle of polymerase chain reaction (PCR). (Image source: Authors' personal collection)

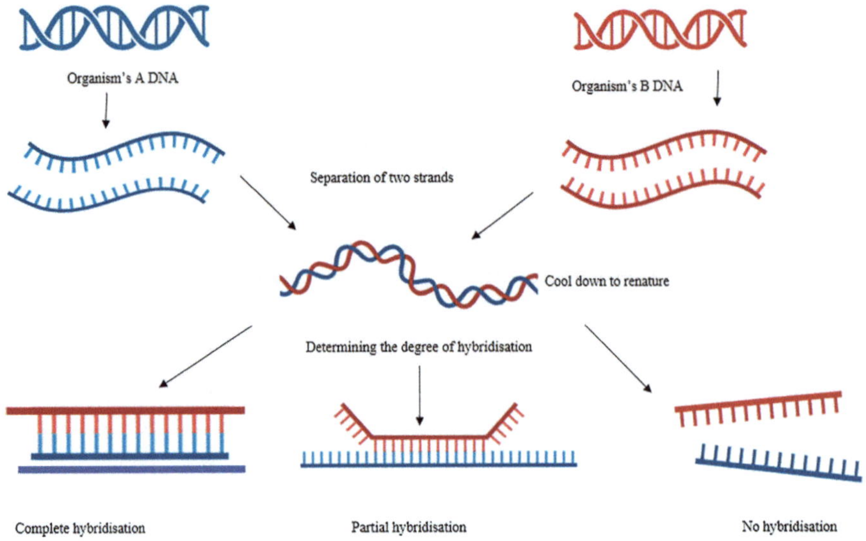

Fig. 2.9 Principle of DNA–DNA hybridisation method. It is used to detect genome similarity between bacteria. (Image source: Authors' personal collection)

DNA Microarrays

These tools consist of a collection of oligonucleotide probes immobilised on a solid surface. They allow for the simultaneous detection and quantification of specific DNA sequences. This is increasingly needed to analyse strain-level differences, gene presence/absence, and gene expression patterns within bacterial communities.

DNA-based molecular approaches provide several crucial advantages for studying bacterial ecology. These offer the following advantages:

1. These tests can perform culture-independent analysis to investigate most bacteria that cannot be cultivated in the lab, leading to a more accurate understanding of natural microbial communities.
2. Techniques like 16S rRNA gene sequencing and metagenomics offer detailed taxonomic information, enabling the identification of microbial strains and a comprehensive assessment of community composition. Metagenomics also provides valuable functional insights by allowing the exploration of the genetic potential within microbial communities and the identification of genes involved in various ecological processes.
3. The rapid data output of modern sequencing technologies facilitates large-scale studies of complex microbial ecosystems, accelerating the pace of discovery.

Despite these significant benefits, DNA-based methods also present certain limitations and challenges. These are:

1. The data complexity arising from the massive datasets, particularly in metagenomics, demands sophisticated bioinformatics tools and expertise for effective processing and interpretation.
2. Technical biases can also be introduced during DNA extraction, where the efficiency may vary across different bacterial species and environmental matrices, potentially skewing results. Similarly, primer selection bias in PCR-based methods can lead to the underrepresentation of certain taxa.
3. A final challenge lies in linking genes to functions. For example, while metagenomics can identify genes, determining their specific roles and how they are expressed and regulated within the environment remains a complex undertaking.

Fluorescence-Based Methods and Stable Isotope Probing

Fluorescence-based methods incorporate fluorescent dyes or proteins to create bacterial detection and visualisation systems that enable the continuous monitoring of bacterial processes (Fig. 2.10). The fluorescent dye SYTO 9, combined with propidium iodide (PI), separates the living from the dead bacteria. SYTO 9 enters the bacterial cells and helps them fluoresce green, while PI enters the damaged cells, enabling them to fluoresce red. Likewise, green fluorescent protein (GFP) allows the tracking of genetically modified bacterial gene expression. Highly sensitive and specific results are attainable with these methods, although equipment expenses and complex sample background fluorescence become limitations to their use.

Microbes under investigation obtain their substrate from isotope-labelled materials as part of stable isotope probing (SIP) analysis. The DNA-SIP, RNA-SIP, and protein-SIP procedures enable researchers to determine active microbial populations while understanding their ecological functions. Direct associations between microbial types and biological functions through SIP are made, including carbon distribution, while studying unmodified natural systems. This method needs expensive isotope instruments, advanced lab equipment, and limited resolution potential when many microbes process equivalent substrates.

Future Directions

Bacterial ecology might face difficulties in the future. However, there are chances for development. The "microbial dark matter," which has remained unexplored for many years, can be addressed through nanofluidic devices, which could help discover bacterial lineages and features. This development can enhance understanding of the roles of microbes in the environment and human health—the sudden rise of antimicrobial resistance (AMR) calls for immediate action. By examining microbial interactions and evolutionary processes that lead to resistance, bacterial ecology can inform the development of targeted

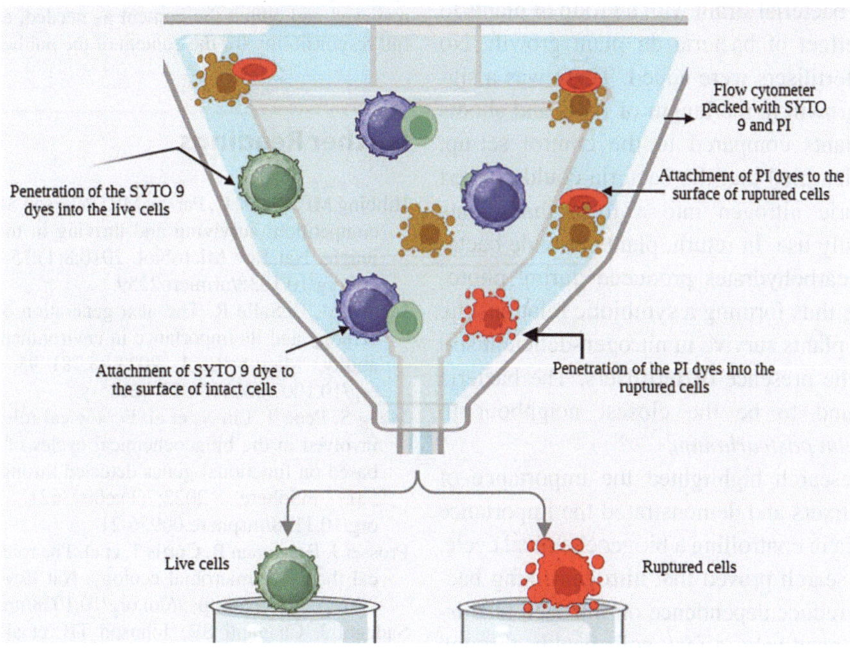

Fig. 2.10 Fluorescence-based method, using SYTO9 and propidium iodide (PI) for differentiating life and dead cells in the laboratory. (Image source: Authors' personal collection)

therapies, such as phage-based treatments, to combat human pathogens.

Combining machine learning and artificial intelligence (AI) can help manage the complexity of omics data. These methods can accelerate metabolic modelling and gene annotation discoveries, decode microbial networks, and predict relationships. Microbial ecosystems are under stress due to pollution and climate change, but bacterial ecology presents chances for sustainability. While artificial communities may improve carbon sequestration or restore ecological balance, microbial bioremediation can detoxify habitats. Improved understanding of microbial resilience will inform strategies to lessen the effects of stress on humans.

Case Study

Umme, a PhD student of Soil Microbiology, isolated a bacterium from the rhizospheric soil region of a marshy wetland with decaying organic matter. The isolated bacterial strain could tolerate up to 950 ppm of cadmium without any nitrogen source. Furthermore, during her study, she inoculated this bacterial strain with a group of plants to see the effect of bacteria on plant growth. No external fertilisers were added. There was a significant growth in the length of roots and shoots in the plants compared to the control set-up. Umme identified that the bacteria could convert atmospheric nitrogen into a form that plants could easily use. In return, plants provide bacteria with carbohydrates produced during photosynthesis, thus forming a symbiotic relationship. It helped plants survive in nitrogen-deficient soil without the presence of fertilisers. The bacteria were found to be the closest neighbour to *Clostridium pasteurianum.*

Her research highlighted the importance of nitrogen fixers and demonstrated the importance of bacteria in controlling a biogeochemical cycle. Umme's search proved that nitrogen-fixing bacteria can reduce dependence on nitrogen fertilisers and positively affect soil health through reduced fertiliser needs, which lower agricultural expenses. The experiment confirmed the critical function of these bacteria in sustainable agriculture practices while improving soil quality.

1. Which biogeochemical cycle is the bacterium involved in? State its role in exhibiting a symbiotic relationship with plants.
2. What type of symbiotic relationship does the bacterium have with the plant? Give some other examples of this symbiotic relationship.
3. What methods did Umme employ to identify this bacterium?

Research Questions

1. How can modern molecular techniques, including 16S rRNA sequencing, FISH, and SIP, be improved to identify the unculturable bacteria?
2. How can bacteria be employed for bioremediation and sustainable energy production?

Declaration During the preparation of this chapter, we have used AI tools, such as QuillBot, in order to improve the grammar and readability of the text and make it more appealing to the readers. After using such tools, we have reviewed and edited the content as needed, and we take full responsibility for the content of the publication.

Further Readings

Hibbing ME, Fuqua C, Parsek MR, Peterson SB. Bacterial competition: surviving and thriving in the microbial jungle. Nat Rev Microbiol. 2010;8(1):15–25. https://doi.org/10.1038/nrmicro2259.

Lemke M, DeSalle R. The next generation of microbial ecology and its importance in environmental sustainability. Microb Ecol. 2023;85:781–95. https://doi.org/10.1007/s00248-023-02185-y.

Meng S, Peng T, Liu X, et al. Ecological role of bacteria involved in the biogeochemical cycles of mangroves based on functional genes detected through GeoChip 5.0. mSphere. 2022;7(1):e0093621. https://doi.org/10.1128/msphere.00936-21.

Prosser J, Bohannan B, Curtis T, et al. The role of ecological theory in microbial ecology. Nat Rev Microbiol. 2007;5:384–92. https://doi.org/10.1038/nrmicro1643.

Sadeghi J, Chaganti SR, Johnson TB, et al. Host species and habitat shape fish-associated bacterial communities: phylosymbiosis between fish and their microbiome. Microbiome. 2023;11:258. https://doi.org/10.1186/s40168-023-01697-6.

Tzamali E, Poirazi P, Tollis IG, Reczko M. A computational exploration of bacterial metabolic diversity identifying metabolic interactions and growth-efficient strain communities. BMC Syst Biol. 2011;5:167. https://doi.org/10.1186/1752-0509-5-167. Published 2011 Oct 18.

Wang R, Wang M, Wang J, Lin Y. Habitats are more important than seasons in shaping soil bacterial communities on the Qinghai-Tibetan Plateau. Microorganisms. 2021;9(8):1595. https://doi.org/10.3390/microorganisms9081595. Published 2021 Jul 27.

Weiland-Bräuer N. Friends or foes-microbial interactions in nature. Biology (Basel). 2021;10(6):496. https://doi.org/10.3390/biology10060496. Published 2021 Jun 2.

Host–Pathogen Relationship and Clinical Manifestations of Bacterial Zoonoses

3

Aditya Kundu, Mallika Sen Gupta, and Ujjala Ghoshal

Learning Objectives

1. To understand the principles of host–pathogen interactions involved in the transmission and pathogenesis of bacterial zoonosis
2. To describe the key clinical manifestations of bacterial zoonosis and diagnostic approaches for bacterial zoonosis
3. To evaluate the role of interdisciplinary strategies, such as the One Health approach, in the prevention and control of bacterial zoonoses

Introduction

The host–pathogen relationship is a crucial component of bacterial zoonoses. It represents a diverse interaction between a host and a bacterial pathogen. Unlike non-zoonotic pathogens, zoonotic bacteria have animal reservoirs and cause infection in humans. These interactions are often at the expense of the host, where the bacteria benefit by utilising the host and replicating within it. Several key factors influence this relationship. Bacterial pathogens can develop mechanisms, such as altering surface antigens, producing toxins, or residing within host cells, to avoid detection and immune recognition, thereby evading immune responses. Concurrently, the hosts develop both congenital and adaptive immune reactions to detect and eliminate these pathogens.

The environment influences the host–pathogen interaction. Climate change, deforestation, intensive cultivation, and urban intervention are altering animal habitats, bringing people into close contact with wildlife and livestock reservoirs, increasing the likelihood of zoonotic infections. Additionally, global journeys and business travel are also spreading zoonotic bacteria beyond their natural habitats. The overuse and misuse of antibiotics apply selective pressure on bacteria, promoting the emergence of resistant strains and increasing the risk of infections caused by these pathogens. As a result, the host–pathogen relationship in zoonoses is not only important for understanding disease pathogenesis and transmission but also an essential aspect of public health. For prompt diagnosis, treatment, and the formulation of preventive strategies, an understanding of host–pathogen interactions is crucial. The "One Health" approach, involving veterinary, human, and environmental health, is crucial in combating these zoonotic infections.

A. Kundu · M. S. Gupta · U. Ghoshal (✉)
Department of Microbiology, All India Institute of Medical Sciences, Kalyani, West Bengal, India

S. C. Parija, A. Poddar (eds.), *Textbook of Bacterial Zoonoses*, Microbial Zoonoses,
https://doi.org/10.1007/978-981-95-6615-0_3

Types of Host–Pathogen Relationship in Bacterial Zoonoses

Commensalism

Commensalism is a one-sided relationship in which the bacteria live within the host without any effect on the host. If the host immunity diminishes, these organisms can become opportunistic pathogens. For example, *Staphylococcus aureus* in the nasal mucosa of cattle or pigs may exist harmlessly. However, it can become an opportunistic zoonotic pathogen, causing abscesses or mastitis, and can infect humans through direct contact or contaminated milk or meat.

Parasitism

Parasitism is a classical disease-causing scenario where the bacterium benefits from the host, resulting in tissue injuries, inflammation, clinical symptoms, and cell death. For example, *Brucella abortus* in cattle causes brucellosis, affecting the reproductive tract and leading to abortion. It is transmitted to humans via unpasteurised dairy or contact with animal fluids. Likewise, *Francisella tularensis* causes tularaemia in small mammals and humans, leading to severe lymphadenopathy and fever.

The extent of infection depends on various factors, including the virulence of the pathogen, host adaptation, and the host's immune response.

Determinants of Host–Pathogen Dynamics

Bacterial Virulence Factors

Bacterial virulence is a crucial molecular factor that enables pathogens to colonise, sustain, and multiply within host organisms. By virtue of their virulence factors, organisms directly affect the host, and these factors also determine the severity, duration, and spread of the infection. In bacterial zoonoses, virulence factors determine whether an interaction results in colonisation, local infection, or systemic disease.

Adhesins and Pili

Adhesins and pili are crucial virulence factors that enable bacteria to attach to host epithelial cells. Following adhesion, some bacteria release proteins that facilitate their entry into host cells. For example, *Yersinia* spp. use invasin proteins to penetrate the intestinal mucosa and infect the deeper cells.

Bacterial toxins are another important factor for virulence. Exotoxins produced by *Clostridium botulinum* or *Bacillus anthracis* damage cellular structures and kill the cell. Endotoxins from Gram-negative bacteria can lead to septic shock.

In addition, bacteria produce various enzymes, such as proteases by *Leptospira* spp. and hyaluronidase by *Pasteurella multocida*, which destroy host structures. These enzymes facilitate the spread of bacteria through tissues and promote the invasion of new sites.

Zoonotic bacteria evade the immune mechanisms through various factors. These include (i) the formation of protective capsules, (ii) antibody recognition, (iii) antigenic variation to evade the secretion system, (iv) interference with immune signalling, and (v) development of resistance to oxidative killing by phagocytes, among others. For example, *Brucella* spp., *Rickettsia* spp., and *Coxiella burnetii* live in the macrophages by preventing phagosome–lysosome fusion, allowing them to persist and spread within the host.

Together, all these factors play a central role in the pathogenicity of zoonotic bacteria.

Host Immune Response

The host immune response, coordinated by multiple molecular and cellular factors, plays a crucial role in combating bacterial zoonoses. The efficiency and uniqueness of the immune system largely determine the outcome of an infection. The immune response is approximately divided into congenital and adaptive immunity.

Innate Immunity

Innate immunity is the body's immediate, non-specific defence. It acts as the first line of defence against bacterial invasion and is crucial in containing transitions in the early stages. Innate immunity components include physical barriers, such as the mucosa and skin membranes, which act as physiological barriers and prevent bacteria from entering the body. Mucosal secretions, including lysozyme and defensins, also possess antimicrobial properties. Other components include pattern recognition receptors (PRRs), such as toll-like receptors (TLRs), that recognise specific molecular patterns responsible for pathogenesis. Phagocytic cells, such as neutrophils and macrophages, cause the lysis of bacteria through a process known as phagocytosis. The complement system promotes opsonisation, lyses bacterial cells by forming membrane-attacking complexes, and helps regulate host immunity.

Adaptive Immunity

The adaptive immune response is slow, but it provides a very specific and long-term effect. It is divided into humoral and cell-mediated immunity. Cell-mediated immunity is primarily mediated by T lymphocytes (CD8+ cytotoxic T cells), which kill intracellular bacteria by activating macrophages through coordinated cellular responses. In contrast, humoral immunity is mediated by antibodies produced by B lymphocytes.

The effect of both innate and adaptive reactions is affected by host-specific factors, including genetics, age, and medical conditions, such as HIV/AIDS, cancer, and diabetes, which can compromise immunity and increase the sensitivity of zoonotic infections.

Host Factor

A series of host-related factors influence the risk of contracting zoonotic bacterial infections and their severity. These determinants influence the hosts' ability to eliminate pathogens and elicit a clinical response to the disease.

Genetic Predisposition

Individual genetic variations are an essential host factor. The genes coding for components, including the major histocompatibility complex (MHC) and TLRs, determine how effectively pathogens are detected and eliminated. For example, TLR mutations can disrupt the recognition of bacterial components. Similarly, the difference in cytokine expression (e.g. IL-1, IL-6, and TNF-α) can change the magnitude and quality of inflammatory reactions. The variation in MHC alleles affects antigen presentation and the initiation of adaptive immunity, which, in turn, impacts the host's ability to defend against pathogens.

Age

Age is a significant factor in immune function. Older adults often experience immunosenescence, characterised by low T-cell activity and decreased antibody production, making them prone to zoonotic diseases, whereas infants and young children have immature immune systems with limited memory cells, making them more susceptible to zoonotic infections as well.

Immunocompromised States

Individuals with a compromised immune system are more susceptible to zoonotic infections. The group includes the recipients of organ transplantation, cancer patients undergoing chemotherapy, HIV/AIDS patients, and patients on long-term immunosuppressive therapy. In these individuals, both innate and adaptive immune responses are deranged, allowing pathogens to establish systemic or opportunistic infections.

Environmental and Occupational Exposure

Some occupations increase the likelihood of exposure to zoonotic bacteria. Farmers, veterinarians, harvesters, and laboratory personnel often meet infected animals or contaminated materials. In addition, individuals who engage in activities such as hunting or butchering are at a high risk of infection.

Understanding these host-related factors is important for risk assessment, targeted prevention, and individual treatment strategies in zoonotic bacteria.

Ecology of Host–Pathogen Relationship

The ecology of host–pathogen interactions in bacterial zoonoses is crucial to understanding how the infection spreads. The ecological structure focuses on the relationship between humans, animals, and the environment. In this interface, the most zoonotic transfers occur. These are dynamic and influenced by environmental changes, human behaviour, agricultural practices, and globalisation. A central concept in this ecological approach is that zoonotic pathogens are not stable in their reservoirs; they constantly interact with the environment and host populations.

Animal Reservoirs

Animal reservoirs are essential components of the ecology of zoonotic bacteria (Table 3.1). A bacterial pathogen can persist over time in these species, often without causing any disease. This asymptomatic carriage allows bacteria to extend over vulnerable populations under favourable conditions from time to time. Many zoonotic bacteria have developed a balanced relationship with their natural animal hosts, which act as long-term reservoirs. *Brucella* spp., for example, infect the livestock, such as animals, goats, and sheep, and reside in these animals, affecting their reproductive systems. *Leptospira* spp. resides in the renal tubules of rodents, especially in the kidneys of mice, and is excreted in the urine, polluting water sources and infecting humans through the mucosa or damaged skin. *Yersinia pestis*, the agent of plague, spreads silently in the rodent population and is transmitted through fleas. However, in these hosts, the bacteria cause minimal infection, yet they can become highly virulent upon transmission to humans. The asymptomatic carriage in animal reservoirs may facilitate unnoticed bacterial persistence, potentially resulting in sudden and unexpected outbreaks in human populations.

Spillover Events

Spillover refers to the process in which a pathogen maintained in an animal reservoir infects a new host. This process is rarely observed, and it occurs due to disruptions in the environment and human activities that increase contact between humans and reservoirs. These include:

Habitat Encroachment

When forests are destroyed for agriculture or urban development, people come into close contact with wildlife, which increases the possibility of direct or indirect contact with zoonotic bacteria. For example, the invasion of houses by rodents increases the risk of exposure to *Leptospira* or *Y. pestis.*

Climate Change

Changing weather patterns affect the distribution and behaviour of hosts and vectors. Hot temperatures and increased rainfall expand the populations of fleas, ticks, and rodents and spread zoonotic bacteria, which then spread into new areas and infect new populations.

Global Travel and Trade

The increase in international movement of humans, animals, and animal products facilitates the spread of zoonotic bacteria across geographical boundaries. For example, trade in infected livestock or contaminated dairy products can introduce *Brucella* or *Coxiella* in areas that were free of disease.

Mechanisms of Transmission

Zoonotic bacteria can spread through various routes (Fig. 3.1). The bacteria spread directly through direct touch, ingestion, inhalation, and vectors. Understanding these mechanisms is crucial for preventing and controlling zoonotic diseases.

Table 3.1 Animal reservoirs of zoonotic bacteria

Bacterial pathogen	Primary animal reservoir(s)	Type of transmission to humans	Remarks
Brucella spp.	Cattle, goats, sheep, pigs	Direct contact, unpasteurised dairy	Occupational risk (farmers, veterinarians); chronic infection possible
Leptospira spp.	Rodents (esp. rats), dogs, livestock	Contact with urine--contaminated water or soil	Common in flood-prone, poor--sanitation settings
Bacillus anthracis	Herbivores (cattle, sheep, goats)	Spores via skin, inhalation, ingestion	Cutaneous, inhalational, gastrointestinal forms
Yersinia pestis	Wild rodents (rats, prairie dogs); fleas as vectors	Flea bites, contact with infected animals	Reservoir in wild rodent populations; can cause epidemics
Francisella tularensis	Rabbits, rodents, deer; ticks as vectors	Tick bites, direct contact, inhalation	"Rabbit fever"; hunters and lab workers at risk
Coxiella burnetii	Cattle, sheep, goats	Inhalation of birthing fluids, dust; unpasteurised dairy	Highly infectious; Q fever outbreaks in farms, abattoirs
Rickettsia spp.	Ticks (vectors); rodents, dogs (hosts)	Tick bites	Endemic in specific geographic and seasonal patterns
Mycobacterium bovis	Cattle, deer, other ungulates	Unpasteurised dairy, inhalation	Causes zoonotic tuberculosis; controlled in many countries
Bartonella henselae	Cats, especially kittens	Scratches, bites	Cat scratch disease; more severe in immunocompromised
Streptobacillus moniliformis	Rats	Bites, contact with rat secretions	Rat-bite fever; also from contaminated food
Campylobacter jejuni	Poultry, cattle, domestic pets	Faecal–oral, undercooked poultry	Leading bacterial cause of gastroenteritis
Salmonella spp.	Poultry, reptiles, cattle	contaminated food/water, contact with animals	Many zoonotic serotypes; risk in the food industry
Listeria monocytogenes	Cattle, sheep, goats	contaminated dairy and meat products	High risk to pregnant women and neonates
Burkholderia pseudomallei	Soil and water	Percutaneous, inhalation, ingestion	Not animal–host dependent; common in South-East Asia
Pasteurella multocida	Cats, dogs	Bites, scratches	Normal flora in pets' mouths; wound infections common
Burkholderia mallei	Horses, mules, donkeys	Contact with infected animals	Glanders; eradicated in many countries; potential bioweapon
Borrelia recurrentis	Humans (epidemic form); rodents (endemic)	Louse or tick bites	Louse-borne relapsing fever during epidemics

Direct Contact

Direct contact transmission occurs when a person comes into physical contact with an infected animal or its biological material (e.g. urine, saliva, and blood). The infection occurs through cuts or scratches in the skin. Cat scratch disease, caused by *Bartonella henselae*, is a classic example of direct contact transmission. Humans contract the disease after contact with an infected cat. Bacteria enter the skin through breaks, causing localised infections and, in some cases, systemic disease. Tularaemia, caused by *Francisella tularensis*, can also be transmitted through direct contact with infected animals or dead bodies of rabbits and rodents. Infected ticks or cuts from infected animals can also transmit the disease. Bacteria usually enter the body through damaged skin or mucous membranes, and the result of the infection is often localised at the site of entry. However, it can also spread systematically,

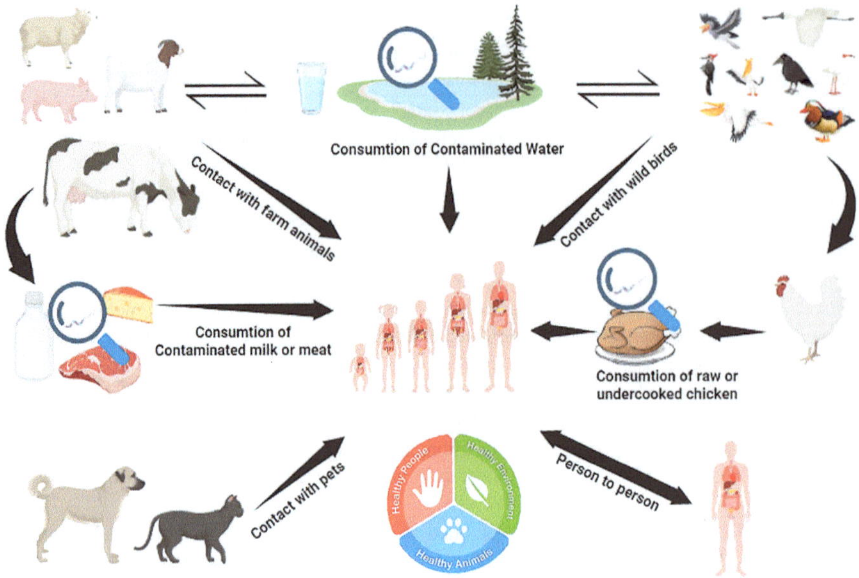

Fig. 3.1 Different transmission routes of zoonotic bacteria. (Image source: Ali S, Alsayeqh AF. Review of major meat-borne zoonotic bacterial pathogens. Front Public Health. 2022;10:1045599. https://doi.org/10.3389/fpubh.2022.1045599. Permitted for reproduction under CC BY license)

depending on the host's immune status and the pathogen's immune response.

Ingestion

Bacteria can be transmitted through the ingestion of contaminated food or water. These bacteria are often present in the intestinal tract of animals and are excreted in the faeces. If not properly handled, faeces can contaminate food products or drinking water; the consumption of contaminated meat or milk can cause diseases such as brucellosis. Bacteria are commonly found in the milk of infected animals, including cattle, goats, and sheep. When humans use these products without proper pasteurisation, bacteria enter the digestive system and cause infection by spreading to different organs. Non-typhoidal *Salmonella*, often transmitted through contaminated meat or eggs, is another example of an ingestion-related zoonotic infection. Infections are usually manifested as gastrointestinal diseases, but more severe systemic diseases can occur if bacteria spread beyond the gut.

Inhalation

Inhalation of aerosolised bacterial particles is an uncommon but crucial transmission route, especially for bacteria that thrive in animal products and contaminate the environment. Q fever, caused by *Coxiella burnetii*, is often transmitted from infected animal products through aerosolised particles. When individuals inhale these particles, especially in areas such as farms or slaughterhouses, they develop respiratory infections, which can then spread to other parts of the body. Tularaemia can also be transmitted by inhalation of aerosolised particles. Inhalation of bacterial aerosols often leads to respiratory symptoms, such as fever and cough. More severe systemic infections can occur in individuals who are immunocompromised.

Vector-Borne Transmission

Certain zoonotic bacteria rely on vectors, such as fleas, ticks, or lice, for transmission to both animals and humans. These vectors are usually blood-sucking arthropods that transmit bacteria from infected animals to humans. Vectors are essential for the multiplication of bacteria and play an important role in host–pathogen relationships.

The plague caused by *Y. pestis* is a prime example of transmission via vectors. Wild rodents or livestock serve as vectors for fleas, transmitting bacteria to humans through bites. *Ixodes* tick bites, which harbour *Borrelia burgdorferi*, cause Lyme disease in humans. This manifests as joint pain, fever, and neurological symptoms. Vector-borne transmission typically involves multiple organs, and untreated cases can result in a fulminant infection.

Host–Pathogen Co-evolution and Adaptation

The interaction between hosts and bacterial pathogens is a dynamic and ever-evolving process. Over time, both hosts and pathogens undergo evolutionary changes that affect host–pathogen conditions. It forms the projection of co-development infection, which affects the severity of the disease and the ability to influence and adapt in both hosts and pathogens.

Pathogen Adaptation

One of the primary methods that develops in response to the host is the development of mechanisms to escape from the immune response within the host. Many zoonotic bacteria have developed different resistant mechanisms against host immune responses. By altering their surface antigens, *Leptospira* and *Brucella* evade the host immune response, including phagocytosis. On the other hand, *C. burnetii* undergoes genetic mutations to evade the immune response. All these factors enable zoonotic bacteria to survive in the host and cause diseases.

Another major adaptation in bacterial pathogens is the development of secretion systems. Many bacteria have developed special protein secretion systems, such as the type III secretion system in *Salmonella* and *Y. pestis*. These systems inject bacterial proteins directly into host cells and manipulate host cell functions to promote the existence and replication of bacteria. These proteins modify the immune signalling pathway or regulate cell death to create a more favourable environment for bacterial development.

Host Adaptation

The hosts, in turn, develop responses to the pressure placed by bacterial pathogens. Immune reactions in the host often undergo selective pressure to become more effective and counteract the pathogen. Adaptive immunity, which includes antibodies and cytotoxic T cells, targets specific antigens and eliminates them. Additionally, the development of immune tolerance in the host enables differentiation between commensals and harmful pathogens, allowing for the avoidance of inflammatory reactions.

Challenges in Host–Pathogen Development

When a bacterium changes to a new, naive host from its natural host, the absence of a long history of co-development often leads to serious illness. However, when pathogens infect a new host with no prior exposure, the resultant effect can be overwhelming, thus highlighting the importance of co-development in the emergence of zoonotic diseases. The absence of co-evolution between the host and the pathogen is a crucial factor in the emergence of new zoonotic diseases in the human population, leading to rapid transmission, high morbidity, and mortality.

Co-development of hosts and pathogens is a complex process. Pathogens have consistently developed various mechanisms to adapt and survive against the host's immune system. At the same time, the host also develops immune mechanisms to fight against the pathogen and maintain immune balance.

Clinical Manifestations of Zoonotic Infections

Zoonotic infections manifest in various ways, depending on the organs involved and the pathogenesis of the disease.

Fever and Systemic Disease

In the acute phase of infection, most zoonotic diseases present with fever and often lead to pyrexia of unknown origin (PUO). It is commonly associated with fatigue, muscle pain, chills and rigors, muscle stiffness, and night sweats. For example, in leptospirosis and Q fever, there are frequent spikes in fever. In some instances, there may be a specific pattern, such as the "undulant" fever seen in brucellosis, where temperature falls and rises like waves.

Localised Infection and Skin Lesions

Many zoonotic bacteria cause localised infections instead of systemic illness. For example, *Bartonella henselae*, which causes cat scratch disease, usually presents as localised lymphadenopathy, where lymph nodes are painful and swollen after being scraped or cut by an infected cat. Similarly, *Francisella tularensis*, which causes tularaemia, causes a painful stomach ulcer instead of an infection (ulceroglandular form). Wounds often occur with inflammatory lymph nodes and, in some cases, progress to a systemic infection.

Multi-organ Involvement

Some zoonotic infections affect many organs and cause systemic disease. In cases of brucellosis, patients may experience inflammation of both the liver and spleen. Other manifestations of systemic involvement include lung problems, such as pneumonia, which are associated with diseases like Q fever. In addition, some zoonotic diseases affect the kidney and liver function, causing jaundice, liver failure, and kidney failure. *Leptospira* is an important example of a bacterium that causes damage to multiple organs, ranging from liver and kidney failure to myocarditis.

Neurological Manifestations

Neurological complications are a significant aspect of certain zoonotic infections, particularly in the chronic stage. For example, *Brucella* infection can cause neurobrucellosis, which is characterised by headaches, memory loss, cognitive dysfunction, and even psychiatric symptoms. Similarly, *C. burnetii* causes fatigue syndrome after the acute phase, during which patients experience long-term fatigue, depression, and other cognitive damage that can last for several months or years.

Serious Systemic Manifestations

In some cases, zoonotic infections spread quickly and cause life-threatening conditions. For example, *Y. pestis* causes septic plague, in which bacteria spread through the bloodstream, causing disseminated intravascular coagulopathy (DIC). This condition is associated with blood clots and gangrene in the extremities. Another potentially fatal condition is seen in the icteric form of leptospirosis, manifesting as severe liver damage, bleeding, and kidney failure.

Chronic Infections

Chronic manifestations of zoonotic infections are also a matter of concern. Many zoonotic bacteria remain in the hosts for extended periods, leading to chronic symptoms and complications. Chronic fatigue is often reported in Q fever and brucellosis. Bone complications, such as spondylitis and brucellosis, occur in chronic brucellosis. Chronic bartonellosis can progress to vascular skin lesions in the immunocompromised due to grave angiomatosis.

Diagnosis

Specimens

The diagnosis of zoonotic infections relies on the collection and analysis of appropriate biological samples from the patient. The choice of specimen is crucial and depends on the suspected pathogen, the clinical presentation, and the affected organ systems.

Blood is the most frequently collected specimen. Serum or plasma is used for serological tests to detect antibodies, indicating the body's immune response to an infection or antigens, the direct components of the pathogen. Whole blood is also used for molecular tests, such as PCR, to directly identify the pathogen's genetic material, such as DNA or RNA, present in the blood.

A stool specimen is essential for diagnosing intestinal zoonoses caused by pathogens, bacteria, or their toxins, which manifest as symptoms such as diarrhoea or abdominal pain. In cases of respiratory involvement, such as zoonotic influenza or Q fever, sputum or mucus swabs are collected from the nose or throat to identify the causative agent. For infections affecting the central nervous system, such as rabies or certain arboviral diseases, cerebrospinal fluid (CSF) may be obtained through a lumbar puncture. Tissue biopsies are invaluable when the infection causes localised lesions, such as cutaneous anthrax, or if an internal organ is involved. Additionally, urine is tested for certain pathogens, such as *Leptospira*, which are excreted in the urine.

The primary specimens of choice for diagnosing different bacterial zoonotic agents are presented in Table 3.2.

Laboratory diagnosis typically includes microscopy, culture, and antigen and antibody detection by serological tests, such as enzyme-linked immunosorbent assay (ELISA), Western blot, immunofluorescence, and point-of-care rapid diagnostic tests. Currently, molecular methods, such as PCR and sequencing, are increasingly used for the specific and accurate detection of pathogens.Conventional techniques, such as blood and tissue culture, remain the gold standard for certain diseases, including non-typhoidal salmonellosis. Culture techniques for certain highly infectious pathogens are quite cumbersome and require high-end biosafety-level (mostly BSL3) facilities, which are often difficult to establish in developing countries. Moreover, trained personnel and equipment are required for these facilities. The sensitivity and specificity of molecular techniques are quite high for zoonotic diseases. Newer techniques, including biosensors, mass spectroscopy, metagenomics, and bioinformatics, are being developed for the early diagnosis of zoonotic diseases.

Microscopy

Different types of microscopy and staining techniques are used for the diagnosis of various zoonotic pathogens. Gram staining is used to detect *Brucella* spp., *Bacillus anthracis*, *Streptobacillus*, *Campylobacter*, *Salmonella*, and other pathogens, with variable sensitivity ranging from 30% to 70% and a specificity of around 90%. Special techniques, such as the McFayden reaction, help in the diagnosis of *B. anthracis* with a sensitivity and specificity of 50–70% and 90%, respectively.

Giemsa stain is helpful in the diagnosis of *Rickettsia*, *Coxiella*, and *Borrelia*, with a sensitivity of 50–70% and a specificity of 90–95%. Other special stains, such as Warthin–Starry, are used for *Bartonella henselae*. Silver staining is used for *Borrelia* and *Leptospira*. Staining is fast, cost-effective, and widely available and, hence, ideal for early screening.

Dark-field microscopy is useful to visualise organisms such as *Leptospira* spp. and *Borrelia* spp. Phase contrast microscopy is also used for the detection of *Leptospira* spp., although it has limited clinical use. Electron microscopy provides high-resolution images and can detect intracellular pathogens such as *Rickettsia* spp. and *C. burnetii*; however, it is primarily used for research purposes due to its complex nature.

In Vitro Culture

Culture remains the gold standard method for diagnosing bacterial pathogens. An automated blood culture system and biphasic media are standard equipment for identifying *Brucella* spp. and *Salmonella* spp. Routine blood and

Table 3.2 Specimens for the diagnosis of bacterial zoonoses

Zoonotic disease	Causative agent	Primary diagnostic sample(s)	Remarks
Brucellosis	*Brucella* spp.	Blood, bone marrow	Bone marrow culture more sensitive; blood used for culture, PCR, and serology
Leptospirosis	*Leptospira interrogans*	Blood, urine	Blood in the first 7 days, urine from the 2nd week; dark-field microscopy/ PCR
Anthrax	*Bacillus anthracis*	Skin lesion swab, blood, sputum, CSF	Sample depends on clinical form; culture and PCR used
Plague	*Yersinia pestis*	Bubo aspirate, blood, sputum	Bubo aspirate ideal; PCR and culture confirmatory
Tularaemia	*Francisella tularensis*	Ulcer swab, lymph node aspirate, blood	Difficult to culture; PCR and serology preferred
Q fever	*Coxiella burnetii*	Blood (serology, PCR)	Culture avoided due to risk; PCR and IgM/IgG ELISA useful
Rickettsial infections	*Rickettsia* spp.	Skin biopsy (rash/ eschars), blood	PCR from rash site or eschar; serology for confirmation
Tuberculosis (zoonotic)	*Mycobacterium bovis*	Sputum, gastric lavage, lymph node biopsy	Similar diagnosis to *M. tuberculosis*; often extrapulmonary
Cat scratch disease	*Bartonella henselae*	Blood, lymph node aspirate or biopsy	PCR or serology useful; culture slow and insensitive
Rat-bite fever	*Streptobacillus moniliformis*, *Spirillum minus*	Blood, synovial fluid, skin lesion swab	Culture or PCR; organism is fastidious
Campylobacteriosis	*Campylobacter jejuni*	Stool	Culture on selective media; PCR for confirmation
Salmonellosis (non-typhoidal)	*Salmonella* spp.	Stool, blood	Blood culture in systemic illness; stool in enteric cases
Listeriosis	*Listeria monocytogenes*	Blood, CSF, placental tissue	Common in pregnant women, neonates; culture and PCR used
Melioidosis	*Burkholderia pseudomallei*	Blood, sputum, pus from abscess, urine	Culture is the gold standard; serology less reliable
Pasteurellosis	*Pasteurella multocida*	Wound swab, aspirate	Culture from the site of an animal bite/scratch
Glanders	*Burkholderia mallei*	Nasal swab, skin lesion, blood	Highly infectious; PCR and culture under BSL-3 required
Relapsing fever	*Borrelia recurrentis*	Blood	Dark-field microscopy during febrile phase; PCR and serology

MacConkey agar are used to culture *B. anthracis*, *Y. pestis*, and *Salmonella* spp.

Selective media have specific compositions that selectively promote the growth of pathogens while inhibiting the growth of other bacteria. Various selective culture media are used to isolate zoonotic pathogens. These include Ellinghausen–McCullough–Johnson–Harris (EMJH) media and Fletcher and Korthof media for *Leptospira* spp., Ashdown's media for *Burkholderia pseudomallei*, polymyxin-lysozyme-EDTA-thallous acetate (PLET) media to isolate *Bacillus anthracis*, polymyxin acriflavine lithium chloride ceftazidime aesculin mannitol (PALCAM) media to isolate *Listeria*, Lowenstein–Jensen (LJ) media for *Mycobacterium bovis*, Skirrow/Campy-BAP agar media for *Campylobacter jejuni*, and buffered charcoal yeast extract agar (BCYA) media to isolate *F. tularensis* with an overall sensitivity and specificity of 50–70% and 95–100%, respectively. Cell culture is used to grow *Rickettsia* and *C. burnetii*.

Serodiagnosis

Serological methods are essential tools for detecting antigen and antibody responses against specific pathogens. These tests are particularly useful for diagnosing infections caused by *Brucella*, *Leptospira*, *Campylobacter* spp., and *B. anthracis*. *Enzyme-linked immunosorbent assay (ELISA)* is widely used, offering a sensitivity of 80–90% and a specificity of approximately 90%. For *Leptospira* spp., the *microscopic agglutination test (MAT)* is considered the gold standard due to its high specificity of around 95%. In the case of *Rickettsia* spp., *B. henselae*, and *C. burnetii*, the *indirect immunofluorescence assay (IFA)* is highly specific. *Latex agglutination* is employed for the detection of *Listeria monocytogenes* infection, while the *Widal test* remains in use for the diagnosis of *Salmonella* infections.

Molecular Diagnosis

Molecular diagnosis, with their high specificity and sensitivity compared to other methods, have revolutionised the diagnosis of infectious diseases. Conventional PCR provides a comparatively faster identification of specific pathogens, such as *Brucella* spp. and *B. anthracis*. Real-time PCR has the advantage of providing quantitative results and a faster turnaround time. They are especially useful for high-risk zoonotic pathogens. They can also detect drug-resistant genes.

An overview of diagnostic modalities is presented in Table 3.3.

Treatment

Bacterial zoonosis includes a variety of pathogens that exhibit different clinical presentations and complications. So a multifaceted approach is required to combat it. Treatment strategies include the eradication of the organisms, along with reducing host damage caused by infections or toxins.

Antibiotic Therapy

The first-line treatment for most bacterial zoonoses includes antimicrobials. They are time- and life-saving when used promptly and in a timely manner. The selection of an antibiotic depends on many factors, including patient factors, the status of the pathogen, and the type of infection. The standard protocol for treating brucellosis is doxycycline 100 mg twice daily, combined with rifampicin 600–900 mg daily, for 6 weeks. Combination therapy is recommended to prevent relapse. In complicated or severe cases, for the initial 2–3 weeks, streptomycin or gentamicin is advised.

In inhalational anthrax, a combination of ciprofloxacin with clindamycin or rifampicin is administered for 60 days, along with an antitoxin. Cutaneous anthrax is treated with ciprofloxacin or doxycycline. Doxycycline or amoxicillin is administered orally in leptospirosis, whereas in severe disease, intravenous penicillin G or ceftriaxone is preferred. Gentamicin, levofloxacin, and doxycycline are used for the treatment of *Y. pestis* and *F. tularensis* infections for a prolonged duration. For *Rickettsia*, doxycycline is preferred. Azithromycin, erythromycin, or fluoroquinolones are used in cases of *Campylobacter* and some non-typhoidal *Salmonella* infections. In endemic regions, the empirical therapy of zoonotic diseases is guided by the local resistance patterns.

Antitoxin Therapy

In the pathogenesis of some zoonotic bacteria, toxins play a crucial role. Antitoxins act as an adjunct to the antimicrobials in severe systemic cases. They neutralise the circulating toxins produced by the bacteria. In cases of severe anthrax, monoclonal antibodies like raxibacumab and obiltoxaximab, or polyclonal immunoglobulin anthrasil, are used as antitoxins to neutralise the protective antigen (PA), a key component of both lethal and oedema toxins. They are indicated in inhalational anthrax and systemic disease, along with antibiotics. Botulism is treated with equine-derived botulinum antitoxin.

Table 3.3 Diagnosis of bacterial zoonoses

Diagnostic technique	Specific tool/method	Example pathogen(s)	Benefits	Remarks (Sensitivity/ specificity)
Microscopy	Light microscopy (Giemsa, Gram, Ziehl–Neelsen)	*Brucella* spp., *Mycobacterium bovis*	Rapid, cost-effective, widely available	Moderate sensitivity, low specificity
	Dark-field microscopy	*Leptospira* spp., *Borrelia* spp.	Visualises live motile organisms	Low sensitivity, moderate specificity
	Phase contrast microscopy	*Leptospira* spp.	Enhanced visualisation without staining	Rarely used clinically
	Electron microscopy	*Rickettsia* spp., *Coxiella burnetii*	High resolution	Not routine, qualitative only
In vitro culture	Blood culture (automated, biphasic media)	*Brucella* spp., *Salmonella* spp.	Definitive diagnosis	High specificity, moderate sensitivity
	Ashdown's agar, selective media	*Burkholderia pseudomallei*	Specific for pathogens	High specificity
	Lowenstein–Jensen (LJ), Microbial growth indicator tube (MGIT)	*Mycobacterium bovis*	Standard for tuberculosis diagnosis	High specificity, variable sensitivity
Serodiagnosis	ELISA	*Brucella* spp., *Rickettsia* spp.	Simple, detects exposure	Variable sensitivity/ specificity
	Microscopic agglutination test (MAT)	*Leptospira* spp.	Gold standard for leptospirosis	High specificity, variable sensitivity
	Indirect immunofluorescence assay (IFA)	*Rickettsia* spp., *Coxiella burnetii*	Highly specific	Gold standard for rickettsial diseases
Molecular diagnosis	Conventional PCR	*Brucella* spp., *Bacillus anthracis*	Rapid, specific	High sensitivity and specificity
	Real-time PCR	*Coxiella burnetii*, *Yersinia pestis*	Quantitative, fast	Very high sensitivity and specificity
	GeneXpert	*Mycobacterium bovis*	Rapid TB detection, resistant genes	High sensitivity for pulmonary TB

Supportive Care

Supportive care is essential in treating zoonotic infections when they are severe and involve multiple organs. Early fluid replacement and the use of vasopressors and ionotropes are helpful in combating sepsis-related zoonotic diseases. Ventilation and oxygen support are crucial in cases of respiratory distress involving multiple organs. Nutritional and psychological support are crucial for long-term effects.

Prevention and Control

Prevention and control of bacterial zoonosis are achieved by reducing human exposure to infected animals, contaminated environments, and disease vectors. Personal protective measures are important, especially for people who have high-risk occupations such as agriculture, veterinary work, and meat processing. Personal protective equipment (PPE), comprising gloves, masks, and gowns, protects against direct contact with high-risk zoonotic pathogens, such as *B. anthracis*, *F. tularensis*, and *Burkholderia pseudomallei*. Food safety practices, including proper food handling and consumption, are crucial safety measures against food-borne pathogens like *C. jejuni*, *Salmonella*, *Listeria*, and *Brucella* spp. Strict adherence to infection control practices, including hand washing and hand hygiene, is recommended in both community and healthcare facilities to prevent zoonotic infections.

Vector and environmental control are equally important. Inspection of rodent population and reducing contact with contaminated soil and water sources help prevent diseases such as leptospirosis, Rickettsial infections, and plague. Maintaining good hygiene, waste management, and drainage systems also reduce zoonotic bacteria in environmental reservoirs. Regular veterinary care, regular vaccination of livestock, and rodent-proof housing significantly reduce the transmission of infections.

Effective monitoring and timely reporting of zoonotic diseases are necessary for initial detection and response. Integrated public health systems will monitor both humans and animals, ensuring rapid investigation and prevention of outbreaks. The coordination between health professionals and veterinary services can intensify outbreak preparation and provide an opportunity for rapid public health intervention.

Ultimately, a One Health approach is essential for sustainable zoonotic control. This strategy emphasises collaboration between the human health, veterinary, and environmental sectors to address shared risks. Through joint monitoring, research, and policy design, a One Health approach ensures a comprehensive response to zoonotic hazards, addresses the causes of disease transmission, and increases global health resilience.

Vaccines

Vaccine plays a dual role in preventing zoonosis by reducing zoonotic reservoirs in animals and as a prophylaxis in high-risk human populations. Vaccination plays a key role in preventing specific zoonotic diseases. However, making human vaccines against most zoonotic pathogens is cumbersome and ineffective due to various factors, including antigenic variations in the pathogen, its intracellular nature, and its tendency to evade host immune responses.

Vaccination strategies are usually integrated as a public health intervention during surveillance or to prevent zoonotic outbreaks. Anthrax vaccine adsorbed (AVA) was licensed in 1970 for the treatment of anthrax. It is used as both pre- and post-prophylaxis in high-risk groups such as laboratory workers, military personnel, and farmers in endemic areas. In the case of brucellosis, live attenuated vaccines from the *Brucella abortus* strain RB51 and *Brucella melitensis* Rev-1 are used in cattle and sheep, respectively. Killed whole-cell and recombinant subunit vaccines against *Y. pestis* have been developed but are currently under evaluation for broader use. For *C. burnetii*, Q-Vax is approved in Australia. BCG vaccine is active against *M. bovis*.

Case Study

A 42-year-old male farmer presented with intermittent fever, intense fatigue, weight loss, and joint pain. In physical examination, he had mild hepatosplenomegaly. On further investigation, close contact with livestock animals, unpasteurised milk consumption, and recent participation in handling aborted foetus was found. The serological test, using the standard agglutination test (SAT), confirmed high titres of *Brucella abortus*. Blood cultures were positive for the organism. Similar results were revealed from livestock testing. Ultrasound in livestock confirmed uterine infection, corresponding to brucellosis.

1. What is the differential diagnosis of the case?
2. Write about the pathogenesis, clinical features, and laboratory diagnosis of the case.
3. What are the various preventive measures to control these conditions?

Research Questions

1. What are the primary host–pathogen interaction mechanisms that facilitate the transmission and pathogenesis of bacterial zoonotic infections?
2. What are the common clinical manifestations of bacterial zoonotic diseases, and what are the effective diagnostic tools for detection?
3. To what extent does the One Health approach improve the prevention, surveillance, and control of bacterial zoonosis?

Declaration During the preparation of this chapter, we used QuillBot (QuillBot, a Learneo Inc. business) and Grammarly (Grammarly Inc.) applications to improve the grammar and readability of the text. After using, we have reviewed and edited the content as needed, and we take full responsibility for the content of the publication.

Further Readings

Azevedo A, Guimarães L, Ferraz J, Whiting M, Magalhães-Sant'Ana M. Pet reptiles—are we meeting their needs? Animals. 2021;11(10):2964. Available from: https://doi.org/10.3390/ani11102964.

Cantas L, Suer K. Review: the important bacterial zoonoses in the "One Health" concept. Front Public Health. 2014;2:144. Available from: https://www.frontiersin.org/articles/10.3389/fpubh.2014.00144/full.

Danon L, Ford AP, House T, et al. Networks and the epidemiology of infectious disease. Interdiscip Perspect Infect Dis. 2011;2011:284909. Available from: https://doi.org/10.1155/2011/284909.

Ebani VV. Domestic reptiles as source of zoonotic bacteria: a mini-review. Asian Pac J Trop Med. 2017;10(7):723–8. Available from: https://doi.org/10.1016/j.apjtm.2017.06.004.

Essack SY. Environment: the neglected component of the One Health triad. Lancet Planet Health. 2018;2:e238–9. Available from: https://doi.org/10.1016/S2542-5196(18)30109-1.

Grudlewska-Buda K, Khamesipour F, Skowron K. Zoonoses and emerging pathogens. BMC Microbiol. 2023;23(Suppl 1):1–2. Available from: https://doi.org/10.1186/s12866-023-02739-3.

Kuria KNJ. Salmonellosis in food and companion animals and its public health importance. In: Salmonella—past, present and future. IntechOpen; 2023. Available from: https://doi.org/10.5772/intechopen.101968.

Rovid Spickler A. Pathogens and host species in zoonoses. In: MSD veterinary manual. 2023. Available from: https://www.msdvetmanual.com/public-health/zoonoses/pathogens-and-host-species-in-zoonoses.

Royce KP, Fu F. Mathematically modeling spillover dynamics of emerging zoonoses with intermediate hosts. PLoS Comput Biol. 2019;15(8):e1007254. Available from: https://doi.org/10.1371/journal.pcbi.1007254.

Santarém VA, Panazzolo GK, Kmetiuk LB, et al. One health approach to toxocariasis in quilombola communities of southern Brazil. Parasit Vectors. 2023;16(1):379. Available from: https://doi.org/10.1186/s13071-023-05776-0.

Stella M, Selakovic S, Antonioni A, Andreazzi CS. Community interactions determine role of species in pathogen spread amplification: the eco-multiplex network model. PLoS Comput Biol. 2017;13(6):e1005567. Available from: https://doi.org/10.1371/journal.pcbi.1005567.

World Health Organization. Critically important antimicrobials for human medicine, 6th ed; World Health Organization: Geneva, 2019; ISBN 978-92-4-151552-8. Available from: https://www.who.int/publications/i/item/9789241515528.

Zhang T, Feng Z. Echinococcosis: from pathogen–host interaction to rapid detection. Trop Med Infect Dis. 2025;10(1):1. Available from: https://doi.org/10.3390/tropicalmed10010001.

Immunology of Bacterial Infections

4

Shampa Anupurba

Learning Objectives

1. To explain the fundamental principles of innate and adaptive immune responses during bacterial infections
2. To elaborate on the immune evasion mechanisms employed by bacteria to survive and persist within the host
3. To describe the detrimental effects of the immune response triggered by bacterial infections, including tissue damage and inflammation

Introduction

Immune response against bacterial pathogen is determined by the location of the bacterium, whether it is extracellular or intracellular (cytosolic or within phagosomes), and by its mode of pathogenicity, whether through direct invasion or toxin production. The key adaptive immune mechanisms involved include (i) the neutralisation of bacterial toxins or enzymes by antibodies, (ii) bacterial lysis via the action of complement and antibodies, (iii) opsonisation by complement-facilitating phagocytosis and bacterial destruction, (iv) intracellular killing of bacteria by activated macrophages, and (v) cytotoxic T lymphocyte (CTL)-driven elimination of infected cells.

The initial entry and establishment of pathogenic bacteria are prevented by physical and chemical barriers that form the first line of defence. Beyond these barriers, host cells utilise various pattern recognition receptors (PRRs) to detect conserved microbial features known as pathogen-associated molecular patterns (PAMPs). These PRRs are classified into intracellular recognition molecules, cell surface and endocytic vesicle receptors, and secreted molecules present in serum and other body fluids. Once activated, these receptors trigger a range of effector responses, including (i) the activation of the complement system, (ii) cytokine production, (iii) phagocyte recruitment, and (iv) release of antimicrobial peptides. Opsonisation, followed by phagocytosis and complement-mediated bacterial lysis, is the shared mechanism for innate and adaptive immune responses.

To survive within the host, bacteria employ strategies to evade immune detection, particularly T-cell-mediated responses. Some bacteria reside within T cells or antigen-presenting cells (APCs) to avoid immune surveillance. Others produce pore-forming toxins that destroy T cells or secrete superantigens (SAgs) that bind non-specifically to T cell receptors. This abnormal activation rewires

S. Anupurba (✉)
Department of Microbiology, Institute of Medical Sciences, Banaras Hindu University, Varanasi, Uttar Pradesh, India
e-mail: shampa@bhu.ac.in

S. C. Parija, A. Poddar (eds.), *Textbook of Bacterial Zoonoses*, Microbial Zoonoses,
https://doi.org/10.1007/978-981-95-6615-0_4

T cells into a hyperactive state, often resulting in anergy or the release of a massive cytokine storm, which can be lethal to the host.

Innate Immune Response

It is the first line of defence against pathogens. Although it responds rapidly and effectively, it is non-specific and lacks immunological memory, making it insufficient to eliminate infections entirely on its own (Table 4.1). The innate system comprises anatomical barriers, soluble molecules, and cellular components that work in coordination to prevent and control infections.

Anatomical Barriers Anatomical barriers form the initial defence against microbial invasion. These include the impermeable epithelial layers of the skin and mucosal surfaces, i.e. the gastrointestinal, respiratory, and urogenital tracts. Additional protective factors include fatty acids secreted by the skin, ciliary movement in the trachea, flushing action in the urinary tract, and acidic environments in the stomach and vagina. Enzymes such as lysozyme also contribute to antimicrobial defence.

Commensal microorganisms on mucosal surfaces play an important role in limiting bacterial colonisation. The endogenous microbiota, especially in the gut, competes with pathogenic bacteria for nutrients, stimulates host immune responses, and produces antibacterial compounds such as colicins. When bacteria penetrate the epithelial surface, they are promptly recognised by the host immune system (Table 4.2).

Table 4.1 Differences between innate and adaptive immunity

Characteristics	Innate immunity	Adaptive immunity
Evolution	The earliest and most primitive line of defence	Emerged later
Response time	Immediate: within minutes to hours	Late: could take days to develop
Specificity	Non-specific	Specific
Immunological memory	Absent	Present
Response to repeated infection	No enhancement	Rapid and more effective
Prior contact with the pathogen	Not required	Required

Soluble Molecules and Cell Surface Receptors The innate immune system can rapidly recognise pathogens using germline-encoded receptors. These receptors detect PAMPs, which are conserved microbial components that are not found in the host cells. This mechanism enables the immune system to differentiate between infectious non-self and self with limited receptors. Bacterial molecules such as peptidoglycan monomers, teichoic acids, lipopolysaccharides (LPSs), porins, mycolic acids, mannose-rich glycans, and flagellins act as PAMPs. These bind to PRRs present on various host defence cells, triggering the synthesis and secretion of cytokines.

Based on their location, the PRRs are broadly categorised into three main types:

Intracytoplasmic Recognition Molecules These are PRRs that are found inside cells, where they recognise microbial structures following phago-

Table 4.2 Anatomical and physiological barriers in innate immunity

Organ/tissue	Innate mechanisms
Skin	Epithelium, antimicrobial peptides (AMPs), fatty acids
Mouth and upper alimentary canal	Enzymes, AMPs, directional fluid flow towards the stomach
Stomach	Epithelium, acidic pH, enzymes, AMPs, fluid flow towards the intestine
Small intestine	Epithelium, enzymes, AMPs, fluid flow towards the large intestine
Large intestine	Epithelium, normal intestinal flora, expulsion of fluid/faeces from the rectum
Airway and lungs	Epithelium, cilia, cough and sneeze reflexes, alveolar macrophages
Urogenital tract	Urinary flushing, low pH, AMPs, vaginal flora
Salivary, lacrimal, and mammary glands	Flushing by secretions, lysozyme, AMPs

cytosis. Once bacteria are internalised, their PAMPs are detected by transmembrane PRRs oriented towards the interior of the phagosome, initiating further immune responses. Bactericidal permeability–increasing proteins (BPI) are located within the primary granules of polymorphonuclear neutrophils. The protein has the ability to bind to lipopolysaccharides (LPSs), contributing to the killing of Gram-negative bacteria. The nucleotide-binding and oligomerisation domain (NOD)–like receptors are cytoplasmic sensors that detect intracellular bacterial components. NOD1, also known as caspase recruitment domain 4 (CARD4), and NOD2, also known as CARD15, recognise LPS and bacterial muramyl dipeptide (MDP), respectively. Their activation triggers the NF kappa B and MAP kinase pathways. These cause the activation of caspases and the production of pro-inflammatory cytokines such as interleukin-18 (IL-18) and interleukin-1 beta (IL-1β). The CARD/NOD protein family shares similarities with Toll-like receptors (TLRs) and CD14, functioning as cytosolic sensors for PAMPs.

Cell Surface and Endocytic Vesicle Receptors Cellular PRRs are typically located on the membranes of host cells, especially antigen--presenting cells. PRRs are critical for the recognition and internalisation of pathogens by phagocytes. They include (i) Toll-like receptors (TLRs), (ii) the type 3 complement receptor, (iii) carbohydrate receptors such as mannose-binding C-type lectins, and (iv) scavenger receptors. TLR4 specifically binds to LPS on the outer membrane of Gram-negative bacteria, facilitating their uptake by macrophages. Both TLR2 and TLR4 also recognise damage-associated molecular patterns (DAMPs). The DAMPs are expressed by apoptotic or necrotic cells, thereby linking infection and tissue damage to immune activation.

Secreted Molecules in Serum and Body Fluids Soluble PRRs present in body fluids bind to microbial targets and facilitate their recognition by cellular receptors or enhance phagocytosis. Several soluble PRRs circulate in body fluids, contributing to the innate immune response. These include acute-phase proteins such as (i) C-reactive protein (CRP), (ii) lysozyme and complement component C3b, (iii) reactive oxygen intermediates (ROIs), (iv) reactive nitrogen intermediates (RNIs), (v) various cytokines, and (vi) soluble CD14. These molecules function by directly neutralising pathogens, promoting opsonisation, and enhancing the recruitment and activation of immune cells.

Lymphocyte-Independent Effector Mechanisms

Lymphocyte-independent effector mechanisms include the following responses:

Complement-Mediated Effector Responses These include the direct lysis of bacteria whose lipid bilayer is susceptible to the membrane attack complex (MAC), i.e. C5b-9. The complement system is activated by alternative or lectin pathways. The opsonisation of bacteria and further interaction with phagocytes are caused by the attachment of cleaved products of complement C3. C3a and C5a are chemotactic and anaphylatoxic. They attract and activate neutrophils and cause degranulation of mast cells, leading to the release of histamines and leukotrienes that increase vascular permeability and extravasate lymphocytes and neutrophils into tissue spaces.

Cytokine-Mediated Effector Responses Tumour necrosis factor (TNF) and IL-1 cause an increase in the adhesive properties of vascular endothelium, by which the circulating phagocytes stick to the walls of capillaries, squeeze out, and enter the inflamed tissue sites. Cytokines, along with chemokines, e.g. CCL2, CCL3, and CXCL8, help in the recruitment of leucocytes. Cytokines, such as IL-1, IL-6, and TNF, initiate an acute response by enhancing complement production. NK cells are stimulated by IL-12 and IL-18, resulting in the release of interferon-gamma (IFN-γ) that further activates macrophages.

Phagocyte-Mediated Bacterial Killing

Elie Metchnikoff first described the role of phagocytosis in immunity. This work resulted in his Nobel Prize in 1908. The type of pathogen and the time since infection influence the composition of phagocytic cells involved in the inflammatory response, either neutrophils or macrophages. For example, acute inflammation caused by *Streptococcus pyogenes* leads to neutrophil-rich lesions, whereas chronic inflammation induced by *Mycobacterium tuberculosis* results in macrophage-dominated granulomas. Neutrophils are known to phagocytose bacteria such as *Haemophilus influenzae, Staphylococcus aureus*, *Streptococcus pneumoniae*, and *Pseudomonas aeruginosa.* In response to bacterial components, cytokines and chemokines are released by macrophages. They initiate and sustain the inflammatory response. These local innate and inflammatory mechanisms are generally beneficial, as they help (i) eliminate pathogens, (ii) remove damaged or dead cells, and (iii) promote tissue repair.

The steps of phagocytosis are:

Binding of Phagocytes with the Pathogen The phagocytes may bind directly to the bacteria by pattern recognition molecules or indirectly through complement receptors.

Uptake of the Pathogen by the Phagocyte After chemotaxis and binding, the uptake of the pathogen takes place. This is enhanced by macrophage-activating cytokines, e.g. granulocyte-macrophage colony-stimulating factor (GM-CSF).

Killing of Bacteria by Phagocytes The pathogen is bound by the host membrane phagosome, which confines it to deliver antimicrobial molecules at high concentration and avoids damage to the host. The killing mechanisms include (i) the acidification of phagosomes, (ii) generation of reactive oxygen intermediates, (iii) promotion of the flux of K+ ions in phagosomes, (iv) activation of microbicidal protease by fusion with lysosomes, and (v) creation of nitric oxide and peroxynitrite from guanidino nitrogen of L-arginine.

Antimicrobial Peptide (AMP)-Mediated Effector Response Various small cationic proteins possess antibiotic-like properties and play a crucial role in the host's innate immunity. They selectively target negatively charged hydroxylated phospholipids such as phosphatidyl serine, cardiolipin, and phosphatidyl glycerol. These phospholipids are commonly found in bacterial membranes but not in the cytoplasmic membrane of mammalian cells. In addition, components like lipopolysaccharides (present in Gram-negative bacteria) and lipoteichoic acid (present in Gram-positive bacteria) contain numerous negative charges, making them ideal targets for these molecules. These antimicrobial peptides (AMPs) are typically produced at anatomical sites where the pathogen invades.

Two major classes of AMPs in most mammals are defensins and cathelicidins.

Defensins They help in the lysis of both Gram-positive and Gram-negative bacteria by integrating within their lipid membranes and disrupting their cellular integrity. They promote phagocytosis and chemotaxis and regulate cytokine production. They have been classified as α-, β-, and θ-defensins (not found in humans). Both α- and β-defensins have been found in azurophilic granules of neutrophils.

Cathelicidins These are another class of antimicrobial peptides found in the specific granules of neutrophils, epithelial cells, and phagocytic cells. They help in bacterial membrane disintegration. LL-37 is the only cathelicidin found in humans.

Adaptive Immune Response

Adaptive immunity complements innate immune mechanisms, where antigen-presenting cells (APCs) serve as a crucial link between them. Dendritic cells (DCs), for instance, present antigens to lymphocytes, activating the adaptive immune response specifically tailored to the invading pathogen. The activation of T cells is the central event in the initiation of adaptive immunity. However, there is typically a delay of 4–7 days before the adaptive immune response becomes effective. During this period, the innate immunity manages the infection and supports the development of the adaptive response.

Cell-Mediated Immune Response T cells use membrane-bound T cell receptors (TCRs) to detect and respond to infected cells expressing foreign antigens. They are classically subdivided based on their expression of major histocompatibility complex (MHC)-restricted αβ or γδ TCRs. Intracellular bacteria that enter the cytosol are processed and presented via major histocompatibility complex class I (MHCI) molecules to naïve CD8+ T cells. In contrast, soluble proteins or proteins derived from phagocytosed organisms are presented to CD4+ T cells that recognise MHCII molecules.

Differentiation and Function of CD4+ Helper T (Th) Cells in Antibacterial Immunity Upon encountering antigen-presenting cells and appropriate cytokine signals, naïve CD4+ T cells differentiate into specialised subsets of (Th) cells. These cells, restricted by MHC class II molecules, are pivotal for orchestrating immune responses. They particularly aid in memory CD8+ T cell development and enhance immunity against bacterial pathogens such as *M. tuberculosis*, *S. aureus*, and *Salmonella enterica* serovar Typhimurium.

Th1 Cells Th1 cells are primarily involved in cell-mediated immune responses. They produce cytokines such as interleukin-1 (IL-1), interferon-gamma (IFN-γ), and tumour necrosis factor-beta (TNF-β). Their hallmark function is the activation of cytotoxic T lymphocytes (CTLs) and macrophages, with IFN-γ being a signature cytokine associated with this subset.

Th2 Cells In response to extracellular bacterial pathogens, CD4+ T cells may differentiate into Th2 cells. These cells aid in humoral immunity by promoting B cell activation and antibody production. Th2 cells secrete cytokines, including IL-4, IL-5, IL-6, IL-9, IL-10, and IL-13. These cytokines are involved in B cell maturation, antibody class switching, and eosinophil recruitment.

Th17 Cells This subset is characterised by the secretion of IL-17A, IL-17F, and IL-22. Th17 cells play a significant role in defence against extracellular bacteria by promoting neutrophil recruitment and enhancing mucosal barrier integrity.

Regulatory T Cells (TREGs) TREG cells are essential for maintaining immune homoeostasis by suppressing excessive T cell responses. Induced TREGs (iTREGs), generated in secondary lymphoid organs under the influence of transforming growth factor-beta (TGF-β), function similarly to natural TREGs (nTREGs). Notably, TREG and Th17 subsets exhibit reciprocal regulation.

Follicular Helper T Cells (Tfh) Tfh cells are specialised in providing help to the B cells within germinal centres and follicles of secondary lymphoid tissues like lymph nodes and tonsils. They secrete IL-21, a key cytokine that facilitates (i) B cell differentiation, (ii) immunoglobulin class switching, and (iii) the generation of long-lived memory B cells.

Apoptosis of Target Cells Induced by Cytotoxic T Lymphocytes (CTLs) Naïve CD8+ T cells, upon antigen exposure and appropriate co-stimulatory signals, become activated and differentiate into cytotoxic T lymphocytes (CTLs). These MHC class I–restricted cells play a vital role in controlling and eliminating intracellular bacterial infections, particularly those caused by *Listeria monocytogenes*, *M. tuberculosis*, *S. enterica serovar* Typhimurium, and *S. aureus*.

Once inside the host, many intracellular bacteria localise within vacuoles following receptor-mediated endocytosis, though some may escape into the cytoplasm. CTLs, through their T cell receptor (TCR)–CD3 complex, recognise bacterial antigens presented on MHCI molecules of infected cells. On recognition, CTLs eliminate the target cells by two major cytotoxic mechanisms. The first mechanism involves the release of perforin and granzymes; perforin forms pores in the target cell membrane, allowing granzymes to enter and activate apoptotic pathways. The second mechanism is mediated by Fas ligand (FasL) expressed on the CTL surface, which binds to Fas receptors on target cells, also initiating apoptosis.

Both cytotoxic pathways ultimately activate intrinsic death programmes already present within the infected cells. After the infection is resolved, most of the effector CTLs undergo apoptosis, while a subset survives as long-lived memory CD8+ T cells. These memory cells, residing in the circulation, lymphoid tissues, and mucosal barriers, mount a more rapid and effective response upon re-exposure to the same pathogen. This phenomenon, known as immunological memory, is a hallmark of the adaptive immune system and involves both T and B lymphocytes. The secondary immune response is typically faster, more intense, and more efficient compared to the primary response.

Humoral Immune Response

Antibody-Mediated Defence by B Lymphocytes Antibodies, or immunoglobulins (Igs), are glycoproteins composed of two identical heavy (H) chains and two identical light (L) chains, which are of the kappa (κ) or lambda (λ) type. The variable (V) regions located at the amino-terminal ends of these chains show high sequence diversity. Within the V region, three hypervariable loops, known as complementarity-determining regions (CDRs), form the antigen-binding site by recognising specific epitopes.

B lymphocytes produce immunoglobulins in both membrane-bound and secreted forms. These antibodies contribute to humoral immunity by neutralising microbial toxins and participating in effector mechanisms that eliminate or destroy pathogens.

Thymus-Dependent (TD) B Cell Activation When B cells encounter soluble protein antigens, they internalise and process them, presenting antigen-derived peptides on their surface using MHC class II molecules. Helper T cells, specifically the Th2 subtype, recognise this peptide–MHC complex through their T cell receptors, leading to their activation. In response, these T cells release cytokines that further stimulate the B cells, inducing their differentiation into plasma cells (which secrete antibodies) and memory B cells. This pathway supports isotype switching, primarily from IgM to other antibody classes such as IgG, IgA, or IgE.

Thymus-Independent (TI) B Cell Activation Certain antigens, particularly polysaccharides and lipopolysaccharides, can activate B cells without T cell involvement. This activation is generally weaker and does not produce memory B cells. The resulting antibody response is mostly restricted to the IgM isotype, with minimal or no class switching.

Effector Mechanisms of Antibodies Antibodies mediate several protective functions through their Fc regions, by the following methods:

Neutralisation of Toxins Antibodies bind directly to bacterial toxins, preventing their interaction with host cell receptors and thereby blocking their entry into the target cells. These neutralised complexes are then cleared by phagocytic cells. High-affinity IgG antibodies are more effective. For example, anti-tetanus toxoid antibodies, generated through vaccination, neutralise *Clostridium tetani* toxins and prevent neurotoxicity.

Opsonisation Opsonisation enhances the uptake of pathogens by phagocytes, such as macrophages and neutrophils. These immune cells express Fc receptors that bind the Fc portion of antibodies and receptors for complement proteins like C3b. A pathogen is more readily engulfed and destroyed when it is coated with antibodies or complement components. This mechanism is particularly effective against Gram-positive bacteria, such as *S. aureus*, which are resistant to complement-mediated lysis.

Complement Activation IgG and IgM are the primary antibody classes that initiate the classical complement cascade. IgM, due to its pentameric structure, is especially potent. Under specific conditions, polymeric forms of IgA or IgE may also activate the complement system through the alternative or lectin pathways. Upon antibody binding to microbial surfaces, a cascade is initiated that results in the formation of the membrane attack complex (MAC), which punctures the membranes of Gram-negative bacteria, leading to cell lysis. However, Gram-positive bacteria are generally resistant to this due to their thick peptidoglycan walls.

Evasion Strategies of Pathogenic Bacteria

Pathogenic bacteria have developed diverse mechanisms to evade detection, attack, and clearance by the host immune system. These strategies allow them to establish infection, persist in the host, and, sometimes, cause severe disease.

Antiphagocytic Mechanisms These include the following mechanisms:

Capsule Formation The polysaccharide capsule acts as a protective layer that inhibits opsonisation and prevents recognition by Fc receptors on phagocytes. This strategy is used by pathogens such as *S. pneumoniae*, *H. influenzae*, *Neisseria meningitidis*, *Salmonella* Typhi (via the Vi antigen), and *Yersinia pestis* (which uses impedins).

M Protein Found in *S. pyogenes*, M protein masks complement-binding sites on the bacterial surface, thereby hindering complement-mediated opsonisation.

Cytotoxin Production *S. aureus* produces haemolysins and leukocidins, which lyse red and white blood cells, thereby impairing the host's immune response.

Molecular Mimicry (Camouflage) The K1 capsule of *Escherichia coli* and the capsule of *N. meningitidis* group B mimic host cell components, such as neural cell molecules, to evade immune recognition by appearing as "self."

Intracellular Survival Strategies Some bacteria have adapted to survive and replicate within host phagocytes by interfering with key cellular processes:

Blocking Phagolysosome Fusion Organisms like *Legionella* spp., *M. tuberculosis*, and *Chlamydia* spp. prevent the fusion of phagosomes with lysosomes, avoiding degradation.

Inhibiting Phagosome Maturation *M. tuberculosis* also delays or halts phagosome progression to a degradative state, allowing intracellular persistence.

Preventing Degranulation *Bordetella pertussis* inhibits the release of lysosomal enzymes, preserving bacterial integrity within immune cells.

Resistance to Lysosomal Enzymes Bacteria such as *S.* Typhimurium, *Coxiella* spp., and *Mycobacterium leprae* withstand the harsh enzymatic conditions of the phagolysosome.

Cytoplasmic Replication or Phagosome Escape *L. monocytogenes*, *Rickettsia* spp., and *Francisella tularensis* escape into the cytoplasm, where they replicate. *Shigella* spp. employs type III secretion systems (T3SSs) to deliver proteins like IpaB, which activate caspase-1 in macrophages, disrupting phagosome integrity.

Inhibiting Respiratory Burst *S. aureus* interferes with the oxidative burst, a major bactericidal mechanism of phagocytes.

IgA1 Protease Production Certain bacteria, including *Neisseria gonorrhoeae*, *N. meningiti-*

dis, *H. influenzae*, and *S. pneumoniae*, secrete IgA1 proteases. These enzymes cleave the hinge region of IgA1, a key antibody in mucosal defence, thereby promoting bacterial colonisation on mucosal surfaces.

Biofilm Formation Biofilms are structured bacterial communities encased in a self-produced extracellular matrix (glycocalyx) that adheres to surfaces. This matrix impedes the penetration of antibiotics, antibodies, and immune cells. Examples include *P. aeruginosa*, *Staphylococcus epidermidis*, and *E. coli* (which produces "curli" fibres for surface attachment and biofilm stabilisation).

Resistance to Antimicrobial Peptides Some bacteria, such as *Brucella* spp., *Burkholderia* spp., and *N. gonorrhoeae*, exhibit intrinsic resistance to host-derived antimicrobial peptides like defensins, allowing survival in hostile environments.

Disruption of Anatomical Barriers *B. pertussis* interferes with mucociliary clearance in the respiratory tract by releasing components that stimulate nitric oxide production, leading to ciliary paralysis (ciliostasis).

Alteration of Pathogen-Associated Molecular Patterns (PAMPs) Some bacteria modify or mask their surface molecules (e.g. lipopolysaccharides or flagellins). These are typically recognised by PRRs, helping them avoid immune detection through stealth. An example is the temperature-dependent change in lipid A of *Y. pestis*.

Immunopathological Effects of the Host Immune Response

While the immune system is essential for defending against pathogens, an uncontrolled or inappropriate immune response can result in tissue damage and severe clinical consequences.

Some key immunopathological effects include:

Endotoxic Shock and Sepsis Endotoxic shock is a life-threatening inflammatory reaction triggered by the presence of lipopolysaccharide (LPS), a major component of the outer membrane of Gram-negative bacteria. LPS stimulates the release of large amounts of pro-inflammatory cytokines such as TNF-α, IL-1β, and IL-6. This overwhelming immune activation leads to clinical manifestations, including high fever, hypotension, widespread disseminated intravascular coagulation (DIC), tissue ischaemia, and, potentially, multi-organ failure.

Sepsis represents a severe, dysregulated immune response against infections. It typically progresses through an initial hyperinflammatory phase, followed by an immunosuppressive stage that increases susceptibility to secondary infections. If untreated, sepsis may culminate in multiple organ dysfunction syndrome (MODS) and death.

T Cell Overactivation by Superantigens (SAgs) Superantigens are a distinct group of bacterial toxins, such as *S. aureus* enterotoxins and *S. pyogenes* pyrogenic exotoxins, that stimulate massive T cell activation. Unlike conventional antigens, superantigens bind directly to the Vβ region of the T cell receptor (TCR) and MHC class II molecules on antigen-presenting cells, without being processed. This leads to the non-specific activation of a large population of T cells. This results in a "cytokine storm" characterised by excessive secretion of cytokines like IL-2, IL-1β, TNF-α, and TNF-β. Clinically, this hyperinflammatory state can cause toxic shock syndrome (TSS), sepsis, and severe respiratory or systemic complications.

Vaccines

Vaccines are defined as substances that stimulate active immunity to a pathogen. Vaccination or immunisation aims to stimulate the immune system to recognise and combat pathogens safely. They are prepared by live modified organisms, inactivated or killed organisms, extracted or cellular fractions, toxoids, or combinations. The route of administration can be oral (e.g. Typhoral), intramuscular (e.g. DPT), or intradermal (e.g. BCG).

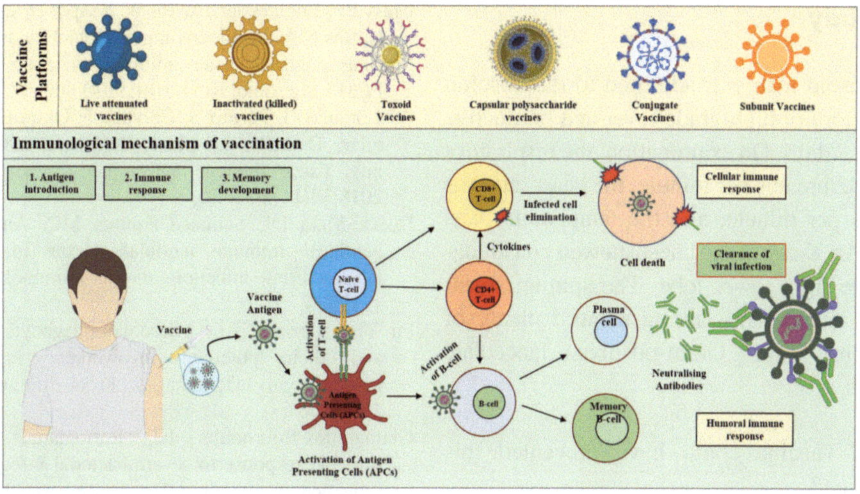

Fig. 4.1 Diagrammatic representation of different types of vaccines and the mechanisms of their modes of action. The vaccine contains an antigen derived from a pathogen in the body, which is recognised and taken up by antigen--presenting cells (APCs), such as dendritic cells. These APCs engulf the antigen and present it on their surface using major histocompatibility complex (MHC) molecules. This activates helper T cells stimulate B cells. B cells produce antibodies and cytotoxic T cells to target infected cells. Finally, memory B and T cells are produced, providing long-term immunity and enabling a rapid and robust response upon subsequent pathogen exposure. (Image source: Authors' personal collection)

Vaccine Types Vaccines are categorised based on their composition and method of inducing immunity (Fig. 4.1). Some of the important vaccine types are discussed below:

Live Attenuated Vaccines These vaccines contain live but virulence lost state of pathogen. As a result, it can replicate but does not cause disease in healthy individuals. Examples are BCG vaccine for *M. tuberculosis* and Typhoral for *S.T* Typhi

Inactivated (Killed) Vaccines These are composed of pathogens that have been killed using heat or chemicals, thus retaining immunogenicity without the risk of causing infection. Examples are the cholera vaccine and the whole-cell pertussis vaccine.

Toxoid Vaccines These vaccines use inactivated bacterial toxins (toxoids) to elicit immunity against the harmful effects of the toxin rather than the bacterium itself. Diphtheria toxoid and tetanus toxoid are examples of such toxoid vaccines.

Capsular Polysaccharide Vaccines These consist of purified polysaccharides from bacterial capsules, inducing immunity primarily through a T cell-independent response. Examples include the pneumococcal polysaccharide vaccine (PPV23), meningococcal vaccine, and Vi polysaccharide vaccine (*S.* Typhi)

Conjugate Vaccines In these vaccines, polysaccharide antigens are chemically linked to protein carriers to enhance immunogenicity, especially in infants and young children. Examples include pneumococcal conjugate vaccine (PCV-7), meningococcal conjugate vaccine, and *H. influenzae* type b (Hib) vaccine.

Subunit Vaccine Acellular pertussis vaccine is an example of a sub-unit vaccine. It contains purified components of *B. pertussis*, such as pertussis toxoid, filamentous haemagglutinin (FHA), and pertactin. This vaccine offers protection with fewer side effects than whole-cell vaccines.

Case Study

A 60-year-old man was admitted to the hospital with a history of high-grade fever and productive cough for 2 days. On examination, the respiratory rate was 42 breaths per minute, the heart rate was 100 beats per minute, and the temperature was 38.2 °C. An X-ray of the chest showed consolidation in the right lower lobe. The sputum smear observed after Gram-staining showed plenty of polymorphs with Gram-positive lanceolate diplococci.

1. Which vaccine could have prevented this condition?
2. Why is there a difference in the immunological responses to protein and polysaccharide antigens?
3. How can vaccines prepared against carbohydrate antigens elicit a T cell response?

Research Questions

1. What are the structural characteristics, mechanism of action, and clinical applications of antimicrobial peptides?
2. How does the internalisation of particulate microbes influence inflammatory cytokine production in macrophages?
3. What are the stages of antigen presentation?
4. How do various vaccine formulations compare in terms of effectiveness and duration of protection?
5. What are pulmonary humoral host responses to a bacterial challenge?

Further Readings

Bülow S, Zeller L, Werner M, Toelge M, Holzinger J, Entzian C, et al. Bactericidal/permeability-increasing protein is an enhancer of bacterial lipoprotein recognition. Front Immunol. 2018:9.

Cress BF, Englaender JA, He W, Kasper D, Linhardt RJ, Koffas MA. Masquerading microbial pathogens: capsular polysaccharides mimic host-tissue molecules. FEMS Microbiol Rev. 2014;38(4):660–97.

De Lorenzo G, Ferrari S, Cervone F, Okun E. Extracellular DAMPs in plants and mammals: immunity, tissue damage and repair. Trends Immunol. 2018;39(11):937–50.

Duarte-Mata DI, Salinas-Carmona MC. Antimicrobial peptides' immune modulation role in intracellular bacterial infection. Front Immunol. 2023;14: 1119574.

Fu YL, Harrison RE. Microbial phagocytic receptors and their potential involvement in cytokine induction in macrophages. Front Immunol. 2021; 12:662063.

Giamarellos-Bourboulis EJ, Raftogiannis M. The immune response to severe bacterial infections: consequences for therapy. Expert Rev Anti-Infect Ther. 2012;10(3):369–80.

Hornef MW, Wick MJ, Rhen M, Normark S. Bacterial strategies for overcoming host innate and adaptive immune responses. Nat Immunol. 2002;3(11): 1033–40.

Ragland SA, Kagan JC. Cytosolic detection of phagosomal bacteria—mechanisms underlying PAMP exodus from the phagosome into the cytosol. Mol Microbiol. 2021;116(6):1420–32.

Reddick LE, Alto NM. Bacteria fighting back – how pathogens target and subvert the host innate immune system. Mol Cell. 2014;54(2):321–8.

Rosales C, Uribe-Querol E. Phagocytosis: a fundamental process in immunity. Biomed Res Int. 2017;2017:9042851.

Roudaire T, Héloir MC, Wendehenne D, Zadoroznyj A, Dubrez L, Poinssot B. Cross kingdom immunity: the role of immune receptors and downstream signaling in animal and plant cell death. Front Immunol. 2020;11:612452.

Shepherd FR, McLaren JE. T cell immunity to bacterial pathogens: mechanisms of immune control and bacterial evasion. Int J Mol Sci. 2020;21(17): 6144.

Twigg HL III. Humoral immune defence (antibodies) recent advances. Proc Am Thorac Soc. 2005;2(5):417–21.

Walch M, Dotiwala F, Mulik S, Thiery J, Kirchhausen T, Clayberger C, et al. Cytotoxic cells kill intracellular bacteria through granulysin-mediated delivery of granzymes. Cell. 2014;157(6):1309–23.

Weigelin B, den Boer AT, Wagena E, Broen K, Dolstra H, de Boer RJ, et al. Cytotoxic T cells are able to efficiently eliminate cancer cells by additive cytotoxicity. Nat Commun. 2021;12(1):5217.

Bacterial Genomics

5

Jacob Abraham, T. R. Arun, and Manoj Nair

Learning Objectives
1. To understand the bacterial genome structure
2. To learn the sequencing technologies in bacterial genomics
3. To find out the applications of genomics in the management of bacterial zoonoses

Introduction

Bacterial genomics has revolutionised the understanding of the microbial world, offering unprecedented insights into the biology, evolution, and interactions of bacteria. Since its inception, bacterial genomics has undergone a remarkable transformation, progressing from early gene sequencing efforts to the current era of high-throughput technologies. These enable researchers to explore bacterial diversity with increasing precision and scale. With the increasing availability of complete bacterial genome collections and advances in computational tools, reference-based genome analysis has become more efficient and accurate. This has enabled high-resolution classification of microbial community composition and dynamics.

J. Abraham (✉) · M. Nair
Institute of Animal Health and Veterinary Biologicals, Palode, Kerala, India

T. R. Arun
ICMR – National Institute for One Health, Nagpur, Maharashtra, India

In zoonotic contexts, bacterial genomics has proven indispensable for identifying critical genomic traits in bacterial pathogens. These studies reveal the genetic mechanisms underpinning virulence, host adaptation, antimicrobial resistance, and interspecies transmission. It is essential in pathogen surveillance, outbreak investigation, antimicrobial resistance monitoring, and vaccine development. Comparative genomics enables the identification of strain-specific virulence factors and resistance genes, thereby illuminating diagnostic and therapeutic strategies. Furthermore, genomics supports studies in taxonomy, the investigation of non-cultivable bacteria, causal agent identification, single-cell sequencing, population genetics, and meta-transcriptomics. All these have contributed to advancing a better understanding of host–pathogen interactions and microbial ecology. This chapter provides an overview of the historical development, genome architecture, technological advancements, and applications of bacterial genomics, with particular emphasis on its transformative impact in the field of zoonoses.

Historical Developments in Bacterial Genomics

The history of bacterial genomics is marked by transformative innovations that have reshaped the understanding of microbial life. The foundation was laid in 1977 with the development of the

S. C. Parija, A. Poddar (eds.), *Textbook of Bacterial Zoonoses*, Microbial Zoonoses,
https://doi.org/10.1007/978-981-95-6615-0_5

Sanger chain-termination method, enabling scientists to read DNA sequences with precision. The sequencing of *Haemophilus influenzae* in 1995 using whole-genome shotgun sequencing demonstrated the power of genomic approaches for studying bacterial biology. Soon after, the genomes of key model and pathogenic bacteria, such as *Escherichia coli* and *Bacillus subtilis*, were completed, providing critical insights into gene regulation, metabolism, and cellular physiology. In the subsequent years, numerous zoonotic and multidrug-resistant bacteria, including *Mycobacterium tuberculosis*, *Salmonella enterica*, and *Staphylococcus aureus,* were sequenced, expanding knowledge on virulence, resistance mechanisms, and environmental adaptation.

The advent of next-generation sequencing (NGS) technologies, including Illumina and 454 pyrosequencing, revolutionised bacterial genomics by enabling massively parallel sequencing at lower costs and faster throughput. Meanwhile, long-read platforms, such as Pacific Biosciences (PacBio) and Oxford Nanopore, have allowed the resolution of complex genomic regions, structural variations, and large plasmids. These technologies facilitated whole-genome sequencing (WGS) for routine pathogen surveillance, antimicrobial resistance (AMR) tracking, and outbreak investigations. Comparative genomics and metagenomics further expanded the field's capabilities, elucidating horizontal gene transfer, microbial diversity, and functional traits relevant to host adaptation and ecosystem roles. The continuous refinement of sequencing technologies and analytical pipelines drives bacterial genomics forwards, positioning it as a cornerstone of microbial research and public health responses. Figure 5.1 explains the historical milestones in bacterial genomics.

Structure and Organisation of Bacterial Genomes

Chromosome

Bacterial genomes exhibit a remarkable diversity in structure and organisation, reflecting the evolutionary pressures and ecological niches they inhabit. The fundamental unit of a bacterial genome is the *chromosome*, which typically consists of a single, circular, double-stranded DNA molecule, a highly condensed entity within the nucleoid region, a non-membrane-bound compartment in the cytoplasm. This circular architecture often facilitates efficient DNA replication and segregation. However, exceptions like the linear chromosomes found in *Borrelia burgdorferi* and certain *Streptomyces* species highlight the evolutionary flexibility within this domain. The size of bacterial genomes exhibits a vast range, from the highly reduced genomes of obligate intracellular symbionts like *Candidatus Carsonella ruddii* (~0.16 Mbp), reflecting their reliance on the host for essential functions, to the expansive genomes of free-living bacteria such as *Sorangium cellulosum* (~14.8 Mbp), which encode a vast repertoire of metabolic pathways for survival in complex environments. This genomic diversity directly correlates with the ecological breadth and metabolic capabilities of different bacterial species.

Plasmid

In addition to chromosomes, many bacteria harbour *plasmids*, which are extrachromosomal DNA molecules that replicate independently of the chromosome. Plasmids often carry genes that provide bacteria with adaptive advantages, such as antibiotic resistance, virulence factors, or metabolic capabilities. The number and types of plasmids harboured by a bacterium can vary widely, contributing to the diversity of bacterial populations and their ability to adapt to changing environments. Plasmids range in size from a few kilobases to hundreds of kilobases and can be present in multiple copies within a cell. Plasmids can be transferred between bacteria through horizontal gene transfer mechanisms, such as conjugation, transduction, and transformation, leading to the rapid spread of antibiotic resistance and other adaptive traits.

Mobile Genetic Elements

Such as insertion sequences, transposons, and integrons, play a crucial role in shaping bacterial genome structure and promoting genetic diversity. Insertion sequences are simple transposable elements that encode only the genes necessary

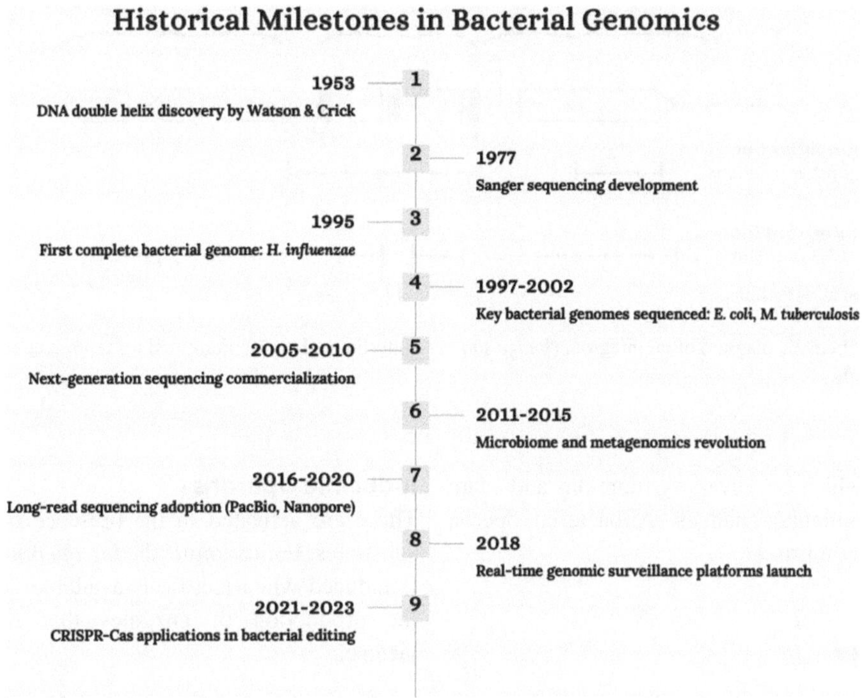

Fig. 5.1 Historical milestones in bacterial genomics. (Image Source: Authors' personal collection)

for their transposition, whereas transposons are more complex elements that carry additional genes, such as those conferring antibiotic resistance. Integrons are genetic platforms that capture and express gene cassettes, often containing antibiotic resistance genes, through a site-specific recombination mechanism. These mobile elements insert themselves into various locations within the bacterial genome, resulting in gene inactivation, gene duplication, or the formation of novel gene fusions. Furthermore, they facilitate the transfer of genes between bacteria, contributing to the spread of antibiotic resistance and other adaptive traits. Many bacteria are intrinsically resistant to certain antibiotics, a classic example of which is the antimicrobial resistance conferred by the chromosomally encoded marRAB locus of *E. coli*. A schematic diagram representing an integron is shown in Fig. 5.2.

Regulatory Elements and Gene Architecture

In bacteria are fine-tuned to control gene expression in response to environmental signals and developmental cues. Bacterial genes are often organised into operons, which are clusters of genes that are transcribed together from a single promoter. The expression of operons is regulated by transcription factors that bind to specific DNA sequences near the promoter, activating or repressing transcription. The precise arrangement of genes and regulatory elements within the bacterial genome is critical for coordinating gene expression and ensuring proper cellular function. The study of regulatory elements and gene architecture provides insights into the mechanisms by which bacteria adapt to changing environments and regulate complex cellular processes.

Operon Model

Bacterial operons are clusters of genes that are co-transcribed into a single polycistronic mRNA, allowing for coordinated expression of functionally related proteins. It evolves through mechanisms like gene duplication, loss, and horizontal

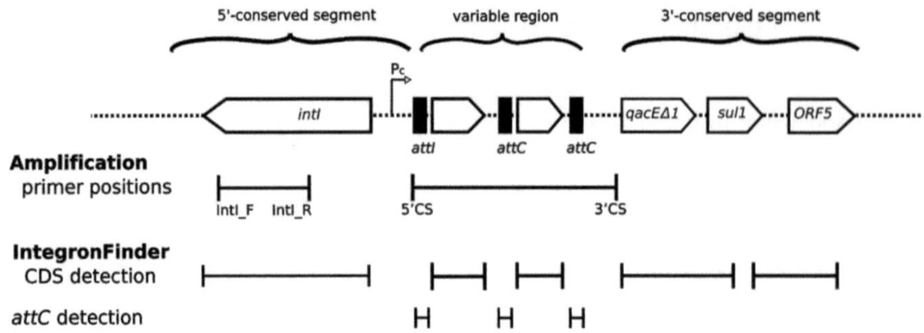

Fig. 5.2 Schematic diagram of an integron. (*Image source*: Sütterlin et al. 2020; Permitted for reproduction under CC BY Lincese)

transfer, which can diversify functions and adapt to environmental changes. A bacterial operon typically comprises:

Promoter

A DNA sequence where RNA polymerase binds to initiate transcription.

Operator

A regulatory DNA sequence that interacts with repressor or activator proteins to control transcription.

Structural Genes

Genes that code for proteins involved in a specific metabolic pathway or function.

Regulatory Gene

A gene that produces a repressor or activator protein, influencing the operon's activity.

Operons allow bacteria to *optimise energy expenditure* by ensuring that enzymes are produced only when needed. This regulation is achieved through mechanisms like induction and repression:

Inducible Operons

These are activated in the presence of specific substrates. For example, the *lac operon* in *E. coli* is induced when lactose is available, leading to the production of enzymes that metabolise lactose.

Repressible Operons

These are typically active but are turned off when the product is abundant. The *trp operon*, responsible for tryptophan biosynthesis, is repressed when tryptophan levels are sufficient.

This regulatory system ensures that bacteria conserve energy and resources by producing enzymes only when the corresponding substrates are present or when the end products are scarce.

Operon in Bacterial Pathogenesis

Beyond metabolism, operons play a crucial role in bacterial *pathogenesis* by regulating genes involved in virulence, antibiotic resistance, and stress responses.

Virulence Gene Regulation

Some operons control the expression of toxins and other virulence factors. For example, the *pho regulon* in *Pseudomonas aeruginosa* indeed activates the production of pyocyanin, a toxin, under phosphate-limited conditions, indicating host tissue damage.

Antibiotic Resistance
Operons like *marRAB* in *E. coli* play a crucial role in conferring multiple antibiotic resistance by reducing porin expression and activating efflux pumps. This operon is induced in the presence of specific antibiotics, enabling the bacterium to survive in hostile environments.

Stress Responses
The *SOS response* operon is activated in response to DNA damage, leading to increased mutation rates (translesion synthesis (TLS)) and the potential emergence of antibiotic-resistant strains.

These operons enable bacteria to adapt to host defences, acquire resistance to antibiotics, and persist under stressful conditions, thereby enhancing their pathogenicity.

Sequencing Technologies in Bacterial Genomics

Sanger Sequencing

This is a chain-termination method, which revolutionised DNA sequencing and became the cornerstone of early bacterial genomics research. This method involves the enzymatic synthesis of DNA fragments of varying lengths, terminated by the incorporation of dideoxy nucleotides that lack the 3′-hydroxyl group necessary for chain elongation. The resulting fragments are then separated by size using electrophoresis, and the DNA sequence is determined by reading the order of the terminating nucleotides. Sanger sequencing played a crucial role in determining the first complete bacterial genome sequence of *H. influenzae* in 1995, marking a significant milestone in the field. Although Sanger sequencing is accurate and reliable, it is also time-consuming and expensive, limiting its application to large-scale genomic studies.

Whole-Genome Shotgun Sequencing

Shotgun sequencing emerged as a powerful alternative to Sanger sequencing, enabling faster and more cost-effective determination of bacterial genome sequences. This method involves randomly fragmenting the entire genome into smaller pieces, sequencing these fragments, and then using computational algorithms to assemble the fragments into a contiguous genome sequence. Whole-genome shotgun sequencing requires a combination of laboratory techniques and computational tools.

Laboratory Techniques
The laboratory component begins with DNA fragmentation, where the bacterial genome is randomly broken into numerous small pieces. Next, these fragments undergo cloning, a process in which they are inserted into vectors and then introduced into host cells, effectively creating multiple copies of each fragment. Finally, the inserted DNA is subjected to sequencing using high-throughput technologies, generating millions of short sequence reads that computational tools later assemble into the complete genome.

Computational Tools
Once the DNA fragments are sequenced, a suite of specialised software tools steps in to assemble and analyse the vast amount of data. Several software tools are employed to assemble and analyse the sequenced fragments. An overview of a few computational tools is presented in Table 5.1.

The shotgun approach eliminates the need for prior mapping or cloning, making it suitable for sequencing large and complex genomes. This approach significantly accelerated the pace of bacterial genome sequencing, allowing researchers to explore the genetic diversity of a wide range of bacterial species.

High-Throughput Sequencing

High-throughput sequencing, also known as next-generation sequencing, has transformed bacterial genomics by simultaneously enabling the rapid and cost-effective sequencing of millions or even billions of DNA molecules. These technologies include platforms such as Illumina,

Table 5.1 Computational tools for genomic data analysis

Tool	Function	Input type	Key features	Use
SPAdes	Genome assembler	Short reads and long reads	Supports hybrid assemblies; uses a multi-k-mer approach	Bacterial genome assembly, hybrid genome assembly
Velvet	Genome assembler	Short reads	Uses de Bruijn graphs; efficient memory usage	*De novo* assembly of small genomes
Phrap	Sequence assembler	Sanger reads	Uses quality scores to improve assembly accuracy	High-quality assembly of Sanger data
Consed	Assembly viewer/editor	Assembled contigs and quality data	Graphical interface facilitates manual finishing of assemblies	Finishing Sanger-based assemblies
Samtools	Post-assembly data processing	Binary Alignment Map (BAM)/Sequence Alignment Map (SAM) files (aligned reads)	Sorting, indexing, variant calling, filtering reads	NGS data analysis pipelines
Deep variant	Variant caller (AI-based)	Aligned reads Binary Alignment Map (BAM)/ Compressed Reference-oriented Alignment Map (CRAM)	Uses deep learning; models variant calling as image classification	High-accuracy SNP and indel calling from Illumina/long reads

454, and Ion Torrent, each with its unique DNA sequencing approach. Illumina sequencing, based on sequencing-by-synthesis chemistry, is widely used for bacterial genome sequencing due to its high accuracy, throughput, and read length. The 454 sequencing, which utilises pyrosequencing technology, was one of the first high-throughput sequencing platforms and offered longer read lengths compared to Illumina, but other technologies have largely superseded it.

Long-Read Single-Molecule Sequencing

This technology, offered by Pacific Biosciences (PacBio) and Oxford Nanopore Technologies, represents a new frontier in bacterial genomics. PacBio utilises single-molecule real-time sequencing, which continuously monitors the incorporation of fluorescently labelled nucleotides as a single DNA polymerase replicates a DNA template. Nanopore technologies employ nanopore sequencing, which involves passing a single DNA molecule through a tiny pore in a membrane and measuring the changes in electrical current as the DNA bases pass through the pore. These technologies produce reads that are tens of thousands or even hundreds of thousands of base pairs long, overcoming the limitations of short-read sequencing technologies. Long-read sequencing facilitates the assembly of complete bacterial genomes, including repetitive regions and complex structural variations that are difficult to resolve with short-read data. Advancements in long-read sequencing technology continue to make whole-genome sequencing more complete, affordable, and accurate. Long-read sequencing offers distinct advantages over whole-genome sequencing, as detailed in the comparative table below. These long-read technologies generate ultra-long sequencing reads and stimulate the development of new tools for assembling highly contiguous genome sequences with unprecedented accuracy across complex repeat regions. Each platform has its strengths and limitations, and the choice of sequencing technology depends on the specific research question and the characteristics of the bacterial genome being studied.

A comparative table demonstrating different sequencing technologies in bacterial genomics is presented in Table 5.2.

Table 5.2 Comparative table demonstrating different technologies

Feature	Sanger sequencing	Whole-genome sequencing (WGS)	High-throughput sequencing (NGS)	Long-read single-molecule sequencing
Read length	500–1000 bp	100–600 bp	100–600 bp	1000–100,000+ bp
Throughput	Low	High	High	Moderate to high
Accuracy	99.99%	High	High (~99%)	Raw: 85–93.8%; Consensus: >99.9%
Cost per base	High	Moderate to high	Low	Moderate to high
Time per run	Hours to days	Days	Hours to days	Hours to days
Sample preparation	PCR amplification, capillary electrophoresis	DNA fragmentation, library preparation	DNA fragmentation, library preparation	DNA extraction, optional fragmentation
Applications	Targeted sequencing, small-scale projects	Comprehensive genome analysis, structural variant detection, *de novo* assembly	Whole-genome/exome sequencing, transcriptomics, metagenomics, targeted panels	Complex genome assembly, structural variant detection, epigenomics, real-time sequencing
Limitations	Low throughput, high cost for large genomes	Short read lengths can complicate the assembly of repetitive regions	Short read lengths can complicate the assembly of repetitive regions, lower accuracy in homopolymer regions	Higher initial error rates, complex data analysis, higher cost per base

Comparative and Evolutionary Genomics

Comparative genomics is a powerful approach for studying the evolutionary relationships, diversity, and adaptation of bacteria by comparing the genomes of different bacterial strains or species. The core genome represents the set of genes shared by all strains of a particular bacterial species, reflecting the essential functions required for survival and reproduction. On the other hand, the pan-genome encompasses all the genes found in all strains of a species, including the core genome and the accessory genome. The accessory genome comprises genes present in some strains but absent in others, often encoding functions related to adaptation to specific environments or lifestyles. Genome plasticity and horizontal gene transfer play a significant role in shaping the evolution of bacterial genomes, enabling bacteria to acquire new genes and adapt to changing environments rapidly.

Phylogenomics is the study of evolutionary relationships among organisms using genome-scale data, providing a powerful approach for inferring the phylogeny of bacteria and understanding their evolutionary history. Strain typing methods based on genomic data have revolutionised bacterial epidemiology and public health, enabling the rapid and accurate identification of bacterial strains and the tracking of outbreaks. By comparing core genomes, researchers can determine the phylogeny of known bacterial species, even those with previously ambiguous phylogenetic positions. The study of bacterial genomes has revealed that horizontal gene transfer (HGT) is a major evolutionary force that has constantly reshaped genomes throughout evolution. This transfer can occur between distantly related species. Through HGT, bacteria acquire novel traits, ranging from antibiotic resistance to metabolic capabilities, augmenting their adaptive machinery in response to environmental shifts.

Bacterial Phylogenomic Tree

A bacterial phylogenomic tree (Fig 5.3) is constructed by analysing the evolutionary relationships among bacterial species based on their genomic data. This tree is built using sequences

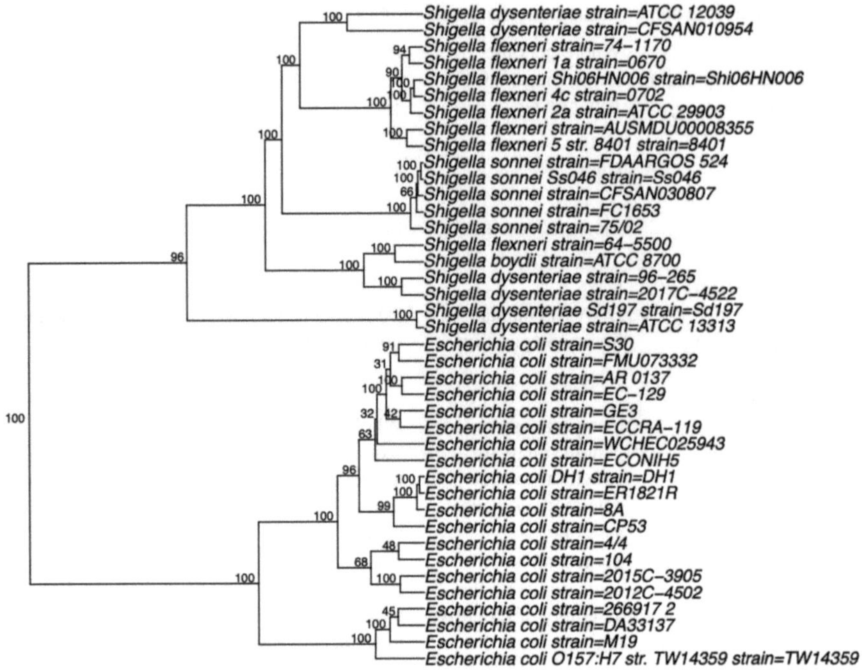

Fig. 5.3 A representative phylogenetic tree containing 20 random sequences from 888 sequences of *E. coli* and 20 random sequences from 92 sequences of *Shigella* based on pentanucleotide frequencies. (Image source: Nakano et al. 2023. Permitted for reproduction under CC BY license.)

from conserved genes across various species. The process involves several key steps.

The process begins with sequence alignment, where genomic sequences from multiple bacterial species are aligned to identify homologous regions, which are portions of the genome that have evolved through relatedness. Following this, a phylogenetic tree is constructed using algorithms such as Maximum Likelihood (ML) or Neighbour-Joining (NJ), which estimate evolutionary distances based on the aligned sequences. These methods help infer the branching patterns and relative relationships between the species under study.

To assess the robustness of the resulting tree, a bootstrap analysis is typically conducted. This involves resampling the data and reconstructing the tree multiple times to determine the consistency of the inferred relationships. Each branch of the tree is then assigned a bootstrap value, indicating the confidence level in that grouping. Furthermore, to orient the tree and infer the direction of evolutionary time, the tree is rooted using an outgroup. These distantly related species serve as a reference point.

The analysis of a bacterial phylogenomic tree provides critical insights into the evolutionary history of bacterial species. It allows researchers to identify common ancestors, trace lineage diversification, and understand how different bacterial groups have evolved over time. Such information is essential not only for bacterial taxonomy but also for advancing research in fields like infectious disease, antibiotic resistance, and industrial microbiology.

Genomics of Some Major Zoonotic Bacteria and Its Role in Pathogenesis

Genomic insights into zoonotic bacteria are key to deciphering their pathogenicity and antimicrobial resistance (AMR). Genomic information helps identify specific genes, genetic elements, and evolutionary pathways that contribute to

Table 5.3 Use of bacterial genomics in the management of zoonotic bacterial pathogens

Zoonotic pathogen	Use of genomic comparison	Conserved target identified	Tools/algorithms used	Application
Brucella spp. (*Brucella melitensis*, *Brucella abortus*)	Comparative genomics of strains from livestock (cattle, goats) in Gujarat, Tamil Nadu, and Punjab	bcsp31, omp31, IS711 (highly conserved)	BLAST, Roary, OrthoFinder, Prokka	Broad-range PCR diagnostics; part of the Brucellosis Control Programme
Leptospira interrogans	Comparison of Indian clinical isolates from Kerala and Gujarat	lipL32, ligB, ompL1 (outer membrane proteins)	MUSCLE, BLAST, Clustal Omega, VaxiJen	Targets for qPCR, ELISA; surveillance of floods/agricultural outbreaks
Bacillus anthracis (anthrax)	Genomic surveillance of strains in Odisha and Chhattisgarh	pagA, capA-C, rpoB (capsule and toxin genes)	Panaroo, SNP-calling (Snippy), PhyML	Design of molecular diagnostic kits used in veterinary labs

their ability to cause disease in both animals and humans, as well as their capacity to resist antibiotic treatments. This genomic information is critical for developing effective surveillance strategies, diagnostics, and targeted interventions to mitigate the threat posed by these important pathogens (Table 5.3). A brief overview of genomic insights into some zoonotic bacteria is described below.

Salmonella enterica

The genome of *S. enterica* is approximately 4.8–5.0 Mb and includes chromosomal regions known as *Salmonella* pathogenicity islands (SPIs), notably SPI-1 and SPI-2, as well as virulence plasmids, such as virulence plasmid of Salmonella enterica serovarTyphinurium strain LT2 (pSLT). These SPIs encode type III secretion systems (T3SSs), syringe-like structures that inject bacterial effector proteins into the host cells. SPI-1 is crucial for host cell invasion, while SPI-2 facilitates intracellular survival, particularly within phagocytic cells. The bacterium's AMR profile is increasingly concerning, with resistance genes such as bla_TEM and qnr often located on plasmids. Efflux pumps and class 1 integrons further enhance resistance to multiple antibiotics, making the treatment of infections challenging.

Brucella spp.

Brucella species possess a compact genome of about 3.3 Mb, distributed across two circular chromosomes. Unlike many pathogens, they lack classic virulence factors, such as toxins, but rely heavily on a type IV secretion system (T4SS), encoded by the virB operon, to facilitate intracellular replication and manipulate host immune responses. *Brucella* has a modified lipopolysaccharide (LPS) structure that contributes to immune evasion by reducing host recognition and immune response. Although *Brucella* exhibits low intrinsic AMR, resistance can develop through chromosomal mutations, such as those affecting rifampin targets, and through efflux mechanisms.

Escherichia coli (STEC/EHEC O157:H7)

Pathogenic strains of *E. coli*, such as Shiga toxin (stx)-producing (STEC) and enterohaemorrhagic (EHEC) variants like O157:H7, have a large and dynamic genome (~5.5–5.8 Mb). This genome is highly plastic, with extensive acquisition of phages, plasmids, and pathogenicity islands. Key virulence factors include Shiga toxin (stx) genes encoded on lysogenic phages and the locus of enterocyte effacement (LEE). This encodes another T3SS, along with adhesion molecules like intimin. *E. coli* also exhibits significant acquisition of AMR, primarily through mobile genetic elements, such as plasmids (e.g. carrying bla_CTX-M, tetA, and sul1). The use of antibiotics in livestock promotes the emergence and spread of these resistance determinants, raising public health concerns.

Applications of Bacterial Genomics

Bacterial genomics has become an indispensable tool in the development of vaccines, diagnostics, and therapeutics. It provides a detailed framework for understanding the complexities of bacterial pathogenesis, thereby guiding the development of targeted interventions. Genomic data allows researchers to identify unique bacterial proteins or epitopes that can serve as targets for antibody development. Furthermore, comparative genomics enables the identification of bacteria-specific sequences that are not present in the host or closely related commensal organisms, minimising the risk of cross-reactivity and enhancing the specificity of the resulting immunodiagnostic assays.

Vaccine Development

Vaccination is known to be one of the most significant public health interventions that has had the greatest impact on global health. Vaccines in use today represent only the tip of the iceberg when considering the number of diseases encountered worldwide over the period. Traditional vaccine development often relies on culturing bacteria and isolating attenuated or inactivated forms of the pathogen. This process is time-consuming and challenging. Nevertheless, the arrival of the genome era has revolutionised the way the identification of vaccine candidates is approached. The advancements in bacterial genomics have significantly transformed vaccine development, resulting in more effective and precisely targeted vaccines. Genomics offers a more rapid and accurate approach by enabling researchers to identify potential vaccine candidates directly from the bacterial genome. For example, in the case of *Neisseria meningitidis* serogroup B, researchers sequenced the bacterial genome and identified over 600 potential antigens. These were tested for immunogenicity, leading to the development of a pentavalent vaccine that induced a bactericidal immune response in mice. This approach has been instrumental in creating vaccines against pathogens with complex surface structures.

In India, there have been large-scale contributions to the comparative genomic studies of *Mycobacterium tuberculosis* strains isolated from Indian patients. These studies have revealed the identification of conserved antigens like ESAT-6, CFP-10, and the *Ag85 complex*, which are present across various Indian lineages. These antigens are used in commercial tests and are being evaluated in *new Indian vaccine candidates*, such as Immuvac, which is in the third phase of its clinical trials.

Development of Diagnostics

Genomic technologies have significantly advanced the development of bacterial diagnostics, enabling the rapid, accurate, and comprehensive detection of bacterial pathogens. Traditional diagnostic methods, such as culturing and biochemical tests, are time-consuming and labour-intensive, often requiring several days to yield results. Genomics-based diagnostics offer a faster and more accurate alternative by directly detecting bacterial DNA in clinical samples. Polymerase chain reaction (PCR) assays, which amplify specific DNA sequences, have become a cornerstone of bacterial diagnostics. DNA sequencing has revolutionised bacterial diagnostics, enabling the identification of virtually any bacterium in a sample, regardless of its culturability or novelty. One such example is Truenat™ for tuberculosis (TB) diagnosis.

Molbio's Truenat™ is a rapid molecular diagnostic platform endorsed by the World Health

Organization (WHO) for the initial diagnosis of tuberculosis and detection of rifampicin resistance. Utilising real-time PCR, Truenat™ provides results in under an hour, facilitating timely treatment decisions. Its integration into India's National Tuberculosis Elimination Programme has improved TB case notification rates by 30% in regions like Andhra Pradesh.

Therapeutics Development

Genomics has fundamentally transformed the landscape of therapeutic development, offering profound insights into bacterial physiology and unveiling novel antimicrobial targets. Historically, the discovery of antibiotics predominantly relied on phenotypic screening, wherein vast compound libraries were empirically assessed for their inhibitory effects on bacterial growth. Genomics, however, provides a more rational and targeted approach by enabling the identification of essential bacterial genes or pathways whose disruption would compromise bacterial viability or virulence. Genomic analysis facilitates the identification of specific enzymes or proteins indispensable for bacterial survival or the manifestation of pathogenicity. These molecular targets can then be exploited with high precision through structure-based drug design. This methodology leverages the three-dimensional structural information of a target molecule to rationally design small molecules or biologics that bind with high affinity and specificity, thereby inhibiting its critical function.

Furthermore, genomics is instrumental in the progression towards personalised antimicrobial therapies. By rapidly sequencing the genome of the aetiologic agent responsible for an infection, clinicians accurately identify the presence of antibiotic resistance genes and precisely ascertain the optimal therapeutic regimen. This genomic intelligence ensures the judicious selection of antibiotics, thereby enhancing treatment efficacy and mitigating the risk of therapeutic failure. Beyond individual patient care, the application of high-throughput genomic tools presents unprecedented opportunities for advanced environmental health monitoring and surveillance.

An example is the treatment of *E. coli* infections, particularly Shiga toxin–producing and enterohaemorrhagic strains. The discovery of the mcr-1 gene, which confers resistance to colistin and is often found on plasmids in zoonotic *E. coli*, has reshaped therapeutic approaches. Genomic screening enables clinicians to identify this resistance gene in patient isolates, leading to the adoption of combination therapies that incorporate colistin and other antibiotics, such as carbapenems or tigecycline. This prevents treatment failure and helps curb the development of further resistance, particularly in cases linked to livestock-origin strains.

In the case of *Brucella melitensis*, genomics has enabled personalised treatment strategies by identifying specific chromosomal mutations linked to resistance, especially to rifampin. By sequencing clinical isolates, healthcare providers avoid using ineffective antibiotics and, instead, tailor regimens to use alternatives, such as doxycycline with streptomycin or quinolones. This is especially valuable in regions where brucellosis is endemic and resistance patterns are not always well-documented. Genomic data, therefore, not only aids in diagnosis but also ensures the efficacy of treatment.

For *S. enterica*, genomic analyses have identified key virulence genes within *Salmonella* pathogenicity islands (SPIs), particularly SPI-2, which encodes a type III secretion system (T3SS) essential for intracellular survival. This knowledge has led to the design of novel anti-virulence compounds, such as Sal-003, which target and inhibit the function of the SPI-2 T3SS. Such targeted therapeutics offer promising alternatives to conventional antibiotics, especially for drug-resistant typhoidal strains of *S. enterica*, and represent a strategic shift towards disrupting bacterial pathogenesis rather than directly killing the bacteria.

Another notable example is the application of genomics in vaccine-linked therapy for *Coxiella burnetii*, the causative agent of Q fever. Genome sequencing of *C. burnetii* revealed immunogenic proteins such as Com1 and ankyrin repeat-containing proteins, which were subsequently used in the development of the Q-Vax vaccine in Australia. While primarily a preventive measure, the vaccine has significantly reduced the incidence of chronic Q fever, thereby lowering the need for prolonged antibiotic treatment in high-risk populations, such as abattoir workers and veterinarians.

Future Directions

Bacterial genomics is poised to further transform the approach to understanding and managing zoonotic bacterial diseases. As sequencing platforms become more portable, affordable, and user-friendly, genomic tools are now being deployed closer to the point of need, including in wildlife surveillance, livestock monitoring, and clinical investigations of zoonotic outbreaks. This enhanced accessibility will allow for more timely and accurate responses to emerging threats, particularly in regions where zoonotic spillovers are most likely to occur. The refinement of long-read and hybrid sequencing technologies continues to improve the capacity to generate highly contiguous and accurate bacterial genome assemblies. This has profound implications for characterising mobile genetic elements, plasmids, and resistance islands that are frequently implicated in zoonotic transmission. These platforms are especially useful in resolving the complex genomes of emerging pathogens, which often evade detection through conventional methods.

Looking forwards, the integration of bacterial genomic data with ecological, epidemiological, and environmental metadata will be essential. Combining these data streams could yield predictive models that map high-risk interfaces between wildlife, livestock, and human populations. Such models could inform targeted surveillance strategies and facilitate early interventions that prevent the escalation of zoonotic outbreaks. Genomic surveillance networks built around the One Health framework will be central to this vision. These networks must prioritise real-time data sharing, interoperability, and equitable access to sequencing infrastructure. International collaboration and capacity building, especially in endemic regions, will be crucial to ensure that genomic innovations translate into local public health benefits.

Finally, bacterial genomics will increasingly inform the development of vaccines and therapeutics tailored to zoonotic pathogens. Genomic insights into host adaptation, immune evasion, and virulence regulation provide a rational basis for identifying novel intervention targets.

Case Study

A 34-year-old male, who recently returned from South-east Asia, presented to the infectious diseases clinic with a two-month history of intermittent fever, fatigue, and abdominal discomfort. Blood cultures confirmed *Salmonella enterica* serovar Typhi. Following local empirical guidelines, he was initially treated with oral ciprofloxacin. Unfortunately, two weeks after completing the ceftriaxone regimen, the patient's symptoms recurred, and *S.* Typhi was again isolated from his blood. It was decided to utilise rapid whole-genome sequencing (WGS) of the *S.* Typhi isolate to gain a comprehensive understanding of the pathogen's genetic profile. The WGS analysis yielded critical information that went beyond conventional susceptibility testing. It confirmed the isolate as *S.* Typhi and, crucially, identified a plasmid carrying multiple resistance genes. These included specific point mutations in the gyrA gene (S83F and D87N), which confer high-level resistance to fluoroquinolones, such as ciprofloxacin, thereby explaining the initial treatment failure. The genome also carried bla_TEM-1B, which confers ampicillin resistance, and catA1, which confers chloramphenicol resistance. Importantly, it confirmed the absence of genes for colistin resistance (mcr-1) and no acquired resistance to azithromycin or ceftriaxone, validating the effectiveness of these drugs. Furthermore, WGS revealed the full complement of classical *S.* Typhi virulence factors, such as the Vi capsule biosynthesis genes and functional

SPI-1 and SPI-2 type III secretion systems, indicating a highly virulent strain. Phylogenetic analysis provided crucial epidemiological context by placing the isolate within the notorious H58 lineage (now Genotype 4.3.1.1), a globally disseminated, multidrug-resistant clone prevalent in South-east Asia. With such a detailed genomic report, the clinical team refined its management strategy. Following the extended azithromycin therapy, the patient's symptoms fully resolved, and subsequent blood cultures were negative. He remained asymptomatic at his 6-month follow-up.

1. How did whole-genome sequencing (WGS) specifically inform the choice and duration of the final antibiotic regimen in this recurrent *S.* Typhi infection?
2. What is the significance of identifying the *S.* Typhi isolate as part of the H58 lineage (Genotype 4.3.1.1) in terms of patient management and public health implications?
3. Beyond antibiotic resistance, what other genomic insights from the WGS analysis helped explain the patient's persistent and recurrent infection?
4. If the WGS had identified resistance to azithromycin and ceftriaxone, what immediate challenges would the clinical team have faced, and what alternative therapeutic strategies might have been considered?

Research Questions

1. Investigate how the genomic transmission dynamics of bacterial pathogens differ between animal and human hosts.
2. How can real-time, point-of-care genomic sequencing technologies be integrated into veterinary and public health surveillance systems to enable immediate outbreak detection and guide rapid, targeted interventions for emerging zoonotic bacterial threats?
3. What novel therapeutic or prophylactic strategies can be discovered and developed by leveraging pan-genomic analyses of zoonotic bacteria to identify universally conserved essential genes, unique virulence factors, or host-specific colonisation determinants?
4. To what extent can longitudinal genomic surveillance of zoonotic bacteria across diverse One Health interfaces (wildlife, livestock, humans, and environment) predict the emergence and evolution and spread of novel antimicrobial resistance genes or highly virulent clones, thereby enabling proactive rather than reactive public health responses?

Declaration During the preparation of this chapter, we have used ChatGPT to improve the readability of the text. After using it, we have reviewed and edited the content as needed, and we take full responsibility for the content of the publication.

Further Readings

Almeida A, Mitchell A, Boland M, Forster SC, Gloor GB, Tarkowska A, et al. A new genomic blueprint of the human gut microbiota. Nature. 2019;568(7753):499. https://doi.org/10.1038/s41586-019-0965-1.

Cameron RA, Bohrhunter JL, Taffner SM, Malek A, Pecora N. Clinical pathogen genomics. Clin Lab Med. 2020;40(4):447. https://doi.org/10.1016/j.cll.2020.08.003.

Dai X, Shen L. Advances and trends in omics technology development. Front Med. 2022;9:911861. https://doi.org/10.3389/fmed.2022.911861.

Fitzgerald JR, Holden MTG. Genomics of natural populations of Staphylococcus aureus. Annu Rev Microbiol. 2016;70(1):459. https://doi.org/10.1146/annurev-micro-102215-095547.

Ghattas M, Dwidedi G, Lavertu M, Alameh MG. Vaccine technologies and platforms for infectious diseases: current progress, challenges, and opportunities. Vaccines (Basel). 2021;9(12):1490. https://doi.org/10.3390/vaccines9121490.

Harvey WT, Ebert P, Ebler J, Audano PA, Munson KM, Hoekzema K, et al. Whole-genome long-read sequencing downsampling and its effect on variant calling precision and recall. bioRxiv [Preprint]. 2023; https://doi.org/10.1101/2023.05.04.539448.

Jung H, Winefield C, Bombarely A, Prentis PJ, Waterhouse PM. Tools and strategies for long-read sequencing and de novo assembly of plant genomes. Trends Plant Sci. 2019;24(8):700. https://doi.org/10.1016/j.tplants.2019.05.003.

McDermott PF, Davis JJ. Predicting antimicrobial susceptibility from the bacterial genome: a new paradigm for one health resistance monitoring. J Vet Pharmacol Ther. 2020;44(2):223–8. https://doi.org/10.1111/jvp.12913.

McGrath C. Breaking genetic barriers: understanding the limits of horizontal gene transfer. Genome Biol Evol. 2023;15(6):evad102. https://doi.org/10.1093/gbe/evad102.

Nafea AM, Wang Y, Wang D, Salama AM, Aziz MA, Xu S, et al. Application of next-generation sequencing to identify different pathogens. Front Microbiol. 2024;14:1329330. https://doi.org/10.3389/fmicb.2023.1329330.

Nakano Y, Domon Y, Yamagishi K. Phylogenetic trees of closely related bacterial species and subspecies based on frequencies of short nucleotide sequences. PLoS One. 2023;18(4):e0268847. https://doi.org/10.1371/journal.pone.0268847.

Sütterlin S, Bray J, Maiden M, Tano E. Distribution of class 1 integrons in historic and contemporary collections of human pathogenic Escherichia coli. PLoS One. 2020;15(6):e0233315. https://doi.org/10.1371/journal.pone.0233315.

Bacterial Proteomics

6

Angeliki I. Katsafadou, George C. Fthenakis, and George Th. Tsangaris

Learning Objectives

1. To gain knowledge regarding the various methodologies that are applied in proteomics
2. To gain knowledge regarding the application of proteomics in bacteria of zoonotic importance
3. To gain knowledge regarding the use of proteomics in antimicrobial resistance mechanisms and food safety

Introduction

Bacteria play a complex and multifaceted role in human and animal health. While they are often associated with infections and diseases, many bacteria are essential for maintaining health. Human and/or animal bodies host trillions of bacteria; hence, understanding the impact of bacteria on health involves both harmful and beneficial aspects.

Beneficial bacteria include probiotics, which restore balance to the gut flora, prevent diarrhoea caused by infections or antibiotics, and reduce the clinical signs of conditions like irritable bowel syndrome. Hence, commensal bacteria live on and within human and/or animal bodies without causing harm, thus providing benefits such as outcompeting harmful pathogens for resources, producing substances that inhibit pathogen growth, and stimulating the immune system.

Pathogenic bacteria on the other hand, enter the body through various routes (food, air, and wounds) and may lead to infections and diseases. Moreover, bacterial infections originating from animals that can be transmitted to humans, known as 'zoonotic infections', represent significant public health risks globally. Other infections, primarily transmitted between humans, can also circulate in animals or have animal reservoirs, potentially leading to severe health emergencies.

The inappropriate and excessive use of antibiotics has accelerated the emergence of bacterial strains that are resistant to conventional antimicrobial therapy, making infections increasingly challenging to manage. Notable examples include methicillin-resistant *Staphylococcus aureus* (MRSA) and vancomycin-resistant Enterococci (VRE), both of which represent serious public health concerns worldwide. These multidrug--

A. I. Katsafadou
Faculty of Public and One Health, University of Thessaly, Karditsa, Greece

Department of Environmental Health Sciences, Yale School of Public Health, New Haven, CT, USA

G. C. Fthenakis
Veterinary Faculty, University of Thessaly,

Karditsa, Greece

G. T. Tsangaris (✉)
Proteomics Research Unit, Biomedical Research Foundation of Academy of Athens, Athens, Greece
e-mail: gthtsangaris@bioacademy.gr

S. C. Parija, A. Poddar (eds.), *Textbook of Bacterial Zoonoses*, Microbial Zoonoses,
https://doi.org/10.1007/978-981-95-6615-0_6

resistant organisms are particularly problematic in healthcare settings, where they can lead to persistent, hard-to-treat infections and heightened morbidity among hospitalised patients. A critical step in safeguarding human populations from zoonotic infections is the control of pathogens at their animal source. Developing and implementing effective strategies to monitor and manage these microorganisms in livestock is essential to prevent transmission between animals and humans. Furthermore, mitigating the onset of antibiotic resistance within animal populations reduces the likelihood of resistant strains entering the human food chain or spreading through environmental routes. The close relationship between animal health, food safety, and human well-being underscores the need for an integrated perspective on disease prevention.

In recognition of this interdependence, the 'One Health' framework, introduced in the early twenty-first century, advocates for a holistic approach that links human, animal, and environmental health. This approach suggests that human and animal health are intertwined and closely linked to the health of the ecosystems they inhabit. Within this context, One Health initiatives focus on several key areas: (i) controlling zoonotic diseases, (ii) combating antimicrobial resistance, and (iii) ensuring food safety. These challenges are further intensified by globalisation, climate change, and shifts in human activity, all of which create new ecological niches for bacterial adaptation and spread.

Overview of Bacterial Proteome

The bacterial proteome represents the complete set of proteins expressed by a bacterial cell, strain, or species under specific conditions. Proteins play vital roles in the metabolism, structural integrity, gene regulation, and virulence of the organism, making proteomic studies essential for understanding bacterial adaptation and biology. The proteome's dynamic nature—shaped by factors like nutrient availability, temperature, and stress—highlights the bacterial ability to thrive in diverse environments.

Advances in mass spectrometry and related techniques have revolutionised proteomics and thus enabled large-scale identification and quantification of bacterial proteins and shed light on their functions in cellular processes, host interactions, and microbial communities. Proteomic research has significant implications across various fields. Studies into infectious diseases have (i) helped identify proteins involved in bacterial pathogenesis and mechanisms of antimicrobial resistance, (ii) offered insights into how pathogens evade antibiotics, and (iii) enabled the development of novel treatments (Fig. 6.1). Proteomics is also valuable in industrial biotechnology, as it has aided in optimising the bacterial production of enzymes, biofuels, and other metabolites. Additionally, it has provided a deeper understanding of bacterial interactions with their environments, including host–pathogen relationships, microbial symbiosis, and the formation of biofilms.

Notably, proteomic studies of bacterial pathogens have focused on species of zoonotic importance, such as *Mycobacterium avium* subsp. *paratuberculosis*, *Mycobacterium bovis*, *Listeria monocytogenes*, *Brucella* spp., *Campylobacter* spp., *Coxiella burnetii*, *Salmonella* spp., *Borrelia* spp., and *Leptospira* spp. Food-borne pathogens, like *Campylobacter jejuni* in poultry, *Francisella tularensis* in hare carcasses, *Salmonella enterica* and *Escherichia coli* O157 in red meat, and *Vibrio parahaemolyticus* in aquatic products, have also been investigated. Additionally, research has targeted bacterial toxins, such as staphylococcal enterotoxins A and B in meat and milk products, *Clostridium perfringens* epsilon toxin (ETX) in milk and water, and mycotoxins in various matrices. By elucidating these proteomic profiles, researchers can advance strategies for food safety, infectious disease control, and industrial applications.

Techniques in Bacterial Proteomics

Advances in proteomics, including sample fractionation and mass spectrometry, can enable bacterial proteome characterisation in inoculated

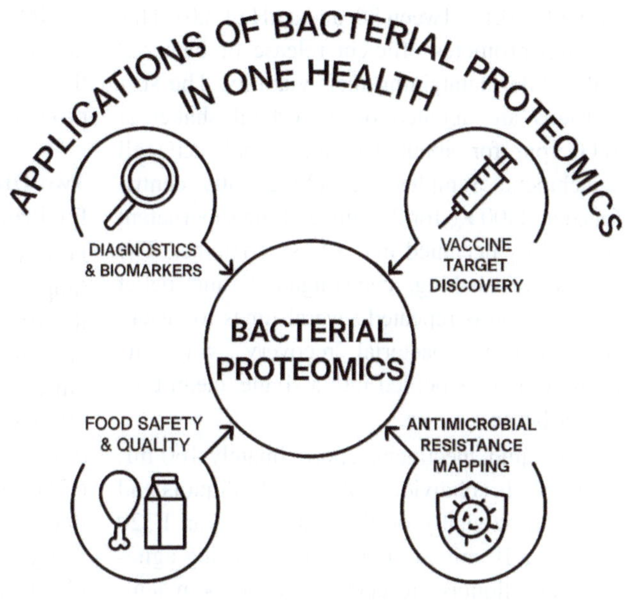

Fig. 6.1 Applications of bacterial proteomics across One Health pillars. The central node ('bacterial proteomics') links to four key application areas: diagnostics and biomarkers, vaccine target discovery, antimicrobial resistance mapping, and food safety and quality. (Image source: Authors' personal collection)

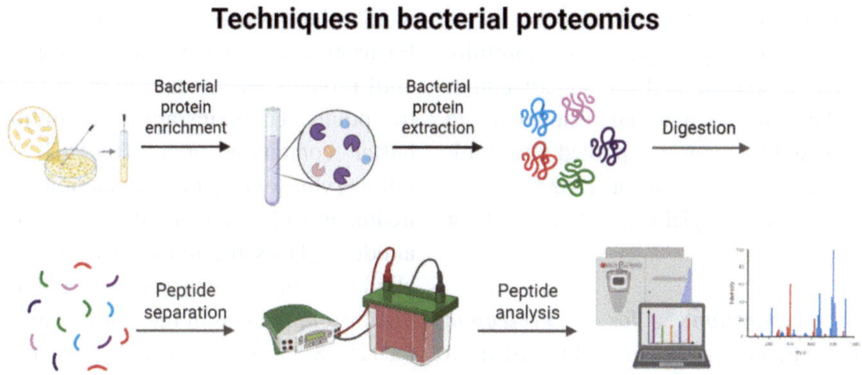

Fig. 6.2 Techniques in bacterial proteomics. The figure was created with BioRender (https://BioRender.com). (Image source: Authors' personal collection)

fluids or tissues, as well as from in vitro cultures. Various approaches, from 'shotgun' sampling to sensitive, quantitative proteomics, are used to analyse bacterial pathogens. Matrix-assisted laser desorption/ionisation time-of-flight (MALDI-TOF) mass spectrometry has been proven quite helpful in identifying and distinguishing bacterial pathogens. Due to the complexity of the data, advanced analytical tools are necessary for biological interpretation. The overall technical flow in the study of bacterial proteomics is described below (Fig. 6.2).

Sample Preparation

Bacterial Fraction Enrichment Enrichment of bacterial fractions typically serves as the initial phase in most proteomic workflows. To preserve the native state and biological activity of the bacterial community, all procedures are conducted at a low temperature (approximately 4 °C), with strict temperature control. For solid biological materials, accurately weighed subsamples (around 0.5 g each) are finely minced and suspended in 15 mL of phosphate buffer (50 mM of

Na_2HPO_4, 0.1% Tween 80, and a pH of 8.0). This solution promotes efficient release of bacterial cells while maintaining their viability. The suspensions are agitated on an orbital shaker at 1600 rpm for about 10 min to enhance cell detachment. Samples are subsequently centrifuged at 2500 ×*g* for 20 min, and the supernatant is carefully decanted into a fresh container. The process of shaking, centrifugation, and pellet resuspension is repeated several times to ensure comprehensive bacterial recovery, with the resulting pellets pooled into a single, clean tube for further use.

For liquid specimens, approximately 160 mL of sample is subdivided into 10 mL aliquots and placed horizontally on the platform of an F320 stirrer for 10 min at 1600 rpm. Following agitation, the aliquots are centrifuged at room temperature (2500 ×*g*, 20 min) to harvest bacterial and cellular material. A faint reddish pellet forms at the base of the tube. The lipid layer is carefully removed with a spatula, and the supernatant is discarded. Pellets with minimal residual liquid are gently mixed by pipetting, pooled into 2 mL tubes, and centrifuged again at 12,000 ×*g* for 20 min at 4 °C to yield the final cellular concentrate.

Sample Preparation for Proteomic Evaluation The bacterial pellets obtained from the enrichment step are resuspended in a lysis buffer composed of 7 M urea, 2 M thiourea, and 4% 3-(3-cholamidopropyl) dimethylammonio-1--propanesulfonate (CHAPS). Mechanical disruption is achieved through six bead-beating cycles of one minute each, separated by one-minute cooling intervals on ice. Each sample is agitated at 4000 rpm using an equal ratio (1:1 v/w) of 0.1 mm zirconium–silica beads. The clarified extract (supernatant) containing soluble proteins—the metaproteome—is collected in a clean microtube for downstream processing. To ensure complete solubilisation, samples are heated at 60 °C for 10 min and centrifuged again (12,000 ×*g*, 20 min, 4 °C). The resulting supernatant, rich in extracted proteins, is quantified using the Bradford assay and stored for subsequent proteomic analysis.

Two-Dimensional Gel Electrophoresis (2-DE) for Protein Separation

Following bacterial enrichment, protein extracts (approximately 1 mg per sample, quantified via the Bradford assay) are prepared for proteomic analysis. Each extract is dissolved in a denaturing buffer composed of 8 M urea, 2 M thiourea, 40 mM Tris-HCl (pH 8.5), 4% CHAPS, 1% dithioerythritol, 0.2% immobilised pH-gradient (IPG) buffer, and 1 mM phenylmethylsulfonyl fluoride (PMSF). Protein separation is carried out using the two-dimensional gel electrophoresis (2-DE) technique. The first-dimension isoelectric focusing (IEF) step employs IPG strips, onto which the protein mixture is evenly loaded. Focusing is conducted in an electrophoresis unit until proteins reach equilibrium at their isoelectric points. The strips are then equilibrated in a buffer containing urea, Tris-HCl, SDS, and dithioerythritol to achieve reduction, followed by treatment with a similar buffer in which iodoacetamide replaces the reducing agent to complete alkylation of cysteine residues. For the second dimension, the equilibrated strips are positioned on 12% SDS–polyacrylamide gels and subjected to electrophoresis to resolve proteins based on molecular weight. Upon completion, proteins within the gels are fixed using a methanol–phosphoric acid solution and visualised with appropriate staining reagents such as colloidal Coomassie Brilliant Blue or other compatible dyes.

Mass Spectrometry (MS) for Protein Identification and Characterisation

Protein identification in proteomic studies is most commonly achieved through peptide mass fingerprinting (PMF). In this process, protein spots of interest are digitally mapped, then carefully excised from the two-dimensional gels, and transferred into microtitre plates. The excised gel pieces are washed with a mixture of 30% acetonitrile and 50 mM ammonium bicarbonate to remove residual stains and impurities, followed by drying under vacuum conditions. Subsequently, proteins within the gel fragments are enzymatically digested, typically using trypsin, and the resulting peptides are extracted with a solution containing acetonitrile and trifluoroacetic acid. The peptide extracts are deposited onto a matrix-assisted laser desorption/ionisation (MALDI) target plate, together with a matrix reagent such as α-cyano-4-hydroxycinnamic acid (CHCA). Internal calibrants, for example, des-Arg-bradykinin and adrenocorticotropic hormone fragment 18–39, are often incorporated to enhance mass accuracy. The samples are then analysed by MALDI–time-of-flight (TOF) mass spectrometry, generating spectra that correspond to the mass-to-charge (m/z) ratios of the peptides. These data are compared with curated databases—such as UniProtKB/Swiss-Prot or the Comprehensive Antibiotic Resistance Database (CARD)—to match the observed peptide masses with known bacterial protein sequences (see Table 6.1).

For direct bacterial identification, the MALDI-TOF workflow involves selecting an isolated colony from a culture plate and transferring a minute portion onto the MALDI target surface. A small volume of CHCA matrix is applied, allowing the mixture to dry and co-crystallise. When introduced into the spectrometer, a laser pulse ionises the proteins, and the generated ions are separated according to their mass-to-charge (m/z) ratios, producing a characteristic spectral profile. The acquired fingerprint is automatically compared with reference spectra stored in the instrument's database, and a matching score quantifies the reliability of identification—higher scores indicate closer similarity to known organisms. Routine calibration and maintenance of the instrument are essential to ensure reproducibility and accuracy of results.

New Techniques

After bacterial enrichment and sample preparation, sample volumes corresponding to 200 ng of total proteins are further subjected to 1-D nano–high-pressure liquid chromatography–tandem mass spectrometry (nano-HPLC–MS/MS). The extraction of proteins and the generation of peptides are performed with HPLC (e.g. Dionex 3000 HPLC system, Thermo Scientific, Rockford, IL, USA) and the generation of mass spectra using a mass spectrometer (e.g. Linear Triple Quadrupole (LTQ)–Orbitrap). The raw data files are analysed using software by searching different databases, which include data from bacteria (e.g. UniProtKB/Swiss-Prot Protein Knowledgebase) (Fig. 6.3 and Table 6.1).

Information Gained from Bacterial Proteomics

Bacterial proteomics, the large-scale study of proteins within bacterial cells, provides information critical to understanding bacterial function, behaviour, and responses to various environmental conditions.

Protein Identification and Quantification

Proteomics allows the identification of the entire set of proteins (proteome) expressed by a bacterial cell or population at a given time. This includes proteins involved in basic cellular functions and stress responses, pathogenesis, or interactions with the environment.

Mass spectrometry (MS) is the most widely used tool for protein identification. It detects proteins based on their mass-to-charge ratio and compares them to known protein databases. Identifying proteins helps link them to specific biological functions or pathways, enabling a deeper understanding of bacterial physiology. In addition to identifying which proteins are present, bacterial proteomics can also quantify their expression levels. This reveals which proteins are

Table 6.1 Commonly used bacterial Proteomic databases and platforms

Database/platform	Primary content and scope	Key features that benefit bacterial Proteomics	Typical use cases	Sources
UniProtKB (Swiss--Prot + TrEMBL)	Curated Protein sequences and functional annotation for >1 million reference proteomes	High-quality, non-redundant sequences; controlled vocabulary for PTMs; cross links to structures, pathways, Antimicrobial Resistance (AMR) ontologies	Building custom search databases; mapping peptide IDs to functions and pathways; comparative proteome annotation	https://www.uniProt.org
PRIDE archive	Public repository of raw and processed MS Proteomics datasets	Minimum Information About a Proteomics Experiment (MIAPE)-compliant metadata; supports LC–MS/MS, Data Independent Acquisition (DIA), and quantitative (iTRAQ/TMT) data; direct submission tool; programmatic API	Re-analyse published spectra, benchmark search pipelines, meta-analyses of bacterial stress responses	https://www.ebi.ac.uk/pride/
ProteomeXchange (PX)	A consortium that federates PRIDE, MassIVE, iProX, jPOST, Panorama, and others	Single ProteomeXchange dataset (PXD) identifier per dataset; Rapid Submission System (RSS) updates; partial and complete submission workflows	Discover multi-species datasets, track new bacterial proteomes, ensure Findable, Accessible, Interoperable, and Reusable (FAIR) data deposition	https://www.Proteomexchange.org
PeptideAtlas	Peptide-centric resource built from reprocessed public MS runs (multi--organism, incl. many bacteria)	Uniform False Discovery Rate (FDR) control; spectral libraries for SRM/DIA; organism-specific 'Atlas' builds (e.g. Borrelia PeptideAtlas)	Targeted-assay design, peptide uniqueness checks, validation of novel bacterial proteins	https://peptideatlas.org
Comprehensive antibiotic resistance database (CARD)	Curated AMR genes, Proteins, mutations, and antibiotic classes	Antibiotic Resistance Ontology (ARO); BLAST/RGI tools for resistome prediction; machine--learning-ready downloads	Annotating AMR determinants in bacterial Proteomes; linking PTMs or expression shifts to resistance	https://card.mcmaster.ca
iMetaLab suite	Web-based pipeline for metaProteomic data processing and visualisation	One-stop workflow (search, taxon mapping, functional enrichment); publishable reports; DIA support	Community- or food-matrix metaProteomics; rapid exploration of complex microbiome samples	https://imetalab.ca

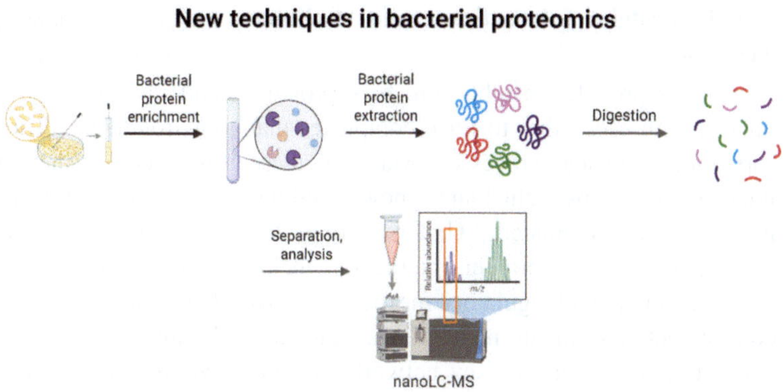

Fig. 6.3 New techniques in bacterial proteomics. The figure was created with BioRender (https://BioRender.com). (Image source: Authors' personal collection)

more or less abundant under different conditions, providing clues about how bacteria adapt to stress (e.g. antibiotics and pH changes) or environmental shifts. Quantitative proteomic techniques, such as label-free quantification or isobaric tagging (isobaric tags for relative and absolute quantitation (iTRAQ) and tandem mass tags (TMT), allow researchers to compare protein levels across different bacterial states.

Currently, MALDI-TOF mass spectrometry has revolutionised bacterial identification by providing a rapid, accurate, and cost-effective method for clinical and research laboratories. Unlike traditional culture-based identification methods, which can be time-consuming, MALDI-TOF can identify bacteria within minutes by analysing their unique protein profiles, primarily ribosomal proteins. This technique has significantly improved patient care by enabling faster diagnosis of infections, leading to timely and targeted antimicrobial treatments. Its ability to process large samples simultaneously has made it invaluable for high-throughput microbial identification, enhancing efficiency in clinical microbiology. Furthermore, MALDI-TOF's impact extends to antimicrobial resistance detection and monitoring, epidemiological studies, and outbreak investigations, as it can facilitate species-level identification of pathogens. Its growing adoption underscores its significance in modern microbiology, improving diagnostic workflows, reducing healthcare costs, and contributing to better patient outcomes.

Post-translational Modifications (PTMs) and Their Functional Significance

Post-translational modifications (PTMs) refer to chemical changes made to proteins after their synthesis. These modifications can regulate protein function, localisation, stability, and interactions with other molecules. Common PTMs in bacteria include phosphorylation, glycosylation, acetylation, methylation, and lipidation. Each of these modifications can dramatically alter a protein's activity or interactions within the cell. PTMs allow bacteria to rapidly adapt to changes in their environment by modulating protein activity and refer to one of the mechanisms for bacteria to develop antimicrobial resistance. For example, phosphorylation can activate or inhibit enzymes and signalling proteins, allowing rapid adaptation to external stimuli like nutrient availability or antibiotic presence. At the same time, glycosylation can affect bacterial pathogenicity by modifying surface proteins involved in immune evasion or host cell attachment.

Protein–Protein Interactions and Cellular Networks

Proteomics provides insights into how proteins interact with each other to form complexes that carry out biological processes. Mapping these interactions helps elucidate how cellular processes are regulated. Understanding the broader network of protein interactions helps reveal pathways and regulatory mechanisms that control bacterial growth, metabolism, virulence, and antibiotic resistance. Key networks include (a) metabolic networks, which refer to interactions between enzymes in bacterial metabolism; (b) signalling networks, which include proteins involved in transmitting signals, especially in response to environmental stress or host interactions; and (c) virulence networks, with proteins being involved in bacterial infection mechanisms, revealing potential targets for antimicrobial therapy.

Applications of Bacterial Proteomics

Applications in Infections

***Mycobacterium avium* subsp. *paratuberculosis* Infection**

Mycobacterium avium subsp. *paratuberculosis* (MAP) is the aetiological agent of paratuberculosis, a chronic enteric disease in ruminants that causes long-term intestinal inflammation. This condition poses major threats to animal health and productivity, resulting in considerable economic losses in the livestock sector worldwide. Moreover, several studies have suggested a possible association between MAP and Crohn's disease in humans, as the bacterium has been detected in numerous—though not all—clinical samples through culture-based and molecular methods. Proteomic profiling of MAP isolates from the ileal tissues of naturally infected sheep, using two-dimensional polyacrylamide gel electrophoresis (2D PAGE), coupled with MALDI-TOF mass spectrometry, revealed a set of ten upregulated proteins. Among these, ArgG and RocA were identified as candidates potentially involved in the pathogen's metabolic adaptation and carbon source utilisation within the host environment. These proteins were further assessed for their immunogenic potential, with promising implications for future vaccine development. Comparative proteomic studies later demonstrated strain-dependent variations between type I/III and type II MAP isolates, highlighting genetic and metabolic diversity within the species. Additionally, a large-scale proteome-wide analysis was carried out to discover antigenic biomarkers that could improve disease diagnosis. Through complementary proteomic and immunological techniques, the first secretome map of MAP was established, identifying 25 candidate diagnostic antigens. Five of these were subsequently validated using an ELISA assay on field serum samples, achieving a sensitivity of 94.7% and a specificity of 97.9%, results comparable to those obtained with established commercial diagnostic kits.

***Mycobacterium* spp. Infections**

Tuberculosis, a chronic disease affecting humans and animals, is caused by various *Mycobacterium* species, with *M. bovis* responsible for zoonotic tuberculosis, particularly in cattle. Although the incidence of *M. bovis* infection in humans is low, transmission from animals remains a concern. Proteomic studies have identified 32 host peptides elevated in infected animals and 16 *M. bovis*-related peptides in cattle, among which vitamin D–binding protein emerged as a key biomarker for diagnosing the infection. Moreover, detecting another protein, polyketide synthase 5, demonstrated the capability to distinguish *M. bovis* from other mycobacteria. Pathogenetic mechanisms were explored by analysing proteomic changes in monocyte cell lines, revealing 61 differentially regulated proteins involved in phagosome maturation and TNF signalling. Further research on the leucocyte proteome in cattle identified 1222 proteins linked to immune defence, particularly in vaccinated animals, with Janus kinase–signal transducer and activator of transcription (JAK-STAT) and protein kinase C

(PKC) pathways implicated in the immune response, offering insights for controlling transmission and improving diagnostics.

***Listeria monocytogenes* Infection**

Listeriosis is a zoonotic disease that can affect both humans and animals, presenting in multiple clinical manifestations. Proteomic investigations have revealed that *Listeria monocytogenes* possesses remarkable temperature adaptability, allowing it to survive and proliferate under cold storage conditions. The bacterium's ability to endure low temperatures is largely linked to changes in protein synthesis and folding, nutrient transport, and oxidative stress response pathways. Gaining insight into these molecular processes is essential for designing strategies that limit bacterial persistence in chilled foods and processing environments. Comprehensive proteomic analyses have demonstrated that cultures maintained at around 4 °C display elevated expression of proteins related to energy metabolism, including those involved in glycolysis and the phosphotransacetylase–acetate kinase (Pta–AckA) pathway. These findings imply that cold exposure increases the organism's energy requirements for sustaining growth and cellular maintenance. Such results underscore the importance of proteomics in deciphering the molecular basis of cold adaptation in *L. monocytogenes.* Further comparative studies among different strains of the bacterium subjected to bile salt stress have shown variable expression of cell wall components, DNA repair enzymes, molecular chaperones, and oxidative stress regulators. These strain-specific differences indicate that the adaptive response of *L. monocytogenes* is influenced by genetic background and environmental conditions, offering valuable clues for improving control and prevention measures in food safety management.

***Brucella* spp. Infections**

Brucella melitensis and *Brucella abortus* are the primary causal agents of brucellosis in small ruminants and cattle, respectively, causing significant economic and health impacts. *B. melitensis* primarily affects sheep and goats, leading to abortions and human infections through direct contact or contaminated food. Proteomic studies using techniques like liquid chromatography–mass spectrometry (LC–MS) have revealed proteins that distinguish between infected and vaccinated animals, assisting in monitoring control efforts. Specific peroxisome protein partners overexpressed exclusively in vaccinated animals were identified, indicating a possible role in boosting antibody production, similar to human findings. Further, immunodominant proteins in naturally infected animals, including heat shock proteins, enzymes, binding proteins, and hypothetical proteins, were discovered, which are thought to help the bacteria evade host immune defences. *B. abortus*, responsible for bovine brucellosis, is linked to abortion, reduced fertility, and milk production loss in cattle. Research has identified antimicrobial resistance and host-specific proteins that contribute to its survival. Proteomic analysis has highlighted immunoreactive proteins in various hosts, aiding in serological detection and vaccine development. Recent studies also identified proteins that are potentially useful as drug targets or vaccine candidates in humans and cattle, offering insights for future treatments and prevention strategies against brucellosis.

***Campylobacter* spp. Infections**

Campylobacter species, including *C. jejuni*, *C. coli*, and others, cause gastrointestinal infections in humans and animals, primarily transmitted through contaminated food. Proteomic studies have been instrumental in understanding the pathogenicity of *Campylobacter* spp. Key findings include identifying differentially expressed proteins between strong and weak *C. jejuni* colonisers and immunogenic proteins in sheep infected with *C. jejuni.* Outer membrane proteins, e.g. the major outer membrane protein (MOMP) and periplasmic enzyme B4 (PEB4), were found to play critical roles in bacterial adhesion, colonisation, and immune response. Studies on *C. jejuni* in poultry and human hosts revealed metabolic adaptations, cytoskeletal changes, and specific host–pathogen interactions, with proteins involved in fatty acid metabolism and cellular processes being upregulated post-infection. These findings offer insights into the molecular mechanisms of infection and potential targets for vaccine development.

Coxiella burnetii Infection

C. burnetii, the intracellular organism causing Q fever, generally leads to mild infections in animals, e.g. abortions during the last stage of gestation. The pathogen is excreted through vaginal excretions, milk, faeces, urine, and aborted materials. Human infections are often linked to consuming contaminated milk, as seen during outbreaks in the Netherlands following infections in goat herds. Proteomic studies have identified bacterial protein fractions, particularly outer membrane proteins, that are exposed to host immune cells. These proteins can be promising targets for vaccine development, potentially reducing bacterial loads in animals and minimising transmission risks to humans.

Salmonella spp. Infections

Salmonella enterica serovars, namely *S. enterica* serovar Typhimurium and *S. enterica* serovar Enteritidis, are major causes of foodborne infections. *S. enterica* serovar Typhimurium commonly affects pigs, while *S. enterica* serovar Enteritidis causes salmonellosis in poultry, which can be transmitted to humans primarily through contaminated eggs. Proteomic studies on infected pigs revealed significant changes in the intestinal mucosa, including damage to keratins and intermediate filaments, and highlighted immune responses and cytoskeleton remodelling. In poultry, infection with Enteritidis serovar triggered increased mucin production and a T cell-mediated inflammatory response in the caecum, the primary site of colonisation. Proteomic analysis of the liver and blood serum of infected chickens showed activation of immune pathways. At the same time, glycolysis and fatty acid metabolism were downregulated. Outer membrane proteins and lymphocyte-specific proteins identified in these studies provide insights into pathogenesis and potential targets for vaccine development.

Borrelia spp. Infections

Lyme disease, caused by *Borrelia burgdorferi sensu lato*, presents a wide range of clinical symptoms in both humans and animals. Wild rodents, resistant to complement-mediated bacterial killing, serve as key reservoirs, while hard ticks act as the primary vectors. Mass spectrometry (MS) identified 30 bacterial proteins, aiding in developing subunit vaccines. *B. burgdorferi* can evade host immune defences by binding to complement regulatory factor H (fH), with varying binding affinities across different species. Studies revealed that BbCRASP-1, a major fH-binding protein, interacts differently with fH in humans, mice, and sheep compared to cattle and pigs. Proteomic network analysis has identified 15 proteins, with erpX, erpL, and erpY proposed as novel vaccine targets for controlling the infection.

Leptospira spp. Infections

Leptospirosis, a widespread zoonotic disease caused by bacteria of the genus *Leptospira*, primarily affects humans through contact with infected animals or contaminated environments, such as water. Rodents and cattle are common reservoirs, and those working with animals are at a higher risk of infection. Clinical manifestations range from jaundice and reproductive issues to uveitis. The large number of serovars of the pathogen and the limited understanding of its pathogenesis have complicated the development of diagnostic tests. Proteomic studies have focused on identifying virulence factors and potential biomarkers across various *Leptospira* species. Key proteins, including OmpL1, LipL32, and LigA/B, which facilitate adhesion and colonisation, are more abundant in pathogenic isolates. These proteins are critical for understanding host adaptation and pathogenicity and hold promise for vaccine development.

Applications in Antibiotic Resistance

The application of proteomics also focuses on understanding antibiotic-resistant bacterial isolates. Proteomic analysis is crucial for exploring differential protein expression and post-translational modifications that contribute to resistance mechanisms, revealing potential therapeutic targets. For instance, studies have identified post-translational modifications like lysine acetylation associated with resistance. A notable example is the proteomic analysis of *C. jejuni*, which identified radical S-adenosylmethionine enzymes linked to chloramphenicol resistance. Similarly, proteomic studies on fluoroquinolone-resistant *S. enterica* isolates have revealed numerous differentially acetylated proteins, many involved in microbial metabolism and antibiotic biosynthesis.

Proteomics has also shed light on antibiotic resistance in *S. aureus* isolates, showing that proteins involved in glycolysis, oxidative stress response, and ATP hydrolysis are more frequently expressed in resistant strains. Comparative proteomic studies have highlighted key surface proteins in *S. aureus*, offering potential therapeutic targets. In addition, *Streptococcus* spp. isolates from cases of bovine mastitis were analysed, identifying virulence and resistance-associated proteins that could inform new treatment approaches.

Proteomic approaches have also been used to understand resistance in other zoonotic bacteria. For example, studies on extended-spectrum β-lactamase (ESBL)-producing *E. coli* from pigs under stress from high ciprofloxacin concentrations revealed overexpression of hydrolase L-asparaginase, involved in ciprofloxacin resistance. Additionally, *Mannheimia haemolytica* isolates treated with chlortetracycline showed a reduced expression of leucotoxin, a key virulence factor, demonstrating how proteomics can identify the impact of antibiotics on bacterial virulence and resistance mechanisms. These findings are essential for developing strategies to counter antibiotic resistance and control zoonotic infections.

Applications in Food-Borne Bacteria

Proteomics plays a critical role in enhancing food safety. It provides in-depth insights into zoonotic agents, quality control, and the identification of potential health risks related to animal food products. As part of the emerging field of '*foodomics*', proteomics also supports monitoring food production quality and safety. It helps detect spoilage and food-borne contamination caused by pathogens. Additionally, proteomics aids in identifying allergenic proteins in food products.

This integration of animal health monitoring into food production processes allows for a comprehensive assessment of potential threats to public health, such as detecting enterotoxins in dairy and meat products.

For example, proteomics has been performed in bacterial strain identification, toxin detection, and the assessment of environmental contamination. Specifically, identifying bacterial isolates from animal-derived food products, e.g. *C. jejuni* in poultry and *L. monocytogenes* in red meat, has provided valuable data on biofilm formation, motility, and stress responses. Proteomics also allows for the rapid detection of harmful pathogens, e.g. *S. enterica* in poultry, *E. coli* O157:H7 in red meat samples, *F. tularensis* in hare carcasses, and *Vibrio parahaemolyticus* in commonly consumed aquatic organisms (e.g. mussels), within hours of contamination, enhancing food safety protocols.

Proteomic techniques are also pivotal in detecting toxins such as staphylococcal enterotoxins in food, including milk and meat products, with high sensitivity. Methods like liquid chromatography–selected reaction monitoring (LC–SRM) and quadrupole–Orbitrap instruments have been developed to detect and quantify various toxins like ricin (a *Ricinus communis* toxin) and *C. perfringens* epsilon toxin (ETX) in food matrices (milk and tap water). Moreover, proteomics has been essential in identifying fungal toxins like mycotoxins (fungal metabolites), which pose significant health risks.

Overall, proteomics contributes to the One Health framework by improving food safety practices, ensuring the quality of animal-derived products, and identifying biomarkers for toxin contamination, all of which are crucial for protecting public health.

Challenges and Future Directions

Bacterial proteomes are dynamic, constantly adapting to environmental conditions, growth stages, and stress, complicating accurate measurement. The diversity of proteins, particularly low-abundance and membrane proteins, poses challenges in comprehensive analysis. High sensitivity and specificity in identifying low-abundance proteins and those with post-translational modifications remain critical. Incomplete or outdated protein databases hinder accurate protein identification and quantification. Maintaining sample purity is essential to prevent contamination, and variability in sample preparation can lead to inconsistent results, affecting reproducibility. The large datasets generated by proteomic studies necessitate advanced computational tools for practical analysis. However, many existing bioinformatic tools are not optimised for bacterial proteomics. Proteomic data integration with genomics, transcriptomics, and metabolomics is also essential for a holistic understanding of bacterial systems.

Future goals include enhancing mass spectrometry techniques for detecting low-abundance proteins and developing innovative sample preparation methods to improve protein extraction. The incorporation of artificial intelligence and machine learning will significantly enhance data analysis and model accuracy. Expanding and updating protein databases for less well-characterised bacterial species will improve functional annotation, providing insights into protein roles. Integrating multi-omics data will further deepen our understanding of bacterial physiology. Additionally, developing rapid proteomics-based methods for detecting bacterial pathogens in clinical and environmental samples is crucial. Bacterial proteomics will advance research and applications in health, industry, and environmental monitoring by overcoming these challenges and pursuing these goals.

Case Study

Recent studies have indicated that listeriosis is primarily food-borne, although animals can occasionally act as sources of the causal organism, *L. monocytogenes*. Outbreaks and most individual cases occur after consuming contaminated foods, such as raw milk and soft cheese, meat and meat products, raw smoked fish, mussels, vegetables, sprouts, and salads. Although pasteurisation generally controls the bacteria, it may not be effective when the initial bacterial load is high, and post-pasteurisation contamination is also a potential concern. Contamination can occur due to poor hygiene practices during food processing, such as contact with soil, dust, or faeces. Milk contamination can also result from *Listeria*-associated mastitis in animals through a haematogenous spread to the mammary gland or direct ascending infection. Proteomic techniques also offer advantages for the early detection of *L. monocytogenes*. Traditional microbiological methods require 4–5 days for accurate detection, whereas the use of MALDI-TOF MS for identification can significantly reduce this time, allowing for diagnosis within 30 h of sample submission.

1. What are some common food sources of *L. monocytogenes* that can lead to listeriosis?
2. Why might pasteurisation not always be effective in preventing *L. monocytogenes* contamination?
3. What advantage does MALDI-TOF MS offer over traditional microbiological methods for detecting *L. monocytogenes*?

Research Questions

1. What could be the involvement of proteins in terms of molecular function, biological process, and subcellular location between the host and bacteria?
2. How could the post-translational modifications of proteins be used by bacteria to develop antimicrobial resistance?

Further Readings

Alcock BP, Raphenya AR, Lau TTY, Tsang KK, Bouchard M, Edalatmand A, et al. CARD 2020: antibiotic resistome surveillance with the comprehensive antibiotic resistance database. Nucleic Acids Res. 2020;48(D1):D517–25.

Chen B, Zhang D, Wang X, Ma W, Deng S, Zhang P, Zhu H, Xu N, Liang S. Proteomics progresses in microbial physiology and clinical antimicrobial therapy. Eur J Clin Microbiol Infect Dis. 2017;36(3):403–13.

Cocolin L, Mataragas M, Bourdichon F, Doulgeraki A, Pilet MF, Jagadeesan B, Rantsiou K, Phister T. Next generation microbiological risk assessment meta-omics: The next need for integration. Int J Food Microbiol. 2018;287:10–7.

Katsarou EI, Billinis C, Galamatis D, Fthenakis GC, Tsangaris GT, Katsafadou AI. Applied proteomics in 'One Health'. Proteomes. 2021;9(3):31.

Lianou DT, Skoulakis A, Michael CK, Katsarou EI, Chatzopoulos DC, et al. Isolation of *Listeria ivanovii* from bulk-tank milk of sheep and goat farms—from clinical work to bioinformatics studies: prevalence, association with milk quality, antibiotic susceptibility, predictors, whole genome sequence and phylogenetic relationships. Biology. 2022;11(6):871.

Liao B, Ning Z, Cheng K, Zhang X, Li L, Mayne J, Figeys D. iMetaLab 1.0: a web platform for metaproteomics data analysis. Bioinformatics. 2018;34(22):3954–6.

Ozma MA, Khodadadi E, Rezaee MA, Asgharzadeh M, Aghazadeh M, Zeinalzadeh E, Ganbarov K, Kafil HS. Bacterial proteomics and its application in pathogenesis studies. Curr Pharm Biotechnol. 2022;23(10):1245–56.

Palma E, Tilocca B, Roncada P. Antimicrobial resistance in veterinary medicine: an overview. Int J Mol Sci. 2020;21(6):1914.

Pérez-Llarena FJ, Bou G. Proteomics as a tool for studying bacterial virulence and antimicrobial resistance. Front Microbiol. 2016;7:410.

Piras C, De Fazio R, Di Francesco A, Oppedisano F, Spina AA, Cunsolo V, Roncada P, Cramer R, Britti D. Detection of antimicrobial proteins/peptides and bacterial proteins involved in antimicrobial resistance in raw cow's milk from different breeds. Antibiotics (Basel). 2024;13(9):838.

Procopio AC, Colletta S, Laratta E, Mellace M, Tilocca B, Ceniti C, Urbani A, Roncada P. Integrated One Health strategies in dengue. One Health. 2024;18:100684.

Roncada P, Modesti A, Timperio AM, Bini L, Castagnola M, Fasano M, Urbani A. One medicine – one health – one biology and many proteins: proteomics on the verge of the One Health approach. Mol Biosyst. 2014;10(6):1226–7.

Smith DGE, Goldstone RJ, Inglis NF, Watson E. Bacterial proteomics in the study of virulence: an overview. In: Tang YW, Sussman M, Liu D, Poxton I, Schwartzman J, editors. Molecular medical microbiology, vol. 1. 2nd ed. Elsevier; 2015. p. 277–90.

Tilocca B, Costanzo N, Morittu VM, Spina AA, Soggiu A, et al. Milk microbiota: characterisation methods and role in cheese production. J Proteomics. 2020;210:103534.

Tsakou F, Jersie-Christensen R, Jenssen H, Mojsoska B. The role of proteomics in bacterial response to antibiotics. Pharmaceuticals (Basel). 2020;13(9):214.

Zubair M, Wang J, Yu Y, Faisal M, Qi M, et al. Proteomics approaches: a review regarding the importance of proteome analyses in understanding the pathogens and diseases. Front Vet Sci. 2022;9:1079359.

Epidemiology of Bacterial Zoonoses

7

Saurabh RamBihariLal Shrivastava

Learning Objectives

1. To define and classify bacterial zoonoses
2. To describe the various modes of transmission of bacterial zoonoses
3. To explain the significance of prevention and control strategies to control the spread of bacterial zoonotic infections in human and animal populations
4. To evaluate the role of One Health approaches in addressing bacterial zoonotic infections

Introduction

Bacterial zoonoses are a group of infectious diseases caused by different pathogenic bacteria and are transmitted between animals and humans in either direction. Many such bacterial zoonotic diseases have the potential to produce major outbreaks in both humans and animals, resulting in high rates of morbidity and mortality. These infections can spread by multiple routes, including (i) direct contact with infected animals, (ii) intake of contaminated food or water, and (iii) vector-borne transmission. These infections tend to account for major public health consequences, including on veterinary health, and this calls for concerted efforts on the global front for the containment of such infections. These infections affect a wide range of hosts, comprising livestock, domestic animals, wildlife, and rodents, which makes prevention and control quite complicated.

From a public health perspective, bacterial zoonoses remain a major cause of food-borne illnesses, occupational diseases, and emerging infections. Some of the bacterial zoonoses have a chronic course (e.g. tuberculosis and brucellosis), while others have a rapid course and fatal outcomes (e.g. plague and anthrax). Similarly, some bacterial zoonotic infections can result in sporadic cases, endemic transmission, or major outbreaks. For instance, *Salmonella* and *Campylobacter* infections are major public health concerns, resulting in the occurrence of millions of cases of gastroenteritis each year, especially in low- and middle-income nations. In contrast, vector-borne bacterial zoonoses like Lyme disease are localised to regions where tick vectors thrive. On the other hand, a group of diseases, such as salmonellosis and listeriosis, is transmitted due to consuming contaminated meat and dairy products.

In veterinary medicine, bacterial zoonoses tend to affect livestock and companion animals, resulting in reduced productivity, restrictions on trade, and financial burdens on farmers. For instance, brucellosis and bovine tuberculosis in cattle essentially require strict containment measures, including culling and strengthening of

S. R. Shrivastava (✉)
Datta Meghe Medical College, Datta Meghe Institute of Higher Education and Research, Nagpur, India

S. C. Parija, A. Poddar (eds.), *Textbook of Bacterial Zoonoses*, Microbial Zoonoses,
https://doi.org/10.1007/978-981-95-6615-0_7

vaccination programmes. On the other hand, anthrax outbreak in livestock is a major threat to humans as well due to the risk of spillover. In fact, bacterial zoonoses have emerged as a major challenge to both public health and veterinary medicine, owing to their impacts on human health, animal welfare, food chain, food safety, and financial stability. In continuation, these diseases not only cause direct morbidity and mortality but also contribute to significant social and financial burdens. At the same time, the emergence of antimicrobial resistance (AMR) in bacterial zoonoses has emerged as a massive threat to the existing therapeutic options, further complicating the domain of the management of diseases.

Moreover, amid rising globalisation, climate changes, deforestation, and urbanisation, the possibility of human–animal interactions has significantly increased, and all this has immensely contributed to the rising trends of bacterial zoonotic infections. The containment of bacterial zoonoses would require a One Health approach, wherein human, animal, and environmental health–related sectors are integrated to develop effective surveillance, prevention, and control initiatives.

Classification of Bacterial Zoonoses

Bacterial zoonotic infections are classified depending on multiple parameters, such as the mode of transmission, reservoir hosts, and geographic distribution (Table 7.1 and 7.2).

Modes of Transmission

Bacterial zoonoses can be transmitted by different ways, and based on these modes, the epidemiology and measures to contain these infections usually vary. Moreover, these modes also determine the extent of the spread of infections among animals and humans, thereby influencing public health, veterinary medicine, and agriculture at large.

These modes of transmissions include the following:

Table 7.1 Classification based on mode of transmission

Mode of transmission	Example	Causative agent
Direct contact	Brucellosis	*Brucella* spp.
	Tuberculosis	*Mycobacterium bovis*
Indirect contact	Leptospirosis	*Leptospira* spp.
	Q fever	*Coxiella burnetii*
Vector-borne	Plague	*Yersinia pestis* (via fleas)
	Lyme disease	*Borrelia burgdorferi* (via ticks)
Food-borne	Salmonellosis	*Salmonella* spp.
	Listeriosis	*Listeria monocytogenes*
Environmental	Anthrax	*Bacillus anthracis* (from soil)
	Melioidosis	*Burkholderia pseudomallei* (from water/soil)

Table 7.2 Reservoirs in bacterial zoonoses

Reservoir type	Example	Causative agent
Wildlife	Plague	*Yersinia pestis* (rodents)
	Lyme disease	*Borrelia burgdorferi* (deer, rodents)
Domestic animals	Brucellosis	*Brucella canis* (dogs)
	Cat scratch disease	*Bartonella henselae* (cats)
Livestock	Bovine tuberculosis	*Mycobacterium bovis* (cattle)
	Anthrax	*Bacillus anthracis* (cattle, sheep, goats)

Direct Contact Transmission

It is one of the major modes of zoonotic disease transmission, which occurs when humans come in close contact with infected animals. This mode of transmission is common among farmers, pet owners, veterinarians, workers in slaughterhouses, etc., who regularly handle animals or their products. For instance, laboratory personnel, butchers, and animal handlers are at a higher risk of infections like tularaemia, which spreads through handling infected rodents or rabbits. Bacterial pathogens can directly enter human tissues via the bites and scratches from infected animals. For example, cat scratch disease occurs because of infection with *Bartonella henselae*, wherein scratches from infected cats transmit bacteria into the human skin. Similarly, handling infected animals or their bodily fluids can also result in zoonotic infections. For example, brucellosis is transmitted through contact with infected livestock during birthing, milking, and/or slaughter.

Vector-Borne Transmission

Vector-borne bacterial zoonoses are transmitted by arthropods such as fleas, ticks, and lice, which act as biological or mechanical carriers of the causative microorganisms. The involved vectors acquire bacteria from the infected animals and subsequently transmit them to humans through their bites or faecal contamination.

- *Fleas*: They are responsible for transmitting plague caused by *Yersinia pestis*. Fleas become infected by feeding on rodents carrying *Y. pestis* and then transmit the bacteria to humans through bites.
- *Ticks*: In the capacity as vectors for bacterial zoonotic infections, ticks are responsible for the transmission of (i) Lyme disease caused by *Borrelia burgdorferi*, (ii) Rocky Mountain spotted fever caused by *Rickettsia rickettsii*, and (iii) tularaemia caused by *Francisella tularensis*. As ticks feed on infected wildlife like deer and rodents, they acquire the infection and subsequently transmit the infection to humans.
- *Lice*: Body lice are involved in the spread of trench fever caused by *Bartonella quintana*, the condition often found among soldiers living in poor hygienic conditions.

Prevention of vector-borne zoonotic infections essentially require implementation of different vector control measures, such as (i) promoting the use of appropriate insect repellents, (ii) encouraging protective clothing, and (iii) environmental interventions to eliminate habitats of vectors. In addition, other measures like animal vaccination, rodent control, and awareness initiatives are also vital to reduce the incidence of such infections.

Food-Borne and Water-Borne Transmission

This kind of transmission is reported when bacteria contaminate food or drinking water, which is subsequently consumed by humans, leading to gastrointestinal and systemic infections. This mode of transmission has been ranked as the most frequent one for bacterial zoonotic outbreaks globally. Some of the common diseases include salmonellosis caused by *Salmonella* spp. from undercooked meat and leptospirosis caused by *Leptospira* spp. from water contaminated by animal urine. Further, unpasteurised milk has been linked with the development of brucellosis and bovine tuberculosis. The containment of such infections requires proper food handling, thorough cooking of meat, pasteurisation of milk, and promotion of access to safe and clean drinking water.

Environmental Exposure

This transmission of infection generally involves indirect contact with bacteria in soil, air, or contaminated surfaces/fomites, which infect humans via inhalation, ingestion, or skin contact. It is noteworthy that humans acquire these infections in the workplace, farms, and recreational settings. For instance, anthrax, a soil-borne bacterial zoonosis occurs because of *Bacillus anthracis*, which can persist in soil for years and act as a major risk to livestock and humans. Aerosol transmission has been reported in cases of Q fever and tularaemia. The preventive strategies include (i) promoting proper ventilation, (ii) advocating the use of personal protective equipment, (iii) disinfecting equipment, and (iv) ensuring safe disposal of animal carcasses. In addition, promoting awareness among masses regarding the environmental risk and occupational safety measures plays a defining role in preventing the acquisition of bacterial zoonoses through environmental modes.

Risk Factors and Vulnerable Populations

The potential risk of acquiring bacterial zoonotic infections depends on a wide range of factors.

The notable vulnerable populations and risk factors include the following:

Occupational Exposure

Occupational exposure has been acknowledged as a significant risk factor for bacterial zoonoses, particularly among individuals who work closely with animals and animal products. Individuals like veterinarians, butchers, and slaughterhouse workers are at a high risk of acquiring these infections. This is because of repeated direct and indirect contact with infected animals, their bodily fluids, and contaminated workplace settings. Farmers who are generally exposed to livestock are at more risk of acquiring infections like brucellosis, leptospirosis, and Q fever. On the other hand, veterinarians are more prone to infections like tularaemia, cat scratch disease, and rabies-related bacterial infections because of exposure to saliva, urine, and faeces of infected animals.

Finally, butchers and slaughterhouse workers are at an increased risk of contracting anthrax, brucellosis, and bovine tuberculosis. This transmission predominantly occurs while handling infected carcasses and inhaling aerosolised bacteria during the processing of meat. To prevent the acquisition of infection, the preventive measures include (i) the use of protective personal equipment such as gloves, masks, and goggles; (ii) maintenance of personal hygiene and safe handling of animals and meat products; and (iii) vaccination, if required. However, this must be equally supported by strengthening workplace policies, including the promotion of periodic health check-ups and awareness activities and strict implementation of biosecurity measures to reduce occupational risks.

Socio-economic and Environmental Determinants

Socio-economic and environmental parameters occupy a vital place in facilitating transmission and augmenting the caseload of bacterial zoonotic infections. Individuals belonging to a low socio-economic status, rural communities, and marginalised communities are at more risk of acquiring these infections because of poor sanitation, overcrowding, and poor access to healthcare services. For instance, limited access to clean water and sanitation facilities significantly augments the risk of acquiring infections like leptospirosis and cholera. Moreover, poor housing conditions, especially in slum communities, act as a predisposing factor for rodent infestations, which further result in increasing the risk of transmission of plague and rat-borne leptospirosis.

Environmental alterations, such as climate change, deforestation, and urbanisation, remain the key factors in promoting the emergence of zoonotic diseases. The urbanisation and expansion of human habitats have destroyed habitats of wildlife animals, bringing them closer to the human habitations. For instance, deforestation has been linked to a rise in the number of Lyme disease patients. At the same time, climate changes have resulted in the expansion of different vector-borne bacterial zoonotic infections like tick-borne relapsing fever. To counter the impact of these factors, there is a need to improve the water quality, sanitation facilities, disposal of wastes, rodent control, and housing standards. These activities must go together with a strengthening of surveillance mechanisms, efforts to improve awareness among community members, and sustainable environmental management.

Immunocompromised Individuals and High-Risk Groups

Individuals with an immunocompromised status are extremely prone to acquiring and developing severe bacterial zoonotic infections because of their reduced ability to fight against diseases. Similarly, people belonging to high-risk groups (viz., elderly, pregnant women, recipients of organ donation, cancer survivors on immunosuppressive therapy, individuals with HIV/AIDS, coexisting diabetes, etc.) are also at a higher risk of acquiring infections. In fact, even minor exposures to bacterial pathogens can result in the development of life-threatening complications.

For instance, tuberculosis has a significantly high probability of progressing to active disease among immunosuppressed individuals, while listeriosis poses a major threat to pregnant women and neonates, resulting in miscarriage, premature birth, and/or neonatal sepsis. To prevent these infections among high-risk groups, there is a definite need to avoid the consumption of raw or undercooked animal products, promote good hygiene around pets, refrain from unnecessary contact with farm animals, and adhere to proper schedules of animal vaccination. In addition, healthcare professionals must educate immunocompromised individuals about safe food handling and infection prevention measures to mitigate the risk of bacterial zoonoses.

Laboratory Diagnosis

Accurate laboratory diagnosis plays a key role in ensuring appropriate patient management, containing infection outbreaks, and performing epidemiological investigations. Laboratory diagnosis can be performed in multiple ways, namely culture, serology, molecular, and genomic techniques, and these methods can detect zoonotic causative pathogens in human, animal, and environmental samples.

The conventional culture methods continue to be the gold standard in promoting the detection of bacteria like *Salmonella*, *Brucella*, and *Yersinia*. However, their main limitation is the requirement of specific growth conditions and a long time span for these organisms to grow in the culture. Serological tests to detect host antibodies or bacterial antigens in the serum and other body fluids is currently the mainstay in the diagnosis of zoonotic bacterial infections.

The introduction of molecular techniques such as polymerase chain reaction (PCR) and whole-genome sequencing (WGS) has revolutionised the process of surveillance and investigation of outbreaks of these diseases. The advantage of PCR is that it offers a rapid and highly specific detection of bacterial DNA, which is crucial for diagnosing infections. On the other hand, WGS enables differentiation of causative strains, profiling of antimicrobial resistance, and tracking of transmissions, which are all crucial in the identification of the source of the outbreaks. Further, advanced techniques like metagenomics and CRISPR-based diagnostics essentially aid in the detection of pathogens in environmental and clinical samples. It is important to integrate molecular epidemiology into public health strategies. This will facilitate the process of real-time surveillance, monitoring of antimicrobial resistance, and detection of early outbreaks. This will, thereby, decrease the magnitude of bacterial zoonotic infections in both human and animal populations.

Prevention and Control Strategies

The prevention and containment of bacterial zoonotic infections essentially require a multipronged approach, wherein interventions targeting human, animal, and environmental health are integrated.
These preventive measures include the following:

Strengthening Surveillance and Outbreak Preparedness

Strengthening surveillance and outbreak preparedness activities would ensure the implementation of a timely response to emerging bacterial zoonoses. Robust surveillance allows for the early detection of unusual disease patterns or increases in animal or human cases, providing crucial lead time to investigate potential zoonotic threats. This early warning enables the prompt implementation of preparedness activities, such as activating rapid response teams, mobilising diagnostic resources, and initiating risk communication strategies.

Vaccination Programmes

Vaccination programmes can significantly reduce disease transmission, safeguard high-risk populations, and avoid financial losses in livestock sectors. Sustained surveillance and monitoring of vaccine coverage are important to assess the efficacy and prevent outbreaks. In fact, the integration of vaccination into the One Health approach would ensure a coordinated global response to bacterial zoonotic infections. Animal vaccination is required to control zoonotic pathogens at the source. For instance, brucellosis vaccines are administered to cattle, sheep, and goats, and this prevents human infections through contaminated dairy products. Human vaccination is essential among high-risk groups like veterinarians, farmers, and abattoir workers. This is because the timely administration of vaccines can prevent diseases like brucellosis and anthrax.

Improvement in the Knowledge

Improvement in the knowledge levels of communities on safe animal handling, proper food preparation, and environmental hygiene through public awareness campaigns can play a defining role in preventing bacterial transmission.

Food Safety Measures

Food safety measures are important in reducing the incidence of *Salmonella*, *Listeria*, and *Campylobacter* infections. These comprise proper cooking of meat, avoiding consumption of raw dairy products, pasteurisation of dairy products, and preventing cross-contamination during food preparation.

Maintenance of High Standards of Hygiene and Sanitation

Maintenance of high standards of hygiene and sanitation is a crucial approach to avoid the acquisition of bacterial zoonotic infections. Personal hygiene includes regular hand washing with soap after handling animals, raw meat, or contaminated surfaces. These measures are vital for the safety of high-risk groups of people, as poor sanitation significantly contributes to the spread of a wide range of food-borne, water-borne, and environmental bacterial zoonoses like leptospirosis, salmonellosis, and cholera. Similarly, water, sanitation, and hygiene (WASH) programmes significantly aid in the prevention of zoonotic diseases in vulnerable populations. In this regard, there is an immense need to facilitate access to clean drinking water, proper sewage disposal, and rodent control measures, as these measures can significantly reduce bacterial contamination in human and animal environments.

Environmental sanitation

Environmental sanitation, including regular cleaning and disinfection of animal shelters, farms, and markets, plays a defining role in reducing the presence of zoonotic bacteria. Further, proper waste management can reduce exposure to contaminated animal faeces, which can harbour *Escherichia coli* and *Brucella* spp.

Strengthening the One Health approach

Strengthening the One Health approach and promoting collaboration between medical, veterinary, and environmental experts will expedite the detection and containment of zoonotic threats as soon as possible.

The One Health Approach

The One Health approach is a comprehensive, multidisciplinary strategy that acknowledges the connection and the need to simultaneously target human, animal, and environmental health in the fight against bacterial zoonoses. Because more than 60% of the emerging infectious diseases are zoonotic in nature, there is enough evidence to advocate for collaboration among medical, veterinary, and environmental professionals to prevent, detect, and control these infections effectively. It becomes crucial to prevent pandemics by facilitating early identification of emerging zoonoses and supplementing them with the implementation of rapid containment measures. The key One Health strategies have been depicted in Fig. 7.1.

However, to ensure effective integration, there is a need to constitute multidisciplinary teams comprising epidemiologists, veterinarians, ecologists, and policymakers. All these stakeholders must work in collaboration to analyse disease patterns, ascertain the potential risks, and ensure the implementation of targeted interventions. Increasing financial investment in One Health research and education is expected to improve our ability to predict, prevent, and control bacterial zoonoses at their source. It is important to understand that the One Health approach not only reduces the over-

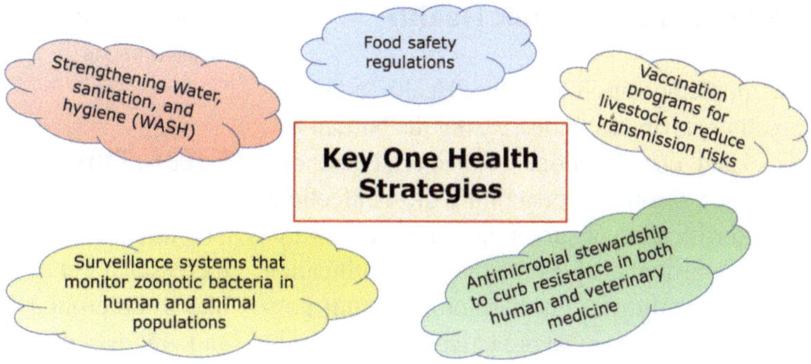

Fig. 7.1 Key One Health strategies

all burden of diseases but even becomes crucial in ensuring food safety and improving pandemic preparedness.

Challenges and Potential Solutions

Various challenges have been identified that can cumulatively contribute to the emergence and spread of zoonotic bacteria. Some of these are described below.

Climate Change and Its Impact on Bacterial Zoonoses

Climate change has been identified as a major contributor to the emergence and spread of bacterial zoonotic infections. This has also impacted their geographic distribution, transmission dynamics, and outbreak frequency. Rising temperatures, changes in precipitation patterns, and the occurrence of extreme weather events aid in the survival and spread of bacterial infections (Fig. 7.1), Similarly, floods have contributed to the spread of leptospirosis cases. In essence, effectively addressing these climate-related risks requires integrated One Health approaches.

Antimicrobial Resistance in Bacterial Zoonoses

Owing to the overuse and misuse of antibiotics in the agricultural, veterinary, and healthcare sectors, there has been a significant rise in the incidence of the emergence of resistant bacterial strains. The available evidence suggests that some zoonotic pathogens, namely *Salmonella*, *Campylobacter*, and *E. coli,* have developed resistance due to the widespread use of antibiotics in livestock. Moreover, antibiotics are often used as growth promoters and preventive treatments in food animals in many nations. Hence, there is a heightened risk of the spread of resistant bacteria to humans via contaminated meat or milk or close contact with animals. Therefore, there is an immense need to strictly enforce antibiotic stewardship programmes, including (i) banning indiscriminate use of antibiotics in livestock, (ii) encouraging the use of alternative options like probiotics, and (iii) improving vaccination coverage.

Emerging Pathogens

The emergence of new bacterial zoonotic pathogens has emerged as a major concern to global public health. There has been a significant rise in the extent of exposure to novel zoonotic infections, primarily due to population growth, global trade, intensive farming, and deforestation practices. For instance, multidrug-resistant zoonotic bacteria, like carbapenem-resistant *Enterobacteriaceae*, are spreading due to antibiotic misuse in livestock. The containment of these threats would require strengthening surveillance activities and widescale implementation of whole-genome sequencing and enabling rapid identification and tracking of new bacterial strains. In addition, mobile health applications and AI-driven predictive models can significantly improve the efficiency of early warning systems by analysing real-time epidemiological data.

Implications for Public Health

The implementation of One Health approach directly contributes to decreasing the burden and associated complications of bacterial zoonotic infections. Central to containing these infections is the fortification of surveillance and early warning systems that enables the timely identification of potential threats. Furthermore, existing governmental policies must proactively address environmental determinants, such as climate change and rapid urbanisation, which significantly contribute to the accelerated spread of zoonotic diseases. Moreover, recognising the escalating challenge of antimicrobial resistance, there is a critical need to formulate robust national and international strategies to curb its development and preserve the efficacy of vital antibiotics.

Governments and international agencies need to substantially enhance investment in the development and deployment of vaccines for both human and animal populations. Simultaneously, comprehensive and robust food safety regulations are urgently required to prevent zoonotic bacteria from entering the food chain, safeguarding public health. At the same time, continuous education and training programmes for healthcare professionals and veterinarians are crucial for effective disease prevention and outbreak containment.

The long-term success of these measures lies in the empowerment of the general population through well-organised public awareness campaigns. These initiatives need to emphasise the significance of hygiene practices and vaccination in mitigating the risk of zoonotic infections. Moreover, policymakers must prioritise building climate-adaptive and resilient health systems to address the increasing risks posed by climate change on the emergence and spread of bacterial zoonoses. Addressing this global challenge necessitates enhanced data sharing and funding for research activities by global leaders to deepen our understanding of various facets of zoonotic diseases and bolster global health security. Finally, sustained efforts are crucial to strengthen research in innovative diagnostics and the application of artificial intelligence to improve outbreak prediction and facilitate timely interventions.

Future Directions

The threat from zoonotic bacteria is expected to increase in future. This is catalysed by factors like antimicrobial resistance, ecological changes, and increased human–animal interaction. To ensure a global health security amid this rising threat, there is an urgent need for an improved future direction in research and public health practices. Some of these future directions are presented in Box 7.1.

Case Study

Mr. Suresh, a 31-year-old farmer from a village in western India, comes to the outpatient department with complaints of intermittent fever, night sweats, fatigue, joint pain, and loss of appetite for the past three weeks. He also reports occasional lower back pain and generalised malaise. Patient was treated by some local practitioner using antipyretics, but the complaints did not subside.

Upon further enquiry, the patient shares that he has been actively involved in milking cattle and assisting in calving over the past few months. He frequently handles raw milk and assists in livestock birthing without using gloves or protective gear. He recalls that one of his cows had a miscarriage recently. On examination, Suresh appears tired but is haemodynamically stable. Mild hepatosplenomegaly is noted. Initial blood investigations reveal anaemia, leucopenia, and raised erythrocyte sedimentation rate (ESR), and subsequently, blood culture and serological tests have been ordered.

1. What are the potential zoonotic diseases transmitted by handling livestock?
2. What precautions should be taken at the individual level to prevent this type of infection?
3. What are the preventive measures needed to stop the transmission of such infections?

Research Questions

1. What are the most effective strategies to integrate real-time zoonotic surveillance across human, animal, and environmental sectors, especially in low-resource settings?
2. What molecular factors contribute to the persistence and relapse of *Brucella* infection despite standard antibiotic treatment regimens?
3. How can we develop affordable, rapid, and accurate point-of-care diagnostic tools for the early detection of bacterial zoonotic infections in rural and remote areas?
4. What environmental thresholds (e.g. rainfall and flooding patterns) most strongly predict *Leptospira* outbreaks in different geographic regions?
5. How do land use changes, such as deforestation and urban sprawl, alter the risk landscape for emerging bacterial zoonotic infections?

Box 7.1: Future Directions for Improved Epidemiology and Infection Control

(a) Integrating genomic bio-surveillance into the existing surveillance mechanisms, as this will aid in tracking the evolution of pathogens, patterns of resistance, and outbreak sources in real time
(b) Advocating for the development of One Health surveillance frameworks, wherein emphasis will be on data sharing and combined investigations in human, animal, and environmental health sectors
(c) Promoting the expansion of global databases for zoonotic bacterial pathogens, thereby creating a platform for open-access data exchange
(d) Conducting population-based cohort studies in endemic areas to gain insights about the long-term burden, transmission dynamics, and risk factors of bacterial zoonoses
(e) Applying artificial intelligence to forecast zoonotic outbreaks depending on the ecological, climate, and host–pathogen interaction data
(f) Streamlining cross-border surveillance and reporting through rapid diagnostics and coordinated outbreak responses
(g) Performing ecological studies to understand bacterial persistence in wildlife, livestock, and the environment
(h) Mapping antimicrobial resistance profiles through integrated surveillance in the human, veterinary, and food sectors
(i) Carrying out research works to understand the impact of climate change on bacterial zoonotic infections, including their emergence and distributions
(j) Strengthening risk communication and public health literacy by using culturally sensitive and accessible approaches
(k) Investing in the development of low-cost, field-deployable diagnostic assays for the rapid detection of bacterial zoonotic infections
(l) Supporting capacity building programmes in low- and middle-income nations to create a sustainable response infrastructure

Further Readings

Almasri M, Ahmed QA, Turkestani A, Memish ZA. Hajj abattoirs in Makkah: risk of zoonotic infections among occupational workers. Vet Med Sci. 2019;5(3):428–34.
Burns JE, Mhapanhar G, Kilabuk E, Penner J. Zoonotic infections of the Canadian Arctic. CMAJ. 2025;197(2):E34–43.
Cascio A, Bosilkovski M, Rodriguez-Morales AJ, Pappas G. The socio-ecology of zoonotic infections. Clin Microbiol Infect. 2011;17(3):336–42.
Cetinkol Y, Enginyurt Ö, Çelebi B, Yıldırım AA, Çankaya S, Aktepe ÖC. Investigation of zoonotic infections in risk groups in Ordu University Hospital, Turkey. Niger J Clin Pract. 2017;20(1):6–11.

Christou L. The global burden of bacterial and viral zoonotic infections. Clin Microbiol Infect. 2011;17(3):326–30.

Ghasemzadeh I, Namazi SH. Review of bacterial and viral zoonotic infections transmitted by dogs. J Med Life. 2015;8(Spec Iss 4):1–5.

Kuitunen I, Renko M. Changes in the epidemiology of zoonotic infections in children: a nationwide register study in finland. Pediatr Infect Dis J. 2022;41(4):e113–9.

Kucharski AJ, Edmunds WJ. Characterizing the transmission potential of zoonotic infections from minor outbreaks. PLoS Comput Biol. 2015;11(4):e1004154.

Kang M, Wang LF, Sun BW, Wan WB, Ji X, Baele G, Bi YH, Suchard MA, Lai A, Zhang M, Wang L, Zhu YH, Ma L, Li HP, Haerheng A, Qi YR, Wang RL, He N, Su S. Zoonotic infections by avian influenza viruses: changing global epidemiology, investigation, and control. Lancet Infect Dis. 2024;24(8):e522–31.

Mason MR, Gonzalez M, Hodges JS, Muñoz-Zanzi C. Protective practices against zoonotic infections among rural and slum communities from South Central Chile. BMC Public Health. 2015;15:713.

Mendoza AP, Muñoz-Maceda A, Ghersi BM, De La Puente M, Zariquiey C, Cavero N, Murillo Y, Sebastian M, Ibañez Y, Parker PG, Perez A, Uhart M, Robinson J, Olson SH, Rosenbaum MH. Diversity and prevalence of zoonotic infections at the animal-human interface of primate trafficking in Peru. PLoS One. 2024;19(2):e0287893.

Mourya DT, Yadav PD, Patil DY, Sahay RR, Rahi M. Experiences of Indian Council of Medical Research with tick-borne zoonotic infections: Kyasanur Forest disease & Crimean-Congo haemorrhagic fever in India with One Health focus. Indian J Med Res. 2021;153(3):339–47.

Diagnosis of Bacterial Zoonoses

8

Sourav Maiti and Abhijit Chaudhury

Learning Objectives

1. To emphasise the importance of appropriate sample collection
2. To understand the various diagnostic modalities in bacterial zoonoses
3. To understand the significance of culture procedures for subsequent genomic and proteomic studies and antibiotic susceptibility tests
4. To emphasise the importance of technological advances in bacterial diagnostics

Introduction

Bacterial zoonoses are infectious diseases caused by bacteria that naturally spread between vertebrate animals and humans. These diseases present enormous threats to public health, agriculture, and the global economy. Zoonoses are responsible for nearly 60% of all infectious diseases in humans and 75% of new infections. Bacterial zoonoses, including anthrax, brucellosis, leptospirosis, and plague, have a particularly large impact. Over 200 zoonotic illnesses have been detected worldwide, with bacterial zoonoses accounting for a significant share. Food-borne bacterial zoonoses like salmonellosis and campylobacteriosis cause major illness and mortality worldwide.

Laboratory diagnosis is important for the early detection, accurate identification, treatment guidance, and surveillance of bacterial zoonoses. Early detection helps prevent outbreaks by identifying the pathogens at an early stage. Accurate identification differentiates between zoonotic and non-zoonotic pathogens, ensuring appropriate treatment and management. Guiding treatment involves suggesting appropriate antimicrobial therapy, while surveillance monitors disease trends to inform public health interventions.

Laboratory Diagnosis

A detailed history, including (i) physical contact with infected animals (e.g. anthrax and brucellosis), (ii) exposure to fleas or ticks (e.g. plague and Lyme disease), (iii) consumption of contaminated animal products (e.g. salmonellosis and listeriosis), and (iv) exposure to animal reservoirs like rodents (*Yersinia pestis*), livestock (*Brucella* species in cattle, sheep, and goats), and companion animals (*Pasteurella* spp. in dogs and cats), often supplements laboratory diagnosis.

S. Maiti (✉)
Diagnostic Division, Consortium of Accredited Healthcare Organizations (CAHO), Kolkata, West Bengal, India

A. Chaudhury
Department of Microbiology, Sri Venkateswara Institute of Medical Sciences University, Tirupati, Andhra Pradesh, India

S. C. Parija, A. Poddar (eds.), *Textbook of Bacterial Zoonoses*, Microbial Zoonoses,
https://doi.org/10.1007/978-981-95-6615-0_8

The geographical information also supplements the diagnosis of zoonotic bacterial infections. For example, leptospirosis is more prevalent in tropical and subtropical climates, brucellosis is endemic in the Mediterranean regions, the Middle East, and parts of Asia, and plague is localised in certain regions of Africa and South America. Occupational history plays an important role, with veterinarians, livestock farmers, abattoir workers, and wildlife researchers being at higher risks of acquiring the zoonotic infections.

The evolution of diagnostic technologies for bacterial zoonoses reflects the rapid scientific progress of the past two centuries. This reflects the deeper understanding of the intricate relationships among animals, humans, and the pathogens that traverse the boundaries between them. During the early days—predating the modern era of microbiology—diagnosis was predominantly based on clinical observation and rudimentary techniques. Physicians and veterinarians relied heavily on patient history, symptomatic presentations, and post-mortem examinations. The advent of microscopy and the pioneering Gram stain initiated the beginning of a more scientific approach to understanding bacterial agents. The early twentieth century saw considerable advancements as culture techniques were refined and serological methods were introduced. Since then, bacterial cultures became more standardised and enabled the growth of pathogens under controlled conditions, albeit often with substantial waiting times and the risk of laboratory-acquired infections when dealing with dangerous agents.

Concurrently, serological assays—ranging from complement fixation tests (CFTs) to agglutination tests and, subsequently, enzyme-linked immunosorbent assay (ELISA)—emerged as pivotal tools in epidemiological investigations. A transformative advancement occurred in the latter part of the twentieth century with the introduction of molecular diagnostic techniques. The invention of the polymerase chain reaction (PCR) revolutionised the field by allowing for the direct amplification and detection of bacterial DNA in clinical specimens. This era of molecular diagnostics provided markedly enhanced sensitivity and specificity and significantly reduced the time needed for diagnosis compared to traditional culture and serology. Complementary techniques, such as immunofluorescence assays and Western blotting, were integrated to confirm the PCR findings, thereby bolstering diagnostic accuracy for hard-to-culture pathogens. Automated systems and matrix-assisted laser desorption/ionisation time-of-flight (MALDI-TOF) mass spectrometry (MS) emerged as game-changers in the identification of bacteria, slashing turnaround times from days to mere hours.

At the same time, next-generation sequencing (NGS) techniques have taken centre stage, providing rapid identification and profound insights into pathogen genomics, including antimicrobial resistance profiles and evolutionary dynamics. This depth of information is critical for surveillance initiatives and for framing targeted public health responses within the increasingly interconnected framework of the One Health approach.

Specimen Collection and Handling

Types of Specimens

Blood

Blood is essential for detecting systemic bacterial infections like brucellosis, leptospirosis, and anthrax. For example, blood cultures are pivotal for isolating *Brucella* spp., particularly in febrile patients. Blood can be collected using a Vacutainer with the appropriate anticoagulant or directly into culture media. It is important to collect blood during the active stage of infection, particularly when the patient has a high fever. Since bacteraemia can occur intermittently, multiple blood samples are collected from different veins within 24 h. Local disinfection needs to be performed diligently to prevent contamination of the specimen.

Serum

Blood is collected in a sterile test tube without an anticoagulant or in a Vacutainer. The tube is kept undisturbed in a slanting position at room temperature for 2–3 h to allow the blood to clot, which helps maximise the amount of serum obtained. Once the blood has clotted, either the serum is collected directly in a sterile vial or the clotted blood is centrifuged at 1500 RPM for 5 min. It is recommended to collect paired serum samples at an interval of 21 days during both the acute and convalescent stages of the disease. Care is taken to avoid haemolysis during and after collection. An adequate volume of serum is collected, depending on the tests to be conducted. Prior to storage, the serum is incubated at 56 °C for 30 min. A preservative such as sodium azide, sodium fluoride (0.1%), or merthiolate (0.01%) is added to minimise bacterial and/or fungal contamination. However, preservatives are not used if the serum is intended for ELISA testing.

Specimen from the Wound

When collecting specimen from wounds, fluids, cells, or tissue are preferred. The samples, preferably, are collected in duplicate—one for aerobic culture and another for anaerobic culture—while following proper local disinfection protocols. A sterile swab may be used for superficial wounds; however, for deeper wounds, aspiration of fluid or a tissue biopsy is recommended over the use of swabs. For example, skin tissue is collected from cutaneous lesions taken with dry swabs and is stored in sterile containers that contain transport media for the diagnosis of anthrax.

Abscess Material

When collecting samples from an abscess, it is important to properly disinfect the area. Dry swabs are used to scrape the wall of the abscess after drainage of pus. Pus obtained from fresh abscesses is optimal for bacterial isolation. The swab is transported on ice or within two hours at room temperature in sterile vials.

Skin and Mucosal Samples

Dry swabs are rubbed gently when raw ulcers are visible after disinfecting the selected area. The material is aspirated using a sterile syringe if pustules or vesicles are present. A biopsy sample of the skin can be taken under general anaesthesia. A moist swab is used for samples collected from mucosal surfaces. In cases of dry skin lesions, skin scrapings might be very helpful. A sterile scalpel blade is used to collect scrapings from the edges of the lesions, along with any hair or scab material that may be present.

Fluids

Cerebrospinal fluid (CSF) is the specimen of choice for the diagnosis of bacterial meningitis. Urine is frequently analysed for leptospirosis detection by the detection of specific antigen or DNA identification.

Swabs

Nasal swabs are collected to identify *Y. pestis* during respiratory outbreaks of plague. Wound swabs also help in diagnosing zoonotic soft tissue infections caused by *Pasteurella* spp.

Respiratory Tract Samples

Aseptically collected bronchoalveolar lavage, tracheal aspirates, bronchial washing, and lung tissue biopsy are usually collected in screw-capped sterile containers. Sputum samples are collected either from a deep cough after brushing of teeth or after induction by saline nebulisation in paediatric or uncooperative patients.

Samples from Animals

Additional specimens from suspected animals include freshly collected faecal samples that are obtained directly from the rectum. It is important to wear protective gloves during the collection process to maintain hygiene and prevent contamination. Samples contaminated with urine or collected from the ground are unsuitable for accurately isolating and identifying pathogenic agents. In the case of small animals, rectal swabs can serve as an effective alternative for sample collection. Once collected, the samples are placed in sterile plastic vials to ensure their integrity.

To ensure accurate testing for mastitis, milk samples are collected immediately upon observing any signs of the condition. Samples from animals currently undergoing antibiotic treatment are usually avoided, as this may interfere with test results. Before collecting milk samples, the teats are disinfected thoroughly using a 70% ethanol solution. After disinfecting, the first stream of milk is discarded, as it may contain bacteria from the teat surface. A small quantity of milk is collected into a sterile bottle specifically designed for sample collection. To preserve the sample's integrity during transportation and storage, a 5% boric acid solution is added to one-tenth of the total volume of the milk sample. This addition helps inhibit bacterial growth and ensures that the sample remains viable for analysis. The sample is labelled clearly with relevant information, including the date, time, and case identification, to facilitate tracing and testing.

Environmental Specimens

Water or soil samples may be analysed for environmental persistence of zoonotic bacteria (e.g. *Leptospira* spp.).When collecting soil samples for brucellosis research, it is necessary to focus on specific locations where a suspected pathogen deposition may occur by aerosols or liquid droplets. Targeting non-vegetated areas is crucial, as these locations are less likely to be contaminated by organic matter that could interfere with the testing results. It is recommended that surface soil samples be collected from a depth of no more than one inch, ensuring that the samples are representative of the area in question. Sterile tools are used to avoid cross-contamination. Subsequently, samples are placed in clean containers, followed by labelling them with the date/time, location, and any relevant observations to facilitate accurate analysis.

Best Practices for Specimen Collection

Proper Technique

Optimal specimen collection is important for accurate laboratory diagnosis, particularly in bacterial infections. Adherence to strict aseptic techniques is crucial to minimise contamination, ensuring that the isolated bacteria reflect the true infectious agent. As such, utilising sterile equipment, such as syringes and containers for blood collection, and employing proper skin disinfection procedures are recommended. The selection of appropriate preservatives or culture media is equally vital, depending on the suspected pathogen and the nature of the specimen. For instance, Cary–Blair transport media is recommended for stool samples suspected of harbouring *Salmonella* spp. and other enteric zoonoses to preserve bacterial viability during transit to the laboratory. Consistent application of best practices ensures reliable results, ultimately facilitating effective patient management and public health interventions. Specimens need to be collected ideally during the acute phase of illness for maximum bacterial load and before initiating antibiotics therapy.

Specimen Transport

All specimens need to be accurately labelled with the patient information, type of specimen, date, time of collection, and suspected diagnosis. The use of waterproof, durable labels helps withstand potential exposure to moisture or contamination during transport. Ambient temperature is suitable for specimens like CSF, which is to be processed immediately. Refrigeration (2–8 °C) is required for most bacterial specimens, such as urine and tissue, to prevent bacterial overgrowth or degradation. Freezing (−20 °C or lower) is recommended for long-term storage of specific specimens like bacterial DNA or frozen serum for serological tests. Portable refrigeration units, ice packs, or cold-chain systems are employed for transport over longer distances. Specialised transport media need to be used during transport to preserve pathogen viability. Some examples of such media include the (i) Cary–Blair medium for stool samples suspected of harbouring enteric pathogens like *Salmonella*, (ii) buffered glycerol saline for urine samples to preserve *Leptospira* spp., and (iii) Stuart or Amies medium for bacterial swabs to maintain viability. Triple packaging protocols for biohazardous materials consist (i) a primary container holding the specimen, securely sealed and leak-proof; (ii) a secondary container, such as a durable, waterproof, and biohazard-labelled bag or plastic container, that houses the primary container; and (iii) an outer container, such as a rigid, insulated box, that provides structural protection and minimises risk during transportation. Packaging needs to comply with international standards like the WHO and International Air Transport Association (IATA) regulations for infectious materials. Prompt transportation to the laboratory is vital to avoid bacterial overgrowth or die-off. Urine specimens should ideally be processed within 1–2 h to ensure accurate diagnosis.

Specimen Storage

Short-Term Storage

Refrigeration at 2–8 °C is needed for specimens such as urine and tissue if immediate processing is not possible. Anthrax spores can persist in environmental samples stored under cool conditions.

Long-Term Storage

Flash freezing or cryopreservation may be necessary for research purposes or for delayed analysis. Examples are the preservation of bacterial DNA in tissue samples for molecular diagnostics.

Biosafety and Handling Protocols

Gloves, gowns, face shields, and biosafety cabinets reduce exposure risk during specimen handling. Work surfaces are disinfected with hypochlorite solutions or approved chemicals after processing potentially infectious specimens. Proper containment and disposal of biohazardous waste are important. For safe laboratory practice, mouth pipetting, eating, drinking, and storing eatables in the laboratory are strictly prohibited. Aerosol-generating activities are performed only inside the biosafety cabinet.

Diagnostic Techniques

The laboratory diagnosis of bacterial zoonoses employs a variety of diagnostic techniques, ranging from traditional methods like microscopy and culture to advanced molecular and serological approaches. Different diagnostic techniques are described, while the advantages and limitations are summarised in Table 8.1.

Microscopy

Microscopy is the initial step in diagnosing bacterial zoonoses due to its simplicity and rapidity. A direct examination assesses the quality of the specimen (sputum mixed with saliva), provides an early indication (e.g. Gram-positive cocci in clusters in an exudate), and ensures better guidance to the benchwork. A situation where three different cellular types are seen on direct Gram stain but only two grow in culture might indicate the presence of an anaerobic bacterium. On the contrary, a positive culture with no organism visible on direct stain would indicate possible contamination. Dark-field microscopy is used to visualise spiral-shaped bacteria like *Leptospira* spp.

Types of Staining in Microscopy

Gram Staining

It differentiates bacteria based on the structure of their cell walls. Gram-positive bacteria have thick peptidoglycan layers that retain the crystal violet stain, appearing purple under the microscope. Gram-negative bacteria with a thin peptidoglycan layer and an outer membrane do not retain the crystal violet colour and appear pink after counterstaining with safranin. For example, *Brucella* spp. appears as Gram-negative coccobacilli and *Bacillus anthracis* (Fig. 8.1) as Gram-positive, spore-forming rods.

Ziehl–Neelsen Staining (Acid-Fast Staining)

It is used to identify acid-fast bacteria, which are resistant to decolourisation with acid alcohol due to the presence of mycolic acids in their cell walls. The primary stain, carbol fuchsin, is used to penetrate the waxy cell wall. After decolourisation with acid alcohol, the bacteria retain the

Table 8.1 Advantages and limitations of different diagnostic techniques

Diagnostic techniques	Types	Advantages	Limitations
Microscopy	Routine microscopic examination of slides	Rapid and cost-effective Useful for high bacterial loads	Low sensitivity for detecting small quantities of bacteria Reliant on skilled interpretation
Culture methods	Use of different culture media	Provides live isolates for further testing (e.g. antimicrobial susceptibility)	Time-consuming, as cultures may take days to weeks Requires stringent biosafety measures for pathogens like *Brucella* spp.
Different bacterial identification systems	VITEK system	Automated biochemical analysis with identification	Reduced sensitivity and specificity for non-fermenters
	API system	User-friendly and cost-effective, numerical profile for computer-aided identification	Cross-contamination, low discrimination scores for atypical strains
	MALDI-TOF system	Rapid diagnosis	Initial capital cost and infrastructure requirement
Serological tests	Enzyme-linked immunosorbent assay (ELISA)	High sensitivity and specificity, suitable for large-scale screening	May not differentiate past infections from acute ones; potential cross-reactivity in endemic settings
	Microscopic agglutination test (MAT)	Gold standard for leptospirosis, serovar-specific diagnosis	Labour-intensive, requires live cultures
	Rose Bengal test	Rapid, simple, cost-effective screening of brucellosis	Low specificity, false positives in vaccinated and cross-reactive infections
	Western blot	Highly specific and confirmatory	Technically demanding, time-consuming, and requires well-characterised antigens
	Indirect immunofluorescence assay (IFA)	High specificity and sensitivity	Requires a fluorescence microscope, subjective interpretation
	Lateral flow assays (LFAs)	Portable, rapid, user-friendly, ideal for field use and point-of-care diagnostics	Often qualitative and provides lower sensitivity than ELISA
	Complement fixation test (CFT)	Useful for the chronic phase of disease	Technically complex, highly affected by antibody class and sample quality
	Radioimmunoassay (RIA)	Highly sensitive, useful in quantification	Uses radioactive isotopes, requires special equipment and disposal protocols
Molecular diagnostics	Polymerase chain reaction (PCR)	Highly specific and versatile, applicable to diverse bacterial pathogens	Requires precision in primer design and thermal cycling equipment
	Loop-mediated isothermal amplification (LAMP)	Rapid, portable, and requires minimal equipment; ideal for resource-limited settings	Lower multiplexing capability compared to PCR
	Next-generation sequencing (NGS)	Comprehensive data and the ability to detect co-infections	Cost-intensive and requires bioinformatic expertise
	Microarray technology	High-throughput and capable of detecting co-infections	Relatively lower sensitivity compared to PCR and sequencing
	Fluorescence in situ hybridisation (FISH)	Direct visualisation of pathogens in clinical specimens	Requires fluorescence microscopy and expertise
	Restriction fragment length polymorphism (RFLP)	Useful for strain typing and epidemiological studies	Labour-intensive and requires sufficient DNA quantity
	CRISPR–Cas Systems	High sensitivity and precision; potential for point-of-care diagnostics	Still in experimental stages for many bacterial zoonoses

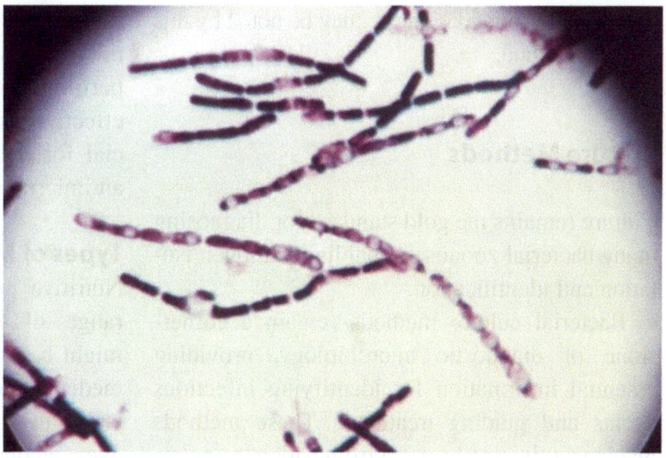

Fig. 8.1 Gram-positive rods of *Bacillus anthracis* with endospores (Gram's stain, ×1000). (Image Source: CDC, PHIL, ID#9826)

red colour of the fuchsin (e.g. *Mycobacterium bovis* appears red).

Giemsa Staining

It is commonly used to detect intracellular bacteria using a mixture of dyes (eosin and methylene blue) that stain cellular components. Bacteria appear blue or purple against a pale background. Rickettsial organisms appear blue or purple in tissue samples. *Anaplasma* species can also be detected in blood smears by this staining.

Wright's Staining

Wright's stain is a combination of eosin and methylene blue, allowing for the differentiation of cellular components and microorganisms. Acidic cellular structures, such as the nuclei of white blood cells and basophilic granules, readily bind to the basic methylene blue, taking on hues of blue to purple. Conversely, basic components like haemoglobin in red blood cells and eosinophilic granules attract the acidic eosin, staining in shades of pink to orange-red. Neutral cellular elements interact with both dyes, resulting in a spectrum of colours that allows for (i) the identification and morphological assessment of different blood cell types, (ii) the detection of malarial parasites, and (iii) the identification of eosinophils in urine, proving its versatility in diagnostic applications.

Silver Staining

Silver stains are used to highlight specific bacterial structures, particularly those that are difficult to visualise with conventional stains. Silver ions are deposited on the bacterial cell walls, making them appear dark black or brown under the microscope. Silver staining highlights the spiral shape of *Leptospira* bacteria in tissue samples. *Francisella tularensis* is small and difficult to visualise, but silver stains can aid in detection.

Methylene Blue Staining

This technique is particularly effective for identifying anthrax bacilli in blood films or tissue sections. The staining process typically involves preparing the tissue sections by removing paraffin and hydrating them, followed by polychrome methylene blue application. This reacts with the capsule of *B. anthracis*, producing a characteristic blue staining (*McFadyean's reaction*).

Fluorescent Staining

Fluorescent staining uses dyes that bind to bacterial cells and emit light when exposed to ultraviolet (UV) light. This is particularly useful for detecting small numbers of bacteria or specific bacterial components. Bacteria are stained with a fluorescent dye (e.g. acridine orange and auramine) that is visible under UV light. *Chlamydia* species are detected using fluorescent antibody techniques.

Haematoxylin and Eosin (H&E) Staining

Haematoxylin stains the nuclei blue, and eosin stains the cytoplasm pink. It is not specific to bacteria but allows for visualising tissue damage and bacterial infiltration. Granulomas in tissues may be visible. In leptospirosis, tissue changes in organs, such as the liver and kidneys, may be noted by this staining method.

Culture Methods

Culture remains the gold standard for diagnosing many bacterial zoonoses, enabling pathogen isolation and identification.

Bacterial culture methods remain a cornerstone of diagnostic microbiology, providing essential information for identifying infectious agents and guiding treatment. These methods involve cultivating bacteria from patient samples, such as blood, urine, sputum, or tissue, in a controlled laboratory setting. The process typically involves inoculating the sample onto or into a suitable growth medium, which provides the necessary nutrients for bacterial proliferation. Different types of media, including agar plates and liquid broths, are used to support the growth of various bacterial species. Selective and differential media are particularly valuable, as they can inhibit certain bacteria's growth while promoting others' growth or differentiate between bacterial species based on their metabolic characteristics. Once sufficient bacterial growth is achieved, the organisms are identified through microscopic examination, biochemical testing, and, increasingly, molecular techniques. Furthermore, antimicrobial susceptibility testing (AST) is often performed on cultured isolates to determine the effectiveness of various antibiotics, which is crucial for optimising patient care and combating antimicrobial resistance.

Types of Media

Nutritive media support the growth of a wide range of non-fastidious microorganisms and might be differential (e.g. blood agar). Selective media support the growth of one group of organisms but not another by adding antimicrobials, dyes, alcohol, or other inhibitory chemicals to a particular medium. MacConkey agar, for example, contains the dye crystal violet, which inhibits Gram-positive organisms. Columbia agar with colistin and nalidixic acid is a selective medium for Gram-positive organisms because the antimicrobials colistin and nalidixic acid inhibit Gram-negative organisms. Chocolate agar is an enrichment medium used in most routine cultures to enhance the growth of fastidious organisms. In some cases, enrichment broths (e.g. thioglycollate broth, brain–heart infusion broth, and tryptic soy broth) are used to increase the yield, particularly in samples containing low microbial load, in the cases treated with antimicrobial agents, and to recover anaerobes. Special media like polymyxin

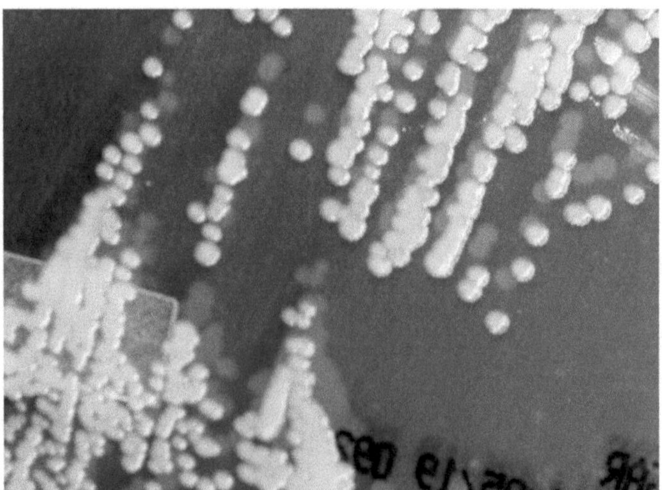

Fig. 8.2 *Bacillus anthracis* cultured on polymyxin B-lysozyme-EDTA–thallous acetate agar (PLET) medium for 24 h at 37 °C (10×). (Image Source: CDC, PHIL/Todd Parker, ID#17096)

Table 8.2 Different media used in diagnosis

Culture media	Type	Key components	Use
Blood agar	Enriched, differential	5–10% sheep blood in nutrient agar	Supports the growth of most bacteria and differentiates between alpha-, beta-, and non-haemolytic colonies
Chocolate agar	Enriched	Lysed red blood cells and nutrients, especially nicotinamide adenine diphosphate (X factor) and haemin (V factor)	To recover *Haemophilus* spp., *Brucella* spp., and *Neisseria* spp.
MacConkey agar	Selective, differential	Bile salts, crystal violet, lactose, neutral red	Inhibits Gram-positive bacteria (crystal violet) and discriminates lactose fermenters (pink colony) from lactose non-fermenters (colourless)
Xylose lysine deoxycholate (XLD) agar	Selective, differential	Xylose, lysine, sodium thiosulphate, ferric ammonium citrate	*Shigella* spp. appears as red colonies; *Salmonella* spp. as red colonies with a central back spot
Salmonella Shigella (SS) agar	Selective, differential	Bile salts, brilliant green, lactose, sodium thiosulphate	Differentiates *Salmonella* spp. (black dot) from *Shigella* spp.
Campylobacter Blood Agar Plate (CAMPY--BAP) medium	Selective	Blood agar with several antibiotics like cefoperazone and vancomycin	To increase the yield of *Campylobacter* spp.
Thiosulphate citrate bile salts sucrose (TCBS) medium	Selective, differential	Bile salts, thiosulphate, citrate, sucrose, bromothymol blue	Isolation of *Vibrio spp.* and differentiation of sucrose fermenters (yellow colonies)
Skirrow's medium	Selective	Blood base, along with vancomycin, polymyxin B, and trimethoprim	Selective for *Campylobacter* spp. (flat, irregular colonies)
Buffered charcoal yeast extract (BCYE) agar	Enriched	Yeast extract, L-cysteine, iron salts, activated charcoal	For fastidious respiratory zoonoses (e.g. *Legionella pneumophila* and *Francisella tularensis*)
Brucella agar/ Castaneda medium	Enriched	Peptone base, dextrose, serum, antibiotics	To cultivate *Brucella* spp.
Tinsdale agar	Selective, differential	Potassium tellurite, serum, cystine	For growing *Corynebacterium pseudotuberculosis*
Tellurite blood agar	Selective, differential	Blood with potassium tellurite	For *Corynebacterium* spp. and *Listeria* spp.
Lowenstein–Jensen (LJ) medium	Enriched, selective	Egg base, malachite green	Growth of mycobacteria, e.g. rough and buff colonies of *Mycobacterium bovis*
Middlebrook 7H10/7H11 medium	Enriched	Defined synthetic medium with oleic acid, albumin, dextrose and catalase (OADC) supplement	For the faster growth of Mycobacteria
Selective enrichment broths (e.g. selenite F and tetrathionate)	Enrichment	Selenite or tetrathionate inhibits normal flora	For *Salmonella* spp.
Thioglycollate broth	Enrichment	Thioglycollate, resazurin	Growth of anaerobes, e.g. *Clostridium* spp.
Cold enrichment medium	Enrichment	Buffered media kept at 4 °C for weeks	For cold-tolerant bacteria (e.g. *Listeria monocytogenes* and *Yersinia enterocolitica)*

B-lysozyme-EDTA–thallous acetate agar (PLET) are sometimes used to culture environmental samples or highly contaminated animal carcasses for anthrax bacilli (Fig. 8.2 and Table 8.2).

Blood Culture Methods

Conventional Blood Culture Methods

It involves inoculating blood samples into bottles or media, followed by incubation and subsequent microbial identification. In the manual blood culture method, blood samples are directly inoculated into culture media such as blood agar or nutrient broth. The samples are then incubated at a suitable temperature (usually 37 °C) for up to 5–7 days. The media used typically contain nutrients and blood, which support the growth of a wide range of bacteria. *Brucella* species are difficult to grow and may require longer incubation periods, sometimes up to 2–3 weeks. A bi-phasic (*Castaneda method*) system is used to prevent contamination and exposure (Fig. 8.3). *Leptospira* spp. can sometimes be cultured from blood, although special media (e.g. Fletcher's medium) are used.

Automated Blood Culture Systems

These are widely used in clinical microbiology to speed up the identification of pathogens and reduce the labour required for manual culture. These systems are especially useful for detecting bacteraemia or septicaemia caused by zoonotic infections, where time is critical. The BacT/ALERT system is an automated blood culture method that involves inoculating blood samples into specialised bottles, which are then placed into an automated incubator. The system continuously monitors the bottles for signs of bacterial growth, typically by detecting CO_2 production or changes in pH. *Brucella* spp. grows slowly but can be detected by the BacT/ALERT system with extended incubation. The BACTEC system uses blood culture bottles inoculated with blood and then incubated. The system measures growth through changes in CO_2 levels, allowing for real-time monitoring and detection of bacterial growth.

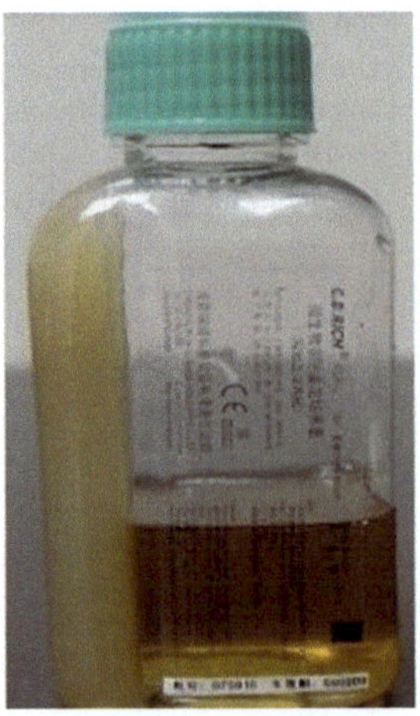

Fig. 8.3 Bi-phasic blood culture medium (Castaneda system) for *Brucella* spp. (Source: Authors personal collection)

Cell Culture Method

Certain bacteria in the *Rickettsia* group, *Ehrlichia*, and *Anaplasma* cannot be grown in an artificial medium. These are cultivated like viruses in cell lines derived from animal or human tissues. These include embryonated egg yolk sacs, Medical Research Council cell strain 5 (MRC 5), Vero cells, canine histiocytic cell lines (DH82), human promyelocytic cell lines (HL60), and others (Fig. 8.4).

Biochemical Tests

Utilisation of a battery of biochemical tests has been fundamental in the phenotypic identification of bacteria, both in the manual and the automated systems. Pure culture remains essential for proper identification. Inoculation with a mixture of bacteria produces mixed and uninterpretable results. The key biochemical tests include enzyme-based tests (e.g. catalase, indole (tryptophanase), L-pyroglutamyl-aminopeptidase, etc.) and metabolic pathway–based tests (e.g. carbohydrate oxidation and fermentation, amino acid degradation, and single substrate utilisations) (Table 8.3).

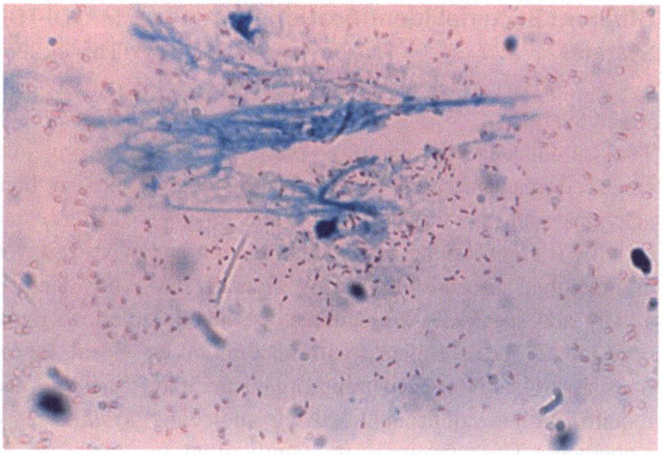

Fig. 8.4 Gimenez-stained yolk sac smear of *Rickettsia* spp. (Image Source: CDC, PHIL/Dr. Charles C. Shepard, ID#20205)

Table 8.3 Different biochemical tests used in laboratory diagnosis

Biochemical test	Principle	Use
Catalase	Detection of enzyme catalase, which breaks down hydrogen peroxide into water and nascent oxygen (seen as bubble formation)	To differentiate catalase-positive (e.g. Enterobacteriaceae) from catalase-negative (e.g. *Brucella* spp.)
Oxidase	Detection of cytochrome c oxidase enzyme using an electron donor chemical, resulting in the development of purple colour within 10 s	To differentiate catalase-positive *Pasteurella* spp. and *Campylobacter* spp. from catalase--negative Enterobacteriaceae
Urease	Detection of the urease enzyme by the hydrolysis of urea to ammonia and the resultant pink colour development due to alkalinity	*Brucella* spp. and *Yersinia enterocolitica* are usually positive, while *Salmonella* spp. is negative
Indole formation	Detection of the ability to degrade tryptophan	To differentiate indole-positive isolates (*Escherichia coli*, *Pasteurella multocida*) from the indole-negatives (*Salmonella* spp., *Yersinia pestis*)
Triple sugar iron agar	Testing of glucose, lactose/sucrose fermentation, along with gas and hydrogen sulphide production	Differentiation of oxidative and fermentative bacteria; hydrogen sulphide production (*Salmonella* spp., *Proteus* spp., etc.)
Citrate utilisation test	Determination of the ability to use citrate as the sole carbon source	To aid in the differentiation between positive (*Salmonella* spp.) and negative (*E. coli*)
Methyl red test	Detection of mixed acid fermentation by measuring stable acid end products from glucose	Differentiation of *E.coli* and *Salmonella* spp. (often positive) from Enterobacteriaceae
Voges–Proskauer test	Detection of acetoin production from glucose fermentation	Identifies *Klebsiella* spp. and *Enterobacter* spp., which are positive for the test
Christie Atkins Munch Peterson (CAMP) test	Detection of synergistic haemolysis with *Staphylococcus aureus* in the form of an arrowhead	To identify *Streptococcus agalactiae*

Different Bacterial Identification Systems

VITEK System

The automated microbiology platform uses enhanced reagent cards to identify metabolic activities such as acid generation and enzyme activity. It combines bacterial identification and antimicrobial susceptibility testing. Overall, sensitivity and specificity are satisfactory, albeit it occasionally faces problems with uncommon or unusual strains. Non-fermenting organisms may require further confirmation using molecular techniques. This system is frequently used in hospitals and diagnostic labs to detect zoonotic pathogens such as *Leptospira* spp. and *Streptococcus suis*.

Analytical Profile Index (API) System

Manual biochemical testing using pre-packaged plastic strips is available with up to 20 tiny reactions. The metabolic activities produce colour changes during incubation, which can be analysed manually or digitally. Sensitivity of the system ranges from moderate to high, depending on pathogen complexity. It has high specificity for common bacterial strains but is restricted to rare zoonotic bacteria. Nevertheless, it is useful for the cost-effective and simple identification of zoonotic pathogens.

MALDI-TOF MS

This system facilitates rapid and precise identification of bacteria using protein profiling (Fig. 8.5). The system generates spectra of cellular proteins and compares them to extensive reference libraries. With high sensitivity, it can identify rare and emerging zoonotic pathogens. Also, a very high specificity ensures minimal misidentifications. The expansive protein library allows for high-accuracy matching. The system is ideal for advanced diagnostics in laboratories tackling zoonotic infections such as *Francisella tularensis*, *Y. pestis*, and *Brucella* spp.

Serological Tests

Serological tests are indispensable tools in the diagnosis of zoonotic bacterial infections. It provides essential information about past or present exposure to pathogens by detecting host antibodies produced against bacterial antigens. Serological assays identify IgM and IgG antibodies in body fluids, signalling an immune response. These tests are vital for diagnosing infections where bacterial isolation is difficult or for retrospective exposure assessments. For example, serological tests are essential in zoonotic diseases such as brucellosis, leptospirosis, and Lyme disease, where bacterial culture is often problematic.

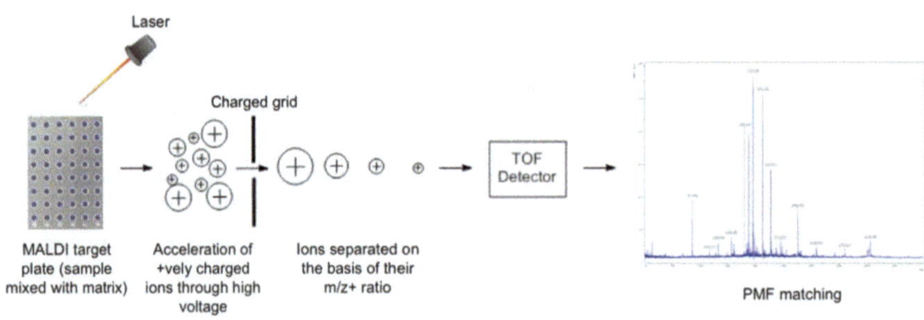

Fig. 8.5 Principle of MALDI-TOF. (Image Source: Singhal et al. 2015. Copyright © 2015. Creative Commons Attribution License (CC BY))

Some of the commonly used serological tests are the following:

Complement Fixation Test (CFT)

CFT detects antigen–antibody complexes by measuring the consumption of complement proteins. Complement binds to antigen–antibody complexes, preventing haemolysis of sensitised red blood cells in the indicator system. The test serum is mixed with a known antigen and complement proteins. If antibodies are present, complement binds to the antigen–antibody complex. In the absence of antibodies, complement remains free and lyses sensitised red blood cells. The test is used, however, less frequently in the diagnosis of Q fever and brucellosis.

Microscopic Agglutination Test (MAT)

MAT uses live antigen strains to detect agglutination in serum samples, which is the gold standard for detecting *Leptospira* serovars. MAT uses heat-killed antigens to improve safety during the testing process. It is suitable for laboratories with limited biosafety capabilities.

Rose Bengal Test

The test detects antibodies using stained antigens in a slide format. It is used for the screening of *Brucella* spp. in humans and livestock. Modified Rose Bengal test enhances sensitivity by adjusting antigen concentrations.

Indirect Immunofluorescence Assay (IFA)

IFA detects pathogen-specific antibodies using fluorescently labelled secondary antibodies. It remains the gold standard for diagnosing Q fever. Dual antigen IFA simultaneously detects antibodies against multiple pathogens.

Enzyme-Linked Immunosorbent Assay (ELISA)

The principle of ELISA is based on the specific interaction between an antigen and its corresponding antibody, utilising an enzyme to produce a detectable signal for quantification. Either the antigen or the antibody is immobilised on a solid surface, such as a microtitre plate. The assay uses enzyme-labelled antibodies or antigens as detection molecules.

Commonly used enzymes include horseradish peroxidase (HRP) and alkaline phosphatase (AP). After the antigen–antibody binding, a substrate specific to the enzyme is added. The enzyme catalyses a reaction that generates a measurable signal—typically a colour change (chromogenic substrates) or fluorescence. The resulting signal, proportional to the target antigen/antibody amount, is quantified using a spectrophotometer or fluorescence reader.

Direct ELISA detects antigens directly using a labelled antibody, for example, to detect *Brucella* antigens in infected samples. Indirect ELISA detects antibodies with a secondary enzyme-labelled antibody and is commonly used for detecting IgM and IgG in *Leptospira* infections. Sandwich ELISA detects antigens captured by a primary antibody and a secondary labelled antibody, enhancing sensitivity for chronic zoonoses like brucellosis. Competitive ELISA measures the presence of specific antibodies competing with a labelled antigen for differentiating antibody subtypes, especially for chronic zoonotic diseases.

Western Blot

In this technique, bacterial proteins are separated by electrophoresis, and specific antibodies are detected. The test is used for leptospirosis and brucellosis. Quantitative Western blot measuring antibody levels for disease staging has also been used for chronic infections such as brucellosis or *Coxiella burnetii* infections.

Lateral Flow Assays (LFAs)

LFA combines the principles of chromatography and immunoassay techniques. It detects antigens or antibodies in a sample using capillary action to transport the sample across a nitrocellulose membrane. The sample is applied to a pad containing labelled antibodies conjugated to coloured particles (e.g. colloidal gold or latex beads). As the sample migrates, antigen–antibody complexes form and bind to immobilised antibodies at the test line. A visible coloured line appears at the test line if the target analyte is present. Antibody detection LFAs are used in anthrax. It has also been used for the field diagnosis of plague.

Radioimmunoassay (RIA)

RIA is based on competitive binding between radiolabelled and unlabelled antigens for antibody binding sites. The concentration of free radiolabelled antigen is inversely proportional to the amount of unlabelled antigen in the sample. Radiolabelled antigens compete with sample antigens for binding to specific antibodies. Bound and free radiolabelled antigens are separated, and radioactivity is measured. Competitive RIA measures antibodies competing with labelled antigens and has been historically used for leptospirosis research. Direct RIA detects antigens directly using radioactively labelled antibodies. It is rarely used in modern diagnostics but is effective for zoonotic studies.

Multiplex Serology

Multiplex serology uses bead-based or microarray platforms to simultaneously detect antibodies against multiple pathogens. It relies on antigen–antibody interactions and signal detection. Antigens are immobilised on beads or microarray spots. Serum antibodies bind to the antigens, and secondary labelled antibodies generate signals (e.g. fluorescence or chemiluminescence). This has been used for high-throughput screening for co-infections in zoonotic diseases like leptospirosis and brucellosis.

Molecular Diagnosis

Polymerase Chain Reaction (PCR)

In PCR, the amplification of DNA sections is contemplated using primers, Taq polymerase, and temperature cycling (denaturation, annealing, and extension). Conventional PCR identifies DNA via electrophoresis, and the results are qualitative. Real-time PCR (qPCR) is a quantitative technique that employs fluorescent probes or dyes to monitor amplification in real time. Reverse transcription PCR (RT-PCR), which amplifies RNA by converting it to complementary DNA (cDNA), has been used to identify *F. tularensis*. Multiplex PCR amplifies several DNA targets at the same time using multiple primer sets, for example, co-infections with *Brucella* spp. and *Leptospira* spp. in endemic regions.

Loop-Mediated Isothermal Amplification (LAMP)

In this test, DNA is amplified at a steady temperature using a strand-displacing polymerase and four to six specifically selected primers. Colorimetric LAMP allows for the visual detection of amplification through colour changes (e.g. pH indicators) and is utilised in field diagnostics for *Leptospira* and *Brucella*. Fluorescence-LAMP detects amplification using real-time fluorescence signals, such as quantitative leptospirosis assays. RT-LAMP combines reverse transcription with RNA detection.

Next-Generation Sequencing (NGS)

This technique involves high-throughput sequencing of complete bacterial genomes or specific sections. Whole-genome sequencing (WGS) generates comprehensive bacterial genomic data, which is essential for detecting uncommon infections and tracking zoonotic outbreaks. Targeted NGS uses specific genomic areas (e.g. 16S rRNA sequencing) to identify bacterial species in mixed clinical samples. Metagenomic sequencing examines all microbial genomes in a sample, making it an ideal method for finding unknown or novel zoonotic infections.

Microarray Technology

This technique involves hybridisation of labelled DNA or RNA, which is accomplished with complementary probes on a chip. The DNA microarray detects DNA sequences using complementary DNA probes. The RNA microarrays assess gene expression levels in bacterial RNA.

Fluorescence In Situ Hybridisation (FISH)

Fluorescently labelled DNA/RNA probes are utilised to recognise complementary sequences in bacterial cells. In direct FISH, probes that hybridise directly with bacterial DNA or RNA are used to rapidly detect *Brucella* spp. in tissue biopsies. RNA-FISH detects *Francisella* spp. and *Rickettsia* spp. more sensitively by targeting bacterial ribosomal RNA.

Restriction Fragment Length Polymorphism (RFLP)

DNA is digested by restriction enzymes, and the resulting fragment patterns are analysed. Conventional RFLP analyses DNA fragment sizes using gel electrophoresis (e.g. differentiating *Brucella* strains). RFLP-PCR combines PCR amplification with RFLP analysis for enhanced sensitivity and is used to differentiate leptospirosis serovars.

CRISPR–Cas Systems

The system utilises CRISPR-associated enzymes (e.g. Cas12 and Cas13) to detect specific nucleic acid sequences. Specific High-sensitivity Enzymatic Reporter unLOCKing (SHERLOCK) uses Cas13 for RNA detection (Fig. 8.6). CRISPR has been used for the rapid detection of zoonotic pathogens like *F. tularensis*.

Metagenomics

Metagenomics involves the unbiased sequencing of all genetic material in a sample, identifying bacterial pathogens and their interactions. Shotgun metagenomics sequences all genetic material, providing comprehensive microbial community profiles, and can help in detecting unknown pathogens in complex infections. Targeted metagenomics focuses on specific loci (e.g. 16S rRNA) for bacterial identification from mixed samples.

Antimicrobial Susceptibility Testing (AST)

Antimicrobial susceptibility testing (AST) is an important laboratory procedure used to determine the effectiveness of various antimicrobial agents against specific bacterial strains. This testing is essential for guiding clinicians in selecting

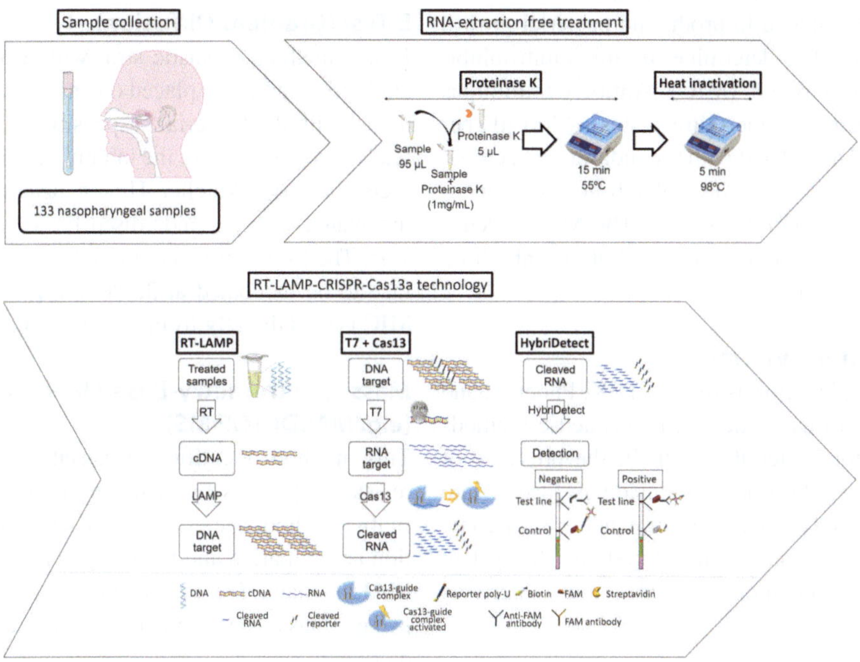

Fig. 8.6 Workflow protocol for infectious disease diagnosis based on CRISPR-Cas13a technology. (Image Source: Ortiz-Cartagena et al. 2022. Copyright © 2022. https://creativecommons.org/licenses/by/4.0/)

the most appropriate antibiotic treatment for patients with bacterial infections.

The different methods of AST used in diagnosing bacterial zoonoses are as follows:

Disc Diffusion Method (Kirby–Bauer Test)

In this method, antibiotic-impregnated discs are placed on Mueller–Hinton agar and inoculated with a standardised bacterial suspension. Antibiotics diffuse radially, creating a concentration gradient. Zones of inhibition are measured to determine susceptibility. Mueller–Hinton agar plates are used with a depth of 4 mm and a pH of 7.2–7.4. The test bacterial suspension is adjusted to 0.5 McFarland standard and swabbed evenly across the agar surface. Antibiotic discs are placed at least 24 mm apart and incubated at 35 °C for 16–18 h. The inhibition zones are measured and interpreted using Clinical and Laboratory Standards Institute (CLSI) or European Committee on Antimicrobial Susceptibility Testing (EUCAST) guidelines.

Broth Microdilution

In this method, serial twofold dilutions of antibiotics are prepared in broth, and bacterial growth is monitored to determine the minimum inhibitory concentration (MIC). Antibiotic dilutions are prepared in microtitre plates (≤200 μL per well) and inoculated with a standardised bacterial suspension (5×10^5 CFU/mL), followed by incubation at 35 °C for 16–20 h. The MIC is determined as the lowest concentration inhibiting visible growth.

Automated Systems

Automated platforms (e.g. VITEK, Phoenix, and MicroScan) use optical or colorimetric methods to monitor bacterial growth in the presence of antibiotics. Bacterial suspension is loaded into pre-prepared reagent cards or panels, followed by insertion into the automated system. The system incubates, monitors growth, and interprets results.

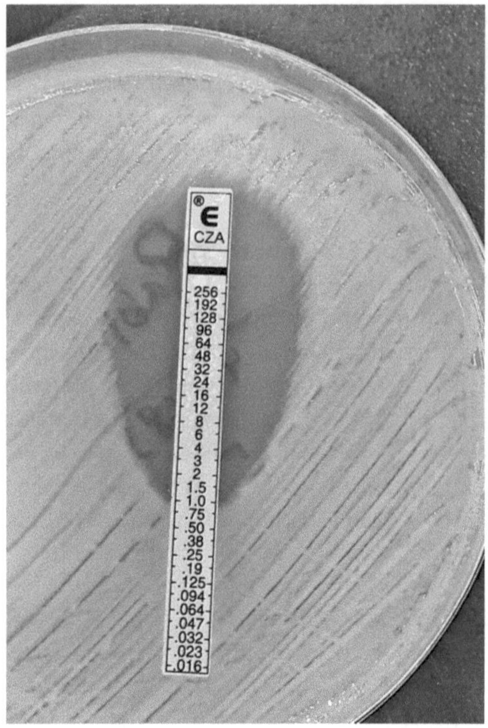

Fig. 8.7 Epsilometer test for antimicrobial sensitivity testing. (Image source: Authors' personal collection)

E-Test (Gradient Diffusion)

In this method, a plastic strip with a predefined antibiotic gradient is placed on an agar plate inoculated with the bacterial suspension. The MIC is read at the point where the inhibition ellipse intersects the strip. Mueller–Hinton agar plates are inoculated with a standardised bacterial suspension. The E-test strips are placed on the agar surface to be incubated at 35 °C for 16–20 h. The MIC is read directly from the strip (Fig. 8.7).

Mass Spectrometry-Based Methods (e.g. MALDI-TOF MS)

This method identifies bacterial species and resistance profiles by analysing protein spectra. In this method, bacterial sample and matrix solution are prepared, and the mixture is applied to a MALDI plate. The system ionises and analyses using mass spectrometry from a pattern dictionary.

Table 8.4 Advantages and limitations of different antimicrobial susceptibility testing (AST) methods

AST methods	Advantages	Limitations
Disc diffusion (Kirby–Bauer)	Simple, cost-effective, widely used	Qualitative results only, less precise for borderline resistance
Broth microdilution	Quantitative MIC results, gold standard for AST	Labour--intensive, requires specialised equipment
Automated systems (VITEK, Phoenix, MicroScan)	Automated, rapid results, standardised testing	High cost, may require confirmation for atypical pathogens
E-test (gradient diffusion)	Combines the simplicity of disc diffusion with MIC determination	Costlier than disc diffusion, requires careful handling
Mass spectrometry (MALDI-TOF MS)	Rapid and precise, effective for rare pathogens	High initial cost, relies on extensive databases
Biosensor--based AST	Potential for point-of-care testing, rapid results	Emerging technology, limited availability

Biosensor-Based AST

Biosensors are used to detect bacterial growth or resistance markers in real time. Specific enzymes or antibodies are immobilised on the biosensor surface. Bacterial samples are introduced, and interactions are monitored. Amperometric or optical signal outputs are analysed to determine susceptibility.

The advantages and limitations of each method are summarised in Table 8.4.

Artificial Intelligence-Based Emerging Technologies

Artificial intelligence (AI) has the potential to change the diagnosis and management of bacterial zoonoses by increasing detection speed, accuracy, and efficiency. AI-powered algorithms use complicated datasets like genomic sequences and protein profiles to accurately identify zoonotic bacteria. Machine learning algorithms, for example, may distinguish between closely similar bacterial strains. Several machine learning methods, including K-nearest neighbours, support vector machines, and artificial neural networks, have demonstrated promising results in detecting *Escherichia coli* and *Staphylococcus aureus* in raw meat emissions. AI systems predict antibiotic resistance trends by analysing genetic markers and phenotypic data, allowing for more tailored treatment techniques for zoonotic illnesses. AI-powered predictive models use environmental and epidemiological data to forecast zoonotic disease outbreaks, assisting in pre-emptive public health actions. A Monte Carlo simulation-based predictive model has been created for *Salmonella* spp. in eggs. AI improves the performance of diagnostic platforms such as MALDI-TOF MS and PCR by optimising data interpretation and shortening the turnaround time. AI systems make real-time diagnostic and therapy suggestions based on patient data, which improves clinical outcomes for zoonotic diseases. AI-powered biosensors and mobile applications enable quick point-of-care diagnostics in remote or resource-constrained environments.

Case Study

A wildlife biologist presented with high fever, painful lymphadenopathy, and a history of working in an area with rodent die-offs. Blood and aspirates from enlarged lymph nodes were obtained. Wright–Giemsa staining revealed bipolar-staining "safety pin" bacteria characteristic of *Y. pestis*. PCR confirmed the genetic identity of *Y. pestis*. The patient received gentamicin and was quarantined. Rodent control measures were introduced in the affected area.

1. What are the advantages of combining traditional methods (microscopy and culture) with modern molecular techniques?
2. How important is the close coordination between field teams and laboratory staff for

timely diagnosis and intervention in bacterial zoonotic infections?
3. For epidemiological purposes, what other investigations are essential in the light of the One Health approach to zoonotic diseases?

Research Questions

1. What are the requirements for developing rapid point-of-care tests that can be used in field conditions, and why is this necessary?
2. How can antibiotic resistance gene targeting be integrated with the existing molecular methods, which are mainly used for identifying organisms?
3. How do we develop economical, easy, reproducible methods for molecular epidemiological studies in bacterial zoonoses?

Further Readings

Oršolić M, Sarač N, Balen TM. Vector-borne zoonotic lymphadenitis – the causative agents, epidemiology, diagnostic approach, and therapeutic possibilities – an overview. Life. 2024;14:1183.

Ortiz-Cartagena C, Fernández-García L, Blasco L, Pacios O, Bleriot I, López M, Cantón R, Tomás M. Reverse transcription-loop-mediated isothermal amplification-CRISPR-Cas13a technology as a promising diagnostic tool for SARS-CoV-2. Microbiol Spectr. 2022;10:e02398–22.

Procop GW, Church D, Hall G, et al. Koneman's Color Atlas and textbook of diagnostic microbiology. 7th ed. Lippincott, Williams & Wilkins; 2017.

Rahman MT, Sobur MA, Islam MS, Ievy S, Hossain MJ, El Zowalaty ME, Rahman AT, Ashour HM. Zoonotic diseases: etiology, impact, and control. Microorganisms. 2020;8(9):1405. https://doi.org/10.3390/microorganisms8091405.

Singhal N, Kumar M, Kanaujia PK, Virdi JS. MALDI-TOF mass spectrometry: an emerging technology for microbial identification and diagnosis. Front Microbiol. 2015;6:791. https://doi.org/10.3389/fmicb.2015.00791.

Tille PM. Bailey & Scott's diagnostic microbiology. St. Louis: Elsevier; 2014.

Wikel SK. Zoonoses: changing, challenging, and increasing global disease threats. Zoonotic Diseases. 2024;4(1):8–10. https://doi.org/10.3390/zoonoticdis4010002.

World Health Organization. Laboratory biosafety manual. 4th ed; 2020. Available from: https://www.who.int/publications/i/item/9789240011311

Yang X, He Z, Zhang G, Lu J, Zhang H, Ren H, Tian Y, Yang H, Chen C, Li L, Fu Y, Allain J-P, Li C, Wang W. Evaluation of reactivity of monoclonal antibodies against Omp25 of *Brucella* spp. Front Cell Infect Microbiol. 2020;10:145.

Bacterial Zoonoses and Antimicrobial Resistance

9

Subham Mookerjee, Shravani Mitra, Sudipto Mandal, and Surojit Das

Learning Objectives

1. To review the current updates on antimicrobial resistance (AMR) and its mechanisms in bacterial pathogens originating from food-producing animals, wildlife, and other non-human sources
2. To understand the contributing factors in the emergence, transmission, and sustainability of antibiotic-resistant zoonotic bacteria
3. To identify the public health implications of zoonotic AMR, focusing on the United Nations Sustainable Development Goals (SDGs)
4. To define global advancement in the surveillance, diagnosis, prevention, control, and treatment of AMR-associated zoonoses

S. Mookerjee
Department of Zoology, Ramakrishna Mission Vivekananda Centenary College, Khardaha, West Bengal, India

S. Mitra
Formerly at Division of Bacteriology, ICMR-National Institute of Cholera and Enteric Diseases, Kolkata, West Bengal, India

S. Mandal
Department of Microbiology, Ramakrishna Mission Vidyamandira, Howrah, West Bengal, India

S. Das (✉)
Department of Biomedical Laboratory Science and Management, Vidyasagar University, Midnapore, West Bengal, India

Introduction

Since the mid-twentieth century, antibiotics have transformed the medical landscape by treating and preventing bacterial infections, saving millions of human and animal lives worldwide. Zoonoses, which are naturally transmissible between animals and humans, cause 2.5 million human cases and 2.7 million deaths annually worldwide. These contribute for 60% of recognised infectious diseases and up to 75% of emerging infectious diseases (EIDs). Antibiotic use in humans, animals, agriculture, and aquaculture climbed by 35% between 2000 and 2010. Globally, BRICS nations such as Brazil, Russia, India, China, and South Africa contribute to nearly a 79% rise in antibiotic resistance. Surprisingly, animal husbandry and production use almost 70% of all antibiotics sold for prophylaxis and growth promotion, which is expected to add an extra 8% increase in antibiotic use by 2030. Irrational use of comparable and related antibiotics from human medicine in animal farming over the years renders them clinically ineffective by creating antimicrobial resistance (AMR) in zoonotic bacteria and even commensal bacteria.

Simple infections frequently become untreatable and fatal due to antibiotic-resistant bacteria (ARB). These bacteria spread AMR as an ever-changing evolutionary phenomenon across human–animal–environmental interfaces. This is

S. C. Parija, A. Poddar (eds.), *Textbook of Bacterial Zoonoses*, Microbial Zoonoses,
https://doi.org/10.1007/978-981-95-6615-0_9

mediated by horizontal gene transmission (HGT) of mobile genetic elements (MGEs) and clonal propagation of mutant types to survive antibiotic--induced selection pressure. This indirectly exposes wildlife, particularly migratory birds, to antibiotic residues and ARB, thereby allowing AMR to spread across borders. The global rise of drug-resistant infections presents a hidden pandemic of AMR. The World Health Organization (WHO) has identified AMR as one of the top three medical emergencies, affecting public health, economic development, and food security. Drug-resistant infections are the third leading cause of death, next only to cardiovascular diseases and cancer. In 2019, the Global Research on AMR Project (GRAM) study estimated 1.27 million attributable and, approximately, five million associated deaths from drug-resistant infections. These infections primarily affect low- and middle-income countries (LMICs) of Asia and Africa. The mortality rate is projected to reach ten million by 2050, exceeding cancer deaths.

History of Antibiotics

Antibiotics are chemical substances that inhibit the growth of or destroy microorganisms, primarily bacteria. They are derived from natural sources such as bacteria and fungi or synthesised chemically to mimic natural compounds. Antibiotics are important in treating bacterial infections as they have saved millions of lives since their introduction in the 1940s.

The early recognition of antimicrobial properties in natural substances existed centuries ago. The modern antibiotic era is largely attributed to Alexander Fleming's groundbreaking discovery of penicillin. In the year 1928, he noted the inhibitory effect of *Penicillium notatum* mould on *Staphylococcus* bacteria. Subsequently, Gerhard Domagk's development of Prontosil in the 1930s marked the arrival of the first commercially available synthetic antibacterial drug. The large-scale production and clinical application of penicillin in the 1940s then ushered in a transformative period in the treatment of infectious diseases. The subsequent decades, particularly the 1950s–1970s, are often celebrated as the "golden era" of antibiotic discovery. This was a period that witnessed the identification of many of the fundamental antibiotic classes still in use today.

However, antibiotic abuse and overuse have resulted in the rise of ARB, posing a substantial global health challenge.

Classification of Antibiotics

Antibiotics are classified based on their chemical structure, mode of action, and spectrum of activity. An overview of the different antibiotics, their classes, mode of action, and spectrum of activity is presented in Table 9.1.

The industrial production of antibiotics primarily relies on sophisticated biotechnological and chemical processes. Fermentation is a key method for naturally derived antibiotics, where carefully cultivated microorganisms in large bioreactors produce the desired antibiotic as a secondary metabolite, which is then extracted and purified. Semi-synthetic antibiotics involve using these fermentation products as starting materials for further chemical modifications in the laboratory. Entirely synthetic antibiotics, on the other hand, are manufactured through a series of carefully designed chemical reactions.

Veterinary Use of Antibiotics

Fluoroquinolones, penicillins, cephalosporins, tetracyclines, sulfonamides, macrolides, and glycopeptides are commonly used antibiotics for treatment and prevention and/or as growth promoters in animals (food-producing, companion, and exotic animals) (Table 9.2).

In 2022, the Office International des Epizooties (WOAH) published the "Terrestrial Animal Code" for the authorisation and administration of veterinary antibiotics through competent authorities and stakeholders. The authorisation process involves "(I) assessment of the potential of antimicrobial agents to select for resistance; (II) establishment of acceptable daily intake (ADI), maximum residue limit (MRL) and withdrawal

Table 9.1 Overview of different antibiotics in terms of actions and activity spectrum

Chemical structure class	Antibiotics	Mechanism of action	Spectrum of activity
Beta-lactams	*Penicillins*: amoxicillin, cloxacillin, piperacillin	Bind to penicillin-binding proteins and inhibit cell wall synthesis	Primarily Gram-positive bacteria; some have extended spectrum against certain Gram-negative bacteria and anaerobes
	Cephalosporins: cephalexin, ceftriaxone, cefepime	Inhibit cell wall synthesis like penicillins	Spectrum varies by generation: early generations primarily Gram-positive, later generations have increasing activity against Gram-negative bacteria, some with activity against *Pseudomonas*
	Carbapenems: imipenem, meropenem	Inhibit bacterial cell wall synthesis; broader spectrum than most beta-lactams and more resistant to beta-lactamases	Very broad spectrum against many Gram-positive and Gram-negative bacteria (including many multi-drug-resistant strains), and anaerobes
	Monobactams: aztreonam	Inhibit bacterial cell wall synthesis; primarily active against aerobic Gram-negative bacteria	Primarily aerobic Gram-negative bacteria, including *Pseudomonas aeruginosa*. Little activity against Gram-positive bacteria or anaerobes
Macrolides	Erythromycin, azithromycin, clarithromycin	Bind to the 50S ribosomal subunit, inhibiting the translocation of peptidyl-tRNA and thus protein synthesis	Primarily Gram-positive bacteria, some activity against atypical bacteria (e.g. *Mycoplasma*, *Chlamydia*, and *Legionella*) and some Gram-negative organisms
Tetracyclines	Tetracycline, doxycycline, minocycline	Bind to the 30S ribosomal subunit, inhibiting the binding of aminoacyl-tRNA and thus protein synthesis	Broad spectrum against Gram-positive and Gram-negative bacteria, as well as atypical bacteria, rickettsiae, and some protozoa
Aminoglycosides	Gentamicin, tobramycin, amikacin	Bind to the 30S ribosomal subunit, causing misreading of mRNA and inhibiting protein synthesis	Primarily aerobic Gram-negative bacteria, including *Pseudomonas*. Often used in combination for serious Gram-positive infections (synergy). Limited anaerobic activity
Fluoroquinolones	Ciprofloxacin, levofloxacin, moxifloxacin	Inhibit bacterial DNA gyrase and topoisomerase IV, blocking DNA replication and transcription	Broad spectrum against many Gram-positive and Gram-negative bacteria, including atypical pathogens. Later generations often have enhanced activity against Gram-positive bacteria and anaerobes
Sulfonamides	Sulfamethoxazole (often combined with trimethoprim: SMX/TMP)	Inhibit bacterial folic acid synthesis by acting as competitive inhibitors of dihydropteroate synthetase	Broad spectrum against some Gram-positive and Gram-negative bacteria, as well as some protozoa. Often used for urinary tract infections, *Pneumocystis jirovecii* pneumonia, and *Nocardia* infections
Glycopeptides	Vancomycin, teicoplanin	Block bacterial cell wall synthesis by attaching to the D-Ala-D-Ala terminus of peptidoglycan precursors	Primarily Gram-positive bacteria. Example: methicillin-resistant *Staphylococcus aureus* (MRSA) and *Clostridium difficile*
Lincosamides	Clindamycin, lincomycin	Bind to the 50S ribosomal subunit, inhibiting peptidyl transferase activity and thus protein synthesis	Primarily Gram-positive bacteria and anaerobes

(continued)

Table 9.1 (continued)

Chemical structure class	Antibiotics	Mechanism of action	Spectrum of activity
Polymyxins	Colistin, polymyxin B	Bind to bacterial cell membranes, increasing their permeability and leading to cell lysis	Primarily Gram-negative bacteria, often used for multi-drug-resistant organisms
Nitroimidazoles	Metronidazole, tinidazole	Undergo reduction in anaerobic bacteria to reactive intermediates that damage bacterial DNA	Primarily anaerobic bacteria and some protozoa
Oxazolidinones	Linezolid, tedizolid	Bind to the 50S ribosomal subunit, preventing the formation of the initiation complex for protein synthesis	Primarily Gram-positive bacteria, including MRSA and vancomycin--resistant enterococci (VRE)
Lipopeptides	Daptomycin	Insert into the bacterial cell membrane, causing rapid depolarisation and cell death	Primarily Gram-positive bacteria, including MRSA and vancomycin--resistant enterococci (VRE)
Rifamycins	Rifampicin, rifabutin	Inhibit bacterial DNA--dependent RNA polymerase, thus blocking RNA synthesis	Broad spectrum, but resistance develops rapidly when used alone. Primarily used for mycobacterial infections (tuberculosis and leprosy) and in combination for some Gram-positive infections

periods in food-producing animals; and (III) post-marketing antimicrobial surveillance." The first-point explains AMR as the "(a) concentration of either active antimicrobial agents or metabolites in the gut of the animal (where the majority of potential food-borne pathogenic agents reside) at the defined dosage level, (b) the pathway for human exposure to antimicrobial--resistant microorganisms, (c) the degree of cross--resistance, and (d) the intrinsic and pre-existing baseline level of resistance in the pathogenic agents of human health concern."

The WHO has ranked the medically important antibiotics as critically important, highly important, and important for developing risk management strategies to control AMR, especially in farm animals. Critically important antibiotic classes are the sole or limited available therapeutic options for serious infections (criterion 1) and are used in ARB infections originating from non-human sources (criterion 2). Highly important classes meet either criterion 1 or 2, while important classes are neither criterion 1 nor criterion 2. Several countries have disapproved of the veterinary use of glycopeptides, polymyxins, chloramphenicol, and nitrofurantoin to limit the spread of ARB in humans.

Antibiotic Resistance in Zoonotic Bacteria

Antibiotics are natural or synthetic compounds that can kill or prevent bacterial growth. AMR is the survival method of bacteria, acquired and accumulated through evolution. Bacteria use AMR as a defence mechanism against these bactericidal (killing) and bacteriostatic (growth-inhibiting) compounds. Suboptimal levels of antibiotics allow the growth of resistant mutants of bacteria to acquire the skills to evade or inactivate one or more antibiotics. These help bacteria proliferate by eliminating the population of microorganisms that are still susceptible. This is called *selection pressure*. The overuse and misuse of antibiotics in the agricultural sector, particularly in poultry and livestock, have contributed to this selection pressure for the emergence and

Table 9.2 Veterinary use of clinically important antimicrobial classes for human medicine in farm animals

Antibiotic class	Antibiotics	Category		Route	Indication (in animals)	Use in poultry			Use in livestock		
		Human	Animal			Th	PrP	GP	Th	PrP	GP
Penicillins	Cloxacillin, flucloxacillin, nafcillin	CI	CI	SC, IM	Mastitis, pneumonia, sepsis	Yes	No	No	Yes	No	No
Sulfonamides	Ormosulfathiazole, phthalylsulfathiazole, sulfaquinoxaline, sulfadimethoxine	HI	CI	Oral, IV	Coccidiosis, colibacillosis, mastitis	Yes	No	No	Yes	Yes	No
Polypeptides	Bacitracin	I	HI	Topical, IM	Skin and eye infection	Yes	Yes	Yes	No	Yes	Yes
Aminoglycosides	Amikacin, gentamicin, neomycin, apramycin, framycetin	CI	CI	Oral, IV, IM	Mastitis, endometritis	Yes	No	No	Yes	No	No
Amphenicols	Florfenicol	HI	CI	Oral	BRD	Yes[a]	No	No	Yes	No	No
Tetracycline	Chlortetracycline, oxytetracycline, doxycycline	HI	CI	IM, IV	Keratoconjunctivitis	Yes	Yes	Yes	Yes	Yes	Yes
Cephalosporins	Second gen. (cefalonium) Third gen. (cefovecin, cefquinome, ceftiofur)	HI (second gen.), CI (third gen.)	HI (second gen.), CI (third gen.)	IM, SC	BRD, UTI	Yes	No	No	Yes	No	No
Polymyxins	Colistin	CI	HI	IV, IM	–	Yes	Yes	Yes	Yes	No	No
Macrolides and ketolides	Gamithromycin, kitasamycin, tilmicosin, tulathromycin, tylosin, tylvalosin, erythromycin	CI	CI	Oral, IV	uRTI, enteritis, UTI	Yes	No	No	Yes	Yes	No
Lincosamides	Pirlimycin, lincomycin	HI	HI	Oral	Wound and abscess	No	No	Yes	Yes	No	No
Streptogramins	Virginiamycin	HI	I	Oral	–	No	No	Yes	No	No	Yes
Glycopeptides	Avoparcin	CI	HI	Oral, IV	–	No	Yes	Yes	No	Yes	Yes
Fluoroquinolones	Danofloxacin, ciprofloxacin, difloxacin, enrofloxacin, bafloxacin, marbofloxacin, orbifloxacin	CI	CI	Oral	Diarrhoea, UTI	Yes	Yes	No	Yes	No	No
Nitrofurantoin	Furazolidone, nitrofurantoin, nifurtoinol, nitrofural	I	HI	Oral	UTI	Yes	Yes	Yes	Yes	No	No
Nitroimidazoles	Metronidazole, tinidazole, ornidazole	I	HI	Oral	Diarrhoea	Yes	No	No	Yes	No	No
Anti--mycobacterial	Rifampicin, cycloserine, ethambutol, ethionamide, isoniazid, pyrazinamide	CI	HI	Oral, IV, IM	zTB	No	No	No	Yes	No	No

Abbreviations: *BRD* bovine respiratory disease, *CI* critically important, *GP* growth promoter, *HI* highly important, *I* important, *IM* intramuscular, *IV* intravenous, *PrP* prophylactic, *SC* subcutaneous, *Th* therapeutic, *UTI* urinary tract infection, *uRTI* upper respiratory tract infection, *zTB* zoonotic tuberculosis

[a]Reserve drug, sparingly used

spread of ARB. Animal-originated ARB can transmit to humans via direct contact or animal products, like milk, eggs, and raw meat. The transmission of zoonotic bacteria takes place due to many factors. These include human–animal proximity, poor hygiene and sanitation, drinking unpasteurised milk, and eating undercooked meats, seafood, and contaminated vegetables.

The WHO, currently, has prepared the bacterial priority pathogen list. The said list includes several zoonotic organisms belonging to critical priority (third-generation cephalosporin-resistant Enterobacterales, carbapenem-resistant Enterobacterales, carbapenem-resistant *Acinetobacter baumannii*, and rifampicin-resistant *Mycobacterium tuberculosis* complex) and high priority (fluoroquinolone-resistant non-typhoidal *Salmonella*, vancomycin-resistant enterococci (VRE), methicillin-resistant *Staphylococcus aureus* (MRSA), and carbapenem-resistant *Pseudomonas aeruginosa*) (Table 9.2). Apart from this, colistin, aminoglycoside, and macrolide–lincosamide–streptogramin (MLS) resistance are also listed in zoonotic pathogens.

The dissemination of epidemic bacterial clones, which already have acquired resistance against important antibiotics and are often armed with several virulence genes, is one of the ways for the rapid spread of AMR. In the family Enterobacterales, the sequence types (STs) of epidemic clones are varied in extended-spectrum beta (β)-lactamase (ESBL)–producing *Escherichia coli* (ST10, ST69, ST93, ST167, and ST410) and *Klebsiella pneumoniae* (ST11, ST37, ST107, ST268, ST274, and ST307) (Table 9.2). Fluoroquinolone-resistant *Salmonella enterica* serovar Kentucky ST198 and *Campylobacter jejuni* ST21, ST206, ST353, ST354, and ST464 are prevalent in poultry and broiler. VRE ST16, ST17, and ST132 are common in both cultivated and wild animal intestinal microbiomes, while MRSA ST398 and ST9 play a major role (Table 9.2). However, most of the studies suggest that clonal dissemination (i.e. vertical gene transmission) is less important in the spread of drug resistance in animals and/or from animals to humans than HGT. These allow the transfer of genes rapidly among bacteria, even across different species, through plasmids and/or other MGEs such as insertion sequence (IS) elements, transposons, integrons, and bacteriophages.

Several antibiotics are reserved for treatment or are not recommended for veterinary use. Nonetheless, AMR in these antibiotics is observed in animals due to the co-transfer of multiple antimicrobial resistance genes (ARGs) through MGEs or exposure to human antibiotic-resistant colonising bacteria.

Mechanisms of Antibiotic Resistance

To defend themselves against antibiotics, bacteria possess a plethora of chromosome-mediated intrinsic and/or plasmid-borne acquired resistance mechanisms.

Chromosome-mediated mechanisms are active efflux, decreased influx, target site alteration, and target bypass (Fig. 9.1). Transmembrane efflux pumps reduce intracellular antibiotic accumulation by exporting them out of bacterial cells, while downregulating porin proteins on bacterial outer membranes decreases the influx of antibiotics. Mutations alter the structures of target proteins or enzymes, reducing the binding of antibiotics. Bacteria can bypass the original targets of antibiotics by synthesising new proteins and continue their metabolic activities without any interference from antibiotics.

Plasmid-borne traits rapidly transmit among bacterial populations, which include enzymatic inactivation of antibiotics by either hydrolysis of functional groups or chemical modifications and protection of antibiotic target proteins, inhibiting their binding (Fig. 9.1). Plasmid-borne mechanisms are mostly investigated in zoonotic ARB. Some of the major antibiotic resistance mechanisms are described.

β-Lactamases

β-lactam antibiotics (e.g. penicillins, cephalosporins, carbapenems, and monobactams) inhibit peptidoglycan cross-linking by targeting penicillin-binding proteins (PBPs), resulting in cell lysis. In response, bacteria resist β-lactams

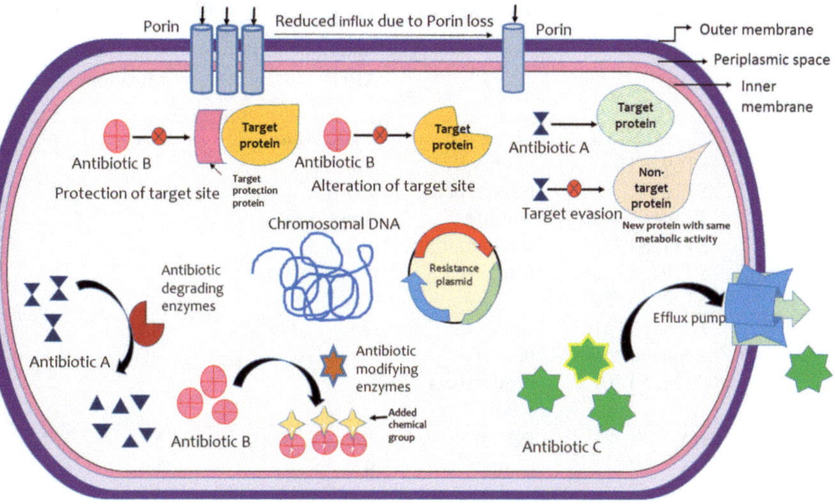

Fig. 9.1 Resistance mechanisms by bacteria to evade the action of antibiotics. (Image source: Authors' personal collection)

by producing β-lactamases, such as AmpC cephalosporinases (*bla*$_{CMY}$), ESBLs (*bla*$_{TEM}$, *bla*$_{SHV}$, and *bla*$_{CTX}$), oxacillinases (*bla*$_{OXA-1}$), and carbapenemases (*bla*$_{NDM}$, *bla*$_{IMP}$, *bla*$_{VIM}$, and *bla*$_{OXA-48}$). β-lactamases reside in the periplasm and cleave the β-lactam ring before it reaches the target. The Ambler classification categorises β-lactamases into four classes (A, B, C, and D). Ambler classes A (ESBLs), C (AmpC cephalosporinases), and D (oxacillinases and carbapenemases) enzymes have serine in their active site. Class B comprises metallo-β-lactamases (MBLs: carbapenemases) having Zn^{2+} ions in their active site. Other mechanisms, such as the modification of PBPs and changes in membrane permeability, are observed in β-lactam-resistant bacteria but are rarely investigated in zoonotic pathogens.

In poultry and livestock, cephalosporins are only used therapeutically and not as growth promoters (Table 9.2). ESBLs and AmpCs are prevalent in Gram-negative bacteria present as gut colonisers or as pathogens causing infections in animals. ESBLs hydrolyse extended-spectrum cephalosporins (e.g. third-generation cephalosporins) and monobactams (aztreonam) and are inhibited by β-lactamase inhibitors, particularly clavulanic acid. The primary ESBLs that spread in animals are *bla*$_{CTX-M-1}$, *bla*$_{CTX-M-14}$, *bla*$_{CTX-M-9}$, and *bla*$_{CTX-M-15}$ (Table 9.3). The IS element IS*Ecp1* has been linked with these genes. Variants such as IS*Ecp1D*-IS*10*, IS*Ecp1*/IS*26*, IS*Ecp1*-IS*903*, and IS*Ecp1*-IS5 have also been reported. In animals, the spread of this gene has been primarily attributed to IncI1, IncF, and IncF1B-type plasmids. However, Inc. K, IncN, IncP, and IncY have also been reported. AmpC cephalosporinases (such as *bla*$_{CMY}$) inactivate first- and second-generation cephalosporins. ESBL inhibitors have no effect on AmpC cephalosporinase activity. In *E. coli*, the resistance is often caused by the overexpression of the chromosomal AmpC gene due to a highly conserved mutation in the promoter/attenuator region.

Filth flies, which reside in the vicinity of animal farms, as well as urban, semi-urban, and rural solid waste collection pits, consume animal and human faeces. They are efficient vectors of AMR pathogens and a hotspot for horizontal gene transfer between different genera. Studies also suggest that filth flies and/or houseflies are responsible for transferring multidrug-resistant (MDR) bacteria from the hospital environment to the community. This problem is acute in lower-middle-income countries and lower-income countries (Fig. 9.2). The prevalence of ESBL-producing *E. coli* in Nigeria, Ethiopia, and Zambia was reported to be 0.8%, 6%, and 13.4%, respectively. The prevalence of the same in highly

Table 9.3 Antimicrobial resistance in the WHO bacterial priority pathogens originating from animals

Pathogen	Clone	Host	Antibiotic resistance mechanisms		
			ARG[a]	Mutations	Efflux activity
Critical priority					
3GCR-E[b]	*E. coli*: ST10, ST69, ST93, ST167 *K. pneumoniae*: ST11, ST37, ST107, ST268, ST274, ST307	Livestock, poultry, broiler, pets, wildlife, sea(fish), filth flies	ESBLs (*bla*$_{TEM-1}$, *bla*$_{SHV-2/12}$, *bla*$_{CTX-M-1/14/15}$); AmpC (*bla*$_{CMY-2}$, *bla*$_{MIR-1}$); penicillinase (*bla*$_{OXA-1}$)	NA	NA
CRE[b]	*E. coli*: ST410 *K. pneumoniae*: ST11, ST307	Livestock, poultry, broiler, pets, wildlife	*bla*$_{NDM-5}$, *bla*$_{VIM-2}$, *bla*$_{IMP-1}$, *bla*$_{OXA-48/181}$	NA	NA
CRAB	NA	Pets, fish	*bla*$_{OXA-23}$, *bla*$_{OXA-58}$	NA	NA
Rifampicin-resistant MBTC[c]	NA	Cattle	NA	NA	*mtrA*, *efpA*
High priority					
Fluoroquinolone--resistant NTS[d]	*Salmonella* Kentucky: ST198	Livestock, poultry, broiler, pets	*qnrB*, *qnrS*, *qnrD*, *aac(6′)-Ib-cr*	*gyrA* (Ser83Glu/Tyr, Ser84Leu, Asp124Glu/Gly); *parC* (Thr57Ser)	NA
VRE	ST16, ST17, ST132	Pets, livestock, broiler	*vanA*, *vanB*	NA	NA
MRSA	ST9 and ST398	Cattle, wildlife, filth flies	*mecA*, *mecC*	NA	NA
CRPsA	ST233	Pets	*bla*$_{VIM-2}$	NA	NA
Others					
Fluoroquinolone--resistant *Campylobacter* spp.[e]	ST21, ST206, ST353, ST354, ST464	Livestock, poultry, broiler	NA	*gyrA* (Thr86Ile)	*cmeABC*
Aminoglycoside--resistant Enterobacterales[b]	NA	Livestock, poultry	*aac(3)-IIa/IV*, *aph(3″)-Ib*, *aph(3′)-Ia*, *aph(6)-Id*, *ant(3″)-Ia*, *strA*, *strB*, *aadA2*	NA	NA
Colistin-resistant *Escherichia coli* and NTS	*Escherichia coli*: ST10, ST101, ST131, ST167, ST354, ST410 NTS: ST14 and ST34	Pigs, broiler	*mcr-1*	NA	NA
MLS-resistant *S. aureus*	ST9 and ST398	Pigs	*lnuB*, *ermA-ermC*, *vgaA*, *vgaE*	NA	NA
ML-resistant *Streptococcus suis* and *S. canis* (Group D)	NA	Pigs, pets	*ermB*, *ermT*, *ermR*	NA	NA
Linezolid--chloramphenicol--resistant *S. aureus*	ST9 and ST398	Pigs	*cfrC*	NA	NA

(continued)

Table 9.3 (continued)

Pathogen	Clone	Host	Antibiotic resistance mechanisms		
Critical priority			ARG[a]	Mutations	Efflux activity
chloramphenicol--resistant organisms	NA	Livestock, poultry	*catA*, *catB*	NA	*cmlA*, *floR*
Trimethoprim--sulfamethoxazole--resistant organisms	NA	Livestock, poultry	*sul1*, *sul2*, *sul3*, *dfrA*, *dfrG*, *dfrK*	NA	NA
Tetracycline-resistant organisms	NA	Livestock, poultry, broiler, pets	*tetM*, *tetO*, *tetT*	NA	*tetA*, *tetB*, *tetC*, *tetD*, *tetK*, *tetL*

Abbreviations: *ARG* antibiotic resistance gene, *3GCR-E* third-generation cephalosporin–resistant Enterobacterales, *CRAB* carbapenem-resistant *Acinetobacter baumannii*, *CRE* carbapenem-resistant Enterobacterales, *CRPsA* carbapenem-resistant *Pseudomonas aeruginosa*, *Gly* glycine, *Ile* isoleucine, *Leu* leucine, *MRSA* methicillin-resistant *Staphylococcus aureus*, *MBTC Mycobacterium tuberculosis* complex, *MLS* macrolide–lincosamide–streptogramin, *NA* not available, *NTS* non-typhoidal *Salmonella*, *Ser* serine, *Thr* threonine, *VRE* vancomycin-resistant enterococci

[a]Major variants of ARGs are represented

[b]Enterobacterales mainly includes *Escherichia coli* and NTS and, rarely, *Klebsiella* spp.

[c]MBTC mainly includes *Mycobacterium bovis*

[d]Major *Salmonella* serovars include Typhimurium, Derby, Weltevreden, Gallinarum, Kentucky, Sofia, and Rissen

[e]Major species are *Campylobacter jejuni* and *Campylobacter coli*

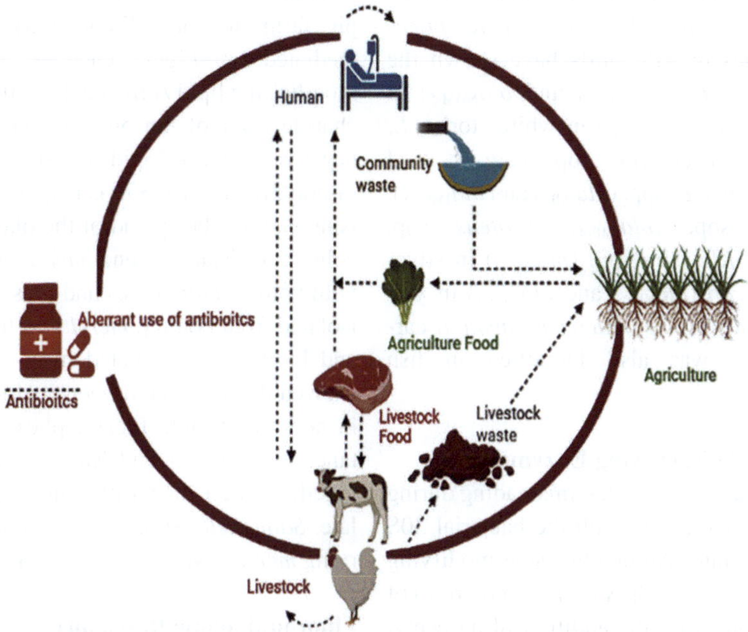

Fig. 9.2 Multidisciplinary approach through One Health to protect the health of humans, animals, plants, and ecosystems. (Image source: Aslam B, Khurshid M, Arshad MI, Muzammil S, Rasool M, et al. Antibiotic resistance: one health one world outlook. Front Cell Infect Microbiol. 2021;11:771510. https://doi.org/10.3389/fcimb.2021.771510. Open Access Article License)

populated countries like China and India was 3% and 11%, respectively.

Carbapenems (e.g. meropenem, ertapenem, and imipenem) are used as last-resort drugs to treat severe ESBL-producing bacterial infections in humans. Notably, it is not used in animals (Table 9.2). Carbapenemase-producing organisms that are resistant to carbapenems are spread to animals due to anthropogenic activities and inactivate all β-lactams except monobactams. Different alleles of bla_{NDM} and/or bla_{OXA-48} carbapenemases are commonly noted through AMR surveillance in food-producing animals. Carbapenemase-producing organisms of critical priority—*E. coli* (bla_{NDM-1}, bla_{NDM-4}, bla_{NDM-5}, bla_{NDM-7}, bla_{NDM-9}, and bla_{NDM-17}), *E. cloacae* (bla_{NDM}), *K. pneumoniae* (bla_{NDM}), *K. oxytoca* (bla_{NDM}), and *Salmonella* spp. (bla_{VIM-1})—have been identified in chicken carcasses or poultry meat (Table 9.3). In bovine meat and milk, the predominant carbapenem-resistant organisms were *E. coli* (bla_{NDM-5}, bla_{OXA-48}, $bla_{OXA-181}$, and bla_{VIM-1}), *K. pneumoniae* (bla_{OXA-48} and bla_{NDM-5}), *Acinetobacter baumannii* (bla_{NDM-1}, bla_{NDM-5}, and bla_{OXA-23}), and *Salmonella* spp.

In wild animals and birds, carbapenem-resistant *Escherichia fergusonii* was identified in baboons. Wild boars in Algeria were infected with bla_{OXA-48}-positive *E. coli* and *K. pneumoniae.* Studies in wild birds have shown the presence of *E. coli* (bla_{NDM-1} and bla_{NDM-5}) in swallows; *E. coli* (bla_{OXA-48}) in white storks; *E. coli* (bla_{IMP-4}), *Klebsiella* spp. (bla_{IMP-4} and bla_{IMP-26}), *Citrobacter* spp. (bla_{IMP-4} and bla_{IMP-38}), *Enterobacter* spp. (bla_{IMP-4}), *Proteus* spp. (bla_{IMP-4}), and *Kluyvera* spp. (bla_{IMP-4}) in silver gulls; and *E. coli* (bla_{KPC-2} and bla_{VIM-1}) in yellow-legged-gulls. *Acinetobacter baumannii* carrying bla_{OXA-23} was also identified in fish (Table 9.3).

Aminoglycoside-Modifying Enzymes

Aminoglycosides cause codon misreading during translation by interacting with the bacterial 30S ribosomal subunit. Aminoglycoside-modifying enzymes modify the hydroxyl or amino group of the antimicrobial with the addition of an acetyl group [aminoglycoside acetyltransferases (AACs)], a phosphate group [aminoglycoside phosphotransferases (APHs)], and a nucleotide [aminoglycoside nucleotidyltransferases (ANTs)]. Genes encoding these enzymes (*aac(3)-IIa/IV*, *aph(3")-Ib/strA*, *aph(3')-Ia*, *aph(6)-Id/strB*, *ant(3")-Ia*, and *aadA*) are commonly found in the *Enterobacterales* isolated from poultry and livestock (Table 9.3). The bifunctional AAC(6')-APH(2") enzyme, found on the common Tn*4001* transposon, is also responsible for significant levels of gentamicin resistance found in strains of MRSA and VRE around the globe.

Colistin Resistance

Colistin, a cationic polypeptide antibiotic of the polymyxin family, disrupts cell membrane permeability by electrostatic interaction between positively charged amino acids of colistin and the negatively charged phosphate group of lipopolysaccharide (LPS) lipid A subunit. The emergence of carbapenem-resistant organisms has reintroduced colistin as a last-resort drug for treating human infections. Long-time indiscriminate use of colistin in animal husbandry has reduced the clinical efficacy of this antibiotic (Table 9.2). Colistin-resistant *E. coli* was first identified at a pig farm in China (Table 9.3). The plasmid-mediated *mcr-1* gene encodes an enzyme that modifies the lipid A molecule of the lipopolysaccharide layer of the outer membrane of Gram-negative bacteria. Lipid A is the target molecule of polymyxins. Tn*6330* is responsible for mobilising *mcr-1*. The spread of the plasmid-mediated colistin resistance gene *mcr-1* was associated with both *E. coli* clones and plasmids. In *E. coli* isolated from poultry, *mcr-1* was found in IncHI2 and IncP. *E. coli* isolated from calves in Spain showed the presence of *mcr-1*, along with *mcr-3*, in non-mobilisable IncHI2 plasmids. The resistance genes *mcr-1* and $bla_{OXA-181}$ were identified together on an IncX4 plasmid in an *E. coli* isolate. *Salmonella* spp. and *Enterobacter* spp. carrying *mcr-1* have also been reported.

Fluoroquinolone Resistance

Fluoroquinolones block DNA gyrase and topoisomerase IV, two enzymes required for bacterial DNA replication. Enrofloxacin usage in chicken production has resulted in the development of fluoroquinolone-resistant *C. jejuni* in poultry. Chromosomal target site mutations in quinolone resistance–determining regions (QRDRs) of gyrase (*gyrA* and *gyrB*) and/or topoisomerase IV (*parC* and *parE*) lead to fluoroquinolone resistance (Table 9.3). Plasmid-mediated quinolone resistance (PMQR) genes (*qnrA/B/S/D*) protect the target sites so that the bacteria can develop the required mutations in the QRDRs to acquire clinical resistance. Efflux pumps (e.g. *cmeABC*) also contribute to fluoroquinolone resistance.

Macrolide–Lincosamide–Streptogramin (MLS) Resistance

MLS are antibiotic classes that prevent protein synthesis by attaching to the bacterial 50S ribosomal subunit. Three different mechanisms cause resistance to these antibiotics. These include (1) target site modification via methylation and/or mutation within the 23S rRNA gene, encoded by *ermA-ermC* genes; (2) enzymatic inactivation of MLS antibiotics by erythromycin esterases (*ereA/B*) genes and lincosamide nucleotidyltransferase (*lnuE*); and (3) efflux pumps that actively pump the MLS out of the cell, reducing their intracellular concentrations (encoded by *msr* genes). The first and second mechanisms have been reported in staphylococcal and streptococcal strains isolated from animals, especially from pigs (Table 9.3). Notably, *vgaA* and *vgaE*, which belong to the ATP-binding cassette (ABC)-F family, are found in *Staphylococcus aureus* and mediate resistance to lincosamides, streptogramin A, and pleuromutilins by displacing them from the ribosome (Table 9.3).

Vancomycin Resistance

Vancomycin, a last-resort glycopeptide antibiotic, inhibits Gram-positive bacterial cell wall synthesis by interfering with the cross-linking of peptidoglycan. High-priority VRE pathogens with their major lineages, ST16, ST17, and ST132, are recovered from food-producing and companion animals (Table 9.3). The *vanA* and *vanB* genes encode enzymes that prevent the binding of glycopeptide antibiotics to the bacterial cell wall by modifying the peptidoglycan precursors. Interestingly, these genes are part of transposons (e.g. Tn*1546* and Tn*1549*) residing in either plasmids or chromosomes.

Chloramphenicol Resistance

Chloramphenicol blocks bacterial transpeptidation during protein synthesis by binding to the peptidyl transferase cavity of the 30S ribosomal subunit. Chloramphenicol resistance is mediated by various mechanisms, including enzyme inactivation and active efflux (Table 9.3). The *catA/B gene*, which encodes chloramphenicol acetyltransferase, is commonly found in food animal–derived *E. coli* and *Salmonella* spp. and prevents the binding of chloramphenicol to the 50S ribosomal subunit by acetylation. The *cmlA* and *floR*, which encode efflux pumps, actively expel chloramphenicol and/or its analogue, florfenicol, respectively, from the bacterial cell, reducing their effective intracellular concentration. These genes are often linked to mobile DNA segments (e.g. Tn*9*, Tn*1696*, Tn*2000*, and Tn*2670*), either in chromosomes or plasmids, contributing to the rapid dissemination of chloramphenicol resistance in animal populations.

Trimethoprim–Sulfamethoxazole Resistance

Trimethoprim and sulfamethoxazole elicit bactericidal action by blocking the folic acid metabolic pathway, which provides precursors of nucleic acid synthesis. Alterations in dihydropteroate synthase (encoded by *sul*) and/or dihydrofolate reductase (encoded by *dfr*) target enzymes hinder the binding of sulfamethoxazole and/or trimethoprim, respectively. The presence of these altered genes in MGEs facilitates their spread under antibiotic stress. *sul1/2/3* and *dfrA/G/K* genes are noted in food-producing animals (Table 9.3).

Tetracycline Resistance

Tetracyclines prevent the attachment of aminoacyl-tRNA to the 30S ribosomal subunit and inhibit bacterial protein synthesis. Tetracycline-resistant organisms have been detected in livestock, poultry, broilers, and pets, carrying resistance genes like *tetA/B/C/D/K/L* (encoding efflux pumps) and *tetM/O/T* (encoding ribosomal protection proteins) (Table 9.3). These genes are generally mobilised through plasmids and transposons.

Zoonotic Resistance Factors

Emerging zoonotic pathogens considerably enhance the worldwide burden of AMR due to the combined effects of multiple factors. Some of these factors are mentioned here:

Community Unawareness Common people, particularly animal handlers, are unaware of the growing threat of AMR and its relationship with bacterial zoonoses due to a lack of community awareness programmes. As a result, they are less concerned with the prudent use of antibiotics in animals for infection treatment and prevention, unknowingly raising the burden of ARB.

High Demand for Animal Proteins Over the past 50 years, the total consumption and production of animal meat has increased across the globe. Livestock and poultry industries are rapidly growing to meet the high demand for animal proteins in overpopulated developing countries, where almost 70% of antibiotics are sold only for the prevention of infections and growth promotion in farm animals. Extensive antibiotic use as a growth promoter to improve the livestock and poultry biomass turns the farms into potential reservoirs of ARB.

Irrational Use of Antibiotics Antibiotics are mis- and overused to treat animal infections without a prescription (Fig. 9.2). Prophylactic and metaphylactic use of broad-spectrum antibiotics, even as growth promoters, in food-producing animals causes an imbalance among the gut-commensal microbes, known as *dysbiosis*, resulting in selective proliferation of ARB. The use of substandard, expired, or falsified veterinary antibiotics often leads to treatment failure and case fatality, aggravating the AMR situation. Over-the-counter availability and oral administration of broad-spectrum antibiotics lead to their rampant use.

Lack of Antimicrobial Stewardship Prescription and administration of antibiotics without any evidence of infections by veterinarians promote AMR in zoonotic pathogens by compromising antimicrobial stewardship (AMS). Poor healthcare infrastructure, lack of rapid diagnostics, and commercial promotion by veterinarians also hinder AMS, especially in LMICs.

Transferrable Antimicrobial Resistance Genes Rapid spread of the zoonotic ARB occurs by HGT of MGEs. The MGEs carry one or more ARGs conferring co-resistance, such as bla_{TEM}, bla_{SHV}, and $bla_{\mathrm{CTX\text{-}M}}$ for ESBL-*E. coli*; *mecA* for MRSA; and *mcr* for colistin resistance.

Human–Animal Proximity Deforestation, consumption of wildlife meat, animal domestication, and farming expose humans to a wide range of zoonotic pathogens, including ARB. Human–animal contact also reverts the pathogens to animals from humans, in a process known as *reverse zoonosis*, which contributes to the long-term survival of ARB via a vicious cycle.

Lack of Surveillance and Vigilance Finally, there is a dearth of properly designed monitoring systems with respect to the genetic changes in bacteria, sanitation, sewage treatment, environmental pollution, climate change, and, most importantly, the usage of antibiotics and their dosage, which adds to the AMR of the bacterial pathogens.

Prevention of AMR in Zoonotic Bacteria

Combating the increasing trend of AMR in zoonotic bacteria requires multimodal approaches, ranging from social vaccines to high-throughput genomics. Some of the approaches are described. Veterinary Use of Antibiotics In 2018, the WHO produced a list of critically important antimicrobials for human medicine in collaboration with the Food and Agriculture Organization (FAO) of the United Nations and the WOAH, in response to the global and animal health issues raised by AMR. It has classified 35 classes of antibiotics as critically important (aminoglycosides, broad-spectrum cephalosporins, carbapenems, glycopeptides, polymyxins, quinolones, and others), highly important (amphenicols, narrow-spectrum cephalosporins, lincosamides, penicillins, tetracyclines, and others), and important (nitrofurans, nitroimidazoles, and others). Antibiotics can be used prophylactically or metaphylactically when there is evidence of infection in other livestock in the same herd, flock, or fish population.

Antimicrobial Stewardship for Veterinarians The American Veterinary Medical Association (AVMA) defines antimicrobial stewardship (AMS) as "the actions veterinarians take individually and as a profession to preserve the effectiveness and availability of antimicrobial drugs through conscientious oversight and responsible medical decision-making while safeguarding animal, public, and environmental health." Continuing medical education promotes the improvement of AMS among veterinary practitioners.

Infection Prevention and Control Practices Infection prevention and control (IPC) are critical in zoonoses because they have the potential to produce worldwide pandemics, which require effective surveillance, early identification, sound governance, and prompt action. These veterinarian services ensure food production, safety, and security, benefitting animal welfare and public health. High-quality vaccinations are effective in preventing infections. The Water, Air, Sanitation, and Hygiene (WASH) intervention programme could strengthen the IPC practices. The WHO, in partnership with the FAO and the WOAH, has released guidelines for the prevention and management of terrestrial and aquatic zoonoses.

Rapid Molecular Diagnostics Commercial and laboratory-developed polymerase chain reaction (PCR) and loop-mediated isothermal amplification (LAMP) rapidly and accurately detect zoonotic pathogens and ARGs in animal samples and products, facilitating epidemiological surveillance and outbreak investigation of zoonotic infections. Major ESBLs (bla_{TEM}, bla_{SHV}, and $bla_{CTX\text{-}M}$), carbapenemases (bla_{KPC}, bla_{NDM}, and $bla_{OXA\text{-}48}$), colistin resistance (*mcr-1*), MRSA (*mecA*), and VRE (*vanA/vanB*) are selected as the AMR diagnostic biomarkers.

Alternatives to Antibiotics (ATAs) The use of ATAs might conquer the dilemma of AMR, improving animal health. Phage therapy targets bacterial pathogens with specific bacteriophages, which is important in MDR infections among food-producing animals. Bacteriocins are produced by probiotic bacteria as antimicrobial peptides with bactericidal or bacteriostatic activity on related bacterial strains extensively used in food preservation. Nisin, an antibiotic family bacteriocin synthesised by *Lactococcus* and *Streptococcus* spp., is used in Gram-positive and Gram-negative zoonotic bacterial pathogens like MRSA, VRE, and ESBL-producing Enterobacterales. Nano-biotics, the next-generation antibiotics, are chemically or naturally derived nanoparticle entities of nanometric size that elicit antimicrobial properties against a broad range of bacteria, preferably by inducing free radical generation. Anti-sense antimicrobial therapeutics include nucleic acid analogues that silence gene expression by binding to complementary RNA of the target organism. It has been tested against several ARBs, including MDR–*Salmonella* spp. and ESBL- and bla_{NDM}-producing *Enterobacterales*.

Microbiota Modulation Probiotics, faecal microbiota transplantation (FMT), and enhanced food additives have all been researched in animal health to reduce the load of intestinal ARB. Intestinal microbiota is regulated by dietary and other factors, like age, stress, medication, and the infection of animals. Advanced genomics may guide the understanding of the dynamics of host–microbe interactions in the success of FMT. Dietary supplementation with *Bacillus subtilis* has been found to increase the frequency of beneficial bacteria, such as Lactobacillaceae and Bifidobacteriaceae, among the intestines of enterotoxigenic *E. coli*-infected weaned pigs.

Case Study

In 1998, a 12-year-old boy, a cattle farmer's son, reported acute abdominal discomfort with fever, followed by diarrhoea. During that time, a few calves on the farm died due to severe diarrhoea following treatment failure with ceftiofur, a third-generation cephalosporin. Stool cultures of both the patient and calves recovered *Salmonella enterica* serovar Typhimurium (STM) isolates that resisted ampicillin, ceftriaxone, chloramphenicol, kanamycin, trimethoprim–sulfamethoxazole, tetracycline, and streptomycin. Molecular characterisation by PCR and Sanger sequencing detected $bla_{\text{TEM-1}}$ (a penicillinase) and $bla_{\text{CMY-2}}$ (a cephamycinase) genes as the mechanisms of resistance to penicillin and broad-spectrum cephalosporins, respectively. Pulsed-field gel electrophoresis (PFGE), a molecular typing method, showed identical genetic fingerprints of the STM isolates designated as clonal. The study concluded that the local salmonellosis outbreak was due to zoonotic transmission of ceftriaxone-resistant STM from infected calves to the child.

1. Which third-generation cephalosporins have applications in both veterinary and human medicine?
2. What specific antibiotic resistance mechanisms were identified in the STM isolates, and how did these mechanisms contribute to the observed treatment failure with ceftiofur in the calves and the ineffective treatment of the child's infection?
3. How did the researchers establish the zoonotic transmission of ceftriaxone-resistant STM from the calves to the child, and what molecular typing method was used to confirm the clonal relationship between the isolates?
4. Considering the context of the case study, what implications does this outbreak have for the use of third-generation cephalosporins like ceftiofur in veterinary medicine, and what potential measures could be implemented to prevent similar zoonotic transmission events in the future?

Research Questions

1. What is the significance of horizontal gene transmission in AMR?
2. What is the diagnostic advancement in the rapid detection of antibiotic residues and resistant bacteria originating from food-producing animals and food chains?
3. What is the significance of surveillance, infection prevention, and antimicrobial stewardship in AMR-associated zoonoses?
4. What is AMR's "global tricycle surveillance system"? Why is the "One Health" approach prioritised to battle the threat of AMR?

Further Readings

Aslam B, Khurshid M, Arshad MI, Muzammil S, Rasool M, et al. Antibiotic resistance: one health one world outlook. Front Cell Infect Microbiol. 2021;11:771510. https://doi.org/10.3389/fcimb.2021.771510.

Biswas U, Das S, Barik M, Mallick A. Situation report on *mcr*-carrying colistin-resistant clones of *Enterobacterales*: a global update through human-animal-environment interfaces. Curr Microbiol. 2024;81(1):12. https://doi.org/10.1007/s00284-023-03521-8.

Caneschi A, Bardhi A, Barbarossa A, Zaghini A. The use of antibiotics and antimicrobial resistance in veterinary medicine, a complex phenomenon: a narrative review. Antibiotics (Basel). 2023;12(3):487. https://doi.org/10.3390/antibiotics12030487.

Cheng G, Fang L-x, Li X, Wu Z, Feng J, et al. Editorial: Antimicrobial resistance in zoonotic bacteria. Front Microbiol. 2024;15:1362380. https://doi.org/10.3389/978-2-8325-4413-6.

Collignon PC, Conly JM, Andremont A, McEwen SA, Aidara-Kane A, et al. The World Health Organization ranking of antimicrobials according to their importance in human medicine: a critical step for developing risk management strategies to control antimicrobial resistance from food animal production. Clin Infect Dis. 2016;63(8):1087–93. https://doi.org/10.1093/cid/ciw475.

Dandachi I, Chabou S, Daoud Z, Rolain J-M. Prevalence, and emergence of extended-spectrum cephalosporin-, carbapenem- and colistin-resistant Gram-negative bacteria of animal origin in the Mediterranean Basin. Front Microbiol. 2018;9:2299. https://doi.org/10.3389/fmicb.2018.02299.

Darby EM, Trampari E, Siasat P, Gaya MS, Alav I, et al. Molecular mechanisms of antibiotic resistance revisited. Nat Rev Microbiol. 2023;21:280–95. https://doi.org/10.1038/s41579-022-00820-y.

EFSA (European Food Safety Authority) and ECDC (European Centre for Disease Prevention and Control). The European Union summary report on antimicrobial resistance in zoonotic and indicator bacteria from humans, animals, and food in 2021–2022. EFSA J. 2024;22(2):e8583. https://doi.org/10.2903/j.efsa.2024.8583. Accessed 15 Aug 2024.

Olaru ID, Walther B, Schaumburg F. Zoonotic sources and the spread of antimicrobial resistance from the perspective of low and middle-income countries. Infect Dis Poverty. 2023;12:59. https://doi.org/10.1186/s40249-023-01113-z.

WHO, FAO, and OIE. Antimicrobial resistance and the United Nations sustainable development cooperation framework: guidance for United Nations country teams. 2021. Available from: https://www.woah.org/app/uploads/2021/10/unsdcf-amr-guidance-en-final-approved.pdf. Accessed 15 Aug 2024.

World Health Organization, Food and Agriculture Organization of the United Nations, United Nations Environment Programme and World Organisation for Animal Health. A one health priority research agenda for antimicrobial resistance. Geneva; 2023. Available from: https://www.who.int/publications/i/item/9789240075924. Accessed 15 Aug 2024.

10 Prevention, Surveillance, and Control of Bacterial Zoonoses

Deepthi Vijay and Pankaj Dhaka

Learning Objectives

1. To describe key principles of zoonotic disease prevention, including reservoir control, transmission interruption, and host protection, and apply them to bacterial zoonoses
2. To explain the components of surveillance systems for bacterial zoonoses, comparing traditional approaches with advanced surveillance tools
3. To discuss integrated control strategies for bacterial zoonoses using a One Health approach for effective prevention and control measures

Introduction

Zoonoses are infectious diseases naturally transmitted between vertebrate animals and humans. Among these, bacterial zoonoses, such as brucellosis, anthrax, salmonellosis, campylobacteriosis, listeriosis, leptospirosis, zoonotic tuberculosis (TB), and Q fever, continue to significantly impact global health and economies. More than 75% of the emerging infectious disease pathogens in recent decades have been zoonotic, with the majority originating from wildlife or domestic animal reservoirs. Each year, zoonotic diseases are responsible for billions of human infections and millions of deaths worldwide, alongside substantial losses in livestock productivity and trade.

Multiple drivers cause the re-emergence of zoonoses, including anthropogenic and environmental changes. Expansion of agriculture, deforestation, and urbanization increase contact between humans, livestock, and wildlife, elevating spillover risks. Climate change and ecosystem disruption further alter pathogen transmission dynamics. Globalization via travel and trade enables localized outbreaks to quickly disseminate internationally. Furthermore, the intensification of livestock production and the rise of antimicrobial resistance (AMR) create new challenges in managing bacterial zoonotic threats. These complex processes mean that no single sector can effectively safeguard health in isolation. Recognizing the interdependence of human, animal, and environmental health, the One Health approach has gained traction as a collaborative framework for combating zoonoses. One Health aims to (i) more efficiently prevent zoonotic diseases at their sources, (ii) improve early detection through coordinated surveillance, and (iii) implement control measures that benefit all sectors.

D. Vijay
Centre for One Health, College of Veterinary Science, Guru Angad Dev Veterinary and Animal Sciences University, Ludhiana, Punjab, India

P. Dhaka (✉)
Centre for One Health, College of Veterinary Science, Guru Angad Dev Veterinary and Animal Sciences University, Ludhiana, Punjab, India

S. C. Parija, A. Poddar (eds.), *Textbook of Bacterial Zoonoses*, Microbial Zoonoses,
https://doi.org/10.1007/978-981-95-6615-0_10

Principles of Zoonotic Disease Prevention and Control

Prevention of bacterial zoonoses requires breaking the chain of infection at critical points, namely (a) the pathogen's animal reservoir, (b) the transmission route, and (c) the susceptible human or animal host. This can be categorized into three major strategies, i.e. (a) reservoir neutralization, (b) reducing contact potential, and (c) increasing host resistance. In addition, supportive measures like public education and policy are essential to ensure that these strategies are effectively implemented.

Reservoir Neutralization

The first principle is to remove or treat the infection at its source, thereby neutralizing the reservoir of the pathogen. In practical terms, this often means identifying and managing infected animals that serve as reservoirs for human infection. For bacterial zoonoses in livestock, this may involve test-and-slaughter programmes to cull positive animals, which is especially effective when a limited number of species are responsible for transmission. For example, bovine tuberculosis and brucellosis control programmes have relied on testing herds and eliminating infected bovines to reduce the animal reservoir. In some settings, mass treatment of animals with antibiotics may be employed to rapidly reduce infection burden within a herd or population, particularly during outbreaks where individual diagnosis are impractical or unavailable. However, such an approach needs to be used judiciously and only under veterinary guidance, as indiscriminate antibiotic use can contribute to the development and spread of AMR, potentially compromising both animal and human health. Environmental interventions are another form of reservoir control. For instance, safely disposing of anthrax-infected carcasses or decontaminating soil removes persistent sources of infection in the environment.

Reducing Contact Potential

Interrupting the transmission pathways of bacterial zoonoses requires the implementation of evidence-based interventions aimed at minimizing opportunities for contact between animal reservoirs and human populations. The underlying principle is to break the chain of transmission at critical control points through a combination of hygiene, environmental management, and animal containment strategies. Some of the important strategies in reducing contact potential include (i) sanitation and personal hygiene, (ii) comprehensive farm biosecurity, and (iii) targeted stray animal and vector control, with overlapping domains that reinforce one another.

Sanitation and Personal Hygiene Measures

Sanitation and personal hygiene represent first-line defences against zoonotic disease transmission, particularly in high-risk occupational settings such as farms, slaughterhouses, and veterinary clinics. Consistent hand hygiene, through soap and water or alcohol-based hand rubs, has proven efficacy in preventing the transmission of pathogens. The correct and consistent use of personal protective equipment (PPE), including gloves, gowns, masks, and eye protection, is crucial for preventing exposure to aerosolized or contact-transmitted pathogens such as *Brucella abortus*, *Coxiella burnetii*, and *Streptococcus suis*. Environmental hygiene, including routine disinfection of equipment and premises, vector-proof waste disposal, and maintenance of clean water supplies, further mitigates indirect transmission risks.

Farm Biosecurity Measures

Farm biosecurity refers to the essential practices and measures implemented on farms to prevent the introduction, spread, and exit of infectious diseases. While primarily protecting the health of animals and the animal handlers, farm biosecurity also ensures food safety and safeguards the economic viability of agricultural operations.

A comprehensive farm biosecurity plan typically encompasses several key areas, with physical control measures representing the first line of defence.

Physical Control Strategies Include isolation and traffic control, especially when introducing new animals into the herds and during an outbreak. Temporary zoning practices, such as establishing infected, buffer, and surveillance zones, are applied to contain the spread, which are essential to contain outbreaks. At the same time, keeping out rodents, insects, and wild animals through integrated pest management strategies (chemicals, traps, fencing, etc.) further prevents unwanted introduction or spread of disease.

Animal Health Management In addition, *animal health management* is another major component that includes clinical surveillance and diagnostic screening. This enables timely identification of pathogens, herd vaccination when available, and prompt reporting of any unusual signs of illness to a veterinarian. It also involves working collaboratively with veterinary professionals to develop and implement tailored herd or flock health plans, including vaccination programmes.

Stray Animal and Vector Management Stray animals and disease-carrying vectors exacerbate zoonotic risk. The transmission becomes a higher risk in densely populated regions and places where animals intimately share space with humans. To reduce the risks of such transmission, focused control measures are essential. For example, the stray dog and cat population is managed through catch–neuter–vaccinate–release (CNVR) programmes to reduce reservoirs of *Brucella canis*, *Leptospira* spp., *Salmonella* spp., etc. Rodent control measures use integrated pest management approaches to prevent *Leptospira* spp., *Yersinia pestis*, and *Streptobacillus moniliformis* transmission. Vector control programmes combine insecticides, biological agents, and habitat modification to reduce exposure to ticks, fleas, and biting flies.

Increasing Host Resistance Strengthening host resistance is a core principle in the prevention of bacterial zoonoses. It aims to protect susceptible individuals so that even upon exposure, infection or disease can be averted or significantly mitigated. This can be achieved through immunization, chemoprophylaxis, and other host-level interventions.

Vaccination and Pre-/Post-exposure Prophylaxis Vaccination remains a critical preventive strategy. Immunizing animal reservoirs, when possible, not only safeguards animal health but also reduces the risk of zoonotic transmission to humans. In contrast, human vaccination is generally reserved for individuals at an elevated occupational or environmental risk. All vaccination efforts, whether in animals or humans, need to be supported by continuous surveillance, outbreak preparedness, and robust biosecurity measures to ensure sustainable disease control. In addition, the timely antibiotic treatment of infected humans and, when feasible, animals is crucial for reducing disease severity, preventing complications, and limiting further transmission to others, including zoonotic spillover. However, treatment protocols should always be followed under the guidance of qualified medical or veterinary professionals, considering AMR patterns, disease stage, and host-specific considerations.

It is important to note that vaccine availability and implementation vary considerably across countries and regions. Moreover, for several bacterial zoonoses, such as those caused by *Campylobacter*, *Listeria*, and *Bartonella*, no effective vaccines are currently available for either humans or animals. In such cases, disease control relies entirely on early diagnosis, strict hygiene and biosecurity practices, vector control, and appropriate antimicrobial therapy.

Table 10.1 presents a summary of the recommended vaccination and treatment options for major bacterial zoonotic diseases.

Table 10.1 Summary of vaccination and treatment options for major bacterial zoonotic diseases

Disease	Causative agent	Affected species	Vaccine	Treatment options
Anthrax	*Bacillus anthracis*	Humans, livestock, wildlife	Live attenuated (*Sterne strain*) for animals; cell-free toxoid for humans (at-risk groups)	Ciprofloxacin or doxycycline; 60 days post-exposure or 7–10 days for cutaneous cases
Brucellosis	*Brucella* spp.	Humans, cattle, sheep, goats, swine	Live attenuated (Strain 19, RB51 for cattle; Rev-1 for goats/sheep); no human vaccine	Doxycycline + rifampicin for ≥6 weeks; not routinely treated in livestock
Leptospirosis	*Leptospira* spp.	Humans, dogs, livestock	Inactivated bacterin (polyvalent); used in animals and selectively in humans	Doxycycline (mild); penicillin/ceftriaxone (severe); 7–10 days
Plague	*Yersinia pestis*	Humans, rodents, cats	Inactivated and subunit vaccines (limited use); live attenuated in endemic countries	Streptomycin or gentamicin; 10–14 days
Tularaemia	*Francisella tularensis*	Humans, rabbits, rodents	Live attenuated vaccine (LVS), investigational only for laboratory workers	Streptomycin/gentamicin (severe); doxycycline/ciprofloxacin (mild); 10–14 days
Q fever	*Coxiella burnetii*	Humans, cattle, sheep, goats	Inactivated whole-cell vaccine (Q-Vax) for high-risk humans; phase I vaccine for livestock	Doxycycline for 14 days (acute); doxycycline and hydroxychloroquine for 18–24 months (chronic)
Tuberculosis (zoonotic)	*Mycobacterium bovis*	Humans, cattle, wildlife	BCG vaccine for humans; experimental use in animals	Isoniazid, rifampicin, ethambutol, pyrazinamide; 6–12 months
Salmonellosis (non-typhoidal)	*Salmonella* spp.	Humans, poultry, livestock	Live attenuated and inactivated for poultry/livestock; no human vaccine	Supportive care; fluoroquinolones/cephalosporins if invasive
Lyme disease	*Borrelia burgdorferi*	Humans, dogs, wildlife	Subunit vaccines for dogs; human vaccine under development	Doxycycline for 10–21 days (early); IV ceftriaxone for late
Listeriosis	*Listeria monocytogenes*	Humans, ruminants	No vaccine available	Ampicillin and gentamicin or trimethoprim/sulfamethoxazole (TMP-SMX); 2–3 weeks
Psittacosis	*Chlamydia psittaci*	Humans, birds	No licensed vaccine available	Doxycycline for 10–14 days
Cat scratch disease	*Bartonella henselae*	Humans, cats	No vaccine available	Azithromycin (mild); rifampicin and others (severe)
E. coli O157 infection	*Escherichia coli* O157:H7	Humans, cattle	Subunit vaccines for cattle (to reduce shedding); no human vaccine	Supportive care only; antibiotics contraindicated

Other Host-Level Interventions Beyond vaccination and chemoprophylaxis, several additional interventions enhance host resistance to bacterial zoonoses. Genetic selection and breeding programmes aimed at improving disease resistance in livestock have demonstrated considerable promise, particularly for endemic bacterial infections such as bovine tuberculosis and mastitis. The advances in genomics and marker-assisted selection are also enabling the development of resilient animal lines that are less susceptible to colonization or clinical disease.

Nutritional management is another key determinant of immunocompetence. Adequate intake of macro- and micronutrients, especially proteins and vitamins A, D, and E, along with trace minerals like zinc and selenium, plays an essential role in supporting robust immune function in both animals and humans. These interventions are particularly relevant in livestock systems where seasonal feed scarcity, suboptimal feed quality, or high production demands may compromise immune responses.

Reducing physiological and environmental stress is equally important, as chronic stress can suppress immune function and increase vulnerability to infection. Stressors such as overcrowding, heat stress, poor ventilation, and frequent transportation elevate the risk of bacterial shedding and disease transmission. The adoption of animal welfare–oriented practices, including improved housing, herd management, and gentle handling, helps mitigate these effects and strengthen natural defences.

Additionally, hormonal and immunomodulatory interventions are being explored to enhance disease resistance, particularly in intensive production systems. These include the administration of immunostimulants such as cytokines, probiotics, and phytogenic feed additives, which can prime or modulate immune responses.

When integrated with vaccination, biosecurity, and surveillance, these host-level strategies contribute to a comprehensive and sustainable approach to reducing susceptibility to bacterial zoonotic infections and minimizing the impact of disease outbreaks.

Public Education and Policy

Health education is essential to inform farmers, animal workers, and the public about zoonotic disease risks and prevention practices. For example, educating communities about the risks of consuming raw milk and how to safely handle birthing materials from animals can lower brucellosis and Q fever transmission. Training and equipping high-risk professionals such as veterinarians, farmers, and slaughterhouse workers with protective practices are equally important. Additionally, legislative measures and public health policy create an enabling environment for disease prevention. For example, regulations on animal movement, compulsory reporting of certain zoonoses, pasteurization requirements, and intersectoral disease control programmes create robust and uniform risk management practices.

Surveillance Systems for Bacterial Zoonoses

Early detection and monitoring of zoonotic disease events depend on effective surveillance systems. Surveillance is defined as the continuous, systematic collection, analysis, interpretation, and dissemination of health data for use in public health action. In the context of bacterial zoonoses, surveillance spans human health, animal health, and, sometimes, environmental monitoring.

Traditional Surveillance

Indicator-Based Surveillance in Human Health Indicator-based surveillance refers to formal systems that rely on confirmed diagnoses of notifiable diseases. In most countries, human cases of bacterial zoonoses such as anthrax, brucellosis, leptospirosis, and plague are legally notifiable. Clinicians or laboratories are required to report confirmed cases to national public health authorities. These systems allow for trend analysis, outbreak detection, and policy formulation. For example, national disease registries in countries like India include mandatory reporting of anthrax and brucellosis. This approach has strong regulatory backing and is effective for known pathogens. Nevertheless, its performance depends heavily on healthcare access, diagnostic capacity, and provider awareness. In low-resource settings, many zoonotic infections remain undiagnosed or unreported, leading to a substantial underestimation of the disease burden. Moreover, these systems often operate independently of animal health reporting, delaying the recognition of cross-species outbreaks.

Traditional Veterinary Surveillance Systems Veterinary surveillance historically includes inspection at abattoirs, sero-monitoring of livestock, and laboratory-confirmed diagnoses reported to animal health authorities. For example, routine bovine tuberculosis or brucellosis surveillance in cattle herds is performed in many countries through annual tuberculin or Rose Bengal testing. However, such surveillance is often focused on economically important diseases or those with trade implications rather than those of public health concern. Furthermore, livestock surveillance tends to exclude companion animals and wildlife, which can serve as reservoirs for zoonotic pathogens. Lack of integration with public health systems limits the ability to detect zoonotic transmission chains early, and an outbreak response may only be triggered after human cases arise.

Passive and Active Surveillance Approaches Passive surveillance, the mainstay of most traditional systems, involves reliance on stakeholders (e.g. physicians, veterinarians, and farmers) to voluntarily report suspected cases. This method is cost-effective but inherently limited by low sensitivity, especially for emerging or subclinical infections. For instance, brucellosis cases in endemic areas often go unnoticed due to non-specific symptoms and diagnostic delays. In contrast, active surveillance involves proactive efforts by authorities to search for cases or detect pathogens. Examples include targeted screening of livestock for Q fever or serosurveys for leptospirosis in flood-affected areas. While active approaches provide more complete data, they are resource-intensive and thus typically limited to priority diseases or specific geographic zones.

Sentinel Surveillance Systems Sentinel surveillance uses strategically selected sites or populations as early warning indicators. In the context of zoonoses, this may involve routine testing of high-risk animal populations. Testing rodents for *Y. pestis* in plague-endemic areas or stray dogs for leptospiral carriage is one such example. Human sentinel sites, such as specialized hospitals, also play a role by providing detailed clinical and laboratory data for diseases like leptospirosis or zoonotic tuberculosis. The value of sentinel surveillance lies in its capacity to detect localized pathogen circulation before large-scale outbreaks occur. However, the representativeness of sentinel sites is a limitation, especially when disease patterns shift or transmission expands to new ecological niches.

Syndromic Surveillance Syndromic surveillance captures and analyses health indicators, such as symptoms, absenteeism, or pharmaceutical purchases, before formal diagnosis. This approach has gained prominence in bio-defence and early outbreak detection frameworks. For bacterial zoonoses, syndromic surveillance might monitor spikes in "fever of unknown origin" or acute febrile illness in high-risk zones, potentially indicating leptospirosis or brucellosis outbreaks. Although syndromic systems enable faster detection, they lack specificity and often result in false alarms or missed signals for atypical presentations. Integration with laboratory confirmation is essential to translate early signals into actionable public health responses.

Environmental Surveillance Environmental surveillance is an increasingly recognized component of zoonotic disease surveillance, especially for pathogens with environmental persistence or indirect transmission routes. Monitoring of water, soil, and fomites can offer early warning signals of pathogen circulation in the absence of clinical cases. For instance, *Leptospira* spp. can be identified in surface waters in flood-prone or peri-urban regions, offering an early indication of leptospirosis risk following heavy rainfall or sanitation breakdowns. *B. anthracis* is known to persist in soil for decades; targeted soil surveillance in historical outbreak zones can support risk mapping and pre-emptive control. Similarly, *Salmonella* and *E. coli* are routinely monitored in abattoir wastewater, irrigation canals, and food-processing environments as part of food safety and One Health surveillance frameworks.

Vector and reservoir surveillance also fall under this domain. Testing ectoparasites (ticks and fleas) or reservoir hosts (rodents and stray animals) for pathogens such as *Rickettsia* or *Y. pestis* provides an ecological description of the zoonotic risk. Wastewater-based surveillance is emerging as a cross-sectoral platform for detecting zoonotic pathogens shed by humans or animals, although applications for bacterial agents are still evolving.

Despite its promise, environmental surveillance remains underutilized due to technical, logistical, and interpretive challenges. It requires intersectoral collaboration between environmental, veterinary, and public health authorities, alongside standardization of sampling and detection protocols.

Laboratory Surveillance and Reference Networks Laboratories are pivotal to confirm the diagnoses of bacterial zoonoses through techniques such as culture, serology, and polymerase chain reaction (PCR). National and international lab networks play a key role in standardization and rapid detection. For example, the US Laboratory Response Network (LRN) includes sentinel and reference labs that handle high-consequence pathogens like *B. anthracis* and *Brucella* spp. Similarly, PulseNet International enables DNA fingerprinting of food-borne pathogens, enhancing traceback and source attribution.

Despite technological advances, many countries still lack adequate laboratory infrastructure, biosafety protocols, or supply chains to support timely diagnosis. Traditional lab surveillance also tends to focus on known pathogens. This limits its utility for early detection of emerging zoonoses or AMR trends.

Advances in Surveillance

In recent years, a paradigm shift driven by the One Health approach, coupled with advances in data science, genomics, and digital technologies, has resulted in advancements in surveillance. Modern systems aim not only to detect known pathogens more efficiently but also to identify emerging threats through integrated, real-time, and predictive approaches.

Event-Based Surveillance (EBS) EBS focuses on capturing informal, unstructured data related to disease occurrence. It supplements traditional indicator-based systems by scanning news media, social media, blogs, and community reports for "signals" of potential outbreaks. EBS has proven particularly valuable for detecting early signs of bacterial zoonoses in underserved or remote areas where formal reporting is delayed. Platforms like ProMED-mail and HealthMap pioneered in the field, aggregating real-time alerts from clinicians, journalists, and the public. ProMED-mail, for instance, was among the first to highlight unusual pneumonia cases linked to a novel pathogen in Wuhan in late 2019. For bacterial zoonoses, EBS can capture early warnings such as reports of sudden livestock deaths (potential anthrax) or unverified brucellosis clusters in rural hospitals. Increasingly, EBS systems use AI-driven web crawlers and multilingual keyword recognition to filter signals and prioritize alerts for further verification.

Digital Epidemiology and Real-Time Data Mining Digital epidemiology refers to the use of Internet-based data streams to detect, monitor, and predict infectious disease activity. For bacterial zoonoses, it offers a dynamic and cost-effective surveillance complement, particularly in low-resource settings. Tools that analyse search engine queries (e.g. "fever + cattle contact"), social media mentions, or online pharmacy purchases provide near-real-time indicators of unusual health trends.

Machine learning and natural language processing models are employed to identify patterns, anomalies, or clusters of disease-like symptoms. These insights can precede formal diagnoses and serve as triggers for active investigation. For instance, spikes in regional Google searches for "leptospirosis symptoms after rain" may suggest an impending outbreak in flood-affected areas. While these systems require careful interpretation due to noise and bias, they are invaluable in augmenting early warning capacity for under-diagnosed bacterial zoonoses.

Predictive Modelling and Geographic Information System (GIS) Applications Geospatial and temporal modelling tools are increasingly used to forecast zoonotic risk and inform targeted interventions. Predictive models integrate data on disease incidence, animal movement, land use, climate, and hydrology to identify areas of elevated transmission risk. For example, rainfall and rodent density data are used to anticipate leptospirosis outbreaks in urban slums or post-disaster zones.

Geographic information system (GIS) enables visualization of hot spots and helps optimize surveillance resource allocation. In anthrax-endemic areas, for instance, mapping of soil types, historical outbreaks, and livestock density helps predict locations with heightened spore reactivation risk. These tools also support dynamic risk maps used in regional surveillance planning for zoonoses like Q fever, tularaemia, and salmonellosis.

Genomic Epidemiology and Molecular Diagnostics Genomic technologies have transformed the speed and precision of pathogen surveillance. Whole-genome sequencing (WGS) allows near-real-time comparison of isolates from humans, animals, and environmental samples. In bacterial zoonoses, this capability confirms outbreak linkage, source attribution, and detection of emerging strains or resistance determinants. For instance, genomic typing of *Brucella* spp. isolates can trace transmission between livestock and humans or identify the introduction of non-native strains. During anthrax outbreaks, sequencing of *B. anthracis* isolates enables authorities to distinguish between a single contamination source and multiple independent spillovers. Global databases like GenBank and Pathogenwatch support data sharing and global strain tracking. Additionally, portable sequencers and multiplex PCR panels now permit field-level molecular testing, reducing the time from sample collection to actionable diagnosis.

Environmental DNA/RNA Surveillance Environmental surveillance is evolving from traditional culture-based detection to modern eDNA/eRNA methods that identify pathogen nucleic acids in water, soil, or air. This non-invasive approach allows early warning detection of pathogens even before symptomatic cases appear. These methods are increasingly supported by PCR-based assays and metagenomic approaches that allow broad-spectrum detection. Environmental surveillance is particularly valuable for bacteria that persist outside of host organisms or are shed asymptomatically by carriers.

Workforce Development and Digital Reporting Tools

Technological advances in surveillance are only as effective as the systems and people implementing them. A key enabler of modern surveillance is the expansion of a cross-sectoral, trained One Health workforce. Initiatives such as the Field Epidemiology Training Programs (FETPs) with One Health tracks, In-Service Applied Veterinary Epidemiology Training (ISAVET), and the One Health Workforce-Next Generation (OHW-NG) are building capacity for integrated zoonotic disease detection. Parallel to workforce development, digital reporting tools (Table 10.2) are closing communication gaps. Mobile applications like DHIS2, Afya Data, and country-specific e-reporting systems allow frontline animal health workers and community health volunteers to report sus-

Table 10.2 Digital surveillance tools for bacterial zoonoses

Tool/platform	Type	Target pathogens	Key features	Application scope
ProMED-mail	Event-based surveillance	All major zoonotic bacteria	Crowd-sourced reports, expert moderation, global alerts	Global (multi-sector)
HealthMap	Event-based surveillance	Bacterial and viral zoonoses	Automated media scraping, geospatial visualization, multilingual search	Global public dashboard
PulseNet International	Genomic surveillance	Food-borne bacteria (e.g. *Salmonella* and *Escherichia coli*)	DNA fingerprinting, cross-border outbreak tracking	Multi-country laboratory network
GLEWS$^+$	Integrated surveillance	Zoonotic diseases with livestock interface	Tripartite coordination (WHO, FAO, WOAH), early warning	Global early warning platform
Go.Data	Digital case management	Outbreak response (incl. bacterial zoonoses)	WHO-supported tool for contact tracing and outbreak visualization	Used in field outbreaks, adaptable to zoonoses
Afya Data	Community reporting tool	Brucellosis, anthrax, leptospirosis	Mobile-based reporting by fieldworkers and veterinarians	East Africa (One Health pilot projects)
Epidemic Intelligence from Open Sources (EIOS)(WHO)	Epidemic intelligence	Multiple zoonoses	AI-assisted filtering of online outbreak signals	WHO and partner institutions
Surveillance Portal, United States Department of Defense (US DOD)	Event-based and strategic monitoring	Select agent bacteria, including *Brucella* spp. and *Bacillus anthracis*	Secure platform for threat detection and coordination	Military/public health interface
Pathogenwatch	Genomic surveillance	*Brucella* spp., *Salmonella*, *E. coli*	Real-time WGS comparison and resistance profiling	Global genomic epidemiology
DHIS2 Tracker	Digital reporting system	Notifiable zoonotic bacteria	Customizable mobile forms, dashboard integration	National and subnational surveillance

pected zoonotic events, such as sudden livestock die-offs or unusual febrile illness in humans, directly into national databases. These tools improve reporting timeliness and enable rapid risk assessment and response, especially in rural or under-resourced regions.

Global Collaboration and Initiatives

As zoonotic pathogens increasingly threaten global health security, coordinated international action has become imperative for timely detection, data sharing, and response. Numerous global initiatives and institutional frameworks have emerged over the past two decades to foster a collaborative surveillance landscape, particularly aligned with the One Health approach.

One Health Coordination Under the One Health umbrella, one of the flagship programmes is the Global Early Warning System for Major Animal Diseases, including Zoonoses (GLEWS$^+$). This is a collaboration between the World Health Organization (WHO), the World Organisation for Animal Health (WOAH), and the Food and Agriculture Organization (FAO). GLEWS$^+$ facilitates real-time sharing of outbreak data between veterinary and public health agencies, particularly for notifiable zoonotic diseases like anthrax, brucellosis, and avian influenza. By combining field reports, epidemiological alerts, and laboratory confirmations, GLEWS+ enables joint risk assessments and synchronized response planning. Although not exclusively bacterial in focus, the system's tripartite structure is designed to support bacterial zoonoses of global concern. Additionally, the One Health Surveillance Codex, supported by the One Health European Joint Programme (OHEJP), has attempted to codify best practices and create harmonized tools for surveillance across sectors. Such frameworks underscore a strategic shift towards surveillance systems that can detect bacterial zoonoses in animals before human spillover occurs—whether via abattoir data, wildlife sampling, or veterinary lab networks.

Early Warning and Rapid Response Systems Early detection of zoonotic threats, especially novel or re-emerging bacterial agents, is critical for preventing widespread outbreaks. Several systems have been established with the specific goal of enabling rapid signal detection and field deployment.

The Global Outbreak Alert and Response Network (GOARN), coordinated by the WHO, is a multinational network of technical institutions and experts that mobilizes rapid response teams during disease outbreaks. GOARN has historically responded to outbreaks of leptospirosis, plague, and other bacterial zoonoses. While it is largely response-focused, GOARN also supports real-time information exchange during investigations, linking local detection to global action.

International Reporting Systems Global coordination relies on international reporting frameworks. The WOAH maintains the World Animal Health Information System (WAHIS), where member states report outbreaks of notifiable animal diseases. Similarly, the International Health Regulations (IHR 2005) require countries to report unusual public health events to the WHO, including zoonotic threats of international concern. While these systems provide essential global visibility, their effectiveness depends on timely, transparent reporting. Underreporting due to trade concerns, limited diagnostic capacity, or weak governance structures remains a significant challenge.

Data Sharing and Interoperability Platforms Real-time sharing and harmonization of surveillance data are fundamental for global surveillance.

Several global initiatives have addressed this by developing digital data-sharing platforms and repositories. For instance, the United States Agency for International Development (USAID) PREDICT project (2009–2019), part of the emerging pandemic threats programme, established wildlife sampling protocols and pathogen discovery pipelines in over 30 countries. Although largely viral in focus, PREDICT also enabled the detection of novel bacterial zoonoses and trained national laboratories to archive data in open-access platforms. Its successor, the STOP spillover initiative, launched in the wake of COVID-19, continues this capacity-building focus, now expanding to incorporate environmental sampling and risk modelling for both viral and bacterial spillover. Moreover, tools like GenBank, Global Initiative on Sharing All Influenza Data (GISAID), and the International Nucleotide Sequence Database Collaboration (INSDC) support genomic data sharing.

For bacterial zoonoses, these platforms enable the comparison of isolates from humans, animals, and food sources across continents. Programmes like Pathogenwatch and EnteroBase have further extended this by offering web-based tools for analysing AMR markers and phylogenetic relationships in zoonotic bacteria. Despite their utility, challenges remain in ensuring equitable access, data standardization, and governance over data use and attribution.

Cross-Border Cooperation and Regional Networks As zoonotic pathogens often emerge and spread irrespective of political boundaries, regional and cross-border surveillance initiatives are essential for coordinated monitoring and control.

The PulseNet International network is one of the most successful examples of cross-border bacterial surveillance globally. It connects public health laboratories across more than 80 countries to enable molecular subtyping and comparison of food-borne and zoonotic bacterial strains (e.g. *Salmonella*, *E. coli*, and *Listeria*). Through standardized DNA fingerprinting and real-time data sharing, PulseNet has enabled the early detection and international tracing of numerous zoonotic outbreaks originating from contaminated animal products. The network has directly contributed to improved food safety measures and outbreak containment across human and animal sectors.

In Africa and Southeast Asia, several regional One Health networks have been piloted to improve surveillance of zoonoses at borders, especially in livestock corridors or transboundary wildlife areas. For example, the South East Asia One Health University Network (SEAOHUN) supports training and surveillance collaboration across veterinary and medical schools, including projects on bacterial threats like leptospirosis and anthrax. Similarly, the East Africa Integrated Disease Surveillance Network (EAIDSNet) promotes joint human–animal surveillance and information exchange across national borders.

Integrated Control Through the One Health Approach

The implementation of One Health often begins with establishing formal coordination between agencies responsible for human and animal health. Many countries have created One Health committees or task forces that bring together public health officials, veterinarians, wildlife experts, and environmental agencies to design and oversee zoonoses control programmes. These coordination mechanisms facilitate information sharing and collective decision-making. Internationally, the importance of such multisectoral collaboration has been recognized through the tripartite alliance of the WHO, FAO, and WOAH, which, in 2019, released a Tripartite Zoonoses Guide (TZG) to help countries operationalize One Health coordination for zoonotic diseases. In 2022, this partnership expanded to a quadripartite with the inclusion of the United Nations Environment Programme (UNEP), reflecting a more holistic approach that includes environmental drivers of zoonoses. The Quadripartite One Health Joint Plan of Action (2022–2026) now guides global efforts in strengthening One Health capacities for disease surveillance, preparedness, and control.

Integrated control of bacterial zoonoses begins with joint risk assessment, where human and animal health sectors collaboratively prioritize diseases. Frameworks such as the five-step One Health model, i.e. Engagement, Assessment, Planning, Implementation, and Monitoring, help structure coordination and operational readiness. A key outcome of such planning is the development of joint contingency plans and standard operating procedures (SOPs), enabling coordinated responses during outbreaks. For example, in a suspected anthrax event, synchronized efforts may include animal vaccination, human prophylaxis, safe carcass disposal, and unified public messaging.

Surveillance integration is central to this approach, involving real-time data sharing across sectors and joint analyses during outbreaks. Platforms allowing mutual access to veterinary and human health data enable faster detection and unified risk communication, enhancing public trust. In outbreak situations like brucellosis, medical and veterinary teams operate concurrently by treating patients, tracing contacts, testing livestock, and vaccinating herds under a common incident command. This coordination reduces redundancy, enhances effectiveness, and ensures that all transmission routes are addressed. During the Rift Valley fever outbreaks, East African countries demonstrated the success of such joint actions by simultaneously implementing animal vaccination, vector control, and human case management.

Integrated prevention measures include synchronized animal vaccination with human health interventions, such as rabies control through dog vaccination and public access to post-exposure prophylaxis. Similarly, environmental interventions like rodent control or managing waste address shared reservoirs of pathogens such as *Leptospira* and anthrax spores. Food safety efforts also benefit from integration. For instance, veterinary services ensure healthy animals, agricultural depart-

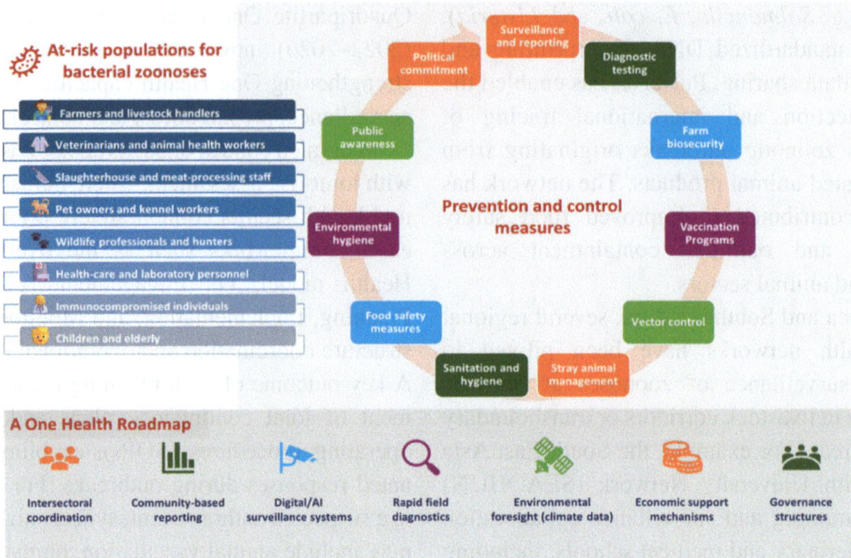

Fig. 10.1 Integrated One Health framework for bacterial zoonoses. (Image source: Authors' personal collection)

ments manage hygienic processing, and health authorities educate consumers, closing the loop from farm to fork.

Political commitment is essential to institutionalize these strategies. Countries like Kenya have embedded One Health into national governance by forming permanent intersectoral units and enacting policies that mandate joint reporting and funding for zoonotic disease control. Such institutionalization ensures that One Health strategies become standard practice rather than temporary projects. Ultimately, One Health integration enables comprehensive planning, surveillance, and response, offering a synergistic advantage critical for controlling bacterial zoonoses that cross sectors and species boundaries. Figure 10.1 illustrates key elements in the prevention, control, and surveillance of bacterial zoonoses through a One Health lens.

Challenges

Despite the availability of structured frameworks and notable success stories, numerous challenges continue to hinder the effective control of zoonoses and the implementation of robust surveillance systems. These challenges encompass technical, operational, and sociopolitical domains, and they often manifest through fragmented coordination, limited resources, and weak community engagement. Recognizing and addressing these constraints is critical to develop context-appropriate, sustainable One Health strategies. Table 10.3 summarizes the key categories of these challenges, outlining their core issues and implications for zoonotic disease prevention, surveillance, and control efforts.

Case Study

A 34-year-old goat herder presented to the district hospital with a two-week history of intermittent fever (39–40 °C), drenching night sweats, profound fatigue, and arthralgias affecting his wrists and knees. He reported assisting with the delivery of multiple goat abortions on his farm 1 month prior. Initial outpatient treatment with broad-spectrum antibiotics yielded no improvement. Physical examination revealed hepatosplenomegaly and tender cervical lymphadenopathy. Routine blood tests showed leucopenia ($3.2 \times 10^3/\mu L$), mild anaemia, and elevated liver enzymes. Blood cultures subsequently grew small Gram-negative coccobacilli,

Table 10.3 Key challenges in controlling zoonoses and implementing surveillance systems

Challenge category	Key issues	Implications
Fragmentation and coordination gaps	Siloed human/animal health systems; unclear roles; lack of joint planning; delayed responses	Delayed outbreak control; inefficient response coordination
Resource and capacity constraints	Limited workforce, diagnostics, and surveillance infrastructure; unsustained vaccination programmes	Continued disease burden; under-diagnosis; weak outbreak response
Technological and methodological challenges	Poor detection of novel pathogens; diagnostic limitations; insufficient laboratory readiness	Delayed recognition of emerging zoonoses; missed intervention windows
Data sharing and interoperability	Non-standardized formats; lack of integrated platforms; hesitancy in international sharing	Missed cross-sector insights; limited global response coordination
Cultural and behavioural challenges	Resistance to behavioural change; stigma and misinformation; weak community engagement	Low public compliance; ineffective outreach strategies
Compliance and enforcement	Weak governance; non-compliance with disease control laws; informal livestock systems	Risk of disease spread via illegal movements or informal trade
Persistence in wildlife and environment	Environmental persistence (e.g. anthrax spores); wildlife reservoirs (e.g. TB in badgers)	Inability to fully eliminate diseases; need for indefinite surveillance
Privacy and ethical issues	Concerns over personal data use in surveillance; fear of punitive measures	Undermined public trust; underreporting of disease events
High cost and sustainability	Funding drops post-crisis; reliance on donors; misalignment with national needs	Interrupted programmes; lack of long-term planning and resilience

and serology (standard agglutination test) was strongly positive for *Brucella* species (titre 1:640).

Simultaneously, two of his co-workers, both involved in milking and birthing assistance, developed similar symptoms and tested positive for brucellosis. A veterinary investigation of the goat herd identified recent abortions in over 30 per cent of does. Milk samples and vaginal swabs from the affected goats yielded *Brucella melitensis* on culture.

In response, the local animal husbandry department instituted a herd vaccination campaign using the Rev. 1 live vaccine, enforced quarantine of affected herds, and introduced a voluntary differentiating infected from vaccinated animals (DIVA) testing protocol to distinguish naturally infected animals from vaccine recipients. Public health teams conducted active case finding among farm families and neighbouring villages, offered free serological screening, and launched community education sessions on safe animal handling, milk pasteurization, and use of PPE during birthing and slaughtering.

1. What is the most likely bacterial pathogen in this case, and which occupational exposures increased the herder's risk?
2. Which surveillance strategies were employed at the human and animal health levels to detect and contain the outbreak?
3. Describe at least three farm biosecurity interventions implemented to interrupt transmission among goats and from goats to humans.
4. Outline recommended personal protective and food safety practices for those handling livestock or consuming dairy products.

Research Questions

1. How can machine learning, genomic sequencing, and other novel tools be most effectively incorporated into integrated surveillance systems?
2. What next-generation vaccine approaches, such as multispecies formulations, oral baits for wildlife reservoirs, or novel human vaccines, can interrupt persistent bacterial transmission?

3. Which community-level factors (e.g. cultural beliefs, economic incentives, and trust in authorities) strongly influence the acceptance of interventions like animal vaccination, culling, or market closures?
4. What adaptive surveillance frameworks and control strategies can be developed to anticipate, monitor, and mitigate these emerging risks?

Declaration During the preparation of this chapter, we have used ChatGPT in order to improve the readability of the text. After using it, we have reviewed and edited the content as needed, and we take full responsibility for the content of the publication.

Further Readings

Bedi JS, Vijay D, Dhaka P, Gill JPS, Barbuddhe SB. Emergency preparedness for public health threats, surveillance, modelling & forecasting. Indian J Med Res. 2021;153(3):287–98.

Erkyihun GA, Alemayehu MB. One Health approach for the control of zoonotic diseases. Zoonoses. 2022;2(1):963.

Ghai RR, Wallace RM, Kile JC, Shoemaker TR, Vieira AR, Negron ME, Shadomy SV, Sinclair JR, Goryoka GW, Salyer SJ, Barton Behravesh C. A generalizable one health framework for the control of zoonotic diseases. Sci Rep. 2022;12(1):8588.

Qiu Y, Guitian J, Webster JP, Musallam I, Haider N, Drewe JA, Song J. Global prioritization of endemic zoonotic diseases for conducting surveillance in domestic animals to protect public health. Philos Trans R Soc Lond Ser B Biol Sci. 1887;2023(378):20220407.

Sharan M, Vijay D, Yadav JP, Bedi JS, Dhaka P. Surveillance and response strategies for zoonotic diseases: a comprehensive review. Sci One Health. 2023;2:100050.

Van Herten J, Bovenkerk B. The precautionary principle in zoonotic disease control. Public Health Ethics. 2021;14(2):180–90.

Part II

Zoonotic Bacterial Infections

Staphylococcal Infections

11

Dhwani Upadhyay, Prasad Andhare,
and Mahendra Pal

Learning Objectives

1. To describe the features of *Staphylococcus aureus*, a potent zoonotic pathogen
2. To understand the human and animal diseases caused by *S. aureus*
3. To comprehend the pathogenesis and response of the immune system against the causative agent
4. To identify the diagnostic methods, prevention, and control strategies for infections

Introduction

Staphylococcus aureus is one of the most prominent and pervasive bacterial pathogens. It is responsible for innumerable chronic skin infections. *S. aureus* presumably causes millions of very serious, invasive infections annually. It is an extensive cause of nosocomial bacteraemia; prosthetic joint, surgical site, and cardiovascular infections; pneumonia; and respiratory tract infections in humans. *S. aureus* is a facultative anaerobic bacterium, which is nonmotile, non-spore-forming, and biochemically positive for coagulase and catalase tests. It is a desiccation-resistant bacterium that may thrive in potentially parched and stressful environments.

The major environmental sources of *S. aureus* are soil, air, and water. The microorganisms are also present as normal flora, mainly the skin and mucous membranes of humans and animals, and can evolve as major opportunistic microbes during immune-compromised situations. The presence of bacteria is also observed in milk and different dairy products that transmit the illness from animals to humans. Multiple diverse diseases in both humans and animals are induced by *S. aureus*. The most common instances of animal documentation are mastitis in dairy cows and bumblefoot in poultry. Pigs with *S. aureus* suffer from exudative dermatitis, sometimes known as greasy pig disease. Several virulence factors of *S. aureus* cause various effects on hosts. These include immune system–interfering substances, extracellular enzymes, and toxins. These factors lead to the induction of tissue necrosis. Surface proteins, however, aid in bacterial adhesion to the host. This adhesion facilitates invasion, which is often followed by chronic infection. These components provide extraordinary potential to cause various infections among humans and animals. *S.*

D. Upadhyay
Department of Life Sciences, Parul Institute of Applied Sciences, Parul University, Vadodara, Gujarat, India

P. Andhare
Department of Life Sciences, School of Sciences, GSFC University, Vadodara, Gujarat, India

M. Pal (✉)
Narayan Consultancy of Veterinary Public Health and Microbiology, Bharuch, Gujarat, India

S. C. Parija, A. Poddar (eds.), *Textbook of Bacterial Zoonoses*, Microbial Zoonoses,
https://doi.org/10.1007/978-981-95-6615-0_11

aureus infections are specifically challenging due to frequently emerging antibiotic resistance.

Methicillin-resistant *S. aureus* (MRSA) is the most significant clinically among other antibiotic-resistant infectious *S. aureus* strains. It is observed that the infection of MRSA has a higher rate of death and morbidity than methicillin-sensitive *S. aureus* (MSSA) infections. Therefore, the infection of *S. aureus* has become a severe threat to humans and animals, affecting public health globally. Additionally, staphylococcosis causes a lack of animal products and other resources that are important for human survival. The current research studies are driving the management of the antibiotic-resistant *S. aureus*, which is becoming an economic risk to mankind worldwide.

History

Despite being around since the beginning of time, staphylococci were discovered to be bacterial pathogens in the nineteenth century. When fluid from a surgical abscess in a knee joint was examined in 1880, Alexander Ogston noticed grape-shaped colonies of bacteria and named them *Staphylococcus*. According to the data from the Centers for Disease Control and Prevention (CDC) and National Nosocomial Infections Surveillance (NNIS) System, *S. aureus* was the most common cause of nosocomial cases of pneumonia and surgical wound infections during the years 1990 through 1992. It was also the second most common cause of nosocomial bloodstream infection. Later, it was found that bacteria cause chronic diseases in animals, such as bursitis, arthritis, dermatitis, synovitis, spondylitis, and omphalitis. Initially, penicillin was reported to be effective against *S. aureus*. However, in a very short time, the bacteria became resistant to the antibiotics and had beta-lactam rings. Later, a new antibiotic, methicillin, was introduced. Methicillin was the first semi-synthetic penicillin to treat infections caused by *S. aureus* strains resistant to penicillin and was made available in 1959 by the Beecham British Pharmaceutical Company. Methicillin was first shown to be effective against *S. aureus*; however, within 2 years of its discovery, the first methicillin-resistant strain of *S. aureus* was found. Methicillin-resistant *S. aureus* (MRSA) strains pose the most risk when they evolve into multi-drug-resistant (MDR) strains. Historical evidence of the emergence of drug-resistant staphylococci is presented in Table 11.1.

Taxonomy

The genus *Staphylococcus*, as part of the *Bacillus–Lactobacillus–Streptococcus* cluster, includes Gram-positive bacteria, with a low G+C content in chromosomal DNA. The genus was transferred to the Staphylococcaceae family of the order Bacillales in the second edition of *Bergey's Manual of Systematic Bacteriology* in 2004. Molecular phylogenetic studies, specifically those utilising 16S rRNA gene sequencing, have partitioned the genus into more than 45 species and 24 subspecies, with some of the most clinically relevant being *Staphylococcus aureus*, *Staphylococcus epidermidis*, *Staphylococcus saprophyticus*, and *Staphylococcus haemolyticus*. Among them are *S. aureus* subsp. *anaerobius* and *S. schleiferi* subsp. *coagulans*. The *Staphylococcus* intermedius group includes *S. delphini*, *S. intermedius*, and *S. pseudintermedius*, each isolated primarily from veterinary infections.

In recent years, whole-genome sequencing (WGS) and multilocus sequence typing (MLST) have been important for improving the taxonomic resolution of *Staphylococcus* spp., extending beyond the constraints of classical phenotypic and biochemical assays. Such methods have identified clonal complexes (CCs) and sequence types (STs) associated with host specificity, virulence potential, and antimicrobial resistance profiles. For instance, *S. aureus* clonal complex CC398 is predominantly associated with livestock but has zoonotic potential, while ST239 is a common cause of nosocomial infections and harbours multi-drug resistance traits. Moreover, a comparison of genomes has also identified mobile genetic elements, including pathogenicity islands, plasmids, and staphylococcal cassette chromosomes (SCCs). These act as drivers of both evolutionary

Table 11.1 Emergence of drug resistance in *Staphylococcus aureus*

Year	Event	Description	Statistical data/impact
1880	Discovery of *Staphylococcus*	Alexander Ogston observed grape-shaped bacterial colonies in surgical abscesses and sepsis	–
1928	Discovery of penicillin	Alexander Fleming discovered penicillin, which became the first antibiotic effective against *Staphylococcus aureus*	–
1940s	Widespread use of penicillin	Penicillin used to treat bacterial infections, including *S. aureus*	Significant reduction in mortality from *S. aureus* infections
1942	First case of penicillin-resistant *S. aureus*	The first penicillin-resistant *S. aureus* strain was reported	Penicillin resistance spread rapidly; within a few years, over 80% of hospital isolates were resistant
1959	Introduction of methicillin	Methicillin, a semi-synthetic penicillin, was introduced to treat penicillin-resistant *S. aureus* strains	Initially effective, methicillin-resistant strains (MRSA) soon emerged
1961	Emergence of MRSA	The first cases of MRSA were documented in the UK	Within a decade, MRSA spread globally, leading to higher morbidity and mortality
1990s	MRSA: a global public health issue	MRSA strains caused severe outbreaks in hospitals worldwide, becoming a leading cause of hospital-acquired infections	By 1992, MRSA accounted for 40–60% of *S. aureus* infections in hospitals across Europe, Asia, and the Americas
2000s	Introduction of new antibiotics	Vancomycin, linezolid, and daptomycin were introduced to combat MRSA	Vancomycin-resistant *S. aureus* (VRSA) strains emerged within a few years
2005	MRSA accounted for most skin infections	In the United States, community-acquired MRSA became the common cause of skin infections	Over 50% of *S. aureus* skin infections were attributed to MRSA by 2005 in the United States
2010	Spread of multi-drug-resistant (MDR) *S. aureus*	MDR strains of *S. aureus* increased, particularly in healthcare settings, making treatment difficult	Global rise in healthcare costs due to prolonged hospital stays and complex treatments
Present	Ongoing resistance issues	MRSA and MDR strains continue to pose clinical challenges, with increasing prevalence of vancomycin- and linezolid-resistant strains	MRSA remains a major cause of morbidity and mortality worldwide, causing over 19,000 deaths annually in the United States alone

adaptability and ecological diversity among *S. aureus* strains. This genomic redefinition taxonomy is vital for tracking infections, containing outbreaks, and developing targeted therapeutics.

Genomics and Proteomics

Genomic Structure

The *S. aureus* genome consists of approximately 2.8 Mbp, encoding around 2600 genes. The percentage of G/C content in the chromosomal DNA of staphylococcal species is approximately 30–40%. The genome shows a high degree of plasticity owing to horizontal gene transfer, which contributes to its adaptability and antibiotic resistance. Comparative genomics between methicillin-sensitive *S. aureus* (MSSA) and MRSA strains have revealed critical genomic islands associated with virulence, such as the *staphylococcal cassette chromosome mec* (SCCmec) element, which harbours the methicillin resistance gene (*mecA*).

Gene Expression and Regulation

Functional genomics has revealed complex regulatory networks in *S. aureus* that control virulence factor expression. For instance, the accessory gene regulator (*agr*) quorum-sensing system plays a pivotal role in regulating the expression of toxins and surface proteins, which modulate adhesion and immune evasion. Transcriptomic analyses of *S. aureus* under various conditions, such as nutrient limitation or antibiotic stress, have provided insights into the genes involved in stress response, virulence, and biofilm formation.

Antibiotic Resistance Mechanisms

Genomic studies have identified several resistance determinants in *S. aureus*. Besides the *mecA* gene, which confers methicillin resistance, other important resistance elements include the *vanA* gene cluster (vancomycin resistance) and the *erm* gene family (macrolide resistance). Whole-genome sequencing (WGS) of resistant strains has revealed point mutations and gene acquisitions that contribute to resistance against multiple antibiotic classes.

Proteomic Techniques

S. aureus proteomics involves high-throughput approaches such as mass spectrometry (MS)–based shotgun proteomics, two-dimensional gel electrophoresis (2-DE), and isobaric tags for relative and absolute quantification (iTRAQ). These techniques have enabled the recognition of proteins that are involved in metabolic adaptation, virulence, and antibiotic resistance.

Virulence Proteins

The secreted protein profile of *S. aureus* encompasses a diverse array of toxins and enzymes that facilitate immune evasion and tissue degradation. Proteomic studies have identified key virulence factors, such as α-toxin, Panton–Valentine leucocidin (PVL), and protein A, all of which are involved in pathogenesis. The agar system and SarA regulon also modulate the production of these virulence proteins in response to environmental signals.

Stress Response and Adaptation

Proteomic analysis under stress conditions, such as antibiotic exposure or oxidative stress, has revealed the upregulation of proteins involved in DNA repair, oxidative stress response, and efflux pump activity. For instance, exposure to vancomycin triggers a distinct proteomic signature, with the upregulation of proteins that enhance cell wall synthesis and stress response mechanisms. These findings directly impact the understanding of how *S. aureus* adapts to hostile environments, including the host immune system.

Biofilm Formation

Biofilm formation is critical in persistent *S. aureus* infections and antibiotic resistance. Proteomic analyses of biofilm-forming *S. aureus* strains revealed proteins associated with extracellular matrix synthesis, including polysaccharide intercellular adhesin (PIA) and teichoic acids. These proteins enhance the ability of bacteria to adhere to surfaces and resist antibiotic treatment, making biofilm-associated infections particularly challenging to treat.

Morphology

The spherical, Gram-positive bacterium *S. aureus* can be seen under a microscope in pairs, short chains, or clusters that resemble grapes. It has a cell wall composed of peptidoglycan and teichoic acid (Fig. 11.1). It develops into lustrous, smooth, elevated, and round colonies on a solid medium. A colony on non-selective media may grow to roughly 4–6 mm in diameter. Pigment production in a colony on agar media can result in various colours. Different colour variations within the colony are possible, ranging from orange to greyish to greyish-white.

Cultivation

The cultivation of *S. aureus* is carried out on various media types. The optimum temperature range for the growth of staphylococci is between 18

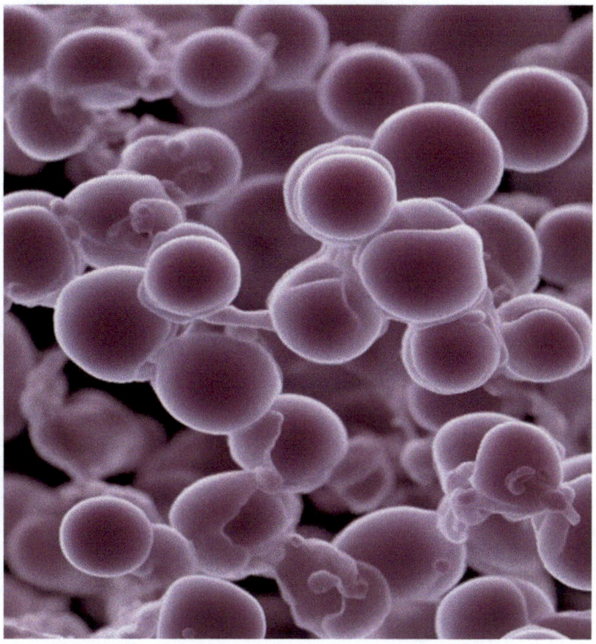

Fig. 11.1 Electron micrograph showing the spherical shapes of *Staphylococcus aureus* arranged in pairs. (Image source: Authors' personal collection)

and 40 °C. Moreover, the bacterium can grow up to 10% NaCl concentration.

S. aureus can grow on a variety of selective, enriched, and differential media. Mannitol salt agar (MSA), *Staphylococcus*-110 agar, and Baird-Parker agar are commonly used for this purpose. These media serve as both selective and differential media for *S. aureus* culture. Each medium offers specific advantages for isolating and identifying *S. aureus* from mixed bacterial populations. Additionally, tryptone soy agar (TSA), Giolitti–Cantoni broth, Luria Bertani (LB) agar, brain heart infusion (BHI) agar, liquid Baird-Parker (LBP), and blood agar media are used as enriched media for the growth of *S. aureus*.

On blood agar, *S. aureus* typically forms medium-to-large, round, golden-yellow-to-creamy-white colonies. This is often surrounded by a clear zone of β-haemolysis due to haemolysin production. When grown on mannitol salt agar (MSA), colonies of *S. aureus* ferment mannitol and appear as yellow colonies due to acid-induced colour change in the medium, distinguishing it from coagulase-negative staphylococci. On Baird-Parker agar, *S. aureus* forms black, shiny colonies with a clear halo, attributable to tellurite reduction and lecithinase activity. On tryptone soy agar (TSA) or LB agar, colonies are smooth, glistening, and slightly elevated. They show a characteristic creamy-to-golden pigmentation depending on the strain and incubation conditions. In Giolitti–Cantoni broth, *S. aureus* growth is indicated by the blackening of the medium due to potassium tellurite reduction.

These distinctive appearances across media are diagnostically valuable for the presumptive identification of *S. aureus* in clinical and environmental specimens.

Animal Models

There are multiple animal models, such as mice, rats, hamsters, Mongolian gerbils, bird-chicken embryos, rabbits, guinea pigs, pigs, dogs, nematodes, insects, and fishes. These are for the strategic selection of animal hosts that mimic specific aspects of *S. aureus* pathophysiology, immune responses, or disease progression in humans or livestock. These models are employed in a targeted manner to (i) investigate organ-specific infections, (ii) evaluate virulence determinants, or (iii) test therapeutic interventions under controlled experimental conditions (Fig. 11.2).

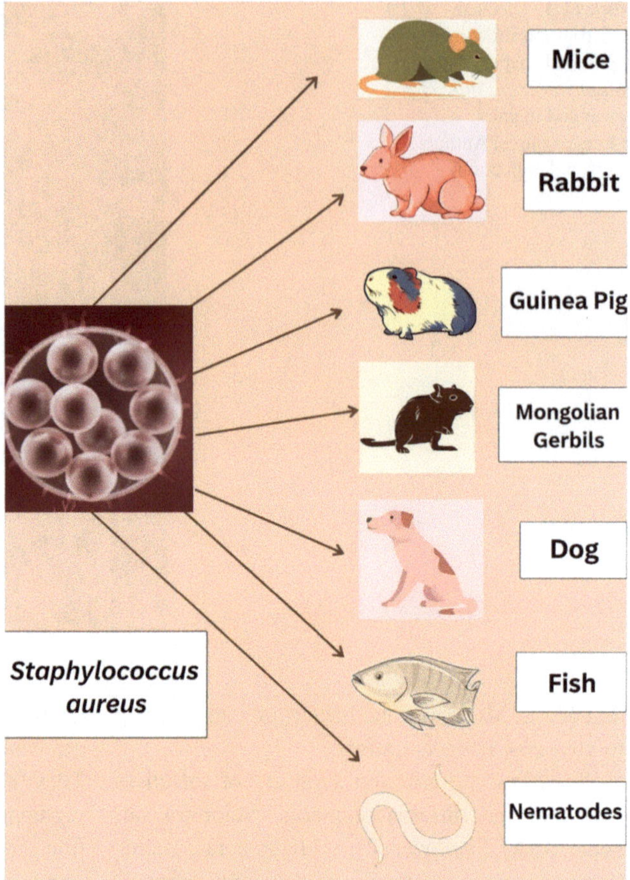

Fig. 11.2 Animal models for experimental trials of *Staphylococcus aureus* infections. (Image Source: Authors' personal collection)

The domestic mouse (*Mus musculus*) model is highly preferred for the study of skin infections, including intradermic and hypodermic. This is because mice have skin like humans and reflect a similar effect of staphylococcal epidermolytic toxins. Additionally, they are used for histopathological studies in cases of arthritis and staphylococcal-specific bone infections. The rat model (*Rattus rattus*) is preferred for endocarditis, osteomyelitis, vaginitis, and biofilm-associated *Staphylococcus* infections. Hamster (*Mesocricetus auratus*) models are used to study drugs, haemorrhagic reactions, and inflammation, especially against MRSA. Another model used for the observation of immunogenic responses, efficacy of immunoglobins, and gastric mucosal layer destruction during the infection of *S. aureus* is the Mongolian gerbil (*Meriones unguiculatus*).

Bird-chicken embryos (*Gallus gallus*) are another potent animal model used to test immunogenic responses during infection and observe drug effects. The next most preferred animal model is the domestic rabbit (*Oryctolagus cuniculus*), which is utilised for clinical trials of diseases such as osteomyelitis, staphylococcal meningitis, enterotoxin-related diarrhoea, endophthalmitis, vaginitis, endocarditis, and other skin infections. Guinea pigs (*Cavia porcellus*) are also important model animals because they possess delicate membranes that allow the easy delivery of *Staphylococcus* strains, This is written in the context of Guinea pigs which are majorly studied for abscesses and wound infection.

Transmission Cycle

S. aureus infection is common among animals and humans. As discussed above, bacteria cause a broad spectrum of diseases in both hosts by targeting various cell types. The major animal hosts

Fig. 11.3 Transmission of *Staphylococcus aureus*. (Image Source: Authors' personal collection)

are goats, sheep, cattle, poultry birds, pigs, and fish.

Disease transmission occurs by both direct and indirect methods (Fig. 11.3). It can be transmitted from human to human and from animal to animal. Direct contact, frequently through the hands, with animals or humans that are colonised or infected, as well as contaminated objects and surfaces, is the usual method of transmission. Typically, the spread of *S. aureus* is caused by handling of or consuming food contaminated with the bacteria, uncontrolled animal traffic between farms, or contact between an infected mammary gland and an uninfected gland through fomites, such as milking equipment or the hands of the milker. People, animals, and other sick or colonised entities shed *S. aureus* from their noses and skin into the environment, suggesting that airborne transmission could be a potential source of infection. Additionally, linked to the spread of *S. aureus* are vectors such as the housefly (*Musca domestica*).

Pathogenesis and Pathology

S. aureus is a systematic human and animal coloniser and potential pathogen that causes invasive infections such as pneumonia, endocarditis, osteomyelitis, mastitis, and vaginitis, as well as diseases of the skin and soft tissues.

Pathogens secrete toxins that can harm the host in two primary ways. Firstly, some toxins directly damage the host membrane by forming pores. Secondly, other toxins interfere with the functions of cell membrane receptors. These mechanisms allow pathogens to disrupt cellular processes and harm the host organism.

Numerous virulence factors, such as pore-forming toxins (alpha-haemolysin and Panton–Valentine leucocidin), phagocytosis inhibitors (polysaccharide capsule and protein A), superantigens (enterotoxins and toxic shock syndrome toxin-1), and immune evasion molecules (chemotaxis inhibitory protein, staphylokinase, and aureolysin), are involved in the pathogenesis of *S. aureus*. Coagulase-positive *Staphylococcus* species have the potential to induce more pathogenicity than coagulase-negative species because they produce the coagulase enzyme, which is another virulence factor.

The pathogenesis can be studied using the following steps:

Adhesion

The first and essential step in pathogenesis is adhesion, wherein virulent bacterial factors adhere to the host cell for biofilm formation. Adhesion is mediated by cell wall–anchored (CWA) proteins as well as extracellular sugar-based polymers. The cell wall–anchored proteins belong to the microbial surface components recognising adhesive matrix molecules (MSCRAMM) family. Some of them are biofilm-associated proteins, surface proteins, iron-regulated surface determinants, collagen adhesins, etc. These are major examples of the MSCRAMM family, which are covalently attached to the cell surface. Additionally, a few noncovalently surface-associated proteins are also part of cell wall–anchored (CWA) proteins. These proteins include autolysin, extracellular matrix–binding protein homologue, elastin-binding protein, and extracellular adherence protein. Some non-proteinaceous adhesins also play a major role in adhesion, such as teichoic acid and lipoteichoic acid, which are present in the walls of bacteria, along with a few polysaccharide adhesins.

Invasion

S. aureus produces various enzymes that not only damage the host cell but also interfere with the host immune system, making it easy to invade the host.

Alpha-haemolysin cytolytic toxins invade various types of cells in humans and animals, such as epithelial cells, endothelial cells, leucocytes, and RBCs. Alpha-haemolysin results in apoptosis, which causes pores in the membrane.

Some proteins, such as chemotaxis inhibitory proteins (CHIPs) and Formyl peptide receptor (FPR) inhibitory proteins, also play an important role in immune evasion as they possess the ability to interfere with leucocyte receptor functioning. The process of invasion is additionally carried out by some enzymes that are directly involved in the degradation of host components. The major enzyme involved in this process is staphylocoagulase. This enzyme converts proactive fibrinogen into active fibrin, leading to the formation of a phagocytosis-impeded fibrin clot (PIFC). This coating acts as a shield around the bacterial surface and prevents the bacteria from host receptors, as it forms clumps that can be further degraded by the enzyme staphylokinase produced by bacteria during the attack on the host. Staphylokinase can convert human plasminogen into plasmin, a fibrinolytic protease that can break apart existing clumps and cause isolated cocci to spread. Sak aids in breaking down the host epidermal barrier. To prevent complement-mediated bacterial death, *S. aureus*'s "toolbox" also contains several specialised proteases, including Aureolysin, glutamyl endopeptidase, and staphopain A and B. These proteases interfere with complement components and prevent bacteria from being killed.

Evasion

The ability of *S. aureus* to escape host innate immunity and resist phagocytosis makes it a potent pathogen for humans and animals. *S. aureus* generates substances that may prevent opsonophagocytosis, whether they are surface-bound or freely released. For instance, complement and antibody deposition is impeded by surface-associated proteins such as SpA. Protein A (SpA) has five immunoglobulin (Ig)-binding domains that bind to the Fab domains of IgG and IgM, as well as the Fc region of IgG. It attaches to the bacterial cell wall envelope. The attachment of SpA to IgG prevents Ab from interacting with neutrophil Fc receptors, which prevents phagocytosis.

In vitro, *S. aureus* strains lacking SpA were more effectively phagocytosed by neutrophils and exhibited reduced virulence in animal infection models. Furthermore, *S. aureus* capsule polysaccharides can make the bacterium immune to phagocytosis. Numerous *S. aureus* strains produce positively charged polysaccharide intercellular adhesion (PIA), which binds to the negatively charged surfaces of bacterial cells to promote the production of biofilms. This polymer can partially shield *S. aureus* from neutrophil phagocytosis and death. All the toxins discussed above, specifically superantigens, toxic schizotomy syndrome toxins, and different types of haemolysins, play a vital role in stimulating a strong immune response.

Biofilm Formation

S. aureus is more likely to remain on prosthetic surfaces when biofilms are formed, and this increases the resilience of bacteria to both human defences and antibiotic treatments. One way to conceptualise biofilm development is as a two-step process: firstly, an initial attachment mediated by CWA proteins and, secondly, a maturation phase involving cell proliferation and the synthesis of the biofilm polymeric matrix.

The biofilm matrix components are extracellular DNA, surface-associated and secreted proteins, host factors, and polysaccharides, including intercellular polysaccharide adhesins. Different methods, such as modifications to the hydrophobicity of the bacterial cell surface and extracellular DNA synthesis, are used by *S. aureus* to adhere to a surface, whether biotic or abiotic. Following adhesion, the bacteria multiply and form microcolonies on the surface. Following the development of microcolonies, an extracellular polymeric substance (EPS) is formed, which eventually matures into a fully formed biofilm. The bacterial cells that make up the biofilm produce specific compounds, such as D-amino acids and EPS-degrading enzymes, such as alginate lyase, to disintegrate and disseminate the biofilm once it has fully developed. These primitive cells are prepared to build a new biofilm by attaching it to a different location or returning to its original location.

Once a biofilm has fully developed, the bacterial cells within it begin to produce specific compounds, including D-amino acids and extracellular polymeric substance (EPS)–degrading enzymes, such as alginate lyase. These compounds facilitate the disintegration and dispersal of the biofilm. Following dispersal, the bacterial cells can initiate the formation of a new biofilm, either by attaching to a new surface or by returning to their original location.

Immunology

To combat common bacterial diseases, including *S. aureus*, humans have evolved several levels of defence mechanisms supplemented with various components, from natural barriers such as mucosal surfaces, skin, and innate immune components, such as neutrophils and complement systems, to adaptive immune components, such as antibodies and effector T cells that participate in *S. aureus* infections.

Neutrophils provide the main innate immune protection against invasive diseases caused by *S. aureus*. Toll-like receptors (TLRs) and C-type lectin receptors (CLRs) are pattern recognition receptors that *S. aureus* can bind to and activate in neutrophils. Such interactions between TLRs and CLRs stimulate the release of neutrophil cytokines and phagocytic activity. Upon activation, neutrophils can harm bacterial invaders via NETosis. This process results in neutrophil death via the ejection of an extracellular trap rich in antimicrobial peptides and DNA. Moreover, activated neutrophils are highly effective in phagocytosing bacteria. They do this by using cytoplasmic granules as well as an intracellular oxidative respiratory burst to produce reactive oxygen species (ROS) that kill phagocytised bacteria. Furthermore, complement receptors and immunoglobulin Fc receptors are expressed by neutrophils, which aid in the phagocytosis of *S. aureus* bacteria coated with complement proteins or antibodies. *S. aureus* has been shown to resist intracellular death and can persist within neutrophils; strong and efficient neutrophil function is essential for the effective killing and clearing of germs. Such intracellular survival may make it easier for live *S. aureus* to spread throughout the body and cause illness.

Another essential element of the host's defence against bacterial infection is complement. *S. aureus* and other Gram-positive bacteria are typically resistant to complement-mediated direct eradication. Complement component 5a (C5a) and complement component 3a (C3a), on the other hand, release complement fragments that alert the adaptive immune system to *S. aureus* infection and cause the recruitment of phagocytic cells. The antibody isotypes are immunoglobulin G and M, which fix complement proteins and aid in the deposition of C3b on the bacterial surface. The process by which neutrophils phagocytose bacteria is facilitated by the deposition of C3b and antibodies onto their surface. The production of functional antibodies against CPs (CP5 and CP8), specifically for *S. aureus*, can aid in opsonophagocytosis. Through the production of T cells and antibodies specific to *S. aureus*, the adaptive immune response can aid in facilitating anti-*S. aureus* neutrophil activity. Strong functional antibody responses against vaccination carrier proteins or staphylococcal protein antigens are necessary to elicit T-helper (Th) responses and encourage class switching. The induction of long-lived memory B cells also depends on T-helper responses. Through opsonisation, the antibody generated by B cells can increase *S. aureus* absorption into neutrophils. It has also been demonstrated that Th1 cells specific to antigens release interferon-γ, which improves phagocyte activity in humans and mice and contributes to *S. aureus* defence. Furthermore, Th17 cells' production of IL-17 contributes to neutrophil recruitment to infection sites, as well as the activation and maintenance of neutrophil survival at inflammatory sites.

Infection in Humans

S. aureus causes multiple diseases in humans. Their severity is based on the patient's immunity and the pathogenicity of the strain. Some important diseases caused by *S. aureus* are shown in Fig. 11.4 and are mentioned below:

Skin Infections

Skin and soft tissue infections are very common inflammatory conditions caused by *S. aureus*. Impetigo is a common bacterial skin illness. Skin and Soft Tissues (SST) infections can include

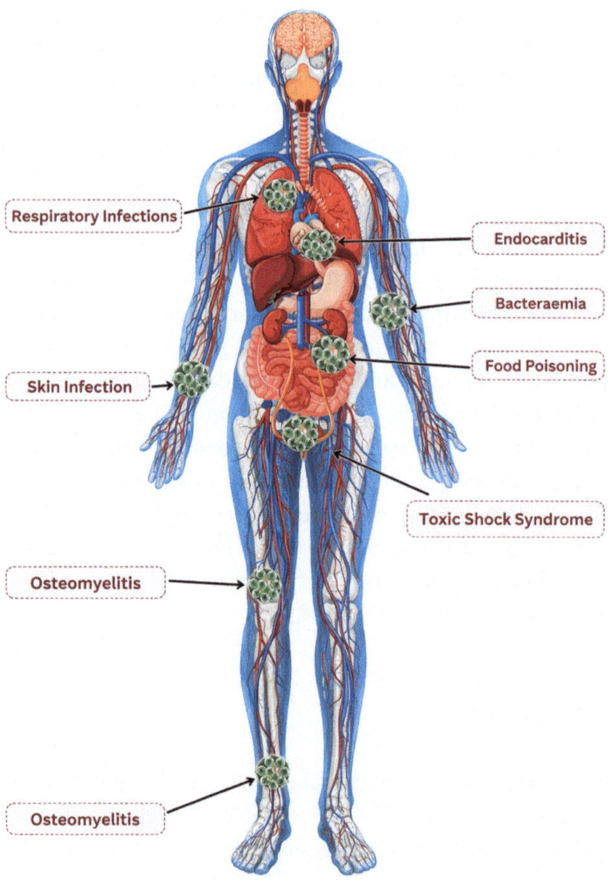

Fig. 11.4 *Staphylococcus aureus* diseases in humans. (Image Source: Authors' personal collection)

folliculitis, wound infections, and hidradenitis suppurativa. The recurrent pyogenic infection of the apocrine sweat glands, and mastitis is a mammary gland infection in nursing women. Localised skin infections lead to the development of cellulitis, necrotising fasciitis, and lymphangitis with lymphadenitis.

Osteomyelitis

Osteomyelitis is frequently caused by *S. aureus*. It might entail a single localised infection of the bone or a more widespread infection involving the marrow, cortex, periosteum, and surrounding soft tissue. The symptoms manifest in a variety of ways, such as local swelling, slow-moving fistula drainage, exposed fractured bone in open wounds, and bone pain without skin lesions. Contiguous osteomyelitis affecting the growth plate usually coexists with septic arthritis. The factors for septic arthritis in adults include age >80 years, diabetes mellitus, rheumatoid arthritis, recent joint surgery, and skin infections.

Endocarditis

S. aureus heart valve infections have significant morbidity and fatality rates. The most common site of infection is the left heart valve, and although the death rate from *S. aureus* endocarditis is minimal in those under 50 years of age, it frequently exceeds 50% in older patients. Individuals who receive haemodialysis, have a medical history of hospitalisation, abuse intravenous drugs, have prosthetic valves, have degenerative valvular disease, or have rheumatic heart disease are at the highest risk of endocarditis.

Toxic Shock Syndrome

Toxic shock syndrome is caused by the advent of superabsorbent tampons for use during menstruation. The disease occurs because of the presence of the superantigen TSS-1. It is categorised as menstrual and nonmenstrual TSS. Menstrual TSS leads to severe skin infection of the cervix, vagina, and other female genital cells due to the colonisation of *S. aureus*. However, nonmenstrual TSS occurs in cases of surgery, insect bites, and post-partum stages.

Food Poisoning

The common reasons for food contamination include inaccurate handling of cooked or processed foods and storage requirements that sustain *S. aureus* growth. Usually manifesting 2–4 hours post-meal, the illness is marked by symptoms such as nausea, vomiting, cramping in the abdomen, diarrhoea, and, maybe, even dehydration. Enterotoxins are the primary cause of food poisoning.

Respiratory Infections

A bloodstream infection or aspiration can cause lung infection with *S. aureus*. Necrotising pneumonia caused by *S. aureus* results in epithelial destruction, compromising ciliary function and bacterial clearance. The illness presents with typical pneumonia symptoms, occasionally accompanied by many cavitary lesions. The gene encoding *Cna* is frequently present in the bacteria that cause this illness. This protein may facilitate *S. aureus* adherence to respiratory epithelial basal cells.

Bacteraemia

Bacteraemia is usually linked to the spread of a local infection, although it can also be temporary. There are three major categories of infected foci: extravascular, intravascular, and assumed intravascular foci. Every year, 20–50 instances of *S. aureus* bacteraemia occur per 100,000 people, and 10–30% of these cases result in mortality. Sepsis is frequently caused by *S. aureus*. Sepsis can result from some illnesses, and the common signs include fever, hypotension, tachycardia, and tachypnoea. Patients receiving chemotherapy or surgery, the elderly, and those with impaired immune systems are among the groups most vulnerable to staphylococcal sepsis. Severe sepsis can result in death, lactic acidosis, disseminated intravascular coagulation, and multiple organ failure.

Infection in Animals

In animals, *S. aureus* infections are the same as in humans. The group of diseases remains the same; however, the infection level varies among host animals. For instance, it causes mastitis, pododermatitis, abscesses, and other skin diseases in rabbits. In poultry animals, it causes osteomyelitis, bumble foot, arthritis, omphalitis (yolk sac infection), tenosynovitis (inflammation of the tendon sheath), abscesses, and necrotic dermatitis. In pigs, *S. aureus* infection can lead to mastitis, castration wounds, and exudative dermatitis. In cattle, these bacteria cause small abscesses, mastitis, and toxic shock syndromes. Sheep and goats show mild folliculitis, mastitis, mammary botryomycosis, and dermatitis. *S. aureus* also infects small ruminants by causing abscesses, mastitis, and septicaemia.

Epidemiology and Public Health

S. aureus is present mainly as normal microflora of the skin and nasal cavity of humans and other warm-blooded animals. They colonise the skin, udders, teats, mucous membranes, and nasal and genital regions of milking animals. These sites are major sources of reservoirs and contaminants. Bacteria survive as opportunistic pathogens in these habitats. Over time, under suitable conditions, they emerge as pathogens and cause several diseases in humans and animals. Pathogenic bacteria produce various toxins, resulting in unusual infections in both hosts.

A significant disease burden is caused by *S. aureus* infections in both community and healthcare structures globally (Fig. 11.5). For many years, it has been the primary source of

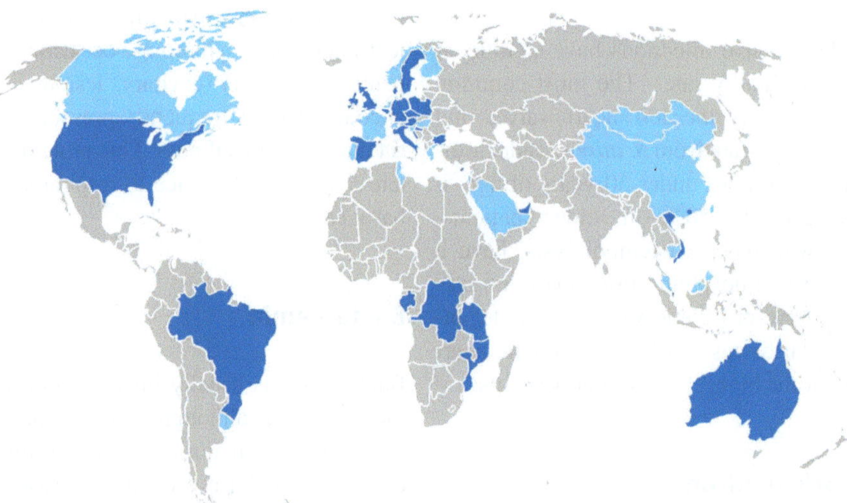

Fig. 11.5 Global distribution of *Staphylococcus aureus* ST45. Countries with ST45 isolates reported in the literature are coloured in blue. The darker shade marks countries with ST45 isolates that were available for this study. (Image source: Effelsberg N, Stegger M, Peitzmann L, Altinok O, Coombs GW, Pichon B, Kearns A, Randad PR, Heaney CD, Bletz S, Schaumburg F. Global epidemiology and evolutionary history of *Staphylococcus aureus* ST45. J Clin Microbiol. 2020;59(1):10–128)

healthcare-associated infections (HAIs). Next to *Clostridium difficile*, *S. aureus* was the second most reported causal pathogen in a recent systematic population-based assessment of nosocomial infections that required antibiotic treatment in hospitals with acute care in the United States. The emergence of antibiotic-resistant *S. aureus* in both community settings and livestock farms has exacerbated this issue.

MRSA infection is known as amphixenosis because it can spread in two ways: from humans to animals and, in close contact, from animals to humans. Consequently, there is a greater probability of livestock-acquired MRSA colonisation among residents in zones with larger livestock populations (cattle, pigs, and veal calves). According to estimates from The European Centre for Disease Prevention and Control (ECDC), MRSA independently causes 171,200 hospital-acquired infections Hospital Aquired Infections (HAIs) and 5400 attributable deaths annually. In the past 15 years, there has been an increase in community-associated *S. aureus* infections globally, particularly in younger patients, which is another significant cause of morbidity. *S. aureus* is associated with other serious clinical syndromes, including sepsis and necrotising pneumonia, and community-associated infections (CAIs).

Diagnosis

Prompt and accurate diagnosis of *S. aureus* infections is critical for initiating appropriate treatment regimens and performing epidemiological surveillance, particularly in the context of antibiotic resistance. Microbiological identification is typically based on a combination of microscopy, classical culture methods, serological assays, and molecular diagnostics, with variable sensitivities, specificities, and scopes of application. The application of these methodologies improves the ability of the laboratory to accurately identify and differentiate between methicillin-sensitive (MSSA) and methicillin-resistant strains (MRSA), as well as toxin-producing isolates.

Microscopy

Microscopy has remained the most important microbiological diagnostic tool, frequently allowing for the rapid presumptive characterisation of a microbe. Due to the nature of its binary fission in multiple planes, *S. aureus* typically appears as Gram-positive cocci in grape-like clusters on Gram staining of clinical specimens. They are purple under light microscopy because they retain crystal violet in their thick peptidoglycan layers.

Electron microscopy techniques have been used for improved structural and ultrastructural visualisations. SEM and TEM provide amphinomid information on surface topography and biofilm architecture, as well as subcellular features, including septal formation and cytoplasmic granules. Confocal laser scanning microscopy (CLSM) with fluorescent probes has been used to visualise live cell imaging of *S. aureus* colonisation in biofilm-associated infections.

In Vitro Culture

Culture-based diagnosis remains the gold standard because it recovers viable organisms and allows phenotypic testing. *S. aureus* grows well on enriched non-selective media, such as blood agar, where colonies are typically β-haemolytic due to haemolysin activity. On mannitol salt agar, it ferments mannitol, resulting in yellow colonies, distinguishing it from coagulase-negative staphylococci. Other selective media include the Baird-Parker agar. On this medium, *S. aureus* produces black colonies with clear halos, indicating tellurite reduction and lecithinase activity. This medium is also used for differentiating between members of the genus *Staphylococcus*.

Confirmatory biochemical tests include coagulase and catalase tests. *S. aureus* is catalase-positive, quickly breaking down hydrogen peroxide into water and oxygen. Owing to the strong correlation between virulence and the production of free or bound coagulase, the coagulase test is of major importance for confirming *S. aureus*.

Serodiagnosis

Serological methods act as complementary methods, particularly when bacterial culture is constrained by previous antibiotic therapy or in the presence of chronic, biofilm-related infections. To assess immunity against *S. aureus*, enzyme-linked immunosorbent assay (ELISA) is used to detect circulating antibodies against *S. aureus* exotoxins (TSST-1, enterotoxins, and PVL), capsular polysaccharides (CP5 and CP8), or structural proteins (for example, protein A).

Western blotting, radioimmunoassay, and passive haemagglutination are also used to quantify host immune responses. These are especially useful in vaccine studies and in assessing a host's exposure history. For instance, the detection of anti-protein A IgG by ELISA has been shown to correspond to an invasive *S. aureus* infection.

However, serodiagnosis has limitations due to cross-reactivity and delayed antibody formation, limiting its utility for acute infection diagnosis. Nevertheless, it is useful for epidemiological and retrospective studies.

Molecular Diagnosis

Molecular diagnostics provide high sensitivity and specificity, allowing rapid and accurate detection of *S. aureus* DNA, resistance genes, and virulence determinants. Polymerase chain reaction (PCR) targeting the *nuc* gene (thermostable nuclease), *mecA* (methicillin resistance), *lukS*/F-PV (PVL toxin), and *tsst*-1 (toxic shock syndrome toxin) are widely used in clinical microbiology.

Quantitative PCR (qPCR) and immuno-PCR improve detection limits and allow for real-time quantification of the bacterial load. Loop-mediated isothermal amplification (LAMP) is emerging as a rapid, equipment-independent alternative with high amplification efficiency, low cost, and great potential for use in point-of-care diagnostics.

Epidemiological monitoring, resistance gene discoveries, and outbreak investigations are being performed in reference laboratories with the increasing use of next-generation sequencing (NGS) and whole-genome sequencing (WGS). These platforms allow for high-resolution genotyping and capture of horizontal gene transfer events that drive the evolution of antibiotic resistance.

Novel label-free rapid detection platforms have strong potential for diagnostic applications with biosensor-based diagnostics, such as electrochemical and optical sensors. Devices based on *S. aureus*–specific aptamers or antibodies (e.g. protein A exotoxins) are sensitive for detecting viable *S. aureus* in clinical samples.

An overview of the diagnostic modalities is presented in Table 11.2.

Treatment

S. aureus has been extremely successful in developing drug resistance. Beta-lactam family antibiotics are the foremost option for the treatment of methicillin-sensitive *S. aureus* (MSSA) infections. An alternative is to use first-generation cephalosporins, such as oral cephalexin and intravenous cefazolin. For many years, vancomycin has been the preferred antibiotic for treating MRSA. However, because of the decreased susceptibility to vancomycin that has emerged in recent years, additional antibiotics have been produced. The various classes of antibiotics that combat MRSA include (1) novel cephalosporins, such as ceftaroline and ceftobiprole; (2) glycopeptides, such as vancomycin and teicoplanin; (3) lipopeptides, such as daptomycin; and (4) oxazolidinones, such as linezolid and more recent medications, including tedizolid and linezolid.

Prevention and Control

To avoid *S. aureus* infections both inside and outside the healthcare context, one of the most effective therapies is practising good hand hygiene. Effective strategies for reducing *S. aureus* infections include bundles of care aimed at enhancing adherence to quality-of-care measures. A research study, for instance, revealed that this approach was linked to a lower death rate and enhanced patient

Table 11.2 Diagnostic approaches for *Staphylococcus aureus* infections

Diagnostic approach	Method	Target	Remarks
Microscopy	Gram staining	Gram-positive cocci in grape-like clusters	Rapid and cost-effective; presumptive identification from clinical specimens; limited specificity
	Electron and confocal microscopy	Biofilm structure and ultrastructural features	Advanced imaging for structural and biofilm analysis; not routine in diagnostic labs
In vitro culture	Blood agar, mannitol salt agar, Baird-Parker agar	Viable *Staphylococcus aureus* colonies	Gold standard; allows isolation, phenotypic characterisation, and antimicrobial susceptibility testing
	Coagulase and catalase tests	Enzymatic activity (virulence markers)	Coagulase-positive test differentiates *S. aureus* from other staphylococci; catalase helps distinguish from streptococci
Serodiagnosis	ELISA, Western blot, passive haemagglutination	Circulating antibodies (e.g. anti-protein A, toxins)	Useful in retrospective diagnosis and immuno-epidemiological studies; limited value for acute diagnosis due to delayed antibody production
Molecular diagnosis	Conventional and real-time PCR	*mecA*, *nuc*, *lukS/F-PV*, *tsst-1* genes	High sensitivity and specificity; useful for detecting MRSA, toxin genes, and confirming identity in culture-negative infections
	Loop-mediated isothermal amplification (LAMP)	Specific genomic sequences	Rapid and cost-effective; minimal equipment required; suitable for field or point-of-care diagnostics
	Whole-genome sequencing (WGS), next-generation sequencing	Resistance genes, virulence islands, phylogenetic data	Offers epidemiological insight; highly accurate but expensive and resource-intensive
Biosensor--based detection	Electrochemical/optical sensors with aptamers or antibodies	Surface proteins (e.g. Protein A), toxins	Emerging platform for rapid, label-free detection; highly sensitive but requires development for widespread diagnostic use

management for patients with *S. aureus* bacteraemia. The best strategy for controlling *S. aureus* mastitis and preventing new infections in dairy farms is to remove any circumstances that could expose the teat ends to bacteria and reduce the chance of cow-to-cow transfer. The best ways to prevent new infections and manage *S. aureus* mastitis in dairy farms are through hygiene measures, such as dipping the teats in an antiseptic solution before and after milking, cleaning and maintaining the equipment properly, and having milkers wear gloves and change them frequently. Another important measure is to reduce the likelihood of cow-to-cow transmission. Public education regarding hand hygiene, the use of gloves during food preparation, and maintaining food at an appropriate temperature are necessary to reduce the growth of pathogens, eradicate them, and reduce toxin production. Maintaining proper hygiene at home and in the healthcare of humans and animals is a crucial preventive and control measure. It has been recommended that traditional (systemic) antibiotics be used as prophylaxis against nosocomial infections caused by *S. aureus*.

Case Study

A 38-year-old male dairy worker was admitted to a primary health centre with a 10-day history of a painful swelling on his right forearm. It started as a pustule and evolved into a fluctuant abscess with surrounding erythema and induration. He had low-grade fever, malaise, and routine exposure to lactating cattle, some recently diagnosed with mastitis. No major trauma was remembered, but minor scratches from milking were common.

His forearm was found to have cellulitis and regional axillary lymphadenopathy. Microbiological testing was performed on pus aspirated from the lesion. Gram stain showed Gram-positive cocci in grape-like clusters. Colony morphology on blood agar and mannitol salt agar (MSA) incubated at 37 °C displayed golden-yellow, β-haemolytic colonies (blood agar) and mannitol fermenting colonies (MSA). Positive catalase and tube coagulase tests identified the isolate as *S. aureus*.

The antibiotic susceptibility testing (Kirby–Bauer method) showed resistance to methicillin and oxacillin. PCR confirmed the *mecA* gene, consistent with methicillin-resistant *S. aureus* (MRSA). Oral linezolid was started, and wound care was recommended. The local veterinary authorities were notified of potential zoonotic transmission. After 10 days, there was significant clinical improvement in the patient, and follow-up cultures were negative.

1. What was the duration and nature of the lesion before the patient presented to the primary health centre?
2. What did the Gram stain of the aspirated pus show?
3. Which gene confirmed methicillin resistance in the isolate?
4. What treatment was given to the patient, and what was the outcome after 10 days?

Research Questions

1. What are the current research gaps and future directions identified in the study of staphylococcal infection?
2. What are the potential evolutionary pathways for staphylococci that could enhance their zoonotic potential in the coming decades?
3. What is the predicted impact of climate change and agricultural practices on the prevalence and transmission of zoonotic *Staphylococcus* infections?
4. What are the potential advancements in diagnostic tools and surveillance methods for the early detection and prevention of future zoonotic *Staphylococcus* outbreaks?

Declaration During the preparation of this chapter, we have used QuillBot in order to improve the grammar and readability. After using, we have reviewed and edited the content as needed, and we take full responsibility for the content of the publication.

Further Readings

Beenken KE, Smeltzer MS. *Staphylococcus aureus* biofilm-associated infections: have we found a clinically relevant target? Microorganisms. 2025;13(4):852.

Brüggemann A, Hailer NP. Never mind the bug: no differences in infection-free survival after periprosthetic joint infections with *Staphylococcus aureus*, Coagulase-negative *Staphylococcus*, or *Streptococcus*. Front Microbiol. 2025;15:1503928.

Chenhsu RY, Hall BA, Tran H, Donnelley MA, Cheema R, Nakra NA. Vancomycin area under the curve to minimum inhibitory concentration ratio for treatment effectiveness in paediatric and neonatal *Staphylococcal* infections: a systematic review. J Pediatr Pharmacol Ther: JPPT. 2025;30(1):52.

Dufresne K, Tuffs SW, Walton NR, Kasper KJ, Mohorovic I, Hasan F, Bentall T, Heinrichs DE, Delport J, Mele TS, McCormick JK. Evidence that *staphylococcal* superantigens promote within-patient bacterial persistence following postoperative surgical site infection. Infect Immun. 2025;93:e00407-24.

Hulme J. Harnessing ultrasonic technologies to treat *Staphylococcus aureus* skin infections. Molecules. 2025;30(3):512.

Jennings MR, Elhaissouni N, Colantuoni E, Prochaska EC, Johnson J, Xiao S, Clark RH, Greenberg RG, Benjamin DK, Milstone AM. Epidemiology and mortality of invasive *Staphylococcus aureus* infections in hospitalized infants. JAMA Pediatr. 2025;179:747–55.

Liu GY, Bradley JS. *Staphylococcal* infections. In: Remington and Klein's infectious diseases of the fetus and newborn infant. Elsevier; 2025. p. 395–419.

Mourad A, Nwafo N, Skalla L, Holland TL, Jenkins TC. Oral versus intravenous antibiotic therapy for *Staphylococcus aureus* bacteremia or endocarditis: a systematic review and meta-analysis of randomized, controlled trials. Clin Infect Dis. 2025;80(1):29–36.

Speziale P, Foster TJ, Arciola CR. The endothelium at the interface between tissues and *Staphylococcus*

aureus in the bloodstream. Clin Microbiol Rev. 2025;38:e00098-24.
Szymczak K, Rychłowski M, Zhang L, Nakonieczna J. Harnessing light-activated gallium porphyrins to combat intracellular *Staphylococcus aureus* using an in vitro keratinocyte infection model. Sci Rep. 2025;15(1):1295.
Xu H, Wang S, Liu X, Li M, Wang X, Chen H, Qu C, Liu Y, Liu J. Strategies for survival of *Staphylococcus aureus* in host cells. Int J Mol Sci. 2025;26(2):720.
Zhang T, Wu W, Zhao Y, Ding Z, Wei B, Yang T, Li J, Wang P, Lan L, Gan J, Yang CG. Structure-guided development of ClpP agonists with potent therapeutic activities against *Staphylococcus aureus* infection. J Med Chem. 2025;68:1810–23.

12 Streptococcal Infection

Shradha R. Gaonkar and Rajeshwar V. Naik

Learning Objectives
1. To describe the pathogens involved in zoonotic streptococcal infections
2. To review the genomic and proteomic characteristics of major zoonotic *Streptococcus* species
3. To know the diagnosis, treatment, prevention, and control of zoonotic streptococcal infections

Introduction

Streptococcus is a large genus that contains both beneficial and harmful bacteria. Some species, such as *Streptococcus thermophilus*, are used in dairy fermentation (e.g. yoghurt and cheese production), while certain oral commensals, like *Streptococcus salivarius*, contribute to both good health and commercial applications. In contrast, harmful streptococci present significant public health challenges and are broadly categorised based on their primary transmission routes. This includes those primarily transmitted from person to person, such as *Streptococcus pyogenes* (causing strep throat and scarlet fever), *Streptococcus agalactiae* (a major cause of neonatal sepsis and meningitis), and *Streptococcus pneumoniae* (responsible for pneumonia and meningitis). Additionally, there are zoonotic streptococci, which are transmitted from animals to humans. As of 2020, several zoonotic streptococci have been documented in people. Notable examples include *Streptococcus suis* (linked to pigs, causing meningitis), *Streptococcus equi* subspecies *zooepidemicus* (from horses, leading to invasive diseases), *Streptococcus canis* (from dogs and cats, often causing wound infections), and *Streptococcus iniae* (from fish, causing cellulitis and meningitis) (Table 12.1).

Each zoonotic *Streptococcus* presents a diverse array of clinical manifestations that is influenced mainly by the specific species involved, the route of transmission, and the host's immune status. For example, *S. suis* commonly causes meningitis, sepsis, and hearing loss in humans through contact with infected animals or raw pork, especially via skin lesions. *S. equi* subspecies *zooepidemicus* causes severe, disseminated infections, including sepsis and meningitis, following direct animal contact. *S. canis*, which is prevalent in companion animals, is often linked to bacteraemia following animal bites. *S. iniae*, a fish pathogen, can cause cellulitis and invasive disease in individuals handling infected fish.

S. R. Gaonkar (✉)
Department of Microbiology, Royal Hospital, Margao, Goa, India

R. V. Naik
Department of Internal Medicine, Royal Hospital, Margao, Goa, India

S. C. Parija, A. Poddar (eds.), *Textbook of Bacterial Zoonoses*, Microbial Zoonoses,
https://doi.org/10.1007/978-981-95-6615-0_12

Table 12.1 Overview of pathogenic *Streptococcus* species

Species/subspecies	Transmission type	Lancefield group	Haemolysis	Reservoir	Common human diseases	Transmission route	Remarks
Streptococcus pyogenes (*Group A strep*)	Human-to-human	A	Beta	Humans	Pharyngitis, impetigo, rheumatic fever	Droplets, direct contact	Highly contagious, common in children
Streptococcus pneumoniae	Human-to-human	None	Alpha	Humans (nasopharynx)	Pneumonia, otitis media, meningitis	Droplets	A coloniser that can become invasive
Streptococcus agalactiae (human genotypes)	Human-to-human	B	Beta	Humans (GI and genital tracts)	Neonatal sepsis, meningitis	Vertical (mother to neonate), sexual	Major cause of neonatal infections
Streptococcus mutans	Human-to-human (indirect)	None	Alpha	Humans (oral cavity)	Dental caries	Not directly contagious; oral flora	Biofilm-forming; no animal reservoir
Streptococcus dysgalactiae subsp. *equisimilis*	Human-to-human	C or G	Beta	Humans	Pharyngitis, soft tissue infections	Droplets, direct contact	Related to *Streptococcus pyogenes*
Streptococcus suis	Zoonotic	D-related (capsular types)	Non-beta	Pigs	Meningitis, sepsis, endocarditis	Contact with pigs/ pork	Most prevalent zoonotic species, esp. in Asia
Streptococcus iniae	Zoonotic	None	Beta	Fish	Cellulitis, endocarditis, meningitis	Handling infected/ raw fish	Common in aquaculture workers
Streptococcus canis	Zoonotic	G	Beta	Dogs, cats	Skin and Soft Tissue Infections (SSTIs), bacteraemia	Bites, scratches, pet exposure	Rare but emerging zoonotic infection
Streptococcus equi subsp. *zooepidemicus*	Zoonotic	C	Beta	Horses, cattle	Meningitis, sepsis, pneumonia	Animal contact, raw milk	Can cause severe outbreaks in humans
Streptococcus dysgalactiae subsp. *dysgalactiae*	Zoonotic	C, L	Alpha-, beta-, or non-haemolytic	Cattle, other animals	Bacteraemia, Skin and Soft Tissue Infection (SSTI) (rare)	Animal contact	Less common in humans
Streptococcus halichoeri	Zoonotic	B	Non-haemolytic	Seals, dogs	Sepsis, wound infections	Animal contact (rare; mechanism unclear)	Emerging, rare zoonotic infections
Streptococcus agalactiae (animal genotypes)	Zoonotic	B	Mostly beta	Cattle, fish, dogs, camels	Rare sepsis, meningitis in humans	Animal contact, milk	Distinct genotypes from human strains
Streptococcus porcinus	Zoonotic (possible)	E, P, U	Usually beta	Pigs	Pharyngitis, cellulitis (rare)	Handling pigs or pork	Rare, poorly documented in humans
Streptococcus gallolyticus subsp. *gallolyticus*	Possibly zoonotic	None	Alpha	Ruminants, birds	Endocarditis, bacteraemia	Possibly food-borne or faecal–oral	Strong link to colorectal cancer; zoonotic role debated

History

Streptococcal infections have been recognised for over a century, with the earliest descriptions of these bacterial diseases dating back to the late nineteenth century. In the 1870s, German surgeon Theodor Billroth observed chain-forming bacteria in surgical wound infections, which were later identified as *Streptococcus* species. Subsequently, in the 1880s, Austrian bacteriologist Anton Weichselbaum and German physician Friedrich Fehleisen independently identified *S. pyogenes* as the pathogen responsible for erysipelas and other suppurative infections.

The development of Lancefield classification in the 1930s by American microbiologist Rebecca Lancefield revolutionised the understanding of streptococcal infections. Lancefield's method classified streptococci based on their carbohydrate antigens, identifying different serogroups (A, B, C, etc.) with distinct pathogenic profiles. This classification system remains a cornerstone in streptococcal research and diagnostics.

In contrast to the recognition of *Streptococcus* as a human pathogen, the historical understanding of its zoonotic nature is a relatively recent development. *S. suis* was first described as causing disease in swine in 1954. However, it was not until 1968 that the first documented case of *S. suis* infection in humans was reported in Denmark. Large-scale outbreaks occurred in China, such as the one in Sichuan Province in 2005, involving hundreds of human cases and dozens of fatalities.

S. equi subspecies *zooepidemicus* is a common pathogen in horses. It has been identified in human infections, often in individuals with close equine contact or those consuming unpasteurised dairy products, with early cases appearing in the late twentieth century.

S. iniae, first isolated from a dolphin in 1972, gained attention as a human pathogen primarily in the 1990s. The first recognised human cases occurred in Texas in 1991 and Ontario, Canada, in 1994. These cases were often associated with individuals handling raw fish, reflecting the growth of aquaculture and increased exposure to risk.

Taxonomy

The genus *Streptococcus* belongs to the phylum Firmicutes, class Bacilli, order Lactobacillales, and family Streptococcaceae.

The genus includes over 100 species, which are classified based on haemolytic patterns (alpha-, beta-, or gamma-haemolysis), serological grouping (Lancefield antigens A–U), and molecular phylogenetics. Taxonomic classification has been refined through 16S rRNA sequencing, multilocus sequence analysis (MLSA), and whole-genome sequencing.

The zoonotic bacterium *S suis* is categorised into 35 distinct serotypes based on the differences in its capsular polysaccharide antigens. These capsular components are critical to the bacterium's ability to cause disease and evade the host's immune responses. In swine populations, a small subset of these serotypes (notably 2, 3, 7, and 9) is most frequently implicated in disease outbreaks. For human infections, serotype 2 is the most isolated strain; however, other serotypes, including serotypes 5 and 14, have also been identified in patients. Furthermore, genetic analysis has revealed certain isolates that diverge from the classic *S. suis* and are often referred to as "*S. suis*-like" organisms. For some of these genetically distinct variants, new species names such as *Streptococcus orisratti*, *Streptococcus parasuis*, and *Streptococcus ruminantium* have been proposed.

Genomics and Proteomics

The genome of the genus *Streptococcus* exhibits a conserved core and a highly variable accessory genome. Members of the genus have a circular chromosome ranging from approximately 1.8 to 2.2 Mb, with moderate-to-low GC content. A comparative overview of genomics and proteomics of zoonotic *Streptococcus* is presented in Table 12.2. A discussion on *S. suis* is presented.

Table 12.2 Comparative genomics of zoonotic *Streptococcus* infections

Species	Genome size (Mb)	Capsular types/ genes	Key virulence genes	Mobile elements/PAI	Notable proteins (proteome)	Proteomic adaptations
Streptococcus suis	2.0–2.3	>35 types (cps locus), esp. type 2	sly, mrp, epf, fbps, cps, orf2	ICEs, prophages, PAIs	Suilysin, MRP, EF, enolase, GAPDH, FBPs	Stress response Superoxide Dismutase ((SOD), thioredoxin), host adhesion, iron uptake
Streptococcus iniae	~2.1	No capsule; capsule gene cluster variable	sagA, cpsD, simA	Fewer mobile genetic elements (MGEs) than *S. suis*	Surface antigen I/II, enolase, cytolysins	Proteins upregulated under fish host conditions
Streptococcus canis	~2.0	Limited data (capsule gene homologues found)	scpA, cbp, fbp	Some ICEs and phage remnants	Collagen-binding protein (CBP), FBP, DNase	Virulence-associated proteins expressed during dog/cat infection
Streptococcus equi subsp. *zooepidemicus*	~2.1–2.2	cps-like locus; capsular synthesis genes present	szm, seeH, lmb, fbpZ	Phage-associated PAIs	SzM (M-like protein), Lmb, fibronectin-binding protein Z (FBPZ), hyaluronidase	Secretes adhesins, invasins in host environments
Streptococcus dysgalactiae subsp. *dysgalactiae*	~2.0	Variable cps-like loci	emm, scpA, sagA (in some strains)	Lower PAI content	Enolase, M-protein (Emm), plasminogen-binding proteins	Proteins adapted to host fibrinogen and immune interactions
Streptococcus agalactiae (animal strains)	~2.1–2.2	Group B (cps operon highly conserved)	cfb, scpB, bibA, pilA-C	ICEs and genomic islands	C5a peptidase, Pili (PilA–PilC), BibA	Pili expression during mucosal colonisation
Streptococcus halichoeri	~2.1	Poorly characterised	Limited known virulence genes	Unknown	Few known proteins; possible surface adhesins	Largely unstudied; low expression variability noted
Streptococcus porcinus	~2.0	Group E/U; homologous cps loci	scpA, emm-like genes	Phages, transposases	M-like proteins, DNase	Closely mimics Group A Strep proteome
Streptococcus gallolyticus subsp. *gallolyticus*	~2.2	No capsule	gallo2178, adhesins, pilB	ICEs and phage elements	PilB, collagen-binding protein, sortases	Expresses biofilm and colon tissue adhesion proteins in vivo

Genome of *Streptococcus suis*

The genome of *S. suis* is relatively compact, typically ranging in size from 2.0 to 2.3 megabases (Mb), with a GC content of approximately 41%. It consists of a single circular chromosome with around 2000–2200 protein-coding genes, depending on the strain. Some strains may possess small plasmids that contribute to antibiotic resistance or metabolic versatility.

The *S. suis* pan-genome is open. This plasticity is evident in the distinction between the core genome (about 1100 conserved genes) and the accessory genome, which includes mobile genetic elements such as prophages, integrative and conjugative elements (ICEs), transposons, and strain-specific virulence and resistance genes. These accessory genes often contribute to the organism's adaptability and virulence, especially in zoonotic contexts.

S. suis virulence is contributed by the capsular polysaccharide (CPS), which has been identified in more than 35 serotypes. Among them, serotype 2 is the most associated with human disease, particularly in South-east Asia. The cps locus, which governs capsule synthesis, varies significantly among serotypes and is a major factor in immune evasion. Other important virulence factors include suilysin (SLY), a cholesterol-dependent cytolysin; muramidase-released protein (MRP); extracellular factor (EF); and fibronectin-binding proteins (FBPs), which facilitate adhesion to host tissues. Highly virulent strains often harbour pathogenicity islands (PAIs) that encode multiple virulence elements, along with secretion systems and stress response genes.

Comparative genomics has shown that *S. suis* strains infecting humans, especially those belonging to sequence type 1 (ST1) and clonal complex 1 (CC1), are genetically distinct from non-zoonotic or low-virulence strains. These zoonotic strains frequently carry PAIs and exhibit unique transcriptional profiles during infection, upregulating genes associated with capsule production, oxidative stress defence, iron acquisition, and tissue adhesion.

S. suis also shows antimicrobial resistance, often due to horizontal gene transfer. Resistance genes for macrolides (erm(B)), tetracyclines (tet(M)), and aminoglycosides have been identified, often located on transposons or ICEs.

Proteomics of *Streptococcus suis*

Based on complete genome sequences, *S. suis* encodes approximately 2000–2200 proteins, depending on the strain. These proteins encompass both housekeeping enzymes and a diverse array of virulence-associated proteins, surface adhesins, immune evasion factors, and stress response proteins. The *S. suis proteome* is characterised *by* the presence of surface-anchored proteins, many of which contain Leu-Pro-X-Thr-Gly amino acid (LPXTG) motifs, allowing them to be covalently attached to the cell wall via sortase enzymes. These include important virulence factors such as muramidase-released protein (MRP), extracellular protein factor (EF), and fibronectin-binding proteins (FBPs). These proteins mediate adhesion to host extracellular matrix, colonisation, and evasion of host immune defences. Suilysin (SLY), a pore-forming cytolysin, is another key virulence-associated protein found in many zoonotic strains. It belongs to the cholesterol-dependent cytolysin family and contributes to host cell lysis and inflammation.

The *S. suis* secretome is especially relevant to virulence. It includes not only toxins like suilysin but also proteases, DNases, and neuraminidases, which assist in tissue invasion and immune modulation. Additionally, moonlighting proteins such as enolase and GAPDH, although traditionally cytoplasmic metabolic enzymes, have been shown to localise on the bacterial surface and bind host molecules like plasminogen, playing roles in invasion and immune evasion.

Comparative proteomic analyses between highly virulent zoonotic strains (e.g. serotype 2 and ST1) and low-virulence or commensal strains reveal significant differences in the expression of these virulence proteins. For instance, serotype 2 strains frequently express higher levels of capsular proteins, suilysin, and MRP/EF, all of which contribute to their pathogenicity in both pigs and humans.

Morphology

Streptococcus species are Gram-positive cocci that typically form chains or pairs (from the Greek word *streptos*, meaning "twisted chain"). They are non-motile and non-spore-forming. Zoonotic *Streptococcus* species are morphologically small, with differences in capsule size, chain formation, and mucoid character that help in laboratory recognition.

S. suis, the most prevalent zoonotic species, typically measures 0.5–1.0 μm in diameter and appears as cocci arranged in short- or medium-length chains (Fig. 12.1). Encapsulated strains, such as the virulent serotype 2, appear larger and more mucoid, with the capsule adding an additional 0.2–0.5 μm in diameter around the cell. These encapsulated forms are particularly noticeable in blood or cerebrospinal fluid (CSF) smears. In contrast, *S. iniae*, primarily found in fish, is slightly smaller, generally 0.6–1.0 μm in size. It appears as short chains or pairs and has a less mucoid surface compared to *S. suis*. Its relatively compact structure supports its adaptation to aquatic environments and the fish host.

Cultivation

Streptococci are facultative anaerobes, capable of growing in both aerobic and anaerobic conditions. Cultures are usually incubated at 37 °C for 18–48 h. *Streptococcus* species are cultivated on enriched media, such as blood agar plates, where they exhibit distinct haemolytic patterns, namely alpha-, beta-, or gamma-haemolysis (Table 12.3). Colonies of streptococci on blood agar are generally small (0.5–2 mm in diameter), round, and often translucent to greyish-white.

Other enriched media, such as trypticase soy agar (TSA), with or without yeast extract, and brain heart infusion broth or agar are also frequently used. Selective media containing specific antibiotics are used to isolate *Streptococcus* from samples with mixed flora.

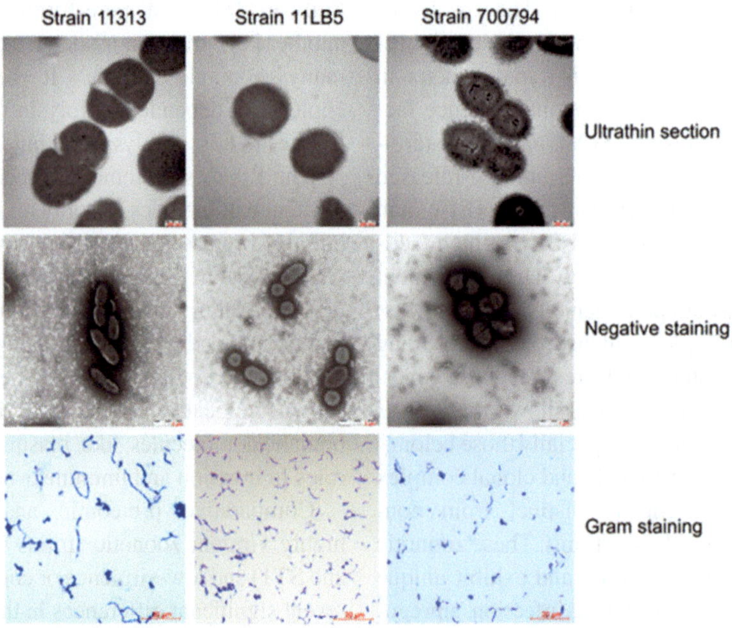

Fig. 12.1 Transmission electron micrograph and Gram-stained smear showing coccoid morphology of *Streptococcus suis* strains arranged in short- or medium-length chains. (Image source: Wang S, Zhang D, Jiang C, He H, Cui C, Duan W, Hu S, Wang J, Cai X. Strain characterization of *Streptococcus suis* serotypes 28 and 31, which harbor the resistance genes optrA and ant(6)-Ia. Pathogens. 2021;10(2):213. https://doi.org/10.3390/pathogens10020213. Permitted for reproduction under CC BY license)

Table 12.3 Classification of *Streptococcus* based on haemolytic properties and representative species

Haemolytic property	Description	*Streptococcus* species (examples)
Alpha (α)	Results in a partial lysis of red blood cells around the bacterial colony on a blood agar plate. This incomplete haemolysis results in a characteristic narrow, greenish or brownish discolouration of the agar due to the reduction of haemoglobin to methaemoglobin	*Streptococcus pneumoniae*, *Streptococcus viridans* group (e.g. *Streptococcus mitis* and *Streptococcus salivarius*)
Beta (β)	Characterised by a complete lysis of red blood cells in the medium surrounding and underneath the bacterial colony. This complete breakdown of red blood cells results in a clear, colourless halo or zone of clearing around the colony on the blood agar	*Streptococcus pyogenes* (Group A *Streptococcus*), *Streptococcus agalactiae* (Group B *Streptococcus*)
Gamma (γ)	Indicates that there is no haemolysis or no apparent lysis of red blood cells occurring around the bacterial colony on the blood agar plate. The agar under and around the colony remains unchanged, with no noticeable discolouration or clearing	*Enterococcus faecalis* (formerly *Streptococcus faecalis*), *Streptococcus bovis* group (now often classified within the *Enterococcus* or other genera)

Animal Models

Animal models such as mice, rabbits, and guinea pigs are generally used to study the pathogenesis and immune response to streptococcal infections. Among zoonotic species, *S. suis* has the most developed models. Piglets and young pigs serve as the most accurate model for human disease. Various inoculation routes can be used, such as intranasal to simulate natural exposure or intravenous to induce systemic infection. These porcine models effectively reproduce clinical signs, such as meningitis, arthritis, septicaemia, and respiratory disease, that are characteristic of *S. suis* infections in both pigs and humans. Other than pigs, mice are widely used for experimental infections, virulence studies, and drug screening. Invertebrate models, such as *Galleria mellonella* (wax moth larvae), provide cost-effective alternatives for high-throughput screening. *S. iniae* is primarily studied in fish models, such as tilapia and zebrafish.

Transmission Cycle

The zoonotic transmission cycle of *Streptococcus* species begins with the pathogen residing in its animal reservoir, which is often an asymptomatic carrier. Depending upon the pathogen, these reservoirs can be livestock, companion animals, or aquatic species. The bacteria are shed into the environment through various routes, making it a potential environmental reservoir. The transmission cycle to humans typically occurs through several pathways.

Direct contact with infected animals or their bodily fluids is a common route, often when individuals have cuts or abrasions that allow bacterial entry. This is particularly relevant for *S. suis*, where pig farmers and abattoir workers are at a high risk through the direct handling of pigs or raw pork, especially when skin lesions are present. Similarly, *S. equi* subspecies *zooepidemicus* can be transmitted through direct contact with infected horses or other livestock.

Indirect contact via contaminated environments, surfaces, or even arthropod vectors (though less common for streptococci) can also play a role. Food-borne transmission is significant for some species; for instance, consuming undercooked pork or contaminated pig products is a major route for *S. suis* infection, and ingesting unpasteurised dairy can transmit *S. zooepidemicus*. For *S. iniae*, handling infected fish, especially with cuts on hands, or consuming raw fish products is the primary mode of transmission.

The infective dose is not always precisely quantified for zoonotic streptococci in humans. However, a higher bacterial load and the presence of predisposing factors (e.g. immunocompromised status or open wounds) increase the risk of disease. Once transmitted, humans are typically incidental hosts and can develop severe disease.

Pathogenesis and Pathology

The pathogenesis mechanisms differ among zoonotic species. However, a common strategy exists among them for adherence, immune evasion, toxin production, and tissue invasion. A major mechanism across many zoonotic streptococci is the production of a polysaccharide capsule. This outer layer, as seen in *S. suis*, has 35 serotypes based on capsular antigens. *S. equi* subspecies *zooepidemicus*, possessing a hyaluronic acid capsule, is critical for evading host phagocytosis and complement-mediated killing, allowing the bacteria to persist and spread within the host. Beyond capsules, various surface proteins play crucial roles in adherence to host cells and tissues. For instance, M-like proteins like SCM in *S. canis* and fibronectin-binding proteins facilitate attachment to host extracellular matrix components.

Once adhered, these bacteria secrete enzymes and toxins to damage host cells, facilitate spread, and modulate the immune response. Haemolysins, such as suilysin, produced by *S. suis*, are pore-forming toxins that can lyse red blood cells and other host cells. It contributes to tissue damage and systemic inflammation. Other enzymes, such as proteases, degrade host proteins, facilitating tissue penetration and immune evasion. Some strains may also produce superantigens. These superantigens which can overstimulate the host immune system, leading to an exaggerated inflammatory response. These potentially contributing to conditions like streptococcal toxic shock syndrome. However, their role varies by species and strain.

Pathology

The specific pathological presentation varies depending on the *Streptococcus* species and the primary sites of infection. In the case of *S. suis*, the major pathological feature is purulent meningitis. Gross examination of affected brains often reveals opaque, thickened meninges with a yellowish, creamy exudate covering the surface of the brain and spinal cord. Histologically, there is a severe neutrophilic infiltration within the subarachnoid space, accompanied by fibrin deposition and abundant bacterial colonies. This meningeal inflammation can extend to affect cranial nerves, particularly the auditory nerve, leading to irreversible hearing loss, frequently observed as a sequela in human survivors.

Beyond the central nervous system, many zoonotic streptococci, including *S. suis* and *S. equi* subspecies *zooepidemicus*, cause septicaemia and multi-organ involvement. In these systemic infections, gross pathological findings may include widespread petechial haemorrhages on various serosal surfaces and within internal organs, indicating microvascular damage. Organs such as the spleen are often markedly enlarged due to reactive hyperplasia, and regional lymph nodes are swollen and oedematous. At the microscopic level, there is evidence of disseminated intravascular coagulation (DIC), widespread inflammation, and microscopic bacterial emboli occluding small blood vessels. These result in focal areas of necrosis and infarction in organs such as the liver, kidneys, and lungs, which contribute to systemic organ failure.

Other common pathological manifestations include septic arthritis and endocarditis. In septic arthritis, typically seen with *S. suis* and *S. zooepidemicus*, the affected joints are swollen and painful, containing copious amounts of purulent synovial fluid. The synovial membrane itself becomes hyperaemic, oedematous, and infiltrated with inflammatory cells. Chronic cases can result in irreversible damage to articular cartilage and subchondral bone.

Immunology

The host initiates a rapid *innate immune response* immediately after *Streptococcus* organisms breach epithelial barriers. Neutrophils and macrophages recognise lipoteichoic acid and peptidoglycan, the pathogen-associated molecular patterns (PAMPs) on streptococcal cell walls by Toll-like receptors (TLR2/TLR6 heterodimers). Stimulation of these receptors triggers NF-κB signalling, leading to the release of pro-inflammatory cytokines (TNF-α, IL-1β, and IL-6) and chemokines (IL-8/CXCL8) that attract additional phagocytes to the infection site. Complement activation through the classical and alternative pathways deposits C3b on the bacterial surface, enhancing opsonophagocytosis. However, streptococcal M-protein and hyaluronic acid capsule can prevent complement deposition, enabling immune evasion.

Adaptive immunity develops subsequently. Antigen-presenting cells (APCs) process streptococcal proteins, such as M-protein, streptolysin O, and Spe exotoxins. They present peptides on major histocompatibility complex (MHC) class II to naïve $CD4^+$ T cells. Differentiation into Th1 cells producing IFN-γ enhances macrophage bactericidal activity. Th17 cells secrete IL-17A/F, which promotes neutrophil mobilisation and epithelial antimicrobial peptide expression, crucial for the mucosal clearance of streptococci. Concurrently, B cell activation gives rise to plasma cells that produce opsonic IgG and IgA directed against surface antigens. These antibodies neutralise toxins, enhance complement fixation, and confer serotype-specific protection. Memory B and T cells generated during the response underlie long-term immunity, although antigenic variability among streptococcal serotypes can limit cross-protection.

In streptococcal infections, dysregulated immunity contributes to pathology. For instance, excessive cytokine release drives tissue damage in severe infections, and molecular mimicry between M-protein epitopes and host cardiac tissue underlies post-streptococcal rheumatic fever. However, some *Streptococcus* species have evolved mechanisms to evade host immunity, such as capsule formation and molecular mimicry.

Infections in Humans

Zoonotic *Streptococcus* infections in humans can be broadly categorised into two distinct types: suppurative and non-suppurative manifestations. While these two are from direct bacterial invasion of the human host, zoonotic *Streptococcus* also presents toxin-mediated manifestations.

Suppurative Manifestations

These infections are characterised by an acute inflammatory response, leading to the formation of pus, rich in neutrophils, at the site of infection. One of the most critical suppurative manifestations is bacterial meningitis, which is prominently seen with *S. suis* infection. Patients typically present with high fever, severe headache, neck stiffness, and altered mental status, reflecting the purulent exudate accumulating within the meninges and cerebrospinal fluid. A notable sequela is often permanent sensorineural hearing loss.

Skin and soft tissue infections, such as cellulitis, are also common suppurative conditions, particularly after skin breaches or animal bites. *S. iniae*, for instance, frequently causes cellulitis in fish handlers, presenting as localised redness, warmth, swelling, and pain. These can progress to abscess formation. Furthermore, septic arthritis, where pus accumulates within joint capsules, causing pain and swelling, can result from infections with *S. suis* and *S. equi* subspecies *zooepidemicus*. Bacterial pneumonia, potentially caused by *S. zooepidemicus* as part of a systemic spread, may also present as a suppurative process with purulent exudates in the lungs.

Non-suppurative and Systemic Manifestations

While not directly characterised by pus formation, these conditions represent severe systemic responses to the infection. Bacteraemia, the presence of bacteria in the bloodstream, frequently progresses to sepsis, a life-threatening systemic inflammatory response syndrome (SIRS). Symptoms include high fever, chills, rapid heart rate, rapid breathing, and septic shock, leading to multi-organ dysfunction, such as acute kidney injury and respiratory distress. This systemic inflammatory cascade is exacerbated by bacterial toxins, such as haemolysins like suilysin from *S. suis* and, potentially, superantigens. These trigger an overwhelming "cytokine storm."

Infective endocarditis**,** particularly associated with *S. suis*, is another severe non-suppurative manifestation. The condition involves the formation of vegetations comprising fibrin, platelets, and bacteria, rather than pus, on heart valves, which can lead to valve damage and embolic events. While toxic shock–like syndrome (TSLS) is more commonly linked to human-adapted streptococci, its potential occurrence in severe zoonotic streptococcal infections underscores the impact of toxin-mediated systemic effects.

Toxin-Mediated Infections

These secreted bacterial toxins directly injure host cells, modulate immune functions, or disrupt vital physiological processes. Among the prominent classes of toxins involved are haemolysins, which are cytolytic toxins capable of lysing various host cells, including erythrocytes, leucocytes, and epithelial cells. A prime example is suilysin (SLY)**,** a potent pore-forming toxin produced by *S. suis*. SLY contributes significantly to the tissue damage, inflammation, and cellular dysfunction observed in severe *S. suis* infections. These facilitate bacterial dissemination and the development of conditions such as meningitis and septicaemia. Similarly, the streptolysins produced by other beta-haemolytic zoonotic streptococci, such as *S. equi* subspecies *zooepidemicus* and *S. canis*, contribute to their ability to cause localised tissue necrosis and systemic spread. These toxins create pores in cell membranes, leading to cell death and the release of pro-inflammatory mediators, further exacerbating the host's inflammatory response.

Some zoonotic streptococci also produce *superantigens* (*SAgs*). These powerful exotoxins act as non-specific immune activators. These bind directly to major histocompatibility complex (MHC) class II molecules and T cell receptors, bypassing conventional antigen presentation. This bypass leads to a massive, polyclonal activation of T cells and an overwhelming release of pro-inflammatory cytokines, often referred to as a "cytokine storm." This uncontrolled systemic inflammation results in *streptococcal toxic shock–like syndrome* (*STSLS*). STSLS is a severe and life-threatening condition characterised by the rapid onset of fever, hypotension, and multi-organ failure. Cases of STSLS have been reported in humans infected with zoonotic species, such as *S. suis* and *S. equi* subspecies *zooepidemicus*.

An overview of human infections by zoonotic *Streptococcus* is presented in Table 12.4.

Infections in Animals

Infections in Pigs

S. suis is a major bacterial pathogen of swine globally. It is considered one of the most significant causes of mortality in piglets, particularly after weaning. While many pigs can be asymptomatic carriers, particularly in their upper respiratory tract (such as the tonsils), virulent strains cause severe systemic disease. Common manifestations include meningitis (neurological signs like tremors, incoordination, paddling, and sudden death), septicaemia (leading to sudden death, with fewer prior signs, like fever, depression, and poor appetite), arthritis (swollen joints and lameness), and endocarditis (inflammation of heart valves). Other less common conditions include pneumonia, polyserositis, and abortion in sows.

Table 12.4 Infections in humans by zoonotic *Streptococcus*

Species	Suppurative manifestations	Non-suppurative manifestations	Toxin-mediated manifestations
Streptococcus suis	Meningitis (often with deafness) Sepsis and septic shock Arthritis Pneumonia endocarditis	Immune-mediated vasculitis (rare) Mild post-infectious phenomena	Toxic shock–like syndrome (TSLS) (especially with serotype 2)
Streptococcus iniae	Cellulitis Abscess Bacteraemia endocarditis Septic arthritis	Fever without localising signs, especially in the elderly	No known specific toxin-mediated syndromes
Streptococcus canis	Necrotising fasciitis SSTI Bacteraemia osteomyelitis (rare)	Not reported	TSLS-like features in immunocompromised cases (rare)
Streptococcus equi subsp. *zooepidemicus*	Sepsis Meningitis Pneumonia Endocarditis Purulent pleuritis or peritonitis	Rare sequelae Inflammatory vasculitis (occasional)	Toxic shock–like syndrome due to superantigen-like proteins
Streptococcus agalactiae (animal strains)	SSTI Bacteraemia Wound infection	Very rare immune sequelae	Beta-haemolysin/cytolysin-mediated tissue damage
Streptococcus dysgalactiae subsp. *dysgalactiae*	Pharyngitis SSTI Endocarditis Wound infection	Post-streptococcal glomerulonephritis (rare, strain-dependent)	Limited data; possible toxin-mediated skin involvement
Streptococcus halichoeri	Wound infections Cellulitis Bacteraemia	Not documented	Not established
Streptococcus porcinus	Pharyngitis Wound infections Vaginal/cervical infections	Not reported	Unclear; may produce M protein–like toxins
Streptococcus gallolyticus subsp. *gallolyticus*	Endocarditis Bacteraemia Hepatic abscess Colonic lesions	Paraneoplastic syndromes	No known toxin but expresses virulence-associated adhesins

Infections in Horses and Other Livestock

S. equi subspecies *zooepidemicus* is a common opportunistic pathogen in horses, where it is often found as commensal flora in the upper respiratory tract. However, under stress or co-infection, it can cause a range of diseases, including upper respiratory tract infections (e.g. pleuropneumonia or "shipping fever"), metritis (a uterine infection in mares), arthritis, mastitis, and septicaemia. It is also increasingly recognised as a pathogen in other species, such as pigs, cattle, sheep, goats, dogs, and cats, causing similar systemic and localised infections.

S. dysgalactiae is another important streptococcal species in livestock, most notably causing mastitis in cattle and small ruminants. This can range from subclinical mastitis, which reduces milk quality and yield, to severe clinical forms, with fever and udder swelling. *S. dysgalactiae* is also a common cause of infectious arthritis (joint-ill) in lambs and can cause septicaemia.

Infections in Companion Animals

S. canis is a typical inhabitant of the skin and mucosal surfaces of the throat and the vagina of healthy dogs and cats. However, it can become an opportunistic pathogen, especially in young, old, or immunocompromised animals or when there are skin wounds. In dogs and cats, *S. canis* causes various infections, including skin and ear infections, pharyngitis, pneumonia, urinary tract infections (UTIs), arthritis, and neonatal sepsis in puppies and kittens. More severe, invasive conditions like necrotising fasciitis and streptococcal toxic shock syndrome (STSS) is characterised by rapid tissue destruction and multi-organ failure. The condition has also been reported in dogs, often following minor trauma.

Infections in Aquatic Animals

S. iniae is a leading bacterial pathogen in aquaculture worldwide. It causes significant economic losses in numerous farmed and wild fish species, including tilapia, rainbow trout, and hybrid striped bass. Infections in fish typically manifest as meningoencephalitis (leading to neurological signs such as erratic swimming, lethargy, and loss of orientation), septicaemia (characterised by lethargy, darkening of the skin, exophthalmia ("pop-eye"), corneal opacity, and internal and external haemorrhages), and skin lesions.

Epidemiology and Public Health

The global epidemiology of zoonotic *Streptococcus* infections presents a diverse distribution. It is influenced by animal reservoirs, occupational exposure, food practices, and regional public health infrastructure. *S. suis* is endemic in nearly all countries with a developed pig industry, including North America, South America, Europe, Asia, Australia, and New Zealand. Outbreaks in Vietnam, Thailand, and China have been linked to the consumption of raw or undercooked pork and occupational exposure to pigs. *S. suis* serotype 2 is globally the most prevalent and pathogenic serotype for both pigs and humans. However, other serotypes such as serotype 14 in Southeast Asia are also important.

S. iniae, primarily affecting fish handlers, is reported from North America, Asia, and the Middle East, often following injuries sustained during the preparation of fish. Species like *S. equi* subsp. *zooepidemicus* and animal strains of *S. agalactiae* have caused outbreaks through unpasteurised dairy or aquaculture products, especially in Europe, China, and parts of Africa. Less common zoonotic agents such as *S. canis*, *S. halichoeri*, and *S. porcinus* have a more sporadic distribution. However, they remain important in immunocompromised individuals and animal-contact professions. *S. gallolyticus* subspp. *gallolyticus*, is not a classic zoonotic pathogen. However, it shows a strong association with colorectal cancer globally, possibly influenced by animal or environmental exposure.

A global epidemiology of zoonotic *Streptococcus* is presented in Table 12.5.

Diagnosis

Specimens

In human patients with suspected meningitis, a cerebrospinal fluid (CSF) sample is obtained by lumbar puncture. In instances of sepsis or bacteraemia, blood cultures collected from the patient are preferred. Localised conditions like skin and soft tissue infections (e.g. cellulitis from *S. iniae* or *S. canis*) necessitate swabs or aspirates from the lesion's leading edge or pus from any formed abscess. Suspected pneumonia cases require the collection of sputum, bronchoalveolar lavage (BAL) fluid, or tracheal aspirates.

In animal hosts, sample collection depends on the animal species. For pigs with suspected *S. suis* infection, samples include brain tissue, ton-

Table 12.5 Global epidemiology of zoonotic *Streptococcus*

Species	Geographic distribution	High-risk populations	Sources of infection	Notable outbreaks/ trends
Streptococcus suis	South-East Asia (Vietnam, Thailand, China), parts of Europe and Latin America	Pig farmers, butchers, pork consumers, abattoir workers	Infected pigs or raw/undercooked pork	Major outbreaks in China (2005, Sichuan); endemic in South-east Asia
Streptococcus iniae	North America, south-East Asia, Middle East, Latin America	Fish handlers, elderly, immunocompromised	Handling or preparing infected freshwater fish (e.g. tilapia)	Repeated cases in Taiwan, Hong Kong, Canada
Streptococcus canis	North America, Europe, sporadically in Asia	Dog owners, veterinarians, elderly or immunosuppressed	Dog bites, scratches, mucosal contact	Rare, sporadic zoonotic cases reported
Streptococcus equi subsp. *zooepidemicus*	Globally distributed; Europe, North America, China	Horse owners, dairy workers, immunocompromised individuals	Horses, unpasteurised milk, close equine contact	Fatal meningitis/ sepsis outbreaks in the UK, China (linked to dairy)
Streptococcus agalactiae (animal strains)	Africa, Asia, South America, parts of Europe	Dairy workers, aquaculture workers	Raw milk, infected fish	Aquaculture outbreaks in south-East Asia, human cases via fish exposure
Streptococcus dysgalactiae	Europe, North America, parts of Asia	Livestock handlers	Cattle contact, contaminated environment	Human cases are rare; more common as an animal pathogen
Streptococcus halichoeri	Europe (Scandinavia), USA, coastal regions	Pet owners, immunocompromised	Contact with seals, dogs, possibly seafood	Emerging; isolated from dogs and a few human cases in the Nordic area
Streptococcus porcinus	Asia (China, Korea), North America	Pig handlers, meat processors	Pigs (oral, urogenital, skin flora)	Rare zoonotic reports; underdiagnosed
Streptococcus gallolyticus subsp. *gallolyticus*	Worldwide, especially in the elderly	Elderly, cancer patients	Possibly food-borne or animal exposure (birds, cattle, rodents)	Strong epidemiological link to colorectal cancer worldwide

sils, lung tissue, spleen, joint fluid, and blood. Nasal or tonsil swabs are also collected for carrier detection in herds. In equine cases suggesting *S. equi* subspecies *zooepidemicus* infection, nasopharyngeal swabs, tracheal washes, bronchoalveolar lavage fluid, joint aspirates, or uterine swabs (for metritis) are frequently taken.

For companion animals, such as dogs and cats, suffering from *S. canis* infections, specimens are collected from wound swabs, ear swabs, pharyngeal or vaginal swabs, blood, or tissue biopsies from affected areas. In aquatic animals infected by *S. iniae*, preferred samples include brain, kidney, and spleen tissues.

Microscopy

Microscopy plays a crucial role, primarily using Gram staining and microscopic examination of clinical specimens. This method enables the initial identification of Gram-positive cocci arranged in chains or pairs, which is indicative of the *Streptococcus* species. This provides rapid preliminary results to guide immediate treatment decisions.

In Vitro Culture

In vitro culture remains the gold standard for definitive diagnosis, involving the isolation and identification of *Streptococcus* organisms from clinical samples. Blood agar is primarily used for streptococcal culture. This not only supports growth but also reveals haemolytic patterns (alpha, beta, or gamma) that offer a preliminary means of differentiation (Table 12.3).

Enrichment broths are often used to increase the bacterial load from samples, especially those with low bacterial counts, before plating on solid media. Brain heart infusion (BHI) broth and Todd–Hewitt broth (THB) are commonly used for this purpose. When grown in these liquid media, zoonotic streptococci produce general turbidity throughout the broth. Due to their non-motile nature and chain formation, they typically settle to form a sediment at the bottom of the tube after a period of incubation. These broths are also vital for preparing bacterial suspensions for subsequent biochemical tests and antimicrobial susceptibility testing.

The biochemical characterisation of zoonotic *Streptococcus* species provides essential clues for laboratory identification and differentiation from non-zoonotic streptococci. While all species are catalase-negative, they exhibit diversity in select biochemical tests that help distinguish them. For example, *S. agalactiae* is the only species that tests positive in the CAMP and hippurate hydrolysis tests. Likewise, *S. gallolyticus* subsp. *gallolyticus* is positive for the bile esculin hydrolysis test, while other species show negative or variable results.

Serotyping

Lancefield Grouping

This test classifies beta-haemolytic *Streptococcus* species based on specific carbohydrate antigens present in their cell walls. This is typically performed using commercial latex agglutination kits. *S. equi* subspecies *zooepidemicus* belongs to Lancefield Group C. *S. canis* is classified as Lancefield Group G. Conversely, *S. suis* generally lacks a definitive Lancefield group antigen. *S. iniae* does not react with common Lancefield grouping antisera (A–G), rendering this method less useful for their direct serological identification.

Capsular Serotyping

It is critical for specific species like *S. suis*. The bacteria possess a diverse polysaccharide capsule, which serves as the basis for their serological classification into at least 29 distinct serotypes. This subtyping is performed on bacterial isolates using specialised antisera through co-agglutination or the capsular swelling test by Quellung reaction. Such serotyping helps in epidemiological tracking, particularly for identifying highly pathogenic strains like *S. suis* serotype 2, which is globally prevalent and virulent in both pigs and humans.

Molecular Diagnosis

Polymerase chain reaction (PCR); its quantitative variant, real-time PCR (qPCR); and multiplex PCR are commonly used for diagnostic purposes. It offers remarkable sensitivity, capable of identifying minute quantities of bacterial DNA, even in challenging clinical samples such as cerebrospinal fluid or blood from patients who have already received antibiotics.

Multiplex PCR further enhances efficiency by simultaneously detecting multiple *Streptococcus* species or even various serotypes in a single reaction. For instance, specific PCR assays for *S. suis* can not only confirm the species (e.g. by targeting 16S rRNA genes) but also identify specific capsular serotypes (e.g. serotypes 2 or 14 via cps genes), which is crucial for epidemiological tracking and understanding virulence.

The diagnostic approaches for detecting *Streptococcus* infections are summarised in Table 12.6.

Treatment

The treatment of zoonotic *Streptococcus* infections in both humans and animals involves targeted antimicrobial therapy, along with supportive care to manage symptoms and complications. The selection of the appropriate antibiotic and the duration of treatment are guided by prompt antimicrobial susceptibility testing (AST) of the isolated pathogen. Beta-lactam antibiotics, particularly penicillins such as penicillin G, ampicillin, and amoxicillin, are often the drugs of choice due to the high susceptibility of *Streptococcus* species to this class. For severe systemic infections, intravenous administration is initially preferred to achieve adequate drug concentrations, especially in difficult-to-penetrate sites like the central nervous system. Supportive measures, including fluid therapy, anti-inflammatory medications, and surgical intervention for abscess drainage or debridement, are vital adjuncts to antibiotic treatment.

An overview of the treatment of different zoonotic streptococci is presented in Table 12.7.

Prevention and Control

Farm Biosecurity and Animal Management

Effective biosecurity protocols are fundamental to controlling zoonotic streptococci in animal populations. This includes strict hygiene measures on farms, such as regular cleaning and disinfection of animal housing, equipment, and transport vehicles. Reducing overcrowding, ensuring proper ventilation, and maintaining stable environmental temperatures minimise stress, which often predisposes animals to infection. Implementing robust quarantine protocols for new animals introduced to a herd or facility prevents the introduction of new strains. Good animal husbandry practices, including optimal nutrition and minimising co-infections with other pathogens, reduce susceptibility to disease. For specific pathogens, such as *S. suis*, strategies in pig farming include ensuring adequate colostrum intake for piglets, optimising gut health during the vulnerable weaning phase, and avoiding the mixing of pigs from different age groups or sources to reduce horizontal transmission.

Vaccination

Vaccination plays a crucial role in preventing and controlling zoonotic streptococcal diseases in animals, although human vaccines are not available. For *S. suis*, commercially available inactivated vaccines are used in pigs. The vaccines induce active immunity in sows and provide passive immunity to piglets through colostrum, thereby reducing mortality and clinical signs, particularly those caused by common serotypes, such as serotype 2. For *S. iniae* in aquaculture, ongoing research into novel vaccine approaches and "green alternatives," such as bacteriophage endolysins, aims to reduce the reliance on antibiotics.

Preventing Human Exposure

Preventing direct human exposure to infected animals or contaminated animal products is paramount. This is particularly relevant for high-risk groups such as pig farmers, abattoir workers, veterinarians, butchers, and fish handlers. This can be achieved by following frequent and thorough hand washing after contact with animals or raw meat and fish and wearing gloves, waterproof adhesive dressings over wounds, and protective clothing (e.g. masks and goggles) when handling animals, carcasses, or raw meat and fish.

To prevent food-borne transmission, particularly relevant for *S. suis* from pork and *S. zooepidemicus* from unpasteurised milk, strict food hygiene practices are critical. This involves purchasing food from reputable sources, separating raw and cooked foods to prevent cross-contamination, thoroughly cleaning surfaces and utensils that have come in contact with raw products, and ensuring that all pork and other animal products are cooked thoroughly to a safe internal temperature before consumption. Avoiding the

Table 12.6 Diagnosis of zoonotic *Streptococcus* infections

Diagnostic technique	Specific tools/methods	Purpose	Examples	Remarks
Microscopy	Gram stain, light microscopy	Rapid preliminary detection; Gram-positive cocci in chains or pairs suggest *Streptococcus*	All zoonotic streptococci (e.g. *Streptococcus suis* and *Streptococcus agalactiae*)	Low sensitivity; requires confirmation; helpful in urgent treatment decisions
In vitro culture	Blood agar for haemolysis; BHI or THB broth for enrichment	Gold standard; isolates viable bacteria; haemolytic pattern assists preliminary ID; enrichment broths aid in low-load samples	*S. suis* (alpha-haemolytic), *S. agalactiae* (beta), *Streptococcus gallolyticus* (gamma)	Essential for antibiotic susceptibility testing and downstream biochemicals
Biochemical tests	CAMP test, hippurate hydrolysis, bile esculin hydrolysis, catalase test	Helps differentiate zoonotic species based on metabolic traits	CAMP+, hippurate+ (*S. agalactiae*), bile esculin+ (*S. gallolyticus*)	All species catalase-negative; multiple tests required for species-level ID
Serodiagnosis	Lancefield grouping (latex agglutination), capsular serotyping (Quellung, co-agglutination)	Classifies beta-haemolytic groups; serotype-specific epidemiological insights	Lancefield C (*Streptococcus equi zooepidemicus*), G (*Streptococcus canis*), serotype 2 (*S. suis*)	Not all species typable by Lancefield (e.g. *S. suis* and *Streptococcus iniae*)
Molecular diagnosis	PCR, qPCR, multiplex PCR (targeting 16S rRNA, cps genes)	High sensitivity and specificity; can detect species and serotype; works on pre-treated samples	*S. suis* detection by 16S rRNA PCR; cps2/cps14 gene for serotype	Useful for surveillance, especially in blood/CSF; unaffected by prior antibiotic use

consumption of raw or undercooked animal products is particularly important in areas where these products are endemic.

Surveillance and Public Health Messaging

Continuous surveillance of zoonotic *Streptococcus* infections in both animal and human populations is crucial for monitoring disease trends, detecting outbreaks early, and identifying emerging resistant strains. Public health campaigns are crucial for raising awareness among at-risk populations about the modes of transmission, symptoms, and preventive measures.

Case Study

A 48-year-old pig farmer, named Ramu, presented to a regional hospital with a sudden onset of high fever (40 °C), excruciating headache, generalised weakness, and confusion. His family reported that he had also developed stiffness in his neck and was unusually irritable. Mr. Ramu's medical history was unremarkable, but his occupation frequently involved handling pigs and their meat, including participating in home slaughtering, often with minor cuts and abrasions on his hands. He also admitted to occasionally consuming traditional pork dishes that might be undercooked. Upon examination, Mr. Ramu was disoriented, exhibited nuchal rigidity, and had a positive Kernig's sign, strongly suggesting men-

Table 12.7 Treatment options for zoonotic *Streptococcus* infections

Species	First-line treatment	Alternative agents	Resistance trends	Special considerations
Streptococcus suis	IV Penicillin G or ceftriaxone	Vancomycin, linezolid, meropenem	Emerging resistance to tetracycline, macrolides	Dexamethasone may reduce hearing loss in meningitis; monitor for sepsis
Streptococcus iniae	Ceftriaxone or penicillin	Clindamycin, vancomycin	Some strains resistant to tetracycline and fluoroquinolones	Treat early to prevent deep soft tissue spread; source control important
Streptococcus canis	Penicillin or amoxicillin	Clindamycin, cephalexin	Rare resistance reported	Manage associated animal bite or wound infection carefully
Streptococcus equi subsp. *zooepidemicus*	Penicillin G, ceftriaxone	Vancomycin, carbapenems	Variable macrolide resistance	Treat as invasive streptococcal disease; severe cases may need ICU support
Streptococcus agalactiae (animal strains)	Ampicillin or penicillin G	Erythromycin, vancomycin	Macrolide and clindamycin resistance possible	Ensure differentiation from human GBS; screen aquaculture/farm sources in outbreaks
Streptococcus dysgalactiae	Penicillin V or G	Clindamycin, cephalosporins	β-Lactam-sensitive; macrolide resistance reported	Clinical presentation may resemble *streptococcus pyogenes*; no need for post-strep prophylaxis
Streptococcus halichoeri	Penicillin, cefazolin	Linezolid, daptomycin	Limited data; generally, β-lactam-susceptible	Rare; treat as an opportunistic infection in immunocompromised or elderly patients
Streptococcus porcinus	Penicillin	Macrolides, cephalosporins	Some macrolide resistance	Infections usually mild; treat based on site and severity
Streptococcus gallolyticus subsp. *gallolyticus*	Penicillin G or ampicillin	Vancomycin, gentamicin (synergy for endocarditis)	High susceptibility to β-lactams; resistance rare	Always screen for colorectal cancer in bacteraemia/endocarditis cases

ingitis. Blood tests revealed elevated white blood cell counts, characterised by neutrophilia. A lumbar puncture was performed, revealing turbid cerebrospinal fluid (CSF), with significantly elevated protein levels, very low glucose levels, and a high leucocyte count, predominantly neutrophils. A Gram stain of the CSF sediment revealed numerous Gram-positive cocci arranged in short chains and pairs. Culture of the CSF on blood agar yielded small, alpha-haemolytic colonies after 24 h of incubation. Empirical intravenous antibiotic therapy with ceftriaxone was immediately initiated. Subsequent biochemical tests of the isolated bacteria were carried out. Species-specific PCR confirmed the isolate as *S. suis*, specifically identified as serotype. He developed significant, irreversible bilateral sensorineural hearing loss, which became apparent after discharge. He is now managing his condition with the help of hearing aids and adjusting his lifestyle.

1. Based on his clinical presentation and the initial laboratory findings, what specific type of infection is highly suspected, and what morphological characteristics of the causative agent support this suspicion?
2. What specific occupational and behavioural risk factors are evident that would have

increased his susceptibility to this zoonotic infection?
3. Considering this case, which actionable public health and animal management measures could be implemented to prevent and control *S. suis* infections in a community?

Research Questions

1. What are the current research gaps in the study of zoonotic streptococcal infections?
2. What are the evolutionary pathways for streptococci that enhance their zoonotic potential?
3. What are the developments in the diagnosis and prevention of zoonotic streptococcal outbreaks?

Declaration During the preparation of this chapter, Gemini was used to improve the grammar and readability of the text. The chapter was reviewed, and the content was edited as needed; therefore, we take full responsibility for the content of the publication.

Further Readings

Botelho AC, Ferreira AF, Fracalanzza SE, Teixeira LM, Pinto TC. A perspective on the potential zoonotic role of *Streptococcus agalactiae*: searching for a missing link in alternative transmission routes. Front Microbiol. 2018;9:608.

Fulde M, Valentin-Weigand P. Epidemiology and pathogenicity of zoonotic streptococci. Curr Top Microbiol Immunol. 2013;368:49–81.

Goyette-Desjardins G, Auger JP, Xu J, Segura M, Gottschalk M. *Streptococcus suis*, an important pig pathogen and emerging zoonotic agent—an update on the worldwide distribution based on serotyping and sequence typing. Emerging Microbes Infect. 2014;3(1):1–20.

Kim M, Heo ST, Oh H, Kim M, Jo J, Kim YR, Lee KH, Yoo JR. Human zoonotic infectious disease caused by *Streptococcus equi* subsp. zooepidemicus. Zoonoses Public Health. 2022;69(2):136–42.

Koh TH, Rahman NB, Sessions OM. Comparative genomic analysis of streptococcus dysgalactiae subspecies dysgalactiae, an occasional cause of zoonotic infection. Pathology. 2020;52(2):262–6.

Lun ZR, Wang QP, Chen XG, Li AX, Zhu XQ. *Streptococcus suis*: an emerging zoonotic pathogen. Lancet Infect Dis. 2007;7(3):201–9.

Ma J, Wu H, Ma Z, Wu Z. Bacterial and host factors involved in zoonotic streptococcal meningitis. Microbes Infect. 2025;27(1):105335.

Richards VP, Zadoks RN, Pavinski Bitar PD, Lefébure T, Lang P, Werner B, Tikofsky L, Moroni P, Stanhope MJ. Genome characterisation and population genetic structure of the zoonotic pathogen, *Streptococcus canis*. BMC Microbiol. 2012;12:1–6.

Zondervan NA, Martins dos Santos VA, Suarez-Diez M, Saccenti E. Phenotype and multi-omics comparison of staphylococcus and streptococcus uncovers pathogenic traits and predicts zoonotic potential. BMC Genomics. 2021;22:1–21.

Anthrax

13

Geetha Lakshmi and Sumetha Suga Deiva Suga

Learning Objectives

1. To describe the epidemiology of anthrax, including transmission routes and risk factors
2. To review the pathogenesis and clinical manifestation of anthrax
3. To understand and interpret the diagnosis of anthrax
4. To review recent developments in the vaccines against anthrax

Introduction

Anthrax, a zoonotic disease caused by *Bacillus anthracis*, poses a significant public health threat. The disease presents high mortality, particularly in its inhalation and gastrointestinal forms, if left untreated. *B. anthracis* is classified as a Category A bioterrorism agent.

Anthrax primarily affects animals and humans, who acquire the disease by contact with infected animals or their products, inhalation of spores, or ingestion of contaminated meat. Rare instances of transmission through direct contact with infected lesions or contaminated objects have also been documented.

Although anthrax remains endemic in numerous regions where livestock farming is prevalent, the incidence of human infection has significantly decreased in many areas, largely because of enhanced animal vaccination programmes and strengthened public health interventions. Despite these advances, sporadic outbreaks persist, particularly in locations where disease control measures are inadequate. Timely administration of antibiotics, along with antitoxins, when necessary, substantially increases the likelihood of survival across the various clinical manifestations of anthrax.

History

The history of anthrax is long and intertwined with human and animal health, dating back to ancient times. Descriptions of diseases resembling anthrax have been found as early as biblical times, with some scholars suggesting it was the fifth plague of Egypt. Ancient Greek and Roman writings also describe illnesses consistent with anthrax, even speculating its potential role in the fall of Rome. Notably, Roman poet Virgil, in his account of the murrain of Noricum, described inflamed papules resulting from exposure to tainted wool as "sacred fire" (*ignis sacer*) or "Persian fire" (*ignis persicus*), further suggesting

G. Lakshmi (✉)
Dr. M.G.R Educational and Research Institute, Chennai, Tamil Nadu, India

S. S. D. Suga
Department of Microbiology, A.C.S Medical College and Hospital, Dr M.G.R Educational and Research Institute, Chennai, Tamil Nadu, India

S. C. Parija, A. Poddar (eds.), *Textbook of Bacterial Zoonoses*, Microbial Zoonoses,
https://doi.org/10.1007/978-981-95-6615-0_13

early recognition of the disease's characteristic skin manifestations.

The first clinical descriptions of anthrax emerged in the eighteenth century. However, it was not until the mid-nineteenth century that significant scientific progress was made. In 1850, rod-shaped bacteria were observed in the blood of infected animals. By the 1870s, Robert Koch identified *B. anthracis* as the causative agent, establishing its life cycle and demonstrating its role in causing the disease. His work with anthrax was fundamental in developing the germ theory of disease.

Anthrax also holds a significant place in the history of vaccines. Louis Pasteur developed the first effective vaccine against anthrax for animals in 1881, a landmark achievement in immunology. Louis Pasteur proposed the attenuation of the virulent pathogen by serial passage to develop a vaccine by using *B. anthracis* in 1881.

During World War II, there was a bioterrorism attack on Gruinard Island with small bombs containing anthrax spores. Aerosols were observed on the skin of sheep that were grazing the grass. The soil remained contaminated with anthrax spores for 36 years from 1979. Decontamination was attained in mid-1980, and the island was finally declared anthrax-free in 1990.

Nineteenth-century British and German wool sorters developed inhalational anthrax, leading to the consideration of anthrax as an occupational disease. In the microbiology facility of the Soviet military in Russia, there was an accidental release of anthrax spores in the year 1979, leading to 70 cases of anthrax due to inhalation. The potential use of anthrax as a biological weapon has raised significant concerns. The 2001 anthrax attacks in the United States, in which anthrax spores were mailed to several media outlets and government offices, demonstrated the potential impact of bioterrorism.

Taxonomy

B. anthracis is classified within the domain Bacteria, phylum Firmicutes, class Bacilli, order Bacillales, family Bacillaceae, and genus *Bacillus*.

The genus currently comprises 110 species. Historically, the genus *Bacillus* was very broad. However, based on phylogenetic and genomic analyses, many species have been reclassified into new genera.

The genus contains species with varying degrees of pathogenicity. *B. anthracis*, causing anthrax, and *Bacillus cereus*, causing food-borne illnesses and opportunistic infections, pose significant health risks. However, most *Bacillus* species are considered non-pathogenic. These non-pathogenic species include the well-studied *Bacillus subtilis* and the industrially important *Bacillus licheniformis* and *Bacillus amyloliquefaciens*. They play crucial roles in environmental nutrient cycling and are widely utilized in biotechnology for enzyme and antibiotic production.

Genomics and Proteomics

B. anthracis genome is stable and highly monomorphic, indicating minimal genetic diversity among different strains. This genetic homogeneity points towards a relatively recent evolutionary divergence from a *B. cereus*-like ancestor.

The genetic material of *B. anthracis* is organized into a large circular chromosome, typically ranging in size from 5.2 to 5.5 million base pairs. This chromosome houses the essential core genes necessary for the bacterium's fundamental survival and metabolic processes. The existence of two virulent plasmids, *pXO1* and *pXO2*, is an important characteristic of *B. anthracis*. These plasmids play an important role in determining host specificity and harbouring the genetic elements responsible for key pathogenic factors.

The *pXO1* plasmid includes the genes *pagA*, *lef*, and *cya*. These encode components of the anthrax toxin: the protective antigen (PA), lethal factor (LF), and oedema factor (EF). In contrast, the *pXO2* plasmid contains the *capB, C, A, D, and E* operons. These are responsible for producing enzymes involved in synthesizing the *B. anthracis* capsule, composed of poly-γ-D--glutamic acid. The genes within the cap operon are arranged in the sequence *capB*, *capC*, *capA*, *capD*, and *capE*, each encoding the respective proteins *CapB*, *CapC*, *CapA*, *CapD*, and *CapE*. These proteins are suggested to facilitate capsule biosynthesis and the transport and attachment of D-glutamic acid residues to the bacterial cell surface. Additionally, the *capA* and *capB* genes, which encode proteins with partially overlapping functions, are also found on the *pXO2* plasmid.

Strains of *B. anthracis* that possess both plasmids are pathogenic to humans and animals. The *pXO1* and *pXO2* plasmids also encode regulatory proteins such as *atxA*. Furthermore, these plasmids influence the expression of chromosomal genes, enhancing the activity of 12 genes while repressing 14 others in minimal media containing CO_2.

Proteomic analyses of anthrax have significantly advanced the understanding of the pathogenesis of the disease. The proteins are classified into three categories: secretome analysis proteins, regulatory network proteins, and spore-specific marker proteins.

Secretome Analysis

Over 400 secreted proteins have been identified, including metalloproteases (*inhA1* and *nprB*) and plasmid-encoded immunogens (pXO1–90 and pXO1–130). These proteins are repressed under host-mimicking conditions to prevent toxin degradation.

Regulatory Networks

Proteogenomic studies detected the global regulator AtxA and the sporulation protein Spo0A, validating their roles in toxin production and spore formation.

Spore-Specific Markers

Comparative proteomics of *B. anthracis*, *B. cereus*, and *B. thuringiensis* identified unique proteoforms, such as surface-layer proteins (Sap and EA1), enabling species-specific detection.

Morphology

B. anthracis is a Gram-positive, rod-shaped bacterium. It is non-motile and capable of forming endospores. It measures approximately 1.5 µm in width and between 3 and 7.5 µm in length. By microscopy, they are observed as elongated chains containing central or para-central oval spores. However, in infected tissues, the bacteria are generally found either as single cells or in short chains of two to three bacilli, with spores being notably absent (Fig. 13.1).

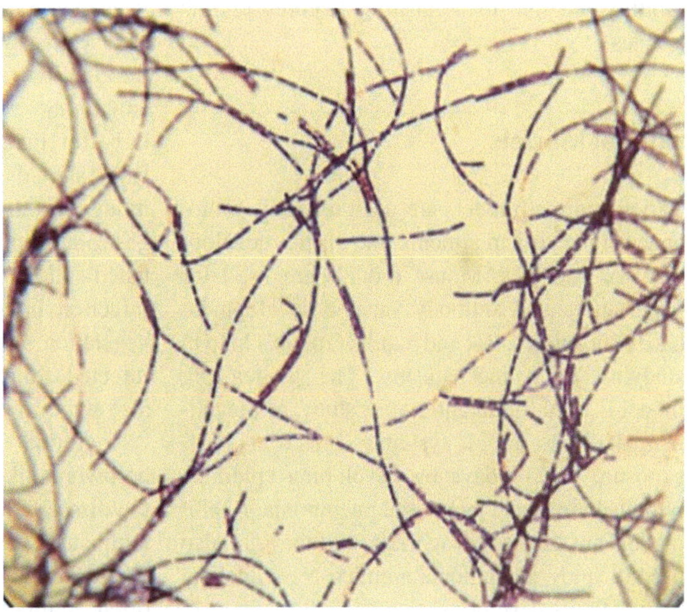

Fig. 13.1 Gram-stained smear of *Bacillus anthracis* showing Gram-positive rod-shaped structures in a bamboo-like arrangement (×1000). (Source: https://www.nc.cdc.gov/eid *Emerging Infectious Diseases* journal)

Cultivation

B. anthracis is a facultative anaerobe, which grows in the presence or absence of oxygen. However, it typically grows best under aerobic conditions. The bacteria are mesophilic, with an optimal growth temperature ranging from 35 to 37 °C, like the body temperature of its mammalian hosts. However, it can grow within a wider range of temperatures, typically from 15 to 45 °C, but growth is less efficient outside the optimal range. The optimal growth occurs around a pH of 7.0–7.5.

The bacteria grow in different bacteriological media. In liquid media, *B. anthracis* typically forms a pellicle, a layer of bacteria, on the surface due to its aerobic preference.

On nutrient agar, the colonies are typically large and flat and have a characteristic "*medusa head*" or "*ground glass*" appearance due to the presence of comma-shaped projections at the colony edges. On the blood agar, the colonies grow rapidly and appear non-haemolytic (gamma haemolysis), resembling a medusa head. These colonies exhibit a strong adherence to the agar, making them difficult to detach for sub-culturing. Hence, they are characterized by a whipped egg-white appearance when a loop is employed to collect them. Catalase positivity and a lack of motility are the distinguishing features of this species.

Animal Models

Various animal models are used to study pathogenesis and test treatments and in the development of vaccines. Mouse models are used for hyper-attenuated antibody vaccine development and toxin studies. Rat and hamster models help in studying toxin mechanisms. The guinea pig model is used for an antibiotics study, as the animal develops rapid systemic infection post-exposure in 2–4 days by developing splenitis, lymphadenitis, and interstitial pneumonia. Rabbit models are used for drug trials, and they are also used to study inhalational anthrax. New Zealand white rabbits develop septicaemia after exposure to 10^6 spores through inhalation. Non-human primate species, including *Rhesus macaques*, *Cynomolgus macaques*, and African green monkeys, serve as valuable models for simulating human conditions in the context of vaccine studies.

Transmission Cycle

The cycle of anthrax transmission is centred on the enduring presence of *B. anthracis* spores in the environment and their relationship with host organisms (Fig. 13.2). There is no evidence of direct transmission between animals or from humans to humans.

B. anthracis spores exhibit remarkable resistance to heat, chemicals, and desiccation, allowing them to remain viable in the soil for many years. When hosts infected with the bacteria die, the vegetative forms released into the environment, such as soil and water, undergo sporulation upon encountering oxygen. This process is generally initiated within 4–24 h under optimal temperature conditions (ranging from 20 to 42 °C), with the ideal conditions being a neutral-to-alkaline pH (between 7.5 and 8.2) and temperatures exceeding 20 °C. This cycle highlights the durability of *B. anthracis* spores, which facilitates their transmission over extended periods.

The transmission of anthrax infection occurs through several key routes (Table 13.1). Aerosols containing spores pose a significant risk, particularly for inhalation anthrax. Ingestion of meat from animals infected with anthrax is another mode of transmission, primarily affecting the gastrointestinal system. Furthermore, contaminated soil harbouring the resistant spores causes infection in grazing animals as they consume vegetation. Animal products derived from infected animals, such as wool and hides, can also serve as vehicles for spore transmission.

Ingestion is a primary pathway for grazing herbivores like cattle and sheep, which contract the disease by consuming spores present in contaminated soil or feed. Cutaneous entry occurs when spores gain access through skin abrasions, often facilitated by insect bites or contact with

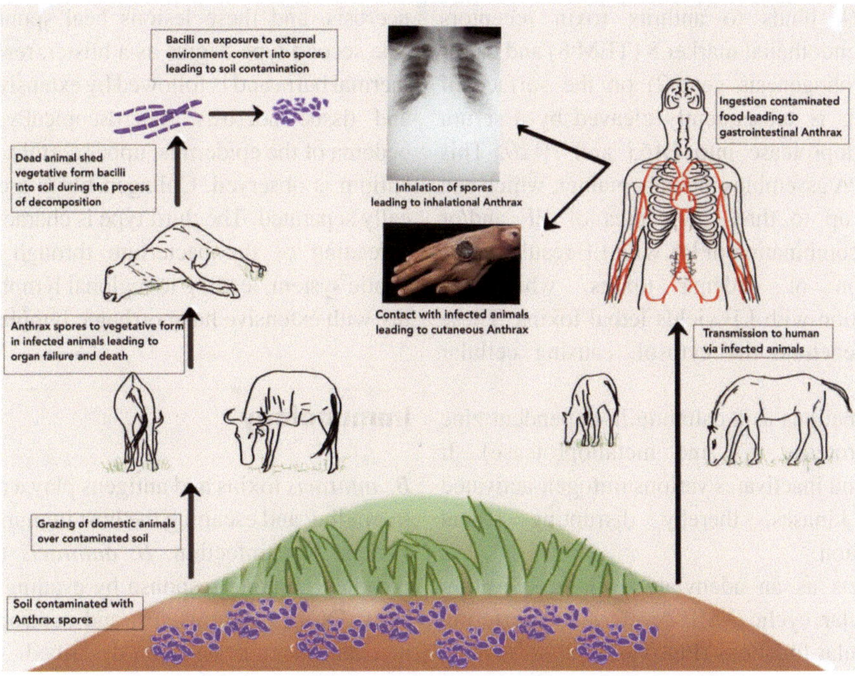

Fig. 13.2 Transmission of *Bacillus anthracis* spores to humans and domestic animals via contaminated soil and infected animals. The illustration was created by Dr. Deva Kumar, ACS Medical College and Hospital, Chennai)

Table 13.1 Route of transmission and types of infection caused by *Bacillus anthracis*

Transmission of infection		Clinical diagnosis
Spores from the hides and skin of infected animals	Skin-to-skin contact	Cutaneous anthrax
Spores	Aerosol entry to the lungs through inhalation	Inhalational anthrax
Vegetative form of *Bacillus anthracis*	Consuming meat of the infected animals	Gastrointestinal anthrax

contaminated materials. Additionally, mechanical transmission can play a role, with biting flies capable of spreading spores between animals during outbreaks (epizootics).

Once the spores enter a host, they germinate into vegetative cells within macrophages or tissues. These actively multiplying bacterial cells then release potent lethal and oedema toxins. These toxins are responsible for causing tissue necrosis, systemic toxaemia, and, ultimately, organ failure. In infected animals, the disease often progresses rapidly, leading to death and the subsequent release of large quantities of vegetative bacteria into the environment through bodily fluids, thus perpetuating the cycle of contamination.

Pathogenesis and Pathology

Pathogenesis

The bacterium exhibits two main virulence factors: an anti-phagocytic capsule composed of poly-γ-D-glutamic acid and a tripartite toxin that includes a protective antigen (PA).

PA is an oedema factor (EF) and a lethal factor (LF), both of which are biologically inactive in their native forms. PA plays a crucial role in facilitating the binding of EF and LF to host cells, enabling their entry and subsequent toxic effects. EF induces oedema, while LF disrupts cellular signalling pathways, ultimately resulting in cell death. PA binds to anthrax toxin receptors (tumour endothelial marker 8 (TEM 8) and capillary morphogenesis gene-2) on the surface of cells. This is subsequently cleaved by a serum furin endoprotease into *PA63* and *PA20*. This form of PA assembles into a heptamer, which can bind to up to three molecules of EF and/or LF. The combination of PA with EF results in the production of oedema toxins, while the combination with LF yields lethal toxins. These toxins penetrate the cytosol, causing cellular damage.

LF functions as a calmodulin-dependent zinc metalloprotease (Cl–zinc metalloprotease). It cleaves and inactivates various mitogen-activated protein kinases, thereby disrupting signal transduction.

EF acts as an adenylate cyclase, elevating intracellular cyclic AMP levels and interfering with cellular functions (Fig. 13.3).

Pathology

The pathological changes in the cellular level present as "vasculitis." This first type is characterized by acellular proliferation of bacilli in the walls of blood vessels, and luminal thrombosis is often observed. A necrotic sore is caused by vascular thrombosis, haemorrhage, and acellular necrosis, and these lesions heal spontaneously. The second type begins as a blister, resembling a thermal burn, and is followed by extensive oedema and tissue necrosis. Microscopically, massive oedema of the epidermis, upper dermis, and interstitium is observed. Collagen fibres are dramatically separated. The third type is characterized by spreading of the bacterium through the lymphatic system, leading to regional lymphadenopathy with extensive haemorrhagic lymphadenitis.

Immunology

B. anthracis toxins and antigens play a major role in evading and escaping the host immune defence system. Post-infection, *B. anthracis* triggers a complex immune response by evading dendritic cells (DCs), macrophages, and neutrophils. The mechanisms of evasion are described.

Targeting Dendritic Cells (DCs)

Lethal toxin (LT), composed of protective antigen (PA) and lethal factor (LF), inhibits dendritic cells (DCs) by cleaving mitogen-activated protein kinases (MEKs). This action obstructs the secretion of pro-inflammatory cytokines such as IL-12 and TNF-α and the upregulation of co-

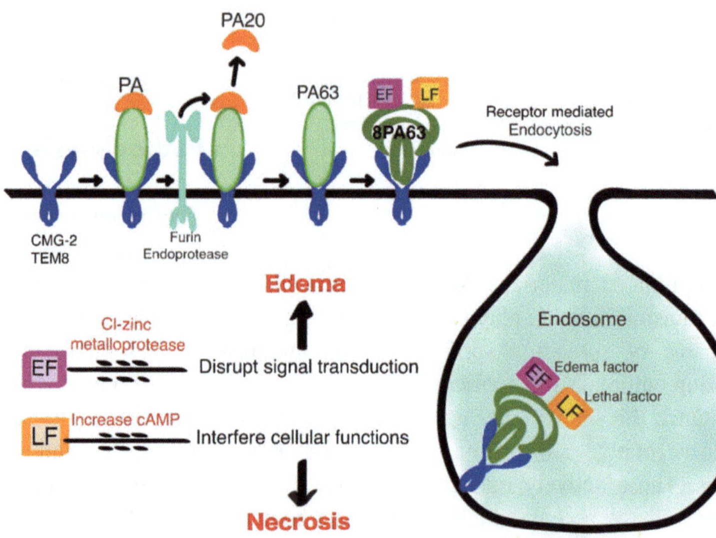

Fig. 13.3 The image illustrates the mechanism of anthrax toxins' action inside the living cell. CMG2, capillary morphogenesis gene-2; TEM8, tumour endothelial marker 8; Cl–zinc metalloprotease, calmodulin-dependent zinc metalloprotease. (The illustration was created by Dr. Deva Kumar, ACS Medical College and Hospital, Chennai)

stimulatory molecules like CD80 and CD86, which are essential for antigen presentation to naïve T cells.

Disabling Macrophages and Neutrophils

LT induces macrophage apoptosis and inhibits phagocytosis. Oedema toxin (ET = PA + EF) elevates cAMP, impairing neutrophil chemotaxis and reactive oxygen species (ROS) production. Neutrophils remain critical for early bacterial clearance via phagocytosis and ROS despite these (Fig. 13.4).

Systemic Immune Paralysis

LT disrupts *MAPK* signalling in innate cells, delaying symptom onset until the pathogen overwhelms host defences. This explains anthrax's rapid progression from non-specific symptoms to fatal shock.

Infection in Humans

Human anthrax is primarily caused by contact with infected animals or the consumption of contaminated animal products, such as wool, bone meal, meat, horns, and hides. Anthrax presents in four main forms.

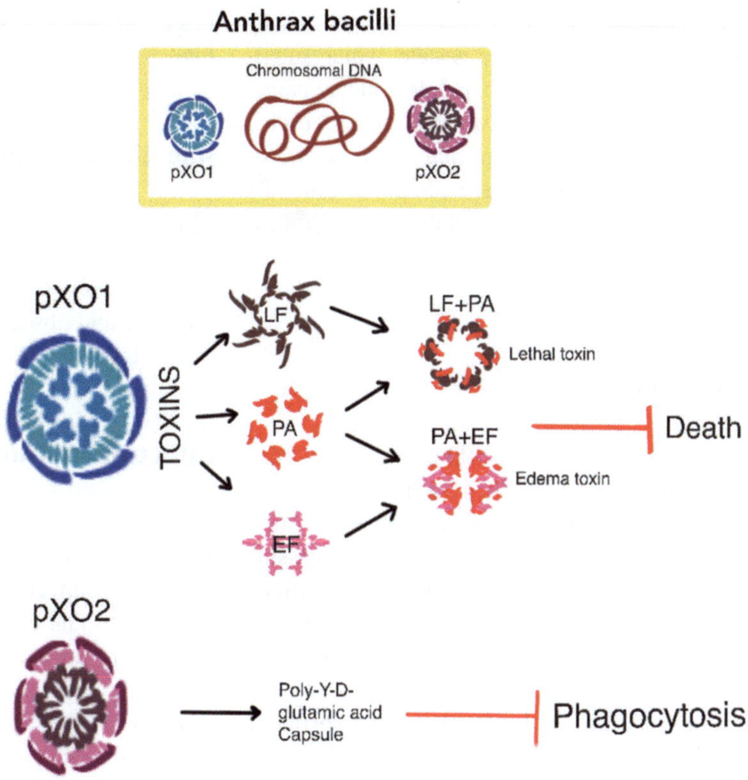

Fig. 13.4 This image illustrates the activation of protective antigen (PA), lethal factor (LF), and oedema factor (EF) by the *Pxo1* gene and the phagocytosis process by the *Pxo2* gene by activating poly-γ-D-glutamic acid capsule. (The illustration was created by Dr. Deva Kumar, A.C.S Medical College and Hospital, Chennai)

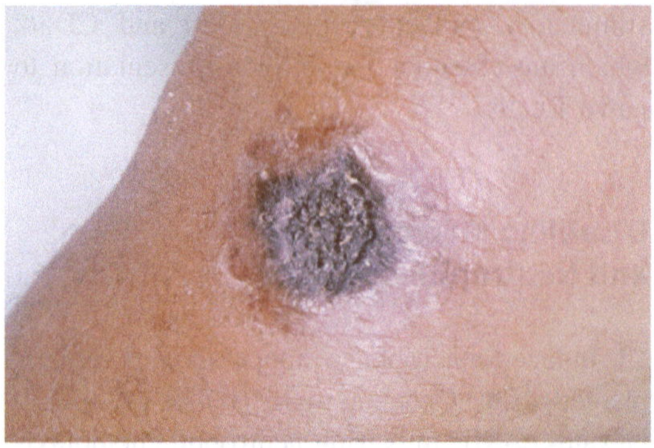

Fig. 13.5 The image shows distinctive coal-black eschar, a characteristic feature of cutaneous anthrax. (Source: https://www.cdc.gov/anthrax/media/images/2024/05/cutaneous-anthrax.png)

Cutaneous Anthrax

Cutaneous anthrax, often referred to as hide porter's disease, primarily affects dock workers who transport bundles of hides and skins on their bare backs, allowing spores to enter through minor skin abrasions. This form predominantly affects dock workers, butchers, abattoir employees, and farmers, constituting about 95% of human cases. The condition typically begins with a small, itchy papule that evolves into a vesicle, which subsequently ruptures to form a painless ulcer characterized by a distinctive coal-black eschar (Fig. 13.5). The term "*ánthrakas*" in Greek translates to coal. Cutaneous anthrax is generally treatable with antibiotics, and the mortality rate is relatively low when treatment is administered promptly.

Inhalational Anthrax

Inhalational anthrax, also known as wool sorter's disease, occurs when spores are inhaled from the wool of infected animals. This form primarily affects workers in wool-processing facilities and is the most severe and potentially lethal variant. Initial symptoms may mimic those of a common cold or flu, including fever, fatigue, and muscle aches. As the illness advances, it leads to bacteraemia, haemorrhagic mediastinitis, haemorrhagic meningitis, severe respiratory distress, shock, and, ultimately, death. Inhalational anthrax is associated with a high mortality rate, even with intensive treatment, and is often linked to bioterrorism incidents.

Gastrointestinal Anthrax

It is associated with the ingestion of contaminated meat. Symptoms include nausea, vomiting, abdominal pain, and bloody diarrhoea. This form results in severe systemic infection and may be fatal.

Injectional Anthrax

This form is associated with the injection of contaminated drugs, particularly heroin. It causes deep tissue infections, often with severe swelling and abscesses. This form is very difficult to treat.

Infection in Animals

Animals are infected by the ingestion of spores while grazing on contaminated pastures. The clinical signs of infection vary, depending on the animal species and the form of the disease. In cattle and sheep, an acute or peracute form is often observed, characterized by a sudden onset of symptoms and rapid progression to death. Affected animals may exhibit a high fever, reaching up to 41.5 °C/107 °F, along with respiratory distress and bloody discharges from body openings. Neurological signs, such as staggering and convulsions, may also be present, culminating in death

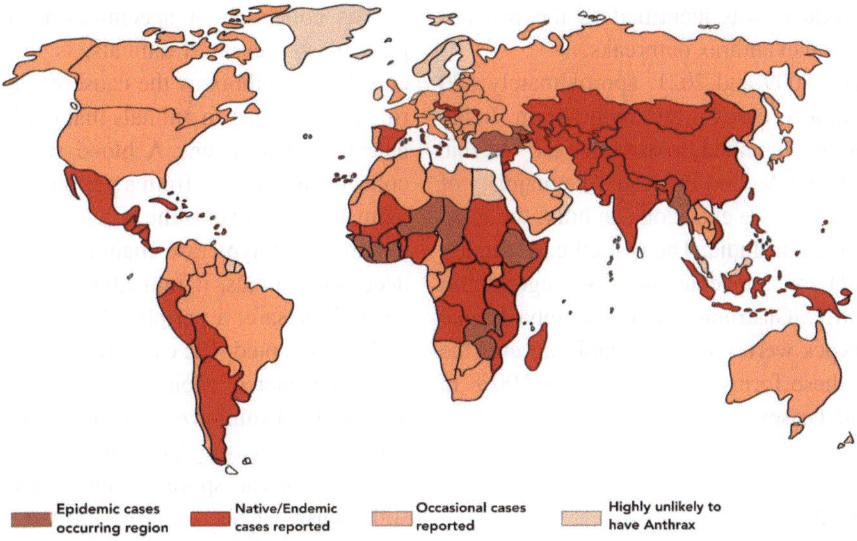

Fig. 13.6 This image illustrates the global epidemiology of *Bacillus anthracis* with respect to epidemic and endemic regions. (The illustration was created by Dr. Deva Kumar, A.C.S Medical College and Hospital, Chennai)

within 2–24 h of symptom onset. In contrast, pigs and carnivores typically develop a subacute or chronic form of the disease, often manifesting as swollen necks or throats (oropharyngeal anthrax). This swelling causes respiratory compromise and poses a significant risk of suffocation.

Epidemiology and Public Health

The annual estimated global incidence of anthrax ranges between ≤1000 cases and 20,000 cases. Primarily, as a zoonotic disease, it affects the herbivores across most continents, except Antarctica (Fig. 13.6). Southern Africa is considered the geographic origin of the pathogen. Additionally, it is prevalent in certain regions of the Middle East, Africa, and Latin America, particularly among grazing herbivores. Zimbabwe reports a higher incidence of anthrax cases in both animals and humans compared to most other countries combined.

Seasonal fluctuations have been documented for many years; significant spring rainfall leads to the accumulation of spores in low-lying areas, which are subsequently followed by a hot, dry period affecting grazing animals. The long-term viability of spores in the soil and on domestic items raises the potential for anthrax outbreaks, especially considering climatic changes. This is evidenced by the significant epizootic event that occurred in the summer of 2016 on the Yamal Peninsula. During the year 2000, a North Dakota ranch anthrax outbreak occurred in the United States. In 2021, the European Union/European Economic Area countries' surveillance system of anthrax outbreak documented seven cases of anthrax, four confirmed and three probable. A recent outbreak of anthrax was documented in Zambia, Africa, which was recorded in the year 2023. Environmental factors such as climate, soil type, and proximity to water bodies also play a role in the persistence and spread of anthrax spores in endemic regions. Additional contributing factors include poverty, lack of awareness, traditional practices, and wildlife–livestock interface.

In India, many cases of human anthrax outbreaks have been documented. In 2004, a paediatric case of inhalational anthrax was reported from Villupuram district, Tamil Nadu. Anthrax meningitis was observed in two separate incidents, one in Tirupati (2007) and another in Assam (2018). In 2015, Puducherry reported cases of gastrointestinal anthrax that progressed to sepsis. Handling or consumption of infected animal products was identified as the primary driver of human anthrax outbreaks.

Between 2009 and 2021, approximately 420 human anthrax cases were reported in India. They were documented in West Bengal, Odisha, Jharkhand, and Andhra Pradesh. The majority of reported cases were cutaneous anthrax, the most common form in India. The overall case fatality rate (CFR) for cutaneous anthrax ranged from 3% to 38%. Gastrointestinal and inhalational anthrax types were rare but notably severe; the CFR for these forms was reported as 100% in documented cases.

Diagnosis

Laboratory diagnosis of anthrax involves a combination of methods and depends on the form of anthrax suspected, cutaneous, inhalational, gastrointestinal, or injectional and on the stage of illness.

There is a high risk of laboratory-acquired infection; hence, it must be processed in biosafety cabinets in a laboratory classified under biosafety level 3.

Specimens

For the diagnosis of anthrax in humans, the types of specimens collected depend on the clinical presentation of the disease. For cutaneous anthrax, the primary specimen is typically a swab or scraping collected directly from the characteristic skin lesion (*eschar*). If there is significant fluid accumulation, an aspirate of the vesicular fluid may also be collected. For inhalational anthrax, the key specimens include blood and pleural fluid. Cerebrospinal fluid (CSF) is collected if there are signs or symptoms of meningitis. For gastrointestinal anthrax, specimens are blood and stool samples. Ascitic fluid is collected if there is evidence of abdominal effusion. For injectional anthrax, specimens include swabs or aspirates from the injection site and blood cultures.

The collection of specimens from animals, particularly deceased animals, is important for confirming anthrax as the cause of death and to rule out infection in animals that could contaminate the environment. A blood sample, ideally collected aseptically from a peripheral vein if the animal is still alive, is the specimen of choice for culture and polymerase chain reaction (PCR). In deceased animals, if immediate necropsy is not feasible or safe, a peripheral blood sample can still be attempted. The ear notch/tip or the earlobe is the specimen of choice from deceased animals. This method minimizes the risk of opening the carcass and releasing large numbers of vegetative bacteria that can sporulate and contaminate the soil. During a necropsy, performed safely and under appropriate biosafety conditions, samples from the spleen and lymph nodes are often collected, as these organs typically harbour high concentrations of the bacteria.

Microscopy

In a Gram-strained smear seen through microscopy, the bacteria appear as Gram-positive rods (Fig. 13.1). Smears prepared from *B. anthracis* exhibit a distinctive bamboo stick appearance, a characteristic feature aiding identification. Gurr's polychrome methylene blue stain is employed to visualize the capsule. This shows an amorphous purple capsule surrounding blue bacilli. Furthermore, spores can be demonstrated using specific staining techniques, such as Ashby's method with malachite green or a modified acid-fast staining procedure utilizing 0.25% sulphuric acid. These methods allow us for better visualization and identification of the spores and presence of spores.

In Vitro Culture

Bacilli are aerobic and non-fastidious and grow in ordinary medium, with temperatures ranging from 12 to 45 °C. Medusa head appearance colonies are seen in the nutrient agar plate. Dry, wrinkled, and non-haemolytic colonies appear on the blood agar plate. If cultured in a solid medium containing penicillin, it shows a "String-of-pearls" formation, reflecting its sensitivity to this antibiotic. While capsule production is primarily visualized microscopically, it can be enhanced on bicarbonate agar under increased carbon dioxide (CO_2). PLET medium is a heart infusion agar consisting of Polymyxin, Lysozyme, EDTA, and Thallous acetate, one of the selective media for *B. anthracis*.

Biochemical tests are essential for confirming the identity of *B. anthracis* isolates obtained through in vitro culture. These biochemical assays help distinguish *B. anthracis* from other *Bacillus* species, particularly the closely related *B. cereus*. Key characteristics include a positive catalase test and a negative oxidase test result. Lecithinase production is typically variable or weak in *B. anthracis*, contrasting with the strong activity often seen in *B. cereus*.

B. anthracis ferments glucose and maltose without gas production, while it is generally negative for arabinose, xylose, and mannitol. Gelatin liquefaction is typically positive, often displaying an inverted "*fir tree*" pattern. Crucially, *B. anthracis* is typically susceptible to lysis by the gamma phage.

Serodiagnosis

During the early stages of infection, the body may produce only minimal antibodies, which could result in false-negative results. For this reason, serological tests should be interpreted cautiously in early disease, and repeat testing after 10 days to 2 weeks may be necessary to detect seroconversion.

Enzyme-Linked Immunosorbent Assay (ELISA)

The anti-PA IgG ELISA method is the most widely accepted and utilized anthrax serological test, detecting IgG antibodies against protective antigens (PAs). This assay has demonstrated an approximate specificity of 78% in patients with cutaneous anthrax. For the anti-PA IgG ELISA specifically developed in India, a specificity of 99.41% has been reported, making it highly reliable for confirming anthrax infection in endemic regions.

The detection of antibodies against lethal factor (LF) by using anti-LF IgG ELISA kit is another valuable diagnostic approach. The test shows a sensitivity of 61% in patients with cutaneous anthrax. When used in combination with anti-PA testing, the anti-LF IgG assessment improves diagnostic sensitivity, as some patients may produce serum antibodies to one antigen but not the other.

Double antibody sandwich ELISA method offers enhanced sensitivity and specificity. Commercial kits utilizing this technique can detect anthrax toxin receptor proteins or antibodies against bacterial antigens with high precision.

Direct Immunofluorescence (DIF) Test

This test is carried out using fluorescent-tagged monoclonal antibodies to target and detect capsular and polypeptide antigens. Following the application and incubation of these labelled antibodies on a fixed sample, any bound antibodies indicating the presence of *B. anthracis* are visualized as a distinct fluorescence under a fluorescence microscope using the appropriate excitation wavelength for the fluorophore.

Key targets for these antibodies include the galactose-N-acetylglucosamine polysaccharide (Gal-NAG-PS), a significant component of the *B. anthracis* cell wall, and the poly-D-glutamic acid (PGA) capsule, a critical virulence factor. Antibodies targeting both Gal-NAG-PS and PGA, each labelled with a distinct fluorophore, are employed simultaneously. This dual-targeting strategy enhances specificity by requiring the simultaneous presence of both antigens for a positive signal. This improves the reliability of the identification, even in cases involving atypical isolates.

Ascoli Precipitin Test

Ascoli precipitin test, also known as the *Ascoli test* or *Ascoli ring test*, is a rapid serological test historically used for the diagnosis of anthrax. This classical serological method involves precipitin reactions between antigen and antibody in solution. It has been used historically for anthrax diagnosis, though it has largely been replaced by more sensitive and specific methods like ELISA.

This test is used particularly in decomposed carcasses or contaminated materials where isolating viable bacteria might be challenging. It depends on the principle of immunoprecipitation, where soluble antigens of *B. anthracis* react with specific antibodies in an antiserum, forming a visible precipitate. The test is described as the best confirmatory precipitation test for anthrax.

Microhaemagglutination Assay (MHA)

This assay is based on the principle of passive haemagglutination. It involves coating red blood cells with specific *B. anthracis* antigens, thereby sensitizing them to react with corresponding antibodies. The patient's serum, potentially containing these antibodies, is serially diluted and then mixed with the antigen-coated red blood cells. Following incubation, specific antibodies in the serum lead to agglutination, a visible clumping of the red blood cells due to antibody–antigen cross-linking. This agglutination, observed as a diffuse pattern in the microtitre well, indicates a positive result, and the highest dilution of serum showing agglutination defines the antibody titre. MHA offers advantages such as simplicity, affordability, and high throughput. However, its sensitivity and specificity may be lower compared to more advanced immunoassays, and it may be susceptible to non-specific agglutination.

Molecular Diagnosis

The molecular diagnosis of anthrax has significantly enhanced detection speed and accuracy, mostly in cases of outbreaks. Polymerase chain reaction (PCR)-based assays, targeting specific virulence genes on the *B. anthracis* plasmids pXO1 and pXO2, remain a cornerstone, with improvements focusing on multiplexing to simultaneously detect multiple targets and real-time quantification for assessing bacterial load.

Isothermal DNA Amplification

It represents a highly promising technique, with various methods being described. These methods provide rapid and field-deployable diagnostic solutions, owing to their straightforward nature and minimal dependence on advanced laboratory apparatus. Loop-mediated isothermal amplification *(LAMP) assays* for *B. anthracis* target unique chromosomal sequences (e.g. the *BA813* gene) to differentiate it from the closely related *B. cereus*. It also targets genes on the pXO1 (e.g. *pag*, encoding protective antigen, and *lef*, encoding lethal factor) and pXO2 (e.g. *capB*, involved in capsule synthesis) plasmids to confirm virulence. *Multiplex LAMP assays* targeting multiple genes can enhance specificity and provide information about the presence of virulence factors. LAMP assays developed for *B. anthracis* have demonstrated high sensitivity and are capable of detecting as low as 10 fg of bacterial DNA or a few colony-forming units (CFUs) per reaction.

Next-Generation Sequencing (NGS)

Despite the overall monomorphic nature of *B. anthracis*, NGS can identify single-nucleotide polymorphisms (SNPs), variable number tandem repeats (VNTRs), and other genetic markers that allow for high-resolution strain typing. This is crucial for epidemiological investigations, tracing the source of outbreaks, and forensic analysis in a bioterrorism incident.

NGS readily identifies the presence of virulence plasmids and specific variants of pXO1 (encoding toxins) and pXO2 (encoding the capsule). Sequencing confirms the presence of key genes like *pag* (protective antigen), *lef* (lethal factor), *cya* (oedema factor) on pXO1, and the *cap* operon on pXO2. Furthermore, it can detect any mutations or variations in these genes that might affect toxin production or capsule formation.

Metagenomic NGS (*mNGS*), or *shotgun sequencing*, is applied to complex samples (e.g. environmental samples and clinical specimens without pure culture) to directly detect *B. anthracis* DNA without prior isolation. This approach involves sequencing all DNA present in the sample and then bioinformatically identifying the *B.*

Table 13.2 Laboratory diagnosis of anthrax

Diagnostic approaches	Methods	Targets	Remarks
Direct microscopy	Gram staining	Vegetative bacilli	Simple and less time-consuming
	Gurr's polychrome methylene blue staining	Capsule	Simple and fast technique, useful to identify the capsule in 30 s. An amorphous purple capsule surrounding the blue bacilli is observed
	Ashby's method with malachite green Or Modified acid-fast staining with 0.25% sulphuric acid	Spores	This method is useful in the identification of spores and their positions within the bacilli
In vitro cultures	*Bacillus anthracis* grows in normal media such as nutrient agar and blood agar PLET medium is a heart infusion selective medium	Vegetative bacilli	*Medusa head* appearance colonies are seen in the nutrient agar plate Dry wrinkled non-haemolytic colonies appear in the blood agar plate
Immunodiagnostics	ELISA method Anti-PA IgG ELISA Anti-LF IgG ELISA	Antibodies against the protective antigen are measured	The positivity rate increases after 1 week of infection
	Direct immunofluorescence test	Capsular and polypeptide antigens	Less time-consuming and easy to perform
	Ascoli's thermos-precipitation test	PA antigen and capsular antigen	If a putrefied specimen needs to be processed, this ring test method is performed
Molecular assays	LAMP, nucleic acid probes, PCR, aptamers, and peptide--nucleic acid chimera probes	Protective antigen and capsular genes	Have high sensitivity and specificity. The BApX01 primer is employed for the detection of the PA antigen. BApX02 primers are utilized to target genes associated with the capsule

anthracis reads. These methods are improving the sensitivity of mNGS for detecting even low levels of *B. anthracis*.

Matrix-Assisted Laser Desorption/Ionization Time-of-Flight (MALDI-TOF) Mass Spectrometry (MS)

It has been utilized for the identification of bacilli and is currently widely adopted in both medical and veterinary diagnostic laboratories. The mass spectrometer generates a mass spectrum, which represents a unique "fingerprint" of the protein composition of the *B. anthracis* isolate. The obtained mass spectrum is compared against a comprehensive database containing reference spectra from well-characterized *Bacillus* species, including *B. anthracis* and its close relatives like *B. cereus*. Sophisticated algorithms are used to match the unknown spectrum to the reference spectra, providing an identification, down to the species level and, potentially, even the strain level in specialized databases.

Laboratory diagnosis of anthrax is summarized in Table 13.2.

Treatment

Prompt commencement of suitable antibiotic treatment is essential for managing anthrax, particularly in severe non-cutaneous instances. Anthrax bacilli exhibit resistance to third and fourth-generation cephalosporins. Nevertheless, they demonstrate a wide sensitivity to various other antimicrobial agents, including tetracyclines, macrolides, aminoglycosides, fluoroqui-

Table 13.3 Antibiotics and dosage for anthrax treatment

Patient category	Initial IV therapy	Oral regimens
Adults	400 mg of ciprofloxacin to be given every 12 h or 100 mg of doxycycline every 12 h	After IV therapy, oral drugs can be initiated if clinical improvements are seen Tablet ciprofloxacin (500 mg) per oral twice daily or tablet doxycycline (100 mg) per oral twice daily
Children	Ciprofloxacin (10–15 mg/kg) every 12 h and not to exceed 200 mg/day	Tablet ciprofloxacin (10–15 mg/kg) per oral every 12 h, not to exceed 1 g/day, or doxycycline can also be given as an alternative

nolones, carbapenems, clindamycin, rifampicin, and chloramphenicol. The recommended dosages and preferred medications are outlined in Table 13.3.

Treatment for Cutaneous Anthrax

Oral penicillin treatment for a duration of 7–10 days is recommended for individuals who have contracted the organism from environmental sources, particularly in scenarios where there is no evidence of intentional spore release, such as in bioterrorism incidents. For adults, the dosage of amoxicillin is 500 mg every 8 h, while for children, it is 80 mg/kg/day administered in divided doses every 8 h, not exceeding 500 mg every 8 h. The Centers for Disease Control and Prevention (CDC) advise using penicillin, in conjunction with other antibiotics, following culture and sensitivity testing despite approximately 10% of *B. anthracis* strains exhibiting resistance to penicillin.

Treatment of Inhalational Anthrax

The treatment period is typically extended to 60 days, particularly for inhalational anthrax, to ensure the complete eradication of any residual spores. In cases of inhalational anthrax, corticosteroids may be utilized as an adjunctive treatment alongside oxygen therapy to manage pulmonary oedema.

Additionally, monoclonal antibodies that target the protective antigen of the anthrax toxin are incorporated into the treatment regimen. Human monoclonal antibodies with a high affinity for protective antigens, such as raxibacumab (40 mg/kg IV) and obiltoxaximab (16 mg/kg IV), have been shown to enhance antibody levels with a single administration.

Prevention and Control

The effective prevention and control of anthrax is based on a multifaceted approach. These include (i) vaccination, (ii) surveillance, (iii) environmental decontamination, and (iv) timely medical intervention, all of which are vital for safeguarding both human and animal health.

Vaccination

For over a century, anthrax vaccines have been utilized to prevent infections in both humans and animals. Historically, three primary anthrax vaccines have been employed. These include the cell-free protein-based BioThrax, or the anthrax vaccine adsorbed (AVA); the anthrax vaccine precipitated (AVP); and the live anthrax vaccine (LAV).

The currently authorized human vaccine is the anthrax vaccine adsorbed (AVA or BioThrax). It is derived from a cell-free culture supernatant that contains protective antigen (PA) from a non-encapsulated, toxin-producing strain of *B. anthracis*. Recently, the US FDA has approved a new vaccine, CYFENDUS (AV7909), which includes an additional adjuvant to enhance its immunogenic response. Furthermore, advancements in vaccine technology have led to the development of plasmid DNA vaccines encoding for PA and lethal factor (LF). A live attenuated vaccine incorporating mutations in genes respon-

Table 13.4 Vaccination schedule for pre-exposure prophylaxis

Schedule	Route of administration	Dosing schedule
Primary series	Intramuscular	0, 1, and 6 months
Booster series	Intramuscular	6 and 12 months after completion of the primary series and at 12-month intervals thereafter

sible for *Bacillus*'s encapsulation has also been developed.

Vaccinating livestock is essential for controlling the disease, as it helps prevent outbreaks among vulnerable animal populations and reduces the risk of human exposure.

BioThrax is administered via the subcutaneous route. For post-exposure prophylaxis, a regimen of subcutaneous injections of the anthrax vaccine at 0, 2, and 4 weeks post-exposure, in conjunction with antimicrobial therapy, is recommended. It is contraindicated for individuals with a previous history of severe allergic reactions to any allergens or the anthrax vaccine. Table 13.4 lists pre-exposure prophylaxis and post-prophylaxis of anthrax.

Surveillance is important for the early identification and management of anthrax outbreaks. Active surveillance initiatives are designed to track animal populations for unusual deaths and symptoms indicative of anthrax. Environmental decontamination is another aspect of reducing the transmission of anthrax spores. The proper disposal of infected animal remains and contaminated materials is essential to prevent further environmental contamination.

Case Study

A 45-year-old farmer was admitted to the hospital presenting with a high fever, headache, and irritability that persisted for 3 days, leading to unconsciousness on the fourth day. The patient's occupational history showed that he is a part-time trader of wool and hides. Ten days prior to his admission, he sustained a small incisional cut on his left hand. Following 5 days post-injury, he noticed discoloration, swelling, pain, and warmth in his finger. Eventually, the swelling progressed to form a pustule and eschar. A fine-needle aspiration cytology (FNAC) was performed, and the aspirated material was sent to the microbiology laboratory for Gram staining, culture, and sensitivity testing. The Gram stain revealed Gram-positive bacilli arranged in chains (bamboo-like appearance), while the culture on blood agar exhibited colonies with a medusa head appearance.

1. What is the probable diagnosis?
2. What is the source of infection, and what may be the infective form?
3. What is the staining method to demonstrate the capsule and spores?
4. What is the choice of antibiotic in this case scenario?

Research Questions

1. What are the comparative efficacies of the existing anthrax vaccines in various populations, including those with different genetic backgrounds?
2. Can novel vaccine strategies, such as recombinant proteins or mRNA vaccines, provide improved protection against inhalational anthrax?
3. How long do *B. anthracis* spores remain viable in various environmental conditions, and what methods are most effective for decontaminating areas exposed to anthrax?
4. What are the implications of soil contamination with anthrax spores for agricultural practices in endemic regions?

Declaration During the preparation of this chapter, we have used Grammarly in order to improve the grammar and/or readability of the text Anthrax. After using, we have reviewed and edited the content as needed and take full responsibility for the content of the publication.

Further Readings

Alam ME, Kamal MM, Rahman M, Kabir A, Islam MS, Hassan J. Review of anthrax: a disease of farm animals. J Adv Vet Anim Res. 2022;9(2):323–34. https://doi.org/10.5455/javar.2022.i599.

Azarkar Z, Bidaki MZ. A case report of inhalation anthrax acquired naturally. BMC Res Notes. 2016;9:141. https://doi.org/10.1186/s13104-016-1955-0.

Badri R, Uwishema O, Wellington J, Thambi VD, Pradhan AU, Adanur I, Onyeaka CVP, Onyeaka H. Anthrax outbreak amidst the COVID-19 pandemic in Africa: challenges and possible solutions. Ann Med Surg. 2022;81:104418.

Baldari CT, Tonello F, Paccani SR, Montecucco C. Anthrax toxins: a paradigm of bacterial immune suppression. Trends Immunol. 2006;27(9):434–40. https://doi.org/10.1016/j.it.2006.07.002.

Beatty ME, Ashford DA, Griffin PM, Tauxe RV, Sobel J. Gastrointestinal anthrax: review of the literature. Arch Intern Med. 2003;163(20):2527–31. https://doi.org/10.1001/archinte.163.20.2527.

Bower WA. CDC guidelines for the prevention and treatment of anthrax, 2023. MMWR Recomm Rep. 2023;72:1–47.

Bower WA, Hendricks KA, Vieira AR, Traxler RM, Weiner Z, Lynfield R, Hoffmaster A. What is anthrax? Pathogens. 2022;11(6):690. https://doi.org/10.3390/pathogens11060690.

Doganay M, Dinc G, Kutmanova A, Baillie L. Human anthrax: update of the diagnosis and treatment. Diagnostics. 2023;13(6):1056. https://doi.org/10.3390/diagnostics13061056.

Ezhova E, Orlov D, Suhonen E, et al. Climatic factors influencing the anthrax outbreak of 2016 in Siberia. Russia EcoHealth. 2021;18:217–28. https://doi.org/10.1007/s10393-021-01549-5.

Fukao T. Immune system paralysis by anthrax lethal toxin: the roles of innate and adaptive immunity. Lancet Infect Dis. 2004;4(3):166–70. https://doi.org/10.1016/S1473-3099(04)00940-5.

Griffith J, Blaney D, Shadomy S, Lehman M, Pesik N, Tostenson S, Delaney L, Tiller R, DeVries A, Gomez T, Sullivan M, Blackmore C, Stanek D, Lynfield R, Anthrax Investigation Team. Investigation of inhalation anthrax case, United States. Emerg Infect Dis. 2014;20(2):280–3. https://doi.org/10.3201/eid2002.130021.

Hendricks K, Person MK, Bradley JS, Mongkolrattanothai T, Hupert N, Eichacker P, Friedlander AM, Bower WA. Clinical features of patients hospitalized for all routes of anthrax, 1880–2018: a systematic review. Clin Infect Dis. 2022;75(Supplement_3):S341–53. https://doi.org/10.1093/cid/ciac534.

Hupert N, Person M, Hanfling D, Traxler RM, Bower WA, Hendricks K. Development, and performance of a checklist for initial triage after an anthrax mass exposure event. Ann Intern Med. 2019;170:521–30.

Kutmanova A, Doganay M, Zholdoshev S. Human anthrax in Kyrgyz Republic: epidemiology and clinical features. J Infect Public Health. 2020;13:1161–5.

Pilo P, Frey J. *Bacillus anthracis*: molecular taxonomy, population genetics, phylogeny, and patho-evolution. Infect Genet Evol. 2011;11(6):1218–24. https://doi.org/10.1016/j.meegid.2011.05.013.

Sushma B, Shedole S, Suresh KP, Leena G, Patil SS, Srikantha G. An estimate of global anthrax prevalence in livestock: a meta-analysis. Vet World. 2021;14(5):1263.

14 Listeriosis

S. B. Barbuddhe, P. Niveditha, Shilpshri V. Shinde, and Deepak B. Rawool

Learning Objectives

1. To learn about the disease listeriosis as an important food-borne zoonotic infection that is caused by both invasive and non-invasive forms
2. To understand the pathogen's capability to traverse the blood-brain, intestinal, and placental barriers, which forms the basis of clinical manifestations in infected hosts
3. To know the disease dynamics, various diagnostic approaches, and prevention and control measures

Introduction

Listeriosis, caused by *Listeria monocytogenes*, is an important food-borne illness. It has been isolated from various sources, including soil, sewage water, and decaying plant material. Also, it has been isolated from food products, including milk and dairy products, meat, fermented sausages, fresh and frozen vegetables, seafood, and fish products. The ability of *L. monocytogenes* to grow at low temperatures and form biofilms and its persistence in food processing plants may lead to contamination of ready-to-eat (RTE) foods, posing a significant food safety challenge. The pathogen can cross the host's blood-brain, placental, and intestinal barriers. The conditions are often associated with severe invasive and non--invasive infections in diverse species, including humans, ruminants, birds, fish, and crustaceans. In ruminants, it is also known as *circling disease* or *silage disease*. The case fatality rate, especially in high-risk populations (immunocompromised individuals, the elderly, and newborns), is as high as 30%, with an overall hospitalisation rate of more than 95%. The US Centres for Disease Control and Prevention (CDC) placed the disease on the list of nationally notifiable diseases in the year 2001. Globally, it is the second most common cause of fatal food-borne illness, followed by *Salmonella* infections.

Globally, the majority of listeriosis cases have been reported either in sporadic or epidemic form. Most often, the infection in animals may be subclinical; however, severe forms may also occur. Listeriosis is characterised by neural, visceral, and reproductive clinical entities, usually manifested with septicaemias, encephalitis, meningoencephalitis, meningitis, rhombencephalitis, abortion, stillbirth, perinatal infections, and gastroenteritis in humans and animals. Meningitis

S. B. Barbuddhe (✉) · D. B. Rawool
ICAR—National Meat Research Institute, Hyderabad, Telangana, India

P. Niveditha
ICAR—National Meat Research Institute, Hyderabad, Telangana, India

Vignan Foundation for Science, Technology and Research, Guntur, Andhra Pradesh, India

S. V. Shinde
Department of Biotechnology, Nagpur Veterinary College, Nagpur, Maharashtra, India

S. C. Parija, A. Poddar (eds.), *Textbook of Bacterial Zoonoses*, Microbial Zoonoses,
https://doi.org/10.1007/978-981-95-6615-0_14

due to *Listeria* has been ranked as the third important bacterial cause of meningitis in humans.

History

In the year 1911, Hülpers isolated bacteria from a case of liver necrosis in a rabbit and named it *Bacillus hepatica*. In the year 1926, E.G.D. Murray and colleagues described *L. monocytogenes* among laboratory guinea pigs and rabbits. It was initially named *Bacterium monocytogenes* due to monocytosis. In 1927, the organism was named *Listerella hepatolytica* by Pirie in honour of Joseph Lister. The first report of *Listeria* infection in humans was confirmed by Nyfeldt in 1929, while the first report in ruminants was reported in 1929 by Gill in sheep (circling disease) in New Zealand. Later, a similar organism was isolated from sick sheep and named *Listerella ovis*. Subsequently, Jones and Little confirmed the first report of listeriosis in cattle in 1934. It was implicated in "typical listeriosis," namely, meningoencephalitis. Its current name, *L. monocytogenes*, came about in the year 1940.

L. monocytogenes was subsequently identified as a causative agent of disease in humans, and the pathogen was isolated in 1949 during an outbreak among newborns. The bacteria were isolated from meconium, blood, and/or organs of 85 stillborn infants or newborns with severe illness and findings of granulomas in various organs. *L. monocytogenes* was recognised as an important food-borne pathogen by the World Health Organization (WHO) in 1986. Also, in 2002, the WHO included *L. monocytogenes* in the list of food-borne pathogens.

Taxonomy

The genus *Listeria* is classified under the phylum Firmicutes, the order Bacillales, the class Bacilli, and the family Listeriaceae. The genus *Listeria* has low G + C content and is closely related to other genera, including *Bacillus*, *Clostridium*, *Enterococcus*, *Streptococcus*, and *Staphylococcus*.

The *Listeria* genus consists of two clades, sensu stricto and sensu *lato. The* genus currently comprises 30 species that were isolated from different habitats (Table 14.1). Among these, *L. monocytogenes* is an important pathogen in humans and animals, while *L. ivanovii* mostly affects animals. *L. monocytogenes* is classified into four evolutionary lineages denoted by I, II, III, and IV and 14 serotypes (1/2a, 1/2b, 1/2c, 3a, 3b, 3c, 4a, 4ab, 4b, 4c, 4d, 4e, 7, and 4h). *L. monocytogenes* serotypes 1/2b and 4b of lineage I are responsible for 95% of cases of listeriosis in humans.

Genomics and Proteomics

As of 2025, genomes for more than 88,000 *Listeria* isolates from different parts of the world are available in the public domain. The size of the *L. monocytogenes* genome is about 3 Mb, encoding nearly 3000 core genes. These are organised within genomic islands (*Listeria* genomic islands (LGIs)) and pathogenicity islands called *Listeria* pathogenicity islands (LIPIs). *L. monocytogenes* harbours various pathogenicity islands, namely, LIPI-1, LIPI-2, LIPI-3, and LIPI-4, and genomic islands such as LGI-1, LGI-2, and LGI-3. The LIPI-1 harbours genes required for the intracellular cycle and plays a major role in the pathogenesis of *L. monocytogenes* infections. Many genes still remain unexplained.

The proteomic study of *L. monocytogenes* has offered a comprehensive view of its protein landscape and its role in the pathogenesis of the disease. Beyond virulence, the proteomic study offers insights into the bacteria's adaptations to diverse and often stressful conditions. A suite of stress response proteins, including cold shock proteins, heat shock proteins, and proteins involved in osmotic stress adaptation, is upregulated under these challenging conditions. Additionally, biofilm contributes to the bacterium's persistence and resistance to sanitisers in food processing facilities.

Table 14.1 Important features of the genus *Listeria*

Species	Habitat/country	Key features
Listeria monocytogenes	Ubiquitous: soil, water, decaying vegetation, animal faeces, silage, sewage, various foods, and food processing environments	Beta-haemolytic (β), CAMP-positive with *Staphylococcus aureus*, motile, psychrotolerant, major human and animal pathogen (listeriosis)
Listeria innocua	Soil, water, vegetation, food, food processing environments (often found with *L. monocytogenes*)	Non-haemolytic, non-pathogenic, genetically very close to *L. monocytogenes*
Listeria welshimeri	Soil, vegetation, animal faeces, and, occasionally, food	Non-haemolytic, generally considered non-pathogenic
Listeria ivanovii	Primarily associated with ruminants, especially sheep and their faeces, silage	Strongly beta-haemolytic (β), CAMP-positive with *rickettsia equi*, primarily an animal pathogen of ruminant abortion, rare human infections
Listeria seeligeri	Soil, vegetation, water, and, occasionally, food	Typically, weakly beta-haemolytic (β) or non-haemolytic, CAMP-negative, rarely associated with infection
Listeria grayi	Soil, decaying vegetation, and animal faeces	Non-haemolytic, considered non-pathogenic, ferments mannitol
Listeria marthii	Soil in forest environments (Finger Lakes National Forest, NY, USA)	Non-haemolytic, non-pathogenic
Listeria costaricensis	Water drainage system of a food processing facility (Costa Rica)	Non-haemolytic, non-pathogenic
Listeria rocourtiae	Food such as leafy greens, food processing environments	Non-haemolytic, non-pathogenic
Listeria fleischmannii	Cheese (e.g. Swiss hard cheese), dairy environments	Non-haemolytic, non-pathogenic can be divided into subspecies
Listeria newyorkensis	Soil samples (New York state, USA)	Non-haemolytic, non-pathogenic
Listeria weihenstephanensis	Widely distributed: soil, water, plants, food, and food processing environments	Non-haemolytic, non-pathogenic, psychrotolerant
Listeria floridensis	Environmental water samples (Florida, USA)	Non-haemolytic
Listeria aquatica	Environmental water samples	Non-haemolytic
Listeria thailandensis	Fried chicken sample, Thailand	Non-haemolytic
Listeria cornellensis	Soil and agricultural run-off water (New York state, USA)	Non-haemolytic
Listeria riparia	Riverbank soil and water	Non-haemolytic
Listeria booriae	Forest soil and stream water (New York state, USA)	Non-haemolytic
Listeria goaensis	Mangrove swamps (Goa, India)	Non-haemolytic
Listeria grandensis	Urban canal water (Colorado, USA)	Non-haemolytic
Listeria valentina	Sheep faeces, irrigation water (Spain)	Non-haemolytic
Listeria farberi	Soil and stream water (New York state, USA)	Non-haemolytic
Listeria portnoyi	Agricultural soil (Colorado, USA)	Non-haemolytic
Listeria cossartiae	Agricultural soil and water	Non-haemolytic
Listeria rustica	Pasture soil	Non-haemolytic
Listeria immobilis	Soil and water near food processing plants (Austria)	Non-haemolytic, non-motile (lacks flagella genes)
Listeria swaminathanii	Stream water (Colorado, USA)	Non-haemolytic
Listeria ilorinensis	Raw milk, dairy farm environments (Ilorin, Nigeria)	Generally considered non-haemolytic and non-pathogenic
Listeria tempestatis	Floodwater (North Carolina, USA)	Non-haemolytic
Listeria rocourtiae subsp. *hofi*	(North Carolina, USA)	Non-haemolytic

One prominent protein family consistently identified in the *L. monocytogenes* biofilm matrix is the biofilm-associated protein (Bap) family. These large, surface-anchored proteins are crucial in the initial attachment to surfaces and intercellular adhesion. Different members of the Bap family have been identified, and their specific contributions to biofilm formation may vary depending on the strain and environmental conditions.

Another important class of proteins found in the biofilm matrix is enzymes. These include extracellular proteases, which may be involved in modifying the matrix composition over time, potentially aiding in nutrient acquisition or dispersal. Enzymes involved in carbohydrate metabolism have also been detected, suggesting a role in synthesising or modifying extracellular polysaccharides. Furthermore, chaperone proteins and stress response proteins are often found in the matrix. They maintain protein stability and protect the biofilm community from environmental stresses.

Morphology

Listeria are small, non-sporulating, Gram--positive (Fig. 14.1), facultatively anaerobic rods measuring 1–2 × 0.5 μm in size. They show characteristic *tumbling motility* at or around 25 °C. *Listeria* are capable of growing at high salt concentrations (10% NaCl) and at a broader range of pH (4.3–9.6) and temperature (0–45 °C; optimum 30–37 °C).

Cultivation

Listeria are ubiquitous organisms with the capability to grow in a broad temperature range (0–45 °C), pH (4.3–9.6), high salt concentrations (up to 10.0% NaCl), and low water activity (A_w to 0.90).

The organism can be cultivated on a number of selective media and chromogenic media in cell cultures. The various selective media used for the isolation of *Listeria* are Oxford agar, polymyxin–acriflavine–lithium chloride (LiCL)–ceftazidime–aesculin–mannitol (PALCAM) agar, McBride agar, and Dominguez-Rodrigues isolation agar. Typical *Listeria* colonies appear black, with a black zone in the surrounding medium on PALCAM agar (Fig. 14.2).

Animal Models

Animal models facilitate better understanding of pathogenesis and host defences in vivo. Animal studies have also helped characterise the cell-mediated immunity in controlling infection and evaluate chemotherapeutic treatments for listeriosis.

Over the past few decades, mice inoculated with *L. monocytogenes* have been the most

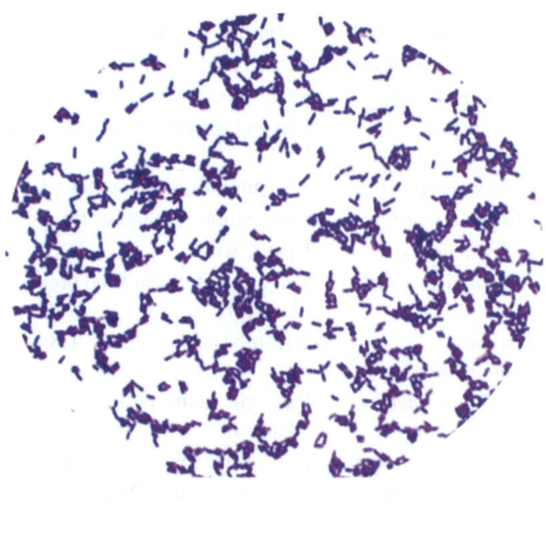

Fig. 14.1 Gram-stained smear of *Listeria monocytogenes* showing Gram-positive rod-shaped structures (×40). (Image source: Authors' personal collection)

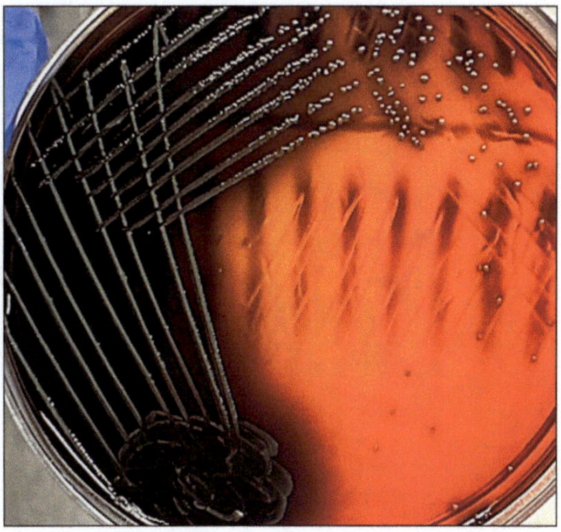

Fig. 14.2 Colonies of *Listeria* on polymyxin–acriflavine–lithium chloride–ceftazidime–aesculin–mannitol (PALCAM) agar. (Image source: Authors personal collection)

widely used small animal model in the study of listeriosis. "*Humanised*" mouse models and human epidemiological data have decoded the mechanisms underlying *L. monocytogenes* crossing of intestinal and placental barriers.

Various other animal models, including gerbils, rats, and guinea pigs, have also been used to study *L. monocytogenes* infection. *Galleria mellonella* (greater wax moth) larvae are used widely as surrogate infectious disease models due to the presence of an innate immune system functionally similar to that of vertebrates. *G. mellonella* has been used in virulence studies of *L. monocytogenes* versus non-pathogenic *Listeria* spp. and validation with wild-type and isogenic mutants in different genetic backgrounds.

Transmission Cycle

L. monocytogenes can survive under unfavourable conditions, which determines its ubiquitous nature in the environment. The organism has been isolated from water, soil, sewage, rotting vegetation, animal feed, and various species of fish, birds, and mammals. The transmission cycle of *Listeria monocytogenes* is presented in Fig. 14.3.

L. monocytogenes infection is usually food-borne, except in the case of mother-to-child transmission. Food-borne transmission is a major route of *L. monocytogenes* infection for humans. Almost all the cases of transmission of *L. monocytogenes* are via contaminated foods. Industrially processed and refrigerated foods are more frequently linked to *L. monocytogenes* outbreaks than raw foods. Ingestion of contaminated fresh and soft cheeses, meat, and fish may be the cause of sporadic cases or outbreaks.

The bacteria bind to surface receptors on gastrointestinal cells through the *inlA and inlB* genes after ingesting contaminated food. It later crosses the intestinal barrier into the bloodstream via endocytosis by endothelial cells. *Listeria* spreads to numerous tissues, especially the central nervous system (CNS), liver, and placenta. Pathogenic *Listeria* species can infect hosts by breaching endothelial and epithelial barriers, including the intestinal, blood-brain, and fetoplacental barriers.

Pathogenesis and Pathology

L. monocytogenes has notable adaptation to intracellular survival and multiplies in professional phagocytic and non-phagocytic cells of vertebrates and invertebrates.

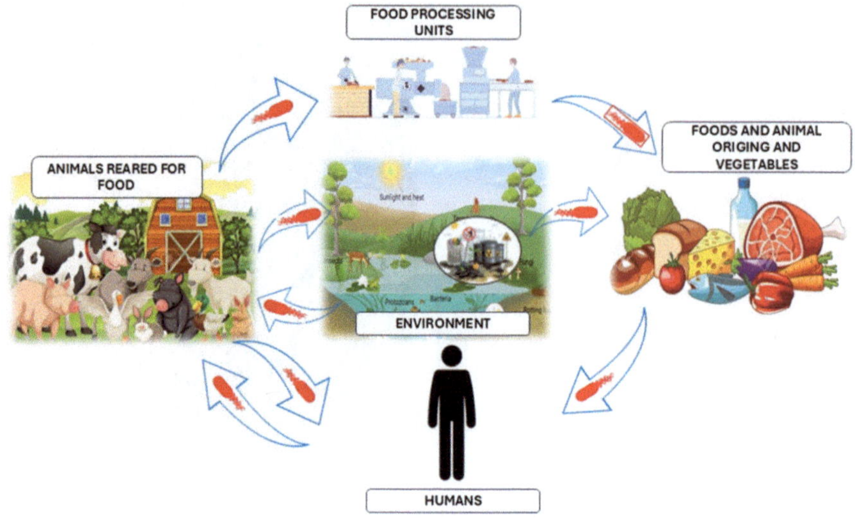

Fig. 14.3 Transmission cycle of *Listeria monocytogenes*. (Image source: Authors personal collection)

Virulence Factors

L. monocytogenes possesses diverse virulence factors with unique functions. A total of 38 virulence factors are identified in *Listeria* spp., particularly in *L. monocytogenes*.

Listeriolysin O (LLO), a haemolysin encoded as the *hly* gene, is a vital virulence factor of *L. monocytogenes*. This, in association with *plc*A (phosphatidylinositol-specific phospholipase C (PI-PLC)) and *plc*B genes (phospholipase C), permits elusion of phagosomes and escape into the cytoplasm, where replication occurs. ActA, the bacterial surface protein, makes the organisation of host cell actin into filaments, forming a "comet tail," allowing the bacteria to push themselves towards the host cell membrane. This cell-to-cell spread and escaping of the extracellular environment allows further evasion of host cell defences.

The six key virulence factors are physically linked in a 9-kb chromosomal island. These are responsible for intracellular parasitism of *L. monocytogenes* (*prf*A, *plc*A, *hly*A, *mpl*, *act*A, and *plc*B). The genes are involved in various cycles, such as the intracellular cycle (*prfA*, *hly*, *plcA*, *plcB*, *inlA*, and *inlB*), stress responses (*sigB*, *gadA*, *caspD*, *clpB*, and *lmo1138*), biofilm formation (*agr* and *luxS*), and disinfectant resistance (*emrELm*, *bcrABC*, and *mdrL*). A number of genes are organised into genomic and pathogenicity islands. LIPI-1 and LIPI-3 contain genes linked to the infectious life cycle and survival in the food processing environment. *Listeria* genomic island-1 (LGI-1) and LGI-2 ensure survival and durability in the production environment.

Pathogenesis of Listeriosis

L. monocytogenes binds to surface receptors on gastrointestinal cells through the *inlA and inlB* genes after ingestion of contaminated food. It later crosses the intestinal barrier into the bloodstream via endocytosis by endothelial cells. *Listeria* spreads to numerous tissues, especially the central nervous system, liver, and placenta.

Pathogenic *Listeria* species infect hosts by breaching endothelial and epithelial barriers, including the intestinal, blood-brain, and fetoplacental barriers. Listeriolysin O (LLO), in association with phosphatidylinositol-specific phospholipase C) and *plc*B gene (phospholipase C), permits elusion of phagosomes and escape into the cytoplasm, where replication takes place. ActA, the bacterial surface protein, makes the organisation of host cell actin into filaments, forming a "comet tail," allowing the bacteria to push themselves towards the host cell membrane.

This cell-to-cell spread and escaping of the extracellular environment allows further evasion of host cell defences.

Immunology

Listeria has been an ideal model for studying the early innate and adaptive immunity molecular mechanisms. The early innate immune responses during systemic *L. monocytogenes* infection in mice are mediated by macrophages, neutrophils, natural killer cells and dendritic cells. These are a significant line of defiance against this pathogen. $CD4^+$ and $CD8^+$ T cell populations are actively involved in the defensive immune response to *L. monocytogenes*. Class I and II MHC-restricted epitopes of listeriolysin O for $CD8^+$ and $CD4^+$ T cells, respectively, have been identified.

Acquired cell-mediated immunity to *L. monocytogenes* is induced by live, replicating bacteria growing in the host cell cytosol. However, killed bacteria or those trapped in a phagosome fail to induce protective immunity. Various antigens produce interferon-gamma after the activation of TH1 cells, a subset of CD^{4+} T helper lymphocytes. Interferon-gamma and tumour necrosis factors are crucial for in vivo immune defence against primary infection with *L. monocytogenes*.

Infection in Humans

Listeria is an opportunistic intracellular pathogen and an important cause of human food-borne infections worldwide. The infection is responsible for the highest hospitalisation rates (91%) among food-borne pathogens. Humans may act as reservoirs and asymptomatic faecal carriers, particularly abattoir and laboratory workers exposed to *L. monocytogenes* organisms.

Clinical Presentations

Human listeriosis often shows non-specific flu-like symptoms during the early stages of infection. The clinical manifestations include chills, fatigue, headache, muscular and joint pain, and gastroenteritis. The condition can develop into septicaemia, meningitis, encephalitis, abortion, and, in some cases, death, in the absence of proper treatment.

Listeriosis causes severe infection in immunocompromised people, such as pregnant women, the unborn, neonates, the elderly, kidney dialysis patients, individuals with prosthetic heart valves or replacement joints, individuals with alcoholism or alcoholic liver disease, and patients undergoing steroid and cytotoxic therapy. Persons with malignant neoplasms are at an increased risk of severe infection. In contrast, the infection may present as febrile gastroenteritis in immunocompetent individuals.

Rare presentations include meningoencephalitis and encephalitis, in association with infections with identifiable foci, such as endocarditis, pneumonia, peritonitis, and deep-seated abscess formation.

Infection in Newborns

L. monocytogenes has the ability to cross the blood-brain and placenta barriers and can also colonise the intestine. An infected mother can infect the newborns in the womb or in the birth canal during childbirth.

Early-onset disease, acquired in utero or during birth, typically presents within the first few days of life. The condition manifests as respiratory distress, fever or temperature instability, irritability or lethargy, poor feeding, and, sometimes, a characteristic skin rash. Jaundice and general signs of sepsis may also be evident. In contrast, *late-onset disease* usually appears a few days to several weeks after birth and commonly manifests as meningitis, with symptoms including fever, irritability, lethargy, poor feeding, vomiting, seizures, and a bulging fontanelle.

Infection in Animals

Animals may get infected by ingestion, inhalation, or direct contact with the pathogen. In animals, *Listeria* is most frequently transmitted by contaminated silage feeding. Asymptomatic carriers may be sources of *L. monocytogenes* infection in the farm environment.

Listeriosis in animals ranges from asymptomatic infection and carriage to a symptomatic condition. Symptomatic infection includes cutaneous lesions, focal infections like conjunctivitis, urethritis, endocarditis, and staggering gait. *L. monocytogenes* causes mastitis, repeat breeding, infertility, and septicaemia in cattle. Abortions and perinatal deaths are usual in ruminants. *Listeria* abortions usually occur in the last trimester of pregnancy and develop in the late winter or early spring.

L. monocytogenes causes invasive and often fatal diseases, including central nervous system (CNS) infection in numerous animal species. Infected farm animals may exhibit encephalitis, septicaemia, and abortions.

L. ivanovii infection is less frequent than *L. monocytogenes* and may cause abortion in cattle and sheep.

Epidemiology and Public Health

L. monocytogenes is considered a pathogen of major public health significance. It is a ubiquitous pathogen with a broad host range, including 40 mammals, 20 birds, crustaceans, ticks, and fish. Many wild and domestic animal species, such as sheep, goats, buffaloes, cattle, pigs, poultry, birds, and fish, are intestinal carriers of *Listeriae*.

The world's largest listeriosis outbreak, attributed to ready-to-eat processed meat products, was reported in South Africa. In this outbreak, a total of 1060 confirmed cases were reported during 2017–2018. In Europe, the incidence of listeriosis appears to have increased since 2000 among those >65 years old. The key global epidemiological facts are given in Fig. 14.4.

Over the years, listeriosis has been reported in many food-borne outbreaks worldwide, e.g. in the Republic of South Africa (ready-to-eat processed meat products, 2017–2018), Spain (chilled roasted pork, 2019), Germany (cooked sausage, 2018), and Italy (soft cheese, 2018). The timeline for major food-borne listeriosis outbreaks is given in Fig. 14.5.

The economic impact caused by listeriosis in humans and animals is immense. This includes significant losses for the meat processing industry and changes in consumer behaviour. For example, a single listeriosis case in the USA can result in legal costs exceeding $1 million. In the

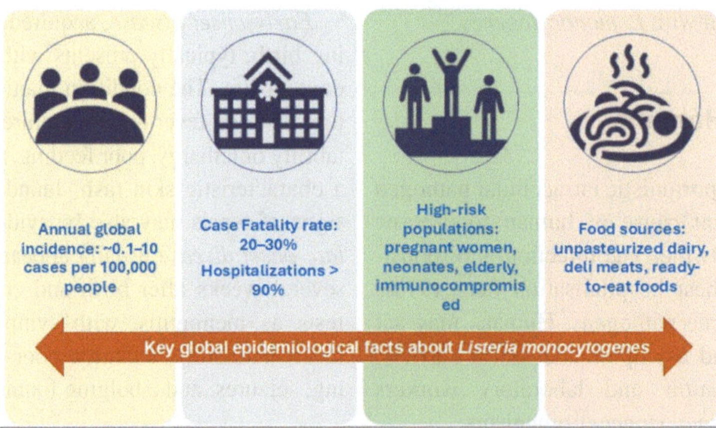

Traditional Foods	Novel Foods
Unpasteurized/Pasteurized milk and dairy products	Cantaloupe melons, stone fruits such as nectarines, peaches, plums, etc.
Cheese Varieties	Asparagus soup
Deli and ready-to-eat meat products	Frozen vegetable (Broccoli, Cauliflower, Onions. Corn, Blueberries, Canberries, Raspberries, Strawberries, etc.)
Refrigerated and smoked seafoods	Mashed potatoes, sprouts, celery, caramel apple
Pates and meat spreads	Package salad mixes, sandwich and wraps

Fig. 14.4 Key global epidemiological facts about listeriosis in humans. (Image source: Authors personal collection)

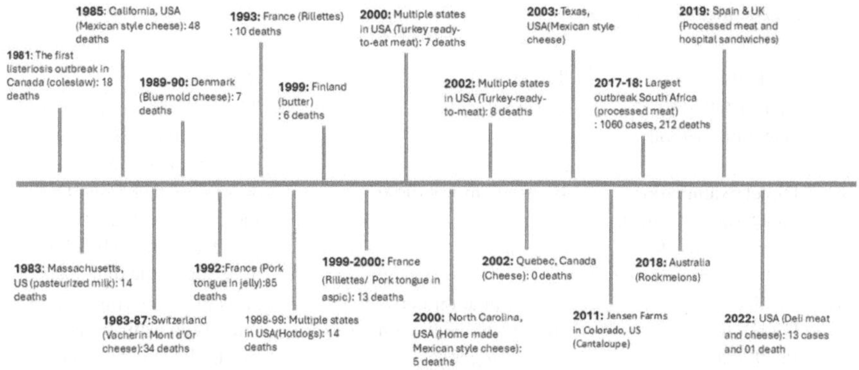

Fig. 14.5 Major food-borne listeriosis outbreaks in humans. (Image source: Authors personal collection)

USA, the estimated cost of treating listeriosis cases is around $2.6 billion annually. The costs include hospitalisation, long-term care for complications, and indirect costs associated with loss of income and productivity.

Diagnosis

The isolation and identification of *L. monocytogenes* and the limitation of its proliferation in foods have been the focus of a significant international effort. This is also useful for regulatory, epidemiological surveillance, and outbreak management purposes. Table 14.2 summarises various methods for the diagnosis of listeriosis.

Specimens

These in human infections include blood, cerebrospinal fluid (CSF), amniotic fluid, placental bits, meconium, lochia, gastric washings, or ear swabs from newborns. Even stool specimens (other than meconium) would be helpful for detection. Similarly, in animals reporting abortion, placental beds and blood would be ideal specimens. CSF, blood, faecal, vaginal, and nasal swabs may be useful for animals with CNS manifestations.

Microscopy

Under a light microscope, *Listeria* spp. appears as Gram-positive short rods (Fig. 14.1). While the presence of Gram-positive rods in a sample can be suggestive of *Listeria*, it is important to note that other bacteria may share similar morphological characteristics. Therefore, microscopy alone is not sufficient for a definitive diagnosis. In the case of *Listeria* meningitis, for instance, Gram staining of CSF may not always reveal the bacteria, especially if the bacterial load is low.

Other staining methods, such as immunofluorescence, offer increased specificity and sensitivity compared to Gram staining. Immunofluorescence involves using antibodies labelled with fluorescent dyes that bind specifically to *Listeria* antigens. This technique helps in rapid identification, especially in samples where the bacterial load is low or mixed with other organisms. Additionally, it is used to visualise *Listeria* within host cells, aiding in understanding the pathogenesis of infection.

In Vitro Culture

The conventional bacteriological methods for the isolation of *Listeria* spp. by culture have been widely used for detection in specimens collected from human and animal cases and in food speci-

Table 14.2 Diagnostic methods for listeriosis

Diagnostic approaches	Methods	Targets	Remarks
Isolation and culture methods	ISO 11290-1:2017 detection method	Bacterial cells	To detect the isolate by enriching and isolating the viable cultures on selective agar plates
Microscopy	Direct examination under 100× eyepiece Motility at 25 °C	Bacterial morphology	A primary technique for morphological identification and to proceed with the biochemical tests, assisting in the diagnosis of infections
Biochemical tests	Catalase CAMP test PI-PLC assay	Discrimination between *listeria* species and pathogenic *Listeria monocytogenes*	Aids in identifying and differentiating microorganisms based on their metabolic and enzymatic activities
Molecular assays	PCR Real-time PCR	*prs* (genus specific) and *isp*, *hlyA*, *plcA* (*Listeria monocytogenes*- and *Listeria ivanovii*-specific)	Detection of multiple virulence--associated genes enables the identification of different serovars and lineages
Molecular subtyping	Restriction fragment length polymorphism (RFLP) Pulsed-field gel electrophoresis (PFGE) Multilocus sequence typing (MLST) Whole-genome sequencing	Restriction endonucleases like *ApaI*, and *AscI* Specific nucleotide sequence	Subtyping, source tracking, and investigation of outbreaks Identifies the structure of the population and gives phylogeny within a population of isolates
Serological tests	ELISA Lateral flow assay of listeriolysin O (LLO) Latex agglutination test	Polyclonal antibodies against pathogenic species	Antigen–antibody-binding enzymatic reactions

mens. Isolation by culture is considered the "gold standard." Culture of the pathogen is useful for treatment, epidemiological surveillance, and outbreak management. Moreover, conventional methods are also sensitive and do not require expensive and sophisticated equipment to isolate and identify the organism, allowing their widespread use.

The various selective media used for isolation of *Listeria* are Oxford agar, polymyxin–acriflavine–LiCL–ceftazidime–aesculin–mannitol (PALCAM) agar, McBride agar, and Dominguez-Rodrigues isolation agar. Typical *Listeria* colonies appear black, with a black zone in the surrounding medium on PALCAM agar (Fig. 14.2) after 3–4 days of incubation at 37 °C.

Some of the biochemical tests used for the identification of *Listeria* spp. are given below.

Catalase Test

It is a quick test that shows a positive reaction by producing visible bubbles with inoculum, which breaks down hydrogen peroxide (H_2O_2) into water and oxygen.

Christie–Atkins–Munch-Petersen (CAMP) Test

This test helps in the differentiation of *L. monocytogenes* from other *Listeria* species. In this test, a β-haemolytic strain of *Staphylococcus aureus* is streaked in a straight line on sheep blood agar, and the *L. monocytogenes* is streaked perpendicular to it. A positive reaction shows the enhanced zone of haemolysis (*arrowhead-shaped clearing*) that appears at the junction of the streaks due to synergistic haemolysis towards *S. aureus*.

Phosphatidylinositol-Specific Phospholipase C Activity (PI-PLC) Assay

The test is performed on the chromogenic selective medium Agar *Listeria*, according to Ottaviani and Agosti (ALOA). An opaque halo zone with blue-green colonies is observed due to the activity of phosphatidylinositol-specific phospholipase C (PI-PLC), an enzyme produced by pathogenic *L. monocytogenes* species only.

Cell Culture Infection/Invasion Assays

These in vitro assays are designed to mimic the initial stages of *Listeria* infection at a cellular level. Cultured mammalian cells, such as Caco-2, HeLa, or J774 macrophages, are exposed to *Listeria* isolates. These assays offer mechanistic insights into *Listeria* pathogenesis, allow for the screening of virulence factors, provide quantifiable results, and are relatively cost-effective for high-throughput analyses.

In Vivo Assays

Chick Embryo Lethality Assay

This in vivo assay evaluates the virulence of *Listeria* isolates by measuring their capacity to cause mortality in developing chick embryos. Typically, 10–12-day-old embryonated chicken eggs are inoculated with a known dose of *Listeria* onto the chorioallantoic membrane (CAM) or into the yolk sac. The eggs are then incubated, and mortality is recorded daily over several days. The Lethal Dose 50 (LD50), representing the bacterial dose that kills 50% of the embryos, is calculated to quantify the isolate's virulence. This assay is relatively inexpensive and easy to perform, providing rapid results and an indication of the systemic virulence of the *Listeria* isolate in a developing organism.

Mouse Inoculation Assay

As a classic mammalian in vivo model, the mouse inoculation assay allows for a comprehensive study of *Listeria* pathogenesis. Mice of specific strains are inoculated with a defined dose of *Listeria* through routes like intravenous, intraperitoneal, or oral administration. Following infection, mice are monitored for clinical signs and mortality, allowing for the determination of LD50. Furthermore, bacterial burden in various organs can be quantified over time, and histopathological analysis of infected tissues can reveal tissue damage and inflammatory responses. However, it is costly, raises ethical considerations, and can be time-consuming, with results potentially varying based on the mouse and *Listeria* strain.

Galleria mellonella Larva Infection Assay

The larvae of the greater wax moth (*Galleria mellonella*) serve as an emerging invertebrate in vivo model for studying bacterial pathogenesis. A known number of *Listeria* bacteria are injected into the haemocoel of the larvae, and their survival is monitored over several days. Mortality curves are generated, and LD50 is calculated. Bacterial burden in the haemolymph can also be quantified, and the melanisation response of the larvae can be observed. This assay offers the advantages of being inexpensive and easy to handle and providing rapid results, with the potential for high-throughput screening.

Rapid Diagnostics

Biosensors are recent rapid diagnostic methods. These include optical (organic light-emitting diode), cell-based (bioelectric diagnostics), and electrochemical (carbon ionic liquid electrode) for the detection of *L. monocytogenes*. These tools are novel, highly sensitive, and specific and produce reproducible results. Also, most of the developed biosensor systems are portable and easy to handle and perform rapid or real-time detection. However, the high cost involved in processing the sample and furthering their validation with cultural methods limits their use in the routine detection of *L. monocytogenes*.

Spectrophotometric methods such as MALDI-TOF MS, VITEK® MS (e.g. VITEK® MS Advanced Expert System™), and MICRO-ID are rapid and highly sensitive, but considering the economics, their use is restricted to clinical settings for the diagnosis of infections in humans only.

Serodiagnosis

Several serological assays detect and quantify antibodies specific to *Listeria* antigens in serum or other bodily fluids:

Antibody-based assays include indirect fluorescent antibody (IFA) and enzyme-linked immunosorbent assay (ELISA).

Indirect Fluorescent Antibody (IFA)

The test is used to detect *Listeria*-specific antibodies. In this assay, *Listeria* antigens are fixed onto a microscope slide. The patient's serum is applied to the slide, allowing any *Listeria*-specific antibodies to bind to the antigens. After washing away unbound serum components, a fluorescently labelled secondary antibody, which recognises human or animal immunoglobulins, is added. The slide is then examined under a fluorescence microscope. If *Listeria*-specific antibodies are present in the patient's serum, the bound secondary antibody will fluoresce, indicating a positive result. IFA provides a visual confirmation of antibody binding but may be more subjective in interpretation than quantitative methods like ELISA.

Enzyme-Linked Immunosorbent Assay (ELISA)

These antibody-based assays employ different antigens derived from the bacteria, namely the somatic (O) antigens, which are associated with the bacterial cell wall; the flagellar (H) antigens, related to the motility structures; cold-extracted or sonicated antigens, representing a broader range of cellular components; and outer membrane protein (OMP) antigens, specific to the bacterial surface. These tests have been evaluated for their potential in screening animal and human cases of listeriosis, aiming to identify individuals who have been exposed to *Listeria*.

Despite the variety of antigens utilised, these antibody-based serological assays often suffered from limitations in their performance characteristics. Specifically, they exhibited poor specificity, which means that they could produce positive results in individuals who were not actually infected with *Listeria* due to cross-reactivity with antibodies against other microorganisms. Additionally, their sensitivity, or the ability to correctly identify true positive cases, was also inadequate.

Antigen-based assays include agglutination tests and indirect ELISA.

Agglutination tests were historically employed in *Listeria* serodiagnosis. These assays rely on the visible clumping (agglutination) that occurs when *Listeria* antigens react with corresponding antibodies in the patient's serum, often in the form of heat-killed bacteria or specific cell fractions. The degree of agglutination is typically assessed visually. For example, latex agglutination tests (LATs) utilise antibody-coated latex particles that clump together in the presence of the target *Listeria* antigen, providing a visible indication of infection. These antigen-based assays hold promise for faster and, potentially, more specific diagnosis of active *Listeria* infections. While relatively simple, agglutination tests generally exhibit lower sensitivity and specificity compared to more modern techniques like ELISA, leading to their less frequent use in contemporary clinical diagnostics.

Indirect Enzyme-Linked Immunosorbent Assays (ELISAs)

Recently, a significant focus of the serodiagnosis of *Listeria* infection has been on the *indirect enzyme-linked immunosorbent assays* (*ELISAs*) designed to detect anti-listeriolysin O (ALLO) antibody. Listeriolysin O is a key virulence factor produced by *Listeria* monocytogenes, and antibodies against it are used to indicate an immune response to infection. This approach has been extensively investigated to diagnose *Listeria* infection in humans and animals. Nevertheless, a major challenge associated with the ALLO-based ELISA is its potential for cross-reactivity with streptolysin O (SLO), a similar haemolysin produced by various species of *Streptococcus*. This cross-reactivity can lead to false positive results, necessitating an additional step of adsorbing the test sera with SLO before performing the ELISA to remove these interfering antibodies.

Molecular Diagnosis

Molecular methods such as polymerase chain reaction (PCR), multiplex PCR (mPCR), real-time/quantitative PCR (qPCR), nucleic acid sequence–based amplification (NASBA), and loop-mediated isothermal amplification have been evaluated in the diagnosis of listeriosis. These tests target different gene targets to detect *L. monocytogenes* in foods and in various clinical specimens obtained from cases in humans and animals.

mPCR is increasingly used to detect multiple virulence-associated genes of *L. monocytogenes*, different species of *Listeria*, different serovars, and the lineage of *L. monocytogenes*. Currently, commercial *qPCR kits* are available to detect *L. monocytogenes* in various clinical specimens.

DNA microarray and next-generation sequencing (NGS) technology are available for the detection and/or identification of *L. monocytogenes*. DNA microarray can simultaneously detect and discriminate *Listeria* spp. and different food-borne bacterial pathogen isolates in food-borne infections.

Typing Methods

Typing methods identify and differentiate closely related *L. monocytogenes* strains. These methods are used to confirm the source of an outbreak, establish a transmission pattern, and determine and monitor the reservoirs of epidemic strains. Two major typing methods are described below.

The Classical Serotyping Method

This is based on somatic O and flagellar H antigens of *L. monocytogenes* strains by serological methods such as ELISA. However, this method has poor discriminatory power. It fails to type certain *L. monocytogenes* isolates recovered from food and environmental sources. The method is also laborious and time-consuming and requires expensive antisera.

Molecular Subtyping Methods

This includes *amplification-based typing methods* such as random amplification of polymorphic DNA (RAPD), PCR-single strand conformation polymorphism (SSCP), PCR-restriction fragment length polymorphism (PCR-RFLP), pulse field gel electrophoresis (PFGE). PFGE is considered the gold standard and was used for subtyping during outbreak investigations until the advent of sequence-based typing methods.

Sequence-Based Typing Methods

Sequence-Based Typing Methods include multiple-locus variable-number tandem repeat analysis (MLVA) and multilocus sequence typing (MLST). MLVA and MLST are being increasingly used for typing outbreak strains because of their high discriminatory power. Recently, *next-generation sequencing (NGS)*, a DNA sequencing–based approach, has been widely used for typing *Listeria* isolates. *NGS* also provides new insights into disease transmission, virulence, and antimicrobial resistance.

Treatment

In Humans

Antibiotics such as penicillin and ampicillin are the first-line drugs for the treatment of *L. monocytogenes* infection in humans. These antibiotics penetrate the infected cell and bind to intracellular targets, i.e. penicillin-binding protein 3 (PBP3), and remain active under an unfavourable intracellular environment. Trimethoprim is the second-line antibiotic used to treat patients who exhibit intolerance to beta-lactam antibiotics.

Rifampin, vancomycin, linezolid, and carbapenems have also been evaluated for the treatment of *L. monocytogenes* infection. Fluoroquinolones such as levofloxacin, clinafloxacin, sparfloxacin, and moxifloxacin are also suggested. Levofloxacin has been found effective against experimental *L. monocytogenes* in mice and a human case of *Listeria* meningitis.

If infected with L. *monocytogenes*, the infection spontaneously clears within 7 days in immunocompetent people. However, in the case of high-risk populations, such as pregnant women, they usually require immediate intravenous treatment of antibiotics to prevent abortion and also to slow down the progression of more severe infection. The duration of antibiotic treatment depends on the severity of the infection; treatment of meningitis is recommended for 3 weeks, while brain abscess treatment may last up to 6 weeks. Table 14.3 denotes the dose and duration of antibiotics for various clinical manifestations of *Listeria* infection.

In Animals

Aggressive treatment with antibiotics such as penicillin, oxytetracycline, ceftiofur, erythromycin, trimethoprim/sulphonamide, and supportive treatment might help the animal recover from the infection. In animals, penicillin at 22,000 U/kg weight is administered every 12 h for 1–2 weeks. It begins with potassium penicillin (IV) followed by procaine penicillin G (IM). Intravenous administration of oxytetracycline at 16.5 mg/kg/body weight has been found effective. Supportive therapy, such as fluid and electrolyte supplements, is recommended for animals having difficulty eating and drinking. In complicated infections, affected animals may exhibit recumbency, abnormal excitability, and/or weak or absent menace reflexes. The prognosis remains poor in such cases, even with the administration of antibiotic therapy and supportive treatment.

Table 14.3 Treatment options for various clinical manifestations of *Listeria* infection

Clinical manifestation of *Listeria* infection	Therapeutic dose	Duration
Bacteraemia	First line: ampicillin@6 g/day IV alone or if the patient is more than 50 years old or has a chronic disease Second line: erythromycin@4 g/day IV or TMP/SMX@200–320 mg tid or Vancomycin@1 g tid	14 days 14 days
Bacteraemia in pregnancy	First line: ampicillin@6 g/day IV Second line: erythromycin@4 g/day IV	7–14 days 7–14 days
Acute meningitis	First line: ampicillin@6 g/day IV alone, or if the patient is more than 50 years old or has a chronic disease, cardiovascular or respiratory compromise, include gentamicin@2.5 mg/kg/day Second line: erythromycin@4 g/day IV or TMP/SMX@200–320 mg tid	21 days 21 days
Brain abscess	First line: ampicillin@14 g/day IV plus gentamicin@2.5 mg/kg/day Second line: TMP/SMX@200–320 mg tid	4–6 weeks 4–6 weeks
Joint/bone infection	Ampicillin@6 g/day IV alone, or if the patient is more than 50 years old or has a chronic disease, cardiovascular or respiratory compromise, include gentamicin@2.5 mg/kg/day	6 weeks if no prosthesis or 2 weeks IV with prosthesis and lifelong oral amoxicillin or TMP/SMX in each case

Prevention and Control

Currently, no effective vaccine is available for listeriosis in humans and animals. Hence, to prevent and control this infection, proper hygienic practices play a key role in controlling the spread of *L. monocytogenes*. There are some general US Food and Drug Administration (FDA) recommendations on preventing an infection with *Listeria* and some additional recommendations, specifically for persons at higher risk and animals.

Preventing *L. monocytogenes* contamination necessitates a comprehensive strategy encompassing safe food handling and meticulous hygiene practices throughout the food chain. This includes rinsing all raw produce under running potable water to remove potential surface contamination. Cross-contamination is avoided by strictly separating raw meats, poultry, and seafood from vegetables, cooked foods, and ready-to-eat items during storage and preparation, utilising dedicated utensils and surfaces, and ensuring proper containment. Maintaining a clean kitchen environment involves regular hand washing, diligent cleaning and sanitisation of all food contact surfaces, and immediate clean-up of spills, especially from raw animal products. Temperature control is vital because of the ability of *Listeria* to grow at refrigeration temperatures. Therefore, refrigerator temperatures should be maintained at or below 4 °C (40 °F) and freezer temperatures at −10 °C (14 °F).

Thoroughly cooking raw animal products effectively eliminates *Listeria*. Ready-to-eat foods need to be consumed promptly and stored according to manufacturer guidelines, with opened packages used within 1 week and unopened ones used within 2 weeks. Leftovers should be cooled rapidly in shallow containers, stored airtight, and consumed within 3–4 days after thorough reheating. Finally, consuming raw (unpasteurised) milk and products should be completely avoided. Pasteurisation is effective for killing of *Listeria* in milk and milk products.

Control Measures in Animals

Controlling listeriosis in animals requires safe animal husbandry and rigorous hygiene practices. Proper silage management, including anaerobic fermentation, providing optimal moisture content, and avoiding the feeding of mouldy or high-pH silage, is most important.

Additionally, good animal welfare practices by maintaining clean and dry housing, effective drainage, and regular bedding replacement contribute significantly to reducing transmissions of infection. Prompt isolation of sick animals is also crucial to limit the spread of the disease.

Routine cleaning and disinfection of animal housing and equipment with effective disinfectants are important. Maintaining clean water sources and practising proper manure management to minimise environmental contamination are also vital. Regular observation for clinical signs of listeriosis and prompt laboratory diagnosis of suspected cases are important for early detection and treatment of cases.

Case Study

A 33-year-old woman presented with a history of repeated abortion, specifically during the third trimester of her pregnancy. She had a history of working on an organised goat farm. On investigation, a cervical swab was collected and subjected to selective enrichment for *Listeria*, followed by plating on selective agar. *L. monocytogenes* isolated from the samples was confirmed by biochemical tests and by PCR for virulence-associated genes. The aborted samples from animals with a history of abortions were collected from the goat firm. *Listeria* isolates were recovered from animal samples. The isolates from the woman and animals were of the same serotype, and PFGE profiles were similar. The woman on specific treatment with ampicillin recovered from the infection.

1. What was the medical history of the woman related to pregnancy?
2. What are the laboratory tests to detect *L. monocytogenes* in clinical samples?
3. How was the zoonotic link between the *Listeria* isolates from the woman and the animals established?
4. What preventive measures can be taken to reduce the risk of *Listeria* infection in pregnant women working in high-risk environments?

Research Questions

1. What are the prevalence rates of listeriosis in the African, Eastern Mediterranean, and South-east Asian WHO regions that are considered to have a significant immunocompromised population?
2. What is the scope for biocontrol approaches such as bacteriophages, bacteriocins, and essential oils for the control of listeriosis in food establishments?
3. What is the significance of increasing surveillance and testing for *Listeria* in previously unsurveyed regions, habitats, foods, and food ingredients?

Further Readings

Barbuddhe SB, Rawool DB, Doijad SP, Vergis J, Malik SS, Chakraborty T. Ecology of *Listeria monocytogenes* and *Listeria* species in India: the occurrence, resistance to biocides, genomic landscape and biocontrol. Environ Microbiol. 2021;24(6):2759–80.

FAO/WHO. *Listeria monocytogenes* in ready-to-eat (RTE) foods: attribution, characterisation and monitoring—meeting report. Microbiological risk assessment series no. 38. Rome: FAO/WHO; 2023.

Končurat A, Sukalić T. Listeriosis: characteristics, occurrence in domestic animals, public health significance, surveillance and control. Microorganisms. 2024;12(10):2055.

Orsi RH, Liao J, Carlin CR, Wiedmann M. Taxonomy, ecology, and relevance to food safety of the genus *Listeria* with a particular consideration of new *listeria* species described between 2010 and 2022. MBio. 2024;15(2):e0093823.

Sarr M, Alou MT, Padane A, Diouf FS, Beye M, Sokhna C, et al. A review of the literature of *Listeria monocytogenes* in Africa highlights breast milk as an overlooked human source. Front Microbiol. 2023;14:1213953.

Erysipeloid

15

Natarajan Vaghesan, Balamurali Venkatesan, and Kesavarajan Govindarajan Karthikeyan

Learning Objectives

1. To provide an overview of the molecular mechanisms of erysipeloid, its effects on endothelial cells, and its clinical manifestations
2. To understand the key concepts of *Erysipelothrix rhusiopathiae* transmission, its association with erysipeloid, and the socio-environmental factors contributing to outbreaks
3. To explore current diagnostic methods, treatment strategies, and challenges in vaccine development

Introduction

Erysipelothrix rhusiopathiae is a Gram-positive bacterium that causes erysipeloid, a rare zoonotic disease in humans. It causes erysipelas in animals such as domestic pigs, poultry, sheep, and fish.

Individuals who handle animals or animal products, such as butchers, farmers, seafood handlers, veterinarians, and meat processors, are usually at a high risk of the infection. Erysipeloid is mostly an occupational disease. Small cuts, abrasions, or puncture wounds on the hands and fingers are typical sites through which bacteria enter the human body and cause infections.

Erysipeloid usually presents as a localised cutaneous infection, which may occasionally progress to a systemic infection or a more widespread cutaneous manifestation. Rarely, *E. rhusiopathiae* causes severe complications, such as septic arthritis, endocarditis, or septicaemia, particularly in individuals with pre-existing medical conditions. Domestic pigs, as well as fish, poultry, sheep, and other animals, are reservoir hosts. The bacteria are excreted in urine, saliva, and faeces, thereby polluting the environment and potentially causing infections in humans. Erysipeloid is frequently misdiagnosed or underdiagnosed since it is generally mild and self-limiting. Hence, laboratory and clinical diagnoses are important for initiating early treatment of the condition. Penicillin is the antibiotic of choice for the effective treatment of the condition.

History

In 1876, Robert Koch identified the first bacterium, later classified under the genus *Erysipelothrix,* during his experiments involving the injection of putrefying blood into mice. In a letter to Dumas in 1882, Louis Pasteur mentioned research on a bacterium that causes "rouget" in pigs. Notably, Pasteur stated that these bacteria

N. Vaghesan · B. Venkatesan
K. Govindarajan Karthikeyan
Department of MicrobiologySri Lalithambigai Medical College and Hospital, Faculty of Medicine, Dr. MGR Educational and Research Institute, Maduravoyal, Adayalampattu, Chennai, Tamil Nadu, India

S. C. Parija, A. Poddar (eds.), *Textbook of Bacterial Zoonoses*, Microbial Zoonoses,
https://doi.org/10.1007/978-981-95-6615-0_15

were quite like those that cause cholera in birds, except for a few unidentified physiological characteristics and their smaller size. He successfully induced infection in pigs, lambs, and rabbits through inoculation but noted that chickens remained unaffected. A few years later, Friedrich Löffler provided a detailed description of the disease in pigs and its causative agent. The bacteria were also recognised as a human pathogen in 1884. In the year 1936, Beaudette and Hudson reported the first documentation of erysipelas in poultry in Turkey. Hausser subsequently reported the disease in hens in 1909, followed by many reports of the infection in different chicken species in the following years.

In 1918, Buchanan formally proposed the name *Erysipelothrix rhusiopathiae*, and in 1920, it was named the type species of the genus. Over the years, the bacterium has been referred to by various names, often differing based on the host species from which it was isolated. Acknowledging that these strains belonged to a single species, Langford and Hansen honoured Trevisan's original nomenclature by proposing the unifying name *Erysipelothrix insidiosa* in 1954. However, due to its extensive historical use, the earlier name *E. rhusiopathiae* was revived in 1966.

Taxonomy

The species *E. rhusiopathiae* belongs to the genus *Erysipelothrix,* family Erysipelotrichaceae, phylum Bacillota, class Erysipelotrichia, and order Erysipelotrichales.

E. rhusiopathiae, with a complex taxonomic structure, is influenced by its wide host range and antigenic variations. Based on traditional phenotypic classification, the species have historically been classified into multiple serogroups and over 28 serovars. The classification depends on the presence of surface protective antigens (Spas), particularly peptidoglycan-associated polysaccharides. Serogroups 1a, 1b, and 2 are the most frequently associated with infections in pigs.

Currently, genomic and proteomic studies play a crucial role in refining the taxonomy of *E. rhusiopathiae* and related species within the genus *Erysipelothrix*. Comparative genomic analyses have helped distinguish *E. rhusiopathiae* from closely related species such as *E. tonsillarum* and other recently identified taxa. These studies are based on core genome comparisons, average nucleotide identity (ANI), and phylogenomic approaches. These provide a more robust taxonomic resolution than traditional phenotypic methods.

Genomics and Proteomics

Genomic and proteomic studies have enhanced understanding of *E. rhusiopathiae*, especially in relation to its virulence, host specificity, and antigenic diversity.

Whole-genome sequencing (WGS) of *E. rhusiopathiae* isolates from different hosts has revealed a relatively small genome, approximately 1.8–2.0 Mb in size, with a high degree of conservation across strains. However, genomic variability exists, particularly in genes associated with surface antigens, such as *spaA*, *spaB*, and *spaC*. These genes encode surface protective antigens and play important roles in serovar differentiation and immune evasion. These antigens are often used as molecular markers in typing schemes and vaccine formulations. Comparative genomic studies have shown that strains from swine and humans commonly possess the *spaA* gene, while strains from fish or marine animals may exhibit different surface antigen gene profiles.

A small genome (1.8 Mbp) was found in the first whole-genome sequence of a pig strain from Fujisawa in 2011, which was in line with the adaptation for intracellular parasitism. A complicated population structure with a clear clade 1 and two less well-defined clades, 2 and 3, as well as intermediate isolates, was later discovered by comparative genomic investigations in swine populations.

In Sweden and Italy, clade 1 isolates were demonstrated in healthy pigs and wild boars, indicating decreased pathogenicity for terrestrial animals, but they were conspicuously absent in poultry and pigs with clinical erysipelas. Pathogenic clones, especially serotype 1a, in pigs in China and Japan have been reported. More genetic variety in pigs and poultry in Europe has been reported. Clonal strains verified by WGS have also been associated with outbreaks among wildlife, including harbour porpoises, muskoxen, and Arctic animals.

Clades 2, 3, and intermediate kinds, on the other hand, appear to be mixing with both mammalian and avian hosts, suggesting the possibility of cross-species transmission between the animal–human interface. Although certain serotypes have been linked to serious pig epidemics, serotyping using PCR or in silico analysis has shown that serotypes are paraphyletic. This means that they do not always match phylogenetic connections. This explains the differences between host specificity and serotype.

A wide range of genes and genomic sites are involved in *E. rhusiopathiae* pathogenicity. The cps gene cluster is essential for the development of a protective polysaccharide capsule, and mutants lacking this capsule exhibit decreased pathogenicity. Additionally, capsule production is associated with serotyping-relevant antigen architectures, which may help explain the variations in virulence among serotypes. Three surface protective antigen (Spa) variations have been found: SpaA (in *E. rhusiopathiae* except clade 1 and *E. amsterdamensis*), SpaB (in *Erysipelothrix rhusiopathiae (ER)* clade 1), and SpaC (in *E. piscisicarius*). Antibodies that target these proteins are protective. Adhesion and phagocytosis resistance are enhanced by SpaA. Haemolysin (hlyIII), neuraminidase (nanH), and genes are other possible virulence factors associated with biofilm development.

Proteomic analyses have been employed to identify immunogenic proteins, virulence factors, and stress response proteins. These studies have highlighted the roles of neuraminidase, hyaluronidase, and capsule-associated proteins in pathogenesis.

Morphology

E. rhusiopathiae is a Gram-positive, rod-shaped bacterium that does not form spores and is often mistaken for *Lactobacillus* species or *Listeria monocytogenes* by microscopy (Fig. 15.1).

Depending on the surroundings, two different morphological forms are demonstrated by microscopy: smooth (S) and rough (R), which measure about 0.8–2.0 μm in length and 0.2–0.4 μm in breadth, respectively. The S-form

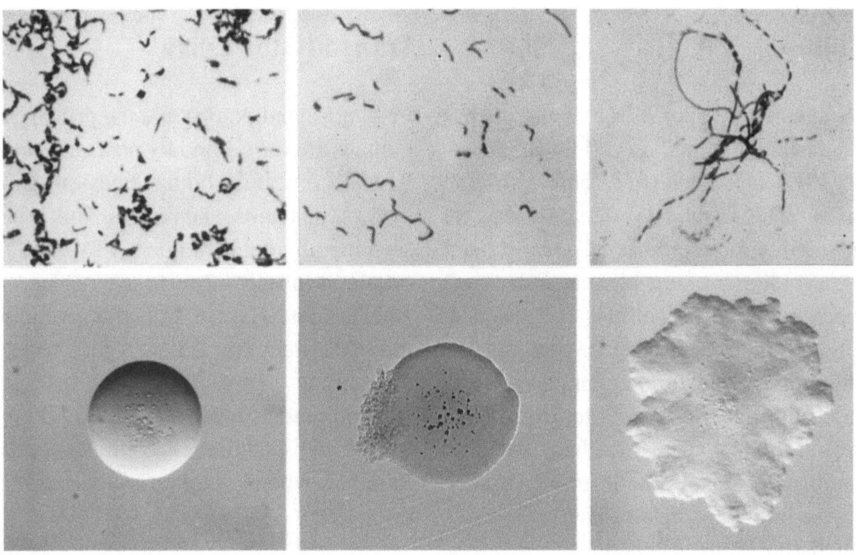

Fig. 15.1 Cellular and colonial morphology of *Erysipelothrix rhusiopathiae*. Upper row: ×1200, crystal violet; lower row: ×32. (Image source: National Animal Disease Center, Ames, Iowa)

appears as small, slender, slightly curved rods with rounded ends. The R-form appears as long filaments that form chains, which may extend longer than 60 μm.

Cultivation

On solid media, *E. rhusiopathiae* exhibits two primary colony types: smooth (S-form) and rough (R-form). The S-form colonies are typically small (0.1–0.5 mm in diameter after 24 h, growing to 0.5–1.5 mm after 48 h), convex, and circular and have a smooth surface with an entire margin. They are usually clear, transparent, and non-pigmented. However, they may develop a pale blue opacity with prolonged incubation or age. These smooth colonies are commonly associated with acute infections. In contrast, R-form colonies are generally larger, flatter, and opaquer than S-form colonies. They have an irregular, rough surface and a fringed or irregular edge.

Lower pH (<7.0) and 37 °C encourage R-form development, whereas a pH range of 7.6–8.2 and 33 °C favour S-form growth. When cultured on blood agar, most strains exhibit a narrow alpha-haemolysis; beta-haemolysis is non-existent, resembling that of viridans group streptococci.

Tissue Cultures

The intracellular survival of *E. rhusiopathiae* has been studied using mammalian cell lines, including RAW 264.7 cells, which resemble murine macrophages. These models demonstrate the bacterium's ability to enter host cells and remain there while avoiding immune clearance. Its virulence depends on intracellular persistence, emphasising the function of surface proteins like SpaA. Gentamicin protection techniques are frequently used in infection tests to distinguish between intracellular and extracellular bacteria. Understanding adhesion, invasion, and resistance to phagocytic killing processes has been made possible by tissue culture methods.

Chick Embryo Inoculation

In the past, strains of *E. rhusiopathiae* have been tested for pathogenicity using chick embryo inoculation. Embryonated chicken eggs are infected through the yolk sac or allantoic cavity, usually between 9 and 11 days of incubation. Because virulent strains cause haemorrhages, distinctive lesions, and embryonic mortality, they are useful parameters of determinants of pathogenicity.

Animal Models

E. rhusiopathiae infection has been widely studied in mouse, rat, and pig models. Mice that are intraperitoneally challenged show signs of septicaemia, which makes virulence evaluations easier. Pigs, the natural host, exhibit diamond-shaped skin lesions, arthritis, and septicaemia, but rats have been used to demonstrate infection imitating endocarditis. A recent study employed fish models, such as tiger barbs (*Puntigrus tetrazona*), to compare several strains of *Erysipelothrix*. These models aid in clarifying patterns of antibiotic resistance, vaccination effectiveness, and host–pathogen interactions.

Transmission Cycle

Fish, pigs, and poultry are the primary hosts, and humans are the secondary hosts. *E. rhusiopathiae* typically enters the human body through breaches in the skin, hands, and fingers. This is particularly seen in individuals frequently exposed to animals or animal products, such as butchers, fishermen, and farmers. Human infection primarily occurs through direct contact with infected animals or their tissues. Transmission can also occur from asymptomatic carrier animals that do not display visible signs of illness. *E. rhusiopathiae* can be found in their blood, tissues, and secretions (Fig. 15.2).

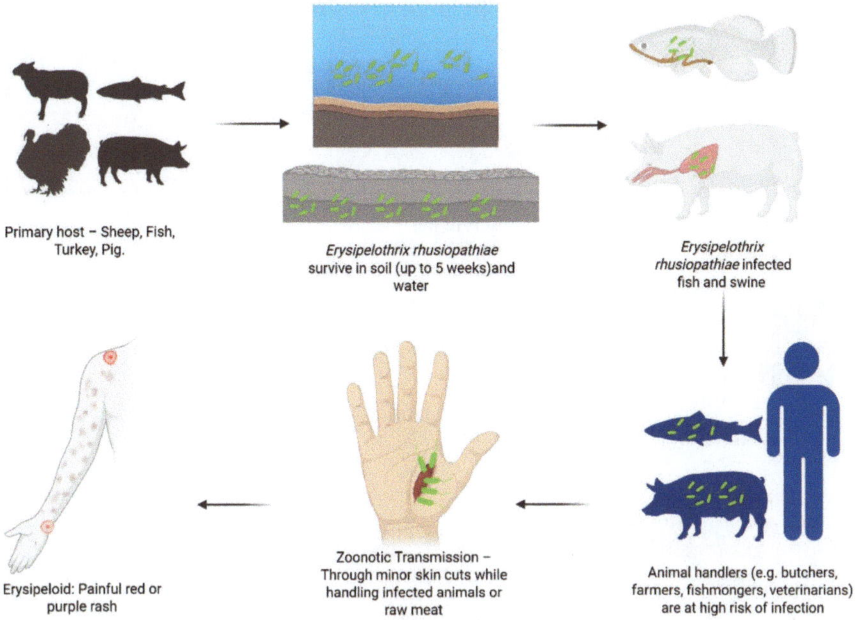

Fig. 15.2 Transmission of infection in humans. (Image source: Authors' personal collection)

Pathogenesis and Pathology

The pathogenicity of *E. rhusiopathiae* is linked to the production of enzymes such as neuraminidase and hyaluronidase. Neuraminidase cleaves sialic acid residues on host cells to facilitate bacterial adherence and invasion. Antibodies against it provide protection, and its expression is correlated with virulence. Bacteria's involvement in vascular disease is supported by their adhesion to desialylated arterial areas. The potential spreading factor hyaluronidase exhibits a patchy correlation with virulence. Therefore, although hyaluronidase may contribute to disease in other hosts, it is not necessary for the disease to be virulent.

E. rhusiopathiae surface proteins are important for host contact and possible pathogenicity. The identity of the trypsin-sensitive, heat-labile surface element essential for adherence remains unknown. Its prominent surface antigens include 43-kDa and 66–64-kDa proteins that are recognised by antiserum. Highly pathogenic strains have higher expression of these antigens. The SpaA protein, a recognised protective antigen, mimics *Streptococcus pneumoniae* choline-binding proteins. Additionally, a 19-kDa surface protein has been discovered. However, it is not haemolytic in and of itself; when produced in *E. coli*, it promotes haemolysis.

A significant non-protein surface component of *E. rhusiopathiae* is its capsular polysaccharide antigen, which has a molecular weight of 14–22 kDa. In contrast to the parent strain, mutants produced by Tn916 transposon mutagenesis that lacked this capsule were avirulent and extremely vulnerable to phagocytosis by polymorphonuclear leucocytes (PMNs). Both phagocytosis resistance and virulence were restored upon capsule reversion, demonstrating the critical function of the capsule in pathogenesis. Although macrophages phagocytose both capsulated and capsular strains, the latter are consumed at noticeably greater rates, according to intracellular survival studies. The primary phagocytes involved are macrophages, rather than PMNs, indicating a capsule-mediated defence against immunological clearance (Fig. 15.3).

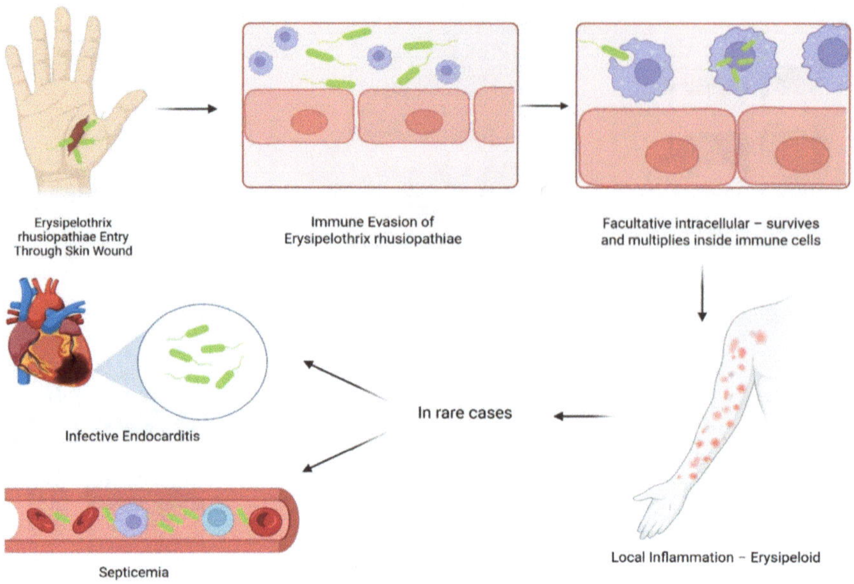

Fig. 15.3 Pathophysiology of *Erysipelothrix rhusiopathiae.* (Image source: Authors' personal collection)

Immunology

The immunology of *E. rhusiopathiae* infections in humans involves both innate and adaptive immune responses.

Once the bacteria have penetrated the epidermal barrier, pattern recognition receptors, particularly Toll-like receptors (TLR2), allow dendritic cells and macrophages to identify the pathogen. Pro-inflammatory cytokines (IL-1, TNF-α, and IL-6) are released in response to this identification and cause redness, swelling, and the migration of neutrophils to the infection site.

However, the ability of *E. rhusiopathiae* to resist phagocytosis and survive within macrophages allows it to evade early immune clearance. Because of its capsule-like exterior, which prevents lysosomal fusion, *E. rhusiopathiae* evades intracellular death when neutrophils and macrophages try to phagocytose the bacterium. Despite this, the spread of microorganisms is restricted by oxidative burst and ongoing phagocytosis in most immunocompetent hosts. Additionally, complement activation takes place, which improves opsonisation and draws in more inflammatory cells.

The adaptive immune response encompasses the activation of B cells and CD4+ T-helper cells. Th1-type responses increase macrophage activity to eradicate intracellular germs, especially through IFN-γ. B cells produce antibodies (IgM and, later, IgG) that neutralise the virulence components of the bacteria and opsonise them. Humoral immunity plays a key role, with antibodies targeting surface antigens, such as the capsule and cell wall components.

Immunity after infection is often incomplete, which may contribute to recurrent or chronic cases, particularly in immunocompromised individuals.

Infection in Humans

E. rhusiopathiae causes *erysipeloid* in humans. The infection usually occurs through skin abrasions after handling infected animals or animal products. The infection occurs even in immunocompetent hosts without occupational exposure and in newborns. It remains an important cause of "crayfish poisoning" among lobster fishermen.

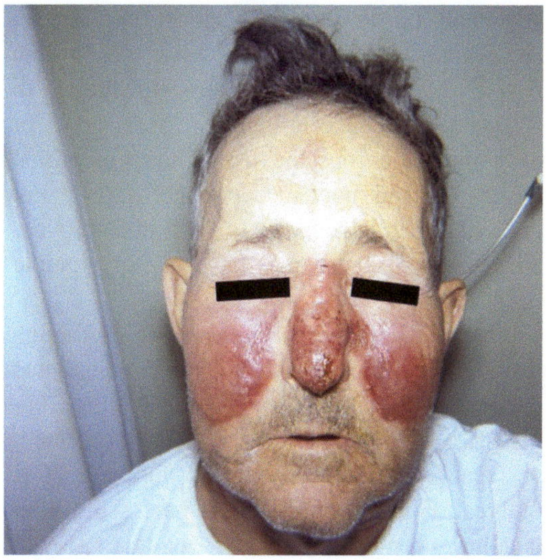

Fig. 15.4 This image depicts an anterior view of a patient's face, which exhibited symptoms of an illness known as erysipelas, a dermatological condition that involves the inoculation of the skin and subcutaneous tissue with streptococcal bacteria, causing oedema and bright red erythema of the affected areas. (Image source: CDC/Emory University, Dr. Thomas F. Sellers)

The *erysipeloid* manifests as a localised cutaneous lesion, usually on the hands or fingers. The presence of a localised, painful, reddish skin lesion, often an elevated zone with a fading centre and peripheral spread, is characteristic. Multiple or distant lesions and bullous eruptions are present. In contrast to staphylococcal infections, it does not cause oedema or suppuration.

In rare cases, the infection may become systemic, leading to endocarditis or septicaemia, especially in immunocompromised individuals. It is frequently associated with endocarditis, even in people with normal valves. Congestive heart failure is a frequent complication. Intracranial abscess, brain infection, and persistent arthritis have also been reported. Erysipeloid does not provide long-lasting immunity, and relapse can occur even after cure (Fig. 15.4).

Infection in Animals

E. rhusiopathiae causes *erysipelas* in animals. Domestic pigs are the main reservoir. Infection in pigs may cause fever, skin lesions, arthritis, etc. The bacteria are frequently found in the tonsils and are excreted in urine, faeces, and nasal secretions, which can contaminate the surroundings. Young pigs that are persistent shedders frequently acquire the bacteria from carrier sows.

Poultry, sheep, and fish are also affected. The bacteria infect the sheep through skin abrasions or wounds after castration or docking and cause post-dipping lameness. The disease can spread from the navel to lambs. Other infected animals and birds may also serve as reservoirs. In aquatic habitats, especially in fish, *E. rhusiopathiae* can remain asymptomatically within the external mucus layer for longer periods.

Epidemiology and Public Health

In humans, the most frequent cases of erysipeloid have been reported in individuals handling pigs, fish, poultry, or animal products. Nearly 38–42% of cases have been reported in butchers, meat processors, veterinarians, or fish handlers. Infection occurs through minor abrasions or puncture wounds, usually affecting the hands and fingers. Ninety per cent of human cases manifest as localised cutaneous infections, whereas 5–10% develop diffuse cutaneous and systemic infections.

Because of underreporting, misdiagnoses, and a lack of standard laboratory confirmation, it is difficult to estimate the global incidence of erysipeloid. It is relatively rare in the USA, Canada, and parts of Europe due to enhanced animal immunisation programmes, better protective gear, and improved workplace cleanliness. *E. rhusiopathiae* was responsible for fewer than 0.02% of occupational skin infections in Germany. Incidence is higher in rural areas or developing countries due to high levels of animal husbandry and poor occupational safety. For instance, in rural China, up to 1.3% of employees who handle fish and pigs have had erysipeloid instances.

Outbreaks of *E. rhusiopathiae*, although uncommon, have been linked to contaminated surfaces or inappropriate food handling. For instance, 16 employees (4.8% of the workforce) in Japan contracted erysipeloid during a 2001 outbreak connected to seafood processing after handling contaminated fish without gloves. Despite being uncommon, the disease's severe systemic form is of concern. The fatality rate for *E. rhusiopathiae*–caused endocarditis is significant, ranging from 30% to 40% in patients that go untreated. People with underlying medical disorders, including alcoholism, diabetes, or immunosuppression, are frequently affected (Table 15.1).

E. rhusiopathiae serovars vary in distribution across host species and geographic regions. Among animals, pigs, the primary reservoir, are associated with serovars 1a and 2. These serovars are also frequently implicated in human infections, typically through occupational exposure. Other reservoirs, including poultry, fish, and marine mammals, may harbour additional serovars with variable pathogenicity. The identification of serovars is important for understanding host–pathogen interactions across different species, epidemiological tracking, and vaccine development.

Table 15.1 Global epidemiology of the genus *Erysipelothrix* causing erysipeloid

Species	Host	Geographical distribution
Erysipelothrix rhusiopathiae	Pigs (primary reservoir), turkeys, sheep, fish, crustaceans	Worldwide; especially prevalent in regions with swine farming—North America, Europe, Asia (notably Japan and China), and parts of South America
Erysipelothrix tonsillarum	Pigs, poultry	Less widely distributed; mainly identified in Europe and Asia
Erysipelothrix piscisicarius	Fish, particularly marine and freshwater species	Asia, notably Japan and China; identified in aquaculture farms
Erysipelothrix larvae	Marine environments, invertebrates	Rare, limited data; isolated cases in East Asia and marine laboratories

Diagnosis

Microscopy

Gram staining serves as an important initial step in indicating a potential *E. rhusiopathiae* infection by revealing the presence of characteristic Gram-positive pleomorphic rods. However, it is never considered a stand-alone diagnostic tool. Its findings must always be correlated with clinical history, culture characteristics, biochemical reactions, and, increasingly, molecular techniques for accurate and definitive diagnosis.

In Vitro Culture

The gold standard for isolating *E. rhusiopathiae* remains in vitro culture, as the bacterium thrives on standard blood culture media commonly used in clinical settings.

Several specialised media have been developed to enhance selective isolation. Erysipelothrix selective broth (ESB) is widely used due to its unique composition. It is a nutritious broth supplemented with serum, tryptose, kanamycin, neomycin, and vancomycin. This formulation leverages the bacterium's natural resistance to these antibiotics, effectively inhibiting the growth of competing microorganisms. While ESB offers superior enrichment, other selective media provide distinct advantages, depending on the sample type and diagnostic needs.

Modified blood azide (MBA) agar, containing sodium azide and horse blood or serum, offers a shorter incubation time than Packer's medium, though its selectivity is not as robust. Packer's medium, also known as sodium azide crystal violet blood agar, was developed by R.A. Packer in 1943. It is particularly useful for processing heavily contaminated samples, providing excellent selectivity through the inclusion of sodium azide and crystal violet. Shimoji's selective broth utilises tryptic soy broth, Tween 80, tris buffer, crystal violet, and sodium azide for improved enrichment. For rapid diagnosis, especially in pigs, combining PCR with Shimoji's broth has proven to be highly effective and quick.

With negative results for catalase, oxidase, indole, and other common reactions, *E. rhusiopathiae* is inactive biochemically. However, a crucial feature that distinguishes it from the *Listeria monocytogenes* and *Corynebacterium* species is the generation of H_2S on triple sugar iron (TSI) agar. The organism does not ferment mannitol, xylose, or maltose, but it does ferment glucose, lactose, galactose, and fructose. *E. tonsillarum* ferments sucrose, in contrast to *E. rhusiopathiae*.

Serodiagnosis

Serodiagnosis involves the detection of specific antibodies in the serum of affected individuals or animals. Enzyme-linked immunosorbent assay (ELISA), fluorescence antibody (FA) test, and agglutination tests are commonly used. These assays are useful in epidemiological studies and herd-level diagnosis. They also allow for the monitoring of infection dynamics and immune responses following vaccination or natural infection. The serological methods are used for confirming exposure, especially in chronic or subclinical cases where bacterial isolation may be challenging. When culture or molecular techniques produce ambiguous findings, serodiagnosis offers a confirmatory way of identifying *E. rhusiopathiae*.

Molecular Diagnosis

Molecular methods, particularly PCR, have significantly advanced the diagnosis and characterisation of *Erysipelothrix* infections.

Early approaches, such as the Makino PCR, offered a genus-specific detection by targeting a conserved region of the 16S rRNA gene. This method is effective for identifying the presence of *Erysipelothrix* species; however, it cannot differentiate between *E. rhusiopathiae* and *E. tonsillarum*. Another early technique, Shimoji's PCR, is devised for greater specificity by targeting virulence-associated sequences unique to *E. rhusiopathiae*. However, its practical application was limited by lower sensitivity, requiring a relatively high bacterial load of over 1000 bacterial cells per reaction.

Significant progress was made with the development of an improved PCR method in 1999. This technique utilises primers derived from sequences covering the entire rRNA gene cluster (including 16S, 23S, and 5S rRNAs, along with non-coding sections). This innovation enabled the distinction of up to four *Erysipelothrix* species, including two potentially novel species, thereby enhancing the accuracy of diagnosis in both human and animal cases.

Multiplex PCR further improved specificity and efficiency. This method employs two distinct primer sets: MO101/ERS-1R and ERY-1F/ERY-2R. These sets amplify species-specific (2210 bp) and genus-specific (719 bp) DNA segments, respectively. The simultaneous amplification of these targets allows for the accurate differentiation and identification of *E. tonsillarum* and *E. rhusiopathiae* in a single reaction.

Spa multiplex real-time PCR test is a more recent and particularly valuable breakthrough. This assay provides information on pathogenicity by differentiating *Erysipelothrix* strains based on the presence of the spaA, spaB, and spaC genes. Specifically, the spaA gene is associated with well-known serotypes such as 1a, 1b, 2, and others. In contrast, spaB and spaC are found in distinct or currently unclassified strains of *E. rhusiopathiae*. To further enhance its epidemiological utility, the spaB gene can be further subtyped into spaB1 and spaB2, providing finer resolution for tracking strain dissemination and understanding disease outbreaks.

Other Tests

The mouse protection test is regarded as the gold standard among conventional techniques. *E. rhusiopathiae* is subcutaneously administered to mice. A test group is given the culture, together with horse hyperimmune antiserum, whereas a control group is given the culture alone. Only the immunised mice survive if the bacteria are present; the non-immunised group perishes after 5–6 days. Despite the test's high reliability, its application is limited due to the moral and practical issues surrounding animal testing.

Table 15.2 summarises the laboratory diagnosis of erysipeloid infection.

Treatment

Antibiotic therapy with penicillin is the mainstay of treatment for *E. rhusiopathiae* infections; it usually takes 48 h for oral penicillin to cure cutaneous erysipeloid infections. Intravenous penicillin is advised for more serious or systemic conditions like endocarditis or septicaemia. Till now, no resistance to penicillin has been documented. Cephalosporins are the best substitute for those who are allergic to penicillin. Clindamycin and erythromycin are not recommended because of their bacteriostatic properties.

Ceftriaxone acts effectively against *E. rhusiopathiae* and has a low minimum inhibitory concentration (MIC). Penicillin is still the first-line treatment, although other antibiotics, such as oxytetracycline, have shown good effectiveness. Despite being found in *E. rhusiopathiae*, plasmids do not seem to have a major role in antibiotic resistance.

Table 15.2 Laboratory diagnosis of erysipeloid in humans

Diagnostic approach	Method	Target	Remarks
Microscopy	Gram staining and morphological observation	Gram-positive rods	S-form: small, slender, slightly curved rods ($0.8–2.0 \times 0.2–0.4$ μm); R-form: long filaments (>60 μm) in chains
In vitro culture	Growth on selective and non-selective media	Viable *Erysipelothrix rhusiopathiae* cells	Grows well on blood culture media. ESB is the most frequently used medium. Packer's medium offers excellent selectivity. Shimoji's and Bohm's media use specific additives to exploit antibiotic resistance and chemical tolerance. Useful for routine isolation and enrichment from clinical and environmental samples
Serodiagnosis	Mouse protection test, fluorescence antibody (FA) tests	Antibodies or antigens of *E. rhusiopathiae*	Mouse test is the gold standard but limited by ethical concerns. Only immunised mice survive if the pathogen is present Direct and indirect FA tests offer ethical, quicker alternatives Useful when culture or PCR results are inconclusive
Molecular diagnosis	Conventional PCR, species-specific and multiplex PCR, spa multiplex real-time PCR	*Erysipelothrix* genus and species-specific DNA sequences	Makino PCR targets 16S rRNA (genus-specific); Shimoji's PCR detects virulence genes (species-specific but lower sensitivity) Improved PCR distinguishes four *Erysipelothrix* species. Multiplex PCR identifies genus and species simultaneously Spa multiplex real-time PCR differentiates spaA, spaB, and spaC types—useful for strain typing, epidemiology, and vaccine evaluation

Prevention and Control

Humans who handle animals, meat, or fish are primarily affected by erysipeloid. Prevention of *E. rhusiopathiae* infection in humans relies on protective measures, such as the use of gloves and proper hygiene when handling animals or animal products. Hand washing and maintaining personal hygiene are of importance. Proper disposal of corpses, rodent control, equipment cleaning, and good farm hygiene are all important. Early outbreak detection and containment are aided by quarantine and regular health examinations. Public health education encourages early symptom detection and treatment by increasing awareness among high-risk groups.

Vaccines

Vaccines for use in humans against *E. rhusiopathiae* are not available, as human infections are relatively rare and usually occupational.

In contrast, vaccines are widely used in animals, particularly in swine and poultry, where they play a critical role in disease prevention and control. Live attenuated or inactivated vaccinations restrict carrier animals and lower the incidence of infection. In pigs, both live attenuated and inactivated vaccines are available and are commonly used to prevent swine erysipelas and reduce the incidence of acute septicaemia, arthritis, and reproductive losses. Poultry vaccines, primarily inactivated formulations, help control outbreaks and improve flock health. Most commercial vaccinations protect for 6–12 months. However, the strain and animal type immunised might affect the effectiveness of the vaccine. Reports of failures in Australia and Japan, where epidemics occurred despite vaccination, highlight the shortcomings of the current vaccines.

Case Study

A 45-year-old butcher from a remote location presented with several painful, glistening, reddish-purple lesions that were well-defined. He had no pets, no history of travel abroad, and no recollection of being bitten by an animal or insect. Antihistamines were used unsuccessfully to treat the first lesion, which had developed on his left fifth finger 2 weeks prior. The lesion became erosive after expanding centrifugally. Additional lesions appeared on the left forearm (annular plaque), right arm (lymphangitic streaking), right thigh (bullae), and posterior thorax (two plaques). Furthermore, the patient reported a fever (38 °C), headache, cough, joint discomfort, and lymphadenopathy. Laboratory tests revealed leucocytosis, neutrophilia, elevated C-reactive protein (CRP), erythrocyte sedimentation rate (ESR), and fibrinogen. Giemsa and Gram stains were negative, as were blood agar cultures. He declined a skin biopsy. Serology for *Toxocara* and *Brucella* was negative. Chest X-ray and echocardiography were unremarkable.

1. What conditions mimic painful, spreading erythematous lesions in *Erysipelothrix rhusiopathiae* infection?
2. What is the significance of the centrifugal spread and erosive evolution of the primary lesion?
3. How does the patient's occupational exposure as a butcher contribute to the possible aetiologies of his condition?

Research Questions

1. What are the diagnostic challenges of zoonotic skin infections of *E. rhusiopathiae* in rural high-risk professionals such as butchers and farmers?
2. How accurate and useful are molecular diagnostics rather than skin biopsy?
3. How does chronic exposure to animal products or pathogens alter the skin's immune response?

Further Readings

Balkhair A, Al Lawati H, Al Riyami M, Alameddine T, Al Amin M, Al Adawi B. Erysipelothrix rhusiopathiae endocarditis diagnosed by broad range 16 s rRNA

PCR gene sequencing. IDCases. 2019;18:e00584. https://doi.org/10.1016/j.idcr.2019.e00584.

Boyd AS, Ritchie C, Fenton JS. Cutaneous Erysipelothrix rhusiopathiae (erysipeloid) immunocompromised child. Pediatr Dermatol. 2014;31:232–5. https://doi.org/10.1111/j.1525-1470.2012.01835.x.

Forde T, Biek R, Zadoks R, et al. Genomic analysis of the multi-host pathogen *Erysipelothrix rhusiopathiae* reveals extensive recombination as well as the existence of three generalist clades with wide geographic distribution. BMC Genomics. 2016;17:461. https://doi.org/10.1186/s12864-016-2643-0.

IJsseldijk LL, Begeman L, Duim B, et al. Harbor porpoise deaths associated with Erysipelothrix rhusiopathiae, The Netherlands, 2021. Emerg Infect Dis. 2023;29(4):835–8. https://doi.org/10.3201/eid2904.221698.

Ingebritson AL, Roth JA, Hauer PJ. *Erysipelothrix rhusiopathiae:* association of Spa-type with serotype and role in protective immunity. Vaccine. 2010;28(13):2490–6. https://doi.org/10.1016/j.vaccine.2010.01.041.

Lukácsi S, Mácsik-Valent B, Nagy-Baló Z, et al. Utilisation of complement receptors in immune cell-microbe interaction. FEBS Lett. 2020;594(16):2695–713. https://doi.org/10.1002/1873-3468.13743.

Miura T, Hashizume K, Ariyoshi T, et al. Active infective endocarditis due to Erysipelothrix rhusiopathiae: zoonosis caused by vancomycin-resistant gram-positive rod. Gen Thorac Cardiovasc Surg. 2013;61:96–9. https://doi.org/10.1007/s11748-012-0105-4.

Pal N, Bender JS, Opriessnig T. Rapid detection and differentiation of Erysipelothrix spp. by a novel multiplex real-time PCR assay. J Appl Microbiol. 2010;108:1083–93. https://doi.org/10.1111/j.1365-2672.2009.04560.x.

Pomorska-Mól M, Markowska-Daniel I, Pejsak Z. Effect of age and maternally-derived antibody status on humoral and cellular immune responses to vaccination of pigs against Erysipelothrix rhusiopathiae. Vet J. 2012;194(1):128–30. https://doi.org/10.1016/j.tvjl.2012.03.009.

Takahashi T, Fujisawa T, Umeno A, Kozasa T, Yamamoto K, Sawada T. A taxonomic study on erysipelothrix by DNA-DNA hybridisation experiments with numerous strains isolated from extensive origins. Microbiol Immunol. 2008;52(10):469–78. https://doi.org/10.1111/j.1348-0421.2008.00061.x.

Tan EM, Marcelin JR, Adeel N, Lewis RJ, Enzler MJ, Tosh PK. Erysipelothrix rhusiopathiae bloodstream infection - a 22-year experience at Mayo Clinic, Minnesota. Zoonoses Public Health. 2017;64:e65–72. https://doi.org/10.1111/zph.12348.

Veraldi S, Girgenti V, Dassoni F, Gianotti R. Erysipeloid: a review. Clin Exp Dermatol. 2009;34(8):859–62. https://doi.org/10.1111/j.1365-2230.2009.03444.x.

Wang Q, Chang BJ, Riley TV. Erysipelothrix rhusiopathiae. Vet Microbiol. 2010;140(3–4):405–17. https://doi.org/10.1016/j.vetmic.2009.08.012.

Dermatophilosis

16

Abhijit Chaudhury

Learning Objectives

1. To describe the causative agent of dermatophilosis and its unique characteristics
2. To explain the pathogenesis and clinical signs associated with dermatophilosis in animals and humans
3. To discuss the various diagnostic methods used to detect dermatophilosis
4. To evaluate the treatment options and control strategies for dermatophilosis

Introduction

Dermatophilosis is a disease of the skin that, though very common among different animals, is a rare entity in humans. It is caused by the actinomycete bacteria *Dermatophilus congolensis*. It causes the disease in various animals, which goes by the common names "lumpy wool" and "strawberry foot rot." Human infection is rare but may be under-reported and occurs through close contact with an affected animal. The clinical manifestations in humans as well as animals are primarily limited to the skin, although systemic infection can occur in animals.

A. Chaudhury (✉)
Department of Microbiology, Sri Venkateswara Institute of Medical Sciences University, Tirupati, Andhra Pradesh, India

History

The condition was first reported in 1915 from the Belgian Congo by Rene Van Saceghem, a military veterinarian stationed in Africa, who described the dermal lesion in cattle. Animal infections have been recorded from many parts of the world. The first human case was described in 1961 in the USA, in which a person handled an infected deer. Reports of human infection have been documented from Africa, Australia, and Asia, including India.

Taxonomy

The organism belongs to the superkingdom Bacteria, Clade Terrabacteria, Phylum Actinomycetota, Class Actinomycetes, Order Micrococcales, Family Dermatophilaceae, and genus *Dermatophilus*. Three species, namely, *Dermatophilus congolensis, Dermatophilus dermatonomus,* and *Dermatophilus pedis*, have been described in the past, but the seminal work by Morris A. Gordon in the 1960s established that they were one and the same species, and *D. congolensis* remains the only valid species at present. *Dermatophilus chelonae*, named by Masters et al. in 1995, was renamed *Austwickia chelonae* in 2011.

S. C. Parija, A. Poddar (eds.), *Textbook of Bacterial Zoonoses*, Microbial Zoonoses,
https://doi.org/10.1007/978-981-95-6615-0_16

Genomics and Proteomics

In keeping with the other actinomycetes, the genome has a high G+C content of 59.4%, and its approximate size is 2.62 Mbp. The genome encodes about 2527 genes, of which at least 48 genes have been found to be associated with virulence and another 23 with antimicrobial resistance. However, the detailed functions of the virulence genes remain unknown. In one study by Branford et al. (2021), the isolates were found in three clusters based on the phylogenetic tree as derived from the whole-genome sequencing data. It corresponded with the three different numbers of Clustered regularly interspaced short palindromic repeats (CRISPR) spacer sequences found in these isolates.

Proteomic studies with this organism from human isolates are lacking because of the rarity of infection.

Morphology

D. congolensis is a Gram-positive bacterium that, on direct smear, exhibits thin branching filaments with septations, along with coccoid bodies in parallel chains. In cultures on solid media, various morphological forms may be observed. These include thin filamentous forms (0.5–1.5 μm) and thick filaments (5–8 μm) with transverse and longitudinal septations, with mucoid capsules. Sporangial forms with parallel chains of packets of coccoid cells may develop. In liquid culture, the coccoid cells are released to form motile flagellated forms or zoospores.

Cultivation

The organism is a facultative anaerobic bacterium that grows well at 37 °C in an atmosphere of 5–10% CO_2. Ordinary culture media like blood agar, tryptose agar, and brain heart infusion (BHI) agar support the growth of *D. congolensis*. On blood agar, after overnight incubation, tiny, pinpoint colonies and creamy white colonies can be seen, which are beta-haemolytic in nature.

Prolonged incubation results in a confluence of the colonies, with pigmentations of various shades of yellow, orange, or grey. These older colonies are characterised by the presence of abundant sporangia. Growth on liquid culture media like tryptose broth results in the formation of motile zoospores with flagella. The organism does not grow on MacConkey's agar and Sabouraud's dextrose agar. Biochemically, it is catalase- and urease-positive. It ferments glucose and fructose but not the other common carbohydrates. It hydrolyses gelatine and starch.

Animal Models

Various animals like sheep, guinea pigs, rabbits, and mice have been used as experimental models to test the efficacy of putative vaccines. The use of a mouse model is convenient for studying the pathogenesis of dermatophilosis. Mouse skin is normally resistant to infection by *Dermatophilus* but may be made susceptible by light abrasion or by prior treatment with ether or methyl alcohol. Mice have also been used to study the infection model in immunocompromised hosts. For this, various types of gnotobiotic mice like athymic (nude) mice, nude mice with beige mutation, and BALB/c mice with nude mutation have been used.

Transmission Cycle

D. congolensis exhibits distinct morphological forms at various stages of its life cycle. The stages of development seen in cultures parallel the morphological forms seen in lesions in experimental animals. The flagellated motile zoospores first germinate into thin branching filaments, followed by the development of rudimentary spores and the beginning of septations. Perpendicular branching and transverse septations, as well as longitudinal segmentations, continue to result in the formation of a branched filamentous structure. Parallel chains of coccoid elements embedded in a gelatine-like matrix develop, which later escape from the gelatinous sheath to form the motile zoospores. These zoospores reinitiate the developmental cycle by germination.

Pathogenesis and Pathology

D. congolensis essentially causes skin lesions involving the epidermis, and rarely, in animals, there is involvement of other tissues. Minor skin trauma or abrasion helps in the entry of the zoospores, and moisture facilitates the process. The epidermis is affected but not the stratum corneum or the dermis layer. The recruitment of neutrophils just beneath the affected epidermis prevents the entry of the organism into the dermis. Bacterial virulence factors like keratinase and ceramidase are involved in overcoming the skin barrier. Once in its favourable niche, the zoospores germinate into the invasive hyphal form and spread along the epidermal layer. This results in an acute inflammatory response at the local site. Normally, the infection subsides in 2–3 weeks of time with healing of the lesion. In chronic infections, the affected foci of infection serve as the nidus and cause intermittent invasion into the healthy epidermis. There is cornification of the epidermal layer, which then separates as scabs. Moisture helps in the production and release of zoospores from the mature hyphae.

Histopathology of the lesion reveals an exudative lesion with an influx of polymorphs and lymphocytes below the stratum corneum. Vesicular changes cause the separation of the cornified layer. The stratum spinosum shows oedema, congestion, and thickening. In chronic cases, continued epidermal regeneration, fresh invasions, and accumulation of neutrophils ultimately result in the formation of thick scabs on the lesion. Apart from skin lesions, involvement of deeper tissues like lymph nodes, subcutaneous tissues, and muscle has been reported in animals.

Immunology

The role of adaptive immune response in dermatophilosis has been studied in various animals, like sheep, cattle, and guinea pigs. A decrease in the prevalence of the disease with age in sheep points towards a natural protective immune response, but there is insufficient evidence and a paucity of studies in this regard. Nevertheless, it has been demonstrated that animals develop resistance to the disease after 4–6 weeks of infection, which indicates the development of acquired immunity. On the other hand, studies have shown that chronic *Dermatophilus* infection induces a weak protective immune response. Studies in animals have shown that bacterial antigens can induce strong antibody responses in the serum as well as inflamed skin in sheep and transfer of IgG1 and IgG2 from serum to the skin.

In essence, an increase in antibody titre is definitely an indication of active humoral response to *Dermatophilus* antigens, but it appears not to confer any protective activity against infection. The antibody response may be directed towards the zoospores rather than the invasive hyphal forms responsible for pathogenesis.

Infection in Humans

Human dermatophilosis is a rare entity, and only a handful of cases have been reported worldwide, mainly from tropical and subtropical countries.

Minor trauma, abrasions, previous skin conditions, immunocompromised state, and, importantly, contact with infected animals are risk factors for developing dermatophilosis. Acute infection may begin as a small skin vesicle, which turns pustular and later becomes erythematous, itchy, and exudative or scaly in nature. Folliculitis and pitted keratolytic lesions have also been described. In chronic cases, maceration and fissuring may occur. In immunocompromised subjects, hairy leucoplakia and subcutaneous nodules may develop. There is a possibility that the condition may be under-reported since the lesions may be confused with fungal infection or non-infectious skin conditions. Careful history regarding recent contact with animals, occupation, and recent travel to tropical countries by the infected person may be suggestive of the infection.

Infection in Animals

The disease is most prevalent in cattle (where it was first demonstrated), sheep, goats, horses, and, less frequently, pigs, dogs, and cats. In animals, the disease is more common in the young, those who are immunocompromised, and those who are exposed to water or high humidity. Colonisation of the skin with *Pseudomonas aeruginosa* has been found to favour the invasion by *Dermatophilus*. The animals may also be infected by biting flies or ticks. In cattle, the disease starts with matting of the hair (paintbrush lesions), followed by scab formation, and, ultimately, hyperkeratinisation due to chronic infection, leading to warty growth on the skin. Prolonged exposure to water in hot climates, as seen among buffaloes in tropical areas, determines the distribution of the lesions in skin folds of the flexor surfaces of the joints, apart from the ventral surface of the body. In sheep, lumpy wool infections can be seen where masses of crusty material are bound to wool on the skin.

Epidemiology and Public Health

Although considered a soil saprophyte, in various studies, the organism has not been isolated from soil. The organism is sensitive to pH and osmotic changes, and it may not be possible for it to survive for long periods in soil. It is possible that animals may get the infection from mud or moist soil, where the infective zoospores can survive. However, it may not be an important source of infection. Some studies suggest that organic matter content protects *D. congolensis* in oxisol soil but not in vertisol, though it retains its pathogenicity. Ponds and dipping tanks may constitute sources of cattle infection as a result of the soil and water environmental mix. Transmission of infection by biting insects like ticks and lice has also been demonstrated (Fig. 16.1).

Asymptomatic animals or those who are having chronic infection may serve as a source of infection for other animals, as well as humans. Climatic conditions play an important role in the seasonality of infection since prolonged periods of rain, high humidity, and temperature favour the germination of zoospores and infection transmission. It has been observed that the advent of a dry climate results in a decline in the incidence of infection among animals. Table 16.1 lists the geographic distribution of *Dermatophilus* infections.

Diagnosis

Clinical diagnosis of dermatophilosis is often supported with laboratory diagnosis, both in animals and humans. In both cases, microscopy and culture remain the mainstay for diagnosis. In recent times, nucleic acid amplification has been used for definitive diagnosis and in doubtful cases. Table 16.2 summarises the laboratory diagnosis of dermatophilosis.

Microscopy

Purulent discharge from the pustules, swabs from the lesion, or fresh crusts or scabs can be used as specimens. A practical method is cytologic examination of fresh crusts and/or impression smears of the underside of freshly avulsed lesions. The material is crushed on a glass slide with a few drops of saline and then stained. Gram stain is not recommended since it is intense Gram-positive, and the details may be obscured. Methylene blue or Giemsa stains demonstrate the details of the filaments better. If biopsy is attempted, then Grocott–Gomori silver methenamine or periodic acid–Schiff stains are useful. In clinical material, the organism commonly appears as branching filaments, but in specimens taken from older lesions like scab material, two to six parallel rows of cocci can be seen, having a *railroad track appearance* (Fig. 16.2). Microscopy can be highly predictive if the characteristic developmental forms can be seen.

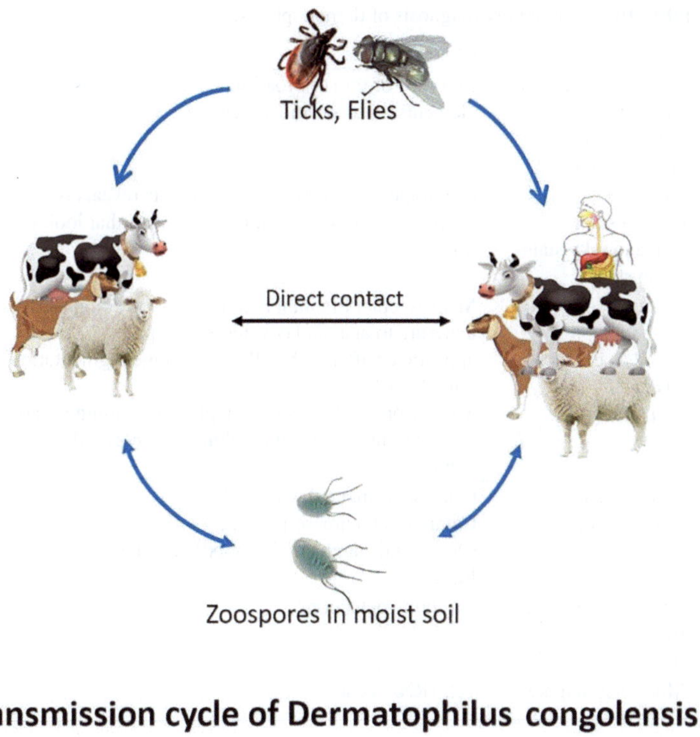

Fig. 16.1 *Dermatophilus congolensis* transmission in humans and animals. (Image source: Authors' personal collection)

Table 16.1 Distribution of *Dermatophilus* infections

Species	Normal hosts	Human infection	Global distribution	Countries from where human infections have been reported
Dermatophilus congolensis	Cattle, goats, sheep, horses, pigs, dogs, cats	Yes	Worldwide in tropical and subtropical countries	Spain, Brazil, USA, Australia, some African countries, and India
Dermatophilus crocodyli	Crocodiles, alligators	No	Australia	Nil

In Vitro Culture

All strains of *Dermatophilus* grow well aerobically at 37 °C in the presence of 5–10% CO_2 on blood agar, BHI agar, or tryptone broth. There may be a wide variation in colony morphology and pigmentation in various media. After 24 h of aerobic incubation on blood agar, tiny (0.5–1 mm) colonies appear greyish-white in colour, which are tough and adherent to the medium, causing pitting with a zone of β-haemolysis. Prolonged incubation results in yellow or orange colonies with a diameter of 4–6 mm after 7 days. Staining on consecutive days during prolonged incubation may be helpful for a correct identification.

Stained preparations from the culture material may reveal various morphological forms. These include branching filaments in any or all stages of segmentation; mucoid or non-mucoid clusters of cocci, either in clumps or lying scattered; free coccoid spore-like structures; or any combination of the above forms (Fig. 16.3). Cultures grown in broth medium may reveal motile zoospores in a hanging drop preparation.

Biochemical identification is useful for definitive identification. Purely fermentative breakdown of glucose is a valuable test to distinguish *Dermatophilus* from other aerobic actinomycetes. The organism is catalase- and urease-positive. Casein and starch are decomposed, and gelatine is liquefied.

Table 16.2 Laboratory diagnosis of dermatophilosis

Test	Characteristics	Remarks
Direct examination: Gram stain Methylene blue Giemsa stain	Presence of thin or broad septate, filamentous elements with coccoid structures	Gram stain causes intense gram positivity and may obscure the details of the hyphal elements
Cytology: Giemsa stain Romanowsky stain Gram stain	Cytologic examination of fresh crusts reveals two to six parallel rows of gram-positive cocci that look like railroad tracks	Considered the most practical method for diagnosis in animals
Culture: Media Blood agar BHI agar Tryptone broth	Macroscopic: Pinpoint beta-haemolytic colonies, adhering to agar and covered with a pitted appearance after 24 h yellow or orange pigmentation after 48–72 h Microscopic: Thin hyaline hyphae, becoming broad with continued incubation along with coccoid elements	Liquid broth may reveal motile zoospores. Can be shown in hanging drop preparation
Biochemical identification	Glucose fermented with acid only Catalase- and urease-positive Casein and starch are decomposed Gelatine is liquefied	Can help in distinguishing from other aerobic actinomycetes
Immunodiagnostics: ELISA IFAT	Antibody detection	Not commercially available. Can be used for serosurveys
Molecular diagnosis: Conventional PCR Sequencing	16S rRNA gene	Sequence analysis of the genes can help in definitive diagnosis

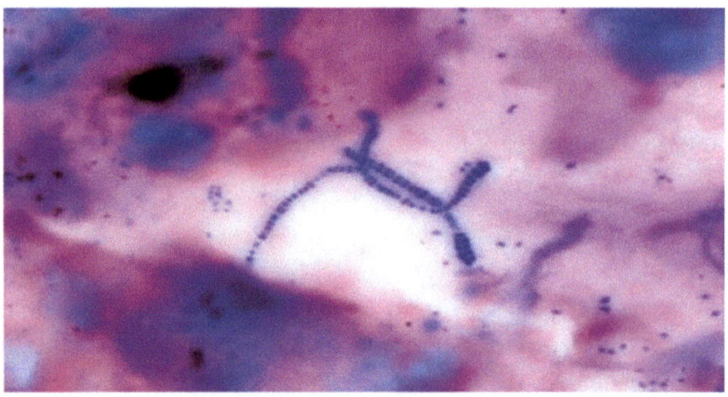

Fig. 16.2 Gram stain of a clinical specimen with characteristic branching filaments of *Dermatophilus congolensis* with "*train track*" form or hypha-like chains that released sporangium (×1000). (Image source: Anchang and Simonsen. Developments and Perspectives in Bryophyte Biotechnology in Sub-Saharan Africa. https://doi.org/10.5772/intechopen.81692, http://creativecommons.org/licenses/by/3.0)

Serodiagnosis

An indirect fluorescent antibody technique and a single-dilution ELISA test have been developed for large serologic and epidemiological surveys among animals. However, they are not commercially available. ELISA using the cell wall- or membrane-enriched preparations is more specific than a crude extract or a soluble antigen.

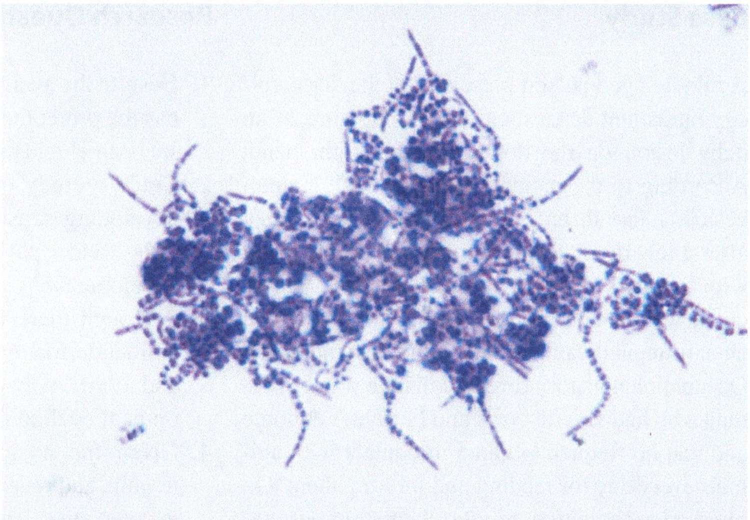

Fig. 16.3 Clustering of *Dermatophilus congolensis* from culture material (Giemsa stain) (×1000). (Image source: CDC/Dr William Caplan, https://phil.cdc.gov/Details.aspx?pid=2986)

Molecular Diagnosis

A nucleic acid amplification test may be employed for identification from culture material. This is carried out by sequential analysis of two overlapped fragments of a partial 16S rRNA gene of 660 and 817 bp. The primers are TGCCGTAAACGTTGGGCGCT and CGTGCAGTGGGTACGGGCAG as the forward primers for the 660-bp and 817-bp fragments, respectively, and TGTTACTTGATCCCCAATCGCCAGT as the reverse primer.

Treatment

Various topical and parenteral antibiotics have been evaluated in animals with variable outcomes. Topical antibiotics may not penetrate the deep epidermis, while parenteral preparations may not reach the avascular upper epidermis. In food-producing animals, topical applications of lime sulphur are a cost-effective adjuvant to antibacterial therapy, while in horses, chlorhexidine can be used. Human infection need not be given any specific treatment since the condition is self-limiting if the affected part is kept dry. In vitro testing of antibiotics has revealed sensitivity to a wide range of agents like penicillin, cephalosporins, aminoglycosides, clindamycin, tetracycline, chloramphenicol, rifampicin, cotrimoxazole, and glycopeptides.

Prevention and Control

Skin wetting is an important factor that predisposes the animals to infection. Hence, keeping the animals dry is important to prevent dermatophilosis. Zinc deficiency in animals should be avoided, as it is associated with outbreaks. For control of infection and in an outbreak situation, isolating clinically affected animals, culling the affected animals, and controlling ectoparasites help in breaking the cycle. Development of an animal vaccine has been attempted using zoospores, hyphal extracts, soluble antigens, serine proteases, and peptides from phage libraries, but all have shown variable efficacy.

Case Study

A middle-aged person presented at the dermatology outpatient department with complaints of an itchy lesion on the dorsum of his right hand. According to the patient, it started with a small vesicle 1 month back, which developed a crust after a few days. The eruptions started recurring with repeated formation of crusty lesions. He had contacted the local physician, who had prescribed an antifungal cream, but there had been no relief. Occupational history revealed that he was a milkman who had several cows and buffaloes at home and that he frequently came in contact with animals every day for feeding and milking them. On physical examination, four erythematous desquamative lesions were found on the dorsum of the hand. Swabs were collected from the lesion, and direct microscopy revealed fragmented hypha-like elements with associated spherical spore-like structures. The sample was inoculated in blood agar, MacConkey's agar, Sabouraud medium with chloramphenicol, and Sabouraud medium with cycloheximide and chloramphenicol. Blood agar was incubated at 37 °C aerobically, while Sabouraud medium was incubated in duplicate at 25 °C and 37 °C. After 24 h, pinpoint β-haemolytic colonies could be seen on blood agar, which became larger and yellowish, with a pitted appearance after 48 h. Gram stain showed Gram-positive branching filaments with some coccoid elements. Sabouraud's medium did not show any growth. A presumptive diagnosis of the isolate as *D. congolensis* was made. The patient was asked not to come in contact with the animals for 1 month and was given ampicillin and advised to use antibacterial soap and keep the area dry. A follow-up after 2 weeks showed a complete resolution of the condition.

1. How was the patient infected, and what was the infective form?
2. The patient was given ampicillin empirically. Is there any drug of choice for treating the condition in animals as well as in humans?
3. Apart from microscopy and culture, what additional laboratory test can be done to confirm the diagnosis?

Research Questions

1. Despite the availability of technological tools, has the proteome of *Dermatophilus congolensis* been elucidated?
2. Could a study of its proteome contribute to developing sensitive and specific serological tests and pave the way for vaccine development?
3. Although there was a surge of research into animal dermatophilosis vaccines in the 1980s and 1990s, why did interest in vaccine development decline following mixed results?
4. Given the advancements in genomic, proteomic, and reverse vaccinology tools, is there a renewed need to identify protective antigen(s) of *D. congolensis* for designing candidate vaccines?
5. Considering the availability of excellent small and large animal models for reproducing experimental dermatophilosis, should these models be utilised to determine the efficacy of different antibiotics and gather relevant information, especially given the absence of a standardised treatment regimen for this condition?

Further Readings

Branford I, Johnson S, Chapwanya A, Zayas S, Boyen F, Mielcarska MB, et al. Comprehensive molecular dissection of *Dermatophilus congolensis* genome and first observation of tet(Z) tetracycline resistance. Int J Mol Sci. 2021;22:7128.

Denman S, Tellam RL, Vuocolo T, Ingham A, Wijffels G, James PJ, et al. Fleece rot and dermatophilosis (lumpy wool) in sheep: opportunities and challenges for new vaccines. Anim Prod Sci. 2022;62:301–20.

Gordon MA. The genus dermatophilus. J Bact. 1964;88:509–22.

Martinez D, Mari B, Aumont G, Vidalenc T. Development of a single dilution ELISA to detect antibody to *Dermatophilus congoensis* in goat and cattle sera. Vet Microbiol. 1993;34(1):47–62.

Tellam RL, Vuocolo T, Denman S, Ingham A, Wijffels G, James PJ, et al. Dermatophilosis (lumpy wool) in sheep: a review of pathogenesis, aetiology, resistance, and vaccines. Anim Prod Sci. 2022;62:101–13.

17 Rhodococcus Infections

Tapashi Ghosh and Vangara Harika

Learning Objectives

1. To describe *Rhodococcus*, primarily a soil-borne organism causing infection in animals, as well as in humans
2. To describe the diagnosis, prevention, and treatment of the infection
3. To understand the epidemiology of infection as a zoonosis and its importance to humans

Introduction

The genus *Rhodococcus* contains Gram-positive to Gram-variable, weakly acid-fast organisms that appear as cocci or coccobacilli with some branching. Members of this genus are primarily considered zoonotic pathogens associated with farm environments, causing rare, potentially fatal, pulmonary and extrapulmonary pyogranulomatous diseases in animals, especially foals. Rhodococcosis is a zoonotic infection in humans, typically occurring in individuals with impaired cell-mediated immunity. Apart from its pathogenic potential, members of the genus *Rhodococcus* are utilised as biocatalytic agents that degrade natural organics and xenobiotic compounds.

T. Ghosh (✉)
School of Tropical Medicine, Kolkata, West Bengal, India

V. Harika
Department of Microbiology, Sri Venkateswara Institute of Medical Sciences University, Tirupati, Andhra Pradesh, India

History

Microbiologists studying soil bacteria and pathogens first noticed *Rhodococcus* species in the late nineteenth to early twentieth century. It was evident that this genus shares traits with other actinomycetes, such as *Mycobacterium*, *Nocardia*, and *Corynebacterium* spp., leading to difficulties in classification. Chester noted the production of reddish pigment while examining some soil bacteria. In 1891, the first proper description of the genus was provided by Zopf, who classified certain pigmented soil organisms under the genus *Micrococcus*. In animals, it was first isolated from foals by Magnusson in 1923 in Sweden. The first reported human case was from a patient with a lung abscess in 1967. A few cases were reported in the next two decades, but the number of human cases started increasing with the beginning of the HIV pandemic.

Taxonomy

Magnusson, in 1923, named the equine isolate *Corynebacterium equi*, and this name continued to be used for several decades. With the advent of chemotaxonomic classification methods and,

S. C. Parija, A. Poddar (eds.), *Textbook of Bacterial Zoonoses*, Microbial Zoonoses,
https://doi.org/10.1007/978-981-95-6615-0_17

later, DNA–DNA hybridisation, the genus has undergone several taxonomic changes.

According to the 2015 classification in *Bergey's Manual of Systematics of Archaea and Bacteria*, these organisms belong to the phylum Actinobacteria, order Corynebacteriales, and family Nocardiaceae. The Nocardiaceae family comprises the genera *Nocardia, Mycobacterium, Corynebacterium, Gordonia, and Rhodococcus,* all of which share certain similarities with one another. The genus *Rhodococcus* comprises more than 50 validly published species, each with a correct name; however, only a few have pathogenic potential. *Rhodococcus equi* is the most common zoonotic pathogen. Other species such as *Rhodococcus fascians, Rhodococcus rhodochrous,* and *Rhodococcus erythropolis* have also been implicated in human infections. *R. fascians* causes leafy gall disease in plants.

Genomics and Proteomics

Members of the genus *Rhodococcus* have a G+C content of 63–73%, with a genome size of 5.04 Mbp. It appears to be genetically stable, with the absence of significant recombination and infrequent insertion sequences, but with a significant propensity for horizontal gene transfer. Comparative genomic studies of this genus have been conducted using a limited number of genomes. A total of 327 *Rhodococcus* genomes have been analysed, and clustering studies have been conducted, revealing 42 phylogenomic groups and 83 species-level clusters.

The genus has a hard-core genome comprising 381 orthologous groups and a soft-core genome comprising 1253 orthologous groups. The hard-core genomes are represented in all genomes. It is expected that the diversity in this genus will increase with the sequencing of further new genomes.

The genomes of 53 *R. equi* strains have been sequenced. They revealed 11,481 pan genes, including 3690 core genes and 7791 accessory soft genes. Six clades of *R. equi* were identified by phylogenetic analysis, unrelated to the host or location. A total of 84 virulence genes were identified in the studied strains of 53 *R. equi* strains, categorised into 20 groups related to metabolic activities, protein secretion, and immune evasion. All strains carried the gene for rifamycin resistance (*rpoB),* apart from six other resistance genes, which were not universal. More than 25 plasmids were also present in the 53 strains.

In addition to the chromosome, virulent strains of *R. equi* harbour a circular or occasionally linear virulence plasmid, which is essential for their pathogenicity. These plasmids vary in size but are typically around 80–90 kilobases (kb). This plasmid contains a specialised region known as a pathogenicity island (PAI), which is believed to have been acquired through horizontal gene transfer. Within PAI are the crucial virulence--associated protein (Vap) genes, which encode a family of surface-associated proteins essential for the bacterium's survival and replication inside macrophages.

The presence of specific Vap plasmids is related to animal host adaptation, with an active exchange occurring across the *Rhodococcus* population. The most extensively studied of these is VapA, found on the pVAPA plasmid type, which is predominantly associated with equine isolates and is indispensable for the intracellular multiplication of *R. equi*. VapB strains produce lesions in the lymph nodes of swine. The last described, VapN, was initially associated with strains collected from the lymph nodes of cattle and sheep. In strains of human origin, the vapN and vapB plasmids were more prevalent than the vapA plasmid (43%, 30%, and 16%, respectively), or they were absent.

Beyond the vap genes, the plasmid's PAI also contains critical regulatory genes, such as virR and virS, which govern the expression of the vap genes and other PAI components. These regulators enable *R. equi* to sense and respond to environmental cues within the host, such as changes in temperature and pH.

Proteomic analysis of *R. equi* has revealed the presence of four distinct types of proteins: virulence-associated proteins, secreted proteins, proteins involved in metabolism, and proteins associated with immunogenicity and immune evasion. An earlier study identified 48 proteins, with cholesterol oxidase being the major secretory protein, and five antigenic proteins were identified by immunoblot. In a recent study with 237 strains of various *Rhodococcus* species, 17,258 orthologous groups were identified. However, this study focused more on the proteins involved in the degradation of organic molecules, a key characteristic of many environmental species. The virulence-associated proteins (VAPs) have been studied extensively, and VAPA, found in equine (and human) strains has been characterised.

Morphology

R. equi is an aerobic, non-endospore-forming, Gram-positive, variable, and weakly acid-fast bacterium showing pleomorphic morphology. As such, specific measurements for length and width are not consistently cited. However, studies indicate that cell length can range from 1.0 to 2.5 μm and cell width from 1.22 to 1.31 μm in some strains. Colonies on solid media can range in diameter from 2 mm to 4 mm.

The bacterium exists as a filamentous, branching structure in the environment and in liquid media. In solid media and tissue, the filaments fragment into pleomorphic coccobacillary forms. It possesses thin cell walls with a type IV composition, the primary and major constituents being glutamic acid, muramic acid, meso-Diaminopimelic acid (DAP), alanine, glucosamine, galactose, and arabinose, along with nocardio-mycolic acids. It is non-motile, and a few numbers of pili may exist.

Cultivation

R. equi is an obligate aerobe and so requires an oxygen-rich environment for growth. Its optimal growth temperature falls between 30 °C and 37 °C. The bacterium also prefers a neutral-to-slightly alkaline pH for optimal growth, typically ranging from 7.0 to 8.5. Visible colonies can often be detected within 48 h of incubation. However, the full development of the characteristic salmon-pink pigmentation may take up to 4–7 days, necessitating prolonged observation of cultures for complete identification.

R. equi has simple nutritional requirements and can grow on a variety of bacteriological media. These include sheep blood agar, brain heart infusion agar, Columbia agar with blood, or sheep blood agar without antibiotics. They can be inhibited, however, by chloramphenicol, penicillin, and streptomycin, which are present in selective media. They also grow on media designed for the recovery of mycobacteria. On sheep blood agar, the colonies are irregular, smooth, and mucoid due to the presence of capsules, and they acquire a salmon-pink colour in 4–5 days.

Biochemically, *Rhodococcus* is catalase- and urease-positive. It produces enzymes such as lipase, phosphatase, DNase, elastase, and lecithinase. It hydrolyses casein, tyrosine, and xanthine.

Animal Models

Considering foals are the primary natural host for severe *R. equi* pneumonia, foal models are considered the gold standard for studying the disease. The models enable researchers to mimic natural infection routes, observe the characteristic pyogranulomatous lung lesions, and evaluate the efficacy of vaccines or antimicrobial treatments. However, foal models are logistically challenging, expensive, and raise ethical considerations, which limit their widespread use.

Alternatively, mouse models have been widely developed and utilised. Researchers commonly employ systemic (intravenous) infection models to assess bacterial burdens in various organs or intranasal challenges to focus on pulmonary infection. Immunodeficient mouse models, such as severe combined immunodeficiency (SCID) mice or nude mice, are particularly valuable. They develop more persistent infections and sometimes granulomas, providing a better reflection of the chronic nature of rhodococcosis seen in susceptible hosts. These models are crucial for (i) initial in vitro and in vivo testing of vaccine candidates, (ii) studying host immune responses (e.g. the importance of Th1 responses and interferon-gamma (IFN-γ)), and (iii) evaluating the efficacy of antimicrobial agents and combinations.

Other animal models, such as pigs, are susceptible to *R. equi* strains carrying the pVAPB plasmid and can develop the disease. This model offers insights into alternative plasmid-mediated virulence. Additionally, there have been case reports and limited studies on *R. equi* infection in other domestic animals, such as goats and cattle, as well as in wild animals, which occasionally serve as natural hosts or models for specific aspects of the disease.

Transmission Cycle

Rhodococcus species is a ubiquitous, soil-borne organism but is also found in aquatic environments, such as rivers and lakes. Soil is contaminated with the manure of herbivorous animals, which is a major source of infection for both animals and humans. This pathogen is harboured by livestock such as equines, bovines, small ruminants, and pigs. It is also present in some wild animals, domestic cats, and dogs, as well as in the gastrointestinal tract of a variety of insects. *Rhodococcus* spp. forms biofilms on soil particles and water surfaces. They thrive at sites contaminated with hydrocarbons, solvents, and heavy metals due to their strong biodegradation abilities, contributing to both natural and engineered environmental clean-up (*bioremediation*).

In humans, inhalation of contaminated dust containing pathogenic strains of R. *equi* is the primary route, especially at horse farms, while contact with contaminated soil through skin wounds is rare. Though it can survive in natural or contaminated waters, soil is the primary reservoir for most *Rhodococcus* species. While soil enriched with manure or decaying organic matter supports their survival and multiplication, they survive well under a variety of soil conditions, such as dry, moist, or acidic. The dusty farm environments, along with animal excreta, secretions, and fomites, serve as important reservoirs for pathogenic *R. equi.* Dry, dusty, and manure-rich conditions favour persistence. Plant surfaces (phyllosphere) and root zones (rhizosphere) can harbour *Rhodococcus* species. *R. fascians* persists in infected plant tissues and surrounding soil.

Pathogenesis and Pathology

Pathogenesis

The metabolically active or dormant live, vegetative bacterial cells of *R. equi* are the infectious form for humans and animals. Virulence factors, such as the VapA plasmid in *R. equi*, are essential for causing disease. Without these factors, environmental *Rhodococcus* is typically harmless.

The exact infective dose of *R. equi* varies depending on the strain's virulence, the host's immune status, and the route of exposure. In experimental settings, a relatively low dose of virulent *R. equi* (e.g. 10^4 colony-forming units (CFU)) induces pneumonia in foals. However, in natural environmental conditions, foals are constantly exposed to varying concentrations of both virulent and avirulent strains. However, the precise infective dose in natural settings is difficult to ascertain.

Once inhaled, *R. equi* is phagocytosed by alveolar macrophages. However, unlike most bacteria that are destroyed within the phagolysosome, virulent *R. equi* possesses a unique mechanism to subvert the normal macrophage-killing process. VapA, encoded on the virulence plasmid, is a key virulence factor. It plays a crucial role in preventing the fusion of the phagosome, the vacuole containing the ingested bacterium, with lysosomes. Such inhibition of lysosome–phagosome fusion within macrophages is responsible for tissue destruction and neutrophilic infiltration.

Bacteria generate a macrophage-killing resistance, which ultimately leads to necrotising pneumonia. This necrotic cell death is highly pro-inflammatory, triggering a cascade of immune responses. The process attracts more phagocytic cells, primarily neutrophils and additional macrophages, to the site of infection. This continuous cycle of bacterial uptake, intracellular replication, macrophage destruction, and the subsequent recruitment of inflammatory cells leads to the characteristic formation of pyogranulomas and abscesses.

The chronic nature of the infection stems from the bacterium's protected intracellular niche and the host's inability to mount an effective cell-mediated immune response, which is crucial for clearing intracellular pathogens. Young foals are particularly susceptible due to their immature cell-mediated immunity.

Pathology

These pyogranulomatous lesions primarily characterise the pathology. These affect various organs, although the lungs are the most common and severely impacted sites in foals.

Multifocal-to-coalescent nodules, which range in diameter from a few millimetres to several centimetres in the lungs, are characteristic. These nodules are typically firm and yellowish-white and may contain a creamy, caseous, or purulent material upon cut section, indicative of abscessation. Extensive consolidation of lung tissue and atelectatic (collapsed) areas may surround these nodules. Histologically, these lesions are characterised by a central core of necrotic debris, neutrophils, and degenerating macrophages, surrounded by a cuff of epithelioid macrophages, lymphocytes, plasma cells, and, often, a fibrous capsule.

Beyond the lungs, *R. equi* disseminates haematogenously or lymphatically to other organs, causing extrapulmonary lesions. The most common extrapulmonary manifestation in foals is suppurative lymphadenitis, particularly affecting the mediastinal and tracheobronchial lymph nodes. These become swollen and may contain abscesses, like those found in the lungs. Intestinal and mesenteric abscesses are also frequently observed, often leading to ulcerative enterocolitis, particularly in the large intestine.

Less common but significant extrapulmonary lesions include septic physitis (inflammation of the growth plate) and osteomyelitis. The latter is debilitating and even leads to pathological fractures of the vertebrae, causing spinal cord compression. In immunocompromised humans, *R. equi* infections often mimic tuberculosis (TB), presenting as cavitary pneumonia and lung abscesses. It can also lead to disseminated disease affecting various organs, including the skin, soft tissues, and even the brain, sometimes resulting in a rare inflammatory condition called malakoplakia.

An overview of the pathology of *R. equi* infection is presented in Table 17.1.

Immunology

R. equi evades immunity through its virulence plasmid VapA, which blocks phagosome maturation. It has plasmid-encoded factors that suppress the host IFN-γ while increasing IL-10.

Host defences against *R. equi* rely on robust cell-mediated immunity, as macrophages themselves are the primary replicative niche. Both innate and adaptive immune responses are engaged—innate sensors (complement and Pattern Recognition Receptors (PRRs)) recognise *R. equi* and trigger inflammation. The $CD4^+$ T cells (Th1/Th17) produce IFN-γ, TNF-α, IL-6, and related cytokines, which activate macrophages to kill the bacteria. Simultaneously, IL-12 released from macrophages drives up IFN-γ production. Opsonisation by complement component C3 is important for efficient binding to macrophages via the CR3 (Mac-1 and CD11b/CD18) receptor.

Neutrophils rapidly influx into the infected lungs, where they phagocytose *R. equi*-infected cells. These cells also secrete chemokines and cytokines that recruit additional leucocytes, with a bias towards Th1 responses. By contrast, Th2 immunity, characterised by the production of IL-4 and IL-5, does not protect against *R. equi.* Dendritic cells (DCs) link innate to adaptive immunity by presenting antigens from *R. equi* to T cells. Epithelial cells in the lungs respond to *R.*

Table 17.1 Pathology of *Rhodococcus equi* infection

Organ/system	Gross lesions	Histological features	Notes
Lungs	Multifocal-to coalescent, firm, yellow-white nodules, creamy/caseous centres,surrounded by consolidation and atelectasis	Central necrotic core with neutrophils and macrophagessurrounded by epithelioid macrophages, lymphocytes, plasma cells,often fibrous capsule	Primary site in foals; most severe impact
Lymph nodes	Swollen mediastinal/tracheobronchial nodes, suppurative abscesses like pulmonary lesions	Similar pyogranulomatous inflammation with abscessation	Common extrapulmonary site; typically, via lymphatic spread
Intestines and mesentery	Intestinal wall and mesenteric abscessesUlcerative enterocolitis, especially in the large intestine	Suppurative inflammation; mucosal ulceration and necrosis	Common in systemic dissemination
Bones (physes and vertebrae)	Lesions in growth plates and vertebraeMay cause pathological fractures and deformities	Septic physitisOsteomyelitis with bone necrosis and inflammatory infiltrates	May result in spinal cord compression
Immunocompromised humans	Cavitary pneumoniaLung abscessesand disseminated lesions in the skin, brain, and soft tissues	Granulomatous inflammationmay show features of malakoplakia (foamy macrophages with Michaelis–Gutmann bodies)	Resembles tuberculosis; opportunistic pathogen in immunocompromised individuals

equi, especially in humans, by secreting IL-6 and IL-8, which recruit neutrophils.

The full role of non-immune cells remains less characterised. High-titre hyperimmune serum or colostrum containing maternal antibodies is needed to protect foals. Natural infection does induce anti-*R. equi* antibodies, especially against surface antigens like VapA or polysaccharides. Nevertheless, these are not sufficient for clearance without cell-mediated immunity.

Infection in Humans

Human rhodococcosis is an opportunistic disease often linked to occupational exposure and the consumption of undercooked pork or beef contaminated with *R. equi* (specifically, strains carrying pVAPB or pVAPN plasmids) during slaughter, when meat comes in contact with infected lymph node material or faeces. Once exposed, approximately 80% of *R. equi* infections in humans manifest as pulmonary disease. Clinical manifestations include subacute onset of fever, cough, and weight loss. Cavitatory pneumonia resembling tuberculosis and lung abscesses with necrotising pneumonia are prevalent in immunocompromised patients. Although rarely reported, in severely immunosuppressed individuals, *R. equi* dissemination beyond the lungs can lead to extrapulmonary manifestations, such as bacteraemia, soft tissue abscesses, lymphadenitis, osteomyelitis, and meningitis.

Infection in Animals

Foals become infected in the first month of life. The clinical signs in animals include severe suppurative bronchopneumonia, characterised by multifocal lung nodules and abscesses, primarily in horses and foals. Tracheobronchial lymphadenopathy, the caseous lymphadenitis, is a common condition. Extrapulmonary abscesses occur in the liver, spleen, and gut. Granulomatous lymphadenitis of various nodes, such as the submandibular or mesenteric nodes, and multifocal visceral abscesses are typical. *R. equi* is often isolated from lymph nodes in infected production animals, such as swine, cattle, and goats. Infections are rare in dogs and cats, but when they do occur, they are mostly granulomatous lesions presenting as either necrotising bronchopneumonia or skin/subcutaneous abscesses, often with draining tracts. Feline cases frequently involve cutaneous lesions of the feet or limbs.

Epidemiology and Public Health

Rhodococcus species are environmental bacteria with a global distribution, predominantly isolated from soil ecosystems. Among them, *R. equi* is the most significant pathogen, particularly in veterinary contexts. It is the primary pathogen, causing severe pneumonia and systemic infection in foals. It is a well-established cause of severe bronchopneumonia and systemic disease in young foals. Transmission primarily occurs by inhalation of dust particles contaminated with *R. equi*, as well as through ingestion of contaminated feed, bedding, or contact with infected faeces and environments such as paddocks and stables. In other domesticated animals, including livestock and companion animals, the pathogen can enter through ingestion or wounds. These infected animals then serve as asymptomatic carriers, continuously shedding the organism and contributing to environmental persistence.

Since its first reported case in 1967, *Rhodococcus* has been reported from Europe, North and South America, Australia, and other countries around the world, except Antarctica. *R. equi* is frequently isolated from soils enriched with faecal matter of herbivores, especially in agricultural and pastoral environments. The bacterium thrives in arid, dusty, and warm conditions, enhancing its ability to become airborne and inhaled by susceptible hosts.

In humans, *R. equi* is considered an opportunistic pathogen, primarily affecting individuals with weakened immune systems. Populations at elevated risk include those with HIV/AIDS, recipients of organ transplants, patients undergoing immunosuppressive therapies (such as chemotherapy or prolonged corticosteroids), and

Table 17.2 Epidemiology and distribution of *Rhodococcus* spp.

Species	Normal host/habitat	Human infection	Global distribution	Countries from where human infections have been reported
Rhodococcus equi	Horses, pigs, many other animals, soil	Yes	Worldwide	Spain, Brazil, USA, Australia, some African countries, India
Rhodococcus corynebacterioides	Soil, water	Yes	Australia	Australia
Rhodococcus erythropolis	Soil	Yes	Worldwide	USA, Korea, Japan, Nigeria, Germany, UK
Rhodococcus rhodochrous	Soil	Yes	Worldwide	China
Rhodococcus fascians	Soil, plants	Yes	USA, India, Russia, Iran, some European countries	USA

individuals with haematological malignancies. Other *Rhodococcus* species, such as *R. erythropolis*, are largely saprophytic and only infrequently associated with opportunistic infections.

The epidemiology and distribution of different *Rhodococcus* spp. are presented in Table 17.2.

Laboratory Diagnosis

Accurate diagnosis requires a high index of suspicion, especially in at-risk populations.

Specimens

The specimens include sputum, bronchoalveolar lavage fluid, or any other body fluid or tissue when applicable for microscopy, culture, or molecular diagnostics such as PCR.

Microscopy

Microscopy, particularly Gram stain and, often, acid-fast stain, serves as a valuable presumptive diagnostic tool. Due to its varying microscopic morphology, depending on the species and stage of growth, care must be taken when identifying *R. equi*.

R. equi exhibits a Gram-positive-to-Gram-variable reaction with pleomorphic coccobacilli or short rods arranged in clusters or diphtheroid (V- or L-shaped) formations (Figs. 17.1 and 17.2). It may exhibit uniform staining or a beaded appearance. Special care is needed during interpretation because the Gram stain of *Rhodococcus* spp. shows pleomorphic Gram-positive coccobacilli that could suggest normal respiratory flora. A BAL fluid analysis can further facilitate diagnosis.

The standard Ziehl–Neelsen method may miss it. The bacterium appears partially acid-fast on modified Ziehl–Neelsen or Kinyoun stain. Modified acid-fast staining of isolates from culture on trypticase soy agar with 5% chocolate agar or blood agar may appear acid-fast negative. This partial acid-fast characteristic is particularly important in human diagnosis, where *R. equi* pneumonia often mimics tuberculosis. The presence of acid-fast bacilli in sputum or other respiratory specimens from an immunocompromised patient, particularly when *Mycobacterium tuberculosis* is not isolated in culture, should raise suspicion for *R. equi* infection.

The histopathology of *Rhodococcus* is unique, characterised by neutrophilic leucocytosis and infiltration of macrophages interspersed with micro-abscesses. In tissues, the organism is often seen intracellularly inside macrophages. In patients with pulmonary *R. equi* infection, a distinctive histological reaction called "*pulmonary malacoplakia*" is found within the macrophages. It is characterised by laminated basophilic concretions (*Michaelis–Gutmann bodies*) (Fig. 17.3). Though macrophages are predominant in the affected tissue, well-formed caseating granulomas are absent. Nodular pul-

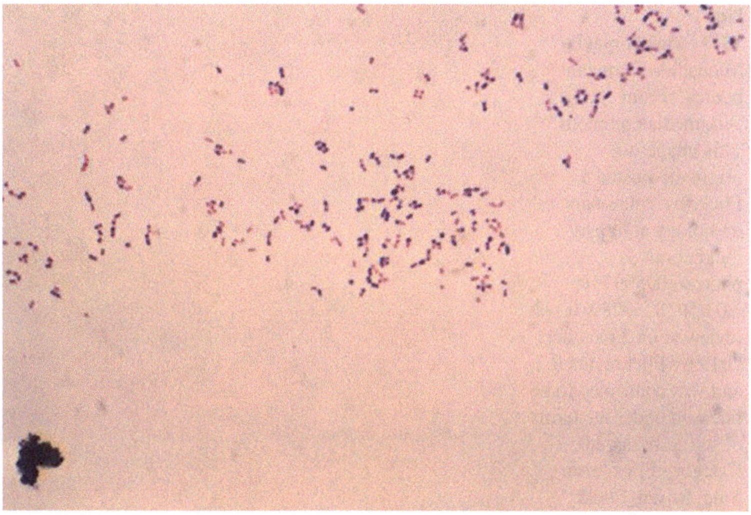

Fig. 17.1 Gram-stained specimen with numerous Gram-positive *Rhodococcus equi* coccobacilli (1000×). (Courtesy: PHIL CDC/ Regina Linder: Hunter College, New York, New York, USA. 1977)

Fig. 17.2 Slide culture sample of tap water agar under a magnification of 400×, with Nocardia-like, Gram-positive organisms: *Rhodococcus* group bacteria. (Courtesy: PHIL CDC/ Dr. David Berd. 1972)

monary infiltrate may sometimes be present; cavitation is common. Histopathologically, micro-abscesses may be conspicuous, accompanied by a marked histiocytic infiltrate and periodic acid–Schiff (PAS)-positive macrophages.

In Vitro Culture

In vitro culture the gold standard for the definitive diagnosis of *R. equi* infection in both animals (especially foals) and humans. *R. equi* is cultured from clinical specimens; however, its identification may be challenging due to its resemblance to other pathogens, such as *M. tuberculosis*.

R. equi grows well on blood agar, chocolate agar, and other routine aerobic media but not on MacConkey agar. Cultures should be incubated for up to 7 days. Visible growth is seen in 48–72 h. Colonies are smooth, mucoid, and non-haemolytic, with salmon-pink or orange pigmentation developing after 2–3 days and becoming rough with ageing (Fig. 17.4). Other species may produce orange, red, buff, or yellow colonies. Mueller–Hinton agar with added ceftazidime and novobiocin is also used for the recovery of *R. equi*.

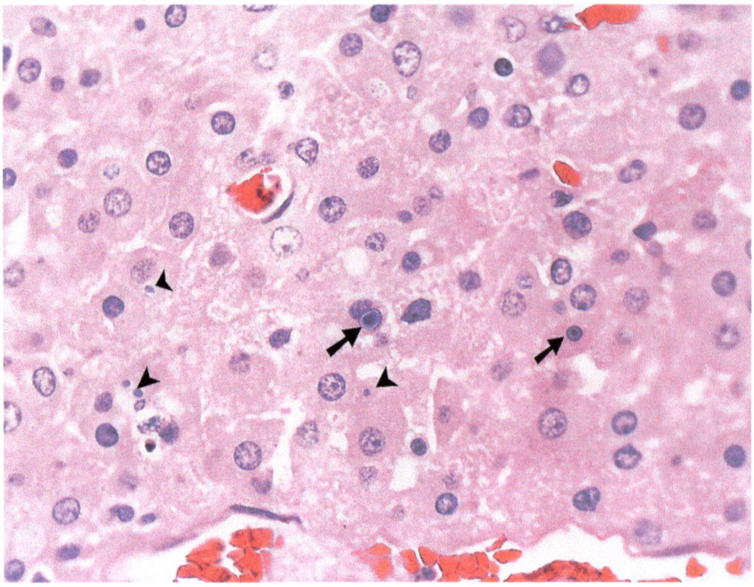

Fig. 17.3 Malakoplakia: Michaelis–Gutmann bodies. (From Wikimedia commons. This image was originally posted to Flickr by Pulmonary Pathology at https://flickr.com/photos/30950973@N03/5938257789. It was reviewed on 3 October 2019 by FlickreviewR 2 and was confirmed to be licensed under the terms of the cc-by-sa-2.0. Category:Files_from_Yale_Rosen_Flickr_stream)

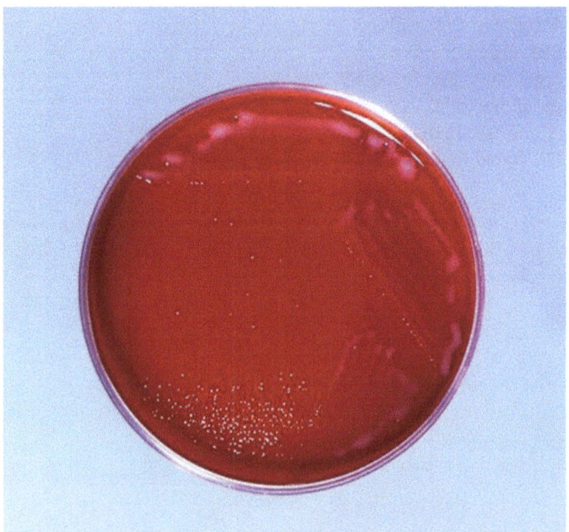

Fig. 17.4 Colonies of *Rhodococcus equi* on blood agar plate. (Courtesy: PHIL CDC/Dr. W.A. Clark. 1977)

Traditional biochemical tests may misidentify organisms of the *Dietzia* and *Tsukamurella* species as *R. equi.*

Serodiagnosis

Currently, there are no validated serological assays available for the diagnosis of *Rhodococcus* infections in humans. In veterinary medicine, antibody-based enzyme-linked immunosorbent assays (ELISAs) targeting virulence-associated protein A (VapA) are being investigated. However, their clinical utility remains limited, and they are not widely adopted for diagnostic purposes.

Molecular Diagnosis

PCR assays targeting conserved chromosomal genes are widely used for the detection and confirmation of *R. equi*. A common target for this purpose is the 16S ribosomal RNA (rRNA) gene. The sensitivity of standard PCR for 16S rRNA is generally high. It can detect very low numbers of bacterial DNA copies in a sample, often in the range of tens to hundreds of colony-forming units (CFU) per millilitre.

Furthermore, PCR is crucial for detecting genes located on its essential virulence plasmid. In equine isolates, the most significant target is the virulence-associated protein A (*vapA*) gene. This type of PCR is paramount in both clinical diagnoses to confirm a virulent infection and in epidemiological studies to track the spread of pathogenic strains on farms or in outbreak investigations.

Quantitative real-time PCR (qPCR) detects *R. equi* DNA and quantifies its initial volume in a sample. qPCR targets either the 16S rRNA gene to determine the total *R. equi* load or, more specifically, the vapA gene to quantify the virulent bacterial burden. The primary application of qPCR is to determine the bacterial load in clinical samples. This is highly valuable for (i) assessing disease severity, (ii) monitoring the patient's response to antimicrobial therapy, and (iii) informing prognostic assessments during treatment.

Specialised PCR-based techniques, such as traA/vapA/vapB (TRAVAP system) typing, are used to type *R. equi* through the presence of *traA*, *vapA*, and *vapB* plasmid genes. The primary application of TRAVAP typing is in molecular epidemiology, enabling detailed strain classification, tracing the origin and spread of specific virulent clones, and gaining a deeper understanding of *R. equi* transmission dynamics within and between animal populations.

Table 17.3 provides a summary of the laboratory diagnosis of *Rhodococcus* infection.

Table 17.3 Laboratory diagnosis of rhodococcosis

Diagnostic methods	Key features	Notes
Microscopy	Gram-positive, weakly acid-fast coccobacilli "Pulmonary Malacoplakia" with Michaelis–Gutmann bodies in pulmonary infection (histopathology)	Intracellular in macrophages, especially *R. equi*
In vitro culture	Blood agar: slow growth (48–72 h) with pink/orange colonies	No growth seen on MacConkey agar Mueller–Hinton agar with ceftazidime and novobiocin can be used for the recovery of *R. equi*
Serodiagnosis	Not useful in humans	Limited to veterinary settings (e.g. VapA ELISA)
Molecular diagnosis	PCR (*choE, vapA* genes) 16S rRNA sequencing	Whole-genome sequencing (WGS) for research

Treatment

Rhodococcus is usually susceptible to macrolides (e.g. azithromycin or clarithromycin), rifampicin, vancomycin, minocycline, and carbapenems. Lipid-soluble agents, such as macrolides like erythromycin, azithromycin, and clarithromycin, usually combined with rifampin, are preferred. They achieve high intracellular concentrations in phagocytes and granulomas. This combination therapy significantly improves outcomes by delivering therapeutic levels directly inside host cells, where *R. equi* resides. While testing for antimicrobial susceptibility, break points for *Staphylococcus aureus* should be used for vancomycin and rifampin. Removal of the catheter is recommended due to the formation of biofilm.

Rhodococcus is resistant to penicillin, ampicillin, and cephalosporins. Susceptibility is variable for fluoroquinolones. Aminoglycosides, such as gentamicin, tobramycin, and amikacin, have high rates of resistance that may preclude use as part of empiric therapy.

Treatment for Humans

The management of rhodococcosis is challenging, as there are no established guidelines regarding the choice and duration of therapy for *Rhodococcus* infections. Commonly employed antimicrobial therapy involves two or three in vitro active antimicrobial agents, typically a combination of macrolides, fluoroquinolones, imipenem, vancomycin, rifampin, linezolid, or aminoglycosides. Rifampin–erythromycin, rifampin–minocycline, erythromycin–minocycline, and imipenem–amikacin have been reported as effective combinations under in vitro conditions.

The chosen regimen is administered for 2–8 weeks based on clinical improvement; longer courses of 2–6 months are generally recommended among immunocompromised hosts or those with severe or disseminated disease. With timely and appropriate therapy, outcomes are favourable. However, mortality rates remain significant, particularly among patients with advanced immunosuppression.

Treatment for Foals/Equines

Erythromycin, along with rifampin, is the antibiotic of choice. Rifampin is often paired with some macrolides for the treatment of foals. The prognosis of *R. equi* pneumonia is poor, even after prolonged treatment. Resistance to macrolides and rifampin in isolates of *R. equi* is increasing.

Prevention and Control

Since the complete elimination of *R. equi* from the environment is not feasible, prevention and control strategies focus on reducing exposure and enhancing host immunity.

Most prevention efforts require effective environmental management. This involves diligent dust control in areas where foals are housed and graze. This is achieved through regular watering of dry paddocks, maintaining adequate pasture cover, and ensuring proper footing in stalls to minimise dust generation. Equally important is effective manure management, as *R. equi* proliferates in horse faeces. Furthermore, good ventilation in foaling stalls and barns, coupled with avoiding overcrowding, helps decrease airborne bacterial concentrations and mitigate infection risk.

For animals, passive immunisation with hyperimmune plasma (HIP) is a common prophylactic practice on farms endemic for *R. equi* pneumonia. This involves administering commercially prepared plasma, rich in *R. equi*-specific antibodies, including those against VapA, intravenously to foals shortly after birth. A second dose is often given a few weeks later. While widely adopted, the efficacy of HIP has been subject to debate, with some studies showing a reduction in disease incidence but not a complete elimination of cases. Controlling *R. equi* in foals depends on early detection, combined with targeted treatment. On farms where *R. equi* pneumonia is endemic, routine screening programmes are often implemented.

For humans, prevention and control strategies require minimising exposure for immunocompromised individuals. This is because *R. equi* is an opportunistic pathogen, predominantly affecting those with weakened immune systems (e.g. HIV/AIDS patients and transplant recipients). Individuals in these risk groups are advised to reduce contact with horses, horse manure, and dusty farm environments where *R. equi* is prevalent. Furthermore, prudent antimicrobial use is a critical aspect of control for both human and animal health.

Case Study

A middle-aged man who was recently diagnosed with HIV infection presented at the emergency department with fever, progressive dyspnoea, and right-sided pleuritic chest pain, 1 month after starting highly active antiretroviral therapy (HAART). He was severely immunocompromised, with a CD4 count of 17/μL. He reported no drug use or recent travel, animal contact, or known TB contact. He worked in an animal harbour, where he came into close contact with containers filled with horse hides. Lung auscultation revealed crepitations on the right side and diffuse scratch-like lesions on the skin. Blood analysis revealed an elevated leucocyte count and C-reactive protein (CRP) level. PCR for CoV-SARS2, influenza A and B, respiratory syncytial virus (RSV), *Chlamydia pneumoniae*, *Legionella pneumophila*, *Streptococcus pneumoniae*, *Mycoplasma pneumoniae*, and *Pneumocystis jirovecii* were negative. A CT scan of the lungs revealed necrotising cavitating pneumonia in the right lower lobe, a smaller cavitating consolidation apically in the left lower lobe, and bilateral hilar necrotising adenopathies. The patient was placed in isolation, and empiric antibiotic therapy with intravenous amoxicillin and clavulanic acid was initiated. A biopsy was done from a tumour-like lesion in the right lower lobe during bronchoscopy. Histological examination revealed chronic bronchitis with a histocyte infiltrate, while Ziehl–Neelsen staining and Grocott staining were negative, and there was no evidence of malignancy. Mycobacterial qPCR in BAL fluid was negative, and *Aspergillus* antigen was not detected. Cultures of BAL and an endotracheal aspirate identified *Rhodococcus* species by matrix-assisted laser desorption/ionization-time of flight mass spectrometry (MALDI-TOF MS), and these were confirmed by 16S rRNA sequencing.

1. What is the empiric antibiotic therapy in this case?
2. Can HAART be continued in conjunction with antibiotic therapy?
3. What is the differential diagnosis for this presentation?

Research Questions

1. How can we implement preventive measures in controlling the transmission of infection due to the role of wild animals in the life cycle of this parasite?
2. How can we develop a safe and effective vaccine by defining an optimum target in the definitive host to interrupt the transmission cycle of the genus *Rhodococcus*?
3. What are the targets needed for newer drug discovery to overcome the limited efficacy of current drugs?

Further Readings

Anastasi E, MacArthur I, Scortti M, Alvarez S, Giguère S, Vázquez-Boland JA. Pangenome and phylogenomic analysis of the pathogenic actinobacterium *Rhodococcus equi*. Genome Biol Evol. 2016;8(10):3140–8.

Bryan LK, Alexander ER, Lawhon SD, Cohen ND. Detection of vaPN in Rhodococcus equi isolated cultured from humans. PLOS One. 15(6):e0235719.

Coulson GB, Agarwal S, Hondalus MK. Characterisation of the role of the pathogenicity Island and vapG in the virulence of the intracellular actinomycete pathogen Rhodococcus equi. Infect Immun. 2010;78(8):3323–34.

Giguère S, Cohen ND, Chaffin MK, Hines SA, Hondalus MK, Prescott JF, Slovis NM. Rhodococcus equi: clinical manifestations, virulence, and immunity. J Vet Intern Med. 2011;25(6):1221–30.

Hondalus MK. Pathogenesis and virulence of Rhodococcus equi. Vet Microbiol. 1997;3–4(56):257–68.

Khurana SK. Current understanding of *Rhodococcus equi* infection and its zoonotic implications. Adv Anim Vet Sci. 2015;3(1):1–10.

MacArthur I, Anastasi E, Alvarez S, Scortti M, Vázquez-Boland JA. Comparative genomics of Rhodococcus equi virulence plasmids indicates host-driven evolution of the vap pathogenicity island. Genome Biol Evol. 2017;9(5):1241–7.

Majidzadeh M, Fatahi-Bafghi M. Current taxonomy of Rhodococcus species and their role in infections. Eur J Clin Microbiol Infect Dis. 2018;37(11):2045–62.

Ranganath N, Mendoza MA, Stevens R, Kind D, Wengenack N, Shah A. *Rhodococcus* infection: a 10-year retrospective analysis of clinical experience and antimicrobial susceptibility profile. J Clin Microbiol. 2024;62(3):e0153723.

Scott MA, Graham BS, Verrall R, Dixon R, Schaffner W, Tham KT. Rhodococcus equi–an increasingly recognised opportunistic pathogen. Report of 12 cases and review of 65 cases in the literature. Am J Clin Pathol. 1995;103(5):649–55.

Silveira BP, Cohen ND, Lawhon SD, Watson RO, Bordin AI. Protective immune response against *Rhodococcus equi*: an innate immunity-focused review. Equine Vet J. 2025;57(3):563–86.

Song Y, Xu X, Huang Z, Xiao Y, Yu K, Jiang M, Yin S, Zheng M, Meng H, Han Y, Wang Y, Wang D, Wei Q. Genomic characteristics revealed plasmid-mediated pathogenicity and ubiquitous rifamycin resistance of *Rhodococcus equi*. Front Cell Infect Microbiol. 2022;12:807610.

Plague

18

Prathyusha Kokkayil

Learning Objectives

1. To understand the zoonotic transmission cycle of *Yersinia pestis*, including its natural reservoirs (rodents), vectors (fleas), and factors contributing to spillover into human populations
2. To describe the mechanisms of pathogenesis of plague and correlate them with the three clinical forms, namely, bubonic, septicaemic, and pneumonic
3. To discuss approaches to diagnosis, treatment, and prevention while addressing emerging challenges such as antimicrobial resistance (AMR) and bioterrorism preparedness

Introduction

Yersinia pestis is a highly virulent Gram-negative bacterium. It is the causative agent of plague, a disease that has historically had a profound impact on human civilisation. Its most infamous manifestation, the "Black Death" in the fourteenth century, devastated populations across Europe, highlighting the profound impact this pathogen can have. The bacterium's discovery by Alexandre Yersin in 1894 marked a significant step in understanding its aetiology.

P. Kokkayil, (✉)
Department of Microbiology, All India Institute of Medical Sciences, Patna, Bihar, India
e-mail: drprathyushak@aiimspatna.org

Plague is primarily a zoonotic disease, maintained in natural cycles involving wild rodents and their fleas. Humans typically become infected through the bite of an infected flea, which regurgitates bacteria during a blood meal. Direct contact with infected animal tissues or fluids, such as during hunting or skinning, can also lead to infection. The most concerning mode of transmission for public health is person-to-person spread, which occurs with pneumonic plague. The pathogenesis of *Y. pestis* is intricately linked to its multiple virulence factors, many of which are encoded on plasmids. These include *Yersinia* outer proteins (Yops) and the plasminogen activator (Pla). Additionally, the F1 capsular antigen forms a protective layer, further inhibiting phagocytosis.

Diagnosis of plague requires a high index of suspicion, especially in individuals with relevant epidemiological exposure and characteristic symptoms like swollen, painful lymph nodes (buboes) in bubonic plague or severe respiratory distress in pneumonic plague. Laboratory confirmation is crucial and involves isolating *Y. pestis* from clinical samples such as blood, bubo aspirate, or sputum. Microscopic examination may reveal the characteristic "safety pin" appearance of the bacilli with bipolar staining. Culture or highly sensitive molecular methods, such as PCR, are used for definitive identification. Early and aggressive antibiotic treatment is vital, as untreated plague can be rapidly fatal.

S. C. Parija, A. Poddar (eds.), *Textbook of Bacterial Zoonoses*, Microbial Zoonoses,
https://doi.org/10.1007/978-981-95-6615-0_18

Streptomycin and gentamicin have historically been considered first-line treatments. However, fluoroquinolones, like ciprofloxacin and levofloxacin, and tetracyclines, like doxycycline, are also highly effective and widely used.

History

History of *Yersinia pestis*

The genus *Yersinia* was named in honour of Alexandre Yersin, who first isolated the bacterium during the 1894 Hong Kong outbreak. However, it was initially misclassified as *Pasteurella pestis* before genomic studies in 1944 confirmed its proper phylogenetic placement.

History of Plague

Y. pestis, the causative agent of plague, has profoundly shaped human history. The first pandemic, the Justinian Plague (541–549 AD), likely originated in Central Asia or East Africa and devastated Constantinople, resulting in the deaths of 25–50 million people. Plague persisted for over two centuries, weakening the Byzantine Empire. The Black Death (1346–1353), originating from Central Asia, swept through Europe via trade routes, killing 30–60% of the population and dismantling feudal structures. Its cultural and social impact was profound, with widespread panic and persecution.

The *Third Pandemic* (1855–1959), which began in Yunnan, China, resulted in the deaths of 12 million people and marked a turning point in plague research. Alexandre Yersin identified the bacterium in 1894. Subsequently, flea-borne transmission of the bacterium was confirmed in 1898 by Paul-Louis Simond. This led to advancements in public health, antibiotics, and vector control.

Although urban plague declined after 1924, *Y. pestis* became endemic in wild rodent populations worldwide. Modern outbreaks have occurred in Madagascar, the United States, and various parts of Africa. Currently, plague remains a Tier 1 bioterrorism agent due to its lethality and historical use as a weapon. Climate change, urbanisation, and genomic studies underscore the importance of a One Health approach in preventing future outbreaks and understanding its evolutionary potential.

History of Plague in India

The history of plague in India is a complex narrative of epidemiological devastation, colonial intervention, and sociocultural transformation. Early descriptions of plague-like illnesses can be found in ancient Sanskrit texts, such as the *Madhava Nidana* (seventh century CE), although their direct link to *Y. pestis* remains uncertain. While the fourteenth-century Black Death likely reached India via Persian trade routes, significant documentation only appears from the seventeenth century onwards, particularly in port cities like Surat and Calcutta, where trade and urban crowding facilitated outbreaks.

The most devastating impact occurred during the Third Pandemic (1896–1921), which began in Bombay, likely via steamships from Hong Kong. The disease spread rapidly through India's cities and transport networks, causing an estimated ten million deaths. The 1994 Surat outbreak reignited national concern. Triggered by flooding and poor sanitation, it caused public panic, a mass exodus, and economic losses exceeding $1.7 billion. Though only 56 official deaths were reported, some questioned whether the outbreak involved true plague or the infection caused by *Yersinia pseudotuberculosis*.

Taxonomy

Y. pestis belongs to the domain Bacteria, phylum Pseudomonadota (formerly Proteobacteria), class Gammaproteobacteria, order Enterobacterales, family Yersiniaceae, and genus *Yersinia*. It is classified within the *Yersinia pseudotuberculosis* complex, being genomically most closely related to *Y. pseudotuberculosis* (sharing ~97% average nucleotide identity) but distinguished by its unique plasmid profile (pPCP1, pMT1, and pCD1) and flea vector adaptation.

Historically, it was known as *Pasteurella pestis* until its reclassification in 1944 based on DNA hybridisation and biochemical characteristics. The species demonstrates limited genetic diversity due to its recent evolutionary emergence (~6000 years ago) from *Y. pseudotuberculosis,* with three biovars recognised (Antiqua, Mediaevalis, and Orientalis) corresponding to major plague pandemics. Its taxonomic position reflects both phenotypic traits (non-motile at 37 °C and urease-negative) and genomic features (loss of ~13% ancestral genes during speciation, including those for enteric survival). The current phylogeny divides *Y. pestis* into four primary branches (0–3), with subspecies-level distinctions based on Clustered regularly interspersed short palindromic repeats (CRISPR) arrays and single-nucleotide polymorphisms (SNPs).

Genomics and Proteomics

Modern genomic analyses have revolutionised the understanding of the emergence of *Y. pestis* from its progenitor *Y. pseudotuberculosis,* approximately 6000–10,000 years ago. This evolutionary transition represents a dramatic pathogenicity shift, transforming a mild enteric pathogen into the causative agent of plague. The transformation was driven by three pivotal genetic events: (i) acquisition of virulence plasmids through horizontal gene transfer, (ii) genome reduction through pseudogenisation, and (iii) adaptation to flea vectors.

The three hallmark plasmids—pCD1 (70 kb), pMT1 (100 kb), and pPCP1 (9.5 kb)—conferred novel pathogenic capabilities. pCD1 encodes the type III secretion system (T3SS) and Yop effector proteins that neutralise phagocytes and suppress cytokine production, enabling systemic invasion. pMT1 carries genes for the antiphagocytic F1 capsule and plasminogen activator (Pla), critical for tissue penetration and dissemination. The smaller pPCP1 enhances fibrinolytic activity through Pla, facilitating haematogenous spread. Concurrently, ~13% of ancestral genes were lost, including metabolic pathways for enteric survival (e.g. inv adhesin), reflecting niche specialisation.

Phylogenetic studies classify *Y. pestis* into four primary branches (0–3), with three historically significant biovars: Antiqua (First Pandemic; capable of fermenting glycerol), Mediaevalis (Black Death; nitrate reductase–positive), and Orientalis (Third Pandemic; glycerol- and nitrate-negative). The avirulent Microtus strains exhibit more than 2000 unique single-nucleotide polymorphisms (SNPs) and lack key virulence factors, prompting proposals for reclassification as *Yersinia pestis* subsp. *caucasica.* Molecular typing tools, such as Multilocus sequence typing (MLST) (targeting aroA, glpD, and dmsA) and CRISPR barcode analysis (spacer arrays YP1–YP8), enable precise strain tracking, as demonstrated during the 2017 Madagascar outbreak, where CRISPR typing linked human cases to specific rodent reservoirs. Ongoing research reveals adaptive mutations in rmpA (biofilm regulation) and ybt (yersiniabactin synthesis) that enhance flea gut colonisation.

Morphology

The morphology of *Y. pestis* reveals a highly adapted bacterial pathogen whose structural features reflect its dual lifestyle in both flea vectors and mammalian hosts.

Y. pestis is a Gram-negative coccobacillus measuring 0.5–0.8 μm in width and 1–3 μm in length. It displays a distinctive short, plump rod shape that often appears oval under light microscopy. The individual cells typically occur singly or in pairs rather than forming extended chains or clusters. It exhibits notable pleomorphism, characterised by a striking *bipolar staining pattern.*

This is often described as a "*safety pin*" appearance.

Y. pestis is a capsulated bacterium. It consists of F1 capsular antigen, a polymer of Caf1 subunits encoded by the pMT1 plasmid, and forms a gel-like envelope (5–10 nm thick) visible by electron microscopy (EM) when expressed at 37 °C. The capsule undergoes phase variation—being repressed at flea gut temperatures (25–28 °C). This variation allows immune evasion in mammals while facilitating biofilm formation in insects. Cryo-EM studies reveal the capsule's lattice structure, which prevents complement deposition and phagocyte receptor engagement.

When examined for motility, *Y. pestis* displays temperature-dependent behaviour. It is non-motile at human body temperature (37 °C) but expresses flagella and demonstrates limited motility at the cooler temperatures (<30 °C) found in its flea vectors, a rare characteristic among *Yersinia* species.

These morphological features exhibit clear distinctions when compared to other *Yersinia* species. *Y. pseudotuberculosis* and *Yersinia enterocolitica* appear as typical rod-shaped bacteria with consistent motility at lower temperatures. In contrast, *Y. pestis* exhibits its characteristic coccobacillus shape, temperature-dependent motility, and prominent bipolar staining, which is not seen in other *Yersinia* species.

Cultivation

Y. pestis is a facultative anaerobe that grows best with 5–10% CO_2 enrichment and maintains optimal growth at a neutral pH of 7.2–7.4. The bacterium grows optimally at two distinct temperature ranges that reflect its dual life cycle in flea vectors and mammalian hosts. *Y. pestis*, at 28–30 °C, which mimics the flea gut environment, exhibits unique characteristics, including biofilm formation and occasional motility. When cultured at 37 °C (human body temperature), these features disappear, while virulence factors such as the F1 capsule and Yops become prominent. Growth stops above 40 °C, and the bacterium is rapidly inactivated by heat or UV exposure.

Animal Models

Animal models are essential for studying *Y. pestis* pathogenesis and host immunity, as well as for evaluating vaccines and therapeutics.

Mice (*Mus musculus*) are the most widely used to study immune responses and test antibiotics (e.g. streptomycin) and vaccine candidates, such as the F1-V subunit. Subcutaneous, intranasal, and intravenous routes are used in various mouse strains (e.g. BALB/c and C57BL/6) to mimic bubonic, pneumonic, and septicaemic plague, respectively.

Rats (*Rattus norvegicus*), natural reservoirs of the plague, are useful for transmission studies. Wild rats are generally resistant, mimicking asymptomatic carriers, whereas lab strains, such as brown Norway rats, can model acute disease through flea bites or injection.

Guinea pigs (*Cavia porcellus*) exhibit human-like disease progression and are sensitive to both bubonic and pneumonic forms. These features make them valuable for vaccine and antibiotic studies, including those involving EV76 and doxycycline.

Non-human primates (NHPs), particularly rhesus macaques and African green monkeys, most closely replicate human plague. Experimental infection shows symptoms such as fever, lymphadenopathy, and respiratory distress. They are useful for late-stage testing of vaccines and therapeutics.

Transmission Cycle

The transmission cycle of *Y. pestis* represents a complex ecological relationship involving animal reservoirs, arthropod vectors, and accidental human hosts (Fig. 18.1).

Wild rodents, particularly ground squirrels, prairie dogs, and gerbils, serve as natural reser-

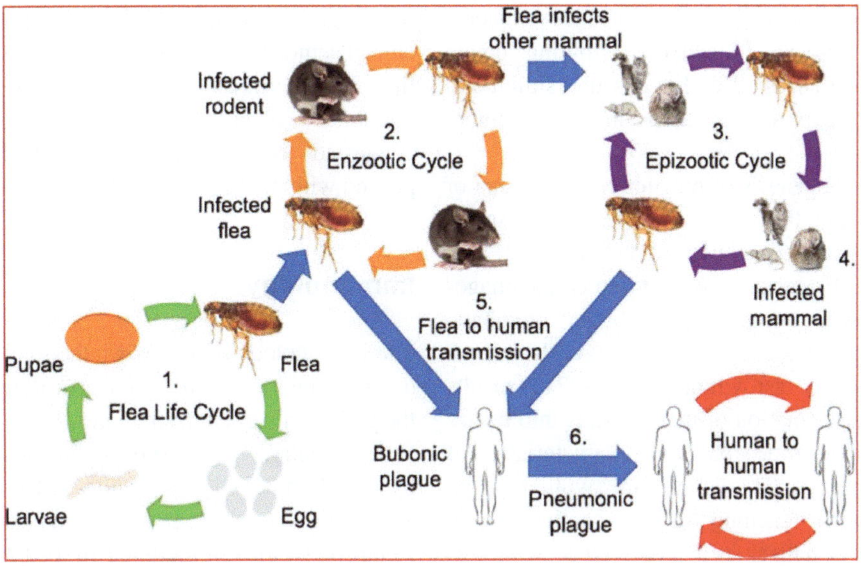

Fig. 18.1 Transmission cycle of *Yersinia pestis.* (Image source: Mackay-Alderson et al. (2020). Permitted for production under CC BY license)

voirs in endemic regions, maintaining the bacterium through persistent, often asymptomatic infections. These enzootic hosts coexist with more susceptible rodent species, such as rats (*Rattus* spp.), which experience high mortality during epizootic outbreaks, creating conditions for rapid bacterial amplification.

The transition from enzootic to epizootic transmission occurs when infected fleas, primarily *Xenopsylla cheopis* (the Oriental rat flea), feed on bacteraemic rodents. They subsequently become blocked due to *Y. pestis* biofilm formation in their digestive tracts. This blockage renders fleas chronically infectious as they repeatedly attempt to feed, regurgitating bacteria into new hosts.

Humans typically enter this cycle as dead-end hosts. They acquire the infection through flea bites that introduce bacteria into the skin, leading to bubonic plague, characterised by swollen, painful lymph nodes (*buboes*). Direct contact with infected animal tissues or fluids also transmits the bacterium, causing primary septicaemic plague without buboes. In contrast, inhalation of respiratory droplets from pneumonic plague cases or infected animals, such as cats, results in the most severe and contagious form of the disease.

Pathogenesis and Pathology

Pathogenesis

Y. pestis produces surface adhesins, including the pH 6 antigen (PsaA), a fimbrial structure that mediates attachment to host cells in the acidic environment of phagolysosomes. It also produces Pla protease, which embeds in the outer membrane to facilitate both tissue invasion and flea gut colonisation.

The bacterium, in addition, produces a wide variety of virulence factors. These include the plasmid-encoded type III secretion system (T3SS) that injects effector proteins (Yops) directly into host cells. YopH dismantles phagocyte signalling networks through tyrosine phosphatase activity, YopE disrupts actin cytoskeletons to block phagocytosis, and YopJ induces macrophage apoptosis, collectively crippling innate immune defences. The F1 capsule simultaneously prevents opsonisation and complement attack.

The pathogenesis of plague represents one of the most rapidly progressing bacterial disease. When transmitted to humans via flea bite, *Y. pestis* uses its F1 capsule to resist immediate phagocytosis while migrating through lymphatic vessels to regional lymph nodes. At the site, it replicates very fast within 2–6 days. This results in the transformation of these tissues into haemorrhagic, necrotic buboes, characteristic of bubonic plague. The *buboes* are swollen lymph nodes, often reaching several centimetres in diameter, with blackened, pus-filled centres. This gave rise to the "Black Death" moniker during medieval pandemics.

Pneumonic plague is the most lethal and contagious form. Inhaled bacteria bypass initial lymphatic defences by being taken up by alveolar macrophages, where they not only survive but replicate intracellularly. They subsequently spread to pulmonary lymph nodes and cause necrotising pneumonia. The latter is marked by alveolar haemorrhage and a frothy, bloody exudate that facilitates deadly person-to-person aerosol transmission.

Pathology

As infection progresses, *Y. pestis* uses its plasminogen activator (Pla) protease to degrade fibrin clots and complement proteins. This facilitates systemic dissemination through the bloodstream, where endotoxic lipopolysaccharide (LPS) triggers a catastrophic cytokine storm dominated by TNF-α and IL-6. All these lead to disseminated intravascular coagulation (DIC), haemorrhagic necrosis in the spleen and liver, and, ultimately, multi-organ failure characteristic of septicaemic plague.

The pathological hallmarks of advanced infection reveal the bacterium's destructive capacity. Bubonic plague shows lymph nodes replaced by necrotic debris and bacterial colonies. Septicaemic plague exhibits widespread petechial haemorrhages and adrenal gland destruction (Waterhouse–Friderichsen syndrome). Pneumonic plague shows the lungs resemble congested, consolidated masses, with alveoli packed with neutrophils and bacteria.

Immunology

The immunology of *Y. pestis* infection involves a dynamic interaction between host defences and the pathogen's evasion strategies.

Upon entry, *Y. pestis* subverts innate immunity by modifying lipid A to evade TLR4 detection and disabling inflammasome activation. Neutrophils, though recruited by IL-8, are functionally paralysed by YopH-mediated dephosphorylation. Macrophages serve as early replication sites but are compromised by YopJ-induced apoptosis and actin disruption via YopE. The complement system is neutralised through Pla-mediated degradation of C3b/C5a and the F1 capsule's antiphagocytic properties.

Adaptive immunity, though often delayed by rapid disease progression, involves protective antibody responses against the F1 capsule. Effective immunity requires high titres of IgG for opsonisation and neutralisation. CD4+ Th1 responses enhance macrophage killing through IFN-γ. However, CD8+ T cells play a limited role. Survivors exhibit durable memory B and T cell responses, offering cross-biovar protection.

Immune evasion is evolved around the type III secretion system (T3SS), with Yop proteins disrupting phagocytosis and inflammatory signalling. Late-stage infection is characterised by cytokine storms (TNF-α, IL-6, and IL-1β), haemophagocytic lymphohistiocytosis, and immunosuppression mediated by IL-10.

Infections in Humans

Y. pestis infection in humans presents a spectrum of clinical manifestations.

Bubonic Plague

Bubonic plague is the most prevalent form of the disease, representing most naturally acquired human infections.

The incubation period ranges from 2 to 6 days following exposure. Clinically, patients present with the abrupt onset of high fever, chills, headache, malaise, and generalised body aches. The hallmark of bubonic plague is the development of painful, swollen lymph nodes, known as buboes. These are most commonly found in the inguinal, axillary, or cervical regions, corresponding to the site of the flea bite. These lymph nodes become tender and erythematous and may eventually suppurate or become necrotic. If left untreated, bubonic plague may disseminate to the bloodstream or lungs, evolving into the septicaemic or pneumonic forms with significantly higher fatality rates.

Septicaemic Plague

Septicaemic plague may develop either as a complication of untreated bubonic or pneumonic plague or as a primary infection in which no buboes are evident. It occurs when *Y. pestis* enters the bloodstream and rapidly multiplies, leading to overwhelming sepsis.

The clinical presentation is abrupt and severe, characterised by high-grade fever, chills, profound weakness, abdominal pain, nausea, vomiting, and, sometimes, diarrhoea. Rapid progression to septic shock is common. It is accompanied by hypotension, multi-organ failure, and disseminated intravascular coagulation (DIC). Cutaneous signs such as petechiae, ecchymoses, and acral gangrene of the fingers, toes, and nose may develop due to endotoxaemia and coagulopathy. These manifestations contributed to the historical reference to plague as the "Black Death." Septicaemic plague can occur without the classical buboes. Hence, diagnosis may be delayed, and mortality rates in untreated cases are extremely high, often exceeding 80%.

Pneumonic Plague

Pneumonic plague is the most severe and contagious form of the disease. It may occur primarily through inhalation of aerosolised droplets containing *Y. pestis* or secondarily when the bacteria spread haematogenously from a bubonic or septicaemic focus to the lungs.

The incubation period is short, typically ranging from 1 to 4 days. Patients present with a sudden onset of high fever, chills, productive cough with purulent or bloody sputum, pleuritic chest pain, and progressive dyspnoea. Tachypnoea, cyanosis, and haemoptysis are common. The disease rapidly progresses to respiratory failure, acute respiratory distress syndrome (ARDS), and circulatory collapse.

Pneumonic plague is a medical and public health emergency due to its capacity for person-to-person transmission via respiratory droplets. It has exceptionally high mortality if not treated within the first 24 h of symptom onset.

Other Clinical Forms of Plague

Meningeal involvement occurs in approximately 6% of cases and is usually secondary to septicaemic dissemination. Patients present with classical signs of meningitis, including headache, photophobia, neck stiffness, vomiting, altered mental status, and seizures. Cerebrospinal fluid (CSF) examination typically reveals neutrophilic pleocytosis, elevated protein levels, and low glucose concentrations. Gram stain of CSF may occasionally demonstrate the characteristic bipolar-staining bacilli. Plague meningitis carries a high risk of mortality and long-term neurological sequelae, particularly when diagnosis and treatment are delayed.

Pharyngeal plague is a rare clinical variant that arises from the direct ingestion of *Y. pestis*-contaminated materials, such as undercooked meat from infected animals, or via inhalation of droplets containing the bacterium, with compatible symptoms and relevant exposure history in endemic regions.

Cutaneous plague is a rare clinical variant that results from direct contact between broken skin and tissues or fluids of infected animals. It is a localised form and is less frequently encountered. The condition begins with the development of a small papule or pustule at the site of bacterial entry, which may evolve into an ulcerative or necrotic lesion.

Infections in Animals

Plague is a multispecies zoonosis that affects a broad range of mammals, with significant clinical, epidemiological, and ecological implications. Rodents serve as the primary reservoirs of *Y. pestis*. In addition, numerous other animal species, including lagomorphs, carnivores, and domestic animals, may also become infected. The clinical manifestations in animals are highly variable, depending on the species, host susceptibility, route of infection, and the virulence of the infecting strain. In many rodent species, infections may be asymptomatic or subclinical, supporting the maintenance of *Y. pestis* in enzootic cycles. In others, particularly in epizootic hosts, infection can cause acute, fatal disease.

Rodents represent the most important natural hosts in the ecology of plague, and their response to infection varies considerably by species. These responses are broadly classified into two categories: enzootic hosts (maintenance) and epizootic hosts (amplifiers).

Enzootic Rodents

Enzootic rodents, such as deer mice (*Peromyscus* spp.), house mice (*Mus musculus*), and certain ground squirrels, typically develop mild or asymptomatic infections. These animals exhibit a degree of natural resistance to *Y. pestis*, permitting chronic, low-level bacteraemia that supports flea infection without causing overt disease. In such species, clinical signs may be absent, and infected individuals may survive long enough to perpetuate the enzootic cycle by serving as a continuous source of infection for flea vectors.

Epizootic Rodents

Epizootic hosts, such as prairie dogs (*Cynomys* spp.), black rats (*Rattus rattus*), and rock squirrels (*Spermophilus variegatus*), are highly susceptible to plague. Infection often leads to acute septicaemia and death within 1–2 days, frequently before the development of overt clinical signs. When signs are observed, they may include fever, lethargy, laboured breathing, incoordination, and swollen lymph nodes. In epizootic situations, mass die-offs of these susceptible rodent populations may serve as an early indicator of plague activity in a region and are often followed by increased transmission to fleas, leading to spillover into other animal and human populations.

Other Animal Hosts

Domestic cats (*Felis catus*) are highly susceptible to *Y. pestis* and are among the few animals that can develop pneumonic plague, capable of transmitting the infection to humans. Zoonotic transmission from cats to humans, especially veterinary personnel, is a recognised risk, particularly in pneumonic cases.

Lagomorphs, such as cottontail rabbits and jackrabbits, are moderately susceptible to plague and can present with systemic illness. Domestic dogs are relatively resistant to clinical disease, often exhibiting mild or asymptomatic infections. Occasionally, plague has been reported in other wild and captive mammals, including camels, llamas, ferrets, monkeys, and zoo carnivores, such as lions and tigers.

Epidemiology and Public Health

Plague remains a serious global health threat, with *Y. pestis* sustaining enzootic cycles in rodents and causing 1000–3000 human cases annually. Madagascar accounts for over 75% of cases, including a 2017 pneumonic outbreak (2417 cases and 209 deaths). Ongoing transmission persists in the Democratic Republic of

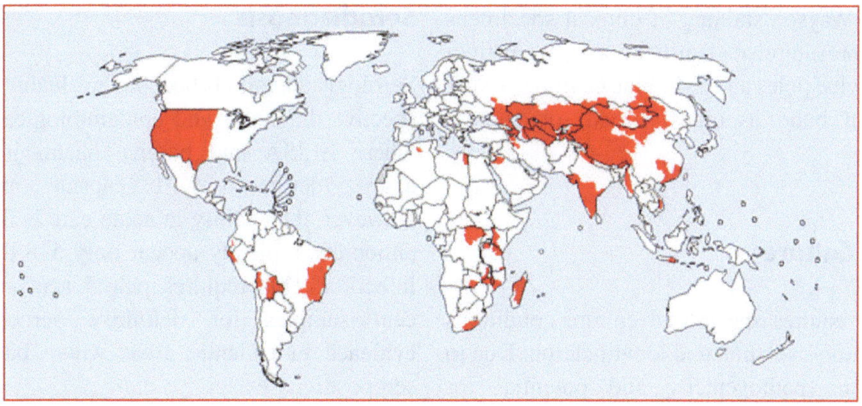

Fig. 18.2 Global epidemiology of plague as of 2016. (Image source: WHO/PED)

Congo and other African foci. Asia reports sporadic cases in China's Inner Mongolia and Mongolia (linked to marmot hunting), as well as in India, which has remained alert since its 1994 outbreak. In the Americas, cases have been reported in Peru, Brazil, and the US South-west, often associated with prairie dog flea bites. Warmer seasons increase flea activity, and climate change threatens to expand plague into new regions, including the former Soviet foci. The global epidemiology of plague is presented in Fig. 18.2.

The efficiency of transmission depends on multiple ecological factors, including rodent population density, flea vector competence, and environmental conditions that allow bacterial persistence in soil or carcasses for extended periods. Urban plague outbreaks historically emerged from commensal rat–flea–human cycles, as seen in the Black Death. At the same time, modern cases more commonly originate from sylvan foci, where humans encounter infected wild rodents or their fleas.

At-risk groups include rural communities, marmot hunters, and urban slum residents. Environmental factors, such as deforestation and extreme weather, increase human exposure. The threat is compounded by *Y. pestis* classification as a Tier 1 bioterrorism agent and the emergence of drug-resistant strains. Innovative solutions—like CRISPR-based diagnostics and next-generation vaccines (e.g. F1-V subunit)—are urgently needed. Public health priorities must emphasise One Health surveillance, community education, and rapid diagnostics to contain this persistent and evolving zoonotic threat.

Public health interventions targeting this transmission cycle emphasise flea control through insecticides, rodent population management, and rapid antibiotic prophylaxis for exposed individuals, particularly in regions where human activities, such as hunting or agricultural expansion, increase contact with enzootic hosts.

Diagnosis

The objectives of laboratory diagnosis of plague are to (i) quickly identify *Y. pestis*, (ii) confirm the clinical diagnosis, (iii) differentiate it from other related infections, and (iv) guide therapeutic interventions.

Microscopy

Microscopy serves as the first-line diagnostic tool in clinical settings, where Gram staining reveals the characteristic Gram-negative coccobacilli with the distinctive bipolar "safety pin" morphology. This appearance is particularly evident when using Wright's or Giemsa stains, especially in lymph node aspirates from patients with bubonic plague. This bipolarity, caused by concentrated nucleic acid material at the poles, serves as an important diagnostic feature in endemic regions.

Rapid Wayson staining of clinical specimens provides presumptive identification by its distinctive dark blue poles and light blue central staining pattern in bubo aspirates, blood, or sputum samples.

In Vitro Culture

Y. pestis requires specialised culture conditions for laboratory isolation and identification. Due to its extreme pathogenicity and potential for aerosol transmission, all culture work must be performed in Biosafety Level 3 (BSL-3) containment facilities.

For primary isolation, blood agar containing 5–10% sheep or horse blood serves as the standard non-selective medium. In this medium, *Y. pestis* produces distinctive small (1–2 mm diameter), grey-white, non-haemolytic colonies after 48–72 h of incubation. These colonies often display a characteristic "*fried egg*" morphology with a raised centre and flat periphery. On MacConkey agar, the bacterium forms colourless colonies because it cannot ferment lactose.

Selective media play a crucial role in isolating *Y. pestis* from contaminated specimens. Cefsulodin–irgasan–novobiocin (CIN) agar produces diagnostic "bull's-eye" colonies with dark red centres and translucent edges due to the inhibitory action of its antibiotic components (cefsulodin, irgasan, and novobiocin). Specialised media like Herellea agar facilitate phage typing for strain identification.

Y. pestis ferments glucose, maltose, and mannose without producing gas but fails to metabolise lactose, sucrose, or rhamnose. Additional tests demonstrate nitrate reduction capability but negative lysine decarboxylase activity.

Y. pestis is positive for catalase but negative for oxidase and urease. It is differentiated from related *Yersinia* species by its inability to produce indole. It shows a unique triple sugar iron (TSI) reaction pattern (alkaline slant/acid butt without H_2S production).

Serodiagnosis

Serological methods become invaluable for retrospective diagnosis and epidemiological studies, where ELISA and passive haemagglutination assays detect anti–F1 capsule antibodies. However, their utility in acute care is limited, as antibodies typically appear only 5–8 days post-infection. This requires paired acute–convalescent samples for definitive seroconversion evidence in endemic areas where background seropositivity exists.

Molecular Diagnosis

Modern molecular techniques have revolutionised plague diagnostics.

Real-time PCR targeting the *pla*, *caf1*, and *lcrV* genes enables sensitive detection within hours, even in antibiotic-treated patients, in whom culture may fail. Plasmid genes (*pla* on pPCP1 and *caf1* on pMT1) or chromosomal markers (*yihN*) are targets for real-time PCR (qPCR) and can detect 1.5–10 CFU/ml in sputum/bubo aspirates within 5 h. Recently, integrated PCR–lateral flow (PCR–LF) assays have been developed, which combine PCR amplification (e.g. yihN gene) with lateral flow visualisation using labelled primers.

Isothermal amplification methods like Loop-mediated Isothermal Amplification (LAMP) are used for field use in resource-limited outbreak zones. LAMP has a sensitivity of up to 0.03–0.2 pg DNA, and targets include caf1A or pla genes.

Whole-genome sequencing (WGS) plays a growing role in outbreak investigation and bioterrorism preparedness, enabling strain typing, resistance gene detection, and transmission tracing through phylogenetic analysis.

A summary of laboratory diagnosis is presented in Table 18.1.

Table 18.1 Laboratory diagnosis of plague

Method	Principle	Key features	Limitations
Microscopy	Staining reveals morphology	Gram stain: Gram-negative coccobacilli with "safety pin" bipolar staining Wayson/Giemsa stains: enhanced bipolar visualisation in bubo aspirates, blood, sputum	Rapid but low sensitivity; requires skilled interpretation
In vitro culture	Isolation on specialised media (BSL-3 required)	Blood agar: grey-white "fried egg" colonies (1–2 mm) after 48–72 h CIN agar: "bull's-eye" colonies MacConkey: colourless colonies (lactose-negative) Biochemical tests: catalase+, oxidase–, urease–; TSI: alkaline slant/acid butt (no H_2S)	Slow (48–72 h); high biohazard risk (BSL-3)
Serodiagnosis	Detection of anti-F1 capsule antibodies	ELISA/passive haemagglutination Useful for retrospective diagnosis/surveillance	Antibodies appear only 5–8 days post-infection; paired acute/convalescent samples needed; low utility in the acute phase
Molecular diagnosis	Nucleic acid amplification test (NAAT)	qPCR: targets *pla*, *caf1*, *lcrV*, *yihN* (sensitivity: 1.5–10 CFU/ml; 5 h) LAMP: targets *caf1A/pla* (sensitivity: 0.03–0.2 pg DNA; 30 min); field-deployable PCR–lateral flow: combines PCR (*yihN*) with lateral flow readout WGS: for outbreak tracing/resistance detection	High sensitivity even post-antibiotics; PCR–LF avoids gel electrophoresis; LAMP ideal for resource-limited settings

Treatment

Early initiation of antimicrobial therapy remains the cornerstone of successful treatment and can drastically reduce mortality rates associated with all forms of plague. Early and effective antimicrobial therapy is important in managing plague, particularly the septicaemic and pneumonic forms. When initiated within the first 24 h of symptom onset, appropriate treatment dramatically reduces morbidity and mortality.

The choice of antimicrobial agents depends on disease severity, clinical presentation, local resistance patterns, availability, and special population considerations, such as pregnancy and age.

First-line antibiotics for plague include streptomycin (15 mg/kg IM twice daily, max 2 g/day for 10 days), gentamicin (5 mg/kg IV/IM once daily for 10 days), doxycycline (100 mg IV/oral twice daily for 10–14 days), and fluoroquinolones like ciprofloxacin (400 mg IV or 500–750 mg oral twice daily for 10–14 days) or levofloxacin (500–750 mg IV/oral once daily for 10–14 days). Gentamicin is often preferred over streptomycin due to its wider availability and better safety profile, although both require toxicity monitoring. Doxycycline and fluoroquinolones offer oral options suitable for outpatient care or prophylaxis but are generally avoided in pregnancy and in young children.

For pneumonic plague, treatment must begin within 24 h of symptom onset using aminoglycosides (gentamicin) or fluoroquinolones, with strict respiratory isolation for at least 48 h after antibiotics are started. In pregnancy, gentamicin or chloramphenicol (25 mg/kg IV q6h) may be used cautiously, while children can receive gentamicin or—in life-threatening cases—doxycycline despite age restrictions. Supportive care (fluid resuscitation, vasopressors, and ICU support) is critical for septic or pneumonic cases. Post-exposure prophylaxis for pneumonic contacts involves doxycycline (100 mg oral twice daily) or ciprofloxacin (500 mg oral twice daily) for 7 days.

Prevention and Control

Effective plague prevention requires an integrated approach targeting rodent reservoirs, flea vectors, human behaviour, and healthcare systems.

Rodent control relies on environmental sanitation, secure food storage, and rodent-proofing, with rodenticide use carefully timed after flea control to prevent the dispersal of infected fleas. Flea control involves the use of insecticides, such as deltamethrin, in burrows and homes, alongside personal protective measures like N-N-Diethyl-meta-toluamide (DEET) repellents and permethrin-treated clothing for individuals at risk. Pets, particularly free-roaming cats, should receive regular flea treatment to prevent zoonotic transmission.

Public health systems encompass active surveillance, using rodent die-offs as early warnings and deploying rapid diagnostics for timely case detection. Prophylactic antibiotics (e.g. doxycycline) are given to contacts of pneumonic cases to halt transmission. Immediate public health notification and droplet precautions are mandatory for suspected cases.

In *healthcare settings*, pneumonic plague cases require airborne precautions with N95 respirators and isolation, while bubonic cases follow standard precautions unless lesions are open. Bioterrorism preparedness includes antibiotic stockpiles, diagnostic surge capacity, and clinician training.

Vaccination is currently limited to high-risk groups due to the limitations of older vaccines; however, recombinant subunit vaccines (F1-V) are in development.

Case Study

In August 2017, Madagascar faced a deadly pneumonic plague outbreak that differed from typical bubonic plague cases by spreading through airborne transmission in urban areas. The epidemic resulted in 2417 suspected cases and 209 deaths, primarily affecting the capital city of Antananarivo and the port city of Toamasina. The outbreak began when a 31-year-old man travelling by bush taxi developed severe respiratory symptoms and died en route, having already infected 31 close contacts. The disease spread rapidly through crowded urban settings, including prisons and hospitals, prompting an aggressive public health response. Authorities implemented contact tracing for over 11,000 individuals, administered prophylactic antibiotics, enforced infection control measures with personal protective equipment, and restricted public gatherings. Treatment relied on streptomycin and doxycycline, while flea and rodent control measures were intensified. The outbreak was contained by November 2017, but it exposed critical challenges such as antibiotic shortages, delayed diagnoses, and the lack of an effective vaccine.

1. What made the August 2017 plague outbreak in Madagascar particularly concerning and different from typical bubonic plague cases?
2. Which two major urban areas were primarily affected by the pneumonic plague outbreak in Madagascar?
3. What were some of the critical challenges exposed by the 2017 pneumonic plague outbreak in Madagascar, even after it was contained?

Research Questions

1. How do climate change and ecological shifts influence the re-emergence of *Y. pestis* in new geographic regions?
2. What genetic adaptations enable *Y. pestis* to transition between flea-borne (bubonic) and airborne (pneumonic) transmission modes?
3. Can novel vaccines or phage therapies overcome the limitations of current antibiotics in treating multidrug-resistant *Y. pestis* infections?

4. What are the roles of emerging biotechnologies to combat antimicrobial resistance (AMR) in plague?

Declaration During the preparation of this chapter, we have used Copy.ai's AI Paragraph Rewriter to improve the readability of the text. After using, we have reviewed and edited the content as needed and take full responsibility for the content of the publication.

Further Readings

Andrianaivoarimanana V, Kreppel K, Elissa N, Duplantier JM, Carniel E, Rajerison M, et al. Understanding the persistence of plague foci in Madagascar. PLoS Negl Trop Dis. 2013;7(11):e2382.

Bertherat E. Plague around the world in 2019. Wkly Epidemiol Rec. 2019;95(19):241–8.

Butler T. Plague gives surprises in the first decade of the twenty-first century in the United States and worldwide. Am J Trop Med Hyg. 2013;89(4):788–93.

Hinnebusch BJ, Bland DM, Bosio CF, Jarrett CO. Infective dose of *Yersinia pestis* in fleas. J Infect Dis. 2017;216(5):612–6.

Mackay-Alderson J, Quastel M, Wilson E, Bellamy D. Factors influencing the re-emergence of plague in Madagascar. Emerg Top Life Sci. 2020;4:423–33. https://doi.org/10.1042/ETLS20200334.

Mead PS. Plague in Madagascar—a tragic opportunity for improving public health. N Engl J Med. 2018;378(2):106–8.

Morelli G, Song Y, Mazzoni CJ, Eppinger M, Roumagnac P, Wagner DM, et al. *Yersinia pestis* genome sequencing identifies patterns of global phylogenetic diversity. Nat Genet. 2010;42(12):1140–3.

Rajerison M, Mel CDC, Ratsitorahina M, Rahalison L, Chanteau S. Use of loop-mediated isothermal amplification for plague detection in Madagascar. Trans R Soc Trop Med Hyg. 2009;103(7):661–6.

Riehm JM, Rahalison L. Plague: a review for clinicians. Infect Dis Clin N Am. 2019;33(4):967–83.

Sebbane F, Jarrett CO, Gardner D, Long D, Hinnebusch BJ. Role of the *Yersinia pestis* plasminogen activator in the incidence of distinct septicemic and bubonic forms of flea-borne plague. Proc Natl Acad Sci USA. 2006;103(14):5526–30.

Stenseth NC, Atshabar BB, Begon M, Belmain SR, Bertherat E, Carniel E, et al. Plague: past, present, and future. PLoS Med. 2008;5(1):e3.

Vogler AJ, Chan F, Wagner DM, Roumagnac P, Lee J, Nera R, et al. Phylogeography and molecular epidemiology of *Yersinia pestis* in Madagascar. PLoS Negl Trop Dis. 2011;5(9):e1319.

Wagner DM, Klunk J, Harbeck M, Devault A, Waglechner N, Sahl JW, et al. *Yersinia pestis* and the plague of Justinian 541–543 AD: a genomic analysis. Lancet Infect Dis. 2014;14(4):319–26.

World Health Organization. Plague [fact sheet]. Geneva: World Health Organization; 2016. Mar 15 [updated 2022 Jul 7; cited 2025 Jun 10]. Available from: https://www.who.int/news-room/fact-sheets/detail/plague.

19 Other *Yersinia* Infections

Prathyusha Kokkayil

Learning Objectives

1. To understand the epidemiology of *Yersinia enterocolitica* and *Yersinia pseudotuberculosis*
2. To describe the pathogenesis of these *Yersinia* infections, focusing on key virulence factors and their mechanisms of immune evasion
3. To learn diagnostic strategies for *Yersinia* infections, including culture techniques, molecular methods, and serological approaches, while recognising their limitations

Introduction

Yersinia enterocolitica and *Yersinia pseudotuberculosis* are emerging zoonotic pathogens of considerable public health concern. These cause a wide range of illnesses varying from self-limiting gastroenteritis to potentially life-threatening systemic infections. These Gram-negative bacteria, although differing in their epidemiology and clinical features, exhibit remarkable adaptations. These allow them to thrive in a variety of hosts, from livestock and wildlife to humans and to persist in environmental reservoirs.

P. Kokkayil (✉)
Department of Microbiology, All India Institute of Medical Sciences, Patna, Bihar, India
e-mail: drprathyushak@aiimspatna.org

Y. enterocolitica, a major cause of food-borne yersiniosis, is closely associated with pork production chains. In contrast, *Y. pseudotuberculosis* typically follows a sylvatic cycle in rodents, occasionally spilling over to humans via contaminated food and water. Their pathogenesis relies on different virulence mechanisms, including plasmid-encoded type III secretion systems. These mechanisms subvert host immune defences. The thermoregulated genes of the bacteria aid survival across diverse ecological settings.

Despite progress in diagnostic techniques, these pathogens remain challenging to detect due to their fastidious growth, non-specific symptoms, and frequent underreporting. All these factors often contribute to delayed or missed diagnoses. The increasing threat of antimicrobial resistance further complicates treatment options, highlighting the urgent need for robust, evidence-based clinical guidelines.

History

The discovery of *Y. enterocolitica* and *Y. pseudotuberculosis* occurred through early twentieth-century investigations into animal diseases. *Y. pseudotuberculosis* was first isolated in 1883 by Malassez and Vignal from tubercular-like lesions in guinea pigs. It was initially misclassified as a mycobacterial species due to its granulomatous

S. C. Parija, A. Poddar (eds.), *Textbook of Bacterial Zoonoses*, Microbial Zoonoses,
https://doi.org/10.1007/978-981-95-6615-0_19

pathology. *Y. enterocolitica* was later identified in New York in 1939 by Schleifstein and Coleman during an outbreak of diarrhoeal illness mimicking appendicitis. However, its taxonomic status remained unclear until the 1960s, when the work of Frederiksen established it as a distinct species within the Enterobacteriaceae family.

The zoonotic potential of these pathogens was increasingly recognised during the 1970s and 1980s. These studies linked human infections to animal reservoirs, particularly pigs, the primary reservoir for *Y. enterocolitica,* and wild rodents for *Y. pseudotuberculosis.* Autoimmune sequelae, such as reactive arthritis and erythema nodosum, were first associated with *Y. enterocolitica* serotype O:3 infections during the 1980s, coinciding with discoveries of molecular mimicry between bacterial lipopolysaccharides and human antigens. Advances in genomic characterisation during the 1990s identified plasmid-encoded virulence factors, such as Yop effectors, shared with *Y. pestis.*

Taxonomy

The genus *Yersinia,* belonging to the family Yersiniaceae within the order Enterobacterales, comprises several species of Gram-negative bacilli.

Y. enterocolitica is highly heterogeneous, with multiple bio groups and over 70 serotypes. Of these, bio group 1B (serotype O:8) and bio group 4 (serotype O:3) are most associated with human disease. *Y. pseudotuberculosis*, more genetically like *Y. pestis*, primarily infects wild rodents and birds but can occasionally cause severe disease in humans.

Beyond these two species, several other *Yersinia* spp. have been isolated from clinical, animal, or environmental sources, though their pathogenic potential is generally lower or uncertain. These include *Yersinia intermedia, Yersinia frederiksenii, Yersinia kristensenii, Yersinia bercovieri*, and *Yersinia mollaretii*, many of which were previously considered bio groups of *Yersinia enterocolitica* but have since been reclassified based on molecular and phenotypic criteria. *Yersinia ruckeri* is unique in that it is a fish pathogen, causing enteric red mouth disease in salmonids, and is not known to cause human disease.

Genomics and Proteomics

Genomic and proteomic studies of *Yersinia* species other than *Yersinia pestis* have revealed evolutionary adaptations that distinguish these organisms from the plague bacillus and underpin their success as enteric and zoonotic pathogens. In particular, *Yersinia enterocolitica* and *Yersinia pseudotuberculosis* retain larger and more diverse genomes than *Y. pestis,* reflecting their ability to survive in environmental reservoirs, animal hosts, and the human gastrointestinal tract. The genome shares a 4.6–4.8 Mb chromosome encoding ~4000 genes and exhibiting significant genomic plasticity. *Y. pseudotuberculosis* possesses a highly conserved genome (97% ANI with *Y. pestis*) with minimal pseudogenisation. *Y. enterocolitica* shows greater diversity across its six biotypes. Pathogenic biotype 1B strains harbour unique pathogenicity islands (e.g. HPI for *Yersiniabactin* synthesis and Ysa encoding a SPI-1-like T3SS) absent in environmental biotypes. Both species carry the 70-kb pYV virulence plasmid encoding the Ysc-Yop type III secretion system (T3SS), but *Y. enterocolitica* additionally possesses chromosomal adaptations like *ail* (attachment invasion locus) and *ystA* (heat-stable enterotoxin), enabling mucosal invasion and diarrheagenic toxicity.

Proteomic analyses reveal a temperature-dependent regulation of virulence and survival factors in Yersinia species. At 37 °C, pathogenic strains upregulate Yersinia outer proteins (YopH, YopE) via the type III secretion system to inhibit phagocytosis and suppress inflammasome activation, facilitating immune evasion during host infection. In contrast, at 25 °C—reflecting environmental conditions—these bacteria predominantly express cold-shock proteins such as CspA and biofilm-associated components like Hms proteins, which enhance environmental persistence and transmission potential. Mass spectrometry-based studies have further characterised species-specific secretomes: *Y. enterocolitica* notably produces Yst enterotoxin during intestinal infection, while *Y. pseudotuberculosis* expresses invasin (Inv), a key adhesin mediating translocation across M cells in the intestinal epithelium. Additionally, antimicrobial resistance genes, including blaA and qnr, are being increasingly identified on mobile plasmids in clinical isolates.

Morphology

Y. enterocolitica and *Y. pseudotuberculosis* are non-spore-forming, facultative anaerobes. They are Gram-negative coccobacilli measuring 0.5–1.0 µm in diameter and 1–3 µm in length. They exhibit pleomorphic morphology that varies from ovoid rods to spherical forms depending on growth conditions. *Y. pseudotuberculosis* often appears more uniformly rod-shaped compared to the slightly curved or straight bacilli of *Y. enterocolitica*. Both species lack prominent capsules, unlike *Y. pestis*. However, they may produce a loose slime layer in vitro.

They demonstrate bipolar staining with Giemsa or Wright stains, giving a characteristic "safety pin" appearance under light microscopy due to concentrated chromatin at the poles. Ultrastructurally, they possess a typical Gram-negative cell wall. The cell wall consists of an outer membrane containing lipopolysaccharide (LPS). The O-antigen of LPS determines serotype specificity. The bacteria exhibit peritrichous flagella at 25–30 °C that confer motility in environmental reservoirs. However, this motility is lost at 37 °C during human infection due to thermoregulation of flagellar genes.

Y. enterocolitica is distinguished from *Y. pseudotuberculosis* by its thicker and more irregular outer membrane invaginations visible via transmission electron microscopy. Additionally, *Y. enterocolitica* has a tendency to form smaller, smoother colonies (1–2 mm diameter) on blood agar compared to the larger and more mucoid colonies of *Y. pseudotuberculosis* after 24–48 h of incubation.

Cultivation

Y. enterocolitica and *Y. pseudotuberculosis* are facultative anaerobes that can grow on standard media. Both species grow on blood agar and MacConkey agar at 28–30 °C and produce small (1–2 mm), non-haemolytic, greyish colonies after 24–48 h of incubation. Nevertheless, *Y. pseudotuberculosis* often appears more mucoid due to its polysaccharide slime layer.

Cefsulodin-irgasan-novobiocin (CIN) agar is the selective medium essential for isolation from stool or food samples. This suppresses gut flora while allowing *Yersinia* to form characteristic "bull's-eye" colonies with red centres and translucent peripheries due to mannitol fermentation. Cold enrichment at 4 °C in phosphate-buffered saline (PBS) or peptone broth for 1–3 weeks significantly improves recovery from low-inoculum specimens. This property is based on exploiting their ability to proliferate at refrigeration temperatures, a trait linked to their food-borne transmission. For environmental monitoring, trypticase-soy-yeast extract broth with prolonged incubation enhances detection in water or soil.

Optimal growth of *Y. enterocolitica* biotype 1B strains occurs at pH 7.0–8.0, requiring calcium-depleted conditions for expression of virulence plasmids (pYV). These are maintained at 37 °C but lost during routine subculture unless the media contain magnesium oxalate or Congo red. *Y. enterocolitica* is urease-positive and ferments sucrose. *Y. pseudotuberculosis* is urease-negative and sucrose-negative but produces a heat-stable enterotoxin detectable in specialised assays.

Animal Models

Animal models are important in elucidating the pathogenesis and host immune responses to *Y. enterocolitica* and *Y. pseudotuberculosis* infections.

Mice are the most widely used models. Oral inoculation with *Y. enterocolitica*, particularly biotype 1B strains like WA-314, replicates human enterocolitis, causing weight loss, diarrhoea, and systemic spread to mesenteric lymph nodes, spleen, and liver. *Y. pseudotuberculosis* infection in mice mimics human pseudoappendicitis and granulomatous inflammation, with strains like IP32953 inducing necrotic foci in Peyer's patches and spleen.

Temperature-sensitive virulence plasmid (pYV) expression is routinely studied in murine models. In these models, infections at 26 °C (permissive for plasmid loss) versus 37 °C (promoting plasmid retention) demonstrate the essential role of Yops (*Yersinia* outer proteins) in immune evasion.

Guinea pigs and rabbits, though less common, develop more severe systemic infections resembling human septicaemia, particularly with *Y. pseudotuberculosis* serotype I strains. It causes tubercle-like lesions in visceral organs, recapitulating the historical observations that led to its initial misclassification as a *Mycobacterium*. Pigs that serve as natural reservoirs for *Y. enterocolitica* are used to study persistent intestinal colonisation and faecal shedding, providing critical data on transmission dynamics. Zebrafish larvae offer a transparent vertebrate model for real-time imaging of neutrophil and macrophage interactions during early infection.

Transmission Cycle

Y. enterocolitica primarily circulates in swine populations, where asymptomatic carriage in tonsils and intestines leads to faecal contamination of meat, especially pork and dairy products. The infective stage appears when psychrotrophic strains multiply in refrigerated foods (4 °C), reaching doses sufficient to cause human infection (>10^4 CFU/g). Human transmission occurs by ingestion of undercooked pork, unpasteurised milk, or contaminated water. Acidic resistance allows the bacteria to survive gastric passage before adhering to ileal M cells. This is followed by translocation to the mesenteric lymph nodes.

In contrast, *Y. pseudotuberculosis* maintains a sylvatic cycle in wild rodents (e.g. voles, rabbits) and birds. The bacteria are excreted in faeces, contaminating soil and leafy vegetables. Its infective stage involves biofilm formation on plant surfaces and cold-adapted growth in stored produce. Human outbreaks linked to raw carrots, lettuce, or untreated spring water.

Transmission is facilitated by the high infectious dose and environmental hardiness, as both species survive for months in moist soils at low temperatures. Zoonotic spillover occurs through direct contact with livestock (*Y. enterocolitica*) or wildlife (*Y. pseudotuberculosis*). Person-to-person transmission is rare but documented in faecal-oral outbreaks.

An overview of the transmission cycle is presented in Fig. 19.1.

Pathogenesis and Pathology

In humans, after ingestion, both *Y. enterocolitica* and *Y. pseudotuberculosis* target M cells within the Peyer's patches of the ileum. They use the surface proteins invasin (Inv) and YadA adhesins to bind β1 integrins on host cells. This initiates actin-mediated endocytosis and enables the bacteria to cross the intestinal epithelium.

Once inside the host, the bacteria deploy a type III secretion system (T3SS), injecting Yop effector proteins into host macrophages and neutrophils. Both species evade complement-mediated killing through plasmid-encoded outer membrane proteins such as Ail and YadA. Their ability to survive within macrophages as small colony variants contributes to chronic or recurrent infections.

Y. enterocolitica primarily causes enterocolitis. It is characterised by neutrophilic infiltration, mucosal ulceration, and oedema in the terminal ileum, often mimicking acute appendicitis ("*pseudo-appendicitis*"). *Y. pseudotuberculosis* typically induces granulomatous inflammation in mesenteric lymph nodes, liver, and spleen. It closely resembles tuberculosis histologically, with central necrosis surrounded by epithelioid macrophages and lymphocytes. Though systemic spread is uncommon, it can occur, particularly in immunocompromised individuals.

Pathologically, *Y. enterocolitica* infection is marked by acute inflammation with microabscess formation in the gut. *Y. pseudotuberculosis* produces caseating granulomas and lymphoid hyper-

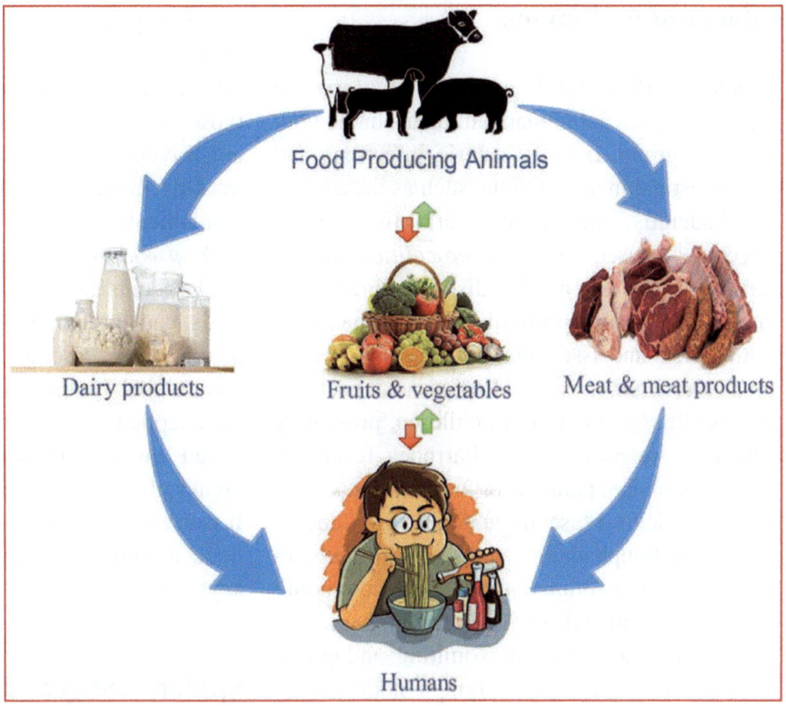

Fig. 19.1 Transmission cycle of Yersiniosis. (Courtesy: Muhammad Shoaib et al., RSC Advances, 2019)

plasia, reflecting its capacity for intracellular persistence.

The differing tissue tropisms and host responses underscore the evolutionary specialisations of these pathogens. *Y. enterocolitica* is better adapted for extracellular colonisation of the gastrointestinal tract, while *Y. pseudotuberculosis* exhibits a more systemic, intracellular mode of infection.

Immunology

Innate immunity in murine typhus begins with recognition by Toll-like receptors (TLR2/TLR4) and NOD-like receptors (NOD1/NOD2), which detect bacterial lipopolysaccharide (LPS) and peptidoglycan. This triggers NF-κB activation and the release of proinflammatory cytokines (IL-1β, IL-6, TNF-α), promoting neutrophil recruitment to the intestinal mucosa. However, *Yersinia* counteracts this response by injecting Yop effector proteins (YopJ, YopP) through its type III secretion system (T3SS). This inhibits Mitogen-Activated Protein Kinase (MAPK) and NF-κB signalling in macrophages and dendritic cells, creating an immunosuppressive environment. The host responds with a Th1-polarised immune response, where IFN-γ activates macrophages to control bacterial replication, and $CD8^+$ T cells target infected cells via MHC-I presentation. *Y. pseudotuberculosis* further evades immune detection by using YopH to downregulate MHC-I expression, impairing $CD8^+$ T cell responses.

Humoral immunity develops against outer membrane proteins (YadA, Inv) and LPS. However, antibodies offer limited protection due to antigenic variation and the bacteria's ability to resist complement-mediated lysis via Ail and YadA.

Postinfectious autoimmune complications like reactive arthritis, linked to HLA-B27, occur due to molecular mimicry between bacterial heat-shock proteins (Hsp60) and host antigens, driving cross-reactive T cell responses. Immunocompromised individuals exhibit severe outcomes due to impaired Th1 responses, highlighting the critical role of IFN-γ in controlling infection.

Infections in Humans

Y. enterocolitica and *Y. pseudotuberculosis* are the most significant human pathogens in the non-*Y. pestis* group, causing enteric infections and extraintestinal manifestations such as mesenteric lymphadenitis and reactive arthritis. Human infections caused by *Y. enterocolitica* and *Y. pseudotuberculosis* exhibit distinct clinical features, although both primarily target the gastrointestinal tract and associated lymphoid tissues.

Y. enterocolitica is a leading cause of bacterial enterocolitis, particularly in children, presenting with acute watery or bloody diarrhoea, fever, and severe abdominal pain. The condition often mimics appendicitis ("pseudo-appendicitis") due to mesenteric lymphadenitis and terminal ileitis. In contrast, *Y. pseudotuberculosis* more commonly causes a systemic febrile illness accompanied by right-sided abdominal pain, vomiting, and granulomatous inflammation of lymph nodes, liver, and spleen. The condition occasionally mimics tuberculosis or Crohn's disease on histological examination.

Both species have the capacity to invade the bloodstream, leading to septicaemia in immunocompromised individuals or those with chronic liver disease, with mortality rates reaching up to 30% in untreated cases. Postinfectious complications include reactive arthritis. It is reported in 10–30% of cases and is associated with HLA-B27. Erythema nodosum is another complication more frequently observed in women. These two complications are suggested to result from molecular mimicry between bacterial lipopolysaccharides (LPS) and host antigens, leading to autoimmune responses. Rarely, glomerulonephritis or myocarditis, resulting from autoimmune responses, are other complications triggered by bacterial antigens.

Infections in Animals

Y. enterocolitica primarily colonises pigs, where it persists asymptomatically in the tonsils and intestines, leading to widespread faecal shedding and contamination of pork products. The infection in pigs is usually subclinical. Nevertheless, outbreaks of enteritis and mesenteric lymphadenitis may occur in young piglets, particularly those infected with virulent biotype 1B strains. *Y. enterocolitica* is infrequently associated with mastitis in dairy cattle, posing a risk of milk contamination.

Y. pseudotuberculosis predominantly infects wild rodents such as voles, mice, rabbits, and birds, causing epizootics. The condition is characterised by caseous granulomas in the liver, spleen, and lymph nodes, often progressing to fatal septicaemia in stressed or immunocompromised animals. It can occasionally cause outbreaks of enteritis and abortion in goats and deer. It causes acute hepatosplenic necrosis and high flock mortality among avian hosts such as turkeys, and pigeons.

Epidemiology and Public Health

Y. enterocolitica and *Y. pseudotuberculosis* exhibit distinct epidemiological patterns shaped by their reservoir hosts, transmission routes, and environmental persistence.

Y. enterocolitica is a leading cause of bacterial gastroenteritis in temperate regions, with an estimated incidence of 3–10 cases per 100,000 population in Europe and North America. Serotyping reveals geographic variations, for example, *Y. enterocolitica* O:3 dominates in Europe/Asia, O:8 in North America, and O:9 in pig-farming regions. The condition is primarily linked to pork consumption, particularly undercooked chitterlings or contaminated processed meats and unpasteurised dairy products. In contrast, *Y. pseudotuberculosis* displays a more sporadic and wildlife-driven epidemiology. Cases are clustered in rural areas where humans encounter infected rodents or consume contaminated vegetables and untreated water, notably in Scandinavia, Japan, and Russia.

Food-borne transmission, primarily through the consumption of contaminated pork and dairy products, has become a significant route of human *Yersinia* infection. Both bacteria thrive in

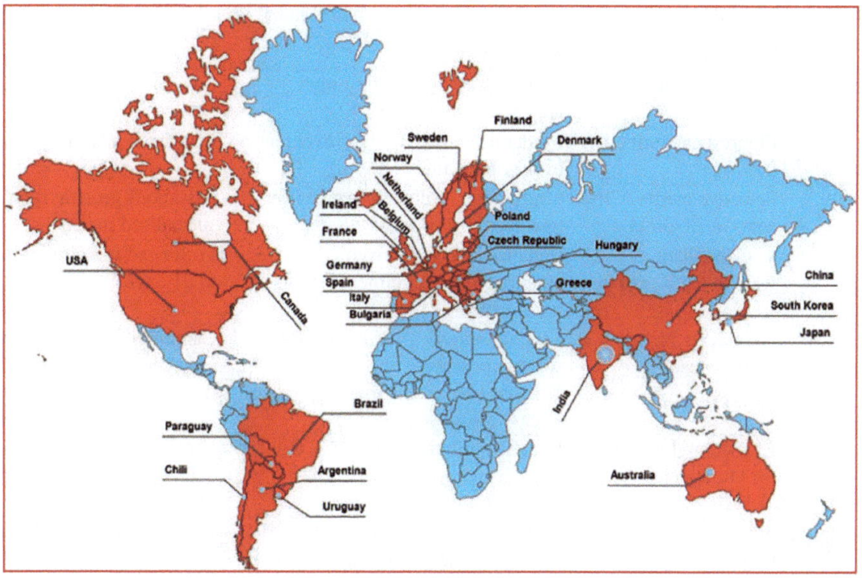

Fig. 19.2 Global epidemiology of Yersiniosis. (Courtesy: Muhammad Shoaib et al., RSC Advances, 2019)

cold climates. *Y. enterocolitica* infections increase in winter, reflecting pork slaughtering cycles, and *Y. pseudotuberculosis* in spring, coinciding with rodent population surges.

Virulence factors like the pYV plasmid and yersiniabactin, essential for systemic spread and iron uptake, are upregulated at 37 °C, enhancing bacterial adaptation within the host following transmission.

Global epidemiology of *Yersinia* infection is presented in Fig. 19.2.

Diagnosis

The diagnosis of *Y. enterocolitica* and *Y. pseudotuberculosis* infections necessitates a multifaceted strategy due to their fastidious growth requirements and clinical overlap with other enteric pathogens (Table 19.1).

Microscopy

Initial microscopy by Giemsa or Wright stains exhibits Gram-negative coccobacilli, giving a characteristic "safety pin" appearance. However, this lacks specificity. Direct fluorescent antibody (DFA) staining marginally enhances sensitivity but remains adjunctive.

In Vitro Culture

Culture methods constitute the diagnostic cornerstone, employing selective media such as cefsulodin-irgasan-novobiocin (CIN) agar. This produces characteristic "bull's-eye" colonies (deep red centres with translucent peripheries) after 24–48 h of incubation at 28–30 °C. Cold enrichment at 4 °C for 1–3 weeks in phosphate-buffered saline or peptone broth significantly improves recovery from paucibacillary specimens such as stool, blood, or food samples. Automated blood culture systems reliably detect bacteraemic cases. Subculture at both 25 °C and 37 °C improves yield, as some virulence factors (e.g. Yops) are temperature-regulated.

Biochemical differentiation is critical. *Y. enterocolitica* is urease-positive and ferments sucrose, whereas *Y. pseudotuberculosis* is urease-negative, sucrose-negative but oxidase-positive.

Table 19.1 Summary of diagnostic methods for *Yersinia enterocolitica* and *Y. pseudotuberculosis*

Method	Principle/specimen	Key features	Limitations
Microscopy	Gram stain of stool/tissue	Bipolar staining ("safety-pin" appearance)	Low specificity; adjunctive only
DFA staining	Fluorescent antibody labelling	Moderate sensitivity improvement over gram stain	Limited availability; rarely used
Culture	Selective media (CIN agar) at 28–30 °C	"Bull's-eye" colonies (red centre/translucent rim) in 24–48 h	Fastidious growth; requires specialised media
Cold enrichment	4 °C incubation (1–3 wks) in PBS/peptone	Enhances recovery from paucibacillary specimens (stool, food, water)	Prolonged turnaround time (up to 21 days)
Biochemical profiling	Urease, sucrose, oxidase tests	*Yersinia enterocolitica*: Urease+/sucrose positive *Yersinia pseudotuberculosis*: Urease-/oxidase positive	Cannot differentiate biotypes/serotypes
Serology	ELISA/agglutination for O-antigens	Detects antibodies (e.g. O:3, O:9 for YE; O:1, O:3 for YP)	Cross-reactivity (e.g. *Brucella*); delayed seroconversion; retrospective only
PCR	*ail*, *yadA*, *virF* gene targets	Rapid (4–6 h); high sensitivity/specificity; detects culture-negative cases	Requires molecular lab facilities
Multiplex real-time PCR	Stool/blood	Simultaneous detection with other enteric pathogens (e.g. *Campylobacter*)	High cost; limited validation
Whole-genome sequencing	Isolates/clinical specimens	Outbreak tracing; resistance gene profiling (e.g. fluoroquinolone resistance)	Research/outbreak settings only; not routine

Serodiagnosis

Serodiagnosis is important for detecting *Y. enterocolitica* and *Y. pseudotuberculosis* infections in humans, especially when culture methods are inconclusive or when samples are collected late in the course of illness. These tests detect specific antibodies against *Yersinia* antigens, such as outer membrane proteins or lipopolysaccharides.

Agglutination tests or enzyme linked immuno sorbent assay (ELISA) are frequently used to detect antibodies against specific O-antigens (e.g. O:3/O:9 for *Y. enterocolitica*; O:1/O:3 for *Y. pseudotuberculosis*). However, the cross-reactivity with *Brucella, Salmonella*, and other Enterobacteriaceae, coupled with delayed seroconversion, limits the utility in acute settings. These tests also have limited sensitivity and specificity.

Molecular Diagnosis

PCR targeting virulence genes (*ail*, *inv*, *yadA*) or plasmid-borne loci (*virF*) is highly sensitive and specific for the detection of non-pestis *Yersinia* infections. This is sensitive for identification even in culture-negative cases. Multiplex real-time PCR panels facilitate simultaneous detection of *Y. enterocolitica* along with common diarrhoeal pathogens. PCR-based methods supplement traditional culture for rapid screening of high-risk samples, addressing the limitations of variable growth rates among strains.

Whole genome sequencing (WGS) is useful for outbreak tracing and resistance profiling, revealing emerging fluoroquinolone-resistant clones. Persistent challenges include the inhibitory effect of sodium polyanethol sulfonate (SPS) in standard blood culture media and limited access in resource-constrained laboratories.

Treatment

The primary objectives of treatment for *Y. enterocolitica* and *Y. pseudotuberculosis* infections are to (i) alleviate clinical symptoms, (ii) prevent complications such as septicaemia or reactive arthritis, and (iii) reduce bacterial shedding to limit transmission.

In most cases, especially mild gastroenteritis, the illness is self-limiting and may not require specific therapy. Antimicrobial treatment is reserved for severe cases, systemic involvement such as bacteraemia or septicaemia, or immunocompromised patients, such as those with diabetes, iron overload, or undergoing immunosuppressive therapy.

First-line antimicrobial agents include fluoroquinolones (e.g. ciprofloxacin 500 mg twice daily) and third-generation cephalosporins (e.g. ceftriaxone 2 g once daily). However, infections caused by *Y. enterocolitica* biotype 1B may require combination therapy due to intrinsic resistance to ampicillin and first-generation cephalosporins. In cases of bacteraemia or disseminated infection, prolonged intravenous antibiotic courses (typically 14–21 days) are preferred. Effective agents include gentamicin (5 mg/kg/day) or trimethoprim-sulfamethoxazole (TMP-SMX, 8–10 mg/kg/day of the TMP component), particularly in patients with endovascular involvement or hepatic abscesses.

The emergence of antimicrobial resistance, such as fluoroquinolone-resistant *Y. enterocolitica* in parts of Asia and β-lactamase-producing strains, underscores the importance of culture-guided therapy where feasible. In refractory cases, susceptibility testing for extended-spectrum penicillins (e.g. ticarcillin-clavulanate) or carbapenems (e.g. imipenem) is warranted.

Prevention and Control

Preventing *Y. enterocolitica* and *Y. pseudotuberculosis* infections requires a One Health approach, integrating food safety, animal reservoir management, and public health education to interrupt zoonotic transmission.

For *Y. enterocolitica*, strict pork hygiene, including thorough cooking (internal temperature ≥63 °C), avoidance of raw pork products (e.g. chitterlings), and prevention of cross-contamination in kitchens, as pigs represent the primary reservoir, is paramount. Pasteurisation of milk and dairy products eliminates the risk of food-borne outbreaks.

In the case of *Y. pseudotuberculosis*, rodent control in agricultural environments and thorough washing of raw vegetables are essential preventive strategies, as wild rodents and birds serve as key carriers. Refrigeration alone is inadequate, since both species can proliferate at 4 °C; therefore, cold-stored foods should be consumed promptly or treated with bacteriostatic agents. PCR-based surveillance of swine herds and slaughterhouse samples enables the early identification of contaminated meat before it enters the food chain.

Public health strategies prioritise high-risk groups, including children, immunocompromised individuals, and occupationally exposed workers (e.g. slaughterhouse staff, veterinarians, farmers), with emphasis on hand hygiene and protective measures. While no human vaccine exists, Yop-based subunit vaccines are under investigation. Rapid diagnostics and antimicrobial stewardship are essential for outbreak control.

Case Study

A 32-year-old farmer presented to a rural clinic with a 5-day history of high-grade fever (39.5 °C), severe right lower quadrant abdominal pain, and watery diarrhoea, following exposure to wild rodents during harvest season. Blood cultures remained negative after 48 h, but cold-enriched stool culture on CIN agar yielded small, lactose-negative colonies later confirmed as *Y. pseudotuberculosis* serotype O:3 by MALDI-TOF MS and PCR targeting the *inv* gene. The patient's occupational rodent exposure and consumption of unwashed field vegetables supported the diagnosis of pseudotuberculosis-associated mesenteric adenitis, distinct from classical appendicitis.

Treatment with intravenous ceftriaxone (2 g daily) led to fever resolution within 72 h. The patient subsequently developed reactive arthritis in both knees, a recognised postinfectious complication linked to HLA-B27 positivity confirmed by genetic testing.

Research Questions

1. How do climate change and agricultural intensification influence the geographic expansion and seasonal prevalence of *Y. enterocolitica* and *Y. pseudotuberculosis* in animal reservoirs and food systems?
2. What genetic and epigenetic mechanisms enable *Yersinia* spp. to transition between environmental survival and host invasion during infection?
3. Can PCR improve the sensitivity and speed of detecting *Yersinia* spp. in clinical and food samples compared to conventional culture?

Declaration During the preparation of this chapter, we have used Copy.ai's AI paragraph rewriter to improve the readability of the text. After using this tool, we reviewed and edited the content as needed. We take full responsibility for the content of the publication.

Further Readings

Bottone EJ, Bercovier H, Mollaret HH. Yersinia enterocolitica: the charisma continues. Clin Microbiol Rev. 1997;10(2):257–76.

European Food Safety Authority, Ricci A, Allende A. Scientific opinion on the public health risks of Yersinia spp. in food. EFSA J. 2013;11(4):3167.

Galindo CL, Rosenzweig JA, Chopra AK. Animal models of Yersinia pathogenesis. Eur J Microbiol Immunol. 2011;1(1):5–12.

Gkouletsos T, Patas K, Lambrinidis G, Neubauer H, Sprague LD, Ioannidis A. Antimicrobial resistance of Yersinia enterocolitica and presence of plasmid pYV virulence genes in human and animal isolates. New Microbes New Infect. 2019;32:100604.

Huovinen E, Sihvonen LM, Virtanen MJ. Symptoms and sources of Yersinia enterocolitica infection: a case-control study. BMC Infect Dis. 2010;10:122.

McNally A, Thomson NR, Reuter S, Wren BW. Genomic evidence for the emergence and evolution of pathogenicity and niche adaptation in Yersinia pseudotuberculosis. Infect Immun. 2016;84(6):1811–22.

Mühlenkamp M, Oberhettinger P, Leo JC. Yersinia enterocolitica targets cells of the innate and adaptive immune system by injection of Yops in a mouse infection model. PLoS Pathog. 2015;11(3):e1004735.

Pham JN, Bell SM, Lanzarone JY. Real-time PCR detection of Yersinia enterocolitica in food. Appl Environ Microbiol. 2011;77(19):6970–4.

Revazishvili T, Kotetishvili M, Stine OC, Kreger AS. Comparative analysis of multilocus sequence typing and pulsed-field gel electrophoresis for characterising Yersinia pseudotuberculosis. J Clin Microbiol. 2004;42(8):3814–22.

Rosner BM, Stark K, Werber D. Yersiniabactin reduces respiratory reactive oxygen species in Yersinia enterocolitica to promote epithelial cell invasion. PLoS Pathog. 2017;13(6):e1006447.

Rosner BM, Werber D, Höhle M. Mouse models for Yersinia enterocolitica infections. Curr Protoc Microbiol. 2010; Chapter 5:Unit 5A.3.

Sabina Y, Rahman A, Ray RC. Overview of Yersinia enterocolitica and molecular methods for its identification. Foodborne Pathog Dis. 2011;8(8):863–72.

Sulakvelidze A, Dalakishvili K, Barry E. Yersiniae other than Y. pestis: the ignored species. Microbes Infect. 2000;2(5):497–513.

Wren BW, Thomson NR, Holden MTG, Crossman LC. The Yersiniae – a model genus to study the rapid evolution of bacterial pathogens. Nat Rev Microbiol. 2003;1(1):55–64.

Brucellosis

20

Akanksha Gupta and Jyotsna Agarwal

Learning Objectives

1. To understand the epidemiology and transmission of brucellosis in the community
2. To describe the clinical features of brucellosis and the diagnostic tests available for it
3. To discuss the available vaccines for preventing brucellosis in animals and to analyse ways to prevent and control brucellosis in humans

Introduction

Brucellosis is a zoonotic infection caused by species of the genus *Brucella*. It primarily infects livestock and wildlife, mainly cattle, sheep, goats, swine, and horses. The major modes of transmission of brucellosis from infected animals to humans include consumption of unpasteurised milk and other dairy products, inhalation of aerosolised products, direct contact with infected body fluids of animals, and unsafe handling of *Brucella* samples in the laboratory. It is of great concern for public health and for economic losses. It is of major importance in countries where farming and livestock handling are very predominant activities and where food safety practices are not ensured.

Brucellosis is endemic in the Mediterranean, South and Central America, the Middle East, and all states of India. In animals, it mainly affects the reproductive system, presenting with infertility, reduced milk production, abortions, etc. Human brucellosis presents as pyrexia of unknown origin, with severity ranging from flu-like symptoms to many serious complications. Diagnosis is based on clinical features and a history of exposure to infected animals. Isolation of *Brucella* from blood/bone marrow/body fluids, serological assays, and molecular-based tests are the frequently used tests for the laboratory diagnosis of brucellosis. Treatment relies on a combination of antibiotics such as streptomycin/rifampicin and doxycycline. Prevention and control of brucellosis focus on eliminating infection from animals and preventing their spread from infected animals to humans.

History

Brucellosis dates to the 1850s, when many cases of the 'mysterious fevers' were seen in Malta during the Crimean War (1853–1856). The local names for this fever were 'Malta fever', 'Gibraltar fever', and 'Cretan fever'. Principal Medical officer to the Indian Army, Jeffery Allen Marston, suffered from similar illness and reported it in himself. It gained wide recognition when the British Royal Army Medical Corps (RAMC)

A. Gupta · J. Agarwal (✉)
Department of Microbiology, Dr. Ram Manohar Lohia Institute of Medical Sciences, Lucknow, Uttar Pradesh, India

S. C. Parija, A. Poddar (eds.), *Textbook of Bacterial Zoonoses*, Microbial Zoonoses,
https://doi.org/10.1007/978-981-95-6615-0_20

started to serve the Mediterranean region after the Crimean War. The causative organism was isolated for the first time by British physician David Bruce in 1887 from four fatal cases. He named it *Micrococcus melitensis*. David Bruce suggested the name 'Mediterranean fever' in 1903.

Maltese scientist Themistocles Zammit made the first causal association of Malta fever with goat milk. In 1905, Zammit demonstrated that the organism causing Malta fever was found in goats and proposed that the consumption of unpasteurised goat milk by the Maltese population and the British military was the cause of Malta fever. Karl Meyer, in 1918, honoured David Bruce and named the organism '*Brucella*'. In 1918, The locally named 'Malta/Undulant/ Mediterranean fever' was replaced with 'Brucellosis', as proposed by Alice Evans, an American microbiologist.

Brucellosis is a well-studied zoonotic infection of major public health importance, with a well-developed laboratory diagnosis. The practice of pasteurising milk before its consumption brought down human brucellosis cases to a significantly lower level.

Taxonomy

The genus *Brucella,* based on 16 s rRNA analysis, belongs to the family Brucellaceae, phylum Proteobacteria, class Alphaproteobacteria, and order Rhizobiales (Fig. 20.1). The genera *Brucella* and *Bartonella* differ from other closely related genera of the order Rhizobiales in being infective to mammals. *Brucella* is differentiated from *Bartonella* in being facultatively intracellular, as opposed to the latter being an obligate intracellular pathogen. Also, the genome size of the former is 50–100% bigger than *Bartonella*.

Currently, it has been proposed to merge the related genera *Ochrobactrum* and *Brucella* into a single genus, *Brucella*. The genus *Brucella* is presently classified into two categories: the formerly *Brucella* genus classified into the category of facultative intracellular organisms and the previously *Ochrobactrum* genus classified into the extracellular category. Based on whole genome sequencing (WGS), the facultative intracellular *Brucella* are further divided into phylogenetic clusters, namely classical intracellular *Brucella,* and non-classical intracellular *Brucella* (Fig. 20.1). 'Classical' refers to the species having >99% DNA similarity with the *B. melitensis* reference genome, while 'non-classical' species are the ones with DNA similarity of >97%. The former comprises the six initially identified species of *Brucella*, which are restricted to mammals. The latter consists of phylogenetically distant species, newer species that are isolated from various vertebrates and not just limited to mammals.

The taxonomic structure of *Brucella* is based on phenotypic characteristics, while speciation is done based on the organism's predilection towards its natural host. Some species, even though phylogenetically very close to each other, with over 97% similarity, differ in their host preference, virulence, and zoonotic potential.

The genus *Brucella* has 12 recognised species, which are known to cause brucellosis in various animals, and some of them can cause zoonotic transmission to humans. These species can be distinguished by host preference, pathogenicity, and genetic differences, although all belong to a highly conserved genus.

These species include *Brucella melitensis* (affecting sheep and goats), which is most pathogenic to humans; *Brucella abortus* (cattle); *Brucella ovis* (sheep); *Brucella canis* (dogs); *Brucella neonatomae* (desert wood rats); *Brucella suis* (pigs); *Brucella pinnipedialis* (seals and other pinnipeds); *Brucella ceti* (dolphins and whales); *Brucella microti* (voles); *Brucella papionis* (baboons); and *Brucella vulpis* (red foxes).

B. melitensis was the first *Brucella* species to be identified. Species are further divided into multiple biovars (Table 20.1).

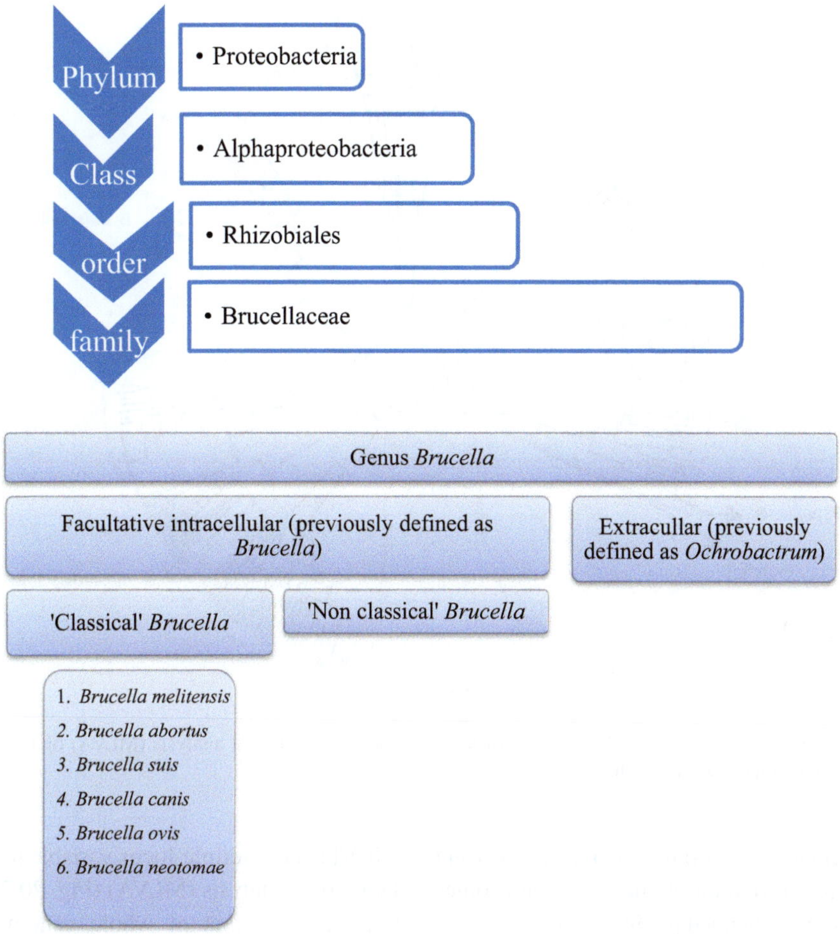

Fig. 20.1 Taxonomic classification of *Brucella*. (Image source: Authors' personal collection)

Table 20.1 Biovars and principal hosts of the six 'classical' *Brucella* species

Name of species	Number of biovars[a]	Principal host
Brucella melitensis	3 (1–3)	Sheep and goats
Brucella abortus	8 (1–7, 9)	Cattle
Brucella suis	5 (1–5)	Swine
Brucella ovis	Not yet defined	Sheep
Brucella canis	Not yet defined	Dogs
Brucellaneotomae	Not yet defined	Wood desert rats

[a]Total numbers of biovars in the given species (parentheses indicate the actual number given to various biovars). For example, *Brucella melitensis* has 3 biovars: biovar 1, biovar 2, and biovar 3. Similarly, *Brucella abortus* has 8 biovars, named biovar 1 to biovar 7 and biovar 9, and *Brucella suis* has 5 biovars, named biovar 1 to biovar 5

Genomics and Proteomics

The genome of *Brucella* is conserved with its G + C content of 57% mol. According to the genomic properties and molecular differentiation, *Brucella* is a monospecific genus. However, it is differentiated into various species based on its preferred natural host, phenotypic characteristics, and biological behaviours. The genome size of the *Brucella* genus is 3.1 Mb, which consists

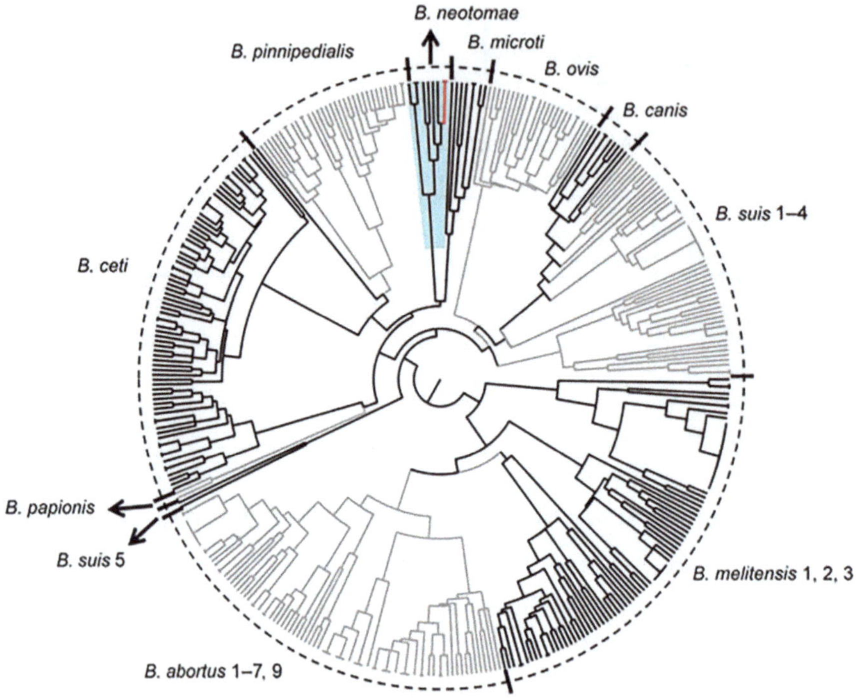

Fig. 20.2 Dendrogram based on multiple-locus variable number tandem repeat analysis (MLVA) of 16 loci panel of *Brucella* species. (Image source: CDC)

of two circular chromosomes of sizes 2.1 Mb and 1.2 Mb. This information about two chromosomes in *Brucella* was obtained in 1990 with the help of pulsed field gel electrophoresis (PFGE). The larger chromosome I (Chr1) spans approximately 2.1–2.2 metabases (Mb) and predominantly houses genes vital for fundamental physiological functions, core metabolic processes, and overall cell viability. In contrast, the smaller chromosome II (Chr2) primarily harbours genes associated with the bacterium's adaptability, virulence mechanisms, and intricate interactions with its host. While this two-chromosome structure is a hallmark of the genus, some variations do exist, such as in *B. suis* biovar 3, where the two chromosomes can fuse into a single replicon. Currently, the genetic diversity of *Brucella* is assessed by studying sequences at multiple genetic loci, using multi-locus sequence type (MLST) and multi-locus variable number of tandem repeats (MLVA) analysis.

It has been demonstrated that the dendrogram generated using multi-locus sequence typing (MLST) and multiple-locus variable number tandem repeat analysis (MLVA) (Fig. 20.2) does not correlate with that of whole genome analysis (WGA). Therefore, currently the most predominant technique used for analysing the genetic diversity of *Brucella* is WGS, using single--nucleotide polymorphism (SNPs).

Various proteomic studies have been carried out on the genus *Brucella* to identify proteins that play a role in virulence, intracellular survival, metabolic pathways, and diagnosis and vaccine candidates. Proteomic studies have shown that lipopolysaccharide (LPS), the type IV secretion system, and the BvrR /BvrS two-component system play an important role in intracellular survival and pathogenicity of *Brucella*.

A study was done to study the proteome of *B. melitensis* strain 16 M and *B. ovis* REO198 and found 274 proteins in the former and 606 in the latter. SDS-PAGE was used to separate the proteins of the above two species. The study found outer membrane proteins N8LTST7 in *B. melitensis* and Q45321 in *B. ovis* as important candidates for improving the diagnosis of brucellosis, which can also be used for vaccination against these species.

Morphology

The genus *Brucella* is non-acid-fast, non-sporing, and non-motile Gram-negative coccobacilli, though its cocci and bacillary forms also exist. *Brucella* species resist decolourisation by weak acids and alkalis. *Brucella* is not completely acid-fast, but it resists decolourisation by 0.5% acetic acid, as demonstrated by Stamp's modified cold Ziehl–Neelsen staining, and stains red. These are short and slender, with a straight axis and parallel sides; the sides may also be convex outside. The bacterium's length and breadth may range from 0.6 to 1.5 μm and 0.5 to 0.7 μm. The cells with smaller lengths appear as oval cocci, which may be arranged in singles or pairs and groups (Fig. 20.3). When grown in liquid media, they may form a short chain of 5–6 cells. When arranged in pairs, their division plane is in line with their long axis, as opposed to *Neisseria*, whose plane of division is perpendicular to its long axis. Morphologically, *B. melitensis* appears typically as a cell-wall–defective bacterium under a light microscope. Even bipolar staining can be seen.

Structural features are quite like other Gram-negative bacteria, like *Escherichia coli*, but some differences are also seen from those of *E. coli*. The outer membrane is made up of an LPS protein complex. This outer membrane is about 9 nm thick and encloses a 3–5 nm thick muramic acid layer; this layer is more predominant than that seen in *E. coli*. On electron microscopy, the LPS protein complex appears as a rugose layer. Beneath the outer membrane lies the periplasmic space. Small vacuoles and polysaccharide-containing granules are seen in the cytoplasm.

Cultivation

Brucella is classified as a biosafety level 2/3 pathogen according to the Indian Risk Group classification. *Brucella* should, therefore, be cultured/handled cautiously in a controlled environment in a BSL2 or BSL3 facility, depending on the nature of the work with the organism. Most *Brucella* species require extra nutrition, which is provided by supplementing culture media with blood, serum, and tissue extracts. Other nutrients added to cultivate different species of *Brucella* include amino acids, thiamine, biotin, nicotinamide, magnesium, etc.

Peripheral blood cultures are recommended to culture *Brucella*, especially during bacteraemia. *Brucella* is cultured from bone marrow, with higher sensitivity than blood cultures. Other samples from which *Brucella* can be cultivated

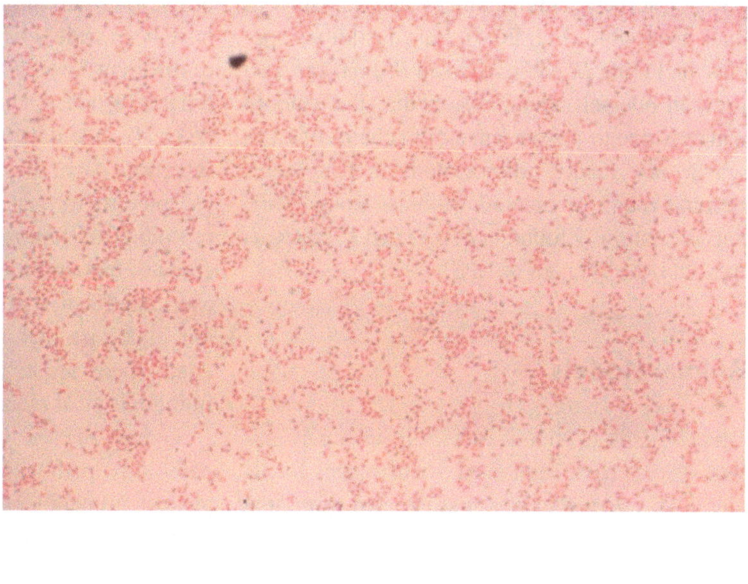

Fig. 20.3 Gram-stained image of *Brucella canis* showing Gram-negative coccobacilli (X1000). (Image source: Public Health Image Library CDC)

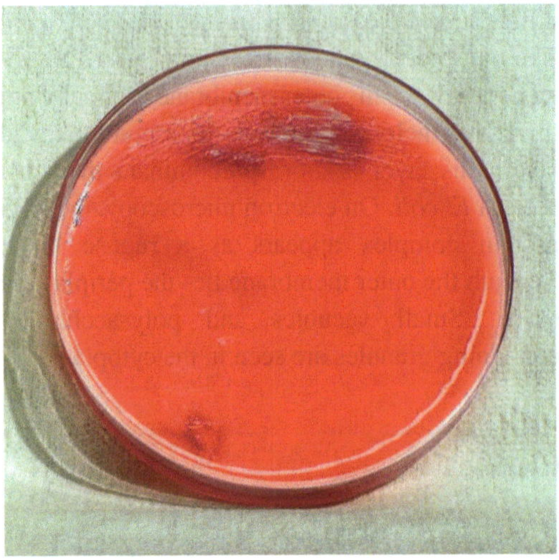

Fig. 20.4 Colony morphology of *Brucella melitensis* on sheep blood agar after 48 hours of incubation at 37 °C, showing smooth, round and moist colonies. (Image source: Public Health Image Library CDC)

include pus, CSF, and body fluids like synovial fluid and pleural fluid with varying sensitivity.

On suitable solid media, *Brucella* shows growth in 2–3 days. The preferred growth conditions are a temperature of 37 °C and 5–10% CO_2. Some strains of *Brucella* are more fastidious, like *B. ovis*, *B. abortus* biogroup 2, and even some strains of biogroups 3 and 4, and these require blood and serum in culture media to grow. *Brucella* grows well on serum dextrose agar, giving convex, circular, translucent colonies. Colonies may be smooth or non-smooth. Smooth ones are soft and emulsifiable, while non-smooth ones are friable (Fig. 20.4). Less fastidious species, including *B. melitensis* and *B. suis,* can also grow in bile salt-containing media. The smooth nature is due to the presence of O chains of LPS, which are readily dissociated, and colonies become rough or mucoid variants. Rough variants are naturally found in *B. ovis* and *B. canis*, as they lack the O chain of LPS.

Animal Models

Brucellosis is a zoonotic infection, and many animal species have been used as surrogates for its study since *Brucella* was first isolated from goat milk.

The various animal models studied include guinea pigs, mice, rats, rabbits, and non-human primates (NHP). The most studied animal species is the mouse model, while the least commonly studied model is that of rats and rabbits, as they are less susceptible to brucellosis. The most common route used for studying disease pathogenesis in mice is via intraperitoneal (i.p.) inoculation, while other routes used are intranasal, intravenous, aerosol, and oral routes. This is one of the most important drawbacks of using mice, as an oral or inhalational route of inoculation is a more biologically significant route of brucellosis transmission than the i.p. inoculation. The mouse model has various other disadvantages, including a lack of development of fever in mice following i.p. inoculation, while fever is a classic sign of brucellosis in humans.

Other models, like the *rhesus macaque*, have also been widely used. It has various advantages over mice, including the development of similar clinical illnesses to those seen in humans. It can be inoculated by routes that mimic natural infection in humans, including inhalation and via *Brucella*-contaminated milk. The major drawback of using NHPs like the *rhesus macaque* is the high cost associated with their care and animal husbandry.

Guinea pigs were the best animal model for studying brucellosis. Many studies have been done on them to discover *Brucella*'s pathogenesis and pathology. Guinea pigs have high susceptibility to different *Brucella* species. Thus, they have been widely used to determine the virulence of different *Brucella* strains and test vaccine candidates' efficacy. Unlike mouse models, they can be easily inoculated by routes mimicking natural infection. Also, they develop a very similar clinical picture to that seen in humans, including reproductive failure, as also reported in some cases of human brucellosis. The major drawback associated with this animal model is its large size and high cost, making its accommodation tough. Also, there are a few reagents used to study immunological reactions.

Guinea pigs are the best models to study the pathology, pathogenesis, and impact of vaccines against brucellosis, but a lack of investment in these animal models has been an important factor contributing to brucellosis being a neglected zoonosis.

Transmission Cycle

Brucellosis is a strict zoonosis in which infection reaches humans directly via infected animals or by consuming their products. *Brucella* species primarily infect mammals, including humans, domestic animals, and wildlife like feral swine. Different species have different principal animal hosts. Even marine mammals and rodents may be infected with *Brucella* species. Humans are susceptible to infection by all *Brucella* species and produce symptoms ranging in severity across various clinical manifestations.

Small animals may become infected by coming into direct contact with the body secretions of other infected animals or mostly by ingestion. Some animals, like sheep and goats, may remain infected for many years and continue to shed the infection in vaginal secretions.

Brucella is transmitted from animals to humans via various routes, including consumption of unpasteurised milk and other dairy products, inhalation of aerosols, and occupational exposure (Table 20.2).

Brucella survives in milk for up to two days if kept at 8 °C. It is stable for up to three weeks in frozen meat and in goat cheese for up to three months. It can survive in damp soil for more than 40 days. *Brucella* is sensitive to most commonly used disinfectants, pasteurisation, heat, and ionising radiation.

Pathogenesis and Pathology

Brucella, being a facultative intracellular microorganism, has the potential to produce chronic intracellular infections and proliferate within macrophages. The LPS is the major virulence factor. The other virulence factors include the type IV secretion system and two components of the regulatory system, BvrR/BvrS. There are four stages in the pathogenesis of brucellosis: adherence, invasion, establishment, and dissemination within a human host. Different species have different virulence in humans, with *B. melitensis* being the most pathogenic. *B. abortus* causes mostly subclinical cases and is less of a clinical infection, while *B. suis* produces infections of intermediate severity.

Brucella enters the human body through the alimentary tract, nasal mucosa, conjunctiva or skin. The first cells to encounter it are mononuclear or polymorphonuclear (PMNL) leukocytes, and they ingest the bacteria. *Brucella* survives within macrophages and phagocytes by releasing certain soluble substances to inhibit phagolysosome formation and inactivating the myeloperoxidase hydrogen peroxide-halide systems, thus inhibiting degranulation. Subsequently, bacteria disseminate to other human cells.

The exact mechanism of *Brucella* invasion and penetration is still to be fully understood. LPS plays an important role in the entry of bacteria into macrophages, immune evasion, and intracellular survival within the macrophages. After phagocytosis, they multiply in the local lymph nodes, which drain the site of infection. As the bacteria are released from the cells, they parasitise and enter the bloodstream, causing bacteraemia, which marks the acute febrile stage of the illness. Bacteraemia persists for some time, after which the bacteria are distributed throughout the reticuloendothelial system (RES), specifically localising within the liver and spleen. The other sites where the bacteria can localise include the central nervous system, genital tract, joints, bones, heart, spleen, lymph nodes, kidneys, and reproductive organs (especially seen in animals). The severity of the bacteraemia and the localised organs may vary among different species.

Immunology

Antibodies may start to develop after the acute phase, or, in subacute cases, antibody production may precede the onset of clinical symptoms. Antibodies function primarily to prevent reinfection. The cell-mediated immune (CMI) response and humoral immune response by production of antibodies are seen in most patients with brucel-

Table 20.2 Different modes of transmission of *Brucella* to humans

Mode of infection	Transmission	Remark
Direct contact	Infected animal or its body fluids, vaginal discharge, semen, products of abortion, etc.	Farmers, veterinarians, livestock handlers are at increased risk
Consumption of contaminated products	Raw or unpasteurised dairy products	Most common mode of transmission. Food safety practices are essential to prevent transmission by this route
Inhalation	Air-borne pathogen	Individuals working in slaughterhouses and meat processing areas are at increased risk
Conjunctiva	Aerosols or by fingers	Individuals handling carcass dressing or those helping in activities of animals may be exposed to body fluids of infected animals and are at increased risk
Intrauterine transmission	Transplacental	Very rare but can be transmitted from infected mother to her unborn child
Traveller's brucellosis	Contact with animal populations like swine, dogs and reindeer	May be infected with *Brucella suis* or *Brucella canis*
Ingestion	Undercooked meat of infected animals	Exposure risk is less likely since the bacterial load in the muscles is very less
person-to-person transmission	Blood transfusion, tissue transplantation, congenital, breastfeeding and sexual contact	Although reported, but very rare

losis. In humans, brucellosis leads to tissue lesions seen in the form of minute granulomas composed of epithelioid cells, PMNL, lymphocytes, and giant cells. Even though the granulomas formed in *B. melitensis* are very small, the toxaemia produced by this species is remarkably high.

Studies done on mouse models have shown the role of a stress-induced protein, Htr-A, in initiating a granulomatous reaction against *B. abortus* infection and suppressing the infection level during the early phase of the disease. However, it failed to reduce the subsequent number of infecting bacteria. Subsequent studies have demonstrated that mice deficient in this protein produced a disease like that seen in the wild-type against *B. abortus*. Many factors that play a role in containing brucellosis remain to be understood.

CMI is the most important component of immunity that helps to recover from brucellosis. This produces activated mononuclear lymphocytes, which have a major role in Th1 lymphocytes and interferon-gamma formation. Interferon-gamma activates the previously inactivated macrophages. Activated macrophages are more effective in phagocytosing and killing *Brucella*. They also present *Brucella* antigens to T cells, thus further aggravating the immune response against *Brucella*. Lymphokines derived from T lymphocytes attract various cells towards the foci of infection and thus lead to granuloma formation. This granuloma formation attracts activated macrophages to the different sites of infection where they are required.

Infection in Humans

Symptoms of brucellosis can range from a mild flu-like illness to a very severe form. Infection by all *Brucella* species except *B. melitensis* has a subacute, insidious onset and is less severe. *B. melitensis* brucellosis is acute and more severe.

Its incubation period ranges from 5 days to 6 months. The duration of the incubation period depends on various factors like the virulence of the infecting strain, inoculum size, host immunity, and the transmission route. *Brucella* can reach various organs via the lymphatic system and can multiply there, leading to infections ranging from localised to systemic infection.

Acute Brucellosis

This is characterised by intermittent fever with a maximum temperature ranging from 38 °C to 41 °C, or sometimes even higher. The temperature usually remains normal during the day while sharply rising by afternoon or evening, along with chills, fatigue, nausea, headache, joint pain, shivering, and malaise. After reaching a peak, the temperature falls rapidly. During the early phase of the disease, the fever follows a typical pattern that changes later in the course of the disease, with longer intervals between the rise and normal body temperature. The characteristic undulatingfever (rises and falls in waves) is seen only in cases that are not treated and have become chronic. Most patients may recover in 1–3 months; some symptoms may last for a longer duration.

Brucella species can involve multiple organs, including bones, joints, the cardiovascular system, the RES, the CNS, and the genitourinary system, during acute or subacute conditions. Bone and joint complications in the form of spondylitis or arthritis of large joints are seen in about 10% of cases. In 3–5% of *B. melitensis* cases, CNS invasion may lead to neurobrucellosis, which is an emergency condition requiring immediate treatment to prevent mortality or permanent disability. Cardiovascular complications include thrombophlebitis and endocarditis. Endocarditis may be the primary cause of death in some patients with brucellosis. Genitourinary complications in males include epididymal orchitis. Brucellosis during pregnancy leads to abortion.

Chronic Infection

In some patients, acute or subacute cases, if left untreated, may recur or persist for more than 12 months, and then they are known as chronic infections. Chronic brucellosis is a very loosely defined term with no clear-cut definition by the WHO or the CDC. Patients with chronic brucellosis may either have objective signs of infection and focal complications, or objective evidence of infection may be missing, with persistent symptoms of infection, and the patients may present with malaise, depression, and anxiety. In endemic regions, brucellosis is considered an important cause of pyrexia of unknown origin (PUO), which is defined as a fever of at least 101 °F (38.3 °C) on several occasions, lasting more than three weeks, and without an obvious diagnosis despite initial investigations.

Relapse

Relapse may be seen in 5–15% of individuals. It usually occurs within 6 months of completion of treatment, but can be seen for up to one year after treatment. It occurs due to inadequate treatment in the form of an improper antibiotic regimen, inaccurate dosage or timing of antibiotics, lack of adherence to treatment, or the presence of localised foci.

Infection in Animals

The most common system involved in animals is the reproductive system, presenting as abortions, stillbirths, or infertility in females and swollen testicles in males. Other symptoms include arthritis, fatigue, skin lesions, decreased milk production, swollen liver and lymph nodes, decreased weight gain, and abdominal pain. Poor reproductive health in animals leads to substantial economic losses in the dairy and farming industries. In horses, swelling of the neck and back may occur, known as poll evil or fistulous withers (oozing lesions on the skin).

B. suis (swine), *B. abortus* (cattle), and *B. melitensis* (sheep and goats) are notifiable diseases, being reported to the World Organisation for Animal Health (WOAH).

Epidemiology and Public Health

Brucellosis is a strict zoonotic disease that is transmitted to humans by the consumption of unpasteurised milk and other dairy products and very rarely by person-to-person transmission via blood transfusion, sexual intercourse, breastfeeding, or direct exposure to contaminated material. Even though it can affect any age group, certain people may be at higher risk, like those involved in handling infected animals or their products, like veterinarians, livestock handlers, abattoir workers, farmers, and laboratory personnel, due to close contact with contaminated samples. Many cases of brucellosis have been seen in wildlife, including feral swine, thus putting hunters at an increased risk of contracting brucellosis.

Brucellosis primarily affects livestock, including sheep, goats, cattle, pigs, and horses. Though present globally (Fig. 20.5), it is of major concern in countries with high populations of livestock and poor safety measures. The condition can affect people of any age group. According to CDC data, 500,000 new cases are reported every year globally, with 2.4 billion people at risk. This may be an underestimation, as most cases might be left undiagnosed due to (i) the lack of a specific clinical picture, (ii) the lack of awareness amongst clinicians, and (iii) a deficiency of well-developed diagnostic laboratories for brucellosis. The most frequently reported case throughout the world is caused by *B. melitensis*, while the most potent source of infection is *B. abortus*.

Human infections are more common in people living in endemic areas or travelling to endemic areas. Brucellosis in sheep, goats, and cattle is endemic in Asia, Central and South America, Africa, and Eastern Europe. Brucellosis in pigs is endemic in North America. Countries with the maximum reported brucellosis cases are Iran, Armenia, Uzbekistan, Tajikistan, and Azerbaijan.

India has shown significant seroprevalence of brucellosis in animals, with about 0.5–8% seroprevalence seen in humans. India is endemic for brucellosis, with the highest number of cases reported from the north-eastern states. Studies have shown seasonal brucellosis trends with maximum cases occurring from March to June, showing a rise during spring, especially in endemic areas.

Effective control of brucellosis is essential, especially in endemic regions. It is an important public health concern, as it can cause severe disabilities and complications in humans.

Diagnosis

Diagnosing human brucellosis relies mainly on microbiological investigations, a very high clinical suspicion, and accurate history taking, including occupational history, travel history, and any history of exposure to infected animals.

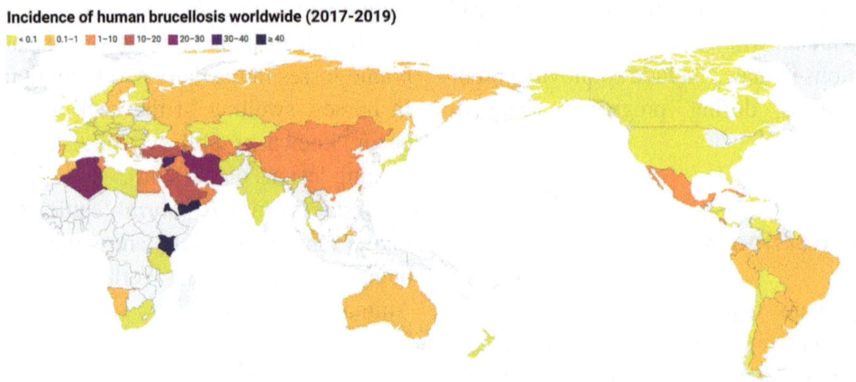

Fig. 20.5 Global epidemiology of human brucellosis. (Image source: Liu Z, Gao L, Wang M, Yuan M, Li Z. Long ignored but making a comeback: a worldwide epidemiological evolution of human brucellosis. Emerging Microbes & Infections. 2024 Dec 31;13(1):2290839. Distributed under Creative Commons Attribution License (http://creativecommons.org/licenses/by/4.0/) permits unrestricted use, distribution, and reproduction in any medium, provided the original work is properly cited)

Table 20.3 Diagnosis of brucellosis in humans

Diagnostic approaches	Methods	Targets	Remarks
Direct microscopy	Gram stain of tissues/aspirates/body fluids	Short Gram-negative coccobacilli	Helps in providing initial clue of brucellosis Not a definitive method of identification
	Stamp's modified ZN staining	Red stained short coccobacilli	Helps in presumptive identification of *Brucella*, not visible by Gram staining
In vitro culture	Blood/bone marrow culture using either manual or automated method Requires media with blood, serum, and tissue extracts	*Brucella* grows as smooth, small round colonies rough; colonies may be formed because of loss of O chain of LPS	Gold standard *Brucella* is a slow-growing bacterium, requires long incubation Important in diagnosing relapses
Serodiagnosis	Standard tube agglutination test (SAT) Rose Bengal plate test Enzyme-linked immunosorbent assay(ELISA) 2-mercaptoethanol agglutination test Radioimmunoassay Coomb's agglutination test and complement fixation test	Detects IgG and/or IgM antibodies Antismooth LPS antibodies	Particularly of importance in areas with high prevalence Not suitable for follow-ups, as the titre of antibodies remains high for a long time even after infection has cleared May cross-react with other bacteria like *Escherichia coli, Francisella tularensis, Vibrio cholerae, Yersinia enterocolitica* O:9 and various serotypes of *Salmonella.* Only indicates exposure, but not recent infection. False negative results during early phase of brucellosis
Molecular diagnosis	PCR, LAMP, multiplex PCR, MLVA at loci 16	16S rRNA is used for diagnosing brucellosis but fails to identify the species	Accurate and rapid diagnosis Can detect the bacteria even in the formalin-fixed tissues Cannot differentiate between active or killed/treated infection

The absence of a specific clinical presentation of the disease, the absence of any characteristic physical symptom, and more subclinical cases pose great difficulty in the accurate diagnosis of brucellosis. Clinically suspected cases show non-specific symptoms, including anaemia, leucopenia, or leucocytosis, often accompanied by thrombocytopenia or thrombocytosis or by the detection of *Brucella* DNA in a sample using molecular methods like PCR (Table 20.3).

The laboratory diagnosis of *Brucella* includes blood cultures, serological assays, and molecular tests.

Specimens

Blood, bone marrow, genital exudates, synovial fluid, cerebrospinal fluid or bone tissue.

Microscopy

Microscopy is not the primary method for diagnosing brucellosis. However, it is used in conjunction with other techniques to confirm the presence of the bacteria in stained smears of infected tissue or secretions. On Gram staining,

Brucella species appear as small Gram-negative coccobacilli or short, slender bacilli. *Brucella* species stain red by Stamp's modification of cold Ziehl–Neelsen staining; however, microscopy alone is often not adequate due to the challenges of staining and the difficulty in differentiating *Brucella* from other similar bacteria.

In Vitro Culture

A definitive diagnosis is established by the isolation of *Brucella* from blood, CSF, tissue, or other body fluids.

Blood Culture

Isolating *Brucella* from blood during acute infection is the mainstay of diagnosing brucellosis. Isolation is possible only during the acute phase of illness caused by *B. melitensis* and *B. suis*, but it is less useful for *B. abortus* and is of no particular use in diagnosing chronic infection. Samples other than blood include bone marrow, any body fluids, or tissues obtained by surgery or after death. The sensitivity of blood culture ranges from 10% to 90%. Multiple peripheral blood samples are needed over a period of 24 hours. It can sometimes be recovered from afebrile or subclinical cases. Blood culture is an important test in cases of suspected brucellosis, even in the absence of fever.

Various blood culture systems are used to culture it manually or automatically. Manual systems include a monophasic and biphasic approach, as well as lysis-based blood culture and blood clot media. BacT Alert, BACTEC, and BACTEC MGIT systems are automated blood culture systems useful for culturing *Brucella*. The automated culture systems have increased culture sensitivity and reduced the turnaround time for cultivating *Brucella* species.

Manual Monophasic Approach

Blood samples are incubated in culture vials at 35 °C for a long duration, followed by blind subcultures in cases of strong clinical suspicion of brucellosis. It may be accompanied by the possibility of fungal contamination.

Manual Biphasic Approach

It consists of biphasic media (both solid and liquid phases in a single medium). The most common amongst biphasic media is the *Castaneda flask*, introduced by Ruiz Castaneda. It was proposed to reduce laboratory contamination, decrease blind subcultures, and prevent media contamination. It is not specific for *Brucella* but is widely used in resource-limited countries because of its affordability and practicality.

Lysis-Based Blood Culture

Brucella is ingested by white blood cells (WBCs), thus decreasing the sensitivity of their culture. In lysis-based blood culture media, WBCs are first lysed and then put on appropriate solid media for culture, thus releasing the ingested *Brucella* into the media and increasing the sensitivity of the culture.

Blood Clot Culture

A blood sample is collected and allowed to coagulate, after which serum is separated by centrifugation and used for serological diagnosis. The clot is broken by shaking and then cultured on appropriate solid culture media. The method is not yet optimised.

Another selective medium widely used for cultivating *Brucella* is Farrell's medium (FM), which is used to culture *Brucella* from veterinary samples. It consists of antibiotics and antifungals, namely vancomycin, bacitracin, natamycin, nystatin, polymyxin B sulphate, and nalidixic acid in basal media like serum dextrose agar.

The sensitivity of blood culture depends on various factors such as the *Brucella* species causing infection, the age of the patient, the presence/absence of systemic symptoms, prior exposure to antibiotic therapy, the type of method used for culturing, the number of samples collected, the timing of collecting the sample, the periodicity of performing blind subcultures, and the incubation conditions of culture.

Bone Marrow Culture

It exhibits higher sensitivity than blood culture and is considered the gold standard test for brucellosis.

Serodiagnosis

Brucella IgM antibodies are formed during the early phase of illness, followed by the development of IgG antibodies. IgM antibodies reach their peak at about the 3rd–4th week of illness and then decline slowly. In chronic brucellosis, IgG may persist for longer and may not show the same pattern as that seen in acute brucellosis.

Various serological tests have been used to test antibodies against *Brucella*. The most commonly used serological assays include ELISA, the standard tube agglutination test (SAT), the Rose Bengal plate test, the 2-mercaptoethanol agglutination test, radioimmunoassay, and the Coombs agglutination test. Demonstrating a fourfold or greater change in serum antibody titres between acute and convalescent-phase serum samples collected at least two weeks apart suggests the diagnosis of brucellosis. A presumptive diagnosis is made based on a total *Brucella* antibody titre of 1:160 or greater in a serum sample taken after the onset of illness, as obtained by the SAT.

Milk Ring Test

It is used to check for the presence of anti-*Brucella* antibodies in sheep milk. The tests used for screening brucellosis in sheep herds include the Rose Bengal plate test, the buffered acidified plate antigen test, and the milk ring test.

Molecular Diagnosis

All PCR-based methods provide a very specific diagnosis of brucellosis. Different amplification-based methodologies include real-time PCR, multiplex PCR, nested PCR, and loop-mediated isothermal amplification (LAMP). Many genes have been targeted. One of the important gene targets for the *Brucella* genus is 16S rRNA, but it is not used for species identification, as all species of the *Brucella* genus have the same 16S rRNA. Insertion sequence IS711 is also another potential target, but it is absent in a few *Brucella* strains and, therefore, not used. MLVA-16, which detects locus 16, is a good tool for diagnosing human brucellosis.

Molecular-based methods offer many advantages in the diagnosis of brucellosis. They provide accurate diagnoses even in doubtful cases. They can detect infection even when the blood culture is negative. However, they detect nucleic acid; hence, they cannot differentiate whether it is an active infection or due to killed bacteria observed in treated brucellosis.

Treatment

Early diagnosis and treatment by a full course of antibiotic therapy are important to prevent chronic infection and complications. Management of symptoms and monitoring for treatment response, as a part of supportive care, play a vital role in comprehensive disease management.

Antibiotic therapy for brucellosis is primarily based on a combination of antibiotics to prevent relapse and ensure complete eradication of the infection. The WHO recommends a regimen of doxycycline given at 100 mg twice daily for six weeks, combined with streptomycin, 1 g daily, for two to three weeks, or alternatively, rifampicin given at 600–900 mg daily for six weeks in place of streptomycin. This dual therapy approach improves efficacy by acting against intracellular *Brucella* species and reducing the likelihood of antibiotic resistance.

Extended treatment duration and additional antibiotics like trimethoprim-sulfamethoxazole or ciprofloxacin may be necessary in complicated cases such as neuroborreliosis, endocarditis, or osteoarticular involvement.

Prevention and Control

Prevention and Control in Animals

Brucellosis has been classified as one of the seven neglected zoonotic diseases that lead to poverty in developing countries. Various approaches are used to prevent animals from getting infected. Infected animals should be prevented from contact with other animals of the herd. In areas with a high prevalence of animal brucellosis, any new animal is introduced into a herd only after proper screening and quarantine of that animal. Proper disinfection is carried out in areas where an infected herd is present, or its body secretions/aborted materials are present. In areas with a high prevalence of brucellosis, vaccination of animals like goats, sheep, and cattle is preferred. While areas with a low prevalence require screening of animals by various serological tests and culling of infected animals. Tracing and controlling the movement of infected animals is carried out to control brucellosis.

Vaccination in Animals Vaccination alone is not sufficient and is not acceptable for controlling brucellosis in any animal species. Various vaccines have been developed for animals with varying efficacies, but no vaccine exists for humans. Vaccine development for animals started with the development of inactivated vaccines, which later progressed to the formation of live attenuated vaccines. The most widely used vaccine for controlling animal brucellosis is *B. abortus* S19; other commonly used live attenuated vaccines include *B. melitensis* Rev1 and *B. suis* 2. A study done in the 1930s found *B. abortus* S19 to be most effective against brucellosis in bovines. According to the WHO, the new vaccine *B. abortus* rough strain RB51 is in use in many American countries. *B. abortus* S19 is effective against *B. abortus* in cattle, and *B. melitensis* Rev1 prevents infection by *B. melitensis* in sheep and *B. ovis* in goats.

The major drawbacks of these live attenuated vaccines include (i) their interference with serological diagnostic tests, (ii) the potential for developing antibiotic resistance in animals, (iii) the residual virulence of the vaccine strain, (iv) abortion in pregnant animals and (v) their pathogenicity to humans. The new approaches for developing newer vaccines include subunit, DNA, and mucosal vaccines with better efficacy that are non-pathogenic to humans.

Prevention and Control in Humans and Zoonotic Transmission

Eliminating infection in animals is the most effective strategy for preventing brucellosis. In areas where both vaccination and elimination of infection in animals are not possible, human brucellosis is prevented by raising awareness amongst people, ensuring food safety, and proper pasteurisation of milk before consumption.

Other preventive measures include safe handling of specimens by laboratory staff and proper disposal of the placenta and aborted materials of infected animals. The use of proper clothing, like impermeable gloves, while handling infected animals, is another method of preventing infection.

No vaccines for humans are currently available, but many approaches for developing human vaccines against brucellosis are in progress.

Case Study

A 6-year-old boy was brought to the emergency department with chief complaints of fever for the past 3 weeks. The fever was of high grade, accompanied by chills and rigors. This pattern was observed 3–4 times a day. He was already on i.v. Ceftriaxone, amikacin, and artemether, but had no relief. There were no other associated findings, such as a skin rash, lymphadenopathy, or splenomegaly. On laboratory testing, leucopenia was found with normal neutrophils and normal lymphocytes. Blood culture and urine culture were negative; the malaria smear and rapid card test were also negative. On further history, it was found that the child had goats at home and had been consuming unpasteurised goat milk for a couple of years. With high suspicion of brucellosis, antibody testing was done, which showed a four times higher titre (1:640) than the normal titre (1:80). Combination therapy of rifampicin 10 mg/kg/day and doxycycline 4 mg/kg/day was started. Within 48 hours, the child became afebrile, and on follow-up after two weeks, no episodes of fever were reported.

1. What in the clinical picture of this boy raises suspicion of brucellosis?
2. What diagnostic tests have confirmed this to be a case of brucellosis?
3. What is the treatment of brucellosis in children?

Research Questions

1. How can the sensitivity, specificity, speed, and applicability of brucellosis diagnostic methods in both human and animal populations be improved, particularly in resource-limited settings, given the limitations of existing approaches?
2. What are the efficacy and challenges associated with current treatment regimens for human brucellosis, specifically addressing concerns such as antibiotic resistance, optimal treatment duration, and persistent relapse rates?
3. What is the current progress, and what are the key obstacles to developing effective and safe brucellosis vaccines for both animal and human use, including the crucial immunological correlates of protection?

Further Readings

Christopher S. Brucellosis: review on the recent trends in pathogenicity and laboratory diagnosis. J Lab Physicians. 2010;2(02):055–60.

Ficht T. *Brucella* taxonomy and evolution. Future Microbiol. 2010;5(6):859–66.

Qureshi KA, Parvez A, Fahmy NA, Abdel Hady BH, Kumar S, Ganguly A, Atiya A, Elhassan GO, Alfadly SO, Parkkila S, Aspatwar A. Brucellosis: epidemiology, pathogenesis, diagnosis and treatment–a comprehensive review. Ann Med. 2023;55(2):2295398. https://doi.org/10.1080/07853890.2023.2295398.

Moreno E. The one-hundred-year journey of the genus *Brucella* (Meyer and Shaw 1920). FEMS Microbiol Rev. 2021;45(1):fuaa045.

Moriyón I, Blasco JM, Letesson JJ, De Massis F, Moreno E. Brucellosis and one health: inherited and future challenges. Microorganisms. 2023;11(8):2070.

Parija SC. Textbook of microbiology & immunology. 4th ed. Springer; 2023.

Poetsch A, Marchesini MI. Proteomics of *Brucella*. Proteomes. 2020;8(2):8.

She R, Anglewicz C, Jerke K, Relich R, Glazier M, Filkins L, Schuetz A. Brucella and Ochrobactrum taxonomic updates for laboratories: frequently asked questions (FAQ) for clinical laboratories.

Silva TM, Costa EA, Paixão TA, Tsolis RM, Santos RL. Laboratory animal models for brucellosis research. Biomed Res Int. 2011;2011(1):518323.

World Health Organization Brucellosis Fact Sheet. https://www.who.int/news-room/fact-sheets/detail/brucellosis

Yang Z, Chai Z, Wang X, Zhang Z, Zhang F, Kang F, Liu W, Ren H, Jin Y, Yue J. Comparative genomic analysis provides insights into the genetic diversity and pathogenicity of the genus *Brucella*. Front Microbiol. 2024;15:1389859.

Salmonellosis

21

Ajanta Sharma

Learning Objectives

1. To study the taxonomy, classification, morphology, cultural characteristics, and genomic features of non-typhoidal *Salmonella* (NTS)
2. To understand the transmission cycle and pathogenesis of non-typhoidal salmonellosis
3. To study the common diseases caused by non-typhoidal *Salmonella* in humans and animals
4. To understand the diagnostic methods and treatment options available for non-typhoidal *Salmonella*
5. To explore the preventive strategies and vaccine development efforts targeting virulence factors of non-typhoidal *Salmonella*

Introduction

"Salmonellosis" is a common food-borne illness caused by bacteria of the genus *Salmonella*, affecting both humans and animals. *Salmonella* is responsible for gastrointestinal illness and fever, collectively termed salmonellosis, and it is among the most prevalent zoonotic food-borne pathogens worldwide. Most human infections are due to *Salmonella enterica* subspecies *enterica*, which includes over 2600 serovars classified based on their somatic lipopolysaccharide (O) and flagellar (H) antigens. These serovars vary in their host specificity and the severity of the diseases they cause. They cause a range of symptoms, from mild or asymptomatic infections to severe gastrointestinal illness and fever.

Certain serovars, such as *Salmonella* Typhi and Paratyphi, are host-restricted to humans. They spread mainly through person-to-person transmission, causing diseases like typhoid and paratyphoid fever without significant animal reservoirs. Most other serovars, classified as non-typhoidal *Salmonella* (NTS), are zoonotic and are typically transmitted from animals to humans through contaminated food, especially poultry, eggs, and dairy products.

NTS is a leading cause of food-borne illness worldwide, causing about 93.8 million cases and 155,000 deaths annually. In India, NTS infections are a growing public health concern, leading to significant illness, deaths, and healthcare costs. While healthy adults usually experience gastroenteritis, it can cause severe invasive disease, such as bacteraemia or meningitis, in young children, the elderly, and immunocompromised hosts.

Animals often act as asymptomatic carriers of *Salmonella*, shedding the bacteria in their faeces and contaminating the food chain. Control in food-producing animals is crucial, but is complicated by the administration of antibiotics in livestock feed, which promotes drug-resistant strains.

A. Sharma (✉)
Department of Microbiology, Gauhati Medical College & Hospital, Guwahati, Assam, India

S. C. Parija, A. Poddar (eds.), *Textbook of Bacterial Zoonoses*, Microbial Zoonoses,
https://doi.org/10.1007/978-981-95-6615-0_21

NTS cause sporadic cases or outbreaks, especially in areas with poor food safety, sanitation, and hygiene. The true extent of the disease burden in developing countries is unclear, and rising antimicrobial resistance is a major global concern.

History

In 1880, German pathologist Karl Joseph Eberth first observed the bacterium now known as *Salmonella* in a patient with typhoid fever—a disease named from the Greek word *typhōdes*, meaning "like smoke", referring to the delirium experienced by sufferers. In tribute to Eberth's discovery, the organism was initially called *Eberthella Typhosa*. Georg Gaffky later isolated it in pure culture, confirming Eberth's findings. Soon after, Theobald Smith, working with Daniel Elmer Salmon, isolated *Salmonella choleraesuis* from swine, mistakenly believing it caused hog cholera. The genus was named "*Salmonella*" by Joseph Lignières to honour Daniel Salmon, after Theobald Smith and Salmon isolated a related bacterium from swine. These discoveries laid the groundwork for understanding *Salmonella*'s role in human and animal disease. This early research established the basis for understanding the many *Salmonella* species that cause a range of diseases in both humans and animals.

Taxonomy

The classification and naming of *Salmonella* are intricate and continue to develop. The genus *Salmonella* is classified within the phylum Proteobacteria, class Gammaproteobacteria, order Enterobacteriales, and family Enterobacteriaceae as per the WHO. There are three recognised species: *Salmonella enterica* (the type species), *Salmonella bongori*, and *Salmonella subterranea* (identified as a distinct species in 2005), which are differentiated based on 16S rRNA analysis. *Salmonella enterica* is divided into six subspecies, and the genus includes about 2659 serovars, identified based on somatic O and flagellar H antigens, as described in the White–Kauffmann–Le Minor scheme. This scheme, first published in 1934, has expanded from 44 to over 2659 serovars until now, reflecting ongoing research by various scientists. *Salmonella enterica* is divided into the following six subspecies: I. *S. enterica subsp. enterica*, II. *S. enterica subsp. salamae*, III. *S. enterica subsp. arizonae*, IV. *S. enterica subsp. diarizonae*, V. *S. enterica subsp. houtenae, and* VI. *S. enterica subsp. indica*. These species and subspecies can be differentiated based on distinct characteristics as described in Table 21.1.

Genomics and Proteomics

Over 340,000 *Salmonella* genomes are publicly available, highlighting the significant diversity within the species. Phylogenetic studies using 16S and 23S rRNA confirm that *Salmonella enterica* and *Salmonella bongori* are distinct species, closely related to *Escherichia coli* and the *Shigella* complex. *Salmonella* DNA has a G + C content of 50–53%. Detailed chromosome maps exist for several *Salmonella* serovars. Serovars like *Salmonella* Typhimurium, *Salmonella* Enteritidis, *Salmonella* Paratyphi B, *Salmonella* Paratyphi A, and *Salmonella* Typhi have similar genome structures, with sizes between 4600 and 4800 kilobases, conserved gene order, and seven *rrn* operons (regions coding for ribosomal RNA) each.

The *Salmonella* genome is made up of a single circular chromosome and, in *S. Enterica* subspecies I, a virulence plasmid (pSLT). Key virulence factors include flagella, fimbriae, type III secretion systems, toxins, two-component regulatory systems, and the Vi capsular antigen found only in *S.* Typhi and *S.* Paratyphi. The *S.* Typhi genome contains about 5 million base pairs and roughly 4000 genes, with over 200 being functionally inactive. NTS genomes usually consist of about 115 DNA segments (contigs), have a GC content of 52%, an L50 value of 17 contigs, and a total genome size of 4.8 million base pairs.

Table 21.1 Differential characteristics of *Salmonella* species and subspecies

Species	*Salmonella enterica*						*Salmonella bongori*
Subspecies	*enterica*	*salamae*	*arizonae*	*diarizonae*	*houtenae*	*indica*	
Dulcitol	+	+	–	–	–	d	+
ONPG (2 h)	–	–	+	+	–	d	+
Malonate	–	+	+	+	–	–	–
Gelatinase	–	+	+	+	+	+	–
Sorbitol	+	+	+	+	+	–	+
Growth with KCN	–	–	–	–	+	–	+
L(+)-tartrate(b)	+	–	–	–	–	–	–
Galacturonate	–	+	–	+	+	+	+
γ-glutamyl transferase	+(*)	+	–	+	+	+	+
β-glucoronidase	d	d	–	+	–	d	–
Mucate	+	+	+	– (70%)	–	+	+
Salicine	–	–	–	–	+	–	–
Lactose	–	–	– (75%)	+ (75%)	–	d	–
Lysed by Phage O1	+	+	–	+	–	+	d
Usual habitat	Warm-blooded animals	Cold-blooded animals and environment					

(a) = *d*-tartrate
(*) = Typhimurium d, Dublin –
+ = 90% or more positive reactions
– = 90% or more negative reactions
d = different reactions given by different serovars

Genome sequencing and comparative genomic analysis show that *Salmonella* genomes have similar core regions. They also exhibit recombination, genomic rearrangements, degradation, and clonal diversity both within and between serotypes. Genomic degradation is linked to host adaptation and increased pathogenicity, especially in host-restricted serotypes (*S.* Typhi, *S.* Paratyphi, and *Salmonella* Gallinarum) and host-adapted serotypes (*S.* Typhimurium and *S.* Enteritidis) and *Salmonella enterica*. Ongoing genetic reassortment in *Salmonella* leads to increased virulence and the development of antimicrobial resistance, which significantly heightens public health risks.

Morphology

Salmonella are straight, rod-shaped, Gram-negative bacilli measuring 0.7–1.5 × 2.0–5.0 μm, in line with the typical features of the Enterobacteriaceae family. They are usually motile with peritrichous flagella, except for non-motile *S.* Gallinarum and *S.* Pullorum. *Salmonella* are facultative anaerobes, non-spore-forming, and most strains have fimbriae, with about 80% showing type 1 fimbriae for mannose-sensitive adhesion.

Cultivation

Salmonellae are facultative anaerobes that grow well in standard and automated blood culture systems and on common laboratory media. On blood agar, they form grey-white, non-haemolytic, non-swarming colonies (2–3 mm in diameter) after 24 hours of incubation. On MacConkey agar, they produce smooth, colourless colonies of similar size. On xylose lysine deoxycholate (XLD) agar, colonies appear red-pink (2–3 mm in diameter), usually with black centres due to H_2S production. Strains like *S.*

Paratyphi A, which do not produce H_2S, form red-pink colonies without blackening. On deoxycholate citrate agar (DCA), colonies are colourless, while on Wilson and Blair bismuth sulphite agar, they form black colonies with a surrounding metallic sheen.

Salmonella species have distinctive biochemical features that help in their identification. Most *Salmonella* species are characterised by their ability to ferment lactose, produce hydrogen sulphide, a negative oxidase test, and a positive catalase test. They can use citrate as their only carbon source, can decarboxylate lysine, and are unable to hydrolyse urea. Key tests for non-typhoidal *Salmonella* (NTS) include glucose fermentation, negative urease and indole reactions, lysine decarboxylase activity, H_2S production, and dulcitol fermentation. Automated methods like MALDI-TOF MS can accurately identify *Salmonella* at the genus level.

Animal Models

The mouse is the most used animal model for studying *Salmonella* pathogenesis, particularly for studies on typhoid fever and enteritis. Currently, susceptible mouse strains like BALB /c are commonly used to study how *S.* Typhimurium infections progress. After oral infection, mice show fever-like symptoms (ruffled fur) but not diarrhoea, helping researchers investigate bacterial spread and immune responses. In mice, *Salmonella* infect by crossing M cells in the Peyer's patches, triggering inflammation and spreading to organs such as the liver and spleen. Streptomycin-pretreated mice are used to study intestinal inflammation similar to human colitis. For modelling human gastroenteritis caused by NTS, non-human primates and bovines are preferred, as they develop diarrhoea and inflammation like humans. Other models, like guinea pigs, rats, and rabbits, are less commonly used, mainly for vaccine research.

Transmission Cycle

The transmission cycle of *Salmonella* involves multiple routes of transmission, reservoirs, and vectors, with both animals and humans playing key roles in the spread of the bacteria. Many animals carry *Salmonella* in their gut or gallbladder, enabling direct and indirect spread to humans. As a result, these animals can contribute to both direct and indirect transmission of the pathogen to humans. Poultry and poultry products are the main sources of *Salmonella* infections.

Transmission occurs through direct or indirect contact in homes, hospitals, or farms. Most cases are food-borne, typically from consuming food or water contaminated with infected animal and human faeces. Common sources include eggs, meat, poultry, milk, and raw fruits and vegetables exposed to contaminated manure or water. Person-to-person spread occurs mainly via the faecal-oral route and is possible, especially in households or childcare centres.

Humans acquire a *Salmonella* infection from contact with animals, including pets, which may carry the bacteria without symptoms and shed them in their faeces. Vertical transmission occurs in birds and reptiles when bacteria from the female reproductive tract infect the eggs, with reptile eggs being more vulnerable due to their thinner and more permeable shells. Indirect transmission can occur through contaminated utensils, surfaces, or via live or inanimate vectors like insects, which can transfer *Salmonella* from contaminated sources to food or humans (Fig. 21.1).

Pathogenesis and Pathology

The incubation period for *Salmonella* infection is typically 5–72 hours. Infection starts when *Salmonella* attaches to host cells and enters the cells either through phagocytosis or by directly invading phagocytic and non-phagocytic cells via

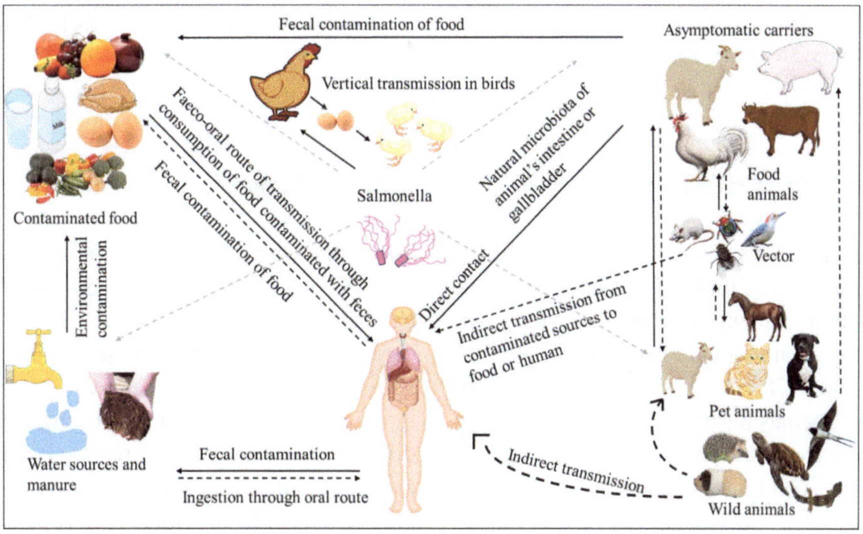

Fig. 21.1 Zoonotic transmission cycle of non-typhoidal *Salmonella*. (Image source: Author's own work using animal, human and other figures from copyright free sources (https://commons.wikimedia.org))

trigger and zipper mechanisms. Host receptors like toll-like receptors (TLRs) recognise bacterial components such as lipopolysaccharide (LPS) and flagellin by activating immune responses.

In NTS infection, the TLR-LPS interaction triggers strong inflammation, sometimes leading to septic shock. Typhoidal strains like *S.* Typhi evade TLR4 detection, reducing immune responses and aiding their survival. Despite these differences, both types induce similar cytokine production in monocytes. The ability of the NTS to invade and survive inside host cells is important to establish infection and disease progression. Major virulence factors include flagella, fimbriae, and type III, type I, and type VI secretion systems, lipopolysaccharide (LPS), and biofilm formation. They are all encoded by genes in *Salmonella* pathogenicity islands (SPIs), virulence plasmids, prophages, and the chromosome. These factors help the bacteria adhere, invade, evade immunity, and survive within host cells.

Major virulence factors of NTS are:

Virulence Plasmids: These are present in some NTS serotypes (e.g. *S.* Typhimurium, *S.* Enteritidis). They carry genes for antibiotic resistance and virulence, such as *spvB* (ADP-ribosylating toxin) and *spvC* (inhibits pyroptosis and inflammation).

Superoxide Dismutase (SOD): Enzymes (especially SodCI) protect against oxidative stress inside macrophages.

Fimbriae: They aid in adherence to host cells and trigger immune responses.

Flagella: They enable motility, adhesion, invasion, and biofilm formation; important for host recognition.

Lipopolysaccharides (LPS): Outer membrane components crucial for structural integrity, immune evasion, and virulence.

Biofilms: They enhance resistance to stress and antibiotics, mainly composed of curli fimbriae and cellulose.

Type I Secretion System (T1SS): This exports adhesins and toxins, e.g. BapA (Biofilm formation), SiiE (Attachment and adhesion), to aid in adhesion, invasion, and biofilm formation.

Type III Secretion System (T3SS): This is encoded by SPI-1 (for invasion and inflammation), which encodes T3SS 1, and SPI-2 (for intracellular survival) and encodes T3SS 2. SPI-1 is found in both *S. enterica* and *S. bongori*; SPI-2 is mainly found in *S. enterica*.

Type VI Secretion System (T6SS): The Type VI secretion system (T6SS) helps NTS kill competing bacteria.

A schematic illustration of the pathogenesis of Non-typhoidal *Salmonella* is presented in Fig. 21.2.

Pathogenesis of NTS Infections

After ingestion, NTS enters the host via contaminated food or water and faces dietary, immune, and microbial challenges in the gut. Many individuals stay asymptomatic because gut microbiota inhibits NTS by producing inhibitory substances such as bacteriocin, lantibiotics, colicins, competing for sites and nutrients, creating low-oxygen conditions, and activating host immunity.

NTS exploits dietary factors (e.g. high-fat diets increasing bile salts) and uses virulence genes on SPIs to outcompete commensals. The type VI secretion system (T6SS) helps NTS kill competitor bacteria. NTS produces siderophores such as enterobactin, salmochelins to acquire iron, giving it an advantage over commensals. If colonization resistance (CR) is breached, NTS proliferates in the large intestine, reaching high densities necessary for invasion. NTS attaches to and invades epithelial cells using fimbrial and non-fimbrial adhesins (e.g. Fim, Pef, Lpf, Agf, MisL, SiiE). Invasion is mediated by both "Trigger" (T3SS-1 effectors causing membrane ruffling) and "Zipper" (outer membrane proteins Rck, PagN) mechanisms.

NTS survives in *Salmonella*-containing vacuoles (SCVs) inside host cells. T3SS-2 effectors inhibit autophagy and suppress immune responses, promoting intracellular survival. NTS thrives on inflammation-induced metabolites such as thiosulfate, tetrathionate, ethanolamine, nitrate, and utilises by-products from other microbes, further supporting its growth. In severe cases, NTS invades deeper tissues, leading to bacteraemia, especially in immunocompromised hosts.

Immunology

The innate immune response is the first line of defence of the body against *Salmonella*, using pattern recognition receptors (like TLRs) to detect bacterial components such as LPS and flagellin. This triggers inflammation and recruits immune cells (neutrophils, macrophages, dendritic cells, monocytes) to release cytokines and antimicrobial molecules, limiting bacterial growth. NK cells and innate lymphoid cells produce IFN-γ to further activate macrophages, while the complement system and phagocytosis help contain the infection.

As the infection progresses, the adaptive immune response becomes essential. CD4+ Th1 cells produce IFN-γ and TNF-α to activate macrophages for killing intracellular bacteria, while CD8+ T cells and B cells (producing mainly IgG antibodies) aid in clearing the infection and providing long-term immunity. Effective adaptive immunity depends on antigen presentation by dendritic cells and B cells, which is crucial for strong T cell activation and lasting immunological memory.

Infection in Humans

Salmonella infections mainly present as two main forms of disease: enteric fever (including typhoid and paratyphoid) and gastroenteritis. Human salmonellosis can manifest in five ways: gastroenteritis, bacteraemia (septicaemia), enteric fever, extraintestinal complications, and a chronic carrier state.

Most *Salmonella* infections cause mild, self-limiting gastroenteritis, but severe illness can develop in infants, young children, the elderly, and immunocompromised individuals.

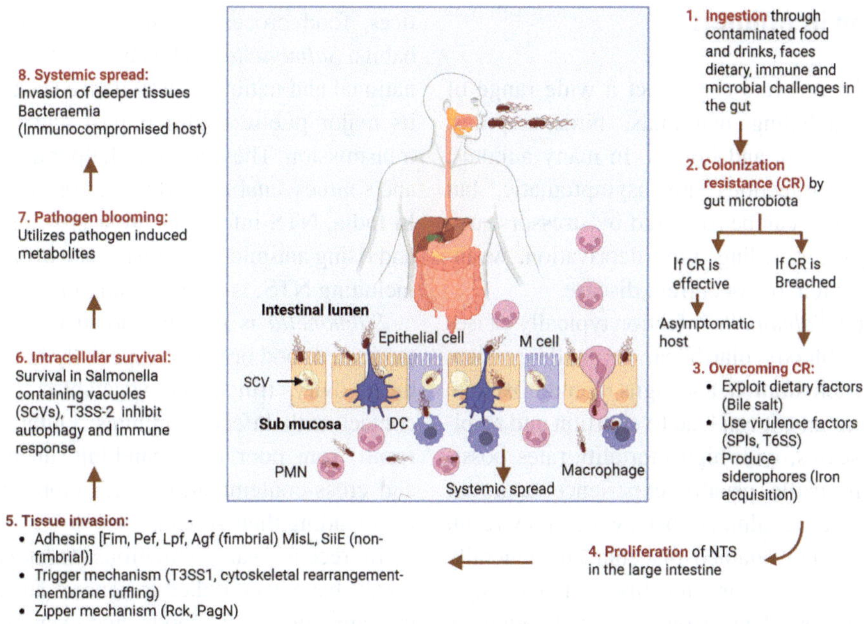

Fig. 21.2 Schematic illustration of pathogenesis of non--typhoidal *Salmonella*. (Image was created with BioRender.com; concept was adopted from Sima et al. (2024)). non--typhoidal *Salmonella* (NTS); *Salmonella* pathogenicity islands (SPIs); type 6 secretion system (T6SS); type 3 secretion system (T3SS-1) encoded by SPI-1; (type 3 secretion system) T3SS-2 encoded by SPI-2; *Salmonella*-containing vacuoles (SCVs); dendritic cell (DC); polymorphonuclear leucocyte (PMN); type I fimbriae (Fim), plasmid-encoded fimbriae (Pef), long polar fimbriae (Lpf), thin aggregative fimbriae (Agf); microbial intestinal surface ligand (MisL); *Salmonella* intestine infection E (SiiE)

Complications may include bacteraemia, meningitis, sepsis, and reactive arthritis. The severity of infection depends on the health of the host, the *Salmonella* serovar, infectious dose, and underlying conditions.

NTS typically cause gastroenteritis, which is usually self-limiting. Symptoms include fever, headache, nausea, vomiting, abdominal pain, non-bloody diarrhoea, and muscle aches. Complications such as cholecystitis, appendicitis, and pancreatitis can occur, especially in vulnerable groups; terminal ileum perforation is also rarely seen. About 6% of patients with NTS enterocolitis develop bacteraemia, particularly infants, young children, the elderly, and immunocompromised individuals. NTS serovars like *S.* Choleraesuis and *S.* Dublin are highly invasive, and some (e.g. Enteritidis, Typhimurium, Dublin) carry *spv* virulence genes that increase invasiveness.

NTS infections can result in both immediate and lasting complications, highlighting the diverse impact on affected individuals. Short-term complications arise from NTS spreading through the blood, causing infections at multiple body sites, whereas long-term complications are linked to NTS persistence in the host, leading to chronic diseases such as neoplasms and autoimmune disorders. The short-term complications include meningitis, mycotic aneurysm, endocarditis, pulmonary and pericardial abscess, osteomyelitis, spondylodiscitis, cellulitis, and urinary tract infection. The long-term complications include colorectal carcinoma, gallbladder cancer, reactive arthritis, and inflammatory bowel disease.

Infection in Animals

Salmonella serovars can infect a wide range of animals, including mammals, birds, reptiles, amphibians, fish, and insects. In many animals, *Salmonella* infections are asymptomatic, but clinical illness can be triggered by stressors such as transport, crowding, food deprivation, weaning, parturition, or concurrent disease.

In cattle, *Salmonella* infection typically causes watery or bloody diarrhoea, fever, depression, anorexia, dehydration, and signs of endotoxemia. Less commonly, it can lead to abortion and respiratory diseases, with high mortality rates possible. Adult cattle usually experience acute or subacute enteric salmonellosis, which may result in abortion and retained placenta but generally does not affect future fertility. Clinical signs include fever, depression, reduced appetite, decreased milk production, and foul-smelling, mucoid diarrhoea with blood or necrotic tissue. Dehydration and colonic congestion are also frequent. In calves, disease most often occurs between 2 and 6 weeks of age and is marked by fever, lethargy, anorexia, and diarrhoea that may contain fibrin, mucus, or blood. Without treatment, rapid dehydration can lead to death within a week, though recovered calves rarely become chronic carriers. Clinical salmonellosis is rare in dogs and cats, and birds are infrequently affected.

Epidemiology and Public Health

Salmonella are primarily intestinal pathogens found in humans and a wide range of animals, including wild birds, domestic pets, and rodents. They can also be present in the blood and internal organs. In the environment, *Salmonella* is commonly detected in sewage, rivers, water, and soil, where it can survive for weeks or even years, although it does not multiply significantly outside a host. These bacteria are frequently isolated from various foods such as fruits, vegetables, and animal protein feed supplements, making them significant food-borne contaminants. The epidemiology of *Salmonella* infections varies greatly by region. It is influenced by factors such as climate, population density, land use, farming practices, food processing methods, and consumer habits. *Salmonella* is closely monitored by international and national surveillance systems due to its major public health impact and food-borne transmission. These systems help track outbreaks and sources, enabling effective control measures. In India, NTS infections are a growing concern, and rising antimicrobial resistance in *Salmonella,* including NTS, is an increasing global threat.

Salmonella is mainly spread by eating contaminated food or water, especially poultry, eggs, meat, dairy, fruits, and vegetables, or through contact with infected animals. Outbreaks often result from poor food handling, undercooking, and cross-contamination, with global trade and travel aiding their spread.

In recent years, numerous *Salmonella* outbreaks have been linked to sources such as poultry, eggs, processed foods, and even pet reptiles and small mammals. Surveillance continues to highlight the persistent challenge of controlling *Salmonella* in the food supply, with *S.* Enteritidis and *S.* Typhimurium most frequently implicated in these outbreaks. Table 21.2 describes epidemiological features of some important zoonotic non-typhoidal *Salmonella.*

Diagnosis

For humans, stool samples are the primary specimens for diagnosing non-typhoidal *Salmonella,* as most cases present with gastroenteritis. In suspected bacteraemia or focal infections, blood, bone marrow, urine, abscess material, or cerebrospinal fluid may be collected. Multiple samples may be needed for accurate detection. In animals, *Salmonella* is typically isolated from faeces, or from blood and bone marrow in cases of systemic infection.

In Vitro Culture

The best way to recover *Salmonella* from faecal samples is by direct plating on selective agar and using enrichment broths. Faecal specimens yield more bacteria than rectal swabs. For isolation, both low-selectivity media, like MacConkey agar

Table 21.2 Epidemiological features of some important Non-typhoidal *Salmonella*

Serovar	Host range	Main reservoir host	Clinical manifestations	Geographical distribution
Salmonella Enteritidis	Humans, poultry	Poultry, eggs	Gastroenteritis, invasive disease-causing septicaemias, spinal meningitis, subcutaneous abscess	Worldwide, most prevalent in Asia, Latin America, Europe, Africa
Salmonella Typhimurium	Humans, domestic, wild animals	Pigs, cattle, poultry	Gastroenteritis, bacteraemia	Most prevalent and widely distributed globally with emergence of multiple drug--resistant strains. Major outbreaks have been reported from USA, UK, Africa, Asia
Salmonella Newport	Humans, animals	Cattle, pig, poultry, eggs	Gastroenteritis, ileocecal lymphadenitis, chest wall abscess, pyosalpinx, spondylitis, osteomyelitis, endocarditis, meningitis, and septicaemias	England, Scotland, Ireland, USA, Africa
Salmonella Kentucky	Humans (rare), poultry, animals	Poultry, eggs	Gastroenteritis, invasive disease bacteraemia	North America, Europe, Asia, Africa, Middle East, USA
Salmonella Hadar	Humans, animals	Poultry, eggs	Gastroenteritis	Prevalent in Latin America, Africa, UK
Salmonella Anatum	Humans, animals	Cattle, pig, poultry	Gastroenteritis	UK, United States, Philippines, Taiwan, Africa
Salmonella Weltevreden	Humans, animals, poultry	Poultry, reptiles	Gastroenteritis, invasive disease	South-East Asaia (including India) and the USA
Salmonella Dublin	Cattle, humans (Rare)	Cattle	Bacteraemia, systemic infection, abscess formation in bones and joints	UK, France, USA
Salmonella Heidelberg	Humans, poultry	Poultry, eggs	Gastroenteritis, blood stream infection	USA, Europe, North America, Asia, Australia, Africa
Salmonella Infantis	Humans, poultry	Poultry, Cattle	Gastroenteritis, invasive disease	Europe (Denmark, Finland, Germany, Ireland, Italy, Luxembourg, Netherlands, Poland, Ukraine, Romania, UK, USA
Salmonella Panama	Humans, animals	Wild reptiles, pig,	Invasive salmonellosis such as septicaemias, meningitis, and osteomyelitis, throat infection, brain abscess, and Bartholin's abscess	UK, USA, Latin America, Asia, Europe
Salmonella Montevideo	Animals, poultry, human (rarely)	Poultry, eggs	Gastroenteritis	UK, USA, Europe, Australia, Asia
Salmonella Virchow	Humans (Rarely), poultry	Poultry	Gastroenteritis	France, Germany, Netherlands, UK, USA, Denmark, Ireland
Salmonella Agona	Human, pigs, poultry	Pigs, poultry	Gastroenteritis	Israel, Canada, USA, UK
Salmonella Thompson	Huaman, poultry	Poultry	Gastroenteritis, intra-cranial infection	Canada, European Union, Norway
Salmonella Stanely	Huamn, poultry	Poultry	Gastroenteritis	Canada, USA, European Union
Salmonella Choleraesuis	Swine, cattle, humans	Swine	Bacteraemia, systemic infection	UK, France, USA

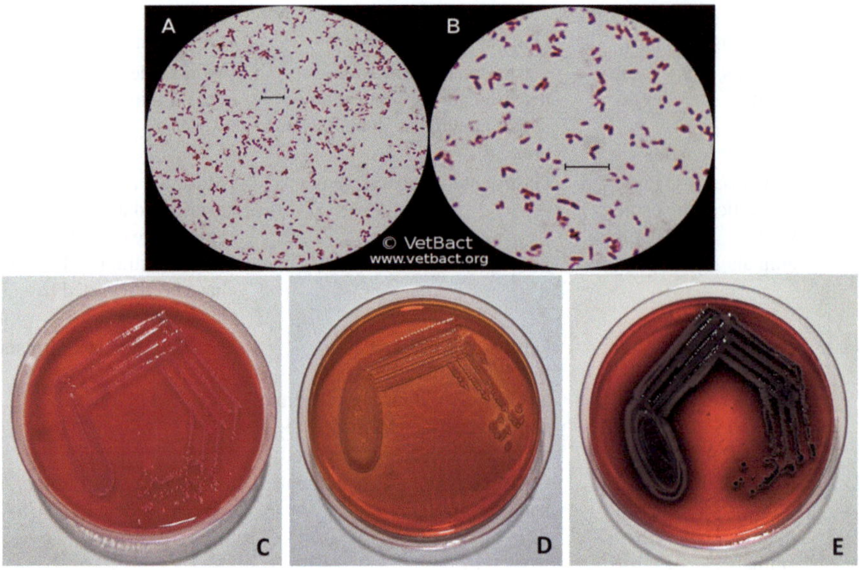

Fig. 21.3 (**a**, **b**) Gram staining of *Salmonella enterica* subsp. *enterica*, serovar Typhimurium. The difference between A and B is the degree of magnification, and the length of the scale bars is equivalent to 5 μm in both images. (Image source: Copyright free; Source: VetBact-Veterinary bacteriology: Information about important bacteria (www.vetbact.org)). Colonies of *Salmonella* Typhimurium on different culture media: (**c**) blood agar, (**d**) MacConkey agar, (**e**) Xylose lysine deoxycholate (XLD) agar. (Image source: Author's personal collection)

and high-selectivity media, such as XLD agar, are used. *Salmonella* forms colourless colonies on MacConkey and DCA, red-pink colonies with black centres on XLD. It produces black colonies with a metallic sheen on Wilson's and Blair bismuth sulphite agar. Enrichment broths like selenite brilliant green (SBG) enhance recovery rates. Chromogenic agars offer easier identification but are not yet routinely used. Figure 21.3 describes the colony characteristics of an NTS, *S. typhimurium,* in different culture media.

Salmonella species are identified by key biochemical features. They ferment glucose but not lactose, produce hydrogen sulphide, grow on citrate, decarboxylate lysine, and do not hydrolyse urea. They are oxidase-negative and catalase-positive. Automated methods like MALDI-TOF MS can accurately identify *Salmonella* at the genus level.

Typing

Primary isolation of *Salmonella* spp. is typically followed by serotyping, which is essential for identifying the specific serovar and guiding further subtyping methods.

Traditional serotyping is performed in specialised reference laboratories using slide or tube agglutination with rabbit polyclonal antisera. This process targets the O (lipopolysaccharide), H (flagellar), and Vi (capsular) antigens, following the White–Kauffmann–Le Minor (WKL) classification scheme. Table 21.3 shows the antigenic formulae of some Non-typhoidal *Salmonellae* of importance for human infection.

Molecular typing is a faster alternative to traditional *Salmonella* serotyping. It uses multiplex nucleic acid-based assays to detect genes encoding key antigens that determine the traditional phenotypic serovars. Commercial kits, like the xMap *Salmonella* serotyping assay, can identify about 85% of the most common serotypes by targeting genes such as sdf (for Enteritidis), Vi (for Typhi Vi antigen), and fljB (second phase flagellar antigen). These molecular assays show high concordance with traditional serotyping, especially for the most common non-typhoidal *Salmonella* (NTS) serovars.

Phage typing was once a valuable method for strain characterisation and played a key role in surveillance, source attribution, and outbreak

Table 21.3 Antigenic formula of some zoonotic non-typhoidal *Salmonella* serovars as according to the White–Kauffmann–Le Minor (WKL) scheme

	O serogroup	Somatic O antigen	Flagellar H antigen	
Serovar			Phasc I	Phase II
Salmonella Enteritidis	O:9 (D1)	1, 9, 12	g,m	–
Salmonella Typhimurium	O:4 (B)	1, 4, [5], 12	i	1, 2
Salmonella Newport	O:8 (C2-C3)	6, 8, 20	e,h	1,2:[z_{67}]
Salmonella Kentucky	O:8 (C2-C3)	8, 20	i	z_6
Salmonella Hadar	O:8 (C2-C3)	6, 8	z_{10}	e,n,x
Salmonella Anatum	O:3,10 (E1)	3, 10 [15] [15, 34]	e,h	1, 6
Salmonella Weltevreden	O:3,10 (E1)	3, 10[15]	r	z_6
Salmonella Dublin	O:9 (D1)	1, 9, 12[Vi]	g,p	–
Salmonella Heidelberg	O:4 (B)	1, 4, [5], 12	r	1, 2
Salmonella Infantis	O:7 (C1)	6, 7, 14	r	1, 5
Salmonella Panama	O:9 (D1)	1, 9, 12	I,v	1, 5
Salmonella Montevideo	O:7 (C1)	6, 7, 14	g,m,[p],s	[1,2,7]
Salmonella Virchow	O:7 (C1)	6, 7	r	1, 2
Salmonella Agona	O:4 (B)	1, 4, 12	f,g,s	[1,2]
Salmonella Thompson	O:7 (C1)	6, 7, 14	k	1, 5
Salmonella Stanley	O:4 (B)	1, 4, [5], 12, 27	d	1, 2
Salmonella Choleraesuis	O:7 (C1)	6, 7	c	1, 5

investigations. However, it has largely been replaced by advanced genomic methods.

Serodiagnosis

In humans, serological diagnostics for non-typhoidal salmonellosis are limited due to low sensitivity and specificity. There is no standardised ELISA for NTS because of its global diversity. Most commercial kits use LPS from *S.* Enteritidis or *S.* Typhimurium, with reported sensitivity around 88.5%. Rapid antigen tests mainly detect *S.* Typhi and *S.* Paratyphi and are less accurate for NTS. Antibody-based tests lack specificity and have high false-positive rates in endemic areas. They cannot distinguish between invasive and gastrointestinal disease, making them unreliable for individual diagnosis.

Serology is more useful for (i) herd or flock screening, (ii) carrier detection in poultry, and (iii) surveillance programmes, but is limited in individual animals due to delayed antibody response and possible presence of antibodies in uninfected animals.

Molecular Diagnosis

Molecular diagnostic tools for non-typhoidal salmonellosis are broadly classified into DNA fingerprinting methods, single-nucleotide polymorphism (SNP) technologies and sequence-based typing.

DNA Fingerprinting Methods Pulsed-field gel electrophoresis (PFGE) is the gold standard for discriminating *Salmonella* strains. Other DNA fingerprinting methods include Clustered Regularly Interspaced Short Palindromic Repeats (CRISPR) typing and multiple locus variable-number tandem repeats analysis (MLVA). All these methods offer higher resolution than traditional serotyping but are less effective for highly clonal serovars such as *S.* Heidelberg and *S.* Enteritidis.

SNP-Based Typing SNP-based typing methods for *Salmonella* include multilocus sequence typing (MLST), ribosomal multilocus sequence typing (rMLST), core genome multilocus sequence typing (cgMLST), and whole genome multilocus sequence typing (wgMLST). MLST analyses variation in seven core genes, *aroC*, *dnaN*, *hemD*, *hisD*, *purE*, *sucA*, and *thrA,* to assign sequence types, correlating well with serovar determination but lacking resolution for highly clonal serovars. rMLST uses 51 ribosomal protein genes encoding the bacterial ribosome protein subunits (*rps* genes), while cgMLST uses the 3002 locus cgMLST v2 scheme, and wgMLST uses the pan genome of *Salmonella*. These offer much higher resolution for distinguishing strains, tracking outbreaks, and identifying resistance and virulence genes.

Sequence-Based Typing 16S rRNA gene sequencing can identify bacteria at the species level and sometimes at the subspecies level. However, analysing only a few genes typically does not provide enough detail to distinguish closely related strains, such as different isolates of the same serotype.

Modern Techniques

Loop-Mediated Isothermal Amplification (LAMP) LAMP is a rapid, highly specific, and sensitive molecular technique for nucleic acid amplification. It uses four to six primers targeting multiple regions of the gene and operates at a constant temperature (60–65 °C), eliminating the need for thermal cycling or post-amplification electrophoresis. Results can be visually detected without specialised equipment. LAMP technology offers a sensitivity that is one to two orders of magnitude higher than conventional PCR techniques. It is suitable for point-of-care testing, especially in resource-limited settings where rapid diagnosis of invasive non-typhoidal *Salmonella* is needed.

Microwave-Accelerated Metal-Enhanced Fluorescence (MAMEF) This assay detects the chromosomal *oriC* locus found in all *Salmonella enterica* subsp. *enterica* serovars. It uses microwave-induced lysis to rapidly release DNA from *Salmonella* cells in bacteriological media. The method is highly sensitive, capable of detecting as little as 1 CFU in 1 ml of medium in 30 seconds. It is also specific, reliably identifying serovars such as *S.* Typhi, *S.* Paratyphi A, B, and C, *S.* Typhimurium, *S.* Enteritidis, and *S.* Choleraesuis.

The various methods for laboratory diagnosis of *salmonellosis* are presented in Table 21.4.

Treatment

Antimicrobial therapy is generally not recommended for immunocompetent patients with uncomplicated non-typhoidal *Salmonella* (NTS) gastroenteritis. According to the 2017 IDSA guidelines, first-line treatment for NTS infection is supportive care, mainly rehydration therapy.

Antibiotics should be considered only for (i) those at increased risk of invasive disease, such

Table 21.4 Different tests for laboratory diagnosis of salmonellosis

Diagnostic approaches	Methods	Targets	Remarks
In vitro cultures	For conventional culture, media used are MacConkey agar (MA), XLD agar, DCA	Colonies are colourless in 2–3 mm diameter, on XLD agar, red-pink colonies with black centre, on DCA and MA, produce colourless colonies	Bacterial colonies exhibit distinct characteristics depending on the agar medium and allows identification of non-typhoidal *Salmonella* species
Phenotypic identification	Key biochemical tests are: fermentation of glucose, urease reaction, lysine decarboxylase, indole test, H_2S production, and fermentation of dulcitol Automated methods such as MALDI-TOF MS	Ferments glucose, maltose, and mannitol with acid and gas production; does not ferment sucrose or lactose. Negative urease reaction, decarboxylate lysine, negative indole test, H_2S production, and ferment dulcitol Automated methods and MALDI-TOF MS can identify on at genus level only	Phenotypic identification has limited role in discriminating different species of NTS
Traditional serotyping for identification	Serovar and serotypes can be identified by serotyping with somatic (O), flagellar (H), and capsular (Vi) antigens	Slide or tube agglutination with specific rabbit antisera is used to identify O and H antigens according to the White–Kauffmann–Le Minor scheme	Allows identification and differentiation of different species of NTS
Sero diagnostics	There are no internationally standardised serological tests like ELISA for NTS	LPS-based ELISA tests have been used to detect antibodies against NTS using *S.* Typhimurium and *S.* Enteritidis LPS	In human, serological diagnostics are limited for non-typhoidal salmonellosis, owing to low sensitivity and specificity of the serological methods
Molecular diagnosis	DNA finger printing (PFGE, CRISPR typing, MLVA) SNP-based typing (MLST, most, cgMLST) Sequence-based typing	PFGE band classes, CRISPR polymorphism, Variable-number tandem repeats (VNTR), unique DNA elements repeated in tandem Seven housekeeping gene loci, *aroC*, *dnaN*,*hemD*, *hisD*, *purE*, *sucA*, and *thrA* in MLST, 51 genes encoding the bacterial ribosome protein subunits (*rps* genes) in ribosomal MLST, 3002 locus cgMLST v2 scheme, pan genome of *Salmonella* in wgMLST 16S rRNA	PFGE is the gold standard and offers higher resolution than traditional serotyping, but is limited for highly clonal serovars like *S.* Heidelberg and *S.* Enteritidis. CRISPR typing and MLVA provide greater discriminatory power. MLST lacks resolution for highly clonal serovars. rMLST rapidly identifies *Salmonella* phylogeny. cgMLST quickly and accurately traces pathogen sources. wgMLST offers the highest resolution and detects antibiotic resistance and virulence genes. 16S rRNA gene sequencing lacks sufficient divergence for reliable differentiation
Newer techniques	Loop-mediated isothermal amplification (LAMP)	LAMP for transcriptional regulator hilA gene of *Salmonella* spp	A sensitive, reliable, and cost-effective molecular test for *Salmonella* that delivers results within 2 hours, requires no specialised equipment, and is suitable for point-of-care use
	Microwave-accelerated metal-enhanced fluorescence (MAMEF)	*Salmonella* specific target *oriC* locus	Ultrafast and highly sensitive; detects *Salmonella* directly from blood

(continued)

Table 21.4 (continued)

Diagnostic approaches	Methods	Targets	Remarks
Molecular typing	Multiplex nucleic acid-based detection of specific genes	Detect genes encoding key antigens that determine the traditional phenotypic serovars *sdf*, Vi, and *fljB* to identify *Salmonella* Enteritidis, *Salmonella* Typhi Vi antigen, and the second phase of the flagellar antigen	Luminex®xMAP® *Salmonella* serotyping assay (SSA) can identify 85% of the most common *Salmonella* serotypes

as infants under 3 months, adults over 50 years with suspected atherosclerosis, and immunocompromised individuals, and (ii) those with cardiac or significant joint disease. Recommended antibiotics include amoxicillin/ampicillin, trimethoprim-sulfamethoxazole, fluoroquinolones, azithromycin, third-generation cephalosporins, and carbapenems. Aminoglycosides are not recommended due to poor clinical efficacy. If bacteraemia or an extraintestinal focal infection is confirmed, the choice of the antibiotics would depend on the susceptibility pattern of the strain and the clinical condition. Options may include ampicillin, trimethoprim-sulfamethoxazole, fluoroquinolones or third-generation cephalosporins, such as ceftriaxone, depending on the results of in vitro susceptibility testing.

Prevention and Control

Effective prevention and control of non-typhoidal salmonellosis requires a comprehensive "farm-to-fork" approach. Key measures include improving sanitation, ensuring safe food handling and storage, and preventing faecal contamination during food production and processing. While typhoidal *Salmonella* cases are declining in some regions due to better sanitation, NTS infections are on the rise.

On farms, strategies such as enhanced hygiene, use of natural antimicrobials, genetic selection for resistance, regular testing, and veterinary vaccines help reduce *Salmonella* levels. New DIVA vaccines, like the attenuated *S.* Typhimurium BBS 866 strain, show promise for broader protection without interfering with disease surveillance.

Integrated surveillance across the food chain is crucial, using rapid and modern molecular detection methods to quickly identify and control outbreaks. While significant progress has been made, ongoing improvements in sanitation, food safety, animal health, vaccine development, and surveillance are essential for effective global control of NTS.

Currently, there are no licensed human vaccines for NTS, but research is ongoing, especially for high-risk groups like children and the elderly in low- and middle-income countries. The NTS vaccine development pipeline includes live-attenuated, subunit-based, and recombinant antigen-based candidates. Promising vaccine candidates include live-attenuated, subunit, and Generalised Modules for Membrane Antigens (GMMA)-based vaccines, which are in Phase I trials.

Case Study

A 24-year-old man working in animal husbandry presented to the emergency department with a one-day history of severe lumbar back pain radiating into the right buttock and lower limb. The pain was accompanied by systemic symptoms including fever, rigors, generalised headache, and vomiting. The patient reported a recent episode of diarrhoea lasting seven days, which had resolved two days prior to admission. On examination, he appeared ill and febrile, with localised tenderness over the lumbar spine. Neurological examination revealed no focal deficits. MRI of the pelvis with contrast showed a small gluteal abscess inferior to the right sacroiliac joint. Laboratory investigations included blood cultures, which subsequently grew *Salmonella* Enteritidis.

1. Which zoonotic *Salmonella* pathogens are associated with livestock exposure?
2. Describe the mode of transmission and pathogenesis of the Non-typhoidal *Salmonella* infection.
3. Describe the laboratory diagnosis of Non-typhoidal *Salmonella* infection.
4. What are the common clinical manifestations of Non-typhoidal *Salmonella* infections in humans?
5. Describe the preventive and control measures for this zoonotic infection.

Research Questions

1. What are the main NTS serovars causing human infections in different regions, and how do their reservoirs and transmission routes differ?
2. How does antimicrobial resistance in NTS affect clinical outcomes and treatment?
3. How effective are new vaccines (e.g. glycoconjugate vaccines, GMMA-based vaccines) in preventing invasive and gastroenteritis forms of NTS?
4. What genetic and pathogenic differences exist between invasive and non-invasive NTS strains, and how do they affect disease severity?
5. What strategies can improve antimicrobial stewardship and surveillance to control multidrug-resistant NTS?

Further Readings

Andrews JR, Ryan ET. Diagnostics for invasive *Salmonella* infections: current challenges and future directions. Vaccine. 2015;33(Suppl 3(0 3)):C8–15. https://doi.org/10.1016/j.vaccine.2015.02.030.

Billah MM, Rahman MS. *Salmonella* in the environment: a review on ecology, antimicrobial resistance, seafood contaminations, and human health implications. J Hazard Mater Adv. 2024;13:100407. https://doi.org/10.1016/j.hazadv.2024.100407.

Diep B, Barretto C, Portmann AC, Fournier C, Karczmarek A, Voets G, et al. *Salmonella* serotyping; comparison of the traditional method to a microarray-based method and an in-silico platform using whole genome sequencing data. Front Microbial. 2019;11:2554.

Grimont PAD, Weill FX. Antigenic formulae of the *Salmonella* serovars. 9th ed. Paris: WHO Collaborating Centre for Reference and Research on Salmonella, Institut Pasteur; 2007. Available at: http://www.pasteur.fr/sante/clre/cadrecnr/salmoms/WKLM_En.pdf

Kurtz JR, Goggins JA, McLachlan JB. *Salmonella* infection: interplay between the bacteria and host immune system. Immunol Lett. 2017;190:42–50. https://doi.org/10.1016/j.imlet.2017.07.006.

Lamichhane B, Mawad AMM, Saleh M, Kelley WG, Harrington PJ 2nd, Lovestad CW, et al. Salmonellosis: an overview of epidemiology, pathogenesis, and innovative approaches to mitigate the antimicrobial resistant infections. Antibiotics (Basel). 2024;13(1):76. https://doi.org/10.3390/antibiotics13010076.

Monte DFM, Sellera FP. Salmonella. Emerg Infect Dis. 2020;26(12):2955. https://doi.org/10.3201/eid2612.ET2612.

Palmer AD, Slauch JM. Mechanisms of *Salmonella* pathogenesis in animal models. Hum Ecol Risk Assess. 2017;23(8):1877–92. https://doi.org/10.1080/10807039.2017.1353903.

Popoff MY, Bockemühl J, Gheesling LL. Supplement 2001 (no. 45) to the Kauffmann–White scheme. Res Microbiol. 2003;154:173–4.

Popoff MY, Le Minor LE. *Salmonell*a. In: Bergey's Manual of Systematics of Archaea and Bacteria online. Wiley & Sons, Inc. in association with Bergey's Manual Trust; 2015. p. 1–79. https://doi.org/10.1002/9781118960608.gbm01166.

Sanchez-Vargas FM, Abu-El-Haija MA, Gomez-Duarte OG. *Salmonella* infections: An update on epidemiology, management, and prevention. Travel Med Infect Dis. 2011;9:263–77. https://doi.org/10.1016/j.tmaid.2011.11.001.

Sima CM, Buzila ER, Trofin F, Paduraru D, Lunca C, Duhaniuc A, et al. Emerging strategies against Non-Typhoidal *Salmonella*: from pathogenesis to treatment. Curr Issues Mol Biol. 2024;46(7):447–7472. https://doi.org/10.3390/cimb46070442.

Stevens MP, Humphrey TJ, Maskell DJ. Molecular insights into farm animal and zoonotic *Salmonella* infections. Philos Trans R Soc B. 2009;364:2709–23. https://doi.org/10.1098/rstb.2009.0094.

Teklemariam AD, Al-Hindi RR, Albiheyri RS, Alharbi MG, Alghamdi MA, Filimban AAR, et al. Human Salmonellosis: a continuous global threat in the farm-to-fork food safety continuum. Foods. 2023;12(9):1756. https://doi.org/10.3390/foods12091756.

Enterohemorrhagic *Escherichia coli* (EHEC) Infections 22

Subhash Chandra Parija, Tuhina Banerjee, and Rakhee Biswas

Learning Objectives

1. To understand the key concepts of enterohemorrhagic *Escherichia coli* (EHEC) transmission, its zoonotic potential, and the factors contributing to outbreaks
2. To provide an overview of the pathogenesis of infection, its major virulence factors, and the clinical manifestations and complications of EHEC infections
3. To explore current diagnostic methods, treatment strategies, and challenges in vaccine development in EHEC

Introduction

Escherichia coli is a Gram-negative bacillus and a commensal in the gut of all warm-blooded animals. Most of these strains reside in the gastrointestinal tract and are usually harmless. However, the diarrhoeagenic strains of *E. coli* are among the most important causes of diarrhoea. There are essentially five pathotypes of diarrheagenic *E. coli*, namely enteropathogenic *E. coli* (EPEC), enterohemorrhagic *E. coli*, also known as Shiga toxin-producing *E. coli* (EHEC/STEC), enteroaggregative *E. coli* (EAEC), enterotoxigenic *E. coli* (ETEC), and enteroinvasive *E. coli* (EIEC). These pathotypes are characterised by their different predilections for colonisation of host-specific sites, their various virulence properties, and the type and severity of diarrhoea they cause. Among these pathotypes, zoonotic transmission has been reported for *E. coli* O157:H7 (EHEC). Humans are infected after exposure to contaminated food, especially of bovine origin, through direct contact with animals shedding and through environmental contamination. In addition to diarrhoea, infection with EHEC often presents with systemic complications, such as haemorrhagic colitis (HC), haemolytic uraemic syndrome (HUS), and others.

History

E. coli was first described by German bacteriologist Theodore Escherich in 1885. It was then called *bacterium coli commune*, which was first isolated from the stool of a healthy infant. With the realisation of the pathogenic potential of diarrhoeagenic strains, several pathotypes of otherwise harmless commensals have been identified.

S. C. Parija (✉)
Formerly at Sri Balaji Vidyapeeth (Deemed to be University), Pondicherry, India

T. Banerjee
Department of Microbiology, Institute of Medical Sciences, Banaras Hindu University, Varanasi, Uttar Pradesh, India

R. Biswas
Department of Microbiology, Jawaharlal Institute of Postgraduate Medical Education & Research, Pondicherry, India

S. C. Parija, A. Poddar (eds.), *Textbook of Bacterial Zoonoses*, Microbial Zoonoses,
https://doi.org/10.1007/978-981-95-6615-0_22

The sporadic case of haemorrhagic colitis was first recorded as a human infection from *E. coli* O157:H7 in the year 1975. Despite this, the bacterium was not formally recognised as a pathogen until 1982. It was in this year that investigations into two outbreaks of haemorrhagic colitis in Oregon and Michigan, both linked to the consumption of undercooked hamburgers, led to the identification of a new category of *E. coli* known as enterohemorrhagic *E. coli* (EHEC). It was distinguished from other pathogenic *E.coli* by its ability to cause severe gastrointestinal illness. Later, in 1983, the bacterium was definitively associated with haemolytic uremic syndrome (HUS), a severe and potentially fatal complication primarily affecting young children and the elderly.

The pathogen gained significant public awareness and became a major public health concern following the infamous 1993 Jack in the Box outbreak. This large-scale incident was caused by eating undercooked hamburgers contaminated with the *E. coli* O157:H7. It resulted in hundreds of illnesses, numerous hospitalisations, HUS cases, and several deaths in the Pacific Northwest. By 1994, it became a nationally notifiable infection in the United States. In the same year, the United States Department of Agriculture (USDA) implemented a ban on the sale of ground beef contaminated with the O157:H7 strain.

Taxonomy

The genus *Escherichia* belongs to the family Enterobacteriaceae within the phylum Proteobacteria, class Gammaproteobacteria, and order Enterobacteriales.

Based on their diarrheagenic potential, *E. coli* is divided into five pathotypes: enteropathogenic *E. coli* (EPEC), enterohemorrhagic *E. coli*, also known as Shiga toxin-producing *E. coli* (EHEC/STEC), enteroaggregative *E. coli* (EAEC), enterotoxigenic *E. coli* (ETEC), and enteroinvasive *E. coli* (EIEC). Of these, zoonotic transmission has been reported to be caused by EHEC.

Genomics and Proteomics

The genome sequence of *E.coli* NCCP15658, an O157:H7 strain isolated from a patient in the Republic of Korea, has revealed a genome composed of 51 contigs, with a total length of 5,462,312 bp and a GC content of 50.45%. This genome encodes 5547 protein-coding genes. The genome has plasmid-borne sequences that are commonly seen in O157:H7 strains. The chromosome and the pO157 virulence plasmid contain genes for the Shiga toxin and a Shiga toxin-like protein that are related to the pathogenesis of infection in EHEC. Genes involved in cell attachment and virulence, including intimin/invasion/adhesion, haemolysis, and pertactin, were also noted.

Interestingly, it has been observed that O islands (OIs), which are unique genomic islands in EHEC O157:H7, comprise 177 isolated genomic features and harbour 26% of the total genes absent in the non-pathogenic *E. coli* K-12 genome. Evolution-based studies of EHEC have revealed that EHEC O157:H7 has acquired an extensive regulatory network for virulence and pathogenicity. In this network, the regulatory proteins serve as the primary pivot for coordinating the expression of virulence-related genes. In total, 16 OI-encoded virulence regulatory proteins, 10 being activators and six being repressors, have been characterised in EHEC O157:H7, like the activators Ler (encoded by the LEE pathogenic island (PAI) OI-148), OvrB (encoded by OI-9), OvrA (encoded by OI-19), Cro (encoded by OI-45), GrvA (encoded by OI-47), RgdR (encoded by OI-51), EtrB (encoded by OI-115), LmiA (encoded by OI-119), AirA (encoded by OI-134), to name a few.

A proteomic analysis of EHEC O157:H7, the prototype strain, revealed 2644 non-redundant proteins, which account for 47% of its predicted proteome. Around 50 proteins, the majority of which are associated with the translation process, have been detected and labelled as the most abundant in the EHEC O157:H7 proteome. This was followed by those involved in energy metabolism, post-translational modification, protein turnover, and chaperones. A KEGG enrichment analysis has shown that glycolysis/gluconeogenesis is the most significant metabolic pathway. This may be responsible for the initial colonisation of the gut by EHEC. On the contrary, the less abundant proteins detected were related to DNA processes, such as replication, recombination, and repair, as well as cell respiration and prophages. Among the proteins constituting the Type III secretion system, the most abundant protein was EspA.

Morphology

E. coli O157:H7 (EHEC O157:H7) is a Gram-negative, rod-shaped bacillus. They are relatively small, with individual cells generally measuring approximately 1 μm in width and 2–6 μm in length (Fig. 22.1).

EHEC O157:H7 is motile and uses peritrichous flagella. The presence and structure of these flagella are key to its "H7" serotype designation. Beyond motility, the bacterium utilises various types of pili (fimbriae), which are critical for the initial stages of infection, facilitating adherence to both host cells and environmental surfaces. Some strains also produce curli fimbriae, which are involved in surface adhesion. While *E. coli* O157:H7 exists as planktonic (free-swimming) cells, it can also form biofilms. However, the ability to form robust biofilms varies significantly among different O157:H7 strains.

Cultivation

E. coli EHEC O157:H7 is a facultative anaerobe. It grows optimally at 37 °C; however, it can grow within a broader temperature range from 7 °C to 50 °C. It can survive in acidic conditions (down to pH 4.4; optimum at 6.8) and low-water-activity environments.

Standard laboratory media, such as Luria-Bertani (LB) broth, are suitable for the routine propagation of EHEC O157:H7. The colonies generally appear as medium to large, typically ranging in diameter from 2 mm to 4 mm. These colonies are characteristically round with an entire (smooth) margin. Their surface is usually smooth and glistening. The colonies are typically off-white to greyish-white or cream-coloured and tend to be opaque.

However, the isolation of EHEC O157:H7 from complex samples, such as food, water, or stool, requires the use of enrichment media. This involves incubating the sample in a selective broth that favours the growth of *E. coli* O157:H7 while inhibiting the growth of competing flora.

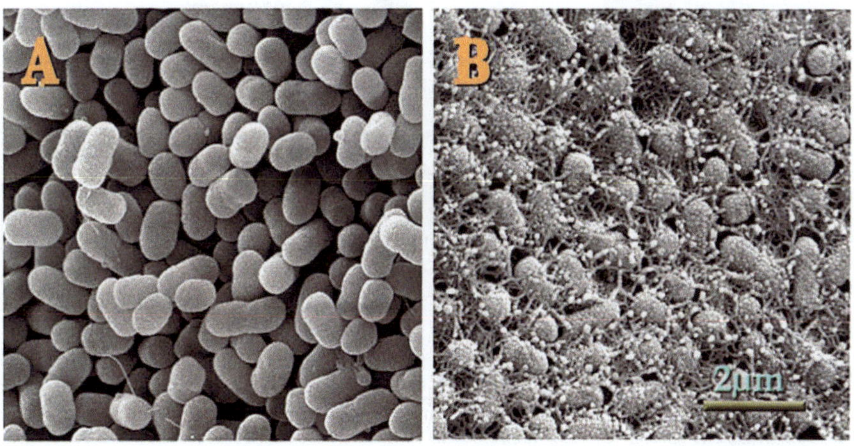

Fig. 22.1 Electron microscopic images of colonies of *Escherichia coli* O157:H7 strains (**a**) 43895OW (curli non-producing) and (**b**) 43895OR (curli-producing) grown on agar for 48 h at 28 °C. (Image source: United States Department of Agriculture)

Common enrichment broths include modified tryptone soya broth (mTSB) with novobiocin and EC Broth with novobiocin. These broths provide essential nutrients, while novobiocin acts as a selective agent to inhibit Gram-positive bacteria and some other Gram-negative bacteria. Enrichment cultures are typically incubated at 42 °C for 15–24 hours, as this temperature is often inhibitory to some background flora but suitable for EHEC O157:H7.

Animal Models

Animal models play an important role in studying EHEC pathogenesis, transmission, and host immune responses. Various models, including mice, rats, rabbits, chickens, pigs, cows, dogs, baboons, and macaques, have been used to mimic aspects of human infection. The cattle model has been extensively used for EHEC rather than the mouse model. Gnotobiotic and germ-free animals are particularly useful for investigating bacterial colonisation and toxin-mediated effects in the absence of competing microbiota. However, no single model fully replicates human disease. Hence, these models facilitate the study of the molecular mechanisms of EHEC virulence, assess the efficacy of potential therapeutics, and evaluate preventive strategies under controlled conditions.

Transmission Cycle

The transmission cycle of *Escherichia coli* O157:H7 (EHEC O157:H7) is primarily zoonotic, involving a cycle between its animal reservoir, environmental contamination, and human infection (Fig. 22.2).

The primary natural reservoir for EHEC O157:H7 is the gastrointestinal tract of ruminant animals, particularly healthy cattle. Infected cattle typically do not show signs of illness but shed the bacteria in their faeces. The infectious dose is low, approximately ingestion of less than 100 CFU, often as low as 10 organisms, resembling the infective dose of *Shigella*. The route of transmission is the faecal–oral route. Generally, two types of transmission occur:

Primary transmission of the disease to humans most commonly occurs through the consumption of contaminated food or water. The most well-known food source is undercooked ground beef, which can become contaminated during the slaughtering and processing of infected cattle. Other significant food sources include unpasteur-

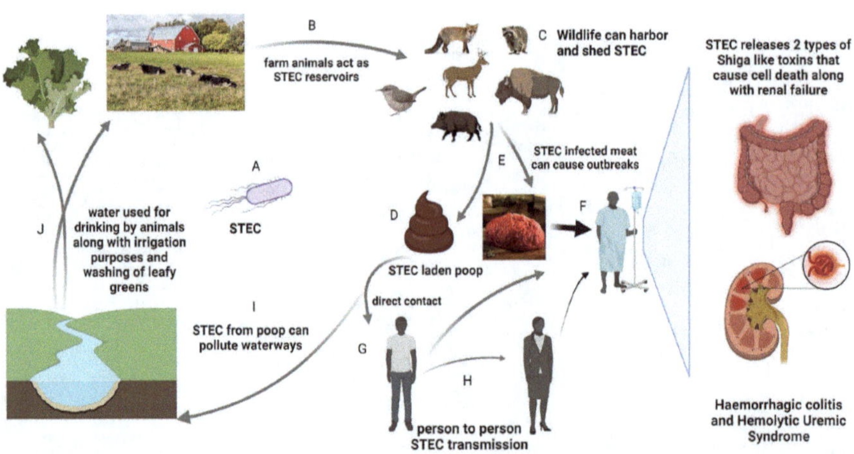

Fig. 22.2 Transmission cycle of *Escherichia coli* O157:H7. (Image source: Ray R, Singh P. Prevalence, and implications of Shiga toxin-producing E. coli in farm and wild ruminants. Pathogens. 2022;11(11):1332. Permitted for reproduction under CC BY license)

ised milk, unpasteurised fruit juices (e.g. apple cider), and fresh produce (e.g. leafy greens, sprouts) that may be contaminated through contact with animal faeces, contaminated irrigation water, or improper handling. Contaminated water can also be a direct source of infection, either through drinking contaminated water or recreational exposure in lakes, rivers, or swimming pools contaminated with animal or human waste.

Direct contact with infected animals (e.g. at farms) or their faeces is another recognised route of primary transmission.

Secondary transmission involves the transfer of bacteria from one person to another. Infected individuals shed the bacteria in their faeces, often for several days to weeks after symptoms resolve. Even asymptomatic carriers can shed the bacteria. This allows for person-to-person spread, particularly in settings with poor hygiene, such as daycare centres, nursing homes, and within families.

Pathogenesis and Pathology

The major virulence factor for EHEC strains is the Shiga toxin (Stx) produced by *stx* genes, which are encoded by bacteriophages. Upon induction with these phages, toxin production occurs. During the colonisation of the gut with EHEC, Stx enters the epithelial cells and subsequently reaches the submucosal immune cells. This causes the release of cytokines, thus ensuring the inflammatory process. There is an increased expression of the Stx receptor globotriaosylceramide (Gb3) on the endothelial linings of the intestine, kidneys, and brain, which mediates endothelial injury. This, in turn, activates coagulation and inhibits fibrinolysis, leading to the accumulation of fibrin. This causes thrombosis. The Stx, with the lipopolysaccharide layer causes the aggregation of platelet-leukocytes, and with the release of tissue factors, it causes a thrombotic state.

Besides this toxin, the outer membrane protein of EHEC, namely intimin, encoded by the *eae* gene, is another virulence factor. EHEC strains that possess the eaeA gene can produce diarrhoea. Intimin plays a role in initial colonisation by attaching to intestinal epithelial cells and causing the effacement of microvilli. The LEE genes of EHEC are regulated by quorum sensing, the expression of which is also responsible for the type 3 secretion system. A few LEE-negative non-O157 EHEC strains might also contain a potent subtilase cytotoxin, the role of which in the disease process is yet to be fully elucidated.

During infection, the major pathological changes in the colon reveal ischaemic changes and pseudomembrane formation, like those seen in *Clostridium difficile* colitis. In HUS, thrombotic microangiopathy progresses to the thrombotic occlusion of small renal arteries and arterioles. Microthrombi are also seen in the glomerular capillaries.

Immunology

Innate Immune Response

As part of the innate response, the gastrointestinal tract's mucosal barrier, comprised of epithelial cells, mucus, and antimicrobial peptides, physically impedes bacterial entry. Furthermore, host cells, such as macrophages and dendritic cells, express pattern recognition receptors (PRRs) that detect conserved bacterial components, including lipopolysaccharide (LPS) and flagellin. This recognition triggers proinflammatory responses, including the production of cytokines (e.g. IL-1β, TNF-α), chemokines, and reactive oxygen species. All these cytokines aim to recruit immune cells, primarily neutrophils, to the site of infection.

Adaptive Immune Response

The adaptive immune system develops a more targeted approach, primarily involving humoral immunity. Infected individuals typically produce antibodies against the O157 lipopolysaccharide (LPS), with IgG and IgM contributing to bacterial clearance. Patients, particularly those who develop severe complications such as HUS, often exhibit strong antibody titers against the translocated intimin receptor (Tir) and intimin proteins. Antibodies are also raised against other Type III Secretion System (T3SS) effectors and, most importantly, against the potent Shiga toxins (Stx1 and Stx2). Neutralising antibodies to Shiga toxins are considered highly protective, as these toxins are the primary drivers of severe systemic disease.

Additionally, a strong mucosal immune response, characterised by secretory IgA (sIgA) in the gut, is developed. This prevents bacterial adherence and colonisation at the intestinal surface.

EHEC Immune Evasion

EHEC O157:H7 has evolved sophisticated strategies to survive, colonise, and cause disease. One of its most significant evasion strategies is its remarkable acid resistance, which allows it to survive the highly acidic environment of the stomach. This capability bypasses a significant innate immune barrier, enabling a very low infectious dose to establish infection in the intestine. Furthermore, EHEC actively modulates the host's inflammatory response. For instance, some of its secreted proteins and Shiga toxins can interfere with host signalling pathways, potentially suppressing crucial proinflammatory and anti-bacterial responses.

EHEC O157:H7 also demonstrates adaptive survival mechanisms within the host. It survives within phagosomes of macrophages for a long period. Its ability to form biofilms provides another layer of protection, contributing to persistence in the gut. Lastly, Shiga toxins themselves can directly damage immune cells, such as macrophages, potentially leading to their death and further disrupting the host's ability to mount an effective defence.

Infection in Humans

EHEC, particularly the O157:H7 serotype, is a pathogenic strain of *E. coli* known to cause severe gastrointestinal infection in humans. The usual presentation of EHEC infection ranges from watery diarrhoea to complications like haemorrhagic colitis (HC) and haemolytic uremic syndrome (HUS).

The initial presentation is self-limiting diarrhoea. This often progresses to haemorrhagic colitis (HC), characterised by the presence of an excess amount of blood in the stool, abdominal pain, low-grade fever, nausea, vomiting, and dehydration. If uncontrolled, this often progresses to complications like intestinal necrosis, perforation, and formation of strictures in the colon.

A minority of those affected with HC can progress to *haemolytic uremic syndrome* (HUS). This is a serious complication that may cause acute kidney failure, especially in children and the elderly. The pathogenicity is largely due to the production of Shiga toxins**,** which damage intestinal and vascular endothelial cells.

Mostly, children are affected by HUS, usually within a week after diarrhoea. HUS includes acute kidney injury, haemolytic anaemia, and thrombocytopenia. Incomplete HUS is an entity in which not all the features are present. However, there may be anaemia without azotaemia, with or without thrombocytopenia. Besides renal involvement, symptoms like irritability, stroke, paralysis, coma, and other features of central nervous system (CNS) involvement might also be present. Respiratory symptoms, such as pleural effusion and adult respiratory distress syndrome (ARDS), may also be encountered.

Infection in Animals

As in humans, EHEC can cause diarrhoea and abdominal pain, especially in ruminants. The infection might progress to HUS, particularly in young animals. Some non-O157 EHEC serogroups, like O26, O111, O118, and O85, can cause diarrhoea in ruminants less than 3 months old, but not in older ones. Neonatal calves aged less than 2 days can also suffer from bloody diarrhoea with these serotypes, sometimes causing high mortality. Naturally infected rabbits and piglets do not show any symptoms of infection.

However, haemorrhagic diarrhoea with symptoms resembling HUS has been reported as an outbreak among domesticated rabbits. Dogs experimentally infected with EHEC O157:H7 have shown self-limiting diarrhoea and vomiting. In contrast, those infected with non-O157 EHEC developed severe disease, showing seizures, cerebral infarction, and finally, death.

Epidemiology and Public Health

There have been several human outbreaks of EHEC worldwide across various countries. Over the decades, more than 1343 outbreaks, impacting over 39,787 individuals, have been documented across four continents and more than 30 countries due to various serotype infections. Some notable outbreaks include the 1996 Japan outbreak in Sakai City, which affected over 8000 cases linked to white radish sprouts served in the school canteen; the 2011 Germany outbreak; and the relatively recent 2020 South Korea preschool outbreak involving 103 cases. Other recent outbreaks include the 2023 Calgary outbreak and the 2024 McDonald's outbreak, among others. Outbreaks from contaminated spinach, hamburgers, and sprout seeds have also been reported.

Most human infections arise from cattle, with studies indicating that 88.5% of human lineages are derived from cattle lineages. Contaminated food products, particularly beef and fresh produce, are primary vehicles for the transmission of pathogens. Transmission commonly occurs through ingestion of contaminated food or water, especially undercooked beef, unpasteurised milk, and fresh produce. Person-to-person transmission is also possible, especially in settings with poor hygiene.

Severe complications include HUS and HC, with hospitalisation rates reaching 50% during outbreaks. The CDC estimates that *E. coli* O157:H7 causes approximately 73,000 illnesses and 61 deaths annually in the United States, underscoring its significant public health impact. While initially prevalent in North America, *E. coli* O157:H7 has emerged in other regions, including Brazil, where it was previously thought to be absent.

Across Asia, *E. coli* O157:H7 exhibits varying characteristics and has been linked to significant outbreaks in different countries. Both South Korea and Japan have reported substantial incidents, with specific food items identified as routes of transmission. For instance, South Korea experienced a large outbreak in June 2020, affecting 103 children at a preschool who presented with symptoms including diarrhoea and HUS. This event was traced to improperly stored food, emphasising the critical role of robust food safety practices. In Japan, a 2016 outbreak linked to uncooked minced meat cutlets resulted in 61 confirmed cases, four of whom developed HUS. Research conducted in China has highlighted a high prevalence of the stx2 gene among *E. coli* O157:H7 strains, along with significant genetic diversity within the pathogen population. A study in Jiangsu province documented 131 cases of *E. coli* O157:H7-related diarrhoea between 1999 and 2006, predominantly affecting elderly farmers in the region. In India, such infections pose a considerable public health challenge due to their association with severe gastrointestinal illnesses.

A meta-analysis of 140 studies involving 220,427 cattle revealed a global prevalence of *E. coli* O157 of 5.68% (Fig. 22.3). Moreover, regional differences in prevalence were observed, with Africa exhibiting the highest prevalence (31.20%) and Latin America and the Caribbean showing the lowest (1.65%). Factors like region, cattle type, specimens used, and pre-enrichment methods were found to influence prevalence estimates.

Studies in India have demonstrated a high prevalence of the strain in bovine populations, with locally isolated strains exhibiting virulence comparable to clinical human isolates. For example, a study conducted in West Bengal, India, found that 2.04% of slaughtered cattle and 7.59% of diarrheic calves tested positive for *E. coli* O157, with all positive isolates carrying Shiga toxin genes. Furthermore, a substantial proportion of these isolates displayed resistance to mul-

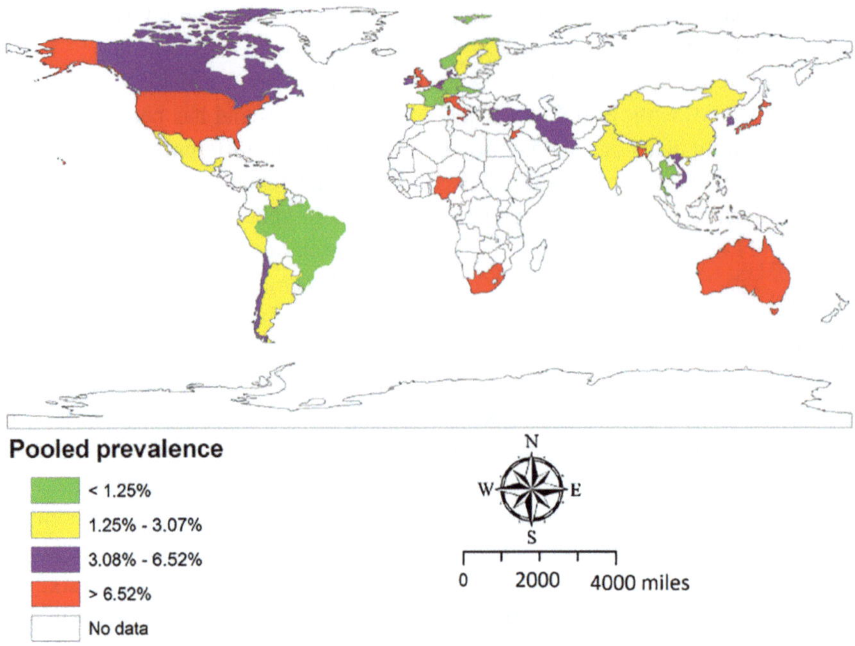

Fig. 22.3 Global epidemiology of *Escherichia coli* O157:H7 in cattle. (Image source: Islam MZ, Musekiwa A, Islam K, Ahmed S, Chowdhury S, Ahad A, Biswas PK. Regional variation in the prevalence of E. coli O157 in cattle: A meta-analysis and meta-regression. PloS One. 2014;9(4):e93299. Permitted for reproduction under CC BY license)

tiple antibiotics, raising significant concerns regarding treatment options and the broader implications for public health.

Diagnosis

The primary specimen used for the diagnosis of EHEC O157:H7 infection in humans is a stool sample. This is because *E. coli* O157:H7 colonises the gastrointestinal tract and is shed in the faeces.

Microscopy

Direct microscopy of stool is not a definitive or specific method for diagnosing *E. coli* O157:H7 infection. It has significant limitations for EHEC O157:H7; hence, it is not considered a primary diagnostic tool. This is because the Gram-negative rod morphology seen by routine light microscopy is indistinguishable from that of other commensal *E. coli* strains and other Gram-negative bacteria that naturally inhabit the human gut and are shed in the stool. Moreover, there are no unique morphological features that would allow for visual differentiation of pathogenic EHEC O157:H7 from non-pathogenic *E. coli* or other enteric bacteria in a stool sample.

In Vitro Culture

The presumptive identification of EHEC strains is performed by culturing the pathogen in specialised media.

Sorbitol MacConkey agar (SMAC) is the most widely used differential medium for the identification of Enterobacteriaceae. Most *E. coli* strains ferment sorbitol, producing acid. However, EHEC O157:H7 typically does not ferment sorbitol, resulting in colourless or transparent colonies on the sorbitol MacConkey agar (SMAC) medium.

Cefixime-tellurite sorbitol MacConkey agar (CT-SMAC) is an enhanced version of SMAC. Cefixime inhibits *Proteus* species, and potassium tellurite inhibits most other Gram-negative bacteria, thereby increasing the selectivity for EHEC O157:H7.

Chromogenic agars (e.g. CHROMagar O157, Rainbow Agar O157) contain chromogenic substrates that are cleaved by specific enzymes produced by *E. coli* O157:H7. This leads to the formation of colonies with distinctive colours (e.g. mauve or purple on CHROMagar O157), making identification easier and faster.

Serodiagnosis

Serodiagnosis is based on detecting antibodies to lipopolysaccharide (LPS) by enzyme-linked immunosorbent assay (ELISA) and immunoblotting techniques. In a study, serology identified 23% of sera with antibodies, providing evidence of infection when bacteriology alone failed to detect cases.

The Shiga toxin is a major virulence factor in EHEC infections and can be detected for diagnostic purposes. Shiga toxin immunoassays (EIA/ELISA) are being employed to detect the toxins (Stx1 and Stx2) directly in stool samples. These assays show sensitivities ranging from 73% to 100% compared to stool culture on sorbitol MacConkey agar (SMAC). The assay, however, can detect only *E. coli* O157 and not the other serotypes.

Due to their clinical significance, the Centres for Disease Control and Prevention (CDC) recommends that all stool samples from patients presenting with acute community-acquired diarrhoea, especially if bloody, or with suspected HUS, be tested for the presence of Shiga toxins.

Molecular Diagnosis

The primary targets for molecular diagnosis include the crucial Shiga toxin genes (*stx1* and *stx2*) and the *eae* gene (encoding intimin, essential for adherence), which are specific to the O157 and H7 serotypes.

Polymerase chain reaction (PCR) and its variants have become essential tools for the detection of *E. coli* O157:H7. Various PCR methodologies, including multiplex PCR, most probable number PCR (MPN-PCR), and quantitative PCR (qPCR), have been developed to enhance sensitivity and specificity in identifying this pathogen in diverse samples.

Multiplex PCR enables the simultaneous amplification of multiple target genes, such as *rfbE* and *fliC*, which are crucial for identifying *E. coli* O157:H7 and its associated virulence factors. This method can detect as few as 5 CFU/g in enriched samples. Studies show that multiplex PCR can effectively differentiate *E. coli* O157:H7 from non-O157 strains without cross-reactivity with other bacteria. *Most probable number PCR (MPN-PCR)* combines statistical methods with PCR to quantify *E. coli* O157:H7 in milk, achieving detection limits as low as 10^1 CFU/mL. *Quantitative PCR (qPCR) and Droplet Digital PCR (ddPCR)* provide accurate quantification of *E. coli* O157:H7, with ddPCR demonstrating high reproducibility and the ability to detect concentrations as low as 1 CFU/mL in environmental samples. These methods are important for monitoring contamination levels in food and water sources.

Loop-mediated isothermal amplification (LAMP) has emerged as a promising method for the rapid and sensitive detection of *E.coli* O157:H7. This technique offers advantages over traditional methods, including higher sensitivity, specificity, and cost-effectiveness, making it suitable for use in resource-limited settings. LAMP assays targeting the *z3276* gene have demonstrated a sensitivity of 100% and specificity of 97.05% in detecting *E. coli* O157:H7. Various studies report detection limits as low as 1.3 CFU/mL, showcasing LAMP's superior sensitivity compared to conventional PCR.

Various diagnostic methods are summarised in Table 22.1.

Table 22.1 Laboratory diagnosis of *Escherichia coli* O157:H7

Diagnostic approach	Method	Target	Advantages	Limitations
In vitro culture	Sorbitol MacConkey agar (SMAC)	*Escherichia coli* O157:H7 (non-sorbitol fermenting)	Cost-effective; provides a live isolate for further characterisation (e.g. antibiogram, molecular typing)	Slower (24–48+ hrs for presumptive ID); misses non-O157 STEC strains; may miss if low bacterial load or non-viable
	Cefixime-tellurite SMAC (CT-SMAC)	*Escherichia coli* (non-sorbitol fermenting)	Increased selectivity over SMAC (inhibits Proteus and other Gram--negatives); provides a live isolate	Slower; misses non-O157 STEC strains; may miss if low bacterial load or non-viable
	Chromogenic agars (e.g. CHROMagar O157, rainbow agar O157)	*Escherichia coli* O157:H7 (specific enzyme activity)	Faster presumptive identification (distinctive colony colours); provides a live isolate	Costlier than SMAC; misses non-O157 STEC strains
	Confirmation steps (post-culture)	O157 LPS & H7 flagellar antigens, biochemical characteristics	Confirms serotype (O157:H7) and *E. coli* species	Does not confirm Shiga toxin production; requires additional time
Serodiagnosis (stool and blood sample)	Shiga toxin immunoassays (EIA/ ELISA)	Shiga toxins (Stx1 and Stx2)	Rapid results (hours); detects both O157 and non-O157 STEC strains (critical for comprehensive diagnosis); can detect toxin even if bacteria are non-viable	Does not provide a live isolate for further studies; may not be as sensitive as PCR in early stages or very low toxin levels
	ELISA/western blot: antibodies (IgG, IgM) to O157 LPS, Shiga toxins (Stx1, Stx2), or virulence factors (e.g. intimin, tir)	Host immune response to infection	Useful for retrospective diagnosis (e.g. HUS cases where stool is negative); epidemiological investigations	Not suitable for acute diagnosis (antibodies take time to develop); cannot distinguish recent from past infection without paired sera; potential for cross-reactivity
Molecular diagnosis	Polymerase chain reaction (PCR) / real-time PCR (qPCR)	*stx1, stx2* genes; *eae* gene; O157-specific genes (*rfbE*); H7-specific genes (*fliC*)	Highly sensitive and specific; very rapid (hours); detects viable/ non-viable organisms; detects both O157 and non-O157 STEC strains (*stx* genes)	Does not provide a live isolate for typing or an antibiogram; detects DNA, not necessarily live, infectious bacteria
	Loop-mediated isothermal amplification (LAMP)	stx genes, eae gene, O157--specific genes	Rapid (under 1 hr); does not require thermal cycler (isothermal); suitable for point-of-care testing	Does not provide a live isolate; specific primer design is crucial

Treatment

The mainstay of treatment for EHEC infection is supportive care. Supportive therapy mainly includes volume repletion and correction of dehydration. Oral rehydration solutions are sufficient for milder cases. However, intravenous fluids become necessary for individuals suffering from significant dehydration, an inability to retain fluids orally, or signs of hypovolemic shock. Beyond fluid management, ensuring proper nutritional support, especially in paediatric patients, is vital for recovery.

Antibiotics have not been shown to reduce any symptoms associated with the carriage of the organism or prevent complications, such as HUS. They are contraindicated in confirmed cases of EHEC O157:H7-mediated infections. This is because of the concern that antibiotics can induce the phages, leading to excessive toxin production and an increased risk of HUS. Similarly, antiperistaltic agents should be avoided because they increase the risk of systemic complications. Newer therapies, such as eculizumab, a C5 complement inhibitor, have been used for the treatment of atypical HUS. However, evidence is still lacking for its successful use.

No specific treatment is available for complications like HUS once it has set in. However, management becomes critical. Patients typically require hospitalisation in an intensive care unit. The primary focus is on managing the multi-organ damage caused by the systemic effects of Shiga toxins.

Prevention and Control

A coordinated and multidisciplinary approach is required for the control and prevention of EHEC infections. Control efforts are focused on reducing the prevalence and shedding of EHEC O157:H7 in cattle, its primary reservoir. Strategies include improving farm hygiene and biosecurity to prevent faecal contamination of the environment and water sources. Vaccinating animals reduces the frequency and intensity of EHEC shedding in bovine faeces is helpful.

To prevent primary transmission, stringent food safety measures throughout the processing chain are paramount. In slaughterhouses, strict hygiene practices are enforced to minimise faecal contamination of carcasses. For ground beef, the risk is higher due to the mixing process, so thorough cooking is essential. Similarly, pasteurisation of milk and fruit juices effectively eliminates *E. coli* O157:H7. For fresh produce, minimising contact with animal waste through improved irrigation practices, proper manure management, and rigorous washing and sanitising procedures during processing is vital.

Vaccines

There are no currently licensed vaccines for EHEC infections in humans. However, for animals, the *E. coli* bacterial extract vaccine with siderophore receptors and porins is available in the USA and Canada. This vaccine targets siderophore receptors and porins (SRPs), thereby interfering with the bacterium's ability to scavenge iron, making the gut environment less favourable for EHEC colonisation. Studies have shown that vaccination with SRP-based vaccines can significantly reduce the prevalence, duration, and magnitude of *E. coli* O157:H7 shedding in cattle faeces. The vaccine typically requires multiple doses, three doses initially, for optimal protection.

Case Study

A 39-year-old female presented to emergency care with a 1-day history of severe diarrhoea and abdominal pain. She had a history of consuming salad from a retail chain market, which was recalled due to suspicion of bacterial contamination. She was given IV fluids and sent home that day. The next day, she presented to the emergency care clinic again with worsening abdominal pain and several episodes of new onset of bloody diarrhoea. She was afebrile with normal liver function tests. A CT scan of the abdomen and pelvis with contrast revealed diffuse wall thickening of the colon, suspicious for colitis. On the second day of hospitalisation, the patient underwent a sigmoidoscopy, which showed severe colitis spanning 15–40 cm from the normal appearing rectum. The colonic mucosa was suggestive of ischaemia. On the fourth day, the bacterial GI panel stool test came back positive for the Shiga toxin gene. The patient's stool sample was negative for *E. coli* O157:H7 on culture. All the antimicrobials were discontinued to avoid the development of haemolytic uremic syndrome. On hospital day 5, the patient improved and could tolerate clear liquids. She was eventually discharged home on the seventh day.

1. What history and presentation suggest EHEC infection?
2. What are the complications of EHEC infection?
3. What is the approach to management?
4. How was EHEC diagnosed in this case?

Research Question

1. How can specific biomarkers be identified to predict the prognosis and assess the severity of EHEC O157:H7 infection?
2. What strategies can be implemented for early screening of at-risk populations and early diagnosis to enhance patient outcomes and prevent complications from EHEC O157:H7 infection?
3. What novel vaccine candidates are currently being investigated for the prevention of EHEC infections in both human and veterinary populations?
4. Which new investigational therapies are most promising for treating EHEC infection?

Further Readings

Beutin L, Fach P. Detection of Shiga toxin-producing *Escherichia coli* from nonhuman sources and strain typing. Microbiol Spectr. 2014;2(5):ehec-0001-2013. https://doi.org/10.1128/microbiolspec.ehec-0001-2013.

Bowser S, Melton-Celsa A, Chapartegui-González I, Torres AG. Efficacy of EHEC gold nanoparticle vaccines evaluated with the Shiga toxin-producing *Citrobacter rodentium* mouse model. Microbiol Spectr. 2024;12(1):e0226123.

Da Silva WM, Bei J, Amigo N, Valacco MP, Amadio A, Zhang Q, Wu X, Yu T, Larzabal M, Chen Z, Cataldi A. Quantification of enterohemorrhagic *Escherichia coli* O157:H7 protein abundance by high-throughput proteome. PLoS One. 2018;13(12):e0208520.

García A, Fox JG, Besser TE. Zoonotic enterohemorrhagic *Escherichia coli*: a one health perspective. ILAR J. 2010;51(3):221–32.

Gomes TA, Elias WP, Scaletsky IC, Guth BE, Rodrigues JF, Piazza RM, Ferreira LC, Martinez MB. Diarrhoeagenic *Escherichia coli*. Braz J Microbiol. 2016;47 Suppl 1(Suppl 1):3–30.

Heo N, Lee J, Kim Y, Lee D, Heo SJ, Park YS, Yun JW, Kim YC. The epidemiological and clinical characteristics of the largest outbreak of Enterohemorrhagic *Escherichia coli* in Korea. J Korean Med Sci. 2023;38(15):e117.

Kehl SC. Role of the laboratory in the diagnosis of enterohemorrhagic *Escherichia coli* infections. J Clin Microbiol. 2002;40(8):2711–5.

Page AV, Liles WC. Enterohemorrhagic *Escherichia coli* infections and the hemolytic-uremic syndrome. Med Clin North Am. 2013;97(4):681–95. xi

Parija Subhash Chandra. Coliforms: in textbook of microbiology & immunology, Springer 4 2023, 495–515.

Reilly A. Prevention and control of enterohaemorrhagic *Escherichia coli* (EHEC) infections: memorandum from a WHO meeting. WHO consultation on prevention and control of Enterohaemorrhagic *Escherichia coli* (EHEC) infections. Bull World Health Organ. 1998;76(3):245–55.

Song JY, Yoo RH, Jang SY, Seong WK, Kim SY, Jeong H, Kang SG, Kim BK, Kwon SK, Lee CH, Yu DS, Park MS, Cho SH, Kim JF. Genome sequence of enterohemorrhagic *Escherichia coli* NCCP15658. J Bacteriol. 2012 Jul;194(14):3749–50.

Stanton E, Wahlig TA, Park D, Kaspar CW. Chronological set of *E. coli* O157:H7 bovine strains establishes a role for repeat sequences and mobile genetic elements in genome diversification. BMC Genomics. 2020;21(1):562.

Wang F, Sun H, Kang C, Yan J, Chen J, Feng X, Yang B. Genomic Island-encoded regulatory proteins in enterohemorrhagic *Escherichia coli* O157:H7. Virulence. 2024;15(1):2313407.

Campylobacter Infections

23

Ujjala Ghoshal, Aditya Kundu,
and Mallika Sen Gupta

Learning Objectives

1. To comprehend the pathogenesis, transmission, signs and symptoms of campylobacteriosis
2. To evaluate various methods for the diagnosis of *Campylobacter* infections
3. To assess antibiotic resistance in *Campylobacter* and its importance
4. To describe prevention and control measures to prevent *Campylobacter* infections

Introduction

Campylobacteriosis caused by *Campylobacter jejuni* and *Campylobacter coli* is emerging as an infection of public health importance. These zoonotic bacteria are found in birds, especially chickens, and other animals like cattle, pigs, and pets. Since the 1970s, significant progress has been made in understanding the epidemiology of campylobacteriosis. Poultry, particularly undercooked chicken, has been identified as a major source of infection, along with contaminated water and raw milk.

Effective control measures, such as better food safety practices, outbreak reporting, and research into epidemiology and transmission pathways, are crucial to reduce their impact and improve public health outcomes. The One Health approach, which integrates people, animal, and environmental health, is important to control *Campylobacter* infections.

History

Early observations of curved or spiral-shaped bacteria in cases of diarrheal illness date back to the late nineteenth century. Theodor Escherich, in 1886, described similar organisms in the stools of infants suffering from "cholera infantum." However, these early sightings were not definitively linked to the currently known bacteria as *Campylobacter*.

The first recognised isolation of these bacteria was documented in 1913 when John McFadyean and Stewart Stockman observed curved organisms in foetal tissues from aborted sheep. They initially classified these as vibrios based on their shape. In 1919, Theobald Smith isolated similar bacteria from aborted bovine foetuses and named the organism *Vibrio fetus*. For several decades, these microaerophilic (requiring low-oxygen levels) vibrio-like organisms were primarily associated with reproductive diseases in livestock populations.

Campylobacter as a distinct genus was reported by Belgian scientists Elisabeth King, along with Sebald and Véron, in 1963. They rec-

U. Ghoshal (✉) · A. Kundu · M. S. Gupta
Department of Microbiology, All India Institute of Medical Sciences, Kalyani, West Bengal, India

S. C. Parija, A. Poddar (eds.), *Textbook of Bacterial Zoonoses*, Microbial Zoonoses,
https://doi.org/10.1007/978-981-95-6615-0_23

ognised significant biological differences between these organisms and true *Vibrio* species by their inability to ferment carbohydrates and their distinct DNA base composition. They proposed the new genus *Campylobacter*, meaning "curved rod," and *Vibrio fetus* was reclassified as *Campylobacter fetus*.

Despite the recognition of the genus, the role of *Campylobacter* as a significant human intestinal pathogen remained largely underappreciated for several more years. This was partly due to the technical difficulties in isolating these fastidious bacteria from faecal samples, which were often overgrown by other, more easily cultured gut bacteria. *Campylobacter*, as a major cause of human gastroenteritis, was reported in the 1970s by Martin Butzler in Belgium in the stools of patients with diarrhoea. This paved the way for many more studies that quickly revealed the high prevalence of *Campylobacter*, particularly *Campylobacter jejuni* and *Campylobacter coli*, in cases of acute diarrheal illness in humans.

Taxonomy

The genus *Campylobacter* belongs to the domain bacteria, phylum Campylobacterota, class "Campylobacteria" (formerly known as Epsilonproteobacteria), order Campylobacterales, and family Campylobacteraceae.

As of November 2024, there are 49 officially described species within the genus *Campylobacter*. While this number reflects the recognised diversity within the genus, not all of these species are known to cause disease in humans. Many *Campylobacter* species are commensal, meaning they live in the gastrointestinal tracts of various animals without causing harm to the host. However, several *Campylobacter* species are significant human pathogens. The majority of human infections are caused by *C. jejuni* and *C. coli*. Other species occasionally implicated in human illness include *Campylobacter lari*, *Campylobacter upsaliensis*, and *Campylobacter fetus*. The study and classification of *Campylobacter* species continue to evolve with the application of advanced molecular techniques, leading to the identification and description of new species and a better understanding of their roles in health and disease.

Genomics and Proteomics

The C. jejuni genome has a small, circular chromosome around 1.6–1.8 Mb in size. A remarkable feature of the *Campylobacter* genome is the high degree of genetic plasticity, which allows these bacteria to quickly adapt to environmental tension, such as temperature, oxygen levels, and acidity.

One of the striking features of the *C. jejuni* genome is its relatively low G + C content, typically around 30%. Unlike many other bacteria, the *C. jejuni* genome is notable for a comparative lack of insertion sequences, prophages, and large repeat regions. However, it does possess a significant number of short, repetitive sequences known as homopolymeric tracts. These tracts are prone to slipped-strand mispairing during DNA replication, leading to frequent phase variation in genes. Gene variations are particularly involved in synthesising and modifying surface structures like flagella, lipooligosaccharide (LOS), and capsule.

While the core genome of *C. jejuni*, representing genes essential for basic survival and found in nearly all strains, is relatively conserved, the species exhibits an "open" pan-genome. This means that the total gene content across all *C. jejuni* strains is much larger than the gene content of any single strain. The accessory genome, comprising genes present in only a subset of strains, contributes significantly to the diversity observed among *C. jejuni* isolates. These accessory genes are often associated with specific host tropisms, environmental adaptation, and virulence.

Several genes and genomic features are recognised as important contributors to *C. jejuni* pathogenesis. These include genes involved in (i) motility and chemotaxis (essential for colonisation), (ii) adherence to host cells (e.g. *cadF*, *peb* genes), (iii) invasion of intestinal epithelial cells, and (iv) the production of toxins, such as the cytolethal distending toxin (CDT). The genes responsible for the biosynthesis of LOS and capsule are also critical virulence factors, influencing immune evasion and the potential to induce post-infectious complications like Guillain-Barré syndrome.

The flagella of the bacterium, coded by specific genes, make it possible to move effectively in a viscous environment, such as the mucus lining in the stomach. The flagella are essential for motility and are important virulence factors in invading gut epithelial cells. Other genomic components include the Type IV secretion system, which transmits bacterial proteins into host cells, helping to evade the immune system and host manipulation. Genomic sequencing has also revealed the presence of several antibiotic resistance genes in *Campylobacter*.

Proteomic studies have identified a number of surface proteins in *Campylobacter*. For example, surface proteins involved in adhesion are important for the initial interaction of bacteria with the intestinal lining. These include glycosylated proteins that help to attach bacteria to host cells. Proteomic studies have also identified stress-related proteins, such as oxidative stress conditions in the acidic stomach or intestine, that help *Campylobacter* survive in a hostile environment. Proteins involved in oxidative stress resistance, such as catalase and superoxide dismutase, are important for the bacteria's ability to survive in the gut and to avoid the host's immune response.

Morphology

Campylobacter spp. are Gram-negative, helical, or curved-shaped, and non-sporing bacteria. Morphologically, they are described as "comma--shaped" or "gull-wing" bacteria with a characteristic "*corkscrew*" motion. With this ability, they traverse the gastric mucosa and colonise the epithelial layer of the gastrointestinal tract of the host.

Campylobacter spp. appear as slender, curved rods, often 0.5–0.8 μm in width and 1.5–5 μm in length. The bacterium has a polar flagellum at one or both ends, contributing to its motility. This morphology is also important for the bacterium's ability to survive in low-oxygen environments, a key feature given the microaerophilic nature of *Campylobacter* species.

Cultivation

Campylobacter are fastidious bacteria that require specific and carefully controlled conditions. *Campylobacter* cultivation requires a microaerobic atmosphere. They grow best in an atmosphere with reduced oxygen concentration, typically between 5% and 10%. They also require an elevated concentration of carbon dioxide, usually around 5% to 10%. The presence of nitrogen and a small percentage of hydrogen helps scavenge residual oxygen. This specific gaseous environment is created using specialised gas mixtures, gas-generating sachets within sealed jars or pouches, or automated incubation systems that precisely control gas levels. Maintaining these conditions throughout the incubation period is essential for optimal growth.

C. jejuni and *C. coli* are thermophilic. The ideal incubation temperature for these species is typically 42 °C. *Campylobacter* species can grow at 37 °C. Thermophilic species require a higher temperature of 42 °C for their isolation from samples that may contain a mixture of bacteria, as it inhibits the growth of many other common flora. *Campylobacter* generally do not multiply at temperatures below 30 °C.

Isolation of *Campylobacter* from stool requires the use of selective media, which can inhibit the growth of other organisms while promoting the growth of *Campylobacter*. Selective media have been supplemented with certain antibiotics and antifungals to inhibit the growth of other organisms. These include Butzler's selective media (containing bacitracin, novobiocin, colistin, cephalothin, actidione), Skirrow's blood agar (vancomycin, polymyxin b, trimethoprim), Preston selective medium (trimethoprim, polymyxin b, rifampin, cycloheximide), Preston blood-free medium (cefoperazone), Campy-BAP (vancomycin, trimethoprim, polymyxin b, cephalothin, amphotericin b), and charcoal-based blood-free selective medium (vancomycin, cefoperazone, cycloheximide).

Chromogenic agars such as RAPID *Campylobacter* agar, CHROM agar, and *Campylobacter* selective chromogenic medium can also be used, and they give different coloured colonies based on enzymatic reactions.

Animal Models

Animal models are important in understanding *Campylobacter* infections. The main animal model used is the chicken since poultry are natural hosts for the *Campylobacter* species and often act as reservoirs of infection. Chickens can also become infected at a very young age, which is often the result of contact with infected water or contaminated feed. They are also able to shed the bacteria in their faeces, which serves as a means of environmental contamination.

Other animal models, such as rodents and non-human primates, have also been used to explore pathogenicity and immune response to *Campylobacter* infections. Most notably, the mouse model is used to study host-pathogen interactions, immunology, and vaccine development. Although rodents are not considered natural hosts of *Campylobacter*, they are artificially infected with the bacteria to study the mechanisms of intestinal colonisation and immune evasion.

Transmission Cycle

Campylobacter species are widely colonised in the stomachs of different animals. These animals excrete *Campylobacter* in the stool, which can then pollute the surroundings.

Humans are infected by direct contact with infected animals or their faeces. This is especially common among pet owners, dog owners, farm workers, individuals who handle livestock and people working in agricultural environments. The faecal–oral transmission is still the most common route for *Campylobacter* transmission. Squamates (lizards, snakes and amphisbaenids) have also been identified as potential reservoirs for *C. fetus* subspecies *testudinum* and *C. fetus* subspecies *fetus*.

The transmission cycle of the *Campylobacter* infection includes mainly animal hosts, especially poultry, which act as a large reservoir. The infectious phase of the bacteria is usually a viable cell, which can be transmitted through contaminated foods, especially undercooked poultry, water or through direct contact with animals or stools, as shown in Fig. 23.1.

Pathogenesis and Pathology

Pathogenesis of campylobacteriosis is determined by a combination of factors that allow these bacteria to survive, colonise and infect their hosts. These include invading host tissue, opposing oxidative stress, producing toxins, acquiring iron, and surviving in unfavourable environmental conditions.

Virulence Factors

There are multiple virulence factors that play a major role in the pathogenicity of *Campylobacter*.

1. *Flagella*: Most *Campylobacter* species have amphitrichous flagella except *Campylobacter gracilis* (non-motile) and *Campylobacter showae* (has multiple flagella). The presence of flagella contributes to their high-speed motility even in viscous environments. It also contributes to biofilm formation, protein export and adhesion and host invasion.
2. *Chemotaxis*: There are certain methyl-accepting chemotaxis proteins (MCPs), also known as transducer-like proteins (Tlps), that can detect the extracellular chemical gradients that control the motility of the bacteria. These extracellular chemicals bind to Tlps and relay a signal to chemotaxis proteins (Che) in the cytoplasm. This initiates a signal that leads to the directional movement of flagella.

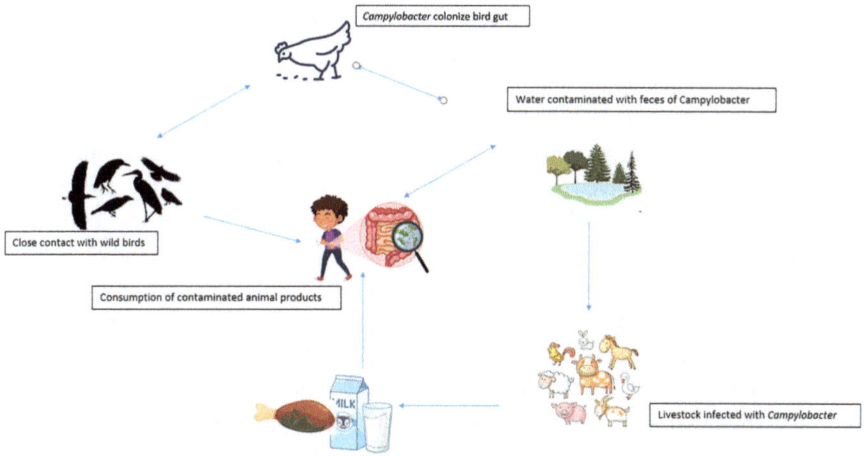

Fig. 23.1 Transmission of *Campylobacter* species. (Image source: Authors' Personal Collection)

3. *Polysaccharide Capsule*: Mutations in the capsular polysaccharide ABC transporter kpsM gene have been associated with decreased invasion, decreased capacity to cause systemic infections and reduced adhesion, thus supporting the role of capsular polysaccharide in virulence. The presence of a polysaccharide capsule may also increase cytokine production by the host immune system and pathogenicity.
4. *Lipooligosaccharide*: It is associated with host immune system evasion, adhesion, invasion, and helps the bacteria survive in hostile environments. The presence of sialic acid in lipooligosaccharide can be associated with a more severe form of disease. Sialylation of lipooligosaccharides have been shown to invade cells more significantly than non-sialylated lipooligosaccharides. Some lipooligosaccharides of *C. jejuni* also mimic the neuronal gangliosides and lead to Guillain-Barré syndrome due to cross-reactive antibody response.
5. *Adhesion*: Several adhesion proteins like CadF, FlpA, Peb1A, Peb3, JlpA, and CapA are associated with pathogenesis and adhesion of *C. jejuni* to host cells. The *Campylobacter* virulence plasmid, pVir, is also associated with adhesion. However, only a few strains carry this virulence plasmid.
6. *Role of Iron*: Iron plays a significant role in colonisation of the host as well as in the virulence of the bacteria and is regulated by the cfrB gene. Iron availability alters the ability of *Campylobacter* in terms of adhesion, invasion, and colonisation of the host's gastrointestinal tract. Inactivation of the cfrB gene results in a reduction and prevention of colonisation by *Campylobacter.*
7. *Cytolethal Distending Toxin (CDT)*: CDT has two binding subunits (CdtA and CdtC) and an active subunit (CdtB). CdtA and CdtC units allow binding of CDT to target cells and deliver CdtB inside the cell. Once CdtB enters the cell, it enters the nucleus and produce DNase I-like activity, resulting in double-strand breaks in chromosomes. This initiates a regulatory process where cell cycle arrest occurs and ultimately results in cell death.

Immunology

The immune response to *Campylobacter* infection is diverse. The gastrointestinal mucosal barrier, antimicrobial peptides, and phagocytic cells act as the primary line of protection that are important for initial defence against *Campylobacter*.

Furthermore, *Campylobacter* directly stimulates Toll-like receptors (TLRs), leading to the activation of inflammatory pathways and the production of cytokines such as interferon-beta. In terms of adaptive immunity, both humoral and cell-mediated responses are involved.

Humoral immunity is characterised by the development of antibodies against *Campylobacter* antigens, with some antibodies specifically targeting flagellin and other protein components. Cell-mediated immunity involves T cells, particularly CD4+ and CD8+ T cells. These contribute to the clearance of infection and the regulation of the inflammatory response. Studies examining B cell responses in individuals with *Campylobacter* infection have revealed alterations in B cell populations, including an increase in transitional B cells and a decrease in mature B cells.

Infection in Humans

Campylobacter infection, also known as campylobacteriosis, is a bacterial illness that primarily affects the human digestive system. While the severity can vary, the infection most commonly manifests as acute gastroenteritis.

The incubation period varies from 2 to 5 days. Nevertheless, the incubation period can vary from as short as one day to as long as ten days. The initial symptoms often include the sudden appearance of watery diarrhoea. A notable feature of campylobacteriosis is the occurrence of bloody stools. The presence of blood in the diarrhoea is a key indicator that differentiates *Campylobacter* infection from other types of gastroenteritis.

Diarrhoea is associated with abdominal pain and cramping. These cramps range from mild to severe and are often a significant source of discomfort. Some individuals may also experience nausea and vomiting. However, vomiting is generally not as prominent compared with some other food-borne illnesses. Fever varying in intensity is another common symptom, often associated with diarrhoea and abdominal pain.

The acute phase of campylobacteriosis typically lasts for about a week. In rare instances, symptoms can be more prolonged or even relapse. Most *Campylobacter* infections are self-resolving. Nevertheless, complications occur in a small percentage of cases.

Guillain-Barré syndrome (GBS) is one of the most serious post-infectious complications. It is a rare neurological disorder where the body's immune system affects the peripheral nerves, leading to muscle weakness and potentially paralysis. *Campylobacter* infection is a significant trigger for GBS. However, the mechanism linking *Campylobacter* to GBS remains an area of active research. Another potential complication is reactive arthritis, which causes painful inflammation of the joints, typically occurring several weeks after the diarrheal illness has resolved. In rare instances, *Campylobacter* infection leads to bloodstream infections (*bacteraemia*), particularly in individuals with weakened immune systems.

Infection in Animals

Campylobacter species are commonly found in a wide variety of animal hosts, both wild and domesticated. While these bacteria cause illness in some animals, their presence often goes unnoticed as the animals serve as asymptomatic carriers. This widespread colonisation in animals, particularly those raised for food, is a primary reason for the transmission of *Campylobacter* to humans.

Poultry is especially important in the epidemiology of *Campylobacter*, because it can carry bacteria in the gut without showing clinical signs. Pets like cattle and wild boar can also harbour bacteria, although they are less often involved in human infections. Symptoms like diarrhoea, dysentery, nausea, vomiting, and loss of appetite are seen in infected animals. Liver disease, less egg production, and other rare complications are seen in poultry, whereas infertility and abortion are seen in cattle.

C. fetus affects a wide range of hosts and causes breeding problems in livestock, while *C. fetus* subgroups *venerealis* are more host-specific and lead to significant financial losses due to infertility and miscarriage in cattle. *C. fetus* subspecies *testudinum* is usually found in reptiles and shows zoonotic capacity. Despite these risks, only bovine genital campylobacteriosis is officially recognised by the World Organisation for Animal Health (WOAH).

Epidemiology and Public Health

Human infection is acquired through intake of contaminated foods especially raw or undercooked chicken, as well as unpasteurised milk or untreated water. Bacteria survive in the environment for a long time, especially under moist conditions. Infected animals, especially poultry, do not show the clinical signs of the disease, making it difficult to identify and control the source of infection. This emphasises the importance of hygiene practices in food handling and animal control to reduce the risk of infection.

Animal Colonisation and Faecal–Oral Route of Transmission

Transmission Sources

Milk The global consumption of raw milk has grown because of increasing health benefits. Though milk is sterile in animals' mammary glands, it becomes contaminated at some stage during handling the milk, particularly if the equipment or animals are infected. *Campylobacter* species like *C. jejuni*, *C. coli*, and *C. lari* may be observed in raw milk from diverse mammals. Consumption of contaminated raw milk has caused campylobacteriosis outbreaks, raising public health issues.

Water Water is an important source for infecting people with *Campylobacter*, especially through contaminated drinking water. This is a significant risk in areas of poor hygiene or insufficient water treatment. Recreational water sources such as lakes and beaches also harbour bacteria.

Meat Contamination occurs during slaughter or processing when faecal material contaminates meat. Improper handling and underwashing also contribute to the infection. While poultry is the primary source, other meats such as pork, beef, and lamb can also carry bacteria.

Other Transmission Sources Cross-contamination occurs in the kitchen, especially when raw meat is not handled properly, or surfaces are not cleaned after contact with contaminated food. Kitchen workers and people with poor hygiene practices are at greater risk. In addition, even though rare, the outbreak of *Campylobacter* can occur in the healthcare system, such as hospitals, where hygiene protocols are not followed.

Campylobacter infection remains one of the most widespread bacterial causes of diarrheal disease across the globe, imposing a significant public health burden in both developed and developing countries (Fig. 23.2). While accurate global incidence figures are challenging to ascertain due to variations in surveillance and reporting, estimates suggest that hundreds of millions of cases occur annually.

Epidemiology in developed countries is often characterised by sporadic cases, though outbreaks linked to contaminated food or water do occur. There is also a distinct seasonality, with a peak in infection typically observed during the warmer summer months. In contrast, *Campylobacter* infection is often endemic in low- and middle-income countries, particularly affecting young children. The incidence in children under the age of five in these regions is remarkably high, and while infections are common, asymptomatic carriage is also widespread in both children and adults.

Certain populations are at higher risk of *Campylobacter* infection globally. Young children, especially those under 5 years old, experience a high incidence of infection, particularly in developing countries. Immunocompromised individuals, the elderly, and travellers to regions with different hygiene standards are also more susceptible to infection and potentially more severe outcomes. While both sexes are affected,

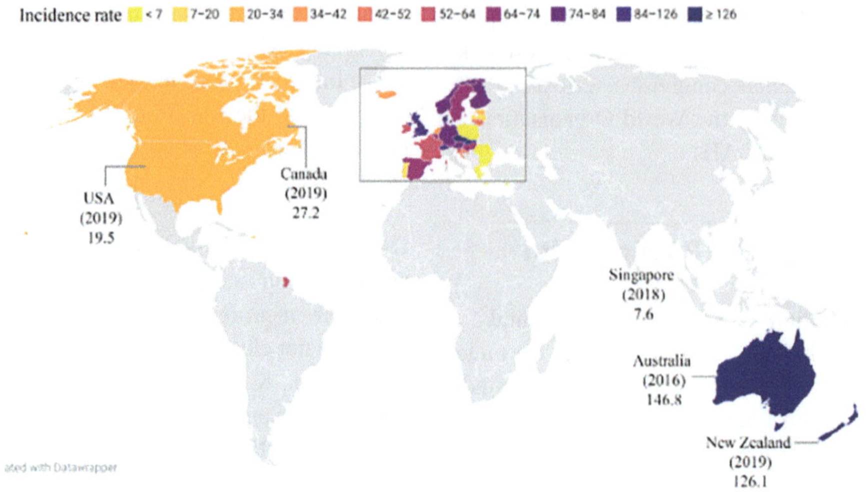

Fig. 23.2 Global epidemiology of campylobacteriosis. (Image source: Liu F, Lee SA, Xue J, Riordan SM and Zhang L (2022) Global epidemiology of campylobacteriosis and the impact of COVID-19. Front. Cell. Infect. Microbiol. 12:979055. https://doi.org/10.3389/fcimb.2022.979055. Open-access article distributed under the terms of the Creative Commons Attribution License (CC BY))

some data from developed countries suggest a slightly higher incidence in males.

Campylobacter infection is less recognised in India as compared with other gastrointestinal pathogens like *Salmonella* and *Escherichia coli*. But nowadays, food-borne infection caused by *Campylobacter* has been increasing. In 2017, the incidence of food poisoning increased from 50 (2008) to 242, making it the second most common cause of outbreaks of infectious disease in the country. A study in North-East India between 2014 and 2016 detected 41/407 (10.1%) *Campylobacter* species in children under 5 years of age presenting with diarrhoea. The highest incidence of *Campylobacter* infection was seen in June and July compared to December and January. In early 2025, Maharashtra experienced an outbreak of Guillain-Barré syndrome (GBS), a rare neurological disorder often associated with *C. jejunii* infections. This outbreak resulted in five deaths and affected 163 people, mainly in Pune. Health authorities recognised *C. jejunii* as a causative agent, which emphasises the need for better food security measures.

In 2018, the US Centers for Disease Control and Prevention (CDC) reported an outbreak of *Campylobacter* infection infecting 113 people in the United States, with 23 patients requiring hospitalisation. This outbreak was then found to be linked to contact with pet store puppies. Multiple antimicrobial resistance genes were also identified from 38 sick patients and 10 puppies using whole genome sequencing. An outbreak of *Campylobacter* infection linked to raw milk consumption was also reported by the CDC in 2014 in Utah.

Diagnosis

Isolation and accurate identification of *Campylobacter* spp. are important to diagnose infections and to study the epidemiology of the bacteria.

Specimens

A stool specimen is the most common and preferred sample type for confirming *Campylobacter* infection. Ideally, stool samples are collected during the acute phase of the illness, when the bacteria are most likely to be present in high numbers. It should also be collected before antibiotic treatment is initiated, as antibiotics can reduce the bacterial load and make detection more difficult. Rectal swabs are also collected, particularly from infants or when a fresh stool sample is difficult to obtain, and should also be placed in a suitable transport medium.

Specimens also include blood for culture to detect bacteraemia or, in rare instances, tissue samples if localised infections in other body sites are suspected.

Microscopy

Microscopic examination helps in the initial assessment and presumptive diagnosis of *Campylobacter* infection, particularly in a clinical laboratory setting. These provide rapid clues based on the characteristic morphology and motility of the bacteria. Different microscopic techniques offer varying levels of detail and are used for specific purposes. The combination of a Gram stain to observe the characteristic curved Gram-negative rods and a wet mount to observe the distinctive motility remains the most common and practical approach for the rapid presumptive diagnosis of *Campylobacter* infection in a clinical setting.

Direct wet mount examination is the simplest and quickest microscopic technique in diagnosis. This involves placing a small amount of fresh stool or a suspension of bacterial culture on a microscope slide, adding a drop of saline, and covering it with a coverslip. The sample is then observed under a light microscope, often using phase contrast or dark-field illumination. The key feature to look for in a wet mount of a *Campylobacter*-positive sample is the characteristic motility. *Campylobacter* bacteria exhibit a rapid, darting, or corkscrew-like motility, which is highly distinctive and can provide a strong presumptive diagnosis, especially in fresh diarrheal stools. The curved or spiral shape of the bacteria may also be visible, but motility is the most striking feature in a wet mount.

Gram staining is most commonly used for observing *Campylobacter* morphology. The bacteria appear as thin, spiral, or curved-shaped Gram-negative bacilli. They often have a specific "Gull wings" appearance shown in Fig. 23.3.

Dark-field microscopy is another technique that is highly effective for visualising thin, motile bacteria like *Campylobacter*. In dark-field microscopy, light is directed onto the

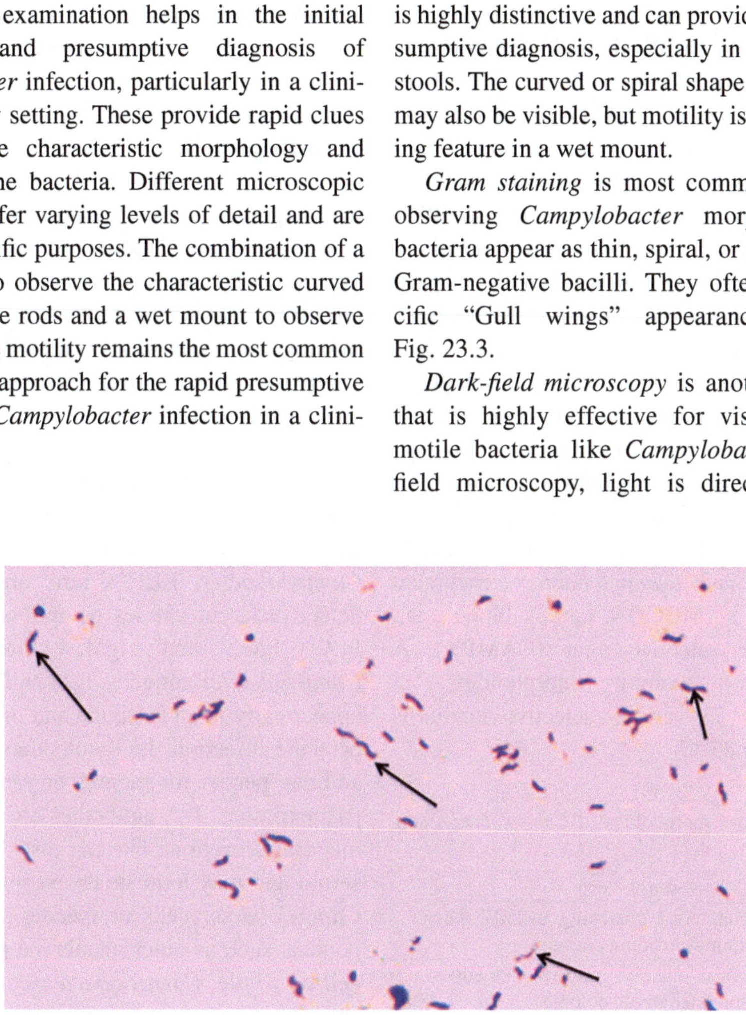

Fig. 23.3 Gram-stained smear of *Campylobacter jejuni* showing thin, spiral, or curved-shaped Gram-negative bacilli with "Gull wing" appearance (X1000). (Mushi, Martha & Paterno, Laurent & Tappe, Dennis & Deogratius, Anna & Seni, Jeremiah & Moremi, Nyambura & Mirambo, Mariam & Mshana, Stephen & Mushi, Martha. (2014). Evaluation of detection methods for Campylobacter infections among under-fives in Mwanza City, Tanzania. Pan African Medical Journal. 19. https://doi.org/10.11604/pamj.2014.19.392.4242. Open-access article distributed under the terms of the Creative Commons Attribution License (CC BY))

specimen from the sides, so only light scattered by the specimen is collected by the objective lens. This creates a bright image of the bacteria against a dark background, significantly enhancing their visibility, especially their motility. The rapid, darting movement of *Campylobacter* is particularly striking and easily observed using dark-field microscopy of a wet mount.

In Vitro Culture

Selective media typically include antibiotics to inhibit the growth of other bacteria. One of the most used selective media is the modified charcoal cefoperazone deoxycholate agar (CCDA) medium. This contains cefoperazone, charcoal, and deoxycholate to inhibit other bacteria. Other selective media include Skirrow's medium, which includes a combination of antibiotics such as polymyxin B, trimethoprim and vancomycin to inhibit other bacteria.

In 1977, Bolton and Robertson developed a selective Preston medium, especially designed to separate *Campylobacter* from food and environmental tests. Since then, many other selective broths and media have been developed for the same purpose. These include Bolton broth, Preston broth and *Campylobacter* enrichment broth, as well as MCCDA, campy blood agar, *Campylobacter* selective agar (CAMPY). A description of colony morphology of *Campylobacter* in various selective media is shown in Table 23.1.

Table 23.1 Colony morphology of *Campylobacter* spp. in various media

Media	Colony morphology
CCDA media	Flat, moist, glistening, metallic sheen, runny/spreading colonies
Skirrow's agar	Moist, mucoid, grey, irregular edges; non-haemolytic colonies
Campy-BAP agar	Round to irregular, glistening, grey wet appearance
Preston agar	Grey, moist colonies, enhanced isolation

Originally, the separation of the *Campylobacter* species was dependent on culture-based techniques, which were necessary to identify bacterial strains responsible for human infections. Despite the progress in culture-based methods, *Campylobacter* species may be difficult to isolate in clinical laboratories due to the sensitivity of bacteria to temperature and oxygen conditions. In addition, the actual processes for isolation can vary between countries and laboratories, which contributes to nonconformity in clinical accuracy.

Biochemical tests are also used to identify *Campylobacter* species, shown in Table 23.2.

Serodiagnosis

Serodiagnosis plays a role by detecting the presence of serum antibodies produced by the host immune system in response to the bacteria. These tests are valuable in certain situations, such as (i) investigating outbreaks, (ii) identifying the cause of post-infectious complications, or (iii) in epidemiological studies.

Enzyme-Linked Immunosorbent Assay (ELISA) ELISA is a widely used serodiagnostic technique for detecting antibodies against *Campylobacter*. ELISA tests are designed to detect different classes of antibodies, typically IgA, IgG, and IgM, directed against *Campylobacter* antigens. IgM antibodies usually appear early in an infection and indicate a recent or acute infection. IgG antibodies develop later and can persist for months or years, suggesting past exposure. IgA antibodies are often found in mucosal secretions but can also be detected in serum and may indicate recent infection. Whole *Campylobacter* cells or specific antigenic components such as outer membrane proteins or flagellin, whole *Campylobacter* cells or specific antigenic components (outer membrane proteins or flagellin) are used as antigens in ELISA. ELISA tests are also designed to directly detect *Campylobacter* antigens in samples like stool.

Table 23.2 Biochemical properties of different *Campylobacter* species

Test	*Campylobacter jejuni*	*Campylobacter coli*	*Campylobacter lari*
Hippurate hydrolysis	Positive	Negative	Negative
Indoxyl acetate hydrolysis	Positive	Positive	Variable
Nalidixic acid susceptibility	Sensitive	Sensitive	Resistant
Cephalothin susceptibility	Resistant	Resistant	Sensitive
Growth at 42 °C	Yes	Yes	Yes
Growth at 25 °C	No	No	Yes

Complement Fixation Test (CFT) CFT is another serological method that has been used to detect antibodies against *Campylobacter*. This test measures the ability of antibodies in a patient's serum to fix complement in the presence of *Campylobacter* antigens. A positive test, indicated by the lack of haemolysis of sensitised sheep red blood cells, suggests the presence of complement-fixing antibodies directed against the *Campylobacter* antigens used in the test. The target antigens in CFT are typically whole-cell preparations or crude antigenic extracts of *Campylobacter*. CFT can detect a rise in antibody titre between the acute and convalescent serum samples, indicating a recent infection.

Western Blot This process, also known as immunoblotting, is a more specific serodiagnostic technique that can identify antibodies in the patient serum that react with specific *Campylobacter* proteins separated by gel electrophoresis. In this method, *Campylobacter* proteins are separated by size, transferred to a membrane, and then incubated with patient serum. Antibodies in the serum that bind to specific protein bands are then detected. Western blot allows for the identification of the specific protein antigens (flagellin, major outer membrane proteins, or other immunogenic proteins) that elicit an antibody response during infection. This technique can be particularly useful in research settings or for confirming serological results by identifying reactivity to known *Campylobacter* virulence factors or other characteristic proteins.

Latex Agglutination Tests (LAT) LAT offer a rapid method for the presumptive identification of *Campylobacter*. These tests typically use latex particles coated with antibodies that bind to specific cell surface antigens. When a suspension of *Campylobacter* bacteria, usually from a culture, is mixed with the sensitised latex particles, the antibodies on the particles bind to the corresponding antigens on the bacterial surface, causing the latex particles to clump together (agglutinate). This visible agglutination indicates the presence of *Campylobacter* with those specific surface antigens. This technique is often used as a quick confirmation test for colonies grown on culture media.

Antigen Detection

Demonstration of specific antigens in stool samples or other clinical samples by enzyme immunoassay (EIA) and immunochromatography test is the most commonly used method for identifying *Campylobacter*. These tests detect *Campylobacter* surface antigen, including *C. jejuni* lipopolysaccharides, and the outer membrane proteins of *C. jejuni* and *C. coli*. These tests are carried out when *Campylobacter* stool cultures are negative or are not feasible. The sensitivity of these tests varies from 88% to 98% whereas the specificity varies from 85% to 95%.

Molecular Diagnosis

In recent years, molecular methods have become increasingly popular for the detection and identification of *Campylobacter* species because of their ability to produce more accurate and quick results.

Polymerase Chain Reaction (PCR) PCR is used for *Campylobacter* detection. This works by amplifying specific regions of the *Campylobacter* genome, allowing for the detection of even small numbers of bacteria in a sample. Primers, short synthetic DNA sequences, are designed to bind to conserved genes within the *Campylobacter* genus or to genes specific to particular *Campylobacter* species, such as *C. jejuni* or *C. coli*. Common genetic targets for *Campylobacter* detection by standard PCR include the 16S rRNA gene. This gene is present in all bacteria but has species-specific variable regions, and genes encoding virulence factors or essential proteins like *cadF* (encoding a fibronectin-binding protein) or *ceuE* (involved in iron uptake). The presence of amplified DNA fragments of the expected size, typically visualised using gel electrophoresis, indicates a positive result.

Real-Time PCR (qPCR) It is an advanced form of PCR that allows for the monitoring of DNA amplification in real-time as the reaction progresses. This provides faster results and can also be quantitative, estimating the initial amount of *Campylobacter* DNA in the sample. qPCR assays for *Campylobacter* utilise fluorescent probes or dyes that bind to the amplified DNA. The increase in fluorescence is measured in real-time, and a positive result is determined based on the threshold cycle value. qPCR assays often target similar genes as standard PCR, such as the 16S rRNA gene or species-specific genes like *hipO* (hippuricase gene, specific for *C. jejuni*) or *gyrA* (encoding a subunit of DNA gyrase, with species-specific sequences). The use of fluorescent probes designed for different targets also allows for the differentiation of multiple *Campylobacter* species in a single reaction.

Multiplex PCR This enables the simultaneous amplification and detection of multiple gene targets in a single reaction. This is particularly useful for identifying and differentiating several *Campylobacter* species or for simultaneously detecting *Campylobacter* and other enteric pathogens in a single test. Multiplex PCR assays for *Campylobacter* use multiple sets of primers, each specific to a different target gene or species. For instance, a multiplex PCR panel might include primers targeting genes specific to *C. jejuni* (*hipO*, *mapA*), *C. coli* (*ceuE*, *glyA*), *C. lari* (*glyA*, *gyrA*), and *C. fetus* (*sapB2*). The different primer sets are designed to produce amplification products of different sizes. These can then be distinguished using gel electrophoresis, allowing for the identification of the *Campylobacter* species present in the sample.

Loop-Mediated Isothermal Amplification (LAMP) LAMP is another molecular technique that offers a rapid, sensitive, and cost-effective approach for detecting *Campylobacter*. The target for *Campylobacter* LAMP assays is typically a conserved gene sequence present in *Campylobacter* species, or specific genes for differentiation of species like *C. jejuni* or *C. coli*, such as the *rplD* gene, *cdtC* gene, or *gyrA* gene. A significant advantage of LAMP is that the reaction can often be visually detected, either by the turbidity caused by the accumulation of magnesium pyrophosphate, a reaction by-product, or by using fluorescent dyes that change colour in the presence of amplified DNA. This eliminates the need for specialised detection equipment like gel electrophoresis, making it suitable for use in resource-limited settings or for on-site testing.

Matrix-Assisted Laser Desorption/Ionisation Time-of-Flight Mass Spectrometry (MALDI-TOF MS) MALDI-TOF MS is a technique increasingly used in clinical microbiology for rapid and accurate identification of microorganisms, including *Campylobacter*. For *Campylobacter*, MALDI-TOF MS rapidly identifies the species from isolated colonies, often within minutes, targeting the characteristic ribosomal protein peaks that differentiate *Campylobacter* species like *C. jejuni*, *C. coli*, and others.

Diagnosis of *Campylobacter* species is described in Table 23.3.

Table 23.3 Diagnosis of *Campylobacter* infections

Method	Description	Target	Turn-around time	Strengths	Limitations
Microscopy Wet mount Gram stain Dark field Phase contrast	Direct visualisation of bacteria and its motility	Bacteria	Minutes	Immediate presumptive result	Low sensitivity, Non-specific, Requires fresh samples
In vitro culture CCDA media Skirrow's agar Campy-BAP agar Preston agar	Growth on selective media under suitable conditions	Bacteria	48–72 hours	Allows antimicrobial susceptibility testing	Time-consuming; requires specific environment; low sensitivity in treated cases and low load cases
Serodiagnosis ELISA LAT CFT	Detection of *Campylobacter* antigen using antibodies and colorimetric reaction	Antibodies and antigens	1–4 hours	Rapid and easy to perform	Low sensitivity; no antimicrobial data
Molecular diagnosis PCR Real-time PCR LAMP	Amplification of *Campylobacter* DNA	Genes like 16S rRNA, hipO, gyrA etc.	2–6 hours	High specificity and sensitivity Fast and detect non-viable bacteria	Expensive; requires molecular lab setup Not widely available in all labs
MALDI-TOF MS	Identifies species based on mass spectroscopy	Protein profile	24–48 hours	Fast species-level identification	Requires culture growth; expensive equipment

Challenges in Diagnosis

Isolation of *Campylobacter* faces various challenges.

Campylobacter tends to die early during the transport and handling of samples. Transport media can prolong the survival of bacteria to some extent, but the exact duration of storage is poorly understood. In addition, special growth media and microaerophilic chambers are needed for the culture of *Campylobacter* and require at least 48–72 hours for growth. Identifying microscopic colonies of *Campylobacter* among the faecal flora is also difficult. Even minor changes from its ideal growth requirements can lead to unsuccessful isolation.

Secondly, the ability of *Campylobacter* to produce biofilms makes isolation more challenging. Biofilms act as a protective shield around the bacteria, making it difficult to isolate them from coexisting microorganisms, and also provide them with protection from external stresses.

Additionally, some *Campylobacter* species, which are viable but non-culturable (VBNC), are a clinical challenge. These species maintain their physical and virulent properties but are unable to grow in culture. This can lead to a false-negative diagnosis, causing the patient to continue on antibiotics that might actually be ineffective. Misuse of antibiotics can lead to antibiotic resistance.

Treatment

First line of therapy to combat *Campylobacter* infection is supportive care. This involves maintaining adequate hydration to offset fluid loss from diarrhoea and vomiting. ORS is often recommended, especially in children and the elderly. For most patients, there is improvement in symptoms within a week without the need for antibiotics.

However, in some cases, such as severe symptoms or immunocompromised individuals, antibiotics can be prescribed. Preferred antibiotics are macrolides or fluoroquinolones. However, resistance to antibiotics, especially fluoroquinolones, is increasing, so the choice of therapy must be based on local resistance patterns and clinical decisions.

In rare cases, *Campylobacter* infection can cause complications such as Guillain-Barré syndrome, an autoimmune condition affecting the nervous system, leading to muscle weakness and paralysis. In such examples, further medical interventions such as immunotherapy may be necessary.

Antibiotic Resistance Antibiotic resistance in *Campylobacter* is associated with increased healthcare costs, long-term illness, increased hospital stays and high mortality and morbidity. Progress in molecular techniques has made detecting resistance genes in *Campylobacter* strains easier from different sources, including food, water, animals, and environmental samples. These studies have identified many resistance genes, such as gyra, gyrb, parc, pare (related to quinolone and fluoroquinolone resistance), bla-oxa-61, and blax-184 (β-lactamase genes). The mechanisms causing *Campylobacter* resistance include target modification (DNA gyrase target modification; Thr-86-lle, Asp-90-Asn), enzyme inactivation, altered membrane permeability (MOMP), efflux pump overexpression (CmeABC) and acquired resistance genes via horizontal gene transfer. These mechanisms reduce antibiotics' effects, enabling *Campylobacter* to survive and grow in the presence of antimicrobial agents.

Prevention and Control

Preventing *Campylobacter* infections involves appropriate practices for food handling. This includes thoroughly cooking poultry at the right temperatures, avoiding cross-contamination in the kitchen and ensuring access to clean, treated water for both household and community use. Personal hygiene should also be maintained through proper hand-washing after handling raw meat, after using the bathroom, and before eating to reduce the chance of infection. Additionally, livestock control against *Campylobacter* infection is essential. This can be achieved by improving hygiene through proper and regular cleaning of farms, facilities, and the living areas of the animals. Biosecurity measures during outbreaks are also important. Regular veterinary supervision to control and detect any *Campylobacter* infection in livestock should be encouraged. Ensuring clean, non-contaminated water and feed for livestock can also reduce infection.

The One Health approach is recommended for effective prevention and control measures. To date, no licensed vaccine is available for *Campylobacter*, mainly due to its pathogenesis, antigenic variations, and limited understanding of infectious complications.

Case Study

In 2023, a 34-year-old woman from a rural agricultural community presented to a local health facility with severe abdominal pain, nausea, vomiting, and bloody diarrhoea. She was a farmer and often handled livestock, including chickens and pigs. The patient also had a history of handling raw chicken meat at home. After reaching the hospital, her initial investigation showed increased white blood cells, and stool cultures were positive for *C. jejuni*. This diagnosis was confirmed through PCR. Clinically, it is a case of acute gastroenteritis. The patient's history of close contact with livestock, poultry consumption, and no recent travel history raised substantial doubt about zoonotic transfer. The patient improved with supportive care with ORS (Oral rehydration therapy).

1. What are the other causes of acute bacterial gastroenteritis?
2. Briefly explain the pathogenicity and laboratory diagnosis of *Campylobacter*.

3. What are the treatment and preventive measures for *Campylobacter* infection?

Research Questions

1. What are the mechanisms causing antibiotic resistance in *Campylobacter* and their variation in different geographical locations?
2. How do climate change and specific environmental factors (e.g. temperature, rainfall, water quality) influence the prevalence and seasonality of *Campylobacter* in animal reservoirs and the incidence of human infections?
3. Are there specific *Campylobacter* species or genotypes that are more strongly associated with severe clinical outcomes or post-infectious complications like Guillain-Barré syndrome or irritable bowel syndrome?

Further Readings

Allos BM. *Campylobacter jejuni* infections: update on emerging issues and trends. Clin Infect Dis. 2001;32(8):1201–6.

Blaser MJ, Engberg J. Clinical aspects of *Campylobacter jejuni* and *Campylobacter coli* infections. In: Nachamkin I, Szymanski CM, Blaser MJ, editors. Campylobacter. 3rd ed. Washington, DC: ASM Press; 2008. p. 99–121.

Butzler JP. *Campylobacter*, from obscurity to celebrity. Clin Microbiol Infect. 2004;10(10):868–76.

Centers for Disease Control and Prevention (CDC). *Campylobacter* (Campylobacteriosis) [Internet]. 2022 [cited 2025 Apr 9]. Available from: https://www.cdc.gov/*Campylobacter*/

Cody AJ, Maiden MC, Dingle KE. Genetic diversity and stability of the *Campylobacter* genome. In: Nachamkin I, Szymanski CM, Blaser MJ, editors. Campylobacter. 3rd ed. Washington, DC: ASM Press; 2008. p. 51–66.

Dasti JI, Tareen AM, Lugert R, Zautner AE, Groß U. *Campylobacter jejuni*: a brief overview on pathogenicity-associated factors and disease-mediating mechanisms. Int J Med Microbiol. 2010;300(4):205–11.

Dearlove BL, Cody AJ, Pascoe B, Méric G, Wilson DJ, Sheppard SK. Rapid host switching in generalist *Campylobacter* strains erodes the signal for tracing human infections. ISME J. 2016;10(3):721–9.

European Food Safety Authority (EFSA). *Campylobacter* [Internet]. 2024 [cited 2025 Apr 9]. Available from: https://www.efsa.europa.eu/en/topics/topic/*Campylobacter*

Humphrey T, O'Brien S, Madsen M. *Campylobacter*s as zoonotic pathogens: a food production perspective. Int J Food Microbiol. 2007;117(3):237–57.

Kaakoush NO, Castaño-Rodríguez N, Mitchell HM, Man SM. Global epidemiology of *campylobacter* infection. Clin Microbiol Rev. 2015;28(3):687–720.

Kwan PS, Barrigas M, Bolton FJ, Whiley A, Elviss N, Fox A, et al. The microbiological safety of ready-to-eat salad vegetables: risk factors associated with contamination by *Salmonella*, *Listeria monocytogenes* and *Campylobacter* spp. Food Microbiol. 2013;34(2):385–93.

Luangtongkum T, Jeon B, Han J, Plummer P, Logue CM, Zhang Q. Antibiotic resistance in *Campylobacter*: emergence, transmission and persistence. Future Microbiol. 2009;4(2):189–200.

Moore JE, Corcoran D, Dooley JS, Fanning S, Lucey B, Matsuda M, et al. Campylobacter. Vet Res. 2005;36(3):351–82.

Olvera-Ramírez AM, McEwan NR, Stanley K, Nava-Diaz R, Aguilar-Tipacamú G. A systematic review on the role of wildlife as carriers and spreaders of *Campylobacter* spp. Animals (Basel). 2023;13(8):1334.

Zhang Q, Beyi AF, Yin Y. Zoonotic and antibiotic-resistant *Campylobacter*: a view through the one health lens. One Health Adv. 2023;1:4.

Vibrio Infections

24

Tuhina Banerjee

Learning Objectives

1. To understand the key concepts of zoonotic transmission of *Vibrio* spp. and the environmental factors contributing to outbreaks
2. To provide an overview of the pathogenesis and the clinical manifestations of infections in humans and marine animals
3. To explore the current diagnostic methods, treatment strategies, and challenges in vaccine development

Introduction

Vibrio spp. are a group of Gram-negative bacilli that inhabit natural freshwater and marine environments as their natural habitat. There are many species of *Vibrio*, with numerous genetic and biological similarities among species. There is genetic diversity among the infections caused by these species, but they all originate in marine and aquatic environments. These *Vibrio* spp. thrive in salty environments, and their numbers tend to increase as the temperature rises in natural reservoirs. Among the prominent species, *Vibrio cholerae*, *Vibrio parahaemolyticus*, *Vibrio vulnificus*, and *Vibrio alginolyticus* are important members. *Vibrio* spp. are the most common pathogens in aquatic environments and seafood, contributing to human illness. They exhibit seasonal variations and are primarily found in estuarine environments from May to October, when the coastal water temperature is high. Infections caused by *Vibrio* spp. often manifest in people when they encounter contaminated water or consume seafood that has not been properly prepared.

T. Banerjee (✉)
Department of Microbiology, Institute of Medical Sciences, Banaras Hindu University, Varanasi, Uttar Pradesh, India
e-mail: drtuhina105@bhu.ac.in

History

V. cholerae was described as the causative agent as early as 1884 by Filippo Pacini. During the same period, John Snow, a physician, linked the causation of cholera to contaminated water, thus laying the groundwork for understanding infectious diseases for the first time. However, these observations were overlooked, and it was Robert Koch who, for the first time, isolated *V. cholerae* in pure culture in 1883 and established it as the causative agent. Later, in 1959, Sambhu Nath De demonstrated the role of the cholera toxin as the potential virulence attribute causing the disease.

V. parahemolyticus was first identified in 1950 by Tsunesaburo Fujino from a seafood outbreak in Japan. Since then, food poisoning caused by this bacterium has been linked to major outbreaks worldwide.

S. C. Parija, A. Poddar (eds.), *Textbook of Bacterial Zoonoses*, Microbial Zoonoses,
https://doi.org/10.1007/978-981-95-6615-0_24

V. vulnificus was first isolated in 1964 by the Centers for Disease Control and Prevention. Initially, it was thought to be a virulent strain of *V. parahemolyticus*. It was labelled as a distinct species in 1970, when it was recognised to cause severe wound infections and septicaemia. The nomenclature of *V. vulnificus* was suggested by J Centres Farmer in 1980. Since its discovery, several outbreaks of *V. vulnificus* have been recorded that have occurred throughout the world.

V. alginolyticus was first identified as a human pathogen in 1973 and labelled as a distinct species in 1974. Otherwise, it was considered a biotype of *V. parahemolyticus*. Found in the marine environment, this organism is an opportunistic pathogen that attacks a compromised immune system when ingested in improperly cooked seafood.

Taxonomy

The genus *Vibrio* belongs to the family Vibrionaceae, phylum Pseudomonadota, class Gammaproteobacteria, and order Vibrionales.

Vibrio cholerae is the type species of this genus. To date, the genus comprises 137 validly published species isolated from various ecosystems.

Based on its potential for zoonotic transmission, *V. vulnificus* is the only true zoonotic Vibrio, because cases of secondary septicaemia in humans transmitted from diseased fish have been reported with this organism. Besides, *V. alginolyticus*, *V. parahaemolyticus,* and occasionally *V. cholerae* have also been linked with zoonotic transmission.

Traditional taxonomy based on phenotypic traits has been significantly improved by whole-genome sequencing (WGS). This provides high-resolution data, enabling accurate classification and phylogenetic mapping. Comparative genomics using average nucleotide identity (ANI) and multilocus sequence typing (MLST) has helped to distinguish between closely related species and subspecies, especially among clinical and environmental isolates.

Genomics and Proteomics

The genome of the *V. vulnificus* is nearly 4.8 Mbp in size and contains two circular chromosomes of approximate sizes of 3.37 Mbp and 1.85 Mbp, respectively. The mean G+C content is 46.8%. It does not consist of any plasmids. In addition, the genome contains around 4617 protein-coding genes, 172 RNA genes, and type I, II, and III secretion systems. On phylogenetic analysis, 5 lineages of *V. vulnificus* could be identified. The L1 lineage consists of a mixture of clinical and environmental isolates, the L2 lineage consists of a mixture of biotypes 1 and 2, the L3 lineage consists of biotype 3, and L4 and L5 consist of biotype 1 Spanish and biotype 1 Israel, respectively.

The *V. alginolyticus* genome size ranges from 4.9 to 5.2 Mbp and consists of two circular chromosomes. These genomes have a G+C content of around 44%, which is typical of *Vibrio* species. The number of coding genes ranges from about 4400 to 4800. Additionally, there can be one or two extrachromosomal plasmids.

Proteomic analysis of *V. vulnificus* has revealed several proteins that are involved in various functions, such as initial attachment to host cells, invasion, and production of toxins. Among the major proteins, OmpU, which is a fibronectin-binding protein, is involved in attachment to host tissues. VvhA, a haemolysin/cytolysin gene, is involved in invasiveness and tissue damage. The RTX toxin, which is a gene cluster encoding the RTX toxin, plays a crucial role in the pathogenesis of *V. vulnificus* infections. Studies have also shown that, of all the proteins in the virulent strains of *V. vulnificus*, approximately 40.34% were acetylated. The acetylated proteins are involved in various metabolic processes, binding functions, the cytoplasm, and multiple central metabolic pathways. In this context, lysine acetylation is widespread in *V. vulnificus,* which can affect the virulence and antibiotic resistance of the strains.

V. alginolyticus has a type III secretion system (T3SS) as one of the important virulence factors, with *vopR* being one of the effector genes. Studies on transcriptomic analysis have revealed that overexpression of *vopR* is associated with the upregulation of 410 genes and the downregulation of 207 genes. Similarly, for this protein, proteomic analysis has shown that 126 proteins were upregulated and 518 proteins were downregulated. Gene Ontology (GO) enrichment analysis has shown that the differential genes were significantly enriched in various biological processes (e.g. cellular processes, metabolic processes, and biological regulation) as well as in cellular components such as cell parts and membranes, and molecular functions such as binding and catalytic activity. These results suggest that *vopR* has a significant impact on the cytoskeleton, metabolism, and immune regulation of host cells during the pathogenic process of *V. alginolyticus*.

Morphology

Vibrio spp. are curved and comma-shaped Gram-negative bacteria (Fig. 24.1). They are typically small, approximately 1–2 μm in length and 0.2–0.4 μm in width by Gram stain. They are comma- or S-shaped when two or more cells are joined end to end.

In diarrhoeal samples of cholera, the stained smears of mucus flakes are often described as having a "*fish in stream*" appearance. They are highly motile bacteria; motility is attributed to a single polar flagellum. Their active motility is often described as darting motility, during which they may appear as a "*swarm of gnats*".

Cultivation

Vibrios are commonly aerobic, with a few species being tolerant of anaerobic conditions. However, growth is slow in the absence of oxygen. They are mesophilic and typically thrive within a temperature range of 16–40 °C, with an optimum of 37 °C. They also have specific pH and salt requirements for growth. They require an alkaline pH within a range of 6.4–9.6 and a salt concentration of 1% for optimum growth.

Vibrios are cultured on ordinary media such as MacConkey agar, where colourless colonies are observed. On prolonged incubation, a reddish colour develops in these colonies due to lactose fermentation. On blood agar, haemolytic colonies are seen. Thiosulfate citrate bile salts sucrose (TCBS) agar is a selective differential medium used for isolating *Vibrio* spp. On TCBS agar, *V. cholerae* produces smooth, round, and yellow colonies due to acid production from sucrose fermentation. Non-sucrose fermenters such as *V. parahaemolyticus* and *V. vulnificus* form green or bluish-green colonies.

Alkaline peptone water (APW) is used as an enrichment broth (pH ~8.5). It enhances the growth of *Vibrio* spp., especially when bacterial counts are low.

Animal Models

The study of animal models has been historical with reference to *V. cholerae* and the disease of cholera. The commonly used models include

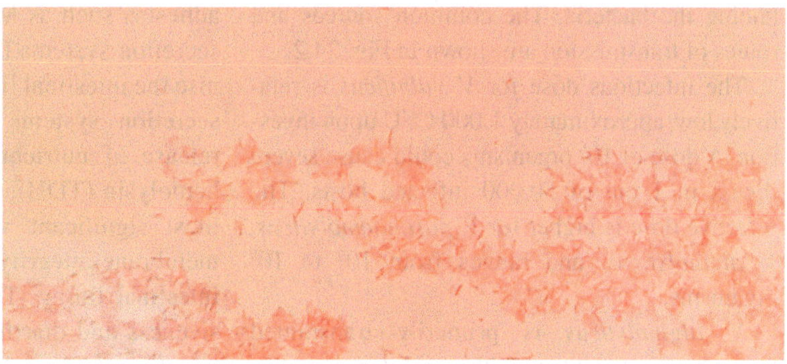

Fig. 24.1 Gram-stained smear showing Gram-negative *Vibrio parahemolyticus* in curved morphology (100X). (Image source: Author's personal collection)

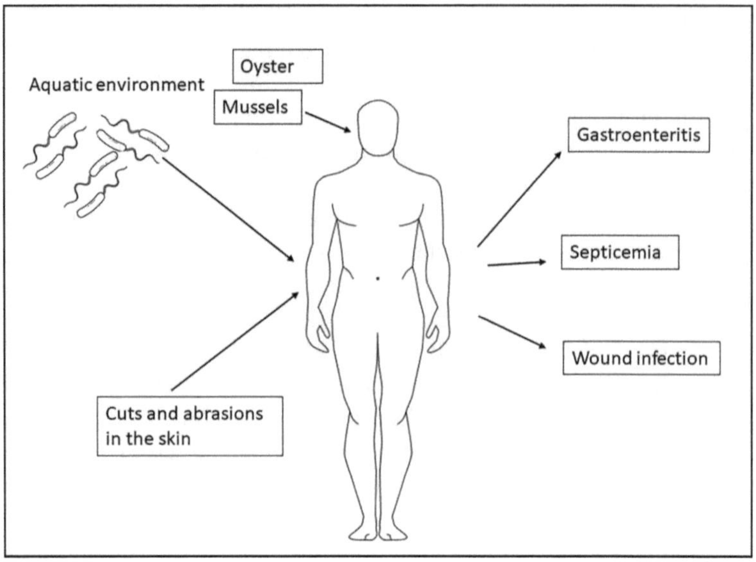

Fig. 24.2 Transmission of zoonotic *Vibrio* species, mainly *Vibrio vulnificus*. (Image source: Author's personal collection)

infant mice and rabbits. Studies on zebrafish (*Danio rerio*) as a natural host have also yielded results.

For *V. vulnificus,* besides mice and zebrafish, the silkworm model has also been used. This model has been used to isolate attenuated *V. vulnificus* mutants. The *Caenorhabditis elegans* model can also be used for *V. vulnificus*. All these models have also been used for *V. alginolyticus*.

Transmission Cycle

V. vulnificus is mainly transmitted through the consumption of raw or undercooked seafood, especially oysters, or through having open wounds exposed to contaminated seawater containing the bacteria. The common sources and routes of transmission are shown in Fig. 24.2.

The infectious dose for *V. vulnificus* is relatively low, approximately 1,000 CFU upon ingestion. A dose of 10^6 organisms could cause severe illness in 1 out of 50,000 infected hosts. The infective dose is higher for *V. parahaemolyticus*, *V. alginolyticus,* and ranges from 10^6 to 10^8 organisms.

V. alginolyticus is primarily transmitted through direct contact with seawater contaminated with the bacteria. It occurs either during swimming in seawater or through cuts, abrasions, and open wounds on the skin. It is also transmitted indirectly through contact with fish, shellfish, and sponges that are colonised by this organism. The infection is also acquired by consuming raw or undercooked seafood, mostly molluscan shellfish such as clams, oysters, and crabs.

V. parahaemolyticus often spreads from one person to another through contaminated seafood, such as oysters. Some outbreaks of *V. cholerae* (non-O1 and non-O139 strains) have been linked to the consumption of contaminated shellfish, including raw oysters and crabs.

Pathogenesis and Pathology

Inside the host, *V. parahaemolyticus* utilises adhesins such as MAM-7, toxins, and type III secretion systems (T3SS1 and T3SS2) to colonise the intestinal lining and cause damage. The secretion systems cause cell lysis and the release of nutrients. The thermostable direct hemolysin (TDH), a pore-forming toxin, is the most significant virulence factor, disrupting membrane integrity and causing damage to the intestinal lining. This contributes to the invasiveness and diarrheal symptoms caused by *V. parahemolyticus*. Additionally, the flagella and capsules of the bacteria also aid in colonisation.

V. vulnificus possesses several virulence factors that account for disease severity and outcome. The *V. vulnificus* haemolysin is a potent haemolysin that can cause cell lysis in diverse types of host cells, thus leading to tissue damage. Once inside the host, *V. vulnificus* undergoes rapid multiplication, often accompanied by severe inflammation. This causes necrosis and severe skin infections. The lipopolysaccharide of the bacteria can induce inflammation and septic shock. Additionally, it has several mechanisms to acquire iron, which is required for its survival. Besides common factors such as capsules, pili, and flagella, various proteolytic enzymes also contribute to its virulence within the host cells. The severity of the infections depends on the host's immune status and other factors, such as chronic liver disease and iron overload, which can exacerbate the disease. The pathogenesis is illustrated in Fig. 24.3.

V. alginolyticus also causes pathogenicity by similar mechanisms, including adhesion, colonisation, and tissue damage mediated through various adhesins, toxins, proteases, and others.

Immunology

The Innate Immunity

For any *Vibrio* infection, the body's initial defence is mounted by the innate immune system, which relies on physical barriers, such as the intestinal mucosa and skin. Furthermore, innate immune cells such as neutrophils release antimicrobial peptides, including calprotectin, which limit bacterial growth by sequestering essential metals. Phagocytes, such as macrophages and neutrophils, are also vital, engulfing and destroying *Vibrio* bacteria, a process enhanced by opsonisation. Toll-like receptors (TLRs), especially TLR4, recognise bacterial components such as lipopolysaccharide (LPS), initiating inflammatory responses and the production of cytokines and chemokines. While inflammation is a crucial defence mechanism, an excessive inflammatory response, especially in systemic infections caused by *V. vulnificus*, can contribute to severe outcomes, such as septic shock.

The Adaptive Immunity

Following the innate immune response, the adaptive immune system provides specific and long-lasting protection. Humoral immunity, mediated by antibodies, plays a pivotal role. Cellular immunity, driven by T cells, also plays a crucial role in protection. T helper (Th) cells, including both Th1 and Th2 subsets, are activated. Th1 responses, which activate phagocytes, are typically associated with intracellular pathogens. In contrast, Th2 responses that promote B cell acti-

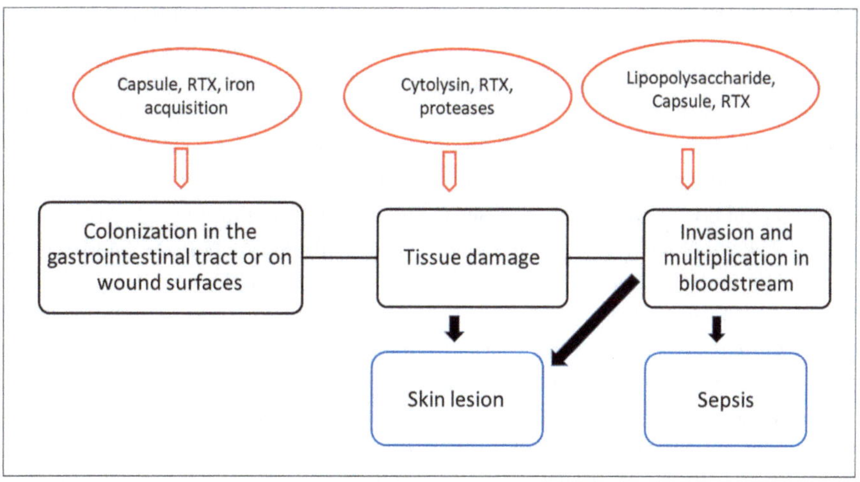

Fig. 24.3 Pathogenesis of *Vibrio vulnificus* infections. (Image source: Author's personal collection)

vation and antibody production are more relevant for extracellular pathogens.

In addition to the general immune responses, immunological mechanisms vary considerably among individual *Vibrio* species, reflecting species-specific host–pathogen interactions and virulence strategies. *V. parahaemolyticus* possesses virulence factors such as haemolysis (thermostable direct hemolysin, TDH; and TDH-related hemolysin, TRH) and type III secretion systems (T3SS). The immune response to *V. parahaemolyticus* involves the recognition of these virulence factors and other bacterial components, leading to inflammation and antibody production.

Infections due to *V. vulnificus* evoke both innate and adaptive immune responses. The severity of the infection and the outcome of the disease are affected by these responses. The major virulence factors, such as the capsular polysaccharide and the RTX toxin, play a crucial role in immune evasion. *V. vulnificus* infection triggers an inflammatory response, thus leading to the release of cytokines such as TNF and IL-8. *V. parahemolyticus* also exhibits comparable immune responses, although acute inflammatory responses are largely self-limiting. Humoral immunity develops particularly against the direct hemolysin (TDH) and lipopolysaccharide (LPS) of the bacteria.

Infection in Humans

Vibrio species cause vibriosis in humans. The infectious dose is relatively low for *V. vulnificus*, approximately 1000 CFU upon ingestion. A dose of 10^6 organisms could cause severe illness in 1 out of 50,000 infected hosts. The infective dose is higher for *V. parahemolyticus*, *V. alginolyticus*, and ranges from 10^6 to 10^8 organisms.

The usual presentation of severe vibriosis caused by *V. vulnificus* is septicaemia, marked by fever, chills, low blood pressure, often accompanied by large blistering lesions and necrotising fasciitis. It also causes wound infections, marked by symptoms of redness, swelling, and pain, along with blisters and discharge. Mild infections cause watery diarrhoea with vomiting and abdominal pain. On the other hand, *V. alginolyticus* typically causes gastroenteritis, characterised by watery diarrhoea, abdominal cramps, nausea, vomiting, and occasionally fever. Wound infections or bloodstream infections are rare.

V. parahaemolyticus causes mild-to-moderate gastrointestinal symptoms like watery diarrhoea, abdominal cramps, nausea, and sometimes vomiting. These symptoms usually appear within a few hours to a few days after consuming contaminated food.

An overview of the clinical characteristics of infections caused by these pathogens is presented in Table 24.1.

Infection in Animals

V. vulnificus, *V. alginolyticus*, and *V. parahaemolyticus* present significant threats to various animal populations, particularly in aquatic environments. These bacteria are part of the natural flora of marine and brackish waters, but they can become opportunistic pathogens, causing disease when animals are stressed or their immune systems are compromised.

V. vulnificus causes vibriosis in fish, especially in farmed eels and tilapia, as well as in crustaceans such as shrimp, lobsters, and crabs. The infection begins when bacteria attach to the fish's gills. Here, the bacteria multiply through biofilm formation, releasing toxins that cause inflammation and ultimately lead to septicaemia. Fish infected with *Vibrio* spp. often exhibit restlessness, skin lesions, exophthalmia, and finally death. Splenic enlargement, abdominal dropsy, intestinal inflammation, epidermal haemorrhage, scale exfoliation, pop-eye, and tail decay have also been reported.

V. alginolyticus is a significant concern in aquaculture, as it is a common pathogen affecting a wide range of marine aquatic animals, including fish and crustaceans. It causes marked economic losses in the aquaculture industry due to disease outbreaks. In marine fish, *V. alginolyticus* can cause severe haemorrhagic septicaemia and other

Table 24.1 *Vibrio vulnificus*, *Vibrio alginolyticus*, and *Vibrio parahaemolyticus* infections in humans

Characteristics	*Vibrio vulnificus*	*Vibrio alginolyticus*	*Vibrio parahaemolyticus*
Transmission	Ingestion of raw/undercooked seafood (esp. oysters); wound exposure to seawater	Wound exposure to seawater; ear infections from seawater	Ingestion of raw/ undercooked seafood (esp. shellfish)
Incubation period	~12–72 h	24–48 h (wound/ear); variable	4–96 h
Main clinical syndromes	Septicaemia (primary), Severe wound infections, necrotising fasciitis, gastroenteritis (less common)	Otitis externa (swimmer's ear), wound infections (mild-to-moderate), rare systemic disease	Acute gastroenteritis, watery or bloody diarrhoea, nausea, vomiting, abdominal cramps
Severity	*High*: life-threatening in immunocompromised or liver disease patients	*Low-to-moderate*: usually self-limiting	*Mild-to-moderate*: self-limiting in most cases
Systemic infection risk	Very high, esp. in liver disease or immunocompromised	Rare	Rare
Mortality rate	Up to 50% in septicaemia cases	Very low	Very low (<1%)
Common risk factors	Liver disease, immunosuppression, diabetes, iron overload, raw oyster consumption	Swimming in warm seawater with open wounds	Consumption of raw/ undercooked seafood, especially shellfish
Notable features	Highly virulent; can cause rapidly fatal infections	Least virulent among the three; common in superficial infections	Major cause of seafood-borne gastroenteritis outbreaks

illnesses. It has been identified as a pathogen in *Penaeus vannamei* (shrimp), *Epinephelus coioides* (grouper), *Crassostrea rhizophorae* (oyster), *Pseudosciaena crocea* (large yellow croaker), and *Sparus latus*. Clinical signs in affected aquatic animals include lethargy, anorexia, dermal ulceration, exophthalmia ("*pop-eye*"), gill necrosis, and darkened skin. In shrimp, it causes shell disease and systemic vibriosis, characterised by lesions on the cuticle, cloudy muscle tissue, and necrosis of organs such as the hepatopancreas.

V. parahaemolyticus is a significant pathogen in various aquatic animals, particularly in the aquaculture industry. It is a major cause of disease in fish, shellfish, and penaeid shrimp. Notably, *V. parahaemolyticus* is considered as the key agent causing acute hepatopancreatic necrosis syndrome (AHPNS), also known as early mortality syndrome (EMS), in shrimp. This devastating disease has caused high mortalities and substantial economic losses in shrimp farming globally. Redness of the body, cuts on the antennae, atrophy of the hepatopancreas, white faeces, and luminescence in the dark are the frequent clinical signs in the shrimp associated with V. *parahaemolyticus* infection.

Epidemiology and Public Health

Vibriosis in humans, caused by zoonotic *Vibrio* spp., mainly presents with gastroenteritis, wound infections, and septicaemia in severe cases. The zoonotic vibrios prefer tropical to subtropical climates, with reports from Asia, North America, and Europe. They are usually isolated from water where temperatures range from 9 °C to 31 °C. Low-to-moderate levels of salinity in the marine environment are associated with the presence of *V. vulnificus*. Consequently, the onset of both systemic and wound infections in humans occurs during the months of April and September in the USA. It is predominantly found along the coastlines of those states that border the Gulf of Mexico. Almost all cases of primary septicaemia and gastroenteritis reported are associated with the consumption of raw shellfish and oysters. The number of people infected worldwide with *V. vulnificus* is low compared with other vibrios. However, in the USA, it is the cause of a substantial percentage of Vibrio-related illnesses, the severity of which makes it the most probable cause of seafood-related mortalities.

There have been several outbreaks of *Vibrio* spp. worldwide, across various countries. In 2005, after Hurricane Katrina, *V. vulnificus* was identified among the evacuees from New Orleans. In 2015, *V. vulnificus* was reported in Florida, causing two deaths among the 8 cases reported. The next hurricane, Ian, in 2022, witnessed many cases of *V. vulnificus* infections. Sporadic cases are on record even in recent times.

Vibriosis is also the oldest bacterial disease among the marine vertebrates. The condition affects aquaculture systems, leading to major fish diseases in both cultured and wild fish. The condition also leads to substantial mortality in fish, shellfish, and crustaceans. The outbreaks of *V. alginolyticus* have been more commonly reported among fish populations, causing significant losses due to high mortality rates. Among fish, vibriosis is typically manifested at a temperature of 220 °C and hence is called warm water vibriosis. These occur as outbreaks or epizootics among tilapia and eel species, even in aquaculture, causing varying rates of mortality. As in human infections, the clinical signs are similar, ranging from abdominal petechiae, haemorrhages at the base of the dorsal and anal fins, and redness in the operculum region.

The global distribution of a few major outbreaks has been shown in Fig. 24.4.

Diagnosis

Specimens

The common specimens for diagnosis of foodborne zoonotic *Vibrio* infections include stool, wound swabs, and blood for blood culture. In addition, food and water samples are also tested for the identification of the source of infection.

Microscopy

Vibrio spp. are curved, comma-shaped Gram-negative bacteria. A hanging drop preparation directly prepared from samples such as watery stool might reveal the active motility of *Vibrio* spp., especially as "darting motility".

In Vitro Culture

In vitro culture stands as the "gold standard" for the definitive laboratory diagnosis of *Vibrio* infections. Specimens such as stool, pus from wounds, or blood are often used for culture.

Vibrio isolation begins with an enrichment step, especially for samples with low bacterial loads or heavy contamination, such as stool.

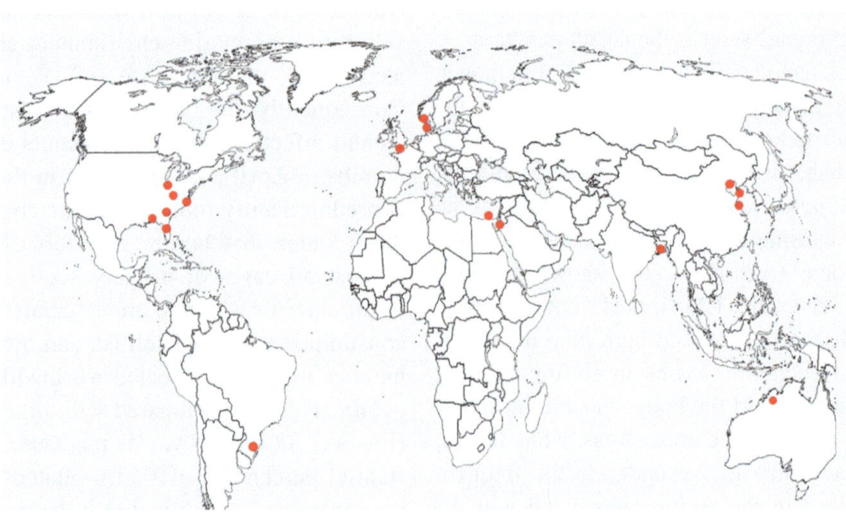

Fig. 24.4 Global distribution of *Vibrio vulnificus* infections causing major outbreaks. (Image source: Author's personal collection)

Alkaline peptone water (APW), with its high pH (typically 8.4–8.6), is the preferred enrichment broth. It selectively inhibits the growth of most other enteric bacteria while promoting the proliferation of *Vibrio*. After a 6–8-h or overnight incubation at 37 °C, the surface pellicle from the APW is subcultured onto selective agar to enhance the likelihood of isolating *Vibrio* species.

Following enrichment, stool specimens are streaked onto specialised selective and differential agar media. Thiosulfate citrate bile salts sucrose (TCBS) agar is the most widely utilised medium for *Vibrio* spp. Its high pH and bile salts selectively inhibit most non-*Vibrio* bacteria, while sucrose serves as a fermentable carbohydrate for differentiation, with pH indicators revealing the colour change. On TCBS, *V. vulnificus* typically forms green or translucent colonies due to its general inability to ferment sucrose. *V. alginolyticus* and *V. parahaemolyticus* both produce large, yellow colonies, indicating that they ferment sucrose. Incubation conditions for *Vibrio* cultures are typically 37 °C under aerobic conditions, as they are facultative anaerobes.

Newer chromogenic Vibrio agars (e.g. CHROMagar™ Vibrio) offer improved differentiation by utilising enzymatic reactions that result in distinct, species-specific colony colours, potentially reducing the need for extensive follow-up biochemical tests.

Once colonies are isolated, further steps are required for definitive identification. A battery of biochemical tests is then used to differentiate between specific *Vibrio* species based on their metabolic profiles (Table 24.2). Slide agglutination is used to rapidly identify *Vibrio* species by observing clumping when bacterial colonies react with known species-specific anti-H or anti-O antisera.

Serodiagnosis

Enzyme-linked immune assay (ELISA) has been used to measure antibody levels in patient serum, thus indicating exposure to or infection with *V. vulnificus*. However, ELISA may not be specific for all strains of *V. vulnificus*, particularly when using polyclonal antibodies. In the case of low levels of *V. vulnificus* in the sample, false negatives may occur. The ELISA has also been used to detect *V. vulnificus* in environmental samples, such as oysters, sediment, and seawater, to identify potential sources of infection.

Immunochromatographic strip tests, also known as rapid diagnostic tests (RDTs), provide quick results for antigen detection. They are widely used for the detection of *V. cholerae* O1/O139 in stool specimens.

Molecular Diagnosis

Simple and quantitative polymerase chain reaction (qPCR) and loop-mediated isothermal amplification (LAMP) have been used to detect and differentiate *V. vulnificus, V. parahaemolyticus*, and *V. alginolyticus*. These techniques target specific genes, such as those encoding haemolysis (*tlh*, *tdh*, *trh*) for *V. parahaemolyticus*, or virulence factors like the T3SS1 and T3SS2 secretion systems, and their corresponding genes (*vcrD*, *vopD*). For *V. vulnificus, rpoD* and *rpoS* genes are targeted for its detection.

For *V. alginolyticus*, virulence genes such as *toxR* and genes associated with pathogenicity islands, such as *VPI*, are commonly targeted. *V. alginolyticus* pathogenic strains are also detected based on isothermal recombinase polymerase amplification (RPA) and lateral flow dipsticks (LFD). The method targets the *toxR* gene for *V. alginolyticus.*

An overview of laboratory diagnosis is presented in Table 24.3.

Treatment

The mainstay of treatment for vibriosis depends on the severity of clinical manifestations. For mild cases, supportive therapy primarily involves volume repletion and correction of dehydration resulting from diarrhoea. Antibiotics are often administered for severe wound infections in the

Table 24.2 Biochemical tests for differentiation of *Vibrio* species

Biochemical test	*Vibrio cholerae*	*Vibrio parahaemolyticus*	*Vibrio vulnificus*	*Vibrio alginolyticus*
Oxidase	Positive	Positive	Positive	Positive
Glucose fermentation	Fermentative (acid only)	Fermentative (acid only)	Fermentative (acid only)	Fermentative (acid only)
Lysine decarboxylase	Positive	Positive	Positive	Positive
Ornithine decarboxylase	Positive	Positive	Variable (often negative)	Positive
Arginine dihydrolase	Negative	Negative	Negative	Negative
Indole production	Positive	Variable	Positive	Positive
H_2S production	Negative	Negative	Negative	Negative
Urease	Negative	Negative	Negative	Negative
Salt requirement	Can grow without NaCl	Requires NaCl (≥1%)	Requires NaCl (≥0.5–1%)	Requires NaCl (≥1%)
Growth in 0% NaCl	Yes	No	No	No
Growth in 6% NaCl	Variable	Yes	Yes	Yes
Sucrose fermentation	Positive (yellow on TCBS)	Negative (green on TCBS)	Variable (often positive)	Positive (yellow on TCBS)
TCBS agar colony colour	Yellow	Green	Green or yellow	Yellow
Haemolysis on blood agar	Variable (often β haemolytic)	β-haemolytic (Kanagawa test positive)	Often β-haemolytic	Often β-haemolytic

Table 24.3 Laboratory diagnosis of *Vibrio* infections

Diagnostic method	Sample type	Target	Remarks
Microscopy	Stool, wound exudate	*Vibrio* spp. (motility and shape)	Gram-negative, curved/comma-shaped rods; darting motility seen in hanging drop preparations; not species-specific; used as a preliminary observation
In vitro culture	Stool, pus, wound swab, blood	*Vibrio* spp. (species-level via colony & biochemistry)	Gold standard. APW used as enrichment medium; TCBS agar differentiates by sucrose fermentation (e.g. yellow colonies for *Vibrio alginolyticus*, green for *Vibrio vulnificus*) CHROMagar™ Vibrio offers improved species differentiation
Biochemical tests	Isolated colonies	Species-specific metabolic traits	Based on tests like indole production, salt tolerance, and sugar fermentation. Used to differentiate *V. vulnificus*, *V. alginolyticus*, *Vibrio parahaemolyticus*, and others
Serodiagnosis	Patient serum, isolated colonies	Antibodies or antigens (e.g. O/H antigens, haemolysins)	ELISA useful for *V. vulnificus* exposure; may lack strain specificity. Slide agglutination with antisera helps in rapid *Vibrio* identification. RDTs are common for *Vibrio cholerae*, less so for other species
Molecular diagnosis	Clinical specimens, environmental isolates	Species-specific genes (*toxR*, *tlh*, *tdh*, *rpoS*, etc.)	PCR, qPCR, LAMP, and RPA used. Targets include: *V. vulnificus*: *rpoD*, *rpoS* *V. parahaemolyticus*: *tdh*, *trh*, *tlh*, T3SS genes; *V. alginolyticus*: *toxR*, *VPI* genes. LFD tests offer rapid, field-applicable detection

form of doxycycline and other tetracycline-like antibiotics, especially for *V. vulnificus*. Alternatively, quinolones or third-generation cephalosporins may be considered for severe infections or in cases where doxycycline is not tolerated. In these cases, if severe wound infections ensue, surgical debridement with the removal of dead tissue, and if uncontrolled, amputation of the infected limbs may be necessary.

Currently, no widely licensed vaccines for *V. vulnificus* infections are available in humans. However, significant developments have been made in vaccine research in this field. Several promising vaccine candidates are being developed. These include the formalin-killed cell (FKC) vaccines, *V. vulnificus* ghost (VVG) vaccines, vaccines targeting specific protein regions, such as RtxA1-C, and vaccines based on TC38-killed *V. vulnificus* cells (TKC). For *V. parahemolyticus*, subunit vaccines targeting outer membrane proteins (OMPs), such as VP0802, have been identified as promising candidates.

Prevention and Control

Vibriosis is prevented by ensuring safe food-handling practices and appropriate care of wounds. Avoidance of the consumption of raw or undercooked seafood, especially shellfish such as oysters, is essential for the prevention of vibriosis. Seafood should be cooked thoroughly to appropriate temperatures. If required, the internal temperature of food should also be verified to check whether it is adequate for killing zoonotic bacterial pathogens. It is advisable to discard the shellfish that do not open during cooking, as this clearly indicates they are unsafe. For oysters, boiling for at least 3 min is essential, with or without prior frying, for the removal of zoonotic vibrios.

Protection of wound surfaces, avoidance of exposure in cases of recent tattoo piercings, open injuries from saltwater, brackish water, and avoidance of raw seafood, including oysters, in these high-risk conditions, are also very important. Wound surfaces should preferably be covered if there are chances of exposure to seawater. If contact with the marine environment is unavoidable, waterproof bandages could be used to protect the wound surfaces from exposure. The use of rubber gloves by persons handling seafood and having open wounds protects the wounds from contamination. However, in the case of inadvertent exposure of the wounds to seawater, thorough washing of the wounds with soap and water should be done immediately. Alternatively, if soap is not available, washing with an antiseptic solution should be done. Immediate medical attention is required if a wound shows any sign of infection.

Thorough hand washing after handling seafood, the use of separate cutting boards and utensils, and proper storage of seafood are also imperative for the prevention of vibriosis. Storage of both raw and cooked seafood in the refrigerator is encouraged, but they should be stored separately. People who are alcoholics, with liver diseases, those with weakened immune systems, or malnourished are at risk of acquiring the infection.

Case Study

A 63-year-old male patient with chronic liver disease and diabetes presented with pain in both lower legs. He had consumed seafood, such as oysters and shellfish, for lunch three days before the presentation. He had taken 800 mg of ibuprofen and sprayed diclofenac sodium on his lower legs, but his pain worsened at night. Within 12 h of arrival at the hospital, he experienced swelling and severe pain with the formation of tense bullae and erythematous plaques. There were noticeable colour changes in both lower legs. Then, the patient rapidly developed primary sepsis with decreased strength and paraesthesia of the legs. Urgent surgical debridement and decompression were performed. Postoperatively, the patient's condition deteriorated rapidly, and he developed multiple organ dysfunction syndrome. He was immediately transferred to the intensive care unit, where intravenous therapy with imipenem/cilastatin sodium, cefoperazone sodium/sulbactam sodium, and levofloxacin was initiated. Bacterial cultures from blood, bullae fluid, and tissues were all positive for *V. vulnificus* by MALDI-TOF. Despite prompt, appropriate antibiotic and surgical treatment, the patient died 16 days after admission.

1. What history and presentation support *V. vulnificus* infection?
2. What are the complications of *V. vulnificus* infection?
3. What is the approach to management?
4. What could have been the preventive measures in this case?

Research Questions

1. How does the interplay of various *Vibrio* species' virulence factors with the human immune response influence the prognosis and severity of infections?
2. What is the relationship between the host immune response to *Vibrio* virulence factors and the assessment of infection severity and prognosis?
3. What are the most promising avenues for investigating potential novel vaccines to prevent vibriosis, particularly in high-risk populations?

Further Readings

Baker-Austin C, Oliver JD, Alam M, Ali A, Waldor MK, Qadri F, Martinez-Urtaza J. Vibrio spp. infections. Nat Rev Dis Primers. 2018;4:8.

Carmona-Salido H, Fouz B, Sanjuán E, Carda M, Delannoy CMJ, García-González N, González-Candelas F, Amaro C. The widespread presence of a family of fish virulence plasmids in Vibrio vulnificus stresses its relevance as a zoonotic pathogen linked to fish farms. Emerg Microbes Infect. 2021;10:2128–40.

Chen Y, Ai X, Yang Y. Vibrio cholerae: a pathogen shared by human and aquatic animals. Lancet Microbe. 2022;3:e402.

Dong Y, Zhao P, Chen L, Wu H, Si X, Shen X, Shen H, Qiao Y, Zhu S, Chen Q, Jia W, Dong J, Li J, Gao S. Fast, simple and highly specific molecular detection of Vibrio alginolyticus pathogenic strains using a visualised isothermal amplification method. BMC Vet Res. 2020;16:76.

Jones MK, Oliver JD. Vibrio vulnificus: disease and pathogenesis. Infect Immun. 2009;77:1723–33.

Kah Sem NAD, Abd Gani S, Chong CM, Natrah I, Shamsi S. Management and mitigation of *Vibriosis* in aquaculture: nanoparticles as promising alternatives. Int J Mol Sci. 2023;24(16):12542.

Manchanayake T, Salleh A, Amal MNA, Yasin ISM, Zamri-Saad M. Pathology and pathogenesis of Vibrio infection in fish: a review. Aquac Rep. 2023;28:101459.

Oliver JD. The biology of Vibrio vulnificus. Microbiol Spectr. 2015;3(3)

Pan J, Sun Y, Yao W, Mao H, Zhang Y, Zhu M. Complete genome sequence of the Vibrio vulnificus strain VV2014DJH, a human-pathogenic bacterium isolated from a death case in China. Gut Pathog. 2017;9:67.

Pang R, Li Y, Liao K, Guo P, Li Y, Yang X, Zhang S, Lei T, Wang J, Chen M, Wu S, Xue L, Wu Q. Genome- and proteome-wide analysis of lysine acetylation in Vibrio vulnificus Vv180806 reveals its regulatory roles in virulence and antibiotic resistance. Front Microbiol. 2020;11:591287.

Parija Subhash Chandra. Vibrio, Aeromonas and Plesiomonas. In: Textbook of microbiology and immunology. 3rd ed. Springer; 2023. p. 563–78.

Pettis GS, Mukerji AS. Structure, function, and regulation of the essential virulence factor capsular polysaccharide of Vibrio vulnificus. Int J Mol Sci. 2020;21:3259.

Tian Z, Yang L, Qi X, Zheng Q, Shang D, Cao J. Visual LAMP method for the detection of Vibrio vulnificus in aquatic products and environmental water. BMC Microbiol. 2022;22:256.

Xu K, Wang Y, Yang W, Cai H, Zhang Y, Huang L. Strategies for prevention and control of *Vibriosis* in Asian fish culture. Vaccines (Basel). 2022;11:98.

Xue M, Huang X, Xue J, He R, Liang G, Liang H, Liu J, Wen C. Comparative genomic analysis of seven Vibrio alginolyticus strains isolated from shrimp larviculture water with emphasis on chitin utilization. Front Microbiol. 2022;13:925747.

Ziarati M, Zorriehzahra MJ, Hassantabar F, Mehrabi Z, Dhawan M, Sharun K, Emran TB, Dhama K, Chaicumpa W, Shamsi S. Zoonotic diseases of fish and their prevention and control. Vet Q. 2022;42:95–118.

25 Melioidosis

Jagadish Mahanta

Learning Objectives

1. To study *Burkholderia pseudomallei* and its potential to infect humans and animals after prolonged exposure to soil and water
2. To study the mechanisms of immune evasion of the host by the bacteria and its ability to reactivate from latency
3. To understand the clinical manifestations and diagnostic modalities in melioidosis
4. To review the treatment modalities, prevention, and control of melioidosis

Introduction

Melioidosis is caused by the opportunistic, environmental bacterium, *Burkholderia pseudomallei*. Humans or animals acquire the infection by inoculation, inhalation, or ingestion.

Patients with melioidosis may present with acute septicaemia with multi-organ involvement, often due to a chronic infection. Mortality in acute septicaemic cases ranges between 20% and 50%.

However, asymptomatic or subclinical manifestations are also observed among many exposed persons.

The relatively large genome of *B. pseudomallei* exhibits significant plasticity, allowing the bacterium to adapt to environmental changes and host defences. Mobile genetic elements, carrying antibiotic resistance and virulence factors, facilitate the spread within bacterial populations. *B. pseudomallei* is inherently resistant to many antibiotics. The "Darwin guidelines" are the "gold standard" for the management of patients with melioidosis.

History

Alfred Whitmore and C. S. Krishnaswami, in 1911, first described melioidosis as a "glanders-like" disease among persons who were addicted to morphine in Rangoon, Burma (Myanmar). In 1917, Fletcher isolated the causative organism from a rubber estate worker. Historically, the disease was known as Whitmore's disease, Nightcliff gardener's disease, pseudoglanders, and the "Vietnam time bomb". The term "melioidosis" was first coined in 1921 from the Greek word "*melis*" (meaning "distemper of asses") and "*eidos*" (similar) by Stanton and Fletcher in 1932, and in 1937, water was indicated as the habitat of the causative bacterium.

Bacteria were known as *Bacillus pseudomallei, Bacillus whitmorii, Malleomyces pseudomallei*, and *Pseudomonas pseudomallei* till 1992, and was then renamed as *Burkholderia pseudo-*

J. Mahanta (✉)
ICMR-Regional Medical Research Centre for Northeast, Dibrugarh, Assam, India

S. C. Parija, A. Poddar (eds.), *Textbook of Bacterial Zoonoses*, Microbial Zoonoses,
https://doi.org/10.1007/978-981-95-6615-0_25

mallei based on the sequences of the 23S rRNA gene. The genus *Burkholderia* was named after Walter H. Burkholder. An outbreak of melioidosis in sheep occurred in North Queensland, Australia, in 1949, followed by human cases in Townsville in 1950. Another outbreak of melioidosis at the Paris Zoo was reported in the 1970s and was suspected to originate from an imported panda or horses from Iran. After the Vietnam War, several American soldiers were diagnosed with melioidosis. In 1990, *B. thailandensis* was identified as a non-virulent "arabinose-positive *Burkholderia* spp". In 2004, the complete genome of *B. pseudomallei* was published.

Taxonomy

The genus *Burkholderia* belongs to the family, Burkholderiaceae; order, Burkholderiales; class, Betaproteobacteria; and phylum: Pseudomonadota (Proteobacteria).

In 1992, *Pseudomonas pseudomallei* was transferred to *Burkholderia pseudomallei* under the genus *Burkholderia.* There are nearly 100 species within the genus *Burkholderia,* and more than 70 species with variable pathogenicity to humans, animals, and plants. They are placed in 3 monophyletic clusters: *Burkholderia cepacia* complex, *Burkholderia pseudomallei* and close relatives, and the other *Burkholderia* phytogenic clade. *B. pseudomallei* causes melioidosis; *B. cepacia* causes lung infections in persons with cystic fibrosis; and *B. xenovorans* affects patients with chronic granulomatous disease. *Burkholderia mallei* causes glanders in horses and related animals, while *Burkholderia glumae, Burkholderia gladioli, B. cepacia,* etc. cause infections in plants.

Genomics and Proteomics

B. pseudomallei has two circular chromosomes (4.07 and 3.17 Mb), demonstrating functional partitioning. The larger one is required for cell wall synthesis, mobility, metabolism, and growth, while the smaller one carries genes required for survival and virulence factors.

A diverse array of genes drives the virulence. First, multiple secretion systems, including type I, II, III (encoded by *bsa, bip,* and *bop* gene clusters), and V, are important for the bacterium to transport effector proteins and other molecules across its membranes and into host cells, thereby facilitating invasion and immune evasion. Second, genes such as *waa, rfb,* and *wcb* are involved in the biosynthesis of lipopolysaccharide (LPS) and capsular polysaccharide. Both play significant roles in protecting the bacterium from the host's immune system and contributing to its structural integrity. Third, the formation of exoproteins, including phospholipase C (*plc*), metalloprotease A, other proteases (*mpr*), and collagenase, is mediated by specific genes that enable the bacterium to degrade host tissues and promote dissemination. In addition, genes responsible for the expression of adhesive surface proteins and mediating host-cell interactions are critical for the initial attachment and subsequent invasion of host cells. Finally, the formation of fimbriae and pili, the filamentous appendages, is involved in adhesion.

Proteomic analyses have been indispensable in dissecting the protein cargo delivered by secretion systems, revealing their diverse roles in the infection process. For instance, the Type II secretion system (TIISS) is known to secrete hydrolytic enzymes such as proteases, phospholipases, and chitinases. These enzymes are believed to contribute to nutrient acquisition and potentially facilitate tissue invasion. The deubiquitinase TssM, a protein implicated in modulating the host immune response, has also been identified as a TIISS substrate.

Beyond the secreted proteins, numerous other bacterial components contribute to the causation of disease. BimA, an autotransporter protein, plays a critical role in actin polymerisation, enabling intracellular motility and dissemination. The capsular polysaccharide and lipopolysaccharide (LPS) are essential for extracellular survival and resistance to the host's complement system. Various secreted enzymes contribute to tissue damage and nutrient acquisition during infection. Proteins involved in the bacterial stress response are often regulated by alternative sigma factors like RpoE. All these proteins are crucial for survival both within the host and in the external environment. Adhesins, such as Sap1, facilitate the initial attachment of the bacterium to host cells.

Morphology

B. pseudomallei is an aerobic, motile, rod-shaped, straight or slightly curved bacterium with rounded ends. It is Gram-negative, showing bipolar staining, with a central vacuole, often described as having a "safety pin" appearance (Fig. 25.1). It measures 2–5 μm in length and 0.4–0.8 μm in breadth. The bacterium has flagella and fimbriae or pili on the surface. It has a thin peptidoglycan cell wall with inner and outer membranes.

Cultivation

B. pseudomallei grows on a large variety of culture media, including blood agar, MacConkey agar, EMB, etc. It takes about 24–48 h to grow in an artificial medium. Optimal growth in vitro is seen at 40 °C in neutral or slightly acidic environments (pH 6.8–7.0). Ashdown's medium containing tryptase soy agar with 4% glycerol, crystal violet, neutral red, and gentamicin (4 mg/L) is suitable for isolation from non-sterile clinical and environmental samples (Fig. 25.2).

Most *B. pseudomallei* strains are non-lactose fermenters. They are oxidase and catalase-positive. It can be distinguished from avirulent *B. thailandensis* by its inability to assimilate L-arabinose.

Animal Models

Inbred mouse strains, chickens, rats, guinea pigs, and Syrian golden hamsters support the growth of *B. pseudomallei.* Of these, BALB/c and C57BL/6 inbred mice are extensively used as animal models. For instance, BALB/c mice are highly susceptible and are used to mimic acute melioidosis in experimental infection, characterised by rapid disease progression and disseminated abscesses, resembling the acute form seen in humans. In contrast, C57BL/6 mice are more resistant and tend to develop a more chronic

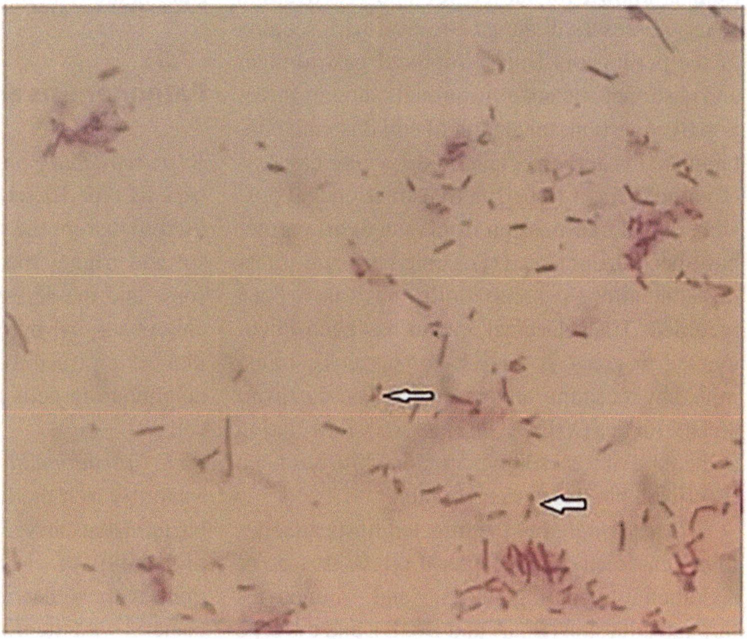

Fig. 25.1 Gram-stained smear of *Burkholderia pseudomallei* showing Gram-positive rod-shaped structures with a safety pin appearance. (Image source: WIKIPEDIA, The free encyclopaedia)

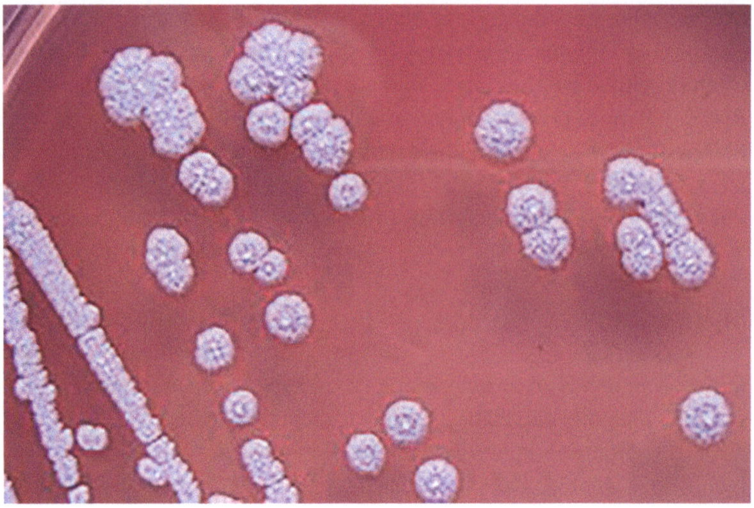

Fig. 25.2 Appearance of *Burkholderia pseudomallei* colonies on Ashdown's medium after 4 days of incubation. (Image source: WIKIPEDIA, The free encyclopaedia)

infection, which is used for studying the long-term aspects of the disease and for vaccine evaluation.

Transmission Cycle

B. pseudomallei infects the human host by inhalation, ingestion, and inoculation (Fig. 25.3). In the bloodstream, it infects antigen-presenting cells to cause bacteraemia and sepsis. The bacterium is known for its latency and reactivation. Micro-abscesses in the gastric wall with seeding of the peritoneum from a ruptured gastric ulcer and gastrohepatic nodes in animals support infection by ingestion. Inhalation of soil dust and contaminated water, especially in drowning cases, is reported. Person-to-person transmission and possible sexual transmission from man to his partner have been documented. Possible perinatal transmission through breast milk has also been recorded. Transplacental spread has been documented in goats. *B. pseudomallei* can be transmitted by mosquito vectors such as *Aedes aegypti* and the rat flea (*Xenopsylla cheopsis*). Inoculation at the time of a snake bite or animal bite has been described, but it is extremely rare.

B. pseudomallei in the infected hosts attaches to protease-activated receptor-1 on the surface of endothelial cells, platelets, and monocytes through adhesion (BoaA and BoaB), and the type IV pilus protein (PilA). It enters the cells through endocytosis. In the cell, the bacterium induces actin polymerisation and uses the process to move forward. This actin-mediated motility is accomplished with the autotransporter BimA. The same process in infected neurons allows the bacterium to travel through nerve roots in the spinal cord and the olfactory sensory nerve to reach the brain. It infects antigen-presenting cells, macrophages and dendritic cells and is used as a vehicle for the dissemination of the bacterium in the lymphatic system.

Pathogenesis and Pathology

B. pseudomallei present in an endocytic vesicle uses its type III secretion system. The system is used to disrupt the vesicle to enter the host cytosol and trigger the formation of the inflammasome and the activation of caspase 1, to induce cell death by pyroptosis. The bacterium also induces protrusions in the cell wall to fuse with neighbouring cells to form multinucleated giant cells.

B. pseudomallei uses the haematogenous route to reach the liver, spleen, and muscle. The bacteria that carry a BimABm allele have a higher probability of causing neurological manifestations with higher mortality in comparison with bacteria with the BimABp variant. *B. pseudomal-*

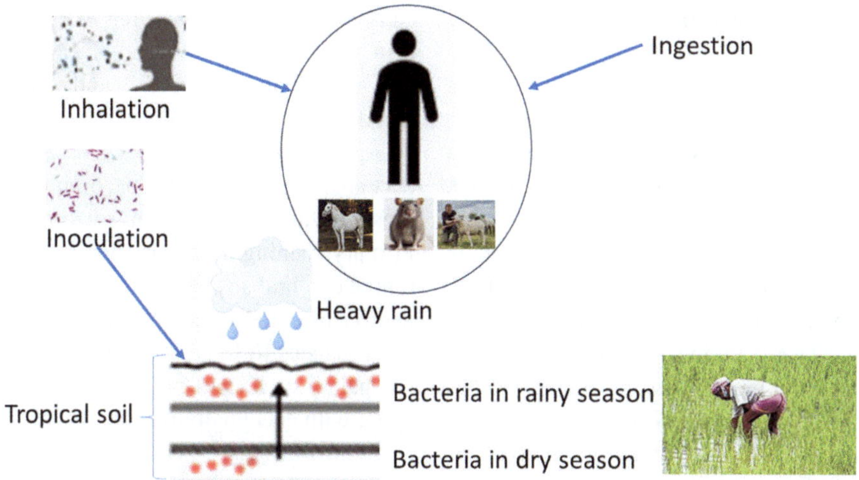

Fig. 25.3 Transmission of *Burkholderia pseudomallei* to vulnerable hosts. (Image source: Authors' personal collection)

lei is one of the first proteobacteria to contain up to six different types of type VI secretion systems contributing to the pathology of the disease. Further, the intrinsic efflux pump of *B. pseudomallei* makes it resistant to many aminoglycosides (*AmrAB-OprA*), tetracyclines, fluoroquinolones, and macrolides (*BpeAB-OprB*).

Evasion of the Immune System

Like other Gram-negative organisms, endotoxins and exotoxins produced by *B. pseudomallei* help in the pathogenesis and evasion of the immune system. Bacterial capsules and LPS help to resist lysosomal degradation and the lethal action of interferon-gamma inside macrophages. They can survive and multiply within macrophages/monocytes and neutrophils, evade phagosome-lysosome fusion, and destroy the phagosome membrane within 15 min after ingestion. Ultrastructural studies of *B. pseudomallei*-infected macrophages suggest less phagolysosome fusion in macrophages among melioidosis patients. Impaired chemotaxis, phagocytosis, oxidative burst, and killing activity among diabetic individuals or chronic alcoholics, those with chronic renal failure and thalassemia help the bacteria to survive inside the host. Intrinsic efflux pumps of *B. pseudomallei* make them resistant to many aminoglycosides (*AmrAB-OprA*), tetracyclines, fluoroquinolones, and macrolides (*BpeAB-OprB*).

Immunology

Innate Immunity

B. pseudomallei is resistant to serum bactericidal components, but the alternative complement pathway activation helps in phagocytosis. Innate immune signalling molecules such as TLR, NOD, MyD88, and the pro-inflammatory cytokines IFN-γ and tumour necrosis factor-alpha (TNF-α) play key roles in infection containment. Neutrophils, monocytes, and macrophages help limit the dissemination of *B. pseudomallei*.

Capsular polysaccharide and lipopolysaccharide (O-PS I and O-PS II), flagellin proteins, and cell wall proteins of *B. pseudomallei* are immunogenic and show strong IgG1 and IgG2 responses. These antibodies even persist in the circulation beyond 3 years. However, repeated natural exposure to *B. pseudomallei* and *B. thailandensis* does not confer protection against primary infection or relapse.

Cell-Mediated Immunity

The type 1 response-prone C57BL6 mice are relatively resistant to *B. pseudomallei* compared with BALB/c mice. IFN-Y plays an important role in stimulating phagocytosis, oxidative burst, and intracellular bacterial killing.

Th1 polarisation, IL-12, is demonstrated in patients with melioidosis and ex vivo blood from mice. Further, high levels of IL-6 (Th1 cytokine) and IL-10 (Th2 cytokine) are associated with poor outcomes in melioidosis. Moreover, mRNA levels for Th1 cytokines (IFN-Y, IL-6, and IL-12) and Th2 cytokines (IL-10) increased in both susceptible BALB/c and resistant C57BL/6 mice.

Toll-like receptors (TLR2, TLR4, and TLR5) recognize LPS and flagella to produce interleukin-1 beta and interleukin-18. IL-18 increases IFN production through natural killer cells, while IL-1 beta reduces IFN production. This leads to nuclear factor κB-induced immune activation by releasing pro-inflammatory cytokines. These immune molecules help in recruiting neutrophils, dendritic cells, B cells, and T cells to the site of infection.

Infection in Humans

Melioidosis in humans is primarily caused by inoculation.

The disease manifests after an initial incubation of 1–21 days (median, 4 days). However, incubation periods of 24–29 years among ex-servicemen in Papua New Guinea and Vietnam have been recorded.

Melioidosis has no pathognomonic clinical manifestations. It can be subclinical, localised, or disseminated. Skin and soft tissue are common primary sites of infection. In half of the chronic cases, infection is localised to a single organ.

Nearly 40–60% of cases have bacteraemia, and 20% develop septic shock in acute melioidosis. More than half of acute cases with pulmonary involvement develop pneumonia and manifest as fever, headache, cough, respiratory distress, and chest discomfort. They clinically and radiologically mimic tuberculosis. Ten to thirty-three percent of cases develop liver, spleen, or para-intestinal abscesses and manifest as jaundice.

Approximately 1–5% of cases develop encephalitis, meningitis, or brain abscess and present with disorientation, seizures, and coma. Primary meningitis occurs possibly due to rupture of cerebral abscesses. Unilateral upper motor neuron type palsy, cerebellar signs, and cranial nerve palsies (VI and VII nerve palsies and bulbar palsy) or flaccid paralysis alone may occur along with encephalomyelitis.

Genito-urinary involvement is seen in 14–28% of cases with pyelonephritis, kidney abscesses, and prostatitis in males. Septic arthritis and osteomyelitis are seen in 4–14% of cases. Mycotic aneurysms, mediastinal infection, thyroid abscesses, corneal ulcers, orbital cellulitis spreading to sinuses, pyogenic pericardial effusion, myocardial abscesses, pancreatitis, and endocarditis are rare in melioidosis. About 10% of patients develop chronic infections with symptoms continuing for more than 2 months.

Less than 5% of melioidosis cases are reactivated from latency by immunosuppression or stress. An Intranuclear site of infection in the cell, slow growth, antibiotic resistance, and genetic adaptation to the host environment may cause the bacterium to remain in latency. Granulomas formed at the site of infection have also been associated with latent human infection.

Melioidosis patients typically reveal a constellation of abnormalities indicative of systemic inflammation and organ dysfunction. These include lymphopenia, elevated serum levels of liver enzymes such as aspartate aminotransferase (AST) and alanine aminotransferase (ALT), as well as increased bilirubin, suggesting involvement of the liver. Renal impairment is indicated by elevated levels of urea and creatinine. Metabolic disturbances such as hypoglycaemia and acidosis are also frequently observed. Inflammatory markers, including increased levels of C-reactive protein (CRP) and procalcitonin, are often elevated.

Infection in Animals

A wide variety of animals, including cattle, goats, pigs, camels, horses, sheep, kangaroos, koalas, alpacas, deer, cats, and dogs, marine animals, fish, and reptiles, are infected by *B. pseudomallei*.

Subclinical infections with no apparent signs of illness are demonstrated in infected animals. In symptomatic infections, the symptoms of melioidosis in animals are highly variable and often mimic other diseases, making diagnosis difficult. The clinical signs can range from acute and severe to chronic infections and vary significantly in different animal species. Fever, weight loss, and depression are common symptoms frequently observed in infected animals. Heavy breathing and nasal discharge, lameness, and swelling of the joints are also frequently reported. In severe acute cases, the infection can lead to sudden death following septicaemia and multiple organ failure.

In certain specific animal species, the clinical manifestations are more distinct and characteristic. For example, sheep often develop severe respiratory disease with the presentation of coughing, nasal and ocular discharge, and may exhibit lameness and nervous signs. Infection in goats presents as mastitis (inflammation of the udder) as well as wasting, lameness, hind leg weakness, and abortions. Adult pigs may demonstrate progressive wasting, neural manifestations such as incoordination, skin ulcers, and diarrhoea. Young pigs develop acute disease with the presentation of fever, anorexia, coughing, and nasal discharge. Camelids such as camels and alpacas develop respiratory distress with coughing and nasal discharge, hind leg weakness, and wasting, and acute infections can progress to sudden death.

Dogs infected by *B. pseudomallei* present with a wide array of symptoms, often referred to as “the great imitator”. These include fever, abscesses under the skin and in internal organs (liver, lungs, testicles), severe diarrhoea, muscle pain, lethargy, and pneumonia. Chronic disease manifests as loss of appetite, muscle pain, limb swelling, and skin abscesses.

Epidemiology and Public Health

Melioidosis is endemic in Southeast Asia and Northern Australia and is a major infection of public health importance. It is increasingly reported from the Indian subcontinent, with more than 52,500 cases per year documented from India. The condition is also reported from Southern China, Hong Kong, and Taiwan. Melioidosis is endemic in Australia, Thailand, Singapore, Malaysia, Burma, Vietnam, Southern China, Hong Kong, Brunei, PDR Laos, Cambodia, Taiwan, Aruba (Netherlands Antilles), and France. Annual incidences of up to 41.7 per 100,000 people have been reported in some countries. More than 50 countries have reported *B. pseudomallei* infection across the globe (Fig. 25.4).

Sporadic cases are reported from Indonesia, Bangladesh, Japan, the Philippines, Pakistan, Sri Lanka, the Americas and the Caribbean Guadeloupe, Martinique, Puerto Rico, Ecuador, Panama, El-Salvador, Haiti, Costa Rica, and Colombia, Venezuela, the Pacific Guam, Fiji, Papua New Guinea, New Caledonia, Africa and the Middle East Iran, Uganda, Sierra Leone, the Gambia, Madagascar, Kenya, Korea, Hawaii, East Timor, Europe, Spain, Germany, the United Arab Emirates, Saudi Arabia, Egypt, South Africa, Turkey, Egypt, Niger, Burkina Faso (Upper Volta), Mexico, and the United States.

Humans and a wide variety of animals residing or visiting endemic areas are susceptible to infection. People with diabetes, thalassemia, alcoholism, and renal impairment are more susceptible to melioidosis. Immunosuppressive therapy or occupational exposure to soil and surface water is associated with an increased risk of melioidosis. High rainfall, humidity, temperature, soil composition, vegetation, use of fertilisers, and recent soil disturbances may influence the distribution of *B. pseudomallei* in the environment (Fig. 25.3). Heavy rains, floods, storms, and cyclones-affected areas show clustering of melioidosis cases.

Relapse after apparently successful treatment and an extended incubation suggests latency and

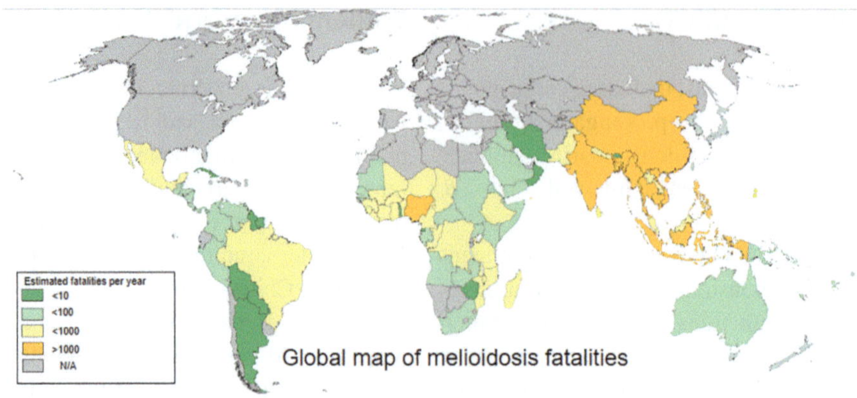

Fig. 25.4 Global case fatalities from melioidosis. (Image source: WIKIPEDIA, The free encyclopaedia)

reactivation. Bacterial latency up to 62 years has been reported.

Diagnosis

Specimens

A diverse range of clinical specimens is collected based on the suspected site of infection. Blood is collected from suspected cases of septicaemia, and multiple blood samples are collected from different sites. Respiratory samples, such as deep cough sputum, are collected from patients with pulmonary involvement. Pus or aspirates from abscesses and swabs from skin lesions offer a high rate of bacterial isolation. Urine is collected and cultured, even in the absence of urinary symptoms. In certain circumstances, throat swabs or other sterile fluids like CSF, pleural, peritoneal, or pericardial fluid may be necessary.

Microscopy

Gram-stained *B. pseudomallei* under a light microscope shows a Gram-negative, straight or slightly curved rod with bipolar staining, central vacuolation, and rounded ends, giving a "*safety pin*" appearance. It shows motility in a hanging drop preparation. Direct fluorescence microscopy is a rapid method for the detection of *B. pseudomallei* in clinical specimens or from blood culture bottles by using monoclonal antibodies against capsular polysaccharides.

In Vitro Culture

Although growth in culture is 100% specific, sensitivity may be as low as 60%, depending on the nature of the sample, the method of collection of samples, and the media used.

B. pseudomallei colonies are cream-coloured with a metallic sheen and earthy odour. Initially, they show smooth colonies, but become dry and wrinkled after incubation for >24 h on blood agar. It also shows small colony variants. On MacConkey agar, *B. pseudomallei* colonies are pale (lactose non-fermenters) and become pinkish, umbonate, or rugose after 48 h. *B. cepacia* agar or Ashdown's medium is employed as the selective medium. The demonstration of a pellicle in Ashdown broth is frequently diagnostic.

B. pseudomallei is typically oxidase-positive, indicating the production of cytochrome c oxidase, and is also catalase-positive. It is generally a non-fermenter or a slow, weak oxidiser of glucose, but does not ferment lactose. It is positive for gelatin hydrolysis, nitrate reduction, arginine dihydrolase, and urease tests. It is negative for indole production, methyl red and Voges-Proskauer tests.

Serodiagnosis

Indirect Hemagglutination Assay (IHA)

IHA is performed using sensitised sheep erythrocytes with crude antigen from cultured bacterial isolates. Serially diluted heat-inactivated patient sera are tested for antibodies. Raised IHA titres reflect an IgM response. A Titre >1:10 is considered positive. Systemic *P. aeruginosa* infections show false-positive, and acute bacteraemia show false-negative results.

Enzyme-Linked Immunosorbent Assay (ELISA)

In ELISA applications for melioidosis, various antigens such as bacterial lipopolysaccharide (LPS), whole-cell preparations, exopolysaccharides (EPS), and affinity-purified EPS antibodies are utilised. Currently, recombinant antigens are preferred, including flagellin proteins derived from *Burkholderia pseudomallei* and *B. thailandensis*. Promising developments include the cloning of genes like *groEL* and *malE* from *B. pseudomallei*. This assay can identify both IgM and IgG antibodies in patient serum.

ELISA formats based on outer membrane proteins (such as Omp3 and Omp85), type VI secretion system component TssD-5, and the serine protease MprA (designated smBpF4) have demonstrated moderate sensitivity for IgG detection. Additionally, monoclonal antibodies have been employed for the detection of exotoxins in culture fluids, as well as for identifying a 40-kDa secreted protein and a 30-kDa cell wall-associated protein. These assay variations have shown acceptable levels of sensitivity and specificity.

Specific Lateral Flow Immunoassay

The test is used to detect the bacterial capsular polysaccharide by high-affinity monoclonal antibodies. The IgG immune-chromogenic test (ICT) has been evaluated to detect IgG and IgM antibodies with 79% sensitivity and 90% specificity.

Latex Agglutination Test

The test using a monoclonal antibody against a 200 kDa protein has been developed for the identification of *B. pseudomallei* culture isolates and from the blood culture supernatant.

Molecular Diagnosis

Polymerase Chain Reaction (PCR) is one of the most widely used molecular techniques. Several PCRs with primers targeting regions in the 23S rRNA, the 16S RNA, and the 16S and 23S RNA junction have been developed. Multiplex PCR has been developed to identify *B. pseudomallei* and differentiate it from closely related species.

Sequencing of the 16S rRNA gene can distinguish *B. pseudomallei* from other *Burkholderia* species. However, a very minor difference (1%) with *B. thailandensis* necessitates *groEL* gene nucleotide sequences of the *B. pseudomallei* strains for improved discrimination. Arabinose assimilation and groEL sequences are better for the identification of *B. pseudomallei*. Multi-locus sequencing appears to have a discriminatory ability like that of PFGE. RAPD, along with PFGE, is more discriminatory than ribotyping.

Current Methods

MALDI-TOF Mass Spectrometry (MALDI-TOF MS)

MALDI-TOF MS is increasingly recognised as an efficient technique for the rapid identification of bacteria by analysing their distinct protein signatures. After the organism is cultured, MALDI-TOF MS produces a protein mass spectrum that is matched against a comprehensive database containing profiles of various bacterial species, including *B. pseudomallei*. Recent studies highlight MALDI-TOF MS's ability to reliably detect *B. pseudomallei*, achieving high sensitivity and specificity. The method also accurately differentiates it from closely related *Burkholderia* species when using refined spectral libraries and defined confidence thresholds.

Metabolomic Profiling Techniques

The use of liquid chromatography-mass spectrometry (LC-MS) to analyse proteins, lipids, and other metabolic byproducts in bacterial culture supernatants provides distinctive biochemical profiles for different bacterial species. Techniques such as principal component analysis (PCA) and partial least squares discriminant analysis (PLS-DA) are commonly used to interpret these profiles. In the case of *B. pseudomallei*, its metabolomic signature clearly diverges from that of related *Burkholderia* strains and other Gram-negative organisms, suggesting the presence of unique metabolic markers that could serve as reliable indicators for this pathogen.

Radiological Diagnosis

Imaging plays an essential role in determining the spread and severity of organ involvement in infectious diseases. In bloodstream infections caused by *B. pseudomallei*, chest radiographs frequently reveal distinctive pulmonary patterns, such as small nodules, patchy infiltrates, and areas of dense necrotising consolidation. A notable radiological feature is the appearance of multiple uneven lobular lesions, sometimes described as a "*honeycomb sign*". For patients presenting with neurological complications like encephalomyelitis, CT imaging may initially show no abnormalities, whereas MRI is more adept at detecting early or subtle signs of central nervous system involvement. MRI scans often show heightened T2 signals in the brainstem and spinal cord, indicative of inflammation and possible tissue injury.

Different diagnostic modalities are summarised in Table 25.1.

Treatment

Melioidosis cases are treated in two phases, covering "intensive" and "eradication" therapy.

Intensive therapy is carried out with six-hourly intravenous ceftazidime (50 mg/kg up to 2 g) or meropenem (25 mg/kg, up to 1 g). However, CNS infection needs a higher dose (40 mg/kg up to 2 g) every 8 h, or six-hourly imipenem (25 mg/kg, up to 1 g) for 2 weeks. An increased dose is given depending on the severity of the disease.

Eradication therapy is prescribed for 3 months, usually with trimethoprim-sulfamethoxazole (TMP-SMX)(6 + 30 mg/kg, up to 320 + 1600 mg) every 12 h, with folic acid. Where trimethoprim-sulfamethoxazole is contra-indicated or not tolerated, amoxicillin/clavulanic acid, every 8 h, or doxycycline, once daily or every 12 h, can be used for eradication therapy.

The revised Darwin melioidosis treatment guideline (2024) provides a minimum dose and duration for therapy. However, these are adjusted based on the foci of infection and the bacteraemia status. Only oral therapy can be considered for a single skin lesion in selected cases. In some cases, a combination of chloramphenicol, doxycycline, and TMP-SMX is tried, but the relapse rate is about 10%. Approximately 20% relapse is seen with short-course (8 weeks) ciprofloxacin-azithromycin or longer-course (20 weeks) quinolone monotherapy.

Six bacteriophages (P37, K96243, E0237, CMS, 365A, and HNBP001) with varying lytic ability (53.5–100%) have been evaluated against *B. pseudomallei* strains.

A higher median lethal dose of *B. pseudomallei* is documented when ciprofloxacin or doxycycline is used as prophylaxis in experimental animals. Prophylaxis with TMP-SMX has not yet been fully evaluated.

Abscesses in the liver, prostate, and brain require drainage, while splenic abscesses often resolve with adequate therapy. At the same time, open drainage and washout are needed in septic arthritis. Osteomyelitis of long bones may require multiple debridements over several months.

Prevention and Control

B. pseudomallei is a resilient organism and is capable of surviving in distilled water and desert soil for more than 16 years. It can tolerate antiseptic and detergent solutions, acidic environments (pH 4.5 for 70 days) and temperatures varying from 24 to 32 °C. The formation of L-form or a non-culturable state may mediate the survival of the bacterium. A biofilm, or "slime" of glycocalyx polysaccharide produced by *B.*

Table 25.1 Different diagnostic tests used in melioidosis

Test type	Tests	Sample used	Target	Sensitivity (%)	Specificity (%)	Remark
Microscopy	Light microscope	Clinical sample	Appearance, Gram stain character	60	70	Economical, ease in doing
	Culture isolate	Culture isolate	Staining character, motility		70–80	Characteristic appearance, motile in hanging drop
	Fluorescence microscopy	Clinical sample	Demonstration of bacteria	60	>90	Quick diagnosis
In vitro Culture	Artificial media	Clinical samples	Colony morphology, staining and biochemical character, motility	60	100	Grows in common media, colony morphology varies with duration of incubation
Serodiagnosis	IHA	Serum	IgG, IgM	62–86	68–100	Purified antigen shows lower sensitivity and specificity
	ELISA	Serum	IgG or IgM	71–87	75–99	
	ELISA (DOT)	Serum	IgG or IgM	72–94	64–99	Culture filtrate as antigen
	Immuno chromogenic card test	Serum	IgG	67–88	69–91	Easy to perform and used as point-of-care test
	Immunofluorescenceassay	Serum		86	99	
		Specimen	Whole cell or cell wall polysaccharide	73	99	Quick and accurate
		Urine	Bacterial sub units	81	96	Easy for sampling
	ELISA	Specimen	Cell wall components	75	98	Use monoclonal antibody (MAb-5F8)
		Serum	Cell wall components	82	95	Use monoclonal antibody (19.5 kDa Ag)
	Latex agglutination	Blood culture supernatant		95–100	96–100	Monoclonal antibody-based latex agglutination
Molecular diagnosis	PCR	Blood	16s rRNA, 23s rRNA sequence	47–100	67–100	Probe-targeting 16s rRNA, 23s rRNA

Note: The sensitivity and specificity of different tests vary with infecting strains in different geographical regions *pseudomallei*, allows for the formation of microcolonies in a protective environment. *B. pseudomallei* may also survive in the environment by infecting *Acanthamoeba* trophozoites.

Personal Protection

Protective wear like boots and gloves, and avoiding outdoor activity during heavy rain and wind, are advisable in endemic areas. Primary prophylaxis with trimethoprim-sulfamethoxazole is used for haemodialysis cases in endemic areas. Chlorination of drinking water can effectively control the spread of *B. pseudomallei*.

Benzalkonium chloride, iodine, mercuric chloride, potassium permanganate, 1% sodium hypochlorite, 70% ethanol, 2% glutaraldehyde, heating to above 74 °C for 10 min, or ultraviolet irradiation can be effectively used to kill *B. pseudomallei*.

Precautions in Clinical Laboratory

Good laboratory practices prevent accidental exposure to *B. pseudomallei*. The organism should be handled within a biosafety level 3 facility. Baseline serum samples should be obtained from all workers handling *B. pseudomallei* and stored at −80 °C.

Immediately after exposure, the site of contamination or inoculation is washed with copious amounts of water, followed by an appropriate cutaneous disinfectant. Oral trimethoprim/sulfamethoxazole for 21 days is the first choice after exposure. Alternatively, amoxicillin/clavulanic acid may be used.

Following laboratory exposure to *B. pseudomallei*, healthcare personnel should be monitored for febrile symptoms, with or without respiratory involvement, and for localised inflammation at any suspected inoculation site. Diagnostic evaluation should include cultures of blood, sputum, throat swab, rectal swab, and urine, ideally using Ashdown medium or *B. cepacia* agar to enhance detection. A chest radiograph should be performed to assess for pulmonary involvement. Serial serologic testing is recommended at weeks 1, 2, 4, and 6 to detect seroconversion, defined by a rising antibody titre between paired samples. In cases of seroconversion, even in the absence of symptoms, a full clinical evaluation and initiation of prolonged antimicrobial therapy are indicated, with follow-up for at least 12 weeks. Patients with culture-confirmed melioidosis should be managed according to established treatment guidelines, including both intensive and eradication phases of therapy.

Vaccine

Burkholderia vaccines have received renewed attention. As humoral antibodies do not confer adequate protection, a cell-mediated response is critical. T-cell responses are critical for vaccine protection against *Burkholderia* infection. Whole cells or different types of antigens have been explored for the development of vaccines.

Live-attenuated B. pseudomallei and outer membrane vesicles (OMVs)–derived vaccines have shown significant (67% at 21 days) protection in animal models, but safety must be evaluated. An OMV with aluminium hydroxide (alum) adjuvant enhances immunogenicity.

Subunit vaccines containing capsular polysaccharide [CPS], cell-associated proteins (LolC, OmpA, OmpW, and FliC), and secreted proteins (BopA, BimA, FlgL, and MprA) are being evaluated. They provide less robust immunity but are safer to administer even in immunocompromised individuals. CPS is structurally conserved and expressed by all known virulent isolates of *B. pseudomallei*. CPS-specific monoclonal antibodies (MAbs) have been used for passive immunisation in mice against lethal challenges of *B. pseudomallei*. However, none of them alone provides complete protection and sterilising immunity. Further, the identification of broadly protective antigens and efficient delivery and adjuvants are a problem.

Purified recombinant *B. pseudomallei* Hcp1 and TssM proteins in C57BL/6 mice with 6-deoxyheptan capsular polysaccharide (CPS CRM197) protect all mice from a lethal inhalational challenge. The recombinant omp85 antigen with Freund's complete adjuvant triggers a Th2-type immune response. It reduces bacterial loads in the lung, liver, and spleen, with 70% survival up to 15 days of lethal challenge.

Combining antigen delivery with nanoparticles, biopolymers, and liposomes with a potent adjuvant is a good vaccine model against melioidosis, specifically for efficient delivery and generation of a broader immune response. Gold nanoparticle glycoconjugate vaccines are currently under trial.

DNA vaccines against the flagellin structural gene and attenuated *B. pseudomallei* mutants have also been tried recently.

Case Study

An adult male presented to a tertiary care hospital with recurrent episodes of fever with dull bodyache, pain in the feet, and big joints for about 15 days. The patient was a farmer, a current smoker, and an alcohol addict. Laboratory tests showed polymorphonuclear leucocytosis and thrombocytopenia. He was diabetic. Empirically, a third-generation cephalosporin was started. However, within 72 h, the patient progressed to septic shock with signs of acute kidney injury. Accordingly, other supportive measures with a change of antibiotics were started.

Blood culture was positive within 48 h of incubation. A Gram stain from the bottle showed bipolarly stained Gram-negative bacilli. Subculture on blood and MacConkey agar showed medium-sized oxidase-positive colonies with dry margins and a metallic sheen. With routine biochemical reactions and resistance to colistin and polymyxin B (300 units), the organism was presumptively identified as *B. pseudomallei*. Colonies were confirmed using the VITEK-I compact system (bioMérieux).

The bacterial isolate was susceptible to ceftazidime, meropenem, imipenem, doxycycline, and cotrimoxazole. An injection of meropenem (3 g/day) was started. Radiological investigations revealed multiple pulmonary nodules, prostatic, and splenic abscesses. He was diagnosed with disseminated septicaemic melioidosis with multiple deep abscesses. The patient improved after initial intensive treatment with intravenous meropenem and was discharged on request after 14 days. He was prescribed oral eradication therapy with cotrimoxazole for 3 months and other ancillary advice.

1. Can early diagnosis of melioidosis with a point-of-care test for *B. pseudomallei* infection prevent progression of cases to bacteraemia and septic shock?
2. Can the development of an antibiotic policy in hospitals for empirical treatment of suspected melioidosis in endemic areas be helpful for a better prognosis?
3. Is personal protection for diabetics, is essential while working in an endemic environment to reduce melioidosis?
4. Do we need to emphasise awareness about relapse and patients' compliance with eradication therapy?

Research Questions

1. What are the survival strategies of bacteria in natural environments, and how do they achieve dormancy within a host during latency?
2. What effective control measures can be developed for *B. pseudomallei* in the environment?
3. How can the burden of melioidosis be accurately estimated, and what screening tools can be developed for latent infection?
4. What point-of-care tests and handy tools can be developed for epidemiological surveys?

Further Readings

Anuntagool N, Naigowit P, Petkanchanapong V, Aramsri P, Panichakul T, Sirisinha S. Monoclonal antibody-based rapid identification of *Burkholderia pseudomallei* in blood culture fluid from patients with community-acquired septicaemia. J Med Microbiol. 2000;49:1075–8.

Appassakij H, Silpapojakul KR, Wansit R, Pornpatkul M. Diagnostic value of the indirect hemagglutination test for melioidosis in an endemic area. Am J Trop Med Hyg. 1990;42:248–53.

Ashdown LR. An improved screening technique for isolation of *Pseudomonas pseudomallei* from clinical specimens. Pathology. 1979;11:293–7.

Cheng AC, Currie BJ. Melioidosis: epidemiology, pathophysiology, and management. Clin Microbiol Rev. 2005;18(2):383–416. https://doi.org/10.1128/CMR.18.2.383-416.2005.

Currie BJ, Fisher DA, Anstey NM, Jacups SP. Melioidosis: acute and chronic disease, relapse, and reactivation. Trans R Soc Trop Med Hyg. 2000;94:301–4.

Currie BJ, Jacups SP, Cheng AC, Fisher DA, Anstey NM, Huffam SE, Krause VL. Melioidosis epidemiology and risk factors from a prospective whole-population study in northern Australia. Trop Med Int Health. 2004;9:1167–74.

Dance D. Treatment and prophylaxis of melioidosis. Int J Antimicrob Agents. 2014;43:310–8.

Dharakul T, Songsivilai S, Smithikarn S, Thepthai C, Leelaporn A. Rapid identification of *Burkholderia pseudomallei* in blood cultures by latex agglutination using lipopolysaccharide-specific monoclonal antibody. Am J Trop Med Hyg. 1999;61:658–62.

Fairley P, Smith S, Maisrikrod S, Henning L. Systematic review and meta-analysis of diagnostic tests for diagnosis of melioidosis. Acta Trop. 2021;214. https://doi.org/10.1016/j.actatropica.2020.105784.

Johnson MM, Ainslie KM. Vaccines for the prevention of melioidosis and glanders. Curr Trop Med Rep. 2017;4(3):136–45. https://doi.org/10.1007/s40475-017-0121-7.

Kang G, Rajan DP, Ramakrishna BS, Aucken HM, Dance DA. Melioidosis in India. Lancet. 1996;347:1565–6.

Phillips ED, Garcia EC. *Burkholderia pseudomallei.* Trends Microbiol. 2024;32(1):105–6. https://doi.org/10.1016/j.tim.2023.07.008.

Phung LV, Han Y, Oka S, Hotta H, Smith MD, Theeparakun P, Yabuuchi E, Yano I. Enzyme-linked immunosorbent assay (ELISA) using a glycolipid antigen for the serodiagnosis of melioidosis. FEMS Immunol Med Microbiol. 1995;12:259–64.

Ramphal R. Infections due to *Pseudomonas*, *Burkholderia* and *Stenotrophomonas* species. In: Loscalzo J, Kasper DL, Longo DL, editors. Harrison's principles of internal medicine. 21st ed. New York: McGraw Hill; 2022. p. 1284–91.

Whitmore A, Krishnaswami CS. An account of the discovery of a hitherto undescribed infective disease occurring among the population of Rangoon. Indian Med Gazette. 1912;47:262–7.

Yamamoto T, Naigowit P, Dejsirilert S, Chiewsilp D, Kondo E, Yokota T, Kanai K. In vitro susceptibilities of *Pseudomonas pseudomallei* to 27 antimicrobial agents. Antimicrob Agents Chemother. 1990;34:2027–9.

Glanders

26

Lenita Ramires dos Santos,
Emanuelle Baldo Gaspar,
Paula Adas Pereira Suniga,
and Flábio Ribeiro de Araújo

Learning Objectives

1. To explore the taxonomy, genomics, and morphology of *Burkholderia mallei*, the causative agent of glanders
2. To comprehend the pathogen, including its hosts, infective stages, and transmission pathways
3. To identify key gaps in current knowledge and propose research questions to advance the understanding and management of glanders

Introduction

Glanders is an infectious disease caused by the Gram-negative coccobacillus *Burkholderia mallei*. It primarily affects solipeds, such as horses, donkeys, and mules, but can also affect camels, bears, small ruminants, and carnivores, including felids and wolves. Pigs and cattle have been shown to be resistant to the infection. The disease has been known in equids since ancient times. However, its connection to human infection, characterised as a zoonosis, was only established in the eighteenth century. Depending on the route of entry of the pathogen, equids can develop nasal, pulmonary, and cutaneous forms of the disease. According to its nature, the disease can be acute or chronic, but some animals remain asymptomatic. Donkeys usually experience the acute form, while horses are more prone to developing chronic glanders and serve as the main reservoir of the disease. Mules usually exhibit intermediary forms. Humans are accidentally infected, and glanders is considered an occupational disease, mainly affecting people in close contact with animals, such as veterinarians, stablemen, or laboratory personnel.

During the twentieth century, with the decline in the use of equines for transportation and the widespread testing and culling of infected animals, the disease was eradicated in several countries. High-income countries are now considered free from glanders. However, glanders remains endemic in South America, Northern Africa, Asia, and the Middle East. Furthermore, it has emerged or re-emerged in areas where it was previously undocumented or had been eradicated. One particular concern is that *B. mallei* has previously been used as a bioterrorism weapon and is classified as a Category B bioterrorism agent by the CDC, primarily due to its high infectivity, degree of incapacitation, and the lack of an available vaccine.

L. R. d. Santos · F. R. d. Araújo
Embrapa Beef Cattle, Campo Grande, MS, Brazil

E. B. Gaspar (✉)
Embrapa Dairy Cattle, Juiz de Fora, MG, Brazil
e-mail: emanuelle.gaspar@embrapa.br

P. A. Suniga
IAGRO – State Agency for Animal and Plant Health Defence, Água Clara, MS, Brazil

S. C. Parija, A. Poddar (eds.), *Textbook of Bacterial Zoonoses*, Microbial Zoonoses,
https://doi.org/10.1007/978-981-95-6615-0_26

History

Hippocrates was the first to describe glanders around 425 BC. Later, Aristotle presented a good, although incomplete, description of the disease and named it "*melis*". In the Latin version, this term became "*malleus*" and was used as a general term for epizootics. In the third century AD, Apsyrtus described several types of malis (epizootic), and among those, two types whose symptoms corresponded to glanders (the pulmonary form of the disease) and farcy (the cutaneous form). His letters were later included in Hippiatrika manuscripts compiled in Byzantium during the tenth century. Vegetius, in the fifth century, also described two varieties of malleus whose symptoms match with glanders and farcy. The fragments of classical authors support the idea that glanders has existed since ancient times, rather than providing unequivocal evidence of an accurate diagnosis. A major challenge in the historical treatment of glanders has been its nonspecific clinical signs and the difficulty of accurate diagnosis. Viborg only demonstrated the common aetiology of glanders and farcy at the end of the eighteenth century.

Solleysel's seventeenth-century description of glanders was remarkably complete. He emphasised the importance of isolating infected animals. He highlighted the risks of transmission through contaminated drinking water and direct contact. Hague (1734) copied Solleysel and reinforced the importance of the isolation and slaughter of infected animals, as well as the disinfection of stables previously used by them. In fact, Apsyrtus (third century) and Vegetius (fifth century) already recognised the contagious nature of the disease and recommended the isolation of affected animals. The infectious nature of the disease was confirmed with the first isolation of the aetiological agent in 1882 by Loeffler and Schütz in Germany. The mallein was developed, inspired by tuberculin, shortly after that and was used as a diagnostic tool as early as 1891. The zoonotic nature of glanders, or its transmission from animals to humans, was first suggested in the eighteenth century. From the beginning of the following century, reports of human cases of farcy and glanders became more common.

Since ancient times, glanders has held special significance during wars, particularly up to World War I, due to the crucial role of horses as military assets. Even the history of modern veterinary medicine has a strong connection with glanders studies. Bourgelat, like the Lafosses, was the pioneer who built the first modern veterinary schools in France and was interested in and wrote about the disease during the eighteenth century.

Taxonomy

B. mallei is an obligate intracellular pathogen with a high tropism for equids, especially mules and donkeys, which are more susceptible to infection. The genus *Burkholderia* belongs to the Burkholderiales order and the Burkholderiaceae family. It consists of over 40 different species, occupying a broad range of ecological niches. While some species are predominantly soil inhabitants and non-pathogenic, *B. mallei* is known for its pathogenicity and ability to infect animals.

Over time, *B. mallei* has been referred to by several names, including *Loefflerella mallei*, *Pfeifferella mallei*, *Malleomyces mallei*, *Actinobacillus mallei*, *Corynebacterium mallei*, *Mycobacterium mallei*, *Pseudomonas mallei*, and *Bacillus mallei*. Its current classification as part of the Burkholderiaceae family is based on phenotypic characteristics, cellular lipid and fatty acid composition, 16S rRNA gene typing, and DNA–DNA homology.

Genomics and Proteomics

The *B. mallei* genome comprises approximately 5.7 Mb arranged on two circular chromosomes. This genome is about 1.5 Mb smaller than that of *B. pseudomallei*. Genome size is the main difference between the two species, as nearly all *B. mallei* genes have an ortholog in *B. pseudomallei*. However, the latter possesses 1200 additional genes. On the other hand, the proportion of variable genes (apart from the core genome) among *B. mallei* strains is higher, ranging from 33% to 41%, whereas in *B. pseudomallei*, it varies between 11% and 22%.

The prevailing hypothesis is that *B. mallei* evolved from a single *B. pseudomallei* ancestor, with an adaptation to the equine host. This evolution reduced genome size, with the loss of genes related to environmental survival, such as antibiotic production and metabolic pathways. As a result, *B. mallei* lost its environmental versatility, becoming an obligate pathogen restricted to a few hosts.

The genome of *B. mallei* contains 5535 open reading frames (ORFs). In terms of replication, it is still not possible to precisely identify the exact start point of the chromosomes, unlike in *B. pseudomallei*.

The flagella *fliP* gene is interrupted by the IS407A element, which results in the immobility of *B. mallei*. This interruption was utilised to develop a PCR-based system specific for *B. mallei*. However, this system recently failed to correctly identify this species, likely due to mutations, loss of the IS407A insertion, or novel genomic recombination events.

The identification and characterisation of *B. mallei* strains are challenging due to their genetic homogeneity and genome complexity. Typing methods such as MLVA (Multiple Locus Variable-number Tandem-repeat Analysis) have been widely used to investigate outbreaks. However, the conservation of certain loci within the species limits their applicability. A more restricted version of MLVA, with 23 loci specific to *B. mallei*, has been effective in differentiating strains. However, more recent methods, such as high-resolution melting HRM, based on single nucleotide polymorphism (SNP) analysis, have also been used to discriminate isolates. Additionally, core genome-based typing techniques (cgMLST) have proven to be efficient, utilising 3328 genetic targets for high-resolution typing, allowing for a clear distinction between strains from different geographical regions. Though more expensive and complex, whole genome sequencing (WGS) is a powerful tool for (i) phylogenetic relationship analyses, (ii) population structure studies, and (iii) transmission events of *B. mallei*, these technological advancements in genomics have provided new perspectives for the molecular investigation of *B. mallei*. These are essential for understanding the genetic characteristics that define the pathogenicity and evolution of this bacterium and enabling a better approach to the diagnosis and control of glanders outbreaks.

Morphology

B. mallei is a Gram-negative, facultative aerobic, and non-motile coccobacillus (Fig. 26.1). Observed under optical microscopy, *B. mallei* appears as Gram-negative coccobacilli with rounded ends, about 2–5 µm long and 0.3–0.8 µm wide. Electron microscopy reveals a capsule composed of neutral carbohydrates that shields the bacteria from harsh environmental conditions. The absence of flagella and subsequent lack of motility are key characteristics for differentiation from other species.

Cultivation

B. mallei grows slowly in culture. After 24 h of incubation, its colonies appear sparse. After 48 h, the colonies evolve into grey, round, punctiform formations, displaying a characteristic translucent and shiny halo without pigment or haemolysis and a viscous appearance. With continued incubation, the growth of *B. mallei* intensifies, and colonies may become confluent and acquire a dark brown colouration while maintaining viscosity. This change in colour and increase in viscosity are important characteristics to be observed during the cultivation of the bacteria. It is a mesophilic microorganism, but some strains can grow at 42 °C. Isolating *B. mallei* from tissue macerates is extremely challenging, as initial samples often result in the accelerated growth of contaminating microorganisms, even in the presence of *B. mallei*. There are several reports of mixed cultures in which the bacteria multiply in association with other microorganisms. For the

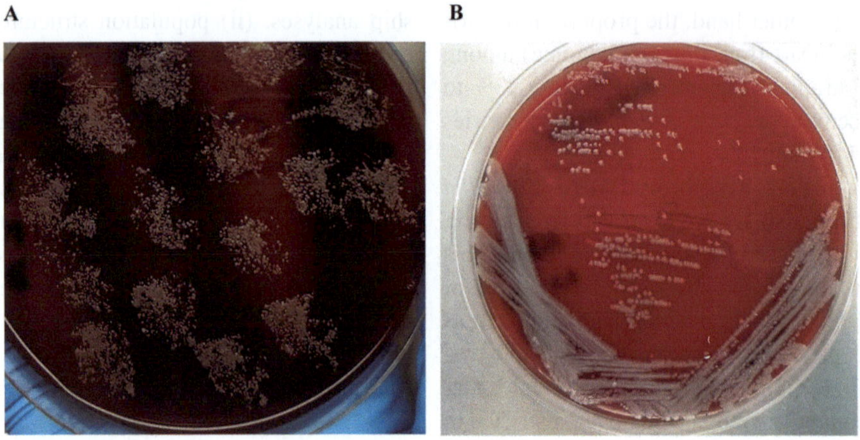

Fig. 26.1 (a) *Burkholderia mallei* growth was observed after imprinting the testicle of an experimentally infected guinea pig (Strauss test) to reactivate the strain. (b) *B. mallei* strain 86/19 (Brazil) after 48 h of incubation in BAG medium

cultivation of *B. mallei,* several media can be used. The agar base, supplemented with 5% defibrinated sheep blood and 2–5% glycerol (blood agar glycerol (BAG)), is one of the most common in the literature. Glycerol plays an important role in increasing bacterial growth.

Animal Models

Animal models play a crucial role in understanding the pathogenesis of *B. mallei* infections and developing potential countermeasures. Various species, including non-human primates, mice, guinea pigs, hamsters, and horses, have been employed to study glanders under different experimental conditions. Among laboratory animals, hamsters and guinea pigs are highly susceptible, while mice exhibit an intermediate susceptibility pattern. BALB/c mice have been widely used to assess the virulence of *B. mallei* strains. Intraperitoneal, aerosol, and intranasal infection models provide valuable insights into disease progression and host responses. Aerogenic infection in mice has been beneficial for mimicking natural exposure routes, offering a reproducible platform to evaluate bacterial dissemination and immune responses. Additionally, subcutaneous and intraperitoneal models have been used to establish sublethal and lethal infection thresholds, aiding in characterising bacterial pathogenesis.

Non-human primates, such as rhesus macaques and common marmosets, have been explored as models for glanders due to their physiological similarities to humans. Comparative studies using aerosol infection in different primate species have helped determine the most suitable models for studying respiratory exposure, which is a primary concern for biodefence research. The common marmoset has been recognised as a relevant model due to its susceptibility to *B. mallei* infection and ability to replicate key clinical features observed in human glanders.

Experimental equine models remain highly valuable. Horses are the natural hosts of *B. mallei*, providing a unique perspective on disease progression, chronic infection, and bacterial shedding. However, there is a notable scarcity of experimental studies and publications involving equine infections. This is largely due to the logistical challenges, high biosafety requirements, and ethical considerations associated with conducting such research. Conducting experimental infections with *B. mallei* in equids presents significant challenges due to its classification as a biosafety level 3 (BSL-3) pathogen. The strict containment requirements, specialised facilities, and rigorous biosafety protocols limit the number of laboratories capable of performing such studies. Additionally, horses' size and handling requirements further complicate experimental procedures, including infection control, monitoring, and sample collection. These constraints make equine glanders models less accessible than small animal or non-human primate models, despite their critical relevance in studying natural disease progression and transmission dynamics.

Transmission Cycle

Although the main hosts for *B. mallei* are equids, several other species can be infected, including felids, camels, bears, wolves, and dogs. In contrast, cattle and pigs are resistant. Humans are considered accidental hosts. Even among equids, the pattern of susceptibility varies: donkeys are the most susceptible, mules show an intermediate level of susceptibility, and horses are relatively less susceptible. This characteristic makes horses prone to developing the disease's chronic and often asymptomatic form, making them the primary reservoir. Chronically infected or asymptomatic horses shed viable bacteria, contaminating food and water troughs through nasal secretions or skin exudates. Infection occurs through ingestion of contaminated water or food, or by contact of mucous membranes or abraded skin with contaminated material. Reports of human infection in the twentieth and twenty-first centuries are rare. There is a possibility of transmission through contact of mucosal surfaces with aerosols generated during laboratory manipulation of the bacteria or through contact of bacterial culture with non-intact skin. Transmission from animals to humans is possible but rare (Fig. 26.2).

Pathogenesis and Pathology

B. mallei infection in horses is marked by severe emaciation, purulent or haemorrhagic nasal discharge, and ulcerative nodules (0.5–1 cm) on the nasal mucosa. Some cases present with cutaneous lesions, forming 1–2 cm nodules on the hindlimbs and ventral abdomen. The lungs typically contain numerous grey, firm miliary nodules (~0.3 cm) and multifocal, coalescing yellowish nodules encased in fibrous tissue. Nodule centres can be caseous or calcified.

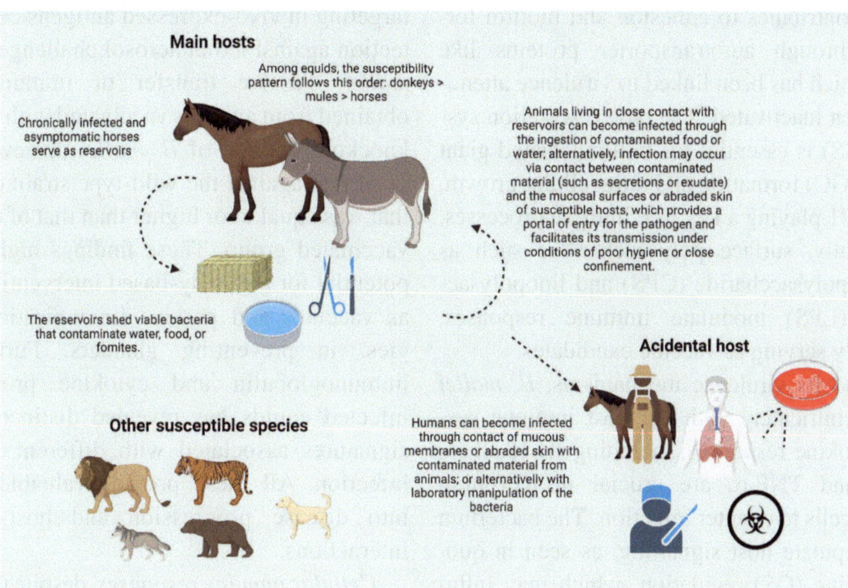

Fig. 26.2 Transmission of *Burkholderia mallei* (The image has been created in BioRender. Image source: Authors' personal collection)

Histopathological findings include pyogranulomas and abscesses, predominantly in the lungs, with occasional involvement of the liver and mediastinal lymph nodes. In the lymphatic system, firm, rounded, and elevated nodules are observed along the course of the lymphatic vessels. The nasal mucosa exhibits necrosis, ulceration, submucosal abscesses, and cartilage destruction, while skin lesions consist of dermal and subcutaneous abscesses. Fibrin exudation is frequently observed in the lungs, nasal mucosa, and skin.

Clinically, affected horses develop ulceration of the nasal conchae and septum, resulting in a thick yellow discharge that is characteristic. The condition also manifests as submaxillary lymphadenopathy, pulmonary nodules with central necrosis, fever, coughing, dyspnoea, progressive debility, diarrhoea, and polyuria. The cutaneous form, known as *farcy,* manifests as enlarged lymphatics with nodular abscesses ("buds") that ulcerate and exude thick, yellow pus.

B. mallei utilises multiple secretion systems to facilitate invasion, intracellular survival, and virulence. The type 3 secretion system (T3SS) is critical for intracellular growth, actin polymerisation, and phagosomal escape, enabling *B. mallei* to spread cell-to-cell. The type 5 secretion system (T5SS) contributes to adhesion and biofilm formation through autotransporter proteins like BpaB, which has been linked to virulence attenuation when inactivated. The type 6 secretion system (T6SS) is essential for multinucleated giant cell (MNGC) formation and intracellular growth, with HCP1 playing a key role in these processes. Additionally, surface polysaccharides such as capsular polysaccharide (CPS) and lipopolysaccharide (LPS) modulate immune responses, potentially serving as vaccine candidates.

Beyond its virulence mechanisms, *B. mallei* interacts intricately with the host immune system. Cytokine responses, including IL-1β, IL-6, IFN-γ, and TNF-α, are crucial in recruiting immune cells to counter infection. The bacterium can manipulate host signalling, as seen in quorum sensing (QS) regulation, which may influence chronic infection dynamics. Additionally, host stress hormones like epinephrine enhance bacterial growth and motility, highlighting metabolic adaptability. Dendritic cells and neutrophils facilitate bacterial dissemination and modulate immune responses, further underscoring the complexity of *B. mallei* pathogenesis. Understanding these interactions is critical for developing targeted therapeutics and vaccines against glanders.

Immunology

The immune response to *B. mallei* infection involves a complex interplay between innate and adaptive immunity. The innate immune system serves as the first line of defence, with pattern recognition receptors (PRRs) detecting bacterial components and triggering pro-inflammatory cytokine production. *B. mallei* can modulate the host immune response by evading complement-mediated lysis, allowing the pathogen to persist within host cells. Additionally, it can survive within macrophages, utilising intracellular mechanisms to avoid immune clearance. This immune evasion strategy contributes to the pathogen's virulence and chronicity in infected hosts.

Humoral immunity plays a crucial role in protection against *B. mallei* infection. Antibodies targeting in vivo-expressed antigens confer protection against lethal aerosol challenges with *B. mallei.* Passive transfer of immune serum obtained from animals vaccinated with a mutant/knockout version of *B. mallei* achieved a survival rate against the wild-type strain challenge that was equal to or higher than that of the initial vaccinated group. These findings highlight the potential for antibody-based interventions, such as vaccines and passive immunisation strategies, in preventing glanders. Furthermore, immunoglobulin and cytokine profiling in infected equids has revealed distinct immune signatures associated with different stages of infection. All these provide valuable insights into disease progression and host-pathogen interactions.

Cellular immune responses, despite the importance of antibody-mediated immunity, are also essential for the adequate clearance of *B. mallei.*

T-cell activation and cytokine secretion are pivotal in controlling bacterial dissemination and promoting macrophage-mediated killing. However, *B. mallei* employs various mechanisms to subvert host immune defences, including modulation of cytokine signalling pathways. Understanding these immune interactions is critical for developing effective therapeutic and prophylactic strategies against glanders, particularly for high-risk populations and biodefence preparedness.

In natural cases of equine glanders, IgG antibody titres are higher than those of IgM, IgG (T), and IgA isotypes. Hcp1 is the most potent antibody inducer among the tested proteins, surpassing TssA, TssB, and others. Elevated levels of IL-1β, MCP-1, IL-17, IL-6, IFN-γ, and TNF-α highlight the crucial role of these cytokines in enhancing cellular defence by recruiting macrophages, neutrophils, and dendritic cells against *B. mallei* infection.

Infection in Humans

Glanders is primarily a disease of solipeds, with human infection occurring accidentally. Although rare, in humans the disease is painful and life-threatening. Approximately 50% of cases present an acute course, with mortality near 100% if untreated. In chronic cases, untreated mortality is about 60%.

As in animals, the disease can be both acute and chronic in humans. The incubation period for acute disease is 1–14 days, but in the chronic form, this can reach up to 12 weeks. Improvement of the clinical signs, followed by a second wave of disease, was reported in accidental cases of infection. Many forms of glanders have been described, including localised, disseminated, pulmonary, and septicaemia.

Localised infections are restricted to a specific region and often feature areas of suppuration, for example, in the skin. Abscesses may ulcerate and drain over extended periods. However, these infections have the potential to spread. Acute glanders is a systemic disease with characteristic septicaemia. Initial symptoms include inappetence, chills, fever, nausea, and severe, widespread pain. Later, granulomas and abscesses can develop in the subcutaneous tissues and muscles. A pustular rash can occur, with papules progressing into pustules and ulcers. Pneumonia, pulmonary abscess, pleuritis, and pleural effusion are common findings in the pulmonary form of the disease.

In a well-documented case of six accidental glanders infections in 1944–1945 at the laboratories of Fort Detrick, the most commonly described symptoms were generalised or localised aches and pain, particularly in the chest (aggravated by breathing), head, abdomen, throat, back, and muscles. Fever, especially in the afternoons, was also common. Fatigue, malaise, chills, and dizziness were reported. At the same time, nausea, diarrhoea, blurring in vision, lacrimation, photophobia, tachypnoea, cough, laryngitis, and mucopurulent sputum production occurred less frequently. Physical examination showed dehydration in two patients. Radiographic imaging revealed lung lesions compatible with abscesses in almost all patients. The leukogram showed lymphopenia in some patients. All patients were treated with sulfadiazine and recovered.

In case descriptions that occurred before the availability of treatment (1909), postmortem analysis of the lungs revealed cystic lesions with softened centres and hyperaemic peripheries or pustular nodules. Pustules in the skin and intramuscular abscesses were also noted. In one chronic case, a patient exhibited destruction of the soft palate. More recently, diagnostic imaging has described diffuse airspace infiltrates in the lungs, heterogeneous patterns of lung attenuation, or mediastinal widening. Small cystic lesions in the liver and slightly larger, more confluent lesions in the spleen were also observed.

Recently, a human case was well documented in Brazil involving a 73-year-old patient from the Northeast region. The patient was hospitalised with severe pneumonia. PCR and genomic sequencing confirmed *B. mallei* in pleural drainage. Genotyping revealed a novel genotype in Brazil, emphasising the need for genetic surveillance in zoonotic infections.

Infection in Animals

Regarding the course of the disease, it can be acute or chronic. The acute form is usually fatal within a few days or weeks. On the other hand, subacute or chronic glanders, while potentially fatal, can persist insidiously for months or years but may exacerbate at any moment. In some cases, animals can recover from clinical signs and become *B. mallei* carriers for life. Subacute or asymptomatic cases are common, especially in horses. Among equids, donkeys are highly susceptible to glanders, usually experiencing the acute form of the disease. Horses are prone to developing chronic cases, while mules show intermediate susceptibility.

B. mallei enters the body through abraded skin, mucous membranes, or inhalation of dried infected particles (aerosol). Depending on the site of entry, equids develop nasal, pulmonary, or cutaneous forms of the disease, the latter also known as *farcy*. However, all three forms can coincide. Pulmonary symptoms are almost always present, with the pulmocutaneous presentation being the most common. The disease is more common in elderly animals or those in poor body condition.

Acute glanders starts with systemic clinical manifestations, such as chills, fever, depression, and inappetence. As specific signs, the animals present bouts of coughing and watery nasal discharge, which progress to mucopurulent or haemorrhagic discharge, leading to laboured breathing (Fig. 26.3). Glottic oedema and enlargement of the jaw lymph nodes may also be present. Respiratory failure due to pneumonia or septicaemia can result in death within a few days or weeks. Some animals may present with limb oedema and polyarthritis in the initial phase of the disease.

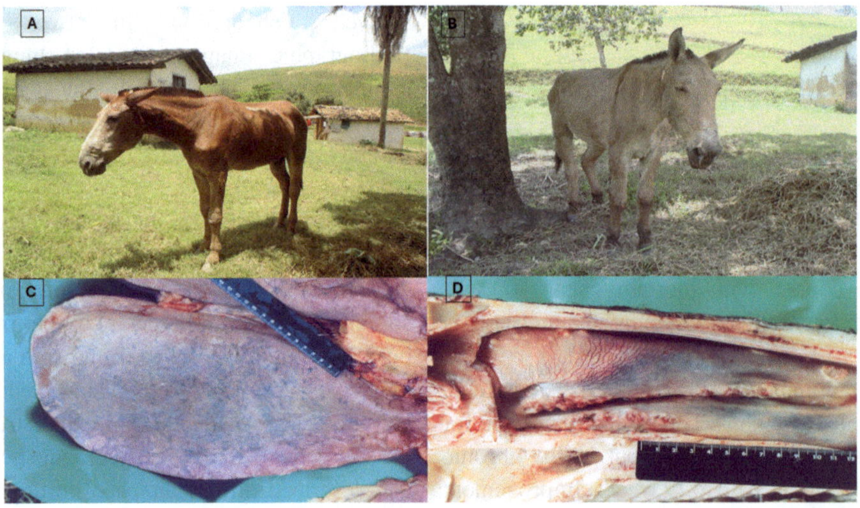

Fig. 26.3 (**A**) A mule with a depressed demeanour: in the left cervical region, there is a thickened lymphatic vessel extending from the nape to the prescapular area, and three circular swellings are observed, as well as on the anterior surface of the right hind limb, just below the patella. In addition, there is nasal discharge with purulent mucus, thickening of the lymphatic vessels in the region above the nostrils, and slight oedema in the hind limbs. (**B**) A mule, which appears depressed, shows purulent nasal discharge. Oedema is observed in all four limbs. In the right costal region and flank, small areas of scarring are found. (**C**)—multifocal pneumonia. Equine lung block. Multiple dark spots are observed distributed across the entire dorsal surface of the right pulmonary lobe. Each of these spots represents a pneumonic focus. (**D**) Glanders—ulcers and hyperaemia. Sagittal section of the head of Muar. In the posterior portion of the nasal cavity, there is intense hyperaemia. In this same region, more dorsally, numerous abscesses are present, seen as yellowish spots. Below, multiple ulcers of varying sizes are found. In general, these ulcers have raised edges and a reddish base. The ulcer located farthest to the right in the image has raised edges, but its base is yellowish, as it is covered by fibrinous exudate. (Courtesy: SANTOS, FERNANDO LEANDRO, UFRPE, PE/Brazil)

Chronic glanders, on the other hand, takes much longer to develop, typically over months, and usually manifests in the pulmonary form, characterised by fever, dyspnoea, and coughing. Pulmonary lesions may remain latent for an extended period with temporary remission of clinical signs. Intermittent fever and dyspnoea may also occur. As lung lesions progress, the bacterium spreads, leading to a decline in the animal's overall condition. Stress can exacerbate the disease and its clinical signs. Ulcers and nodules develop in the nasal mucosa, accompanied by purulent nasal discharge. Enlargement of the submandibular or submaxillary lymph nodes is also observed. Septal perforation and stellate scars in the nasal mucosa are frequent. Fibrosis and adhesion of lymph nodes are also present in chronic cases. As a laboratory finding, the leukogram of some animals reveals leukocytosis with neutrophilia and elevated fibrinogen levels.

Pulmonary glanders can remain asymptomatic for extended periods. According to our experience, it is common for animals with positive serology but no clinical signs of the disease to exhibit abscesses, particularly in the lungs, liver, and spleen, from which *B. mallei* DNA can be detected.

Experimental glanders in equines have been demonstrated to be possible, with intratracheal inoculation of *B. mallei* producing clinical signs like natural infection. The animals already exhibited signs of illness as early as 1 day after inoculation. For humanitarian reasons, the animals had to be euthanised on day seven, preventing the disease from being observed through to the chronic phase.

Felids and other carnivores can become infected after consuming meat from animals with glanders. Lions and tigers exhibit clinical signs, including depression, anorexia, vomiting, skin ulcers, dyspnoea, and respiratory disorders.

Epidemiology and Public Health

Equids are the main hosts of *B. mallei,* especially those living in crowded and unsanitary environments. The spread of glanders is facilitated by the use and transport of horses. These factors made this epizootic particularly significant during wars, up to WWI, when horses played a crucial role in military operations. Besides equids, felids have also experienced outbreaks. Camels, bears, wolves, and dogs are susceptible, while cattle and pigs are resistant. Carnivores can become infected by consuming contaminated meat.

Among equids, overcrowding, poor hygiene, and stress are risk factors for infection. Subclinical or asymptomatic carriers play a greater role in disease transmission than clinical cases. Chronically or subclinically infected equids can shed *B. mallei* intermittently or continuously into food and water troughs through nasal secretions or skin exudates. Thus, chronically infected or asymptomatic horses are the main reservoirs of the disease.

Natural transmission from sick animals to humans is rare, even with frequent and close contact with diseased animals. Transmission occurs through contact with contaminated material from animals or bacterial cultures via non-intact skin or nasal, oral, or conjunctival mucous membranes. Transmission from humans to humans is extremely rare. Inhalation with deep lung deposition can also occur, but the transmission through ingestion of contaminated water and food is improbable. The aerosol route is believed to cause human glanders, particularly in laboratory contamination cases.

The bacteria can also be mechanically transmitted by flies and fomites. Venereal transmission, as well as vertical transmission from the mare, can occur. Recently, *B. mallei* was isolated from a mare's milk, and transplacental transmission was also demonstrated.

Currently, the disease is no longer widespread. The decline in the use of horses for transport and military purposes, together with systematic testing and culling of positive animals, led to the eradication of the disease, especially in high-income countries, such as the USA, Australia, Canada, and Western Europe, during the twentieth century. However, the disease has recently emerged and re-emerged in areas in which it was previously unknown or had been eradicated.

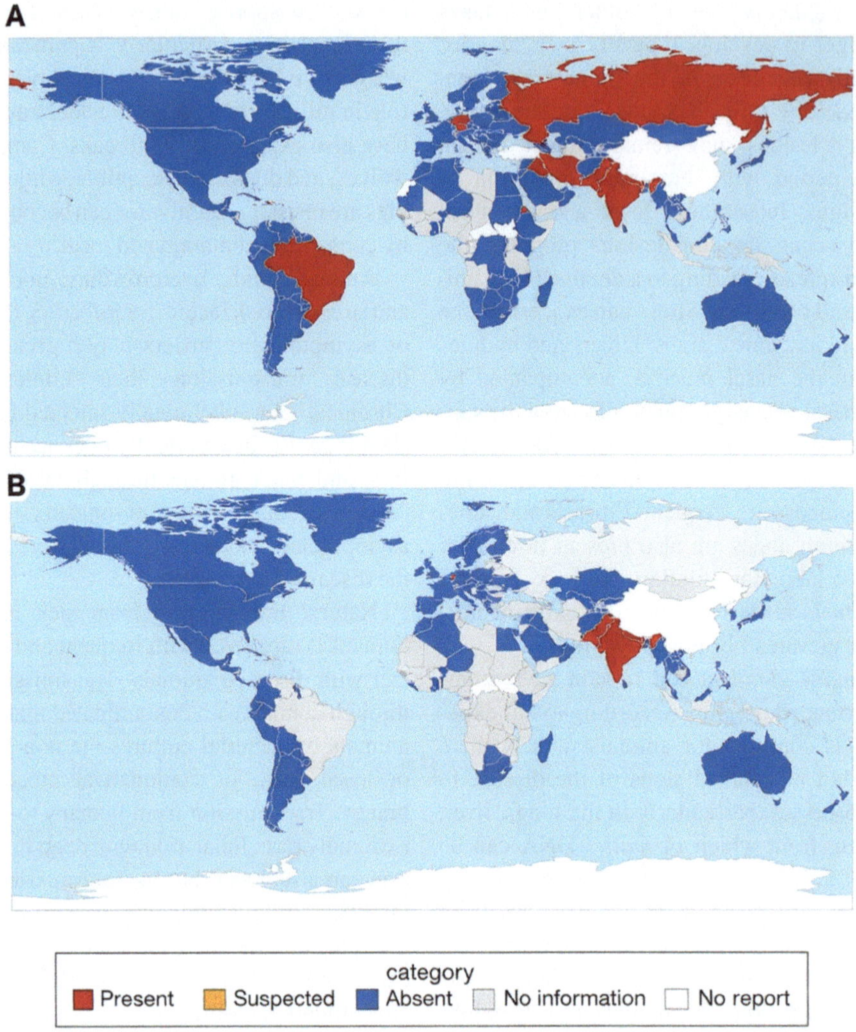

Fig. 26.4 Global distribution of glanders (2015–2025) according to the World Organisation for Animal Health (WOAH) in (**a**) domestic or (**b**) wild animals. Information accessed in April 2025. Map generated in RStudio. (Courtesy: CAMPOS, FRANCELLY GERALDA. MG/BRAZIL)

Glanders is still endemic in South America, Northern Africa, and parts of Asia and the Middle East. Focal cases/outbreaks were reported in several countries, including Brazil, India, Turkey, Eritrea, Ethiopia, Iran, Iraq, Bahrain, Kuwait, the United Arab Emirates, Pakistan, Mongolia, and others, and have been considered a re-emerging disease (Fig. 26.4).

Since the correlation between animal and human disease was established in the eighteenth century, reported cases have highlighted its occupational nature in humans, affecting mostly people with close contact with equines, such as farmers, grooms, horse keepers, veterinarians, or stable workers and laboratory staff.

Human epidemics have never been reported, but isolated cases in humans are still reported, even with the drop in glanders over the last 100 years. *B. mallei* is a potential bio-weapon, being considered a Category B biological agent for the Centres for Disease Control and Prevention (CDC). This threat is due to its high infectivity, degree of incapacitation, and unavailability of vaccines. Furthermore, the disease is relatively unknown, and its symptoms are nonspecific, making it hard to identify promptly. In fact, *B. mallei* is suspected to have been used in bioterrorist attacks during the First and Second World Wars, and possibly during the war in Afghanistan in the 1980s.

Diagnosis

Controlling and preventing glanders necessitates a multifaceted approach due to its zoonotic potential. The diagnosis of glanders in equids cannot rely solely on clinical signs, as they are insufficient for accurately identifying the disease. Moreover, the clinical presentation may resemble that of other illnesses, making laboratory tests essential for confirmation, along with appropriate biosafety and containment procedures.

The bacteria can multiply in association with other microorganisms, making successful isolation and in vitro culture challenging. The identity of suspected isolates can be confirmed using molecular methods, such as PCR, qPCR, and/or whole genome sequencing, enabling lineage determination. Carrying out interlaboratory studies is essential to validate the robustness of PCR and qPCR.

Microscopy

Direct microscopy of nasal or cutaneous smears may reveal Gram-negative, beaded rods suggestive of *B. mallei*. However, this organism is morphologically indistinguishable from many other bacilli and isolating sufficient organisms from clinical material is notoriously difficult. As a result, smear-based diagnosis has very low sensitivity and specificity. When microscopic examination is used, it should be paired with immunohistochemistry targeting *B. mallei* antigens, as with laser scanning confocal microscopy (LSCM) or molecular methods, to improve diagnostic accuracy.

In Vitro Culture

Various culture media, both liquid and solid, are used to isolate *B. mallei* and maximise the chances of detecting the bacteria, overcoming competition from other microorganisms present in the biological samples.

BAG is a selective medium that provides a suitable nutritive environment for bacterial growth. It is used either alone or supplemented with antibiotics to prevent the growth of contaminants. Ticarcillin disodium (32 μg/mL), ampicillin(32μg/mL),andtrimethoprim-sulfamethoxazole (50–10 μg/mL) can be included in BAG to inhibit the growth of contaminants.

The semi-selective BM (modified BHI) medium is also used. BM contains cycloheximide (50 μg/mL), ticarcillin disodium (16.7 μg/mL), sodium phosphomycin (197.6 μg/mL), polymyxin B and crystal violet (3 mg/L). These substances provide a rigorously selective environment that favours the proliferation of *B. mallei* and prevents the growth of other competing microorganisms, ensuring greater specificity in the isolation of the bacteria. In addition, *B. mallei* also grows in BHI broth supplemented with 2% glycerin. This provides an environment conducive to the growth of *B. mallei* in suspension, allowing greater recovery of the bacteria compared to traditional solid culture methods. Antibiotics such as polymyxin B (50 U/mL) and penicillin G (100 U/mL) can be added to the liquid medium to inhibit the growth of contaminating bacteria, further increasing the specificity. These semi-selective culture media, combined with the strategic use of antibiotics and selective agents, allow the effective isolation of *B. mallei* in 93.3% of the samples, both from symptomatic and asymptomatic animals (data from our group).

B. mallei are non-fermenters of sugars, non-producers of H_2S, indole negative, metabolise glucose through the oxidative pathway, oxidase variable (most positive), and catalase positive.

Serodiagnosis

The recommended tests for diagnosing glanders in equids include the complement fixation test (CFT), which should be confirmed by a second test with equal or higher sensitivity and greater specificity, such as the enzyme-linked immunosorbent assay (ELISA) or the *mallein test* (a cellular immunity test with animal welfare concerns). Although the CFT is an internationally recognised serological test for glanders, its use has been criticised due to limitations, such as variable sensitivity and specificity, technical complexity, challenges in result interpretation, and other factors that impact its reliability.

Molecular Diagnosis

Multiple PCR-based assays have been developed to detect *B. mallei*, targeting species-specific genes, insertion sequences, clusters, and intergenic regions such as *fli*P, IS407A, *bim*A, and the type III secretion system (T3SS). However, most of these assays have been validated using limited panels of strains, and genetic variation among isolates may affect primer-binding regions, potentially compromising assay sensitivity and specificity.

Currently, the World Organisation for Animal Health (WOAH) (formerly OIE) recommends a PCR assay targeting the *fli*P gene as a validated method for molecular diagnosis. The *fli*P-based TaqMan real-time PCR has shown high sensitivity and specificity, with a detection threshold as low as 60 femtograms (fg) of DNA per reaction. This makes it a powerful tool for rapid and reliable detection of glanders, particularly in cases where culture is difficult or when fast confirmation is critical for outbreak control.

Additionally, whole genome sequencing (WGS) offers definitive identification and subtyping of *B. mallei*, facilitating outbreak investigations and monitoring emerging variants. Its clinical use is still limited by cost and turnaround time. Nevertheless, advancements in sequencing technologies and more accessible bioinformatics tools are increasingly enabling its application in public and animal health laboratories.

Diagnosis data are summarised in Table 26.1.

Treatment

No veterinary-approved drug treatments are available for animals. On the other hand, treating human disease remains challenging due to the lack of clear recommendations and *B. mallei*'s resistance to many common antimicrobial drugs. In Fort Detrick, accidental contamination in the 1940s, all patients were successfully treated with

Table 26.1 Laboratory diagnosis for glanders

Diagnostic approaches	Methods	Targets	Remarks
Direct microscopy	Direct examination	Coccobacilli	Economical and fast technique; low sensibility and specificity; material needs to be pure or enriched
	Immunohistochemistry (IHC) and LCMS	Specific antigen	Time-consuming technique and high-cost equipment (LSCM). IHC—moderate throughput and turnaround time (1–2 days)
In vitro cultures	Under optimised conditions to maximise *Burkholderia mallei* detection/identification	Purified coccobacilli	Although these techniques are expensive and time-consuming, they enable biochemical screening and provide material for other assays
Indirect immunodiagnostics	Complement fixation ELISA Malein skin test Western blotting	Specific antigens, LPS and specific antibodies	Many tests are not validated yet; dependence on antigen specificity; need for confirmatory testing
Molecular assays	PCR Real-time PCR Genome sequence	*fli*P sequence difference to *B. pseudomallei*	High sensitivity and specificity. Genomic analysis also allows classification into variant subtypes

sulfadiazine. However, several antibiotics have been developed since then, and this is no longer the first therapeutic choice. In 2000, one patient accidentally contaminated in a research laboratory only presented improvement in symptoms after being treated with imipenem and doxycycline, but the *B. mallei* strain had been previously phenotyped as sensitive to these two antibiotics. More recently, published case reports from Asia have provided valuable insights into treatment approaches in the past 2 years. In China, initial therapy with levofloxacin and cefotaxime/sulbactam failed, but successful treatment was achieved by adding meropenem, doxycycline, and TMP/SMX. In Iran, effective management involved imipenem and doxycycline. In Turkey, meropenem was chosen for a 30-day treatment in a child. For definitive eradication, oral ciprofloxacin was administered based on antibiotic susceptibility testing.

Our group observed and analysed the recovery of a critically ill patient in Brazil as part of an investigative research effort. Initial treatment with ceftriaxone, followed by meropenem and azithromycin, failed to improve symptoms. However, the medical decision to use a combination of meropenem, linezolid, TMP/SMX, and levofloxacin resulted in clinical improvement.

Prevention and Control

Effective prevention and control of glanders rely on early detection, rigorous surveillance, and strict biosecurity measures. Once detected, immediate quarantine and humane euthanasia of positive cases are recommended to prevent further transmission. Implementing strict movement controls and biosecurity measures within and between equine facilities further mitigates the risk of spread. For countries without diagnosed cases of glanders, specific guidelines are in place for equine importation.

Preventive strategies also focus on minimising exposure to potential sources of *B. mallei.* Maintaining proper hygiene, disinfecting equipment, and ensuring clean water and feed supplies are essential practices. Educating handlers and veterinarians about glanders' clinical signs enhances early recognition and response.

While no licensed vaccine for glanders currently exists, research into vaccine development is ongoing, aiming to provide a viable prophylactic tool in the future. Inactivated whole-cell vaccines like *B. mallei* heat-killed strain ATCC23344, irradiation-inactivated, and *B. mallei* capsule-negative mutant were tested in BALB/c mice by the intraperitoneal (i.p.) or subcutaneous route using 10^8 CFU for immunisation, but with a low protection range. The most promising outcomes were achieved with live-attenuated vaccines. These vaccines conferred high protection against *B. mallei* challenge at 21 and 28 days post-immunisation using the ΔtonB and ΔtonBΔhcp1 mutant strains, respectively. Advancements in research using genetically attenuated *B. mallei* should focus on ensuring the safety of attenuation and evaluating and controlling the persistence of the vaccine strain to induce a memory immune response effectively. Other vaccine platforms, such as subunit protein, glycoconjugate, and DNA vaccines, may offer safer and more reproducible alternatives while providing high protection. In these cases, the selection of effective adjuvants and strategies for sustaining immune memory should be carefully considered to enhance long-term protection.

Given the zoonotic nature of *B. mallei*, protecting human health is crucial to glanders control programmes. Individuals in close contact with equines, such as veterinarians, stable workers, and laboratory personnel, should employ personal protective equipment and adhere to biosafety protocols to reduce infection risk. To prevent occupational exposure, public health initiatives need to emphasise awareness and education in endemic regions. Collaborative efforts between the veterinary and public health sectors are crucial for effective surveillance, reporting, and management of glanders outbreaks, ensuring that animal and human health are safeguarded.

Case Study

A 73-year-old male patient was admitted to a hospital in Northeast Brazil, presenting with cough, fever, malaise, chest pain, and arterial hypertension. Clinical examination, chest radiography, and computed tomography (CT) revealed pneumonia with pleural effusion. On the third day of hospitalisation, the patient reported a history of contact with a horse that had shared facilities with a mare that tested positive for glanders. Clinical deterioration led to pleural drainage, performed on the seventh day of hospitalisation. The pleural fluid culture was negative for acid-fast bacilli, *Mycobacterium tuberculosis*, but PCR and whole genome sequencing were positive for *B. mallei*. The initial treatment with ceftriaxone followed by meropenem and azithromycin did not improve symptoms. However, combining meropenem, linezolid, TMP/SMX, and levofloxacin resulted in improvement in the patient's condition, leading to his hospital discharge on the eighteenth day of hospitalisation. The genotypic analysis of *B. mallei* revealed a novel strain, emphasising the importance of genomic surveillance in tracking emerging infectious diseases.

1. What were the patient's presenting symptoms on admission?
2. Which diagnostic methods confirmed *B. mallei* infection?
3. Which combination of antibiotics ultimately led to the patient's improvement?

Research Questions

1. What are the historical significance and impact of glanders on human and animal populations globally?
2. How is the current prevalence and geographical distribution of glanders, including factors contributing to its persistence or re-emergence in certain regions?
3. What could be the potential future research directions for glanders, focusing on advancements in diagnostics, treatment strategies, and preventive measures, considering factors like antimicrobial resistance and zoonotic potential?
4. How to address the socio-economic implications of glanders outbreaks in modern times?

Further Readings

Appelt S, Rohleder AM, Jacob D, Von Buttlar H, Georgi E, Mueller K, et al. Genetic diversity, and spatial distribution of *Burkholderia mallei* by core genome-based multilocus sequence typing analysis. PLoS One. 2022;17:e0270499.

Gaspar EB, Santos LRD, Egito AAD, Santos MGD, Mantovani C, Rieger JDSG, et al. Assessment of the virulence of the *Burkholderia mallei* strain BAC 86/19 in BALB/c mice. Microorganisms. 2023;11:2597.

Kettle AN, Wernery U. Glanders and the risk for its introduction through the international movement of horses. Equine Vet J. 2016;48:654–8.

Laroucau K, Aaziz R, Vorimore F, Varghese K, Deshayes T, Bertin C, et al. A genetic variant of *Burkholderia mallei* detected in Kuwait: consequences for the PCR diagnosis of glanders. Transbound Emerg Dis. 2021;68:960–3.

Luz KG, Bezerra FR, Sicolo MA, Silva AA, Egito AA, Suniga PA, et al. Clinical and molecular characterisation of human *Burkholderia mallei* infection, Brazil. Emerg Infect Dis. 2024;30:2400–3.

Nierman WC, DeShazer D, Kim HS, Tettelin H, Nelson KE, Feldblyum T, et al. Structural flexibility in the *Burkholderia mallei* genome. PNAS. 2004;101:14246–51.

Pal M, Gutama KP. Glanders: a potential bioterrorism weapon disease. Am J Infect Dis Microbiol. 2022;10:98–101.

Suniga PAP, Mantovani C, Santos MG, Egito AA, Verbisck NV, Santos LR, et al. Glanders diagnosis in an asymptomatic mare from Brazil: insights from serology, microbiological culture, mass spectrometry, and genome sequencing. Pathogens. 2023;12:1250.

Suniga PAP, Mantovani C, Santos MG, Rieger JSG, Gaspar EB, dos Santos FL, et al. Molecular detection of *Burkholderia mallei* in different geographic regions of Brazil. Braz J Microbiol. 2023;54:1275–85.

Waag DM, Chance TB, Trevino SR, Rossi FD, Fetterer DP, Amemiya K, et al. Comparison of three non-human primate aerosol models for glanders, caused by *Burkholderia mallei*. Microb Pathog. 2021;155:104919.

WHOA. Listed disease – glanders. Available in: https://www.woah.org/en/disease/glanders/. Accessed Mar 2025.

WHOA Terrestrial Animal Health Code. Chapter 12.10. Infection with *Burkholderia mallei* (Glanders). 2024.

Available in: https://www.woah.org/en/what-we-do/standards/codes-and-manuals/terrestrial-code-online-access/?id=169&L=1&htmfile=chapitre_glanders.htm. Accessed Mar 2025.
WOAH Terrestrial Manual. Chapter 3.6.11: glanders and melioidosis. Manual of diagnostic tests and vaccines for terrestrial animals. 2024. Available in: https://www.woah.org/fileadmin/Home/fr/Health_standards/tahm/3.06.11_GLANDERS.pdf. Accessed Mar 2025.
Wilkinson L. Glanders: medicine and veterinary medicine in common pursuit of a contagious disease. Med Hist. 1981;25:363–84.
Woods DE. The use of animal infection models to study the pathogenesis of melioidosis and glanders. Trends Microbiol. 2002;10:483–4.

27 Pasteurellosis

Srujana Mohanty
and Tribhuban Mohan Mohapatra

Learning Objectives

1. To learn about pasteurellosis and its zoonotic transmission
2. To discuss clinical presentations in humans and animals, and prevention and control measures
3. To describe the pathogenesis and global epidemiology of the disease
4. To elaborate on the diagnostic methods, including microscopy, culture, and molecular methods

Introduction

Pasteurellosis, caused primarily by *Pasteurella multocida,* is an uncommon infectious disease in humans. The disease denotes a generic term indicating various infections associated with this bacterium. This facultatively anaerobic Gram--negative bacterium is an integral part of the commensal oral, nasopharyngeal, upper respiratory tract, and gastrointestinal flora of various animals and birds. These include companion and common livestock animals such as cats (70–90%), dogs (50–66%), pigs (51%), and cattle (20–60%). Wild animals harbouring *Pasteurella* include tigers, leopards, lynxes, rats, opossums, lions, horses, rabbits, boars, and panthers.

Pasteurella species cause fatal zoonotic infections in humans, primarily associated with close contact with domestic and pet animals. Infection is acquired from these sources primarily through scratches, licks, and bite wounds. Non-traumatic respiratory exposure through oral and nasal secretions also contributes to the transmission. Vertical transmission from mother to fetus is an essential route in the neonatal population. The characteristics of human *Pasteurella* infections range from the commonly reported localised infection of a bite wound to invasive infections. These invasive infections, which include bacteraemia, meningitis, and infective endocarditis, especially in immunocompromised patients, are associated with higher mortality rates.

History

P. multocida was discovered by Pasteur in 1880. Perroncito of Italy, in 1878, was the first to observe a bacterium with a rounded form in the tissues of birds affected with fowl choléra, occurring singly or in pairs. Kitt first isolated the organism during an epidemic disease among wild hogs in 1878, and Toussaint isolated it in 1879 from a chicken. Toussaint shared his work with Louis

S. Mohanty
Department of Microbiology, All India Institute of Medical Sciences, Bhubaneswar, Odisha, India

T. M. Mohapatra (✉)
Institute of Medical Sciences, Banaras Hindu University, Varanasi, Uttar Pradesh, India

S. C. Parija, A. Poddar (eds.), *Textbook of Bacterial Zoonoses*, Microbial Zoonoses,
https://doi.org/10.1007/978-981-95-6615-0_27

Pasteur, who further advanced the understanding of *P. multocida*. Pasteur isolated it and described it as the causative agent of "*choléra des poules*" (fowl choléra), in whose honour the bacterium was named *Pasteurella*. However, it was not until 1930 that the first true human infection was identified by Kapel and Holm, following a cat bite.

Taxonomy

The genus *Pasteurella* belongs to the family Pasteurellaceae in the phylum Pseudomonadati, class Gammaproteobacteria, order Pasteurellales, and the kingdom pseudomonadati.

The type species is *Pasteurella multocida* (Trevisan, 1887).

The family Pasteurellaceae currently also includes the genera *Haemophilus*, *Actinobacillus*, *Mannheimia*, *Aggregatibacter*, and other minor genera. Recent taxonomic studies have led to significant changes, particularly concerning species previously classified within the genus *Pasteurella*. Many of these species, which were previously misclassified, including *Pasteurella haemolytica*, based on recent comparative genomic and phylogenetic analyses, have been reclassified into new genera, such as *Mannheimia* and *Actinobacillus*.

Currently, a total of 22 species of *Pasteurella* have been described. *P. multocida* remains the most isolated species in human clinical infections. Other species, such as *Pasteurella dagmatis*, *Pasteurella aerogenes*, *Pasteurella canis*, *Pasteurella stomatis, and Pasteurella pneumotropica,* can be pathogenic in humans. In 1985, three subspecies were formally recognised. These include *P. multocida* subsp. *multocida*, subsp. *gallicida*, and subsp. *septica*. *P. multocida* subspecies *multocida* is the most frequently reported subspecies in human infections. It is the main agent associated with the anthropozoonosis of numerous warm-blooded animals. *P. multocida* subsp. *septica* is associated with injuries due to the feline species, while subsp. *gallicida* is rarely encountered in human infections.

Genomics and Proteomics

Early whole genome sequencing (WGS) of *P. multocida* strain Pm70 (serotype F:3) has revealed a relatively compact genome size, typically ranging from 2.05 to 2.70 Mb, with a GC content of 36.9–44.0%. The genome contains 2014 predicted protein-coding regions.

Genomic studies have characterised a wide array of virulence factors. In part, the differentiation of *P. multocida* into its five distinct capsular serogroups (A, B, D, E, and F) is based on variations within specific gene clusters responsible for the biosynthesis of their unique capsular polysaccharides (CPS). These crucial genetic determinants are located within a designated region of the bacterial chromosome, often referred to as the capsule biosynthetic locus or *cap* locus, that is organised into three main functional regions.

Region 1 (R1) typically contains genes involved in the initial steps of polysaccharide synthesis. It often includes components of an ATP-binding cassette (ABC) transport system, which facilitates the import of necessary precursor molecules. Region 3 (R3), at the other end, encodes proteins responsible for the export of the fully synthesised polysaccharide to the bacterial surface and its proper attachment. The critical distinction between serogroups, however, lies in region 2 (R2). This segment is the serogroup--specific region. It houses the genes that encode the unique glycosyltransferases and other enzymatic machinery responsible for the polymerisation of the specific monosaccharide units that characterise each serogroup's capsule (Table 27.1).

P. multocida toxin (PMT), a key virulence factor in porcine atrophic rhinitis, is encoded by the *toxA* gene. It is frequently located within an intact prophage, suggesting that it was acquired via bacteriophage-mediated horizontal gene transfer.

Other critical virulence factors identified through genomics include various adhesins (e.g. PfhB1, PfhB2, PtfA). These are crucial for colonisation. The sophisticated iron acquisition sys-

Table 27.1 Genetic basis for serogroups in *Pasteurella multocida*

Serogroup	Capsular polysaccharide (CPS) composition and key features	Genetic basis in cap locus (region 2)
A	Hyaluronic acid (HA). This is a repeating disaccharide of N-acetyl-β-D-glucosamine and β-D-glucuronic acid, chemically like mammalian hyaluronic acid	Genes-encoding HA synthases and other enzymes for HA biosynthesis (e.g. *hya* genes, often part of the hay cluster)
B	Primarily mannose. It also contains arabinose and galactose. It forms a branched polysaccharide structure	Genes-encoding specific glycosyltransferases are responsible for the polymerisation of mannose, arabinose, and galactose into the serogroup B-specific polysaccharide
D	Polysaccharides susceptible to chondroitin sulfatases A and C and heparinase. This forms a chondroitin-like structure	Genes-encoding enzymes for the synthesis of chondroitin-like polysaccharide units, including specific glycosyltransferases and sugar transferases
E	A distinct, antigenically unique polysaccharide. Its exact chemical composition has been less characterised compared to other serogroups	A unique set of genes within region 2–encoding specific enzymes for the biosynthesis of the serogroup E-specific polysaccharide
F	Like chondroitin. It shares structural similarities with chondroitin	Genes-encoding enzymes involved in the synthesis of a chondroitin or chondroitin-like polysaccharide, genetically distinct from serogroup D's related structure

tems (e.g. *exbB, exbD, tonB, hgbA, hgbB*, and *fur*) are essential for survival in iron-limited host environments. These, along with outer membrane proteins (OMPs) and extracellular enzymes like neuraminidases and hyaluronidases, facilitate host interaction and tissue degradation. Furthermore, genes for superoxide dismutases (*sodA, sodC*) have been found to facilitate neutralising host immune responses.

Genomic sequencing has also revealed the presence of mobile genetic elements (MGEs), such as integrative and conjugative elements (ICEs) and plasmids. These are key vehicles for the horizontal transfer of multiple antibiotic resistance genes. Specific resistance genes identified include tet(H) and tet(Y) for tetracyclines; *strA, strB, aadA31*, and *aph(3')-Ia* for streptomycin; *catA2* and *floR* for chloramphenicol/florfenicol; *sul2* for sulfonamides; *blaTEM-1B* for beta-lactams; and *mef(C), mph(G)* for macrolides.

Morphology

P. multocida is a small Gram-negative coccobacillus measuring 0.3–1.0 μm in width and 1.0–2.0 μm in length (Fig. 27.1). It is facultatively anaerobic and non-motile; the organism demonstrates typical bipolar-staining with methylene blue in smears obtained from wounds or tissues.

Cultivation

P. multocida strains grow well on enriched media, such as tryptic soy agar, chocolate agar, or brain heart infusion agar, supplemented with 5–10% fetal calf serum or horse or sheep blood, under aerobic conditions at 37 °C. Colonies are typically small, often ranging from 0.5 to 1.0 mm in diameter after 24 h of incubation. It grows slightly larger (1.0–2.0 mm) with extended incubation. They are generally round or oval with an entire, smooth, and unbroken edge. The colonies appear greyish-yellow or greyish-white. The colonies on blood agar are non-haemolytic. Direct exposure to sunlight, even for 10 min, kills the organism; however, survival in soil and water can extend for up to 21 and 25 days, respectively.

Animal Models

Animal models for *P. multocida* research include mice, sheep, goats, and pigs. Mice are used for studying haemorrhagic septicaemia and other infections. Sheep and goats are valuable models

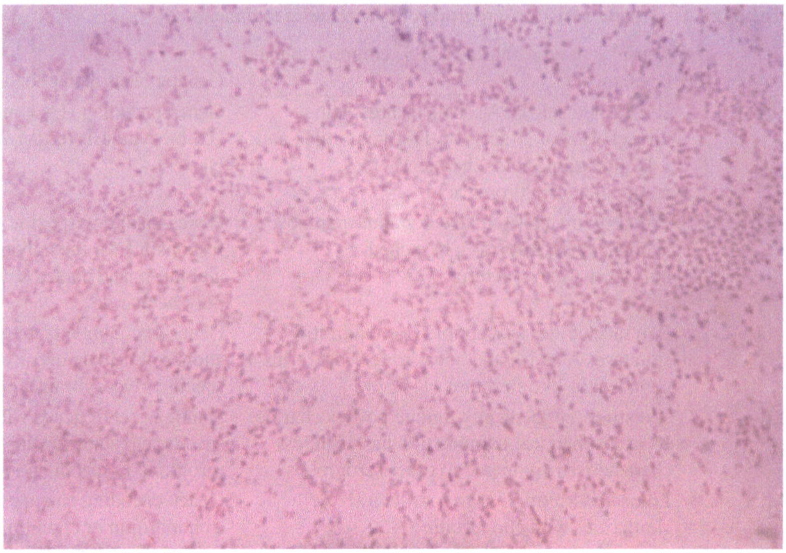

Fig. 27.1 Gram-stained smear showing coccobacillus forms of *Pasteurella multocida* (X1125). (Image source: CDC)

for studying respiratory diseases due to their similar lung anatomy and immune systems. Pigs are used to model atrophic rhinitis and other infections.

Transmission Cycle

P. multocida is a common commensal or opportunistic pathogen found in the upper respiratory tracts of most livestock, domestic, and wild animals. These include chickens, turkeys, and other wild birds, cattle and bison, swine, rabbits, dogs, cats (domestic house cats as well as large wild cats, such as tigers, leopards, cougars, and lions), goats, chimpanzees, and even marine mammals (seals, sea lions, and walruses).

Animal-to-human transmission occurs through biting, scratching, or licking, as well as by the respiratory route. Human infections generally occur after cat or dog bites, scratches, or licks. Respiratory tract infections caused by *Pasteurella* occur in individuals who are occupationally exposed to animals or involved in the processing of animal tissues. No food or water-borne illness has been reported.

Human-to-human transmission has been reported to occur both horizontally and vertically. Horizontal transmission occurs via contaminated blood products or from colonised humans via nasopharyngeal secretions, faeces, and urine, especially in persons suffering from chronic obstructive pulmonary disease. Vertical transmission to infants from mothers positive for *Pasteurella* infection has been reported in many instances.

An overview of the transmission of *Pasteurella multocida* is shown in Fig. 27.2.

Pathogenesis and Pathology

P. multocida by using Carter's and Heddleston's methods has been serologically classified into five commonly isolated serogroups (A, B, D, E, and F). The classification is based on the composition of the polysaccharide capsule and 16 lipopolysaccharides (LPS) serovars (serovar 1 through serovar 16).

The organism is capable of invading human bronchial epithelial cells and evading host defences through various mechanisms. These include capsule formation, lipopolysaccharide (LPS) activity, iron acquisition systems, a battery of key enzymes, and P. *multocida* toxin (PMT).

The hydrophilic capsule helps resist phagocytosis and contributes to mucosal colonisation, particularly in respiratory infections. The capsule is one of the major determinants of the pathotype, host specificity, local epidemiology, pathogenicity,

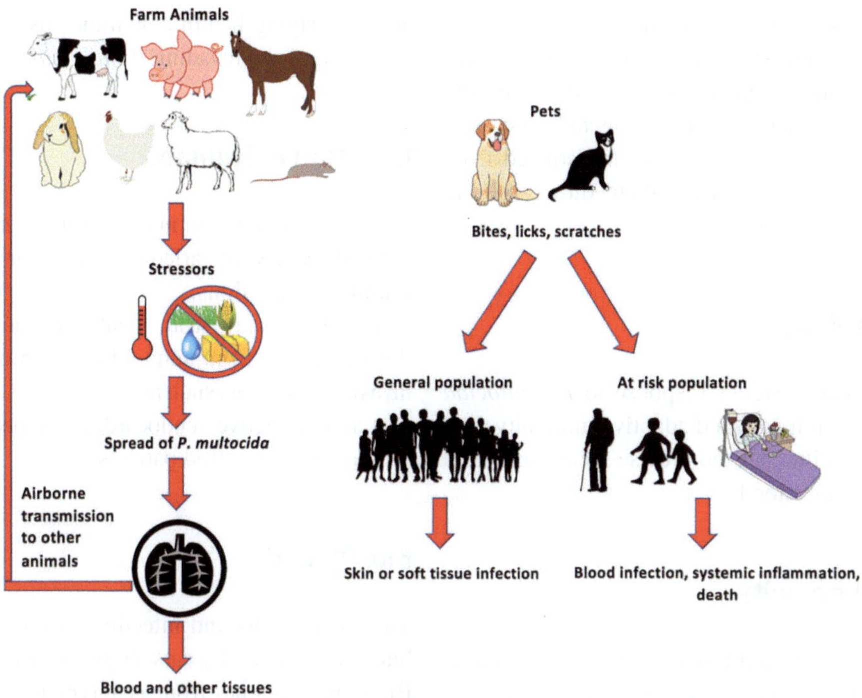

Fig. 27.2 Transmission of *Pasteurella multocida*. (Image source: Garofalo & Kalaska. https://mechpath.com/2017/12/18/pasteurella-multocida/. Copyright free)

and immunogenicity of *P. multocida* diseases. Capsule components of serogroups A, D, and F belong to the glycosaminoglycan (GAG) family. It specifically comprises hyaluronic acid in serogroup A, heparin in serogroup D, and chondroitin in serogroup F. Capsule components of serogroups B are non-GAG–like and are composed explicitly of arabinose, mannose, and galactose in a certain structural form. In type E, the specific composition and chemical structure remain uncertain.

Lipopolysaccharide (LPS) is a critical component that contributes significantly to its virulence. It plays a role in immune evasion by resisting complement-mediated lysis and inhibiting phagocytosis. LPS also stimulates a strong inflammatory response by activating host innate immune receptors such as Toll-like receptor 4 (TLR4), leading to the release of proinflammatory cytokines. Variations in the LPS structure among strains influence their pathogenic potential and interaction with the host immune system.

Iron acquisition systems by *P. multocida* are important for survive and proliferate within the iron-limited environment of the host. These systems include siderophore-independent mechanisms that extract iron directly from host sources such as transferrin, lactoferrin, or haemoglobin. Efficient iron acquisition not only supports bacterial growth but also enhances virulence by promoting colonisation and evasion of host immune defences.

Key enzymes such as hyaluronidase, neuraminidase, and sialidase facilitate tissue invasion and immune evasion. The bacterium also produces several outer membrane proteins (OMPs), such as OmpA, OmpH, Oma87, and PlpE, which act as adhesins and immune targets. Fimbrial adhesins (FhaB1 and FhaB2) contribute to biofilm formation and colonisation of the respiratory tract.

P. multocida toxin (PMT) is an important virulence determinant. It is a 146-kDa protein encoded by the *toxA* gene, often carried by a lysogenic phage. PMT is osteolytic and associated with atrophic rhinitis in animals. It modulates host cell signalling through G protein deamidation, influencing immune responses and potentially contributing to oncogenic processes. While PMT-producing strains are isolated from human respiratory cases, their direct role in human disease remains unclear.

Immunology

The immune system's response to *P. multocida* involves both innate and adaptive immunity. It is serovar-specific, but the precise mechanisms are yet to be understood.

Innate Immunity

Upon entry through a wound, *P. multocida* triggers an inflammatory response involving neutrophils, macrophages, and other immune cells. These innate immune cells attempt to phagocytose and destroy the bacteria. *P. multocida*, being capsulated, resists phagocytosis and survives within the host. The complement system is also activated, resulting in opsonisation. The formation of membrane attack complexes (MACs) causes damage to the bacterial cell membrane.

Adaptive Immunity

Adaptive immunity comes into play when antigens from *P. multocida* are presented to T cells, which can then initiate an adaptive immune response. T cells, particularly CD4+ helper T cells, play a crucial role in activating macrophages and other immune cells to combat the infection. B cells produce antibodies against *P. multocida* antigens, which help to neutralise the bacteria and enhance phagocytosis. The adaptive immune response also generates memory B and T cells, eventually providing long-term protection against future infections.

Nevertheless, the severity of the infection is influenced by factors such as age, immune status, and underlying health conditions, as well as the virulence of the invading *P. multocida* strain.

Infection in Humans

P. multocida infection in humans is a serious systemic disease with varied manifestations and disseminated infections.

Infections vary from localised wound infections resulting from animal bites to more severe invasive infections such as bacteraemia, meningitis, and infective endocarditis, especially in immunocompromised patients.

Bite Wound

The localised wound infections occur following bites or scratches by cats, dogs, or other animals. Bites and scratches from a cat typically affect the upper limbs and face.

The progression of symptoms in *P. multocida* bite wound infections is quite rapid, with an incubation period usually less than 24 h. This contrasts with the slower rate of progression observed with wound infections due to other organisms. Local wound site manifestations include swelling, erythema, tenderness, and purulent exudate. It is accompanied by fever, chills, and lymphadenopathy, which are also observed in healthy individuals. In these cases, infection is most often limited to the skin and subcutaneous tissue. Complicated wound infections may also occur due to the spread of soft tissue infection or direct penetration, developing into abscesses, tenosynovitis, osteomyelitis, and septic arthritis.

Respiratory Infection

Chronic pulmonary infections usually result from exposure to mucous secretions from animals or pets that are colonised or infected with the bacteria. *Clinical manifestations of respiratory infection with P. multocida* are varied. The condition manifests as pneumonia, bronchitis, and tracheitis, and can progress to more serious complications, including empyema and lung abscess. The risk factors include old age, chronic respiratory diseases such as chronic obstructive pulmonary disease, bronchiectasis, and malignant tumours.

Disseminated Infections

P. multocida causes a variety of disseminated infections, including bacteraemia, septicaemia, meningitis, endocarditis, and ocular infections. *Pasteurella* bacteraemia is more likely in patients with liver disease, organ transplants, and other immunodeficient states, as well as in neonates and the elderly. Older age, diabetes, rheumatoid arthritis, immunosuppressive therapy, malignancy, and arthroplasty revision have been identified as risk factors.

In almost 90% of cases, meningitis is caused by direct contact with an infected animal, typically through licking, sniffing, kissing, or touching the animal. Some reports on cardiovascular infections include *P. multocida* bacteraemia causing aortic valve dysfunction with endocarditis; myopericarditis has also been reported. These conditions have been reported following a dog bite, pericardial tamponade after a cat bite, and endocarditis in an intravenous drug abuse patient.

Epidemiological studies have shown that different serovars of *P. multocida* strains are associated with distinct types of diseases (Table 27.2). For example, strains associated with fowl choléra are most frequently determined to be of serovar A:1, A:3, or A:4; haemorrhagic septicaemia (HS) is commonly associated with serovars B:2 and/or E:2; bovine respiratory diseases are associated with serovar A:3; progressive atrophic rhinitis in swine is generally induced by toxigenic serogroup D and/or A strains; and pasteurellosis in rabbits is generally associated with serovar A:3, A:4, or A:12.

Infection in Animals

P. multocida-associated diseases are of great economic importance in global animal husbandry. These include fowl choléra in avian species,

Table 27.2 Serovars/serogroups of *Pasteurella multocida* and associated diseases

Serovar/ serogroup	Associated disease(s)	Infected group(s)	Remarks
Serovars 1, 3, 4	Fowl choléra	Poultry (chickens, turkeys, ducks, geese, etc.)	Serotype 1 is particularly prevalent in North American poultry
Serogroup B	Bovine haemorrhagic septicaemia (BHS)	Cattle, buffalo	Also implicated in mass mortality of Saiga antelopes (specifically serotype B)
Serogroup E	Bovine haemorrhagic septicaemia (BHS)	Cattle, buffalo	Primarily found in tropical and subtropical regions
Serogroup D	Atrophic rhinitis	Pigs	Specifically, toxigenic strains (producing *Pasteurella multocida* toxin, PMT) are responsible for this disease
Serogroup A	Pneumonia/bovine respiratory disease (BRD) complex	Cattle, sheep, goats, pigs	Often, a secondary bacterial invader, following viral infections or stress. Serotype A:1 is common in cattle pneumonia
Various serovars	Snuffles	Rabbits	Multiple serovars can cause snuffles; it is a very common *P. multocida* infection in rabbits
Various serovars	Wound infections, abscesses, respiratory infections, systemic infections	Humans	Primarily, acquired from bites/scratches from domestic animals (cats, dogs), which carry diverse *Pasteurella multocida* strains
Various serovars	Localised or systemic infections (e.g. abscesses, mastitis, septicaemia)	Cats, dogs, other various mammals, and birds	*Pasteurella multocida* is a common commensal in the oral cavity and upper respiratory tract of these animals and can cause disease when conditions are favourable

haemorrhagic septicaemia (HS) and bovine respiratory disease in ruminant species, progressive atrophic rhinitis and pneumonic pasteurellosis in swine, and snuffles in rabbits. In some cases of septicaemic pasteurellosis in cattle, once clinical signs appear, mortality is typically 100%.

Upper respiratory disease, characterised by rhinitis (irritation and inflammation of the nasal mucosa and nasal secretions), and lower respiratory disease, the condition is also referred to as bovine respiratory distress syndrome in cattle are predominant syndromes of pasteurellosis in wild and domestic animal populations. It manifests as pneumonia. Mild symptoms include sneezing, copious mucous secretions, mild rhinitis, mild pneumonia with laboured breathing, and fever. The condition can progress to disseminated disease (haemorrhagic septicaemia) and/or atrophic rhinitis (atrophy of nasal mucosa, seromucous glands, and turbinate bones) associated with toxinogenic strains. Pasteurellosis pneumonia without symptoms of atrophic rhinitis is most often caused by nontoxigenic capsular type A strains of *P. multocida*. The organism can also cause bovine pneumonia without septicaemia, commonly called shipping fever.

Haemorrhagic septicaemia (HS) is a severe, acute, and highly fatal disease in livestock, particularly cattle and buffalo, in tropical regions worldwide, including Asia, Africa, southern Europe, and the Middle East. HS is caused primarily by *P. multocida* serotypes B:2 and E:2 and is thought to occur at the later stages of pasteurellosis disease. Symptoms usually begin with fever, lethargy, and oedema with copious salivation, lacrimation, and nasal discharge. It is rapidly followed by respiratory distress, septic shock with widespread haemorrhaging, and death within 1–3 days. HS is observed less commonly in swine, sheep, goats, deer, and elk.

Fowl choléra caused by *P. multocida* subsp. *multocida* is the most predominant cause of fowl choléra worldwide in a variety of avian species. However, *P. multocida* subsp. *septica* and *P. multocida* subsp. *gallicida* are also sometimes isolated. Fowl choléra is an asymptomatic or mild, chronic condition characterised by sinusitis, conjunctivitis, or pneumonia-like pasteurellosis. However, it can suddenly and rapidly develop into a fatal, disseminated disease. Capsular serotype A (mainly A:1, A:3, and A:4) is highly correlated to disease predilection for strains associated with fowl choléra. Serotype B:3 is often isolated from avian disease cases that manifest as sinusitis.

Epidemiology and Public Health

Zoonotic transmission of pasteurellosis to humans occurs following animal bites, scratches, or contact with nasal secretions, leading to skin or soft tissue infections or, in rare cases, systemic infections such as pneumonia or meningitis.

Pasteurellosis in humans has a widespread geographic distribution. There has been an increase in cases of human pasteurellosis, particularly over the past 20 years. The crude incidence of *Pasteurella* infections has increased in some regions, indicating a higher rate of infection per 100,000 population. This increase has been observed in various regions, including Szeged, Hungary, and tropical Australia. The crude incidence of *Pasteurella* spp. infections in a tertiary care hospital in Australia increased from 1.5 per 100,000 population in 2000 to 11.4 per 100,000 population in 2021.

P. multocida infection in animals is commonly found in wild birds, particularly during migratory periods in North America. Outbreaks have been reported in waterfowl from western and central states, such as Texas, California, and Nebraska. Outbreaks in wild animals, including wild boar in Spain and saiga antelopes in Kazakhstan, have been linked to pasteurellosis. Reports of pasteurellosis in wildfowl and wild animals have been documented in Europe, Africa, and Australia.

Animal migration, food management practices, environmental factors, and zoonotic transmission are some of the factors suggested for an increase in pasteurellosis in both humans and animals. Migratory birds, particularly lesser snow geese, are known to carry *P. multocida* and are associated with outbreaks along their flyways in North America.

Management practices, such as food supplementation and the translocation of captive animals, can increase the risk of pasteurellosis outbreaks. Outbreaks in waterfowl are often associated with the concentration of birds in wetlands during spring migrations, where the bacteria may linger in the environment.

Diagnosis

Pasteurella spp. are suspected in a case of soft tissue infection resulting from scratching, biting, or licking by a cat or dog.

Specimens

These include wound swabs, sputum, cerebrospinal fluid, pleural fluid, synovial fluid, and other specimens, depending on the nature of the infection.

Microscopy

Gram stain, Wright's, or Giemsa stains of exudates from wounds, respiratory secretions, or tissues are useful for rapid presumptive identification of *P. multocida* infections, particularly in acute cases. Gram staining shows small, Gram-negative coccobacilli, exhibiting typical bipolar-staining, giving them a characteristic "safety pin" appearance. This morphological feature is especially prominent in Wright's or Giemsa-stained smears.

Microscopy is always followed by culture and molecular methods for confirmatory diagnosis, as similar morphology may be shared by other organisms.

Culture

A definitive diagnosis of *P. multocida* infection is made by culture.

P. multocida grows well on 5% Columbia agar with sheep's blood, chocolate agar, and casein–sucrose–yeast medium. 5% Columbia agar with sheep's blood is recommended for culture of *P. multocida* at a temperature of 37 °C for 24–72 h, in an atmosphere of 10% CO_2. Fresh cultures growing as non-haemolytic colonies on blood agar show small, pleomorphic, Gram-negative coccobacillus with bipolar-staining. Further identification of *P. multocida*, once isolated, is confirmed through biochemical tests or molecular techniques such as PCR.

P. multocida is catalase, oxidase, indole, and ornithine decarboxylase-positive. Most isolates also ferment sucrose, glucose, and maltose. The identification of *P. multocida* is often carried out using rapid biochemical strips such as Oxi/Ferm strips. However, it has limited accuracy, since *P. multocida* can be confused with *Mannheimia haemolytica*, *Haemophilus influenzae*, or other Pasteurellaceae species.

The VITEK-2 microbial identification system is the newer method used for *P. multocida*.

Serodiagnosis

Serological tests can aid in the diagnosis of *P. multocida* infections, particularly in chronic or systemic cases where culture may be challenging. However, their application in human diagnostics is limited due to cross-reactivity with other bacterial species and variability in immune response.

Various antibody detection methods, including ELISA and indirect immunofluorescence, have been developed for their application, especially in veterinary settings for surveillance and herd health management.

Molecular Diagnosis

Molecular diagnostic methods are highly useful for the accurate and rapid detection of *P. multocida* infections.

PCR is a fast, sensitive, specific, and highly effective method for identifying *P. multocida* and its subspecies by the identification of species-specific genes such as *kmt1*, *ompH*, and *toxA*. These molecular targets help differentiate *P. mul-*

tocida from other closely related pathogens. Real-time PCR and multiplex PCR further allow simultaneous detection and typing of strains, including capsular and virulence gene profiling.

Conventional ribotyping based on PCR amplification alone is a reliable and discriminative method for characterising clinical isolates of *P. multocida* at species levels. Whole genome sequencing and molecular typing tools such as MLST (multilocus sequence typing) contribute to epidemiological studies and strain tracking. These approaches not only improve diagnostic accuracy but also support surveillance and outbreak investigations.

Molecular Typing

The most frequently used molecular typing methods are capsular genotyping, LPS genotyping, multilocus sequence typing (MLST), and/or virulence genotyping based on different virulence gene profiles.

The capsular genotyping method is based on PCR detection of *cap* genes, using primers highly specific for different serogroups: hyaD–hyaC for serogroup A, bcbD for serogroup B, dcbF for serogroup D, ecbJ for serogroup E, and fcbD for serogroup F. The genotypes determined by this multiplex capsular PCR-based typing system reliably and accurately discriminate among the capsular serogroups.

A multiplex PCR assay based on the genetic organisation of the LPS outer core biosynthesis loci is employed for LPS genotyping. This assay is based on the genetic organisation of the LPS outer core biosynthesis loci, enabling the differentiation of *P. multocida* into eight genotypes (L1–L8) corresponding to the 16 Heddleston serovar strains as follows: L1(1, 14); L2 (1, 2, 5); L3 (3, 4); L4 (6, 7); L5(9); L6(10, 11, 12, 15); L7 (8, 13); and L8 (16).

The diagnosis of pasteurellosis is summarised in Table 27.3.

Treatment

Surgical debridement, removal of debris, and, in older wounds, removal of abscesses and dead tissue, along with the prompt use of antibiotics, are essential in treating the infection.

Antimicrobial therapy is crucial for *P. multocida* infections, with early intervention being crucial to prevent complications. The bacterium is sensitive to a broad range of antimicrobials, including penicillins, tetracyclines, fluoroquinolones, and cephalosporins. Penicillin or amoxicillin-clavulanic acid is considered a first-line therapy in both human and veterinary medicine.

Parenteral administration of extended-spectrum cephalosporins or fluoroquinolones is suggested for the treatment of severe or systemic infection. Antimicrobial susceptibility testing is important to guide effective therapy of recurrent, chronic, or non-responding cases, since resistance patterns can occasionally emerge. Supportive care and surgical intervention, such as drainage of abscesses, may be required depending on the clinical presentation and severity of infection.

Prevention and Control

Avoiding bites, scratches, and direct contact with animal saliva is recommended to prevent pasteurellosis. Children should be supervised to prevent bites and scratches from cats and dogs. Individuals should be educated about the proper handling of domestic animals and the associated risks. Prophylactic antibiotics are administered, particularly for individuals who are immunocompromised or diabetic, with high-risk bites such as those to the hands or face.

Occupational pasteurellosis is effectively prevented and controlled in veterinary clinics, farms, and research laboratories by maintaining high standards of hygiene and sanitation. Strict implementation of safety protocols is crucial for prevention. Robust and validated procedures for handling animal waste and disinfecting surfaces and equipment can prevent the spread of bacteria.

Case Study

A 75-year-old woman was admitted to the infectious disease ward of a hospital with a two-day history of productive cough and shortness of

Table 27.3 Laboratory diagnosis of pasteurellosis

Diagnostic approaches	Methods	Targets	Remarks
Direct microscopy	Direct examination by Leishman's stain, methylene blue, or Giemsa stain of clinical specimens (wound swab, sputum, cerebrospinal fluid, synovial fluid, and others, depending on the location of the infection)	Bipolar-staining Gram-negative bacilli or coccobacilli	Difficult to distinguish from other Gram-negative bipolar-staining bacteria such as *Francisella tularensis*, *Yersinia* species, and other Pasteurellaceae species
In vitro cultures	The recommended medium for growing *Pasteurella multocida* is 5% Columbia agar with sheep's blood, maintained at a temperature of 37 °C for 24–72 h, in an atmosphere of 10% CO_2 Other medium—chocolate agar with the same incubation conditions Specimens to be considered—wound swab, sputum, blood, cerebrospinal fluid, synovial fluid, and others, depending on the location of the infection	Non-hemolytic colonies demonstrating bipolar-staining Gram-negative bacilli or coccobacilli Pathogen identification is to be performed by methods such as Vitek2, VITEK MS, Bruker biotype MS tests, and traditional biochemical tests	Definitive method of diagnosis; easy technique; good growth occurs on blood agar and chocolate agar medium
Molecular diagnosis	PCR amplification of 16S rRNA gene sequence-based ribotyping and subtyping using non-16S rRNA genes such as those associated with virulence traits of *P. multocida* isolates	16S rRNA gene and non-16S rRNA gene amplification	A fast, sensitive, specific, and highly effective method for identifying *P. multocida* and its subspecies

breath, preceded by a 5-day history of left leg redness, hotness, and pain. The patient provided a history of playing with her daughter's cat, which had scratched her legs multiple times over the last year, with the most recent incident occurring almost a week prior to her presentation. Repeat blood cultures on the 10th day of admission yielded no growth, confirming clearance of the bacteria from her blood. The patient's clinical status and laboratory parameters showed an excellent response to treatment. She was treated with amoxicillin-clavulanic acid 625 mg oral tablets three times per day to complete a total of 14 days. Two weeks of follow-up in the OPD setting showed complete resolution of symptoms.

1. Which other bacterial diseases in humans are associated with animal bites?
2. Which morphological and cultural characteristics of *Pasteurella multocida* were obtained from clinical specimens?
3. Which methods could be used to prevent pasteurellosis?

Research Questions

1. How can molecular diagnostic techniques be developed to confirm the presence of infection by *P. multocida*, since other species as well as related genera may be involved as causative agents of infection?
2. Why has there been a lack of information about the pathophysiology and immunology of *P. multocida*, and how can this be addressed?
3. Which immunological factors are responsible for protection in humans?

Declaration During the preparation of this chapter, we, *Dr Srujana Mohanty and Dr Tribhuban Mohan Mohapatra*, used *ChatGPT* in order to improve the grammar and/or readability of the text *Pasteurellosis*. After using it, we reviewed and edited the content as needed and take full responsibility for the content of the publication.

Further Readings

Boyce JD, Seemann T, Adler B, Harper M. Pathogenomics of *Pasteurella multocida*. Curr Top Microbiol Immunol. 2012;361:23–38.

Harper M, Boyce JD. The myriad properties of *Pasteurella multocida* lipopolysaccharide. Toxins (Basel). 2017;9(8):254.

Hurtado R, Maturrano L, Azevedo V, Aburjaile F. Pathogenomics insights for understanding *Pasteurella multocida* adaptation. Int J Med Microbiol. 2020;310(4):151417.

Kehrenberg C, Schulze-Tanzil G, Martel JL, Chaslus-Dancla E, Schwarz S. Antimicrobial resistance in pasteurella and mannheimia: epidemiology and genetic basis. Vet Res. 2001;32(3–4):323–39.

Kubatzky KF. *Pasteurella multocida* toxin – lessons learned from a mitogenic toxin. Front Immunol. 2022;13:1058905.

Nakwan N, Nakwan N, Atta T, Chokephaibulkit K. Neonatal pasteurellosis: a review of reported cases. Arch Dis Child Fetal Neonatal Ed. 2009;94(5):F373–6.

Mogilner L, Katz C. Pasteurella multocida. Pediatr Rev. 2019;40(2):90–2.

Peng Z, Wang X, Zhou R, Chen H, Wilson BA, Wu B. *Pasteurella multocida*: genotypes and genomics. Microbiol Mol Biol Rev. 2019;83(4):e00014–9.

Piorunek M, Brajer-Luftmann B, Walkowiak J. *Pasteurella multocida* infection in humans. Pathogens. 2023;12:1210.

Wilson BA, Ho M. *Pasteurella multocida*: from zoonosis to cellular microbiology. Clin Microbiol Rev. 2013;26(3):631–55.

28 Tularemia

Fatih Buyuk, Yaren Ersoy, and Derya Karatas Yeni

Learning Objectives

1. To gain insights into the eco-epidemiology of *Francisella tularensis*
2. To understand the infection mechanism of the *F. tularensis*
3. To understand the clinical manifestation and management of tularemia infection in humans

Introduction

Tularemia is a multisystemic disease of humans and animals caused by a pleomorphic, facultative intracellular Gram-negative bacterium, *Francisella tularensis*. Although the first descriptions of the disease date back to earlier years, the causative agent was first identified in 1911 in ground squirrels by George McCoy and Chapin in California. *F. tularensis* has been isolated from humans as well as animals.

According to the epidemiology, biochemical features, and virulence, *F. tularensis* is taxonomically divided into four subspecies, including *F. tularensis* subsp. *tularensis* (Type A), *F. tularensis* subsp. *holarctica* (Type B), *F. tularensis* subsp. *Mediasiatica,* and *F. tularensis* subsp. *novicida*. Although all subspecies are associated with human disease, *F. tularensis* subsp. *tularensis* and *F. tularensis* subsp. *holarctica* have been reported to be of primary clinical and epidemiological relevance. *F. tularensis* subsp. *tularensis* is highly infectious and restricted almost exclusively to North America. It is associated with infections in wild rabbits, ticks, and flies. *F. tularensis* subsp. *holarctica* infection exists throughout the Northern Hemisphere, Asia and Europe. It causes mild infection and low mortality in humans, but high mortality in wildlife.

F. tularensis is regarded as a potential biological warfare agent. Its low infectious dose, potential for easy dissemination through aerosols, and higher virulence properties make the bacterium ideal as a bioweapon. Nevertheless, antimicrobials such as gentamicin, doxycycline and ciprofloxacin are effective for the treatment of the infection caused by *F. tularensis*, which is classified as a category A biological weapon by the Centers for Disease Control and Prevention (CDC).

This chapter describes the history, ecology, microbial and pathogenomic characteristics, epidemiology, treatment and prevention of tularemia, which often has been called a chameleon due to its adaptation to a wide variety of hosts and habitats.

F. Buyuk (✉) · Y. Ersoy
Department of Microbiology, Faculty of Veterinary Medicine, Kafkas University, Kars, Turkey

D. Karatas Yeni
Department of Microbiology, Faculty of Veterinary Medicine, Necmettin Erbakan University, Konya, Turkey

S. C. Parija, A. Poddar (eds.), *Textbook of Bacterial Zoonoses*, Microbial Zoonoses,
https://doi.org/10.1007/978-981-95-6615-0_28

History

The idea that *F. tularensis* can be used as a biological warfare agent by infecting enemy bases dates back to 1300 BC. However, the first record of the disease was a case of Homma Soken in 1837, which resembled a typical case of tularemia and developed a high fever and lymphadenopathy after eating rabbit meat. The first definitive case was described by Charles Chapin and George McCoy, who isolated the causative agent from squirrels around a lake in the Tulare region of California. The pathogen was named '*Bacterium tularense*' at the time. The first verified human case occurred in Ohio in 1914. In the following period, cases related to rodents were also reported. In the 1920s, Dr Edward Francis documented the causative agent of tularemia in the blood of patients with deer fly fever. In 1924, Dr Ohara published a human case report from Japan in which a febrile systemic illness developed after rabbit skinning. In 1924, Parker, Spenser and Francis also announced that the disease could be transmitted by ticks. Towards the 1950s, the bacterium was taxonomically classified as '*Pasteurella tularensis*' with technological advances that enabled fluorescent antibody staining. The genus name was changed to '*Francisella tularensis*' in 1959 in honour of Dr Francis.

In later periods, it was used as a weapon of biological warfare due to its easy transmission with a small number of bacteria and its potentially lethal effects.

Taxonomy

F. tularensis is a bacterium belonging to the Proteobacteria phylum, the Gammaproteobacteria class, the Thiotrichales order, the Francisellaceae family and the *Francisella* genus. There are two main genetic clones within the genus *Francisella*. One comprises *F. tularensis*, *Francisella novicida* and *Francisella hispaniensis* species, and the other includes *Francisella philomiragia* and *Francisella naotunensis* species.

Based on the biochemical characteristics, the virulence, and the geographical distributions, *F. tularensis*, responsible for tularemia in humans and animals, is composed of four subspecies. These include *Francisella tularensis* subsp. *tularensis* (type A), *Francisella tularensis* subsp. *holarctica* (type B), *Francisella tularensis* subsp. *mediasiatica* and *Francisella tularensis* subsp. *novicida*. Among these, *F. tularensis* subsp. *tularensis* and *F. tularensis* subsp. *holarctica* are known as the major subspecies in humans. *F. tularensis* subsp. *novicida* is very rarely associated with human infections, while *F. tularensis* subsp. *mediasiatica*-based infection has never been reported. Recent studies on the *Francisella* genus have described new species, such as *Francisella noatunensis*, *Francisella asiatica*, *Francisella hispaniensis*, *Francisella halioticida*, *Francisella guangzhouensis*, and *Francisella persica*.

Genomics and Proteomics

The genome sequences of different *F. tularensis* strains, starting with the SCHU S4 strain, contribute to understanding the biology of this agent, revealing its evolution, elucidating its adaptation to different environmental niches and developing vaccine and diagnostic methods. Comparative genomic and molecular methods have shown that the four subspecies of *F. tularensis* are genetically distinct, and there are numerous DNA rearrangements of IS elements. The subspecies of *F. tularensis* appear to have evolved from a common ancestor, probably *F. tularensis* subsp. *novicida*.

Proteomic studies have revealed different virulence factors among strains, and several candidate virulence genes have been identified. When the proteomes of *Francisella* strains with different virulence were compared, similarities and differences were detected in the profile of expressed proteins. Proteomic studies have also identified reactive proteins in sera from naturally infected or vaccinated subjects. Thus, the comprehensive study of protein clusters involved in immunity has been termed '*immunoproteomics*', and findings have been made regarding

Francisella immunodominant antigens. The proteomes of *F. tularensis* subsp. *tularensis* and *F. tularensis* subsp. *mediasiatica* strains were found to be more similar than the proteomes of *F. tularensis* subsp. *holarctica*.

Morphology

F. tularensis is a microscopic, non-spore–forming, facultative intracellular Gram-negative coccobacillus bacterium with dimensions of 0.2–0.5 μm × 0.7–1.0 μm (Fig. 28.1). Gram staining shows the presence of small and punctate Gram-negative bacteria. For the staining of the agent, which stains weakly by Gram staining, it was recommended to stain with dye compounds with the addition of carbol fuchsin and crystal violet for approximately 1.5 h. The Giemsa method was also recommended for histopathological examination and staining of cultures. Slides prepared in this way may show the staining of bacteria from pinkish-purple to purplish--blue. Bacteria from fresh cultures stain pink, whereas in old cultures, they appear dark blue.

Cultivation

F. tularensis is a bacterium with high nutrient requirements, and culture media should be enriched with cysteine. Blood or chocolate agar (CHAB) enriched with glucose and cysteine are used for culture. Cysteine agar with glucose, thioglycolate liquid medium, modified Thayer--Martin medium, charcoal–yeast extract agar, and cysteine agar with 9% sheep blood and heart infusion are also used. On CHAB medium, *F. tularensis* grows in a 5% CO2 atmosphere at 35 °C for 2–5 days and forms small (2–4 mm in diameter), smooth, round mucoid colonies (Fig. 28.2). The causative agent, which classically has a coccobacilli morphology, varies as bacilli or cocci depending on the agent's host type, culture time and infectiousness. Bacillus morphology continues to dominate in fresh cultures, susceptible hosts, and highly infected species. Antibiotic-added cysteine media (CHAB-A and CHAB–PACCV) are used to isolate *F. tularensis* from samples containing contaminants.

Animal Models

Mice are the most widely used animal model for *F. tularensis* infections due to their low cost, availability of different genetic variants and susceptibility to all *F. tularensis* subspecies. However, since species that are avirulent in humans are virulent in mice, there is a need to use more resistant species, such as rabbits and guinea pigs, in infection modelling. Apart from these, primates and rats are also used. Especially in monkeys, the immunity obtained with the *F. tularensis* live vaccine strain (LVS) vaccine provides immunity similar to that in humans.

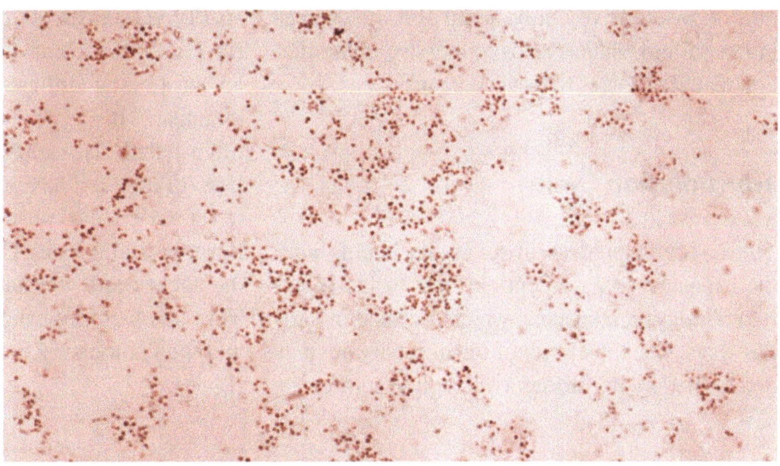

Fig. 28.1 *Francisella tularensis* is a small (0.2 to 0.5 × 0.7 to 1.0 μm), pleomorphic, Gram-negative coccobacillus that stains weakly in Gram staining and is mainly observed as single cells. (Accessed from https://www.cdc.gov/tularemia/hcp/diagnosis-testing/index.html)

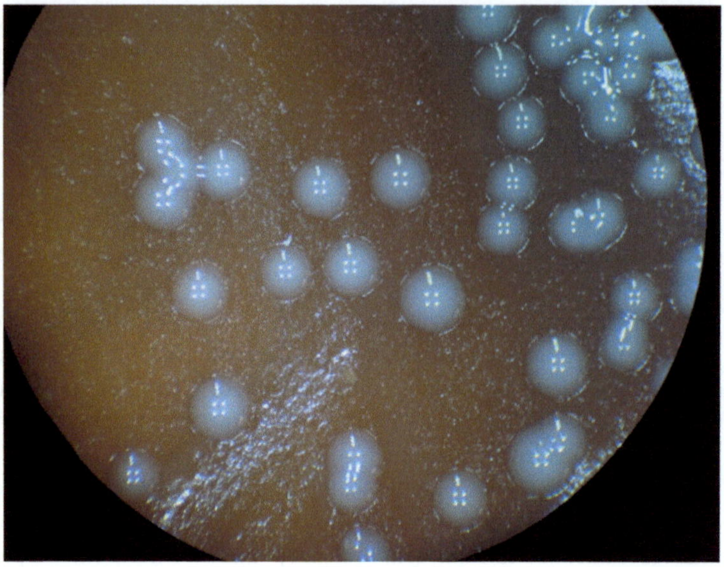

Fig. 28.2 Colonies of *Francisella tularensis,* 2–3 mm in diameter, whitish-grey and bluish-grey, flat, entire-edged, smooth and shiny surface after 72 h of aerobic culture on Francis Medium at 37 °C. (Accessed from the culture collection of Dr Derya Karatas Yeni)

In mice, different disease manifestations occur depending on the route of entry. Twenty-four hours after inhalation, the bacterium is found predominantly within alveolar macrophages. By the third day after infection, 50–80% of neutrophils are infected. Regardless of the route of entry into the body, the bacteria spread rapidly to the spleen and liver. Apart from phagocytic cells, the agent can also proliferate in alveolar epithelial cells and hepatocytes. Infection with the LVS strain is limited by granuloma formation in the liver. Mice infected with SCHU S4 die rapidly (within 5 days) without an immune response. The prognosis of *F. novicida* infection in mice varies depending on the route of entry. Furthermore, intradermal injection of the causative agent is followed by the production of an effective antibacterial IFN-γ, whereas intranasal inoculation leads to massive inflammation that contributes to tissue damage.

Transmission Cycle

The infection process in tularemia begins with the exposure of a susceptible host to the infectious *Francisella* species, which are shed in various ways. The infectivity of the microorganism is determined by its various virulence structures, its number and stability outside the host, and the route and mode of transmission. As a facultative intracellular and lean bacterium, *F. tularensis* cannot tolerate heat, desiccation and chemical irritants for long periods like spore-bearing bacteria, nor does it need to be spread by the exchange of absolute body fluids or tissues as intracellular bacteria do. An important factor in the ecology of tularemia is that the agent can contaminate surface waters, stream beds and pond mud and survive at infectious levels for more than 1 month.

Francisella species cause infection in a host through an epithelial surface, such as the mucosal surfaces of the skin, respiratory, gastrointestinal, and urogenital tracts. *F. tularensis* subsp. *tularensis* (Type A) is highly contagious, with 10 bacteria being sufficient for skin infection and 25 bacteria being sufficient for aerosol infection in humans. This type is associated with severe, often lethal disease. *F. tularensis subsp. holarctica* (Type B) has a higher infectious dose (>100 CFU) and causes progressive disease with lower mortality rates. Once contact is established, the agent needs to establish a focus of infection. This involves adhering to the epithelial surface and then colonising or penetrating to multiply in tissues.

Pathogenesis and Pathology

Virulence Factors The contact-dependent type VI secretion system provides the passage of *F. tularensis* from the cytosol to the target cells. In addition, there are other secretion systems, such as the TolC-dependent efflux system and the type IV pilus (Tfp). Lipid A in the *F. tularensis* cell wall provides the ability to resist natural immunity. The capsule-like structure, defined as a polysaccharide identical to the O-antigen subunit of LPS, carries out an antiphagocytic function. Francisella virulence is controlled by various gene expression programmes. In addition, virulence factors such as KatG, SodB, DsbA, Hfq, and AcrAB play a role in the agent's immunity and adaptation to stress.

As an intracellular agent, *F. tularensis* proliferates within phagocytic cells. Phagocytosis of *F. tularensis* is enhanced by opsonisation mediated by antibody or complement. The cellularised agent resides in a *Francisella*-containing phagosome (FCP). Following an early and late endosomal process that starts shortly after entry into the cell and peaks at approximately 4 h, *F. tularensis* escapes into the host cell cytosol. The passage of the agent into the cytosol is thanks to many genes located in the genomic region called the Francisella Pathogenicity Island (FPI). Biotin synthesis, which provides nutrition to the microorganism, plays an important role in the escape of the agent from the phagosome. In addition, glutamate, arginine permease, isoleucine permease and four acid phosphatases (AcpA, AcpB, AcpC and HapA) have been reported to contribute to the escape of *F. tularensis* into the cytosol by inhibiting phagosomal NADPH oxidase assembly and reactive oxygen species production. The resistance of *F. novicida* to oxidative stress is mediated by lysine decarboxylase through the production of cadaverine, which scavenges oxygen radicals. Citrulline ureidase in *F. tularensis* type A strains may support the survival of the agent by limiting nitric oxide (NO) production in infected macrophages through various metabolic products mediated by citrulline ureidase or by contributing to bacterial growth in macrophages.

F. tularensis infects macrophages and spreads haematogenously throughout the body. Bacteraemia is rare in the early stage of infection. Bacteria and cell debris reaching the capillary endothelium interact with thrombotic foci and cause necrotic foci in organs such as the liver, spleen, and lymph nodes. Lymphocytes, macrophages and epithelioid cells, primarily neutrophils, respond. Occasionally, caseating granulomas are mistaken for tuberculosis. Postmortem examination reveals hepatomegaly, splenomegaly or both. There are usually numerous small foci of greyish, yellow or white necrosis in the spleen, liver and lungs. Lymph node enlargement, hepatic necrosis, and multifocal fibrinonecrotic pneumonia may also be observed.

Immunology

Innate Immune Responses

Due to a specialised envelope composed of several components, *F. tularensis* resists cationic antimicrobial peptides and the bactericidal activity of serum components. On first encounter with host cells, it is recognised by TLR2s rather than TLR4s. Following escape into the host cytosol, *Francisella* is recognised by the cGAS/STING pathway as a first step in so-called cytosolic responses, leading to the production of type I IFNs. Type I IFNs control multiple genes during infection. Type I IFNs block the production of IL-12 released by dendritic cells and IL-17 released by gamma delta T cells. They also induce a pro-inflammatory response and activate IRF1 to induce guanylate-binding proteins (GBPs). GBPs induce bacteriolysis, allowing bacterial DNA to pass into the host cytosol and activate the cytosolic DNA sensor Aim2. Activated Aim2 leads to the activation of caspase-1, the release of pro-inflammatory cytokines such as IL-1β and IL-18, and rapid cell death, termed pyroptosis.

Furthermore, GBPs promote the activation of the non-canonical caspase-4 inflammasome. IFNγ also upregulates macrophage metabolism by inducing IRG-1 and producing itaconate to control intracellular replication. Although neutrophils appear to be important cells in the fight against infection, their inability to produce reactive oxygen species suggests that the causative agent exerts a blockade on these cells. Despite the latent nature of *Francisella* and its ability to reduce immune responses, the late stages of fatal infections are characterised by an immune response leading to severe sepsis.

Adaptive Immune Responses

Although *Francisella* spp. are intracellular agents, antibodies were found to be protective against tularemia. Especially, anti-LPS IgMs were found to be protective in passive immunisation experiments. Both CD4+ and CD8+ T lymphocytes play an important role in the immune response against *Francisella*, especially through IFN-γ and TNF production. IFN-γ shows high bactericidal activity in infected macrophages. Macrophages require both TNF and IFN-γ to control infection.

Infection in Humans

The incubation period is 3–5 days, followed by flu-like symptoms with nonspecific signs such as fever, malaise, chills and headache. Symptoms vary according to the subtype of the agent and its route of entry. In humans, tularemia presents in six clinical forms: ulceroglandular, glandular, oculoglandular, oropharyngeal, pulmonary, and typhoid. The most common forms of tularemia are ulceroglandular and glandular, accounting for approximately 60–70% of cases.

Ulceroglandular or glandular tularemia occurs when the agent enters through the skin or mucous membranes. Transmission occurs after bites from haematophagous arthropods (such as ticks, mosquitoes and horseflies) or after contact with a contaminated environment during various outdoor activities. Ulceroglandular forms are characterised by an inoculation ulcer with one or more satellite lymphadenopathies. It is distinguished from the ulceroglandular form by the absence of ulcers in the glandular form. These are the dominant forms caused by *F. tularensis* subsp. *holarctica*, but other subspecies can also cause these forms. Even if patients receive appropriate treatment, the disease causes lymph node abscesses that require surgical intervention in some patients with adenopathy.

Respiratory tularemia is rare and is transmitted by ingestion of infected aerosols. Respiratory infection caused by *F. tularensis* subsp. *holarctica* is not fatal. However, infection caused by *F. tularensis* subsp. *tularensis* causes an acute, severe infection characterised by high fever, malaise, chills, and cough and can be life-threatening. In the period when effective treatment options were not available, the mortality rate from *F. tularensis* subsp. *tularensis* was >50%, but this rate has decreased to less than 2% today.

Oculoglandular tularemia is the other rare form of tularemia, which involves the eye and usually presents with unilateral conjunctivitis, marked swelling of the eyelids, photophobia, a purulent discharge and swelling of the periauricular lymph nodes.

Oropharyngeal tularemia, which occurs because of ingestion of contaminated water and food and presents with pharyngitis, digestive discomfort and cervical lymphadenopathy, and *typhoidal tularemia*, which is of unknown origin but presents with septicaemia, fever, fatigue and neurological or digestive system disorders.

Infection in Animals

Small and medium-sized mammals are considered the main natural reservoirs of tularemia. These include rabbits, aquatic rodents (beavers, muskrats and water voles), field mice, squirrels, lemmings and hamsters. All of these species are highly susceptible to *F. tularensis*, and infection usually results in severe and fatal disease. Other mammals such as monkeys, dogs, cats, sheep, and cattle, as well as predators such as bears, foxes, coyotes, and various bird species, as well as fish and amphibians, are considered accidental hosts for tularemia. Among domestic animals, cats are more susceptible to tularemia and the clinical picture is well described. In cats, mild infection with lymphadenopathy and fever, as well as subclinical infections or severe infections leading to death, can be observed. Clinical signs seen in other domestic species are related to the route of infection, virulence, and infectious dose. Clinical signs may include local inflammation, ulceration, and enlarged regional lymph nodes. Deaths are rare, and the disease is self-limiting. In more susceptible hosts, *F. tularensis* can persist for long periods.

Clinical signs of tularemia in wildlife are not always apparent. Infected wild animals are usually found moribund or dead. Subclinical infection or carriage in these hosts predispose them to outbreaks of tularemia. Reports of natural infection in dogs are rare. Dogs fed on infected tissues present with fever, mucopurulent nasal and ocular discharge. After intradermal inoculation, pustules at the inoculation site and regional lymphadenopathy are visible. The disease resolves spontaneously with supportive treatment. In addition, there are numerous studies of seroconversion in dogs. Outbreaks of tularemia in sheep, causing significant morbidity and mortality, have been reported in Montana and Idaho. Cases in sheep have been associated with increased stress factors and intense tick infestation. High rectal temperature, low body weight, regional lymphadenopathy, and diarrhoea have been reported in affected sheep. There is ample serological evidence of natural infection in cattle, but no definitive clinical syndrome has been described. Infection in horses has also been associated with tick infestation, with fever, dyspnoea, depression, and incoordination in infected horses, and deaths in foals.

Epidemiology and Public Health

Host

Among the zoonotic disease agents, few agents have been reported to have a wider host range than *F. tularensis*. Species in Lagomorpha, Rodentia and Sciuromorpha are the primary hosts of *F. tularensis*. Snowshoe hare, black-tailed jackrabbit, white-tailed jackrabbit, and cottontail rabbit belonging to the Leporidae family; beaver belonging to the *Castoridae* family; mouse, white-footed mouse, rat, muskrat, woodrat, Norway rat, meadow vole, and vole belonging to the *Muridae* family; chipmunk, woodchuck, yellow-bellied marmot, ground squirrel, white-tailed prairie dog, black-tailed prairie dog, and squirrel belonging to the *Sciuridae* family are the most common species that are sources of human infections.

Sheep are the most commonly associated hosts with tularemia and have been a source of human infections in the USA, Canada and Russia. Recently, domestic cats have become an increasing source of infection for human cases. Infection in pigs, foals and dogs is relatively common, but clinical disease has rarely been reported. Clinical disease is more common in cattle, camels and cats. Birds are highly resistant to tularemia. However, natural infections have been reported in several species, some of which are fatal. Human cases have been associated with the preparation for consumption of game birds such as pheasants, partridges, quail and prairie chickens.

Invertebrates such as mites, biting flies, fleas and lice, especially ticks, have important roles in both the survival of *F. tularensis* in nature and the transmission of the disease. Hard ticks are biological vectors of *F. tularensis* and act as both hosts and reservoirs for some species. Ticks belonging to the genera *Amblyomma* and *Dermacentor* are generally responsible for human cases. In contrast, those belonging to the genera *Ixodes* and *Haemaphysalis* are responsible for the persistence of enzootic foci in wildlife. Various other animal species, such as frogs, turtles, fish, snakes, spiders, etc., have also been naturally or experimentally infected and are therefore referred to as incidental species.

Arthropod Vectors

Various arthropod vectors such as ticks, mosquitoes and biting flies play an important role in the transmission of the infection to mammalian hosts.

While stinging flies are common vectors in regions of the USA such as Utah, Nevada and California, ticks are the most important vectors around the Rocky Mountains. In the former Soviet Union, both mosquitoes (*Aedes*, *Culex* and *Anopheles*) and ticks (*Ixodes*) were reported to transmit the infection. Most cases reported in Japan between 1972 and 1998 were of arthropod origin and were predominantly reported from spring to fall. In Central Europe, *Dermacentor reticulatus* and *Ixodes ricinus* ticks have been reported as important vectors. In some outbreaks, mosquito bites are assumed to be the main risk factor for the disease.

Transmission of Infection

F. tularensis is one of the most infectious agents known to humans. As few as 10–50 bacterial cells can initiate infection in mice and humans. Several transmission routes have been identified. Tularemia is transmitted to humans by (i) handling infected animals, (ii) consumption of contaminated water or food, vector (tick, fly) bites, (iii) contact with the aquatic environment and (iv) inhalation of aerosols. Therefore, the disease is mostly observed in areas where people live and work close to animals and in agricultural areas. In cases of tularemia where human-to-human transmission is not observed, clinical forms vary depending on the mode of transmission and route of entry.

Laboratory-acquired F. tularensis infections have also been reported because of exposure to the agent. It has also been reported that the deliberate release of the aerosolised form of the agent can cause significant morbidity and mortality, indicating that *F. tularensis* has a high potential to be used as a biological warfare agent or in bioterrorist attacks.

In wildlife, tularemia has two cycles: terrestrial and aquatic (Fig. 28.3). In the terrestrial cycle, rabbits serve as mammalian hosts and ticks, stinging flies, or mosquitoes serve as arthropod vectors. The aquatic cycle is mediated by the secretions or carcasses of muskrats or field mice. Various wild predators also contribute to the spread of tularemia.

Water contaminated with the urine and faeces of primary hosts such as beavers, muskrats, or lemmings, or with the carcasses of deceased animals are important source of infection for the transmission of tularemia. Depending on the characteristics and type of contamination, water that harbours *F. tularensis* in a viable and infective form for 3 weeks to 4 months has become an important medium for disease transmission. In fact, outbreaks of waterborne tularemia have been reported worldwide in recent years. Waterborne transmission in humans can occur not only through water consumption but also through the use of contaminated water and various water-based activities. Cases linked to drinking water have increased globally. Countries like

Fig. 28.3 Transmission cycle of *Francisella tularensis*. (Image source: Authors' personal collection)

Türkiye, Kosovo, Bulgaria, Georgia, Macedonia, Norway, Sweden, Italy and Germany have become more susceptible to the infection. In Türkiye, human infections are related to drinking water consumption. Large outbreaks occurred in Kosovo and Bulgaria and were found to be linked to water sources.

Tularemia has been reported all around the world (Table 28.1). The incidence has changed from 0.5 to 5 per one million cases per year in the USA. *F. tularensis* subsp. *tularensis* is responsible for approximately 90% of human cases in North America. It can be said that tularemia, which is also common in European countries, has followed a fluctuating course in the last three decades. The total number of confirmed cases per 100,000 inhabitants varied between 0.1 and 0.06 from 2014 to 2018. Tularemia is endemic in countries such as Finland and Sweden, and cases have been linked to mosquito bites. Tularemia is also seen in Eastern Europe and Russia, with incidence increasing after World War II. Kosovo has the highest annual incidence in Europe, with a rate of 5.2 per 100,000, and its prevalence is still high. Increased recreational outdoor activities in July–November in these regions lead to a seasonal pattern of the disease. Among the European countries, Albania, Greece, Iceland, the UK, Ireland, and Liechtenstein are known to be free from human tularemia. Tularemia is seen in some countries in Asia. China and Japan, Iran and Türkiye in West Asia, Armenia and Azerbaijan in the Caucasus, and Kazakhstan, Uzbekistan and Turkmenistan in Central Asia are the first countries where tularemia cases have been reported. In Türkiye, the main route of transmission is waterborne, and cases are mostly reported in late summer or early fall. Currently, Africa and South America are considered free of tularemia. Only a few cases have been reported in Australia. Limited cases were noted from Sudan and Kenya, but these need to be confirmed by standard methods.

Tularemia is a potential agent of bioterrorism for intentional use, especially when used as an aerosol challenge, because it can cause a disabling and sometimes fatal infection. Due to some of its superior properties, *F. tularensis* has been designated as a risk group 3 pathogen and a category A high-priority bioterrorism agent. Rapid diagnosis and specific treatment are required to prevent a fatal outcome in patients presenting with primary pneumonia. There is currently no vaccine available, and an outbreak could pose a significant public health burden. A collaborative and interdisciplinary One Health perspective is required because tularemia can affect humans and animals and is transmitted via vectors or environmental sources.

Diagnosis

Specimens

For the laboratory diagnosis of tularemia, the type of specimen submitted is critically dependent on the clinical presentation of the disease. In cases of ulceroglandular or glandular tularemia, the most appropriate samples include scrapings or swabs collected directly from the base of any skin ulcers, as well as needle aspirates obtained from enlarged lymph nodes. Similarly, for oculoglandular tularemia, conjunctival scrapings or swabs of the eye surface are recommended. In patients presenting with oropharyngeal tularemia, throat swabs or washings from the pharyngeal area, particularly from any visible lesions, should be collected. For individuals with pneumonic tularemia, sputum samples, if available, are valuable. However, in more severe cases or when sputum is not readily produced, bronchoalveolar lavage (BAL) fluid may be necessary for effective sampling of the lower respiratory tract. Finally, in cases of typhoidal or systemic tularemia, blood samples are essential for both blood culture and molecular detection methods such as polymerase chain reaction (PCR).

Since *F. tularensis* is a biohazardous pathogen, handling of the specimen and diagnosis should be performed in specially equipped laboratories, such as biosafety level 3 laboratories.

Table 28.1 The global distribution of tularemia (all notifications in humans, animals, reservoirs and vectors)

Continent	Country/region						Distribution
Africa	Algeria Angola Benin Botswana Burkino Faso Burundi Cabo Verde Cameroon Central African R. Chad	Comoros Congo Côte d'Ivoire Djibouti Egypt Equatorial Guinea Eritrea Eswatini Ethiopia	Gabon Gambia Ghana Guinea Guinea Bissau Lesotho Liberia Libya	Madagascar Malawi Malaysia Mali Mauritania Mauritius Morocco Mozambique Namibia	Niger Nigeria Réunion Rwanda São Tomé and Príncipe Senegal Seychelles Sierra Leone	Somalia South Africa South Sudan Tanzania Togo Tunisia Uganda Zambia Zimbabwe	Absent; no presence record; no information
	Kenya	Sudan					Present
Asia	Afghanistan Bahrain Bangladesh Bhutan Brune Cambodia	Chinese Taipei India Indonesia Iraq Jordon Kuwait	Kyrgyzstan Laos Lebanon Maldives Mongolia Myanmar	Nepal North Korea Oman Pakistan Palestine Philippines	Qatar Singapore South Korea Sri Lanka Suudi Arabia	Taiwan Thailand Timor Leste Vietnam Yemen	Absent; no presence record; no information
	Armenia Azerbaijan China	Georgia Iran Israel	Japan Kazakhstan	Syria Tajikistan	Türkiye Turkmenistan	United Arab Emirates Uzbekistan	Present
Europe	Albania Andorra Belarus	Faroe Islands Greece Iceland	Ireland Isle of Man Jersey	Liechtenstein Luxembourg	Malta Moldovia	Montenegro San Marino	Absent; no presence record; no information
	Austria Belgium Bosnia and Herzegovina Bulgaria Croatia	Cyprus Czechia Denmark Estonia Finland France	Germany Hungary Italy Latvia Lithuania	Netherlands Macedonia Norway Poland Portugal	Romania Russia Serbia Slovakia Slovenia	Spain Sweden Switzerland Ukraine United Kingdom	Present
North America	Bahamas Barbados Belize Bermuda British Virgin Islands	Cayman Islands Costa Rica Cuba Curacao Dominica	Dominican Republic El Salvador Greenland Grenada	Guatemala Haiti Honduras Jamaica Mexico	Nicaragua Panama Saint Kitts and Nevis	Saint Lucia St. Vincent and Grenadines Trinidad and Tobago	Absent; no presence record; no information
	Canada	United States					Present

Oceania	Cook Island Fiji French Polynesia	Kiribati Marshall Islands Micronesia	New Caledonia New Zealand	Palau Papua New Guinea	Samoa Tonga	Tuvalu Vanuatu	Absent; no presence record; no information
	Australia						Present
South America	Argentina Aruba Bolivia	Brazil Chile Colombia	Ecuador Falkland Islands	French Guiana Guyana Paraguay	Peru Solomon Islands	Suriname Uruguay Venezuela	Absent; no presence record; no information

Microscopy

The Gram staining method has low diagnostic value because *F. tularensis* agents, which have a very small Gram-negative pleomorphic or coccobacillus morphology with a size of 0.2–0.5 and 0.7–1 μm, take the stain lightly. However, the agent can be presumptively identified with fluorescence microscopy conjugated with mono- or polyclonal antibodies.

Alternatively, methods such as methylene blue, the Kozlowski method, and the Burry method can be used to stain the agent. The Kozlowski method is also used to distinguish *F. tularensis*, which has bipolar staining, from *Brucella* agents.

In Vitro Culture

In vitro culture of *F. tularensis*, a facultative intracellular bacterium, can be carried out especially in media supplemented with sulphydryl compounds and cysteine. For this purpose, enriched and selective media such as CHAB (cysteine heart agar with blood), BCYE (buffered charcoal–yeast extract), McCoy and Chapin, Francis, and Cystine Heart Agar are frequently used. Colonies of 2–4 mm in diameter are obtained after 48–72 h of incubation. The colony is white, grey or greenish-grey, or opaque. On the surface, it appears smooth and slightly mucous. In the extraction of this slowly growing agent from environmental samples, the addition of antimicrobials such as penicillin, nystatin, polymyxin B, and cycloheximide to the media increases the chance of isolation. For this purpose, CHAB–PACCV medium is used. While the agent can be grown in blood culture systems such as BACTEC in samples taken from sterile areas, in cases where direct culture fails or in the isolation of the agent from samples taken from contaminated areas, isolation can be attempted following inoculation into experimental animals. Since the biochemical activities of the agent are weak, these methods are not used very often in diagnosis and require confirmation with other tests. Nevertheless, in this context, catalase, beta-lactamase (cefinase test), oxidase, urease, nitrate, etc., can be used for preliminary diagnosis.

Serodiagnosis

Due to the bacterium's high infectivity and the difficulties associated with culture, serological tests are frequently preferred. The primary target of serological tests for *F. tularensis* is the detection of antibodies produced by the human body in response to the infection. The main antigen that these tests aim to detect antibodies against is the lipopolysaccharide (LPS), a major component of the outer membrane of F. tularensis. Some tests may also use whole-cell bacterial antigens. Several types of serological tests are used for the diagnosis of tularemia:

Microagglutination Test (MAT) This is a traditional and widely used test that detects total antibodies against F. tularensis. It involves mixing serial dilutions of the patient's serum with a suspension of formalin-inactivated F. tularensis bacteria. Agglutination (clumping) indicates the presence of antibodies. A fourfold rise in titer between acute and convalescent samples or a single titer of ≥1:160 is generally considered diagnostic.

Enzyme-Linked Immunosorbent Assay (ELISA) ELISA is a highly sensitive method that can detect and differentiate between different classes of antibodies, specifically IgM and IgG. IgM antibodies typically appear early in the infection, while IgG antibodies indicate a later stage or past infection. ELISAs use *F. tularensis* antigens (often LPS extracts) bound to a solid surface to capture specific antibodies in the patient's serum, which are then detected using enzyme-linked secondary antibodies.

Indirect Immunofluorescence Assay (IFA) IFA involves fixing *F. tularensis* bacteria onto a slide and then incubating them with the patient's serum. If specific antibodies are present, they bind to the bacteria. These bound antibodies are then detected using a fluorescently labelled secondary antibody, and the slide is examined under a fluorescence microscope.

Immunochromatographic Assays (ICT) These are rapid, point-of-care tests that can detect F. tularensis-specific antibodies in serum within minutes. They are often based on lateral flow technology and typically detect IgG antibodies. While convenient, they may have varying sensitivity and specificity compared to other methods.

Molecular Diagnosis

Polymerase chain reaction (PCR) variants can be used for agent analysis from clinical or environmental samples and for confirmation of suspicious cultures. Gene regions commonly used in the identification of microorganisms and the distinction of subtypes/variants with PCR are *fopA*, *tul4*, *pdpD*, *ISFtu2*, *lpnA*, *RDs* and *16S rRNA*. Genes like *tul4*, *fopA*, and the 23 kDa protein gene are frequently used targets due to their presence and relative conservation within *F. tularensis*. Moreover, the insertion sequence ISFtu2 is a multicopy element in *F. tularensis* and is often targeted to enhance the sensitivity of detection.

Loop-mediated isothermal amplification (LAMP) is an innovative molecular diagnostic technique that offers rapid, sensitive, and specific detection of *F. tularensis* under isothermal conditions (i.e. at a constant temperature), typically between 60 and 65 °C. This eliminates the need for the thermal cycling equipment required for traditional PCR, making it a potentially valuable tool, especially in resource-limited settings or for point-of-care applications for tularemia.

Matrix-assisted laser desorption/ionisation time-of-flight mass spectrometry (MALDI-TOF MS) is a rapid and increasingly popular proteomic method used for the identification of microorganisms, including *F. tularensis*. It analyses the protein profile of an organism to generate a unique '*fingerprint*' that is then compared against a database for identification. MALDI-TOF MS provides results within minutes to hours, significantly faster than traditional biochemical methods that can take days. This rapid turnaround time is crucial for timely diagnosis and appropriate management of tularemia. When coupled with a well-curated database, MALDI-TOF MS offers high accuracy in identifying *F. tularensis*. Studies have shown its performance to be comparable to or even better than traditional phenotypic methods and some molecular methods.

Protocols for these methods are presented in detail in Table 28.2.

Treatment

Antimicrobials such as gentamicin, doxycycline and ciprofloxacin are used in the treatment of tularemia. The treatment regimen and duration vary according to the type of antimicrobial, severity of the disease, and the patient's age, pregnancy, etc. Thus, tularemia treatment is grouped as standard treatment and post-exposure treatment. In the event of an unexpected fever or flu-like symptoms occurring within 14 days following presumptive exposure, standard treatment should be applied. The duration of standard treatment may vary between 10 and 21 days. The duration of gentamicin (5 mg/kg, IM or IV, daily) and ciprofloxacin (400 mg, IV or 500 mg PO, twice daily) in adults is 10–14 days, while this period has been stated as 21 days for doxycycline (100 mg, IV or PO, twice daily).

In children, the duration of administration is the same, and gentamicin (2.5–5 mg, IV or PO, twice or three times/day) is preferred. Treatment of tularemia in pregnancy in clinical cases or suspected exposure is similar to standard treatment. Gentamicin, ciprofloxacin and doxycycline can be used in such cases because they do not cause birth-related anomalies. Streptomycin is also preferred instead of gentamicin. In post-exposure prophylaxis, doxycycline and ciprofloxacin treatment are used for 10–14 days. This dose, which is 100–500 mg PO twice daily in adults, is used as 2.2–15 mg PO twice daily in children.

Resistance to antimicrobials used in standard or prophylactic treatment of tularemia has not yet been detected. However, antibiotic resistance continues to be a concern due to the potential use

Table 28.2 Diagnosis of *Francisella tularensis* infections

Diagnostic approaches	Methods	Targets	Remarks
Direct microscopy	Gram staining	Short rods or coccoid forms of bacteria	Cheap and fast technique, but has a low diagnostic value due to the faintly staining feature of *Francisella tularensis* Smears prepared from cultures or tissues can be stained *F. tularensis* appears as small (0.2–0.5 and 0.7–1 μm), Gram-negative pleomorphic bacilli or coccobacilli
	Staining alternatives		The Kozlowski method, a methylene blue staining method, can be used to stain the agent Distinguish *F. tularensis* (stained green) from *Brucella* spp. (stained red) Additionally, *F. tularensis* is characterised by its bipolar feature in staining
Cultural analysis	Basic culture media	Viable *F. tularensis*	*F. tularensis*, which has a very low infective dose and therefore poses a high biohazard, should be cultured in specially equipped laboratories such as Biosafety Level 3 laboratories Culture is considered the 'gold standard' for *F. tularensis*. However, these techniques are expensive, time--consuming, and tedious Enriched and selective media such as CHAB (blood cysteine heart agar), BCYE (buffered charcoal–yeast extract), McCoy and Chapin, Francis, and Cystine Heart Agar are often used. Growth on CHAB provides presumptive identification of *F. tularensis* because of the characteristic features (green, opalescent, raised, shiny colonies in 24–48 h) Antibiotic-supplemented media (CHAB-A) are recommended for contaminated samples
	Blood culture media	Viable *F. tularensis*	The BACTEC system or an equivalent system is recommended for blood culture
	Mouse inoculation	Viable *F. tularensis*	Mouse inoculation can be used in cases where direct culture fails, in the isolation of the agent from samples taken from contaminated areas, and in the recovery of isolates or for pathogenicity tests
	Phenotypic identification	*F. tularensis* subspecies	The organism is relatively inert biochemically; only a few tests (oxidase, urease, nitrate, glucose, maltose, sucrose and glycerol) are needed for confirmation, although these tests are used

(continued)

Table 28.2 (continued)

Diagnostic approaches	Methods	Targets	Remarks
Antigen detection	Direct fluorescent antibody (DFA)	Whole killed *F. tularensis* cells	It is used for direct identification of *F. tularensis* in clinical specimens or for confirmatory identification of isolates obtained in culture with a high diagnostic reliability It is a direct fluorescent antibody staining test using FITC-labelled rabbit antibodies directed against killed *F. tularensis* cells A positive DFA result appears as bright, intense green staining over the entire surface of the bacterial cell
	Slide agglutination		A positive reaction results from the binding of high titer anti-*F. tularensis* antibodies with pathogen-specific antigen. It results in immediate clustering of cells visible to the naked eye Suspected isolates should be processed in parallel with the positive and negative controls
	Immunohistochemical staining	*F. tularensis*	Using a monoclonal antibody against LPS, *F. tularensis* can be detected in formalin-fixed tissues
	Capture enzyme-linked immunosorbent assay (cELISA)	LPS	This test is specific for all subspecies of *F. tularensis* and has a sensitivity rate of approximately 10^4–10^3 cfu/mL This test is used for the confirmation of suspect cultures and the testing of animal tissues
Immunodiagnostics	Microagglutination (MAT)	Antibodies against *F. tularensis*	Despite their practical and widespread use, the most important problem is that they cause cross-reactions due to their antigenic structural similarities (with Q fever, psittacosis, *Brucella* and *Yersinia*) Either microagglutination or tube agglutination is the standard serological test for the detection of *F. Tularensis*--specific antibodies
	Enzyme-linked immunosorbent assay (ELISA)		It is a more specific and sensitive test that determines the titers of anti--*Francisella* IgG and IgM An ELISA (for LPS) combined with Western blot (against antigens extracted from whole killed cells) shows very good sensitivity and specificity for the diagnosis of infection

(continued)

Table 28.2 (continued)

Diagnostic approaches	Methods	Targets	Remarks
Molecular assays	*Polymerase chain reaction* (PCR)	Species/ subspecies of *F. tularensis* and genus *Francisella*	Valuable and highly accurate diagnostic tool when organisms cannot be cultured PCR can be used for direct analysis of the agent from clinical or environmental samples and for confirmation of suspicious cultures Gene regions commonly used in the identification of species/subtypes with conventional PCR are *fopA*, *tul4*, *pdpD*, *ISFtu2*, *lpnA*, *RDs* and *16S rRNA* A multitarget real-time TaqMan PCR assay has also been developed for *F. tularensis*, with greater specificity, sensitivity and speed compared with conventional PCR. These assay targets include the *ISFtu2* element, the *23 kDa* and *tul4* genes
	Matrix-assisted laser desorptionionization-time-of--flight mass spectrometry (MALDI-TOF MS)	*F. tularensis*	It has high specificity and reliability It can be used to identify agents based on the database's repertoire

of *F. tularensis* as a biological weapon. In this context, new antimicrobial alternatives such as ketolides, tigecycline and some fluoroquinolone types (grepafloxacin, sparfloxacin, etc.) and antimicrobial peptides such as LL-37 have proven effective.

Prevention and Control

Tularemia is a zoonosis that can infect more than 100 wild and domestic vertebrate species and more than 100 invertebrate species, posing a significant threat to public health. Collaboration between veterinary, medical, animal health and public health professionals is essential to elucidate the complex epidemiology of the disease. High exposure risks and even human outbreaks often occur before the disease is observed in animals. Therefore, timely sharing of animal cases and raising awareness in outbreak situations may facilitate early diagnosis or prevention of human cases. In cases in humans, identifying and eliminating sources of infection through anamnesis is important to prevent subsequent transmission.

In addition to gathering information on the prevalence of ticks, mosquitoes, and possibly other insects that can transmit *Francisella*, it is also important to track the sources of infection in cases or outbreaks and investigate their prevalence in various animal and environmental samples. This requires strong collaboration between animal health and human health experts. International synergy, data and expertise, and internationally supported research projects appear important to further advance tularemia prevention and control.

Considering the very low infective dose of the agent, it is essential for those at risk to be aware of the high contagiousness of contaminated materials. Many useful protective practices can be applied to professional activities or daily life to minimise exposure to the agent. These include: (i) avoiding the consumption and use of water of unknown origin for various purposes, (ii) avoiding polluted water contaminated with animal faeces or carcasses or rodent contact in terms of various water activities, (iii) paying attention to sanitation, hygiene and cooking methods in the consumption of meat obtained from wild animals in hunting, (iv) minimising contact with insects or arthropods, (v) taking maximum protective measures when handling wild animals, especially rabbits, providing protective equipment to prevent possible aerosol contamination in breeding stations and (vi) minimising contact between animals in case of possible infections in farm animals.

Although the effectiveness of immunotherapy has been proven in various animal studies, it has not been applied in humans. Although a live-attenuated vaccine obtained from a low-virulent strain of *F. tularensis* subsp. *holarctica* in the 1940s provided prophylactic success; its use in humans or animals is not permitted. Apart from this, efforts to develop killed, subunit, recombinant vector-based and DNA vaccines that have been developed since the 1920s and provide partial protection continue. These vaccine development efforts should also aim to induce cell-mediated immunity to provide long-term protection against all subtypes of *F. tularensis*.

Case Study

A 64-year-old female patient applied to the infectious disease clinic with complaints of high fever, sore throat and swelling in the left neck area. The sore throat complaint started 2 months ago and continued despite the 14-day amoxicillin-clavulanate treatment prescribed by an external medical centre. The patient had no chronic disease or history of drug use in her medical history. She was engaged in cattle farming, used spring water and consumed cheese made from raw milk. In the systemic examination, it was learned that the high fever occurred 1–2 times a day, increased with chills and shivering, was accompanied by weakness, fatigue and headache, and she had lost 5 kg involuntarily in the past month. Physical examination revealed that the oropharynx was hyperemic, and there was a painful, hard and non-fluctuating lymphadenopathy of approximately 8 × 6 cm in diameter extending from under the left ear to the left side of the trachea. No other features were observed in her systemic examination. Laboratory examination revealed leukocytosis ($14.58 \times 10^3/\mu L$) and elevated C-reactive protein (8 mg/L). Superficial tissue ultrasonographic examination of the neck region revealed lymphadenopathy with cystic necrotic areas within hypoechoic areas of approximately 80 × 65 mm starting from the left neck zone 2 and extending to zone 3. There were increased echoes consistent with inflammation in the surrounding fatty tissue. The patient was interpreted in terms of granulomatous lymphadenitis or lymphoproliferative disease. The patient was initially diagnosed with glandular tularemia, tuberculous lymphadenitis, infectious mononucleosis, acute retroviral syndrome, deep neck infection, malignancies and brucellosis. After excluding differential diagnoses and with the tularemia microagglutination test being 1/1280, a 10-day gentamicin (5 mg/kg) treatment was planned with the diagnosis of glandular tularemia. After the treatment, the lymphadenopathy turned into an abscess, and the patient underwent abscess drainage and was discharged with recovery (Approved by Associate Professor Zeynep Ture Yuce, Faculty of Medicine, Erciyes University, Türkiye).

1. How do you track the transmission of tularemia in such a case?
2. What other microbiological methods would you use in a similar case to diagnose tularemia?
3. What kind of tests do you perform as part of a differential diagnosis, and what do you expect?

Research Questions

1. What kind of epidemiological approach should be used to demonstrate environmental contamination with *F. tularensis*?
2. What role can water play in the epidemiology and transmission of tularemia?
3. What should be the approach to developing a protective immunity against *F. tularensis* and a viable vaccine?

Further Readings

Feldman KA. Tularemia. J Am Vet Med Assoc. 2003;222:725–30.

Friend M. Tularemia. Reston: U.S. Geological Survey, Circular, 1297; 2006. 68 p.

GIDEON. Infectious diseases of the world. Los Angeles: GIDEON Informatics; 2019. https://www.gideononline.com/ebooks/country/infectious-diseases-of-the-world/

Hennebiquea A, Boisset S, Maurin M. Tularemia as a waterborne disease: a review. Emerg Microbes Infect. 2019;8:1027–42.

Hestvik G, Warns-Petit E, Smith LA, Fox NJ, Uhlhorn H, Artois M, et al. The status of tularemia in Europe in a one-health context: a review. Epidemiol Infect. 2015;143:2137–60.

Hopla CE, Hopla AK. Tularemia. In: Beran GW, Steele JH, editors. Handbook of zoonoses. 2nd ed. Boca Raton: CRC Press, Inc.; 1994. p. 113–26.

Karatas Yeni D, Büyük F, Ashraf A, Shah MSUD. Tularemia: a re-emerging tick-borne infectious disease. Folia Microbiol. 2021;66:1–14.

Maurin M. *Francisella tularensis*, tularemia and serological diagnosis. Front Cell Infect Microbiol. 2020;10:512090.

Mörner T. The ecology of tularaemia. Rev Sci Tech. 1992;11:1123–30.

Olsufév NG. Tularemia. In: Pavlovskii EN, editor. Natural foci of human infections. Moscow: Foreign Languages Publishing House; 1963. p. 127–63.

Petersen JM, Schriefer ME, Gage KL, Montenieri JA, Carter LG, Stanley M, et al. Methods for enhanced culture recovery of Francisella tularensis. Appl Environ Microbiol. 2004;70:3733–5.

Sjöstedt A. Family XVII. Francisellaceae, genus I. In: Francisella: Bergey's manual of systematic bacteriology. 2nd ed. New York: Springer; 2005.

Sjöstedt A. Tularemia: history, epidemiology, pathogen physiology, and clinical manifestations. Ann N Y Acad Sci. 2007;1105:1–29.

Svensson K, Larsson P, Johansson D, Byström M, Forsman M, Johansson A. Evolution of subspecies of Francisella tularensis. J Bacteriol. 2005;187:3903–8.

World Health Organization (WHO). Guidelines on tularaemia. World Health Organization; 2007. https://iris.who.int/handle/10665/43793

Mycobacterium tuberculosis Complex

29

Deepjyoti Kalita and Sangeeta Deka

Learning Objectives

1. To describe zoonotic *Mycobacterium* species' taxonomy, morphology, and genetics
2. To explain zoonotic tuberculosis transmission and host–pathogen interactions
3. To understand the challenges of diagnosing and treating zoonotic tuberculosis in India
4. To evaluate One Health control and research techniques

Introduction

Zoonoses significantly contribute to the global burden of infectious diseases. Due to their chronicity, diagnostic challenges, and public health impact, mycobacterial infections are major bacterial zoonoses. Although *Mycobacterium tuberculosis* (Mtb) kills millions annually, other species in the *Mycobacterium tuberculosis* complex (MTBC), including *Mycobacterium bovis*, can also spread zoonotic diseases. Zoonotic tuberculosis (zTB) is transmitted through unpasteurised milk, direct contact with diseased animals, or aerosol inhalation.

Zoonoses contribute significantly to the global infectious disease burden. Mycobacterial infections, particularly those caused by members of the *Mycobacterium tuberculosis* complex (MTBC), represent a major bacterial zoonotic threat due to their chronicity, diagnostic challenges, and profound public health impact. While *M. tuberculosis* (Mtb) is the predominant cause of human TB, killing millions annually, several other species within the MTBC are bona fide zoonotic agents.

The zoonotic MTBC species include *M. bovis*, the primary cause of bovine tuberculosis, *Mycobacterium caprae*, primarily found in goats, and *Mycobacterium orygis*, found in oryx and other antelopes. Other members of the MTBC have documented cases, although they are rarely reported. These include *Mycobacterium africanum* (predominant in West Africa), *Mycobacterium microti* (associated with rodents), *Mycobacterium canettii* (found in Africa), *Mycobacterium pinnipedii* (from marine mammals), *Mycobacterium mungi* (from mongooses), and *Mycobacterium suricattae* (from meerkats).

Zoonotic tuberculosis (zTB) spreads to humans through various routes, including the consumption of unpasteurised milk and dairy products, direct contact with diseased animals, especially in occupational settings like farming or slaughterhouses, or aerosol inhalation from infected ani-

D. Kalita (✉)
Department of Microbiology, All India Institute of Medical Sciences, Guwahati, Assam, India
e-mail: deepjyotikalita@aiimsguwahati.ac.in

S. Deka
Department of Microbiology, Nagaon Medical College & Hospital, Nagaon, Assam, India

S. C. Parija, A. Poddar (eds.), *Textbook of Bacterial Zoonoses*, Microbial Zoonoses,
https://doi.org/10.1007/978-981-95-6615-0_29

mals. Effective veterinary tuberculosis control programmes have significantly reduced zTB incidence in many developed countries. Nevertheless, they remain largely underdiagnosed in high-burden regions due to the conspicuous absence of such comprehensive control programmes. Coexisting human and bovine tuberculosis, cultural customs, insufficient food safety, and the emergence of agents like *M. orygis* complicate the condition. This requires molecular diagnosis and a comprehensive surveillance programme for animals and zTB.

History

Historically, *Mycobacterium* is traced back as far as 150 million years ago. An early progenitor of *M. tuberculosis* may have infected early hominids in East Africa as far back as 3 million years ago. Archaeological findings, including skeletal deformities typical of tuberculosis in Egyptian mummies dating back to 2400 BC and in Peruvian mummies, demonstrate its ancient presence across continents. In ancient Greece, Hippocrates described the disease as "phthisis" or "consumption", noting its fatal nature, especially in young adults. During the Middle Ages, active TB of the neck and lymph nodes, known as "scrofula" or "king's evil", was widely believed to be curable by a "royal touch".

For centuries, the exact cause of tuberculosis remained a mystery. While some early thinkers, such as Benjamin Marten (1720), hypothesised about "wonderfully minute living creatures" causing the disease, it was not until the nineteenth century that scientific breakthroughs revolutionised the understanding. In 1865, the French physician Jean-Antoine Villemin demonstrated the contagious nature of TB by transmitting it to rabbits. However, it was on March 24, 1882, that Robert Koch announced his groundbreaking discovery of the causative agent of tuberculosis, a bacillus he later named *Mycobacterium tuberculosis*. This finding helped him earn the Nobel Prize in Medicine in 1905.

The recognition of zoonotic TB began in the late eighteenth century, after it was observed that TB also affected animals, particularly cattle. In 1898, Theobald Smith differentiated between human and bovine tubercle bacilli, setting the stage for understanding zoonotic transmission. *M. bovis*, primarily associated with bovine tuberculosis, was a significant cause of human illness, especially in the nineteenth and early twentieth centuries. This prompted extensive public health measures in Europe and North America. Measures such as milk pasteurisation, test-and-slaughter programmes, etc., contributed to a significant reduction in zoonotic tuberculosis.

Taxonomy

The genus *Mycobacterium*, within the phylum Actinobacteria, includes over 190 species. Broadly, medically important mycobacteria are categorised into two major groups. *Mycobacterium tuberculosis* complex (MTBC) is considered an obligate pathogen of humans and animals, while the non-tuberculous mycobacteria (NTM), identified as environmental agents with few of them having pathogenic potential, are largely opportunists.

Species under the *Mycobacterium tuberculosis* complex (MTBC) belong to the domain bacteria, phylum Actinomycota, class Actinomycetes, order Mycobacteriales, family Mycobacteriaceae, and genus *Mycobacterium*. This "complex" status acknowledges that despite sharing over 99.9% nucleotide sequence identity and high 16S rRNA gene similarity, these organisms exhibit distinct host preferences, geographical distributions, and sometimes varying pathogenic behaviours.

Zoonotically significant MTBC species are presented in Table 29.1.

Genomics and Proteomics

Members of the MTBC exhibit differences in host adaptability, pathogenicity, immunological response, and treatment sensitivity, largely influenced by minor genomic changes. MTBC show a genetic identity of 99.9%. Comparative genomic

Table 29.1 Different species of *Mycobacterium tuberculosis* complex (MTBC)

Species	Characteristics
Mycobacterium tuberculosis	Primary cause of human TB globally
Mycobacterium bovis	Causes bovine TB, zoonotic, resistant to pyrazinamide
Mycobacterium africanum	Endemic in West Africa, causes TB in humans
Mycobacterium orygis	Emerging member, found in south Asia, zoonotic, in cattle, deer, and humans
Mycobacterium canettii	Rare, smooth-colony variant, found mostly in the Horn of Africa
Mycobacterium caprae	Affecting goats and cattle, zoonotic, found in Europe
Mycobacterium microti	Infecting voles and other rodents, occasional human TB, especially in immunocompromised
Mycobacterium pinnipedii	Seals and sea lions, Zoonotic transmission to humans reported

and proteomic studies have yielded a clear understanding of these differences, especially in zoonotic situations.

The reference strain *M. tuberculosis* H37Rv has a 4.4 Mb genome with approximately 4000 genes. *M. bovis* possesses a marginally reduced genome (4.3 Mbp) with deletions, such as *RD4*, and mutations in *pncA*, which reflect resistance to pyrazinamide. *M. orygis* has distinctive SNPs and deletions signalling a diverse evolution.

Three crucial elements are responsible for the virulence of MTBC. These include (i) the RD1 region, (ii) the expansive PE/PPE gene family, and (iii) the critical DosR regulon. The RD1 area is a genomic segment present in virulent strains but absent in attenuated ones like the BCG vaccine, and is essential for pathogenicity. It notably encodes the genes for ESAT-6 and CFP-10, two proteins that form a complex and are secreted by the ESX-1 system. These proteins are known to disrupt the phagosomal membrane, allowing the bacteria to escape into the host cell cytoplasm and evade degradation, effectively causing a localised "cell lysis" within the host cell compartment. Beyond this, ESAT-6 and CFP-10 intricately modulate host immune responses and interfere with antigen presentation and influence cytokine production. These play an important role in "immunological regulation" that ultimately facilitates bacterial survival and dissemination within the host.

The PE/PPE gene family is characterised by its repetitive nature and sequence diversity. While their functions are still being actively researched, a leading hypothesis is their involvement in antigenic variation. These proteins are believed to help the bacterium evade recognition and attack by the host's adaptive immune system, thereby contributing to the establishment of a chronic infection. Moreover, some PE/PPE proteins are also implicated in modulating host immune cell responses, influencing cell wall integrity, and facilitating the acquisition of essential nutrients.

The DosR regulon allows pathogens to survive in challenging host environments, particularly during periods of dormancy. This regulatory system, activated by stressors such as hypoxia and nitric oxide encountered within the host granuloma, initiates the expression of numerous genes. These genes enable the bacterium to shift its metabolism to a low-energy, non-replicating state, allowing it to persist in a dormant or latent phase for extended periods.

Whole genome sequencing studies from India have validated *M. orygis* in humans and cattle, suggesting its incorporation into diagnostic procedures. Proteomics studies indicate the role of unique stress-induced protein expression in this infection. Prominent antigens comprise the Ag85 complex, HspX, GroEL1, and LprG, all potential candidates for vaccines and diagnostic tools. The ESX-1 and ESX-5 secretion systems are found to influence host immune responses. Differences in protein profiles between *M. bovis* and *M. tuberculosis* are found to affect disease phenotype and immune response.

Morphology

Zoonotic *M bovis*, *M. caprae*, and *M. orygis* share the typical morphology of the MTBC. They are rod-shaped, acid-fast, non-motile, and non-spore–forming bacteria. The typical lipid-rich cell walls provide resistance to environmental

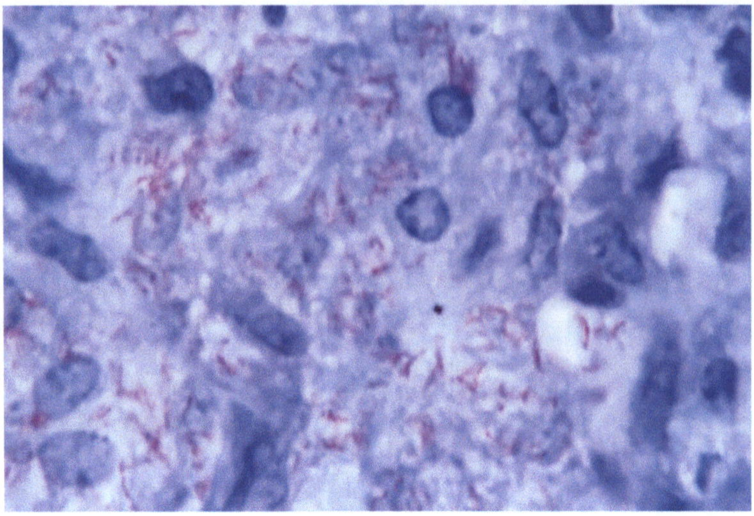

Fig. 29.1 Photomicrograph showing a close view of a murine lung tissue specimen, revealing the presence of a granulomatous inflammatory lesion containing intact, magenta-stained *Mycobacterium bovis*. (Image source: CDC ID#:21806)

stress, immune responses, and numerous antibiotics.

MTBC bacilli are slender rods (2–4 μm × 0.2–0.6 μm) under light microscopy (Fig. 29.1). The increased mycolic acid concentration enables the retention of carbol fuchsin following acid-alcohol decolourisation, which is the basis of Ziehl–Neelsen (ZN) staining. Fluorescence staining techniques, known for their improved sensitivity, are also increasingly used in many laboratories. The "Auramine O" and "Auramine O-Rhodamin B" stains are commonly used, especially for paucibacillary forms such as TB mastitis or lymph node tuberculosis.

Electron microscopy reveals a complex envelope comprising an inner membrane, a peptidoglycan–arabinogalactan–mycolic acid layer, and an outer lipid-rich capsule containing lipoarabinomannan (LAM) and trehalose dimycolate (TDM).

Cultivation

All members of the MTBC are obligate aerobes that grow optimally at 37 °C, with doubling times ranging from 15 to 20 h. These are cultivated in the LJ medium (preferably glycerol-free for *M. bovis*), MGIT 960 liquid culture (which detects growth within 7–14 days and is utilised in national and international reference laboratories), and Middlebrook 7H9 broth (for research, diagnostic susceptibility testing, and viability assessment).

However, their growth characteristics can differ. For example, *M. tuberculosis* develops coarse, desiccated colonies on glycerol-enriched Lowenstein–Jensen (LJ) media within 3–6 weeks. *M. bovis*, dysgonic on glycerol media, exhibits enhanced growth on pyruvate-supplemented LJ or Middlebrook medium, requiring 6–8 weeks for proliferation. In contrast, *M. orygis*, akin to *M. bovis*, may remain unnoticed in the absence of specific media or molecular techniques.

Animal Models

Animal models are essential for zoonotic tuberculosis research, facilitating the advancement of diagnostics, vaccines, and therapeutics in both human and veterinary fields. Given the largely strict host specificity displayed by the MTBC, particularly *M. bovis* and *M. orygis*, the selection of appropriate models is essential.

BALB/c and C57BL/6 mice are commonly preferred in experimental models due to their cost-effectiveness and well-characterised immune systems. In *M. tuberculosis* and *M. bovis* infections, they do not produce caseating granulomas. Hence, they are valuable for examining lesion architecture. Guinea pigs, however, consistently produce necrotic granulomas similar to those found in human tuberculosis, and are therefore used in vaccination and pharmacological research. Rabbits are resistant to *M. tuberculosis* and are useful for studies linked to latent TB, cavitation, and reactivation. Zebrafish infected with *M. marinum* provide benefits for the real-time visualisation of granuloma dynamics. Non-human primates (NHPs) closely resembling humans are also being used.

Transmission Cycle

Zoonotic mycobacteria, such as *M. bovis* and *M. orygis*, have complex transmission cycles involving multiple host species and environmental conditions. Successful One Health initiatives to manage zoonotic tuberculosis require an understanding of these mechanisms.

Zoonotic tuberculosis (zTB) is transmitted through several distinct pathways (Fig. 29.2). Aerosol inhalation remains the primary mechanism for transmission among cattle themselves and between infected animals and humans, particularly in confined spaces such as barns, cattle markets, or abattoirs, where airborne particles can be readily disseminated. This route poses a significant occupational hazard for those working near infected livestock.

Dietary habits play a crucial role, especially in regions where the consumption of raw or underboiled milk remains prevalent. This practice is a well-documented cause of extrapulmonary forms of zTB in humans, frequently leading to conditions such as cervical lymphadenitis or gastrointestinal tuberculosis. Furthermore, individuals in specific professions face heightened risks due to direct exposure. Veterinarians, dairy workers, and meat processors, for instance, are susceptible to infection. They acquire the infection through direct contact with infected animal tissues, fluids, or carcasses, which can occur via skin lacerations or exposure through mucosal membranes.

The environment also serves as a reservoir for the transmission of diseases. Environmental exposure can occur when humans or susceptible animals encounter bacilli present in contaminated bedding, soil, or feed, underscoring the importance of proper hygiene and sanitation in animal husbandry.

The interface between domestic animals and wildlife presents an additional transmission route. This is observed particularly near forested areas, where wild reservoirs may exist, significantly increasing the risk of spillover events. Adding another layer of complexity is the phenomenon of reverse zoonosis, where human-adapted species like *M. tuberculosis* are transmitted from humans to animals.

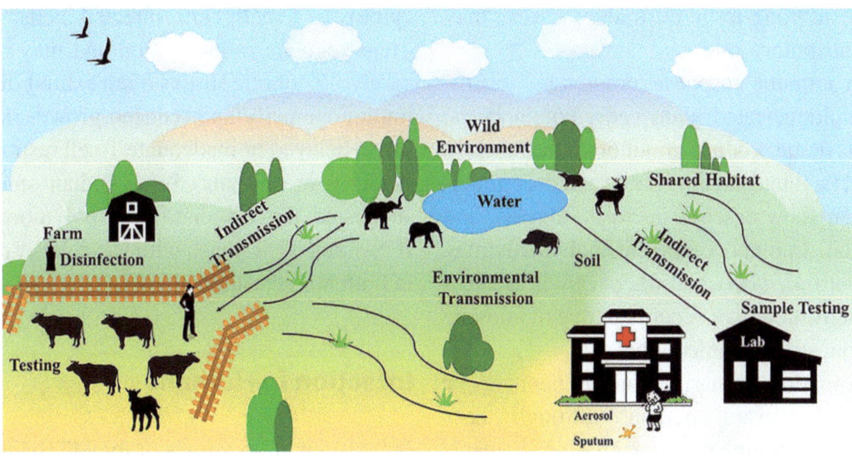

Fig. 29.2 Transmission cycle *of Mycobacterium tuberculosis* complex (MTBC). (Image source: Zhang et al. 2022. Permitted for reproduction under CC BY license)

Documented cases in diverse species, such as elephants, dogs, and caged monkeys, highlight this bidirectional transmission potential, suggesting that the flow of mycobacterial infections is not strictly unidirectional from animals to humans.

Pathogenesis and Pathology

M. bovis and *M. orygis* share a basic pathogenic mechanism with *M. tuberculosis*. However, they differ in terms of host reactions, lesion shape, and disease outcomes among different animals. The infection outcome, whether it remains localised, latent, or disseminated, is dependent on the interplay between bacterial pathogenicity and the host's immune response.

Infection is initiated upon the infiltration of bacilli into alveolar or intestinal macrophages, depending on the respiratory or intestinal route of entry of the pathogen. Once inside macrophages, bacilli avoid elimination by preventing phagosome–lysosome fusion, neutralising ROS/RNS, and surviving in acidic, nutrient-deficient vacuoles. Zoonotic MTBC strains produce virulence proteins, such as ESAT-6, CFP-10, and the PE/PPE family molecules. These enhance intracellular survival by modulating apoptosis and promoting protective cytokine production. According to some studies, intracellular bacterial survival inside mesenchymal stem cells (CD271+ cells), especially in long-term dormancy cases, may have a contributory role.

As the immune response progresses, macrophages, multinucleated giant cells, lymphocytes, and fibrous tissue produce granulomas that enclose bacilli and facilitate latency. Cattle and deer granulomas often show caseous necrosis, fibrosis, and calcification. Immunocompetent individuals may harbour both *M. bovis* and *M. orygis* in a latent form. Reactivation may occur later due to immune suppression, stress, coinfections, etc.

Dissemination primarily impacts neonates or immunocompromised persons through haematogenous or lymphatic transmission. It manifests as miliary tuberculosis or infection of remote organs, including the brain, bones, liver, and spleen. Histologically, human lesions display caseating granulomas with Langhans giant cells and fibrosis. In animals, analogous structures are frequently observed, typically showing significant calcification.

Immunology

The immune response to zoonotic mycobacteria, such as *M. bovis* and *M. orygis*, resembles that of *M. tuberculosis*. However, host-related differences can alter the resolution, latency, or progression of infection to active illness.

During the innate immune phase, bacilli are identified by pattern recognition receptors (PRRs), including TLR2, TLR4, and NOD-like receptors, on macrophages and dendritic cells. This activates pro-inflammatory cytokines such as TNF-α, IL-12, and IL-1β, which attract neutrophils and monocytes, facilitate granuloma formation, and stimulate adaptive immunity. Zoonotic strains circumvent these defences by various mechanisms: (i) inhibiting phagosome–lysosome fusion, (ii) neutralising reactive oxygen and nitrogen species, and (iii) downregulating antigen presentation via MHC class II. Moreover, virulence proteins such as ESAT-6 and CFP-10 interfere with immunological signalling and may induce macrophage death. Th1-type immunity drives adaptive responses, with IFN-γ and IL-2 promoting macrophage activation and granuloma stability. CD8+ cytotoxic T cells kill infected cells; however, Tregs regulate inflammation and may help them survive. Zoonotic strains often exhibit diminished immunogenicity and reduced growth rates, leading to delayed or inadequate T-cell responses, particularly in animals. Some Indian studies have suggested the association of polymorphisms in IFN-γ and TLR genes with variable susceptibility in both humans and animals.

Infection in Humans

Infection with members of the MTBC in humans causes a spectrum of clinical outcomes, ranging from asymptomatic latent infection to severe, life-threatening active disease. While *M. tuberculosis* is the most common cause of human TB, zoonotic MTBC species, especially *M. bovis* and *M. caprae*, also pose a significant threat.

Latent TB Infection

Upon initial exposure to MTBC bacteria, most individuals develop latent TB infection (LTBI). During this stage, the bacteria remain alive but inactive in the body, walled off by the immune system, typically within granulomas. Individuals with LTBI do not experience symptoms, cannot transmit the bacteria to others, and their chest X-rays are usually normal. However, there is a lifetime risk (5–10%, higher in immunocompromised individuals) of progression to active TB disease if the immune system weakens.

Active TB disease

Active TB disease occurs when the immune system fails to contain the infection, allowing the bacteria to multiply and cause overt illness. Active TB affects any part of the body, but it most commonly impacts the lungs, known as pulmonary TB. Symptoms of pulmonary TB include a prolonged cough lasting 3 weeks or longer, sometimes with sputum or blood, chest pain, fever, night sweats, unexplained weight loss, loss of appetite, and general fatigue or malaise. These symptoms are mild initially, leading to delays in seeking care and increasing the risk of transmission.

Extrapulmonary TB

MTBC infections also spread to other organs, leading to extrapulmonary TB. The manifestations vary depending on the affected site of the infection.

Lymph node TB (lymphadenitis) is presented as swollen, sometimes ulcerated, lymph glands, particularly in the neck (scrofula). This is a common form of *M. bovis* infection in children, often due to the ingestion of contaminated milk. Pleural TB is the inflammation of the membranes surrounding the lungs, causing chest pain and shortness of breath. Bone and joint TB affects the spine (Pott's disease), causing back pain and deformity, or other joints. Meningitis is an infection of the membranes surrounding the brain and spinal cord, leading to severe headache, confusion, and neurological symptoms. This is particularly dangerous, especially in young children.

Genitourinary TB affects the kidneys, bladder, or reproductive organs, potentially causing blood in the urine or fertility issues. Abdominal TB involves the intestines, peritoneum, or abdominal lymph nodes, leading to abdominal pain, diarrhoea, and weight loss. This is also relatively common in *M. bovis* cases, especially from ingestion. Miliary or disseminated TB is a severe, widespread form of the disease in which bacteria enter the bloodstream and spread to multiple organs throughout the body. This condition is often seen in individuals with compromised immune systems.

Infection in Animals

The clinical presentation of TB in animals is often chronic and insidious, making early detection challenging. *M. bovis* is the primary animal pathogen within the complex; other species also play crucial roles in their respective animal hosts.

Infection in Cattle

Infection in cattle, caused by *M. bovis* and *M. caprae*, is slow, often taking months or years for clinical signs to appear, making detection difficult. Symptoms, when present, are non-specific and may include gradual emaciation, lethargy, weakness, anorexia, a low-grade, fluctuating fever, and a chronic, moist cough in cases of pulmonary involvement. Enlarged superficial lymph nodes, particularly in the head and neck region, are common, as are tuberculous lesions in the lungs and associated lymph nodes, such as the bronchial and mediastinal nodes. In advanced cases, infected animals may also exhibit dyspnoea or experience alternating diarrhoea and constipation. *M. caprae* also causes tuberculosis in cattle, with similar clinical signs.

Infection in Other Animals

Infections in goats are caused by *M. caprae*. Clinical signs are like those in cattle, including weight loss, reduced appetite, and enlarged lymph nodes. Mammary gland infections are encountered in animals, whereas clinical mastitis is less common in humans than in cattles. Infections in wild boar and deer are caused by *M. bovis*, *M. caprae*, and *M. microti*. These wild animals are important reservoirs for *M. bovis* and *M. caprae* in many regions, including Europe. They often exhibit lesions in the lymph nodes and lungs. *M. microti* has also been reported in wild boar and red deer, causing pyogranulomatous pleuropneumonia and granulomas. Infections in rodents are caused by *M. microti*. Field voles, wood mice, and shrews are considered primary reservoirs for *M. microti*. Infection results in granulomatous lesions in various organs.

Epidemiology and Public Health

In endemic areas, cattle and buffalo are *M bovis's* main reservoirs. Many dairy and draught animals are known to be asymptomatic carriers of these bacteria. *M. orygis* has been found in cattle, deer, and zoo elephants in South Asia, suggesting a broader host range. During the infectious stage, live bacilli are released through respiratory droplets, milk in cases of TB mastitis, urine, faeces, and lesion exudates. These bacilli remain as infectious aerosols or environmental contaminants. Wildlife such as deer, elephants, and captive felines serve as reservoirs, throwing light on cross-species transmission at the interface between humans and animals.

Zoonotic tuberculosis (zTB) is a much-neglected aspect of the global tuberculosis epidemic, particularly in places characterised by close human–animal interactions, such as agricultural zones and forest peripheries. The disease has a heterogeneous global distribution influenced by socio-economic factors, livestock management practices, the presence of wildlife reservoirs, and the implementation of public health interventions such as milk pasteurisation and animal surveillance.

The World Health Organization estimates 140,000 new cases of TB and 12,000 fatalities each year. However, these figures are likely understated due to inadequate species-level diagnostics and frequent misclassification as *M. tuberculosis*. In many countries, pasteurisation, test-and-slaughter initiatives, and stringent animal transportation regulations have significantly diminished zTB. Nonetheless, in low- and middle-income countries (LMICs), the disease persists as endemic and is not well documented.

zTB is endemic in sub-Saharan Africa and constitutes a significant proportion of extrapulmonary TB in humans. The condition is seen particularly in pastoral communities with close animal contact and widespread consumption of raw milk. Countries such as Uganda, Nigeria, and Ethiopia report a high burden of bovine TB in cattle and occasional spillover into human populations. Similar challenges are evident in parts of South and Southeast Asia, including India, Bangladesh, Nepal, and Pakistan, where *M. bovis* circulates extensively in cattle and buffalo. The practice of consuming unpasteurised dairy products and the limited implementation of systematic bovine TB testing contribute to the persistence of human cases in these regions. In Latin America, zTB is moderately prevalent, with countries like Mexico, Brazil, and Peru reporting both animal and human infections. In certain Mexican regions, *M. bovis* accounts for up to 30% of TB cases, often linked to the artisanal production and consumption of raw milk cheese.

The Middle East and North Africa (MENA) region also shows moderate endemicity, with reports from Egypt, Iran, and Algeria pointing to ongoing transmission cycles among livestock and vulnerable human populations, particularly in areas with informal milk supply chains. Western Europe has largely controlled bovine TB in domestic herds through rigorous surveillance and pasteurisation policies. In Eastern Europe, the inconsistent implementation of control programmes in countries such as Romania and

Ukraine has allowed the disease to persist, despite limited comprehensive human surveillance. In North America, the burden of zTB is low overall. However, the United States continues to report sporadic cases, particularly among immigrant communities that consume imported unpasteurised dairy products. Wildlife reservoirs, including white-tailed deer in Michigan and bison in the western states, represent a localised source of infection for both livestock and potentially humans. New Zealand remains a notable exception, where wildlife (particularly the brushtail possum) plays a key role in the maintenance of bovine TB, despite the country's success in eradicating the disease from domestic cattle. Australia, on the other hand, has been officially declared free of bovine TB and maintains rigorous surveillance to prevent reintroduction.

The global distribution of MTBC is shown in Fig. 29.3.

Diagnosis

Diagnosis of zoonotic tuberculosis continues to pose challenges to scientists and doctors. Accurate identification is challenging due to the shared phenotypic characteristics within the MTBC and the infrequent availability of species-specific techniques.

Specimens

In humans, the type of specimen depends on the site of the disease. Sputum is the most common diagnostic specimen for pulmonary tuberculosis. However, zTB frequently presents in an extrapulmonary form, necessitating aspirates or biopsies from lymph nodes or fluids such as cerebrospinal fluid, pleural effusion, or urine.

In veterinary settings, milk samples are routinely analysed for bovine tuberculosis mastitis, and tissues from lymph nodes or lungs are obtained post-mortem. Tracheal aspirates, nasal swabs, or biopsies may be utilised in wildlife and zoo animals when practicable.

Decontamination is essential to reduce contamination and protect slow-growing bacilli. Human samples generally receive NALC-NaOH treatment, whereas oxalic acid is frequently employed for veterinary specimens, especially for milk.

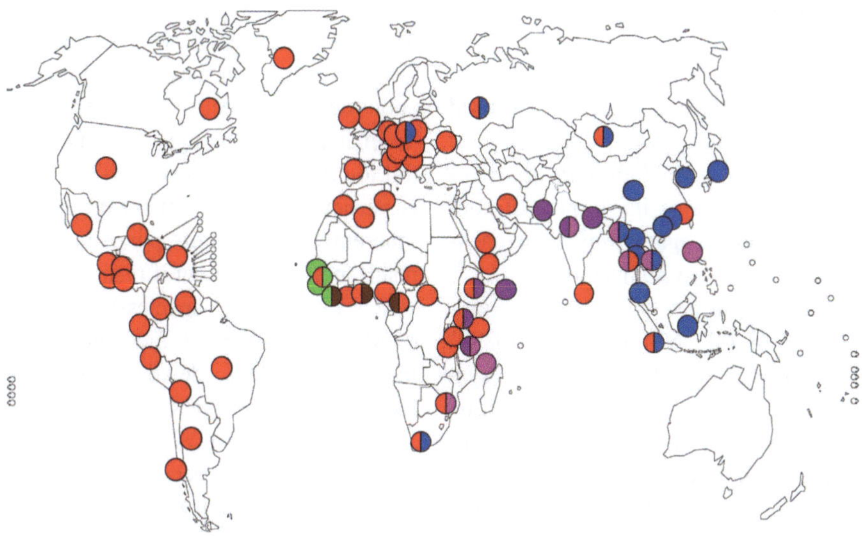

Fig. 29.3 Global distribution of the six main lineages of human MTBC. (Image source: Comas and Gagneux 2009. Permitted for reproduction under CC BY license)

Microscopy

Two principal microscopic techniques are employed for diagnosis.

Ziehl–Neelsen (ZN) staining is a traditional and widely adopted method. Under an oil-immersion microscope, MTBC bacilli appear as red, rod-shaped structures against a blue backdrop. ZN staining is cost-effective, relatively simple, and provides rapid results. However, its main drawback is its low sensitivity, which requires a substantial bacterial load of thousands of bacilli per millilitre for a positive result. It cannot differentiate between *M. tuberculosis*, other zoonotic MTBC species such as *M. bovis*, or non-tuberculous mycobacteria (NTM).

Fluorescence microscopy offers an enhanced alternative that uses fluorescent dyes such as auramine O, which bind to the mycolic acids in the cell wall. Under a fluorescence microscope, the acid-fast bacilli appear as brightly fluorescing yellow or green organisms against a dark background, making them significantly easier to detect. This method has a higher sensitivity than ZN microscopy. However, it cannot differentiate between various MTBC species or NTM.

In Vitro Culture

Lowenstein–Jensen (LJ) medium and Middlebrook 7H10/7H11 agar are widely used solid media for culture-based diagnosis. LJ medium, an egg-based formulation, incorporates malachite green to suppress the growth of contaminants. Middlebrook agar offers clear backgrounds for easier colony visualisation.

Liquid media systems, such as the automated BACTEC MGIT 960 and BacT/ALERT, are widely used for their efficiency. These systems typically employ broth-based media, often modified Middlebrook 7H9 broth, coupled with sensors that detect oxygen consumption as bacteria proliferate. The principal advantages of liquid cultures include a significantly faster time to detection. Positive results often appear within 1–3 weeks (median 10–14 days for MGIT). They also demonstrate higher sensitivity. They detect lower bacterial loads, making them particularly valuable for smear-negative pulmonary TB cases or extra-pulmonary specimens where bacterial numbers are minimal.

Serodiagnosis

The single intradermal tuberculin test (SITT), commonly employed in cattle, exhibits insufficient specificity owing to cross-reactivity with non-tuberculous mycobacteria (NTM). This test, however, fails to distinguish between latent and active tuberculosis (TB). Interferon-gamma release assays (IGRAs) are another test that provides enhanced accuracy for the diagnosis. However, it is not commonly utilised outside research environments due to its cost and complexity.

Despite decades of research and the availability of numerous commercial kits, serodiagnosis for active TB disease has faced significant limitations. The tests are generally not recommended by the WHO for routine diagnostic use.

Molecular Diagnosis

Molecular diagnostics have transformed tuberculosis detection, but they are still underutilised in zoonotic tuberculosis. GeneXpert MTB/RIF and Truenat MTB, although extensively used, even fail to differentiate between MTBC species. This is particularly important for *M. bovis*, which is inherently resistant to pyrazinamide and may result in poor treatment if misidentified.

Mycobacterium species-level identification requires sophisticated methodologies, such as line probe assays, deletion typing (e.g. RD4, RD9), and whole genome sequencing (WGS). WGS has effectively identified *M. orygis* in human and cattle specimens. The necessity of integrating molecular technologies into standard diagnostics, particularly in high-risk areas, is strongly recommended.

An overview of laboratory diagnosis is presented in Table 29.2.

Table 29.2 Diagnosis of *Mycobacterium tuberculosis* complex (MTBC)

Diagnostic technique	specific tools/ approaches	Features	Target	Advantages	Limitations
Microscopy	Ziehl–Neelsen (ZN) staining	Acid-fast staining using carbol fuchsin and methylene blue under oil immersion	Acid-fast bacilli (AFB)	Low cost, rapid, simple	Low sensitivity; requires high bacillary load; no species level identification
	Fluorescence microscopy (auramine O)	Fluorescent dye binds mycolic acids; visualised under fluorescence microscope	Acid-fast bacilli (AFB)	Higher sensitivity than ZN; easier detection	Specialised microscope needed; no species-level identification
In vitro culture	Solid media: LJ, Middlebrook 7H10/7H11	Slow growth of colonies on solid media	Viable MTBC organisms	Enables downstream drug susceptibility testing	Time-consuming (weeks); risk of contamination
	Liquid media: BACTEC MGIT 960, BacT/ALERT	Broth-based automated systems monitor growth via oxygen sensor	Viable MTBC organisms	Faster detection (1–3 weeks); more sensitive	Expensive, biosafety needs
Serodiagnosis	SITT (tuberculin skin test)	Measures delayed-type hypersensitivity	Host immune response to MTBC proteins	Simple, low-cost	Non-specific (NTM cross--reactivity); cannot distinguish latent/active TB
	IGRA (e.g. QuantiFERON)	Measures IFN-γ response to MTBC antigens in blood	Host cellular immune response (IFN-γ)	More specific than SITT	High cost; limited use in animals
Molecular diagnosis	GeneXpert MTB/RIF, Truenat MTB	Cartridge-based PCR for detection of MTBC DNA and rifampicin resistance	rpoB gene (MTBC + RIF resistance)	Rapid (~2 h); detects drug resistance	Species indistinguishable (e.g. *M. bovis* vs. *M. tuberculosis*)
	Line probe assays, deletion typing (RD4, RD9)	DNA hybridisation--based detection of MTBC markers	Species--specific genomic regions	Can identify MTBC species	Requires molecular lab
	Whole genome sequencing (WGS)	Sequencing of complete genome	Full genome of MTBC isolate	Precise species ID (e.g. *Mycobacterium orygis*); high resolution	High cost; advanced infrastructure required

Treatment

Treatment for human zTB is typically similar to the treatments for *M. tuberculosis*. However, some distinctions, particularly with *M. bovis*, call for attention to pharmacological susceptibility and species-specific diagnosis.

M. tuberculosis, M. orygis, and *M. bovis* are susceptible to isoniazid, rifampicin, and ethambutol. *M. bovis* exhibits inherent resistance to pyrazinamide owing to a pncA mutation, for which conventional first-line HRZE (isonazid, rifampicin, pyrazinamide, ethambutol) regimens are inadequate. *M. bovis* patients, ideally, should receive HRE therapy for a minimum of 9 months, although a global consensus has yet to be reached. No resistance pattern has been recorded yet for *M. orygis*.

Without drug susceptibility testing or a known history of clinical failure, traditional first-line therapy is used. Instances of multidrug-resistant (MDR) and extensively drug-resistant (XDR) tuberculosis in zoonotic strains are rare but documented. In such cases, second-line medications such as fluoroquinolones, linezolid, and bedaquiline are usually preferred. Difficulties arise in determining the appropriate duration of treatment, managing drug toxicity, and ensuring effective monitoring, especially in immunocompromised individuals or those with extrapulmonary disease.

Prevention and Control

Preventing zTB requires a comprehensive strategy targeting human and animal reservoirs. *M. tuberculosis* is generally handled through individual-level interventions. In contrast, zTB requires population-based solutions to control animal health, food safety, environmental cleanliness, and occupational protection within a One Health framework.

Milk pasteurisation is considered an effective prevention method for *M. bovis* infection. Public understanding of safe dairy operations and strict enforcement of pasteurisation standards are crucial. Occupational exposure among veterinarians, dairy workers, and abattoir personnel is at times overlooked. Regular use of gloves, masks, and protective attire, along with regular health checks in high-risk zones are essential. Timely species-level identification of the MTBC is important.

Regular tuberculin testing, isolating positive cases, improving farm hygiene, and restricting interactions between wildlife and livestock are the key components of controlling zTB in livestock. Integrating the One Health concept is critical.

Case Study

A 42-year-old Patna dairy worker had persistent cervical lymphadenopathy and moderate fever. He had long been exposed to unpasteurised milk and livestock in his backyard dairy. Fine-needle aspiration cytology revealed granulomatous inflammation, and GeneXpert testing confirmed the presence of the *M. tuberculosis* complex. Insufficient symptom relief was achieved with the standard HRZE. Following complete genome sequencing, a reference lab categorised the strain as *M. bovis*. After 9 months on a regimen without pyrazinamide, the patient recovered.

1. What was the most likely reason for the initial insufficient symptom relief with the standard HRZE regimen, given the final diagnosis?
2. What were the two most probable routes of transmission for this *M. bovis* infection?
3. Why was whole genome sequencing (WGS) crucial for the successful management of this patient, even when GeneXpert had already confirmed the presence of *M. tuberculosis* complex?

Research Questions

1. What are the actual prevalence and geographic spread of zoonotic tuberculosis caused by *M. bovis* and *M. orygis* in human and cattle populations?
2. What is the efficacy of contemporary molecular diagnostic methods in distinguishing zoonotic MTBC species under field conditions?

Further Readings

Comas I, Gagneux S. The past and future of tuberculosis research. PLoS Pathog. 2009;5(10):e1000600. https://doi.org/10.1371/journal.ppat.1000600.

Couto RM, Santana GO, Ranzani OT, Waldman EA. One Health and surveillance of zoonotic tuberculosis in selected low-, middle- and high-income countries: a systematic review. PLoS Negl Trop Dis. 2022;16(6):e0010428.

Duffy SC, Srinivasan S, Schilling MA, Stuber T, Sreevatsan S, Waters WR, et al. Reconsidering *Mycobacterium bovis* as a proxy for zoonotic tuberculosis: a molecular epidemiological surveillance study. Lancet Microbe. 2020;1(2):e66–73.

Duffy SC, Marais BJ, Kapur V, Behr MA. Zoonotic tuberculosis in the 21st century. Lancet Infect Dis. 2024;24(3):339–41.

Hugh BT, Sim EM, Crighton T, Sintchenko V. Emergence of *Mycobacterium orygis*: novel insights into zoonotic reservoirs and genomic epidemiology. Front Public Health. 2025;13:1568194.

Islam MR, Sharma MK, Khun Zinchenko R, Singh N, Chatterjee S, Dutta P, et al. Whole genome sequencing-based identification of human tuberculosis caused by animal-lineage *Mycobacterium orygis*. J Clin Microbiol. 2023;61(4):e00260–23.

Kock R, Michel AL, Yeboah-Manu D, Azhar EI, Torrelles JB, Vilaplana C, et al. Zoonotic tuberculosis – the changing landscape. Int J Infect Dis. 2021;113(Suppl 1):S68–72.

Luciano SA, Hattendorf J, Madhivanan P, Pathak R, Bergmann J, Gagneux S, et al. Human zoonotic tuberculosis and livestock exposure in low- and middle-income countries: a systematic review identifying challenges in laboratory diagnosis. Zoonoses Public Health. 2020;67(2):97–111.

Ministry of Health and Family Welfare (India). India TB report 2023. New Delhi: Central TB Division, Government of India; 2023.

Ramanujam H, Palaniyandi K. Bovine tuberculosis in India: the need for One Health approach and the way forward. One Health. 2023;16:100495.

Rani I, Kumar R, Singha H, Sharma P, Das B, Mahajan NK, et al. *Mycobacterium orygis* and its unseen impact: re-evaluating zoonotic tuberculosis in animal and human populations. Front Public Health. 2025;13:1505967.

Refaya AK, Ramanujam H, Ramalingam M, Vimalraj P, Baskaran K, Raja A, et al. Tuberculosis caused by *Mycobacterium orygis* in wild ungulates in Chennai, South India. Transbound Emerg Dis. 2022;69(6):e3327–33.

Sharma M, Mathesh K, Dandapat P, Roy S, Pal D, Sinha P, et al. Emergence of *Mycobacterium orygis*–associated tuberculosis in wild ruminants, India. Emerg Infect Dis. 2023;29(3):661–3.

Singh AV, Yadav VS, Chauhan DS, Singh SV. *Mycobacterium bovis*-induced human tuberculosis in India: status, challenges, and opportunities. Indian J Med Res. 2022;156(1):21–30.

Thapa J, Gordon SV, Nakajima C, Suzuki Y. Threat from *Mycobacterium orygis*-associated tuberculosis in South Asia. Lancet Microbe. 2022;3(9):e641–2.

WHO. Global Tuberculosis report 2023. Geneva: World Health Organization; 2023.

Zhang H, Liu M, Fan W, Sun S, Fan X. The impact of Mycobacterium tuberculosis complex in the environment on one health approach. Front Public Health. 2022;10:994745. https://doi.org/10.3389/fpubh.2022.994745.

Other Mycobacterial Infections

30

Deepjyoti Kalita and Sangeeta Deka

Learning Objectives

1. To comprehend the taxonomy, morphology, and genetic properties of nontuberculous mycobacteria and *Mycobacterium leprae*
2. To elucidate pathogenesis, clinical signs, and diagnostic techniques of zoonotic mycobacterial infections
3. To evaluate therapeutic and public health measures for zoonotic mycobacterial infections, particularly in India

Introduction

Non-tuberculous Mycobacteria (NTM) are a heterogeneous group of environmental mycobacteria (other than *Mycobacterium tuberculosis* complex and *Mycobacterium leprae*) that include a varied collection of environmental mycobacteria. NTMs are a varied collection of environmental mycobacteria, including species such as *Mycobacterium avium* complex, *Mycobacterium kansasii*, *Mycobacterium abscessus*, and *Mycobacterium fortuitum*, which are increasingly linked to opportunistic infections in humans. These are widely distributed in soil, water, and aerosols. Unlike *M. tuberculosis*, which is primarily spread from person-to-person, NTM infections are predominantly acquired from environmental sources. However, the intersection of humans and animals in shared environments, such as contaminated water sources or soil, leads to zoonotic transmission. Animals, including livestock, pets, and wildlife, are also infected with NTMs. In some cases, these infections pose a risk to human health, especially for immunocompromised individuals. NTM infections in humans manifest in various ways, from pulmonary disease to skin and soft tissue infections, lymphadenitis, and disseminated disease.

Mycobacterium leprae, the causative agent of leprosy, remains a pathogen of enduring historical and contemporary relevance. It causes a persistent granulomatous inflammation of the skin and peripheral nerves. Notwithstanding the success of multidrug therapy (MDT) and a major worldwide reduction, leprosy continues to persist in many locations, with India playing a substantial role in the global burden. While it is primarily transmitted from person-to-person through prolonged close contact, compelling evidence suggests a zoonotic component in some geographical regions. The nine-banded armadillo (*Dasypus novemcinctus*) in the Americas has been definitively established as a natural reservoir for *M.*

D. Kalita (✉)
Microbiology Department, All India Institute of Medical Sciences, Guwahati, Assam, India
e-mail: deepjyotikalita@aiimsguwahati.ac.in

S. Deka
Nagaon Medical College & Hospital, Nagaon, Assam, India

S. C. Parija, A. Poddar (eds.), *Textbook of Bacterial Zoonoses*, Microbial Zoonoses,
https://doi.org/10.1007/978-981-95-6615-0_30

leprae, and transmission to humans has been documented, often through direct handling or the consumption of armadillo meat. More recently, *M. leprae* DNA has also been detected in red squirrels (*Sciurus vulgaris*) in the British Isles and even in wild chimpanzees in West Africa, indicating a broader zoonotic potential than previously understood.

History

History of Nontuberculous Mycobacteria (NTM)

Following Robert Koch's discovery of *M. tuberculosis* in 1882, Alvarez and Tavil began to describe other mycobacterial varieties, some of which were found in animals and the environment. These were initially considered "non--disease-causing organisms" or as distinct from human tuberculosis.

One of the earliest significant observations came in 1868 with the description of "tuberculosis" in chickens that mimicked human disease. By 1890, this avian bacterium, later identified as *Mycobacterium avium*, was recognised in the laboratory as distinct from *M. tuberculosis*. Crucially, these organisms did not cause characteristic diseases when injected into guinea pigs, a common diagnostic test for *M. tuberculosis* at the time. This confirmed the idea that they were different and perhaps not relevant to human health.

However, until the 1930s, NTMs were not recognised as human pathogens. Recognition of human infection stemmed from a landmark case in 1943, which involved the recovery of *M. avium* from a patient with chronic lung disease and silicosis. By the 1950s, pulmonary disease caused by NTMs became more commonly recognised, especially in tuberculosis sanatoriums in the southeastern United States, accounting for 1–2% of admissions.

Recognition of NTM infections as distinct illnesses began in the 1950s with Runyon's categorisation, which grouped them based on pigment synthesis and developmental rate. They include photochromogens, scotochromogens, nonchromogens, and rapid growers.

The 1980s saw a significant increase in interest in NTM, particularly with the emergence of HIV/AIDS, as immunocompromised patients frequently developed disseminated NTM infections, especially due to *M. avium* complex (MAC).

History of *Mycobacterium leprae*

Leprosy is one of the oldest recorded diseases, with archaeological evidence (skeletal remains) dating back 4000 years in India. Ancient texts from India, China, and Egypt (circa 600 BCE) describe conditions consistent with leprosy. However, the terminology used for skin diseases was often broad. Historically, leprosy was often seen as a divine punishment, a curse, or a hereditary ailment, leading to severe social stigma and the isolation of affected individuals in "leper colonies" or "lazarets".

Modern-day genetic analysis of *M. leprae* suggests it may have originated in East Africa or the Near East around 100,000 years ago, spreading with human migrations along trade routes and through military campaigns (e.g. Alexander the Great's army). It reached Europe, Africa, and eventually the Americas via European colonists and the slave trade. Medieval Europe saw a significant incidence of leprosy, with thousands of dedicated hospitals. The decline in Europe from the fourteenth to the sixteenth centuries is not fully understood, but it may be linked to improved living conditions or the widespread presence of *M. tuberculosis*, which confers some cross-immunity.

Zoonotic Potential of *M. leprae*

For centuries, the prevailing view was that leprosy was an exclusively human disease, spread through close and prolonged person-to-person contact. However, in 1971, scientists Kirchheimer and Storrs achieved a breakthrough by successfully infecting the nine-banded armadillo (*Dasypus novemcinctus*) with *M. leprae* in a laboratory setting. This discovery not only provided a desperately needed animal model for leprosy research but also laid the groundwork for the subsequent, crucial realisation of natural infection.

The definitive proof of zoonotic transmission came in 1975. Walsh and colleagues, including Kirchheimer and Storrs, reported the presence of natural *M. leprae* infection in wild, free-living armadillos in Louisiana, USA. This demonstrated that armadillos were not only susceptible to laboratory infection but also actively harboured the bacterium in their natural habitats. Subsequent epidemiological investigations in the southern United States and Brazil began to draw connections between human leprosy cases and individuals who had a history of contact with armadillos, such as through hunting, skinning, or consuming them, even in the absence of known human-to-human exposure.

More recently, the understanding of *M. leprae*'s zoonotic potential has expanded beyond the armadillo. In the 2010s, it was discovered that wild red squirrels (*Sciurus vulgaris*) in the British Isles were naturally infected with *M. leprae*, with strains genetically similar to those found in medieval human cases in Europe. Further broadening this perspective, *M. leprae* was detected in wild chimpanzees (*Pan troglodytes*) in West Africa in 2021, indicating a potentially wider circulation of the bacterium in wildlife than previously conceived.

Taxonomy

The genus *Mycobacterium* belongs to the family Mycobacteriaceae, order Mycobacteriales, class Actinomycota, phylum Actinomycota, and domain Bacteria.

Within the genus, species are broadly categorised into three significant groups: (i) the *Mycobacterium tuberculosis* Complex (MTC), which is the primary causative agent of tuberculosis; (ii) the *Mycobacterium leprae* complex, responsible for leprosy; and (iii) the nontuberculous mycobacteria (NTM), which encompass all other mycobacterial species not belonging to the MTC or the *M. leprae* complex.

Non-tuberculous Mycobacteria (NTM) contain over 200 recognised species. Historically, their classification relied on the Runyon classification system. The system used a phenotypic method that grouped them based on observable characteristics, such as growth rate (slow vs. rapid growers) and pigment production (photochromogens, scotochromogens, and nonchromogens).

However, at present, molecular methods like 16S rRNA gene sequencing and whole genome sequencing have improved phylogenetic classification and resulted in the identification of numerous new species. NTMs now contain complex groups such as the *Mycobacterium avium* complex (MAC) and the *Mycobacterium abscessus* complex, both of which are highly significant in human disease due to their prevalence and often challenging drug resistance profiles.

In contrast, the taxonomy of *M. leprae* is more straightforward. A more recently identified species, *Mycobacterium lepromatosis,* is often grouped with *M. leprae* in the *Mycobacterium leprae* complex.

Genomics and Proteomics

The genomics of NTMs exhibit genomic diversity and adaptability, with genome sizes ranging from 4.5 Mb to 7.0 Mb and a high G+C content of 65–70%. Their genomes encode a variety of metabolic and pathogenicity features. Essential traits include Type VII secretion systems (ESX-1–ESX-5), which facilitate protein transport across the complex mycobacterial cell wall and are critical to immune modulation and pathogenicity.

Intra-species diversity, particularly within *Mycobacterium avium* complex (MAC), correlates with variation in virulence and drug resistance. Rapid growers, such as *M. abscessus*, contain genes associated with biofilm formation and antibiotic resistance. Proteomic investigations have identified antigenic targets, such as heat shock proteins (HSP65, HSP70) and the antigen 85 complex, which are pertinent to diagnostics and vaccine development.

M. leprae comprises a significantly shortened 3.27 Mb genome, with roughly 50% pseudogenes, i.e. genetic material rendered non-functional. This reduction reflects the strict dependence of *M. leprae* on its host for survival and its inability to grow in artificial media. Despite this, *M. leprae* retains critical genes associated with host-cell invasion, such as those encoding the Phenolic Glycolipid-I (PGL-I), which mediates adhesion to Schwann cells. Proteomic research suggests that structural lipoproteins promote immune evasion.

Morphology

Nontuberculous Mycobacteria (NTMs) are acid-fast bacilli distinguished by lipid-rich cell walls composed of mycolic acids, arabinogalactan, and peptidoglycan. Morphologically, they appear as slender, curved rods (0.2–0.6 μm in width; 1–10 μm in length), which are optimally identified using Ziehl–Neelsen staining. Their thick cell walls confer resistance to desiccation, disinfectants, and many antibiotics.

M. leprae is acid-fast, manifesting as straight or curved rods, frequently in clusters ("*globi*") among histiocytes or Schwann cells. Measuring 0.2–0.5 μm in width and 1–8 μm in length, it is excellently stained in tissues using modified staining techniques, like the Fite–Faraco stain. The stain, which preserves the weakly acid-fast bacteria embedded in host cells. Electron microscopy indicates the presence of a cell wall and the absence of any organelles associated with motility, aligning with its intracellular existence.

Cultivation

Most NTMs are cultured on Lowenstein–Jensen (LJ), Middlebrook 7H10/7H11 agar, and MGIT 960 liquid media. Growth rates vary: rapid growers produce colonies within a week, whereas slow growers may take several weeks. The optimal temperature for growth also varies by species. Most NTMs grow best at 35–37 °C; species such as *M. marinum* and *M. ulcerans* prefer lower incubation temperatures, ranging from 30 to 32 °C.

Rapid growers, such as M. *fortuitum* and M. *abscessus*, develop colonies within 2–7 days. Slow-growing species, like *M. avium* and *M. kansasii,* may require up to 8 weeks. Colonies could be smooth, rough, pigmented, or non-pigmented. Certain NTMs, such *as M. marinum* and *M. ulcerans*, prefer growth at lower temperatures (around 30 °C), a reflection of their aquatic ecological niches.

M. *leprae* differs from most other mycobacteria, including the NTMs, in its unique biological characteristics. It is an obligate intracellular parasite that cannot be cultured on artificial laboratory media. The mouse footpad model, where limited multiplication occurs, and the nine-banded armadillo (*Dasypus novemcinctus*), are alternative methods for propagation (Fig. 30.1). These models are essential for research and drug testing.

Animal Models

Animal models have substantially enhanced our understanding of NTMs and *Mycobacterium leprae*.

Murine models, especially immunocompromised strains like SCID mice, are employed to evaluate pulmonary and widespread NTM infections. The zebrafish (*Danio rerio*) has shown value in imitating *M. marinum* infection and studying granuloma biology.

The mouse footpad model developed by Shepard in the 1960s remains a gold standard for studying *M. leprae,* pathogenesis, and drug efficacy. This model allows limited multiplication of *M. leprae* at the cooler temperatures found in mouse footpads, but does not produce full clinical disease. The nine-banded armadillo has been invaluable for larger-scale studies. Armadillos naturally develop disseminated leprosy following experimental infection, closely resembling human disease both clinically and histopathologically.

Transmission Cycle

NTMs are primarily environmental organisms found in both natural and man-made reservoirs. They are found in soil, freshwater, marine environments, dust, and man-made water systems,

Fig. 30.1 Nine-banded *Armadillo*. (Image source: Authors' personal collections)

such as plumbing, hospital water lines, and biofilms on medical equipment. Humans usually acquire infections through direct environmental exposure—that is, from skin trauma, surgery, cosmetic procedures, or contaminated medical equipment for cutaneous and soft tissue infections, or from inhalation of aerosolised particles resulting in pulmonary infections. Invasive procedures and immunosuppression also facilitate opportunistic infections. While most NTM infections are sporadic, breaches in sterilising techniques have been documented, leading to healthcare-associated outbreaks. Zoonotic transmission of NTMs, though rare, has been documented, particularly in aquatic animals such as fish and amphibians. In such circumstances, *Mycobacterium marinum* and *M. fortuitum* serve as potential sources of infection for humans, especially for those in close contact with aquaculture or aquarium settings. Although the exact infectious doses for NTMs are unknown, high environmental loads, coupled with host sensitivity, seem to be essential for the emergence of disease.

M. leprae primarily depends on human-to-human transmission. The primary route is through nasal droplets from multibacillary patients; close and prolonged contact facilitates transmission of infection. Animal reservoirs have been increasingly recognised, for example, the nine-banded armadillo (*Dasypus novemcinctus*) in the southern United States and red squirrels in the United Kingdom, which naturally harbour *M. leprae* and *M. lepromatosis.* There are reports of confirmed zoonotic transmission to humans through the handling or consumption of infected animals. Additionally, the environmental persistence of *M. leprae* DNA in water and soil sources suggests possible, though untested, indirect transmission routes. Although the exact infectious dose in humans is not yet unknown, disease establishment seems to depend on prolonged exposure and compromised immunity.

An overview of zoonotic transmission of *M. leprae* is presented in Fig. 30.2.

Pathogenesis and Pathology

NTM infections result from an interplay between environmental exposure, pathogen pathogenicity, and host immunity. Upon entry, NTMs are phagocytosed by alveolar macrophages or tissue macrophages, where they survive and replicate by inhibiting phagosome–lysosome fusion and resisting reactive oxygen species.

Granuloma formation is an important host defence in mycobacterial infections, representing the host's attempt to contain the organisms. However, in immunocompromised people, NTMs may escape containment, producing progressive lung disease, widespread infections, or

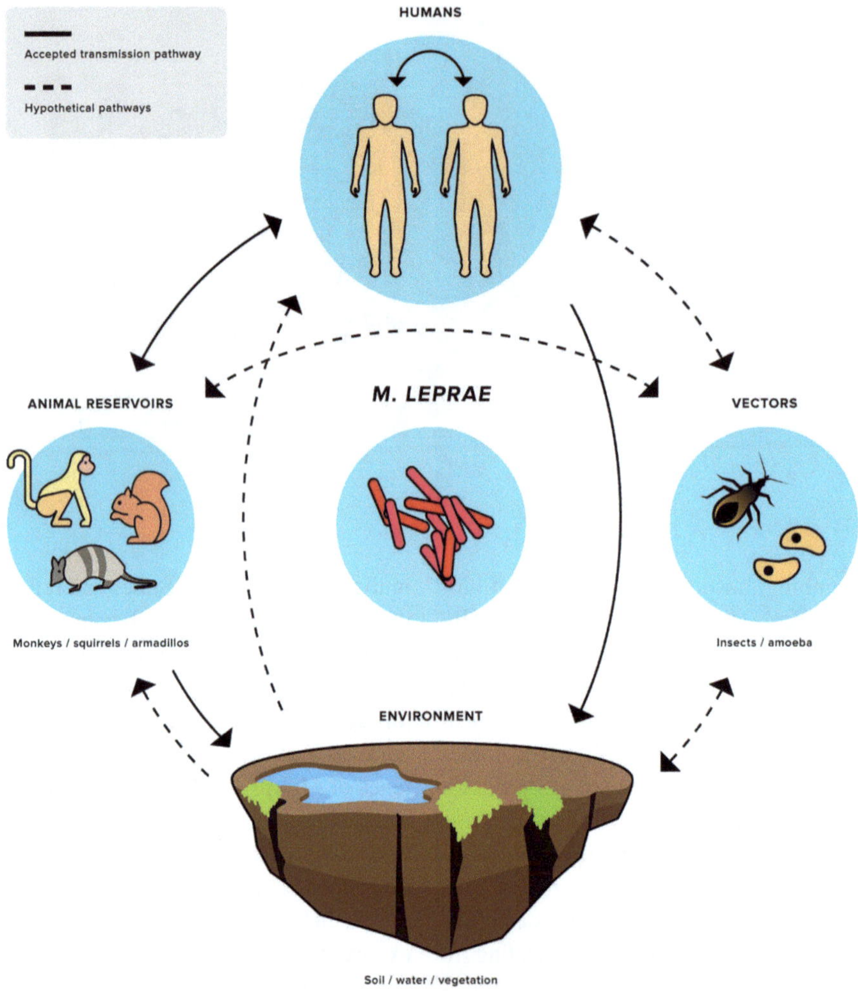

Fig. 30.2 Zoonotic transmission cycle of *Mycobacterium leprae.* (Image source: Ploemacher et al. 2020. Permitted for reproduction under CC BY licence)

chronic skin lesions. Biofilm formation, particularly by rapid growers such as *M. abscessus*, enhances persistence in tissues and medical devices, complicating therapy. Disease severity varies with species and host status—*M. Avium* promotes diffuse disease in advanced HIV, whereas *M. marinum* produces localised dermal infections following aquatic exposure.

M. leprae demonstrates a specialised pathogenic mechanism, marked by a preference for infecting Schwann cells in peripheral nerves and dermal macrophages. The Phenolic Glycolipid-I (PGL-I), surface molecule of the pathogen, enables adhesion to laminin-2 on Schwann cells, enhancing bacterial internalisation. Upon entering Schwann cells, *M. leprae* manipulates host-cell–signalling, resulting in demyelination, nerve injury, and disability.

The host immune response to *M. leprae* determines the clinical manifestations of leprosy, which vary from the localised, self-limiting tuberculoid form to the widespread, progressive lepromatous form. In tuberculoid leprosy, robust cell-mediated immunity leads to granulomatous inflammation with a limited presence of bacilli. Lepromatous leprosy is characterised by a deficient cellular immune response, facilitating rapid bacterial growth, widespread skin and nerve damage, and the development of lesions rich in foam cells.

Tuberculoid lesions exhibit epithelioid granulomas and lymphocytes, often accompanied by substantial nerve injury. Lepromatous types produce sheets of foamy macrophages filled with abundant acid-fast bacilli ("globi"), minimal lymphocytic infiltration, and relative preservation of the nerve structure until late stage.

Leprosy reactions cause acute morbidity in two forms. Type 1 (reversal) reactions involve intensified cellular responses resulting in lesions and nerve irritation. Type 2 (ENL) reactions are immune complex-mediated, causing systemic symptoms and boosting the probability of disability.

Immunology

Host immunity has a profound influence on the outcome of NTM and *M. leprae* infections. Inside the host, NTMs are recognised by innate immune receptors, notably Toll-like receptors (TLRs). These receptors, such as TLR2 and TLR4, initiate the activation of macrophages and dendritic cells. This leads to the release of pro-inflammatory cytokines—TNF-α, IL-12, and IFN-γ—crucial for granuloma formation and pathogen control.

An extensive Th1-type adaptive response, characterised by the production of IFN-γ and IL-2, is essential for macrophage activation and intracellular eradication. Impaired IFN-γ pathways, such as in advanced HIV, predispose individuals to disseminated NTM illness. NTMs evade immunity by inhibiting phagosome–lysosome fusion, modifying antigen presentation, and forming immune-protective biofilms.

The clinical spectrum reflects the host immunological status in *M. leprae* infection. At the tuberculoid end of the spectrum, a robust cell-mediated immune response dominated by Th1 cytokines (e.g. IFN-γ, IL-2) leads to effective control of bacillary multiplication. It is characterised by the formation of well-organised granulomas and a limited bacterial load. Conversely, in lepromatous leprosy, a Th2-predominant response is characterised by high levels of IL-4, IL-5, and IL-10. This results in impaired macrophage activation, allowing unchecked proliferation of *M. leprae.*

These immunological extremities determine the appearance of disease and influence the probability of leprosy reactions. Type 1 (reversal) reactions comprise sudden Th1 changes, inflaming existing lesions and nerves. Type 2 reactions (ENL) are mediated by immune complexes, resulting in systemic inflammation characterised by nodules, fever, and vasculitis.

Infection in Humans

Non-tuberculous mycobacteria (NTMs) cause a range of clinical syndromes, primarily affecting individuals with pulmonary disorders or immunosuppression.

Pulmonary disease is most frequent, mimicking tuberculosis with chronic cough, sputum production, haemoptysis, weight loss, and fatigue. *M. avium* complex (MAC), *M. kansasii*, and *M. abscessus* are common species. Risk factors include bronchiectasis, chronic obstructive pulmonary disease (COPD), prior tuberculosis (TB), cystic fibrosis, and the use of immunosuppressive medications.

Extrapulmonary infections include dermatological and soft tissue conditions, lymphadenitis, and disseminated infections. Skin infections frequently occur in conjunction with trauma, surgical procedures, or exposure to water. *M. marinum* induces nodular lesions resulting from fish tank-related injuries. Rapid-growing species such as *M. abscessus* and *M. fortuitum* are related to wound infections and healthcare-associated outbreaks.

Cervical lymphadenitis in children is another condition associated with MAC. Disseminated NTM disease in advanced HIV (CD4 <50 cells/μL) manifests with fever, weight loss, hepatosplenomegaly, and anaemia. Antiretroviral medication has considerably lowered such incidence.

M. leprae causes leprosy, a chronic infectious disease that primarily affects the skin, peripheral nerves, mucous membranes, and eyes. The disease has several borderline forms, representing immunological intermediates.

The "Ridley-Jopling classification" describes the clinical spectrum of leprosy as ranging from tuberculoid to lepromatous forms. Patients with tuberculoid leprosy show a strong cell-mediated immune response, producing a few well-defined, hypopigmented or erythematous skin patches, with notable sensory loss resulting from early nerve involvement. Often, lesions are anaesthetic, dry, and hairless. Usually, localised peripheral nerve thickening is prominent. On the other hand, the lepromatous leprosy response shows a weak cell-mediated immune response accompanied by significant bacillary proliferation. Often, patients exhibit characteristic facies, characterised by numerous symmetrical skin nodules, plaques, or diffuse infiltration. Progressively, bilateral peripheral nerve involvement causes glove-and-stocking sensory loss, muscle weakness, and deformities, including claw hand and foot drop.

With asymmetrical and more numerous lesions than tuberculoid cases but less than lepromatous leprosy, borderline forms—borderline tuberculoid, borderline all, and borderline lepromatous—show varied features depending on the immunological balance. Sensory loss in these forms is erratic and patchy.

Lepra Reactions

In addition to the primary lesions, patients may experience lepra reactions. Type 1 (reversal) reactions cause sudden inflammation of existing lesions and nerves, thereby risking acute neuritis and nerve damage. Type 2 reactions (erythema nodosum leprosum) cause systemic symptoms, including fever, painful nodules, arthritis, iritis, and lymphadenopathy. Delayed disease diagnosis and recurrent lepra reactions cause severe deformities, sensory loss, trophic ulcers, contractures, blindness, and major disability, contributing to social stigma and psychological stress. Initiating appropriate multidrug therapy and avoiding permanent disability depend on the early identification of clinical forms.

Infection in Animals

NTMs are major causes of diseases in animals, generally contracted by environmental exposure. In livestock, *M avium* subspecies *paratuberculosis* causes Johne's disease—chronic granulomatous enteritis in cattle, sheep, and goats. It results in diarrhoea, weight loss, and mortality, leading to severe economic harm.

Aquatic animals are very vulnerable to NTMs, such as *M. marinum*, *M. fortuitum*, and *M. chelonae.* These cause skin lesions and systemic illness in fish, affecting both natural ecosystems and aquaculture industries. Aquatic NTM infections also pose a potential zoonotic risk to humans, particularly to aquarium workers and fish handlers. Domestic pets, particularly dogs and cats, may develop localised or disseminated NTM infections, especially if immunocompromised. Disseminated disease in dogs has been attributed to infections with *M. avium* complex and *M. fortuitum.* Granulomatous lesions in multiple organs mimic neoplastic conditions and complicate diagnosis.

M. leprae demonstrates a very restrictive host range but plays a unique role in zoonotic transmission. The nine-banded armadillo (*D. novemcinctus*) in the southern United States has been identified as a natural reservoir of *M. leprae*. Armadillos develop disseminated leprosy, similar to the human lepromatous form, with an established association between handling or consuming armadillos and the development of leprosy in humans.

Recent findings have demonstrated the presence of *M. leprae* and *M. lepromatosis* DNA in red squirrels in the United Kingdom. Although zoonotic transmission from squirrels to humans has not been demonstrated, it raises new questions about potential environmental reservoirs.

Epidemiology and Public Health

With improved diagnostic techniques, increased awareness, and potentially rising environmental exposure, such as to household water systems,

NTMs are now recognised as emerging public health threats. These bacteria pose challenges due to their environmental ubiquity, intrinsic drug resistance, and the increasing recognition of person-to-person transmission for certain species, such as *M. abscessus*, in specific settings such as cystic fibrosis.

NTM infections have been documented increasingly due to ageing populations, increased diagnostic capabilities, and a growing number of immunocompromised individuals. Despite the ubiquity of NTMs in the environment, disease occurs in a minority of exposed persons. Pulmonary NTM infections have markedly increased in developed countries, with reported rates ranging from 1.8 to 15 per 100,000 individuals. MAC is generally the most common NTM species isolated globally, particularly in North America, East Asia (e.g. Japan, South Korea), and Australia. Other species like *M. kansasii* are more prevalent in specific regions (e.g. parts of Europe and the central United States). *M. xenopi* is notably more common in certain parts of Europe (e.g. Hungary, Belgium, the UK). These regional differences are likely influenced by environmental factors such as climate and soil composition, as well as human population density and underlying risk factors. Healthcare-associated NTM outbreaks, linked to contaminated equipment, catheters, and water systems, are increasingly recognised. Biofilm-forming nontuberculous mycobacteria obstruct disinfection techniques.

Hospital-based studies in India indicate the isolation of NTM in patients with chronic pulmonary disease, particularly concerning *M. fortuitum*, *M. abscessus* and MAC. Misdiagnosis of tuberculosis continues to constitute a substantial public health concern, resulting in inadequate therapy and postponed diagnosis.

Leprosy epidemiology is better defined. The global epidemiology of leprosy has undergone a dramatic transformation over the past few decades, largely due to the success of multidrug therapy (MDT). Nevertheless, leprosy remains a public health problem in several regions, and the detection of new cases continues. Following the introduction of MDT by the World Health Organisation (WHO) in the 1980s and its widespread provision free of charge since 1995, the global prevalence of leprosy has declined by over 95%. The WHO declared leprosy "eliminated as a public health problem" (defined as a prevalence of less than 1 case per 10,000 population) globally in 2000. India achieved the WHO's eradication criterion in 2005. India's National Leprosy Eradication Programme (NLEP) prioritises early case diagnosis, access to multidrug therapy (MDT), disability prevention, and community education.

Despite the remarkable progress, new cases of leprosy continue to be detected annually, indicating ongoing transmission (Fig. 30.3). In 2023, approximately 182,815 new cases were reported globally by 184 countries, areas, and territories. Most new cases are concentrated in the WHO South-East Asia Region. Brazil, in the Region of the Americas, and Indonesia, in the South-East Asia Region, also report significant numbers (over 10,000 new cases each in 2023). African and South-East Asia Regions, including India, Bangladesh, Democratic Republic of the Congo, Ethiopia, Madagascar, Mozambique, Myanmar, Nepal, Nigeria, Philippines, Somalia, Sri Lanka, and the United Republic of Tanzania, have also reported cases in 2023.

Diagnosis

Specimen

Sputum is the preferred specimen for NTM pulmonary disease, while aspirates, tissue biopsies, or drainage material may be collected for skin, soft tissue, or lymph node infections. Optimal decontamination protocols using N-acetyl-L-cysteine and sodium hydroxide are necessary for sputum samples to eliminate competing flora without compromising NTM viability. Decontamination of clinical specimens, especially sputum, is critical to prevent overgrowth by non-mycobacterial flora. N-acetyl-L-cysteine (NALC) combined with sodium hydroxide is commonly used. It effectively digests mucus and kills non-mycobacterial organisms without significantly harming mycobacteria.

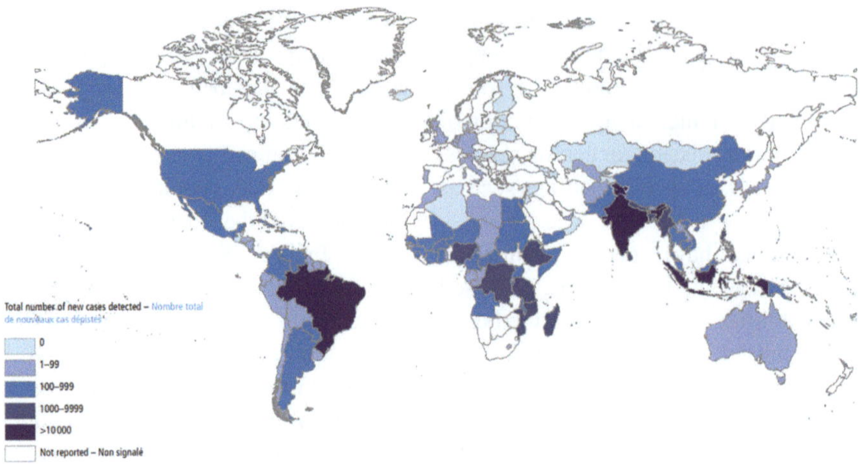

Fig. 30.3 Global epidemiological landscape of *Mycobacterium leprae* as of 2016. (Image source: WHO)

For leprosy, slit–skin smears are taken from multiple sites, particularly active skin lesions, ear lobes, and elbows. Nerve biopsy may be performed in cases of pure neuritic leprosy.

Microscopy

For NTMs, microscopy primarily serves as a rapid preliminary screening tool for the presence of mycobacteria in clinical specimens such as sputum, bronchoalveolar lavage, or tissue biopsies. The *Ziehl–Neelsen (ZN) stain and fluorescence microscopy (e.g. auramine-rhodamine stain)* are commonly employed. These methods offer the advantages of speed and cost-effectiveness, particularly in resource-limited settings. However, their major limitation is the inability to differentiate between pathogenic *M. tuberculosis* complex (MTC) and the diverse range of NTM species.

Microscopy for *Mycobacterium leprae* is a cornerstone in the diagnosis and classification of leprosy. The *Fite–Faraco (Fite) stain* is the preferred method for tissue sections (e.g. skin biopsies), utilising an oil-based deparaffinising solvent to preserve the bacilli's delicate acid-fastness. Modified Ziehl–Neelsen or fluorochrome stains are also used, particularly for slit–skin smears collected from standard body sites.

Beyond mere detection, microscopy in leprosy is vital for quantifying the bacterial load using the Ridley logarithmic scale. The *bacteriological index (BI)*, a logarithmic scale that helps classify the disease as paucibacillary (low BI) or multibacillary (high BI), directly informs treatment regimens. *The morphological index (MI)***,** which assesses the proportion of solid-staining (viable) bacilli, is also valuable for monitoring treatment response.

In Vitro Culture

Solid media, such as Lowenstein–Jensen and Middlebrook 7H10/7H11, and liquid culture systems like MGIT 960, are routinely used. Colony morphology, growth rate, and pigment production offer preliminary species identification, but definitive identification now relies largely on molecular methods.

M. leprae is not cultivable on artificial laboratory media.

Serodiagnosis

For NTM infections, serodiagnosis is not routinely used.

Serodiagnosis plays a more defined role in the diagnosis and monitoring of leprosy. *Phenolic*

Glycolipid-I (PGL-I) is the most extensively studied and utilised antigen for leprosy serodiagnosis. It is an *M. leprae*-specific antigen that elicits a strong IgM antibody response in MB patients. ELISA or rapid lateral flow assays based on PGL-I demonstrate high sensitivity for diagnosing MB leprosy. These tests are valuable for confirming diagnosis, disease classification, and potentially for active case finding in highly endemic areas. However, a significant limitation of PGL-I serology is its *low sensitivity in paucibacillary cases*, where the bacterial load and, consequently, the antibody response are minimal or absent. Therefore, a negative PGL-I test does not rule out PB leprosy.

Molecular Diagnosis

Nucleic acid amplification tests (NAATs), including PCR targeting the 16S rRNA gene, *hsp65, rpoB,* and the internal transcribed spacer (ITS) regions, facilitate rapid identification at the species level. Line probe assays (e.g. GenoType NTM-DR) allow simultaneous detection of species and common drug resistance mutations. Sequencing methods, particularly 16S rRNA sequencing, are considered the reference standard for species identification.

In the case of leprosy, diagnosis is primarily clinical, supported by laboratory findings. The cardinal signs include hypoesthetic skin patches, thickened peripheral nerves, and demonstration of acid-fast bacilli in slit–skin smears or biopsies. Advanced methods, including whole genome sequencing directly from clinical specimens, are being evaluated to better understand transmission dynamics and drug resistance in *M. leprae.*

An overview of laboratory diagnosis is presented in Table 30.1.

Treatment

Managing NTM infections is complex due to varying drug sensitivity, intrinsic resistance mechanisms, and the lengthy treatment requirements of the bacteria.

For lung infections caused by MAC, the typical regimen includes a macrolide (clarithromycin or azithromycin), ethambutol, and a rifamycin (rifampicin or rifabutin). Treatment generally continues for at least 12 months after sputum culture conversion. In cases of severe or fibrocavitary disease, an injectable aminoglycoside, such as amikacin, is given during the initial intensive phase. Macrolide resistance, often due to 23S rRNA mutation, significantly influences outcomes.

M. kansasii is frequently sensitive to isoniazid, rifampicin, and ethambutol, and treated for at least 12 months beyond culture conversion. In contrast, rapid growers like *M. abscessus, M. fortuitum*, and *M. chelonae* often acquire substantial drug resistance. Their treatment frequently involves a macrolide, combined with antibiotics such as amikacin, cefoxitin, imipenem, or tigecycline, selected based on in vitro susceptibility patterns. These regimens are limited by medication toxicity or intolerance.

Extrapulmonary NTM infections, including skin and soft tissue disease, lymphadenitis, and disseminated infections, are treated similarly with combination regimens, although duration may vary based on clinical response.

The WHO regimen for leprosy treatment is based on multidrug therapy (MDT). Multibacillary cases are treated by rifampicin, clofazimine, and dapsone for a period of 12 months. Paucibacillary leprosy is treated with rifampicin and dapsone for six months. MDT is extremely successful, and recurrence rates are low if therapy is completed. In endemic regions like India, treatment is provided free under national leprosy efforts.

Leprosy reactions necessitate further care. Type 1 reactions are treated with corticosteroids, generally prednisolone 40–60 mg/day with slow tapering. Type 2 reactions (ENL) may require corticosteroids, thalidomide (where permitted), or other immunomodulators. MDT is continued during responses unless contraindicated. Ongoing monitoring for adverse pharmaceutical responses, management of nerve dysfunction, and access to rehabilitation programmes are critical to preventing long-term disability and better patient outcomes.

Treatment regimes are detailed in Tables 30.2 and 30.3.

Table 30.1 Laboratory diagnosis of nontuberculous mycobacteria (NTM) and *Mycobacterium leprae*

Diagnostic method	Target organism	Sample type	Technique	Remarks
Microscopy	NTM	Sputum, BAL, tissue biopsy	Ziehl–Neelsen (ZN) stain; auramine--rhodamine fluorescence stain	Rapid and inexpensive, useful for preliminary screening; cannot distinguish NTM from MTC
	Mycobacterium leprae	Skin biopsy, slit–skin smear	Fite–Faraco stain; modified ZN stain; fluorochrome stains	Essential for diagnosis and classification; supports BI and MI determination; vital due to inability to culture *M. leprae*
In vitro culture	NTM	Sputum, BAL, tissue biopsy, etc.	Lowenstein–Jensen; Middlebrook 7H10/7H11; MGIT 960	Allows growth and preliminary species identification; molecular tools needed for confirmation
	M. leprae	–	Not applicable	*M. leprae* is not cultivable on artificial media; microscopy and molecular tools remain primary diagnostics
Serodiagnosis	NTM	Serum	Not routinely recommended	Limited diagnostic utility; not included in major guidelines (e.g. ATS/IDSA)
	M. leprae	Serum	PGL-I antigen-based assays (ELISA, lateral flow)	High sensitivity for MB leprosy; poor sensitivity for PB forms; helpful for classification and monitoring, not reliable for diagnosis of PB leprosy
Molecular diagnosis	NTM	Sputum, BAL, biopsy	PCR (16S rRNA, hsp65, rpoB, ITS), Line probe assay, sequencing	Rapid species level identification; detects resistance; 16S rRNA sequencing is the reference method
	M. leprae	Slit–skin smear, biopsy, tissue	PCR, whole genome sequencing (WGS)	Supports clinical diagnosis; PCR detects uncultivable bacilli; WGS increasingly used for resistance profiling and epidemiological tracking

Prevention and Control

Controlling infections caused by NTMs and *M. leprae* poses significant challenges due to their diverse transmission routes, reservoirs, and epidemiological characteristics.

For NTMs, prophylaxis primarily focuses on restricting environmental exposure, especially in high-risk categories such as those with structural lung disease or immunosuppression. Since NTMs are not typically transmitted person-to-person, conventional public health strategies, such as contact tracing, isolation, etc., are not applicable. In hospital contexts, strict infection control procedures are essential. This involves thorough cleaning of surgical equipment, maintenance of hospital water systems, and effective disinfection of tools such as bronchoscopes.

While no vaccinations are available for NTM infections, rapid diagnosis and species-specific treatment are critical for disease management.

The prevention and control of leprosy have reached substantial success via global public health programmes. The invention of MDT has been highly successful in reducing the prevalence, and measures such as mass treatment, active case detection, and disability prevention have been widely implemented. The BCG vaccine, primarily administered to prevent tuberculosis, provides partial protection against leprosy. Contemporary candidates, such as the *Mycobacterium* indicus pranii (MIP) vaccine,

Table 30.2 WHO treatment regimen for leprosy

Clinical category	Treatment regimen	Duration
Paucibacillary leprosy (PB) (≤5 skin lesions, negative slit–skin smear)	Rifampicin 600 mg once monthly (supervised) Dapsone 100 mg daily (self--administered)	6 months
Multibacillary leprosy (MB) (>5 skin lesions or positive slit–skin smear)	Rifampicin 600 mg once monthly (supervised) Clofazimine 300 mg once monthly (supervised) Clofazimine 50 mg daily (self--administered) Dapsone 100 mg daily (self--administered)	12 months

Table 30.3 Treatment of lepra reactions

Type of reaction	Treatment	Duration
Type 1 reaction (Reversal reaction)	Prednisolone 40–60 mg/day initially, then tapered gradually based on clinical response	3–6 months or longer depending on severity
Type 2 Reaction (Erythema nodosum leprosum, ENL)	Prednisolone (as above) Thalidomide (100–300 mg/day, where permitted) Clofazimine (in higher anti--inflammatory doses if needed) Supportive care (analgesics, antibiotics if ulcerated nodules present)	Varies depending on severity and recurrence; may require prolonged courses

have demonstrated potential in both therapeutic and preventive applications. Recent reports of zoonotic reservoirs—such as armadillos in the southern United States—highlight the importance of coordinated surveillance systems.

Case Study

A 65-year-old man with a 15-year history of chronic obstructive pulmonary disease (COPD) and a previous episode of pulmonary tuberculosis presented to the outpatient clinic with complaints of persistent productive cough, intermittent low-grade fever, fatigue, weight loss, and occasional haemoptysis for the past four months. Chest radiography revealed fibrocavitary lesions predominantly in the upper lobes with bilateral nodular infiltrates. Initial sputum smear examination was positive for acid-fast bacilli (AFB), raising concern for tuberculosis reactivation. However, the GeneXpert MTB/RIF assay was negative for *M. tuberculosis*. Given the persistent symptoms and radiographic findings, sputum samples were cultured using the MGIT 960 system, which yielded growth of *Mycobacterium avium* complex (MAC). Species identification was confirmed by 16S rRNA gene sequencing. Drug susceptibility testing showed sensitivity to clarithromycin, with no macrolide resistance detected. The patient was initiated on a combination regimen consisting of azithromycin, ethambutol, and rifampin. After 18 months of therapy, the patient showed significant clinical improvement, radiological resolution of cavities, and consecutive negative sputum cultures.

1. What underlying condition increased this patient's susceptibility to NTM infection?
2. Why was GeneXpert MTB/RIF helpful in excluding tuberculosis in this case?

3. Which diagnostic technique confirmed the species of NTM?
4. What is the standard recommended duration of therapy for *Mycobacterium avium* complex pulmonary disease?

Research Questions

1. What genetic and immunological factors predispose certain individuals to develop clinically significant pulmonary infections caused by nontuberculous mycobacteria?
2. How can rapid, affordable, and species-specific molecular diagnostic methods for nontuberculous mycobacteria be deployed in resource-limited settings?
3. What are the mechanisms underlying antimicrobial resistance in *Mycobacterium abscessus*?
4. In the context of zoonotic reservoirs of *Mycobacterium leprae*, what are the ecological, genetic, and epidemiological factors influencing cross-species transmission?

Declaration During the preparation of this chapter, we used Microsoft Word (with integrated Copilot) to improve the readability of the text. After its use, we reviewed and edited the content as needed and took full responsibility for the content of the publication.

Further Readings

Britton WJ, Lockwood DN. Leprosy. Lancet. 2004;363(9416):1209–19.

Cole ST, Eiglmeier K, Parkhill J, James KD, Thomson NR, Wheeler PR, et al. Massive gene decay in the leprosy bacillus. Nature. 2001;409(6823):1007–11.

Cowman S, van Ingen J, Griffith DE, Loebinger MR. Non-tuberculous mycobacterial pulmonary disease. Eur Respir J. 2019;54(1):1900250.

Daley CL, Iaccarino JM, Lange C, Cambau E, Wallace RJ Jr, Andrejak C, et al. Treatment of non-tuberculous mycobacterial pulmonary disease: an official ATS/ERS/ESCMID/IDSA clinical practice guideline. Clin Infect Dis. 2020;71(4):e1–36.

Falkinham JO 3rd. Environmental sources of non-tuberculous mycobacteria. Clin Chest Med. 2015;36(1):35–41.

Fedrizzi T, Meehan CJ, Grottola A, Giacobazzi E, Fregni Serpini G, Tagliazucchi S, et al. Genomic characterisation of Non-tuberculous Mycobacteria. Sci Rep. 2017;7:45258.

Griffith DE, Aksamit TR, Brown-Elliott BA, Catanzaro A, Daley CL, Gordin FM, et al. An official ATS/IDSA statement: diagnosis, treatment, and prevention of non-tuberculous mycobacterial diseases. Am J Respir Crit Care Med. 2007;175(4):367–416.

Haworth CS, Banks J, Capstick T, Fisher AJ, Gorsuch T, Laurenson IF, et al. British Thoracic Society guidelines for the management of non-tuberculous mycobacterial pulmonary disease (NTM-PD). Thorax. 2017;72(Suppl 2):ii1–ii64.

Kumar B, Kar HK, Dogra S, editors. IAL textbook of leprosy. 3rd ed. Jaypee Brothers: New Delhi/London; 2023.

Monot M, Honoré N, Garnier T, et al. On the origin of leprosy. Science. 2005;308(5724):1040–2.

National Leprosy Eradication Programme (India). National Strategic Plan and Roadmap for Leprosy 2023–2027. New Delhi: Central Leprosy Division, DGHS, Ministry of Health & Family Welfare; 2023.

Ploemacher T, Faber WR, Menke H, Rutten V, Pieters T. Reservoirs and transmission routes of leprosy: a systematic review. PLoS Negl Trop Dis. 2020;14(4):e0008276. https://doi.org/10.1371/journal.pntd.0008276.

Prevots DR, Marras TK. Epidemiology of human pulmonary infection with non-tuberculous mycobacteria: a review. Clin Chest Med. 2015;36(1):13–34.

Rodrigues LC, Lockwood DN. Leprosy now: epidemiology, progress, challenges, and research gaps. Lancet Infect Dis. 2011;11(6):464–70.

Scollard DM, Adams LB, Gillis TP, Krahenbuhl JL, Truman RW, Williams DL. The continuing challenges of leprosy. Clin Microbiol Rev. 2006;19(2):338–81.

Sharma SK, Upadhyay V. Epidemiology, diagnosis & treatment of non-tuberculous mycobacterial diseases. Indian J Med Res. 2020;152(3):185–226.

Stinear TP, Seemann T, Pidot S, et al. Reductive evolution, and niche adaptation inferred from the genome of Mycobacterium ulcerans, the causative agent of Buruli ulcer. Genome Res. 2007;17(2):192–200.

Tortoli E. Microbiological features and clinical relevance of new species of the genus Mycobacterium. Clin Microbiol Rev. 2014;27(4):727–52.

van Ingen J, Boeree MJ, van Soolingen D, Mouton JW. Resistance mechanisms and drug susceptibility testing of non-tuberculous mycobacteria. Drug Resist Updat. 2012;15(3):149–61.

World Health Organization. Treatment of Mycobacterium ulcerans disease (Buruli ulcer): guidance for health workers. Geneva: WHO; 2012.

World Health Organization. Global Leprosy (Hansen's disease) Strategy 2021–2030: towards zero leprosy. Geneva: WHO; 2021.

Leptospirosis

31

Diptanu Majumder and Nonika Rajkumari

Learning Objectives

1. To understand the nature of the pathogen belonging to the genus *Leptospira*
2. To learn about the modes of transmission and risk factors in the epidemiology of leptospirosis
3. To understand the disease spectrum and laboratory diagnosis of the disease
4. To describe the prevention and control modalities for the use of this pathogen

Introduction

Leptospirosis is a zoonotic disease caused by pathogenic spirochetes of the genus *Leptospira*, which are characterised by a broad spectrum of clinical manifestations. Before its formal identification, the disease was referred to by various regional names: in China as "rice field jaundice", in Japan as "autumn fever" or "seven-day fever", and occupationally as "cane-cutter's disease", "swine-herd's disease", and "Schlammfieber" (mud fever). It has a worldwide distribution but is more common in the tropics, where conditions for its transmission are particularly favourable. In recent years, leptospirosis has increasingly been recognised among adventure travellers involved in water-based activities, as well as among holiday-makers in developed nations and sporadically among inner-city populations.

Transmission primarily occurs through abraded skin or mucous membranes when they meets the environment contaminated with the urine of infected animals. Consequently, leptospirosis constitutes a significant occupational hazard for individuals engaged in animal handling, including livestock farmers, abattoir workers, veterinarians, hunters, game managers, animal control personnel, and researchers. Although not classified as a biological weapon, the clinical presentation of leptospirosis may resemble that of viral haemorrhagic fevers, necessitating its consideration in differential diagnoses in bioterrorism scenarios.

History

Leptospira was first mentioned in 1812 by D.J. Larrey among Napoleon's troops at Cairo. However, German physician Adolph Weil in 1886 described it as a new disease characterised by splenomegaly, jaundice, and nephritis of kidney tissues. Subsequently, the infection was named after his discovery as "Weil's disease". Additional synonyms for the condition included Weil–Vasiliev disease, Fort Bragg fever, waterborne fever, nanukayami fever, swamp fever, Mgunda fever, Stuttgart disease, and canicola

D. Majumder · N. Rajkumari (✉)
Department of Microbiology, All India Institute of Medical Sciences, Kalyani, Nadia, West Bengal, India

S. C. Parija, A. Poddar (eds.), *Textbook of Bacterial Zoonoses*, Microbial Zoonoses,
https://doi.org/10.1007/978-981-95-6615-0_31

fever. Arthur Stimson (1907) used Levaditi silver deposition and differentiated it as leptospires from the kidney tissue of a person who died of yellow fever, indicating that their shape resembled a question mark. The Weil's disease outbreak among coal miners in Japan was reported between 1914 and 1915. During that time, simultaneous efforts by Noguchi and Inada in Japan, and Hübener, Uhlenhuth, and Fromme in Germany, successfully transmitted leptospirosis to guinea pigs and isolated the pathogen at Kyushu University Medical School in Fukuoka.

In 1918, Noguchi proposed the reclassification of *Spirochaeta interrogans* to *Leptospira interrogans*. This led to the creation of the genus *Leptospira* (from Ancient Greek *leptos-* "fine, thin, narrow, etc." and Latin *spira* "coil") that differentiated the Weil's disease spirochete from others known at the time. Subsequently, in 1940, *Leptospira* species were identified as the aetiological agents of cattle yellow fever. By the 1980s, leptospirosis was well documented as a significant veterinary disease affecting dogs, cattle, swine, horses, and potentially sheep. Recent developments in genome sequencing have helped to understand the organism and infection at the molecular level.

Taxonomy

Leptospira is a genetically and serologically highly heterogeneous genus. It is currently classified within the family Leptospiraceae, order Leptospirales, class *Spirochaetia,* phylum Spirochaetota, subkingdom Negibacteria, and kingdom Bacteria, according to the International Code of Nomenclature of Prokaryotes (ICNP).

Initially, antigenically distinct strains were assigned the status of species like *Leptospira icterohaemorrhagiae*, *Leptospira hebdomadis*, *Leptospira canicola* and *Leptospira. biflexa*. However, Wolf and Broom (1954) proposed that serologically distinct strains should not be allocated to different species but rather as serotypes. These serotypes could be clustered further into serogroups. These numerous serotypes were defined by agglutination after cross-absorption with homologous antigen—the cross-agglutinin absorption test (CAAT). Subcommittee on the taxonomy of *Leptospira* in the "Technical Screening Criteria (TSC) meeting" in Montreal in 1962 recommended the division of the genus *Leptospira* into two species by phenotypic characterisation. These species were *Leptospira interrogans*, representing the pathogenic strains, and *Leptospira biflexa*, the saprophytic strains. In 1973, the term "serotype" was replaced by the designation "serovar" (serologic variety). The subcommittee on the "TSC meeting" of 1986 approved the serovar list of *Leptospira*. As per the list, pathogenic *L Leptospira interrogans* sensu lato complex consisted of 202 serovars distributed across 23 serogroups. The *Leptospira biflexa* sensu lato complex consisted of 63 serovars contained in 38 serogroups.

This classification remains under continuous revision, with more than 300 serovars identified to date. Key milestones in the taxonomic development of *Leptospira* are illustrated in Fig. 31.1.

Genomics and Proteomics

Genomics

Classical serotyping, based on the similarity of the surface-exposed epitope, is still useful for identifying reservoirs and generating a control strategy. However, serotyping provides limited insight into the genetic diversity of *Leptospira* species. In 1987, the International Committee for Systematic Bacteriology recommended that complete deoxyribonucleic acid (DNA) sequences should serve as the reference standard for determining phylogeny, and that taxonomy and nomenclature should align with genomic information.

The introduction of rapid analysis of the 16S rRNA sequence subsequently redefined prokaryotic taxonomy, including *Leptospira.* The 16S rRNA-based framework of *Leptospira* strains revealed three clades based on pathogenicity status: pathogenic, saprophytic, and intermedia test strains. The intermedia group contains strains of uncertain pathogenicity that have been recovered

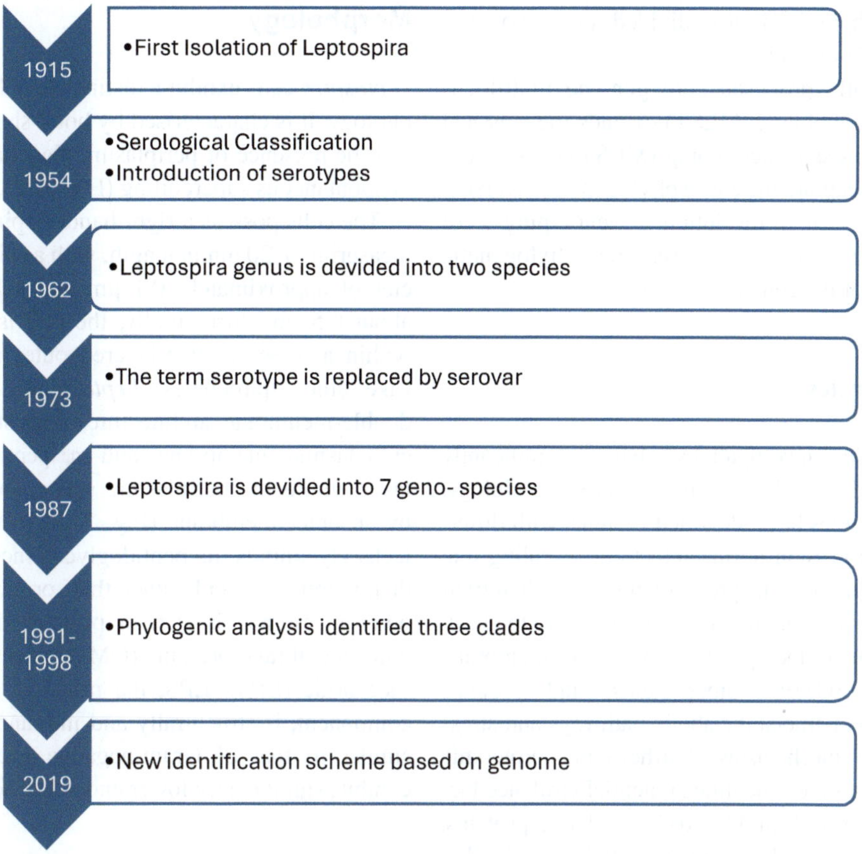

Fig. 31.1 Key events in *Leptospira* taxonomical evolution. (Image source: Authors' personal collection)

from patients but do not cause death. They induce clinical symptoms or establish renal infections in animal models.

Whole genome sequencing (WGS) comparisons soon highlighted the limitations of 16S rRNA sequence analysis and its limited taxonomic resolution. This is because it represents only 0.05% of an average 3 Mbp prokaryotic genome and is often highly conserved. The availability of complete bacterial genome sequences has increased due to next-generation sequencing (NGS) technologies. Comparative tools like average nucleotide identity (ANI) and average amino acid identity (AAI) have significantly improved taxonomic resolution. Comparative ANI analyses between novel isolates and reference genomes identified 45 new *Leptospira* species. These species include 24 pathogenic and 21 saprophytic, bringing the total recognised *Leptospira* species to 68.

Leptospira genomes are 3.73–4.99 Mb in size and are located on two circular chromosomes. Between 2003 and 2008, the first sequencing of the whole genome of *Leptospira* strains was published for three of the most common species. These are *L. interrogans, L. borgpetersenii*, and the saprophyte *L. biflexa*. The recent increase in environmental isolations of *Leptospira* has further enriched the understanding of the genus's diversity.

Phylogenetic analysis of described species has identified two major clades. "Pathogens" contain the species responsible for animal and human infections, and "Saprophytes" contain species isolated from the environment. The clade "Pathogens" is subdivided into two subclades: P1—containing species formerly referred to as pathogens, and P2—formerly known as an intermediate group. Similarly, the second clade,

"Saprophytes", is also divided into two subclades, S1 and S2.

Recently proposed core genome multilocus sequence typing (cgMLST) extends the classical multilocus sequence typing (MLST) to 545 core genes to capture the entire phylogenetic diversity, allowing high-resolution genotyping of *Leptospira* isolates across the phylogenetic breadth of the genus.

Proteomics

Proteomic investigations have significantly advanced the understanding of *Leptospira* biology. These studies have identified proteins with differential expression during infection, revealing that *Leptospira* alters its protein landscape when transitioning from an in vitro environment to the host. This includes the upregulation of outer membrane proteins (OMPs), lipoproteins, and proteins involved in crucial metabolic pathways and stress response mechanisms. Furthermore, proteomic approaches have identified potential virulence factors, such as LipL32, Loa22, and Lig proteins. These proteins play major roles in host cell adhesion, invasion, and host immune system evasion.

Resources like LeptoDB, an integrated database, support these efforts by providing comprehensive genomic and proteomic information for various *Leptospira* species.

Morphology

Leptospira cells exhibit a distinctive helical morphology. It is characterised by hook-shaped ends and the presence of periplasmic flagella capable of spontaneous supercoiling (Fig. 31.2).

The cells possess a right-handed spiral shape, measuring 6–20 μm in length, with a spiral diameter of approximately 0.1 μm and a length of about 0.5 μm. Structurally, the cell is encased within a three- to five-layered outer envelope. Like other spirochetes, *Leptospira* exhibits a double-membrane architecture, consisting of a cytoplasmic membrane and a peptidoglycan layer within the periplasmic space, surrounded by an outer membrane (Fig. 31.3). Although it lacks glycolipids, its peptidoglycan incorporates diaminopimelic acid rather than ornithine. The outer membrane is rich in proteins, including outer membrane proteins (OMPs) and lipopolysaccharide (LPS). LPS, the principal antigenic component, is structurally and immunologically similar to that of Gram-negative bacteria but exhibits significantly lower endotoxic activity.

Motility

Leptospira shows dynamic motility, transitioning between spiral and hook-like shapes. Motility is mediated by two periplasmic flagella, each origi-

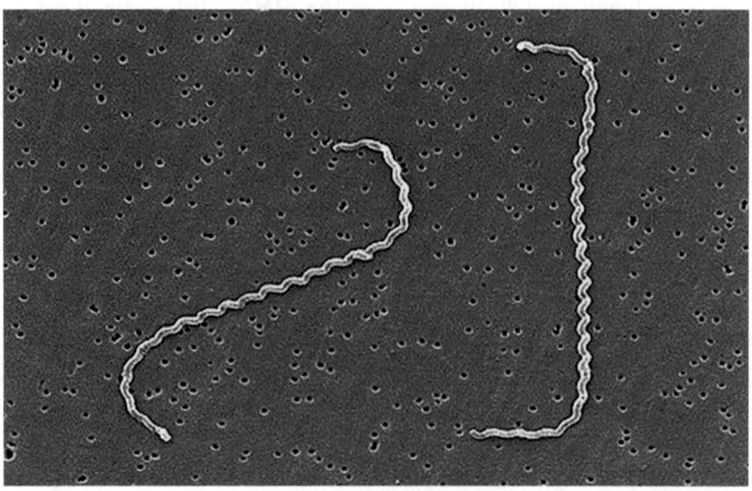

Fig. 31.2 Scanning electron microscope (SEM) showing spiral-shaped *Leptospira interrogans*. (Image source: ID#1220 Public Health Image Library/ Centers for Disease Control and Prevention), USA (with permission))

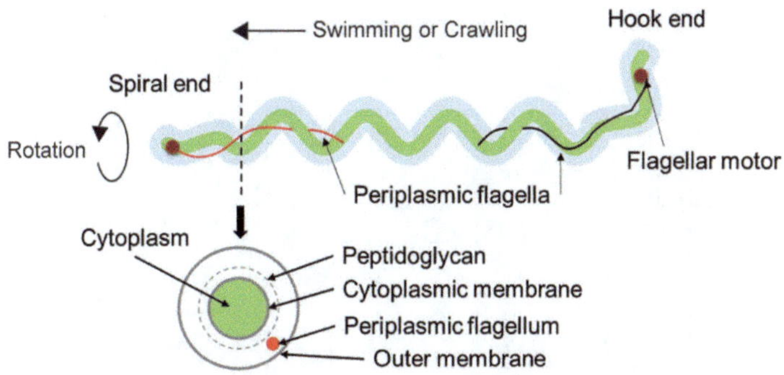

Fig. 31.3 Schematic diagram of the *Leptospira* cell structure. (Image source: Xu J, Koizumi N, Nakamura S. Crawling. Motility on the host tissue surfaces is associated with the pathogenicity of the zoonotic spirochete *Leptospira*. Front Microbiol. 2020;11:1886 (with permission))

nating at opposite poles and extending toward the centre without overlapping. This feature distinguishes *Leptospira* from other spirochetes. The flagellar structure comprises flagellin A (FlaA) and flagellin B (FlaB) proteins, which form the sheath and core, respectively. Electron microscopy of FlaB mutants has revealed the absence of endoflagella and the loss of motility. Importantly, *Leptospira* pathogenicity is closely linked to its motility, and reduced motility correlates with attenuated virulence.

Cultivation

Leptospira are obligate aerobes that grow readily at a temperature of 28–30 °C. They do not resist drought or hypertonicity but support alkalinisation to pH 7.8. They grow in simple media enriched with vitamins B1 and B12, long-chain fatty acids, and ammonium salts. Long-chain fatty acids are utilised as the sole carbon source and are metabolised by β-oxidation.

A biosafety level 2 laboratory is always recommended while handling *Leptospira* isolates. When cultivated at 30 °C in a pH 7.4 medium, their average generation time is approximately 12 hours. The growth is often slow in primary isolation, particularly in late-phase infections, and cultures are kept for about 13 weeks before being discarded. Aeration is required for maximal growth. In semisolid media (0.1–0.2% agar), growth reaches a maximum density in a discrete zone beneath the surface of the medium. On further incubation, the growth becomes increasingly turbid, which is known as a Dinger's ring or Dinger's disc (Fig. 31.4). The cultures are typically maintained through repeated subculture or by storage in haemoglobin-containing semisolid agar. Long-term storage in liquid nitrogen is preferred to maintain virulence.

Animal Models

Various animal models, such as mice, guinea pigs, gerbils, and hamsters, have been employed to understand the pathophysiology and virulence of *Leptospira* species. The choice of model is based on the study's goal, the serovar involved, and host-pathogen dynamics.

The mouse model (*Mus musculus*) is very useful in providing insights, particularly into both sublethal and chronic leptospirosis, especially in elucidating the host immune mechanisms that control pathogenic *Leptospira* species. They are genetically modifiable and are very useful in molecular pathogenesis and immunological studies, showing subclinical, acute, or chronic infections. Moreover, the murine models are appropriate for testing new vaccines and drugs to treat leptospirosis. They are used to establish infection by pathogenic *Leptospira* spp. in reservoir hosts without causing disease. This is in contrast to the immune response developed by humans (the "accidental" host) who succumb to the pathogenic effect of the agent.

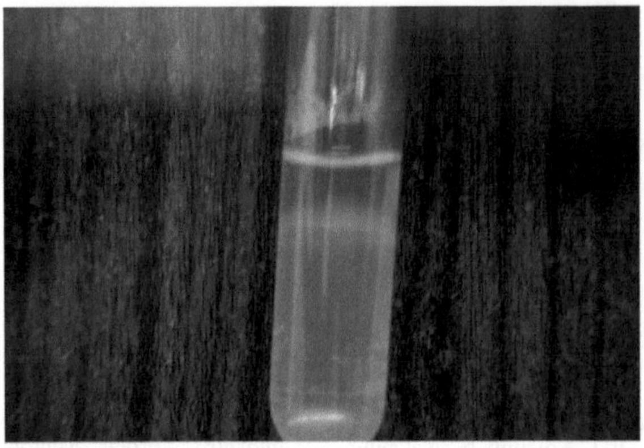

Fig. 31.4 Dinger's ring (the opaque ring just below the surface of the semisolid Ellinghausen-McCullough-Johnson-Harris (EMJH) media in the tube) seen on the ninth day of culture. (Image source: Isa SE, Onyedibe KI, Okolo MO, Abiba AE, Mafuka JS, Simji GS, Nathan SY, Udoh UA, Awang SK, Egah DZ, Banwat EB. A 21-year-old student with fever and profound Jaundice. PLoS Negl Trop Dis. 2014;8(1):e2534 (with permission))

The Golden Syrian hamster (*Mesocricetus auratus*) is highly susceptible to pathogenic *Leptospira*. The experimental infection mimics severe, acute, and often fatal leptospirosis in humans and is helpful for pathogenic studies.

Guinea pigs (*Cavia porcellus*) develop systemic infection with fever and organ lesions. They are helpful for vaccine trials, as well as histological and serological studies.

Rabbits are useful for chronic infection and carrier state modelling, as they develop renal colonisation with minimal disease. This is helpful in leptospiruria studies (urine shedding), as well as serological monitoring and long-term colonisation and transmission studies.

Zebrafish, as a model of leptospirosis, an alternative to mammal use, show promise for studies in innate immunity since the NF-κB and interferon signalling pathways in zebrafish are conserved through evolution.

Transmission Cycle

The bacteria are maintained in the environment and animal reservoirs, whereas humans are accidental hosts (Table 31.1). The transmission of leptospirosis is intricately maintained through a complex interplay of the reservoir–environment–host triad (Fig. 31.5).

Table 31.1 Different types of hosts of *Leptospira* spp.

Host type	Description	Examples
Reservoir hosts	Naturally infected animals that carry and shed *Leptospira* without showing severe illness	Rodents, especially, rats, pigs, cattle, dogs, wild animals
Accidental hosts	Becomes infected through exposure but does not play a significant role in the ongoing transmission; may develop a severe disease	Humans, dogs when infected with non-host-adapted serovars, horses, sheep
Amplifying hosts	Some animals can transiently shed large amounts of *Leptospira* after acute infection	Livestock, also some wildlife during outbreaks

Motile, spiral-shaped, live *Leptospira* organisms are shed in the urine of infected animals, especially reservoir hosts. They are present in contaminated water, soil, or vegetation and occasionally via foetal membranes, placenta, or aborted material in livestock. They survive in moist, alkaline environments for weeks to months.

Animal-to-human (zoonotic) transmission is the primary way humans contract leptospirosis. Carrier animals shed the bacteria in their urine, often without showing signs of illness, leading to

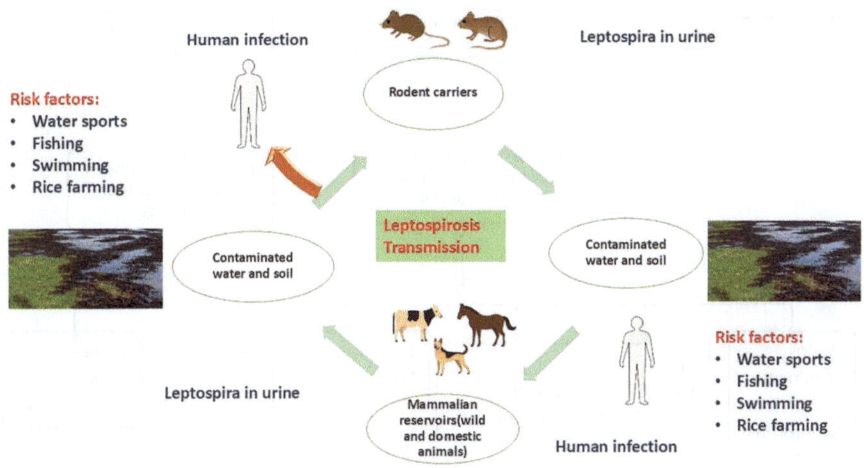

Fig. 31.5 Transmission cycle of leptospirosis. (Image source: Authors' personal collection)

environmental contamination. Humans typically are infected through direct or indirect contact with the urine of these infected animals or with environments (water, soil) contaminated with their urine. The bacteria enter the human body through breaks in the skin (cuts, abrasions), mucous membranes (eyes, nose, mouth), or even through prolonged immersion in contaminated water that softens the skin.

Pathogenesis and Pathology

After entering the body, the bacteria reach the bloodstream rapidly, leading to leptospirosis. They resist phagocytosis by macrophages and neutrophils until specific antibodies appear in the blood. They may survive inside macrophages by escaping apoptosis. The organism disseminates to multiple organs such as the liver, kidneys, lungs, and central nervous system (CNS) (Fig. 31.6). The primary damage occurs in the cell membrane, mediated by *leptospiral* proteins and/or toxic cellular components. This may result in severe consequences for the host, such as loss of vascular integrity, ischaemia, and necrosis, leading to significant organ dysfunction. *Leptospira* OMP inhibits Na^+-K^+ ATPase and Na^+-K^+-Cl^- cotransporter, leading to hypokalaemia and hyponatraemia. A host immune response following this marks the onset of the second phase of symptoms.

The most severe form of leptospirosis (Weil's disease) is characterised by haemorrhage and hepatic and renal dysfunction. Acute renal injury manifests as tubule-interstitial nephritis that may progress to fibrosis in chronic stages. Pulmonary damage occurs through the direct action of toxins and an autoimmune process. Vascular injury is characterised by segmental lesions on the coronary artery's walls and their branches. Autoimmune reactions and re-infection trigger uveitis and Guillain-Barré syndrome.

Immunology

The innate immune system plays a crucial role in the early recognition and elimination of leptospires. The acquired immune response involves the production of specific antibodies against LPS and the activation of the classical complement pathway.

Following invasion, leptospires are recognised by pathogen recognition receptors (PRRs), which detect the molecular motifs known as pathogen-associated molecular patterns (PAMPs). PRRs are classified into three categories such as endocytic, signalling, and secreted receptors. Among these, two main families of signalling PRRs—

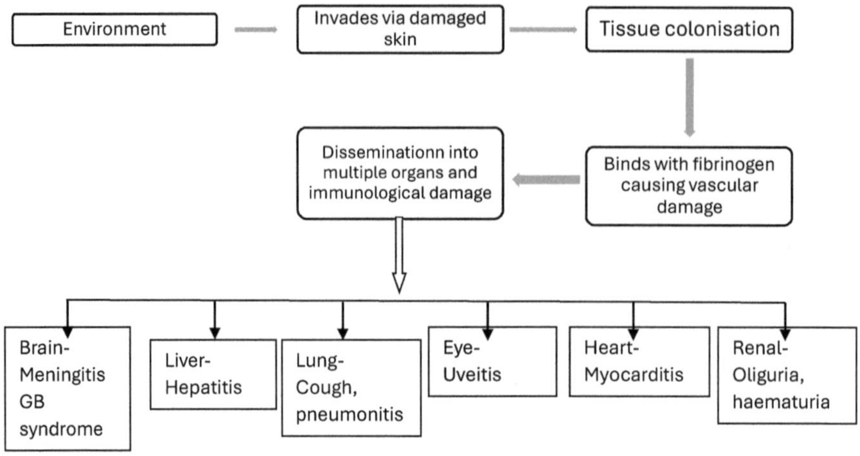

Fig. 31.6 Pathogenesis of human leptospirosis. (Image source: Authors' personal collection)

toll-like receptors (TLRs) and Nod-like receptors (NLRs) play a pivotal role in pathogens' recognition. *Leptospira* LPS activates human macrophages via TLR2, in contrast to TLR4 in Gram-negative organisms. Being the largest population among the intravascular phagocytes, polymorphonuclear neutrophils (PMNs) play an important role in the bacteria. Hence, passive immunisation with monoclonal or polyclonal anti-LPS antibodies may confer protection against the disease.

Infection in Humans

Symptoms often manifest abruptly, following an incubation period of 2–20 days.

Leptospirosis typically presents with non-specific symptoms such as fever, myalgia, headache, malaise, abdominal pain, vomiting, diarrhoea, or constipation, photophobia, and arthralgia, cough, affecting both paediatric and adult populations. The majority of patients (80–90%) present with mild or subclinical disease. The remaining 10–20% present with more severe illness, characterised by jaundice, acute renal failure, bleeding (Weil's disease), or pulmonary haemorrhage syndrome.

Typically, the clinical course is biphasic.

Septicaemic Phase

It generally lasts 4–7 days, during which *Leptospira* can be detected in the patient's blood and cerebrospinal fluid. The non-specific symptoms during this phase include fever, headache, chills, muscle aches, abdominal pain, and conjunctival suffusion. It is followed by a 1–3-day period of defervescence, in which the fever and other symptoms temporarily subside.

Immune Phase

Following the acute septicaemic phase, anti-*Leptospira* antibodies appear in the immune phase, characterised by inflammatory infiltration of affected organ systems. Aseptic meningitis and renal dysfunction are hallmarks of the immune phase. Fever and earlier symptoms recur. This may persist for 6 days to more than 4 weeks, with a mean duration of 14 days, and leptospires start shedding in the urine.

Severe Leptospirosis

Approximately 10–15% of patients develop severe forms of leptospirosis, which sometimes

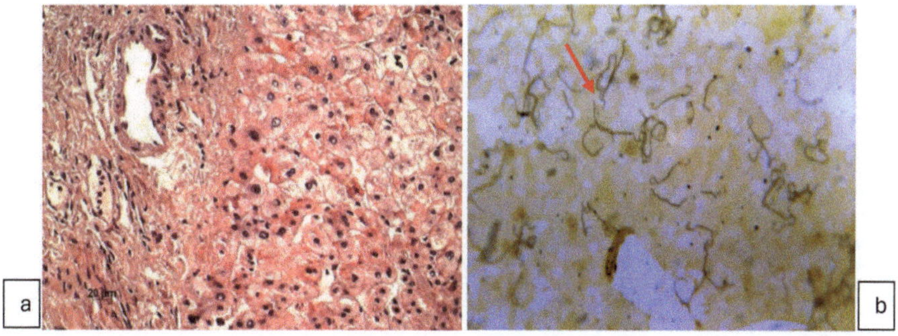

Fig. 31.7 (**a**) Liver biopsy in human leptospirosis. Hepatic cells show slight variation in size and occasional, more marked hyaline cytoplasmatic change. Portal space is unremarkable. (Image source De Brito T, Silva AM, Abreu PA. Pathology and pathogenesis of human leptospirosis: a commented review. Rev Inst Med Trop Sao Paulo. 2018;60:e23 (with permission)). (**b**) Liver tissue, showing the presence of *Leptospira*, with the silver impregnation method. (Image source: ID#22374 Public Health Image Library/CDC, USA (with permission))

progress rapidly to death. The clinical manifestation may closely resemble septic shock, presenting with hypotension, multi-organ dysfunction, and a high risk of mortality. The severe manifestations include jaundice that is primarily attributed to hepatocellular dysfunction rather than fulminant hepatic necrosis (Fig. 31.7). Acute renal failure occurs in 16–40% of cases. Pulmonary haemorrhagic syndrome is often associated with fatal outcomes in severe leptospirosis. Importantly, severe and fatal cases of leptospirosis can also occur in the absence of jaundice.

Weil's Disease Approximately 10% of individuals diagnosed with leptospirosis progress to a classical severe form of the infection, known as Weil's disease. Classically, it is characterised by the triad of jaundice, acute renal failure, and pulmonary haemorrhage. The condition is associated with a mortality rate of more than 10%, even with intensive care management.

Infection in Animals

Leptospirosis affects animals, including pets (dogs, cats), farm animals (cattle, pigs, sheep, goats), rodents, and various wild animals (feral swine and marine mammals) (Fig. 31.8). They acquire the infection directly through contact with urine or body fluids of infected animals or indirectly by exposure to contaminated water or soil. The signs and symptoms vary widely depending on the species and the severity of the infection. Some animals may also remain asymptomatic.

In cattle, leptospirosis often manifests as reproductive disorders, including abortion, stillbirth, and infertility. There may be a sudden drop in milk production accompanied by fever and sometimes bloody or discoloured milk, known as "milk drop syndrome". In pigs, reproductive losses are common, with frequent abortions, mummified foetuses, and weak growth of piglets with signs of jaundice and haematuria in more severe cases. Dogs are more susceptible to leptospirosis than cats, and the disease can be acute and very severe.

Leptospirosis was first documented in marine mammals in 1970, with initial cases reported in California sea lions (*Zalophus californianus*) along the coasts of California, Oregon, and Washington. Since then, seasonal outbreaks have been observed in sea lion populations, with more severe epizootics occurring at intervals of approximately 4–5 years. In addition to sea lions, other marine mammals like the northern fur seal (*Callorhinus ursinus*), northern elephant seal (*Mirounga angustirostris*), and Pacific harbour seal (*Phoca vitulina richardsi*) are also infected by *Leptospira* spp.

Fig. 31.8 Leptospirosis symptoms in animals. (Illustration inspired by publicly available educational material from Centers for Disease Control and Prevention [www.cdc.gov], redesigned and redrawn by the authors)

Epidemiology and Public Health

Animal-to-animal transmission of *Leptospira* occurs through several routes. Direct contact with bodily fluids from infected animals, including urine, placental fluids, and milk, is an important mode of transmission. This can happen through licking, sniffing, or other forms of physical interaction, especially in environments where animals are in proximity. In some animal species, transplacental transmission, where the bacteria pass from an infected mother to her offspring during pregnancy, and venereal transmission, occurring during mating, have also been documented as routes of infection. Additionally, direct transmission can result from bite wounds inflicted by an infected animal or through ingesting infected tissues, such as during predation or scavenging.

A crucial factor in the animal-to-animal transmission of *Leptospira* is the role of reservoir hosts. These animals can become chronically infected and shed leptospires in their urine for extended periods, ranging from months to even years. This long-term, often asymptomatic shedding contaminates the environment and serves as a continuous source of infection for other susceptible animals. Direct contact can occur through handling infected animals or their tissues. Indirect contact is more common and happens through exposure to contaminated water sources such as rivers, lakes, and floodwaters, or through contact with contaminated soil, especially after heavy rainfall or flooding events that can spread the bacteria widely.

Leptospirosis in humans is a re-emerging infectious disease and is a classic example of a One Health issue with substantial public health and economic implications. It is suggested to be the most widespread zoonosis in the world, with an estimated more than a million human cases worldwide, including almost 60,000 deaths annually (Fig. 31.9). It is an under-reported infection with no reliable global incidence figures due to non-specific symptoms and limited diagnostic facilities in endemic regions.

It is mostly prevalent in tropical regions, including South and Southeast Asia, Oceania, the Caribbean, sub-Saharan Africa, and Latin America. It poses a significant public health challenge, especially in developing countries and disaster-prone regions, affecting low-income populations and exacerbating poverty and inequality. Outbreaks are common after natural disasters (e.g., hurricanes, floods) due to increased human contact with contaminated water. Urban outbreaks are common in slums

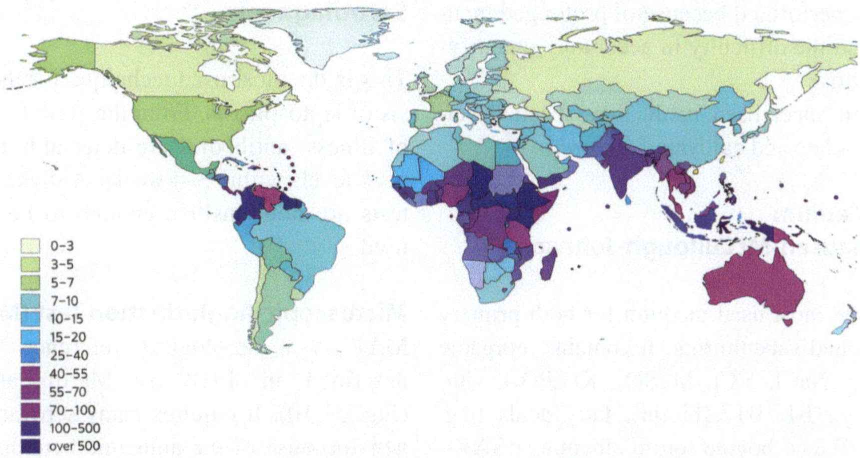

Fig. 31.9 Global burden of leptospirosis estimated in terms of disability adjusted life years (DALY). (Image source: Adapted from Torgerson PR, Hagan JE, Costa F, Calcagno J, Kane M, Martinez-Silveira MS, Goris MG, Stein C, Ko AI, Abela-Ridder B. Global burden of leptospirosis: estimated in terms of disability adjusted life years. PLoS Negl Trop Dis. 2015;9(10):e0004122 (with permission). Visual styling and formatting were modified for clarity and consistency)

with poor sanitation, high rodent populations, and inadequate drainage. In developed countries, it is sporadic and is often related to occupational exposure, recreational water activities, or imported cases. Farmers, veterinarians, sewer workers, and those involved in outdoor recreational activities such as swimming or kayaking in freshwater carry a higher risk of exposure.

Despite being preventable, leptospirosis continues to be the cause of significant morbidity and mortality in vulnerable populations.

Diagnosis

Specimens

Blood or cerebrospinal fluid (CSF) is collected for culture in the early stage (first 7–10 days) of the disease, and urine (>10 days) is collected in the second stage of the disease.

Microscopy

Direct microscopic examination is very useful for detecting the organism in body fluids and assessing culture growth. By dark-field or phase-contrast microscopy, the organisms appear as thin, bright, actively motile structures exhibiting a characteristic, rapid spinning and jerking motility. The sensitivity of dark-field microscopy is approximately 10^7 leptospires/L, but sensitivity may decline from 100% to 90.9% as the duration of infection increases.

Histopathological stains like Warthin–Starry and silver stains have been used to detect leptospires in clinical specimens. Immunostaining is employed in addition to dark-field microscopy to enhance diagnostic sensitivity. Pathogenic and saprophytic *Leptospira* species are morphologically indistinguishable using this technique.

In Vitro Culture

Blood, CSF, and urine are collected for culture before administering antibiotics.

The sample is inoculated aseptically in tubes or special culture bottles and incubated at 28–30 °C in the dark. As *Leptospira* is slow-growing, cultures are kept for up to 13 weeks and are examined weekly under dark-field microscopy for motile spirochetes. There may not be any growth if the patient has already taken antibiotics. Susceptibility testing is not routinely performed because of prolonged incubation and the difficulty in accurately quantifying growth.

Several specialised media have been developed to isolate and cultivate *Leptospira*.

EMJH Medium (Ellinghausen-McCullough-Johnson-Harris)

This is the most used medium for both primary isolation and subculturing. It contains inorganic salts (e.g. NaCl, KCl, $MgSO_4$, KH_2PO_4), vitamins (e.g., B1, B12, biotin), fatty acids (e.g. Tween 80), and bovine serum albumin (BSA)—essential for binding fatty acids and detoxifying the medium. Once growth is visible, fresh EMJH media is used for subculture to avoid nutrient depletion. To obtain pure growth, multiple passages may be needed, along with a final passage in an antibiotic-free medium to ensure no residual effects.

Modified EMJH Medium supplemented with 5-fluorouracil or nalidixic acid is used to prevent contamination. It contains rabbit serum or foetal bovine serum to enhance primary isolation. Supplementation of charcoal may absorb toxic substances from serum or additives.

Korthof Medium It is enriched with rabbit serum and peptone water, which have been historically used for isolation. It is now largely replaced by EMJH due to standardisation issues. It is still used in animal inoculation studies.

Fletcher's Medium It is a semisolid medium (0.1–0.2% agar) that is often used for primary isolation. It allows observation of the diffuse zone of growth under dark-field microscopy, but is nutritionally less rich than EMJH.

Other Media

These include Hornsby-Alt-Nally (HAN) medium for the faster growth of *Leptospira interrogans* and Tchan's medium for *Leptospira biflexa* and other saprophytic strains.

Serodiagnosis

This is the most used technique for the diagnosis of leptospirosis. From the sixth to tenth day of illness, antibodies are detectable and reach peak levels within 3–4 weeks. Antigen detection tests are not sensitive enough to be routinely used.

Microscopic Agglutination Test (MAT)

MAT is a serological reference test first described in 1918 by Martin and Pettit (Fig. 31.10). It requires many serovars as antigens because of the antigenic heterogeneity of *Leptospira* spp. It has a sensitivity of 41% during the 1st week, 82% during the 2nd to 4th week, and 96% beyond the 4th week of illness. A positive result is defined as a fourfold rise in titre between acute and convalescent specimens. A single titre exceeding 1:200 or serial titres exceeding 1:100 suggests leptospirosis, but the findings are not definitive. Additionally, MAT cannot distinguish between active, recent, or past infections, as it detects both IgM and IgG antibodies.

Enzyme-Linked Immunosorbent Assay (ELISA)

ELISA detects IgM antibodies reacting with a broadly reactive genus-specific antigen, but it is not suitable for detecting the causative serovar or serogroup. The diagnostic cut-off point is determined by the same considerations as for the MAT. Most commercial ELISA kits use the non-pathogenic *Leptospira biflexa Patoc* strain as an antigen.

Additional tests like indirect haemagglutination assay (IHA), latex agglutination (LA), complement fixation test, sensitised erythrocyte lysis, macroscopic slide agglutination, microcapsule agglutination, and the Patoc slide agglutination tests are also in use.

Rapid Screening Tests

Rapid diagnostic tests offer ease of use. These can be performed on whole blood samples and are

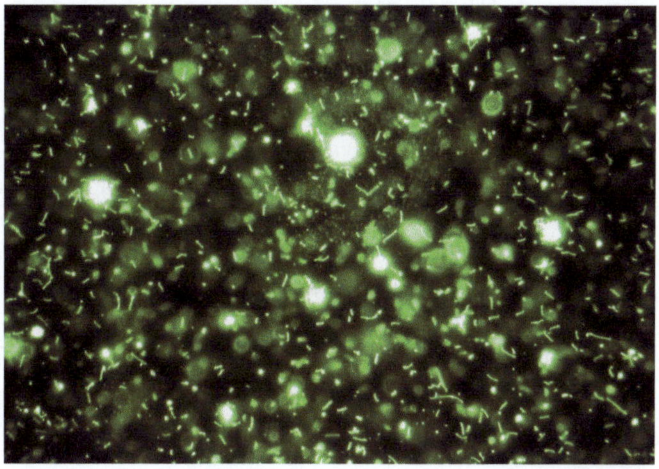

Fig. 31.10 *Leptospiral* microscopic agglutination test (MAT) (100×) highlights the use of live antigen using dark-field illumination microscopy. (Image source: ID#29921Public Health Image Library/ Centers for Disease Control and Prevention (CDC), Atlanta (with permission))

suitable for storage at ambient temperatures without the need for standard laboratory equipment. They predominantly detect IgM antibodies. However, as IgM is generally not detectable until the second week of illness, the sensitivity of these tests is limited during the acute phase, when most patients first seek medical attention.

Molecular Diagnosis

Polymerase Chain Reaction (PCR)

PCR is based on detecting genes like *gyrB*, *rrs*, *secY*, or genes like *lipL32*, *lfb1*, *ligA*, *ligB2*, which are restricted to the pathogenic *Leptospira* spp. Real-time quantitative PCR (qPCR) and multiplex assay using fluorescent probe technology, such as TaqMan probes, have excellent sensitivity and specificity over conventional PCR. PCR can detect DNA in blood within the first 5–10 days of symptom onset, up to the 15th day. Moreover, after effective antimicrobial treatment, it can detect leptospires in culture-negative blood.

Loop-Mediated Isothermal Amplification (LAMP)

LAMP assays designed for the diagnosis of leptospirosis rely on the amplification of specific DNA sequences within the *Leptospira* genome.

The selection of the target gene is a critical determinant of the assay's performance, influencing both its sensitivity and specificity. Several genes have been successfully employed as targets in LAMP-based detection of *Leptospira*. One prominent target is the *lipL32* gene, which encodes a major outer membrane lipoprotein highly conserved among pathogenic *Leptospira* species. Its high copy number and association with virulence often contribute to the good sensitivity observed in assays targeting this gene. Similarly, *lipL41*, encoding another major outer membrane lipoprotein specific to pathogenic *Leptospira*, and the 16S rRNA gene (*rrs*) are frequently used in LAMP. Housekeeping genes such as *secY*, encoding a preprotein translocase subunit, and *gltA* (citrate synthase) have also been utilised as targets, primarily aiming for broad detection of *Leptospira* species.

LAMP assays typically employ a set of four to six primers that recognise six to eight distinct regions on the chosen target gene, significantly enhancing specificity. Some assays further incorporate loop forward (LF) and loop backward (LB) primers that bind to regions between the inner and outer primer binding sites, which can accelerate the amplification process.

Next-Generation Sequencing (NGS)

Next-generation sequencing (NGS) employs various targets depending on the specific diagnostic or research objectives, which can range from simple detection to detailed genomic characterisation. Like its use in PCR and LAMP, a common target is the 16S rRNA gene. Amplicon sequencing of this gene using NGS can identify *Leptospira* at the genus level and, based on the analysis of variable regions, sometimes provides species-level identification.

Whole genome sequencing (WGS) is employed for a more comprehensive genetic analysis when sufficient *Leptospira* DNA is available. WGS aims to sequence the entire genome of the infecting strain, providing the most extensive genetic information. This also allows for precise species and serovar identification, the detection of genes associated with virulence and antibiotic resistance, and detailed phylogenetic analysis that is crucial for investigating outbreaks.

Typing Methods

Identifying the serovar is not required for clinical care. However, it is of particular interest from the public health perspective, indicating the sources of infection and reservoirs and thus contributing to the choice of methods for prevention and control. Given the complexity of serotyping, several alternative molecular techniques have been developed. These include DNA–DNA hybridisation, restriction fragment length polymorphism (RFLP) analysis, pulsed-field gel electrophoresis (PFGE), ribotyping, amplification with specific primers, and PCR-based typing.

The limited concordance between PCR, MAT, and IgM ELISA reflects the different diagnostic windows associated with each method, highlighting the complementary roles of molecular and serological assays across the various stages of infection.

Other Laboratory Tests

These include an elevated erythrocyte sedimentation rate, a raised white blood cell (WBC) count, and thrombocytopenia. Liver function tests show elevated serum aminotransferases, bilirubin, and alkaline phosphatase. Renal involvement is often indicated by raised serum creatinine. In cases with central nervous system involvement, an increased cerebrospinal fluid (CSF) pressure with a predominance of lymphocytes and polymorphs is noted. Urinalysis may reveal pyuria, proteinuria, microscopic haematuria, and the presence of hyaline and granular casts.

Commonly used laboratory diagnostic methods for *Leptospira* are summarised in Table 31.2.

Treatment

Leptospirosis is self-limiting in most patients. Antimicrobials have a role in shortening the duration of illness and reducing the shedding of the organism in the urine.

Mild disease is treated with doxycycline 100 mg orally twice daily for 7 days for adults, or 2 mg/kg per day in two equally divided doses [not to exceed 200 mg daily] for 7 days for children. Alternatively, oral azithromycin 500 mg once daily for 3 days is used for adults and for children 10 mg/kg orally on day 1, followed by 5 mg/kg/day orally once daily on subsequent days. Azithromycin is preferred in pregnant women with mild disease.

For hospitalised adults with severe disease, the treatment of choice is penicillin (1.5 million units intravenously every 6 hours), IV doxycycline 100 mg twice daily, and ceftriaxone (1–2 g IV, once daily) for a duration of 7 days.

The Jarisch–Herxheimer reaction is an acute inflammatory response characterised clinically by fever, rigours, and hypotension. This may occur following antimicrobial therapy, with less than 3 days between symptom onset and antibiotic therapy. The condition is also seen in infection by the *L. interrogans* serogroup Australis strain.

Prevention and Control

The primary control measures for leptospirosis involve avoiding potential sources of infection, such as stagnant water, runoff from animal farms, and areas contaminated by rodents. Protective clothing such as boots and gloves is recommended when the exposure is unavoidable.

Table 31.2 Laboratory diagnosis of leptospirosis

Diagnostic approaches	Methods	Targets	Remarks
Direct microscopy	Dark-field microscopy	*Leptospira* morphology and motility	Rapid and inexpensive; requires fresh urine or blood; low sensitivity and specificity
Culture methods	EMJH medium	*Leptospira* spp.	Takes weeks to grow; not practical for early diagnosis; used for epidemiology and research
	Fletcher's medium		Allows to observe diffuse zone of growth under dark-field microscopy
Serology	Microscopic agglutination test (MAT)	Antibodies (IgM/IgG)	MAT is the gold standard but complex
	ELISA	*Leptospira* IgM antibody	ELISA is faster and suitable for screening Not suitable for the detection of serovar or serogroup
Molecular diagnosis	PCR	*Leptospiral* DNA (targeting flaB, lipL32, or 16S rRNA genes)	Highly sensitive and specific Detect early phase of infection before antibody response
Immuno-histochemistry	Staining of tissue sections	*Leptospira* antigens	Useful in fatal cases or biopsy samples; requires trained personnel
Rapid diagnostic tests	Lateral flow or dipstick-based IgM tests	IgM antibodies	Quick and easy to perform; useful in low-resource settings; variable accuracy

Human vaccines are serovar-specific, but not widely available. Vaccination of the pets and livestock, and personal hygiene after handling animals or potentially contaminated materials, is important to reduce the risk of transmission. Surveillance programmes are essential for monitoring cases in high-risk areas to detect early outbreaks. Antimicrobial prophylaxis is recommended at 200 mg once weekly to travellers and individuals with a high risk of exposure. These combined efforts help manage and reduce morbidity due to leptospirosis (Fig. 31.11).

Coordinated control strategies and multidisciplinary approaches, including those of veterinarians, public health professionals, environmental scientists, and communities, can help reduce both animal and human cases.

Case Study

A 32-year-old man was brought to the emergency services. His family has a history of exposure to floods during the ongoing rainy season. The patient had symptoms of fever, severe myalgia, and haemoptysis. The patient was successfully treated with veno-venous extracorporeal membrane oxygenation (VV-ECMO) and plasma exchange. The treatment resulted in improvement of the patient's condition, with an increased ejection fraction of the left ventricle, increased urine output, and clearing of lung opacities.

1. What is the possible diagnosis and etiological agent?
2. What is this patient's laboratory diagnosis?
3. Mention some points of prevention of this disease.

Research Questions

1. What are the underlying pathophysiological mechanisms contributing to the chronic conditions in leptospirosis in different patient populations?
2. How do different environmental factors in endemic regions influence the survival,

Fig. 31.11 Preventive measures and control measures of leptospirosis. (Image source: Authors' personal collection)

distribution, and virulence of specific *Leptospira* serovars?

3. What are the comparative effectiveness and cost-effectiveness of novel diagnostic tests versus traditional diagnostic methods for the detection of leptospirosis?
4. What are the key immunological responses, both protective and pathogenic, elicited during *Leptospira* infection in humans and animal reservoir hosts?

Declaration During the preparation of this chapter, we, Diptanu Majumder and Nonika Rajkumari, used ChatGPT (OpenAI) to improve the grammar and readability of the text. After using, we reviewed and edited the content as needed and take full responsibility for the content of the publication.

Further Readings

Ahmad SN, Shah S, Ahmad FM. Laboratory diagnosis of leptospirosis. J Postgrad Med. 2005;51(3):195–200.

Arent Z, Pardyak L, Dubniewicz K, Plachno B, Kotula-Balak M. *Leptospira* taxonomy: then and now. Med Weter. 2022;78(09):6694–2022.

Costa F, Hagan JE, Calcagno J, Kane M, Torgerson P, Martinez-Silveira MS, Stein C, Abela-Ridder B, Ko AI. Global morbidity and mortality of leptospirosis: a systematic review. PLoS Negl Trop Dis. 2015;9(9):e0003898.

Gomes-Solecki M, Santecchia I, Werts C. Animal models of leptospirosis: of mice and hamsters. Front Immunol. 2017;8:58.

Haake DA, Levett PN. Leptospirosis in humans. In: Leptospira and leptospirosis. Springer; 2014. p. 65–97.

Leptospirosis in Animals. https://www.cdc.gov/leptospirosis/pets/index.html.

Rajapakse S. Leptospirosis: clinical aspects. Clin Med. 2022;22(1):14–7.

Terpstra WJ. Supplement-historical perspectives in leptospirosis. Indian J Med Microbiol. 2006;24(4):316–20.

Relapsing Fever

32

Amita Jain

Learning Objectives
1. To understand the taxonomy, morphology, and transmission cycle of *Borrelia* species causing relapsing fever
2. To describe the pathogenesis, clinical manifestations, and immune responses associated with relapsing fever
3. To discuss diagnostic methodologies, treatment protocols, and preventive strategies
4. To evaluate the epidemiological significance and public health impact of relapsing fever

Introduction

Relapsing fever is an acute febrile illness caused by several species of *Borrelia* spirochetes and transmitted by the bite of infected soft ticks (*Ornithodoros* species), more recently recognised hard-bodied ticks (*Ixodes* spp.) and body lice (*Pediculus humanus corporis*). It is endemic in various regions, particularly Africa, the Americas, and Asia. In the twentieth century, louse-borne relapsing fever (LBRF) caused epidemics, but currently it is reported only in Africa. Most cases of relapsing fever are tick-borne (TBRF), occurring either sporadically or in small outbreaks. LBRF is an anthroponotic disease caused by *B. recurrentis*, and TBRF is a zoonotic disease caused by various *Borrelia* species. The disease manifests in recurrent febrile episodes due to antigenic variation occurring in the pathogen to escape attacks by antibodies formed by the host's immune response. Many *Borrelia* are detected in the blood at the time of fever. Manifestations can be mild, such as flu-like illness, or severe and life-threatening, such as sepsis. Complications involve central nervous system (CNS) involvement and disordered haemostasis. The host immune response forms antibodies to each of the several variants. Diagnosis is easily made by microscopy, serological tests, and molecular assays. Management is challenging, although treatment with antibiotics is successful. However, a risk of moderate to severe Jarisch–Herxheimer reaction remains.

History

The first clinical descriptions of relapsing fever date back to the eighteenth century. Spiral filaments and motile bacilli, with corkscrew motility, were first observed in the blood of patients with relapsing fever in the 1880s. They were named spirochaetes and assigned to the genus *Borrelia*. In the nineteenth century, its association with lice was recognised, and in the twentieth century, the role of *Borrelia* species was elucidated. Since 1904, TBRF has been recognised in Africa. Advances in micros-

A. Jain (✉)
All India Institute of Medical Sciences,
Raebareli, Uttar Pradesh, India

S. C. Parija, A. Poddar (eds.), *Textbook of Bacterial Zoonoses*, Microbial Zoonoses,
https://doi.org/10.1007/978-981-95-6615-0_32

copy and molecular techniques have further enhanced our understanding of the disease. Currently, because of the decline of Lice infestation LBRF, it has become a rare disease, and TBRF has become the common variety of disease.

Taxonomy

The genus *Borrelia* is placed within the phylum Spirochaeta, class Spirochaetia, order Spirochaetales, and family Spirochaetaceae. Within the genus *Borrelia*, a considerable diversity of species exists. These are broadly categorised into three major groups based on their genetic makeup and the clinical presentations of the diseases they cause.

Relapsing Fever Group

This group, responsible for relapsing fever, comprises species *B. recurrentis*, *Borrelia hermsii* and *Borrelia parkeri*, all implicated in tick-borne forms of the disease. *Borrelia miyamotoi* is also part of this group, notorious for being a tick--borne relapsing fever spirochete that can be transmitted by *Ixodes* ticks, the vector responsible for Lyme disease.

Lyme Disease Group

The group, also known as the *Borrelia burgdorferi* sensu lato complex, encompasses the various species responsible for Lyme borreliosis, including *Borrelia burgdorferi* sensu stricto, *Borrelia afzelii* and *Borrelia garinii*. Other species like *Borrelia bavariensis*, *Borrelia spielmanii*, and *Borrelia mayonii* also contribute to the global incidence of Lyme disease.

Echidna-Reptile Group

This includes *Borrelia* species primarily associated with reptile and echidna hosts, such as *Borrelia turcica*.

Genomics and Proteomics

Borrelia has a linear chromosome (~900 kb) and multiple plasmids. The members of the genus have exceptionally low GC content that typically falls in the range of 27–29%. For example, *B. burgdorferi* B31, a well-studied strain, has a chromosomal GC content of approximately 28.6%. Similarly, *B. recurrentis* has a mean GC content of around 27.5%.

Plasmids in *Borellia* range widely in size (from 5 to over 200 kb) and number. Their plasmids encode variable major proteins (VMPs) responsible for antigenic variation. Antigenic variation is facilitated by an extensive repertoire of variable surface proteins encoded by gene cassettes. Comparative genomics has provided invaluable insights into the diversity and evolution of *Borrelia*. While the core chromosome is generally well conserved across species in terms of gene content and organisation, the plasmid content is highly variable. This plasmid diversity drives phenotypic differences, including host tropism and disease manifestations. For example, comparisons between Lyme disease-causing Borrelia (e.g. B. *burgdorferi*, *B. afzelii*, *B. garinii*) and relapsing fever Borrelia (e.g. *B. recurrentis*, *B. duttonii*) have revealed differences in plasmid content and gene sets that correlate with their distinct transmission cycles and clinical outcomes.

Proteomic studies have identified key virulence factors involved in immune evasion and pathogenesis. There are several proteins that are found on the surface as well as the subcellular level of bacterial cells. Outer surface proteins (OSPs) like OspA, OspB, OspC, OspD and others are expressed on the outer membrane. These are involved in various aspects of infection, including attachment to host cells and tissues, immune evasion, and nutrient acquisition. Lipoproteins like VlsE, OspC, DbpA and DbpB are present in *Borellia* and are involved in binding to host cell proteins and promoting adhesion. Other proteins include DNA-binding proteins like EbfC, moonlighting proteins like enolase and HtrA, and members of the paralogous gene family 54 (PFam54) like BBA64 and BBA66, etc., which play a significant role in pathogenesis, colonisation of host tissues and evasion of the immune system.

Morphology

B. recurrentis has a helical or spiral shape. However, compared to other members of the genus, its coils are typically loose, irregular, and relatively open, distinguishing it from the more tightly coiled *Treponema* species.

The organism typically measures around 8–20 micrometres (μm) in length and 0.2–0.4 μm in width (Fig. 32.1). This larger size, along with its looser coiling, makes it generally easier to visualise in blood smears compared to other spirochetes. Unlike Gram-positive and Gram-negative bacteria, *Borrelia* has a double-membrane structure but lacks classical lipopolysaccharides (LPS). Its outer membrane contains lipoproteins and variable surface proteins (VSPs), which play a role in immune evasion.

B. recurrentis is highly motile, exhibiting a characteristic corkscrew-like or undulating motion. This movement is facilitated by endoflagella, also called axial filaments. These flagella are in the periplasmic space, between the outer membrane and the peptidoglycan layer, and wrap around the protoplasmic cylinder of the bacterium. There are typically 15–20 endoflagella per cell, which generate the rotational and translational forces required for movement through viscous environments like blood.

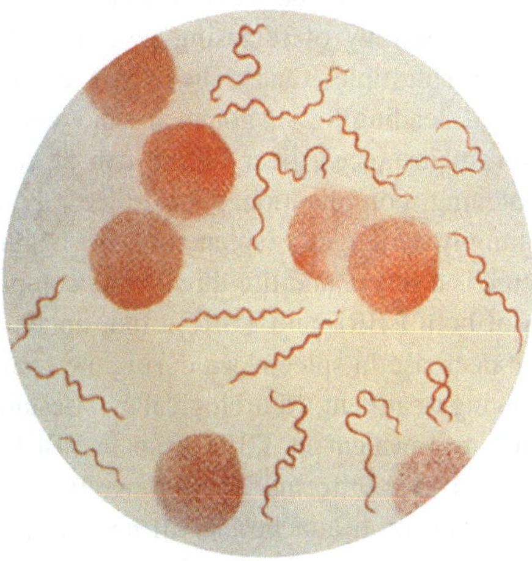

Fig. 32.1 Light microscopic image of a specimen revealing the presence of numerous *Borrelia recurrentis* bacteria with spiral morphology that caused European relapsing fever. (Image source: Centers for disease control and prevention's public health image library (PHIL), with identification number #14351)

Cultivation

Borrelia is cultured in liquid media enriched with blood, serum, or tissue. Serial passaging in culture media makes the strain avirulent. *Borrelia* species are difficult to culture, requiring specialised media such as the Barbour-Stoenner-Kelly (BSK) medium. Optimal growth occurs at a temperature range of 33–35 °C. Furthermore, *Borrelia* are microaerophilic, requiring low concentrations of oxygen to grow. Due to their inherently slow metabolic rate, *Borrelia* require prolonged incubation times, typically ranging from several days to weeks, to grow.

In vitro growth of *Borrelia* on artificial culture media is slow, which limits its use in routine diagnosis. Multiplication is rapid in the chorioallantoic membrane of the chick embryo. At 4 °C, the organisms survive for several months in the presence of organic material. Spirochetes may be passed vertically from generation to generation, known as transovarian transmission.

Animal Models

Rodent models, particularly mice, are the most frequently utilised to study relapsing fever. Immunocompetent mice are particularly well-suited for studying tick-borne relapsing fever (TBRF) species such as *B. hermsii*, *B. parkeri*, and *B. turicatae*, as these *Borrelia* species naturally use rodents as reservoirs.

However, for *B. recurrentis*, the causative agent of louse-borne relapsing fever (LBRF), immunocompetent rodents generally do not develop a stable infection. In these cases, immunodeficient mice are used. Severe combined immunodeficiency (SCID) mice, which lack functional B- and T-cells, and SCID BEIGE mice, deficient in B-, T-, and natural killer (NK) cells, have proven successful in establishing persistent infections with *B. recurrentis*.

Furthermore, the humanised NSG (NOD. Cg-Prkdcscid Il2rgtm1Wjl/SzJ) mice, engineered to engraft human haematopoietic stem cells and develop human myeloid and lymphoid lineages, offer an even more human-relevant model. These advanced models are highly promising for studying *B. hermsii* and potentially *B. recurrentis* infection.

Non-human primates such as grivet monkeys also serve as experimental models for relapsing fever research. However, there are significant ethical considerations and costs associated with their use.

Transmission Cycle

Louse-borne relapsing fever (LBRF) is caused by *B. recurrentis* and spread by the human body louse (*Pediculus humanus corporis*), and possibly the head louse (*Pediculus capitis*). Humans serve as the sole reservoir. Transmission does not occur through the louse bite itself. However, contamination of the bite site with infected louse faeces, often introduced by scratching or through contact with mucous membranes or open wounds, does occur.

Tick-borne relapsing fever (TBRF) is caused by *Borrelia hermsii*, *Borrelia turicatae*, *Borrelia duttonii*, *Borrelia parkeri*, and many other species. All known relapsing fevers other than LBRF are caused by *Borrelia* species other than *B. recurrentis*. These species are transmitted primarily by soft ticks of the *Ornithodoros* genus. The infection reservoirs are usually small- to medium-sized mammals, especially rodents. Pigs and other domestic animals near human dwellings may also serve as hosts. Primarily, it is a tick-borne infection.

Human-to-human transmission is only rarely suspected but has not been confirmed. The exception is *Borrelia duttonii* in sub-Saharan Africa, where humans are the reservoir. It is primarily transmitted between humans via ticks. The soft tick vectors feed quickly, usually taking less than 30 minutes, often while the person is asleep. Because ticks can transmit the infection transovarially, risk can persist in a location long after mammalian hosts are gone.

Borrelia miyamotoi is an emerging spirochaete that causes relapsing fever and is transmitted by hard-bodied ticks of the *Ixodes* genus.

The transmission cycle of LBRF and TBRF is shown in Fig. 32.2.

Pathogenesis and Pathology

The pathogenesis of relapsing fever is caused by *Borrelia* spirochetes evading the host immune system through antigenic variation. After transmission, spirochetes rapidly multiply in the bloodstream, doubling every 6 hours to very high numbers of 10^6–10^7/mL or more, leading to systemic inflammation. *Borrelia* species are extracellular pathogens. After phagocytosis, dead bacteria may appear intracellular. The febrile episodes correspond to waves of spirochetaemia (Fig. 32.3). This results in recurrent fever cycles. Severe cases can lead to disseminated infection involving the central nervous system, myocardium, liver, spleen, and other organs.

The pathology of relapsing fever reveals its impact on multiple organs. The liver is frequently involved, leading to hepatomegaly and jaundice. In severe instances, liver failure can be a life-threatening complication. The spleen is also commonly enlarged (splenomegaly). Splenic rupture is a severe and life-threatening complication of both LBRF and TBRF, often linked to a rapid decrease in spleen size during the crisis or following treatment. Neurological manifestations are more prevalent in TBRF than in LBRF but can occur in both forms. These range from meningitis, causing headaches, a stiff neck, and light sensitivity, to more severe conditions like encephalitis, cranial nerve palsies (e.g. facial droop), focal neurological deficits, and even subarachnoid haemorrhage, as *Borrelia* has the capacity to directly invade the CNS. The heart can also be affected by myocarditis, which may lead to

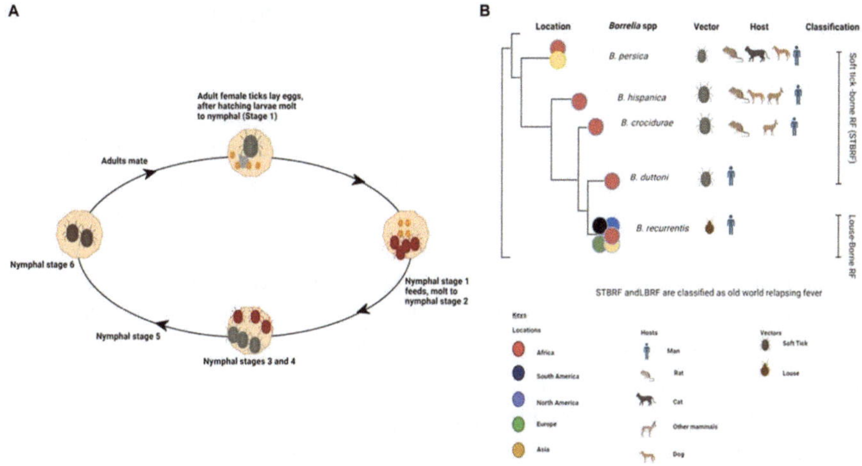

Fig. 32.2 Transmission cycle of louse-borne and tick--borne relapsing fever. (Image source: Adamu A, Reyer F, Lawal N, Hassan AJ, Imam MU, Bello MB, Kraiczy P. Aetiologies of bacterial tick-borne febrile illnesses in humans in Africa: diagnostic limitations and the need for improvement. Front Med. 2024;11:1419575. Distributed under CC BY license for use)

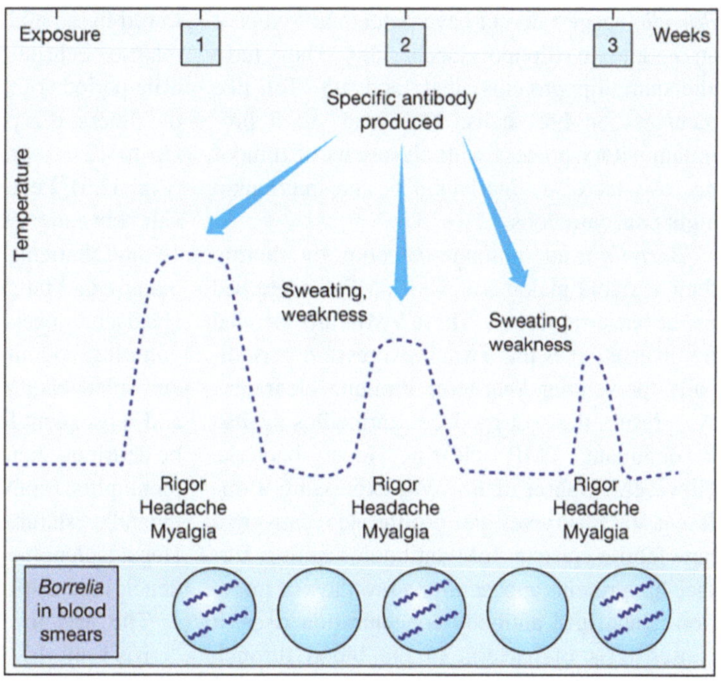

Fig. 32.3 Natural progression of relapsing fever, time lapse, spirochaetemia, and febrile episodes. (Image source: Parija SC. *Treponema*, *Borrelia* and *Leptospira*. In: Textbook of microbiology and immunology. Singapore: Springer Nature Singapore; 2023. p. 617–39)

arrhythmias and conduction abnormalities. Cutaneous manifestations, including rashes (macular, papular, petechial, or purpuric), also appear, especially in TBRF. Ocular complications, such as iritis, uveitis, and iridocyclitis, are also observed.

Jarisch–Herxheimer Reaction (JHR) After Antibiotic Treatment

It is a systemic inflammatory reaction that occurs within 2–4 hours of antibiotic therapy due to massive bacterial lysis and release of endotoxin-like compounds. Symptoms include high fever, tachycardia, hypotension, rigours, and shock-like symptoms. This is managed with supportive care, including fluids, antipyretics, and close monitoring.

Immunology

Humoral immunity is key, with antibody responses targeting variable surface proteins. However, immune evasion via antigenic variation complicates long-term immunity. The innate immune system, particularly macrophages and dendritic cells, plays a role in early pathogen recognition.

Antigenic Variation and Immune Evasion

Borrelia species do not have potent exotoxins or an endotoxin (lipopolysaccharide). They have abundant lipoproteins that activate Toll-like receptors on host cells. This leads to a pro-inflammatory process with elevations of tumour necrosis factor α, interleukin 6, and interleukin eight concentrations.

Borrelia evade immune detection by altering their variable major proteins (VMPs) expressed on the bacterial surface. These VMPs are encoded by diverse genes that switch expression periodically, preventing long-term immune clearance. As a result, the host produces antibodies against a dominant VMP, clearing most bacteria. However, a subset of *Borrelia* expressing a different VMP survives and proliferates, causing a new febrile episode. IgM antibodies against surface lipoproteins appear after a few days of infection. Once IgM antibody concentration rises, it causes lysis of bacteria in the blood through either direct bactericidal action or opsonisation. The release of bacterial lipoproteins and other bacterial products provokes a "crisis", marked by increased temperature, hypotension, and other signs of shock.

Systemic Inflammation and Febrile Episodes

Each wave of spirochaetaemia triggers a pro-inflammatory immune response, leading to fever and systemic symptoms. Fever episodes last 3–5 days, followed by an afebrile period, and relapse occurs when antigenic variants emerge. The inflammatory response is primarily mediated by cytokines such as IL-6, TNF-α, and IL-1β, which contribute to fever, headache, myalgia, and malaise.

Infection in Humans

The incubation period of TBRF is 4–18 days.

Relapsing fever presents as a high fever (39–41 °C), chills, headache, muscle pain, and nausea. Each febrile episode lasts 3–5 days, followed by an afebrile period of 7–10 days before relapse. Febrile periods are punctuated by afebrile periods of a few days. Within the course of the illness, this pattern occurs at least two times. Up to 12 recurrent febrile episodes may occur (Fig. 32.3). Fever has a sudden onset. The interval between fever episodes ranges from 4 to 14 days. Patients may have milder symptoms like malaise and fatigue during a period of remission. Headache, neck stiffness, arthralgia, myalgia, vomiting, abdominal pain, non-productive cough, and spleenohepatomegaly may occur in the first and subsequent febrile episodes. The patient may be delirious. Neurological manifestations such as meningitis, cranial nerve palsies, and altered mental status occur in some cases. Hepatosplenomegaly, jaundice, and bleeding tendencies are also seen in severe cases.

The first fever episode may present with a "crisis" of 15–30 minutes, manifesting with rigours, increased body temperature, pulse rate and blood pressure. The fever is often followed by abnormal sweating and a fall in body temperature and blood pressure, lasting for several hours. LBRF usually has only two episodes. The first episode is more severe and more prolonged than the second episode. Acute respiratory distress syndrome may occur during TBRF. Epistaxis, petechiae, and ecchymoses are common during LBRF but not in TBRF. Insects like body lice or hard ticks may be present on the body.

Infection in Animals

While humans are the primary focus of relapsing fever disease, domestic animals can also be infected. Dogs can be infected with various TBRF *Borrelia* species, including *Borrelia turicatae* and *Borrelia persica*. Clinical signs include lethargy, inappetence, pyrexia (fever), and sometimes neurological signs like lameness, ataxia, abnormal posture, and cranial nerve deficits. While less frequently reported than in dogs, tick-borne relapsing fever has been noted in horses, with occasional reports of equine abortion. Cases of relapsing fever spirochaetemia have been reported in cats with clinical signs that include fever, lethargy, and anaemia.

Epidemiology and Public Health

Historically, LBRF caused massive epidemics across Europe, Asia, and Africa, especially during wars and periods of displacement. With improved sanitation, personal hygiene, and the widespread use of insecticides like DDT in the mid-twentieth century, LBRF dramatically declined in most parts of the world. Currently, Ethiopia and neighbouring countries in the Horn of Africa are considered the primary endemic foci, where an estimated 10,000 cases occur annually. The condition predominantly affects homeless populations in unhygienic and crowded conditions. However, the disease has seen resurgences or has been identified among migrant populations from these endemic regions arriving in Europe. The global epidemiology of LBRF is shown in Fig. 32.4.

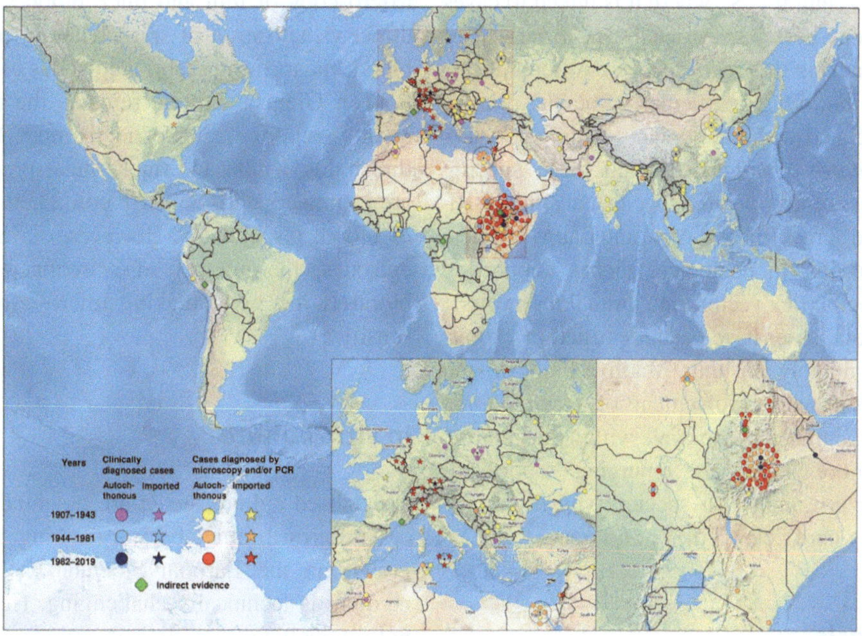

Fig. 32.4 Global epidemiology of louse-borne relapsing fever. (Image source: Kahlig P, Paris DH, Neumayr A. Louse-borne relapsing fever—a systematic review and analysis of the literature: part 1—epidemiology and diagnostic aspects. PLoS Negl Trop Dis. 2021;15(3):e0008564. Distributed under CC BY license for use)

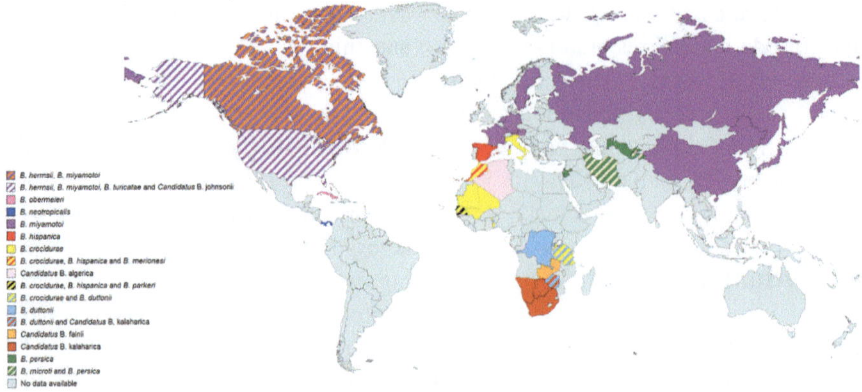

Fig. 32.5 Global epidemiology of tick-borne relapsing fever. (Image source: Jakab Á, Kahlig P, Kuenzli E, Neumayr A. Tick borne relapsing fever-a systematic review and analysis of the literature. PLoS Negl Trop Dis. 2022;16(2):e0010212. Distributed under CC BY license for use)

In contrast, TBRF is endemic worldwide (Fig. 32.5). In North America, for instance, *B. hermsii* is predominantly found in the mountainous regions of the western United States and Canada. It is frequently associated with rustic, rodent-infested cabins that serve as a common source of human exposure. Another significant species in the region, *B. turicatae*, is prevalent in the south-central U.S., where it is linked to caves and rodent burrows. Africa hosts a variety of TBRF-causing *Borrelia* species. Among these, *Borrelia duttonii* is particularly notable, primarily associated with the human-dwelling soft tick *Ornithodoros moubata*. Beyond these specific examples, other *Borrelia* species causing TBRF are found across Europe, Asia, and Latin America, each thriving in ecological niches supporting their specific tick vectors and animal reservoirs.

In India, cases have been identified in various regions, notably Kashmir, Madhya Pradesh, and Karnataka, with involvement of *Ornithodoros* ticks in the transmission cycle within the country. Furthermore, there are emerging reports of *Borrelia miyamotoi*—a species typically transmitted by hard ticks of the *Ixodes* genus.

Diagnosis

The primary clinical specimen for diagnosing relapsing fever is blood, particularly collected during a febrile episode, since this is when spirochete concentrations in the bloodstream are highest, often exceeding 10^6 spirochetes/mL in LBRF. An overview of the diagnosis is presented in the Table 32.1.

Microscopy

Microscopy of thin and thick blood smears by direct visualisation of spirochetes in peripheral blood smears during febrile episodes using dark-field or Giemsa staining remains the most frequently reported diagnostic method. Positivity thresholds of thin and thick smear preparations are estimated at 10^5 and 10^4 spirochetes per mL of blood, respectively. However, during febrile episodes, the number of spirochaetes in the blood is too high, making microscopy highly sensitive.

In Vitro Culture

Specialised culture media are essential for successful growth in a laboratory setting. However, growth in the Barbour–Stoenner-Kelly (BSK) medium is technically challenging. Hence, it is not routinely used in clinical settings. Culture is restricted to very few laboratories worldwide and has primarily been used for research.

The BSK medium has undergone considerable evolution and modification over the years

Table 32.1 Laboratory diagnosis of relapsing fever

Name of test	Specimen	Target	Comment
Microscopy by dark-field or Giemsa staining	Thin and thick blood smears	Bacteria	Most frequently reported diagnostic method Very sensitive during febrile episode Positivity thresholds of thin and thick smear preparations are estimated at 10^5 and 10^4 spirochetes per mL of blood respectively
In vitro culture Barbour-Stoenner--Kelly (BSK) medium Modified Kelly–Pettenkofer (MKP) medium	Blood	Bacteria	Culture is challenging due to fastidious nature Specialised media and controlled environmental conditions are essential
Serodiagnosis (ELISA)	Serum	Bacterial antigen or antibodies against it	Detects anti-*Borellia* antibodies; antibodies develop late in illness; cross-reactivity is an issue
Molecular diagnosis (Real time PCR)	Blood, serum, body fluids and tissues	Conserved genes like 16S rRNA, *glpQ*	Highly sensitive and specific, front-line test in current time; can differentiate between LBRF and TBRF; good probability of being used as point-of--care test

to enhance its efficacy. Extensive research and development have led to widely used modified media such as BSK-II and BSK-H. These are complex liquid media that provide a rich array of amino acids, vitamins, nucleotides, salts, glucose (as an energy source), and buffering agents (such as HEPES) crucial for maintaining the precise pH that *Borrelia* requires. More importantly, these media contain supplements of bovine serum albumin (BSA) and rabbit serum.

The modified Kelly-Pettenkofer (MKP) medium is another medium of importance. This formulation is often favoured for the primary isolation of *Borrelia* from challenging clinical and environmental samples, especially when the initial bacterial load is very low.

Serodiagnosis

ELISA and immunofluorescence assays are available to detect anti-*Borrelia* antibodies, but serology plays a limited role in diagnosis. In endemic areas, seroprevalence is high, which may result in false-positive results. The seroconversion also takes too long to yield a positive result in the acute stage. Moreover, cross-reactivity of assays may confuse TBRF with Lyme disease.

Molecular Diagnosis

The most widely used molecular diagnostic technique for *Borrelia* is the polymerase chain reaction (PCR), especially real-time PCR (qPCR). These assays targeting *Borrelia* DNA provide high sensitivity and specificity. These assays are mainly helpful in low-density infections.

PCR is the second most frequently reported diagnostic method after microscopy. PCR is sensitive and specific and can differentiate between TBRF and LBRF. Borrelia has ~99% genetic similarity at the 16S rRNA gene sequence. Hence, the development of discriminatory PCR assays will not always be able to provide species identification. PCR as a point-of-care routine diagnostic method has been reported.

The *glpQ* gene and its protein product are particularly important targets for distinguishing relapsing fever from Lyme disease. GlpQ is present in all relapsing fever *Borrelia* species but absent in Lyme disease *Borrelia* and *Treponema pallidum.*

Treatment

Tetracyclines, doxycycline, macrolides, and penicillins are effective against *Borrelia* species. Although for LBRF, a single dose of intramuscular procaine penicillin is effective, TBRF requires prolonged and high-dose penicillin treatment to prevent relapse. The lack of resistance to penicillin in spirochetes makes penicillin still an option. However, the restricted availability of procaine penicillin and the need for intravenous doses several times per day restrict its use in clinical practice. *B. miyamotoi* is resistant to amoxicillin but sensitive to ceftriaxone.

Since tetracycline is preferred over penicillin, concerns remain about doses, duration, and treatment success rate. Antibiotic levels in the CNS should be adequate. In the absence of CNS involvement, oral or parenteral treatment with tetracycline (500 mg every 6 hours for 10 days or doxycycline 100 mg every 12 hours for 7–10 days) is recommended for adults. In children, due to side effects, doxycycline is preferred for up to 21 days (doxycycline 4.4 mg/kg body weight/day divided into two doses (max. 200 mg/day). If β-lactam or tetracycline is not recommended, erythromycin (500 mg or 12.5 mg/ kg body weight every 6 hours for 7–10 days) is recommended. Azithromycin is equally effective. Treatment with tetracycline appears to be associated with a higher rate of Jarisch–Herxheimer reaction.

Prevention and Control

Vector control, improved sanitation, and public health interventions are critical in reducing the disease burden. Antibiotic post-exposure prophylaxis with doxycycline following tick bites within endemic areas may be given. Even a single dose of doxycycline is adequate. Preventive measures include reducing exposure to tick-infested areas, implementing lice control programmes in crowded settings, and implementing surveillance programmes to monitor and contain outbreaks.

Targeted delousing measures are vital for LBRF. This includes physical delousing, where washing clothes and bedding in hot water (above 55 °C or 130 °F) and drying on high heat effectively eliminates lice and their eggs. In outbreak scenarios or for individuals with heavy infestations, chemical delousing using insecticides (pediculicides) like permethrin dust, applied to clothing and bedding by trained personnel, or even oral ivermectin in specific contexts, can be highly effective.

Case Study

An elderly woman was hospitalised with a history of a 2-day fever. She had an episode of fever 2 weeks ago and recovered following symptomatic treatment. On examination, she was febrile (temperature, 40 °C), her blood pressure was 110/70 mm Hg, and her pulse was 70/min. She had petechial rash and haematuria. No abnormality was detected in other systems, including the cardiorespiratory system and the central nervous system. Hepatosplenomegaly and lymphadenopathy were absent. An examination of the peripheral blood smear revealed the presence of spirochetes morphologically resembling *Borrelia* species. A diagnosis of relapsing fever was made. The platelet count was low. The patient was given oral doxycycline, 100 mg b.i.d. The pyrexia subsided within 2 days of therapy, and the temperature remained normal thereafter. The PBF was negative after 1 week of treatment. The patient was followed up for a few weeks, after which she had no recurrence of fever.

1. What are the clinical features of relapsing fever?
2. What are the diagnostic criteria for relapsing fever?
3. What empirical antibiotic therapy is given for relapsing fever?

Research Questions

1. How does antigenic variation contribute to the chronicity of relapsing fever?
2. How do host immune responses influence disease severity and relapse rates?
3. Can novel molecular diagnostics improve early detection and management?

Declaration During the preparation of this chapter, we have used ChatGPT in order to improve the readability of the text. After using it, we reviewed and edited the content as needed and took full responsibility for the content of the publication.

Further Readings

Cutler SJ. Relapsing fever–a forgotten disease revealed. J Appl Microbiol. 2010;108(4):1115–22. https://doi.org/10.1111/j.1365-2672.2009.04598.x.

Cutler SJ, Abdissa A, Trape JF. New concepts for the old challenge of African relapsing fever borreliosis. Clin Microbiol Infect. 2009;15(5):400–6. https://doi.org/10.1111/j.1469-0691.2009.02819.

Dworkin MS, Anderson DE Jr, Schwan TG, Shoemaker PC, Banerjee SN, Kassen BO, et al. Tick-borne relapsing fever in the United States: a clinical review of 62 cases. Clin Infect Dis. 1998;26(2):395–400. https://doi.org/10.1086/516302.

Elbir H, Henry M, Diatta G, Mediannikov O, Sokhna C, Tall A, et al. Multiplex real-time PCR diagnostic of relapsing fevers in Africa. PLoS Negl Trop Dis. 2013;7(1):e2042. https://doi.org/10.1371/journal.pntd.0002042.

Jawetz E, Melnick JL, Adelberg EA. Chapter on Spirochetes and Borrelia. In: Medical microbiology. 28th ed. New York: McGraw-Hill Education; 2019.

Mandell GL, Bennett JE, Dolin R. Chapter 246: Relapsing fever. In: Mandell, Douglas, and Bennett's principles and practice of infectious diseases. 9th ed. Philadelphia: Elsevier; 2020.

Murray PR, Rosenthal KS, Pfaller MA. Chapter 37: Spirochetes. In: Medical microbiology. 9th ed. Philadelphia: Elsevier; 2020.

Parija SC. Section on Spirochetes. In: Textbook of microbiology and immunology. 3rd ed. New Delhi: Elsevier; 2020.

Ryan KJ, Ray CG. Chapter 33: Borrelia and relapsing fever. In: Sherris medical microbiology. 7th ed. New York: McGraw-Hill Education; 2018.

Talagrand-Reboul E, Boyer PH, Bergström S, Vial L, Boulanger N. Relapsing fevers: neglected tick-borne diseases. Front Cell Infect Microbiol. 2018;8:98. https://doi.org/10.3389/fcimb.2018.00098.

Lyme Disease

33

Debadatta Dhar Chanda
and Barnamoy Bhattacharjee

Learning Objectives

1. To learn the taxonomy, morphology, and genomics of *Borrelia burgdorferi*
2. To acquire knowledge regarding the history, pathogenesis, and pathology, clinical features, diagnosis, and treatment of Lyme disease
3. To understand the epidemiology and public health importance of Lyme disease

Introduction

Lyme borreliosis (Lyme disease) is a vector-borne zoonosis of public health importance caused by the spirochete *Borrelia burgdorferi*. Transmission to human hosts occurs primarily through the bite of infected ticks of the genus *Ixodes*. The disease was first formally recognised in 1975. The pathogen exhibits genetic heterogeneity, forming the *B. burgdorferi* sensu lato complex. This includes three primary pathogenic species: *B. burgdorferi* sensu stricto (predominant in North America), *Borrelia afzelii* and *Borrelia garinii* (prevalent across Europe and Asia). Lyme borreliosis is common in temperate regions, particularly in the United States. However, its global incidence is rising due to ecological and environmental factors. These factors include increasing reforestation, which expands vector habitats, suburban encroachment on newly wooded areas, which increases human contact with infected ticks, and dynamic shifts in the geographic range of the vector.

D. D. Chanda (✉) · B. Bhattacharjee
Department of Microbiology, Silchar Medical College & Hospital, Silchar, Assam, India

History

Lyme disease is named after the town of Lyme, Connecticut, where the illness was first recognised in 1975 following an investigation into an outbreak of juvenile arthritis. This epidemiological study ultimately led to the identification of the spirochete bacterium *Borrelia burgdorferi* as the etiologic agent in 1982. Lyme borreliosis is currently endemic across specific regions of North America and Europe, favouring the geographic habitats of its vector, the *Ixodes* tick. These tick species play a critical role in facilitating the zoonotic transmission of the spirochete from animal reservoirs to human hosts. Furthermore, the disease is expanding its known geographical range, with increasing reports of its incidence in new territories, including parts of Asia.

Taxonomy

The genus *Borrelia* is classified under the family Borreliaceae. It belongs to the phylum Spirochaetota, class Spirochaetia, and order Spirochaetales in the domain Bacteria.

S. C. Parija, A. Poddar (eds.), *Textbook of Bacterial Zoonoses*, Microbial Zoonoses,
https://doi.org/10.1007/978-981-95-6615-0_33

Initially, all spirochetes causing both Lyme disease and relapsing fever were grouped into the genus *Borrelia.* However, the advent of advanced molecular techniques, such as 16S rRNA gene sequencing, multilocus sequence analysis (MLSA), and whole-genome comparisons, has shown significant genetic and phenotypic differences between the Lyme disease-causing species of *Borrelia* and those responsible for relapsing fever. Based on these findings, Adeolu and Gupta (2014) proposed a taxonomic revision in which the Lyme disease-causing species, collectively referred to as the *Borrelia burgdorferi* sensu lato complex, would be placed in a newly established genus, *Borreliella.* Meanwhile, species responsible for relapsing fever would remain within a redefined *Borrelia* genus.

This proposal was supported by phylogenomic analyses showing significant evolutionary divergence and distinct molecular signatures (such as conserved indels and proteins) between the two groups. Despite its strong scientific rationale, the proposed genus *Borreliella* has not gained wide acceptance across the microbiological community, and most microbiologists still use the traditional genus name *Borrelia* for both groups.

Genomics and Proteomics

B.burgdorferi was one of the first bacteria to have its genome fully sequenced, with the strain B31 published in 1997. Its genome is atypical, consisting of a ~1000 kb linear chromosome and numerous extrachromosomal elements, including 9 circular and 12 linear plasmids. The genome has a low G + C content of about 30 mol%. About 150 genes code for lipoproteins, underscoring their importance in the survival and pathogenesis. Comparative analysis with *B. garinii* (strain PBi) shows high conservation, with ~92.7% identity at DNA and protein levels across the chromosome.

Plasmids in *B. burgdorferi* carry genes critical for host interaction and immune evasion. Relapsing fever species such as *B. hermsii* and *B. turicatae* use plasmid-encoded variable major proteins (Vmps) for antigenic variation, a strategy mirrored in *B. burgdorferi* through recombination of Vmp-like cassettes. This process generates diverse epitopes, aiding immune escape and affecting serodiagnosis. Key plasmid--encoded proteins include OspA and OspB on lp54, and OspC on cp26. OspC shows high sequence variability across strains, reflecting gene exchange that enhances antigenic diversity and promotes persistence in the host.

Morphology

The genus *Borrelia* consists of spirochetes characterised by their slender, helically coiled morphology, typically 20–30 µm in length and 0.2–0.5 µm in diameter. The bacteria exhibit irregularly coiled spirals that are optically distinct.

Due to their weak affinity for dyes, *Borrelia* stain poorly with the Gram stain, often appearing as weakly Gram-negative bacilli. Consequently, they are optimally visualised using specialised techniques such as dark-field microscopy (Fig. 33.1). Ultrastructurally, *Borrelia* share the foundational architecture common to other members of the Spirochaetaceae family.

A defining feature of the spirochetes is the presence of periplasmic flagella (endoflagella). Each cell of *Borrelia* possesses 7–11 of these flagella, which are subterminally inserted at both poles of the protoplasmic cylinder and overlap centrally. These flagella are present between the outer membrane sheath and the cell cylinder. This unique arrangement is essential for both the characteristic rotational motility of the organism and the maintenance of its helical shape. Mutants lacking these flagella lose their wave-like structure and present a rod-shaped morphology. Crucially, the flagella are indispensable for *Borrelia* survival within both its tick vector and its mammalian hosts.

B. burgdorferi also exhibits the following morphological variations:

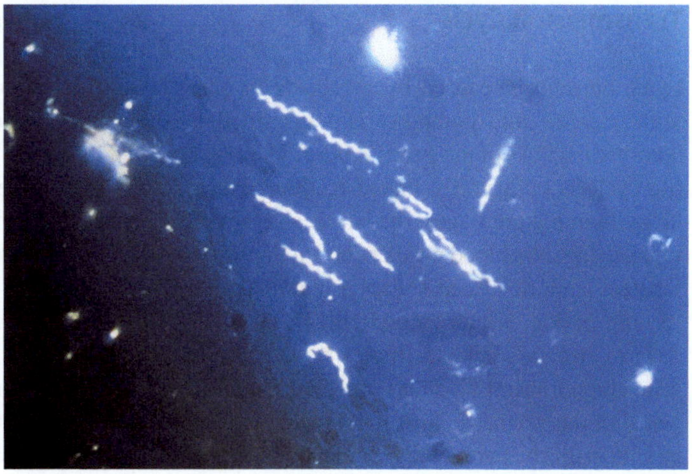

Fig. 33.1 Dark-field microscopic image of *Borrelia burgdorferi* showing "corkscrew-shaped" morphology (×400). (Image source: CDC's Public Health Image Library (PHIL), with ID #6631)

Pleomorphic Forms *B. burgdorferi* can exist as Spirochetes, Blebs (Detached Outer membrane Vesicles), and spherical, round bodies (RBs). Such morphologically variable forms have been thought to aid in the persistence, immune evasion, antimicrobial tolerance, and antimicrobial resistance in *B. burgdorferi.*

Aggregates At higher cell densities, *B. burgdorferi* can form aggregates in culture.

Cultivation

Borrelia species are characterised physiologically as microaerophilic, slow-growing spirochetes. Optimal growth in vitro is achieved using a complex liquid medium (e.g. Barbour-Stoenner-Kelly (BSK) formulations) incubated within a temperature range of 30–35 °C, typically yielding high cell densities of 10^7–10^8 cells per millilitre. Growth is also supported in semi-solid media, sucH as modified Kelly medium (BSK-II, BSK-H, or MKP supplemented with 1% agar). When cultivated in semi-solid matrices, the resultant colonies develop below the surface and exhibit non-differentiating macroscopic properties.

The nutritional profile of *Borrelia* species is like that of other spirochetes, particularly the genus *Treponema.* They possess stringent growth requirements, notably the need for unsaturated and saturated fatty acids. Required supplementary nutrients include glucose, amino acids, and, uniquely, N-acetylglucosamine. Serum proteins are a critical medium constituent, as they serve to solubilise and provide a non-toxic source of the required fatty acids.

Animal Models

Animal model studies have greatly contributed to the understanding of Lyme borreliosis. Rodents and rhesus monkeys are the main experimental models. These models help explain both the host immune response and the way the bacteria adapt in living systems.

Research has shown that proteins derived from the *ospA* gene can protect against infection. When presented in lipoprotein form, they stimulate a strong immune defence. Such OspA-based candidates are now being evaluated in clinical trials as potential vaccines.

Inbred laboratory mice have proven effective for (i) characterising *B. burgdorferi* infection kinetics and (ii) studying the host immune responses in pathogen and inflammatory pathology. In susceptible mouse strains (C3H mice), infection leads to subacute arthritis and carditis characterised by joint swelling and histopathological manifestations like those of human disease, including leukocyte infiltration.

Transmission Cycle

Lyme borreliosis, caused by the *Borrelia burgdorferi sensu lato* complex, is transmitted by hard-bodied ticks of the genus *Ixodes*. The *Ixodes* life cycle is obligatorily linked to three distinct hosts. In immature stages, the larvae and nymphs typically acquire blood meals from small rodents, whereas adult ticks feed on larger mammals. The feeding period, which can extend from several days to over a week, facilitates both pathogen acquisition/transmission and the geographic dispersal of the vector via host movement. Notably, migratory birds, particularly seabirds, contribute to the long-range transportation of *Ixodes* ticks (e.g. *Ixodes uriae*), thus promoting the global spread of *Borrelia* species, most notably *B. garinii.*

In unfed ticks, *B. burgdorferi* sensu lato mainly resides in the midgut. When feeding begins on a mammalian host (e.g. human, mouse), the spirochete undergoes key transcriptional shifts. Outer surface protein A is downregulated, while Outer Surface Protein C is expressed, allowing migration from the midgut lumen to the salivary glands. Although the precise role of OspC in tick dissemination remains debated, its expression is essential for establishing mammalian infection. The migration timeline depends on the vector species. In *Ixodes scapularis*, movement takes about 36 hours, whereas in *Ixodes ricinus*, transmission may occur within 17 hours of feeding.

The *Ixodes* life cycle usually takes about 2 years. Adults lay eggs in spring, and larvae emerge in summer. After feeding, larvae moult into nymphs, which then develop into adults by autumn, following another blood meal. Although adult ticks show nearly twice the infection prevalence of nymphs, most human cases are linked to nymphal ticks. This is because nymphs are abundant in summer, coincide with peak human outdoor activity, and are small enough to escape detection, allowing longer attachment. Transmission depends on feeding duration: nymphs typically need 36–48 hours, while adults require 48–72 hours (Fig. 33.2).

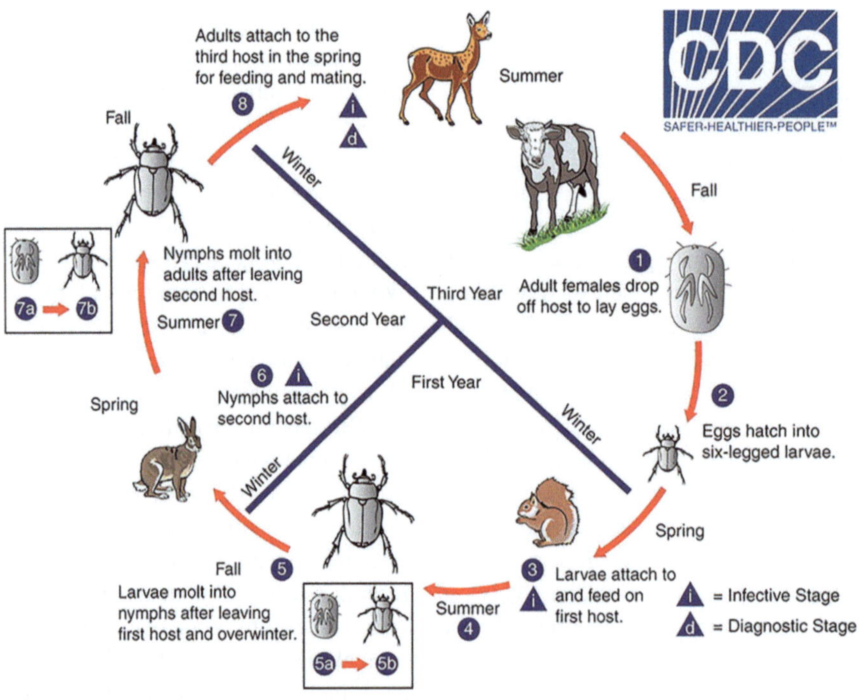

Fig. 33.2 Transmission cycle of Lyme disease. (Image source CDC)

Pathogenesis and Pathology

B. burgdorferi exhibits significant phenotypic plasticity, transitioning from its typical motile cellular state into various alternative morphological forms, including round bodies, stationary phase cells, persisted cells, and structured biofilms. The capacity for this pleomorphism allows *B. burgdorferi* to sustain viability while entering a non-culturable state, a characteristic that contributes substantially to its inherent antimicrobial tolerance and treatment resistance.

Biofilm formation by *Borrelia* represents a complex survival strategy. Within biofilms, *Borrelia* cells exist in a protected state. The extracellular matrix restricts antibiotic penetration, while slow-growing or metabolically dormant subpopulations—often termed "persister cells"—are less susceptible to antimicrobial action. Quorum-sensing within the biofilm community further enhances survival strategies. These factors explain why *Borrelia* biofilms may persist despite adequate antimicrobial therapy and are thought to contribute to the pathogenesis of post-treatment Lyme disease syndrome (PTLDS).

The pathogenesis of Lyme disease reflects the ability of *B. burgdorferi* to adapt to varied host and environmental conditions. Within 3–30 days, the spirochetes can disseminate via the blood or lymphatics, producing systemic disease in nearly half of untreated cases. The organism shows tissue tropism for the skin, joints, heart, and nervous system, and may also localise to the bone marrow, liver, and spleen. Spirochetemia and early central nervous system (CNS) involvement can occur even when only cutaneous symptoms are present.

A major mechanism of persistence is the organism's capacity for intracellular localisation, which supports immune evasion and may drive chronic infection. Even after appropriate therapy, some patients continue to experience symptoms. Around 20% of treated individuals may develop what is known as post-treatment Lyme disease syndrome (PTLDS). This condition is marked by persistent fatigue, pain, and cognitive difficulties despite clearance of the active infection. Proposed mechanisms for PTLDS include ongoing autoimmunity, residual persistent infection, and antibiotic tolerance.

Immunology

B. burgdorferi actively interacts with the host immune system, eliciting both cellular and humoral immune responses.

A key survival strategy of the spirochete in both arthropod and mammalian environments is the modulation of outer surface protein (Osp) expression. This is best illustrated by the lipoproteins OspA and OspC. OspA, a 31-kDa lipoprotein, is strongly expressed on the outer membrane while the organism resides in the tick midgut. During the blood meal and subsequent transmission, OspA is downregulated. In contrast, OspC, a 23-kDa lipoprotein, is upregulated during feeding, enabling migration to the tick salivary glands and ensuring transmission to the mammalian host.

In early Lyme disease, the host immune response is mainly directed against two major antigens: outer surface protein C (OspC) and the 41-kDa flagellar antigen (FlaB).

Clinical Relevance of OspA and OspC

OspA plays a crucial role in vaccine development. A recombinant OspA-based vaccine was once licensed and shown to induce protective antibodies that blocked transmission by targeting spirochetes within the tick midgut before entry into the mammalian host. Although no OspA vaccine is currently in routine use, research continues into multivalent formulations covering diverse *Borrelia* genotypes.

OspC is important in diagnostics and early disease detection. Antibodies against OspC appear early during infection, making it a reliable marker for acute Lyme disease. However, OspC exhibits high antigenic variability among *Borrelia* strains, which limits the sensitivity of single-antigen assays. Panels including multiple OspC variants, along with FlaB and other antigens, improve diagnostic accuracy.

Infection in Humans

Clinically, the disease is divided into the following three stages:

Early Localised Disease (Stage I)

Erythema chronicum migrans is the hallmark clinical sign of early localised infection. It is a circular, outwardly spreading red rash. The lesion may be warm, but it is usually painless. Usually appears 7–10 days (ranging from 3 to 30 days) after the bite of a tick. The single annular lesion can expand to a diameter of ≥10 cm. Half of the patients develop multiple smaller erythematous lesions at sites distant from the bite within several days.

Classically, erythema migrans is described as a "*bull's-eye*" rash because the innermost portion is dark red and indurated, the outermost portion is red, but the portion in the middle clears. Peripheral oedema with central clearing (average diameter, 15 cm) of the lesion has been demonstrated.

Flu-like presentations such as fever, malaise, headache, and muscle soreness may also be present. In most adult patients, systemic symptoms are associated with regional lymphadenopathy.

Early Disseminated Disease (Stage II)

In this stage, the spirochete disseminates in the body. EM can develop at sites other than the original tick bite site. *Lymphocytoma cutis* may develop in a certain group of patients, who typically affect the face, ear lobe, nipple, or scrotum.

Common symptoms include acute asymmetric, oligoarticular joint inflammation and neuroborreliosis manifestations. In meningitis, cerebro-spinal fluid (CSF) may appear purulent. Patients with neuroborreliosis may also have symptoms of meningoradiculoneuritis and mild encephalitis. Altered mental status may be the only symptom reported in some cases of early central nervous system (CNS) infection. In addition, infection may also cause acute hepatitis, arthritis, acute myocarditis and infection of the eye.

Late Persistent Disease (Stage III)

This stage represents the chronic phase of Lyme disease in American patients. It is characterised by chronic arthritis and, less commonly, persistent nervous system symptoms. European patients infected with *B. afzelli* or *B. garinii* are more likely to produce chronic skin lesions, chronic neurological symptoms, and chronic arthritis. The recurrence of arthritis diminishes in frequency and severity each year, but chronic synovitis and permanent disability develop in some patients.

Congenital Transmission of Lyme Disease in Children *Borrelia* species exhibit the capacity for transplacental transmission to the foetus, which may occur as early as the first month of gestation. This early transmission window contrasts with that of *Treponema pallidum*, the agent of syphilis, in which transmission typically occurs in the second trimester.

Infection in Animals

B. burgdorferi sensu stricto is the main species that causes Lyme borreliosis in animals. The infection is spread through the bite of hard-bodied ticks belonging to the genus *Ixodes*.

In the United States, different species of *Ixodes* ticks act as vectors in different regions. *Ixodes scapularis*, commonly called the deer tick, is the major vector in the Northeast and Midwest. On the Pacific coast, transmission mainly occurs through *Ixodes pacificus*. In Europe and Asia, the primary vectors are I. ricinus and I. persulcatus.

Symptomatic disease has been reported mostly in dogs and horses, though many of them are seropositive but asymptomatic. Dogs can display several forms of disease. The most common symptoms are joint swelling, swollen lymph nodes, loss of appetite, lameness, and fatigue. Severe cases may cause serious kidney complications. In horses, clinical signs include neuroborreliosis, uveitis, cutaneous lymphoma, shifting-leg lameness, generalised stiffness, hypersensitivity to touch, weight loss, and poor performance.

Epidemiology and Public Health

Lyme borreliosis is one of the prevalent tick-transmitted diseases in the Northern Hemisphere, particularly in the temperate regions, with reported cases spanning over 50 countries. Major endemic areas include the eastern and Midwestern United States, northwestern, central, and eastern Europe (with the highest incidence in Scandinavian countries, Germany, Austria, and Switzerland), and forested regions of Asia (Fig. 33.3).

In India, cases exhibiting Lyme disease-associated pathologies, such as monoarthritis and neuroretinitis, have been documented across numerous states, including Bihar, Uttar Pradesh, Uttarakhand, Haryana, Maharashtra, Himachal Pradesh, parts of South India, and the states of Northeastern India. Epidemiologically, the lack of a definitive correlation between clinical symptoms and IgG seropositivity in certain regions emphasises the critical need for a comprehensive patient history and a high degree of clinical suspicion for diagnosis in endemic settings.

The maintenance of the *B. burgdorferi* sensu lato complex in nature depends on a complex enzootic cycle involving arthropod vectors (*Ixodes* ticks) and a diverse range of vertebrate hosts that serve as pathogen reservoirs. Human infection with Lyme borreliosis occurs through ticks of the *Ixodes ricinus* complex. These ticks act as regional vectors, with different species predominating in specific areas. On the western coast of the United States, *Ixodes pacificus* is the main vector. In the central and eastern parts of the country, *Ixodes scapularis* is more common. In Europe, transmission is mainly linked to *Ixodes ricinus*, while in Asia, *Ixodes persulcatus* plays the major role.

The prevalence of Lyme disease and the life cycle dynamics of its tick vectors are influenced by climatic conditions and host availability. Crucially, the distribution of *B. burgdorferi* sensu lato does not perfectly over-

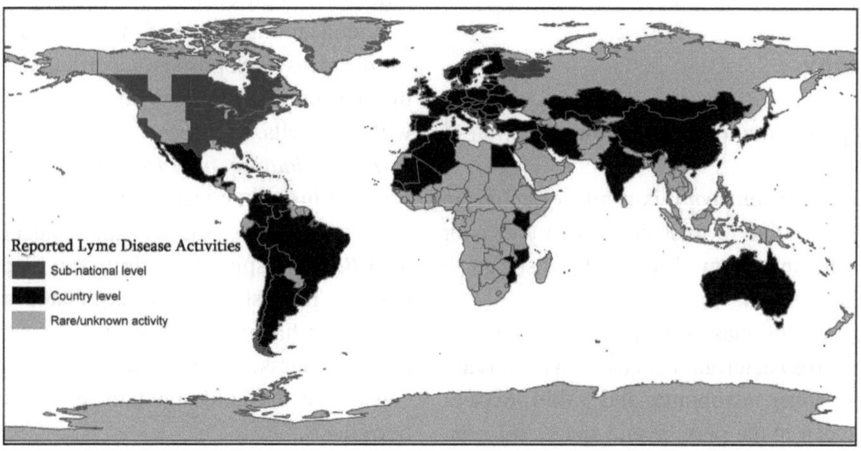

Fig. 33.3 Global epidemiology of Lyme disease. (Image source: Ozdenerol E. GIS and remote sensing use in the exploration of Lyme disease epidemiology. Int J Environ Res Public Health. 2015;12(12):15182–203. Copyright to Ozdenerol E under CC BY License)

lap with the geographic range of *Ixodes* ticks alone. The behavioural interaction between ticks and humans, alongside the diverse vertebrate hosts, is a pivotal factor governing pathogen maintenance and disease incidence. For example, if a competent tick vector rarely feeds on humans, the disease prevalence will be low despite the presence of the *Borrelia* species. Specific host-pathogen links have been established in Europe, such as the association of *B. afzelii* with small rodents and *B. garinii* and *B. valaisiana* with avian hosts. Furthermore, a significant correlation exists between Lyme borreliosis incidence and moisture levels in the two preceding years.

Diagnosis

Specimens

These include blood, cerebro-spinal fluid (CSF), synovial fluid, or synovial biopsy specimens. In a patient suspected of neuroborreliosis, both CSF and serum samples are collected. Punch biopsy specimens are collected from EM or acrodermatitis chronica atrophicans lesions. These are placed in sterile saline or Barbour-Stoenner-Kelly (BSK) medium for isolating *B. burgdorferi*, especially in untreated cases.

Microscopy

Microscopy has limited value in the detection of Lyme disease.

Bright-field microscopy is rarely effective in detecting them due to their small size and poor staining, making them difficult to distinguish from cellular debris.

Gram staining classifies *B. burgdorferi* as a Gram-negative bacterium because it has an outer membrane, inner membrane, and a thin layer of peptidoglycan in the periplasmic space. *Borrelia* are very slender and do not easily retain the crystal violet during Gram staining and often appear very pale pink or are not visible at all against the background tissue or blood components. Hence, Gram staining is not a reliable method for its direct detection of *B. burgdorferi* in clinical samples.

Dark-field microscopy was once used to identify *B. burgdorferi*. However, it is no longer recommended due to its low sensitivity and specificity. *B. burgdorferi* is often scarce and may not be readily detectable even when present. In addition, other microorganisms and cellular artefacts exhibit similar morphology and motility, giving rise to false-positive results.

Fluorescence microscopy combined with specific fluorescent antibodies provides enhanced sensitivity and specificity. Antibodies tagged with fluorescent dyes bind to *Borrelia* antigens, making the spirochetes visible under UV light. This method is more reliable than a bright-field or dark-field microscope and is frequently used to confirm the presence of *Borrelia* in culture rather than for direct diagnosis from patient specimens. However, it fails to detect infection with low bacterial loads and requires specialised equipment and technical expertise.

Electron microscopy, by its high resolution, can demonstrate the ultrastructure of *B. burgdorferi*. Scanning electron microscopy and transmission electron microscopy provide highly detailed images of the spirochete. Despite its precision, the electron microscope is impractical for routine diagnosis because of the complex sample preparation, the high cost of the equipment, and the time-consuming nature of the analysis. It remains a tool for research purposes, particularly in studying the morphology and interaction of *Borrelia* with host cells.

Focus floating microscope (FFM) is a new immune-histochemical method that has been designed to detect the rare and sparsely distributed *Borrelia* spirochetes in tissue biopsies, particularly from skin lesions associated with Lyme disease. It has proven to be more sensitive than the polymerase chain reaction (PCR). This is considered the gold standard test for cutaneous borreliosis.

The process involves two steps. First, tissue sections are treated with specific antibodies against *B. burgdorferi* antigens, making spirochetes potentially visible under a microscope.

Additionally, FFM uses a bright red chromogen like aminoethylcarbazol (AEC) for staining, which enhances contrast against the typically yellowish collagen in the tissue, further aiding in the identification of the spirochetes.

In Vitro Culture

Borrelia species can be successfully propagated in vitro using specialised liquid media, such as modified Kelly medium (e.g. BSK II, BSK-H, or MKP). Optimal growth conditions require a temperature range from 30 to 33 °C and microaerophilic conditions. Under these conditions, the organisms exhibit a relatively slow growth rate, with a generation time of approximately 7–20 hours. Positive cultures can occasionally be obtained in as early as 4 days from clinical samples such as erythema migrans (EM) biopsies, synovial fluid/blood/ Cerebrospinal Fluid.

Most primary isolates, however, require prolonged incubation. Hence, negative cultures must be monitored by dark-field microscopy for a minimum of 6 weeks, with blind passage into fresh media recommended every 2 weeks to ensure adequate nutrient supply.

The culture of *Borrelia* is inherently a slow and time-consuming method and hence is generally not utilised for routine clinical diagnosis. This technique is typically reserved for research applications and specialised reference laboratories.

Serodiagnosis

The center for disease control, Atlanta (CDC) recommends a two-tier testing (TTT) strategy. Step one uses enzyme linked immunosorbent assay (ELISA) or immunofluoroscence assay (IFA). Step two involves an immunoblot, which is performed only if the first test is indeterminate or positive.

For higher sensitivity, IgM detection by immunofluorescence or immunoblot is advised. Yet, in patients ill for more than 1 month, IgM positivity alone does not confirm active disease. In suspected early lyme disease (LD) with negative serology, paired serum testing offers better evidence. Patients with disseminated or late-stage LD usually show a strong IgG response to *B. burgdorferi* antigens. Most infected individuals mount a protective immune response, lowering the chance of reinfection with a similar strain.

Serological confirmation is not needed for Erythema migrans. Detailed serological criteria have been established and formally recommended by European concerted action on lyme borreliosis (EUCALB) (Table 33.1).

Molecular Diagnosis

Polymerase chain reaction (PCR) is widely used for the direct detection of *B. burgdorferi* DNA in clinical specimens.

In *Borrelia*, certain genetic loci are commonly targeted for molecular studies. The ribosomal genes *rrs* and *rrl* are frequently used because they are conserved and provide reliable markers for bacterial identification. The *ospA* and *fla* genes are also important, as they encode surface and flagellar proteins that show useful variations among species and strains. In addition, the ribosomal intergenic spacers *rrf–rrl* and *rrs–rrl* are examined for their polymorphisms, which help in distinguishing closely related genotypes. These loci are therefore valuable for species identification, strain differentiation, and phylogenetic analysis in Lyme borreliosis research.

The diagnostic sensitivity of PCR is highly dependent on the sample type. In skin biopsy samples (e.g. from erythema migrans lesions), DNA detection sensitivity typically ranges from 50% to 90%. Conversely, in cases of neuroborreliosis, the sensitivity of CSF PCR is highly variable, ranging from 12% to 80%.

Nucleic acid sequence-based amplification (NASBA) serves as an isothermal alternative to PCR. NASBA is not as commonly used as PCR in routine Lyme diagnostics due to the challenges associated with RNA stability and handling.

Loop-mediated isothermal amplification (LAMP) is another isothermal method for DNA

Table 33.1 European Concerted Action on Lyme Borreliosis (EUCALB) guidelines for serodiagnosis

Clinical manifestations	According to EUCALB: diagnostic guidelines
Erythema migrans	It is reported that in only 40–60% of untreated cases, specific IgG and/or IgM antibodies are present. Thus, serology is not considered reliable for establishing the clinical diagnosis of erythema migrans (EM)
Early neuroborreliosis	The demonstration of intrathecal antibody production is necessary A significant alteration in the titer of specific IgG and/or IgM antibodies observed between sequentially drawn (paired) blood samples provides strong supportive evidence for the diagnosis of Lyme borreliosis. However, it is critical to recognise that serum antibodies may be absent during the early stages of infection, thus limiting the utility of serology in acute presentations
Lymphadenosis benigna cutis	A significant change in the titer of specific IgG and/or IgM between paired sera and a high IgG titer is significant for the diagnosis
Chronic neuroborreliosis	Simultaneously, CSF and serum samples are collected to determine the titers of specific intrathecal antibody production Furthermore, a serological profile exhibiting isolated IgM positivity in the absence of corresponding IgG antibodies is of no significance in the diagnosis
Lyme arthritis	High levels of specific IgG antibodies Similarly, like chronic neuroborreliosis, the demonstration of specific IgM antibodies without IgG antibodies is of no significance in the diagnosis
Lyme carditis	The laboratory confirmation requires the detection of high titers of IgM or IgG antibodies or, alternatively, the demonstration of a change in the paired serum samples for IgG antibody titer
Acrodermatitis chronica atrophicans (ACA)	Specific IgG antibodies (usually high levels). IgM antibodies may or may not be present A positive IgM test in the absence of IgG antibodies argues against the diagnosis of ACA

amplification, which is characterised by high sensitivity and specificity. It targets specific *B. burgdorferi* DNA sequences, including the flaB and ospA genes, as well as other unique genomic regions linked to Lyme disease. Due to its simplicity and speed, LAMP shows point-of-care applications. However, it is yet to be used widely in routine Lyme diagnosis.

Next-generation sequencing (NGS) is used for whole-genome sequencing of *Borrelia*, targeted sequencing of specific virulence genes or ribosomal RNA genes, and metagenomic sequencing to identify *Borrelia* within a complex microbial community. However, it is not yet used routinely for the diagnosis of the infection.

Xenodiagnosis

Xenodiagnosis involves using uninfected ticks, typically *I. scapularis,* to feed on a patient, and after which, the ticks are tested for infection. This technique is not currently used for routine diagnosis.

Different diagnostic methods in Lyme disease are summarised in Table 33.2.

Treatment

The treatment of early-stage Lyme borreliosis primarily depends on oral antimicrobial therapy.

Doxycycline is generally considered the preferred first-line agent for both prophylaxis and treatment due to its efficacy against *B. burgdorferi* and its concurrent activity against common co-infections, such as babesiosis and human granulocytic anaplasmosis (which occur in 10–15% of patients). Amoxicillin and cefuroxime axetil serve as effective alternatives. Newer-generation cephalosporins and azithromycin may be considered for patients with contraindications to doxycycline. Efficacy is highest when treatment is initiated during these early localised stages.

Severe or disseminated infections necessitate parenteral antimicrobial regimens. The standard intravenous agents include ceftriaxone, cefotax-

Table 33.2 Laboratory diagnosis of Lyme disease

Diagnostic approaches	Characteristics	Remarks
Direct microscopy Dark-field Immunostaining Focus floating microscope	Organisms are twisted irregularly with easily visible spirals	*Borrelia burgdorferi* is not visible by light microscopy. Being a Spirochete, it is best visualised using dark-field or phase-contrast microscopy, or after staining with silver (Warthin–Starry) or immunofluorescence
In vitro cultures Modified Kelly medium, e.g. BSK-II, BSK-H, (Barbour, Stoenner and Kelly) or MKP	*Borrelia* species are characterised physiologically as microaerophilic, slow-growing spirochetes. They can be cultivated in vitro using specialised liquid and semi-solid formulations, such as modified Kelly medium (BSK media), which are known to support their propagation. When grown in semi-solid matrices, the resulting colonies are typically subsurface and lack distinguishing macroscopic characteristics	While positive cultures of *Borrelia* may occasionally be obtained in as early as 4 days from highly cellular clinical specimens, such as erythema migrans (EM) biopsies or synovial fluid/blood/CSF, most primary isolates necessitate several weeks of Incubation. Due to the slow growth rate of these spirochetes, cultures deemed negative must be rigorously monitored using dark-field microscopy for a minimum of 6 weeks. This extended monitoring period requires a crucial procedural step: blind passage into fresh culture medium must be performed every 2 weeks to replenish nutrients and optimise growth conditions for potentially viable, but low-density, organisms
Immunodiagnostics A two-tiered testing approach. —Typically, an ELISA or an IFA test—Followed by a confirmatory immunoblot when the primary screening results are either positive or equivocal/indeterminate	Targets of IgM immunoblot assay: 24 kDa or 21 kDa: OspC1 39 kDa: BmpA 24 or 21 kDa: OspC 41 kDa: Fla	The diagnosis of localised early-stage Lyme borreliosis, as defined by the presence of erythema migrans (EM), is established clinically and does not necessitate serological confirmation. However, serological testing is consistently recommended across both the USA and European diagnostic guidelines to corroborate the diagnosis of disseminated Lyme borreliosis (LB)
Molecular diagnosis PCR NASBA LAMP NGS	The most frequently targeted genetic loci utilised in nucleic acid amplification assays for *Borrelia* detection and identification include the highly conserved *rrs* and *rrl* ribosomal genes, the outer surface protein A *ospA* and flagellin (*fla*) genes, and the ribosomal spacer regions designated as rrf–rrl and rrs–rrl	PCR-based detection of *Borrelia burgdorferi* sensu lato DNA demonstrates variable diagnostic sensitivity across different clinical specimen types. In skin biopsy samples, for instance, the sensitivity typically ranges from 50% to 90% The success of PCR depends upon the duration of the disease. For example: In neuroborreliosis, if the disease duration is less than 14 days, the rate of PCR-positivity is significantly high; it indicates that the early-stage infection is associated with a greater pathogen DNA burden in the CSF Furthermore, the synovial fluid constitutes a crucial diagnostic tool for confirming Lyme arthritis. The ospA gene is frequently utilised as a molecular target for this specific diagnostic application

ime, or high-dose penicillin G. Patients exhibiting complex late manifestations, such as refractory arthritis or symptomatic neurological or cardiac disease, may require prolonged intravenous therapy.

An essential preventive and immediate therapeutic measure is the proper removal of attached ticks. Using fine-tipped forceps to remove the tick close to the skin surface without compressing the body significantly reduces the risk of pathogen transmission. Removal within 24 hours of attachment is critical for preventing disease progression.

Prevention and Control

Preventive strategies remain an important area of focus for managing Lyme borreliosis. The most effective method for preventing this vector-borne zoonotic disease is the avoidance of exposure to the *Ixodes* arthropod vector, which is particularly crucial for at-risk populations. Patient education is paramount to reduce the risk of tick attachment.

Currently, there is no approved vaccine for Lyme disease in active market circulation. Historically, a vaccine known as LYMErix was approved by the US FDA. This vaccine utilised recombinant OspA of *B. burgdorferi* as its target antigen and was recommended for individuals aged 16 years and older. It demonstrated a reported efficacy of 70–80% after the primary three-injection series; however, this vaccine is no longer commercially available.

Case Study

A 28-year-old male from forest areas presented to the Medicine OPD with complaints of sudden onset, painless rashes migrating from the thigh to the lower legs following an insect bite 1 week earlier. It was associated with fever, malaise, and pain in the knee joints. On examination, an erythematous annular rash associated with inguinal lymphadenopathy was found. A mild swelling of the left knee joint and a lymphangitic streak on the inner aspect of the thigh were also noted. Systemic examination was unremarkable. Laboratory investigations showed a raised polymorphonuclear (PMN) leukocyte count and a raised C-reactive protein of 106 mg/L. Left knee ultrasonography showed mild effusion with synovial thickening. Serological enzyme linked immunosorbent assay (ELISA) was positive for IgM antibodies against *B. burgdorferi* sensu lato.

1. What is the provisional diagnosis?
2. What investigations can be done to support the provisional diagnosis?
3. What could be the possible differential diagnosis for this clinical scenario?
4. What are the different guidelines formulated for treatment of this condition?

Research Questions

1. Is there any correlation between the prevalence of Lyme disease and farming practices, or the use of antimicrobials in pesticides and livestock?
2. How does the morphological variability of *B. burgdorferi* (e.g. biofilm-like forms) impact its pathogenicity and resistance to antibiotics?
3. What is the impact of climate change on the geographic distribution and prevalence of Lyme disease, and how should public health strategies adapt?

Further Reading

Cabello FC, Embers ME, Newman SA, Godfrey HP. Borreliella burgdorferi antimicrobial-tolerant persistence in Lyme disease and post treatment Lyme disease syndromes. mBio. 2022;13:e03440-21. https://doi.org/10.1128/mbio.03440-21.

Di Domenico EG, Cavallo I, Bordignon V, D'Agosto G, Pontone M, Trento E, Gallo MT, Prignano G, Pimpinelli F, Toma L, Ensoli F. The emerging role of microbial biofilm in Lyme neuroborreliosis. Front Neurol. 2018;9:1048. https://doi.org/10.3389/fneur.2018.01048. PMID: 30559713; PMCID: PMC6287027.

Eisendle K, Grabner T, Zelger B. Focus floating microscopy: "gold standard" for cutaneous borreliosis? Am J Clin Pathol. 2007;127(2):213–22.

Kliegman RM, Behrman RE, Jenson HB, Stanton BM. Nelson textbook of pediatrics e-book. Elsevier Health Sciences; 2007.

Lyme disease: diagnosis and management [M]. Evidence review for person-to-person transmission NICE guideline intervention evidence review. 2017.

Mahajan VK. Lyme disease: an overview. Indian Dermatol Online J. 2023;14:594–604.

Parija SC. Textbook of microbiology & immunology. 4th ed. Springer; 2023.

Qiu WG, Martin CL. Evolutionary genomics of *Borrelia* burgdorferi sensu lato: findings, hypotheses, and the rise of hybrids. Infect Genet Evol. 2014;27:576–93. https://doi.org/10.1016/j.meegid.2014.03.025. Epub 2014 Apr 3. PMID: 24704760; PMCID: PMC4299872.

34 Rat-Bite Fever

Bhaskar Thakuria

Learning Objectives

1. To explain the epidemiology of rat-bite fever (RBF) caused by *Streptobacillus moniliformis* and *Spirillum minus*
2. To comparatively evaluate the pathogenesis of *S. moniliformis* and *S. minus*, linking their virulence factors to clinical manifestations
3. To evaluate diagnostic challenges in the specific diagnosis of rat-bite fever
4. To discuss evidence-based treatment strategies for uncomplicated and disseminated rat-bite fever

Introduction

Rat-bite fever (RBF) is an uncommon but clinically significant bacterial zoonosis caused by the Gram-negative *Streptobacillus moniliformis* and *Spirillum minus*. *S. moniliformis* grows only under very strict conditions; however, *S. minus* cannot be cultured in vitro. The transmission dynamics of the disease reflect complex interactions between humans and both wild and domestic rodents. The infection is acquired by humans primarily through direct inoculation through bites and scratches caused by infected rats, accounting for 50% of *S. moniliformis* cases. The infection is also acquired through contaminated food and water, such as Haverhill fever outbreaks from unpasteurised dairy.

The causative agents of relapsing fever exhibit antigenic variation, leading to clinical presentations that can resemble malaria or borreliosis. Infections by these pathogens commonly present with a clinical triad comprising recurrent fever, a maculopapular rash—often affecting the palms and soles, and migratory polyarthralgia. Diagnostic challenges persist due to the demanding culture conditions required for *S. moniliformis* and the inability to culture *S. minus* using standard methods. Blood cultures require extended incubation in enriched media. Molecular techniques such as PCR targeting the 16S rRNA or *groEL* genes offer more rapid and specific detection. Retrospective diagnosis may be supported by serological evidence, particularly the presence of antibodies against Yersinia outer proteins (Yops). Penicillin remains the drug of choice for the treatment of *S. moniliformis* infections. However, the emergence of β-lactamase-producing strains has highlighted the need for alternative agents such as doxycycline or ceftriaxone. For *S. minus* infections, streptomycin is considered the antibiotic of choice.

B. Thakuria (✉)
Department of Microbiology, All India Institute of Medical Sciences, Patna, Bihar, India
e-mail: drbhaskart@aiimspatna.org

S. C. Parija, A. Poddar (eds.), *Textbook of Bacterial Zoonoses*, Microbial Zoonoses,
https://doi.org/10.1007/978-981-95-6615-0_34

History

The history of rat-bite fever (RBF) dates back centuries, with early descriptions of illnesses following rodent bites appearing in ancient Indian and Chinese medical texts. In Japan, a condition resembling RBF, later termed *sodoku* (from *so*, meaning "rat", and *doku*, meaning "poison"), was documented as early as the second century but was not formally characterised until the nineteenth century.

The disease gained medical attention in the West during the 1800s, when physicians began documenting cases of fever, rash, and arthritis following rat bites. In 1839, Wilcox described a relapsing fever in a patient bitten by a rat, which may have been an early account of RBF. Subsequently, Japanese physicians identified *S. minus* as the causative agent of sodoku. The microbiological basis of RBF was further elucidated in the early twentieth century, when Schottmüller isolated *S. moniliformis* from a German patient in 1914.

Concurrently, Carter first observed *S. minus* microscopically in 1887; later, its role in RBF in the early 1900s was confirmed by Japanese researchers. A Haverhill fever outbreak in Massachusetts in the United States, involving 86 cases, was documented by Blake in 1925. This was confirmed as food-borne transmission of *S. moniliformis* without a rat bite. During the mid-twentieth century, two distinct forms of RBF were recognised. These included *streptobacillary RBF*, caused by *S. moniliformis*, which is more common in Western countries, and *spirillary RBF (sodoku)*, caused by *Spirillum minus*, predominantly found in Asia.

Taxonomy

S. moniliformis and *S. minus*, the causative agents of rat-bite fever (RBF), belong to two distinct bacterial taxa with markedly different biological characteristics.

S. moniliformis, the primary agent in Western countries, is classified within the domain Bacteria, phylum Fusobacteriota (formerly Fusobacteria), class Fusobacteriia, order Fusobacteriales, and family Leptotrichiaceae. This fastidious, Gram-negative bacterium exhibits its pleomorphic morphology, often appearing as filamentous chains of bead-like structures, which explains its species epithet "moniliformis". Unlike typical members of its phylum, which are often commensals in oral and gastrointestinal flora, *S. moniliformis* has adapted to become an invasive pathogen. Its fastidious growth requirements initially led to taxonomic confusion, with historical classifications under *Haemophilus* and temporary designations like *Actinomyces muris*, before modern genomic analysis confirmed its current taxonomic position.

In contrast, the Asian variant of RBF is caused by *S. minus*, a tightly coiled, spiral-shaped bacterium belonging to the Proteobacteria phylum, class Betaproteobacteria, order Nitrosomonadales, and family Spirillaceae. This motile organism, possessing characteristic bipolar flagella, has resisted cultivation on artificial media, requiring microscopic identification or animal inoculation for detection.

The taxonomic divergence between these two pathogens is profound. *S. moniliformis* shares closer genetic relationships with oral commensals, while *S. minus* shows evolutionary links to environmental Proteobacteria.

Morphology

The bacterial agents causing rat-bite fever exhibit distinct morphological characteristics crucial for their identification and differentiation.

S. moniliformis is a Gram-negative, pleomorphic, non-motile bacillus. It typically forms long, filamentous chains resembling beaded strings, a feature reflected in its species name, "*moniliformis*". Dark-field microscopy reveals its distinctive filamentous, beaded structure. Under microscopic examination, these bacteria appear as slender, straight, or slightly curved rods measuring 0.1–0.5 μm in width and 1–5 μm in length, often arranged in clusters or tangled masses in liquid media (Fig. 34.1). The organism stains poorly with Gram staining but is better visualised

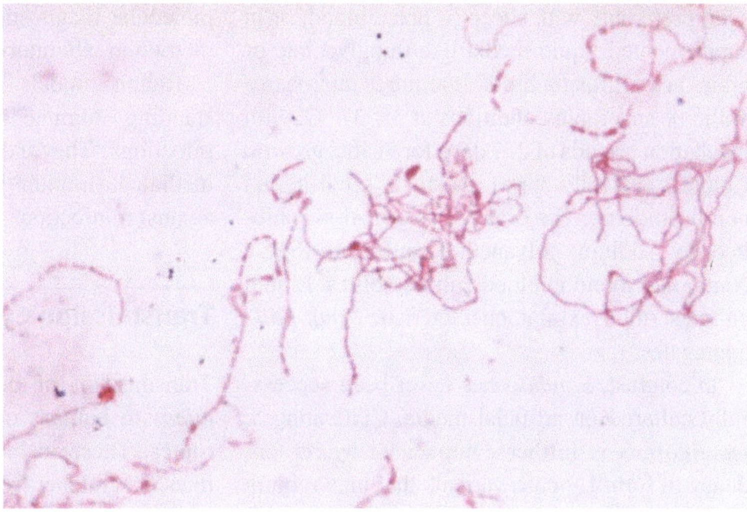

Fig. 34.1 Light microscopic image of Gram-stained smear showing Gram-negative *Streptobacillus moniliformis* morphology arranged in chains (×100). (Image source: CDC)

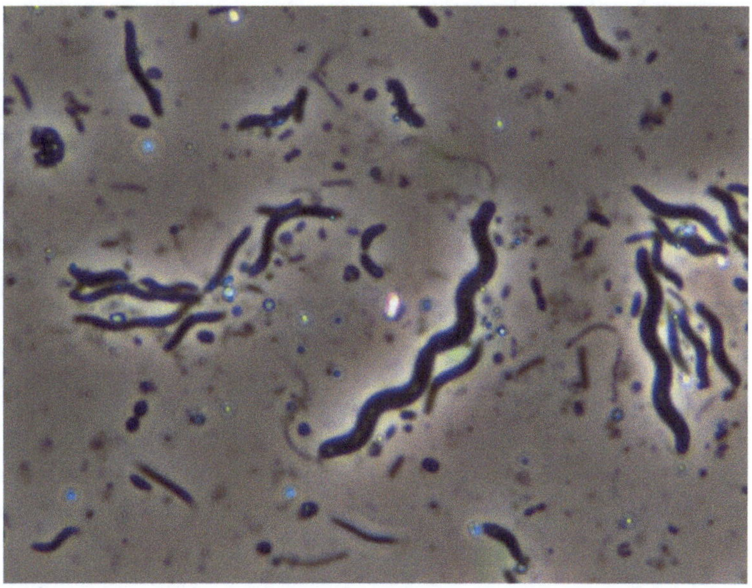

Fig. 34.2 Microscopic morphology of *Spirillum minus* showing coiled, spiral-shaped bacteria with bipolar flagella. (Image source: Wikipedia, Copyright free)

using Giemsa or Wright stains, particularly in blood smears or joint fluid.

In contrast, *S. minus* is a tightly coiled, spiral-shaped, Gram-negative bacterium exhibiting rapid, corkscrew-like motility due to its bipolar tufts of flagella (Fig. 34.2). It measures 0.2–0.5 µm in width and 3–10 µm in length. It is best demonstrated using dark-field or phase-contrast microscopy, where its rigid helical morphology and darting movements serve as key diagnostic features. *S. minus* requires specialised staining techniques, such as silver impregnation (e.g. Warthin-Starry stain), to be visualised in tissue samples.

Cultivation

The cultivation of rat-bite fever pathogens presents distinct challenges due to the differing growth requirements of *S. moniliformis* and *S. minus*.

S. moniliformis is a fastidious bacterium that can be cultured on enriched media such as blood agar, preferably with sheep or horse blood, or in supplemented liquid media like thioglycolate or brain–heart infusion broth. It requires microaerophilic or anaerobic conditions at 35–37 °C, with incubation periods of 2–7 days for visible growth. Colonies typically appear as small, greyish, and non-haemolytic. The organism's growth is inhibited by sodium polyanethol sulfonate (SPS), commonly found in blood culture bottles. Liquid cultures often exhibit characteristic "*puff-ball*" aggregates.

In contrast, *S. minus* has never been successfully cultured on artificial media. Cultivating *S. moniliformis* is further complicated by its tendency to form L-phase variants that may require hypertonic media. At the same time, its sensitivity to environmental conditions demands careful specimen handling and prompt processing.

Animal Models

Murine models, notably Swiss Webster and BALB/c mice, have been extensively used for experimental infection by *S. moniliformis*. Intraperitoneal or intravenous inoculation produces a systemic infection characterised by bacteraemia, arthritis, and splenomegaly within 3–7 days, closely mimicking human disease progression. Immunocompromised mice develop more severe manifestations, including endocarditis and abscess formation, similar to human cases in vulnerable populations. These murine models have demonstrated that virulence factors such as the haemolysin and outer membrane proteins contribute to the pathogenicity of *S. moniliformis*.

Guinea pigs and rats have also been employed to study the natural oral transmission of *S. moniliformis*. It has been revealed that *S. moniliformis* can colonise the oropharynx and be shed in saliva, providing insights into zoonotic transmission dynamics.

The historical use of mouse inoculation remains the gold standard for the isolation and propagation of *S. minus*. Intraperitoneal injection leads to detectable bacteraemia within 5–14 days, with organisms visible in peritoneal exudates by dark-field microscopy. However, currently, molecular diagnostics have largely replaced this method in laboratories.

Rabbit models have contributed to understanding immune responses against *S. minus* infections. They exhibit both humoral and cell-mediated immunity, with partial protection against reinfection.

Transmission Cycle

Transmission of both *S. moniliformis* and *S. minus* to humans occurs through three primary routes. These include (i) direct inoculation via bites or scratches from infected rodents, accounting for most *S. minus* cases and many *S. moniliformis* infections; (ii) indirect contact with rodent secretions through mucous membranes or breaks in the skin, especially relevant for laboratory workers and children handling pet rodents; and (iii) ingestion of food or water contaminated with rat excreta, responsible for *S. moniliformis* outbreaks like Haverhill fever.

For *S. moniliformis*, additional transmission routes include handling rodent carcasses, exposure to rodent urine, and rare cases of person-to-person transmission through saliva. At the same time, *S. minus* appears more strictly dependent on direct inoculation through bites.

The persistence of these bacteria in rodent populations is facilitated by vertical transmission from mothers to offspring and horizontal spread through social behaviours in the community. Overcrowding and poor sanitation amplify transmission risks by increasing bacterial loads in rodent habitats and the likelihood of human exposure, particularly in urban settings with large rat populations.

Pathogenesis and Pathology

The bacteria enter the human host through skin breaches or mucosal surfaces. *S. moniliformis* demonstrates tissue tropism for synovial membranes and cardiac tissue, while *S. minus* shows a predilection for lymphatic spread and systemic dissemination.

S. moniliformis initiates infection by adherence to epithelial cells and extracellular matrix components via surface adhesins. This is followed by the invasion of subcutaneous tissues at the inoculation site. At the site of inoculation, it evades phagocytosis through its antiphagocytic capsule and induces local inflammation characterised by neutrophil infiltration, oedema, and tissue necrosis. The bacteria then disseminate hemaetogenously, facilitated by endotoxin (LPS) and haemolysin production, leading to systemic symptoms such as fever, chills, and myalgias within 3–10 days post-exposure.

The development of septic arthritis in 50% of cases is a key pathological feature, caused by bacterial colonisation of synovial membranes that trigger intense inflammatory responses. This results in joint swelling, fibrin deposition, and eventual cartilage destruction if untreated. *S. moniliformis* also demonstrates tropism for cardiac tissues, where it can cause endocarditis by colonising damaged valves, leading to vegetations composed of platelets, fibrin, and bacterial clusters that may embolise to distant organs.

Severe complications such as abscesses in multiple organs, including the liver, brain, and lungs, occur more frequently with *S. moniliformis*. These occur particularly in immunocompromised hosts, due to their ability to form biofilm-like aggregates and persist as L-form variants resistant to antibiotic treatment.

S. minus pathogenesis involves local multiplication at the bite site, producing a characteristic indurated ulcer with regional lymphangitis and lymphadenopathy. It is followed by systemic spread via lymphatic and haematogenous routes that induce recurrent febrile episodes due to antigenic variation and cyclic bacteraemia. Histopathological examination of infected tissues reveals neutrophilic infiltrates, microabscess formation, and vasculitis in both forms of rat-bite fever. However, *S. minus* infections often show more pronounced perivascular inflammation and endothelial damage.

Immunology

The Innate Immune Response

Following infection with *S. moniliformis* or *S. minus*, the innate immune system responds through pattern-recognition receptors (TLRs 2 and 4). These receptors detect bacterial lipoproteins and LPS, triggering neutrophil recruitment and pro-inflammatory cytokine release (IL-1β, IL-6, TNF-α) that mediate initial fever and acute phase responses.

S. moniliformis evades early immune clearance through multiple mechanisms. These include its antiphagocytic capsule, L-form transformation that avoids cell wall-targeting immune responses, and suppression of complement-mediated killing. At the same time, *S. minus* employs antigenic variation of surface proteins to escape antibody recognition during its characteristic relapsing bacteraemia.

The Adaptive Immune Response

The adaptive immune response develops within 7–10 days following with *S. moniliformis* infection. Th1-type cellular immunity (IFN-γ production) for intracellular clearance and robust antibody production against outer membrane proteins are produced. However, these antibodies often fail to prevent reinfection due to antigenic heterogeneity among strains.

S. minus infection generates predominantly humoral immunity. It is characterised by cyclical fever episodes corresponding to waves of bacteraemia that stimulate new antibody production against variant surface antigens. This explains the relapsing nature of spirally rat-bite fever.

Both pathogens can trigger autoimmune-like phenomena. *S. moniliformis* causes molecular mimicry-induced arthritis through cross-reactive antibodies targeting joint epitopes. At the same time, *S. minus* infection may lead to immune complex deposition in small vessels, contributing to vasculitic manifestations.

Protective immunity appears strain-specific and short-lived, as demonstrated by animal models showing only partial protection against heterologous challenge, explaining the absence of a natural vaccine.

Infections in Humans

The incubation period differs between pathogens, typically 3–10 days for *S. moniliformis* but often longer (7–21 days) for *S. minus*, reflecting their distinct pathogenic strategies.

These infections typically manifest with a characteristic triad that includes recurrent episodes of fever, a maculopapular rash, frequently involving the palms and soles, and migratory joint pains (polyarthralgia). Patients develop an abrupt onset of high fever, often >39 °C, chills, and severe migratory polyarthralgia affecting large joints in 50% of cases. A maculopapular or petechial rash also develops, which typically spreads centrifugally to involve the palms and soles. The rash is a hallmark feature distinguishing RBF from other febrile illnesses.

S. moniliformis infections frequently progress to septic arthritis, mainly in the knees, elbows, and wrists. At the same time, *S. minus* classically causes indurated ulceration at the inoculation site with regional lymphadenitis and a relapsing fever pattern resembling malaria or borreliosis.

Severe complications include endocarditis, particularly in those with valvular abnormalities; meningitis, and abscess formation, such as hepatic, splenic, and cerebral abscesses. The overwhelming sepsis, with mortality rates, has been observed in 10–13% of untreated cases and in immunocompromised hosts. These individuals exhibit more severe disease due to impaired neutrophil function and delayed antibody responses, particularly with *S. moniliformis*.

Table 34.1 highlights variations in clinical presentation in rat-bite fever caused by *S. moniliformis* and *S. minus*.

Infections in Animals

Rats rarely show symptoms; occasional cases of cervical lymphadenitis, otitis media, or pneumonia have been reported in heavily colonised animals. Other species, like mice, gerbils, and guinea pigs, acquire the infection through contact with infected rats but typically remain asymptomatic carriers. Notably, *S. moniliformis* causes more severe disease in non-rodent species, including "*streptobacillary fever*" in laboratory primates. It causes fatal septicaemia in turkeys. Dogs and cats may develop a mild febrile illness after consuming infected rodents.

Table 34.1 Comparison of clinical features caused by *Streptobacillus moniliformis* and *Spirillum minus*

Feature	*Streptobacillus moniliformis*	*Spirillum minus* (Sodoku)
Incubation period	3–10 days	7–21 days
Fever pattern	Abrupt onset (>39 °C)	Relapsing (malaria-like)
Joint involvement	Migratory polyarthralgia (50% cases); septic arthritis (knees/elbows/wrists)	Mild or absent
Characteristic rash	Maculopapular/petechial; centrifugal spread to palms/soles	Rare or localised
Inoculation site	Heals normally	Indurated ulcer with lymphangitis
Lymphadenopathy	Mild generalised	Marked regional
Common complications	Septic arthritis Endocarditis Meningitis	Disseminated granulomas; hepatosplenic involvement
Mortality (untreated)	10–13%	5–10%
High-risk populations	Children with pet rodents; lab workers; immunocompromised	Similar; less documented

Epidemiology and Public Health

Both *S. moniliformis* and *S. minus* are primarily maintained in nature through asymptomatic colonisation of rodents, particularly rats. Both wild *Rattus* species and laboratory/pet rats serve as the principal reservoir hosts. These harbour the bacteria in their nasopharynxes, middle ear, and respiratory tract without showing clinical signs. *S. moniliformis* is particularly prevalent in the oropharynx of healthy-appearing rats. These commensal bacteria are shed in saliva, nasal secretions, or the urine of infected rodents. The bacteria persist in rodent populations through vertical transmission to offspring and horizontal spread via social activities. Traditional *S. minus* transmission routes follow rat bites in agricultural settings. Unique non-bite transmission has also been reported. Food-borne outbreaks of *S. moniliformis*, also known as Haverhill fever, occur through contaminated milk or water.

Rat-bite fever (RBF) is a significant yet underrecognised public health concern. Its complex epidemiology reflects zoonotic and environmental transmission dynamics. The disease has a worldwide distribution *S. moniliformis* infections are predominant in North America and Europe, while *S. minus* cases cluster predominantly in Asia. Most cases occur in Japan, where it is traditionally called "sodoku". They are also reported in the United States, Europe, Australia, and Africa.

Epidemiologic studies suggest substantial underreporting, with an estimated incidence of 0.1–2 cases per million people annually in developed countries. High-risk populations include children with pet rats, accounting for 50% of US cases, 20% of cases of laboratory workers handling rodents, and residents of impoverished communities with poor rodent control. Rat-bite fever (RBF) continues to be reported, particularly among laboratory workers, pet owners, and children exposed to rodents, underscoring the ongoing relevance of this zoonosis.

Public health challenges include the rising popularity of pet rats (with 30–50% colonised asymptomatically), increasing laboratory rodent use in research, and climate change expanding rodent habitats.

Diagnosis

Specimens

Wound exudates, blood, serum, synovial fluid, and tissue exudates are essential for microscopy and serum for serodiagnosis and molecular diagnostics.

An overview of diagnosis is presented in Table 34.2.

Microscopy

Microscopy of wound exudates, blood, or synovial fluid stained by Gram stain shows *S. moniliformis* as pleomorphic, Gram-negative rods in chains. *S. minus* is seen as a tightly coiled, spiral organism. Both species stain poorly and require specialised staining and techniques.

Giemsa or Wright staining enhances the detection of *S. moniliformis*, visualised as blue-purple filaments with characteristic bulbous swellings, and *S. minus* as deep violet spirals adhering to erythrocytes. Wayson stain is used as a rapid field screening stain, visualising *S. moniliformis* with bipolar "safety-pin" morphology and *S. minus* as distinct blue coils. Silver impregnation techniques, such as Warthin-Starry stain, are used to identify *S. minus* in tissues as blackened spirals against a yellow background and delineate *S. moniliformis* filaments. Acridine orange fluorescence staining detects *S.moniliformis* as orange fluorescent threads and *S. minus* as glowing spirals at low concentrations.

Due to poor inherent staining, *S. minus* diagnosis depends heavily on dark-field microscopy. It is important for visualising corkscrew motility and tight spirals (2–6 coils) of *S. minus* in blood smears or lesion exudates.

In Vitro Culture

In vitro culture of *S. moniliformis* remains a challenge due to its fastidious nature. The bacterium requires blood agar or thioglycolate broth supplemented with 10–20% serum for optimum growth.

Table 34.2 Diagnosis of rat-bite fever

Diagnostic technique	Specific tool/method	Pathogen	Sample type	Advantages	Disadvantages	Remarks (sensitivity/specificity)
Microscopy	Gram stain	*Streptobacillus moniliformis*, *Spirillum minus*	Wound exudate, blood, synovial fluid	Basic initial identification	Poor staining; may miss pathogen	Low sensitivity
	Giemsa/Wright stain	*S. moniliformis*, *S. minus*	Blood, tissues	Enhances detection, shows morphology	Needs good smear quality	Moderate sensitivity
	Wayson stain	*S. moniliformis*, *S. minus*	Field samples	Rapid screening	Poor specificity	Useful in low-resource settings
	Warthin-Starry stain	*S. moniliformis*, *S. minus*	Tissue biopsies	Sensitive visualisation	Technically demanding	Detects organisms in tissue
	Acridine orange fluorescence stain	*S. moniliformis*, *S. minus*	Exudate, blood	High sensitivity in low bacterial load	Requires fluorescence microscopy	*S. minus* glows spirally
	Dark-field microscopy	*S. minus*	Blood, lesion exudate	Best for visualising motility and spiral shape	Cannot differentiate species	Critical tool for *S. minus*
In vitro culture	Blood agar/thioglycolate broth + serum	*S. moniliformis*	Blood, synovial fluid	Confirms viable organism	Slow, fastidious, may fail	3–7 days incubation; microaerophilic
	Animal inoculation	*S. minus*	Blood, tissue	Only isolation method	Ethically and logistically limited	Historical; now rarely used
Serodiagnosis	ELISA (PIM protein, LPS)	*S. moniliformis*	Serum	High specificity (>85%), IgM/IgG detection	Cross-reactivity, not for acute phase	Paired sera improve accuracy
	Microagglutination test (MAT)	*S. moniliformis*, *S. minus*	Serum	Useful in convalescence	Needs high titer (≥1:80)	Culture-negative cases
	Indirect immunofluorescence assay (IFA)	*S. moniliformis*	Serum	Rapid, specific	Requires expertise	Interpretation dependent
	Passive hemagglutination	*S. minus*	Serum	Still used in Asia	Low sensitivity (50–60%)	Limited utility
Molecular diagnosis	PCR (16S rRNA)	*S. moniliformis*, *S. minus*	Synovial fluid, tissue, blood	High sensitivity/specificity, culture-independent	Requires lab setup	Preferred method
	qPCR (16S rRNA)	*S. moniliformis*	Same as above	Quantification possible	More costly	Rapid turnaround
	Nested PCR (16S/flagellin *flaB*)	*S. minus*	Blood, tissue	Enables *S. minus* detection	Contamination risk	Crucial for *S. minus* diagnosis
	Multiplex PCR panels	*S. moniliformis*, *S. minus*	Mixed samples	Simultaneous detection	Need standardisation	Useful for syndromic panels
	MALDI-TOF MS	*S. moniliformis*, *S. minus*	Culture isolates	Rapid ID	Needs culture growth	Promising adjunct

It grows best under microaerophilic conditions and typically requires 3–7 days of incubation. Even under these specialised conditions, growth may be scant and slow, making laboratory identification difficult.

Biochemical identification of *S. moniliformis* is similarly limited, as it exhibits minimal reactivity in standard biochemical tests. It is oxidase- and catalase-negative and non-motile. Fermentation reactions are weak or inconsistent, often requiring molecular confirmation for definitive identification.

S. minus cannot be cultivated on artificial media, historically necessitating animal inoculation in mice for isolation.

Serodiagnosis

Serological diagnosis of rat-bite fever in humans is primarily used for retrospective confirmation or in culture-negative cases. Specific serum antibodies typically develop 1–3 weeks post-infection.

The indirect enzyme-linked immunosorbent assay (ELISA) using *S. moniliformis*-specific antigens such as the recombinant PIM protein or LPS is the most sensitive method. The test detects IgM (acute phase) and IgG (convalescent) antibodies with >85% specificity. However, cross-reactivity with *Leptospira* or streptococci may occur.

Microagglutination tests (MAT) using whole-cell antigens, demonstrating antibody titers ≥1:80 in convalescent sera, are diagnostic. Indirect immunofluorescence assays (IFA) visualise pathogen-specific antibodies binding to fixed bacterial smears, offering rapid results but requiring experienced interpretation. Passive hemagglutination remains a historically important test in Asia for serodiagnosis of *S. minus* infection, but suffers from low sensitivity (50–60%).

Serology is unreliable in acute illness, where antibodies are often undetectable. In addition, due to antigenic diversity, paired acute and convalescent sera (collected 2–4 weeks apart) showing a fourfold titer rise are essential for confirmation. Commercial tests are not widely accessible, limiting routine use outside reference laboratories.

Molecular Diagnosis

Molecular diagnostic techniques have become increasingly valuable in diagnosing rat-bite fever, particularly due to the limitations of conventional culture methods. Polymerase chain reaction (PCR) is crucial in identifying *S. moniliformis* and *S. minus*, as they are challenging to culture with routine methods.

For *S. moniliformis*, species-specific PCR targeting, the 16S rRNA gene has high sensitivity and specificity. The method has the advantage of rapid and accurate detection of the bacterium directly in synovial fluid, tissue samples, and sometimes blood. Real-time PCR (qPCR) further enhances diagnostic utility by quantifying bacterial load and reducing turnaround time. The molecular diagnosis of *S. minus* is even more important given its inability to grow on artificial media. Nested PCR targeting the 16S rRNA gene or flagellin genes (e.g., *flaB*) has been used successfully to detect *S. minus* DNA in blood or tissue samples.

Currently, multiplex PCR panels with the simultaneous detection of various fastidious pathogens, including those responsible for zoonotic infections and rat-bite fever, are also standardised.

MALDI-TOF MS is increasingly being utilised for the rapid identification of rat-bite fever pathogens. This technique is especially valuable in cases where clinical suspicion is high, but culture methods fail due to the fastidious nature of *S. moniliformis* and *S. minus*.

Treatment

Treatment of rat-bite fever (RBF) requires prompt antibiotic therapy to prevent severe complications. Penicillin remains the gold standard for both *S. moniliformis* and *S. minus* infections. Intravenous penicillin G (1.2–2.4 million units daily) or high-dose oral penicillin V (500 mg four times daily) for 7–14 days is recommended for uncomplicated cases. Endocarditis or other invasive infections require extended IV therapy (4–6 weeks), with ceftriaxone often added for enhanced CNS penetration. Doxycycline (100 mg twice daily) and streptomycin (15 mg/kg daily) are effective alternatives for penicillin-allergic patients. However, tetracyclines are avoided in pregnant women and young children due to potential adverse effects.

S. moniliformis exhibits intrinsic resistance to vancomycin, trimethoprim-sulfamethoxazole, and aminoglycosides, necessitating careful antibiotic selection. Treatment challenges include increasing reports of β-lactamase-producing *S. moniliformis* strains in some regions, prompting consideration of adding β-lactamase inhibitors in severe or non-responsive cases.

Prophylactic antibiotics after rodent bites remain controversial. However, it may be considered for high-risk patients such as immunocompromised individuals or those with pre-existing valvular disease, by using a 3–5-day oral penicillin or doxycycline course.

Prevention and Control

Primary prevention focuses on minimising contact with wild and pet rodents. Proper handling methods include using gloves when cleaning cages, avoiding face-to-face contact with pet rats, and supervising children during interactions with rodents. Strict personal protective equipment (PPE) protocols should be followed by high-risk groups such as laboratory workers and pet store employees when handling rodents or their bedding, including masks, gowns, and eye protection.

Control focuses on awareness, surveillance, the quarantine of new animals, and the treatment of infected breeding stocks. However, complete eradication is challenging because these pathogens are common in rodent populations. Environmental control measures include maintaining rodent-proof housing to prevent wild rat infestations, proper food storage to avoid attracting rodents, and community-level sanitation programmes in urban areas with high rat populations.

Case Study

A 7-year-old boy presented with a 5-day history of intermittent high-grade fever (39–40 °C), migratory polyarthralgia affecting both knees and wrists, and a generalised maculopapular rash. The rash progressed to involve his palms and soles following a pet rat bite to his left index finger 10 days prior, which had initially healed without intervention.

Blood cultures drawn in SPS-free bottles grew thin, Gram-negative, pleomorphic bacilli forming "fried egg" colonies on blood agar after 72 hours of incubation. This was later confirmed as *S. moniliformis* by MALDI-TOF mass spectrometry and 16S rRNA gene sequencing. Synovial fluid analysis from his swollen right knee demonstrated 12,000 WBC/μL with 90% neutrophils. It was PCR-positive for *S. moniliformis*, while echocardiography, fortunately, showed no valvular vegetations, leading to a definitive diagnosis of rat-bite fever complicated by septic arthritis. The patient showed dramatic clinical improvement within 48 hours of initiating intravenous penicillin G (250,000 units/kg/day), transitioning to oral amoxicillin after 1 week for a total 14-day course, with complete resolution of systemic symptoms and arthritis at follow-up.

1. What are the key clinical differences between *S. moniliformis* and *S. minus* infections in rat-bite fever?
2. Why are conventional culture methods often inadequate for diagnosing rat-bite fever, and how do molecular techniques overcome these limitations?
3. What are the potential complications of untreated rat-bite fever, and how can timely diagnosis improve patient outcomes?
4. What public health measures can help prevent rat-bite fever, especially in high-risk populations or occupational settings?

Research Questions

1. What are the emerging antimicrobial resistance patterns in *S. moniliformis*, and how might they influence future treatment guidelines for rat-bite fever?
2. How do host immune responses differ between *S. moniliformis* and *S. minus* infections?
3. What role do urbanisation and changing rodent ecology play in the epidemiology of rat-bite fever?

Declaration During the preparation of this chapter, we used Copy AI's AI Paragraph Rewriter to improve the readability of the text. After using it, we reviewed and edited the content as needed and took full responsibility for the publication's content.

Further Readings

Adam JK, Varan AK, Pong AL, McDonald EC. Notes from the field: fatal rat-bite fever in a child – San Diego County, California, 2013. MMWR Morb Mortal Wkly Rep. 2014;63(50):1210–1.

Boot R, Bakker RH, Thuis H, Veenema JL. An enzyme-linked immunosorbent assay (ELISA) for monitoring rodent colonies for *Streptobacillus moniliformis* antibodies. Lab Anim. 1993;27(4):350–7.

Boot R, Oosterhuis A, Thuis HC. PCR for the detection of *Streptobacillus moniliformis*. Lab Anim. 2002;36(2):200–8.

CDC. Fatal rat-bite fever – Florida and Washington, 2003. MMWR Morb Mortal Wkly Rep. 2005;53(51):1198–202.

Dendle C, Woolley IJ, Korman TM. Rat-bite fever septic arthritis: illustrative case and literature review. Eur J Clin Microbiol Infect Dis. 2006;25(12):791–7.

Eisenberg T, Ewers C, Rau J, Akimkin V, Nicklas W. Approved and novel strategies in diagnostics of rat bite fever and other *Streptobacillus* infections in humans and animals. Virulence. 2016;7(6):630–48.

Elliott SP. Rat bite fever and *Streptobacillus moniliformis*. Clin Microbiol Rev. 2007;20(1):13–22.

Freels LK, Elliott SP. Rat bite fever: three case reports and a literature review. Clin Pediatr (Phila). 2004;43(3):291–5.

Gaastra W, Boot R, Ho HT, Lipman LJ. Rat bite fever. Vet Microbiol. 2009;133(3):211–28.

Graves MH, Janda JM. Rat-bite fever (*Streptobacillus moniliformis*): a potential emerging disease. Int J Infect Dis. 2001;5(3):151–5.

Khatchadourian K, Ovetchkine P, Minodier P, Lamarre V, Lebel MH, Tapiéro B. The rise of the rats: a growing paediatric issue. Paediatr Child Health. 2010;15(3):131–4.

Kimura M, Tanikawa T, Suzuki M, Koizumi N, Kamiyama T, Imaoka K. Detection of *Streptobacillus moniliformis* and *Spirillum minus* by PCR from wild rat samples in Japan. J Vet Med Sci. 2008;70(6):637–40.

Schachter ME, Wilcox L, Rau N, et al. Rat-bite fever, Canada. Emerg Infect Dis. 2006;12(8):1301–2.

Shanks N, Greek R, Greek J. Are animal models predictive for humans? Philos Ethics Humanit Med. 2009;4:2.

Torres L, López AI, Escobar S, Torres M, Ugalde E. Rat-bite fever caused by *Streptobacillus moniliformis* in a child. Arch Argent Pediatr. 2019;117(4):e404–7.

Spotted Fevers 35

Archana Keche, Girish Patil, Kumar Vikram, and Yogendra Keche

Learning Objectives

1. To know causative agents of spotted fevers, including reservoirs
2. To discuss the pathogenicity of spotted fever
3. To describe the diagnosis, prevention, and control of spotted fever

Introduction

Rickettsial diseases are emerging and/or re-emerging zoonotic infections transmitted to humans and animals through the bite of ticks and mites. The genus *Rickettsia* comprises small, obligate intracellular, Gram-negative bacteria, which are broadly categorised into the typhus fever group (TFG) and the spotted fever group (SFG).

Rickettsial diseases are broadly categorised into the typhus fever group (TFG) and the spotted fever group (SFG) based on a combination of their molecular, biological, and immunological characteristics. The SFG consists of over 30 species that can be found worldwide. Clinical symptoms include fever, eschar, rash, and lymphadenopathy. It can progress to serious complications, and death may occur if left untreated. Rocky Mountain spotted fever (RMSF), caused by *Rickettsia rickettsii,* is recognised as the most severe infection among the SFG.

All spotted fevers are zoonotic. Most rickettsial bacteria are transmitted directly to humans by infected arthropod vectors (i.e. fleas, lice, mites, or ticks) during feeding. Initial clinical manifestations typically consist of fever, headache, muscle pain, rash, local lymphadenopathy, and a characteristic inoculation eschar. The timely diagnosis of SFG is challenging due to the intracellular nature of the bacteria. This cannot be grown in cell-free media like viruses. Molecular methods (polymerase chain reaction (PCR)) and loop-mediated isothermal amplification (LAMP)) should be performed in the first 5 days of infection. Serological methods like Weil-Felix test, immunofluorescence assay (IFA), and enzyme-linked immunosorbent assay (ELISA) yield better results after seroconversion, 5–7 days post-infection. Treatment mainly includes tetracyclines, which help to prevent fatal consequences.

A. Keche (✉) · G. Patil
Department of Microbiology, All India Institute of Medical Sciences, Raipur, Chhattisgarh, India

K. Vikram
Department of Microbiology, ESIC Medical College and Hospital, Joka, Kolkata, West Bengal, India

Y. Keche
Department of Pharmacology, All India Institute of Medical Sciences, Raipur, Chhattisgarh, India

S. C. Parija, A. Poddar (eds.), *Textbook of Bacterial Zoonoses*, Microbial Zoonoses,
https://doi.org/10.1007/978-981-95-6615-0_35

History

The first description of spotted fever is known to have occurred in 1899, when Maxcy described Rocky Mountain spotted fever (RMSF). Initially recognised in 1896 in Idaho's Snake River Valley, RMSF was locally known as "black measles" due to its characteristic, often severe, rash. This mysterious disease frequently led to death.

Dr. Howard T. Ricketts, an American pathologist, embarked on groundbreaking research that would revolutionise the understanding of RMSF. By 1906, he definitively demonstrated that the disease was transmitted by the bite of the Rocky Mountain wood tick, *Dermacentor andersoni*. His meticulous work, conducted between 1907 and 1909, culminated in the identification of the causative bacterial agent, which was later named *Rickettsia rickettsii,* in his honour. Tragically, he succumbed to typhus, another rickettsial disease, in 1910 while studying an outbreak in Mexico City.

In 1919, Simeon Burt Wolbach provided the first detailed description of *R. rickettsii* within endothelial cells, clarifying the disease's pathology. The subsequent decades saw the development of early preventive measures, such as the Spencer-Parker vaccine in 1924. The said vaccine, while effective in reducing mortality, was eventually superseded by the advent of antibiotics. The mid-twentieth century marked a revolutionary turning point with the discovery and widespread availability of broad-spectrum antibiotics, particularly tetracycline-class drugs like doxycycline. This transformed the prognosis for spotted fevers, drastically reducing fatality rates from as high as 20–30% to significantly lower levels with prompt treatment.

A historical timeline showing major events in the study of spotted fever is presented in Table 35.1.

Table 35.1 History of spotted fever group rickettsia

Discovery	Year	Author
First description of RMSF	1899	Maxcy
Role of tick in RMSF	1908	McCalla
First description of RMSF	1909	Conor
Isolation of *R. rickettsii*	1911	Ricketts
Weil-Felix test	1916	Weil-Felix
Isolation of *R. conorii*	1932	Brumpt
Description of Siberian tick typhus	1935	Shmatikov
Isolation of *R. sibirica*	1938	Krontovuka
Description of *Rickettsial pox*	1946	Huebner
Isolation of *Rickettsia akari*	1946	Huebner
Isolation of *Rickettsia australis*	1946	Plotz and Smadel
Queensland tick typhus	1946	Plotz and Smadel
Isolation of *Rickettsia slovaca*	1968	Brezina
Culture of *Rickettsia conorii Israel*	1974	Goldwasser
Isolation of *Rickettsia helvetica*	1979	Burgdofer and peter
Japanese spotted fever	1984	Mahara
Culture of *Rickettsia japonica*	1989	Udida and Walker
Isolation of *Rickettsia africae*	1990	Kelly
First case of *Rickettsia africae*	1992	Kelly and Raoult
First case of flea-borne spotted fever	1994	Schriefer and Azad
First infection by *R. slovaca*	1997	Raoult
First case of acute infection by *Rickettsia helvetica*	2000	Fournier and Raoult

Taxonomy

Genus *Rickettsia* belongs to the class Alphaproteobacteria, order Rickettsiales and family Rickettsiaceae. The universal gene coding for 16S rRNA played a crucial role in the classification of *Rickettsia* and other related genera.

The genus *Rickettsia* is classified into the typhus fever group (TFG) and the spotted fever group (SFG) based on a combination of their molecular, biological, and immunological characteristics. Briefly, a key distinction lies in their cellular localisation: SFG rickettsiae reside in both the cytoplasm and the nucleus of host cells, while TFG rickettsiae are confined to the cytoplasm. Optimal growth temperatures also differ, with SFG typically thriving at 32 °C (except for *Rickettsia felis,* which thrives at 28 °C), and TFG at 37 °C. Furthermore, the G + C content of their DNA varies, being 32–33% for SFG and 29–30% for TFG. Immunologically, TFG and *R. rickettsii* show cross-reactivity with the OX19 strain of non-motile *Proteus vulgaris*, whereas other SFG members react with the OX2 strain.

Genomics and Proteomics

Rickettsial species such as *Rickettsia prowazekii* and *Rickettsia conorii* are among the first 50 sequenced bacterial genomes, with sizes ranging from 1.1 to 2.3 Mbp and a G + C content of 29–33%. Rickettsial genomes have a large fraction of non-coding DNA (up to 24%).

There is progressive gene degradation resulting from the endosymbiotic lifestyle that has allowed them to discard genes involved in metabolism supplied by their eukaryotic host cells. Degraded genes include mostly those coding for amino acids, ATP (adenosine triphosphate), LPS (Lipopolysaccharide), and cell wall component biosynthesis. In addition, rickettsial genomes exhibit a paradoxical expansion of genetic elements, including plasmids, short palindromic elements named Rickettsia palindromic elements (RPEs), ankyrin and tetratricopeptide repeats, toxin–antitoxin modules, ADP–ATP translocases, type IV secretion systems (T4SS), as well as *sca*, *spoT*, *proP*, and *ampG* genes.

Gene family duplication is a frequent occurrence in rickettsial genomes and is believed to facilitate adaptation to environmental changes in the host. The two most duplicated genes encode ADP/ATP translocases, often found in several copies, and enable the exploitation of energy produced by host cells. *SspoT* genes found in 4–14 copies are involved in the microbial response to environmental stress. Other duplicated gene families include proline/betaine transporters, toxin/antitoxin modules, T4SS, sca, and ampG, involved in rickettsial pathogenesis. Surface cell antigen (sca) proteins are a family of 17 orthologous autotransporters diversely detected in all rickettsial genomes. They were demonstrated to be localised at the surface of bacteria and play roles in mammalian cell infection as well as in infecting their arthropod hosts' cells, notably by promoting actin-based motility.

In addition, many repetitive elements (RPEs) are distributed in intergenic regions (tandem repeats) or both intergenic and coding regions. RPEs are at least five times more numerous in SFG than in TFG. Small non-coding RNAs (sRNAs) are identified in intergenic sequences, depending on the species. A total of 1785 sRNAs were detected from 13 species spanning all rickettsial groups. In addition, single-nucleotide polymorphisms (SNPs) are identified in many species. In *R. rickettsii*, 29 SNPs have been identified, whereas in *R. japonica*, there are 112 SNPs.

Morphology

Rickettsia are small intracellular Gram-negative bacilli, measuring about 0.3–0.5 μm in diameter and 1–2 μm in length. The composition of the cell wall is typical of that of proteobacteria, including peptidoglycan, lipopolysaccharide, and an outer membrane with only a few identified proteins. A family of five Omp (the Sca family) has been identified. The spotted fever group rickettsiae have both Omp A and Omp B, while the typhus fever group has only Omp B. The Giménez technique is commonly used for microscopy of tissue sections and infected cell lines (Fig. 35.1). They do not stain well with the Gram stain.

Cultivation

Rickettsia, being an obligate intracellular bacterium, cannot be grown in cell-free media like viruses. Several methods are used for cultivation, including the inoculation of embryonated eggs, cell lines, and experimental animals. Cultivation is essential for obtaining a pure bacterial culture to study its virulence and pathogenicity, as well as its genetic and proteomic characteristics, and its antibiotic susceptibility.

Embryonated Egg

Specimens are injected into the yolk sac. This may result in the death of the embryo after sev-

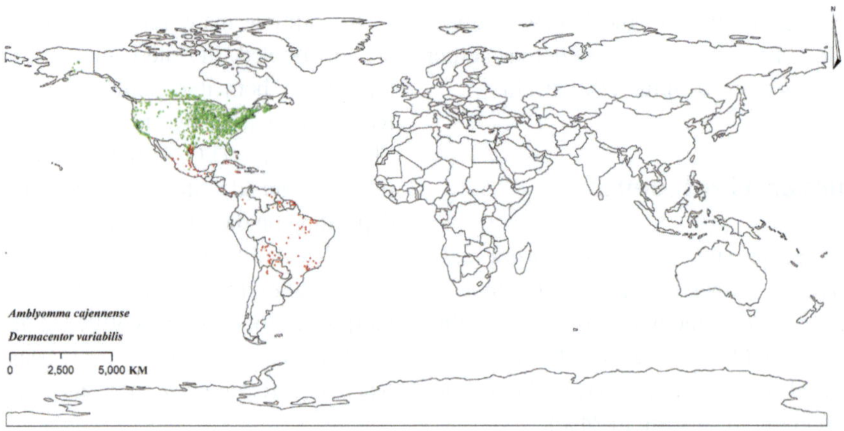

Fig. 35.1 Global distribution of Rocky Mountain Spotted Fever vectors (*Dermacentor variabilis* and *Amblyomma cajennense*). (Image source: Ji H, Wei X, Ma D, Wang X, Liu Q. Predicting the global potential distribution of two major vectors of Rocky Mountain Spotted Fever under conditions of global climate change. PLoS Negl Trop Dis. 2024;18(1):e0011883. Permitted for reproduction under CC BY license)

eral days. The presence of rickettsiae can be confirmed by Giménez staining of the yolk sac. This method was also used to assess the antibacterial activity of commonly used antibiotics. Penicillin and aminoglycosides were not effective, whereas tetracyclines and chloramphenicol were considered rickettsiostatic.

Cell Culture

Different cell lines have been used for initial isolation tests, including Vero (ATCC-CCL81, *Cercopithecus aethiops* kidney) 14, L929 (*Mus musculus* fibroblasts) 15, and HEL (human embryonic lung). Vero CCL-81 and L929 allow for the rapid isolation of highly infected samples. Tick cell lines have also been used to study differences in the proteomic and genomic expression between the vector and the host. Penicillin–streptomycin 0.2% and 1% Fungizone (amphotericin B) effectively limit contamination. The optimal growth temperature is 32 °C with 5% CO_2.

Isolation Techniques

The shell vial technique has been the gold standard for *Rickettsia* isolation. In the shell vial technique, only a small volume of cells and inoculum are required, which are brought into close contact through centrifugation to enhance the penetration of the bacteria into the cells.

Animal Models

The guinea pig (*Cavia porcellus*) has been used for rickettsial research. Ticks naturally feed on guinea pigs and transmit spotted fever rickettsioses (SFR). This was used for identifying *R. rickettsii* as the agent of Rocky Mountain spotted fever and for evaluating protective immunity and treatment for SFR. The development of genetically modified murine strains has resulted in the expansion of murine-specific reagents and assays.

The advantage of the guinea pig model for studying tick-borne rickettsioses is a function of its basic biology. The longer lifespan, ease of handling (e.g. guinea pigs rarely bite), and larger blood volume present the opportunity to follow the same animals with repeated sample collection. The ability of guinea pigs to develop clinical signs of tick-borne rickettsiosis, including changes typical of those observed in human disease, is likely driven by the genetic and functional similarities in the guinea pig and human immune systems.

Transmission of Infection

The transmission dynamics of SFG rickettsiae involve various arthropods, predominantly ticks, and a range of vertebrate reservoir hosts. Rickettsiae are considered *endosymbionts* (living in endocellular symbiosis) because of their regular association with ticks. Ticks are the main reservoir and vector for most SFG rickettsiae. The list of vectors of different pathogens is given in Table 35.2.

Two families of ticks are of medical importance: *Ixodidae* (hard ticks) and *Argasidae* (soft ticks). Various routes of pathogen transmission among ticks have been described. In transstadial transmission, the pathogen is maintained through the tick's developmental stages (larva to nymph to adult). Through transovarial transmission, infected female ticks can pass *Rickettsia* species vertically to their offspring via eggs. This facilitates the long-term maintenance of the bacteria in tick populations, even in the absence of vertebrate hosts. In co-feeding transmission, bacteria are directly spread from an infected tick to an uninfected tick due to feeding at closely situated bite sites. Sexual transmission of *R. rickettsii* has also been described.

Pathogenesis and Pathology

Rickettsia species exhibit a distinct affinity for endothelial cells lining the microcirculation, particularly capillaries (microvascular endothelium in the case of infection by *R. conorii* and both microvascular and macrovascular endothelium in the case of *R. rickettsii*). The resulting vascular injury underlies the pathological and pathophysiological characteristics observed across rickettsial diseases. Progressive endothelial cell injury leads to the generation of a characteristic erythematous rash, disseminated vasculitis, cutaneous necrosis, pneumonitis, meningoencephalitis, and multiorgan failure.

Transmission and Initial Infection

Regardless of their mode of entry, rickettsiae invade the capillary vascular endothelium and frequently cause focal swelling and a mixed inflammatory infiltration at the site of inoculation. Eschars have been reported in cases of infection with *Rickettsia parkeri*, *Rickettsia philipii*, *Rickettsia conorii*, and *Rickettsia africae.*

Attachment and Invasion

Seventeen surface cell antigens (sca) proteins have been identified, out of which only five [Sca0 (outer membrane protein A {OmpA}), Sca1, Sca2, Sca4, and Sca5 (OmpB)] are active and functional in the spotted fever group. Following attachment, *Rickettsia* species enter the host cells by phagocytosis. After internalisation, the phagocytic vacuole is lysed, and the rickettsiae escape phagocytic digestion to multiply freely in the cytoplasm and the nucleus.

Endothelial Dysfunction

Infections with pathogenic SFG *Rickettsia* species induce increased vascular permeability. This is associated with rickettsial replication and the disruption of vascular endothelial cells, characterised by perivascular infiltration of T cells and macrophages. The mechanisms underlying early host cell responses to rickettsial infection are mediated by the activation of nuclear transcription factor NF-κB, MAP kinase pathway(s), and phosphorylation–dephosphorylation mechanisms leading to the activation of protein kinase pathways. Infection of endothelial cells with *R. rickettsii* or *R. conorii* (i) stimulates the secretion of high levels of IL-6 and IL-8, (ii) increases the expression of adhesion molecules on the cell surface, and (iii) increases adhesion of leukocytes in an IL-1α pathway-dependent manner. As the endothelial cells die, necrosis of the intima and media of the blood vessels leads to the formation of hyaline thrombi composed of fibrin and cellular debris. These hyaline thrombi cause

Table 35.2 Classification of spotted fever group (SFG) rickettsiosis

Disease	Pathogen	Vector	Animal hosts	Geographic distribution
Indian tick typhus	*Rickettsia conorii subsp. indica* (proposed)	Ticks (*Rhipicephalus*)		South Asia (India)
Mediterranean spotted fever (Boutonneuse fever)	*Rickettsia conorii* subsp. *conorii* (proposed)	Ticks (*Rhipicephalus*)	Dogs; rodents	Southern and western Asia (India); Africa; southern Europe
Rickettsial pox	*Rickettsia akari*	Mites (*Allodermanyssuss sanguineus*)	House mice; wild rodents	Asia (Korea); South Africa; North and South America; southern and eastern Europe
North Asia tick typhus	*Rickettsia sibirica*	Ticks (*Haemophysalis, Dermacentor*)	Rodents	China; Mongolia; Russia
Japanese spotted fever	*Rickettsia japonica*	Ticks (*Dermacentor, Ixodes*)	Rodents	Japan
Far Eastern spotted fever	*Rickettsia heilongjiangenisis*	Ticks (*Dermacentor*)	Rodents	East Asia; Northern China; far East Russia
Cat flea rickettsiosis	*Rickettsia felis*	Fleas (*Ctenophalides felis*)	Domestic cats, opossums, rodents	Asia, Africa, Europe; North and South America
Lymphangitis associated rickettsiosis	*Rickettsia sibrica mongolotimonae*	Ticks (*Haemophysalis*)	Rodents	Asia (China); Europe (Southern France, Portugal)
An eruptive fever	*Rickettsia helvetica*	Ticks (*Ixodes*)	Rodents	Asia; central and northern Europe
African tick-bite fever	*Rickettsia africae*	Ticks (*Amblyomma*)	Domestic and wild ruminants	Sub-saharan Africa; West Indies
Astrakhan spotted fever	*Rickettsia conorii subsp. caspiae* (proposed)	Ticks (*Rhipicephalus*)		North Caspian region of Russia
Flinders Island spotted fever (Thai tick typhus)	*Rickettsia honei*	Ticks (*Rhipicephalus, Ixodes*)	Reptiles; rodents	Australia; Thailand
Israeli tick typhus	*Rickettsia conorii subsp. israelensis* (proposed)	Ticks (*Rhipicephalus*)		Southern Europe; Middle East
Maculatum infection	*Rickettsia parkeri*	Ticks (*Amblyomma*)	Rodents	North and South America
Rocky Mountain spotted fever	*Rickettsia rickettsii*	Ticks (*Dermacentor, Amblyomma*)	Rodents	North, Central, and South America
Mediterranean spotted fever-like illness	*Rickettsia monacensis*	Ticks (*Dermacentor*)	Lizards; birds	North Africa; Europe
Queensland tick typhus	*Rickettsia australis*	Ticks (*Ixodes*)	Rodents	Australia; Tasmania
Tick-borne Lymphadenopathy (TIBOLA) Dermacentor-borne necrosis and lymphadenopathy	*Rickettsia slovaca*	Ticks (*Dermacentor*)	Asia; southern and eastern Europe	European boar, lagomorphs, rodents

microinfarcts and extravasation of blood, which are manifested grossly by the petechial lesions that are the hallmark of rickettsial diseases.

Systemic Spread and Organ Involvement

As rickettsiae disseminate via the bloodstream, systemic vasculitis affects multiple organs. Capillary leak syndrome, hypotension, and multiorgan dysfunction may occur in severe infections.

Pathology

Pulmonary Involvement

SFG rickettsioses can result in interstitial pneumonitis, although the frequency and severity of pulmonary involvement vary among cases. The underlying vascular damage leads to increased vascular permeability, resulting in alveolar oedema, fibrin deposition, macrophage infiltration, and haemorrhage. Clinically, these manifestations may present as respiratory symptoms, such as coughing, pneumonia, and pulmonary oedema, a hallmark of pulmonary dysfunction that is occasionally observed in RMSF.

Central Nervous System (CNS) Involvement

Neurological complications are significant in RMSF. All regions of the brain and spinal cord can be affected. Nevertheless, *Rickettsia* species display a predilection for the midbrain and the inferior olivary nuclei. The CNS manifestations of RMSF are largely ischaemic, with microinfarcts that often affect the white matter, contributing to neurological deficits.

Cardiac Involvement

A notable inconsistency exists between the presence of myocarditis and the severity of cardiac insufficiency in RMSF. Hypotension and shock are predominantly caused by increased capillary permeability, leading to decreased intravascular volume and systemic vascular resistance.

Renal and Urinary Tract Involvement

The kidneys are frequently affected by perivascular interstitial nephritis in RMSF, although overt renal symptoms are uncommon. Renal dysfunction, when present, is typically caused by hypovolaemia and reduced perfusion rather than direct bacterial injury.

Other Organ System Involvement

Histopathological involvement of the testes, epididymis, scrotal skin, and adrenal glands has been documented in epidemic typhus, scrub typhus, and RMSF. These findings further reflect the systemic nature of rickettsial infections. Although rickettsial organisms affect various parts of the gastrointestinal tract, including the stomach and the small intestine, lesions in this system are relatively rare and usually not clinically significant.

Immunology

The immune response against rickettsiae primarily relies on cytokine-mediated intracellular killing and the action of cytotoxic CD8+ T lymphocytes. In mice, interferon-gamma (IFN-γ) and tumour necrosis factor-alpha (TNF-α) work together to activate nitric oxide synthase-2, facilitating the destruction of infected cells. Antibodies targeting outer membrane proteins, especially rOmpB, play a crucial role in protection by preventing bacterial entry into vascular endothelial cells.

The clearance of *Rickettsia* species begins with innate immune mechanisms. Dendritic cells internalise free bacteria and become activated through Toll-like receptor 4 (TLR4) binding to lipopolysaccharide (LPS) and TLR2 binding to peptidoglycan. These activated dendritic cells release interleukin-12 (IL-12), which stimulates natural killer (NK) cells and secrete IL-1, IL-6, and TNF-α. These cytokines are responsible for inducing fever, increasing vascular permeability, and promoting the liver's production of acute-phase proteins. TNF-α and IFN-γ act on endothelial cells to stimulate intracellular bacterial killing. Antigen presentation by dendritic cells to CD4+ helper T cells, combined with IL-12 signalling, supports a Th1-type immune response. These enhance the cytotoxic activity of CD8+ T cells against infected cells.

Humoral immunity also contributes significantly, particularly through antibodies against surface antigens on *Rickettsia* species. These antibodies, especially IgG and IgM, block the phagocytic uptake of the bacteria into host cells. These also enhance opsonisation, making the bacteria more recognisable to phagocytes and NK cells via Fc receptors. Additionally, IgG and IgM activate the classical complement pathway. Anti-Rickettsial IgG is commonly detected using an indirect immunofluorescence assay, which confirms the presence of infection. Seroconversion usually occurs about 4 weeks after symptom onset. However, early antibiotic treatment may prevent the development of detectable antibody levels.

Infection in Humans

The incubation period of spotted fever group rickettsial diseases generally ranges from 7 to 21 days.

Most diseases are of moderate severity, except for RMSF, which is life-threatening in cases of delayed treatment. Initial clinical manifestations typically consist of fever, headache, muscle pain, rash, local lymphadenopathy, and a characteristic inoculation eschar (*'tache noir'*) at the bite site.

Rocky Mountain spotted fever, caused by *R. rickettsii*, typically begins with high-grade fever and headache. It is associated with non-specific signs and symptoms, including malaise, nausea, myalgias, anorexia, vomiting, and diarrhoea. The classical triad of fever, headache, and rash is present in only 44% of cases. Although the rash is the hallmark of rickettsial infection, only 50% of patients with RMSF develop it in the first week of the illness. The rash begins on the wrists and ankles, and later becomes generalised, involving the palms and soles in 80% of patients. The rash initially appears macular and blanching, and later develops into petechial and purpuric lesions. The rash is absent in a substantial proportion of RMSF, i.e. 'spotless' fever.

In severe RMSF, patients suffer from oedema, hypovolemia, hypoalbuminemia, and hypotension, leading to shock. Renal failure also results from hypovolemia and shock. Neurological manifestations include confusion, stupor, delirium, seizures, and coma. Cerebrospinal fluid examination reveals pleocytosis and elevated protein levels. Pulmonary involvement typically results in cough and radiologic changes, including alveolar infiltrates, interstitial pneumonia, and pleural effusions. In cases of delayed antimicrobial drug treatment, the case fatality rate ranges from 20% to 40%.

Mediterranean spotted fever, also known as boutonneuse fever, is caused by *R. conorii.* It is characterised by the onset of fever, myalgia, and headache after an incubation period of 7 days. A characteristic inoculation eschar ('*tache noir*') is seen at the bite site in two–thirds of patients. The disease is severe in diabetics, alcoholics, the elderly, and patients with G6PD deficiency. It is manifested by more frequent petechial rashes, gastrointestinal symptoms, obtundation, dehydration, tachypnea, hepatomegaly, leukocytosis, coagulopathy, acute renal failure, hyperbilirubinemia, and elevated serum transaminase and creatine kinase levels. This infection has been designated by various geographic names, including Marseille fever, Mediterranean spotted fever, Kenya tick typhus, Israeli tick typhus, Astrakhan spotted fever, and Indian tick typhus.

African tick-bite fever is caused by *R. africae* and is prevalent throughout sub-Saharan Africa. It is characterised by the appearance of multiple inoculation eschars, frequently associated with regional lymphadenopathy in the groin. A rash is absent in more than half of the patients, and may be vesicular. Patients typically have fever, headache, and myalgia.

Major clinical characteristics of SFG rickettsial diseases are listed in Table 35.3.

Table 35.3 Clinical characteristics of spotted fever group (SFG) rickettsiosis

Bacteria	Disease	Rash	Specificity	Inoculation eschar	Multiple eschars	Draining lymphadenopathy
Rickettsia rickettsii	Rocky Mountain spotted fever	Yes	Purpuric (50%) cases	Very rare	No	No
Rickettsia conorii subsp. conorii (proposed)	Mediterranean spotted fever (Boutonneuse fever)	Yes	Purpuric (rarely)	Yes	No	Rare
Rickettsia conorii subsp. caspiae (proposed)	Astrakhan spotted fever	Yes	No	Rare	No	No
Rickettsia conorii subsp. israelensis (proposed)	Israeli tick typhus	Yes	Rarely purpuric	No	No	No
Rickettsia conorii subsp. indica (proposed)	Indian tick typhus	Yes	Maybe purpuric	Rare	No	No
Rickettsia africae	African tick-bite Fever	Rare (<50%)	Vesicular	Yes	Yes	Yes
Rickettsia sibirica	Siberian tick	Yes	No	Yes	No	Yes
Rickettsia japonica	Japanese or oriental spotted fever	Yes	No	Yes	No	No
Rickettsia heilongjanhensis	Far Eastern spotted fever	Yes	Maculopapular	Yes	No	Yes
Rickettsia felis	Cat flea rickettsiosis	Yes	–	Yes	Possible	Maybe
Rickettsia helvetica	Aneruptive fever	No	–	No	–	–
Rickettsia mongolotimonae	Lymphangitis associated rickettsiosis	Yes	Few spots Lymphangitis	Yes	Yes	Yes
Rickettsia australis	Queensland tick typhus	Yes	Vesicular	50% cases	No	Yes
Rickettsia honei	Flinders Island spotted fever (Thai tick typhus)	Yes	–	Yes	No	Yes

Infections in Animals

Most animals infected with SFG rickettsiae are asymptomatic. However, clinical disease has been observed in dogs infected with *R. rickettsii* and *R. conorii.* Clinical manifestations include fever, lethargy, petechiae, conjunctivitis, lymphadenopathy, weight loss, and haematological abnormalities (anaemia, thrombocytopenia, leukocytosis). However, many animals do not develop symptoms despite infection with rickettsiae pathogenic to humans.

Epidemiology and Public Health

Humans are typically accidental dead-end hosts for *Rickettsia* species and play no role in the maintenance of these bacteria in nature. Most rickettsial bacteria are transmitted directly to humans by infected arthropod vectors such as fleas, lice, mites, or ticks during feeding. Inhaling bacteria or inoculating the conjunctiva with infectious material may also initiate an infection for some rickettsial pathogens.

Although rickettsial diseases have a worldwide distribution, there are endemic and hyperendemic areas. The spotted fever group of Rickettsia (SFG) encompasses more than 30 distinct species with a global distribution. Among these, *R. rickettsii*, the causative agent of Rocky Mountain spotted fever (RMSF) in North America, stands out as one of the most extensively described infections. In northern Australia, *R. australis* and *R. honei* are prevalent. *R. conorii*, the causative agent of Mediterranean spotted fever (MSF), is prevalent in various regions across Europe, Africa, and Asia. Notably, *Rickettsia felis*, known as the agent responsible for flea-borne spotted fever, is recognised as an emerging infectious disease. Initially, it was identified in the USA and is now observed worldwide. It also accounts for human infections that occur through bites from infected ticks when humans enter the natural infection cycle.

Tick-borne rickettsioses of the spotted fever group exhibit a distinct geographical division:

Western Hemisphere

Rocky Mountain spotted fever (RMSF), also known as Brazilian spotted fever in some regions, is caused by the bacterium *R. rickettsii*. This illness occurs in many parts of the Western Hemisphere and is transmitted by vectors such as *Dermacentor variabilis* and *Amblyomma cajennense* (Fig. 35.1). The condition has been reported from Canada, Mexico, the United States, and numerous countries in Central and South America, including Argentina, Brazil, Colombia, Costa Rica, and Panama. Sometimes, more people in the same family or area may get sick at the same time. Spending time with dogs in the countryside or cities and engaging in outdoor activities such as camping, fishing, hiking, or hunting increases the likelihood of contracting this infection.

Eastern Hemisphere (Tick Typhus)

Rickettsia sibirica causes Siberian tick typhus. It is mostly predominant in the former Soviet Union and parts of the Indo-Pakistan subcontinent. *R. australis* causes Queensland tick typhus, which is geographically distributed in eastern and northern Australia. *R. honei* is responsible for spotted fever in the region of northern Australia. Travel-associated cases of Mediterranean spotted fever (also known as Boutonneuse fever) caused by *R. conorii* are reported less frequently but occur across a substantial geographical range. This area encompasses the Mediterranean Basin, including southern France and Italy, as well as regions within Africa, much of Europe, India, and the Middle East.

R. felis is the etiological agent of cat flea rickettsiosis, exhibiting a global distribution across diverse invertebrate hosts. It has been documented as a significant contributor to febrile illnesses in specific African nations. *R. akari* is transmitted via house mouse mites and primarily circulates within urban centres in the Balkan states, Korea, South Africa, Ukraine, and the United States. Outbreaks of rickettsial pox frequently arise following contact with infected peridomestic rodents and their mites. These outbreaks occur particularly during periods of natural rodent mortality or extermination efforts that compel the mites to seek alternative hosts, including humans. While urban rodents appear to be more commonly linked to human cases, the agent has also been identified in a limited number of wild rodent populations.

Diagnosis

The variability and often the severity of clinical symptoms in individual patients, coupled with the potential absence of an eschar, lead to diagnostic challenges and subsequent delays in initiating appropriate antibiotic therapy. Recognising less common findings such as hyponatraemia, neurological issues, elevated liver enzymes, and low platelet counts can increase clinical suspicion for SFG infections. A detailed travel and social history can offer diagnostic clues.

Specimens for *rickettsiae* are handled in biosafety level 3 laboratories. *Rickettsiae* are known to cause laboratory infections, typically when they are cultured. The diagnostic modalities available for *Rickettsia* infections are summarised in Table 35.4. Different diagnostic modalities applied based on the pathogen's transmission cycle are presented in Fig. 35.2.

Table 35.4 Laboratory diagnosis of spotted fever

Diagnostic technique	Target	Sample type	Remarks
Microscopy (Giemsa/ Gimenez stain)	Intracellular *Rickettsia* spp.	Skin biopsy from rash lesions	Low sensitivity; requires expert personnel. Rarely used for acute diagnosis. Supportive role only
Immunohistochemistry (IHC)	*Rickettsia* antigens (species/group level)	Tissue sections (e.g., rash biopsy)	High specificity; allows visualisation of organism in situ. Confirmatory but needs specialised laboratory facilities
In-vitro culture	Viable *Rickettsia* spp.	Blood or tissue	Not routinely performed; requires BSL-3 lab and host cell lines due to obligate intracellular nature of the pathogen
Weil-Felix test	Cross-reactive antibodies to Proteus OX19, OX2, OXK	Serum	Obsolete, but still used in some regions for preliminary screening. Low sensitivity and specificity
Indirect immunofluorescence assay (IFA)	Anti-Rickettsial IgM/ IgG antibodies	Serum (acute and convalescent)	Gold standard for serological diagnosis. Requires four-fold rise in titers between paired samples. Antibodies detectable during 7–10 days
ELISA (IgM capture)	Anti-Rickettsial antibodies (IgM)	Serum	High-throughput and sensitive. Needs high-purity antigens. Less useful in sporadic SFG infections unless included in a broader febrile illness panel
PCR (conventional/ real-time)	*Rickettsia* DNA (*gltA*, 17 kDa, 16S rRNA, *ompA*, *ompB*, *sca4*, *rplP*)	Whole blood, skin biopsy, eschar, tissue	Useful in early disease before antibody development. Gene target selection is crucial. Sequencing recommended for species confirmation
LAMP (loop-mediated isothermal amplification)	Rickettsia DNA	Whole blood, tissue, or eschar	Rapid, simple, and field-friendly. Useful for point-of-care testing, though species resolution is limited compared to PCR with sequencing

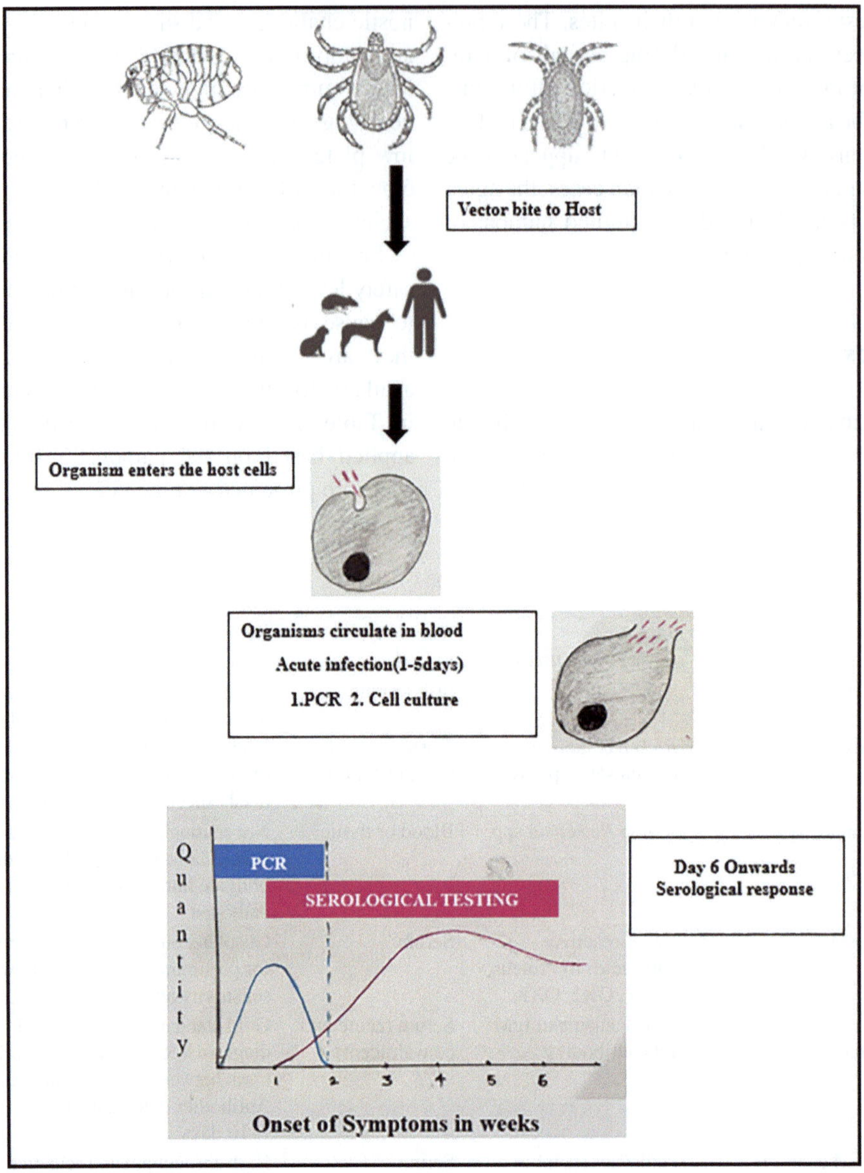

Fig. 35.2 Timeline of laboratory diagnosis of spotted fever with respect to transmission cycle. (Image source: Authors personal creation)

Microscopy

Microscopy plays a supporting role in the laboratory diagnosis of spotted fever. Direct visualisation of *Rickettsia* in clinical samples is challenging due to their small size, intracellular nature, and often low numbers. Hence, specialised staining techniques like Giménez or Giemsa are used to demonstrate them in tissue biopsies, particularly from rash lesions. This method, however, requires highly experienced personnel and offers limited sensitivity.

Immunohistochemistry (IHC) is a useful microscopic technique in this context. IHC involves using specific antibodies that bind to *Rickettsia* antigens in tissue sections, which are then visualised with a colourimetric or fluorescent tag under a microscope. This method offers high specificity and directly confirms the presence of the pathogen in infected tissues.

In Vitro Culture

In vitro culture is not a routine diagnostic method for the laboratory diagnosis of spotted fever. This is primarily due to the unique characteristics of *Rickettsia* species, which are obligate intracellular bacteria, and the need for a BSL3 laboratory.

Serodiagnosis

The Weil-Felix test aids in establishing a preliminary diagnosis for diseases caused by both the typhus and spotted fever groups of rickettsiae. In rickettsial infections, this serological reaction relies on the patient developing antibodies that can agglutinate specific non-motile strains of *Proteus* bacteria, namely *Proteus vulgaris* OX19 and OX2, and *Proteus mirabilis* OXK.

Indirect immunofluorescent antibody (IFA) test remains the gold standard for the serological diagnosis of spotted fever group (SFG) rickettsial infections. This test detects specific IgM and IgG antibodies produced by the patient in response to the infection. While highly sensitive and specific, a key consideration for spotted fever diagnosis is the timing of sample collection. Antibodies typically become detectable only 7–10 days after the onset of symptoms. Therefore, a single positive result in an early acute sample may not be definitive. A fourfold rise in antibody titer between acute- and convalescent-phase samples (collected 2–4 weeks apart) is often required for a conclusive diagnosis, particularly in regions where background antibody levels are present.

Enzyme-linked immunosorbent assay (ELISA), particularly IgM capture assays, offer high sensitivity for rickettsial diagnosis but require large quantities of purified antigens, which are difficult to obtain commercially. ELISAs suit high-throughput screening. This is less useful for sporadic rickettsioses, except for epidemic typhus and Q fever. However, the sensitivity of ELISA could be valuable when relevant rickettsial antigens are included in the fever panel caused by multiple pathogens.

Molecular Diagnosis

Polymerase chain reaction (PCR)-based detection is the primary method for detecting SFG, particularly for the early detection of infection before the development of detectable antibodies. A positive PCR test result is confirmatory, but a negative test does not rule out the possibility of infection. Essentially, PCR is a precise method for both studying how SFG spreads and directly detecting infections in people.

The effectiveness and accuracy of PCR for *Rickettsia* species depend significantly on the selection of appropriate gene targets. For broad detection at the genus or species family group (SFG) level, conserved gene targets are employed. Commonly used examples include genes for citrate synthase (*gltA*), 16S ribosomal RNA (16S rRNA), and 17 kDa lipoprotein outer membrane antigens (17 kDa). Detecting these targets generally confirms the presence of SFG or typhus Group (TG) rickettsiae.

For more specific identification down to the species level, other gene targets such as outer membrane protein A (*ompA*), outer membrane protein B (*ompB*), *sca4*, and *rplP* are often used. These targets offer higher specificity for identifying SFG rickettsiae in both human and animal samples.

Following PCR amplification, sequencing of the resulting product is typically performed to identify the specific *Rickettsia* species.

Loop-mediated isothermal amplification assay (LAMP) presents an alternative to conventional PCR and is particularly advantageous for field-based diagnostics. This technique is a relatively simple and rapid molecular method for detecting the pathogen. This technique involves the amplification of DNA at a constant temperature of 60–65 °C and the stopping reaction at 80 °C. The LAMP has been reported to be more sensitive than nPCR to detect ompB from SFG, with 73% sensitivity and 100% specificity. Add a few lines.

Treatment

Tetracyclines can be lifesaving in rickettsial infections, such as RMSF, scrub typhus, Rickettsial pox, and murine typhus. Doxycycline is the drug of choice recommended to treat suspected rickettsial disease in children, including those under 9 years of age. It is given orally or intravenously, depending on the severity of the illness. The use of antibiotics other than doxycycline increases the risk of severe illness and death. Beta-lactam antibiotic coverage does not treat RMSF.

Treatment options for different spotted fevers are presented in Table 35.5.

Prevention and Control

Prevention and control of spotted group fever caused by *Rickettsia* species in humans and animals primarily focus on minimising contact with infected ticks, the main reservoir, and vectors. Additionally, avoiding direct contact with tick-infested animals and their habitats contributes significantly to reducing human infection risks. Preventive measures, especially in endemic areas, include routine inspection for and prompt removal of ticks, protective clothing, and the use of insect repellents. Regular use of acaricides and tick control measures reduces the risk of transmission in animals, particularly domestic and companion animals.

Currently, no vaccine is available to prevent infections caused by spotted fever-causing *Rickettsia* or whole rickettsial organisms.

Case Study

A 29-year-old male presented with fever, headache, rash, and was also associated with altered mental status and hypertensive crisis. Elevated WBC were noted with thrombocytopenia. Improvement occurred with empiric antibiotics. Keeping a tick bite in mind due to the history of exposure to areas with ticks, Immunoglobulin M (IgM) IFA assays using *R. rickettsii* antigen were performed, which showed positive results.

1. What are the key features from the above clinical history?
2. What is the differential diagnosis?
3. What could be the species involved?
4. How would you confirm the diagnosis?

Table 35.5 Treatment of Rickettsial diseases

Disease	Pharmacological treatment	Non-pharmacological measures
Rocky mountain spotted fever	Doxycycline: 100 mg twice daily for 4–10 days Children ≤45 kg: 2.2 mg/kg/dose twice daily (max 200 mg/day) Chloramphenicol: 50–75 mg/kg/day in 4 divided doses (oral/IV), alternative in pregnancy Continue treatment for ≥3 days after defervescence	Use of protective clothing and tick repellents Frequent tick checks and prompt removal Community tick control efforts No prophylaxis after tick bites
Rickettsial pox	Doxycycline: 200 mg/day orally for 7 days Usually self-limited, but severe cases may need hospitalisation	Eliminate rodents (esp. mice) from homes Use insecticides in infested areas
Tick typhus (Rickettsial fever)	Doxycycline: 200 mg/day orally for 7–10 days Chloramphenicol: 50–75 mg/kg/day in 4 divided doses Azithromycin + Rifampin: Safe and effective in pregnancy Avoid ciprofloxacin (linked to poor outcomes)	Wear protective clothing and use tick repellents. Regular inspection and removal of ticks
Cat flea typhus (*Rickettsia felis*)	Same treatment as other rickettsial fevers: Doxycycline-based therapy (e.g. 200 mg/day for 7 days)	Vector control (fleas). Minimise contact with flea-infested animals

Research Questions

1. What are the major hindrances to the control of infection?
2. What are the challenges in developing a safe and effective vaccine?
3. What are the new targets to develop new drugs?
4. What are the new molecular targets or antigens which can be targeted during the development of a rapid test?

Declaration During the preparation of this chapter, we, Dr. Keche Archana, Dr. Patil Girish, Dr. Kumar Vikram, and Dr. Keche Yogendra, used AI tools to improve the readability of the text. After using them, we reviewed and edited the content as needed and took full responsibility for the content of the publication.

Further Readings

Ammerman NC, Beier-Sexton M, Azad AF. Laboratory maintenance of *Rickettsia rickettsii*. Curr Protoc Microbiol. 2008;Chapter 3:Unit 3A.5. https://doi.org/10.1002/9780471729259.mc03a05s11.

Brunton LL, Knollmann BC. Goodman & Gilman's: the pharmacological basis of therapeutics. 14th ed. McGraw-Hill Education; 2023.

Centers for Disease Control and Prevention. Rickettsial diseases. n.d. Retrieved from https://wwwnc.cdc.gov/travel/yellowbook/2024/infections-diseases/rickettsial-diseases.

Clark EH, Ashley P, Shandera WX. Spotted fevers. In: Papadakis MA, McPhee SJ, Rabow MW, McQuaid KR, Gandhi M, editors. Current medical diagnosis & treatment 2024. 63rd ed. McGraw-Hill Education; 2024. p. 1442–6.

Jia N, Liu H-B, Zheng Y-C, Shi W-Q, Wei R, Chu Y-L, Ning N-Z, Jiang B-G, Jiang R-R, Li T, Huo Q-B, Bian C, Liu X, Sun Y, Li L-F, Wang Q, Wei W, Wang Y-W, Jongejan F, Jiang J-F, Song J-L, Wang H, Cao W-C. Cutaneous immunoprofiles of three spotted fever group rickettsia cases. Infect Immun. 2020;88:e00686-19. https://doi.org/10.1128/IAI.00686-19.

Kim HK. Rickettsia-host-tick interactions: knowledge advances and gaps. Infect Immun. 2022;90:e00621-21. https://doi.org/10.1128/iai.00621-21.

Pradeep J, Anitharaj V, Sangeetha B. Human rickettsial infections in India – a review. J Vector Borne Dis. 2024;61(1):5–22. https://doi.org/10.4103/0972-9062.392255.

Raoult D, Dumler JS. *Rickettsia and Orienta*. In: Murray PR, Funke G, Borriello SP, editors. Topley and Wilson's microbiology and microbial infections (bacteriology). 10th ed; John Wiley & Sons, Ltd; 2010.

Stokes JV, Walker DH, Varela-Stokes AS. The guinea pig model for tick-borne spotted fever rickettsioses: a second look. Ticks Tick-Borne Dis. 2020;11(6):101538. https://doi.org/10.1016/j.ttbdis.2020.101538.

Tello-Martin R, Dzul-Rosado K, Zavala-Castro J, Lugo-Caballero C. Approaches for the successful isolation and cell culture of American *Rickettsia* species. J Vector Borne Dis. 2018;55(4):258–64. https://doi.org/10.4103/0972-9062.256560.

Zhang Y-Y, Sun Y-Q, Chen J-J, Teng A-Y, Wang T, Li H, Hay SI, Fang L-Q, Yang Y, Liu W. Mapping the global distribution of spotted fever group rickettsiae: a systematic review with modelling analysis. Lancet Digit Health. 2022;5:e5–e15.

African Tick-Bite Fever

36

Shashikala and Anil Jacob Purty

Learning Objectives

1. To understand the aetiology, transmission, and clinical presentation of African tick-bite fever
2. To analyse the taxonomy, genomics, and morphology of *Rickettsia africae*
3. To evaluate diagnostic methods and treatment options for African tick-bite fever
4. To assess public health strategies for the prevention and control of tick-borne diseases

Introduction

International travel, one of the largest and most rapidly evolving industries worldwide, has a significant impact on healthcare in many developed countries. One example of such a recently recognised travel-associated disease is African tick-bite fever (ATBF), a tick-borne rickettsial infection endemic to sub-Saharan Africa and some Caribbean regions caused by *Rickettsia africae*. ATBF is an emerging concern for both local populations and international travellers. In the wake of the rapidly expanding international safari tourism to southern Africa, ATBF has emerged as a common cause of imported fever in Europe and elsewhere. Unlike more virulent rickettsioses, ATBF is typically mild but can cause considerable discomfort and diagnostic challenges due to overlapping symptoms with other febrile illnesses.

History

African tick-bite fever was first described in 1930, but it gained prominence in the 1990s following reports of febrile illnesses in European travellers returning from Africa. The early description of African tick-bite fever, likely a very old disease in sub-Saharan Africa, is related to that of Mediterranean spotted fever. In 1911, 1 year after the first description of Mediterranean spotted fever, a similar flu-like disease was described in Mozambique and South Africa. This illness was also transmitted by ticks and was named tick-bite fever. The epidemiology and clinical manifestations of the two conditions were investigated in South Africa in the 1930s. Patients with Mediterranean spotted fever typically acquired the disease in urban areas, whereas most patients with tick-bite fever became ill after contact with cattle ticks in the bush. Also, the clinical manifestations of tick-bite fever were milder, with few complications and frequently no cutaneous rash. The causative *Rickettsia* of the tick-bite fever was isolated in guinea pigs and

Shashikala (✉)
Department of Clinical Microbiology, Pondicherry Institute of Medical Sciences, Pondicherry, India

A. J. Purty
Department of Community Medicine, Pondicherry Institute of Medical Sciences, Pondicherry, India

S. C. Parija, A. Poddar (eds.), *Textbook of Bacterial Zoonoses*, Microbial Zoonoses,
https://doi.org/10.1007/978-981-95-6615-0_36

was shown that the organism was different from *Rickettsia conorii*. However, as subsequent workers could not confirm the findings because the earlier isolate was lost, tick-bite fever in Africa became erroneously considered synonymous with Mediterranean spotted fever for several decades.

In August 1992, a spotted fever rickettsia was isolated from a patient presenting to a hospital with tick-bite fever in Zimbabwe. This isolate, by PCR and restriction endonuclease fragment length polymorphism, was found to be distinct from all other spotted fever group rickettsiae, including *R. conorii*. The name African tick-bite fever was assigned to this disease, along with its causative organism, *Rickettsia africae*. This nomenclature was made official in 1996.

The identification of *R. africae* as the causative agent and its classification within the spotted fever group of rickettsiae facilitated more targeted research and diagnostic efforts. Historical studies also underscored the role of *Amblyomma* ticks, particularly *Amblyomma variegatum* and *Amblyomma hebraeum*, in transmission.

Taxonomy

Rickettsia africae belongs to the Domain Bacteria, Phylum Pseudomonadota, Class Alphaproteobacteria, Order Rickettsiales, Family Rickettsiaceae, and the Genus *Rickettsia*.

Phylogenetic analyses based on the sequencing of genes such as 16S rRNA, *gltA, ompA, ompB,* and *sca4* have refined the internal classification of the genus *Rickettsia*. Four common groupings have been identified:

Spotted Fever Group (SFG)

This is the largest and most diverse group, comprising numerous species, many of which are pathogenic to humans. Examples include *Rickettsia rickettsii* (Rocky Mountain spotted fever), *Rickettsia conorii* (Mediterranean spotted fever), *R. africae* (African tick-bite fever), *Rickettsia parkeri, Rickettsia sibirica, and Rickettsia akari* (rickettsial pox). These are predominantly tick-borne, though some are mite--borne. The differential characteristics of different spotted fevers are presented in Table 36.1.

Typhus Group (TG)

This group comprises the classic human pathogens *Rickettsia prowazekii* (epidemic typhus) and *Rickettsia typhi* (murine typhus). These are typically transmitted by lice and fleas, respectively.

Bellii Group

This group includes *Rickettsia bellii* and *Rickettsia canadensis*. These are often considered phylogenetically older lineages within the genus. These are generally less associated with human disease.

Transitional Group Rickettsiae (TGR)

Some classifications include a transitional group that contains species like *Rickettsia felis*, which has some characteristics that bridge the typhus and spotted fever groups.

Table 36.1 Comparison of African tick-bite fever, Mediterranean spotted fever and Rocky Mountain spotted fever

Feature	African tick-bite fever	Mediterranean spotted fever	Rocky mountain spotted fever
Causative agent	*Rickettsia africae*	*Rickettsia conorii*	*Rickettsia rickettsii*
Vector	*Amblyomma variegatum*	*Rhipicephalus sanguineus*	*Dermacentor* spp.
Eschar frequency	Common (often multiple)	Common	Rare
Rash	Variable, less common	Prominent	Prominent
Severity of infection	Mild	Moderate	Severe
Endemic Regions	Sub-Saharan Africa, Caribbean	Mediterranean basin	Americas

Genomics and Proteomics

The genome of *R. africae* consists of a single circular chromosome with an estimated size of 1248 kb and encodes approximately 1200 proteins. Based on initial genotypic studies, *R. africae* could be differentiated from all other spotted fever group rickettsiae except *R. parkeri*, a species of unknown pathogenicity transmitted by *Amblyomma maculatum* ticks in the USA. Sequencing studies of the 16S rRNA gene sequences have shown that *R parkeri*, *R. akarai* and *R. sibirica,* the agent of Siberian tick typhus are the closest (homology ≥ 99.6%). *R akarai* and *Rickettsia australis,* the agents of rickettsial pox and Queensland tick typhus, respectively are the most distant relative (homology ≥ 97.9%).

Comparative genomic studies have revealed similarities with other SFG rickettsiae, as well as unique features, such as genes potentially involved in tick colonisation and survival. Proteomic analyses have identified immunogenic antigens useful for serological diagnosis and the exploration of vaccine candidates. Whole-genome sequencing has further illuminated the gene regulatory networks and virulence factors that may facilitate iñfection and host specificity.

Morphology

R. africae is a small, Gram-negative, obligate intracellular bacterium that measures 0.3–0.5 × 0.9–0.6 μm. It appears as short rods or coccoid forms and has an outer slime layer and a trilaminar cell wall, as observed under electron microscopy. It primarily resides within the cytoplasm of endothelial cells, forming characteristic vacuoles (Fig. 36.1).

Within the cytoplasm of endothelial cells, *R. africae* can be observed moving by polymerising host cell actin, forming "*actin tails*" that propel them through the cytoplasm and facilitate cell-to-cell spread.

Cultivation

Cultivating *R. africae* requires specialised biosafety level 3 (BSL-3) facilities. However, it is not used for routine diagnostics; rather, it is explored for experimental studies and antigen production.

R. africae is a strictly obligate intracellular pathogen. As such, it cannot be grown on standard artificial culture media; rather, it requires

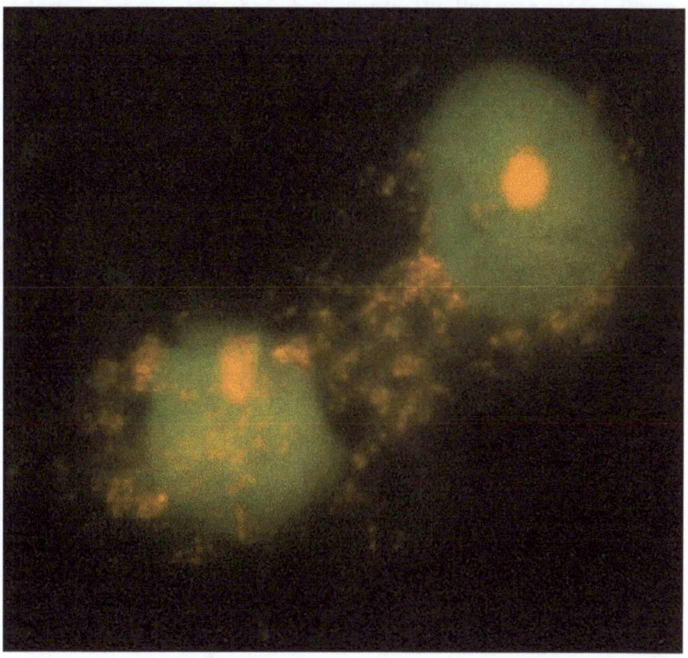

Fig. 36.1 Acridine orange-stained *Rickettsia africae* isolated in Vero E6 cells from an eschar biopsy specimen from a patient with African tick-bite fever. (Image source: CDC)

living eukaryotic host cells for its growth and replication. Historically, the organism was cultivated in the yolk sac of developing chicken embryos after Ernest William Goodpasture pioneered the technique in the 1930s.

In recent times, cell culture systems have been the most common and versatile method for cultivating *R. africae*. Vero cells are widely used due to their susceptibility to rickettsial infections and ease of maintenance. Other cell lines, such as L929 mouse fibroblast cells, are also utilised. Growth is slow, taking several days to weeks, under optimal conditions (34–37 °C with 5% CO_2) and is monitored by cytopathic effects or immunofluorescence.

Animal Models

Mouse models, particularly immunocompromised strains, have been employed to study the pathogenesis of African tick-bite fever. Guinea pigs and non-human primates are also used for research on immune responses and vaccine testing. These models help elucidate mechanisms of endothelial infection, immune evasion, and the effectiveness of antimicrobial therapy. Limitations remain due to species-specific immune responses and differences in disease progression compared to humans.

Transmission Cycle

R. africae is transmitted by the hard ticks of the members of the genus *Amblyomma. A. variegatum*, the tropical bont tick and *A. hebrareum*, the southern African bont tick, are recognised as principal vectors. *A. variegatum* and *A. hebrareum* are also important reservoirs in which infection is maintained through *transovarial transmission* (from adult female to offspring) and *transstadial transmission* (from larva to nymph to adult). The rates of R. *africae* infection are reported to exceed 30% in *A. variegatum* and up to 70% in *A. hebrareum.*

A. variegatum and *A. hebrareum*, like other hard tick species, have three active stages: larvae, nymphs, and adults. Larvae are minute and may be difficult to detect on skin and clothing, whereas a fully engorged adult female may approach the size of a grape. A blood meal is required at each stage for development to proceed to the next. In places with a single rainy season, one generation occurs per year, but where there are two rainy seasons, shorter generations are possible. These ticks are notoriously aggressive, and many ticks may attack a single host, or several hosts may be attacked simultaneously.

Humans are accidental hosts and acquire the infection through the bite of infected nymphs or adult ticks, usually on the legs. However, once on the skin, amblyommas may crawl around for hours before attaching, typically behind the knee, in the groin, or the axilla, where the skin is thin, moist, and warm. The nymphal stage is considered most dangerous due to its small size, which facilitates unnoticed feeding. Human activities such as farming, hunting, and ecotourism increase exposure risk.

Pathogenesis and Pathology

Once introduced into the skin, *R. africae* infects endothelial cells, leading to an increased secretion of endothelium-associated mediators, such as von Willebrand factor and soluble E-selectin. Subsequent damage to the endothelium is followed by perivascular infiltration of T cells and macrophages, resulting in lymphohistiocytic vasculitis, the hallmark of rickettsial disease.

The clearance of *R. africae* is believed to involve complex interactions between T cells, macrophages, natural killer cells, B lymphocytes, and antibodies, characterised by increased circulating concentrations of cytokines (tumour necrosis factor alpha, interleukin 6, and interleukin 10).

Immunology

Infection triggers both humoral and cell-mediated immune responses. Antibodies against outer membrane proteins and heat shock proteins are

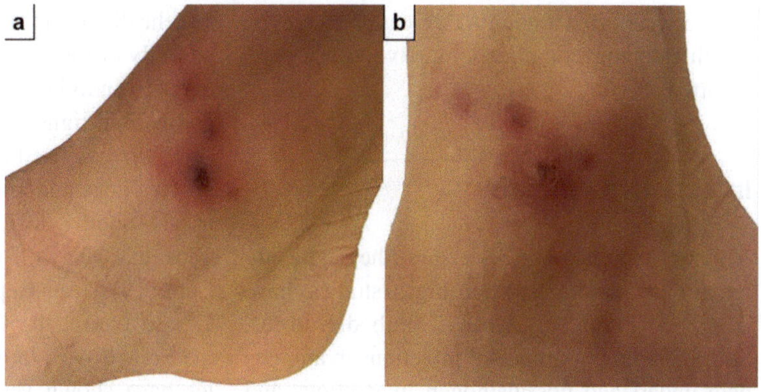

Fig. 36.2 The rickettsial eschar on the right ankle of the patient. (**a**) During the first visit before treatment was initiated, and (**b**) at the follow-up 1 week later after treatment with doxycycline. (Image source: Harrison et al. (2016). Permitted for reproduction under CC-BY license)

common. The immune response causes tissue damage, resulting in the formation of eschar, rash, and lymphadenopathy. Systemic infection can lead to fever and myalgia, but severe complications are rare. Histological findings typically include perivascular infiltration, endothelial swelling, and leukocyte accumulation. Inflammatory cytokines, such as TNF-α and IL-6, are often elevated.

Memory T-cell responses confer partial immunity, which may explain milder symptoms in individuals with repeated exposure. Studies also indicate cross-reactivity with other SFG rickettsiae, complicating serodiagnosis. The development of immunity in endemic regions is poorly understood and requires further research.

Infection in Humans

Any person who enters the habitat of tick vectors in endemic areas is at risk of contracting African tick-bite fever. Reported cases represent a wide range of travellers, including leisure safari tourists, backpackers, hunters, sports competitors, students, foreign aid workers, film crew staff, and deployed soldiers. Game hunting (an activity typically characterised by heavy exposure to ground vegetation, ungulates, and their hides) travel to southern Africa (where *A. hebraeum* is the principal vector) and travels between November and April (when tick activity peaks in many endemic areas) have been identified as independent risk factors of ATBF.

The incubation period of ATBF following tick bite to symptom onset is usually 5–7 days but may be as long as 10 days. Most patients present with abrupt flu-like symptoms, such as fever, nausea, fatigue, headache, and myalgia, particularly neck muscle myalgia, with subjective neck stiffness also occurring. An inoculation eschar, a black crust surrounded by a red halo at the site of the tick bite (Fig. 36.2). It is present in most cases but may easily be overlooked, particularly in dark skin, in the hair, or in the anogenital region. Regional lymphadenitis is a common finding and may be present even in the absence of a frank eschar. A generalised cutaneous rash, sometimes vesicular and usually best seen near the eschar, is present in some patients.

Human ATBF include aphthous stomatitis and lymphangitis. During the first 10 days of illness, when most patients are likely to present, increased concentrations of serum C-reactive protein and moderate lymphopenia are seen in most cases. Elevated serum liver enzymes and thrombocytopenia are also common.

Less common clinical manifestations of ATBF in humans may include vesicular or pustular skin eruptions, regional lymphadenopathy, and myalgia without fever. Atypical presentations, such as multiple eschars or prolonged fatigue, can also occasionally occur.

Complications seem to be rare in ATBF; however, when assessing an international traveller with a fever of more than 3 weeks' duration after travel to an endemic region, this condition needs to be considered. Reactive arthritis has occasionally been reported among ATBF cases. However, life-threatening complications have never been reported.

Infection in Animals

Although animals are reservoirs, they typically remain asymptomatic. Serologic studies have confirmed a high prevalence of antibodies in cattle and goats. Experimental infections in animals help understand host-pathogen interactions and the role of vertebrate reservoirs in transmission. Persistent infections in ticks and transient bacteraemia in animals underscore the complexity of the epidemiological cycle.

Epidemiology and Public Health

Humans are accidental hosts and acquire the infection through the bite of infected nymphs or adult ticks, usually on the legs. However, once on the skin, amblyommas may crawl around for hours before attaching, typically behind the knee, in the groin, or axilla, where the skin is thin, moist, and warm. The nymphal stage is considered the most dangerous due to its small size, which facilitates unnoticed feeding. Human activities such as farming, hunting, and ecotourism increase the exposure risk.

Cattle are the most important domestic hosts for adult *A. hebrareum* and *A. variegatum*, followed by sheep, horses, donkeys, and pigs. Adult ticks also feed readily on a range of wild animals, such as giraffes, buffalo, and antelopes, which are important in maintaining the species in areas where domestic hosts are reduced or absent. Travellers who go on safari, especially those travelling to national parks, game hunters, and ecotourists in sub-Saharan Africa, are at risk of African tick-bite fever. *R. africae* is also endemic to several islands of the Caribbean West Indies, and imported cases have been described from this region *R. africae* remains the most frequently reported rickettsial infection acquired during travel. Commonly, cases of African tick-bite fever cluster among people travelling together, and the diagnosis of the disease in one member of a family or tourist group can alert other similarly exposed individuals to seek care if they develop compatible signs and symptoms.

R. africae has been detected in Amblyomma species ticks and humans from 15 African countries: South Africa, Botswana, Zimbabwe, Tanzania, Kenya, Burundi, Sudan, Ethiopia, the Central African Republic, Gabon, Mali, Niger, Côte d'Ivoire, the Gambia, and Guadeloupe in the French West Indies. In addition, several studies have shown very high seroprevalence rates (up to 70%) for spotted fever group rickettsial infections in areas of sub-Saharan Africa where *A. variegatum* or A. *hebraeum* are prevalent, and where cattle breeding is intense. Thus, direct and indirect epidemiological data suggest that *R. africae* occurs widely throughout sub-Saharan Africa and the eastern Caribbean (Fig. 36.3).

Despite the wide geographical distribution of *R. africae*, reports on ATBF among Indigenous African populations are unexpectedly scarce. Studies have shown that Indigenous people are generally infected at a young age. The disease might be very mild or subclinical, and many patients do not seek medical attention. It is also difficult to see inoculation eschars in pigmented skin, and this important clue might not be visible unless an extensive physical examination is performed. Moreover, the definitive diagnosis of ATBF requires sophisticated diagnostic tests that are not available in many African countries. Increased tourism and ecotourism have resulted in a higher number of travel-associated cases. Surveillance is limited, and the true incidence is likely underestimated. Public health strategies emphasise awareness, tick control, and early diagnosis. Occupational risk groups include veterinarians, farmers, soldiers, and outdoor workers. Climate change and land-use changes may influence tick distribution and disease incidence.

Diagnosis

The laboratory diagnosis of African tick-bite fever (ATBF), caused by *Rickettsia africae*, requires a combination of clinical suspicion and

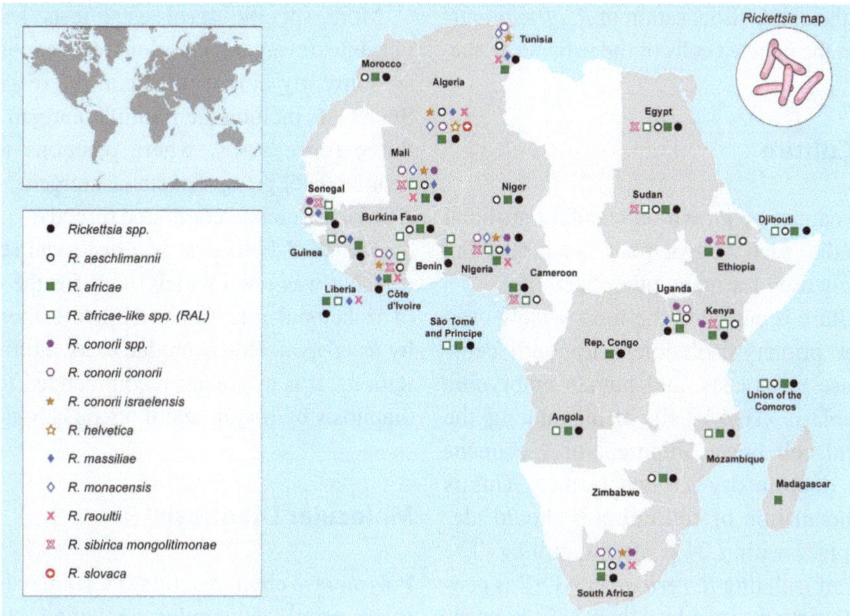

Fig. 36.3 Geographical distribution of *Rickettsia* species across Africa. (Image source: Onyiche et al. (2022). Permitted for reproduction under CC BY license)

confirmatory laboratory techniques, as the symptoms are often nonspecific.

Specimens

Serum is collected for various serological tests. Whole blood is typically used for molecular detection methods, such as polymerase chain reaction (PCR), although its sensitivity may be lower during early infection. The skin biopsy from an eschar or rash is often the most sensitive specimen for direct detection of *R. africae* DNA by PCR and for immunohistochemistry (IHC), given that the eschar is the primary site of bacterial replication.

Microscopy

Microscopy is not a primary method for routine clinical diagnosis of *R. africae* due to the organism's small size and obligate intracellular nature. However, it is indispensable in research and specialised laboratory settings. While being Gram-negative, they stain poorly with the conventional Gram stain in tissues. As such, specialised stains like Giemsa or Gimenez are used. The bacteria appear as reddish-purple or bright red/purple coccobacillary forms within the host cell cytoplasm. Acridine orange, a fluorescent stain, can also be employed for visualisation in cell cultures.

Light microscopy may reveal infected cells with intracellular bacteria, though it cannot provide species-level identification. More powerfully, fluorescence microscopy is used, particularly with the indirect immunofluorescence assay (IFA), to visualise rickettsial antigens in infected cells or tissues, offering higher specificity. A unique and important application is the visualisation of "actin tails" using fluorescently labelled phalloidin, demonstrating the bacterium's ability to polymerise host cell actin for intracellular movement and cell-to-cell spread.

Immunohistochemistry, while not a primary screening tool, is a highly valuable confirmatory test, especially in cases where a skin biopsy (particularly from an eschar or rash) is available. It allows for the direct visualisation of *R. africae* antigens within the infected cells of the patient's tissue.

In Vitro Culture

R. africae cannot be grown on standard artificial culture media, such as agar plates or broth, that are routinely used for many other bacteria.

Cell culture is currently the most widely used system for primary isolation, with Vero cells, L929 mouse fibroblasts, and human embryonic lung fibroblasts (HEL) cells being among the most useful cell lines. Shipment of specimens should be done in dry ice and be fast. This is because inoculation of cell cultures should ideally take place within 24 h after sampling. The sensitivity of isolating *R. africae* in ATBF is generally poor, with rates of less than 15% in heparinised blood and 44% in skin biopsy specimens.

Serodiagnosis

The indirect immunofluorescence assay (IFA) is considered the reference standard for serological diagnosis of rickettsial diseases, including African tick-bite fever. IFA assays typically measure both IgM (indicating a more recent infection) and IgG (indicating current or past infection) antibody titers.

For a definitive diagnosis, paired serum samples are highly recommended: an acute-phase sample collected early in the illness (ideally within the first week of symptom onset) and a convalescent-phase sample collected 2–4 weeks later (or up to 4 weeks for *R. africae*, as per some guidelines). A four-fold or greater rise in antibody titer between the acute and convalescent samples is considered conclusive evidence of a recent *Rickettsia* infection.

Other serological methods include enzyme-linked immunosorbent assays (ELISA) and Western blotting (WB). Western blotting provides a more detailed picture by separating rickettsial proteins and allowing for the detection of antibodies to specific protein bands (e.g., OmpA, OmpB).

More specific serological tests for African tick-bite fever are laborious and expensive and are currently only available at reference centres. Such tests include the multiple-antigen immunofluorescence assay, where reactions to several spotted fever group rickettsial antigens, including *R. africae*, can be compared directly.

The Weil-Felix test is a historical serological test that was once widely used for the diagnosis of rickettsial infections, including those caused by *R. africae*. However, due to its significant limitations, it is no longer recommended for routine diagnosis by major health organisations.

Molecular Diagnosis

Polymerase chain reaction (PCR) may be applied to an array of samples, including blood, skin biopsy samples, and arthrodial tissues. Detection strategies based on recognition of sequences within the genes encoding the 16S rRNA, outer membrane proteins rOmpA, rOmpB and PS120 have been effectively used. With standard PCR, *R. africae* DNA are detected in 48% of specimens. The sensitivity approaches 100% when suicide PCR, a nested PCR using single-use primers targeting a gene never amplified previously in the laboratory, is used.

A summary of the various methods used for diagnosing ATBF is presented in Table 36.2.

Treatment

There is limited data on the optimal antibiotic treatment of African tick-bite fever, including indications, drug of choice, and proper dosing. Doxycycline 100 mg twice daily for 7 days or until 48 h after defervescence can be recommended in cases of ATBF with high fever, intense headache, or other pronounced symptoms. Alternatives include azithromycin for pregnant women and children. Early initiation of therapy shortens disease duration. No long-term sequelae have been reported with appropriate treatment. Antibiotic resistance is not currently a concern for *R. africae*.

Table 36.2 Laboratory diagnosis of African tick-bite fever

Diagnostic method	Specimen type(s)	Target	Advantages	Limitations
Immunohistochemistry (IHC)	Skin biopsy (eschar/rash), other tissue biopsies	*Rickettsia africae* antigens in host cells	Direct visualisation of bacteria in tissue; Confirmatory; Pathological evidence of infection	Invasive (requires biopsy); limited to specialised labs; not for routine screening
Light microscopy (e.g., Giemsa, Gimenez stain)	Infected cell cultures, rarely direct clinical samples	*Rickettsia africae* morphology	Visualisation of intracellular bacteria in culture	Poor sensitivity/ specificity in clinical samples; cannot differentiate species; a research/lab tool
Serodiagnosis				
Indirect immunofluorescence assay (IFA)	Serum (acute and convalescent phase)	Patient antibodies (IgG, IgM) to Rickettsia antigens	Reference standard; high sensitivity/ specificity for rickettsial group	Delayed seroconversion (negative early); significant cross-reactivity within SFG; paired sera needed for confirmation
ELISA / Western Blot	Serum	Serum antibodies (IgG, IgM) to Rickettsia antigens	High-throughput (ELISA); more specific banding patterns (WB); can be more sensitive than IFA	Like IFA regarding delayed seroconversion and cross-reactivity
Weil-Felix test	Serum	Patient antibodies cross-reacting with Proteus antigens	Historical; inexpensive	Very low sensitivity and specificity; frequent false-positives/negatives; no longer recommended for routine diagnosis
Molecular diagnosis Polymerase chain reaction (PCR)	Skin biopsy (eschar/rash), whole blood (EDTA), tick samples, CSF	*R. africae* DNA (*gltA, ompA, ompB,* 16S rRNA)	Early detection (before seroconversion); high specificity (species-level identification with sequencing) Direct pathogen detection	Variable sensitivity (especially blood); bacterial load can be low; requires specialised equipment/ expertise

Prevention and Control

Preventive measures include the use of tick repellents, protective clothing, and inspecting the skin after exposure to tick-prone environments. Tick control in livestock and public health education campaigns is also essential. Personal protective measures are critical for tourists and field workers. Integrated tick management and community education programmes can significantly reduce exposure.

Travellers to sub-Saharan Africa should be aware that they may encounter the tick vectors of African tick-bite fever if they visit rural areas. The ticks may be locally very abundant on vegetation; they are highly aggressive feeders and are usually active 24 h a day. During bush walks, travellers are recommended to wear protective clothing, preferably impregnated or sprayed with acaricides. They should also use topical insect repellents on exposed skin and regularly examine themselves for ticks. Most commercially available insect repellents, based on diethyl-3-methylbenzamide (DEET), piperidine compounds, or citronella oil, have only short-lasting efficacy against *Amblyomma* species ticks, typically lasting less than 2 h.

Case Study

Ms Anuja, a 32-year-old female, presented to a travel clinic in Mumbai with a 5-day history of persistent fever, headache, fatigue, and generalised muscle aches. She had returned from a two-week safari trip to South Africa approximately 1 week prior to the onset of symptoms. She had no significant past medical history. On physical examination, her temperature was 102.5 °F. A distinctive finding was a single, tender, ulcerated lesion, approximately 1 cm in diameter, with a black necrotic centre (eschar) surrounded by an erythematous halo, located on her left ankle. She also exhibited mild regional lymphadenopathy, with palpable and slightly tender lymph nodes in her left inguinal area. No other rashes were noted. Her neurological examination was unremarkable. Blood tests showed a normal white blood cell count and slightly elevated inflammatory markers (CRP 45 mg/L). Serological tests for *R. africae* were sent, along with a PCR test from a swab of the eschar. The serology later showed a significant rise in IgG and IgM titers against *Rickettsia africae*, and the PCR from the eschar was positive, confirming the diagnosis. Ms Anuja was started on oral Doxycycline, 100 mg twice daily, immediately upon presumptive diagnosis. Within 24 h of initiating treatment, her fever began to subside, and her headache and muscle aches significantly improved. She continued the doxycycline for 7 days. At a follow-up visit 1 week after completing the antibiotic course, she was completely asymptomatic. The tache noire was healing well, forming a small crust. She experienced no long-term complications from the infection.

1. What was Ms Anuja's recent travel history that was relevant to her diagnosis?
2. What is the specific name of the characteristic skin lesion found on Anuja, and where was it located?
3. Which bacterium is the primary causative agent of African tick-bite fever?
4. Which diagnostic tests were necessary to confirm the pathogen?
5. What was the initial antibiotic prescribed and why?

Research Questions

1. What genomic features of *R. africae* contribute to its vector specificity and pathogenicity?
2. How does repeated exposure to *R. africae* affect long-term immunity in endemic populations?
3. Can novel vaccine candidates be developed using proteomic data?
4. What role do asymptomatic animal reservoirs play in sustaining *R. africae* transmission?

Further Readings

Angelakis E, Mediannikov O, Parola P, Raoult D. *Rickettsia felis*: the complex journey of an emergent human pathogen. Trends Parasitol. 2016;32(7):554–64.

Binder WD, Gupta R. African tick-bite fever in a returning traveler. J Emerg Med. 2015;48(5):562–5.

Frean J, Grayson W. South African tick-bite fever: an overview. Dermatopathology. 2019;6(2):70–6.

Harrison N, Burgmann H, Forstner C, et al. Molecular diagnosis of African tick-bite fever using eschar swabs in a traveller returning from Tanzania. Wien Klin Wochenschr. 2016;128:602–5. https://doi.org/10.1007/s00508-016-1047-0.

Jensenius M, Fournier PE, Kelly P, Myrvang B, Raoult D. African tick bite fever. Lancet Infect Dis. 2003a;3(9):557–64.

Jensenius M, Fournier PE, Vene S, Hoel T, Hasle G, Henriksen AZ, Hellum KB, Raoult D, Myrvang B, Norwegian African Tick Bite Fever Study Group. African tick bite fever in travelers to rural sub-Equatorial Africa. Clin Infect Dis. 2003b;36(11):1411–7.

Jensenius M, Fournier PE, Helium KB, Wesslén L, Caruso G, Priø T, Løhne K, Vene S, Raoult D, Myrvang B. Sequential changes in hematologic and biochemical parameters in African tick bite fever. Clin Microbiol Infect. 2003c;9(7):678–83.

Jensenius M, Ueland T, Fournier PE, Brosstad F, Stylianou E, Vene S, Myrvang B, Raoult D, Aukrust P. Systemic inflammatory responses in African tick-bite fever. J Infect Dis. 2003d;187(8):1332–6.

Kernif T, Leulmi H, Raoult D, Parola P. Emerging tick-borne bacterial pathogens. Emerg Infect. 2016;4:295–310.

Lepidi H, Fournier PE, Raoult D. Histologic features and immunodetection of African tick-bite fever eschar. Emerg Infect Dis. 2006;12(9):1332.

Ndip LM, Biswas HH, Nfonsam LE, LeBreton M, Ndip RN, Bissong MA, Mpoudi-Ngole E, Djoko C, Tamoufe U, Prosser AT, Burke DS. Risk factors for

African tick-bite fever in rural central Africa. Am J Trop Med Hyg. 2011;84(4):608.
Onyiche TE, Labruna MB, Saito TB. Unraveling the epidemiological relationship between ticks and rickettsial infection in Africa. Front Trop Dis. 2022;3:952024. https://doi.org/10.3389/fitd.2022.952024.
Owen CE, Bahrami S, Malone JC, Callen JP, Kulp-Shorten CL. African tick bite fever: a not-so-uncommon illness in international travelers. Arch Dermatol. 2006;142(10):1312–4.
Roch N, Epaulard O, Pelloux I, Pavese P, Brion JP, Raoult D, Maurin M. African tick bite fever in elderly patients: 8 cases in French tourists returning from South Africa. Clin Infect Dis. 2008;47(3):e28–35.
Silva-Ramos CR, Faccini-Martínez ÁA. Clinical, epidemiological, and laboratory features of *Rickettsia africae* infection, African tick-bite fever: a systematic review. Infez Med. 2021;29(3):366–77.
Tsai YS, Wu YH, Kao PT, Lin YC. African tick-bite fever. J Formos Med Assoc. 2008;107(1):73–6.

Epidemic Typhus

37

Natarajan Vaghesan, Balamurali Venkatesan,
and Kesavarajan Govindarajan Karthikeyan

Learning Objectives

1. To understand the key concepts of *Rickettsia prowazekii* transmission, its association with epidemic typhus, and the socio-environmental factors contributing to outbreaks
2. To provide an overview of the molecular mechanisms of infection, their effects on endothelial cells, and the clinical manifestations of epidemic typhus
3. To explore current diagnostic methods, treatment strategies, and challenges in vaccine development

Introduction

Rickettsia prowazekii is the causative agent of epidemic typhus, one of the earliest pestilential ailments in the evolution of humanity. In the past, typhus outbreaks frequently accompanied hunger, war, climatic extremes, and social unrest because typhus spreads in cluttered and unhygienic environments. In the rural highlands of Asia, Africa, and South America, epidemic typhus persists as an imminent threat. In regions of the globe where civil unrest and inadequate public health initiatives continue, epidemic typhus may reappear as a severe transmissible disease. Outbreaks of typhus have occurred in parts of Morocco, Burundi, Russia, and Andean Peru within the last 20 years.

R. prowazekii is a Gram-negative, obligate intracellular bacterium belonging to the Rickettsiaceae family. The disease is transmitted by the human body louse (*Pediculus humanus*) (Fig. 37.1).

Additionally, Brill-Zinsser disease (BZD) is a relapse variant of epidemic typhus. Some infected individuals carry the bacteria in a latent form for the remainder of their lives due to an unidentified process that occurs after the initial infection and clinical cure. Latent microbes may reactivate and cause disease again in some individuals due to weakened immunity over time.

History

Henrique da Rocha Lima discovered the genus *Rickettsia*, which consists of intracellular pathogens in lice. It bears the name of Howard Taylor Ricketts, who investigated these microbes in his work on Rocky Mountain spotted fever. The cause of the infection is yet unknown. According to some historians, the disease originated in ancient Europe. It may have been connected to Thucydides' account of the Athenian epidemic.

N. Vaghesan (✉) · B. Venkatesan ·
K. Govindarajan Karthikeyan
Department of Microbiology, Sri Lalithambigai Medical College and Hospital, Faculty of Medicine, Dr. MGR Educational and Research Institute, Maduravoyal, Adayalampattu, Chennai, Tamil Nadu, India

S. C. Parija, A. Poddar (eds.), *Textbook of Bacterial Zoonoses*, Microbial Zoonoses,
https://doi.org/10.1007/978-981-95-6615-0_37

Fig. 37.1 Microscopy of lice. (Image source: Authors personal collection)

The word "typhus" comes from the Greek word "*tuphos*", which means "smoke". and refers to the delirium it induces. Although its source remained unknown, the disease may have been identified in Spain as early as 1083. Doctors were able to differentiate typhus from other fevers by the sixteenth century due to its abrupt onset and rash. Its prevalence among military ranks was recorded by Zavorziz in 1676. Before John Huxham distinguished it from typhoid fever in 1739, it was frequently mistaken for the latter. William Gerhard further distinguished "exanthematic typhus" from typhoid disease in 1836 after Boissier de Sauvages coined the phrase in 1760.

Twenty per cent of troops died from typhus, which decimated armies throughout the Napoleonic Wars. It reappeared during World War I, infecting 25 million Russians. Outbreaks happened in concentration camps and throughout Europe during World War II. Outbreaks also happened in Ethiopia (1995), Burundi, Russia, and Peru (1997) despite a post-war drop in cases. The Centers for Disease Control (CDC) has identified epidemic typhus as a possible bioterrorism agent, and it continues to pose a concern.

Taxonomy

The organism *R. prowazekii* belongs to the *Domain Bacteria*, *Phylum Pseudomonadota* (formerly Proteobacteria), *Class Alphaproteobacteria*, *Order Rickettsiales*, *Family Rickettsiaceae*, and *Genus Rickettsia.*

The genus, currently containing 29 validly published names, has been divided into two groups: the spotted fever group (SFG) and the typhus group (TG). *Rickettsia prowazekii* is a member of the TG. Gillespie et al. recently introduced an ancestral group (AG) and a transitional group (TRG). The spotted fever group has outer membrane proteins A and B, whereas the typhus group lacks them (OmpA).

Genomics and Proteomics

The circular chromosome of the *R. prowazekii* strain Madrid E covers 1,111,523 base pairs and has an average G+C content of 29.1%. Its genome has 834 full open reading frames, each of which is 1005 bp long on average. 75.4% of the genome comprises genes that code for proteins, whereas 0.6% is made up of stable RNAs. While the GvF12 strain originated from a flying squirrel in the USA, the Madrid E, Breinl, and Cairo strains came from patients in Spain, Poland, and Egypt, respectively. Madrid E is avirulent, but Breinl and Cairo are virulent. According to Nagarajan et al., the genomes of Madrid E. and Breinl, which were sequenced at the Naval Medical Research Centre (NMRC), were first put together into 197 and 60 contigs, respectively, and then resolved into a single contig using in silico gap closure. The Cairo and GvF12 genomes were assembled from scratch, yielding 752 and 46 contigs, respectively. Each genome was around 1.1 Mb in size and had a sizeable percentage of noncoding regions (~25–26%), as was to be expected. A GC level of roughly 29% and 33 tRNA genes were also present in each genome.

Only two flying squirrel isolate genomes are now publicly accessible. This limits the ability to make inferences about the traits of strains in contrast to the flying squirrel-arthropod life cycle, isolated as a component of the human body louse cycle. Thus, the genomes of the flying squirrel strains GvV257 and GvF24, as well as the previously reported Madrid E genome (NCBI accession number AJ235269.1), were compared with the genome of the flying squirrel isolate GvF12. There were differences between the draft GvF12 genome and the GvV257 and GvF24 genomes at 226 and 11 locations, respectively. In contrast, the GvF12 and Madrid E genomes were found to vary at 869 positions. On the other hand, 292 locations in the genomes of Madrid E and Breinl were found to vary.

Morphology

R. prowazekii exhibits a characteristic morphology as an obligate intracellular bacterium. These microorganisms are typically small, ranging in size from 0.3 to 0.6 μm in width and 0.8 to 2.0 μm in length. Their shape is generally described as pleomorphic, meaning they can appear as short rods (bacilli) or sometimes as coccobacilli (oval to almost spherical). The organism has a slime layer or a loosely associated surface layer composed of proteins and carbohydrates. This layer is believed to be involved in adherence to host cells and potentially in protection from the host's immune system.

Cultivation

R. prowazekii is an obligate intracellular bacterium, and so, it cannot be cultivated on artificial, cell-free media. It requires living eukaryotic host cells for growth and replication. Therefore, cultivation methods rely on providing these host cells in a controlled environment (Fig. 37.2). Briefly, cultivation begins with separating blood into a clot and serum. The clot, containing blood cells and potentially the *Rickettsia* organisms, is used to inoculate guinea pigs, tissue cultures, and the yolk sac of embryonated chicken eggs. Guinea pigs develop symptoms, and their tissues (such as the spleen) are harvested to propagate and study the bacteria.

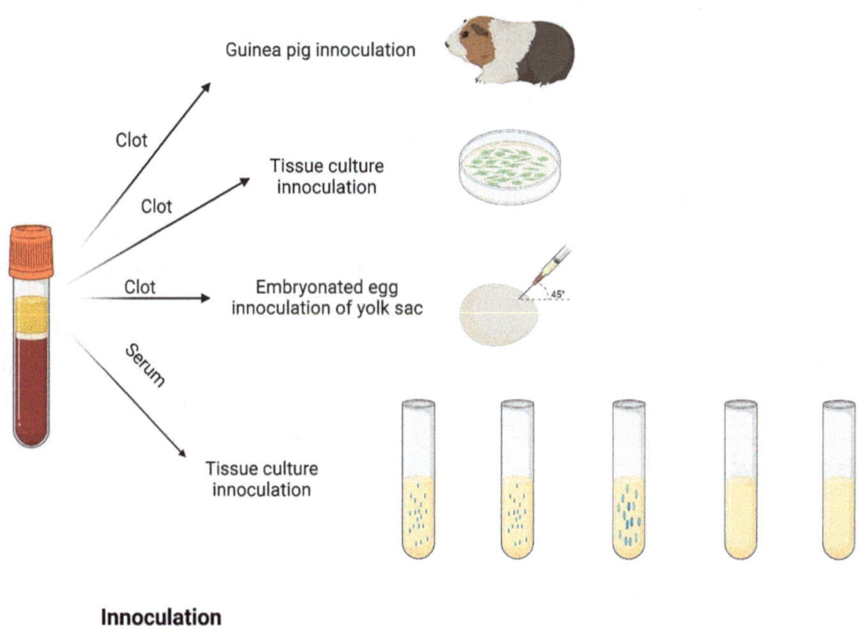

Fig. 37.2 Cultivation of Rickettsial organisms. (Image source: Authors personal collection)

Tissue cultures represent early attempts at in vitro cultivation using host cells. The yolk sac of embryonated chicken eggs is used for *Rickettsia* propagation, particularly for vaccine production. *Rickettsia*-containing material from infected sources (such as guinea pig spleen) is introduced into the yolk sac, providing a nutrient-rich environment. Serum is also used as a potential inoculum for tissue culture, possibly to provide growth factors or study the interaction of *Rickettsia* with host cells. The outcome of these inoculation methods is assessed by observing turbidity or cellular growth in test tubes, indicating the propagation of *Rickettsia* within the different host systems.

Rickettsial culture is mainly employed for antigen production, pathogen stock maintenance, and diagnostics. Although there is increasing support for permitting certain rickettsial cultures in biosafety level 2 (BSL2) facilities for low-concentration amplification, these organisms are usually handled in biosafety level 3 (BSL3) laboratories due to biosafety concerns.

Animal Models

In 1906, Dr H. T. Ricketts conducted research in Missoula, Montana, using guinea pigs as animal models to study typhus. He first used blood from a patient with Rocky Mountain spotted fever to inject two guinea pigs. In further tests, he examined both unfiltered and filtered serum and extended the research to include blood from more individuals. Unless filtered serum was utilised, the guinea pigs constantly developed fever and clinical symptoms of disease. Most of the afflicted guinea pigs eventually perished, demonstrating their susceptibility and significance as a model for investigating rickettsial diseases.

Transmission Cycle

R. prowazekii, a bacterium that becomes infectious within the body louse, is the cause of epidemic typhus in humans. The louse's faeces include *R. prowazekii*, which grows in its intestines.

There are three phases in a louse's life cycle: larvae, nymphs, and adults. A louse needs its host's blood meal, which supplies the energy it needs to advance to the next stage. *R. prowazekii* is acquired by the louse while feeding, making it a link in the epidemiological cycle of disease transmission. Transstadial transmission (TS), in which the acquired pathogen stays in the tick throughout moulting, is made possible by this mechanism, which is referred to as horizontal transmission. Infection can also occur when an individual scratches a bite wound, introducing infectious fluids or faeces into the skin, eyes, or mucous membranes. An overview of the transmission cycle is presented in Fig. 37.3.

Pathogenesis and Pathology

R. prowazekii enters the body through the epidermis, travels through the circulation, infects the endothelium, and occasionally attacks vascular smooth muscle cells (Fig. 37.4). Certain Rickettsia species infiltrate target cells, multiply in the cytosol, and disrupt cells. Although it lacks intracellular mobility, *R. prowazekii* multiplies by binary fission. As a result, unless the host cell membrane is mechanically lysed, it builds up inside the host cells to reach large numbers without harm. The load of intracellular microorganisms is proportionate to the amount of cell damage. Bacteria get nutrients in the cytoplasm through transmembrane exchange processes, as encoded by the ATP/ADP translocase. To spread and infect nearby endothelial cells, infectious bacteria are discharged into the surrounding tissues.

Cellular injury results in the pathological hallmarks of all rickettsial infections, in addition to symptoms unique to the afflicted organ, such as (i) skin rashes, (ii) damage to the pulmonary microcirculation with non-cardiogenic pulmonary oedema, and (iii) brain lesions with seizures and coma. The condition is associated with extensive vasculitis accompanied by oedema, hypovolemia, elevated vascular permeability, and the activation of coagulation and humoral immune systems. Vascular dysfunction becomes

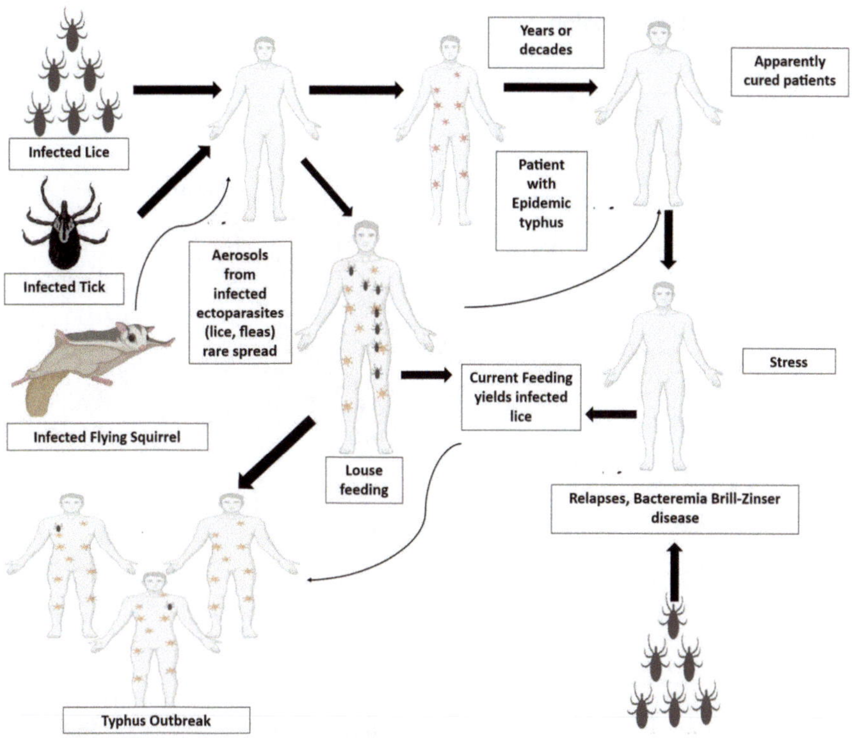

Fig. 37.3 Transmission of infection in humans. (Image source: Authors personal collection)

prevalent as the disease progresses due to increased endothelial degradation. Furthermore, lymphocytes and macrophages gather near tiny blood arteries because of Rickettsia-induced cell injury. Permeability variations and the flow of plasma and plasma proteins from intravascular spaces to the interstitium are caused by endothelial damage in cases of severe infection. Moreover, endothelial damage results in compromised vessel integrity, which shows up as both microscopic and macroscopic haemorrhage foci (Fig. 37.5).

Immunology

The components of the immune system, including dendritic cells, innate immune signalling, macrophages, NK cells, CD8+ T cells, CD4+ T cells, endothelial cells, antibodies, cytokines, and inflammatory chemokines, are crucial in regulating *Rickettsia* infection through mechanisms that facilitate its eradication, based on numerous studies done on both humans and animals. However, a highly severe and deadly illness will emerge if these immuno-protective components fail to function properly in the host response or if *Rickettsia* alters or manipulates the immune system's reaction. In other conditions, the pathophysiology and progression of *Rickettsia* are influenced by immune system mutations, including inducible regulatory (TR) T cells.

The highly conserved cellular immune response known as autophagy was initially identified as a means of nutrient-deficient cell survival. It targets and engulfs invasive microorganisms in double-membrane vesicles before merging with lysosomes to remove them. To create intracellular niches for reproduction, intracellular pathogens have evolved a variety of immune escape techniques and mechanisms to either avoid or take advantage of autophagosomes. According to recent studies, several *Rickettsia* species have adapted to these intricate mechanisms to survive and spread disease.

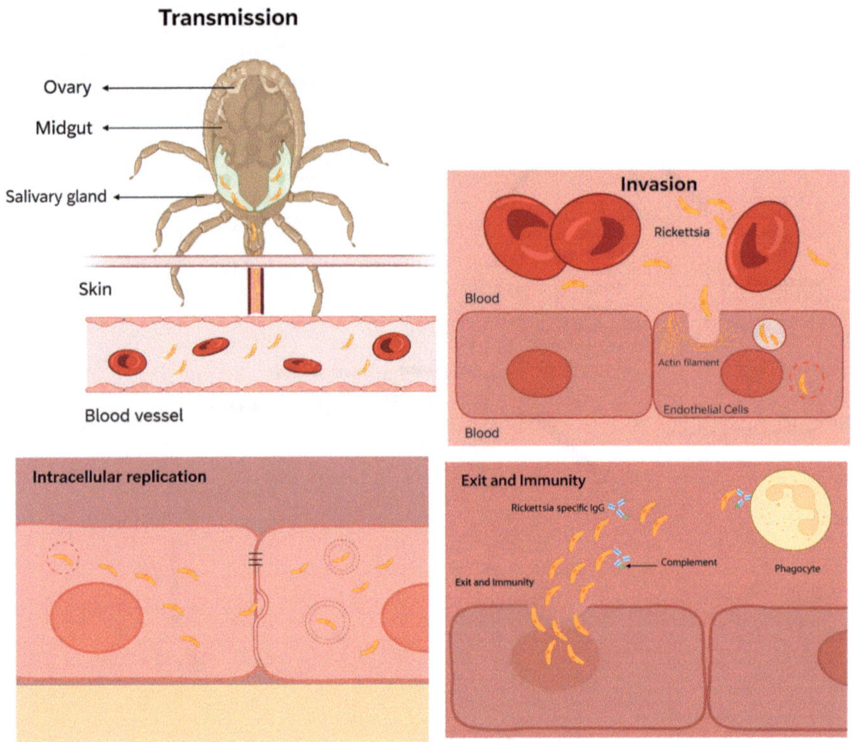

Fig. 37.4 Transmission cycle of *Rickettsia prowazekii.* (Image source: Authors personal collection)

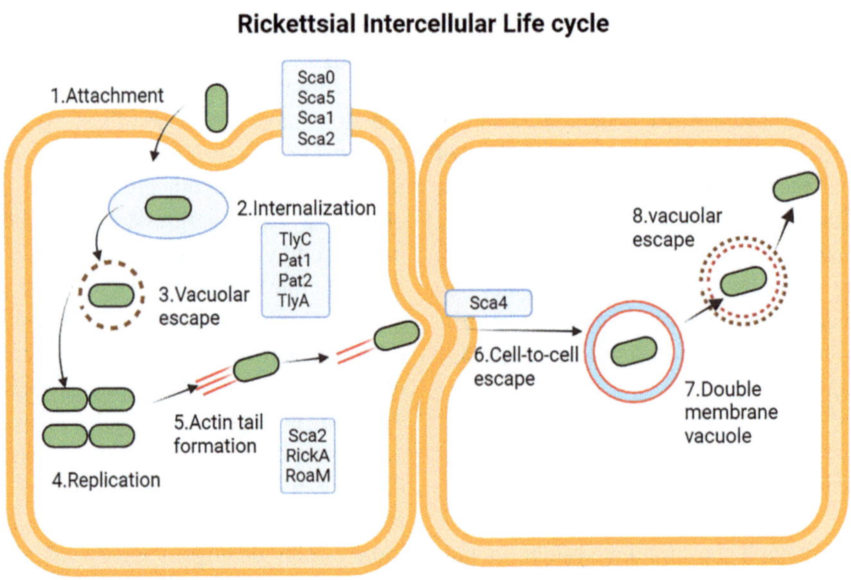

Fig. 37.5 Pathophysiology of *Rickettsia prowazekii.* (Image source: Authors' personal collection)

Infection in Humans

During infection, *R. prowazekii* spreads to endothelial cells, leading to dysfunction and loss of integrity. This results in microvascular thrombosis and increased permeability, causing oedema. Studies using human umbilical vein endothelial cells (HUVECs) have shed light on rickettsial pathophysiology. It has been found that spotted fever group Rickettsia enters endothelial cells via fibroblast growth factor receptor-1. Infected HUVECs express tissue factor and release von Willebrand factor multimers, contributing to thrombosis. *Rickettsia*-infected HUVECs promote neutrophil adhesion through E-selectin, leading to inflammation and tissue damage. These findings help explain the severe vascular effects observed in human rickettsial infections.

Clinically, certain rashes are still essential for diagnosing rickettsioses, and patients usually also experience headaches and a high temperature (39–40 °C). Due to delayed diagnosis, rickettsioses continue to claim the lives of patients. Epidemic typhus typically takes 10–14 days to incubate. Before the sudden onset of severe headaches and fever, patients typically experience malaise for 1–3 days.

Rashes, severe myalgias, arthralgias, and general constitutional symptoms are among the other symptoms, which could vary (chills, anorexia, and malaise); in as many as 80% of cases, central nervous system symptoms (seizures, delirium, and coma) may appear. Rashes were universally present in the original histological description of epidemic typhus, but new data indicate that only 20–40% of individuals may have rashes. This is most likely due to most of the research being conducted in Africa, where it is difficult to identify rashes on dark skin as serious medical issues.

When rashes initially appear as non-confluent erythematous and blanching patches, over one-third of people go on to develop non-blanching macules, petechiae, and even purpura. Most skin lesions are found on the trunk, starting in the axillae and possibly spreading to the extremities in a centrifugal pattern. They are not seen on the face, palms, or soles, although they are occasionally seen on the conjunctiva and soft palate. There are no scars.

Infection in Animals

R. prowazekii infection in natural animal reservoirs is the prevalence of subclinical infections, where the animals harbour the bacteria without exhibiting any obvious signs of illness. This differs sharply from the significant morbidity seen in human cases. In laboratory animals, while symptoms are more apparent and studied to understand the pathogenesis of the infection, they generally lack the characteristic prominent rash and widespread systemic involvement that defines human epidemic typhus.

In experimental infection of guinea pigs, which have historically served as a significant experimental model for *R. prowazekii* research, a consistent finding upon infection is fever. Male guinea pigs may also exhibit a classic sign known as *Kerandel's sign*, characterised by swelling and redness of the scrotum. Other observed symptoms in guinea pigs include lethargy, a general reduction in activity levels, loss of appetite, and a subsequent decrease in body weight. Upon necropsy, splenomegaly, or enlargement of the spleen, is a common finding.

Epidemiology and Public Health

Humans are the primary hosts. *Pediculus humanus* (body louse) acts as a vector. The body louse has a long, membranous, seven-segmented abdomen, a compact thorax, and a small, narrow head with two antennae (Fig. 37.1). Its host's skin tone is frequently reflected in the colour of its outer cuticle. Other than their hosts, body lice have no natural predators. The lifespan of mature lice is around 20 days. The male-to-female ratio of local louse populations varies, ranging from around 1:6. Extreme instances involve thousands of lice per host, whereas infections typically involve just a handful. Two hundred lice might theoretically be produced in a month by a single mating pair. Evans and Smith calculated that the population might rise by 11% per day, demonstrating the rapidity with which epidemics may intensify.

Reports of *R. prowazekii*-caused epidemic typhus in visitors are uncommon. Epidemic

Table 37.1 Epidemiology of louse-causing epidemic typhus

Louse species	Specific host	Country/region distribution
Pediculus humanus corporis	Human	Ethiopia, Burundi, Rwanda, Nepal, Peru, Afghanistan, refugee camps worldwide
Pediculus humanus capitis	Human	Global—high prevalence in school-age children (USA, India, Brazil, UK, Australia)
Pthirus pubis	Human (pubic region)	Global—more common in sexually active populations
Polyplax spinulosa	Rattus species	Urban areas in Asia (China, India), parts of Europe, Latin America
Polyplax serrata	Mice, rats	Central Europe, North America, Eastern Asia
Haematopinus suis	Pigs	USA, China, Philippines, Vietnam, Eastern Europe
Linognathus setosus	Dogs	Worldwide (noted in Brazil, USA, parts of Africa)
Linognathus vituli	Cattle	North America, Europe, parts of sub-Saharan Africa

typhus outbreaks have occurred locally in refugee groups when body lice are common (Table 37.1). These outbreaks happen in the colder months. Depending on the clinical context, epidemic typhus mortality varies. The fatality rate from an untreated initial infection can reach 60%, with older and undernourished individuals having the greatest mortality rates. The death rates for patients suffering from BZD and sylvatic typhus are significantly lower. BZD is seldom lethal and manifests with less severe symptoms.

Historically, the disease exhibited a widespread distribution across Europe, Asia, Africa, and the Americas. The condition was associated with devastating epidemics occurring during periods of conflict and social upheaval, such as the World Wars, when louse infestations were rampant in vulnerable populations. A significant typhus outbreak during the civil war in Burundi was documented in 1997; an estimated 100,000 individuals were afflicted, with a 15% case mortality rate. Since 1997, there have also been reports of minor outbreaks and isolated cases in several countries, including the USA, France, Peru, Russia, and North Africa. Because of the uncertain role of non-human reservoirs and the potential for *R. prowazekii* to reappear in infected individuals, which is linked to BZD, eradicating epidemic typhus may be challenging.

Despite a substantial reduction in global burden, epidemic typhus has not been entirely eradicated and continues to persist in specific endemic foci. The infection is primarily concentrated in regions characterised by poverty, social disruption, and inadequate public health infrastructure. These areas include the highland regions of Central and South America, such as Peru, Bolivia, and Ecuador. In these countries, cold climates may favour louse survival, and access to hygiene resources is limited. Certain parts of Africa also harbour endemic foci, particularly in areas experiencing conflict or hosting refugee populations, where louse infestation can be prevalent. While significantly diminished, sporadic cases or localised outbreaks may still occur in specific regions of Asia facing similar socio-economic challenges.

Several factors influence the current epidemiological landscape of *R. prowazekii* infection. Socio-economic conditions, including poverty, overcrowding, poor hygiene practices, and limited access to clean clothing and washing facilities, remain the primary drivers of louse infestation and subsequent typhus transmission. Conflict and displacement resulting from war, civil unrest, and natural disasters can lead to mass population movements, creating environments conducive to louse proliferation and increasing the risk of outbreaks. Climate, particularly low temperatures, favours the survival of the human body louse, potentially contributing to the persistence of typhus in highland regions.

An additional aspect of *R. prowazekii* epidemiology is the phenomenon of Brill-Zinsser disease(BZD), a recrudescence of a latent *R. prowazekii* infection that can occur years after the initial episode of epidemic typhus. Individuals experiencing BZD can serve as a source of infection for newly acquired lice, potentially initiating new outbreaks if environmental and social conditions are conducive to louse infestation.

Diagnosis

Key sample types for laboratory diagnosis of *R. prowazekii* infections include whole blood and serum for molecular and serological testing. Paired serum samples are crucial for serological confirmation. Skin biopsies (if a rash is present) are preferred for polymerase chain reaction (PCR) and immunohistochemistry.

Microscopy

Microscopic diagnosis of *R. prowazekii* is not typically the primary method for routine clinical diagnosis due to the intracellular nature of the bacteria. However, various microscopic techniques, such as direct detection in specific sample types and visualising the bacteria in cell culture, are used for research purposes.

Under *oil immersion microscopy*, the distinct colouration and morphology (small, rod-shaped or coccobacillary) of the *Rickettsia* can be observed within the host cells. *R. prowazekii* stains poorly with the standard Gram stain. The organisms appear faintly pink or may not be clearly visible against the background. This is because they are obligate intracellular bacteria and have a cell wall structure that does not readily retain the crystal violet dye during the decolourisation step of the Gram staining procedure.

Due to this poor staining with the Gram method, other specialised staining techniques are preferred for the microscopic visualisation of the bacteria. One of the key staining techniques used for visualising Rickettsia is the *Gimenez stain*. This differential stain employs a basic fuchsin dye that preferentially stains the bacteria a bright red or reddish-purple colour, while a counterstain, usually malachite green, stains the background and host cell cytoplasm a light blue or greenish colour. The Gimenez stain is particularly useful for examining smears from cell cultures infected with *R. prowazekii* or impression smears from tissues of infected animals.

Giemsa staining is also used to visualise the bacteria within infected host cells, typically in cell culture or tissue smears. When stained with Giemsa, *Rickettsia* organisms typically appear as small, bluish-purple coccobacilli or short rods within the cytoplasm of the host cells. The host cell nucleus is stained darker purple, while the cytoplasm appears light blue or pink, depending on the cell type and staining intensity.

Another staining method that has been historically used is the *Macchiavello stain*. This technique involves staining with basic fuchsin, followed by decolourisation with citric acid and counterstaining with methylene blue. It stains *Rickettsia* red, however, with a diffusely stained background and host cells.

Immunofluorescence microscopy offers a more specific and sensitive approach for visualising *R. prowazekii*. This technique utilises antibodies that are specifically raised against *R. prowazekii* antigens and are conjugated to a fluorescent dye (e.g. fluorescein isothiocyanate—FITC). When a sample containing *R. prowazekii*-infected cells or tissue is incubated with these fluorescently labelled antibodies, the antibodies bind to the bacteria. The *Rickettsia* appear brightly fluorescent against a dark background, allowing for specific identification and localisation within the host cells.

In Vitro Culture

Early techniques that used designed solutions for tissue cultures and chick embryo inoculation played a significant role in the initial isolation and study of *R. prowazekii* before the widespread adoption of established mammalian cell lines. This technique was used to isolate, propagate, and study these obligate intracellular bacteria, leading to a better understanding of their biology, pathogenesis, and the development of diagnostic tools and vaccines.

Chick Embryo Inoculation

The yolk of eggs that had been incubated at 39.4 °C for 5–6 days was injected, and the eggs were then incubated at 35 °C. Every 5–9 days, transfers took place. Cultures were treated, centrifuged, and then injected intraperitoneally or subcutaneously into chick embryos or guinea pigs for testing. The presence of *Rickettsia* was assessed after the final supernatants were titrated.

Mammalian Tissue Culture

The in vitro cultivation of rickettsial organisms is commonly conducted using methodologies adapted from viral culture systems, wherein tissue culture flasks containing adherent host cell lines provide the necessary environment for bacterial growth. As rickettsiae are obligate intracellular bacteria, the use of antibiotic-free culture media is essential; antibiotics typically used in viral systems would otherwise inhibit rickettsial replication.

Successful propagation requires host cell lines that are both permissive to infection and capable of sustaining rickettsial growth at their optimal temperature. Cell lines derived from mammals (e.g. Vero), amphibians (XTC-2), insects (C6/36), and ticks are frequently employed, with selection criteria based on species-specific thermal preferences and cellular susceptibility. An additional consideration is the potential of the cell line to exhibit cytopathic effects (CPE), which serve as visual indicators of infection. However, CPE is not consistently observed during primary isolation, necessitating secondary confirmation through immunofluorescence assays or polymerase chain reaction (PCR) for the detection of rickettsial DNA.

The temperature sensitivity of different *Rickettsia* species influences the choice of host cell: species favouring lower temperatures (~28 °C) are typically cultured in XTC-2, C6/36, or tick-derived lines, while those requiring higher temperatures (32–35 °C) are better supported by mammalian cells such as Vero. To maximise the likelihood of successful isolation, multiple cell lines are often tested concurrently.

In earlier experimental systems, *modifications of Tyrode's and Baker's solutions* were evaluated for tissue culture media. Optimal results were obtained using Baker's solution supplemented with 50% ascitic fluid or filtered human ascitic fluid. Historically, cultures also employed chick embryo tissues, incubated for 9–10 days at 39.4 °C. Inocula, prepared by Berkefeld filtration of spleens from infected guinea pigs, were added to Erlenmeyer flasks containing minced embryonic tissue in suspension media. Subculturing was performed at 1:10 dilutions every 8–14 days, with cultures maintained at 37 °C.

Serodiagnosis

Since 1916, with the development of the Weil-Felix test, serology has been used as the primary method for diagnosing rickettsial illnesses.

Weil-Felix Test

It is one of the oldest serological tests for rickettsial infections. It is based on the principle of cross-reactivity between *Rickettsia* antigens and different *Proteus* serotypes (*Proteus vulgaris* OX-19 antigen (Ag)-TG antibody (Ab); *Proteus vulgaris* OX-2 and OX-19 Ag-SFG Ab). In the case of *R. prowazekii* (typhus group), the patient's serum is tested for agglutinating antibodies against *P. vulgaris* strain OX19. A rise in agglutination titre against OX19 in paired sera can suggest epidemic typhus. However, the Weil-Felix test has limitations in terms of sensitivity and specificity, as false positives can occur with other infections, and false negatives can occur early in the disease or in treated cases. Due to these limitations, it has largely been replaced by more specific tests.

Complement Fixation Test (CFT)

It was another historically significant serological assay for diagnosing rickettsial infections, including epidemic typhus. This test detects the presence of antibodies in the patient's serum that can bind to *Rickettsia* antigens and "fix" complement, preventing it from lysing sensitised red blood cells. A lack of haemolysis indicates a positive result. While more specific than the Weil-Felix test, the CFT has been largely replaced by immunofluorescence assays and enzyme-linked immunosorbent assays.

Indirect Immunofluorescence Assay (IFA)

It is the most widely employed serological method for the detection of antibodies, such as IgG antibodies, against rickettsial antigens. The assay utilises fluorescein-labelled conjugates to detect antibodies bound to rickettsial antigens fixed onto glass slides. From the second week of illness onward, IgG-based IFA demonstrates high diagnostic performance for both typhus group (TG) and spotted fever group (SFG) rickettsioses, with reported sensitivities ranging from 83% to 100% and specificities from 91% to 100%.

The diagnostic accuracy of the assay is influenced by the selection of antigenic targets; commercially available tests often incorporate only a limited number of well-characterised species, such as *R. rickettsii* or *Rickettsia conorii*. While cross-reactivity among SFG rickettsiae permits group-level identification, definitive species-level discrimination requires more specialised methods such as microimmunofluorescence,

An alternative to IFA is the *indirect immunoperoxidase assay (IIP)*, which substitutes peroxidase for fluorescein, enabling visualisation via standard light microscopy. This method is particularly suitable for laboratories lacking fluorescence microscopy infrastructure.

Enzyme-linked Immunosorbent Assay (ELISA)

For SFG and TG rickettsial infections, IgM or IgG is frequently utilised and is more appropriate for low-resource environments since it can be tested in batches and requires less technical knowledge. All rickettsial groups have shown high sensitivity and specificity. The drawback of optical density (OD) readers is that they only provide qualitative data, that is, whether they are reactive or not. However, they do reduce reader bias. A diagnostic cut-off of OD > 0.5 is commonly employed; nevertheless, there is a lack of independent validation. For SFG rickettsial infections, a quantitative ELISA has recently been created, eschewing arbitrary OD cut-offs in favour of directly computing cut-offs from negative controls. The results are presented in titles. Combined, these developments make it possible to evaluate data in a more objective and accurate way.

Historically, early ELISA protocols often utilised crude preparations of *R. prowazekii*, such as whole-cell lysates or sonicated bacterial suspensions, as the source of antigenic material. While these complex mixtures contain a broad array of bacterial proteins and components capable of eliciting an antibody response, they also carry the risk of including antigens that may cross-react with antibodies produced in response to other, unrelated bacteria. This potential for cross-reactivity can compromise the specificity of the assay, leading to false-positive results.

To enhance the diagnostic accuracy, subsequent efforts have focused on employing purified or partially purified antigens derived from *R. prowazekii*. Among these, outer membrane proteins (OMPs) have garnered significant attention due to their surface exposure and high immunogenicity. Key OMPs, such as OmpA and OmpB, are prime candidates for ELISA targets. The production of these proteins in recombinant forms ensures a high degree of purity and batch-to-batch consistency. Additionally, other surface-exposed proteins and lipoproteins, such as the well-characterised 17-kDa lipoprotein, have also been successfully utilised in rickettsial serology.

Western blotting, or cross-adsorption assays, is typically available only in reference laboratories.

Molecular Diagnosis

Nucleic acid amplification tests (NAAT) have successfully identified *Rickettsia* species from various clinical specimens, including blood, buffy coat, plasma, tissue (fresh, frozen, or paraffin-embedded), and ulcer base swabs. Several molecular markers, such as the *16S rRNA* gene, *gltA*, *ompA*, *sca0*, *sca4*, *sca5*, *HSP60*, and the 17-kDa antigen gene, are used to detect *Rickettsia* DNA. While PCR is more sensitive than culture-based methods, detection in the blood is often limited due to transient bacteraemia, as *Rickettsia* mainly infect endothelial cells. Sensitivity is significantly higher in tissue samples, particularly in eschars, where PCR sensitivity can reach 92%. Although grouped and nested PCR assays effectively distinguish *Rickettsia* groups, species-level identification remains challenging.

Metagenomic techniques, including next-generation sequencing (NGS) and nanopore sequencing (NS), have revolutionised the detection and diagnosis of rickettsial diseases. These methods enhance our understanding of *Rickettsia* infections in ticks and improve diagnostic accuracy in humans. Unlike conventional methods, metagenomic approaches offer several advantages. They enable pathogen detection without prior targeted knowledge and allow for simultaneous identification of multiple pathogens, reducing time and cost. Additionally, they provide insights into the functional and phylogenetic characteristics of *Rickettsia*. However, challenges such as high costs, lack of standardisation, and limited availability of trained personnel hinder widespread adoption.

Table 37.2 presents the diagnostic approaches used for detecting rickettsial infection.

Table 37.2 The diagnostic approaches used for detecting Rickettsial infection

Diagnostic approaches	Method	Target	Remarks
Microscopy	Giemsa Staining	Rickettsia directly visible in diseased tissues	Requires certain staining; the small dimensions of the bacterium limit its use
In vitro culture	Vero, XTC-2, C6/36, tick cell lines	Isolation of live *Rickettsia*	Requires BSL-3; takes a long time to cultivate
Serodiagnosis	Weil-Felix test (WFT)	*Proteus vulgaris* OX-19 antigen (Ag)-TG antibody (Ab); *P. vulgaris* OX-2 and OX-19 Ag-SFG Ab	Considering its low specificity, WFT is still utilised in environments with limited resources
	Indirect immunofluorescence assay (IFA)	Fluorescein-labelled conjugates on rickettsial antigens and IgG detection for rickettsial infection	High specificity and sensitivity; costly equipment are required
	Microimmunofluorescence (MIF)	Simultaneous detection of multiple rickettsial antigens and species-level identification	Potential for cross-reactions
	Western blot	Differentiation of non--specific and species-specific antigens, and confirmation	improves diagnostic accuracy but test performed only in reference labs
	ELISA (IgM/IgG)	Antibody detection against the typhus group (TG) and the spotted fever group (SFG)	Ideal for low-resource batch testing; qualitative outcomes
Molecular diagnosis	PCR detection of rickettsial DNA using markers like 16S rRNA, *gltA*, *ompA*	Pathogen detection in clinical specimens	Better outcomes with tissue samples like eschars; more sensitive than culture
	Metagenomic analysis: Sequencing of all DNA/RNA in a sample	Identification of multiple pathogens	High accuracy; expensive; requires advanced equipment and trained personnel

Treatment

Tetracycline and chloramphenicol continue to be the mainstays of the treatment for epidemic typhus. These antibiotics act by inhibiting the manufacture and proliferation of rickettsial proteins. Before the diagnosis is confirmed, every suspected case of fever of unknown cause, even in children, should be treated with antibiotics as soon as possible. It is considered clinical evidence to rule out epidemic typhus if there is no response within 48–72 h after beginning empirical therapy. To prevent relapses, therapy has been advised for 5 days or 2–4 days following defervescence.

In outbreak settings, a single oral dose of 200 mg doxycycline may be administered to patients as a treatment option. This regimen is associated with a low incidence of relapse. Doxycycline can also be given for a short duration (5–10 days) and may be administered intramuscularly or intravenously, depending on clinical needs. Although its use is limited by potential side effects, chloramphenicol serves as an alternative treatment, particularly for patients with severe hypersensitivity to tetracyclines.

Vaccines

The first inactivated vaccine was developed by Weigl in Poland (1920–1930). He used *R. prowazekii*-infected lice fed on convalescent patients to allow bacterial multiplication before formalin inactivation. However, production and standardisation were challenging. In 1939, an attenuated murine brain tissue vaccine was introduced but later abandoned due to excessive reactivity. Castaneda developed a formalin-inactivated vaccine from the lungs of infected mice, which provided protection. The attenuated Madrid E strain of *R. prowazekii*, developed in the early 1950s, was widely used in research and by volunteers. This vaccine prevented epidemic typhus but occasionally caused moderate to severe reactions, possibly due to reversion to a virulent phenotype.

The Cox vaccine, produced in chicken egg embryos, utilised purified, inactivated bacterial suspensions and was effective, especially with high antigen concentrations. The Durand vaccine, derived from inactivated rat lungs, showed promise but was tested on a limited number of volunteers. Attempts in the 1970s to develop vaccines with formalin-killed *R. prowazekii* faced toxicity and standardisation challenges. The possibility of DNA vaccines offers promise, with 24 *R. prowazekii* genes identified as potential candidates.

Prevention and Control

The primary control strategy during epidemic typhus outbreaks is to prevent louse infestation. Delousing techniques include washing the patient, replacing, boiling infected clothing, and eliminating all lice. The easiest method of delousing, however, is to remove and leave potentially infected items unworn for a week. This is because body lice only consume blood and perish within 5 days when deprived of blood. Using a pesticide (10% DDT, 1% malathion, or 1% permethrin) to eradicate lice is also a successful strategy. However, because infected people carry *R. prowazekii* throughout the rest of their lives, eradicating epidemic typhus may not be conceivable. Furthermore, relapses in treated individuals and persistent asymptomatic bacteraemia might start new epidemics at any moment if the louse infestation returns. Travellers visiting *Rickettsia* endemic regions are advised to avoid animal reservoirs, especially dogs and rats, and to reduce their exposure to infectious lice.

Travellers can reduce their risk of infection by properly applying insect repellents to both skin and clothing, wearing protective clothing, and conducting thorough self-examinations after visiting areas where disease vectors are prevalent. Additionally, the effectiveness of public health infrastructure—including surveillance systems, vector control programmes, and access to healthcare services—plays a crucial role in the prevention and management of outbreaks.

Case Study

A 17-year-old male who was living in a hostel had a high-grade fever for 9 days. Four days later, he developed rashes on his trunk that spread to his limbs, sparing his face, palms, and soles, and eventually turned into purpura fulminans. On the sixth day, he experienced syncope and seizures, and on the ninth day, he was admitted to the intensive care unit due to altered sensorium, including left facial paralysis without neck rigidity. The patient was treated empirically with cefoperazone-sulbactam, doxycycline, and acyclovir, and the condition was de-escalated later. The results of the investigation revealed Hb 13.6 gm/dL, WBC 14900/mm^3 (Neutrophil 79%), and platelets 26,000/mm^3. CSF and other blood tests were found to be normal. As determined by tube agglutination, the Weil-Felix test was positive for the *Proteus* antigens OX19 (1:640), OXK (1:640), and OX2 (negative). Both the brain MRI and CT scans were normal.

1. What clinical features distinguish severe *R. prowazekii* infections from other febrile illnesses?
2. How effective is the Weil-Felix test in diagnosing *R. prowazekii* infections, and what are its limitations?
3. What role do empirical treatments like doxycycline, cefoperazone-sulbactam, and acyclovir play in managing suspected *R. prowazekii* infections?
4. How can differential diagnoses be refined to prioritise *R. prowazekii* in cases of neurological symptoms like syncope and seizures?

Research Questions

1. What specific biomarkers can be identified to predict prognosis and assess disease severity in severe cases of *R. prowazekii* infection, and how does the immune response differ across varying levels of severity?
2. Beyond doxycycline, what alternative or adjunct therapies, such as combination therapies, can enhance patient outcomes in the treatment of *R. prowazekii* infections, and what is their comparative effectiveness?
3. What are the potential novel vaccine candidates for preventing *R. prowazekii* infections, and how effective are they in providing protection, particularly in high-risk and endemic areas?

Further Readings

Andersson SGE, Zomorodipour A, Andersson JO, Sicheritz-Pontén T, Alsmark UCM, Podowski RM, Näslund AK, Eriksson A-S, Winkler HH, Kurland CG. The genome sequence of *Rickettsia prowazekii* and the origin of mitochondria. Nature. 1998;396:133–43.

Angelakis E, Botelho E, Socolovschi C, Sobas CR, Piketty C, Parola P, Raoult D. Murine typhus as a cause of fever in travelers from Tunisia and Mediterranean areas. J Travel Med. 2010;17:310–5.

Bechah Y, Capo C, Mege JL, Raoult D. Epidemic typhus. Lancet Infect Dis. 2008;8:417–26.

Bozeman FM, Masiello SA, Williams MS, Elisberg BL. Epidemic typhus rickettsiae isolated from flying squirrels. Nature. 1975;255:545–7.

Centers for Disease Control (CDC). Epidemic typhus associated with flying squirrels—United States. MMWR Morb Mortal Wkly Rep. 1982;31:555–6.

Clifton DR, Rydkina E, Huyck H, Pryhuber G, Freeman RS, Silverman DJ, Sahni SK. Expression, and secretion of chemotactic cytokines IL-8 and MCP-1 by human endothelial cells after *Rickettsia rickettsii* infection: regulation by nuclear transcription factor NF-kappaB. Int J Med Microbiol. 2005;295:267–78.

Damås JK, Davì G, Jensenius M, Santilli F, Otterdal K, Ueland T, Flo TH, Lien E, Espevik T, Frøland SS, Vitale G, Raoult D, Aukrust P. Relative chemokine and adhesion molecule expression in Mediterranean spotted fever and African tick bite fever. J Infect. 2009;58:68–75.

Houhamdi L, Fournier PE, Fang R, Lepidi H, Raoult D. An experimental model of human body louse infection with *Rickettsia prowazekii*. J Infect Dis. 2002;186:1639–46.

Lutwick LI. Brill-Zinsser disease. Lancet. 2001;357:1198–200.

McQuiston JH, Knights EB, DeMartino PJ, Paparello SF, Nicholson WL, Singleton J, Brown CM, Massung

RF, Urbanowski JC. Brill-Zinsser disease in a patient following infection with sylvatic epidemic typhus associated with flying squirrels. Clin Infect Dis. 2010;51:712–5.

Medina-Sanchez A, Bouyer DH, Cantara-Rodriguez V, Mafra C, Zavala-Castro J, Whitworth T, Popov VL, Fernandez-Salas I, Walker DH. Detection of a typhus group *Rickettsia* in *Amblyomma* ticks in the state of Nuevo Leon, Mexico. Ann N Y Acad Sci. 2005;1063:327–32.

Mokrani K, Fournier PE, Dalichaouche M, Tebbal S, Aouati A, Raoult D. Reemerging threat of epidemic typhus in Algeria. J Clin Microbiol. 2004;42:3898–900.

Raoult D, Roux V. Rickettsioses as paradigms of new or emerging infectious diseases. Clin Microbiol Rev. 1997;10:694–719.

Raoult D, Woodward T, Dumler JS. The history of epidemic typhus. Infect Dis Clin N Am. 2004;18:127–40.

Stein A, Purgus R, Olmer M, Raoult D. Brill-Zinsser disease in France. Lancet. 1999;353:1936.

Rickettsial Pox

38

Natarajan Vaghesan, Balamurali Venkatesan, and Kesavarajan Govindarajan Karthikeyan

Learning Objectives

1. To understand the epidemiology, transmission, and risk factors associated with *Rickettsia akari* infection
2. To recognise the clinical features and diagnostic methods used to identify rickettsial pox
3. To describe the treatment options and preventive measures for managing rickettsial pox effectively

Introduction

Rickettsia akari, a member of the spotted fever group of *Rickettsia* species, is the aetiologic agent of rickettsial pox, an acute, self-limiting disease. Humans contract the disease through the bite of an infected mite, *Liponyssoides sanguineus*, which typically parasitises the common house mouse (*Mus musculus*). The disease was first discovered during an outbreak in New York City in the 1940s. The prevalence is low, with 5–15 annual cases reported in New York City (2012–2017). In 2020, Canada reported one case of possible local transmission. The seroprevalence of *R. akari* antibodies has been found in Albania, India, Sri Lanka, and parts of Africa, indicating broader exposure despite limited confirmed cases.

A vesicular rash and the development of a distinctive eschar at the site of the mite bite are the typical manifestations of rickettsial pox (Fig. 38.1), which clinically mimics chickenpox. Initial symptoms include fever, chills, myalgia, and headache. Due to its non-specific early symptoms and similarity to viral exanthems, the condition is frequently misdiagnosed. The infection usually begins and resolves on its own or with antibiotic treatment. Endothelial cells are the primary target of the obligatory intracellular bacterium. *R. akari* spreads throughout the body after entering the bloodstream, causing vascular damage and an inflammatory reaction. Rickettsial pox is not frequently reported. Nevertheless, it is possibly still underdiagnosed, especially in metropolitan areas where people frequently in contact with rodent-infested surroundings.

History

R. akari, the causative agent of rickettsial pox, was discovered in New York in 1946. The species name *akari* is derived from the Greek word for "mite", reflecting its transmission by mites. *R. akari* is the only known spotted fever group Rickettsia transmitted by mites. Dr. Benjamin Shankman treated an 11-year-old child in Kew

N. Vaghesan (✉) · B. Venkatesan · K. G. Karthikeyan
Department of Microbiology, Sri Lalithambigai Medical College and Hospital, Faculty of Medicine, Dr. MGR Educational and Research Institute, Maduravoyal, Adayalampattu, Chennai, Tamil Nadu, India

S. C. Parija, A. Poddar (eds.), *Textbook of Bacterial Zoonoses*, Microbial Zoonoses,
https://doi.org/10.1007/978-981-95-6615-0_38

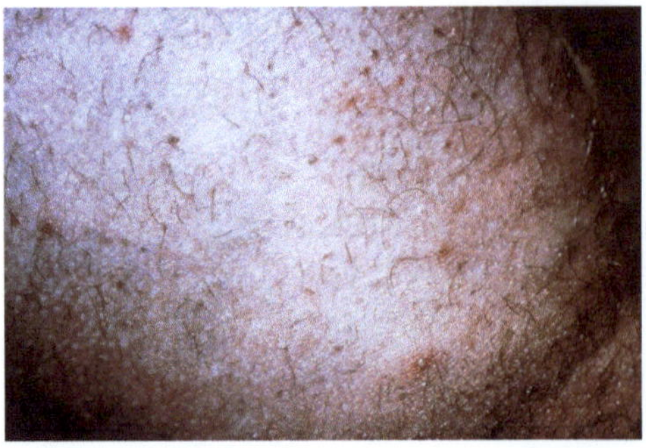

Fig. 38.1 The maculopapular rash seen on this patient's abdominal skin surface was attributed to a case of rickettsial pox. (Image source: CDC/Dr. R. Huebner, NIH)

Gardens, Queens, in February 1946, marking the first recorded observation. The child had a high fever, lymphadenopathy, and a papulovesicular lesion. After penicillin therapy, he was hospitalised for around a week before recovering successfully. Over the subsequent three months, approximately 20 additional patients from the same residential subdivision presented with similar symptoms. In the early summer, he and other medical professionals discovered about 100 case reports of so-called "Kew Gardens Mystery Fever".

Within five months of the initial outbreak, the vector responsible for transmitting *R. akari* was identified by Charles Pomerantz, an amateur entomologist and pest control specialist. The clinical syndrome was subsequently termed "rickettsial pox" by Huebner and colleagues due to its resemblance to varicella (chickenpox). Further studies established that the mite *L. sanguineus*, which parasitises the common house mouse (*Mus musculus*), served as the vector in the zoonotic transmission cycle.

Over the subsequent five decades, interest in rickettsial pox declined significantly. From 1969 to 1977, only twelve serum samples were tested for *R. akari*, and between 1980 and 1999, just 45 cases were officially recorded. However, in 2001, the heightened concern over potential bioterrorism, particularly following the anthrax letter attacks and the focus on eschar-associated febrile illnesses, led to renewed surveillance efforts. This resurgence highlighted the continued relevance of rickettsial pox in the differential diagnosis of eschar-associated illnesses in urban settings.

Taxonomy

The species, *Rickettsia akari*, belongs to the genus, *Rickettsia* in the Family Rickettsiaceae, in the Phylum Pseudomonadota, Class: Alphaproteobacteria and Order: Rickettsiales.

Within the genus *Rickettsia*, species are categorised into distinct groups based on their antigenic properties and the diseases they cause. The primary groups include the spotted fever group (SFG) and the typhus group (TG). The scrub typhus group, which was initially included, has since been reclassified into a separate but related genus, *Orientia*, although it still falls within the order Rickettsiales and is clinically grouped with rickettsial diseases. *R. akari* belongs to the spotted fever group (SFG) of *Rickettsia*.

Recent phylogenetic analyses have further refined the classification within *Rickettsia*. It is proposed to divide *Rickettsia* into five main groups: the Belli group (considered the oldest or ancestral group), the Canadensis group, the Typhus group, and two distinct Spotted Fever groups (Spotted Fever Group I and Spotted Fever Group II).

Genomics and Proteomics

R. akari (strain Hartford) showed a single circular chromosome with 1,231,204 base pairs (bp) and an estimated 1285 open reading frames (ORFs). Its coding capacity is 75.2%. It possesses a low guanine-cytosine (G+C) content (32.3%), similar to other SFG rickettsiae. In addition to three ribosomal RNAs (5S, 16S, and 23S), the genome encodes 33 transfer RNAs (tRNAs) and three non-coding RNAs: tmRNA (*ssrA*), the RNA component of RNase P (*rnpB*), and the 4.5S RNA of the signal recognition particle (*srp*). Comparative genomic investigations show that *R. akari* and other *Rickettsia* species exhibit significant conservation in gene content and local gene order.

Remarkably, compared to other SFG or typhus group rickettsiae, *R. akari* and *Rickettsia felis* contain more preserved short DNA elements. Based on BLAST analysis, about 63.8% of the *R. akari* ORFs had the highest similarity to *R. felis* proteins. PCR studies using primers designed to amplify conserved open reading frames (ORFs) between *Rickettsia prowazekii* and *Rickettsia conorii* yielded findings for only 50 out of 60 targets, suggesting sequence variation, particularly at gene termini, rather than gene absence. For years, it was thought that *R. akari* did not have the rompA gene since its 5′ terminus was missing. However, strain Hartford lacks the tandem repeat found in strain MK, possibly due to strain differences or genomic cloning artefacts. Its whole genome analysis confirmed that *rompA* is a split gene.

In addition to *rompA,* the genome encodes nine other surface cell antigens (Sca proteins). While four (*sca3, sca7, sca8,* and *sca10*) are fragmented genes, five (*sca1, sca2, sca4, sca5* [also termed *rompB*], and *sca6*) are encoded by complete ORFs. The expression patterns and antigenic roles of these proteins remain mostly unclear.

Morphology

R. akari is a Gram-negative, obligate intracellular, coccobacillary bacterium measuring about 0.3–0.5 µm in diameter and 0.8–2.0 µm in length. It is non-motile in its extracellular form and does not produce spores. Conventional techniques, such as Gram or eosin-azure staining, do not stain *R. akari* well. Nonetheless, it may be identified using Giménez or Macchiavello's stains and is visualised as rods or coccobacilli in infected tissues.

The vector, *Linognathus sanguineus*, is a small mite that ranges in size from 400 to 700 micrometres (Fig. 38.2). It is frequently found close to rodent breeding sites, including burrows, nests, and adjacent cracks or fissures.

Cultivation

Rickettsial culture is primarily used for antigen production, pathogen stock maintenance, and diagnostic purposes.

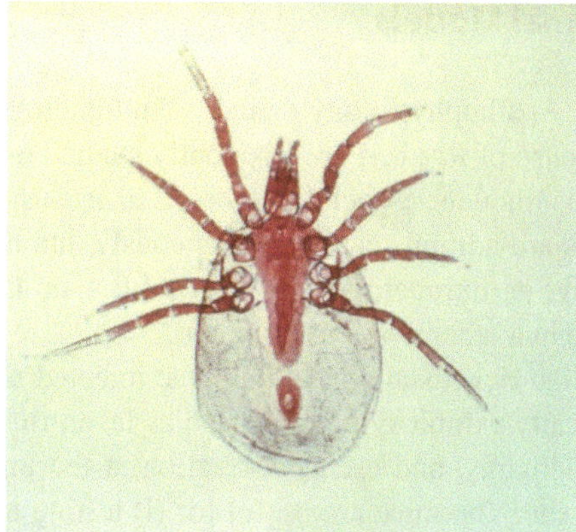

Fig. 38.2 A ventral view of the house mite, *Liponyssoides sanguineus*, formerly known as *Allodermanyssus sanguineus*, the vector that transmits to humans, the bacterium *Rickettsia akari*, the cause of rickettsialpox. (Image source: CDC, 1958)

Tissue Cultures

R. akari is an obligatory intracellular bacterium. It is effectively cultured using tissue culture methods. Frequently used cell lines include HeLa (), L929 (mouse fibroblasts), and Vero (monkey kidney). In a CO_2 incubator, these cells are maintained at 37 °C while being cultivated in RPMI-1640 or DMEM supplemented with foetal bovine serum. After inoculating the monolayers with *R. akari,* the infection is confirmed by examining the cytopathic effects.

Chick Embryo Inoculation

It is traditionally cultivated via chick embryo injection. Usually, 5–7-day-old, fertilised, pathogen-free chicken eggs are employed. The yolk sac is recommended for inoculation because it facilitates effective bacterial multiplication. A minimal amount of the *R. akari*-infected suspension is injected through the eggshell into the yolk sac cavity using sterile procedures. Following inoculation, the eggs are sealed and incubated for five to seven days at 35–37 °C. To determine if an infection is effective, embryos are monitored for symptoms of disease or death.

Animal Models

The pathophysiology and immunological response of *R. akari* are frequently studied using animal models, especially mice. *R. akari* suspensions are administered subcutaneously, intradermally, or intraperitoneally to BALB/c or C3H mice in a laboratory environment.

Like rickettsial pox in humans, infected mice typically exhibit symptoms such as fever, ruffled fur, lethargy, and eschar formation at the injection site. The mice are useful for (i) testing antimicrobial drugs such as doxycycline, (ii) assessing host-pathogen interactions, and (iii) researching histopathological alterations. Mouse lesions frequently exhibit necrosis, vasculitis, and perivascular infiltration, which are characteristics of cutaneous disease in humans. Although mice are most frequently utilised, rabbits and guinea pigs have also been used, albeit with differing degrees of vulnerability.

Transmission Cycle

R. akari has been isolated from house mouse mites, house mice, and other wild rodents. *Mus musculus* (House mouse) is the primary host. *L. sanguineus* (House mouse mite) acts as a vector. *R. akari* becomes infectious within the *L. sanguineus*.

The mite *L. sanguineus* facilitates the transmission of *R. akari,* which is mostly maintained in nature through a cycle involving house mice and other rodent species. Accidental human infection usually happens when an infected mite bites a human after failing to find its native rodent host. The bacterium present in the mite's saliva enters the body during the bite. Therefore, ecological factors influencing rodent and mite populations are directly linked to the introduction of rickettsial pox into human communities. The fact that *R. akari* may spread both transovarially and transcardially within *L. sanguineus* suggests that this mite may be a major reservoir for the disease, in addition to acting as a vector (Fig. 38.3).

Pathogenesis and Pathology

R. akari enters the skin through a mite bite. It invades dermal endothelial cells, causing localised inflammation and necrosis that leads to the development of a distinctive black eschar at the site of the bite. Widespread focal vasculitis leads to vascular necrosis, endothelial cell enlargement, and perivascular infiltration by macrophages and lymphocytes.

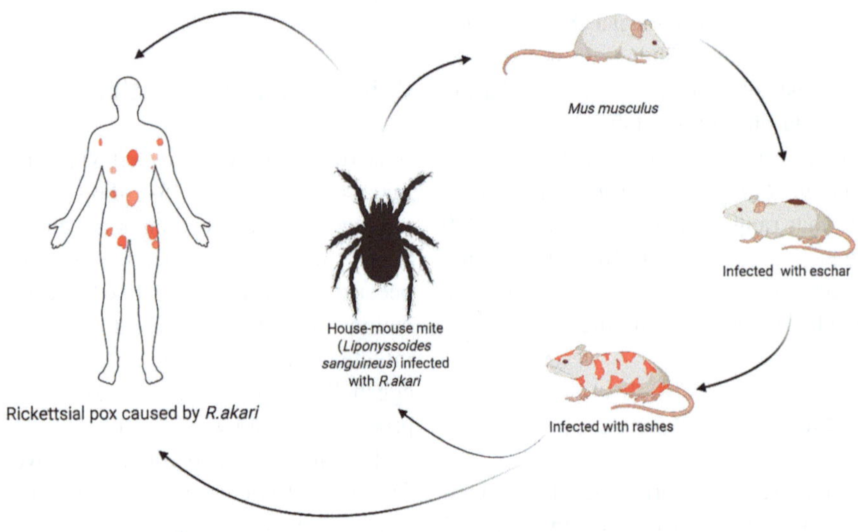

Fig. 38.3 Transmission of infection in humans. (Image source: Authors' personal collection)

A systemic vasculitis reaction is the outcome of the hematogenous dissemination of the pathogen that infects vascular endothelium throughout the body. Fever, chills, headache, and the development of a widespread papulovesicular rash occur two to three days after the onset of fever. The face, limbs, and trunk are frequently affected by the rash. Histologically, skin lesions exhibit enlarged endothelial cells, perivascular lymphocytic infiltration, and dermal oedema. Vascular necrosis, endothelial cell damage, and inflammatory cell infiltration around afflicted arteries are the main pathological hallmarks of vasculitis.

Regional lymphadenopathy is frequently caused by reactive hyperplasia and sinus histiocytosis in the lymph nodes that drain the infection site. In rare instances, mild involvement of the liver and spleen may manifest as activation of Kupffer cells or hepatic enlargement. Although it also contributes to inflammatory diseases, the host immune response, specifically T-cell activation and cytokine production, is crucial for the clearance of the bacteria from the circulation (Fig. 38.4).

Immunology

The immunology of the intracellular pathogen *R. akari* infection encompasses both innate and adaptive immune responses.

Innate Immune Responses

R. akari enters the host by a mite bite, and it mostly infects the endothelial cells. When dendritic cells and macrophages identify the pathogen via pattern recognition receptors (PRRs), including Toll-like receptors (TLRs), the innate immune system reacts initially. Pro-inflammatory cytokines, such as TNF-α, IL-1β, and IFN-γ, are produced as a result, drawing immune cells to the infection site and triggering an inflammatory response. IFN-γ stimulates macrophages and enhances their capacity to eliminate intracellular pathogens. Natural killer (NK) cells also contribute to the early phase. By living inside host cells, *R. akari* circumvents some of the intrinsic defences and elude detection outside the cell.

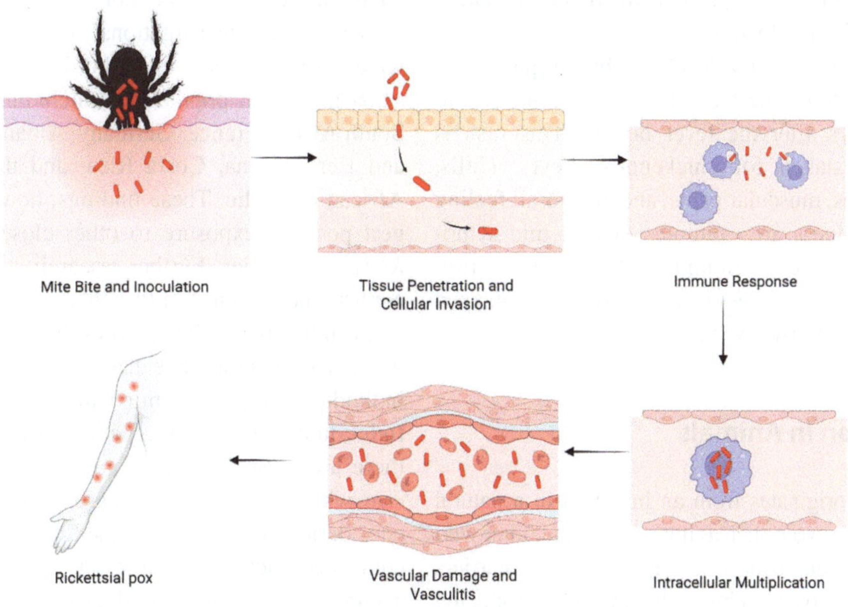

Fig. 38.4 Pathophysiology of *Rickettsia akari*. (Image source: Authors' personal collection)

Adaptive Immune Response

An adaptive immune response is essential to eradicate the infection. CD4+ helper T cells facilitate the activation of macrophages and the stimulation of antibody production, whereas CD8+ cytotoxic T lymphocytes target and destroy infected endothelial cells. Th1 responses predominate. IFN-γ is an important cytokine for regulating intracellular development of bacteria. Though humoral immunity is less successful than cellular responses in rickettsial infections, B cells help neutralise and opsonise *R. akari* by producing specific antibodies.

Infection in Humans

The majority of *R. akari* infections in humans are mild and self-limiting. The bacteria penetrate the skin following the mite bite and begin to multiply inside the local endothelial cells. A small, red papule appears at the bite site within three to seven days, and it eventually ulcerates, forming a distinctive black eschar. Regional lymphadenopathy close to the eschar is formed and is characteristic of the infection.

Patients typically develop a broad, papulovesicular rash on the face, trunk, and limbs two to three days after the fever begins. This rash is often mistaken for chickenpox. Fever, chills, headaches, muscular pains, and an overall feeling of unwellness are examples of systemic symptoms. Despite resembling other feverish rash conditions, the illness usually resolves on its own within one to two weeks.

Infection in Animals

R. akari originates from an infection in a natural animal reservoir, such as house mice and other wild rodents. Rats behave as asymptomatic carriers because the bacteria infect them without producing overt manifestations of the disease. *R. akari* is mostly found in the skin and blood of these hosts, where it can be obtained by feeding *L. sanguineus*. By biting infected rodents, the mites contract the infection, which allows the bacteria to grow within the mite's tissues, including its salivary glands. This maintains the rodent–mite transmission cycle in nature. Through bites, the infected mites subsequently spread *R. akari* to other individuals. Since the infection in rats is typically subclinical, meaning it does not cause obvious illness, it allows the bacteria to survive in the environment.

Epidemiology and Public Health

Rickettsial pox is a zoonotic disease in urban settings, where house mice (*Mus musculus*) and their ectoparasites, especially the house mouse mite (*L. sanguineus*), cohabit. Since its discovery in New York City in 1946, the disease has been intermittently reported in various regions around the world, including the USA, Ukraine, Croatia, Turkey, South Korea, Mexico, and the Netherlands (Table 38.1). During 2012–2017, 5–15 cases of rickettsial pox were recorded annually in North America, especially in New York City. The possibility of local transmission was highlighted in 2020 when a 47-year-old woman in Ontario, Canada, suffered from rickettsial pox despite having no recent travel history.

Serological investigations conducted worldwide have shown serum antibodies reactive to *R. akari* in several populations, including those in South Africa, France, Germany, Albania, Bosnia and Herzegovina, Costa Rica, and the Central African Republic. These findings, however, suggest possible exposure to other closely related *Rickettsia* species. Further research is needed to confirm the presence of true infections.

In India, as per the seroprevalence studies, *R. akari* serum antibodies have been detected in individuals in the Kashmir Valley and Northeast India, indicating that the pathogen is present in these areas. Spotted fever group (SFG) rickettsiae, which include *R. akari*, are responsible for a considerable percentage of the 249 serologically confirmed rickettsial illnesses in Sri Lanka, according to a study conducted between 2018 and 2023 (Fig. 38.5).

People of all ages are susceptible to rickettsial pox; instances have been documented in people as elderly as 92 years old and newborns as little as six months old. The condition is more prevalent in

Table 38.1 Epidemiology of mites causing rickettsial pox

Louse species	Specific host	Country/region distribution
Liponyssoides sanguineus	*Mus musculus* (house mouse)	USA (New York City, Boston, Philadelphia), Ukraine, Turkey, Mexico, Netherlands, South Korea
Ornithonyssus bacoti	*Rattus* spp., *Mus musculus*	Worldwide, especially in tropical/subtropical regions; common in rodent-infested urban areas
Laelaps echidninus	*Rattus* spp., *Mus musculus*	Africa, Asia, Europe, and some parts of the Americas
Blankaartia acuscutellaris	Small mammals, rodents	Southeast Asia, India, Australia
Trombicula spp. (chiggers)	Rodents, reptiles, birds	Global; especially rural and forested areas in Asia and North America
Dermanyssus gallinae	Birds, occasionally rodents	Global; primarily affects poultry, but can bite humans and rodents

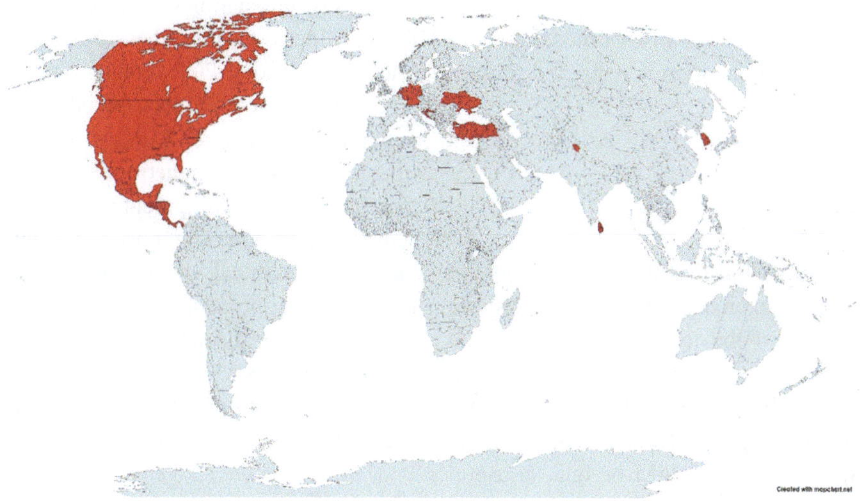

Fig. 38.5 Worldwide epidemiology of Rickettsial pox. (Image source: Authors' personal collection)

metropolitan areas, especially among the homeless and heroin injectors, who are more likely to be exposed to rodent-infested conditions. Because of its vague symptoms and propensity for misinterpretation as other feverish infections, rickettsial pox may remain unreported.

Diagnosis

Rickettsiae are typically handled in Biosafety Level 3 (BSL-3) laboratories due to biosafety concerns.

Microscopy

Rickettsia are small, pleomorphic, Gram-negative bacteria that typically appear as short rods or coccobacilli. Microscopy plays a limited role in the routine diagnosis of rickettsial pox due to the intracellular nature of *R. akari*. Direct visualisation using light microscopy is challenging. Nevertheless, Giemsa or Giménez staining of tissue samples or skin lesions may reveal intracellular organisms in endothelial cells. Electron microscopy demonstrates characteristic rickettsial morphology but is mainly used for research purposes.

In Vitro Culture

The cultivation of *R. akari* is technically demanding due to its obligate intracellular nature. *R. akari* does not grow on conventional bacteriological media.

Isolation typically involves the use of embryonated chicken eggs or specific cell lines such as Vero (African green monkey kidney) or L929 (mouse fibroblast) cells. These cell lines support intracellular replication and facilitate further characterisation of the organism. After inoculating the monolayers with *R. akari*, the infection is confirmed by examining the cytopathic effects. Routine culture, however, is rarely performed in diagnostic laboratories because of biosafety concerns. Hence, its application is restricted primarily to research settings.

Serodiagnosis

Serological testing remains the primary tool for the diagnosis of rickettsial pox. Traditional methods like the Weil–Felix test are largely obsolete due to low sensitivity, especially as *R. akari* typically does not induce agglutinating antibodies against *Proteus* antigens (OX-19, OX-2, OX-K).

The ELISA, detecting IgM or IgG, is useful for the diagnosis of rickettsial pox. It is used for low-resource settings and batch testing, with high sensitivity and specificity. While conventional ELISAs are qualitative, newer quantitative versions provide more reliable thresholds based on negative controls rather than arbitrary OD cut-offs.

The indirect immunofluorescence assay (IFA) is currently the standard, detecting IgG antibodies with high sensitivity (83–100%) and specificity (91–100%) from the second week of infection. However, its accuracy is species-dependent based on the antigen used. Advanced techniques like microimmunofluorescence, Western blot, and cross-adsorption tests, available in reference labs, aid in species-level identification.

Molecular Diagnosis

Polymerase chain reaction (PCR) and its modifications have been employed to detect *R. akari*. Nested PCR targeting the 17-kDa outer membrane protein gene, conserved within the *Rickettsia* genus, has been developed. The test has been successfully used to amplify DNA from eschar biopsy samples of infected individuals. In certain studies, PCR results have correlated with immunohistochemical findings, supporting its diagnostic potential. Additionally, nested PCR assays targeting the 120-kDa outer membrane protein B gene (*rompB*) have identified *R. akari* DNA in serum samples from patients with acute febrile illness. Despite these advances, the routine clinical use of PCR remains limited.

In addition to PCR-based methods, the whole genome sequencing of *R. akari* enables the identification of unique genetic markers that can be targeted for the development of more specific and sensitive molecular diagnostics for detecting the bacteria in clinical specimens. However, these molecular approaches remain valuable for research and confirmatory diagnostics in specialised laboratories.

Table 38.2 presents the diagnostic approaches used for detecting rickettsial infection.

Treatment

Doxycycline, a tetracycline antibiotic, is the antibiotic of choice for rickettsial pox. Clinical reports involving 28 patients showed resolution of fever and systemic symptoms within 24 h of initiating doxycycline or tetracycline, with complete recovery by 48 h. The standard oral doxycycline regimen is 100 mg twice daily, continued for at least three days, and is effective.

Current evidence indicates that a short doxycycline course does not cause dental discolouration or enamel hypoplasia in children under eight. For patients who do not tolerate doxycycline, chloramphenicol is an alternative, administered

Table 38.2 The diagnostic approaches used for detecting Rickettsial infection

Diagnostic approaches	Method	Target	Remarks
Microscopy	Gram stain, Giemsa stain	Intracellular Gram-negative coccobacilli	*Rickettsia* appears as pleomorphic, Gram-negative rods or coccobacilli. Require host cells to grow; not visible in routine stains unless tissue is properly processed
In vitro culture	Tissue culture in host cell lines	Viable *Rickettsia akari* organisms	Requires specialised facilities. Growth in Vero, XTC-2, C6/36, or tick cells at optimal temperature. Cytopathic effects (CPE) may not be evident, so PCR or immunofluorescence is needed to confirm infection. Antibiotic-free media must be used
Serodiagnosis	Complement Fixation, IFA, ELISA, WFT, IIP	Host antibodies (IgG/IgM) to *R. akari* antigens	IFA is gold standard. Titers ≥64 after 3 weeks are diagnostic. Weil-Felix test is usually negative. Cross--reactivity among SFG *Rickettsia* can occur. Western blot and cross-adsorption are used for species-level confirmation. ELISA is suitable for resource-limited settings
	IFA and Micro-IFA	Serum IgG (occasionally IgM)	High specificity and sensitivity from 2nd week of illness. Cross-reactivity limits species-level diagnosis. IgM is less specific. Micro-IFA, Western blot, and cross--adsorption help improve accuracy
	IgG/IgM-based antigen detection	Host antibodies	Batch testing possible. Quantitative ELISA yields titers; OD-based qualitative tests may be biased. Diagnostic cut-off: OD >0.5. High sensitivity/specificity but limited standardisation
Molecular diagnosis	PCR	17-kDa and 120-kDa (*rompB*) genes	PCR from eschar or serum; confirmatory via sequencing. Limited studies; potential inconsistencies in results. High specificity but low availability in clinical settings
Metagenomic analysis	NGS, Nanopore Sequencing	Whole genome/ pathogen DNA	Detects multiple pathogens simultaneously. No prior target knowledge required. Useful for co-infections and surveillance. Limited by high cost, lack of standardisation, and requirement for skilled personnel

in a dosage of 12.5 mg/kg every six hours for 5–10 days. The availability of chloramphenicol is limited due to discontinued oral formulations and restricted intravenous access in some regions. Additionally, *Rickettsia* species, including *R. akari*, are highly susceptible to rifampin in vitro.

Prevention and Control

The mainstay of prevention is the efficient management of house mice in buildings and homes. The mite populations are reduced by vacuuming and dusting mite-infested areas with diatomaceous earth. Acaricide treatments are applied to the walls, ceilings, runways, and mouse nests in affected areas in addition to rodent eradication. Pyrethrin and pyrethroid dust and sprays have been shown to be successful against parasitic mite species transmitting rickettsial pox.

Currently, there is no vaccine to prevent rickettsial pox.

Case Study

A 19-year-old woman developed a painful cheek lesion and asymptomatic facial and neck bumps following travel to a tourist place, where she reported frequent spider sightings. Initially treated with cephalexin for a presumed infected bite, she later experienced fever, sweats, and headaches upon returning to her home state. A new rash prompted a switch to doxycycline. Serologic testing via indirect immunofluorescence during the convalescent phase revealed a *R.*

akari titer of 1:128. Her symptoms fully resolved after a 10-day course of doxycycline.

1. How did the patient's travel history and environmental exposure contribute to the diagnosis?
2. Why was doxycycline chosen as the treatment, and what was the outcome?
3. What diagnostic test confirmed the infection with *R. akari* in this patient?

Research Questions

1. What are the genetic variations of *R. akari* strains isolated from different geographic regions, and how do these affect disease severity?
2. How does environmental exposure to rodent mites correlate with the incidence of rickettsial pox in urban versus rural settings?
3. What role do public health interventions targeting rodent control play in reducing the incidence of rickettsial pox in endemic areas?

Declaration During the preparation of this chapter, no AI tool has been used.

Further Readings

Brouqui P, Parola P, Fournier PE, Raoult D. Spotted fever rickettsioses in southern and eastern Europe. FEMS Immunol Med Microbiol. 2007;49(1):2–12. https://doi.org/10.1111/j.1574-695X.2006.00138.x.

Brown LD, Macaluso KR. *Rickettsia felis*, an emerging flea-borne rickettsiosis. Curr Trop Med Rep. 2016;3:27–39. https://doi.org/10.1007/s40475-016-0070-6.

Choi YJ, Lee SH, Park KH, Koh YS, Lee KH, Baik HS, Choi MS, Kim IS, Jang WJ. Evaluation of PCR-based assay for diagnosis of spotted fever group rickettsiosis in human serum samples. Clin Diagn Lab Immunol. 2005;12(6):759–63. https://doi.org/10.1128/CDLI.12.6.759-763.2005.

Choi YJ, Lee EM, Park JM, et al. Molecular detection of various rickettsiae in mites (acari: trombiculidae) in southern Jeolla Province, Korea. Microbiol Immunol. 2007;51(3):307–12. https://doi.org/10.1111/j.1348-0421.2007.tb03912.x.

Huebner RJ, Jellison WL, Armstrong C. Rickettsialpox; a newly recognised rickettsial disease; recovery of Rickettsia akari from a house mouse (Mus musculus). Public Health Rep (1896). 1947;62(22):777–80.

Kass EM, Szaniawski WK, Levy H, Leach J, Srinivasan K, Rives C. Rickettsialpox in a New York City hospital, 1980 to 1989. N Engl J Med. 1994;331(24):1612–7. https://doi.org/10.1056/NEJM199412153312403.

Krusell A, Comer JA, Sexton DJ. Rickettsialpox in North Carolina: a case report. Emerg Infect Dis. 2002;8(7):727–8. https://doi.org/10.3201/eid0807.010501.

Madison G, Kim-Schluger L, Braverman S, Nicholson WL, Wormser GP. Hepatitis in association with rickettsialpox. Vector Borne Zoonotic Dis. 2008;8(1):7–10. https://doi.org/10.1089/vbz.2007.0135.

Paddock CD, Zaki SR, Koss T, Singleton JR, Sumner JW, Comer JA, Eremeeva ME, Dasch GA, Cherry B, Childs JE. Rickettsialpox in New York City: a persisting urban zoonosis. Ann N Y Acad Sci. 2003;990:36–44. https://doi.org/10.1111/j.1749-6632.2003.tb07334.x.

Paddock CD, Tesse T, Eremeeva ME, Dasch GA, Zaki SR, Sumner JW. Isolation of *Rickettsia akari* from eschars of patients with rickettsialpox. Am J Trop Med Hyg. 2006;75(4):732–8.

Quevedo-Diaz MA, Song C, Xiong Y, Chen H, Wall BM, Radulovic S, Eremeeva ME. Involvement of TLR2 and TLR4 in cell responses to *Rickettsia akari*. Leukoc Biol. 2010;88(4):675–85. https://doi.org/10.1189/jlb.1009674.

Radulovic S, Feng HM, Morovic M, Dejiabala B, Popov VL, Crocquet-Valdes P, Walker DH. Isolation of *Rickettsia akari* from a patient in a region where Mediterranean spotted fever is endemic. Clin Infect Dis. 1996;22(2):216–20. https://doi.org/10.1093/clinids/22.2.216.

Radulovic S, Price PW, Beier MS, Gaywee J, Macaluso JA, Azad A. Rickettsia–macrophage interactions: host cell responses to *Rickettsia akari* and *Rickettsia typhi*. Infect Immun. 2002;70(5):2576–82. https://doi.org/10.1128/iai.70.5.2576-2582.2002.

Weyer F. The behavior of *Rickettsia akari* in the body louse after artificial infection. Am J Trop Med Hyg. 1952;1(5):809–20. https://doi.org/10.4269/ajtmh.1952.1.809.

Zavala-Castro JE, Zavala-Velázquez JE, Peniche-Lara GF, Sulú Uicab JE. Human rickettsialpox, Southeastern Mexico. Emerg Infect Dis. 2009;15(10):1665–7. https://doi.org/10.3201/eid1510.081507.

Zavala-Castro JE, Zavala-Velázquez JE, del Rosario Garcia M, Arias Leon JJ, Dzul-Rosado KRA. Dog naturally infected with *Rickettsia akari* in Yucatan, Mexico. Vector Borne Zoonotic Dis. 2009;9(3):345–7. https://doi.org/10.1089/vbz.2008.0189.

Murine Typhus

39

Kuntal Kumar Sinha and Vibhor Tak

Learning Objectives

1. To understand the epidemiology and clinical presentation of murine typhus
2. To review the zoonotic transmission and immunology of the infection
3. To describe the modalities to diagnose, prevent, and control murine typhus

Introduction

Murine typhus, also known as endemic typhus and flea-borne typhus, is a zoonosis caused by the *Rickettsia typhi*. The bacteria belong to the family Rickettsiaceae, comprising two genera of small, obligately intracellular bacteria: *Rickettsia* and *Orientia*. The geographic distribution is worldwide, especially in warm coastal areas throughout the tropics, with high rat infestations. Recently, it has been increasingly reported from Southeast Asia and the Western Pacific. Transmission of infection occurs by autoinoculation, following the rubbing or scratching of abraded skin or mucosa contaminated by insect faeces or by aerosol inhalation of dried louse or flea faeces.

Murine typhus typically manifests as fever, rash, and general constitutional symptoms. Serological tests form the cornerstone of laboratory diagnosis, while polymerase chain reaction (PCR) assays can identify *R. typhi* DNA in blood, plasma, or tissue samples during the early stages of infection. Timely administration of doxycycline is vital, often resulting in swift clinical improvement. Due to its non-specific symptoms, murine typhus is frequently misdiagnosed as other febrile illnesses, contributing to under--recognition. In endemic regions, inadequate sanitation and frequent human–animal interactions, heightened clinical awareness, and early suspicion are important for appropriate management and control of potential outbreaks.

History

The disease has been known for centuries, as mentioned in various scripts. However, *R. typhi* received its recognition as an important pathogen in 1920 by Wolbach and Todd, who first described the pathogen as *Dermacentroxenus typhi*. Subsequently, in 1921, Hermann Mooser identified *R. typhi* as the causative agent, using guinea pigs. Stuart published the first study on murine typhus in 1924. In 1926, Maxcy successfully isolated *R. typhi* from the blood and also proposed the hypothesis regarding the reservoir and vector of this disease. Dyer and Mooser, in 1930, established the role of fleas and rats as vectors.

K. K. Sinha · V. Tak (✉)
Department of Microbiology, All India Institute of Medical Sciences-Jodhpur, Jodhpur, Rajasthan, India

S. C. Parija, A. Poddar (eds.), *Textbook of Bacterial Zoonoses*, Microbial Zoonoses,
https://doi.org/10.1007/978-981-95-6615-0_39

Murine typhus, characterised by fever, rash, and systemic symptoms, was historically associated with significant morbidity and elevated mortality, particularly among the elderly. Prior to the 1950s, outbreaks were more common, especially in urban areas with poor hygiene. However, after that, improvements in sanitation, housing conditions, and vector control, especially targeting fleas and the rodent population, led to a marked decline in the incidence.

Taxonomy

The genus *Rickettsia* belongs to the Family Rickettsiaceae in the Phylum Proteobacteria, Class Rickettsiales, Order Rickettsiales. *Rickettsia typhi* (Hermann Mooser, 1921) is the type species. The genus *Rickettsia* is classified into two groups based on clinical manifestations: the typhus group and the spotted fever group. The typhus group includes *Rickettsia typhi*, which causes murine typhus (endemic typhus), and *Rickettsia prowazekii*, which causes epidemic typhus (louse-borne typhus). The species does not contain any subsets or subgroups. However, genetic variations, in the form of single--nucleotide polymorphisms (SNPs), are present among isolates of *R. typhi* from different geographical locations.

Genomics and Proteomics

The genome of *R. typhi* has been fully sequenced, providing significant insights into its biology, evolution, and pathogenic mechanisms. The *R. typhi* genome is relatively small, approximately 1.11 million base pairs (bp) in size. It consists of a single circular chromosome. The genome encodes approximately 838 proteins, along with rRNA and tRNA genes, and has a low G+C content of 28.9%. A notable feature is the presence of numerous pseudogenes (non-functional genes), including the entire cytochrome c oxidase system. This reflects the bacterium's adaptation to an intracellular lifestyle, where it relies on the host cell for many metabolic functions.

Comparing the *R. typhi* genome with those of other *Rickettsia* species, such as *R. prowazekii* (epidemic typhus) and *Rickettsia conorii* (Mediterranean spotted fever), reveals a core set of shared genes, as well as unique genes specific to *R. typhi*. Genomic analysis supports the close phylogenetic relationship between *R. typhi* and *R. prowazekii*, placing them within the typhus group of rickettsiae. These studies also contribute to understanding the evolutionary processes that have shaped the genomes of obligate intracellular bacteria. Analysis of single-nucleotide polymorphisms (SNPs) and insertion–deletion (INDEL) sites within the *R. typhi* genome allows for the genetic typing of different isolates from various geographical locations. This is valuable for epidemiological studies and for tracing the global spread of the bacterium.

Proteomics, the large-scale study of proteins, complements genomics by examining the expressed proteins of *R. typhi*. Key findings in *R. typhi* proteomics include a family of *sca* genes (*sca1* to *sca5*) that encode autotransporter proteins localised on the bacterial surface. These Sca proteins are believed to play a significant role in adherence, invasion, and survival within the host. Proteomic approaches have identified key antigens of *R. typhi* that are recognised by the host immune system. For example, the GroEL heat shock protein has been identified as a major immunodominant antigen in typhus group rickettsiae. The components of genomics and proteomics are depicted in Fig. 39.1.

Morphology

R. typhi is a small, Gram-negative coccobacillus, measuring 0.5 μm in width and 0.8 to 2.0 μm in length. The bacteria occur singly, in pairs, or occasionally in short chains or filaments. It is non-motile. Like other Gram-negative bacteria, *R. typhi* possesses a thin peptidoglycan layer and stains poorly with the standard Gram stain. The bacteria are non-capsulated, although some sources mention a loose slime layer. They lack flagella and do not form spores.

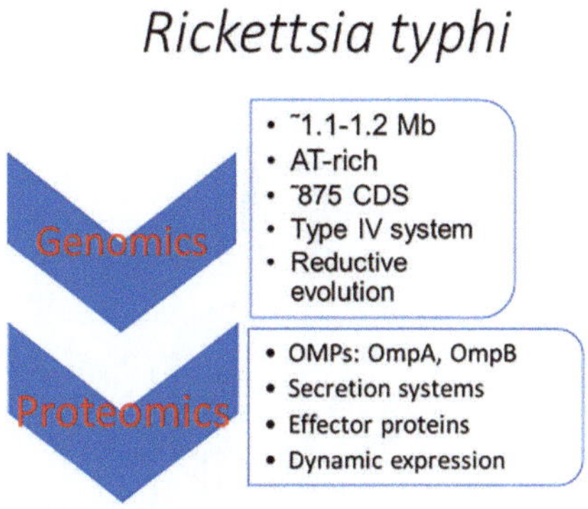

Fig. 39.1 Components of genomics and proteomics of *Rickettsia typhi.* (Image source: Authors' personal collection)

Table 39.1 Morphology components of *Rickettsia typhi*

Feature	Description
Shape	Pleomorphic coccobacillus (rod-like to oval forms)
Size	Approximately 0.3–0.5 μm in width and 0.8–2.0 μm in length
Cell envelope	Gram-negative structure: outer membrane (with LPS and OMPs), thin peptidoglycan layer, inner membrane
Motility structures	Lacks flagella and pili; relies on surface proteins for host interactions
Intracellular niche	Adapted for survival within host cell cytosol; often observed in clusters

Electron Microscopy

R. typhi is an obligate intracellular bacterium that can only survive and multiply within the cytoplasm of living eukaryotic host cells, primarily the endothelial cells lining blood vessels in mammals and the midgut epithelial cells in flea vectors. Electron microscopy studies of related *Rickettsia* species show a dense cytoplasm and a regular cell wall–cytoplasmic membrane complex. The cell wall comprises sugars such as glucose, galactose, and glucuronic acid, as well as amino sugars like glucosamine, muramic acid, and diaminopimelic acid. The components of the cell of the organism are tabulated in Table 39.1.

Cultivation

Cell culture methods are the preferred and most widely used methods for cultivating *R. typhi.* Mammalian cell lines are susceptible to infection and can support the growth of *R. typhi*. Common cell lines used include *Vero cells, L-929 cells, human embryonic lung (HEL) fibroblasts*, MRC-5, BHK-21, and endothelial cells.

Dulbecco's modified Eagle medium (DMEM) or RPMI 1640 supplemented with 5–10% heat-inactivated foetal bovine or calf serum and L-glutamine are used for cell culture maintenance. The optimal growth temperature is around 34–35 °C. Growth is slower below 32 °C and above 40 °C. A 5% CO_2 atmosphere is required during incubation to support bacterial proliferation.

Embryonated chicken eggs were a common method for cultivating *Rickettsia* species, including *R. typhi, in the past.* The bacteria are typically inoculated into the yolk sac of 5- to 8-day-old embryos and incubated. Bacterial growth is then assessed by examining the yolk sac fluid or tissues. However, cell culture methods have largely replaced this approach due to their greater ease of use and reduced risk of contamination.

Animal Models

R. typhi can be propagated in susceptible laboratory animals, such as guinea pigs or mice. However, this method is less common for routine cultivation due to ethical considerations and the complexity of animal maintenance.

Transmission Cycle

The transmission cycle is a classic example of an enzootic cycle where small mammals and their parasitic fleas play the leading role (Fig. 39.2).

The primary reservoir hosts are rodents, commonly the black rat (*Rattus rattus*) and the brown rat (*Rattus norvegicus*). In these animals, *R. typhi* persists without causing severe disease, allowing the pathogen to maintain a low-key, long-term presence in the environment. The Oriental rat flea (*Xenopsylla cheopis*) is the principal vector. When fleas feed on the blood of an infected rodent, they ingest the bacteria that cause the disease. It survives and replicates within the flea and can be transmitted *transstadially* (from one life stage to the next) and *transovarially* (from parent

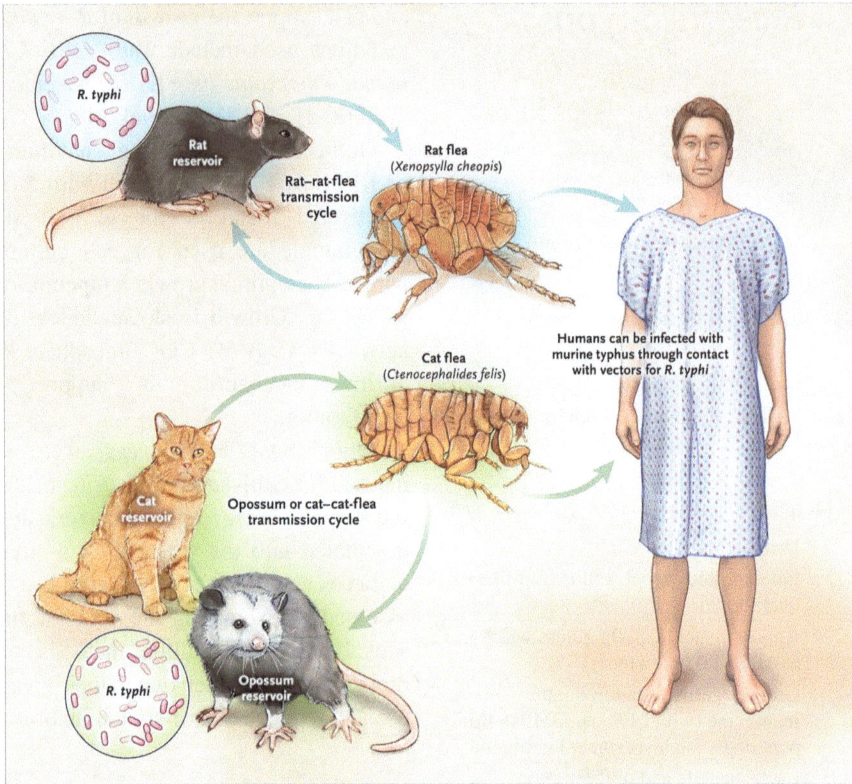

Fig. 39.2 Transmission cycle of murine typhus. (Image source: Stern et al. 2018)

flea to offspring). Once the flea lineage is infected, it remains a potential threat across generations.

The flea, during its blood meal on a human, defecates. The faeces contain *R. typhi*, and if these contaminated droppings get introduced to the skin through scratches, often following an itch from a bite, the bacteria can invade the body. Also, proximity to rodent populations, poor sanitation and warmer climates that support rodent and flea populations are important in disease transmission.

Pathogenesis and Pathology

After entering the host, *R. typhi* spreads via the lymphatic system and bloodstream, resulting in a transient rickettsiosis. The primary target cells for *R. typhi* are the *endothelial cells* that line the small blood vessels throughout the body. *R. typhi* actively invades endothelial cells through a process of induced phagocytosis. Once inside the cytoplasm, the bacteria escape the phagosome and multiply freely. This intracellular growth is essential for their survival and pathogenesis. The replication of *R. typhi* within endothelial cells leads to *cell damage and death*. This damage triggers a localised inflammatory response.

The widespread infection of endothelial cells results in a generalised *vasculitis*. It is characterised by increased vascular permeability, causing oedema and hypovolaemia. Activation of the coagulation cascade results in microthrombi in the affected vessels, leading to localised ischaemia and organ dysfunction. Lymphocytes, macrophages, and other inflammatory cells also infiltrate the perivascular spaces. The infected endothelial cells and the host immune response contribute to the release of various cytokines, such as tumour necrosis factor-alpha (TNF-α) and interleukins. These cytokines contribute to the systemic symptoms of murine typhus, including fever, headache, and myalgia.

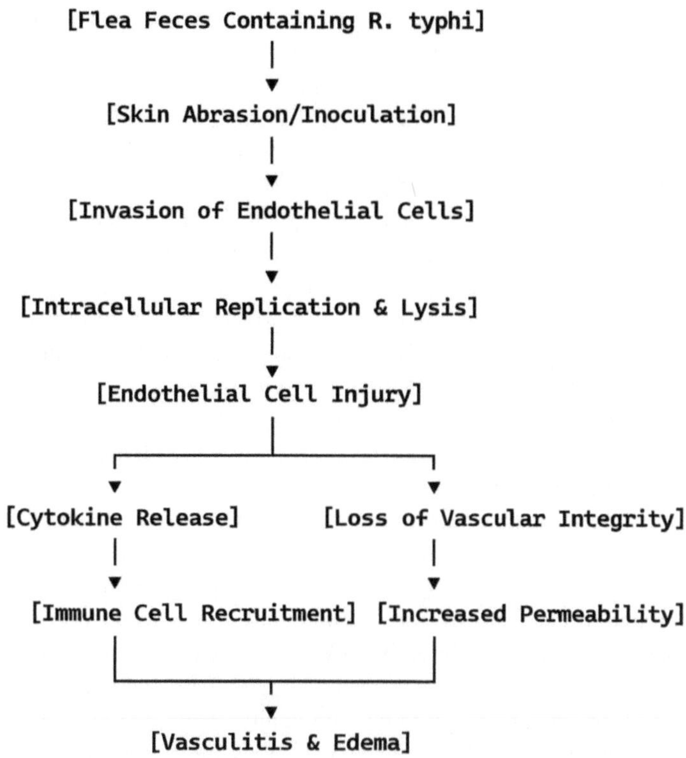

Fig. 39.3 Pathogenesis of *Rickettsia typhi* infection. (Image source: Authors' personal collection)

While the primary pathology involves the vasculature, various organs can be affected due to the widespread vasculitis and reduced blood flow. The severity of the pathological changes generally correlates with the clinical severity of the disease. In most cases, murine typhus is a self-limited illness with mild to moderate pathological findings. However, in rare cases, particularly in the elderly or immunocompromised individuals, more severe vasculitides and multi-organ involvement can occur, potentially leading to complications and even death. Autopsy findings in fatal cases have revealed haemorrhagic pneumonia, acute tubular necrosis, and ischaemic necrosis in various organs, including the liver, adrenals, and brain. The pathogenesis of *R.typhi* infection is depicted in Fig. 39.3.

Immunology

Innate Immune Response

The infection triggers the innate immune system through the recognition of pathogen-associated molecular patterns (PAMPs) by pattern recognition receptors (PRRs) on host cells, such as Toll-like receptors (TLRs). Infected endothelial cells and innate immune cells, such as macrophages and monocytes, release pro-inflammatory cytokines. These cytokines include TNF-α, IL-1β, and IL-6, as well as chemokines like MCP-1 and IL-8. These mediators contribute to local inflammation, recruit other immune cells to the site of infection, and contribute to the systemic symptoms. Macrophages and neutrophils, in conjunction with *natural killer (NK) cells*, produce NO and ROS, which are important for the intracellular killing of bacteria. The production of NO is enhanced by interferon-gamma (IFN-γ) produced by T cells.

Adaptive immune response

The adaptive immune response involving T and B cells is crucial for the clearance of *R. typhi* and the development of long-term immunity.

Cellular Immunity

T helper cells, particularly Th1 cells, produce IFN-γ, which is critical for activating macrophages to kill intracellular *R. typhi* through the production of NO and ROS. TNF-α produced by Th1 cells also contributes to macrophage activation and inflammation. The *CD8+ Cytotoxic T cells* recognise and kill *R. typhi*-infected host cells by releasing cytotoxic granules containing perforin and granzymes, inducing apoptosis of the infected cells. CD8+ T cells also contribute to the production of IFN-γ. Interestingly, T cells have been shown to mediate cross-protective immunity between different *Rickettsia* species, including those from the spotted fever group and the typhus group.

Humoral Immunity

B cells differentiate into plasma cells that produce antibodies against *R. typhi* antigens. These antibodies contribute to bacterial clearance through mechanisms like opsonisation and complement activation. IgM antibodies are typically the first to appear, followed by IgG antibodies, which provide more sustained immunity. The presence of a fourfold rise in IgG antibody titer between acute and convalescent serum samples is a key serological diagnostic criterion for murine typhus. Recovery from *R. typhi* infection results in a robust immune response and the persistence of IgG antibodies for several months to years. This indicates the development of at least short- to medium-term immunity. However, it is not definitively known if lifelong immunity is conferred.

Infection in Humans

Murine typhus has an incubation period of 7–14 days, typically around 10 days.

Symptoms begin abruptly, characterised by shaking, chills, fever, and headache. The fever is usually high, reaching 102–104 °F (39–40 °C) and can last for several days to a couple of weeks if untreated. Additional common symptoms include myalgia, nausea, vomiting, abdominal pain, cough, and anorexia. A maculopapular rash develops in about 40–50% of patients, typically appearing 5–7 days after symptom onset. The rash typically begins on the trunk and spreads peripherally, sparing the palms and soles, and lasts 1–4 days.

Neurological symptoms such as confusion, stupor, seizures, or imbalance occur in up to 45% of cases, especially in severe illness. The disease is generally mild and self-limiting, with low mortality. However, severe complications like pneumonitis, hepatitis, meningoencephalitis, and multi-organ failure may occur, particularly in the older or immunocompromised individuals.

Infection in Animals

The bacteria primarily affect rodents, mainly rats (*Rattus rattus* and *Rattus norvegicus*), which serve as the main reservoirs. These animals typically do not show obvious clinical signs of the infection, meaning they can carry the bacteria without appearing sick. This allows them to maintain the *R. typhi* in the environment and transmit it to fleas.

However, *R. typhi* can also infect other animals.

Dogs

Evidence suggests dogs can be infected with *R. typhi*, as antibodies and the bacterium's DNA have been detected in some studies. Some infected dogs may be asymptomatic or exhibit mild, non-specific signs, such as fever. However, their role in the transmission cycle is still being investigated.

Cats

Domestic cats are infected and may play a role in the suburban transmission cycle, acting as hosts for infected cat fleas. While some studies suggest that cats may not exhibit overt clinical signs, others indicate that they could develop a mild, self-limiting illness characterised by fever and lethargy in some cases.

Opossums

These marsupials have been identified as important reservoirs in some regions, particularly in the United States. Like rats, they may not show clear signs of illness.

Epidemiology and Public Health

The epidemiology and public health impact of murine typhus vary by region and are influenced by ecological, social, and climatic factors. Murine typhus occurs worldwide, especially in temperate, tropical, and subtropical coastal port regions where rodent reservoirs and flea vectors are common. It is endemic in parts of the United States (notably southern California and Texas), Mexico, Central and South America, Asia, Australia, and parts of Europe, such as Spain. In India, cases of murine typhus have been reported from Shimla, Kashmir, Mumbai, Jabalpur, Lucknow, and Pune. The worldwide hotspots are depicted in Fig. 39.4.

The primary reservoirs are rodents, with opossums, cats, and dogs also implicated in some cycles. Peri-domestic cycle involving the transmission of murine typhus is found in urban areas. In urban and suburban areas, cats, dogs, opossums, and sheep, along with their fleas (e.g. *Ctenocephalides felis*), act as significant reservoirs. Although direct transmission from cats to humans has not been definitively demonstrated, high rates of infection in cats correlate with human cases of murine typhus in the same regions.

Fleas on these animals can transmit *R. typhi* to humans similarly through faeces or bites. Humans are incidental or "accidental" hosts; they are not necessary to perpetuate the cycle, but they become infected when they are in contact with the infected flea. Fleas, particularly rat and cat fleas, serve as vectors transmitting the bacteria to humans. Cases tend to rise during warmer months, possibly due to increased flea populations during this time.

Endemic typhus affects all age groups and both sexes equally, though most cases occur in adults. People living in rural or disadvantaged areas, as well as those with close contact with reservoir animals or flea vectors, are at a higher risk. The incidence of endemic typhus is likely underestimated globally because of its non-specific clinical presentation and limited diagnostic testing in many endemic areas. Surveillance is limited in many endemic regions, contributing to the underdiagnosis.

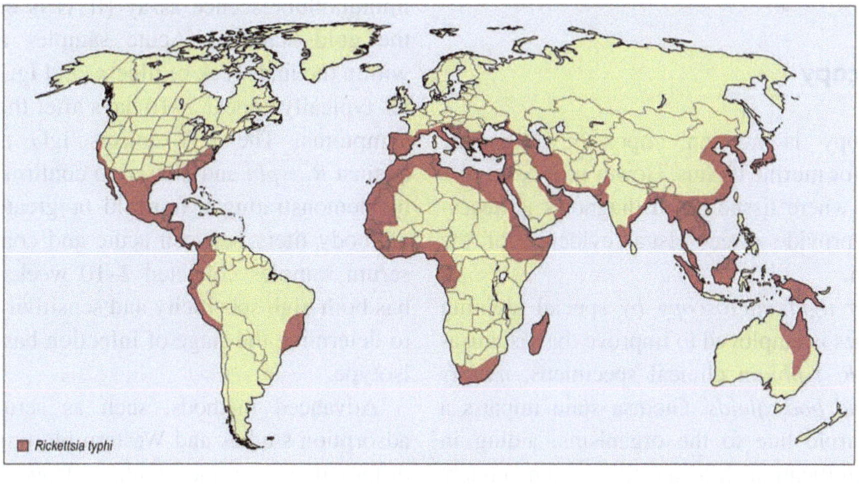

Fig. 39.4 Global epidemiology of murine typhus. (Image source: https://www.houstonchronicle.com/local/prognosis/article/More-about-that-ancient-tropical-disease-found-in-6361671.php)

Diagnosis

Laboratory diagnosis continues to depend primarily on serological methods, with polymerase chain reaction (PCR) assays enabling early detection of *R. typhi* DNA in clinical specimens. *R. typhi*, the causative agent of murine typhus, is generally handled at biosafety level 2 (BSL-2) for routine diagnostic procedures. However, propagation or culture of *R. typhi* should be conducted under BSL-3 conditions due to the increased risk of laboratory-acquired infection.

Specimens

Blood is collected as early as possible in the course of illness to diagnose murine typhus. For isolation by cell culture methods, blood is collected in a sterile heparin, EDTA-, or sodium citrate-anticoagulated vial. Blood for serologic diagnosis is collected as early as possible. A second specimen is collected after 1 or 2 weeks, and if a fourfold rise in antibody titer has not occurred, a third sample is collected 3 or 4 weeks after onset. PCR provides a higher diagnostic yield when applied to whole blood, plasma, buffy coat, or biopsy specimens of Rickettsia-infected lesions.

Microscopy

Microscopy is not an important diagnostic method for murine typhus. However, it has a role in cases where tissue-based diagnosis is necessary to provide direct visual evidence of the organism.

Direct light microscopy by special staining techniques is employed to improve the visualisation of *R. typhi* in clinical specimens, *mainly blood and body fluids*. Giemsa stain imparts a bluish-purple hue to the organisms, aiding in their identification. Machiavellian stain renders the bacteria bright red on a blue background, facilitating clear differentiation from surrounding tissue elements. Giemsa stain highlights *R. typhi* in red against a bluish-green background, offering good contrast. These techniques are valuable tools in the histopathological diagnosis of rickettsial infections.

Direct fluorescent antibody (DFA) technique and immunohistochemistry (IHC) are applicable to ante-mortem tissue specimens or cell lines and post-mortem samples, respectively. Antibodies labelled with fluorescent dyes specifically bind to rickettsial antigens in infected cells.

Transmission electron microscopy (TEM) demonstrates the ultrastructure of *R. typhi* within host cells. Its high cost, need for specialised equipment, and labour-intensive nature limit its routine application mainly to research.

In Vitro Culture

The culture of *R. typhi* is possible using cell lines, as *R. typhi* is an obligate intracellular bacterium. However, it is rarely performed in clinical settings due to biosafety concerns and is also slow and labour-intensive.

Serodiagnosis

The laboratory diagnosis of murine typhus primarily relies on serological testing. The indirect immunofluorescence assay (IFA) is considered the gold standard. Acute samples are taken within the first week of illness, and IgG antibodies typically appear 7–10 days after the onset of symptoms. The IFA detects IgG antibodies against *R. typhi* and is used to confirm infection by demonstrating a fourfold or greater rise in antibody titers between acute and convalescent serum samples collected 2–10 weeks apart. It has both high specificity and sensitivity, helping to determine the stage of infection based on the isotype.

Advanced methods, such as serum cross-adsorption studies and Western blotting, help to differentiate murine typhus from epidemic typhus. Cross-reactivity between *R. typhi* and other rickettsiae, such as *R. prowazekii* and *Rickettsia felis*, occurs, complicating serological interpretation.

Table 39.2 Diagnostic modalities available for *Rickettsia typhi*

Method	Specimen	Advantages	Limitations	Ideal timing
In vitro culture	Tissue, blood	Definitive when positive	Requires BSL-3 facilities; time-consuming and not routine	Rarely used; mostly research settings
Serodiagnosis indirect immunofluorescence (IFA)	Serum	High specificity and sensitivity with paired samples	Delay in antibody response; requires convalescent sample	After 7–10 days; convalescent phase crucial
Molecular diagnosis Polymerase chain reaction	Whole blood, serum, tissue	Rapid, specific detection of bacterial DNA	May be less sensitive due to low bacterial load; needs specialised set-up	Early acute phase

Molecular Diagnosis

PCR is used to detect *R. typhi* DNA in blood, plasma, or tissue specimens during the acute phase of illness, especially within the first week before antibiotic treatment begins. However, PCR sensitivity is variable, and a negative PCR result does not exclude infection; therefore, treatment should not be delayed if murine typhus is suspected clinically.

Real-time PCR (qPCR) is the preferred method. This method offers faster results, higher sensitivity, and specificity, and the ability to quantify the bacterial load. Common gene targets for *R. typhi* PCR include *gltA* (the citrate synthase gene), the 17-kDa protein gene, and *ompB* (the outer membrane protein B gene).

The diagnostic modalities currently in use are listed in Table 39.2.

Treatment

Murine typhus is effectively treated with antibiotics, and prompt treatment usually leads to a full recovery. Treatment is initiated immediately following clinical suspicion, even before laboratory confirmation. In cases of fever with a duration of five days or more where malaria, dengue, and typhoid have been ruled out, therapy with antibiotics is started.

In adults, doxycycline is administered in a dosage of 200 mg/day in two divided doses for individuals above 45 kg for a period of seven days. Alternatively, azithromycin 500 mg in a single dose is given for five days. If the clinical signs and symptoms persist, an alternative diagnosis should be considered. *In children*, doxycycline is administered in a dosage of 4.5 mg/kg body weight/day in two divided doses for children below 45 kg. Alternatively, azithromycin is given in a dosage of 10 mg/kg body weight for five days. *In pregnant women,* doxycycline is contraindicated. Azithromycin is the drug of choice; it is given in a single dose of 500 mg for five days.

Prevention and Control

Prevention is by rodent and flea control, which involves reducing contact with infected fleas and their animal hosts. Implementation of effective flea control measures for pets, such as cats and dogs, using appropriate treatments, is important. Steps are also needed to prevent rodents, such as rats and mice, from entering homes and surrounding areas by sealing holes, storing food properly, and removing potential nesting sites.

Individuals can significantly reduce their risk of infection by adopting personal protective measures that minimise direct exposure to fleas. The application of EPA-registered insect repellents containing active ingredients such as DEET or picaridin on exposed skin and clothing deters fleas from biting. Practising good personal hygiene, including regular bathing and washing clothes and bedding, especially in homes with pets or after potential flea exposure, can also be beneficial.

Surveillance programmes are crucial for monitoring human cases and detecting potential outbreaks, enabling targeted interventions in high-risk areas. This often includes surveillance of rodent and flea populations. Public education plays a critical role in reducing the incidence of *R. typhi* infection.

Case Study

A 43-year-old man with type 2 diabetes, with a chief complaint of a sore chest for a few days, visited his own hospital emergency department, where he works as a janitor. He had not been feeling well for approximately 1 week. The patient, who lived alone, had two cats as pets. His vital signs in the emergency department included a temperature of 39.1 °C, a heart rate of 140 beats/min, a blood pressure of 70/50 mmHg, and a respiratory rate of 30 breaths per minute. His oxygen saturation was 89% while breathing room air.

Laboratory studies showed a white blood cell count of 6900 cells/μL, with a differential of 42% neutrophils, and significant renal dysfunction, characterised by a blood urea nitrogen level of 90 mg/dL and a serum creatinine level of 7.1 mg/dL. The patient was given intravenous saline, vasopressors, and intravenous antibiotics. Routine blood cultures and sputum cultures grew no significant microorganisms. Serology for dengue and malaria antigens was negative. His serum sample was tested for rickettsial diseases by PCR, which showed the presence of *R. typhi* DNA. The patient was treated with doxycycline for 10 days and ultimately discharged. However, the doctors were unsure of the source of the infection, whether it was due to contact with rats from his job or if his pets were infected and acted as a source of infection.

1. What are the differential diagnoses in this case?
2. Describe the risk factors for acquiring this infection in this case.
3. Mention the diagnostic methods available for the diagnosis of this agent.
4. Mention three preventive measures to prevent this infection.

Research Questions

1. Due to climate change, mainly global warming, is there a shift in the hotspots of the disease, leading to a change in the pattern of transmission in the primary host?
2. Do we need to develop a vaccine that could be administered to susceptible populations?
3. Do we need to develop diagnostic modalities for the detection of this agent in animals?

Declaration During the preparation of this chapter, we have used ChatGPT to improve the readability of the text. After using, we reviewed and edited the content as needed and take full responsibility for the content of the publication.

Further Readings

Blanton LS. Murine typhus: a review of a re-emerging flea-borne rickettsiosis with potential for neurologic manifestations and sequalae. Infect Dis Rep. 2023;15(6):700–16. https://doi.org/10.3390/idr15060063.

Condit ME, Jones E, Biggerstaff BJ, Kato CY. Procedure for spotted fever group *Rickettsia* isolation from limited clinical blood specimens. PLoS Negl Trop Dis. 2022;16(10):e0010781. https://doi.org/10.1371/journal.pntd.0010781.

Dhawan S, Robinson MT, Stenos J, et al. Selection of diagnostic cutoffs for murine typhus IgM and IgG immunofluorescence assay: a systematic review. Am J Trop Med Hyg. 2020;103(1):55–63. https://doi.org/10.4269/ajtmh.19-0818.

Kato CY, Chung IH, Robinson LK, Eremeeva ME, Dasch GA. Genetic typing of isolates of *Rickettsia typhi*. PLoS Neglect Trop Dis. 2022;16(5):e0010354. https://doi.org/10.1371/journal.pntd.0010354.

La Scola B, Rydkina L, Ndihokubwayo JB, Vene S, Raoult D. Serological differentiation of murine typhus and epidemic typhus using cross-adsorption and Western blotting. Clin Diagn Lab Immunol. 2000;7(4):612–6. https://doi.org/10.1128/CDLI.7.4.612-616.2000.

McLeod MP, Qin X, Karpathy SE, et al. Complete genome sequence of Rickettsia typhi and comparison with sequences of other rickettsiae. J Bacteriol. 2004;186(17):5842–55. https://doi.org/10.1128/JB.186.17.5842-5855.2004.

Ming DK, Phommadeechack V, Panyanivong P, Sengdatka D, Phuklia W, Chansamouth V, Roberts T, Blacksell SD, Newton PN, Robinson MT. The isolation of Orientia tsutsugamushi and Rickettsia typhi from human blood through mammalian cell culture: a descriptive series of 3,227 samples and outcomes in the

Lao People's Democratic Republic. J Clin Microbiol. 2020;58 https://doi.org/10.1128/jcm.01553-20.
Osterloh A. Immune response against rickettsiae: lessons from murine infection models. Med Microbiol Immunol. 2017;206(6):403–17. https://doi.org/10.1007/s00430-017-0514-1.
Sahni A, Fang R, Sahni SK, Walker DH. Pathogenesis of rickettsial diseases: pathogenic and immune mechanisms of an endotheliotropic infection. Annu Rev Pathol. 2019;14:127–52. https://doi.org/10.1146/annurev-pathmechdis-012418-012800.
Stern RM, Luskin MR, Clark RP, Miller AL, Loscalzo J. A headache of a diagnosis. N Engl J Med. 2018;379(5):475–9. https://doi.org/10.1056/nejmcps1803584.
Tello-Martin R, Dzul-Rosado K, Zavala-Castro J, Lugo-Caballero C. Approaches for the successful isolation and cell culture of American Rickettsia species. J Vector Borne Dis. 2018;55(4):258–64. https://doi.org/10.4103/0972-9062.256560.

Scrub Typhus

40

Mukesh Kumar, Subarnarekha Chowdhury, Sumukh Shourya, and Namrata Kumari

Learning Objectives

1. To describe the aetiological agent *Orientia tsutsugamushi*, including its morphology, classification, and ecological niche
2. To explain the transmission cycle, pathogenesis, and clinical manifestations of scrub typhus in humans
3. To discuss the laboratory diagnosis of scrub typhus
4. To evaluate current treatment protocols, preventive measures, and strategies for controlling scrub typhus outbreaks

M. Kumar
Department of Biophysics, All India Institute of Medical Sciences, New Delhi, India

School of Allied Health Science, Galgotias University, Greater Noida, India

S. Chowdhury
Department of Biophysics, All India Institute of Medical Sciences, New Delhi, India

S. Shourya
Department of General Medicine, Bharati Vidyapeeth Medical College, Pune, India

N. Kumari (✉)
Department of Microbiology, Indira Gandhi Institute of Medical Sciences, Patna, Bihar, India

Introduction

Scrub typhus, also known as bush typhus or tsutsugamushi disease, is a neglected, vector-borne zoonotic disease caused by the Gram-negative bacterium *Orientia tsutsugamushi*. It is transmitted by larval stage trombiculid mites ("chiggers") and is endemic to the Tsutsugamushi Triangle, which covers South and East Asia, parts of the Pacific Rim, and northern Australia. This region puts over a billion people at risk, with nearly one million cases each year. The disease mainly occurs in semi-arid areas and regions with scrub vegetation.

Among the known strains of *O. tsutsugamushi*, the prototype Karp strain and Karp-like strains are most prevalent in endemic regions. Other notable strains include Gilliam, Kato, Kawasaki, and TA763. These strains exhibit geographical variations and strain-specific pathogenic properties, which are crucial for understanding disease dynamics and developing diagnostic methods.

Clinically, scrub typhus manifests as an acute febrile illness, ranging from mild symptoms to severe complications such as ARDS, meningoencephalitis, multi-organ failure, and death, with untreated mortality rates reaching 30%. In India, it accounts for roughly 30% of fever-related hospitalizations during the monsoon season. Insufficient diagnostic tools and limited clinical awareness often lead to misdiagnosis. The intri-

S. C. Parija, A. Poddar (eds.), *Textbook of Bacterial Zoonoses*, Microbial Zoonoses,
https://doi.org/10.1007/978-981-95-6615-0_40

cate ecology of scrub typhus involves over 70 *O. tsutsugamushi* strains, showing marked variability in virulence that affects clinical outcomes.

Despite its historical presence and increasing disease burden, significant gaps remain in understanding vector biology, human risk factors, and environmental dynamics. No licensed vaccine exists, and diagnosis largely depends on serological testing, which lacks sensitivity in the early stages of infection. Treatment primarily involves antibiotics, with doxycycline and azithromycin as first-line therapies.

History

Scrub typhus has been documented since 313 AD in ancient Chinese medical texts. The Chinese clinical manual *Zhouhofang* described an illness remarkably like scrub typhus. This was followed by further detailed accounts in 610 AD and 1596 AD, which even linked the disease to mite bites in specific environments. In Japan, the disease was similarly recognized, with Hakuju Hashimoto providing a medical description in 1810 of "tsutsuga." It was recognized as a harmful disease in the Niigata prefecture, and the term "tsutsugamushi" itself reflects an early understanding of the mite's involvement.

The twentieth century marked a pivotal moment in the scientific understanding of scrub typhus. Although clinicians across the Asia--Pacific region had been reporting cases of what was likely scrub typhus in the early 1900s, it was not until 1920 that Naosuke Hayashi first identified a causative agent, initially mistaking it for a protozoan. However, by the 1930s, further research by scientists like Nagayo, Ogata, and Kawamüra reclassified the organism as a Rickettsia, leading to the proposed names *Rickettsia orientalis* and *Rickettsia tsutsugamushi*.

The disease gained widespread attention during World War II when it caused significant illness and death among Allied forces operating in the Pacific theater, especially in the jungle and rural areas. This devastating effect prompted intensive research into its epidemiology, prevention, and treatment. The development of effective antibiotic treatments, such as chloramphenicol in 1948 and later doxycycline, greatly improved patient outcomes.

Taxonomy

Orientia tsutsugamushi belongs to the domain Bacteria, phylum Pseudomonadota, class Alphaproteobacteria, and order Rickettsiales. The genus *Orientia* currently includes *O. tsutsugamushi* as the primary species responsible for scrub typhus, along with *Orientia chuto* and potentially other Candidatus species.

O. tsutsugamushi was initially classified within the *Rickettsia* genus due to shared traits, including being an obligate intracellular parasite and its transmission through arthropods. However, genetic and molecular studies, especially 16S rRNA sequencing and whole genome analysis, later revealed significant differences (Table 40.1). As a result, it was reclassified into its own genus, i.e., *Orientia*, in the 1990s.

Genomics and Proteomics

O. tsutsugamushi possesses a large and complex genome, ranging from 2.0 to 2.7 megabases with an average G + C content of 30–31%. The genome has an exceptional proportion of repetitive DNA, which constitutes up to 59% of it. This high level of repetitiveness is mainly caused by the proliferation of mobile genetic elements, especially the Rickettsiales amplified genetic element (RAGE), which is found in numerous copies and various states of degradation.

Furthermore, the genome shows a high number of genes linked to the Type IV secretion system (T4SS), indicating a unique method of host interaction and possibly lateral gene transfer. Unlike many bacteria, *Orientia* mainly diversifies its gene functions through gene duplication and subsequent divergence rather than by taking up exogenous DNA.

O. tsutsugamushi produces a distinctive 56--kDa type-specific protein that is not present in

Table 40.1 Comparison between *Rickettsia* spp. and *Orientia tsutsugamushi*

Feature	*Rickettsia* spp.	*Orientia tsutsugamushi*
Taxonomic family	Rickettsiaceae	Rickettsiaceae
Genus	Rickettsia	Orientia
Genome size	~1.1–1.5 Mb	~2.0–2.3 Mb (larger)
GC content	Higher (32–42%)	Lower (~30%)
Peptidoglycan layer	Present (though reduced)	Absent or minimal
LPS (Lipopolysaccharide)	Present	Absent
Outer membrane proteins (OMP)	OmpA, OmpB (highly conserved)	Lacks OmpA/OmpB; has 56-kDa TSA proteins
Antigenic diversity	Less diverse	Highly antigenically variable
Replication site	Cytoplasm	Cytoplasm
Vector	Ticks, fleas, lice	Chigger mites (*Leptotrombidium*)
Disease caused	Typhus, spotted fevers	Scrub typhus
16S rRNA phylogeny	Close similarity within *Rickettsia* species	~80–85% similarity with *Rickettsia* spp. (significant divergence)

other bacteria or Rickettsiae. This protein, unique in its sequence, contains cross-reacting epitopes that contribute to the bacterium's genetic diversity. Variations within this protein sequence are important in defining the diversity of O. *tsutsugamushi* strains, which affect the pathogen's virulence and clinical features. Because of its uniqueness, this 56 kDa protein has been extensively examined as a potential target for vaccine development.

Morphology

O. tsutsugamushi is a Gram-negative pleomorphic bacterium that appears as short rods or coccoid forms, typically measuring 0.5–0.8 μm in width and 1.2–3.0 μm in length. Unlike other Rickettsiae, it lacks a traditional peptidoglycan layer and lipopolysaccharide, which contribute to its distinctive cell wall structure. Consequently, despite being classified as a Gram-negative bacterium, *O. tsutsugamushi* does not stain well with the standard Gram stain method.

Cultivation

O. tsutsugamushi is difficult to cultivate because of its obligate intracellular nature. It cannot grow in artificial media and requires living host cells such as mouse fibroblasts (L929 cells), Vero cells, HeLa cells, and BHK cells. These cells are cultured as monolayers in flasks or plates, providing the necessary environment for the bacteria. *Orientia* replicates within the cytoplasm of these host cells, which sets it apart from *Rickettsia* species that typically reside in the nucleus. After infection, the bacteria multiply by binary fission, often showing a "budding" process as they gather on the host cell surface before being released to infect new cells. The cultivation is usually carried out at 35–37 °C in a humidified atmosphere with 5% CO_2.

Additionally, it can be grown in the yolk sacs of developing chicken embryos. This method offers a living biological system for the bacteria to reproduce.

Transmission Cycle

Small mammals, especially rodents (e.g., rats, voles, field mice), act as amplifying and maintenance hosts. They can become infected and have *Orientia* circulating in their blood, but primarily they serve as a blood meal source for the chiggers. This helps to increase mite populations and may contribute to local transmission. Importantly, infected animals can carry the bacterium without showing any symptoms. Humans are accidental hosts. They are not part of the natural maintenance cycle of the bacterium but become infected when an infected chigger bites them.

Trombiculid mites, mainly of the genus *Leptotrombidium*, commonly known as "chiggers" in their larval stage, serve as the primary hosts (Fig. 40.1). Key species include *Leptotrombidium deliense, Leptotrombidium akamushi,* and other *Leptotrombidium* species. Only the six-legged larval stage of the mite, known as the chigger, feeds on vertebrates, including humans. In contrast, the nymph and adult stages are free-living and feed on small arthropods or plant matter. The bacterium is maintained within the mite population primarily through transovarial transmission, in which infected female mites transmit the pathogen directly to their offspring via the eggs. Additionally, transstadial transmission allows the bacteria to persist through the mite's developmental stages, from larva to nymph to adult. An overview of transmission cycle is presented in Fig 40.2.

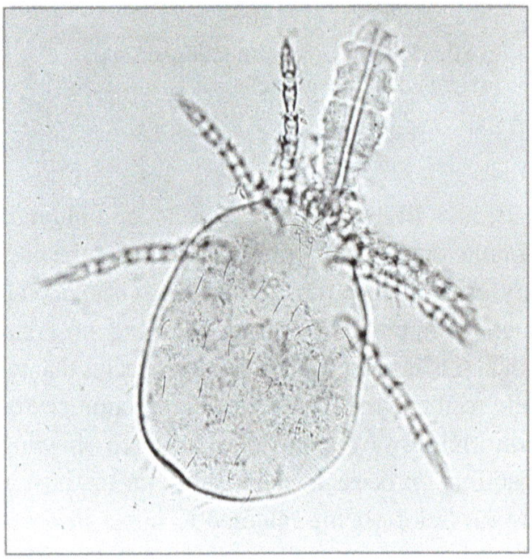

Fig. 40.1 Trombiculid mite larvae. (Image Source: Alan R Walker (talk) 09:27, 28 January 2014 (UTC) - CC BY-SA 3.0, https://commons.wikimedia.org/w/index.php?curid=27942833. Permitted for reproduction)

Infection occurs when an infected chigger bites a human, injecting the bacteria into the skin while feeding on tissue fluids. The mite typically needs to remain attached for 36–72 h to effectively transmit the pathogen. Outbreaks often increase after the monsoon season, when elevated humidity and temperatures create favorable conditions for mite activity.

Pathogenesis and Pathology

The progression of the disease involves complex interactions between the host's immune system and the pathogen's intracellular life cycle.

After entry through a bite from an infected mite, *O. tsutsugamushi* infects dendritic cells, endothelial cells, and macrophages, where it evades immune defenses by replicating inside cells. It uses a Type IV secretion system (T4SS) to manipulate host cellular pathways, ensuring its

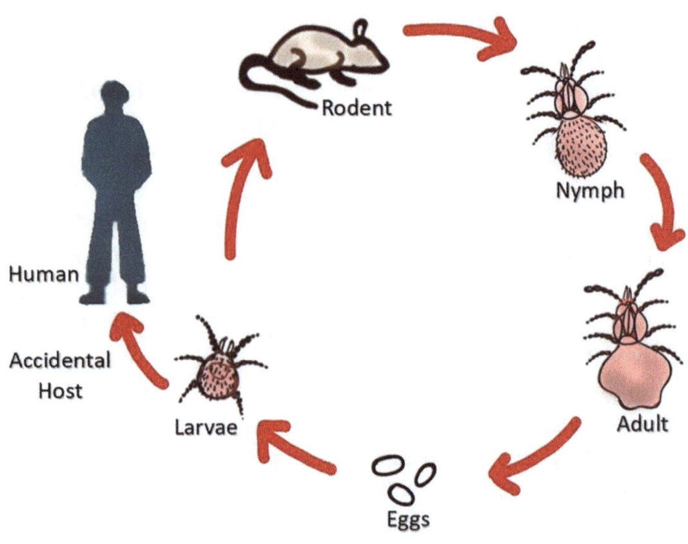

Fig. 40.2 Transmission cycle of Trombiculid mite. (Image source: Authors' personal collection)

survival and spread. The bacterium mainly targets endothelial cells lining blood vessels, causing vasculitis, increased vascular permeability, and systemic inflammation. This widespread endothelial dysfunction contributes to clinical signs such as fever, rash, an eschar at the inoculation site, and lymphadenopathy. Spread through the blood results in multi-organ involvement, with severe complications including acute respiratory distress syndrome (ARDS), myocarditis, renal failure, and meningoencephalitis.

The host's immune response involves activating Toll-like receptor (TLR) pathways, which induce the production of pro-inflammatory cytokines such as TNF-α, IL-6, and IFN-γ. However, the pathogen evades immune clearance by altering autophagy processes and delaying antigen presentation, resulting in persistent and recurring infections.

A key feature of scrub typhus pathophysiology is widespread vasculitis, leading to vascular damage in vital organs such as the liver, lungs, kidneys, and brain. This causes complications like disseminated intravascular coagulation (DIC), vascular leakage, shock, and multi-organ failure.

Immunology

Innate Immunity

Although *Orientia* lacks typical bacterial markers like Lipo polysacharide (LPS) and peptidoglycan, it provokes a strong innate immune response. Studies indicate that C-type lectin receptors (CLRs), such as Mincle and Clec5a, are involved in detecting *Orientia* and activating inflammatory pathways. This process triggers downstream signaling cascades, including Syk-dependent responses, and leads to the production of pro-inflammatory cytokines and chemokines. These mediators are essential for attracting other immune cells, like neutrophils and natural killer (NK) cells, to the infection site and promoting a Type 1-skewed immune response, characterized by cytokines such as IFN-γ, TNF-α, IL-1β, and IL-6.

Adaptive Immunity

The adaptive immune response, involving T cells and B cells, is crucial for clearing *O. tsutsugamushi* infection and for forming memory. Studies in both human and animal models show a strong Th1-skewed response, characterized by the activation of CD4+ T helper 1 cells and CD8+ cytotoxic T lymphocytes (CTLs). CD4+ Th1 cells produce IFN-γ, which activates macrophages and other cells to destroy intracellular bacteria. CD8+ CTLs are essential for directly killing infected host cells, thus eliminating the intracellular pathogen.

B cells produce antibodies against *O. tsutsugamushi*. The main antibody targets are often the major outer membrane proteins, especially the 56-kDa type-specific antigen (TSA), which is highly immunodominant. Antibodies contribute to immune defense by neutralizing extracellular bacteria, facilitating opsonization for phagocytosis, and activating the complement system. Although *Orientia* is primarily an intracellular pathogen, it does have an extracellular phase during its spread between host cells, during which antibodies can exert their effects.

Infection in Humans

The incubation period lasts from 6 to 21 days, after which patients usually develop nonspecific symptoms such as fever, headache, myalgia, and malaise.

When bitten by an infected chigger, bacteria from the salivary glands of the mite are introduced into the skin. The bacteria then multiply locally, often causing an eschar at the bite site (Fig. 40.3). Eschar is a necrotic lesion surrounded by an erythematous halo at the site of the chigger bite. It is a key clinical feature; however, its occurrence varies and is often absent in individuals with darker skin tones.

A high-grade fever lasts up to 2 weeks if not treated, often accompanied by rash, lymphadenopathy, and hepatosplenomegaly. Severe cases may lead to systemic complications caused by

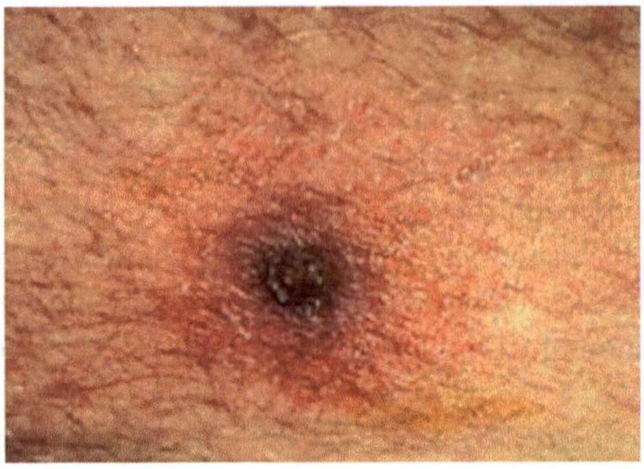

Fig. 40.3 Eschar at the bite site of an infected mite. (Image source: CDC)

vascular endothelial damage, such as disseminated vasculitis and multi-organ dysfunction. Common complications in advanced cases include pneumonia, acute respiratory distress syndrome (ARDS), myocarditis, hepatitis, acute kidney injury, and meningoencephalitis. Neurological symptoms may involve meningitis, encephalitis, cerebral venous thrombosis, Guillain-Barré syndrome, and transient parkinsonism (Fig. 40.3).

Thrombocytopenia and disseminated intravascular coagulation (DIC) rarely cause hemorrhagic events. Critical illness often involves multi-organ failure, especially respiratory, cardiovascular, and hepatic failure. Mortality rates in untreated cases can range from 7% to 30%, depending on host factors and strain virulence.

Infection in Animals

The infection in animals is usually asymptomatic or results in very mild, undetectable disease. This enables them to carry the bacterium for long periods without displaying obvious signs of illness.

Domestic animals, especially dogs, are increasingly recognized as potential hosts. Studies from Japan and other endemic regions have identified bacterial DNA in the blood of dogs, including asymptomatic ones. This indicates that dogs can be naturally infected with various *Orientia* serotypes (e.g., Karp, Kato, Kuroki, Gilliam, Kawasaki). Unlike humans, infected domestic animals usually remain asymptomatic or show very mild, non-specific symptoms such as lethargy, anorexia, or low-grade fever. The classic human symptoms, like an eschar or rash, are generally not observed or are difficult to detect due to fur. Elevated liver and kidney enzymes may be seen on blood tests in some symptomatic cases.

Epidemiology and Public Health

The zoonotic transmission pathway of *O. tsutsugamushi* to humans is indirect. The cycle begins with the bacterium being maintained within mite populations through transovarial transmission. These infected chiggers then live on vegetation, especially in areas with dense scrub, brush, tall grass, and forest habitats, often shared with small mammals. As chiggers attach to and feed on these small mammals, with rodents being common hosts, they help increase mite populations.

Scrub typhus is a significant yet often overlooked global public health issue, with its epidemiology marked by both traditional endemicity and worrying signs of re-emergence and spread (Fig. 40.4). Human infection happens accidentally when people enter mite-infested areas, known as "mite islands" or "tsutsugamushi habitats." This occurs usually during outdoor activities like farming, military exercises, or recreation. An overview of the transmission cycle is presented in Fig 40.2. It is estimated that more than one billion people worldwide are at risk of infection, with around one million new cases each year. Without suitable treatment, the case fatality rate can be quite high, ranging from 3% to as much as 30% or more in some untreated severe cases, highlighting its potential severity.

The classical geographical distribution of scrub typhus is famously known as the "Tsutsugamushi Triangle." This extensive region covers over eight million square kilometers and stretches from northern Japan and far-eastern Russia in the north to Pakistan and Afghanistan in the west, extending south into northern Australia. Countries within this traditional triangle where the disease is highly endemic include South and Southeast Asia (e.g., India, China,

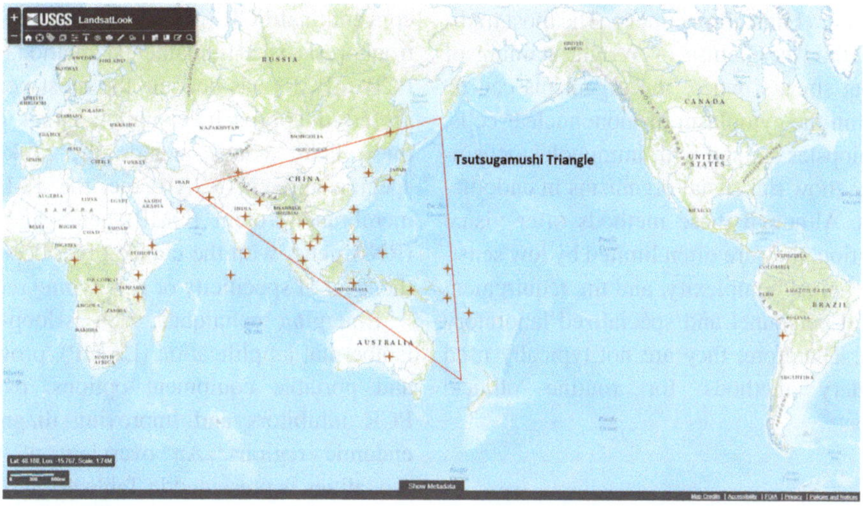

Fig. 40.4 Global epidemiology of scrub typhus. (Image source: Xu G, Walker DH, Jupiter D, Melby PC, Arcari CM (2017) A review of the global epidemiology of scrub typhus. PLoS Negl Trop Dis 11(11): e0006062. https://doi.org/10.1371/journal.pntd.0006062, Permitted for reproduction under CC BY license)

Thailand, Malaysia, Vietnam, Laos, Indonesia, Sri Lanka, Myanmar, Nepal), East Asia (e.g., Korea, Japan, Taiwan), and parts of the Pacific Islands and northern Australia. Countries such as India, China, Thailand, Korea, and Malaysia report significant numbers of cases.

Currently, the global epidemiology of scrub typhus shows a concerning shift, suggesting that its distribution is extending beyond the traditional Tsutsugamushi Triangle. Recent reports have documented cases and evidence of endemicity in formerly non-endemic regions, including parts of Africa, the Arabian Peninsula, and South America (e.g., Chile, Peru). This expansion is due to several factors, such as climate change, environmental encroachment, changes in animal reservoirs, and increased travel and globalization. Scrub typhus is now recognized as one of the top 10 causes of acute febrile illness among travelers returning from tropical regions. Despite its considerable burden, scrub typhus remains classified by the WHO as a neglected tropical disease.

Within endemic areas, most cases occur in rural settings where abundant vegetation harbors mites. However, there is a rising trend of cases being reported from peri-urban and even urban areas, linked to activities such as weekend farming, hiking, and increased human–rodent–mite contact in green spaces within cities. The seasonality of the disease varies with the climate. In temperate zones, it often peaks in autumn and sometimes in spring, correlating with mite activity. In tropical and subtropical regions, transmission can happen year-round, with peaks influenced by rainy and dry seasons.

Diagnosis

Specimens

Blood samples are the most versatile and frequently used specimens. Whole blood is employed for molecular detection. Serum, on the other hand, serves as the gold standard for serological tests. Biopsy samples, particularly from the characteristic eschar, are useful for direct detection via PCR, immunohistochemistry (IHC), or even challenging bacterial culture.

Microscopy

Microscopic diagnostic techniques for scrub typhus mainly involve directly visualizing *O. tsu-*

tsugamushi in clinical specimens. The most traditional method is Giemsa or Wright staining of buffy coat smears, where the organisms can be seen within the cytoplasm of mononuclear cells. Eschar biopsies stained with immunohistochemistry may show rickettsial organisms in endothelial cells. Although these methods offer visual confirmation, they are often limited by low sensitivity, technical complexity, and the requirement for skilled personnel and specialized laboratory facilities. Therefore, they are not typically used as primary methods for routine clinical diagnosis.

In Vitro Culture

Cell culture is the gold standard for isolation, using cell lines like L929 and HeLa. However, it is impractical for routine use due to the need for biosafety level III facilities and lengthy incubation periods.

Serodiagnosis

Serodiagnosis remains the primary method of diagnosis. The indirect immunofluorescence assay (IFA) is the gold standard for detecting IgM and IgG antibodies, with confirmation requiring a fourfold increase in titer between acute and convalescent sera. Enzyme-linked immunosorbent assay (ELISA) provides a reliable and widely used alternative, especially for detecting IgM antibodies, which indicate acute infection. Rapid immunochromatographic tests (ICTs) are useful in resource-limited settings but show variable sensitivity and specificity. Antigen detection methods, such as immunohistochemistry, are currently limited to research applications and have low sensitivity.

Molecular Diagnosis

Nucleic acid amplification tests (NAATs), primarily PCR and its variants, are used for molecular diagnosis of *O. tsutsugamushi.* While conventional PCR (cPCR) followed by gel electrophoresis is a fundamental method, real-time PCR (qPCR) has become the most popular and preferred approach. Common target genes for these PCR assays include the 56-kDa type-specific antigen (TSA) gene, the 47-kDa outer membrane protein (OMP) gene, and the 16S rRNA gene, with the choice often depending on the level of specificity or genotyping required.

Emerging techniques, such as loop-mediated isothermal amplification (LAMP), provide rapid and portable equipment options, overcoming PCR inhibitors and improving diagnostics in endemic regions. An overview of treatment modalities is presented in Table 40.2.

Treatment

Doxycycline remains the primary treatment, recommended at a dose of 100 mg twice daily for 7–14 days. It is highly effective in both adults and children over 8 years of age, with early initiation significantly reducing mortality. For severe cases, intravenous administration followed by oral therapy is preferred. Azithromycin is an alternative for pregnant women and children under 8 years of age, with a dose of 500 mg daily for 3–5 days, owing to its safety and efficacy, including against doxycycline-resistant strains. Chloramphenicol, although effective, is less frequently used due to its risk of bone marrow suppression.

Rifampicin shows promise in resistant cases but is avoided in tuberculosis-endemic regions. Emerging therapies, including telithromycin, offer potential alternatives, although their roles are still under investigation. Supportive therapy and symptomatic treatment are essential in managing severe cases.

An overview of treatment modalities is presented in Table 40.3.

Prevention and Control

Prevention and control of scrub typhus involve minimizing exposure to infected chiggers, especially in endemic rural and forested areas. Protective mea-

Table 40.2 Diagnosis of scrub typhus

Diagnostic technique	Specific tools/tests	Sample type	Advantages	Limitations	Remarks
Microscopy	Giemsa/Wright stain of buffy coat; IHC on eschar biopsy	Buffy coat, eschar biopsy	Visual confirmation; useful in research or specialized labs	Low sensitivity; technically complex; requires skilled personnel	Rarely used in routine diagnosis; adjunct tool
In vitro culture	Isolation in L929 or HeLa cell lines	Blood, eschar tissue	Gold standard; confirms viable organism	Requires BSL-3 lab; slow turnaround (days to weeks); not suitable routinely	Used in reference labs for confirmation and research
Serodiagnosis	IFA (IgM/IgG), ELISA (IgM), rapid ICTs	Serum (acute and convalescent)	Mainstay in clinical practice; ELISA suitable for LMICs	IFA: Needs paired sera and fluorescence microscope; ICTs have variable accuracy	IgM ELISA widely used in endemic regions for acute diagnosis
Molecular diagnosis	Conventional PCR, qPCR (56 kDa, 47 kDa, 16S rRNA); LAMP	Blood, eschar biopsy	High specificity; early detection; LAMP is field-friendly	PCR: Expensive, equipment-heavy; false negatives possible	qPCR and LAMP are promising for point-of-care and early diagnosis

Table 40.3 Treatment options for scrub typhus

Category	Drug/approach	Recommended dosage	Target group	Remarks
First-line therapy	Doxycycline	100 mg orally or IV twice daily for 5–7 days	Adults, children >8 years	Rapid defervescence; avoid in pregnancy and < 8 years due to teeth discoloration
	Azithromycin	500 mg orally/IV once daily for 5 days	Pregnant women, young children	Safe in pregnancy; also effective in doxycycline--resistant cases
Alternative therapy	Chloramphenicol	500 mg orally/IV four times daily for 7–14 days	Adults (only if no other options)	Risk of aplastic anemia; generally avoided
	Rifampicin	600–900 mg orally once daily for 5–7 days	Adults with doxycycline resistance	Effective against resistant strains; caution in co-infections (e.g. TB)
	Tigecycline	100 mg IV loading, then 50 mg IV twice daily	Severe or complicated hospitalized cases	IV only; reserved for refractory infections
	Minocycline	100 mg orally twice daily for 5–7 days	Adults	Similar efficacy to doxycycline
Supportive care	Antipyretics	Paracetamol 500–1000 mg orally every 6–8 hrs	All groups	Controls fever and improves comfort
	Fluid and electrolyte balance	As needed based on dehydration status	All groups	Important in severe illness with hypotension or renal involvement
	ICU support	Oxygen, vasopressors, organ support	Severe cases (MODS, ARDS)	Required in advanced complications
	Monitoring	LFTs, renal function, platelet count	Hospitalized patients	To assess recovery and detect complications early

sures include wearing long-sleeved clothing, using insect repellents containing N,N-Diethyl-meta-toluamide (DEET) or permethrin, and avoiding sitting or sleeping on bare ground. Environmental control through vegetation clearance and rodent control near human dwellings also reduces the risk. Public health education is vital for raising awareness about early symptoms and effective prevention strategies. No licensed vaccine is currently available, making early diagnosis and prompt antibiotic treatment essential. Surveillance programs and research into vaccines, as well as improved diagnostics, are vital for long-term control.

Case Study

A 45-year-old male patient, a teacher by profession, presented to the outpatient department with a history of intermittent fever for 8 days, accompanied by profuse sweating, headache, and myalgias. On general examination, he was noted to have suffused conjunctivae, mild dehydration, and a characteristic eschar measuring 1.8 cm × 1 cm located in the right infra-axillary region of the chest. An ELISA test confirmed scrub typhus infection. Systemic examinations, including chest X-ray, ECG, and neurological assessments, were within normal limits. Based on the clinical and laboratory findings, a diagnosis of scrub typhus with early-stage acute kidney injury was established. The patient was promptly started on empirical oral doxycycline 100 mg twice daily for 7 days at the primary health center, along with intravenous antipyretics and crystalloids to manage fever and dehydration. The patient showed clinical improvement and recovered without any adverse or unexpected outcomes.

1. What epidemiological detail might suggest how he acquired the infection?
2. What was the underlying pathogenesis?
3. What specific benefit did this rapid empirical treatment likely provide for this patient?

Research Questions

1. What are the genetic and antigenic variations of *O. tsutsugamushi* strains across different endemic regions, and how do they affect diagnostic accuracy and vaccine development?
2. What are the immunological mechanisms underlying severe complications of scrub typhus, such as acute respiratory distress syndrome (ARDS) and multi-organ failure?
3. What is the actual burden and distribution of scrub typhus in non-endemic or newly emerging regions, and how is climate change influencing its geographic spread?

Further Readings

Al Amin MM, Paul SK, Aung MS, Paul A, Aziz MA, Khan NA, et al. Molecular characterisation of *Orientia tsutsugamushi* causing scrub typhus among febrile patients in north-Central Bangladesh. New Microbes New Infect. 2019;32:100595. https://doi.org/10.1016/j.nmni.2019.100595.

Banerjee S, Latif A, Ghosh S, Chakraborty B, Ra R, Ghosh S. Case study of a 14-year-old male patient with scrub typhus infection and acute meningitis. Med Case Rep Rev. 2023;6 https://doi.org/10.15761/MCRR.1000178.

Barathan M. From fever to action: diagnosis, treatment, and prevention of acute undifferentiated febrile illnesses. Pathog Dis. 2024;82:ftae006. https://doi.org/10.1093/femspd/ftae006.

Bhandari M, Singh RK, Laishevtcev A, Mohapatra TM, Nigam M, Mori E, et al. Revisiting scrub typhus: a neglected tropical disease. Comp Immunol Microbiol Infect Dis. 2022;90–91:101888. https://doi.org/10.1016/j.cimid.2022.101888.

Cassisa A, Cima L. Cutaneous vasculitis: insights into pathogenesis and histopathological features. Pathologica. 2024;116:119–33. https://doi.org/10.32074/1591-951X-985.

Dasgupta S, Asish PR, Rachel G, Bagepally BS, Chethrapilly Purushothaman GK. Global seroprevalence of scrub typhus: a systematic review and meta-analysis. Sci Rep. 2024;14:10895. https://doi.org/10.1038/s41598-024-61555-9.

Devamani C, Alexander N, Chandramohan D, Stenos J, Cameron M, Abhilash KPP, et al. Incidence of scrub typhus in rural South India. N Engl J

Med. 2025;392:1089–99. https://doi.org/10.1056/NEJMoa2408645.

Devasagayam E, Dayanand D, Kundu D, Kamath MS, Kirubakaran R, Varghese GM. The burden of scrub typhus in India: a systematic review. PLoS Negl Trop Dis. 2021;15:e0009619. https://doi.org/10.1371/journal.pntd.0009619.

Druey KM, Arnaud L, Parikh SM. Systemic capillary leak syndrome. Nat Rev. Dis Primers. 2024;10:86. https://doi.org/10.1038/s41572-024-00571-5.

Fisher J, Card G, Liang Y, Trent B, Rosenzweig H, Soong L. *Orientia tsutsugamushi* selectively stimulates the C-type lectin receptor Mincle and type 1-skewed proinflammatory immune responses. PLoS Pathog. 2021;17:e1009782. https://doi.org/10.1371/journal.ppat.1009782.

Gautam R, Parajuli K, Tadepalli M, Graves S, Stenos J, Sherchand JB. Scrub typhus and molecular characterization of *Orientia tsutsugamushi* from Central Nepal. Pathogens. 2021;10:422. https://doi.org/10.3390/pathogens10040422.

Ghosal S, Pradhan R, Singh S, Velayudhan A, Kerketta S, Parai D, et al. One health intervention for the control and elimination of scrub typhus, anthrax, and brucellosis in Southeast Asia: a systematic review. Lancet Reg Health Southeast Asia. 2024;30:100503. https://doi.org/10.1016/j.lansea.2024.100503.

Kala D, Gupta S, Nagraik R, Verma V, Thakur A, Kaushal A. Diagnosis of scrub typhus: recent advancements and challenges. 3 Biotech. 2020;10:396. https://doi.org/10.1007/s13205-020-02389-w.

Kore VB, Mahajan SM. Recent threat of scrub typhus in India: a narrative review. Cureus. 2022;14:e30092. https://doi.org/10.7759/cureus.30092.

Lu C-T, Wang L-S, Hsueh P-R. Scrub typhus and antibiotic-resistant Orientia tsutsugamushi. Expert Rev. Anti-Infect Ther. 2021;19:1519–27. https://doi.org/10.1080/14787210.2021.1941869.

Q Fever

41

Pradeep Jothimani and Mashila Arasan

Learning Objectives

1. To define Q fever and identify the causative organism
2. To describe the epidemiology of Q fever, including transmission routes and risk factors
3. To interpret the diagnostic test results for Q fever, including serology and polymerase chain reaction (PCR)
4. To promote occupational safety practices for at-risk groups (farmers, veterinarians, abattoir workers)

Introduction

Q fever (QF) is an emerging zoonotic infectious disease caused by the pathogen *Coxiella burnetii*. It is transmitted from animals to humans. The bacterium is extremely resistant to environmental stresses and can survive harsh conditions for extended periods. The disease primarily affects domestic animals, including cattle, sheep, and goats. Moreover, it infects a wide range of wild and domesticated animals and humans. The bacteria persist in the environment for long durations due to their ability to withstand desiccation, heat, and disinfectants. Inhalation of airborne particles contaminated with *C. burnetii* is the primary mode of transmission. Infection can also occur through ingestion of contaminated food and direct contact with infected animals or their bodily fluids.

QF is a significant public health concern due to its ability to spread rapidly among both animal and human populations. The disease is highly infectious, and even small amounts of the bacterium can cause infection. Outbreaks of QF have been reported worldwide, affecting both rural and urban communities. The disease poses a risk to veterinarians, farmers, slaughterhouse workers, and laboratory personnel who are in close contact with ill animals and their products. Economic implications include loss of livestock productivity, healthcare costs, and the need for strict biosecurity measures. QF outbreaks often lead to substantial economic losses in livestock industries due to reproductive disorders, including abortions, infertility, and weak offspring in infected animals. Furthermore, controlling outbreaks requires costly interventions such as mass vaccination programmes, improved sanitation, and strict surveillance measures.

History

QF was first identified in Australia in 1935 by Edward Holbrook Derrick, who was investigating an outbreak of an unknown febrile illness in

P. Jothimani (✉) · M. Arasan
Mahatma Gandhi Medical Advanced Research Institute, Sri Balaji Vidyapeeth (Deemed-to-be-University), Pondicherry, India

S. C. Parija, A. Poddar (eds.), *Textbook of Bacterial Zoonoses*, Microbial Zoonoses,
https://doi.org/10.1007/978-981-95-6615-0_41

abattoir workers. The bacterium *C. burnetii* was isolated subsequently in 1935 at Rocky Mountain Laboratory, Montana, USA. The bacteria were isolated from the ticks *Dermacentor andersoni*, which were obtained near Nine Mile Creek (Davis and Cox 1938) and were identified as the causative agent. Rickettsiae and viruses were characterised and differentiated by Burnet, Freeman, Davis, and Cox. Dyer (1939) established that both *R. burnetii and R. diaporica* are the same organisms in honour of Burnet.

Early outbreaks of QF were observed among individuals who worked closely with animals, particularly in slaughterhouses and farms. Research efforts in the mid-twentieth century focused on understanding the disease's transmission, epidemiology, and pathogenesis. The development of diagnostic tools and effective treatments further advanced the understanding of QF. Large outbreaks in Europe, particularly the Netherlands, from 2007 to 2010, highlighted the importance of vaccination and surveillance. Several studies have examined QF in military personnel, as the bacterium has been associated with biological warfare due to its high infectivity and environmental resistance. Research on *C. burnetii* has also contributed to the development of biosafety protocols for handling infectious agents.

Taxonomy

The genus *Coxiella* belongs to the family Coxiellaceae, within the order Legionellales, class Gammaproteobacteria, phylum Pseudomonadota (formerly Proteobacteria), and domain Bacteria.

C. burnetii is the only species within the genus *Coxiella* and is officially recognised and validly published according to taxonomic standards. However, research continues to uncover "Candidatus" species and various "*Coxiella*-like endosymbionts." These closely related organisms are frequently found within different hosts, particularly ticks. While these organisms are genetically similar to *C. burnetii*, they are distinguished by their lack of virulence factors and their inability to infect humans.

Traditionally, *C. burnetii* was considered the sole species of clinical relevance; however, comparative genomics and phylogenetic analyses have revealed considerable genetic diversity within the genus. This has led to the identification of *Coxiella*-like endosymbionts (CLEs) in ticks and other arthropods. These CLEs, though genetically related to *C. burnetii*, differ significantly in pathogenic potential and ecological niche. Recent comparative genomics and phylogenetic analyses have revealed that *Coxiella* encompasses a broader spectrum of organisms than previously recognised.

Genomics and Proteomics

The genome of *C. burnetii* is relatively small and consists of a single circular chromosome. It contains genes encoding virulence factors, metabolic pathways, and mechanisms for intracellular survival. The genome also shows evidence that horizontal gene transfer and genomic reduction are common in intracellular pathogens.

The *C. burnetii* genome Nine Mile phase I RSA493 is approximately 1,995,275 base pairs (bp) long. In addition to this main chromosome, *C. burnetii* strains often carry plasmids, which are crucial for their classification and for their virulence factors. The GC content of the main chromosome typically ranges from 42.4% to 42.9%. The plasmids found in *C. burnetii,* such as QpH1, have a similar GC content, for instance, around 39.3% in strain 109.

These plasmids vary in size, typically ranging from 36 kilobases (kb) to 42 kb. These are categorised into different groups (I to VI) based on their sequence. Groups I to III are associated with the QpH1 plasmid (around 36 kb). Strains possessing these plasmids are frequently isolated from domestic livestock, ticks, and human acute Q fever cases, thus often termed "acute strains." The QpH1 plasmid is notably present in fewer than three copies per cell.

Groups IV and V are characterised by the QpRS plasmid (approximately 39 kb). These strains have been isolated from sources like aborted goat fetuses and are strongly linked to chronic Q fever, particularly endocarditis, earning them the designation of "chronic strains."

Group VI isolates carry the QpDG plasmid (around 42 kb), initially identified in rodents from Dugway; however, its specific role in *C. burnetii* pathogenesis remains largely unknown. Another distinct plasmid, QpDV (about 33 kb), has also been identified in *C. burnetii* strains from French patients suffering from endocarditis.

Genomic predictions for *C. burnetii* Nine Mile phase I RSA493 indicate that it encodes 2134 coding sequences (CDSs). Of these, a substantial portion, 719 (37.3%), are classified as hypothetical proteins. As such, their exact functions have not yet been experimentally determined or confidently inferred from homology to other known genes.

A major focus of *C. burnetii* proteomics is the identification and characterisation of its virulence factors, particularly the effector proteins. These proteins, secreted into the host cell via the Type IV secretion system (T4SS), are also known as the Dot/Icm system. *C. burnetii* is known to translocate over 150 different effector proteins into the host cytoplasm to manipulate host cell processes and establish its unique intracellular role.

Morphology

C. burnetii, a small, pleomorphic, and Gram--negative obligate intracellular bacterium. It is also known to produce spore-like structures. It exists in two distinct morphological forms: the small cell variant (SCV) and the large cell variant (LCV) (Fig. 41.1).

Small Cell Variant (SCV) These are compact, electron-dense cells, measuring up to 0.2 to 0.5 μm in length. These have a robust structure, characterised by a thickened peptidoglycan layer, a dense cell envelope, and condensed DNA. These morphological features make the SCV highly stable in harsh environmental conditions.

Large Cell Variant (LCV) The LCV represents the active metabolic and replicative phase of the bacterium. It is considerably larger than the SCV and exceeds 1 micrometre in length. The LCV displays a more pleomorphic, or variable, shape. Structurally, the LCV possesses a more typical Gram-negative cell envelope. It has clearly defined inner and outer membranes and a dispersed nucleoid, within the protective confines of the specialised *Coxiella*-containing vacuole (CCV), that resembles a lysosome-like compartment.

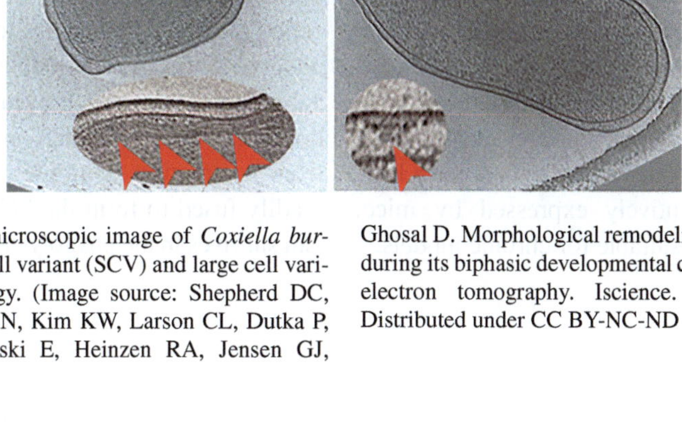

Fig. 41.1 Electron microscopic image of *Coxiella burnetii* showing small cell variant (SCV) and large cell variant (LCV) morphology. (Image source: Shepherd DC, Kaplan M, Vankadari N, Kim KW, Larson CL, Dutka P, Beare PA, Krzymowski E, Heinzen RA, Jensen GJ, Ghosal D. Morphological remodeling of *Coxiella burnetii* during its biphasic developmental cycle revealed by cryo--electron tomography. Iscience. 2023 Jul 21;26(7). Distributed under CC BY-NC-ND 4.0 license)

Cultivation

Initially, *C. burnetii* was cultivated in embryonated chicken eggs or mammalian cell cultures. Various cell lines, like Vero cells (African green monkey kidney fibroblasts) and macrophage cell lines (e.g. J774A.1), were frequently employed.

Axenic (host cell-free) cultivation of *C. burnetii* is a major breakthrough in cultivation. The introduction of acidified citrate cysteine medium (ACCM), and specifically its refined version ACCM-2 around 2009, revolutionised the ability to grow this obligate intracellular bacterium outside its host cell. This semi-defined medium was formulated to mimic the acidic, nutrient-rich environment found within the *Coxiell*a--containing vacuole, supporting bacterial growth under microaerophilic conditions.

This pathogen is not suitable for traditional bacterial culture methods. It requires biosafety level III (BSL-3) containment due to the high infectivity of the organism.

Animal Models

Animal models, including mice, guinea pigs, and nonhuman primates, are used to study QF pathogenesis, immune responses, and vaccine development.

Group I isolates are persistent at elevated levels of cytokine production throughout infection in a mouse infection model. Group IV and V isolates produce only moderate secretion during the elevation of clinical disease. In animal models, the guinea pig shows that group IV isolates are unsuccessful in producing fever. In contrast, group I isolates are able to cause mortality at an increased challenge dose. An acquired immune response is not possible in a severe combined immunodeficiency (SCID) mouse. Group I isolates cause extreme cachexia and death, while group IV and V isolates cause only weight loss. IL-10 was constitutively expressed by mice, which is said to be suitable for chronic models.

Transmission Cycle

C. burnetii has a unique biphasic life cycle involving complex host interactions and unique survival mechanisms (Fig. 41.2). Domestic livestock, like sheep, goats, dogs, cats, reptiles, wild animals, and birds, are reservoirs for the transmission of infection to humans. The infection occurs by inhalation and/or ingestion of contaminated aerosols or through animal products like blood, aborted products, and milk, etc. Other modes of transmission of infection can be either by vector or vertical, i.e. the bite of infected ticks in the pregnant animals, which can pass *C. burnetii* to their offspring through the placenta.

C. burnetii can attach to and enter into the host cells, including phagocytic cells, by utilising specific receptors like $\alpha V\beta 3$ integrin. *C. burnetii* primarily infects macrophages and monocytes in humans and various animal species. The bacterium enters host cells via phagocytosis and replicates itself within an acidic vacuole called the *Coxiella*-containing vacuole (CCV). The specialised compartment allows *C. burnetii* to manipulate host cell pathways and avoid destruction while creating a favourable environment for its replication. The Type IV secretion system (T4SS) facilitates the infection process, which conveys effector proteins to the host body to subvert immune responses and support intracellular survival. *C. burnetii* may persist in tissues such as the heart, liver, and blood vessels in chronic infections, leading to conditions like endocarditis and vascular infections.

The extracellular form of the SCV is fused to a cell membrane and engulfed into a phagosome. *C. burnetii* actively modifies the phagosome, transforming it into a spacious, acidic vacuole called the CCV, and begins to multiply. This can undergo extensive remodelling, similar to the autophagosome, and prevents fusion with the host cell lysosome, which protects the bacterium from destruction. Multiple phagolysosomes are readily fused to form the LCV. How this organism survives in the phagolysosome is still unclear.

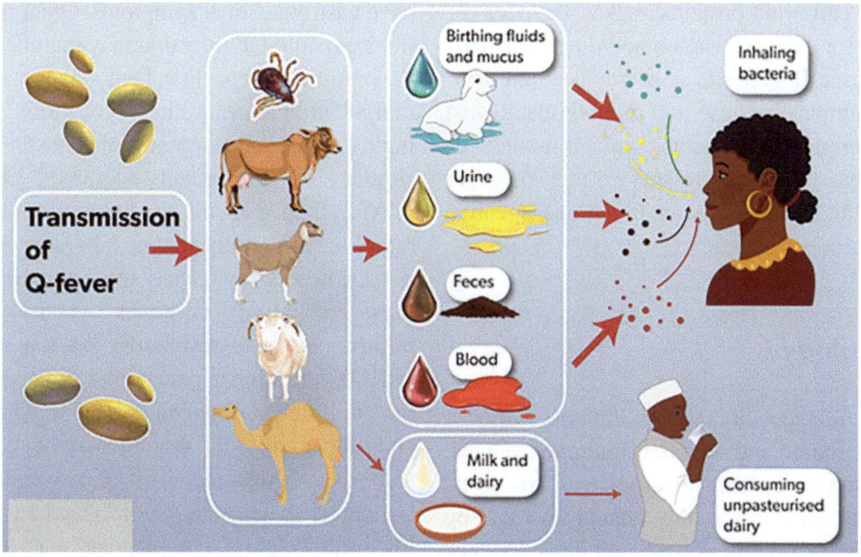

Fig. 41.2 Transmission cycle of *Coxiella burnetii*. (Image source: International Livestock Research Institute, ILRI—Nairobi, Kenya)

In some cases, *C. burnetii* can persist in host tissues for months or years, leading to chronic infections such as endocarditis, hepatitis, or vascular complications.

Pathogenesis and Pathology

C. burnetii possesses a T4SS and lipopolysaccharide (LPS) as two virulence factors. These factors either manipulate or establish the infection in the host cell and evade immune responses. The T4SS helps the bacterium inject effector proteins into host cells, allowing it to manipulate host cell functions and create a niche for replication. The LPS plays an important role in phase variation and virulence.

Proteomics of *C. burnetii* has been identified based on the key proteins involved in stress resistance and immune evasion. These proteins enable the bacterium to survive in harsh intracellular environments, such as the acidic and oxidative conditions within host phagolysosomes. Heat shock proteins (e.g. GroEL and DnaK) help in protein folding and protect against cellular stress. Antioxidant enzymes like superoxide dismutase and catalase neutralise reactive oxygen species produced by the host immune system. Additionally, some effector proteins secreted via the Dot/Icm T4SS interfere with host immune signalling pathways, reducing immune detection and response.

C. burnetii exhibits a unique phenomenon known as phase variation. This significantly impacts its virulence, host interaction, and diagnostic detection. This variation is primarily associated with changes in the bacterium's lipopolysaccharide (LPS) structure.

Phase I Antigen It is extremely contagious in the transmission of infection to humans. LPS acts as a virulence factor, and it prevents the immune system from quickly reacting to the proteins on the bacterial wall. The smooth type of *Coxiella* LPS lacks the binding and activation of complement system components, which can delay or prevent the lysis of bacterial cells. LPS reactivity is the basis for the historical serologic distinction amongst *Coxiella* strains.

Phase II Antigen *C. burnetii* has the ability to grow in cell cultures or cells that are not immunocompetent, such as embryonated eggs. LPS synthesis is usually incomplete, and forms as an intermediate, and sometimes, the protection of the cell wall may be suppressed. The accessibility of many cell wall proteins makes them easily available for characterisation and diagnostic purposes. Phase II *Coxiella* is not considered virulent for humans because it is quickly inactivated by the complement system in animal models. *C. burnetii* has the ability to survive at 1–6 °C for 6 weeks and persists as an infectious agent in donated blood products.

Immunology

C. burnetii triggers a complex immune response involving innate and adaptive immune systems. The bacterium infects macrophages, survives, and reproduces inside the phagolysosome by altering its environment. Infected cells produce pro-inflammatory cytokines, such as TNF-α, IL-6, and IFN-γ, that aid in the control of bacterial replication. The immune system initiates the activation of macrophages, neutrophils, and dendritic cells, which help recognise pathogens and present antigens. B cells generate specific antibodies against antigens in phases I and II of LPS. CD4+ T cells stimulate macrophages, while CD8+ T cells target infected cells for destruction. These factors may cause alterations in phase variation, prevent immune-mediated clearance, and delay programmed cell death for intracellular survival.

Infection in Humans

There is no specific genotype linked to QF infections in humans that manifests as acute, chronic, or subclinical forms. *C. burnetii* is highly infectious to humans. Nearly 1–10 viable organisms are sufficient to induce an infection via the acrogenic route to humans.

Acute QF Clinical symptoms like fever, chills, vomiting, abdominal pain, and pneumonia are the common manifestations of acute QF. It is self-limited, but sometimes it can progress. A non-productive cough and skin rashes are common in both adults and children. About 50% of children have gastrointestinal symptoms. Other manifestations of acute QF infection may include hepatitis, pericarditis, encephalitis, hemophagocytosis, haemolytic uremic syndrome, myocarditis, and meningitis. Signs and symptoms of acute QF are more similar to community-acquired pneumonia (CAP). It is difficult to differentiate clinically, radiologically, or by routine laboratory investigations. Chest X-ray reveals some uncommon findings, including infiltrates (>96%), patchy infiltration, acute respiratory distress syndrome (ARDS), and lobar pneumonia in acute QF. QF infection during pregnancy is quite complicated, and proper antibiotic therapy may be required. In some cases, pregnancy with QF may lead to abortion, preterm delivery, stillbirth, and a foetus with low birth weight. QF in pregnant women requires proper follow-up and antibiotic therapy.

Chronic QF Endocarditis, granulomatous hepatitis, or chronic fatigue syndrome are common manifestations of chronic QF. Endocarditis is the leading cause of up to 60–78% of all cases worldwide. Other nonspecific symptoms like fever, fatigue, chest pain, pulmonary embolism, and congestive cardiac failure may likely occur. A transesophageal echocardiogram (TEE) or a transthoracic echocardiogram (TTE) is not useful for detecting chronic QF endocarditis. For serological confirmation, the QF Phase I IgG antibody titre should be ≥1:1024. The occurrence of complications has led to a range of clinical manifestations, including osteomyelitis in children, chronic hepatitis or endocarditis among animal farm workers, and aneurysms in pregnant women.

Post-QF Fatigue Syndrome Only 20% of patients may experience this syndrome after acute disease, which is rare. It is more likely to develop in those who have severe acute illness. Many patients who develop QFS commonly experience persistent fatigue with severe symptoms. Severe involvement of acute QF leads to the complexity of other symptoms like headache, myalgia, stress, night sweats, short-term memory loss, alcohol intolerance, etc. Elevation of antibodies against *C. burnetii* antigens could be observed. The condition may last up to a year or span an entire lifetime. There was no evidence-based therapy for post-QF-fatigue syndrome.

Infection in Animals

Coxiella is associated with a vast array of infections in animals, including ruminants, dogs, cats, reptiles, wild animals, and birds. It primarily infects cattle, sheep, and goats, leading to reproductive disorders such as infertility, neonatal mortality in affected herds, abortions, decreased milk production, and stillbirths. In many cases, infected herds or flocks shed the bacteria without any history showing signs of abortions or other abnormal reproductive events.

The infection in ruminants is subclinical, and it causes anorexia and abortions with a low bacterial load in reproductive fluids. In cattle, *C. burnetii* shedding in the vagina occurs quickly, usually within 14 days. However, it sheds milk for a more extended period. Approximately 65% of cows have PCR-positive milk samples that remain negative. Goats' milk-shedding patterns are similar to vaginal patterns. However, there are fewer pathogens, and they last longer (8 weeks or more). After pregnancy, the vaginal shedding level usually decreases within 2–3 weeks. Symptoms are not usually present in infected cats, but pregnant queens may have abortions or give birth to weak kittens. The placenta and aborted foetuses need to be examined when investigating the cause of abortion.

Cats that are experimentally infected experience a temporary fever, malaise, and anorexia that last for several days. Shedding of the bacteria routinely occurs in birth fluids, vaginal mucus, milk, and faeces. However, the relative number of bacteria that can be shed in these bodily fluids is not known.

Epidemiology and Public Health

QF cases have been documented across 51 countries on five continents, except New Zealand (Fig. 41.3). Most developing countries in the

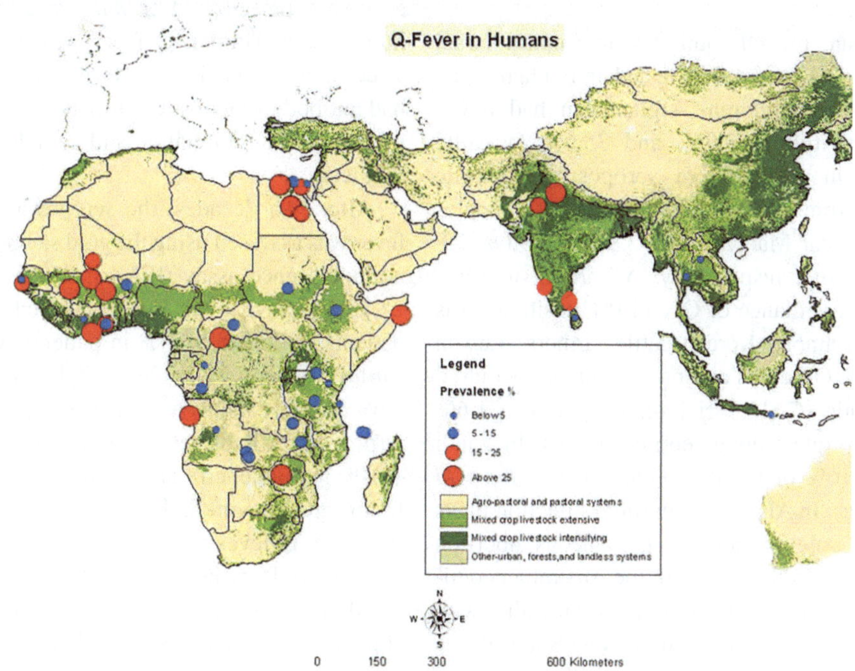

Fig. 41.3 Global prevalence of Q fever in humans. (Image source: Grace D, Mutua FK, Ochungo P, Kruska RL, Jones K, Brierley L, Lapar ML, Said MY, Herrero MT, Phuc PM, Thao NB. Mapping of poverty and likely zoonoses hotspots. Zoonoses Project 4: Report to Department for International Development, UK)

world rely on seropositivity since their cultures require level III containment facilities. For the first time in the history of the Netherlands, over 4000 cases of QF were reported due to a major outbreak of QF between 2007 and 2009. In South American states such as Trinidad, QF was detected among slaughterhouse workers (4.4%) and 3.3% of febrile patients, and an outbreak was observed among military cadet training officers in Brazil. The North American continent exhibited a 3.1% seropositivity rate for *C. burnetii* antibodies in healthy individuals in the USA. The prevalence of community-acquired pneumonia was reported to be 24.4% among patients in French Guiana. In Greece, the prevalence was 2.9%, while Austria reported a seroprevalence of 6% among European countries. The seropositivity rate for stock breeding staff in the Diyarbakir region of Turkey was 6.6%. Between 2000 and 2015, nearly 904 cases of acute QF were recorded in England. The prevalence of QF has been reported in Spain, with a rate of 39.4% in older adults aged 65 years or older. In Italy, QF prevalence was observed in 10.8% of prisoners, 16.5% of prison guards, and only 3.2% of residents.

In Asia, the QF outbreak in Central Israel affected 144 out of 164 tested individuals, primarily schoolchildren. Afghanistan had a QF seropositivity of 63.9%, and 35.6% of cattle farmers in China were seropositive. Febrile patients from Bhutan and the Orang Asli peoples of Peninsular Malaysia had 2.8% and 9.6% QF seropositivity, respectively. According to serological surveillance of QF, 11.0% positivity was observed among Korean cattle farmers. Taiwan had 7.2% QF cases among community-acquired pneumonia (CAP). Sri Lanka reported a 1.6% positivity rate among febrile patients. In Saudi Arabia, 16% of febrile patients had *C. burnetii* antibodies. In Algiers, *C. burnetii* infection plays a major role in human spontaneous abortion cases with febrile illness. In the Australian continent, the cases occur throughout Australia, with varying positivity in Northern New South Wales (3.1 per 100,000 per annum), Southern Queensland (6.3 per 100,000 per annum), South Australia (1.1 per 100,000 per annum), and < 1.0 per 100,000 per annum in the other states and territories.

Epidemiology in India

The first human QF was reported in 1954 based on antibody titre in follow-up blood samples. After 2 years, a human case of QF was reported from Jammu province based on clinical and radiological findings, which were later confirmed by the patient's serum. A first serological survey conducted in the Nilgiris and Coimbatore districts of Madras state found that seven individuals had QF antibodies with titres ranging from 1:16 to 1:64.

For the first time in India, the camel was identified as one of the primary sources of human QF infection in a northern state. The seroprevalence of human Q fever in Jammu and Kashmir was found to be 20.7%. In Rajasthan, *C. burnetii* was isolated from human breast milk. The prevalence of QF using the complement fixation test (CFT) varied from 0.4% to 23.5% in the same state. Sero-epidemiological studies of Q fever revealed a 10.85% presence of *C. burnetii* antibodies in human sera. Human Q fever was estimated to affect approximately 12.3% of individuals who had antibodies against C. burnetii, as determined by capillary agglutination and complement fixation tests.

After two decades, the seroprevalence of Q fever was assessed using the gold standard immunofluorescence assay (IFA). In Uttar Pradesh, a 25.7% positivity rate was observed in febrile patients, followed by 8% in patients with endocarditis in Tamil Nadu, 26.2% in Puducherry, and 2.6% in Karnataka. Animal handlers and cattle farm workers had prevalence rates of 28.2% and 46.4% in Puducherry and Izatnagar, respectively. Using IgG enzyme-linked immunosorbent assay (ELISA), Q fever infection was recorded at 4% among febrile patients in Tamil Nadu.

QF has been reported in patients with different clinical manifestations like human abortions (21.62%), neonatal sepsis, and endocarditis. In India, PCR has been used as a diagnostic tool for identifying cases of QF since 2008.

Diagnosis

Specimens

Blood is the most frequently collected sample for direct detection, when bacterial loads are typically higher. Other specimens include biopsy specimens from affected organs, such as valve tissue (for endocarditis), liver biopsies (for Q fever hepatitis), or bone marrow. In pregnant women suspected of having Q fever, placental tissue or amniotic fluid is also collected for diagnosis.

Serum is the primary sample for the presence and levels of both antibodies, IgM as well as IgG, against Phase I and Phase II antigens.

Microscopy

Microscopy is not commonly utilised as a primary tool for QF due to the intracellular nature of *C. burnetii*. It is Gram-negative but often poorly stained. However, histopathological examination of tissue samples provides indirect evidence of infection. In tissue biopsies from chronic QF patients, immunohistochemical staining may reveal the presence of *C. burnetii* antigens. Additionally, Giménez staining and modified Ziehl-Neelsen staining are used to visualise *C. burnetii* in infected cells.

In Vitro Culture

C. burnetii does not grow in conventional agar medium, and it requires axenic media like acidified citrate cysteine medium (ACCM). This pathogen needs an acidic pH, cysteine, and a CO_2 environment for its growth. Colony morphology of ACCM in solidified agar is pinpoint, translucent, and dome-shaped, often visible under a microscope.

They are catalase positive but negative for oxidase, urease, and nitrate reduction tests. The tissue culture is used to isolate *C. burnetii* from clinical materials. The application of shell vial culture was successful in isolating this pathogen. Several cell lines have been used to isolate *C. burnetii*, like HeLa, Vero, human embryonic lung and L929 mouse fibroblasts, to observe the cytopathogenic effect (CPE) of this pathogen. The continuous flow high salt technique is a simple method for the purification of *C. burnetii* and other rickettsiae.

Fertile Hen's Eggs Placental bits or blood specimens are homogenised and incorporated with antibiotics (gentamicin or penicillin 50–100 μg/ml and streptomycin 100–200 μg/ml) in PBS solution. Supernatant fluid is inoculated into 6- to 7-day-old chick embryo eggs and incubated for up to 2 weeks, following which the yolk sacs are harvested. Smears are prepared with Giemsa/Machivallo/Giménez stains.

Serodiagnosis

The conventional methods like the anti-globulin sensitisation test, the microscopic slide agglutination test, Luoto's capillary agglutination test, slide agglutination test, Fiset's microagglutination test and complement fixation test (CFT) were some of the important landmarks in the serodiagnosis of *C. burnetii* infection in humans and animals. Many of these tests are less sensitive but more specific.

The direct and indirect peroxidase test has been applied for the detection and differentiation of *C. burnetii* and *R. conorii* in tissue smears. It is also used to detect antibodies against *C. burnetii*. The development of fluorescent antibody techniques for the direct identification of this pathogen in tissues has proved to be yet another useful diagnostic tool. Enzyme-linked immunosorbent assay (ELISA) and indirect immunofluorescence assay (IFA) are being increasingly used (Fig. 41.4).

Serological tests require paired sera samples (acute and convalescent) to confirm seroconversion and establish a definitive diagnosis. IFA and ELISA for the detection of QF show 95% sensitivity and 100% specificity. Unlike other infections, serum IgM phase II antibodies to *C. burnetii* last for several months.

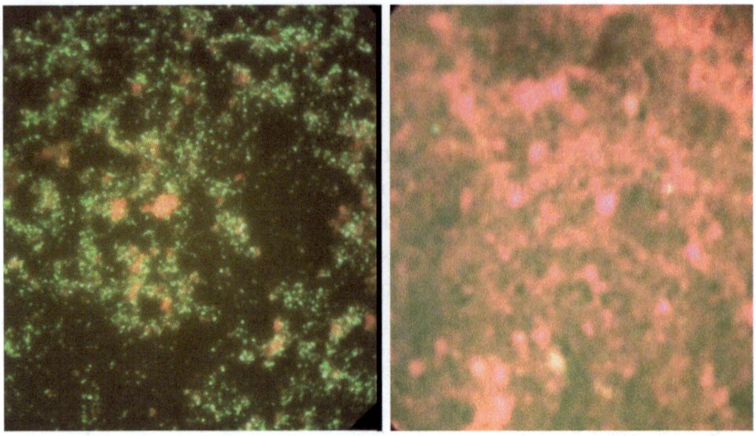

Fig. 41.4 Indirect immunofluorescence assay (IFA) showing (**a**) Q fever positive (apple fluorescence); (**b**) Q fever negative (red) samples. (Image source: Authors' personal collection)

Molecular Diagnosis

Molecular methods are useful for the diagnosis of QF during the early stage of the illness.

Polymerase chain reaction (PCR) for *C. burnetii* DNA uses different gene targets for the diagnosis of the infection. *Real-time PCR (qPCR)* is a highly sensitive and quantitative method that allows detection in blood, serum, or tissue samples within hours.

Loop-mediated isothermal amplification (LAMP) is easier to perform and does not require sophisticated equipment. LAMP exhibits high diagnostic sensitivity and specificity, with outer and inner primers designed to detect specific DNA in human clinical samples and in animals. A few reports of the application of QF LAMP are available for the diagnosis of *C. burnetii*.

Whole genome sequencing (WGS) is highly helpful in research and outbreak investigations to analyse the genetic variations among *C. burnetii* strains. This technique deciphers the entire DNA sequence of an organism. During disease outbreaks, WGS plays a crucial role in epidemiological tracking by identifying the related *C. burnetii* strains.

Multi-locus sequence typing (MLST) is particularly useful for tracking the spread of *C. burnetii* in populations and across regions. It is often a practical and cost-effective way to study the genetic relationships among isolates, monitor epidemiological trends, and map the distribution of strain types over time.

An overview of laboratory diagnosis of Q fever is presented in Table 41.1.

Treatment

Treatment for Q fever primarily involves antibiotic therapy. The choice and duration of treatment depend on whether the infection is acute or chronic and on the presence of complications. Early diagnosis and prompt initiation of appropriate antibiotics are crucial for preventing progression to chronic Q fever, which carries a higher mortality rate.

Acute Q Fever

For acute Q fever, doxycycline is the antibiotic of choice. It is highly effective against *C. burnetii* and is generally well-tolerated. The recommended dosage for adults is 100 mg orally twice daily. The duration of treatment for uncomplicated acute Q fever typically ranges from 10 to 14 days, or until the patient has been afebrile for at least 5 days. For patients who cannot tolerate doxycycline, alternative options include azithromycin or fluoroquinolones, though these are generally considered less effective. Pregnant women

Table 41.1 Laboratory diagnosis of Q fever

Diagnostic approaches	Methods	Targets	Remarks
Histopathological/ microscopy	Immunohistochemistry (IHC)	Antigen in tissue biopsies	Helps to confirm chronic QF in tissues (e.g. heart valves, liver)
	Giemsa stain	Intracellular coccobacilli	Visualises organisms inside macrophages; used in tissue/blood smears
	Giménez stain	Tissues	Red-staining bacteria on blue background; more sensitive than Giemsa stain
	Macchiavello stain	Bright red intracellular	Used in older pathology protocols
	Hematoxylin and eosin (H&E)	Inflammatory lesions	May show fibrin ring granulomas (liver) in QF hepatitis
In vitro culture	Shell vial culture (in BSL-3)	Viable *Coxiella burnetii*	Time-consuming; rarely performed due to biosafety risks
	Embryonated egg yolk sac inoculation	*C. burnetii*	Historical method; discontinued from routine use
Serolodianosis	Indirect immunofluorescence assay (IFA) and ELISA	Phase I and II IgM and IgG antibodies	Gold standard serological method to detect QF phase II IgM/ IgG → acute, QF phase I IgG → chronic phase
	CFT, LAT	Antibodies	Lacks high sensitivity and specificity
	Western blot/ immunoblot	Specific proteins of *C. burnetii*	It is used as confirmatory test for *C. burnetii* infection
Molecular diagnosis	PCR/qPCR (real-time PCR)	IS1111, com1, icd gene, 16S rRNA	High sensitivity/specificity; used on blood, tissues, serum
	Loop-mediated isothermal amplification (LAMP)	IS1111 or 16S rRNA	Rapid, useful in resource-limited settings
	Metagenomic NGS	Whole genome sequences	For atypical cases; unbiased pathogen detection

with acute Q fever are a special consideration, as doxycycline is contraindicated due to its potential effects on fetal bone and tooth development. In such cases, co-trimoxazole (trimethoprim-sulfamethoxazole) is the preferred treatment throughout pregnancy, continuing for the entire duration of the pregnancy.

Chronic Q Fever

The standard regimen for chronic Q fever, particularly endocarditis, involves a combination of doxycycline and hydroxychloroquine. Hydroxychloroquine is included not for its direct antimicrobial effect but because it alkalinises the phagolysosomes within host macrophages, where *C. burnetii* replicates, thereby enhancing the intracellular activity of doxycycline. The dosage for doxycycline includes 100 mg orally twice daily, while hydroxychloroquine is typically given at 200 mg orally three times daily. This combination therapy is typically administered for at least 18 months and, in many cases, for several years, often guided by monitoring of serological markers and clinical response.

Monitoring and Special Considerations

During and after treatment, patients, especially those with chronic Q fever, require careful monitoring of clinical symptoms, liver function tests, and serological titers for *C. burnetii*. A sustained decrease in antibody titers indicates effective treatment, while persistently high or rising titers may suggest ongoing infection or treatment failure. In cases of chronic Q fever endocarditis, surgical intervention (e.g. valve replacement) may be necessary in addition to prolonged antibiotic therapy. Children with Q fever are generally treated with doxycycline, with studies indicating that short courses do not cause tooth staining.

Prevention and Control

Preventing and controlling QF requires a combination of public health interventions and vaccination strategies.

Individuals in close occupational contact with animals, such as farmers, veterinarians, slaughterhouse workers, and butchers, face an increased risk of exposure to diseases through direct contact with the handling of diseased animals and their products, including carcasses, hair, and wool. Bacteria may be transmitted through the consumption of unpasteurised milk. Handling infected blood samples using protective clothing, masks, and gloves prevents the spread of disease. Researchers, veterinarians, and workers need to use adequate personal protective equipment (PPE) to protect against exposure to small droplets and aerosols during high-risk occupational or medical procedures. Initiatives on public awareness and education programmes targeting high-risk populations play a critical role in controlling the spread of disease.

Vaccination is an effective measure against QF, particularly for high-risk populations. It remains one of the most effective approaches for QF, especially in high-risk occupational groups. The Q-Vax vaccine, a formalin-inactivated whole-cell vaccine, is licensed in Australia for individuals at high risk (e.g. veterinarians, farmers, abattoir workers). Live-attenuated vaccines are under investigation. Several vaccines, such as Coxevac, are available for livestock to reduce bacterial shedding and prevent the transmission of diseases. The disease can be controlled by regularly vaccinating animals in endemic areas and properly disposing of animal reproductive products to prevent environmental contamination.

Case Study

A 48-year-old adult male presented with acute febrile illness and abdominal pain for more than 7 days of duration and developed chills after a few hours. The patient complained of severe fatigue, headache, myalgia, dry cough, and malaise. No history of loose stools or weight loss was observed. The chest X-ray showed subtle interstitial infiltrates. Blood culture was negative after 5 days of incubation. CBC was normal, with slight thrombocytopenia. Liver enzymes were slightly raised. He reported prolonged exposure to domestic livestock, including cattle, goats, and dogs, which may have served as potential sources of infection. He also works as a butcher in a private shop. The serological QF ELISA Phase II IgG investigation was done to confirm the diagnosis.

1. What is the causative agent for this disease?
2. What is the definitive diagnosis for this disease?
3. Name the gold standard serological assays available for the diagnosis.
4. Name the confirmatory tests available for this disease.

Research Questions

1. To what extent PCR is more reliable than serological testing for the detection of acute Q fever during outbreaks under current surveillance systems?
2. What is the impact of physician awareness on the timely diagnosis of Q fever in non-endemic regions?
3. What are the challenges in diagnosing and managing asymptomatic Q fever cases in rural or outbreak-prone areas?
4. How effective are One Health strategies and public health communication in the prevention of Q fever?

Declaration In this chapter, no AI tool has been used.

Further Readings

Angelakis E, Raoult D. QF. Vet Microbiol. 2010;140(3–4):297–309.

Arricau-Bouvery N, Rodolakis A. Is QF an emerging or re-emerging zoonosis? Vet Res. 2005;36(3):327–49.

CDC and WHO Reports on QF—Periodic reports from the Centers for Disease Control and Prevention (CDC) and the World Health Organization (WHO) provide updates on surveillance, outbreak management, and control measures.

Davis GE, and Cox HR. A filter-passing infectious agent isolated from ticks: Isolation from Dermacentor andersoni, reactions in animals, and filtration experiments. Public Health Reports, 1938;53:2259–67.

Eldin C, Mélenotte C, Mediannikov O, Ghigo E, Million M, Edouard S, et al. From QF to *C. burnetii*Infection: a paradigm change. Clin Microbiol Rev. 2017;30(1):115–90.

Marrie TJ. QF: the disease—this book provides a comprehensive overview of QF, covering clinical symptoms, diagnosis, and treatment options. 1990.

Maurin M, Raoult D. QF. Clin Microbiol Rev. 1999;12(4):518–53.

Omsland A, Beare PA, Hill JCockrell DC, Howe D, Hansen B, Samuel JE, Heinzen RA. Isolation from animal tissue and genetic transformation of Coxiella burnetii are facilitated by an improved axenic growth medium. Appl Environ Microbiol. 2011;77(11):3720–5.

Pradeep J, Kumar S, Stephen S, Kamboj DV, Gunasekaran D, Hanifah M. Detection of acute Q fever human cases by indirect immunofluorescence & real-time polymerase chain reaction in a tertiary care hospital in Puducherry. Indian J Med Res. 2018;148(4):449–52.

Shepherd DC, Kaplan M, Vankadari N, Kim KW. Morphological remodeling of *Coxiella burnetii* during its biphasic developmental cycle revealed by cryo-electron tomography. iScience. 2023;26(7):107210.

Psittacosis

42

Aditya Kundu and Nonika Rajkumari

Learning Objectives

1. To comprehend the transmission, pathogenesis, and clinical manifestations of psittacosis
2. To understand various laboratory methods for the diagnosis of *Chlamydia psittaci* infection
3. To describe prevention and control measures to prevent psittacosis

Introduction

Chlamydia psittaci, previously known as *Chlamydophila psittaci*, is the causative agent of the zoonotic disease psittacosis. It is also called ornithosis or parrot fever. Primarily, humans become infected through aerosolised droppings or secretions from infected birds, which contain this obligate intracellular pathogen. The disease in humans can range from mild flu-like symptoms to severe pneumonia. Due to its zoonotic potential and capacity to cause serious health outcomes, psittacosis is regarded as a significant public health concern, especially in occupational environments such as bird breeding, poultry farming, and animal husbandry.

Globally, psittacosis occurs sporadically, with cases reported in both developed and developing countries. Although the overall incidence has decreased in some regions due to likely improvements in diagnosis and increased awareness of underreporting, the issue persists, particularly in areas with limited monitoring. The disease is often linked to exposure to infected birds, and multiple outbreaks are associated with the trade of birds from poorly regulated sources. human-to-human transmission is rare, and most cases result from direct or indirect contact with infected birds.

In recent years, the challenge of local outbreaks has increased. From 2023 to early 2024, several European countries (Austria, Denmark, Germany, Sweden, and the Netherlands) reported an unusual rise in psittacosis cases. This increase, which began around November 2023, resulted in at least five deaths and was often associated with both wild and domestic birds. The surge in cases has prompted public authorities in these nations to conduct epidemiological investigations to identify clusters and possible causes of infection. Diagnostic efforts have expanded to include birds, especially those already sampled for avian influenza surveillance, to assess the wider prevalence of *C. psittaci*.

A. Kundu · N. Rajkumari (✉)
Department of Microbiology, All India Institute of Medical Sciences, Kalyani, West Bengal, India

S. C. Parija, A. Poddar (eds.), *Textbook of Bacterial Zoonoses*, Microbial Zoonoses,
https://doi.org/10.1007/978-981-95-6615-0_42

History

The first documented details of psittacosis date back to 1879, when J. Ritter published a case report describing a mini-outbreak. In this outbreak, three of seven patients died, and Ritter made several major observations. He identified the infected parrot as the source of infection, estimated the incubation period, and notably noted the lack of human-to-human transmission, an insight that remains largely accurate today.

Further scientific progress was achieved during 1893, when Edmund Nocard, a French veterinarian and microbiologist, isolated a Gram-negative bacterium from a parrot that had died from the disease. He named it *Bacillus psittacosis*, although this initial identification was not yet connected to the comprehensive understanding of the disease in humans.

The term "psittacosis" was first introduced in 1895, originating from the Greek word *Psittacus* (meaning "parrot"), which highlights a strong connection between the disease and psittacine birds. Between 1929 and 1930, there was a significant global epidemic affecting approximately 750–800 people, with several deaths reported. This outbreak was mainly linked to the importation of parrots from Argentina. The severity of this incident led to a ban on parrot imports and the enforcement of strict regulations for the international trade in birds in various countries. Currently, psittacosis remains a concern, although it is uncommon and often overlooked.

Taxonomy

The genus *Chlamydia* is classified under the domain Bacteria, phylum Chlamydiota, class Chlamydia, order Chlamydiales, and family Chlamydiaceae. Historically, there was a proposed division into two genera, *Chlamydia* and *Chlamydophila*, based on 16S rRNA gene sequences in 1999. However, this split was challenged by subsequent DNA analysis using more extensive genomic data, resulting in a proposal in 2009 to consolidate all species within a single genus, *Chlamydia*. Formal emendation of the family Chlamydiaceae was carried out in 2015, placing all species under one genus, *Chlamydia*. As of today, there are now 14 validly recognised species. Among these, *Chlamydia abortus, Chlamydia psittaci, Chlamydia felis, Chlamydia caviae, Chlamydia suis,* and *Chlamydia gallinacean* are zoonotic.

C. psittaci, being the most important zoonotic agent within the genus *Chlamydia*, exhibits diversity following the sequencing of the ompA gene, which encodes the major outer membrane protein (MOMP). Based on ompA gene sequencing, C. psittaci has been divided into at least ten, and now up to 17, recognised genotypes, including A to G, E/B, M56, Cattle (WC), and others like 1 V, 6 N, Mat116, R54, YP84, and CPX0308. Each of these genotypes often shows a particular association with specific bird species or, in some cases, mammalian hosts. For example, genotype A is widely prevalent in psittacine birds such as parrots, genotype B is commonly found in pigeons, genotype C in geese, genotype D in turkeys, and genotype E/B in various bird species, including ducks. The genotype WC has been identified in cattle, while M56 occurs in muskrats.

Although a degree of host preference is observed, all *C. psittaci* genotypes are regarded as potentially transmissible to humans, capable of causing psittacosis. Genotypes A and D are more frequently associated with severe human infections, while others tend to produce milder diseases.

Genomics and Proteomics

The genome of *C. psittaci* spans approximately 1.1 megabases and encodes about 1000 genes. Consistent with other *Chlamydia* species, it demonstrates a highly reduced genome, a hallmark of obligate intracellular pathogens. It lacks key enzymes required for essential biosynthetic and metabolic pathways, including amino acid, nucleotide, and lipid synthesis. As a result, it relies on the host cell for survival and replication. This genomic reduction reflects a long-standing co-evolution and adaptation to intracellular parasitism.

Advancements in second-generation sequencing technologies have enabled the complete sequencing of over 100 *Chlamydia* genomes. These include 28 strains of *C. psittaci* that represent nearly all known serotypes and host species. These genomic data offer significant insights into the genetic structure, evolutionary relationships, and variability within *C. psittaci*. Each sequenced strain of *C. psittaci* features a conserved single chromosome (about 1.1 Mb) and a plasmid (approximately 8 kb), both containing 911 core coding sequences (CDS) that are consistent across the strains. *C. psittaci*, compared to closely related species such as *C. abortus, Chlamydia pneumoniae,* and *Chlamydia trachomatis,* shares 736 genes, indicating a substantial level of conservation.

Specific genomic regions, including the plasticity zone (PZ), inclusion membrane proteins, and polymorphic membrane protein (PMP) genes, show divergence linked to host adaptation and virulence. Polymorphic membrane proteins (PMPs) assist in protein adhesion, immune evasion, and antigenic variation. The PZ region of *C. psittaci* features distinct characteristics, including a notable toxin/adhesin gene and a guaAB--add cluster, which are believed to be involved in purine salvage and adherence mechanisms. Importantly, these traits are primarily seen within the '6 BC clade'. This clade is a group of strains characterised by low genetic diversity and a broad host range, likely resulting from a horizontal gene transfer event or selective gene loss. Furthermore, the pmp gene family, especially the G/I subfamily, has significantly expanded within *C. psittaci*. This potentially improves its adaptability to various hosts by aiding immune evasion or enhancing adhesion. Despite the presumed genetic isolation of intracellular bacteria, the unexpectedly high levels of homologous recombination in *C. psittaci* contribute notably to genomic variation over point mutations. Around 50% of recombinant DNA fragments appear to originate from either *C. abortus* or unidentified *C. psittaci* strains. This indicates that a substantial amount of genomic diversity remains to be uncovered in this species.

Table 42.1 Various genes of *Chlamydia psittaci* and their functions

Gene category	Examples	Function
Housekeeping genes	*rpoB, dnaK, groEL*	Basic cellular processes such as transcription, replication, etc.
Major outer membrane protein	*ompA*	Structural integrity, immune evasion
Type III secretion system (T3SS)	*sctN, sctL, sctC*	Host cell invasion and effector protein translocation
Polymorphic membrane proteins	*pmpA to pmpI*	Host interaction, immune evasion
Plasmid-encoded genes	*pgp1–pgp8* on plasmid pCp	Virulence regulation, possibly host adaptation
Metabolic remnants	*gltT, tpiA*	Limited energy metabolism
Stress response proteins	*htrA, clpB*	Intracellular survival, protein quality control
Hypothetical proteins	Various	Related to virulence

A plasmid known as pCp is often found in several *C. psittaci* strains and is thought to play an important role in the disease within the organism. Although its exact function is not fully understood, current evidence suggests that this host may influence the regulation of genes involved in immune evasion and intracellular survival systems.

Proteomic analysis has identified many significant protein families involved in pathogenesis. These include the most crucial outer membrane protein (MOMP), encoded by the ompA gene, which plays a vital role in structural integrity and immune recognition. Additionally, the Type III secretion system (T3SS), a molecular syringe used to inject bacterial components into host cells, is important for cell entry and modulation of host cell processes. The functions of various genes are described in Table 42.1.

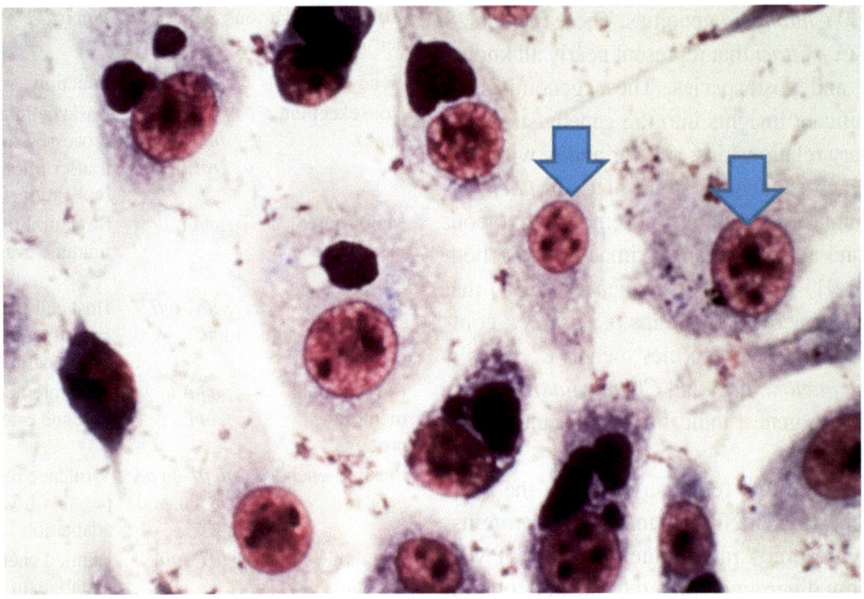

Fig. 42.1 Light microscopy image of *Chlamydia psittaci* inclusion bodies showing characteristic inclusion bodies appear as dark basophilic structures of size 0.25–0.5 μm in diameter within the infected cells. (Reproduced with permission)

Morphology

C. psittaci exhibits a bipolar life cycle composed of two morphologically and functionally distinct forms: the elementary body (EB) and the reticulate body (RB).

EBs are extracellular and contagious, about 0.3 μm in diameter, and are characterised by a dense, electron-opaque appearance. They are metabolically inactive and highly capable of surviving outside the host.

Conversely, the RB is large (0.5–1.5 μm), highly metabolic and divides within host cell inclusions. The transitions between these forms are crucial for the pathogen's life cycle and its ability to cause disease.

A description of the light microscopy image of *C. psittaci* inclusion bodies is shown in Fig. 42.1.

Cultivation

Historically, *C. psittaci* has been cultured in chicken embryos by inoculating the yolk sac of 6–8-day-old embryos. This provides an auxiliary, nutritious environment that allows the bacteria to replicate within host cells. After several days of incubation, infected embryos either show symptoms of illness or die, and the egg yolk can be dissected for further analysis. Although effective, this technique is labour-intensive and, therefore, mainly restricted to large-scale research settings.

Recently, the cell culture system has become a preferred method for laboratory diagnosis. Commonly used cell lines include HeLa (human cervical epithelial cells), McCoy (fibroblasts), and L929 (mouse connective tissue). The bacteria are introduced into these cells, leading to the formation of characteristic intracellular inclusion bodies. All laboratory procedures involving live bacteria are carried out in a Biosafety Level 3 (BSL-3) containment facility. This is due to the zoonotic and aerosol-transmissible nature of the pathogen. The infection can occur through the inhalation of aerosolised particles.

Animal Models

Many animal models are developed to research *C. psittaci* pathogenesis, host immune responses, and transmission. Murine models, particularly BALB/c and C57BL/6 mice, are commonly used. They replicate key features of human psittacosis, such as pneumonia and systemic dissemination, and effectively stimulate the immune system, including strong Th1 responses. Guinea pigs are also valuable models for examining lung infections and mucosal processes.

A bovine model, specifically calves, is used to study respiratory tract infections caused by *C. psittaci*. The structure and function of the bovine lungs closely mimic those of humans. These enable accurate evaluation of lung pathology and functionality through non-invasive tests. Additionally, the large size of the calves offers opportunities for frequent sampling, which aids in understanding disease progression and tracking immune responses.

Avian models, particularly pigeons, parrots, and turkeys, are utilised to investigate natural infections, host conversion, and zoonotic transmission.

Transmission Cycle

Birds act as reservoirs and primary hosts, while humans typically become accidental, terminal hosts. The disease's persistence in nature is mainly due to the widespread presence of *C. psittaci* in various bird populations, covering over 460 species across at least 30 avian orders. Common carriers include popular pet birds such as parrots, parakeets, cockatiels, and budgerigars, as well as poultry like turkeys, ducks, chickens, geese, and pigeons. A key aspect of this cycle is that many infected birds remain asymptomatic carriers.

Infected birds shed *C. psittaci* via various routes, including their faeces, respiratory secretions (e.g., nasal discharge), and feather dust. This shedding is intermittent and is often worsened by stressors like transportation, overcrowding, or reproductive activities. The bacteria are highly resilient in the environment. The EBs are contagious and survive in dried form for several weeks when protected within organic debris such as cage litter or dried faeces.

Bird-to-bird transmission usually happens through the inhalation of infectious particles such as dried droppings, respiratory secretions, or feather dust that become aerosolised. Eating contaminated material, including carcasses, can also help spread the disease among bird populations. Close contact, especially in crowded situations, further speeds up transmission within bird groups. The migration of wild birds usually spreads *C. psittaci* to new areas and vulnerable populations.

Human-to-human transmission of psittacosis is extremely rare, although isolated cases have been documented. The transmission cycle is presented in Fig. 42.2.

Pathogenesis and Pathology

After inhaling aerosolised elementary bodies (EBs), the bacteria target epithelial cells in the upper and lower respiratory tract linings. They attach to host cells via specific targets, such as heparan sulphate-like proteoglycans and bacterial adhesins. Subsequently, they are internalised by endocytosis into the host cells.

Pathogenesis

In the host cell, EBs rapidly differentiate into metabolically active, replicative-form reticulate bodies (RBs). This infection takes place within inclusion bodies that stay separate from lysosomes because *C. psittaci* actively prevents pha-

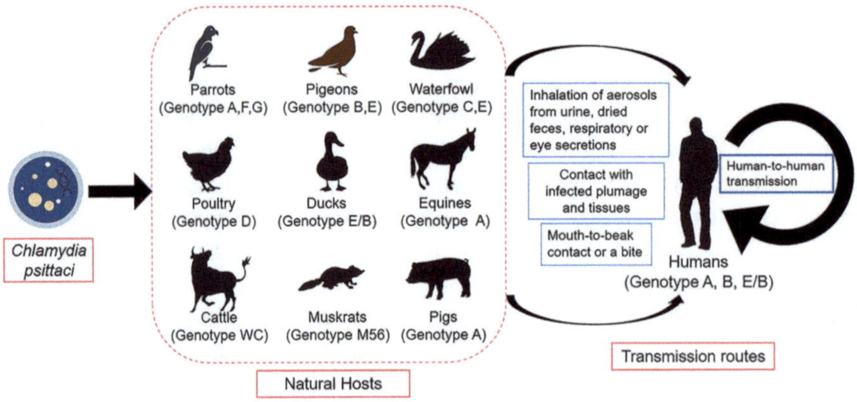

Fig. 42.2 The natural hosts and transmission routes of *Chlamydia psittaci.* (Image source: Wang J, Wang B, Xiao J, Chen Y, Wang C. *Chlamydia psittaci*: A zoonotic pathogen causing avian chlamydiosis and psittacosis. Virulence. 2024 Dec 31;15(1):2428411. Permitted for reproduction under CC BY license)

golysosomal fusion. RBs undergo binary fission and then divide into EBs within the inclusion bodies. After replication, EBs are released by the host cell through lysis and infect neighbouring cells, with the infection persisting.

The Type III secretion system (T3SS) is essential during the transition process and delivers effector proteins, such as IncA and IncB. These Inc. proteins interact with host molecules like G3BP1 and Din, affecting processes such as the inhibition of apoptosis and the movement of intracellular components. The cytoskeletal system, composed of actin, tubulin, kinesin, and dynein, maintains the structural integrity of the inclusion and supports bacterial growth.

Ultimately, the separation of reticulate bodies from elementary bodies results in the lysis of host cells and the release of infectious particles. The mechanisms behind lysis are still not fully understood. However, it is believed that effector proteins from *Chlamydia* may trigger apoptosis, which would help spread the bacteria and continue the infection cycle.

Besides local replication, *C. psittaci* can systemically disseminate through infected mononuclear phagocytes (such as macrophages) to extra-pulmonary sites like the liver, spleen, heart, and central nervous system. This systemic spread results in multi-organ involvement in severe cases of psittacosis.

Pathology

The direct cytopathic effect and the host immune response damage the host cells. Inflammatory cell infiltration, cytokine release (especially TNF-α, IL-6, and IFN-γ), and the production of reactive oxygen species contribute to tissue damage. Granulomatous inflammation, necrosis, and vascular thrombosis may occur in the affected tissue. Histopathological findings may include alveolar damage, lymphoid hyperplasia, hepatic cell necrosis, hepatomegaly, splenomegaly, as well as meningoencephalitis or myocarditis in severe systemic infections.

Chlamydia, when exposed to stressful conditions, enters an "endurance phase," during which the organism remains viable but non-cultivable. This phase enables the bacteria to evade the host's immune response. The analysis of gene expression during this endurance phase shows a reduction in key chlamydial functions, particularly those related to IFN-γ, which decreases the expression of apoptosis-related chlamydia protein associating with death domains (CADD) genes.

An overview of the pathogenesis of *Chlamydia* infection is presented in Fig. 42.3.

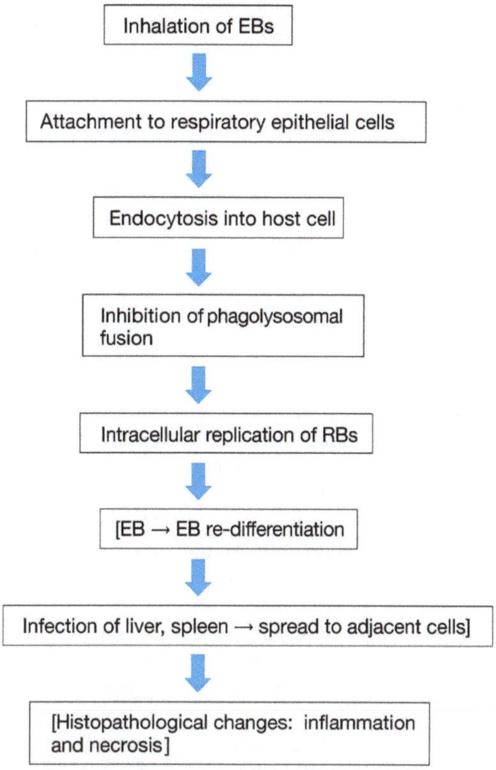

Fig. 42.3 Flow chart showing pathogenesis of *Chlamydia psittaci*. (Image source: Author's personal collections)

Immunology

Both innate and adaptive immunity contribute to the immune response.

Innate Immunity

The innate immune system plays a vital role in the early detection of *C. psittaci* infections. *C. psittaci* is more effective at managing early inflammatory reactions than other chlamydial species. Genes like *Inclusion membrane protein A (IncA)*, *ftsW*, *groEL*, and *cpaf* are expressed at high levels, aiding the bacteria's spread. The complement system is also activated, mainly through the C3a/C3aR pathway. This enhances host defence by increasing inflammation, activating B cells, and supporting T-cell activity. Experiments using a mouse model suggest that the absence of C3aR weakens the immune response to *C. psittaci*.

Adaptive Immunity

The adaptive immune response involves $CD4^+$ T-helper (TH1) cells, macrophages, and cytotoxic $CD8^+$ T cells, which activate IFN-γ and IL-2. $CD8^+$ T cells directly kill infected cells, while B cells produce antibodies against surface proteins like MOMP, aiding in opsonisation and neutralisation of the bacteria. However, *C. psittaci* can evade apoptosis by altering cytokine secretion and entering a metabolically inactive state observed in both animals and humans.

Dendritic cells (DCs) serve as important antigen-presenting cells, playing a crucial role in both innate and adaptive immune responses. When infected, DCS utilise autophagosomal and endovacuolar enzymes to break down chlamydial elements, facilitating antigen presentation through major histocompatibility complex (MHC) class I and triggering the activation of cytotoxic $CD8^+$T cells. This process is significant because *C. psittaci* hosts cell inclusions and avoids lysosomal fusion. Additionally, the release of IFN-γ from T cells expressing MHC promotes bacterial spread and autophagy in affected epithelial cells.

Pattern recognition receptors (PRRs), such as Toll-like receptors (TLR2 and TLR4), recognise *C. psittaci* cell components, including lipopolysaccharide (LPS) molecules and heat shock proteins. They activate signalling pathways, such as NF-κB, leading to the production of pro-inflammatory cytokines, including TNF-α, IL-1β, and IFN-γ. IFN plays an important role in activating infected cells, enhancing antigen presentation, and killing bacteria.

Infection in Humans

The incubation period for psittacosis usually ranges from 5 to 14 days. *C. psittaci* typically presents as an acute fever illness, like the flu. Common symptoms include fever, chills, headaches, myalgia, and a dry cough. These symptoms often appear suddenly and vary in severity depending on the individual's health and immune status. In more severe cases, the infection may affect multiple organs. Complications include atypical pneumonia, endocarditis, hepatitis, and even encephalitis.

Although rare, *C. psittaci* infections can have atypical manifestations. These include reactive arthritis, meningitis, myocarditis, and neurological conditions such as Guillain-Barré syndrome. In pregnant women, psittacosis has been linked to miscarriage and other unwanted pregnancy outcomes, although such cases are unusual.

People in high-risk groups include poultry workers, veterinarians, bird enthusiasts, and staff in pet shops. Seventy percent of reports are linked to contact with birds. Other species, such as *C. gallinacea* and *C. abortus*, can also transmit the disease to humans.

While human-to-human transmission is rare, it has been documented. The outbreaks in Sweden and China confirmed secondary and tertiary spread, as well as direct transmission. The incubation period can range from 5 to 14 days, sometimes extending up to 39 days. Symptoms often resemble community-acquired pneumonia (CAP) and may include fever, dry cough, headache, fatigue, chills, and muscle pain. Less common symptoms may involve light sensitivity, throat inflammation, diarrhoea, and neurological issues such as encephalitis and myelitis. Additionally, a facial rash called '*Horder spots*' can be observed.

Infection in Animals

C. psittaci infects various mammals and birds, including sheep, goats, and domestic cats. In livestock, symptoms such as conjunctivitis, pneumonia, and miscarriage raise concerns about zoonotic risks. *C. psittaci* infections in non-human mammals are rare.

Table 42.2 Difference between human and animal *Chlamydia psittaci* infections

Aspect	Humans	Birds
Causative agent	*Chlamydia psittaci*	*Chlamydia psittaci*
Common symptoms	Fever, chills, dry cough, myalgia, headache	Conjunctivitis, nasal discharge, diarrhoea
Severe manifestations	Pneumonia, hepatitis, endocarditis, encephalitis	Wasting, green urates, respiratory distress
Incubation period	5–14 days	Days to weeks (variable)
Diagnostic findings	Leukopenia, high liver enzymes, X-ray changes	Positive PCR, serology, histopathology
Transmission	Inhalation of aerosolised bird droppings or secretions	Direct contact, aerosols, faeces
At-risk populations	Bird handlers, veterinarians, poultry workers	All avian species, especially psittacines
Complications	Encephalitis, myocarditis, miscarriage (rare)	Death, chronic carriers
Case fatality (untreated)	Up to 15%	Can be high in untreated or stressed birds

C. psittaci infection in birds is referred to as avian chlamydiosis or ornithosis. Symptoms vary from mild to severe systemic disease. Clinical presentations in birds include lethargy, anorexia, ruffled feathers, conjunctivitis, diarrhoea, and respiratory distress. In some cases, infected birds may experience dehydration, green droppings, or sudden death. Latent infections are common and can be reactivated by various stressors. The condition is likely underdiagnosed due to its vague symptoms and limited diagnostic tools. There have been reports of outbreaks at major aviary centres. They can still cause a variety of symptoms.

Differences between infections in humans and animals are listed in Table 42.2.

Table 42.3 Global epidemiology of psittacosis

Region/country	Human prevalence	Bird prevalence	Key occupational groups at risk	Epidemiological patterns
United States	~10–50 confirmed cases/year (likely underreported)	High in pet birds (esp. parrots, cockatiels); poultry also affected	Pet shop workers, poultry workers, veterinarians	Psittacosis is nationally notifiable; past outbreaks linked to poultry processing and pet bird exposure
Europe (e.g. Germany, UK, Netherlands)	Sporadic to small outbreaks; ~100 cases/year in Germany	Wild pigeons, ducks, captive psittacines	Bird owners, zoo staff, abattoir workers	Multiple outbreaks; seroprevalence studies show underreporting; human cases often misdiagnosed
Australia	~20–50 cases/year (underdiagnosed)	Common in wild parrots and pigeons	Farmers, bird handlers	Notable for localised outbreaks in New South Wales and Victoria
China	Increasing detection in poultry & wild birds	Seroprevalence high in ducks, chickens, pigeons	Poultry farm workers, bird sellers	Surveillance improving; some zoonotic transmission clusters observed
Japan	Rare, sporadic cases	Detected in domestic and imported birds	Bird breeders, pet industry	Case reports mainly from urban bird owners and breeders
Middle East	Limited data; occasional zoonotic cases	Moderate in feral birds	Poultry handlers, pigeon racers	Underreporting likely due to limited diagnostic capacity
South America (e.g. Brazil, Argentina)	Rarely reported human cases	High seroprevalence in wild birds	Pet traders, wildlife traffickers	Illegal bird trade contributes to risk
Africa	Very limited data	Unknown; suspected presence in wild birds	Live bird market workers	Diagnostics limited; potential public health blind spot
India	Few documented outbreaks; likely underreported	Moderate-to-high in pigeons, parrots	Bird market workers, poultry staff	No national reporting system; misdiagnosis as atypical pneumonia common

Epidemiology and Public Health

Human infection, a zoonotic event, primarily occurs through the inhalation of aerosolised *C. psittaci* particles. This takes place when dried droppings, respiratory secretions, or feather dust from infected birds become airborne. Less common routes of human acquisition include direct mouth-to-beak contact with an infected bird, bird bites, or handling plumage or tissues from infected birds. Certain individuals face a higher risk of exposure, including pet bird owners, particularly those with psittacine birds, pet shop employees, poultry workers, veterinarians, zoo workers, and wildlife handlers.

Human-to-human transmission of psittacosis is extremely rare, although isolated cases have been documented. This usually involves exposure to infectious, aerosolised droplets from a severely ill patient experiencing forceful coughing. It is not regarded as a significant public health issue for widespread transmission. Several factors influence the psittacosis transmission cycle.

Psittacosis is responsible for outbreaks that are international, localised, and sporadic in nature

(Table 42.3). Actual occurrences are often underestimated as they are underdiagnosed or misclassified as other types of respiratory diseases. In the United States, psittacosis is classified as a notifiable disease according to the centers for disease control and prevention (CDC). The highest percentage of outbreaks occurs in Europe at 40.3%, followed by Asia at 29.7%, North America at 25.8%, and Oceania at 24.7%. Variations in these rates may stem from differences in surveillance systems, the necessity for disease alerts, and the quality of public health infrastructure in the regions.

Globally, the outbreak timing is inconsistent. However, sporadic spikes occur in Europe during winter, possibly linked to factors such as bird migration, which increases shedding and transmission of *Chlamydia*.

Diagnosis

C. psittaci is classified as a BSL-3 pathogen, requiring containment facilities.

Specimens

These include tissue biopsies and respiratory samples such as sputum, bronchoalveolar lavage, and nasopharyngeal swabs. For zoonotic investigations, cloacal swabs or bird faeces are also used. Ensuring the integrity of samples through proper storage and transportation is essential, especially for molecular analyses.

Microscopy

Microscopic tests, such as Giemsa or Gimenez, demonstrate basophilic, intracytoplasmic inclusions within epithelial cells or macrophages in respiratory tract samples. However, this method has low sensitivity and specificity. It cannot reliably distinguish *C. psittaci* from other chlamydial species or intracellular pathogens. It is now mostly outdated as a standalone diagnostic tool; nonetheless, it may still be employed for initial assessment in resource-limited settings.

In Vitro Culture

Culture is the definitive standard for diagnosis; however, it is rarely achieved due to significant biosafety risks. The organism is typically isolated in cell cultures such as McCoy cells or embryonated bird eggs, where it produces functional cytopathic effects. Culture is useful for antimicrobial susceptibility testing and full genomic analysis, but it is slow, takes days to weeks, and is technically demanding.

Given these challenges, in vitro culture for *C. psittaci* is not a common diagnostic method for psittacosis in most clinical or veterinary laboratories. It is mainly carried out in specialised reference laboratories for confirming difficult cases, research purposes, or epidemiological investigations.

Serodiagnosis

Complement fixation (CF), microimmunofluorescence (MIF), and microscopic agglutination tests (MAT) are frequently used serological tests. A fourfold increase in antibody titres between acute and convalescent samples is diagnostic. However, these tests are affected by cross-reactivity with different *Chlamydia* spp., particularly *C. pneumoniae*, which can lead to false positives. MIF is more specific than CF but requires skilled personnel and access to paired sera. Nonetheless, MAT is considered obsolete due to its low sensitivity and tendency for cross-reactivity. Serodiagnosis is often utilised for retrospective confirmation rather than early diagnosis.

ELISA is widely utilised to detect *C. psittaci* immunoglobulin M (IgM) and immunoglobulin G (IgG). ELISA is suitable for mass screening, although some cross-reactivity with different species can occur. Lateral flow assays have been developed but currently lack validation.

Molecular Diagnosis

Polymerase Chain Reaction (PCR)

Is a popular technique for detecting *C. psittaci* due to its high sensitivity, specificity, and rapid turnaround time. Conventional, nested, and real-time PCR assays detect *omp*A (outer membrane protein A), 16S rRNA, and *groEL*. Real-time PCR offers quantification and faster results, while nested PCR is advantageous for samples with low bacterial loads. Sequencing PCR products enables genotyping and tracking epidemiological outbreaks. PCR can be performed on various medical samples, including respiratory swabs, tissues, and blood.

Loop-Mediated Isothermal Amplification (LAMP)

Is a nucleic acid amplification technique designed for use in resource-limited settings. LAMP assays are developed to target specific genes such as ompA, groEL, and 16S rRNA, ensuring high specificity and minimising cross-reactivity with other species. These analyses are highly sensitive, capable of detecting as few as 10–100 DNA copies, and provide rapid results under isothermal conditions.

Next-Generation Sequencing (NGS)

Allows for precise identification of C. *psittaci* in mixed or unusual infections. It plays a crucial role in outbreak investigations and offers detailed strain-level analysis. Furthermore, it can also identify antimicrobial resistance patterns.

Techniques such as *ompA* gene sequencing, multilocus sequence typing (MLST), and single nucleotide polymorphism (SNP) analysis are vital for differentiating strains and understanding transmission routes. This is essential in public health research, allowing the detection of outbreaks and tracking the development of resistance in pathogens.

An overview of the diagnosis is presented in Table 42.4.

Treatment

C. psittaci infection is treated with tetracyclines, especially doxycycline. The usual dose is 100 mg twice daily for 10–14 days. In severe cases such as pneumonia or endocarditis, the duration may be extended to 21 days or longer to ensure complete eradication and reduce the risk of recurrence. Doxycycline is highly effective because of its ability to penetrate cells well. Patients generally see significant clinical improvement within 48 to 72 h after beginning treatment.

For those unable to tolerate tetracyclines, macrolides are suitable alternatives. In such cases, azithromycin is frequently prescribed. The recommended dose is 500 mg on the first day, followed by 250 mg daily for the next 4 days, or alternatively, 500 mg daily for 3 days. Macrolides, such as erythromycin and clarithromycin, have also demonstrated effectiveness, although they are not commonly prescribed. Long-term antibiotic treatment is recommended for treatment of persistent or relapsing infections, especially among immunocompromised individuals or those with pre-existing lung conditions. Monitoring for symptom recurrence and radiological changes is essential for evaluating treatment success.

Although antimicrobial resistance in *C. psittaci* is uncommon, there are reports of decreasing susceptibility, especially in birds exposed to antibiotics. Therefore, antimicrobial stewardship and routine surveillance in both human and veterinary settings are vital.

In veterinary medicine, similar antibiotic approaches are used for infected birds, with doxycycline typically given through drinking water (e.g., 25–50 mg/kg/day) or directly in the feed. Treatment usually lasts for at least 45 days to prevent recrudescence in asymptomatic birds.

Prevention and Control

Prevention measures for *C. psittaci* infection depend on reducing the risk of transmission from birds to humans, especially in high-risk environ-

Table 42.4 Laboratory diagnosis of psittacosis

Tests	Types	Sensitivity	Specificity	Time to result	Remarks
Microscopy	Giemsa stain Gimenez stain	Low	Low	Same day	Inexpensive; poor diagnostic value; used mainly for preliminary screening
In vitro culture	Cell lines Eggs	high	high	7–21 days	Gold standard; allows further analysis; labour-intensive and risky. Not routinely used in diagnosis
Serodiagnosis	Complement fixation (CF) Microimmunofluorescence (MIF)	Moderate	Moderate-to--high	5–10 days	Requires paired sera; cross-reactivity is common; good for retrospective diagnosis
Molecular diagnosis	Conventional PCR	high	high	1–2 days	highly sensitive; may require confirmation via sequencing
	Real-time PCR	Very high	Very high	Few hours to 1 day	Fast, specific, and quantitative; widely used in modern diagnostics
	Nested PCR	Very high	high	1–2 days	Extremely sensitive; useful for low-load samples; higher contamination risk
	Amplicon sequencing	N/A	Very high	2–5 days	Enables genotyping and source tracking; requires PCR-positive sample

ments such as aviaries, pet stores, poultry farms, and veterinary clinics.

A vital step involves quarantining and examining imported birds, especially psittacines such as parrots, parakeets, and macaws. Imported birds are quarantined for a minimum of 30 to 45 days, during which they are observed for any signs of illness and may receive preventive antibiotics. This reduces the bacterial load, preventing contamination of unexposed birds. Routine examinations are essential for early diagnosis and management.

Environmental hygiene is equally important. Proper cleaning of bird cages, feeding areas, and water containers with suitable disinfectants such as quaternary ammonium compounds and bleach solutions greatly reduces environmental contamination. Cleaning should be carried out in a way that minimises the generation of dirt and aerosols.

The One Health approach, including public training and occupational health education for veterinarians, plays a vital role in long-term prevention. Pet owners, breeders, and aviary workers need to be educated about the symptoms of psittacosis, its modes of transmission, and key preventive measures. Raising awareness through these strategies is crucial to reducing the zoonotic risk of *Chlamydia psittaci*.

Case Study

A 41-year-old male zookeeper from a wildlife rehabilitation centre presented to the district hospital with a high-grade fever, chills, dry cough, severe headache, and myalgia. He reported fatigue, shortness of breath, and recent weight loss. There was no recent travel history or contact with individuals who were ill. His occupational history revealed close daily contact with rescued birds, especially parrots and pigeons, many of which were sick or stressed. Chest X-ray revealed bilateral patchy infiltrates consistent with atypical pneumonia. Laboratory tests showed leukopenia, elevated liver enzymes, and mildly raised inflammatory markers. Based on clinical suspicion of zoonotic pneumonia and occupational exposure to birds, a nasopharyngeal swab was sent for PCR testing, which confirmed *C. psittaci* DNA. Serology later demonstrated a fourfold increase in antibody titre, consistent with psittacosis. He was treated with oral doxycycline 100 mg twice daily for 14 days, leading to rapid symptomatic improvement.

1. What are the other causes of atypical pneumonia?
2. Write about the transmission, pathogenicity and laboratory diagnosis of *C.psittaci?*
3. What are the treatment and preventive measures of psittacosis?

Research Questions

1. What is the mechanism of pathogenesis and transmission of psittacosis?
2. How does the One Health approach help to prevent psittacosis?
3. What are the various diagnostic methods for the diagnosis of psittacosis?
4. What are the newer preventive strategies, including vaccines?

Further Readings

Balsamo G, Maxted AM, Midla JW, Murphy JM, Wohrle R, Edling TM, et al. Compendium of measures to control Chlamydia psittaci infection among humans (Psittacosis) and pet birds (avian chlamydiosis), 2017. J Avian Med Surg. 2017;31(3):262–82.

Beeckman DS, Vanrompay DC. Tyrosine phosphorylation of host and bacterial proteins during Chlamydia psittaci infection of epithelial cells. Front Cell Infect Microbiol. 2012;2:40.

Beeckman DS, Vanrompay DC. Zoonotic Chlamydophila psittaci infections from a clinical perspective. Clin Microbiol Infect. 2009;15(1):11–7.

De Gier B, Baan B, Hogerwerf L. Animal-to-human transmission of Psittacosis: a systematic review. One Health. 2018;6:27–33.

Gaede W, Reckling KF, Dresenkamp B, Kenklies S, Schubert E, Noack U, et al. Chlamydophila psittaci infections in humans during an outbreak of Psittacosis from poultry in Germany. Zoonoses Public Health. 2008;55(4):184–8.

Hogerwerf L, De Gier B, Baan B, Van Der Hoek W. Chlamydia psittaci (Psittacosis) as a cause of community-acquired pneumonia: a systematic review and meta-analysis. Epidemiol Infect. 2020;148:e139.

Kalmar ID, Dicxk V, Vanrompay D. Zoonotic infection with Chlamydia psittaci at chicken slaughterhouses. Emerg Infect Dis. 2014;20(1):110–3.

Matsui T, Nakashima K, Ohyama T, Kobayashi J, Arima Y, Kishimoto T, et al. An outbreak of Psittacosis in a bird park in Japan. Epidemiol Infect. 2008;136(4):492–5.

Mitchell SL, Wolff BJ, Thacker WL, Ciembor PG, Gregory CR, Everett KD, et al. Genotyping of Chlamydophila psittaci by real-time PCR and high-resolution melt analysis. J Clin Microbiol. 2009;47(1):175–81.

Szymanska-Czerwinska M, Niemczuk K, Mitura A, Jodełko A, Zaręba K, Gaweł J. Chlamydia psittaci in poultry—review of the current knowledge. J Vet Res. 2017;61(3):279–85.

Wolff BJ, Morrison SS, Winchell JM. Development of a real-time PCR assay for detection of chlamydia psittaci in human clinical specimens. J Clin Microbiol. 2015;53(7):2112–8.

Bartonellosis

43

Rakesh Singh and Mahalakshmi Kumaresan

Learning Objectives

1. To gain knowledge of various clinical conditions or disease manifestations and sequelae for bartonellosis
2. To describe the various modes of transmission and the importance of different vectors in the spread of infection
3. To review the current laboratory diagnostic tests available and their limitations
4. To discuss the treatment and preventive measures of bartonellosis

Introduction

Bartonellosis is a zoonotic disease caused by bacteria of the genus *Bartonella*, which infect a wide range of mammalian hosts. Within the family Bartonellaceae, the genus *Bartonella* comprises facultative intracellular Alphaproteobacteria. *Bartonella henselae*, *Bartonella quintana,* and *Bartonella bacilliformis* infect and cause Cat scratch disease (CSD), Trench fever, and Carrion's disease in humans, respectively. Emerging evidence suggests a link between *Bartonella* infections and chronic conditions such as autoimmune disorders and neuropsychiatric symptoms, emphasising the need for further research.

Bartonella spp. is transmitted primarily by arthropod vectors such as fleas, ticks, lice, and sand flies. The cat flea (*Ctenocephalides felis*), the most common flea species in central Europe, is host-opportunistic and thus a common source of infection. Potential vertical transmission and transmission through blood transfusions have been suggested, emphasising the need for continued vigilance in recognising all possible routes of infection. The bacterium has developed unique mechanisms to evade the host immune system, enabling it to persist for extended periods within erythrocytes and endothelial cells. The clinical symptoms of *Bartonella* infections in humans vary greatly. These range from asymptomatic carriage to systemic diseases, including endocarditis, neuroretinitis, and bacillary angiomatosis, especially in immunocompromised patients.

Despite advances in molecular diagnostics, *Bartonella* infections continue to be underdiagnosed because of their nonspecific clinical signs and limitations in conventional laboratory detection methods. Due to its zoonotic characteristics and widespread presence, *Bartonella* infections remains a significant public health issue.

History

The discovery of *Bartonella* spp. dates back thousands of years, demonstrating its long-standing influence on both human and animal health. DNA analysis of a 4,000-year-old human tooth from France offered the earliest evidence of *B. quintana* bacteraemia. *B. henselae* was found in the teeth of cats approximately 800 years ago, emphasising the ancient link between cats and this bacterium.

R. Singh (✉) · M. Kumaresan
Department of Microbiology, Jawaharlal Institute of Postgraduate Medical Education & Research (JIPMER), Pondicherry, India

S. C. Parija, A. Poddar (eds.), *Textbook of Bacterial Zoonoses*, Microbial Zoonoses,
https://doi.org/10.1007/978-981-95-6615-0_43

In 1531, the first medical description of bartonellosis was documented in pre-Columbian Peru, linked to the death of the Inca ruler Huayna Capac. During Napoleon's 1812 invasion of Russia, *B. quintana*, transmitted by lice, played a significant role in historical events by contributing to widespread illness and death among French soldiers, and notably also during World War I in 1916. Its impact on troop health and combat readiness highlights the importance of vector-borne diseases in shaping military and historical outcomes.

Oroya fever was first reported in 1870 when a railway line was being constructed in Peru from Lima to Oroya. It resulted in an outbreak that killed thousands of workers, and the disease was named after the place of the outbreak. In 1885, Daniel Alcides Carrion, a medical student, demonstrated *B. bacilliformis*, as the cause of Oroya fever, by self-inoculating fluid from a verruga lesion. He developed Oroya fever and meticulously documented his disease progression, which ultimately led to his death. Consequently, Oroya fever is also known as Carrion's disease in honour of the medical student who sacrificed his life. In 1889, Henri Parinaud identified an association between *B. henselae* and oculoglandular syndrome. Alberto Barton first observed the causative bacterium of Carrion's disease during a Peruvian outbreak in 1905. Subsequent confirmation by researchers at Harvard led to its classification and naming as *B. bacilliformis*.

A breakthrough occurred in 1990 when *B. henselae* was identified through DNA technology in an AIDS patient, marking the first time a pathogen was detected directly from tissue without culture. Subsequently, *Bartonella* spp. was linked to prolonged bacteraemia in cats, establishing them as major reservoirs.

Taxonomy

The genus *Bartonella* belongs to the phylum Pseudomonadota (formerly Proteobacteria), class Alphaproteobacteria, order Hyphomicrobiales, and family Bartonellaceae. It is the only genus within the family Bartonellaceae.

There are currently over 40 recognised species within the genus, with *B. bacilliformis* serving as the type species. Several species, including *B. henselae*, *B. quintana*, and *B. bacilliformis*, are significant human pathogens. *Bartonella elizabethae*, *Bartonella vinsonii*, *Bartonella koehlerae*, and *Bartonella clarridgeiae* are emerging zoonotic pathogens.

Genomics and Proteomics

Bartonella spp. genomes generally range from 1.4 to 2.6 megabases (Mb). The genomes have highly dynamic genomic islands (GIs) and regions of high plasticity. These segments of DNA are often acquired through horizontal gene transfer events involving bacteriophages or plasmids. These are responsible for the bacteria's adaptability to various hosts and their capacity to cause disease. Within these plastic regions, crucial virulence factors such as adhesins (e.g., *Bartonella* adhesin A (BadA), in *B. henselae*) are present. Adhesins facilitate attachment to host cells, and genes related to secretion systems, notably Type IV secretion systems (T4SS), like the VirB/D4 and Trw systems. The variation in the content and presence of these GIs across different *Bartonella* spp. contributes to distinct host specificities and diverse clinical manifestations.

B. koehlerae is a new species isolated from cats and has been associated with human endocarditis. Researchers used comparative genomic hybridisation (CGH) against a *B. henselae* microarray for this species. Results indicated that *B. koehlerae* has a genome size of approximately 1.7–1.8 Mb, slightly smaller than that of *B. henselae*. Notably, *B. koehlerae* lacks sequences in genomic islands and prophages, similar to those found in *B. quintana*. Variability was also seen in the chromosome II-like region, suggesting independent gene deletions rather than a uniform gene loss. This shows that host specificity is not solely determined by genes in genomic islands.

Simultaneously, studies on *Bartonella* spp. in rodents and bats from Mexico uncovered new genomic diversity. *Bartonella* strains were isolated from the rodent (*Peromyscus yucatanicus*) and bats (*Desmodus rotundus* and *Pteronotus parnellii*), and their genomes were sequenced. The rodent-associated strain is closely related to

B. vinsonii. Conversely, the two bat-associated strains formed a distinct new clade positioned between lineages 3 and 4, potentially ancestral to rodent-associated *Bartonella* of lineage 4. These three genomes exhibited less than 95% similarity (OrthoANI) to known *Bartonella* spp., indicating they are likely novel species.

Furthermore, the findings suggest a revision of the *B. vinsonii* complex, recommending that subspecies be recognised as full species. The absence of the Vbh/TraG conjugation system in the bat-associated strains, which are part of the T4SS system, supports their classification into a newly proposed *Bartonella* lineage, named lineage 5 (L5).

Morphology

Bartonella spp. primarily exhibit a bacillary morphology, although they show pleomorphism. These bacteria are relatively small and rod-shaped, typically measuring between 1.0 and 1.7 micrometre (μm) in length and between 0.3 and 0.6 μm in width.

Bartonella spp. are Gram-negative bacteria, but they stain weakly. Therefore, specialised silver-impregnated stains like the Warthin-Starry stain are used to visualise these bacteria more clearly in tissue samples.

Motility within the *Bartonella* genus varies. Many species, including notable ones like *B. henselae* and *B. quintana*, are considered non-flagellated. Despite this, they can execute twitching motility, a type of movement facilitated by pili. In contrast, other species, such as *B. bacilliformis, B. clarridgeiae*, and *B. schoenbuchensis*, possess flagella at one pole of the cell (polar flagella) (Fig 43.1).

Cultivation

Bartonella spp. is cultured on enriched blood-containing agar, usually 5–10% defibrinated sheep blood agar or chocolate agar plates. Alternatively, *Bartonella* alphaproteobacteria growth medium (BAPGM) is utilised. The medium is prepared with a base of insect cell culture media and supplemented with various components, including foetal bovine serum (FBS), hemin, yeast extract, L-glutamine, and pyruvate, along with additional amino acids, nicotinamide adenine dinucleotide (NAD), nicotinamide adenine dinucleotide phosphate (NADP), and adenosine triphosphate (ATP). The addition of blood, often in the form of lysed erythrocytes, to liquid media significantly enhances the growth of *Bartonella* spp.

The bacteria need a moist atmosphere enriched with 5% CO_2 to grow. The ideal temperature for most medically important *Bartonella* spp., such as *B. henselae* and *B. quintana*, usually ranges from 35°C to 37°C. However, some species, including *B. bacilliformis* and *B. clarridgeiae*, grow at slightly cooler temperatures, between 25°C and 30°C.

Animal Models

The clinical diversity and complexity of bartonellosis, combined with the bacterium's ability to adapt to various reservoir hosts, have made it difficult to establish a reliable, standardised animal model. Currently, no animal model fully satisfies the criteria of an ideal model.

Cats often serve as models for chronic bacteraemia but rarely show pathology, while dogs may develop disease, though with less predictable bacteraemia. Certain mouse strains, such as Swiss Webster mice, display pathological signs in various tissues; however, their low infection rates limit their use in therapeutic trials.

Transmission Cycle

The transmission cycle of *Bartonella* spp. is intricate, involving various mammalian hosts, arthropod vectors, and specific infective stages that facilitate persistence and dissemination.

Mammals, especially reservoir hosts like cats, rodents, and canines, act as primary carriers, often sustaining chronic or asymptomatic bacteraemia. Several *Bartonella* spp. mainly infect specific mammalian hosts and show strong host specificity. For instance, *B. henselae*, *B. clarridgeiae*, and *B. koehlerae* are frequently found

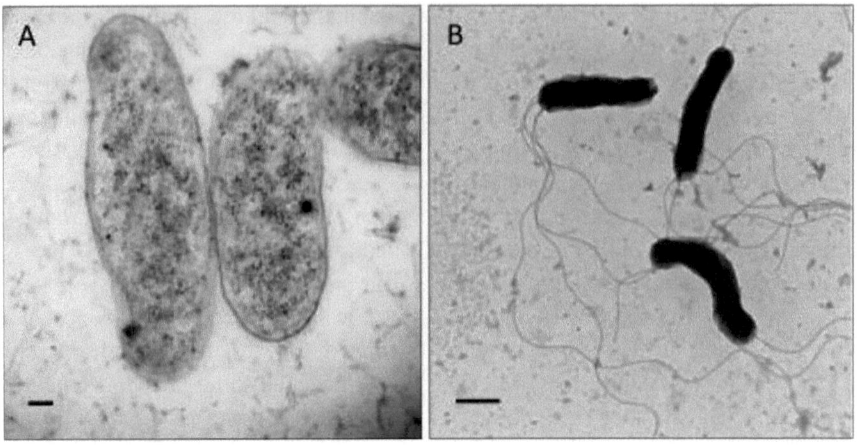

Fig. 43.1 Transmission electron micrographs showing rod-shaped morphology of *Bartonella bacilliformis* with polar flagella. Scale bars represent 100 nm in (**a**) and 500 nm in (**b**). (Image source: Minnick MF, Anderson BE, Lima A, Battisti JM, Lawyer PG, Birtles RJ. Oroya Fever and Verruga Peruana: Bartonelloses unique to South America. PLoS Neglected Tropical Diseases. 2014 Jul 17;8(7):e2919. Permitted for reproduction.)

in domestic cats, while *B. vinsonii* subsp. *berkhoffii* infects dogs and coyotes, and *B. alsatica* is linked to wild rabbits. Cats are notable for their ability to host multiple *Bartonella* spp. These are considered major reservoirs for human infections, particularly in flea-infested regions where seroprevalence is considerably higher.

Humans are generally incidental hosts and acquire infection through contact with infected vectors or reservoir hosts. The infective stage of *Bartonella* spp. involves its ability to localise within erythrocytes and endothelial cells, facilitating both intracellular survival and immune evasion.

Vectors such as cat fleas (*Ctenocephalides felis*), human body lice (*Pediculus humanus corporis*), sand flies (*Lutzomyia verrucarum*), and ticks have been linked to the transmission of *Bartonella* spp., depending on the specific species (Table 43.1). For example, *B. henselae* is mainly transmitted through the cat flea, while *B. bacilliformis* is spread by sand flies, and *B. quintana* is spread by human body lice.

Transmission occurs (i) primarily through vector bites, (ii) contamination of broken skin with vector faeces, or (iii) direct inoculation via scratches or bites from infected animals. After entering the host, *Bartonella* spp. undergo an initial endothelial phase, in which they infect vascular tissues, followed by a homeotropic phase characterised by colonisation of erythrocytes. This biphasic infection cycle allows *Bartonella* spp. to establish prolonged bacteraemia, facilitating uptake by blood-feeding arthropods and thus continuing the transmission cycle. Persistent infection in reservoir hosts ensures ongoing opportunities for vector acquisition and environmental dissemination.

An overview of the transmission cycle of *B. henselae* is shown in Fig. 43.2.

Pathogenesis and Pathology

Bartonella spp. utilise a variety of sophisticated strategies to invade host cells, evade immune responses, and establish chronic infections. These mechanisms allow the bacteria to persist within the host for long periods, often leading to recurrent or persistent disease manifestations.

Bartonella spp. has evolved the ability to invade and persist within both endothelial cells and erythrocytes, enabling it to evade immune detection and sustain long-term bacteraemia. The

Table 43.1 Overview of zoonotic *Bartonella* spp.

Bartonella spp.	Associated human disease(s)	Primary reservoir(s)	Typical vector(s)
Bartonella henselae	Cat scratch disease (CSD), bacillary angiomatosis, peliosis hepatis, endocarditis, neuroretinitis, various disseminated infections	Domestic cats especially kittens, wild felids	Cat fleas (*Ctenocephalides felis*)
Bartonella quintana	Trench fever, bacillary angiomatosis, chronic bacteraemia, endocarditis	Humans (main reservoir in historical context), possibly other primates	Human body louse (*Pediculus humanus corporis*)
Bartonella bacilliformis	Carrion's disease (Oroya fever, verruga peruana)	Humans	Sand flies (*Lutzomyia* spp.)
Bartonella elizabethae	Endocarditis	Rodents such as rats	Fleas, potentially other arthropods
Bartonella vinsonii spp. *arupensis*	Endocarditis, bacteraemia	Canids such as dogs, coyotes, foxes	Fleas, ticks
Bartonella koehlerae	Endocarditis	Domestic cats	Fleas
Bartonella clarridgeiae	Cat scratch disease-like illness, bacteraemia	Domestic cats	Cat fleas
Bartonella washoensis	Myocarditis, bacteraemia	Ground squirrels	Fleas
Bartonella rochalimae	Lymphadenopathy, bacteraemia	Canids such as dogs, possibly other wild animals	Fleas, potentially other arthropods
Bartonella tamiae	Lymphadenopathy	Rodents such as rats, shrews	Mites, fleas
Bartonella grahamii	Neuroretinitis, ocular infections	Rodents	Fleas, possibly ticks
Bartonella schoenbuchensis	Lymphadenopathy, bacteraemia	Ruminants such as deer	Keds (louse flies)
Bartonella doshiae	Endocarditis (rare)	Rodents	Fleas
Bartonella bovis	Bacteraemia (rare in humans)	Cattle	Ticks, biting flies
Bartonella chomelii	Endocarditis (rare)	Dogs, livestock	Fleas, ticks
Bartonella weissii	Bacteraemia (rare)	Rodents	Ticks, fleas

bacteria adhere to and invade the endothelial cells of blood vessels. This invasive interaction and the ability of intracellular survival, contributes to vascular abnormalities, such as angio-proliferative lesions, which leads to bacillary angiomatosis.

Unlike many other intracellular pathogens, *Bartonella* spp. enters and persists within red blood cells (RBCs) without triggering immune clearance. Once inside erythrocytes, *Bartonella* spp. avoids immune detection because mature RBCs lack major histocompatibility complex (MHC) molecules and do not support conventional immune responses. This strategy allows the bacteria to establish prolonged bacteraemia, ensuring their transmission via blood-feeding arthropods. *B. henselae* produces the autotransporter adhesin *Bartonella* adhesin A (BadA), which mediates bacterial attachment to host cells and promotes endothelial invasion. BadA also contributes to immune evasion by interfering complement activation.

Type IV secretion systems (T4SS) are specialised molecular machines that *Bartonella* spp. uses to inject virulence factors (effector proteins) into host cells, altering cellular processes to promote bacterial survival and pathogenesis. The VirB/D4 T4SS delivers *Bartonella* effector proteins (Beps)

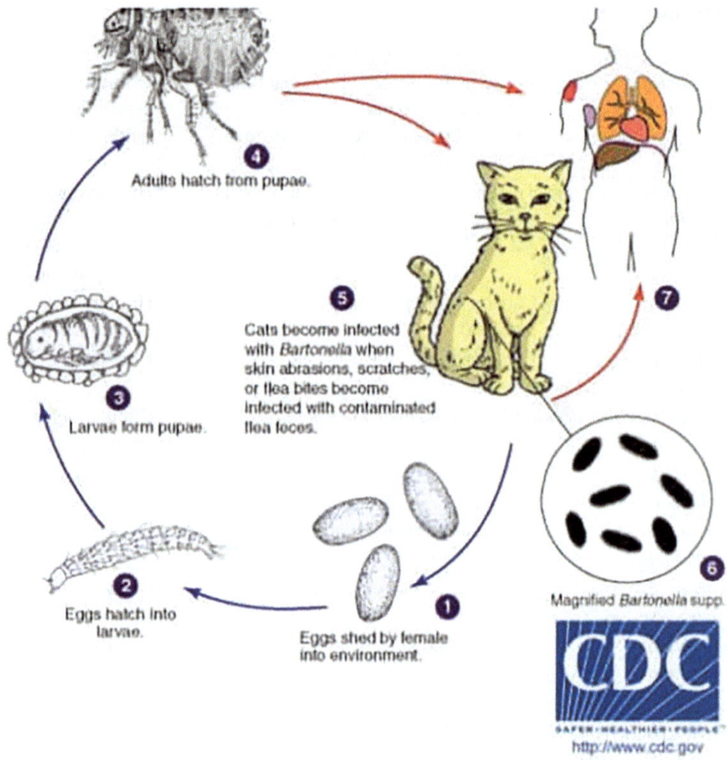

Fig. 43.2 Overview of *Bartonella henselae* transmission cycle. (Image Source: CDC (2015) https://www.cdc.gov/bartonella/index.html)

into host cells. These effector proteins manipulate host signalling pathways, suppress apoptosis, and support intracellular survival. The T4SS plays a vital role in mediating bacterial adhesion to and invasion of erythrocytes, ensuring efficient bacterial colonisation and sustained bacteraemia.

Bartonella infections are characterised by the formation of abnormal blood vessels, leading to angio-proliferative disorders such as bacillary angiomatosis and verruga peruana, seen in Carrion's disease. The bacteria promote excessive blood vessel growth by upregulating host vascular endothelial growth factor (VEGF) and hypoxia-inducible factor-1α (HIF-1α). BadA stimulates the secretion of pro-inflammatory cytokines, which contribute to endothelial cell proliferation and the development of vascular lesions. The hyperactivation of endothelial cells results in tumour-like blood vessel growth, contributing to the pathology of bacillary angiomatosis.

Immunology

The immune response to *Bartonella* spp. involves both innate and adaptive components, with the pathogen employing various evasion strategies. Initially, the innate immune system detects *Bartonella* spp. through pattern recognition receptors (PRRs) that recognise pathogen-associated molecular patterns (PAMPs), such as Lipopolysaccharides (LPS). Macrophages and dendritic cells phagocytose the bacteria and secrete cytokines such as IL-6, IL-8, and IL-10. The adaptive immune response, particularly Th1-type cellular immunity involving IFN-γ and IL-12, is crucial for controlling infection, while humoral responses also contribute via antibody production.

The host immune response to *Bartonella* infections involves a complex interaction between innate and adaptive immunity, often including both pro- and anti-inflammatory reactions. The host's immune status is a key factor in determining disease outcome. In individuals with a healthy immune system, a robust cellular response usually contains the infection, causing localised effects such as cat scratch disease. However, in immunocompromised hosts like those with HIV, ineffective immune responses result in persistent bacteraemia, allowing *B. henselae* to exploit host cells and promote pathological angiogenesis. This leads to severe vascular proliferative conditions, such as bacillary angiomatosis and peliosis hepatis. Importantly, the immune system not only tries to control the infection but also plays a role in disease development. This is because immune cells activated by the pathogen may release pro-angiogenic cytokines that encourage these abnormal vascular lesions. This dual function of the host immune response, as both a defence mechanism and a contributor to disease, emphasises its vital role in the progression and severity of *B. henselae* infections.

Bartonella spp. employs various immune evasion strategies, including suppressing apoptosis through specific effector proteins delivered via the T4SS. These prevent programmed cell death in infected host cells and thus prolong bacterial survival. *Bartonella* spp. also alters cytokine production, diminishing the host's ability to mount an effective immune response. It can suppress pro-inflammatory responses while fostering conditions that support bacterial persistence. Additionally, the bacterium modifies its surface proteins over time, making it more difficult for the immune system to recognise and eliminate it.

Due to its ability to persist in erythrocytes and endothelial cells, *Bartonella* spp. often causes chronic and relapsing infections.

Infections in Humans

Bartonella infections cause a diverse range of clinical signs and symptoms (Table 43.2).

Cat Scratch Disease (CSD)

B. henselae is the primary causative agent of Cat scratch disease (CSD), which typically presents with regional lymphadenopathy. It is observed after a scratch or bite from an infected cat or bite from an infected cat flea. It may manifest as a disseminated disease, affecting the liver, spleen, or central nervous system. Immunocompromised individuals may develop severe systemic complications, such as bacillary angiomatosis (BA), peliosis hepatis, and endocarditis. Bacillary angiomatosis appears as red to purple vascular skin lesions resembling Kaposi's sarcoma, while peliosis hepatis leads to blood-filled cavities in the liver.

Trench Fever

B. quintana causes Trench fever, a louse-borne infection that has historically affected soldiers in wartime conditions. It results in recurrent fever, headache, and severe bone pain, particularly in the tibia. It is also known as *5-day fever* or *quintan fever* (*quintan* means five in Spanish). The fever lasts for 4–5 days. Chronic bacteraemia is common among homeless individuals and HIV-positive patients, with some cases progressing to endocarditis. It is also linked to bacillary angiomatosis.

Carrion's Disease

B. bacilliformis is the causative agent of Carrion's disease, which is endemic to certain regions of South America. This disease has two distinct stages: Oroya fever, characterised by severe haemolytic anaemia, and verruga peruana, which presents with cutaneous angio-proliferative lesions and lasts for months. Oroya fever is the most deadly condition among all forms of bartonellosis and is associated with a 40% mortality rate.

Other Disease Manifestations

B. henselae, *B. quintana*, and other *Bartonella* spp. are associated with culture-negative endocarditis, which often requires specialised diag-

Table 43.2 Overview of Cat scratch disease, Trench fever, and Carrion's disease

Feature	Cat scratch disease	Trench fever	Carrion's disease
Causative agent	*Bartonella henselae*	*Bartonella quintana*	*Bartonella bacilliformis*
Mode of transmission	Cat scratches/bites, fleas	Body lice (*Pediculus humanus corporis*)	Sandfly (*Lutzomyia* spp.)
Incubation period	3–10 days (lesion); 1–3 weeks (systemic)	5–20 days	3–12 weeks (acute), months to years (chronic)
Initial lesion	Papule or pustule at scratch site	None	None (acute phase)
Fever	Low-grade, intermittent	Relapsing (5-day fever)	High fever (*Oroya fever*)
Lymphadenopathy	Prominent, tender	Rare	Rare in acute phase
Pain	Painful nodes, malaise	Shin, head, back pain	Bone/joint pain (chronic)
Rash	Sometimes (maculopapular/pustular)	Sometimes	Verruga peruana (chronic)
Other systemic symptoms	Fatigue, headache, anorexia	Myalgia, conjunctivitis, hepatosplenomegaly	Anaemia, jaundice (acute), bleeding nodules (chronic)
Neurologic complications	Rare (e.g., encephalopathy)	Rare	Rare
Duration of illness	Weeks to months	Weeks to months, may relapse	Acute: weeks; chronic: months to years
Commonly affected groups	Children, young adults	Homeless, immunocompromised	Endemic in Andes, all ages

nostic techniques such as serology or polymerase chain reaction (PCR). The condition may cause severe valvular destruction, often necessitating surgical intervention. This is because of their intracellular presence in erythrocytes and endothelial cells, leading to chronic infection. *B. elizabethae*, *B. vinsonii*, *B. koehlerae*, and *B. clarridgeiae* are emerging pathogens also linked to endocarditis, bacteraemia, and other systemic infections in humans.

Severe neurological complications have been reported, including *B. henselae* encephalitis. These comprise seizures, confusion, altered mental state, and neuroretinitis, characterised by optic disc swelling and macular star formation. Meningitis and myelitis are especially common in immunocompromised individuals or following significant systemic infection. A potential link between *Bartonella* infections and certain vascular tumours, such as haemangiosarcoma, has been suggested in animals, particularly dogs. The *Bartonella* infection has possible associations with Kaposi-like lesions in immunocompromised patients.

Infection in Animals

In animals, *Bartonella* infections are often asymptomatic, especially in naturally infected cats and dogs, which makes assessing their clinical significance more difficult.

In cats, *B. henselae* infection usually causes no illness, although experimental studies have occasionally observed mild signs such as fever, lymphadenopathy, lethargy, myalgia, and even behavioural or neurological abnormalities. Reproductive problems and myocarditis have also been reported, especially with flea exposure. Case reports have linked *B. henselae* to endocarditis, pyogranulomatous myocarditis, gingivostomatitis, uveitis, and potential urinary tract disease, although consistent causality remains uncertain.

In dogs, although clinical signs are rare, *Bartonella* spp. has been associated with a broad range of syndromes, including fever, pyogranulomatous lymphadenitis, granulomatous hepatitis, endocarditis, myocarditis, vasculitis, polyarthritis, peliosis hepatis, ocular signs and neurological signs.

In cattle, *B. bovis* is commonly present without obvious illness, although it has been linked to rare cases of endocarditis. Herd-level studies indicate minimal clinical impact, with no significant connection between bacteraemia and productivity measures. Horses experimentally infected with *B. henselae* or *B. bovis* exhibited mild injection-site reactions, regional lymphadenopathy, limb oedema, or urticaria.

Epidemiology and Public Health

Bartonellosis occurs worldwide, although its prevalence varies considerably depending on the distribution of vectors, interactions with reservoir hosts, environmental conditions, and socioeconomic factors. The global burden of *Bartonella* spp. is affected by climate, geography, human and animal migration, and the ecology of vectors. An overview of the global distribution of *Bartonella* spp. is shown in Table 43.3.

Table 43.3 Global distribution of zoonotic *Bartonella* spp.

Region	Prevalent *Bartonella* spp.	Primary animal reservoirs	Main vectors	Human infections	Remarks
North America	*Bartonella henselae*, *Bartonella quintana*, *Bartonella vinsonii*, *Bartonella elizabethae*	Cats, dogs, rodents	Fleas (Ctenocephalides felis), lice, ticks	Yes, especially *B. henselae* (Cat scratch disease) and *B. quintana* (Trench fever)	High prevalence in stray cats; *B. quintana* common among homeless individuals
South America	*Bartonella bacilliformis*, *Bartonella henselae*	Humans (for *B. bacilliformis*), cats	Sandflies (*Lutzomyia* spp.), fleas	Yes, *B. bacilliformis* (Carrion's disease) endemic in Andes	Carrion's disease restricted to high-altitude valleys in Peru, Ecuador, and Colombia
Europe	*Bartonella henselae*, *Bartonella quintana*, *Bartonella grahamii*	Cats, rodents, dogs	Fleas, lice, ticks	Yes, mostly *B. henselae* and *B. quintana*	*B. grahamii* isolated from rodents, possible emerging human pathogen
Africa	*Bartonella henselae*, *Bartonella quintana*, *Bartonella elizabethae*	Cats, monkeys, rodents	Fleas, lice	Yes, often underdiagnosed	Coinfections with HIV; vector exposure in poor sanitary conditions
Asia	*Bartonella henselae*, *Bartonella clarridgeiae*, *Bartonella bacilliformis* (sporadically), *Bartonella tribocorum*	Cats, rodents, dogs	Fleas, ticks	Yes, especially in Japan, China, Thailand	Limited surveillance in many countries; high seroprevalence in cats
Australia	*Bartonella henselae*, *Bartonella clarridgeiae*	Cats, dogs	Fleas	Yes, mostly mild cases of Cat Scratch Disease	Native marsupials may harbour novel *Bartonella* spp.

B. henselae is widely distributed and commonly found in domestic cats, which serve as natural reservoirs. Infection in cats is often asymptomatic and can last for months. Fleas, especially the cat flea, *C. felis*, are crucial in the intraspecies transmission among felines. These fleas occasionally cause infections in humans by contaminating scratches or wounds with flea faeces. High prevalence rates of *B. henselae* are reported in warm, humid regions where flea infestations are more frequent. Cat scratch disease is estimated to occur at a rate of 6.4 cases per 100,000 adults worldwide, with a higher incidence of 9.4 cases per 100,000 among children aged 5–9 years.

B. quintana, which causes Trench fever, is closely associated with *Pediculus humanus corporis* (human body lice) and is endemic in Russia and Eastern Europe. It is also common in populations with poor hygiene, overcrowding, and limited access to healthcare, such as homeless people, alcoholics, refugees, and those in conflict zones. This species is historically known for transmitting Trench fever among soldiers during World Wars I and II. *B. quintana* remains a public health issue in urban areas with poor sanitation. It has also been linked to chronic bacteraemia and endocarditis, especially in immunocompromised individuals.

Carrion's disease, caused by *B. bacilliformis,* is endemic in specific high-altitude regions of South America, particularly in Peru, Ecuador, and Colombia. It is transmitted by the female sandfly of the *Lutzomyia* spp., which thrives in the warm, humid valleys of the Andes Mountains. Carrion's disease is a significant concern in these areas due to its high mortality rate during the acute phase (Oroya fever) and its potential to cause outbreaks.

While fleas, lice, and sandflies are well-established vectors, recent studies suggest that ticks (*Ixodes*, *Rhipicephalus*, and *Dermacentor* spp.) may also harbour *Bartonella* spp. and could play a role in transmission. However, the role of ticks remains debated due to conflicting evidence regarding their ability to effectively transmit the bacteria to mammalian hosts. Some studies indicate that *Bartonella* DNA has been detected in *Ixodes* spp., and a subset of Lyme disease patients have tested positive for *Bartonella* spp.; however, causality remains unproven.

Laboratory Diagnosis

Microscopy

Microscopy serves as an adjunctive diagnostic tool for *Bartonella* infections. It is used for the examination of stained tissue samples and, to a lesser extent, blood smears.

The Warthin-Starry (WS) silver staining technique is frequently used on tissue biopsies taken from suspected lesions, such as those seen in bacillary angiomatosis, peliosis hepatis, or lymphadenitis related to Cat scratch disease. This method reveals *Bartonella* spp. bacilli as dark brown or black structures, often forming chains or clumps, which can be visualised within endothelial cells, macrophages, or necrotic areas. However, this staining lacks specificity, as other bacteria, such as *Helicobacter pylori* and *Legionella,* can also produce positive results.

Blood smear microscopy is another method, although its usefulness is limited, mainly for Oroya fever caused by *B. bacilliformis*. Blood smears used to diagnose bartonellosis show variable results, with up to 90% erythrocyte parasitisation linked to disease severity. Giemsa-stained peripheral blood smears may show *Rocha-Lima* inclusions. These are intra-erythrocytic cytoplasmic inclusions that represent *Bartonella* organisms within red blood cells. This technique requires skilled microscopists for accurate interpretation. It is generally not regarded as definitive and often needs confirmation through more sensitive molecular or serological tests.

In Vitro Cultures

In vitro culture remains a crucial definitive diagnostic method, particularly in reference or research contexts. Positive cultures are obtained in only about 50% of patients with endocarditis and 6% of those with CSD. *B. bacilliformis* is more easily isolated, with positive cultures in 71–83% of Oroya fever cases, using enriched media such as Columbia blood agar.

The long incubation period of 3–4 weeks, specific growth conditions (35–37°C, 5% CO_2, high humidity), and the organism's fastidious nature make culture labour-intensive and unsuitable for routine diagnostics.

Serodiagnosis

Detection of *Bartonella*-specific antibodies using enzyme-linked immunosorbent assay (ELISA) or indirect immunofluorescence assays (IFA) is helpful. However, serology can sometimes cause cross-reactivity with other infections, thereby limiting its specificity. Most laboratories measure separate Immunoglobulin M (IgM) and Immunoglobulin G (IgG) antibody levels for *B. henselae* and *B. quintana*.

Antibody levels usually become detectable within 1–2 weeks after symptoms start, with IgM peaking at 4 weeks and IgG peaking at 7–8 weeks. IgG levels may either persist or decline over time. IFA remains the most common test for diagnosing *B. henselae* infection, with a titer of 256 or higher is considered positive. Recent studies suggest that ELISA can achieve a sensitivity of 83% for IgM detection.

For *B. bacilliformis*, IFA has 81% sensitivity for acute bartonellosis, while a new ELISA test reported 94% positivity in chronic cases. A positive IgM titer of ≥ 1:16 indicates acute disease, whereas an IgG titer of ≥ 1:256, coupled with strong clinical suspicion, indicates current or past infection. IFA for anti-*Bartonella* antibodies is genus specific and cross-reactivity between *B. henselae* and *B. quintana* can reach up to 95% in IFA testing.

Molecular Diagnosis

Molecular techniques have considerably advanced the diagnosis of *Bartonella* spp. infections. PCR testing has demonstrated sensitivities ranging from 43% to 100%, depending on specimen type.

PCR is recommended for lymph node specimens because it is more sensitive than pathological analysis and is minimally invasive sample. Various PCR techniques, including ERIC-PCR and PCR-RFLP, are employed to identify *Bartonella* spp. Most PCR tests target the bacterial 16S ribosomal RNA or the variably expressed outer membrane protein (*vomp*) genes. PCR is highly specific; however, a negative result does not rule out infection due to potentially suboptimal sensitivity.

Blood, tissue, and bodily fluids are tested using PCR. PCR for bacterial endocarditis is recommended for whole blood, plasma, or serum samples. In cases of endocarditis, PCR detection of *Bartonella* deoxyribonucleic acid (DNA) in cardiac tissue has a variable sensitivity of 38% to 98%. Tissue samples are preferred for bacillary angiomatosis. However, routine PCR testing of blood or serum for CSD is discouraged due to lower sensitivity compared to serology.

Next-generation sequencing (NGS) of microbial cell-free DNA is an emerging tool for diagnosing *Bartonella* infections, particularly in cases with negative serology and PCR results. However, the clinical sensitivity and specificity of NGS for *Bartonella* spp. have not yet been fully established.

An overview of diagnostic approaches is presented in Table 43.4.

Treatment

Treatment of *Bartonella* infections depends on the severity of the disease and the patient's immune status.

Mild cases of CSD may resolve spontaneously, but azithromycin can shorten the duration of symptoms. Tablet azithromycin 500 mg on the first day and 250 mg from the second to the fifth day is recommended. In severe cases, either oral doxycycline 100 mg twice daily or oral rifampin 300 mg twice daily for 4–6 weeks may be considered. Oral doxycycline 200 mg daily for 4 weeks, combined with intravenous gentamicin at 3 mg/

Table 43.4 Diagnosis of *Bartonella* infections

Method	Sample type	Target	Remarks
Microscopy Warthin-Starry (WS) silver stain Giemsa stain Histopathology	Blood smear, tissue biopsy	Whole bacilli, cytoplasmic inclusions, tissue changes	WS stain detects clumps/chains of bacilli but lacks specificity; Giemsa stain shows erythrocyte inclusions (Rocha-Lima) in *Bartonella. bacilliformis*; Histopathology reveals syndrome-specific patterns
In vitro culture Columbia blood agar bartonella alphaproteobacteria growth medium (BAPGM)	Blood, tissue	Viable *Bartonella* organisms	Not routinely used in diagnosis for fastidious nature
Serodiagnosis ELISA indirect immunofluorescence assay (IFA)	Serum	IgM and IgG antibodies against *Bartonella* spp.	IgM ≥1:16 (acute); IgG ≥1:256 (current/past infection) Cross-reactivity up to 95% between *B. henselae* and *Bartonella quintana* IFA most used; IgM peaks at 4 weeks, IgG at 7–8 weeks
Molecular diagnosis PCR (standard, ERIC-PCR, PCR-RFLP) DNA microarrays Next-generation sequencing	Blood, tissue, serum, FFPE, body fluids	*Bartonella* DNA (16S rRNA, *vomp* genes)	High specificity; variable sensitivity (43–100%) depending on specimen Fresh tissue is best; blood PCR is discouraged for CSD PCR preferred for lymph nodes (CSD) and cardiac tissue (endocarditis) NGS is promising but costly.

kg for 2 weeks, is used to treat Trench fever. Prolonged therapy may be required for persistent bacteraemia. Bacillary angiomatosis and peliosis hepatis may be treated with oral erythromycin 500 mg, administered four times daily, for 3–4 months.

Ciprofloxacin 500 mg twice daily for 10–14 days or chloramphenicol 50 mg/kg/day for the first 3 days, followed by 25 mg/kg/day for 14 days, is recommended for the treatment of Carrion's disease during the acute phase. Treatment is especially necessary in severe cases with anaemia and during the chronic phase. Azithromycin 500 mg orally once daily for 7 days or rifampin (10 mg/kg/day) is commonly used to treat cutaneous lesions. Combination therapy with doxycycline and rifampin is preferred, particularly in immunocompromised patients, for bacillary angiomatosis. Treatment typically lasts at least 3 months to prevent relapse.

Bartonella-associated endocarditis requires extended antibiotic treatment. The suggested regimen includes oral doxycycline 100 mg for 6 weeks and intravenous gentamicin at a dose of 3 mg/kg/day for 14 days in immunocompromised patients. Persistent *Bartonella* infections might need longer therapy or combination regimens, including macrolides, fluoroquinolones, or beta-lactams, depending on susceptibility patterns.

Prevention and Control

Veterinarians, animal handlers, and pet owners are at increased risk of exposure to *Bartonella* spp. Enhanced surveillance and diagnostic testing are essential for understanding the complete clinical range of *Bartonella* infections and preventing future outbreaks.

Preventing *Bartonella* infections mainly depends on effective flea control, proper pet management, and good personal hygiene practices. Avoiding scratches or bites from cats, managing flea infestations, and maintaining hygiene in high-risk groups help lower the risk of transmission.

Dogs should be regularly treated with approved oral or topical flea and tick preventives such as fipronil, methoprene, imidacloprid, or permethrin. However, permethrin is never used on cats due to its toxicity. Immunocompromised individuals are particularly vulnerable and should avoid close contact with kittens or free-roaming cats. Additionally, rough play with cats should be discouraged to prevent scratches, which are a known route of transmission. Any scratches sustained should be promptly washed with soap and water, and hands should always be cleaned after handling cats.

There is currently no approved or available protective vaccine for *Bartonella* infections in humans or animals. The flagellin protein is a potential vaccine candidate.

Case Study

A 31-year-old male veterinary doctor presented to the medicine outpatient department of a hospital with complaints of neck swelling, which had been present for 1 month, intermittent high-grade fever for 1 month, and generalised weakness for 1 week. He reported a history of a cat scratch on the right elbow and neck that occurred while handling a cat 2 months ago. On examination, multiple, firm, non-tender, immobile lymphadenopathy was noted in the right axillary and right posterior cervical regions, but without discharging sinuses.

1. How do you confirm the diagnosis?
2. What is the pathogenesis of Cat scratch disease?
3. What are the preventive measures for Cat scratch disease?

Research Questions

1. What is the current understanding of the pathophysiology and immunology of *Bartonella* infections, particularly regarding their neuropsychiatric presentation?
2. What are the epidemiological and clinical studies being conducted on emerging zoonotic *Bartonella* pathogens such as *B. elizabethae, B. vinsonii, B. koehlerae,* and *B. clarridgeiae*?
3. Are there ongoing efforts or developments towards creating human or animal vaccines for bartonellosis?

Declaration During the preparation of this chapter, we have used Quillbot to improve the readability of the text. After using, we reviewed and edited the content as needed and take full responsibility for the content of the publication.

Further Readings

Bartonella infections. CDC Yellow Book 2024. Accessed 26 Apr 2025. Available from: https://wwwnc.cdc.gov/travel/yellowbook/2024/infections-diseases/bartonella-infections.

Bullard RL, Olsen EL, Cheslock MA, Embers ME. Evaluation of the available animal models for Bartonella infections. One Health. 2024;18:100665.

Cat Scratch Disease and Other Zoonotic Bartonella Infections. Accessed 24 June 2025. Available from: http://www.cfsph.iastate.edu/Factsheets/pdfs/cat_scratch_fever.pdf

Dong Y, Alhaskawi A, Zou X, Zhou H, Ezzi SHA, Kota VG, Abdulla MHAH, Olga A, Abdalbary SA, Lu H. Post-COVID reactivation of latent Bartonella henselae infection: a case report and literature review. BMC Infect Dis. 2024;24:422. https://doi.org/10.1186/s12879-024-09336-7.

Gonçalves-Oliveira J, Gutierrez R, Schlesener CL, Jaffe DA, Aguilar-Setién A, Boulouis HJ, Nachum-Biala Y, Huang BC, Weimer BC, Chomel BB, Harrus S. Genomic characterisation of three novel Bartonella strains in a rodent and two bat species from Mexico. Microorganisms. 2023;11:340. https://doi.org/10.3390/microorganisms11020340.

Król N, Krügel M, Kempf VAJ, Pfeffer M, Obiegala A. Emerging rodent-associated Bartonella: a threat for human health? Parasit Vectors. 2022;15:113. https://doi.org/10.1186/s13071-022-05162-5.

Okaro U, Addisu A, Casanas B, Anderson B. Bartonella species, an emerging cause of blood-culture-negative endocarditis. Clin Microbiol Rev. 2017;30:709–46. https://doi.org/10.1128/CMR.00013-17.

Puri K, Kreppel AJ, Schlaudecker EP. Bartonella Osteomyelitis of the acetabulum: case report and review of the literature. Vector-Borne Zoonotic Dis. 2015;15:463–7.

Québatte M, Dehio C. Bartonella gene transfer agent: Evolution, function, and proposed role in host adaptation. Cell Microbiol. 2019;21:e13068. https://doi.org/10.1111/cmi.13068.

Resto-Ruiz S, Burgess A, Anderson BE. The role of the host immune response in pathogenesis of Bartonella henselae. DNA Cell Biol. 2003;22:431–40. https://doi.org/10.1089/104454903767650694.

Rivera-Asencios D, Espinoza-Culupú A, Carmen-Sifuentes S, Ramirez P, García-de-la-Guarda

R. Design of a multi-epitope vaccine candidate against Carrion disease by immunoinformatics approach. Comput Biol Med. 2025;184:109397.

Siewert LK, Korotaev A, Sedzicki J, Fromm K, Pinschewer DD, Dehio C. Identification of the Bartonella autotransporter CFA as a protective antigen and hypervariable target of neutralising antibodies in mice. Proc Natl Acad Sci USA. 2022;119:e2201288119.

Yen TS, Freeman K, Baird R. Clinical manifestations associated with Bartonella henselae infection in a tropical region. Am J Trop Med Hyg. 2021;104:198–206.

Zuckerman EE. The history of Bartonella: from antiquity to the present. National Veterinary Laboratory. 2012. Accessed 26 Apr 2026. Available from: https://www.natvetlab.com/PDF/newsletters/NVLNewsLtr-Summer-2012-Vol11-No3.pdf.

44 Ehrlichiosis

Kokila Banerjee

Learning Objectives

1. To understand the epidemiology and risk factors of ehrlichiosis
2. To identify the clinical presentation of ehrlichiosis in humans and animals
3. To learn treatment options and prevention strategies of ehrlichiosis in humans and animals

Introduction

Ehrlichiosis is an emerging tick-borne zoonosis caused by several species of obligate intracellular bacteria under the family Anaplasmataceae and the order Rickettsiales. Three agents, namely, *Ehrlichia chaffeensis*, *Ehrlichia ewingii*, and *Ehrlichia muris eauclairensis,* formerly known as *Ehrlichia muris*-like agents, are responsible for causing ehrlichiosis in humans. They are transmitted through the bites of various ticks, mainly the Lone Star and black-legged ticks. All these Gram-negative coccobacillary intracellular bacteria cause disease with nearly similar clinical symptoms. However, each of these agents differs in geographic distribution and the severity of infection.

E. chaffeensis causes human monocytic ehrlichiosis (HME). It is the commonest agent and causes more severe infections compared with the other two agents. The disease is self-limiting in most patients, except in patients who are immunocompromised or have had prior exposure to certain drugs. Though the infection is easily treatable with timely antimicrobial therapy, diagnostic confirmation is difficult due to the lack of availability of kits for serological and molecular testing at peripheral centres. A delay in diagnosis can lead to a fatal infection. A knowledge of epidemiology and a high degree of suspicion based on the history of travel and tick bite/exposure are essential for timely diagnosis and prompt treatment delivery to save lives.

History

The history of ehrlichiosis is relatively modern compared to some other infectious diseases, as it was not fully recognised until the late twentieth century. Tick-borne diseases have been recognised for centuries, with some older descriptions of symptoms fitting what we now understand as ehrlichiosis. However, it was not distinguished from other illnesses, and its specific bacterial cause was not identified. In 1910, *Anaplasma marginale*, an agent of cattle disease, was first identified by Theiler. Subsequently, more strains infecting animals such as *Cowdria ruminantium, Ehrlichia canis,* and *Ehrlichia phagocytophila* were identified in 1925, 1935, and 1940, respectively. The genus *Ehrlichia* was named in honour

K. Banerjee (✉)
Drs. Tribedi and Roy Diagnostic Laboratory, Kolkata, West Bengal, India

S. C. Parija, A. Poddar (eds.), *Textbook of Bacterial Zoonoses*, Microbial Zoonoses,
https://doi.org/10.1007/978-981-95-6615-0_44

of the German microbiologist Paul Ehrlich. *Ehrlichia sennetsu*, identified in Japan in 1945, was considered a rickettsia for several years. *Ehrlichia chaffeensis* was first identified in 1986, in the blood of an infected 51-year-old male from the United States of America. This case was earlier linked to *Ehrlichia ewingii*. However, the researchers began investigating the disease in more detail. In the early 1990s, *E. chaffeensis*, the bacterium responsible for HME, was identified as the causative agent. It was discovered that *E. chaffeensis* is primarily transmitted by the Lone Star tick (*Amblyomma americanum*). The agent was later identified for the first time in Europe, in Portugal, in 1991.

Public health awareness increased as more cases were documented, especially in the south--eastern and south-central United States. In the 2000s, reports of the disease spread beyond the original geographic range, and its recognition grew internationally. In the following decades, other species of *Ehrlichia* were discovered to cause similar diseases in humans, including *E. ewingii* (causing a condition with a different clinical presentation) and *E. muris* (which have been linked to infections in certain regions). Ehrlichiosis cases have been reported in several countries, with tick species such as *Ixodes* ticks (the same ones that spread Lyme disease) also identified as potential vectors for some *Ehrlichia* species.

Taxonomy

The genus *Ehrlichia* is part of the family Anaplasmataceae, which includes three other genera, *Anaplasma*, *Wolbachia*, and *Neorickettsia*, and belongs to the order Rickettsiales. Earlier, morphology and intracellular existence were the criteria for classifying the organisms, resulting in the grouping of several unrelated genera. Recently, molecular methods have been used to reclassify the members of the genus *Ehrlichia*. A novel molecular-based taxonomy that uses two highly conserved genes, *rrs* (16S ribosomal RNA gene) and *groESL*, has been used for classification. This has resulted in a smaller amended *Ehrlichia* genus consisting of just five species, namely, *E. canis*, *E. chaffeensis*, *E. muris*, *E. ruminantium*, and *E. ewingii*. *E. chaffeensis* and *E. ewingii*, are recognised as human and veterinary pathogens, and *E. canis* and *E. ruminantium*, infect animals alone.

Genomics and Proteomics

The genomes of *Ehrlichia* are small, ranging from 1.2 to 1.6 Mb, with a high degree of gene conservation despite evolutionary changes and a low GC content (~30%). The genome coding ratio is small due to the presence of long tandem repeat (TR) sequences and non-coding regions in the genome. The TRs are actively created and deleted through DNA slippage, which assists these intracellular pathogens in adapting to the hostile host environment. The *E. chaffeensis* genome contains 1105 protein genes and 43 RNA genes.

The obligate intracellular members of the Anaplasmataceae family possess multiple genes belonging to the Pfam01617.8 family. Seven major proteins have been identified by the Western blotting technique in *E. chaffeensis*, the most common and deadly pathogen of ehrlichiosis. These are 22,28, 29, 44, 58, 66, and 120 kDa in size. Amongst them, 22, 28, 29, and 120 kDa are surface-exposed proteins responsible for inter-strain antigenic variation. The 120 kDa protein containing a region of 80 amino acid tandem repeat units is responsible for host cell adhesion.

Morphology

The genus *Ehrlichia* is a small Gram-negative coccoid or oval bacterium that exists in two morphological variants or stages. A large reticulate intracellular stage (0.4–0.6 μm X 0.7–19 μm) and a small infectious dense core cell (0.4–0.6 μm in diameter). The reticular cells have uniformly dispersed nucleoid filaments and ribosomes, whereas the dense-cored cells have condensed nucleoid filaments and ribosomes. They form morulae or microcolonies within the cytoplasm of host cells, such as monocytes, macrophages, or neutrophils. Morulae range from 1 to 6 μm in size and contain 1–40 organisms of either reticulate or dense-cored variety. Each infected host cell may contain 1 to >400 morulae. The cell walls of the bacteria, in both reticulate and dense-cored forms, lack peptidoglycan and contain outer and inner walls separated by a periplasmic space.

Cultivation

E. chaffeensis is an obligate intracellular bacterium, meaning it requires host cells for survival and replication. Therefore, it cannot be cultivated in conventional cell-free media. It can be propagated in mouse embryos, Vero, and L929 cells. The agent induces a cytopathic effect with different morphological characteristics in each of these cell lines. A cytopathic effect is seen within 4 days of inoculation in untreated and irradiated mouse embryo cells. As the infected foci gradually expand, the centre develops necrotic cells with pyknotic nuclei, and the morulae degenerate. In irradiated L929 cells, *E. chaffeensis* produces cell lysis with the formation of distinct, round macroscopic plaques at 18 days post-inoculation. In untreated and irradiated Vero cells, *E. chaffeensis* produces a reticular pattern with interspersed foci of loosely interwoven necrotic cells, spaces of detached cells, cells with morulae, and uninfected cells. However, no cytopathic effects are produced in DH82 and Buffalo Green Monkey (BGM) cell lines.

Other *Ehrlichia* spp., such as *E. canis,* can be cultivated in the canine macrophage cell line DH82. Tick cell lines such as ISE6, IDE8 and AAE2, which require a specialised medium and an incubation temperature of 34 °C without carbon dioxide gas, are also used to cultivate *Ehrlichiae.*

Animal Models

Several animal models have been employed to study *E. chaffeensis* infection. Each model has its own unique advantages and limitations in mimicking human monocytic ehrlichiosis (HME). Models of immunocompetent mice such as BALB/c, C3H/HeJ, C.B-17, C3H/HeN, DBA, and C57BL/6 have been used to diagnose *E. chaffeensis*. Several knockout models of mice that lack one or the other arm of the immune response have also been studied. C57BL/6 mice models are great options for producing histological evidence of pathogenicity by *Ehrlichia.*

Larger animal models, such as dogs and white-tailed deer, also play a role in understanding *E. chaffeensis* infection. Dogs, as natural hosts for *E. chaffeensis*, can develop canine monocytic ehrlichiosis (CME), which shares some clinical similarities with HME, including fever, lethargy, lymphadenopathy, and thrombocytopenia. Dogs serve as a relevant model for studying the natural host–pathogen interaction and for tick transmission studies. However, their size and maintenance costs can be limiting factors. White-tailed deer, another natural host crucial for the pathogen's ecological cycle, typically develop subclinical or mild infections. Consequently, while less suitable for studying severe disease pathogenesis, they are invaluable for understanding the natural maintenance and transmission dynamics of *E. chaffeensis* in the environment. However, these large animal models are not ideal for mechanistic evaluations of analysing immunity and genetic knockout studies.

Transmission Cycle

During the spring and summer months, the adult and nymphal stages of the Lone Star ticks, *Amblyomma americanum*, tend to bite humans involved in outdoor activities. The white-tailed deer, *Odocoileus virginianus*, serves as the final host for the life cycle of ticks (Fig. 44.1). Both *E. chaffeensis* and *E. ewingii* are transmitted by the bite of the Lone Star tick, which is the most common mode of transmission of ehrlichiosis (Fig. 44.2).

E. chaffeensis has also been isolated from *Dermacentor variabilis*, the American dog tick.

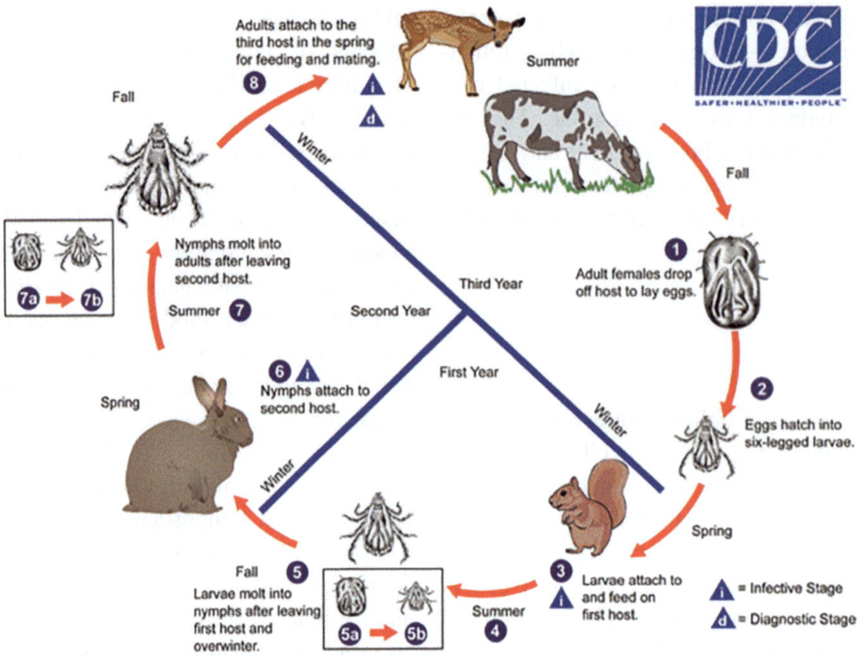

Fig. 44.1 Lifecycle of a Lone Star tick. (Image source: CDC)

Fig. 44.2 A Lone Star tick with a characteristic white spot on its back. (Image source: CDC)

E. muris eauclairensis is spread by *Ixodes scapularis*, commonly called the black-legged tick.

In rare cases, patients have been infected through blood transfusion and stem cell or solid organ transplantation. *E. chaffeensis* can survive for 11 days in refrigerated blood. If a patient presents with an undiagnosed febrile illness and with leukopenia and thrombocytopenia, and the donor is from an endemic area, ehrlichiosis should be kept in mind. The infected blood or blood components need to be quarantined or recalled from the blood bank if the donor or recipient develops symptoms of ehrlichiosis. Leucoreduction or irradiation of blood components does not eliminate the chance of transmission of pathogens.

Pathogenesis and Pathology

Despite their small genomes, *Ehrlichia* have evolved complex, elaborate mechanisms to enable them to adapt to intracellular life inside the bodies of vertebrate and invertebrate hosts. *E. chaffeensis* gets phagocytosed within immune effector cells such as monocytes and tissue macrophages, and thrives as well as multiplies within the membrane-bound cytoplasmic vacuoles. The pathogen survives by blocking the fusion of infected phagosomes with lysosomes. Interferon γ mediates the killing of *E. chaffeensis* by macrophages. Organs of the reticuloendothelial system, such as lymph nodes, spleen, liver and bone marrow, are the preferred organs where these agents proliferate within white blood cells. Bone marrow examination reveals myeloid hyperplasia, megakaryocytosis and the presence of granulomas. The myeloid hyperplasia occurs in the bone marrow due to increased peripheral consumption of cells caused by sequestration and destruction in the spleen and liver. Cholestasis, necrosis, granulomas and diffuse mononuclear hyperplasia are noted in the liver. Necrosis and diffuse mononuclear hyperplasia are seen in the spleen and lymph nodes. Other organs, such as the kidney, meninges, brain and heart, show lymphohistiocytic perivascular infiltrates. Lungs develop interstitial pneumonitis with infiltration by mononuclear cells. There is, however, no vasculitis, endothelial injury, or thrombosis as seen in other rickettsial diseases.

The main pathology of the disease process in HME is not due to the direct effect of bacteria on the host cells but to the infected host's inflammatory response.

Immunology

Human monocytic ehrlichiosis (HME) immunology involves a complex interplay of innate and adaptive immune responses. The host's immune system aims to eliminate the pathogen while also contributing to the clinical manifestations of the disease.

Innate Immune Response

The initial response to *E. chaffeensis* infection involves the innate immune system. Monocytes and macrophages, the primary target cells of *E. chaffeensis*, play a crucial role. Upon infection, these cells recognise pathogen-associated molecular patterns (PAMPs) through pattern recognition receptors (PRRs) like Toll-like receptors (TLRs). This recognition triggers the activation of intracellular signalling pathways, leading to the production of pro-inflammatory cytokines such as tumour necrosis factor-alpha (TNF-α), interleukin-1 (IL-1), and interleukin-6 (IL-6). These cytokines contribute to the early flu-like symptoms observed in HME.

Natural killer (NK) cells also participate in the early immune response. They directly lyse infected cells and produce interferon-gamma (IFN-γ), a key cytokine in controlling intracellular pathogens. The role of the complement system in HME is not fully elucidated, but it may contribute to bacterial clearance or inflammation.

Evasion Mechanisms

E. chaffeensis has evolved several strategies to evade the host's innate immune defenses. First, it resides within a phagosome and inhibits phagolysosomal fusion, preventing its destruction by lysosomal enzymes. Second, the bacterium can also modulate host cell gene expression to promote its survival and replication, including inhibiting apoptosis and interfering with antigen presentation. Finally, *E. chaffeensis* can suppress the transcription of certain early innate immune response cytokines, potentially acting as a "*stealth*" organism within the macrophage.

Adaptive Immune Response

The adaptive immune response, involving T and B cells, is critical for the long-term control and clearance of *E. chaffeensis*.

T Cells

CD4+ T helper 1 (Th1) cells are crucial for effective immunity. They produce IFN-γ, which activates infected macrophages, enhancing their ability to kill intracellular bacteria. IFN-γ also promotes the production of antibodies that play a role in cell-mediated cytotoxicity. CD8+ cytotoxic T lymphocytes (CTLs) directly kill *E. chaffeensis*-infected cells, contributing to pathogen clearance. However, in severe cases, an overproduction of pro-inflammatory cytokines by CD8+ T cells, such as TNF-α and IL-10, has been implicated in immunopathology.

B Cells and Antibodies

Antibodies play a role in immunity against *E. chaffeensis*. They neutralise extracellular bacteria, block bacterial attachment to host cells, and enhance phagocytosis and killing of infected cells through opsonisation and antibody-dependent cell-mediated cytotoxicity (ADCC). Specific antibodies against outer membrane proteins of *E. chaffeensis* have been shown to inhibit infection through both extracellular neutralisation and intracellular mechanisms involving the TRIM21 protein.

Immunopathology

While the immune response is essential for controlling *E. chaffeensis* infection, it can also contribute to the pathogenesis of HME. The excessive production of pro-inflammatory cytokines leads to systemic inflammation, endothelial damage, and potentially severe complications. These complications include acute respiratory distress syndrome (ARDS), acute kidney injury, and multiorgan failure. The balance between protective immunity and immunopathology is important in determining the outcome of *E. chaffeensis* infection.

Inflammasomes and Autophagy

Recent research highlights the involvement of inflammasomes and autophagy in the host response to *E. chaffeensis*. Inflammasomes are intracellular multiprotein complexes that can be activated upon detection of pathogens or cellular stress, producing pro-inflammatory cytokines like IL-1β and IL-18. Autophagy is a cellular process that degrades and recycles cytoplasmic components, including intracellular pathogens. The interplay between inflammasome activation and autophagy influences the outcome of *E. chaffeensis* infection, with both protective and potentially detrimental roles depending on the context.

Infection in Humans

Human monocytic ehrlichiosis (HME) varies from asymptomatic infection to severe infection in immunocompetent patients. The condition shows an incubation period of 7–14 days after a tick bite and a mean duration of illness of 23 days. In immunocompromised patients, the disease has a fulminating course with a septic shock-like syndrome. A complete blood count reveals leucopenia (1300–4000cells/μL) and thrombocytopenia (50,000–1,40,000 cells/μL). Anaemia is seen later in the course of the illness. Relative or absolute lymphocytosis is seen during the recovery phase. Hyponatraemia and elevated liver enzymes, especially transaminases, are the other findings.

Initial symptoms are flu-like manifesting with fever, chills, myalgia, headache, lethargy, and malaise. Later, diarrhoea, nausea, vomiting, loss of appetite and hence weight loss could follow. Rashes are seen in only half of the patients and are more common among children. They could be maculopapular or petechial and are seen mostly on the extremities. Adults with severe illness can have generalised as well as system-specific symptoms. Gastrointestinal, respiratory, renal, haematologic and neurologic involvements with features such as diarrhoea, acute respiratory distress syndrome, acute renal failure, gastrointestinal bleeding and meningoencephalitis, confusion and even death are some of the features of fulminating disease. A case fatality rate of 3% has been noted. Those who receive prompt treatment recover early.

E. ewingii ehrlichiosis infects granulocytes and has clinical features like those of HME; however, rash and gastrointestinal symptoms are less common. Signs and symptoms of *Ehrlichia muris*-like agent ehrlichiosis are like those of *E. ewingii* ehrlichiosis.

Whenever a person presents a history of exposure to ticks, use of insect repellants, travel to a wooded area in an endemic region, during spring or summer months, ehrlichiosis should be suspected in the differential diagnosis for flu-like illnesses that present with or without a rash. Patients with any immunocompromised state, such as retroviral infection, malignancies, or elderly patients or even those on a prior sulfonamide treatment, tend to have overwhelming infections. Hence, an early diagnosis is key to saving lives.

Infection in Animals

E.canis infection presents in two forms: the acute form is seen in warm months, and the chronic form is seen throughout the year. The reticuloendothelial system is affected just as in humans. Symptoms include anorexia, loss of stamina, reluctance to walk, depression, cough, dyspnoea, stiffness, oedema of limbs/scrotum. On blood investigations, normocytic normochromic anaemia, leukopenia/mild leukocytosis, and thrombocytopenia are seen. Death is rare during the acute phase. Spontaneous recovery may occur. There is an asymptomatic phase, which is followed by chronic disease.

The chronic phase may be seen in any breed of dog. It is, however, most frequently noted in the German Shepherd and Siberian Husky. During the chronic phase, the bone marrow is hypoplastic, and organs are infiltrated with lymphocytes/plasmacytes. Marked splenomegaly, glomerulonephritis, renal failure, interstitial pneumonitis, anterior uveitis, meningitis, severe weight loss, and marked thrombocytopenia, evidenced by epistaxis, melena, petechiae and ecchymosis, are seen.

Infections by *E.chaffeensis* and *E.ewngii*, are clinically similar but milder in severity and self-limiting. The hosts, dogs and white-tailed deer, serve as reservoirs of infection, and the pathogen can persist for more than 2 years. Rarely, chronic infections can lead to kidney and liver involvement.

Epidemiology and Public Health

The disease's epidemiology largely depends on the tick population's prevalence. The Lone Star tick is widely distributed across the eastern, south-eastern, and south-central United States (Fig. 44.3). Its range extends as far west as Texas and includes the southern parts of Iowa, Illinois, Indiana, Ohio, and Pennsylvania. The northernmost extent of its established range reaches New Jersey, Rhode Island, and New York, and has been recently documented in Connecticut and possibly Maine. These ticks thrive in wooded areas with dense underbrush, as well as along the edges of meadows and streams, habitats often frequented by white-tailed deer, a primary host for adult ticks. In contrast, the black-legged tick is primarily found across the eastern half of the United States and extends into southern portions of Canada (Fig. 44.4). Its distribution is concentrated in the Northeast and upper Midwestern states, with established populations also present in the Mid-Atlantic and some south-eastern states. These ticks inhabit areas with leaf litter, shrubby plants, and forest grasses, often found in wooded environments.

HME infection is commonly seen in the summer months from May through July, associated with outdoor recreational or occupational activities in rural and semi-urban areas. More than three-fourths of those infected are males. As the Lone Star ticks mature from larva to nymph and then to the adult forms, the bacteria, *E.chaffeensis*, are transmitted from one stage to the other transstadially. However, no transovarial transmission has been reported in ticks. The deer maintain the bacteria in the environment by transmitting the ticks to uninfected deer. The disease has been noted in the south-central states such as Oklahoma, Missouri, and Arkansas. Cases have

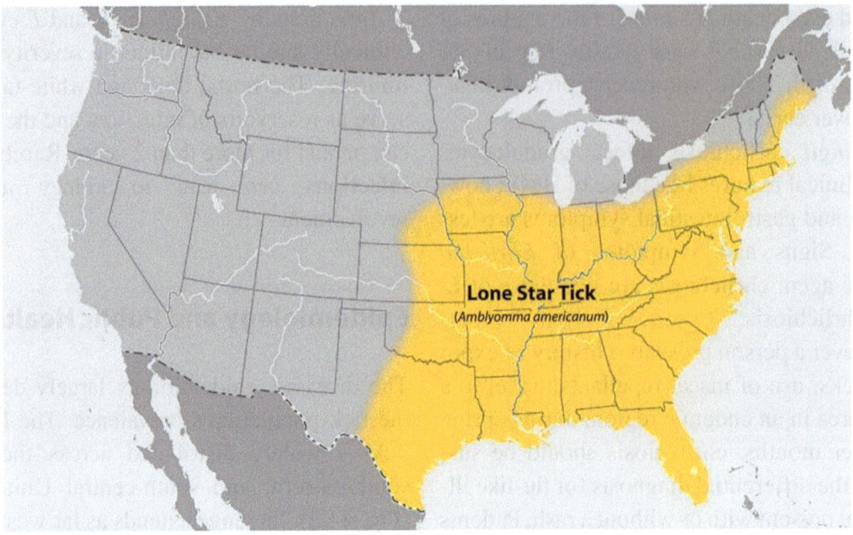

Fig. 44.3 Epidemiology of the Lone Star tick. (Image source: CDC)

Fig. 44.4 Geographical distribution of black-legged ticks. (Image source: CDC)

also been reported from the Northeast and north-central states of the USA. Additional vectors, such as *Dermacentor variabilis,* have also been reported to spread infection. More than 30 American states have reported an increase in cases, from 338 in 2004 to 7309 between 2013 and 2017. The global epidemiology of human ehrlichiosis is presented in Fig. 44.5.

A case fatality rate of 2% in adults and 4% in children under 5 years has been reported for HME. The rate of hospitalisation is 49% for HME. By the 2000s, cases of human ehrlichiosis were reported not just in the United States, but in Europe, Asia, and parts of Africa, where *Ehrlichia* species were found in wildlife and their ticks. Some of these regions had zoonotic connections,

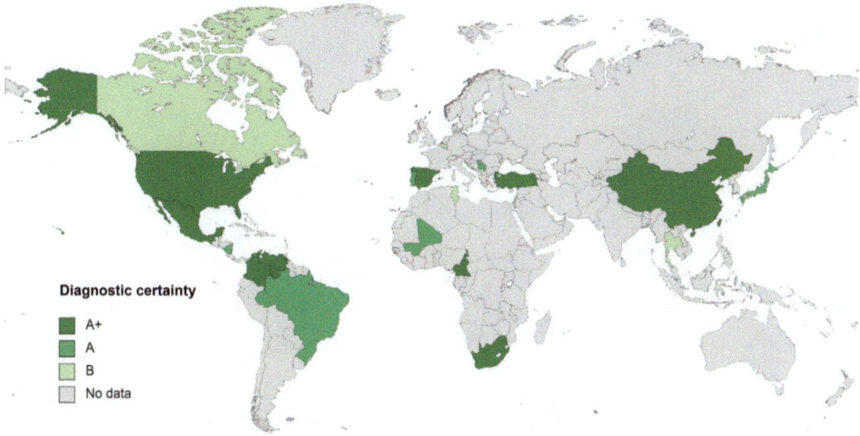

Fig. 44.5 Human ehrlichiosis cases by country. A+, cases diagnosed by PCR, culture and/or immunostaining of biopsy/autopsy tissue; A, cases diagnosed by paired IgG IFA serology; B, cases diagnosed by single IgG IFA or ELISA serology. (Image source: Gygax L, Schudel S, Kositz C, Kuenzli E, Neumayr A. Human monocytotropic ehrlichiosis—A systematic review and analysis of the literature. PLOS Neglected Tropical Diseases. 2024 Aug 2;18(8):e0012377)

with humans being infected through contact with domestic animals or wildlife. Ehrlichiosis in dogs in India has been reported in various regions, at different prevalence rates. Punjab has reported 0.35%, Nagpur, 18.9%, Bhubaneswar, 31.5% and Chennai, 46.9%.

Diagnosis

The diagnosis of ehrlichiosis is a comprehensive approach that requires a proper understanding of the disease's epidemiology, meticulous history taking, and a heightened level of clinical suspicion. Several key clues from a patient's history can strongly suggest a diagnosis of ehrlichiosis, including a documented history of a tick bite. Furthermore, exposure to tick-infested environments, such as engaging in treks into the woods or participating in outdoor recreational or professional activities in rural and semi-urban areas, significantly increases the likelihood of exposure. A history of travel to regions endemic for ehrlichiosis is another important epidemiological factor to consider. Finally, the presence of similar infections in family members, co-workers, or pets might indicate a common exposure source or a shared risk environment, further raising suspicion for ehrlichiosis.

Specimens

The primary clinical sample crucial for the diagnosis of ehrlichiosis is whole blood. This is because *Ehrlichia* bacteria exhibit a predilection for infecting white blood cells, specifically monocytes in the case of *E. chaffeensis* and granulocytes in *E. ewingii*, which circulate within the bloodstream. Consequently, whole blood serves as the most readily accessible and informative specimen for direct pathogen detection and characterisation.

Microscopy

While *Ehrlichia* species are classified as Gram-negative bacteria based on their cell wall structure, the Gram stain is not a practical or sensitive diagnostic tool for identifying them in clinical blood samples.

During the first week of illness, a meticulous microscopic examination of a peripheral blood

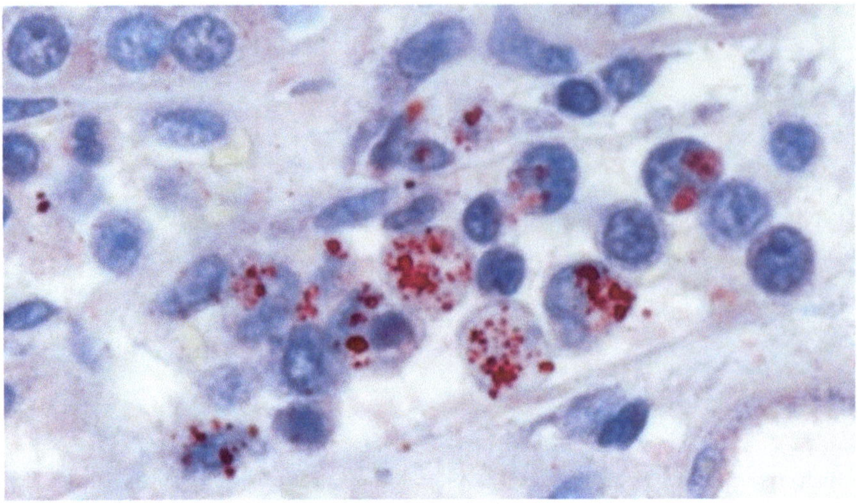

Fig. 44.6 Morulae seen within inflammatory cells. (Image source: CDC)

smear might reveal morulae within the cytoplasm of leukocytes. They are seen as dark blue or purple intracytoplasmic inclusions on Romanowsky stains (Fig. 44.6). Specifically, *E. chaffeensis* is observed within the cytoplasm of monocytes, while *E. ewingii* is found within granulocytes. For *E. muris eauclairensis*, the specific target cell remains unidentified.

A buffy coat examination increases the yield of morulae. They should, however, be differentiated from other structures such as intracellular bacteria, Dohle bodies, overlying platelets, toxic granules, and other cytoplasmic inclusions.

Rarely, morulae can also be visualised within cerebrospinal fluid (CSF) and bone marrow leukocytes. However, despite a strong suggestion that ehrlichiosis may be based on these microscopic findings, confirmatory tests are needed to clinch the diagnosis. Merely observing morulae in a particular leucocyte can neither conclusively identify the *Ehrlichia* species nor confirm the diagnosis of ehrlichiosis. A molecular method is required for specific diagnostic confirmation.

In Vitro Culture

Culture isolation is difficult and insensitive, and hence not a widely used method for the diagnosis of ehrlichiosis. Any lab with a BSL-3 pathogen facility and experience in cell culture techniques can culture these pathogens. Culture samples should be collected before the start of antibiotics and should not be frozen.

Serodiagnosis

Indirect Fluorescent Antibody Assay (IFA)

Commercially available serological tests, such as the IFA for IgG detection, are used to diagnose ehrlichiosis only in the later phases of illness. Paired serum samples are collected during the acute and convalescent phases. The acute phase sample is collected within the first 2 weeks of illness, and the convalescent phase sample is collected 2–10 weeks later. If there is a fourfold rise in titer between the two samples, it gives a retrospective diagnosis of the disease. Ehrlichiosis cannot be confirmed using a single acute antibody result. A single antibody titer equal to or less than 1:64 is considered negative, and that of 1:128 or greater is considered presumptively positive.

Immunoglobulin M antibody testing is not a reliable or specific test. It should not be used as the sole diagnostic test to confirm a recent infection. Antibodies against *Ehrlichia* species might remain elevated for several months to years after the disease has resolved and do not necessitate treatment in the absence of the development of new symptoms. Cross-reactivity exists between closely related species of the *Ehrlichia* and *Anaplasma* genera. In areas endemic to ehrlichiosis and anaplasmosis, IFA using antigens from both genera should be performed simultaneously.

Enzyme-Linked Immunosorbent Assay (ELISA)

ELISA represents another widely employed serological technique for ehrlichiosis, valued for its potential for standardisation and automation, making it suitable for processing a larger number of samples. In this assay, specific *Ehrlichia* antigens are coated onto the surface of microtiter wells. When the patient's serum is added to these wells, any *Ehrlichia*-specific antibodies present in the serum will bind to the immobilised antigens. After washing away unbound serum components, an enzyme-linked secondary antibody that specifically binds to human antibodies is added. Following another washing step, a substrate for the enzyme is introduced. The enzyme acts on the substrate to produce a colour change, and the intensity of this colour reaction is directly proportional to the amount of *Ehrlichia*-specific antibodies present in the patient's serum. The optical density, reflecting the colour intensity, is then measured spectrophotometrically, providing a quantitative or semi-quantitative measure of the antibody level.

Western Blotting

This technique offers a more refined approach to serodiagnosis by allowing for the detection of antibodies directed against specific *Ehrlichia* proteins. In this technique, *Ehrlichia* proteins are first separated based on their molecular weight using gel electrophoresis and then transferred ("blotted") onto a membrane. The membrane is then incubated with the patient's serum, allowing *Ehrlichia*-specific antibodies to bind to their corresponding protein bands. Bound antibodies are subsequently detected using a labelled secondary antibody, as in ELISA. The resulting pattern of protein bands that react with the patient's serum can provide valuable information, potentially aiding in the differentiation of antibody responses to different *Ehrlichia* species and helping to reduce the likelihood of cross-reactive results.

Point-of-care Tests

These tests represent a category of rapid serological assays, often utilising principles of ELISA or lateral flow immunochromatography. These tests are designed to provide quick results, often within minutes. They can be performed outside a traditional laboratory setting. They typically involve applying a small volume of the patient's blood or serum to a test strip or device containing *Ehrlichia* antigens. If *Ehrlichia*-specific antibodies are present, they will bind to the antigens, and this binding is usually indicated by a visible colour change or signal on the test device. While these tests offer the advantage of speed and ease of use, their sensitivity and specificity may sometimes be lower compared with more complex laboratory-based assays like IFA and ELISA, and positive results often require confirmation with a more established method.

One prominent example of a point-of-care test in the veterinary field is the IDEXX SNAP 4Dx Plus Test. This rapid immunoassay detects antibodies against *E. canis*, *E. chaffeensis*, and *E. ewingii*, along with other tick-borne pathogens like *Anaplasma* and *Borrelia burgdorferi* (Lyme disease). It utilises ELISA technology in a user-friendly format, providing qualitative results within minutes using a small blood sample. While specifically labelled for canine use, detecting antibodies to the same *Ehrlichia* species that infect humans highlights the potential for adapting similar rapid antibody detection platforms for human diagnostics.

Another example is the Zoetis Vetscan canine *Ehrlichia* rapid test. This test is designed for the qualitative detection of antibodies to *E. canis, E. chaffeensis,* and *E. ewingii* in canine whole blood, serum, or plasma at the point-of-care. It is a visual rapid test that provides results quickly, aiding in the immediate assessment of potential *Ehrlichia* exposure in animals. The simplicity and speed of such tests underscore the feasibility of developing similar rapid antibody tests for human use in clinics or as potential over-the-counter diagnostics. However, rigorous validation for human application would be essential.

Molecular Diagnosis

Polymerase Chain Reaction (PCR)

PCR-based methods are the most sensitive and specific techniques for the early diagnosis of ehrlichiosis. These assays directly detect the presence of *Ehrlichia* DNA in clinical samples, typically whole blood.

Different PCR techniques, including conventional PCR and real-time PCR (qPCR), can be employed. Real-time PCR often includes fluorescent probes that allow for quantifying the bacterial load. For *E. chaffeensis*, common target genes include the 16S rRNA gene, dsb gene, groEL gene (heat shock protein), and genes encoding major outer membrane proteins like the omp-1 multigene family. For *E. ewingii*, target genes often include the 16S rRNA and groEL genes. Similarly, for *E. muris eauclairensis*, diagnostic PCR assays often target the 16S rRNA gene, gltA gene (citrate synthase), and the groEL gene. Multiplex PCR assays can simultaneously detect DNA from multiple *Ehrlichia* species and other tick-borne pathogens.

Nested PCR

Nested PCR involves two sets of primers used in sequential PCR reactions. The first PCR amplifies a larger region of the target gene, and then the product is used as a template for a second PCR with primers that anneal to sequences within the first amplicon. This method can enhance sensitivity and specificity, particularly when the initial bacterial load is low. Target genes are like those used in conventional PCR for each *Ehrlichia* species, such as the 16S rRNA gene for *E. chaffeensis* and *E. ewingii*.

Quantitative Real-time PCR (qPCR)

qPCR allows for the real-time monitoring of PCR amplification, enabling the quantification of the target DNA. This technique often utilises fluorescently labelled probes or intercalating dyes. For *Ehrlichia* diagnosis, qPCR assays targeting genes like the 16S rRNA, dsb, groEL, and species-specific genes within the omp-1 family for *E. chaffeensis* have been developed. For *E. ewingii,* qPCR assays are used to target the 16S rRNA and *groEL* gene. Similarly, for *E. muris eauclairensis*, qPCR assays targeting the *P13*, *gltA*, and *groEL* gene have been described. Quantification can be useful for monitoring treatment response or understanding the kinetics of infection.

Loop-Mediated Isothermal Amplification (LAMP)

LAMP is an isothermal nucleic acid amplification technique that can amplify DNA with high specificity and rapidity under constant temperature. LAMP assays have been developed for the detection of *Ehrlichia* species, often targeting the 16S rRNA gene. This method can be particularly useful in resource-limited settings as it does not require sophisticated thermal cycling equipment.

Next-Generation Sequencing (NGS)

While not typically used for routine primary diagnosis, NGS can be valuable in identifying novel or unexpected *Ehrlichia* species, especially in cases with unusual clinical presentations or negative results from standard PCR assays. NGS can provide sequence information for various genomic regions, allowing for comprehensive characterisation of the infecting agent. For known *Ehrlichia* species, NGS can be used for research purposes, such as genomic comparisons and phylogenetic studies.

Multiplex Assays

Several molecular diagnostic platforms allow for the simultaneous detection of multiple tick-borne pathogens, including different *Ehrlichia* species, *Anaplasma, Borrelia*, and *Babesia*. These multiplex assays often utilise PCR with different primer sets and detection methods (e.g., fluorescent probes with distinct emission wavelengths) to identify multiple targets in a single reaction. Target genes for each *Ehrlichia* species within these multiplex assays are typically the same as those used in single-target PCR.

The different diagnostic approaches are summarised in Table 44.1.

Table 44.1 Diagnosis of ehrlichiosis

Diagnostic approach	Method	Target	Remark
Microscopy	Direct examination of stained smears by Romanowsky stains	Morulae within white blood cells, in whole blood, bone marrow, and CSF	Low sensitivity, low specificity
Culture	In vitro cell culture	Gram-negative coccobacilli	Difficult technique
Serology	IFA, ELISA	IgG antibodies, a fourfold rise in paired sera collected 2-4 weeks apart	Antibodies can persist for several months to years. Cross-reactivity seen. IgM antibody and single sera are not reliable
Molecular	PCR on whole blood sample, CSF, solid tissue, and bone marrow.	16S rRNA gene, p28 gene family, *groEL* gene, *dsb* gene, *VLPT*, *trp* genes (*trp19*, *trp36*), *gltA* gene	Laboratory developed assays, varying sensitivities, and specificities. Real-time PCR is the most sensitive method

Treatment

Doxycycline is the drug of choice for the treatment of ehrlichiosis in patients of all age groups, including children. The treatment of ehrlichiosis is started immediately and never delayed while awaiting diagnostic confirmation. The dosage is 100 mg twice daily orally or intravenously in adults. For children, it is 2.2 mg/kg of body weight twice daily if the body weight is below 45 kg. Patients with mild disease who have a caregiver at home, do not have comorbid conditions, and are not immunosuppressed can be treated on an outpatient basis. Others will need hospitalisation. The duration of treatment is usually between 5 and 7 days. If the fever does not subside within 24–48 h, an alternative diagnosis needs to be considered. Patients with multiple systemic involvement may take longer to respond to treatment and require a longer duration of therapy. Of note is that children below 8 years of age can be safely administered doxycycline without fear of tooth staining. It is extremely important to treat children promptly, as mortality is much higher in them compared with adults.

Alternative antibiotics for treatment in the case of severe doxycycline allergy include rifampicin, 300 mg twice daily for adults and 10 mg/kg body weight in children, not exceeding 300 mg/dose. Other broad-spectrum antibiotics are ineffective, including fluoroquinolones, beta-lactams, aminoglycosides, and macrolides. Trimethoprim-sulfamethoxazole can cause severe complications and must be avoided if suspicion of a tick-borne infection is high.

Treatment of complications such as renal impairment needs to be managed with fluid and electrolyte balance and vasopressors. ARDS needs to be recognised early and not confused with cardiac failure or pneumonia.

There is no role for prophylactic antibiotic treatment in patients with recent tick bites who have no signs or symptoms. Treatment of asymptomatic seropositive individuals is not warranted, as antibodies can persist for years.

Prevention and Control

Meticulous screening of people and pets to identify and remove ticks promptly needs to be performed regularly, as well as when going for outdoor activities. Areas to be specially scanned for tick attachment are the scalp, axilla, groin, abdomen, behind belt lines, and under socks on the feet. After returning from outdoor activities, bathing is important to remove ticks promptly. This is helpful as ticks do not immediately transfer the infectious pathogen after attaching. Fingers should never be used to remove attached ticks, as the body fluids of ticks could contain infectious agents. Instead, tweezers or forceps must be used. Tick repellents containing DEET, IR3535, picaridin (1-piperidinecarboxylic acid, 2-[2-hydroxyethyl], 1-methlypropyl ester), or other EPA-registered products should be used when travelling outdoors. Pets need to be protected from tick bites by regularly applying ectoparasite control products, such as acaricidal tick collars, shampoos, or topical applications. Exposure to tick-infested habitats can be limited by creating safe zones such as clear walking trails, avoiding walking amongst shrubs and vegetation, and wearing protective clothing such as pants, closed shoes, boots, and permethrin-impregnated clothing.

Case Study

A 72-year-old gentleman, a retired professional living in India, was admitted to the intensive care unit with altered sensorium for 1 day, fever, myalgia, lethargy, and nausea for 3 weeks. His wife informed them that they had consulted a local physician who treated him with oral analgesics and a proton pump inhibitor. His blood counts revealed anaemia, leukopenia, and thrombocytopenia. His liver enzymes were raised. Serum Sodium was 129 mEq/L, creatinine-200 mmol/L. He was treated with steroids with no improvement. His CSF and bone marrow tap revealed intracytoplasmic inclusions within mononuclear cells. Travel history was elicited. The couple had visited their son living in the USA for the summer break to be with their grandchildren in Tennessee. On further probing, it was known they had gone camping on a weekend in the nearby woods along with their grandchildren and pet, a six-year-old German Shepherd. On learning this, the doctors promptly initiated doxycycline, suspecting a particular tick-borne infection. However, his condition deteriorated. He developed septic shock and multiorgan failure and succumbed to his illness on day 5 of admission.

1. What could be the causative agent of this illness?
2. What are the modes of transmission of infection and the pathology of the disease?
3. What are the confirmatory tests available?
4. What preventive strategies should be taken?

Research Questions

1. What are the current gaps in the understanding of the global prevalence of ehrlichiosis in human and animal populations?
2. Why are comprehensive epidemiological studies crucial to address these gaps, particularly considering the expanding tick population?
3. Is there adequate awareness amongst clinicians about the mode of transmission and epidemiology of the disease, considering global travel?
4. Can vaccines be designed for animals to prevent them from serving as reservoirs?

Further Readings

Chakraborty A, Rath PK, Panda SK, Mishra BP, Dehuri M, Biswal S, Jena MK, Sahu BP, Paital B, Sahoo DK. Molecular confirmation, epidemiology, and pathophysiology of *Ehrlichia canis* prevalence in Eastern India. Pathogens. 2024;13(9):803. https://doi.org/10.3390/pathogens13090803.

Gygax L, Schudel S, Kositz C, Kuenzli E, Neumayr A. Human monocytotropic ehrlichiosis-a systematic review and analysis of the literature. PLoS Negl Trop Dis. 2024;18(8):e0012377.

Ismail N, Bloch KC, McBride JW. Human ehrlichiosis and anaplasmosis. Clin Lab Med. 2010;30(1):261–92.

Kondethimmanahalli C, Ganta RR. Proteome analysis of *Ehrlichia chaffeensis* containing phagosome membranes revealed the presence of numerous bacterial and host proteins. Front Cell Infect Microbiol. 2022;12:1070356. https://doi.org/10.3389/fcimb.2022.1070356.

Loftis AD, Massung RF, Levin ML. Quantitative real-time PCR assay for detection of *Ehrlichia* chaffeensis. J Clin Microbiol. 2003;41(8):3870–2. https://doi.org/10.1128/JCM.41.8.3870-3872.2003. PMID: 12904406; PMCID: PMC179802

Rochlin I, Toledo A. Emerging tick-borne pathogens of public health importance: a mini-review. J Med Microbiol. 2020;69(6):781–91.

Schudel S, Gygax L, Kositz C, Kuenzli E, Neumayr A. Human granulocytotropic anaplasmosis-a systematic review and analysis of the literature. PLoS Negl Trop Dis. 2024;18(8):e0012313.

Walker DH, Dumler J. Emergence of the Ehrlichioses as human health problems. Emerg Infect Dis. 1996;2(1):18–29. https://doi.org/10.3201/eid0201.960102.

Corynebacterium Infections 45

Aditi Gupta, Kumar S. Abhishek, and Vibhor Tak

Learning Objectives

1. To understand the pathology and pathogenesis of infections caused by zoonotic *Corynebacterium* species
2. To identify common infections caused by these organisms and the diagnostic tools used for their detection
3. To discuss various strategies for control and prevention of zoonotic infections in humans and animals

Introduction

The genus *Corynebacterium* includes bacteria characterised by their distinctive "club-shaped" appearance. This genus comprises several species that can cause infections in humans. Among them, *Corynebacterium diphtheriae* is the most significant pathogen, historically associated with numerous outbreaks and epidemics. It is also the type species of this genus. Hence, other terms such as diphtheroid and coryneform bacteria are used for those organisms that do not cause human infection and are commonly found as commensals or contaminants.

Organisms responsible for the clinical manifestation of diphtheria include toxin-producing *C. diphtheriae*, *Corynebacterium ulcerans*, and *Corynebacterium pseudotuberculosis*. The latter two are known to be zoonotic, whereas *C. diphtheriae* demonstrates person-to-person transmission.

New additions to this tox-bearing group include *Corynebacterium silvaticum* and *Corynebacterium rouxii*. Other clinically relevant species include *Corynebacterium bovis*, *Corynebacterium jeikium*, *Corynebacterium minutissimum*, *Corynebacterium pseudodiphtheriticum*, *Corynebacterium striatum*, *Corynebacterium urealyticum*, and *Corynebacterium xerosis*. These organisms have been found to spread through both zoonotic and nosocomial routes. While the *Corynebacterium* genus is associated with various infections, notable conditions include upper respiratory tract diseases, such as diphtheria and exudative pharyngitis, as well as cutaneous infections, including ulcers and granulomatous lymphadenitis. Diagnosis of these infections can be performed using phenotypic methods and MALDI-TOF MS, while 16S rRNA gene sequencing remains the gold standard for species identification. Prompt diagnosis and timely treatment play a crucial role in reducing morbidity and mortality associated with these infections. Widespread vaccination in both humans and animals can lower the prevalence of these diseases. Implementing proper precautions when handling domestic animals can further reduce the risk of transmission.

A. Gupta · K. S. Abhishek · V. Tak (✉)
Department of Microbiology, All India Institute of Medical Sciences, Jodhpur, Rajasthan, India

S. C. Parija, A. Poddar (eds.), *Textbook of Bacterial Zoonoses*, Microbial Zoonoses,
https://doi.org/10.1007/978-981-95-6615-0_45

History

Corynebacterium as a genus was first described by Lehmann and Neumann in 1896, deriving the name from the Greek "*Kyrone*", indicating club and "*bacterion*", meaning little rod. *Corynebacterium diphtheriae* is the representative organism of the group, whose earliest written records date back to the fifth century BC (Table 45.1). Diphtheria outbreaks and epidemics have been recorded in Spain and New England, which caused massive morbidity and mortality. The disease diphtheria was coined by Bretonneau, meaning *leather-hide* in Greek. Klebs isolated the bacilli, while Loeffler proved it to be the etiologic agent. Evidence of passive immunity, as discovered by von Behring, paved the way for the treatment and prevention of this disease.

The discovery of another toxigenic *Corynebacterium* followed the isolation of *Corynebacterium ulcerans* in 1926. Zoonotic organisms, such as *Corynebacterium pseudotuberculosis*, were isolated in 1891 from sheep; however, the first clinical case was not reported until 1966. Other organisms, such as *C. urealyticum* and *C. straitum*, were recently reclassified from CDC group organisms. Differentiation of *Corynebacterium* species from other coryneform bacteria has been further enhanced with the evolution of molecular techniques, resulting in the expansion of the genus.

Taxonomy

The genus *Corynebacterium* belongs to the family Corynebacteriaceae in the Phylum Actinomycetota, Class Actinomycetes, and Order Mycobacteriales. The genus *Corynebacterium* comprises 88 published species, of which 53 species have been shown to cause human infections, while 35 species have only been recovered from animal sources. The CDC published group names and identification schemes for *Corynebacterium*-like species, all of which have been almost entirely reclassified into species or other genera.

The taxonomy of the genus *Corynebacterium* has been evolving through newer genomic and sequencing methods established through 16S rRNA sequencing or multilocus sequence typing. Another widely used method is *rpoB* gene sequencing, where taxa with ≥95% identity are considered members of the same species and those with ≥98% identity are deemed members of the same subspecies. New species of *Corynebacterium* continue to be isolated from sites such as the female urine microbiome and horse conjunctiva.

Genomics and Proteomics

The genome of *Corynebacterium* contains a single, circular chromosome with a genome size typically between 2.3 and 2.5 Mbp. For example, *C. diphtheriae* NCTC 13129 has a genome size of approximately 2.49 Mbp, and *C. ulcerans* strains also fall within the range of approximately 2.4–2.5 Mbp. Similarly, *C. pseudotuberculosis* P54B96 possesses a genome of approximately 2.34 Mbp. The species has a high G + C content, typically ranging from 52% to 54%.

The genomes of these pathogens are densely packed with coding regions. Each species typically contains over 2000 protein-coding sequences (CDSs). *C. diphtheriae* NCTC13129, for instance, has 2320 CDSs, while *C. ulcerans* P69 has 2291. *C. pseudotuberculosis* FRC41 is annotated with 2110 protein-coding regions. Beyond protein-coding genes, all three species contain the necessary ribosomal RNA (rRNA) and transfer RNA (tRNA) genes for protein synthesis. For example, *C. ulcerans* P69 has 51 tRNA and 12 rRNA genes, and *C. pseudotuberculosis* FRC41 has 49 tRNA genes and four rRNA operons.

C. diphtheriae and *C. ulcerans* are identified by the presence of the diphtheria toxin *(tox)* gene. This gene is located on a temperate bacteriophage (*corynephage*) integrated into the bacterial chromosome. The acquisition of this phage through horizontal gene transfer is a critical event in the evolution of toxigenic strains. While *C. diphtheriae* is classically known for producing diphtheria toxin, *C. ulcerans* strains can also harbour and express this tox gene, leading to

Table 45.1 Key features of different *Corynebacterium* species

Species	Gram stain	Morphology	Key features	Associated infections	Normal habitat/ reservoir
Corynebacterium diphtheriae	Positive	Club-shaped rods, palisades	Produces diphtheria toxin (if toxigenic strain)	Diphtheria (respiratory and cutaneous), systemic complications (myocarditis, neuropathy)	Human nasopharynx, skin
Corynebacterium ulcerans	Positive	Small irregular rods with clubbed ends	Produces diphtheria toxin (if toxigenic strain)	Diphtheria (respiratory and cutaneous)	Cats, dogs, monkeys
Corynebacterium pseudotuberculosis	Positive	Pleomorphic coccoid morphology, palisades	Produces diphtheria toxin (if toxigenic strain)	Lymphadenitis (human)	Sheep, goats
Corynebacterium jeikeium	Positive	Pleomorphic rods	Lipophilic, often multidrug-resistant	Bacteraemia, endocarditis, prosthetic device infections, skin infections	Human skin
Corynebacterium urealyticum	Positive	Pleomorphic rods	Urease-positive, associated with alkaline urine	Urinary tract infections (often with encrustation), bacteraemia	Human skin, urinary tract
Corynebacterium striatum	Positive	Pleomorphic rods	Increasingly antibiotic-resistant	Respiratory tract infections, bacteraemia, wound infections, prosthetic infections	Human skin, nasopharynx, cats, and dogs
Corynebacterium pseudodiphtheriticum	Positive	Slender rods	Non-lipophilic	Respiratory tract infections (pharyngitis, bronchitis, pneumonia), endocarditis	Human nasopharynx
Corynebacterium amycolatum	Positive	Pleomorphic rods	Lipophilic, common skin commensal	Opportunistic infections (bacteraemia, skin infections, prosthetic infections)	Human skin
Corynebacterium minutissimum	Positive	Small rods	Lipophilic, produces porphyrins (coral-red fluorescence under Wood's lamp)	Erythrasma (superficial skin infection)	Human skin
Corynebacterium kroppenstedtii	Positive	Pleomorphic rods	Lipophilic	Granulomatous mastitis, breast abscess	Human skin
Corynebacterium bovis	Positive	Irregular rods with clubbed ends, occasional coccobacillary forms	Lipophilic	Sepsis in humans, mastitis in cows	Cows
Corynebacterium canis	Positive	Club-shaped rods, occasional filamentous, and occasional branching	Facultative aerobic, non-lipophilic	Wound infection in humans after dog bite	Dogs

(continued)

Table 45.1 (continued)

Species	Gram stain	Morphology	Key features	Associated infections	Normal habitat/reservoir
Corynebacterium freiburgense	Positive	Club-shaped rods, singly, pair, or cluster	Non-lipophilic, >5-day-old colony may have a spoke-wheel appearance	Wound infection in humans after dog bite	Dogs
Corynebacterium falsenii	Positive	Club-shaped rods, singly, pair, or cluster	Non-lipophilic, yellowish pigment after 72 h of incubation	Sepsis	Mouth of an eagle
Corynebacterium glucuronolyticum	Positive	Rods with coryneform morphology	Non-lipophilic, growth in 5% CO_2	Prostatis and urethritis in humans	Urogenital tract of pigs
Corynebacterium jeikeium	Positive	Pleomorphic club-shaped rods with V forms or palisades	Obligate aerobe	Endocarditis, sepsis, prosthetic device infections	Human body surfaces and clinical specimens
Corynebacterium kutscheri	Positive	Irregular stained rods, metachromatic granules present	Reduces potassium tellurite, large number of fimbriae present	Skin and soft tissue infection	Mice and rats

diphtheria-like illnesses in humans, often with zoonotic origins. Comparative analyses have shown that the tox gene in *C. ulcerans* can exhibit high sequence similarity (up to 95%) to that found in *C. diphtheriae*, although the specific prophages carrying the gene can differ in their sequences.

In contrast, *C. pseudotuberculosis* does not typically carry the diphtheria toxin gene. Its primary virulence factor is the phospholipase D (pld) gene. This enzyme, also known as sphingomyelinase D, is crucial for the bacterium's ability to spread within host tissues. The enzyme is also a significant contributor to the pathology of diseases such as caseous lymphadenitis in animals. *C. ulcerans* also produces phospholipase D. This suggests a shared virulence mechanism and evolutionary relatedness between these two species, which are sometimes grouped in the "diphtheriae complex" or "C. pseudotuberculosis complex" due to shared pathogenicity characteristics.

Morphology

All species in the *Corynebacterium* genus are Gram-positive rods arranged in either pairs or palisades, often with club-shaped swellings at the poles. This arrangement is due to the snapping type of division that occurs in the members of this genus. They are non-motile, non-spore-forming, non-capsulated, and non-acid-fast. The key features of different *Corynebacterium* species have been summarised in Table 45.1.

The presence of meso-diaminopimelic acid characterises the cell walls of these species, with arabinose and galactose being the most prevalent sugars. Short-chain mycolic acids, called coryno-mycolates, of 22–36 carbons in length, are present. However, some species, such as *Corynebacterium amycolatum*, *Corynebacterium atypicum*, and *Corynebacterium kroppenstedtii*, lack mycolic acids altogether. Long-chain fatty acids are present with very few or absent branched-chain fatty acids. Phospholipids in this group comprise basic types such as phosphatidylinositol, phosphatidylinositol dimannosides, phosphatidylglycerol trehalose dimycolates, and various other glycolipids. Notably, phosphatidylethanolamine is generally absent, except for *C. bovis* and *C. urealyticum*.

C. diphtheriae bacilli are pleomorphic, slender, non-capsulated and non-sporing Gram-positive rods. Length may range from 0.5 to 1 μm in ovoid forms to Gram-variable rods of 1.5–5 μm. The morphology varies depending on the age of the culture, the degree of subculturing, and the type of media on which it is grown. The bacilli contain polyphosphate granules, also known as volutin, Babes-Ernst, or metachromatic granules. When stained, these bacilli form a *Chinese-letter appearance* due to their tendency to lie in pairs at obtuse angles. Certain biotypes of *C. diphtheriae*, such as intermedius, may not form volutin granules.

As the type species of this genus, the ultrastructure of *C. diptheriae* has been described in detail. Common cellular inclusions found include concentrated lipids in the polar areas of the cell. Volutin granules, a characteristic feature of this species, are composed of inorganic phosphates. They are characteristically present only in slow-dividing organisms, with the absence of these inclusion bodies noted in actively multiplying organisms.

Cultivation

Corynebacterium species are considered fastidious and are facultative anaerobes that typically require the presence of blood or serum for optimal growth. Preferred media include potassium tellurite agar containing lysed horse blood or Loeffler's serum slope. The optimal growth temperature is typically within the mesophilic range, around 35–37 °C, and the pH of the culture media is usually around 7.0.

These organisms are recovered from clinical samples such as throat swabs, pseudomembranes, and abscesses using standard incubation protocols. Lipophilic species, such as *C. afermentans*, *C. jeikeium*, and *C. urealyticum*, can also be cultivated by supplementing lipid-free media, such as Mueller-Hinton agar, with 0.1–1% Tween 80.

These bacteria display varied metabolic profiles, ranging from fermentative to oxidative, or in some cases, lacking both.

Animal Models

Guinea pigs are the preferred animal model for *C. diphtheriae*, primarily because their cells possess a functional diphtheria toxin receptor (HB-EGF), which renders them susceptible to the toxin's cytotoxic effects. Experimental infection in guinea pigs typically mimics human diphtheria, resulting in the characteristic formation of a pseudomembrane and systemic symptoms. Furthermore, the nematode *Caenorhabditis elegans* has emerged as a high-throughput, cost-effective invertebrate model, useful for initial screens of bacterial colonisation, survival, and the impact of virulence factors.

In the study of *C. ulcerans*, given its significant zoonotic potential, animal models often focus on understanding its natural reservoirs and the mechanisms of transmission. Companion animals, such as dogs and cats, are frequently observed with natural *C. ulcerans* infections, making them crucial natural reservoirs. Research in these animals typically involves epidemiological studies to investigate the prevalence, molecular epidemiology, and transmission pathways, providing invaluable insights into the zoonotic cycle.

For *C. pseudotuberculosis*, which is primarily an animal pathogen, the animal models directly involve its natural hosts. Sheep and goats are the most relevant and widely used models for studying caseous lymphadenitis (CLA), the chronic disease caused by this bacterium. Experimental infections in these small ruminants enable the precise replication of CLA, including the formation of abscesses in lymph nodes and internal organs.

Transmission Cycle

Transmission of zoonotic pathogen *C. ulcerans* from animals to humans primarily occurs through direct or indirect contact. Direct contact often involves exposure to animal secretions or exudates, particularly from companion animals. For example, humans are infected through contact with respiratory secretions from a coughing or sneezing animal, wound exudates, or skin lesions. Activities like petting, grooming, or caring for sick animals significantly increase this risk. Less commonly, animal bites or scratches contaminated with *C. ulcerans* can directly inoculate the bacteria into a human. Indirect transmission also occurs when bacteria are shed into the environment (e.g., soil, animal bedding) through animal secretions or faeces, allowing humans to acquire the infection by touching contaminated surfaces and then transferring the bacteria to their mouth, nose, or open wounds. Historically, consuming unpasteurised milk or dairy products from infected cows was a significant route, underscoring the importance of food safety practices.

There is limited, though debated, evidence suggesting the potential for human-to-animal or human-to-human spread. A few documented cases have shown genetically indistinguishable *C. ulcerans* strains in humans and their household pets, raising the possibility of human-to-animal transmission. Unlike *C. diphtheriae*, human-to-human transmission of *C. ulcerans* is considered very rare and poorly documented, with most human cases clearly linked to animal contact.

Pathogenesis and Pathology

Corynebacterium diphtheriae

Adhesion is mediated by pili or fimbriae, which help the organism attach to the host cell wall. They are encoded by pili gene clusters, which encode pilin proteins that each play a specific role in adherence to the pharynx. Other factors are also postulated to play a role in adhesion, including hemagglutinins and non-fimbriae-associated proteins.

Diphtheria toxin (DT), the major virulence factor of these species, is encoded by the tox gene carried on corynebacteriophages. The bacterium becomes toxigenic when the phage genome integrates into specific attB sites within Arg-tRNA2 genes through lysogenic conversion, with some highly virulent strains carrying multiple copies of the tox gene. Although of phage origin, toxin expression is regulated by the bacterial DtxR gene, a metal-activated repressor that controls both iron metabolism and DT production. In low iron conditions, DtxR repression is lifted, resulting in increased toxin expression.

DT is secreted as a 58 kDa exotoxin composed of two subunits, the A-subunit (catalytic) and the B-subunit (responsible for receptor binding and membrane translocation) linked by a disulfide bond. The A-subunit catalyses the ADP-ribosylation of elongation factor 2 (leading to inhibition of protein production and cell death). The B-subunit binds to the heparin-binding epidermal growth factor (HB-EGF) receptor on human cells and is internalised via receptor-mediated endocytosis (Fig. 45.1). Notably, many non-toxigenic *C. diphtheriae* and related species harbour functional DtxR genes. They, by phage lysogeny, may become toxigenic, posing a potential reservoir for emerging pathogenic strains.

Corynebacterium ulcerans

C. ulcerans causes disease through a variety of virulence factors, including the diphtheria toxin and phospholipase D. DT was initially thought to be exclusive to *C. diphtheriae*. Although both toxins share a 95% similarity, there are certain key differences: (1) Toxigenic strains are lysogenised by a bacteriophage, which differs from the phages found in *C. diphtheriae*. (2) Additionally, toxin production in *C. ulcerans* is lower than in *C. diphtheriae*.

Genomic comparisons show *C. ulcerans* is more closely related to *C. pseudotuberculosis* than to *C. diphtheriae*. PLD is conserved across *C. ulcerans* strains and disrupts host cell membranes, promoting bacterial dissemination. Additional candidate virulence factors include (i) a Shiga-like ribosome-inactivating protein, (ii) neuraminidase NanH, and (iii) protease Cpp. All of these are potentially acquired via horizontal gene transfer.

Corynebacterium pseudotuberculosis

Diphtheria toxin, as well as phospholipase D, is considered major virulence factor of *C. pseudotuberculosis*. Encoded by the *pld* gene, it facilitates tissue invasion and dissemination. PLD is conserved across most *C. pseudotuberculosis* strains, although some exhibit mutations that may affect virulence. Genomic studies have also identified other key virulence-associated proteins, including Spa surface proteins, corynebacterial protease CP40, serine proteases, neuraminidase NanH, and nitric oxide reductase.

New Diphtheria Toxin-Producing Species

Corynebacterium silvaticum sp. nov. is a recently recognised species first isolated from wild boars with caseous lymphadenitis. It is closely related to *C. ulcerans* and *C. pseudotuberculosis*. It also exhibits distinct genetic, biochemical, and phenotypic traits, including unique fermentation patterns and lipid profiles. It has been isolated from both wildlife and livestock, indicating potential zoonotic transmission. Proteomic and transcriptomic investigations have identified key virulence-associated proteins, supporting their pathogenic role.

Additionally, *Corynebacterium rouxii* sp. nov. is another species, closely related to *C. diphtheriae*. It was isolated from both humans and animals and shows similarities to the Belfanti biovar. Molecular and genomic analyses confirm it as a distinct species, characterised by high genetic homogeneity and unique biochemical markers.

Immunology

The immunology of *Corynebacterium* infections involves both innate and adaptive immune responses. These eliminate the pathogen and neutralise its potent virulence factors, particularly toxins. While there are overarching similarities in the host's defensive strategies across the genus, the specific immune responses elicited often

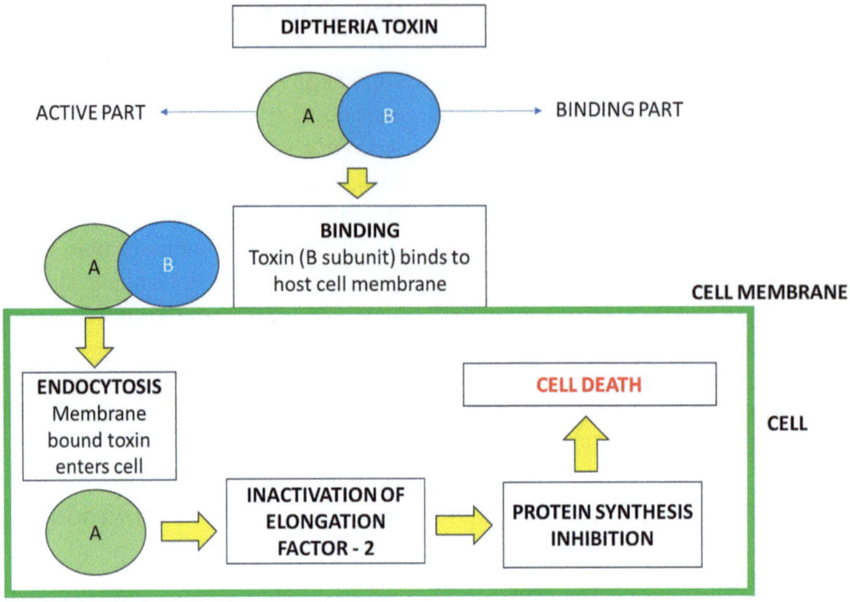

Fig. 45.1 Mechanism of action of diphtheria toxin. (Image source: created by authors)

diverge. It is based on the infecting *Corynebacterium* species and its unique pathogenic mechanisms.

Corynebacterium diphtheriae

The immune response is dominated by the host's reaction to the diphtheria toxin (DT). The innate immune system forms the initial line of defence, with epithelial barriers providing physical protection. Upon infection, immune cells like macrophages and neutrophils are rapidly recruited. These phagocytes recognise bacterial pathogen-associated molecular patterns (PAMPs), such as components of the *C. diphtheriae* cell wall, through pattern recognition receptors (PRRs) like Toll-like receptors (TLR2), initiating an inflammatory cascade.

However, *C. diphtheriae* exhibits strategies to evade innate immunity. The distinctive pseudomembrane formation is a complex outcome of bacterial proliferation and intense host inflammation leading to tissue necrosis.

The adaptive immunity against diphtheria is humoral immunity, specifically the generation of antitoxin antibodies (IgG). These antibodies are crucial, as they bind to and neutralise diphtheria toxin, thereby preventing its attachment to host cells and inhibiting protein synthesis.

Corynebacterium ulcerans

The immune response against *C. ulcerans* shares significant commonalities with *C. diphtheriae* when toxigenic strains are involved. Beyond the toxin, the immune system mounts responses to various bacterial surface components and secreted factors. However, the precise contributions of cellular versus humoral immunity are subjects of ongoing investigation.

Corynebacterium pseudotuberculosis

Immunology primarily focuses on controlling an intracellular pathogen that typically establishes chronic abscesses in its animal hosts. Protection against *C. pseudotuberculosis*, which causes caseous lymphadenitis (CLA), necessitates a complex interplay between both humoral and cell-mediated immune responses. Antibodies, particularly IgM and IgG, are produced in response to bacterial antigens, including the crucial virulence factor, phospholipase D (PLD). Antitoxin antibodies targeting PLD are considered important for neutralising its tissue-damaging effects and preventing bacterial dissemination.

However, given the organism's facultative intracellular location within host macrophages, a potent cell-mediated immune response (Th1-type) is triggered to control the infection. Cytokines such as IFN-γ and TNF-α, produced by Th1 cells and other immune cells, are crucial. They activate macrophages (i) to effectively eliminate intracellular bacteria and (ii) facilitate the formation and maintenance of granulomas, which serve to wall off the infection.

Infection in Humans

Exudative Pharyngitis

Causative agents include *C. diphtheriae*, *C. ulcerans* and *C. pseudodiphtheriticum*. Signs and symptoms of infection typically become apparent within 4–5 days, with involvement of the tonsillar pillars and pharynx. Clinical manifestations include fever and throat pain, accompanied by the formation of a pseudomembrane 72 h after the first clinical symptom. The membrane is initially white and may become greyish-green or black as it extends to the soft palate, nasopharynx, or laryngopharynx. Cervical lymph node enlargement and submandibular oedema lead to a "*bull-neck*" appearance in affected individuals. Complications include stridor and respiratory distress secondary to airway obstruction. Cardiac and neurological toxicity can also occur secondary to pharyngeal diphtheria.

Cutaneous diphtheriae

It is caused by non-toxigenic *C. diphtheriae* and *C. ulcerans*. The condition manifests as an ulcer, which later progresses to a pseudomembrane. Widespread immunisation has led to a decrease in respiratory tract diphtheria, and consequently, cases of cutaneous diphtheriae have increased.

Granulomatous lymphadenitis can occur in farm and veterinary workers secondary to *C. pseudotuberculosis*. Erythrasma, secondary to infection by *Corynebacterium minutissimum*, is a superficial skin infection that causes red, scaly patches. Diagnosis is made by observing coral-red fluorescence under a Wood's lamp and culturing skin scrapings. Breast abscess and granulomatous mastitis can occur secondary to infections caused by *Corynebacterium kroppenstedtii* and *Corynebacterium tuberculostearicum*.

Other Infections

Organisms such as *Corynebacterium glucuronolyticum* and *C. urealyticum* have been implicated in causing urinary tract infections and chronic prostatitis. *Corynebacterium propinquum* is a causative agent of native valve endocarditis. Some species, such as *Corynebacterium straitum*, are also causes of native joint infections.

Infection in Animals

C. pseudotuberculosis is a common cause of caseous lymphadenitis in sheep and goats. It is a chronic disease characterised by abscesses in the skin, lymph nodes and internal organs. Due to its propensity to become endemic in a flock and poor response to treatment, the dreaded infection can have a significant economic impact on the farmer.

C. ulcerans infects both wild and domestic animals, including cattle, causing skin and soft tissue infections. It is also found as a routine commensal in a wide range of animal hosts. *C. bovis* is an important cause of bovine mastitis, with complications ranging from septicaemia to necrosis. *C. amycolatum* and *C. urealyticum* are associated with urinary tract infections in dogs and cats. Additionally, otitis externa in dogs is caused by *Corynebacterium auriscanis*. *C. renale* is associated with cystitis and pyelonephritis in cows.

Epidemiology and Public Health

Zoonotic species of *Corynebacterium*, particularly *Corynebacterium ulcerans* and *C. pseudotuberculosis*, have emerged as important but underrecognised pathogens capable of causing diphtheria-like illness in humans. While *C. diphtheriae* remains the primary agent in classical diphtheria, *C. ulcerans*—which can also carry the *tox* gene—has increasingly been identified in both sporadic and clustered human cases across Europe, the United Kingdom, and Japan. Notably, surveillance data from France and the UK indicate that human infections with *C. ulcerans* have, in some years, surpassed those caused by *C. diphtheriae*, with domestic pets and raw milk products implicated as common sources of exposure. In Japan, a retrospective review over two decades showed a steady increase in *C. ulcerans* cases.

C. pseudotuberculosis predominantly affects livestock such as goats and sheep, causing caseous lymphadenitis. Human infections are infrequent but typically occur in individuals with occupational or environmental exposure to infected animals. Though widely reported in veterinary literature, human cases are rare. In South and Southeast Asia—including India—the true burden of zoonotic *Corynebacterium* may be significantly underestimated. Zoonotic Corynebacterium infections likely represent a considerable proportion of cases that go undetected due to under-reporting and limited clinical recognition. In India, *C. pseudotuberculosis* has been identified in livestock, notably in the Andaman and Nicobar Islands. The expanding role of zoonotic *Corynebacterium* in both developed and resource-limited settings underscores the urgent need for integrated "One Health" surveillance strategies that bridge human, animal, and environmental health systems.

The newly discovered tox-bearing species, *Corynebacterium silvaticum*, is isolated from wild boars, while *Corynebacterium rouxii* has been found in the peritoneum of dogs. Hosts of other zoonotic *Corynebacterium* infections include dogs for *Corynebacterium canis*, *Corynebacterium freiburgense* and *C. urealyticum*; cows for *C. bovis* and *C. xerosis*; and rats for *C. kutscheri*. Many of these bacteria are not exclusive to one animal origin. As diagnostic methods improve, the global spread of zoonotic Corynebacterium species is becoming better understood, with many countries, including India, now reporting infections caused by species other than *C. diphtheriae*.

Diagnosis

Microscopy

Corynebacterium are Gram-positive, non-sporing strains, sometimes club-shaped and occasionally with volutin granules.

In Gram-stained smears, *C. diphtheriae* is characterised by a pleomorphic arrangement in a V or L shape, also known as the Chinese-letter pattern (Fig. 45.2). Albert's stain is used to stain volutin or storage granules. The bacilli stain green, while the granules stain bluish-black; hence, they are also called *metachromatic granules*, characteristic of *C. diphtheriae* and *C. xerosis*. Although this appearance on microscopy is influenced by the type of medium and the age of the culture. This method also serves as a means of rapid identification from clinical specimens, such as throat swabs and pseudomembranes.

In Vitro Culture

Corynebacterium species typically grow well on enriched media containing serum or blood, such as horse or sheep blood agar.

On blood agar, *Corynebacterium* species produce convex, greyish, translucent colonies with a granular texture and more opaque centres. Colony morphology varies across species and biotypes. *C. diphtheriae* biotype gravis forms non-haemolytic, dull grey to black colonies with a matt surface and friable consistency on Hoyle's medium. *C. diphtheriae* biotype mitis, by contrast, is beta-haemolytic, producing glossy, smooth colonies. *C. diphtheriae* biotype intermedius yields smaller, translucent colonies with a

Gram's Stain: Gram-positive, non-sporing bacilli, sometimes club shaped and occasionally with volutin granules. Characterised by a pleomorphic arrangement in V or L shaped also named as Chinese-letter pattern.	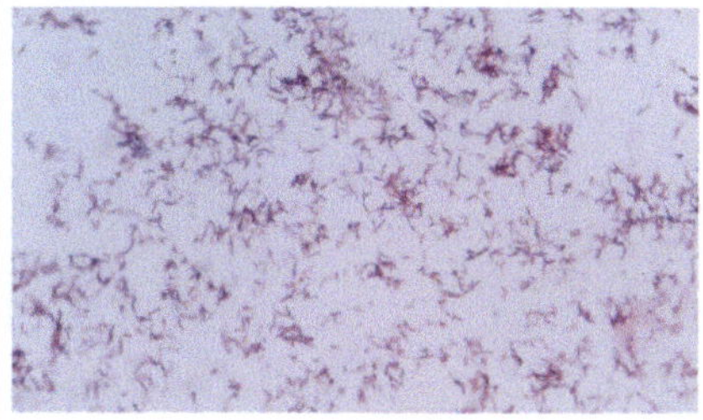
Albert's Stain: The bacilli stain green while the granules stain bluish black with characteristic Chinese-letter pattern.	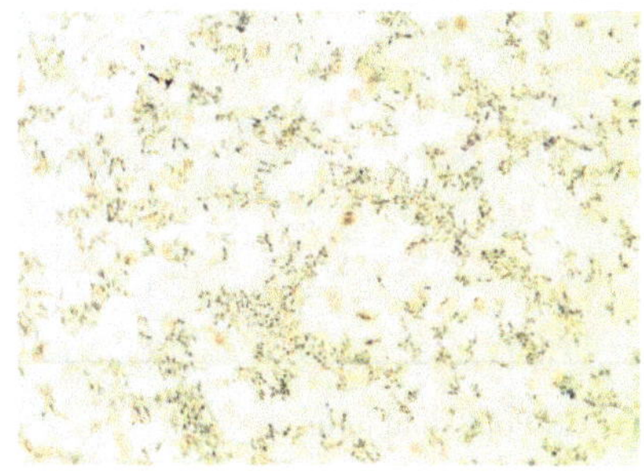

Fig. 45.2 Gram stain and Albert stain of *Corynebacterium diphtheriae* showing characteristic Gram-positive bacilli and metachromatic granules. (Image source: Department of Microbiology, AIIMS Jodhpur)

shiny surface. *C. diphtheriae* biotype belfantii has now been reclassified into a separate species, *C. belfantii*.

Hoyle's tellurite blood agar is employed for the selective isolation and preliminary identification of *Corynebacterium*. Tellurite suppresses background flora and aids in the presumptive identification of *C. diphtheriae* by the characteristic black colour of its colonies (Fig. 45.3). Both *C. ulcerans* and *C. pseudotuberculosis* produce beta-haemolytic, dry, opaque colonies with black pigmentation on tellurite media.

Loeffler's serum slope is an enriched medium used for enhancing the growth of *C. diphtheriae*, while Tinsdale agar, which detects the cystinase enzyme, is useful in differentiating potentially toxigenic corynebacteria. Toxigenic *Corynebacterium* species, such as *C. diphtheriae*, *C. ulcerans*, and *C. pseudotuberculosis*, typically produce black colonies with a surrounding brown halo.

These distinct morphological traits, in combination with selective and differential media, support accurate identification and biotyping of *Corynebacterium* isolates.

Confirmation of *Corynebacterium* isolates can be achieved by biochemical reactions, such as the nitrate reduction test, urea hydrolysis, pyrazinamidase testing, and carbohydrate fermentation. The urease test differentiates urease-positive *C. ulcerans* and C. *pseudotuberculosis* from that of urease-negative *C. diphtheriae*. *C. diphtheriae* and *C. ulcerans* are typically nitrate-positive. *C. pseudotuberculosis* shows variability in the nitrate reduction test, with *biovar equi* generally positive and *biovar ovis* usually negative, aiding in biovar differentiation.

Once the isolate is confirmed to be *C. diphtheriae*, *C. pseudotuberculosis*, or *C. ulcerans*, the demonstration of toxin can be detected by Vero cell cytotoxicity assay and Elek's gel precipitation test.

The Elek test is an *in vitro* immunodiffusion method that detects toxigenic strains by antigen-antibody interactions. In this test, a strip of antitoxin-impregnated filter paper is embedded in a proteose peptone agar medium, and the test strain is streaked perpendicular to it alongside known positive and negative controls. If the isolate produces diphtheria toxin, it

A. Loeffler's serum slope	
B. Blood agar	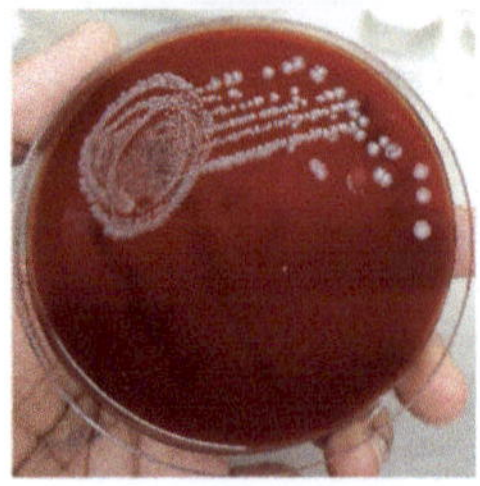
C. Potassium tellurite blood agar	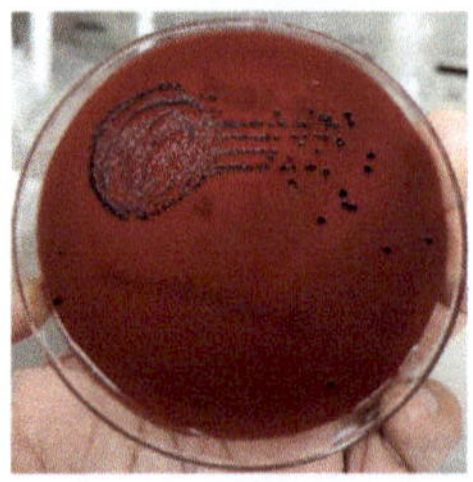
D. Elek's gel precipitation reaction showing line of precipitation for toxigenic strains.	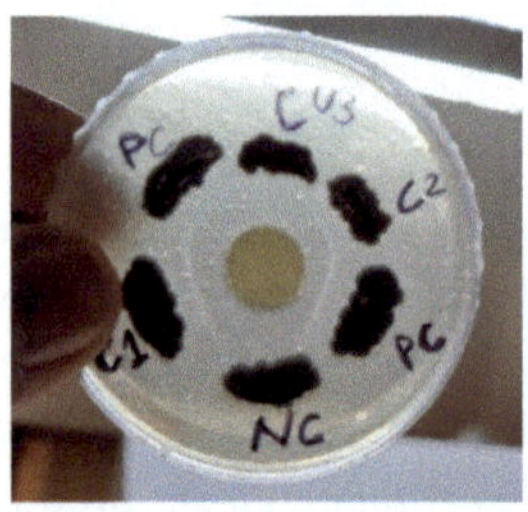

Fig. 45.3 Culture characteristics of *Corynebacterium diphtheriae* and demonstration of toxic production. (**a**) Loeffler's serum slope. (**b**) Blood agar. (**c**) Potassium tellurite blood agar. (**d**) Elek's gel precipitation reaction showing a line of precipitation for toxigenic strains. (Image source: Department of Microbiology, AIIMS Jodhpur)

diffuses into the agar and reacts with the antitoxin to form visible precipitin lines. The appearance of a line of identity between the test and the positive control confirms toxigenicity. A novel advancement in the detection of toxigenic *Corynebacterium* species is the modified Elek test, which offers a significant improvement over the conventional method. This assay shortens the time to result while maintaining high sensitivity and specificity. This method has optimised variables such as agar thickness, antitoxin concentration, and inoculum placement, thus reducing misinterpretation.

Vero cell cytotoxicity assay offers a cell-based alternative that assesses the biological activity of the toxin. Serial dilutions of culture supernatant are added to a monolayer of Vero cells, which are known to be highly sensitive to diphtheria toxin. After incubation, cytopathic effects such as cell rounding and detachment are observed under the microscope. The test well is compared with the controls to determine the presence of antitoxin. When required, neutralisation with antitoxin can be included to confirm specificity.

Other automated methods include the Vitek 2 Anaerobe/Corynebacterium card, BD Phoenix, and the API Coryne panel.

Serodiagnosis

Serodiagnosis of toxigenic *Corynebacterium* species is made by detecting toxins through enzyme immunoassay or latex agglutination. These tests use monoclonal antibodies coated with antitoxin.

Enzyme-linked immunosorbent assay (ELISA) is the most widely used method. ELISA kits are designed to detect human IgG antibodies against diphtheria toxoid. They are quantitative, allowing for the determination of antibody concentrations, typically expressed in International Units (IU) per millilitre (mL). A protective level is generally considered to be 0.1 IU/mL or higher. In contrast, a level of ≥0.01 IU/mL is considered the minimal protective level.

Toxin neutralisation tests (TNTs) are considered the "gold standard" for measuring functional antitoxin antibodies. They involve mixing patient serum with diphtheria toxin and then inoculating the mixture into susceptible cell lines (e.g., Vero cells) or laboratory animals (e.g., guinea pigs). The highest dilution of serum that neutralises the toxin's cytotoxic effect indicates the antibody titre.

Schick test is the in vivo skin test that was used historically to assess diphtheria immunity. A small amount of diphtheria toxin was injected intradermally. A positive reaction (localised inflammation) indicated susceptibility (lack of protective antitoxin), while a negative reaction indicated immunity. Due to safety concerns and the availability of in vitro tests, it is rarely used currently.

Molecular Diagnosis

Since the 1990s, *polymerase chain reaction (PCR)* has been widely employed to detect the diphtheria toxin gene (*tox*), particularly the biologically active fragment A. Various primer sets targeting different regions of this gene have been successfully developed and are used to detect *tox* in clinical isolates and direct specimens.

Real-time PCR has since emerged as a more rapid alternative, eliminating the need for post-amplification gel electrophoresis. However, early assays had limitations in detecting certain *C. ulcerans* strains, prompting the development of more inclusive assays targeting conserved regions across *C. diphtheriae*, *C. ulcerans*, and *C. pseudotuberculosis*. A notable advancement includes the quadruplex PCR developed by De Zoysa et al. (2016) which simultaneously detects *tox* and species-specific *rpoB* targets, thereby improving diagnostic accuracy and enabling the rapid triaging of potential toxigenic isolates before phenotypic confirmation.

For epidemiological investigations, *multilocus sequence typing (MLST)* is used to track genetic lineages of *C. diphtheriae* and *C. ulcerans*, based on housekeeping genes. Molecular tools, such as whole-genome and RNA sequencing, can further aid in outbreak analysis.

An overview of laboratory diagnosis of *Corynebacterium* infections is presented in Table 45.2.

Treatment

The cornerstone of treatment of diphtheria caused by *C. diphtheriae* involves prompt administration of diphtheria antitoxin (DAT) to neutralise circulating toxin, alongside antibiotic therapy to halt further bacterial growth. The treatment is initiated on clinical suspicion without waiting for laboratory confirmation.

Diphtheria antitoxin is administered to neutralise existing antitoxin and is administered to all cases. Since the toxin is equine-derived, hypersensitivity prior to testing is recommended.

Table 45.2 Laboratory diagnosis of *Corynebacterium* infection

Diagnostic method	Target	Remarks	*Corynebacterium diphtheriae* specifics	*Corynebacterium ulcerans* specifics	*Corynebacterium pseudotuberculosis* specifics
Direct detection methods					
Microscopy: Gram stain	Gram-positive, non-sporing rods, pleomorphic, club-shaped, V/L/ Chinese-letter pattern.	Provides rapid preliminary identification based on characteristic shapes and arrangements	Characteristic pleomorphic arrangement in V or L shapes ("Chinese-letter pattern")	Similar pleomorphic, club-shaped rods, "Chinese-letter" patterns	Gram-positive, pleomorphic rods, often in "Chinese-letter" or palisades from abscess material
Albert's stain	Volutin (metachromatic) granules	Stains bacilli green and granules bluish-black	Characteristic appearance due to prominent metachromatic granules	Granules may be present but potentially less prominent	Not typically associated with prominent metachromatic granules
In vitro culture	Bacterial growth and isolation on enriched/ selective media	Requires enriched media (serum/blood). Tellurite suppresses background flora and aids presumptive ID	Grows well on blood agar, Loeffler's serum slope, and tellurite media (black colonies)	Grows on enriched media; produces black colonies with a brown halo on Tinsdale agar	Colonies on blood agar show convex, greyish, translucent colonies and often beta-haemolytic, black colonies on tellurite media with brown halo
Colony morphology on tellurite media	Black colonies with brown halos (Tinsdale agar) or specific features on Hoyle's	Presumptive identification on the basis of characteristic black pigmentation	Biovar gravis: non-haemolytic, dull grey to black colonies, matt surface, friable on Hoyle's. Biovar mitis: beta-haemolytic, glossy, smooth colonies Biovar intermedius: smaller, translucent, shiny colonies	Beta-haemolytic, dry, opaque colonies with black pigmentation on tellurite media; brown halo on Tinsdale	Beta-haemolytic, dry, opaque colonies with black pigmentation on tellurite media; brown halo on Tinsdale
Biochemical tests	Metabolic reactions (e.g., enzyme production, carbohydrate fermentation)	Differentiates species based on unique metabolic profiles. Automated methods are available (Vitek 2, BD Phoenix, API Coryne)	Urease-negative. Nitrate-positive. Ferments glucose and maltose but not sucrose	Urease-positive. Nitrate-positive. Ferments glucose, maltose, often sucrose and starch/glycogen	Urease-positive. Nitrate reduction variable (biovar equi positive, biovar ovis negative)
Toxin detection assays	Presence and activity of diphtheria toxin	Vero cell cytotoxicity assay (gold standard). Elek's gel precipitation test (precipitin lines with antitoxin strip)	Vero cell cytotoxicity assay. Elek's gel precipitation test	Not applicable (does not produce diphtheria toxin)	Crucial for clinical management of toxigenic *Corynebacterium diphtheriae* and *Corynebacterium ulcerans* to determine the need for antitoxin

Serodiagnosis (antibody/toxin detection)					
Toxin detection (EIA, latex agglutination)	Diphtheria toxin antigen	Detects diphtheria toxin directly using monoclonal antibodies	Detects diphtheria toxin directly	Not applicable (does not produce diphtheria toxin)	Rapid detection of circulating toxin
ELISA (antibody detection)	Human IgG antibodies against diphtheria toxoid	Quantitative detection of antitoxin levels (IU/mL). ≥0.1 IU/mL is considered protective	Detects antibodies against diphtheria toxoid, indicating immunity or exposure to toxigenic strains	Detects antibodies against diphtheria toxoid, indicating immunity or exposure to toxigenic strains	Not routinely used for direct *Corynebacterium pseudotuberculosis* infection diagnosis in humans
Toxin neutralisation tests (TNTs)	Functional antitoxin antibodies	"Gold standard" for measuring functional antitoxin antibodies	Measures functional antitoxin antibodies by neutralising toxin in cell lines or animal	Measures functional antitoxin antibodies if toxigenic *C. ulcerans* exposure occurred	Not applicable
Schick test	In vivo diphtheria immunity (historical)	Assessed skin reaction to intradermal diphtheria toxin	Positive reaction (inflammation) indicated susceptibility	Historically applicable if *C. ulcerans* exposure is considered	Not applicable
Molecular diagnosis					
PCR (conventional and real-time)	Diphtheria toxin gene (tox), species-specific genes (e.g., rpoB)	Detects tox gene (Fragment A) in isolates and direct specimens	Detects tox gene (Fragment A) in isolates and direct specimens	Detects *C. pseudotuberculosis*-specific genes, often pld	Rapid and sensitive detection of specific genes, especially the tox gene. Real-time PCR eliminates post-amplification electrophoresis
Quadruplex PCR (De Zoysa et al.)	*Tox gene* and species-specific rpoB targets	Simultaneously detects tox and rpoB targets	Simultaneously detects tox and rpoB targets, improving accuracy	Simultaneously detects rpoB for species ID	Advanced method for rapid triaging of potential toxigenic isolates and species identification
Multilocus sequence typing (MLST)	Housekeeping genes	Tracks genetic lineages of *C. diphtheriae* for epidemiology	Tracks genetic lineages of *C. ulcerans* for epidemiology	Used for tracking genetic lineages for epidemiological investigations	Useful for outbreak analysis and understanding genetic relatedness of strains
Whole-genome sequencing (WGS) and RNA sequencing	Entire genome sequence; gene expression	Aids in comprehensive outbreak analysis and understanding strain diversity	Aids in comprehensive outbreak analysis and understanding strain diversity	Aids in comprehensive outbreak analysis and understanding strain diversity	Provides high-resolution epidemiological data and insights into virulence and resistance

The dosage of antitoxin is determined based on the severity and duration of symptoms: 20,000 units for laryngitis, 40,000 units for extensive pseudomembrane, and 80,000 units for symptoms involving respiratory distress. It should be administered as soon as possible to prevent further complications.

Antibiotic therapy is given to both confirmed and suspected cases of diphtheria to reduce the risk of organ damage and transmission. The WHO recommends the use of macrolides in preference to penicillin G, with a dosage of 10–12 mg/kg of azithromycin daily for the paediatric population and 500 mg for adults.

Treatment of other non-toxin-producing ***Corynebacterium*** *species* is guided by antimicrobial susceptibility testing, with nosocomial isolates typically displaying susceptibility to vancomycin and linezolid. Some infections caused by *C. pseudotuberculosis* are treated through the excision of caseous lymphadenitis, along with concurrent antibiotics.

Prevention and Control

Vaccines

Diphtheria toxoid remains the foundational component of diphtheria vaccines. While Émile Roux and Alexandre Yersin discovered the diphtheria toxin in 1888, it was Gaston Ramon who made the groundbreaking discovery of the diphtheria toxoid in 1923. Diphtheria toxoid is either administered alone or in combination with other antigens, such as tetanus toxoid (in Td or DT formulations) or pertussis (as in DTwP or DTaP vaccines). In addition to these, multivalent formulations, including combinations with vaccines targeting poliovirus, *Haemophilus influenzae* type B, and hepatitis B, are available. Diphtheria toxoid potency is measured in flocculation (Lf) units, which are used to assess the antigen content. Paediatric vaccines typically contain between 7.5 and 25 Lf per dose. In contrast, formulations for older children and adults include lower doses (2–3 Lf) and are usually administered without the pertussis component.

The WHO currently advocates for six doses of diphtheria vaccine: three primary and three booster doses. The first dose is recommended at a minimum of 6 weeks of age, with subsequent doses spaced 4–8 weeks apart from each other. The booster schedule incorporates the first booster at 12–23 months with a DPT-containing vaccine, the second at 4–7 years and the final third at 9–15 years with Td.

Vaccines for other *Corynebacterium* species like *C. pseudotuberculosis* and *C. urealyticum* are currently under research.

Active surveillance of close contacts of diphtheria is conducted to prevent the further transmission of the disease. Chemoprophylaxis is administered either with a macrolide or penicillin in conjunction with the completion of the immunisation series. Diphtheria is a notifiable disease to the World Health Organisation, with active surveillance being conducted through the vaccine-preventable diseases programme. Additionally, many countries have implemented their own diphtheria prevention programmes, further enhancing the global efforts to curb diphtheria.

Case Study

A 55-year-old woman sustained a finger laceration while handling raw chicken. Within 24 h, the wound became red and painful, and she presented to the emergency department a week later with ascending cellulitis. She had multiple comorbidities, including end-stage heart failure, diabetes, chronic kidney disease, and a recent SARS-CoV-2 infection. Cultures grew *C. ulcerans*, confirmed by MALDI-TOF and PCR as toxigenic. Susceptibility testing showed intermediate sensitivity to penicillin and clindamycin, with full susceptibility to other antibiotics. Follow-up nasopharyngeal carriage swabs were negative in the patient and five household contacts. Both household dogs tested positive for toxigenic *C. ulcerans*, and genomic analysis showed near-identical isolates between the patient and her pets, supporting zoonotic transmission.

The patient initially received intravenous cefazolin and then switched to oral flucloxacillin.

After confirmation of toxigenic *C. ulcerans*, she was recalled and started on a 7-day course of oral doxycycline. DAT was not required, as there were no systemic signs. Household contacts received prophylactic antibiotics and booster vaccinations as needed. Both pet dogs were treated with amoxicillin-clavulanate, and one required a second course with enrofloxacin after continued carriage was detected.

1. What other diseases can be transmitted by dogs to humans?
2. What are the methods for the detection of toxins?
3. What are the strategies for preventing and controlling such bacterial zoonoses?

Research Questions

1. What is the current need for vaccine development against non-toxigenic *C. diphtheriae* (NTCD) and other zoonotic *Corynebacterium* species?
2. Given the diverse existence of *Corynebacterium* species, how can the exact role of various species in human and animal diseases be more precisely defined?
3. What further studies are needed to understand the pathogenesis and transmission of non-diphtheria *Corynebacterium* species?

Acknowledgement Dr Vidhi Jain, Associate Professor, AIIMS Jodhpur, and Dr Harsh Vardhan Goswami, Junior Resident, AIIMS Jodhpur, for contributing high-quality laboratory pictures.

Further Reading

Barksdale L. *Corynebacterium diphtheriae* and its relatives. Bacteriol Rev. 1970;34(4):378–422.

Bernard K. The genus *Corynebacterium* and other medically relevant coryneform-like bacteria. J Clin Microbiol. 2012;50(10):3152–8.

Bernard KA, Funke G. Corynebacterium. In: Bergey's manual of systematics of archaea and bacteria. Hoboken: Wiley; 2015. p. 1–70.

Bonmarin I, Guiso N, Le Flèche-Matéos A, Patey O, Patrick AG, Levy-Bruhl D. Diphtheria: a zoonotic disease in France? Vaccine. 2009;27(31):4196–200.

Burkovski A. Proteomics of toxigenic corynebacteria. Proteomes. 2023;12(1):2. https://doi.org/10.3390/proteomes12010002.

Clarke KE, MacNeil A, Hadler S, Scott C, Tiwari TS, Cherian T. Global epidemiology of diphtheria, 2000–2017. Emerg Infect Dis. 2019;25(10):1834.

Cutler SJ. Bacterial zoonoses: an overview. In: Molecular medical microbiology. San Diego: Elsevier; 2015. p. 1771–80.

De Zoysa A, Efstratiou A, Mann G, Harrison TG, Fry NK. Development, validation and implementation of a quadruplex real-time PCR assay for identification of potentially toxigenic corynebacteria. J Med Microbiol. 2016;65(12):1521–7.

Hacker E, Antunes CA, Mattos-Guaraldi AL, Burkovski A, Tauch A. *Corynebacterium ulcerans*, an emerging human pathogen. Future Microbiol. 2016;11(9):1191–208.

Hillan A, Gibbs T, Weaire-Buchanan G, Brown T, Pang S, McEvoy SP, Parker E. Zoonotic transmission of diphtheria toxin-producing *Corynebacterium ulcerans*. Zoonoses Public Health. 2024;71(2):157–69.

Kolybo DV, Labyntsev AA, Korotkevich NV, Komisarenko SV, Romaniuk SI, Oliinyk OM. Immunobiology of diphtheria. Recent approaches for the prevention, diagnosis, and treatment of disease. Biotechnol Acta. 2013;6(4):43–62.

Oliveira A, Oliveira LC, Aburjaile F, Benevides L, Tiwari S, Jamal SB, Silva A, Figueiredo HC, Ghosh P, Portela RW, De Carvalho Azevedo VA. Insight of genus *Corynebacterium*: ascertaining the role of pathogenic and non-pathogenic species. Front Microbiol. 2017;8:1937.

Parija SC. *Corynebacterium* and coryneform bacteria. In: Text book of microbiology and immunology. 4th ed. Singapore: Springer; 2023. p. 395–418.

Peel MM, Palmer GG, Stacpoole AM, Kerr TG. Human lymphadenitis due to *Corynebacterium pseudotuberculosis*: report of ten cases from Australia and review. Clin Infect Dis. 1997;24(2):185–91.

Sharma NC, Efstratiou A, Mokrousov I, Mutreja A, Das B, Ramamurthy T. Diphtheria. Nat Rev Dis Primers. 2019;5(1):81. https://doi.org/10.1038/s41572-019-0131-y.

WHO. Clinical management of diphtheria: guideline, 2 February 2024 (WHO/DIPH/Clinical/2024.1). Geneva: World Health Organization; 2024. Licence: CC BY-NC-SA 3.0 IGO.

Capnocytophaga Infection

46

Geetha Lakshmi and Sukanya Sugumaran

Learning Objectives

1. To identify and describe the morphology and cultural characteristics of the bacterium *Capnocytophaga*
2. To explain the pathogenesis and laboratory diagnosis of the zoonotic infection caused by the bacteria
3. To describe the treatment and prevention of zoonotic infections caused by *Capnocytophaga*

Introduction

Capnocytophaga species, fastidious and slow-growing Gram-negative bacilli characterised by their distinctive fusiform to spindle-shaped morphology and motility, are prevalent members of the oral microbiota in humans, cats, and dogs. While the genus encompasses eight recognised species, six are typically commensals within the human oral cavity, primarily associated with periodontal health in immunocompromised individuals. In contrast, the two species found in the oral flora of cats and dogs pose a more significant risk of zoonotic transmission through bites, scratches, or close contact. Although human infections with these animal-associated *Capnocytophaga* are rare, they can lead to severe and even fatal systemic diseases such as bacteraemia, endocarditis, and meningitis, even in immunocompetent individuals. Due to their fastidious nature and slow growth, *Capnocytophaga* infections are likely underreported, highlighting the need for increased awareness and improved diagnostic strategies.

History

Capnocytophaga was first described by Prevot, a French Bacteriologist at the Pasteur Institute, in 1956. Initially, they were the fermentative strains of the species *Fusobacterium nucleatum.* So, they were named *F. nucleatus* var. *ochraceus*. It was renamed *Ristella ochracea* on further studies since it shared common characteristics with organisms belonging to the genus *Ristella.* Loesche isolated this organism from the oral cavity along with *Bacteroides oralis*. So, he classified this new organism under Bacteroides and called it *Bacteroides oralis* var. *elongatus* due to its fusiform nature. This name was later changed to *B. ochraceus*. The bacterium was further studied in detail by the CDC in the 1960s and was designated as CDC group DF-1. The abbreviation of DF is Dysgonic Fermenter, indicating its inability to properly ferment sugars without serum supplementation. A series of articles was published by Boston's Forsyth Dental Centre and the University of Massachusetts after extensive

G. Lakshmi (✉)
Dr. M.G.R. Educational and Research Institute, Chennai, Tamil Nadu, India

S. Sugumaran
Department of Microbiology, ACS Medical College & Hospital, Dr. M.G.R. Educational and Research Institute, Chennai, Tamil Nadu, India

S. C. Parija, A. Poddar (eds.), *Textbook of Bacterial Zoonoses*, Microbial Zoonoses,
https://doi.org/10.1007/978-981-95-6615-0_46

research proving that periodontal infections are caused by a fusiform bacterium, which exhibits gliding motility and requires 5–10% CO_2 for its growth. *Capnocytophaga* was named because of its capnophilic nature. Wadsworth VA hospitals and the University of California, Los Angeles (UCLA) compared that *Bacteroides ochraceus*, CDC DF-1, and strains from the Boston study. In comparison, they concluded *Bacteroides ochraceus*, CDC DF-1 and the strains from the Boston study are the same organism. They named the organism *Capnocytophaga ochracea* and the related organisms *Capnocytophaga sputigena* and *Capnocytophaga gingivalis*. In 1994, human dental plaques yielded *Capnocytophaga granulosa* and *Capnocytophaga haemolytica*. In 2008, two more species were added to the list of *Capnocytophaga* isolated from the oral microbiome of human beings—*Capnocytophaga leadbetteri* and *Capnocytophaga* genospecies AHN8471.

In 1976, a similar organism was identified, which was taken from the blood culture and cerebrospinal fluid culture of a patient who was bitten by a dog, and it was classified under the CDC group DF-2. From 1976 to 1987, multiple isolates and strains like CDC group DF-2 were identified and sent to CDC to be examined at their special bacteriology section. All the organisms were grouped under *Capnocytophaga* based on their phenotypic properties and similarities in DNA. CDC group DF-2 is renamed *Capnocytophaga canimorsus* (dog bite in Latin). The CDC group DF-2-like strains are now classified into *C. cyanodegmi* (dog bite in Greek). *Capnocytophaga canimorsus* is cytotoxic following phagocytosis, whereas *Capnocytophaga cyanodegmi* does not demonstrate cytotoxic properties.

Taxonomy

Based on 16S RNA analysis, *Capnocytophaga* is taxonomically classified into Phylum Bacteroidetes, Class Flavobacteria, Order Flavobacteriales, Family Flavobacteriaceae and Genus *Capnocytophaga*. Other genera belonging to the family Flavobacteriaceae include *Flavobacterium*, *Elizabethkingia*, *Myroides*, *Chryseobacterium* and *Weeksella* species.

Genus *Capnocytophaga* is a fusiform Gram-negative bacillus that comprises a total of 8 species. The species that are isolated from human sources include *C. haemolytica*, *Capnocytophaga sputigena*, *Capnocytophaga gingivalis*, *Capnocytophaga ochracea*, *Capnocytophaga leadbetteri* and an unnamed taxon (AHN8471). The species that are isolated from animal sources include *Capnocytophaga cynodegmi* and *Capnocytophaga canimorsus*.

Among the 8 species, human infections are caused by all except *C. leadbetteri* and AHN8471.

Genomics and Proteomics

C. canimorsus genome consists of a circular, single chromosome which measures 2,571,406 bp. Gram-negative bacteria consist of an array of surface polysaccharides that play a role in their pathogenicity. These include lipopolysaccharides (LPS), exopolysaccharides (EPS) and capsular polysaccharides (CPS). Lipopolysaccharide is an essential part of the outer membrane. It has three parts, namely the lipid A, which typically initiates inflammation, the O-antigen and the core oligosaccharide.

Compared to the LPS of *Escherichia coli*, the LPS from *Capnocytophaga* is significantly less endotoxic, 100 times less so. The chemical structures of *Capnocytophaga*'s core oligosaccharide and lipid A have been determined. Instead of the usual LPS, it contains a lipooligosaccharide (LOS) in addition to a capsular polysaccharide (CPS). The CPS, rather than the LOS, enhances resistance to being recognised by macrophages and being engulfed by human polymorphonuclear leukocytes, reducing proinflammatory cytokines. Additionally, it can remove sugar moieties from mammalian proteins, such as IgM, IgG, and glycoproteins found on the surface of the phagocytes.

An amino acid and sugar analysis on the peptidoglycan layer indicated that the cell walls majorly comprise alanine, glucosamine, diaminopimelic acid, D-glutamic acid and muramic acid. The peptidoglycan is A1γ-type, and the cell wall does not contain serine and glycine.

The genome was discovered to code for various lipoproteins located within 13 polysaccharide utilisation loci (PULs), which are characteristic of the Bacteroides-Flavobacteria group of organisms. PULs encode surface complexes that resemble the classic starch utilisation system (Sus). At least nine PUL products were identified within the surface proteome, and eight accounted for over 50% of the peptides found in the surface proteome. Systematic deletion of the 13 PULs showed that around 50% of Sus-like complexes are responsible for the growth of animal cells.

Morphology

Capnocytophaga are straight or slightly curved Gram-negative bacilli that measure 1–4 μm in length (Fig. 46.1). They are *capnophilic* (from Greek: carbon dioxide-loving), facultative anaerobes. These bacteria are fusiform or spindle-like and exhibit gliding motility. Cells from older cultures show pleomorphism. They have similarities with species from *Fusobacterium* and *Bacteroides*. They do not possess flagella and exhibit gliding motility. It is often not possible to demonstrate the gliding motility on culture media. *Capnocytophaga* has 35% G+C in its DNA.

Cultivation

Capnocytophaga are facultative anaerobes and are capnophilic in nature. They require 5–10% of carbon dioxide for their growth. They are fastidious in nature. They grow slowly on blood (5% sheep blood agar) or chocolate agar under 5–10% carbon dioxide. Media must be kept in a BOD incubator for a minimum of 5 days at 37 °C. Colonies are not clearly visible during the initial 48 h on blood agar. The initial colonies are pinpointed. The colonies later become large, smooth, and convex with a moist and mottled centre. Colonies are non-haemolytic in nature with a typical shiny spreading corner resembling a thin film over the centre and gliding motility seen as finger-like projections (Fig. 46.2). The colour of the colonies varies from pink to yellow. However, when the colonies are scraped from the culture plate, they invariably appear yellow. MacConkey agar and heart infusion media do not support the growth of the organism.

Capnocytophaga is resistant to colistin, vancomycin and trimethoprim. It grows on a selective medium such as Thayer-Martin medium. It also grows on kanamycin-vancomycin laked blood medium (KVLB agar) and Martin-Lewis's medium. An incubation temperature of 35 °C in a

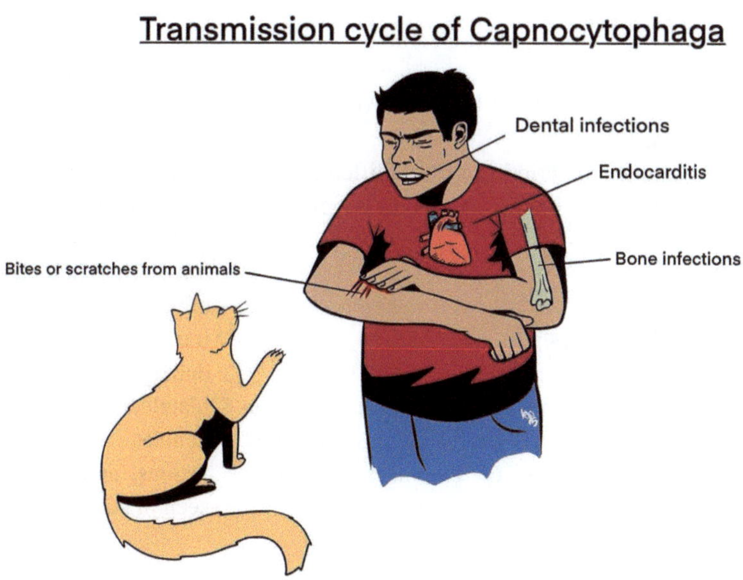

Fig. 46.1 Transmission cycle of *Capnocytophaga*. (Image source: Author's personal collection)

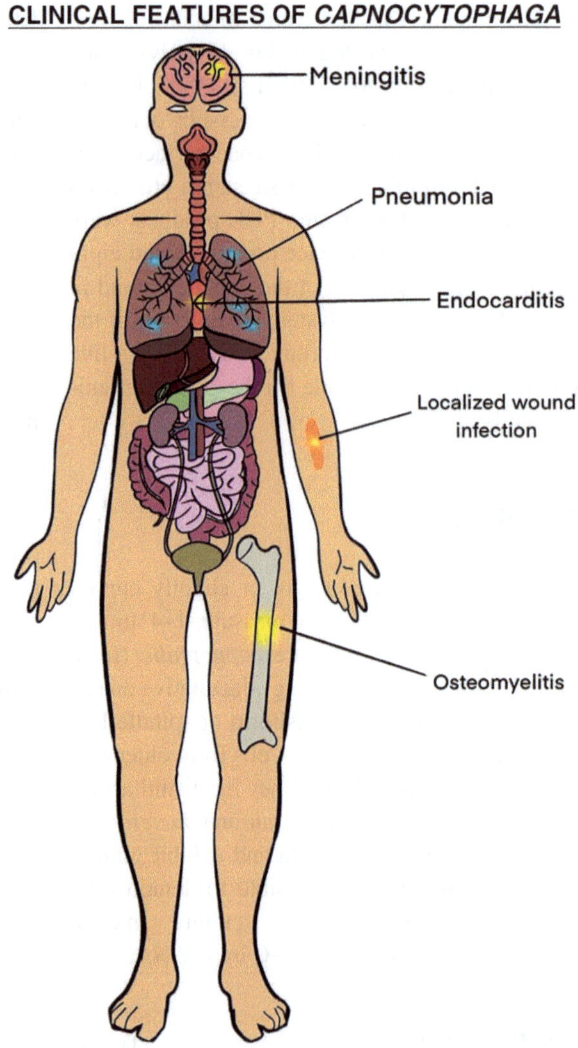

Fig. 46.2 Clinical features of *Capnocytophaga* infection. (Image source: Author's personal collection)

BOD incubator is ideal for the growth of the organism.

A detailed description of the differentiating features between the various species is mentioned in Table 46.1.

Animal Models

Capnocytophaga are commensals that are usually found in the oral cavity of cats and dogs. Some studies show *C. canimorsus* exhibits cytotoxic properties on phagocytosis when cultured in mouse macrophages.

In experimental infection of rabbits, they demonstrate symptoms like septicaemia, disseminated intravascular coagulopathy and necrosis, mimicking human infection. They exhibited a clinical syndrome that included disseminated intravascular coagulation, skin gangrene, cellular damage in the kidneys and adrenals, decreased platelet counts, low blood pressure, bleeding tendencies marked by petechiae and purpuric lesions, and renal failure, mirroring symptoms seen in humans.

Table 46.1 Growth characteristics and biochemical reactions of *Capnocytophaga* species

Character	*Capnocytophaga ochraceus*	*Capnocytophaga gingivalis*	*Capnocytophaga sputigena*	*Capnocytophaga haemolytica*	*Capnocytophaga granulosa*	*Capnocytophaga leadbetteri*	*Capnocytophaga genospecies* AHN8471	*Capnocytophaga canimorsus*	*Capnocytophaga cynodegmi*
Haemolysis on blood agar	–	–	–	+	–	–	–	–	–
Growth in MacConkey agar	–	–	–	–	–	–	–	–	–
Gliding motility	+	+	+	+	+	+	+	+	+
Catalase	–	–	–	–	–	–	–	+	+
Oxidase	–	–	–	–	–	–	–	+	+
Nitrate reduction	–	–	–	+	–	–	–	–	–
Indole	–	–	–	–	–	–	–	–	–
Urease	–	–	–	–	–	–	–	–	–
Arginine	–	–	–	–	–	–	–	+	+
Ornithine	–	–	–	–	–	–	–	–	–
Glucose	+	+	+	+	+	+	+	+	+
Lactose	+	+	+	+	+	+	+	+	+
Sucrose	+	+	+	+	+	+	+	–	+
Maltose	+	+	+	+	+	+	+	+	+
Xylose	–	–	–	–	–	–	–	–	–
Mannitol	–	–	–	–	–	–	–	–	–

+ indicates positive reaction, and – indicates negative reaction

Transmission Cycle

Most *Capnocytophaga* species are found in the oral cavity of humans and are causative organisms of periodontal infections in the immunocompromised. Humans are the only susceptible hosts, and *C. canimorsus* and *C. cyanodegmi* infection transmission occurs through scratches or dog and cat bite wounds (Fig. 46.1).

The transmission cycle of *C. canimorsus* begins with its natural presence as a commensal bacterium residing in the oral cavity of healthy dogs and, less frequently, cats. These animals serve as the primary reservoir for the bacteria, typically carrying them without exhibiting any signs of illness.

The bacterium is transmitted to humans primarily through direct contact with these animal reservoirs. The most common routes of transmission include bites, which introduce the bacteria directly into the bloodstream or tissues. Although less frequent, scratches can also facilitate transmission if the animal's saliva contaminates the wound. In some instances, transmission can occur through licks on open wounds or broken skin, allowing the bacteria to enter the human body. Rarely, close contact with animals, even without a bite or scratch, might lead to transmission, possibly through contact with saliva and mucous membranes.

Pathogenesis and Pathology

Certain *Capnocytophaga* species, such as *C. ochracea* and *C. gingivalis* from human oral flora, and *C. canimorsus* and *C. cynodegmi* from canine sources, are associated with periodontal diseases and opportunistic infections. Isolates from both human and animal oral lesions express different virulence factors that facilitate tissue invasion and immune evasion.

These virulence factors include the production of proteolytic enzymes such as collagenases and trypsin-like proteases, which degrade the extracellular matrix and contribute to periodontal damage. Some strains also secrete cytotoxins that attack epithelial and immune cells, increasing their pathogenic potential. Lipooligosaccharides (LOS) enable resistance to complement-mediated lysis and modulate inflammatory responses.

Sialidases, by cleaving sialic acids from host glycoproteins, support nutrient acquisition and mucosal adherence. Iron acquisition mechanisms from host proteins further enhance survival in nutrient-limited environments. Particularly in animal-derived strains like *C. canimorsus*, a polysaccharide capsule and surface proteins help resist neutrophil-mediated killing.

Other virulence-related enzymes, such as aminopeptidases, elastases, and neuraminidases, contribute to tissue degradation and modulate plaque-mediated inflammation. These enzymes promote vascular permeability, leukocyte recruitment, and pain through molecules like bradykinin. Neuraminidase activity also facilitates co-infection by enhancing the virulence of other microbes.

Furthermore, *Capnocytophaga* species can secrete low-molecular-weight factors that impair neutrophil chemotaxis and morphology, particularly impacting immunocompromised individuals. The extracellular polysaccharides suppress T-cell responses to antigens and mitogens. In addition, lipopolysaccharides (LPS) confer resistance to serum bactericidal activity.

Immunology

Capnocytophaga possesses a group-specific, heat-labile protein antigen that is susceptible to degradation by trypsin and sodium dodecyl sulfate.

C. canimorsus are resistant to phagocytosis. When in contact with phagocytes, bacterial cells show a significant increase in growth. This occurs due to the presence of a sialidase enzyme on its surface. The sialidase enzyme plays an important role in obstructing the bactericidal action of macrophages. The enzyme also prevents the macrophages from releasing nitric oxide. The enzyme interferes with phagocytosis and also inhibits nitric oxide secretion in most infections.

In addition, *C. canimorsus* can avoid interaction with the TLR-4 receptor of the host and escape the immune response. The receptors undergo down-regulation that could induce an immune reaction in the presence of a pathogenic organism. The sialidase complex gets activated if the bacterium is in contact with phagocytes. It also enables the bacteria to access internal amino sugars present in the glycoproteins of host cells from the glycan chains.

In a murine infection model, it has been shown to release carbohydrates from the surface mucosa, thereby promoting commensalism. It has also been demonstrated that *C. canimorsus* cells can ingest deglycosylated glycans present in hosts and consume PMNs in mice. Macrophages that were infected with either heat-killed or live *Capnocytophaga* strain derived from a patient were incapable of secreting interleukin-1α (IL-1α), tumour necrosis factor, IL-8, IL-6, nitric oxide, macrophage inflammatory protein 1b or gamma interferon, thereby leading to serious septicaemia. In response to CC5, toll-like receptor 4 fails to get activated, leading to the absence of a proinflammatory response.

Infection in Humans

Certain species, such as *C. ochracea* and *C. gingivalis*, present in human oral flora, and *C. canimorsus* and *C. cynodegmi,* present in canine oral flora, are associated with periodontal disease and other opportunistic infections in humans.

Following the infection, the incubation period varies from 1 to 8 days, with a maximum of up to 3 weeks in some instances. *C. canimorsus* infection in humans, following bites, scratches, or close interactions with dogs or cats, remains relatively uncommon. However, if infection does occur, the clinical manifestations vary significantly. These range from mild, localised wound infections characterised by redness, swelling, pain, blisters, and pus to severe, life-threatening systemic infections (Fig. 46.2).

C. canimorsus infections in immunocompetent hosts are frequently asymptomatic. However, the risk of infection is significantly elevated in individuals who lack a spleen. Pet owners, particularly dog owners, as well as professionals like veterinarians, dog breeders, and animal shelter staff, are at an increased risk of contracting this infection.

Clinical symptoms are more commonly observed in immunocompromised populations, including individuals with liver cirrhosis, alcohol use disorder, corticosteroid therapy, neutropenia, chronic pulmonary conditions, and haematologic malignancies. Older adults, especially those over 50 years or more, also demonstrate increased susceptibility. Patients with conditions such as beta-thalassaemia major, hemochromatosis, and chronic alcohol consumption, or those who smoke heavily, may exhibit elevated iron levels in bodily tissues. This iron overload provides an environment conducive to the multiplication of *C. canimorsus.*

Prodromal Phase

Patients within 8–12 h of a dog bite may present with localised lesions showing minimal inflammation. As the infection progresses, it may manifest as pain at the bite/scratch site, lymphangitis, and localised lymphadenopathy. Early symptoms may include fever, chills, muscle aches, vomiting, diarrhoea, abdominal discomfort, general weakness, breathing difficulties, confusion, and headaches.

Rapid and severe infections can develop in those with immune deficiencies. In such instances, the infection might present as sepsis, osteomyelitis, meningitis, endocarditis, peritonitis, pneumonia, or purulent and septic arthritis, as well as fulminant purpura. Disseminated intravascular coagulation (DIC) can occur in both immunocompromised and previously healthy individuals (Fig. 46.5).

Skin and Soft Tissue Infections

The condition manifests as cellulitis. It is characterised by redness, pain at the bite site, and purulent leakage from the wound, often accompanied by lymphangitis and focal lymphadenopathy.

Sweet's syndrome is a rare complication, characterised by acute neutrophilic dermatitis accompanied by fever. The skin rash associated with Sweet's syndrome typically occurs on the back, neck, limbs, and chest.

Bloodstream Infections

Disseminated intravascular coagulation (DIC) caused by *Capnocytophaga* can lead to gangrene in the lips, ears, extremities, and nose, frequently requiring amputation of the gangrenous area. Purpura may appear as tiny pinpoint bruises (*petechiae*) or larger blue marks (*ecchymosis*) under or on the skin. Waterhouse-Friderichsen syndrome and haemolytic uremic syndrome are also known to occur. *C. canimorsus* causes infectious endocarditis in individuals with pre-existing heart damage, often ending in paravalvular abscesses, fistulas, myocardial infarctions, or myocarditis.

Central Nervous System (CNS) Manifestations

Meningitis caused by *C. canimorsus* typically presents as meningitis (photophobia, altered mental status, and meningeal signs), associated with fever, observed in only 32% of reported cases. Other less common clinical presentations include hearing impairment, cholecystitis and paravertebral abscesses.

Septicaemia resulting from *C. canimorsus* infection is associated with high mortality and substantial morbidity, such as gangrene-induced amputations, renal dysfunction, and myocardial infarction. In contrast, meningitis due to *C. canimorsus* has a lower mortality rate of approximately 5%. However, mortality from *C. canimorsus* endocarditis, as reported in case studies, is relatively high (25%). Generally, outcomes of infections are poorer among immunocompromised patients compared to those with normal immune function.

Comparative characteristics of infections arising from *Capnocytophaga* species of human versus animal origin are summarised in Table 46.2.

Table 46.2 Distinguishing features between *Capnocytophaga* species of human and animal origin

Characters	Human origin	Animal origin
Common in children	Yes	No
Common in adults	Yes	Yes
Leukaemia/Lymphoma as risk factor	Yes	No
Asplenia as risk factor	No	Yes
Ethanol abuse as risk factor	No	Yes
Neutropenia as risk factor	Yes	No
Transmits following dental procedure	Yes	No
Transmits following animal bite	No	Yes

Infection in Animals

There is no evidence of *C. canimorsus* infections in cats following a bite. However, there has been a case report where a cat developed rhinitis and chronic sinusitis because of infection with *Capnocytophaga*. This organism showed a similarity of 97% to *C. canimorsus* based on its DNA sequencing. The frequency of bite wounds caused by cats and dogs is not promptly documented. It is suggested that out of all the emergencies linked to trauma, these injuries account for about 15%. Like human bite wounds, a lot of these injuries go unreported, and pet owners may not seek veterinary care. In dogs and cats, skin and bite injuries can be less obvious or penetrate deeper tissues, making them hard to detect. Additionally, cats and dogs might not be very prone to *C. canimorsus* infection, which is why a minimal number of cases are reported in animals.

Only two cases have been documented in other animals, such as rabbits, caused by *C. canimorsus*.

Epidemiology and Public Health

C. canimorsus is the most common species causing zoonotic infections in humans (Fig. 46.3).

Animal keepers, pet owners, veterinarians, and animal breeders are high-risk populations for *Capnocytophaga* infections. The transmission of

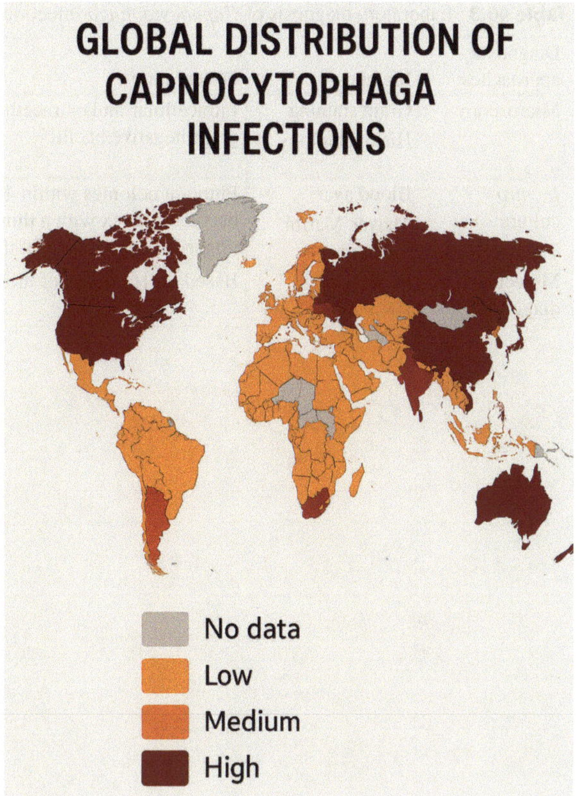

Fig. 46.3 Global distribution of *Capnocytophaga* infection. (Image source: Author's personal collection)

this bacterium can occur through close contact between pet owners and their pets. Transmission does not always lead to infection.

Most humans over 40 years of age are affected. Males are three times more prone to infection than females. Of those infected with this organism, 60% of people had pre-existing health conditions. Individuals who have had their spleens removed previously are particularly at risk, facing a 30–60 times higher likelihood of severe septicaemia. Frequently, cases of overwhelming post-splenectomy infections (OPSI) can lead to a swift decline in health within a day of the onset of symptoms, resulting in multiple organ failure and possibly death.

Globally, lakhs of individuals are getting bitten or scratched by animals annually; approximately 50% of the population will experience injury due to an animal at any stage in their lives. Most of these bites, about 90%, involve dogs and cats. Fewer than 50% of bites go unreported, and only 18% of those affected seek medical help. Most bites are not severe and typically occur on the hands, feet, head, and neck. They were able to isolate the bacteria in 80% of the wounds resulting from dog bites on humans. A higher incidence has been noted among males compared to females (M/F ratio = 3.75:1).

Both dogs and cats can carry *C. canimorsus* in their oral cavity, but it is more commonly found in dogs. Incidents involving scratches, bites or close contact with cats account for less than 10% of reported cases. The sharp and deep punctures from cat bites likely create more unfavourable conditions for the bacteria to grow compared to the superficial and bigger wounds caused by dog bites. Additionally, patients receiving timely antibiotic treatment are unlikely to develop an infection with *C. canimorsus*.

Table 46.3 Laboratory diagnosis of *Capnocytophaga* infections

Diagnostic approaches	Methods	Observation	Remarks
Microscopy	Gram staining	Intracellular and extracellular fusiform gram-negative bacilli	Cost-effective and fast technique to visualise the bacteria directly from the samples or culture
	Histopathology		
In vitro culture	Blood agar	Pinpoint colonies within 48 h, which later become convex with a thin film on top and finger-like projections at their periphery	Laborious method
	Thayer-Martin medium		
Molecular diagnosis	PCR	Based on 16S RNA of the bacteria	Currently widely used
	DNA sequencing		

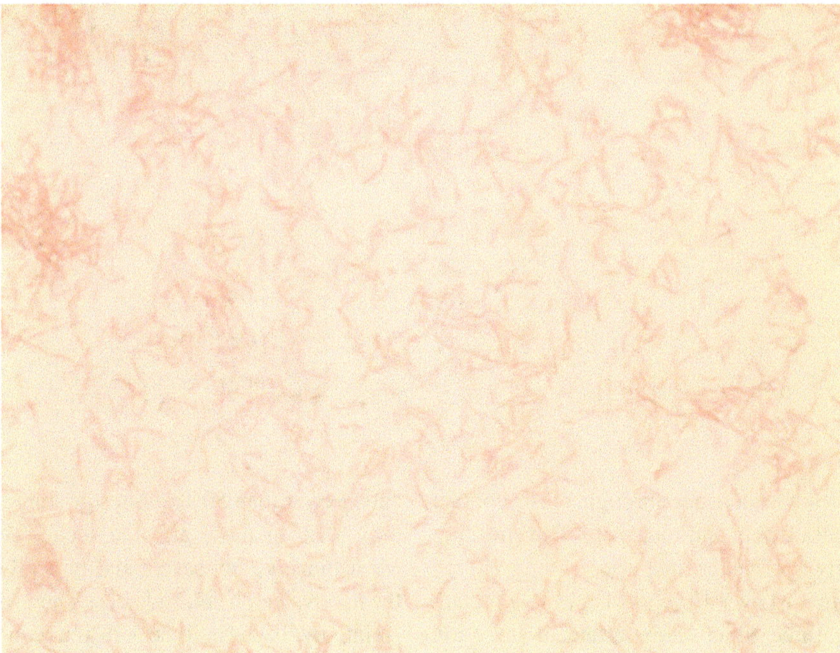

Fig. 46.4 Gram-stained smear of *Capnocytophaga* showing fusiform Gram-negative bacilli. (Image source: Copyright free image: Brichacek et al. 2012)

Diagnosis

Different laboratory diagnostic approaches are summarised in Table 46.3.

Specimens

Blood is the specimen of choice. Other samples include cerebrospinal fluid and tissue or pus specimens from bite and scratch wounds. Polyanethol-sulfonate, a commonly used anticoagulant in automated blood culture bottles, inhibits the growth of *C. canimorsus*. Mouth swabs from the dogs that bit the patients have been collected in several instances, but only half of the samples showed growth of *C. canimorsus*.

Microscopy

Currently, the culture of *C. canimorsus* from blood or other samples remains the gold standard for diagnosis. Buffy coat is obtained from blood samples by centrifugation and examined using Gram's stain under a bright field microscope.

Gram staining shows multiple extracellular and intracellular fusiform Gram-negative bacilli, which become pleomorphic on repeated subcultures (Fig. 46.4).

Phase-contrast microscopy and dark-field microscopy, however, offer no specific advantage over Gram staining for routine identification.

In Vitro Culture

Blood agar and modified Thayer-Martin medium are used for culture. Specimens are inoculated in 5% sheep or horse blood agar and are incubated in a candle jar (with 5–10% CO_2) at 37 °C. Pinpoint colonies are formed after 5–7 days of incubation, which later become yellow circular colonies with a pinkish-purple tint and irregular edges. The centre of the colonies shows a film-like layer on top, and the edges show finger-like projections due to gliding motility (Fig. 46.5). *Capnocytophaga* does not grow on MacConkey agar.

Human *Capnocytophaga* isolates are negative for catalase and oxidase. *C. canimorsus* and *C. cyanodegmi* are positive for catalase and oxidase. They produce acid but no gas. They ferment glucose, sucrose, mannose, and maltose. They do not ferment mannitol, ribose, and xylose. They are indole negative (except *C. leadbetteri*) and urease negative. They cannot reduce nitrates to nitrites, except *C. haemolytica*. They are negative for amino acids—arginine does not get dihydrolased, and lysine and ornithine do not get decarboxylated.

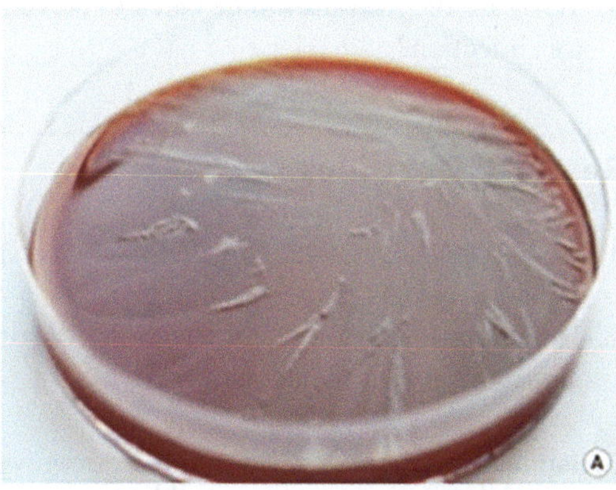

Fig. 46.5 Colonies of *Capnocytophaga* on a blood agar plate showing gliding motility. (Copyright free image-https://openi.nlm.nih.gov/imgs/512/240/4071192/PMC4071192_alm-34-325-g001.png?keywords=)

Serodiagnosis

Serodiagnosis has a limited role in the routine diagnosis of *C. canimorsus* infections.

Molecular Diagnosis

Molecular diagnostic techniques are increasingly used for the rapid and accurate detection of *C. canimorsus*. These techniques primarily depend on amplifying and detecting specific *C. canimorsus* DNA sequences.

Polymerase Chain Reaction (PCR)

PCR is widely used for the detection of *C. canimorsus*. This technique involves the enzymatic amplification of a specific target DNA sequence, allowing for the detection of even small numbers of bacteria in clinical samples like blood or cerebrospinal fluid. Various target genes have been utilised by PCR for *C. canimorsus* detection. One commonly targeted gene is the 16S rRNA, which contains both conserved and variable regions.

Quantitative Real-Time PCR (qPCR)

This depends upon the principles of conventional PCR by enabling the real-time monitoring of DNA amplification. Real-time PCR assays incorporate fluorescent probes or dyes and allow for the simultaneous amplification and detection of the target DNA. This provides rapid results and quantitative information. Multiplex qPCR assays, which can simultaneously detect multiple targets, have also been employed to differentiate *C. canimorsus* from other related species present in the sample.

Nucleic Acid Sequencing

It is the best method for definitive identification of *C. canimorsus* at the species level. Sequencing is particularly useful when culture results are negative or ambiguous. This is also useful when there is a need to differentiate *C. canimorsus* from other related species. Sanger sequencing is traditionally used. However, recent next-generation sequencing (NGS) technologies have yet to be used as routine diagnostic tools for *C. canimorsus*.

Treatment

Penicillin G is the antibiotic of choice for *C. canimorsus* infections, although some isolates have demonstrated resistance. *Capnocytophaga* strains that show resistance to the penicillins produce beta-lactamase enzymes that also grant resistance to extended-spectrum cephalosporins. These beta-lactamase-producing strains show significant resistance to penicillin, cefazolin, and amoxicillin. Compared to beta-lactamase-negative strains, these resistant strains demonstrate lower susceptibility to cefuroxime, cefotaxime, and ceftazidime. Imipenem appears effective against isolates that produce beta-lactamase and against isolates that do not produce beta-lactamases. Generally, *Capnocytophaga* species exhibit resistance to aminoglycosides, colistin, trimethoprim, and vancomycin, though some strains may show susceptibility.

Prevention and Control

Capnocytophaga infections, although uncommon, pose significant risks, especially in immunocompromised individuals and following animal bites. Preventive strategies are based on both human and veterinary public health measures.

In humans, prompt and thorough wound care after animal bites or scratches, particularly from dogs or cats, is essential. Prophylactic antibiotic therapy may be recommended in high-risk individuals (immunocompromised patients, people with asplenia and people with advanced liver disease and for deeper injuries that involve bones) to prevent infection. These patients should be advised to avoid close contact with animals known to harbour *Capnocytophaga* species. In animals, regular veterinary check-ups and maintenance of oral hygiene reduce the bacterial load in the oral cavity, limiting zoonotic transmission. Public education on responsible pet ownership and bite prevention also plays a key role in reducing infection risks.

Case Study

A 62-year-old patient arrived at the emergency OPD with complaints of fever for 1 day and chest pain that had persisted for 3 days. She has had a previous history of splenectomy following a road traffic accident. Four days before her visit to the hospital, she had suffered a bite injury on the right hand from her dog, resulting in four bite wounds. Diagnostic laboratory tests revealed a white blood cell count of 17,000 per cubic millimetre (the normal range is 4000–10,000). Her platelet count was 29,200 per cubic millimetre (normal range is 1,50,000–3,70,000). Her procalcitonin levels exceeded 150 μg per litre (normal value is ≤0.05). Blood samples were collected and sent for culture. She was started on broad-spectrum antibiotics, along with norepinephrine and intravenous fluids. A peripheral blood smear was taken, which indicated the presence of both extracellular and intracellular bacillary forms of bacteria.

1. What is the causative agent?
2. What are the possible risk factors predisposing to this infection?
3. How will you treat the condition?
4. What are the measures to be taken to prevent this infection?

Research Questions

1. How to develop a culture medium that allows rapid growth of the organism?
2. Do animals apart from cats and dogs contain *Capnocytophaga* in their oral cavity?

Declaration During the preparation of this chapter, we used *GRAMMARLY* in order to improve the grammar and/or readability of the text. After using it, we reviewed and edited the content as needed and take full responsibility for the content of the publication.

Further Readings

Bobo RA, Newton EJ. A previously undescribed gram-negative bacillus causing septicemia and meningitis. Am J Clin Pathol. 1976;65:564–9.

Brichacek M, Blake P, Kao R. *Capnocytophaga canimorsus* infection presenting with complete splenic infarction and thrombotic thrombocytopenic purpura: a case report. BMC Res Notes. 2012;5:1–6.

Brenner D, Hollis D, Fanning R. *Capnocytophaga canimorsus* sp. Nov. (formerly CDC group DF-2), a cause of septicemia following dog bite, and *C. cynodegmi* sp. Nov., a cause of localised wound infection following dog bite. J Clin Microbiol. 1989;27:231–5.

Butler T. *Capnocytophaga canimorsus*: an emerging cause of sepsis, meningitis, and post-splenectomy infection after dog bites. Eur J Clin Microbiol Infect Dis. 2015;34(7):1271–80.

Chesdachai S, Tai DBG, Yetmar ZA, Misra A, Ough N, Abu Saleh O. The characteristics of *Capnocytophaga* infection: 10 years of experience. Open Forum Infect Dis. 2021;8(7):ofab175. https://doi.org/10.1093/ofid/ofab175.

Gaastra W, Lipman LJA. Capnocytophaga canimorsus. Vet Microbiol. 2010;140(3–4):339.

Jolivet-Gougeon A, Buffet A, Dupuy C, Sixou JL, Bonnaure-Mallet M, David S, Cormier M. In vitro susceptibilities of *Capnocytophaga* isolates to beta-lactam antibiotics and beta-lactamase inhibitors. Antimicrob Agents Chemother. 2000;44(11):3186–8. https://doi.org/10.1128/AAC.44.11.3186-3188.2000.

Leadbetter ER. The genus *Capnocytophaga*. In: Dworkin M, Falkow S, Rosenberg E, Schleifer KH, Stackebrandt E, editors. The prokaryotes. Springer; 2006. https://doi.org/10.1007/0-387-30747-8_28.

Popiel KY, Vinh DC. 'Bobo-Newton syndrome': an unwanted gift from man's best friend. Can J Infect Dis Med Microbiol. 2013;24(4):209–14.

Sardo S, Pes C, Corona A, Laconi G, Crociani C, Caddori P, Luisa Boi M, Finco G. The great pretender: the first case of septic shock due to *Capnocytophaga canimorsus* in Sardinia. A case report and review of the literature. J Public Health Res. 2022;11(4):22799036221133234. https://doi.org/10.1177/22799036221133234.

Spentzouri D, Baliou S, Ioannou P. Infective endocarditis by *Capnocytophaga* species-a narrative review. Medicina (Kaunas). 2024;60(3):382. https://doi.org/10.3390/medicina60030382.

Suzuki M, Kimura M, Imaoka K, Yamada A. Prevalence of *Capnocytophaga canimorsus* and *Capnocytophaga cynodegmi* in dogs and cats determined by using a newly established species-specific PCR. Vet Microbiol. 2010;144(1–2):172–6.

Zajkowska J, Król M, Falkowski D, Syed N, Kamieńska A. *Capnocytophaga canimorsus* – an underestimated danger after dog or cat bite – review of literature. Przegl Epidemiol. 2016;70(2):289–95.

Zangenah S, Abbasi N, Andersson AF, Bergman P. Whole genome sequencing identifies a novel species of the genus *Capnocytophaga* isolated from dog and cat bite wounds in humans. Sci Rep. 2016;6(1):22919. https://doi.org/10.1038/srep22919.

Trueperella pyogenes Infections 47

Ajanta Sharma

Learning Objectives

1. To study the taxonomy, classification, morphology, cultural characteristics, and genomic features of *Trueperella pyogenes*
2. To understand the transmission cycle and pathogenesis of *T. pyogenes*
3. To study the common diseases caused by *T. pyogenes* in humans and animals
4. To understand the diagnostic methods and treatment options available for *T. pyogenes*
5. To explore the preventive strategies and vaccine development efforts targeting virulence factors of *T. pyogenes*

Introduction

Trueperella pyogenes, previously referred to as *Arcanobacterium pyogenes, Actinomyces pyogenes,* and *Corynebacterium pyogenes*, is a widely distributed Gram-positive bacterium known for its opportunistic pathogenicity. It is frequently found on the skin, oropharynx, and mucous membranes of the upper respiratory, gastrointestinal, and urogenital tracts in animals. This bacterium is associated with various purulent infections in livestock, such as cattle, swine, and small ruminants, leading to significant economic losses in animal husbandry. *T. pyogenes*-induced infections affect multiple tissues and organs, such as pneumonia, liver abscesses, mastitis, metritis, endocarditis, and osteoarthritis. Human infections due to *T. pyogenes* are uncommon but can occur sporadically, often linked to specific occupational exposures. These infections are primarily associated with individuals working closely with farm animals and their surroundings, underscoring the zoonotic potential of this bacterium. Such cases highlight the importance of awareness and preventive measures for agricultural workers.

History

T. pyogenes, originally named *Bacillus pyogenes* by Glage in 1903, has undergone multiple taxonomic revisions over the years. It was reclassified as *Corynebacterium pyogenes* (Eberson 1918), *Actinomyces pyogenes* (Collins et al. 1982), and later as *Arcanobacterium pyogenes* (Ramos et al. 1997). Finally, phylogenetic and chemotaxonomic analyses led Yassin et al. (2011) to propose the genus *Trueperella*, honouring German microbiologist Hans Georg Trüper. The name *pyogenes*, derived from Latin and Greek roots (*pyum*, pus; *genes*, producing), reflects its role as a pus--inducing bacterium. The analyses of phylogeny and chemotaxonomy revealed that the genus referred to as *Arcanobacterium* is not monophyletic, necessitating taxonomic revision, as reported by Yassin et al. (2011). The data sup-

A. Sharma (✉)
Department of Microbiology, Gauhati Medical College & Hospital, Guwahati, Assam, India

S. C. Parija, A. Poddar (eds.), *Textbook of Bacterial Zoonoses*, Microbial Zoonoses,
https://doi.org/10.1007/978-981-95-6615-0_47

ported the revision of the classification of five species—*Arcanobacterium abortisuis*, *Arcanobacterium bernardiae*, *Arcanobacterium bialowiezense*, *Arcanobacterium bonasi*, and *Arcanobacterium pyogenes*—into a new genus, for which the new genus name "*Trueperella*" was proposed.

Taxonomy

The genus *Trueperella* is placed within the family Actinomycetaceae in the phylum Actinomycetota (previously known as Actinobacteria), class Actinomycetia, and order Actinomycetales. Species within the genus share a close relationship with several pathogenic genera, such as *Mycobacterium, Corynebacterium, Actinomyces, Rhodococcus, Dermatophilus,* and *Nocardia*. This genus currently includes five species: *Trueperella pyogenes*, *Trueperella abortisuis*, *Trueperella bonsai*, *Trueperella bernardiae*, and *Trueperella bialowiezensis*. Among these, *T. pyogenes* and *T. bernardiae* have been identified in clinical specimens from humans, highlighting their significance as opportunistic pathogens capable of causing both human and animal infections. *Trueperella pecoris* was introduced as a novel species of the genus *Trueperella* in 2021.

Genomics and Proteomics

The complete genome of *T. pyogenes* was sequenced and published for the first time in 2014, isolated from bovine mastitis. It provides valuable insights into its pathogenicity and genetic composition. The NCBI genome database contains 25 complete genome sequences of *T. pyogenes*, 19 of which are derived from different animal species.

The genome of *T. pyogenes* comprises one circular chromosome, with genome sizes ranging from 1.9 to 2.4 Mb based on the strain. It is characterised by a high guanine and cytosine (GC) percentage of 59.6, 2081 genes and 2074 proteins. Among RNA types, tRNA is the most abundant, while rRNA constitutes 3–9% of the total nucleic acid content. Additionally, several non-coding RNAs, ranging from one to three per strain, are present. These genomes exhibit a relatively high number of pseudogenes and encode multiple factors contributing to virulence, such as pyolysin, collagen-binding protein, neuraminidases, and four fimbrial proteins. In addition to virulence-associated genes, the genome of *T. pyogenes* can also harbour multiple antibiotic--resistance determinant genes. These include genes conferring resistance to tetracyclines (*tetW*, *tetM*, *tetA(33)*, *tetK*, and *tetL*), macrolides (*ermX* and *ermB*), aminoglycosides (*aadA1*, *aadA9*, *aadA11*, and *strA-strB*), trimethoprim (*dfr2a*), and β-lactams (*blaP1*). The presence of these resistance determinants highlights the genetic adaptability of *T. pyogenes* and its potential role in the emergence of antimicrobial resistance.

Morphology

T. pyogenes is a Gram-positive, pleomorphic, non-spore-forming, non-motile, non-acid-fast, and non-capsulated organism ranging from coccobacilli to short rods (0.5–2 μm). These bacteria are typically found singly or in pairs (forming V, T, or palisade arrangements). Short diphtheroid shapes with small club-like structures are also observed, especially in young cultures resembling coryneform bacteria. In older cultures, they may display variability in their Gram reaction. The peptidoglycan layer of the cell wall is composed of lysine, rhamnose, and glucose and does not contain mycolic acid.

Cultivation

T. pyogenes exhibits distinct microbiological and biochemical features that aid in its identification. Its growth requirements are relatively simple; however, the culture of this bacterium requires blood- or serum-enriched culture media. On sheep blood agar, this bacterium forms pinpoint colonies (0.2–0.5 mm) with a distinct zone of β-haemolysis after 24 h at 37 °C. Subsequently,

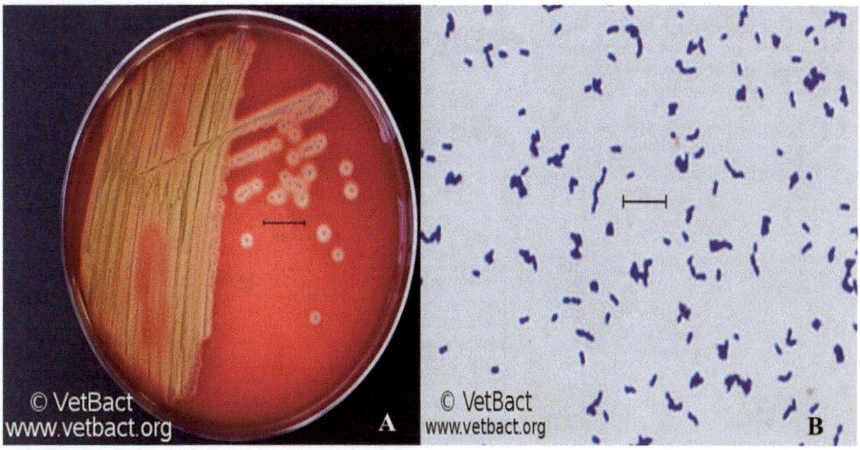

Fig. 47.1 (**a**) β haemolytic colonies of *Trueperella pyogenes* on blood agar medium; (**b**) Gram-stained smear showing gram-positive rod-shaped *T. pyogenes* bacteria (X 40). (Image source: Copyright free; Source: VetBact-Veterinary bacteriology: Information about important bacteria (www.vetbact.org))

these colonies enlarge to 0.5–1.5 mm, becoming convex and circular in shape, with a translucent appearance and smooth margins. The haemolysis zone is usually two to three times larger than the colony's diameter. *T. pyogenes* are facultative anaerobes, they metabolise proteins (hydrolyse gelatin, casein), coagulate serum and ferment carbohydrates such as lactose, predominantly yielding lactic acid. Their growth is enhanced in the presence of higher CO_2 concentrations. Figure 47.1 describes the colony characteristics and microscopic morphology of the bacteria.

Animal Models

Most of the research focused on eliciting protective immune responses to *T. pyogenes* infection has utilised mouse models, with limited research on ruminants or swine. Early attempts to protect ruminants from *T. pyogenes* focused on treatments using whole bacterial cells or culture supernatants, which largely proved ineffective. However, certain therapeutic benefits in the progression of mastitis were observed. Mouse models have demonstrated the ability to induce specific immunity that effectively protects animals against challenges involving *T. pyogenes* and *Clostridium perfringens*.

Transmission Cycle

The dissemination and transmission of *T. pyogenes* infections remain poorly understood. Most infections are believed to arise from an endogenous source, as the bacterium is a frequent constituent of the skin and mucous membrane flora in animals. However, exogenous infections are also possible, with transmission occurring through contaminated husbandry equipment or direct animal-to-animal contact. Contaminated natural habitats serve as a major source of *T. pyogenes*, particularly linked to mammary and foot infections in domestic and wild animal populations. Establishing infection routes is often challenging, but mechanical injuries to the skin and mucous membranes are primary contributors. Inadequate surgical practices, including castration and intramuscular injections, can also facilitate bacterial invasion. Additionally, micro-injuries caused by arthropods like flies or ticks may serve as entry points for *T. pyogenes*, highlighting their potential role in transmission. *T. pyogenes* infections in humans usually occur through zoonotic transmission, as they are transmitted from animals to humans, particularly those with occupational exposure to farm animals. The zoonotic transmission cycle of *T. pyogenes* is depicted in Fig. 47.2.

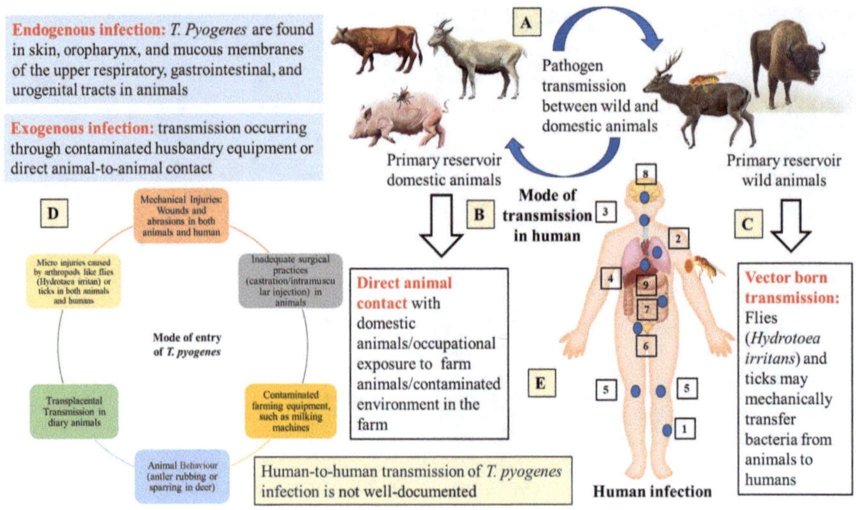

Fig. 47.2 Zoonotic transmission cycle of *Trueperella pyogenes*; (**a**) Transmission between wild and domestic animals; (**b**) Human infection through direct animal contact; (**c**) Human infection through vector-borne transmission; (**d**) Mode of entry of *T. pyogenes*. (Author's own work using animal and human figures from copyright free sources (https://commons.wikimedia.org)); (**e**) Sites and types of human infection (1: skin ulcer/abscesses; 2: Pneumonia; 3: Pharyngitis; 4: Endocarditis; 5: Arthritis; 6: Urethritis; 7: Intra-abdominal infection; 8: Intracranial infection; 9: Sepsis)

Pathogenesis and Pathology

The mechanisms of interplay between *T. pyogenes* and host immune response are less understood. A range of extracellular products and surface-exposed proteins produced *by T. pyogenes* play key roles in attachment, colonisation, and inducing host tissue damage. Several virulence factors of *T. pyogenes* contribute to its pathogenicity, such as (i) cell wall macromolecules, (ii) pyolysin (PLO), the sole known toxin of *T. pyogenes,* (iii) neuraminidases (NanH and NanP), (iv) fimbriae, (v) extracellular matrix binding proteins like collagen-binding proteins (Cbp), (vi) various exoenzymes, such as serine proteases with gelatinase and caseinase activity, as well as DNases, and (vii) its ability to form biofilms, further enhancing its capacity for adhesion, colonisation, and persistence within host tissues. The virulence factors and pathogenesis of *T. pyogenes* are presented in Fig. 47.3.

Cell Wall Macromolecules: They include lipoteichoic acids and peptidoglycan, which interact with macrophages, inducing the release of the pro-inflammatory cytokines. This interaction is crucial for initiating an immune response against the bacterium.

Pyolysin: It is the major virulence factor and is a pore-forming exotoxin classified as a cholesterol-dependent cytolysin (CDC). It exhibits haemolytic, cytolytic, dermonecrotic, and lethal effects on various cell types. It is haemolytic for sheep, horse, human, rabbit, and guinea pig erythrocytes. It is also cytolytic for immune cells and others such as polymorphonuclear neutrophils (PMNs), macrophages, epithelial cells, fibroblasts, and endometrial stromal cells. The cytolytic activity of PLO is attributed to its ability to bind to cholesterol in the plasma membrane and form transmembrane pores. These pores disrupt the cell membrane's integrity, allowing ions and other molecules to leak out, ultimately leading to cell lysis and death. PLO can also induce pyroptosis (a type of programmed cell death) along with the secretion of IL-1β, an inflammatory cytokine, in macrophages, further contributing to the inflammatory response.

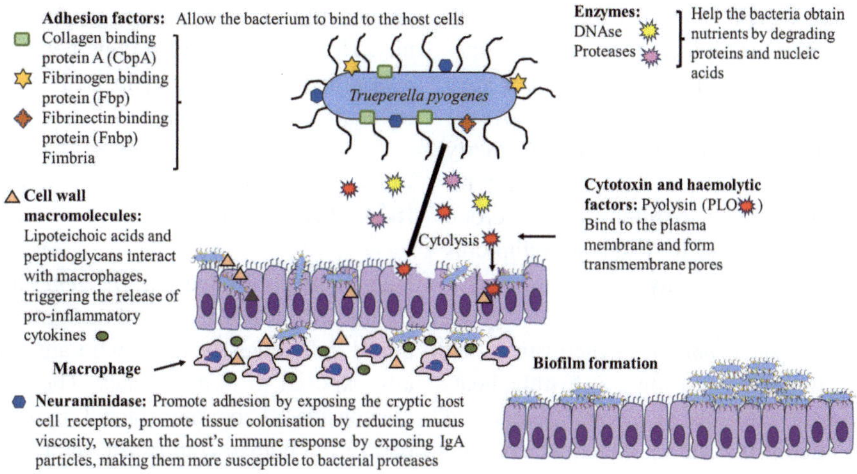

Fig. 47.3 Schematic representation of different virulence factors and their functions in *Trueperella pyogenes* infection. (Image source: Authors' personal collection)

Neuraminidases: Also known as sialidases, these enzymes assist *T. pyogenes* in attaching to the host's alimentary and respiratory epithelial mucosa by cleaving terminal sialic acid residues from host carbohydrates or glycoproteins on cell receptors. By increasing the mucosal viscosity, these enzymes facilitate tissue colonisation and impair the host's immune response by exposing immunoglobulin A (IgA) to bacterial proteases.

Fimbriae: While fimbriae are rarely present in Gram-positive bacteria, *T. pyogenes* stands out as a notable exception. In *T. pyogenes*, five distinct fimbrial types have been recognised—FimA, FimB, FimC, FimE, and FimG. These structures play a crucial role in enabling the bacterium to attach to host cells. However, the exact molecular processes involved in this attachment are yet to be completely clarified.

Extracellular Matrix Protein: The surface of *T. pyogenes* contains several proteins that attach to components of the host's extracellular matrix. This matrix is composed of important glycoproteins such as collagen, fibronectin, fibrinogen, laminin, and elastin. Collagen-binding protein (Cbp) is essential for attachment to collagen-rich tissues.

Other Enzymes: *T. pyogenes* produces virulence factors such as serine proteases and calcium-dependent proteins with proteolytic activity, contributing to tissue invasion and damage of host tissues, evasion of host defences, and modulation of the host immune system during infection and inflammation.

Biofilm Formation: It is a well-documented feature of microorganisms, enhancing their resistance to adverse conditions, including antimicrobial agents. Biofilm formation also increases adherence capabilities and effectively protects against the host's immune response. Biofilm formation creates an optimal environment for *T. pyogenes* during colonisation, promoting bacterial growth and reproduction by providing essential nutrients and moisture. To acquire nutrients, *T. pyogenes* utilises proteases and DNases that degrade proteins and nucleic acids. These mechanisms are critical for its survival and pathogenicity within the host environment. The multifaceted strategies employed by *T. pyogenes* highlight its adaptability and potential for causing infections under various conditions.

The ability of *T. pyogenes* to invade deeper tissues usually depends on pre-existing tissue injury or stress. Such conditions may result from wounds or prior viral or bacterial infections. In many cases, this organism is also associated with mixed infections involving other bacterial species. Particularly Gram-negative anaerobes such as *Fusobacterium necrophorum*, *Bacteroides* spp., and *Peptoniphilus indolicus* (formerly *Peptostreptococcus*). These infections are typically characterised by purulent and necrotic lesions, which can lead to systemic symptoms and result in animal mortality. An important synergistic relationship is observed between *T. pyogenes* and *F. necrophorum*. This synergistic effect involves *T. pyogenes* releasing a diffusible, heat-sensitive factor that lowers oxygen levels and redox potential at the site of infection, creating favourable conditions for *F. necrophorum*. In turn, *F. necrophorum* produces leukotoxins that protect *T. pyogenes* from phagocytosis by lysing or inducing apoptosis in leukocytes. Furthermore, the lactic acid generated by *T. pyogenes* acts as an energy source for *F. necrophorum*.

Immunology

Infections caused by *T. pyogenes* primarily affect immunocompromised individuals and are characterised by suppurative conditions such as mastitis, endometritis, and abscess formation. Pyolysin, the major virulence factor, plays a critical role in immune evasion. Pyolysin (i) triggers the activation of the complement system, (ii) stimulates cytokine production, and (iii) suppresses both the oxidative burst and the bactericidal activities of neutrophils and monocytes. These mechanisms allow *T. pyogenes* to modulate the host immune response and evade immune defences.

Additionally, the neuraminidases, NanH and NanP, contribute to pathogenesis by increasing mucus viscosity, facilitating bacterial colonisation, and impairing the host's immune response. This involves compromising a primary line of defence, specifically the degradation of IgA, a crucial component of mucosal immunity, by bacterial proteases. The initial immune response to infection involves inflammation. This attracts neutrophils, monocytes, and macrophages to the infection site for bacterial clearance through phagocytosis and stimulation of adaptive immunity. However, the interactions between *T. pyogenes* and host phagocytic cells significantly influence disease outcomes.

Infection in Humans

Human infections due to *T. pyogenes* are rare but significant, primarily affecting immunocompromised individuals or those who are occupationally exposed to farm animals. These infections are considered zoonotic since *T. pyogenes* is not part of the human commensal microbiota and is transmitted through contact with infected animals or contaminated environments.

Current literature is restricted to case reports only. Available literature regarding human infections caused by *T. pyogenes* is scarce. It encompasses conditions such as pharyngitis, urethritis, sepsis, endocarditis, pneumonia, skin ulcers, arthritis, various purulent lesions, intra-abdominal infections, abscesses, and erythema multiforme associated with *T. pyogenes* bacteraemia.

Infection in Animals

T. pyogenes functions as a commensal bacterium as well as an opportunistic pathogen in livestock. It infects a wide range of animal species, including antelopes, bison, camels, chickens, deer, elephants, gazelles, horses, macaws, reindeer, turkeys, wildebeest, cattle, swine, sheep, and goats, with rare cases in horses and birds. These infections often lead to significant economic losses in animal husbandry due to reduced productivity and increased healthcare costs. In cattle, *T. pyogenes* mainly leads to infections of the reproductive system and mammary tissue, along with pneumonia and hepatic abscesses. It is a significant cause of metritis and endometritis in dairy cows, with clinical manifestations occurring in 23–52% of animals following parturition. These conditions negatively impact fertility and milk production.

Mastitis is one of the most common infections caused by *T. pyogenes* in cattle, affecting lactating cows, dry cows, and heifers. A notable form of this infection is summer mastitis, a polymicrobial disease primarily occurring in pastured cows during warmer months. It is often transmitted by biting flies, such as *Hydrotaea irritans*. *T. pyogenes* can also cause various suppurative infections in cattle, including pneumonia, encephalitis, pyelonephritis, kidney abscesses, lymphadenitis, endocarditis, and abscesses in different locations. Additionally, the bacterium has been isolated from cases of septicaemia and abortion.

Summer mastitis typically involves mixed infections with anaerobes like *F. necrophorum*, with synergistic interactions between the pathogens exacerbating the condition. Liver abscesses linked to *T. pyogenes* are most common in feedlot cattle and are associated with the presence of bacteria in the rumen wall. Primary lesions in the rumen allow bacterial penetration via the hepatic portal venous system to the liver, where secondary infection sites develop into abscesses. Rupture of these abscesses can result in systemic symptoms and potentially fatal outcomes. The formation of liver abscesses is often connected to ruminal acidosis (acute or subacute), which damages the protective surface of the rumen wall, facilitating bacterial invasion.

Epidemiology and Public Health

Environmental factors, such as high humidity and mild temperatures, may play a major role in the development of *T. pyogenes* infections and significantly contribute to the occurrence of *T. pyogenes* infections. In cases of summer mastitis, transmission between animals can occur via biting flies like *Hydrotaea irritans*. However, *T. pyogenes* can invade tissues via micro-injuries inflicted by ectoparasites such as ticks, which is a plausible route of infection. Studies have highlighted the potential risk of bacterial transmission between wild and domestic animals. This is particularly seen in agricultural areas like pastures and meadows, where wildlife and livestock frequently co-exist. These shared environments may act as reservoirs for *T. pyogenes*. Furthermore, animal behaviour may play an important role in pathogen transmission and susceptibility to cranial or intracranial abscess infections.

T. pyogenes infections are widespread among domestic and wild animals globally but are rare in humans. There is variation in the prevalence of *T. pyogenes* isolation based on host species and geographic region. Most research on *T. pyogenes* infections focuses on food animals, particularly in regions such as Europe, China, Japan, Brazil, and the United States. In livestock, *T. pyogenes* infections lead to considerable economic losses, particularly in cattle and swine breeding. These losses stem from reduced meat and milk production, diminished breeding effectiveness, and the necessity to cull affected animals. The clinical severity of these pyogenic infections varies, with mortality rates rising in cases of misdiagnosis or inadequate treatment.

Although rare, *T. pyogenes* has also been associated with human infections, including endocarditis, sepsis, pneumonia, arthritis, and leg ulcers. Notably, it was identified in leg ulcers among 495 schoolchildren in Thailand, underscoring its zoonotic potential and global distribution.

The epidemiological features of different *Trueperella* species are presented in Table 47.1.

Diagnosis

For the diagnosis, the clinical specimens include swabs, tracheal aspirate samples, lung tissue, blood, milk, umbilical cord, or other affected tissues/fluids.

Microscopy

T. pyogenes is a Gram-positive, pleomorphic bacterium characterised by non-spore-forming, non-motile, non-acid-fast and non-capsulated coccobacilli or short rods (0.5–2 μm). These bacteria are typically found singly or in pairs, forming V, T, or palisade arrangements. Short

Table 47.1 Epidemiological features of different *Trueperella* species

Trueperella species	Reservoir host	Disease susceptibility across animal species	Diseases caused	Geographical distribution
Trueperella pyogenes	Livestock (cattle, swine, sheep, horses), companion animals (dogs, cats), wildlife reservoirs (deer, European bison, etc.), ticks (Dermacentor reticulatus)	Cattle, pigs, goats, sheep, European bison, antelope, buffalo, camels, horses, chickens, elephants, dogs, gazelles, macaws, reindeer, turkeys, reptiles, pigeons	Mastitis, metritis, abscesses, pneumonia, balanoposthitis, urogenital tract infection, lymphadenitis, septicaemia, encephalitis, pyometra, prostatitis, orchitis, seminal vesiculitis, pericarditis, omphalitis	Worldwide *Southeast Asia* (China, India, Japan), the *Middle East* (Iran, Jordan) *Europe* (Poland, Germany, United Kingdom, Austria, Denmark, Norway) *North America* (USA, Canada), *South America* (Brazil, Argentina) *Africa* (Egypt), *Oceania* (New Zealand)
	Domestic livestock, companion animals, wildlife reservoir	Human (Rarely)	Sepsis, leg ulcer, pharyngitis, pneumonia, endocarditis, urethritis, intra-abdominal infection, septic arthritis, intracranial infection, wound infection	*Asia* (India, Thailand, Malaysia) *Europe* (France, Spain, Ireland, Denmark) *North America* (Canada) *South America* (Brazil)
Trueperella abortisuis	Swine, boar, companion animals	Swine, cats, dogs, goats, boars	Reproductive tract infection, urinary tract infection, abscess	*North America* (USA, Canada), *Europe* (Germany, United Kingdom), *Asia* (Japan, Vietnam)
Trueperella barnardiae	Human skin, oropharynx, pigs	Humans (mainly immunocompromised), rarely pigs, ducks	Bloodstream infection, osteoarticular infection, urinary tract infection, skin and soft tissue infection, prosthetic joint infection, brain abscess, wound infection, septic thrombophlebitis, eye infection, necrotising fasciitis, leg ulcer, abscess, osteomyelitis	*Europe* (France, Denmark, Spain, United Kingdom), *North America* (USA) *South America* (Brazil) *Asia* (Korea)
Trueperella bialowiezensis	European bison	European bison	Balanoposthitis	*Europe* (Poland)
Trueperella bonasi	European bison	European bison	Balanoposthitis	*Europe* (Poland)
Trueperella pecoris	Dairy cattle	Camel, pig	Necrotic vestibulitis, mastitis, abortion, suppurative bronchopneumonia	*Europe* (Switzerland, Germany) *Africa* (Egypt)

diphtheroid shapes with small club-like structures are also observed, especially in young cultures resembling coryneform bacteria.

In Vitro Culture

T. pyogenes grows well on blood agar, typically exhibiting pinpoint colonies (0.2–0.5 mm) with a distinct zone of β-haemolysis after 24 h at 37 °C. Brain heart infusion agar (BHI) supplemented with 5% sheep blood can also be used for initial enrichment. By 48–72 h, colonies enlarge to 0.5–1.5 mm, becoming convex, circular, and translucent with smooth edges. The haemolysis zone is usually two to three times larger than the colony's diameter. Their growth is enhanced in microaerophilic conditions. To improve the isolation rate, neomycin (0.6–0.7 units/ml), polymyxin B sulphate (25 units/ml) or 0.1% Tween 80 can be added to the medium.

Phenotypic identification of *T. pyogenes* isolates is based on colony characteristics on blood agar, microscopic morphology and a negative catalase test. Christie Atkins Munch Peterson (CAMP)-like reactions have been observed in *T. pyogenes* when tested with *Rhodococcus equi* as the indicator strain or, to some extent, with β-haemolytic *Staphylococcus aureus*. The isolates also hydrolyse gelatine and hippurate. Biochemical profiles can also be studied using commercially available systems (e.g. API Coryne) and automated identification platforms (e.g. VITEK 2 Compact).

Serodiagnosis

There is no specific serological test for *T. pyogenes* in humans; diagnosis primarily relies on bacteriological culture and identification of the organism, along with molecular methods. Serological tests remain underutilised but could complement molecular methods in surveillance or diagnosis of chronic cases.

Enzyme-linked immunosorbent assay (ELISA) tests have been used to detect antibodies against *T. pyogenes* and its fimbrial proteins, particularly FimC, FimE and pyolysin (PLO) in animals. Indirect immunodiagnosis can be achieved by analysing the patient's immune response to the infection, which could involve measuring inflammation markers or cytokines.

Molecular Diagnosis

Molecular methods include *T. pyogenes*-specific polymerase chain reaction (PCR). These targets include several genes, including the 16S–23S rRNA intergenic spacer region. This also targets the *nanH* gene, *rpo*B encoding the β-subunit of bacterial RNA polymerase, *plo* encoding the putative virulence factor pyolysin, and the *tuf* gene encoding translation elongation factor. Multiplex PCR has been developed for *T. pyogenes* with other bacteria such as *Pasteurella multocida* and *Mannheimia haemolytica*.

Apart from that, single enzyme amplified fragments length polymorphism (SE-AFLP), BOX-PCR, the superoxide dismutase A gene (*sodA*) sequencing, random amplification polymorphic DNA (RAPD-PCR), and multilocus sequence analysis (MLSA) have been used for the identification of *T. pyogenes*.

Real-time PCR has also been developed for the detection of 16S rRNA of *T. pyogenes*. Other molecular techniques include 16S rRNA gene sequencing, loop-mediated isothermal amplification (LAMP) assay targeting the *cpn*60 gene using a real-time fluorometer, targeting the *plo* gene and next-generation sequencing (NGS), which can be used for the detection of *T. pyogenes*.

Matrix-assisted laser desorption/ionisation time-of-flight (MALDI-TOF) mass spectrometry (MS) is emerging as a reliable method for the identification of *T. pyogenes* based on protein profiling. Fourier transform infrared (FT-IR) spectroscopy analyses the total composition of components of the bacterial cell (infrared spectrum). Fluorescent in situ hybridisation (FISH) uses fluorescence-labelled oligonucleotide probes specific for *T. pyogenes*, which bind specific regions of the target RNA/DNA of *T. pyogenes*.

The various methods for laboratory diagnosis of *T. pyogenes* are presented in Table 47.2.

Table 47.2 Laboratory diagnosis of *Trueperella pyogenes* infection

Diagnostic approaches	Methods	Targets	Remarks
Direct microscopy	Gram stain of swab samples, blood	Gram-positive, pleomorphic, non-spore-forming, non-motile, non-capsulated coccobacilli and short rods (0.5–2 μ) found in single or in pairs or clusters	Economical and fast technique
In vitro cultures	Culture media used for isolation are sheep blood agar and brain heart infusion agar supplemented with 5% sheep blood	Colonies are pinpoint, β-haemolytic colonies (0.2–0.5 mm) after 24 h at 37 °C, appearing white, opaque, and glistening	Allows identification and differentiation of different species
Phenotypic identification	Catalase test, CAMP-positive (requires *Staphylococcus aureus* as an indicator), hydrolysis of gelatine and hippurate, commercially available systems (e.g. API Coryne), and automated identification platform (e.g. VITEK 2 Compact)	catalase-negative, CAMP-positive, hydrolyses gelatine and hippurate, and demonstrates enzymatic activity or fermentation of carbohydrates using commercially available systems (e.g. API Coryne), a battery of biochemical tests measuring carbon source utilisation and enzymatic activities (VITEK 2 Compact)	Allows identification and differentiation of different species Commercially available systems can identify coryneform bacteria within 24 h Automated ID system accurately gives results within 6 h
Serodiagnostics	There is no specific serological test for *Trueperella pyogenes* in humans	ELISA tests have been used to detect antibodies against *T. pyogenes* and its fimbrial proteins, particularly FimC and FimE, in animals	There is no specific serological test for *T. pyogenes* for diagnosis of human infection
Molecular assays	PCR, multiplex PCR, real-time PCR, targeted gene sequencing, SE-AFLP, BOX-PCR, *sodA* gene sequencing, RAPD-PCR, MLSA and NGS	16S rRNA, 16S–23S rRNA intergenic spacer region, PCR for detection of target genes *cpn*60, *tuf gap, rpo*B, nanH and *plo* genes, and *sodA* gene	High sensitivity and specificity Used to confirm isolates and assess pathogenicity
Newer techniques	Loop-mediated isothermal amplification (LAMP)	LAMP for *cpn*60 (Chaperonin 60) using real-time fluorometer and for detecting *plo* gene	A sensitive, reliable, and cost-effective way for the molecular identification of *T. pyogenes*, with no need for specialised equipment
	Matrix-assisted laser desorption/ionisation time-of-flight (MALDI-TOF) mass spectrometry (MS)	Protein profile	Fast, reliable method to distinguish *T. pyogenes* from other closely related species within the *Trueperella* genus and other genera like *Arcanobacterium* and *Actinomyces*, comparable to molecular methods
	Fourier transform infrared (FT-IR) spectroscopy	Total composition of components of the bacterial cell (infrared spectrum)	Rapid, reliable and does not require extraction, amplification, labelling, or staining, simplifying the process. Aids in distinctly differentiating this species from other *Trueperella* species as well as from species of the genera *Arcanobacterium* and *Actinomyces*
	Fluorescent in situ hybridisation (FISH)	Specific region of target RNA/DNA	A molecular diagnostic approach that enables direct identification of pathogens in clinical specimens without the need for culture, provided the organisms are present in high numbers

Treatment

Beta-lactams, tetracyclines, and macrolides are commonly used antibiotics for treating *T. pyogenes* infections. However, antimicrobial resistance in *T. pyogenes* is emerging as a significant issue due to the widespread agricultural use of these drugs. *T. pyogenes* is generally susceptible to benzyl penicillin, ampicillin, gentamicin, and macrolides but resistant to cotrimoxazole, streptomycin, and tetracyclines.

Reports indicate substantial resistance to antimicrobials like tetracycline, erythromycin, and clindamycin in *T. pyogenes*, with variations across regions. This raises concerns about the efficacy of current therapies and highlights the importance of monitoring resistance patterns to guide effective treatment strategies. The use of antimicrobials to treat *T. pyogenes* infections can disrupt the natural bacterial flora, potentially leading to the overgrowth of harmful microorganisms and the development of new infections, a phenomenon commonly referred to as superinfection.

Prevention and Control

Preventing *T. pyogenes* infections requires an effective fly control programme alongside maintaining clean and dry environments, especially in calving areas. Fly control measures include regular manure removal, maintaining clean, dry pens, cleaning waterers and feed bunk areas, and using insecticides or fly traps. Proper sanitation is critical, including removing organic fly-breeding materials like spilt feed and wet litter, composting manure to destroy fly eggs, and ensuring proper drainage to eliminate wet areas.

Various strategies have been explored in vaccine development against *T. pyogenes*. A key approach involves targeting the bacterium's virulence factors, such as pyolysin (PLO), fimbriae, and collagen-binding proteins, as antigenic targets to elicit a strong immune response and prevent infection.

These include attenuated vaccines created through sequential passage and antibiotic treatment to reduce pathogenicity while mimicking natural infection. Recombinant protein vaccines combining virulence factor genes from *T. pyogenes* and *C. perfringens* have shown partial protective efficacy. Additionally, encapsulating DNA vaccines within chitosan nanoparticles enhances host resistance to *T. pyogenes*, setting benchmarks for future vaccine innovations.

Case Study

A 54-year-old male farmer with a history of aortic stenosis, alcoholic cirrhosis, and poorly controlled type 2 diabetes mellitus presented with symptoms of fever, chills, and an abscess located on his left foot. The patient reported a history of raising pigs and cattle on his farm. Five days after symptom onset, he developed slurred speech and left-sided weakness, prompting hospital admission. On physical examination, the patient was found to have an aortic murmur, neurological abnormalities, and a body temperature of 39.8 °C. Laboratory results showed a white blood cell count of 18,000/mm^3 and a haemoglobin concentration of 11.2 g/dL. Additionally, urinalysis detected 5–10 red blood cells per high-power field. CSF analysis showed a white blood cell count of 12/mm^3, an RBC count of 430/mm^3, a protein level of 32 mg/dL, and a glucose level of 62 mg/dL. Blood cultures and swabs from the leg ulcers were collected. Initial antibiotic therapy was started with ceftriaxone. Cultures grew pinpoint, whitish, moderately β-hemolytic colonies, and microscopy revealed Gram-positive pleomorphic bacilli. The organism fermented glucose, maltose, and xylose; hydrolysed gelatin; and was CAMP test positive. It was identified as *T. pyogenes* via MALDI-TOF MS and 16S rDNA sequencing. Susceptibility testing showed sensitivity to penicillin, cephalothin, erythromycin, clindamycin, vancomycin, and trimethoprim-sulfamethoxazole. On day seven, ceftriaxone was replaced with high-dose penicillin G (two million IU every 4 h). Despite treatment, the patient's condition deteriorated rapidly. Repeat transthoracic echocardiography revealed calcific aortic stenosis with vegetations of size 3–5 mm. Severe respiratory distress ensued, and the patient succumbed shortly thereafter.

1. Describe the mode of transmission and immunopathogenesis of the *T. pyogenes* infection.
2. Describe the laboratory diagnosis of *T. pyogenes* infection.
3. Describe the preventive and control measures for this zoonotic infection.

Research Questions

1. How can we understand the genetic diversity among *T. pyogenes* isolates from different animal species, humans, and geographic locations?
2. What are the primary reservoirs and transmission routes of *T. pyogenes* in livestock populations?
3. How to address the gap regarding the pathophysiology and immunology of *T. pyogenes?*

Further Readings

Galán-Relaño Á, Gómez-Gascón L, Rodríguez-Franco A, Luque I, Huerta B, Tarradas C, et al. Search of potential vaccine candidates against *Trueperella pyogenes* infections through proteomic and bioinformatic analysis. Vaccines (Basel). 2020;8(2):314. https://doi.org/10.3390/vaccines8020314.

Huang T, Song X, Jing J, Zhao K, Shen Y, Zhang X, et al. Chitosan-DNA nanoparticles enhanced the immunogenicity of multivalent DNA vaccination on mice against *Trueperella pyogenes* infection. J Nanobiotechnol. 2018;16(1):8. https://doi.org/10.1186/s12951-018-0337-2.

Jost BH, Billington SJ. *Arcanobacterium pyogenes*: molecular pathogenesis of an animal opportunist. Antonie Van Leeuwenhoek. 2005;88:87–102.

Kreitlow A, Ningrum SG, Lämmler C, Erhard M, Hoffmann C, Plötz M, et al. Identification of the novel potential pathogen *Trueperella pecoris* with interspecies significance by LAMP diagnostics. Sci Rep. 2023;13:14005. https://doi.org/10.1038/s41598-023-40787-1.

Lämmler C, Sammra O, Glaeser S, Kämpfer P. *Trueperella*. In: Trujillo ME, editor. Bergey's manual of systematics of Archaea and Bacteria. Online © 2020 Bergey's Manual Trust. John Wiley & Sons, Inc. in association with Bergey's Manual Trust. https://doi.org/10.1002/9781118960608.gbm01515.

Liu N, Shan Q, Wu X, Xu L, Li Y, Wang J, Wang X, Zhu Y. Phenotypic characteristics, antimicrobial susceptibility and virulence genotype features of *Trueperella pyogenes* associated with endometritis of dairy cows. Int J Mol Sci. 2024;25:3974.

Machado VS, Bicalho RC. Complete genome sequence of *Trueperella pyogenes*, an important opportunistic pathogen of livestock. Genome Announc. 2014;2:e00400–14. https://doi.org/10.1128/genomeA.00400-14.

Nagaraja TG. Trueperella. In: Scott McVey D, Kennedy M, Chengappa MM, Wilkes R, editors. Veterinary microbiology. Chapter 27. 5th ed. Willey Blackwell publishers; 2022. p. 252–6.

Quinn PJ, Markey BK, Leonard FC, Fitzpatrick ES, Fanning S, Hartigan PJ. Veterinary microbiology and microbial disease. Wiley-Blackwell; 2011. p. 245–57.

Ribeiro MG, Risseti RM, Bolaños CAD, Caffaro KA, de Morais ACB, Lara GHB. *Trueperella pyogenes* multispecies infections in domestic animals: a retrospective study of 144 cases (2002 to 2012). Vet Q. 2015;35:82–7.

Rzewuska M, Kwiecień E, Chrobak-Chmiel D, Kizerwetter-Świda M, Stefańska I, Gieryńska M. Pathogenicity and virulence of *Trueperella pyogenes*: a review. Int J Mol Sci. 2019;20:2737.

Stuby J, Lardelli P, Thurnheer CM, et al. *Trueperella pyogenes* endocarditis in a Swiss farmer: a case report and review of the literature. BMC Infect Dis. 2023;23:821.

Wen X, Cheng J, Liu M. Virulence factors and therapeutic methods of *Trueperella pyogenes*: a review. Virulence. 2025;16(1):2467161.

Yassin AF, Hupfer H, Siering C, Schumann P. Comparative chemotaxonomic and phylogenetic studies on the genus *Arcanobacterium* Collins et al. 1982 emend. Lehnen et al. 2006: proposal for *Trueperella* gen. nov. and emended description of the genus *Arcanobacterium*. Int J Syst Evol Microbiol. 2011;61:1265–74.

Zhang W, Wang P, Wang B, Ma B, Wang J. A combined *Clostridium perfringens/Trueperella pyogenes* inactivated vaccine induces complete immunoprotection in a mouse model. Biologicals. 2017;47:1–10.

Zhang Z, Liang Y, Yu L, Chen M, Guo Y, Kang Z, Qu C, Tian C, Zhang D, Liu M. TatD DNases contribute to biofilm formation and virulence in *Trueperella pyogenes*. Front Microbiol. 2021;12:758465.

Arcobacter Infections

48

Ashoka Mahapatra

Learning Objectives

1. To comprehend the morphology, growth characteristics, and taxonomic position of *Arcobacter* spp.
2. To recognise the clinical relevance of *Arcobacter* as an emerging zoonotic pathogen
3. To understand the reservoirs and transmission modes in animals and humans
4. To identify diagnostic methods culture, biochemical identification and molecular diagnosis and outline treatment and prevention strategies

Introduction

Arcobacter spp. are recognised as emerging zoonotic, food-borne, and water-borne pathogens linked to gastroenteritis and bacteraemia in humans. These Gram-negative, spiral-shaped, motile bacteria are closely related to *Campylobacter* but display aerotolerance. Among the numerous *Arcobacter* species identified, three are most implicated in zoonotic transmission and human disease. These are *Arcobacter butzleri*, *Arcobacter cryaerophilus*, and *Arcobacter skirrowii*.

A. Mahapatra (✉)
Department of Microbiology, All India Institute of Medical Sciences, Bhubaneswar, Odisha, India

The *Arcobacter* species are frequently found in the gastrointestinal tracts of a wide variety of animals, both domesticated and wild, which serve as significant reservoirs. They are associated with gastroenteritis, bacteraemia, and reproductive tract infections in livestock. In humans, they cause a wide range of infections varying from self-limiting gastroenteritis to severe conditions such as endocarditis, peritonitis, and prolonged diarrhoea. Transmission occurs through contaminated water, food, and direct animal contact. Pathogenicity is driven by virulence factors enabling adhesion, invasion, and cytotoxicity. Diagnosis involves culture on selective media, biochemical identification and molecular diagnosis. Treatment primarily involves chemotherapy with ciprofloxacin and tetracycline, although resistance is increasing. Prevention focuses on food safety, water sanitation, and proper animal handling.

History

Arcobacter was first reported in 1977 with the isolation of these bacteria from aborted bovine and porcine foetuses. Initially, these organisms were referred to as "aerotolerant *Campylobacter*-like organisms" due to their morphological similarities to bacteria in the genus *Campylobacter*. The bacteria, however, were able to grow in the presence of oxygen, unlike the typically microaerophilic *Campylobacter* species. The formal recognition of these bacteria as a distinct genus,

S. C. Parija, A. Poddar (eds.), *Textbook of Bacterial Zoonoses*, Microbial Zoonoses,
https://doi.org/10.1007/978-981-95-6615-0_48

Arcobacter, was reported in 1991 by Vandamme et al. (1991) based on ribosomal RNA genetic similarity (95% cut-off). This classification, based on detailed taxonomic studies, separated them from the *Campylobacter* genus. The genus name *Arcobacter* is derived from the Latin term "*arcus*" (bow) and the Greek word "*bacter*" (rod), symbolising its arch-shaped morphology.

The understanding of *Arcobacter* as a human pathogen gained attention during the 1990s and 2000s. Studies began to associate *Arcobacter* species, particularly *A. butzleri*, with cases of gastroenteritis and diarrhoea in humans. While often causing self-limiting illness, *Arcobacter* infections have also been implicated in more severe extra-intestinal manifestations like bacteraemia, although these are less common.

Taxonomy

The genus *Arcobacter* is placed within the domain Bacteria, the phylum Campylobacterota, the class "Campylobacteria", the order Campylobacterales, and the family Arcobacteraceae. The genus *Arcobacter* was established in 1991 with three species, *A. butzleri*, *A. cryaerophilus*, and *A. skirrowii*, separating it from the genus *Campylobacter*. This distinction was primarily based on their physiological differences, particularly the ability of *Arcobacter* species to tolerate aerobic conditions and grow at lower temperatures, traits not typically found in *Campylobacter*. By 2021, 33 *Arcobacter* species had been identified. The evolution and emergence of the new genus *Arcobacter* and new species over the timeline, along with the source and year of some species, have been depicted in Fig. 48.1.

The taxonomic classification of *Arcobacter* at the genus level is currently a subject of scientific discussion and proposed changes. While traditionally recognised as a single genus encompassing a growing number of species, recent comprehensive studies utilising advanced genomic and phylogenetic analyses have challenged this structure. One significant proposal, supported by detailed analyses of core genomes and other genetic markers, suggests that the genus *Arcobacter* should be divided into multiple distinct genera, including *Arcobacter* (in a more restricted sense) and several newly proposed genera such as *Aliarcobacter*, *Pseudoarcobacter*, *Haloarcobacter*, *Malacobacter*, *Poseidonibacter*, and "Candidatus Arcomarinus". This perspective is based on evidence of significant genetic divergence among groups of species currently classified under *Arcobacter*, suggesting they represent separate evolutionary lineages deserving of generic status.

However, this proposed reclassification is not universally accepted within the scientific community. A counterpoint, supported by extensive comparative analyses of genomic and phenotypic characteristics, argues against the division of *Arcobacter* into multiple genera. Proponents of retaining the single genus *Arcobacter* contend that, despite some genetic variations, the group exhibits sufficient phenotypic and phylogenetic coherence to remain classified as a single genus. They also raise concerns that splitting the genus might lead to practical difficulties in identification and communication, particularly in clinical and environmental settings. This ongoing debate underscores the dynamic nature of bacterial taxonomy, which is continuously refined as new data and analytical approaches emerge.

Regardless of the ongoing discussion surrounding the genus-level classification, there is a clear recognition of the increasing diversity of species within *Arcobacter* (sensu lato). Since its initial description with just a few species, the number of recognised *Arcobacter* species with validly published names has expanded considerably, now exceeding 30. These species have been isolated from a diverse array of environments, highlighting their adaptability and widespread distribution. They are found in various water sources, associated with a wide range of animals, and detected in food products. Furthermore, several *Arcobacter* species, including *A. butzleri*, *A. cryaerophilus*, and *A. skirrowii*, have been identified as important emerging pathogens capable of causing illness in both humans and animals, making their accurate identification and taxonomic placement crucial for public health and veterinary medicine.

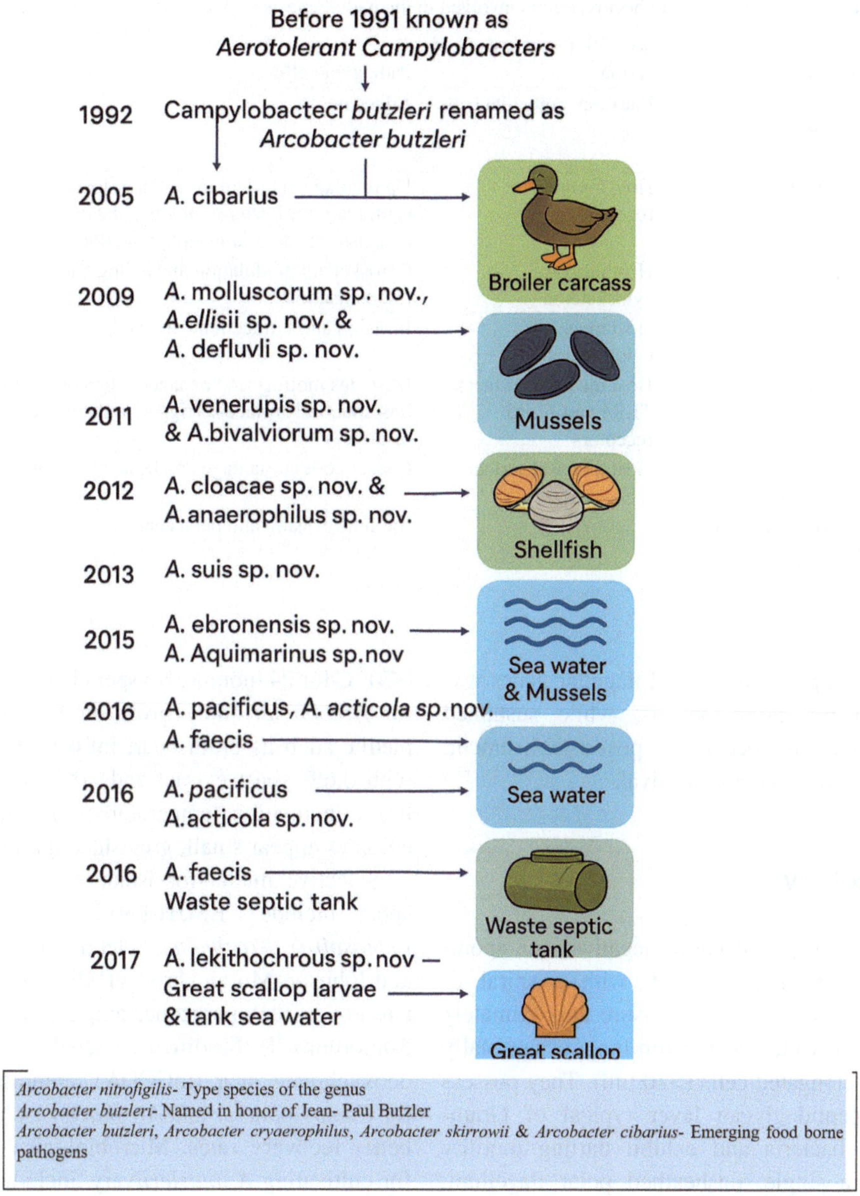

Fig. 48.1 Year-wise evolution and the emergence of the new genus *Arcobacter* beginning from 1991 and the addition of new species over the timeline along with the source of some species. (Image source: Author's personal collection)

Genomics and Proteomics

Genomic and proteomic analyses reveal key virulence factors contributing to *Arcobacter* spp. pathogenicity and antibiotic resistance. Whole-genome sequencing has facilitated the identification of genes associated with adhesion, invasion, and toxin production. Proteomic studies highlight outer membrane proteins and stress-response factors aiding host colonisation and immune evasion (Table 48.1). Medina et al. (2019) demonstrated the adaptive response of *A. butzleri* through RT-PCR transcriptional analysis of flagellar and virulence genes, identifying its biphasic transcriptional pattern during endocytobiosis with *Acanthamoeba castellanii*. Upregulation of

Table 48.1 Bacterial ligands and host receptors involved in the pathogenesis of Arcobacter infection

Virulence determinants/ bacterial ligands	host cell receptors/site of action	Pathogenic effect
Outer membrane proteins (OMP21, OMP31, and OMP50)	Intestinal epithelial cells	Adhesion
Fibronectin-binding protein (CadF protein)	Host fibronectin receptors	Up regulates interleukin-8 expression, which is significant for bacterial binding and colonisation of host tissues and establishing infection
MreB protein	Host mucosal cytoskeleton	Cytoskeletal modulation and aiding bacterial internalisation
CiaB	Host mucosal cytoskeleton	Invasion and intracellular survival
Flagellin (FlaA/FlaB)	Host mucosal surfaces, TLR4 and TLR5 receptors	Mediates motility and enhances chemotaxis towards host mucosal surfaces. Triggers inflammatory responses
Toxins—phospholipases and haemolysins, CDT	Host mucosal surfaces	Causes cellular damage and host cell apoptosis
Lipo-oligosaccharides (LOS) and biofilms	–	Immune evasion and persistence

ciaB, *cadF*, *pldA*, *mviN*, and flagellar genes was observed pre-endocytobiosis, while sustained downregulation occurred post-establishment, enabling intra-amoebic survival.

Morphology

Arcobacter spp. are Gram-negative, non-spore-forming, and non-acid-fast, with a spiral or curved rod shape. Cells measure approximately 0.2–0.9 μm wide and 1–3 μm long, occasionally forming elongated cells (>20 μm). They possess a thin peptidoglycan layer typical of Gram-negative bacteria and exhibit darting motility through a single, unsheathed polar flagellum. This motility aids in membrane filtration, allowing their isolation. Phase-contrast or dark-field microscopy is used to visualise this characteristic motion.

Cultivation

Arcobacter are aerotolerant, growing under both microaerophilic and atmospheric oxygen levels at 15–37 °C and pH 6.8–8.0. Unlike *Campylobacter*, they cannot grow at 42 °C. They tolerate freezing at −20 °C for 6 months and −70 °C for 24 months. No specific medium exists for *Arcobacter*; they grow on *Campylobacter* media, such as brain heart infusion agar (BHI) with 0.6% yeast extract and 10% blood, but at lower incubation temperatures. On blood agar, colonies appear small, greyish, and translucent.

Selective media for isolation of *Arcobacter* spp. include EMJH-P80 (originally for *Leptospira*), *Arcobacter* Selective Broth (ASB), and Johnson-Murano broth (JMB), enriched with antibiotics (cefoperazone, amphotericin B, and 5-fluorouracil). Modified charcoal cefoperazone deoxycholate agar (mCCDA), commonly used for *Campylobacter* and *Arcobacter* isolation, has better recovery rates. Microbiological methods for cultivating *A. butzleri* vary, including differences in nutrient media and incubation conditions. Merga et al. tested five cultivation techniques and found that *Arcobacter* broth-mCCDA-Columbia agar, a selective medium with antibiotics, was the most sensitive and specific for detecting *A. butzleri*.

Animal Models

Various animal models are used to study *Arcobacter* pathogenicity. *A. butzleri* is the most invasive species, capable of colonising piglet intestines. However, results vary in chickens and turkeys. Mouse and rat models reveal enhanced adherence of *A. butzleri* in intraperitoneal infections. In BALB/c mice, oral and intramammary administration causes severe inflammation, abscesses, and organ enlargement. Due to its transparency and ease of genetic screening, zebrafish is increasingly used as a model for real-time visualisation. Gnotobiotic IL-10-deficient mice infected with *A. butzleri* are used to study the animal immune response. Studies reveal increased secretion of pro-inflammatory cytokines—IFN-γ, TNF-α, and IL-6—in both small and large intestines, indicating intestinal immune activation. This immune activation is not only limited to the intestines. High levels of a cytokine called MCP-1 were also found in the blood, suggesting that *A. butzleri* can trigger a systemic immune response, even when there is no infection outside the intestines. This suggests that *A. butzleri* can trigger widespread inflammation in the body.

When compared with *Campylobacter jejuni*, *A. butzleri* induces less distinct pro-inflammatory sequelae, indicating its relatively low pathogenic potential. Cytotoxicity has also been demonstrated in Caco-2, HT-29, porcine intestinal tissues, and Hep-2 cell lines.

Transmission Cycle

The transmission cycle of *Arcobacter* species is primarily indirect, with contaminated food and water serving as the principal vehicles for infection in both animals and humans (Fig. 48.2). The cycle is intricately linked to various reservoirs and environmental pathways.

Arcobacter spp. is widely distributed in animal and environmental reservoirs, including livestock (pigs, cattle, sheep, and poultry), wild birds, reptiles, rodents, and marine animals. These animals often carry the bacteria asymptomatically in their intestinal tracts and shed them in their faeces. Wild birds and other animals can also contribute to the environmental load of *Arcobacter*.

Humans primarily acquire *Arcobacter* infections through the consumption of contaminated food and water. Drinking water that has not been adequately treated or has become contaminated after treatment poses a risk. Similarly, consuming raw or undercooked meat and seafood is a direct pathway for infection. While less common, contact with infected animals or their faeces, and potentially even person-to-person transmission, cannot be entirely ruled out, though these routes

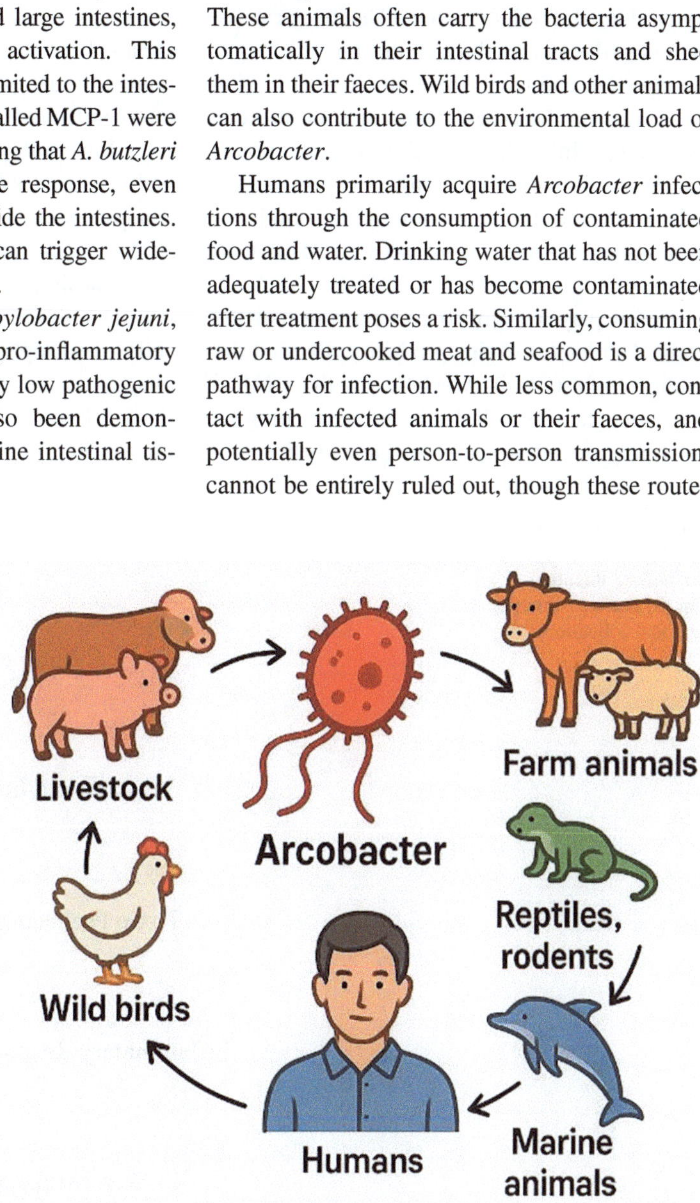

Fig. 48.2 Zoonotic transmission cycle of *Arcobacter* species. (Image source: Author's personal collection)

are considered less significant compared to food- and water-borne transmission.

Pathogenesis and Pathology

Pathogenesis

Arcobacter infection, mainly caused by *A. butzleri*, *A. cryaerophilus*, and *A. skirrowii*, follows several steps typical of gut pathogens (Fig. 48.3). Infection usually starts after eating or drinking contaminated food or water. The bacteria use their flagella to move through the mucus lining of the intestine and reach the epithelial cells. They then attach to these cells using specific proteins like CadF, helping them to colonise. Some strains can invade the intestinal lining, allowing them to hide from the immune system and multiply. Genes such as *ciaB*, *mviN*, *pbpA*, and *tlyA* are involved in this process.

Toxins play a major role in causing damage. Some *Arcobacter* strains produce cytolethal distending toxin (CDT), which harms the DNA of host cells, causing cell death and weakening the gut lining. Other toxins, like haemolysins (e.g. from *tlyA*), also contribute to cell damage. To survive in the host, *Arcobacter* depends on systems like iron uptake proteins (e.g. IrgA) to obtain essential nutrients. It may also evade the immune system and form biofilms to persist longer. Invasion, toxin production, and inflammation cause symptoms like diarrhoea and abdominal pain. However, virulence can vary between different species and strains.

Pathology

Detailed studies of *Arcobacter* infection in human tissues are limited because the disease is often mild and self-resolving, making biopsies rare. Most information comes from its known pathogenesis, symptoms, related bacteria (like *Campylobacter*), and animal studies.

Arcobacter mainly affects the gastrointestinal tract, causing inflammation of the intestinal lining. Histology typically shows many immune cells (neutrophils, lymphocytes, and macrophages) in the lamina propria (the tissue below the gut lining). This inflammation causes swelling (oedema) of the intestinal wall.

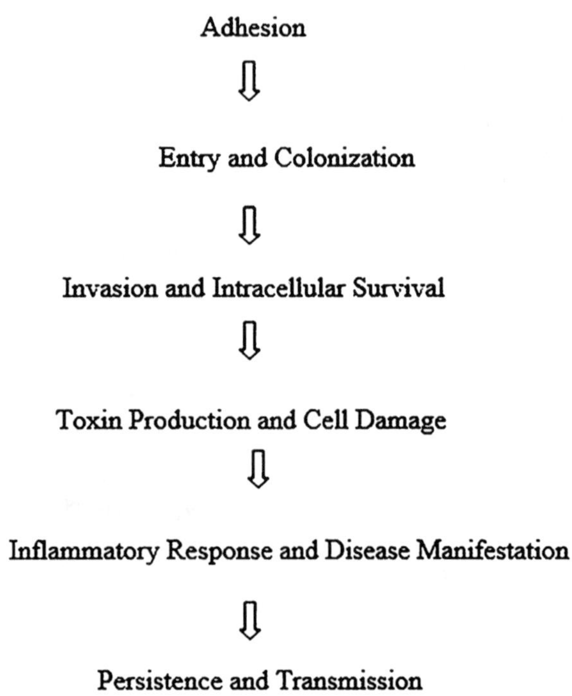

Fig. 48.3 Steps for pathogenesis of *Arcobacter* infection. (Image source: Author's personal collection)

Damage to the gut lining occurs due to bacterial invasion and toxins like CDT. Changes may include surface erosion, increased death of epithelial cells (apoptosis), shortening or blunting of intestinal villi, and inflammation or pus-filled abscesses in the colon crypts (cryptitis and crypt abscesses). These changes disrupt normal gut function, leading to diarrhoea and abdominal pain. Overall, the pathology resembles the acute inflammation seen in *Campylobacter* infections. However, virulence can vary between different species and strains.

Immunology

The interaction between *Arcobacter* species and the host immune system is a developing area of research. The precise mechanisms are yet to be fully elucidated. Nevertheless, studies have shown that *Arcobacter* can trigger both innate and adaptive immune responses in infected hosts, although the outcome of infection can vary, ranging from asymptomatic carriage to overt disease. Despite the observed immune responses, the ability of *Arcobacter* to establish persistent colonisation in some animals and cause asymptomatic carriage suggests potential mechanisms of immune modulation or evasion. However, specific immune evasion strategies employed by *Arcobacter* are not yet well characterised and represent an important area for future research.

The innate immune system serves as the first line of defence against *Arcobacter*. Research, particularly in animal models, has highlighted the crucial role of pattern recognition receptors, such as Toll-Like Receptor 4 (TLR4). These receptors recognise *Arcobacter* components such as lipooligosaccharide (LOS) or lipopolysaccharide (LPS). This recognition triggers downstream signalling pathways that lead to the induction of inflammatory responses. Studies using gnotobiotic mice, for instance, have demonstrated that TLR4 is essential for mediating intestinal and systemic immune responses to *Arcobacter butzleri*. These influence the recruitment of immune cells like macrophages, monocytes, and lymphocytes, and the production of pro-inflammatory cytokines such as tumour necrosis factor alpha (TNF-α), Interleukin-6 (IL-6), Interferon-gamma (IFN-γ), and monocyte chemoattractant protein-1 (MCP-1). Macrophages, as key phagocytic cells, are involved in the initial recognition and response to *Arcobacter*. Studies have indicated that *Arcobacter* infection can modulate miRNA expression in macrophages, suggesting a layer of fine-tuning in the innate immune signalling.

In contrast, specific details on the adaptive immune response, including the roles of T and B lymphocytes and antibody production, are less extensively documented. The presence of immune cell infiltrates in infected tissues in animal models further supports the activation of adaptive immune components.

Arcobacter species possess potential virulence factors that likely play a role in their interaction with the host immune system. These include factors associated with adhesion to and invasion of host epithelial cells. This disrupts the intestinal barrier and exposes deeper tissues to the bacteria, thereby influencing the nature and intensity of the immune response. Some studies also suggest the potential production of toxins or enzymes that could contribute to tissue damage and inflammation, further impacting the immune landscape. The induction of apoptosis in intestinal epithelial cells, observed with certain *Arcobacter* strains, can also influence the host's defence mechanisms.

Infection in Humans

The International Commission on Microbio logical Specifications for Food (ICMSF) identifies *Arcobacter* spp. as a significant food-borne pathogen. The commission ranks it as the fourth most common *Campylobacter*-like organism causing diarrhoeal illness. In humans, *Arcobacter* spp. cause acute diarrhoea, abdominal pain, and fever in immunocompetent individuals. In immunocompromised patients, it can lead to bacteraemia, peritonitis, endocarditis, and septic shock. Occasionally, it causes reproductive tract infections. *A. butzleri*, *A. skirrowii*, and *A. cryaerophilus* are the primary species linked to human disease, with *A. butzleri* identified as a significant public health concern.

Infection in Animals

The primary *Arcobacter* species associated with animal infections are *A. butzleri, A. cryaerophilus, A. skirrowii*, and *Arcobacter thereius*. These pathogens mainly cause gastroenteritis and reproductive disorders in cattle, sheep, pigs, dogs, and poultry. Gastrointestinal symptoms include acute and chronic diarrhoea (sometimes haemorrhagic), colitis, dehydration, reduced feed intake, and weight loss. *A. cryaerophilus* is particularly linked to reproductive issues in livestock, causing abortion, stillbirth, and foetal loss. It may also lead to infertility due to metritis and endometritis. Systemic infections, including septicaemia, are more common in calves, lambs, and piglets, presenting with fever, lethargy, and anorexia.

Epidemiology and Public Health

Arcobacter spp. are emerging zoonotic pathogens associated with gastrointestinal diseases in humans and animals. However, the precise incidence data for human infections remain limited due to challenges in routine diagnostics. However, several studies from various regions worldwide indicate a significant presence of these bacteria in key reservoirs and a clear association with human illness (Tables 48.2 and 48.3).

As per the data of a systematic review and meta-analysis published in 2021, the pooled global prevalence of *Arcobacter* in the food chain was 30% (95% CI: 26.2–34.2%), ranging from 0.5% to 99.4%. The highest prevalence was in Belgium (57.5%) and the lowest in Ireland (3.1%). Oceania (40.9%) and high-income countries (34.2%) showed the highest prevalence in subgroup analyses.

Infection occurs through ingesting or direct contact with contaminated water, food, or animal

Table 48.2 Global epidemiology of *Arcobacter* infection

Species	Host	Geographical distribution	Mode of presentation	Outcome	Year and reference
Arcobacter lanthieri	Animal (cattle, water sources)	Canada, USA	No confirmed human cases yet	N/A	2017 Douidah et al. (2017)
Arcobacter cryaerophilus	Human, animal	Europe, North America, Australia	Diarrhoea, abdominal cramps	Mild to moderate illness	2014 Ferreira et al. (2014)
Arcobacter cloacae	Animal (water samples)	Europe, USA	No confirmed human infection	N/A	2013 Levican et al. (2013)
Arcobacter butzleri	Human, animal	Worldwide (notably Europe, North America, Asia, Africa)	Gastroenteritis (diarrhoea, abdominal pain), bacteraemia	Often self-limiting; severe cases in immunocompromised	2011 Collado and Figueras (2011)
Arcobacter thereius	Animal (pigs, poultry)	Europe (mainly Belgium and Spain)	No clear human infection confirmed yet	N/A	2011 De Smet et al. (2012)
Arcobacter skirrowii	Human, animal	Europe (few cases in Asia and North America)	Diarrhoea, rare systemic infections	Generally mild; rare systemic spread	1998 Atabay and Corry (1998)

Table 48.3 Indian epidemiology of *Arcobacter* infection

Species	Host	Geographical distribution	Mode of presentation	Outcome	Year and reference
Arcobacter butzleri	Human	Pune, Maharashtra	Diarrhoea in hospitalised patients	Mostly mild; supportive care	2019 Sharma (2019)
Arcobacter cryaerophilus	Animal (livestock, milk samples)	Tamil Nadu, Kerala	Asymptomatic in animals	Public health concern via food chain	2017 Anandhi and Blocksome (2017)
Arcobacter butzleri	Human	Delhi	Diarrhoea, abdominal pain	Mostly self-limiting	2017 Rajalakshmi et al. (2017)
Arcobacter butzleri	Human	Vellore, Tamil Nadu	Acute diarrhoea	Dehydration in some cases	
Arcobacter butzleri, Arcobacter cryaerophilus	Animal (poultry, livestock)	Gujarat	Asymptomatic carriage	Reservoirs; zoonotic potential	
Arcobacter skirrowii	Animal (poultry)	Punjab, Gujarat	No confirmed human infection	N/A	

faeces. Human transmission routes include faecal-oral, food-borne, direct animal contact, or person-to-person spread, commonly via undercooked meat, unpasteurised milk, or contaminated produce. In animals, it spreads through faecal-oral routes, contaminated feed, or direct contact with aborted foetuses and placentas. *Arcobacter* survives in biofilms and resists disinfectants in food processing environments, facilitating transmission.

Environmental sources include water (irrigation and environmental), raw foods, and animal-origin products (milk, meat, fish, and seafood). Food contamination is another major route. During the slaughter and processing of livestock and poultry, *Arcobacter* can be transferred from the intestinal contents to the meat. Raw or undercooked meat, particularly poultry, is frequently identified as a source of human infection. Contamination can also occur during food handling and preparation through contact with contaminated water or surfaces. Shellfish from contaminated waters can also harbour *Arcobacter* and serve as a source of infection when consumed raw or undercooked.

Arcobacter was most frequently detected at the processing stage (33.5%) compared to the primary and retail stages. Poultry products had the highest contamination rate (40.4%), with *A. butzleri* being the most common species. The findings emphasise the widespread distribution of *Arcobacter* in the food chain, highlighting the need for preventive measures to reduce its public health impact. A 2024 meta-analysis reported a pooled prevalence of 21.9% across various sources, with human stool samples accounting for 1.8% and animal food products 25.9%. However, the prevalence varies globally. For instance, in Thailand, a culture-based detection method revealed a prevalence of 2.4%. In contrast, based on molecular detection techniques, South Africa reported a considerably higher prevalence of 12.9%. In India, a study reported 4% from human diarrhoeal stool compared to a high percentage from animal sources (25.33%—pigs,17.33%—poultry).

A. butzleri is the most widespread species globally, with seasonal peaks in summer and early autumn. Recent outbreaks (2020–2023) have included severe cases of bacteraemia and sepsis, highlighting their clinical significance. These bacteria are found in food products (poultry, beef, dairy, seafood, and vegetables) and water sources (rivers, lakes, and drinking water), demonstrating their public health risk. Occasionally, *A. butzleri* may cause severe enteritis and septicaemia, and the disease progression may be dose-dependent. The isolation of this bacterium from varied human, animal, and environmental sources suggests its transmission should be viewed with the background of the One Health concept.

Diagnosis

Different diagnostic methods for the detection of *Arcobacter* are available.

Specimen

Laboratory diagnosis of *Arcobacter* infection relies on the appropriate collection and processing of various sample types, depending on the suspected source of infection (clinical or non-clinical). The selection of samples is guided by the clinical presentation in humans or animals or by the focus of environmental or food safety investigations.

For suspected human infections, the most common clinical sample submitted for *Arcobacter* testing is stool. In cases of suspected systemic infection, such as bacteraemia, blood samples are crucial for diagnosis. In veterinary diagnostics, samples from animals, those showing signs of illness (e.g. diarrhoea, reproductive issues) and potentially asymptomatic carriers, are important. Animal faeces are frequently tested, as are tissue samples from affected organs in cases of systemic disease or abortion.

Beyond clinical cases, *Arcobacter* is commonly looked for in non-clinical samples as part of source attribution, epidemiological studies, and food and water safety monitoring. Food products, particularly raw or undercooked meats (poultry, pork, beef), seafood (especially shellfish), and raw milk, are frequently tested due to their role in transmission. Environmental water samples from various sources, including rivers, lakes, groundwater, and wastewater, are also important matrices for *Arcobacter* detection, reflecting their widespread presence in the environment. Samples from food processing environments, such as slaughterhouses, can also be collected to identify potential points of contamination.

Microscopy

By microscopy examination, Gram staining reveals the bacteria as small, curved, or spiral-shaped Gram-negative rods with tapered ends, closely resembling *Campylobacter* (Fig 48.4). To improve visibility, especially in clinical specimens, a modified Gram stain technique can be used, where prolonged counterstaining helps enhance contrast and clarity of the organisms.

Additionally, fluorescent staining methods, including fluorescent dyes or fluorescent in situ hybridisation (FISH), offer increased sensitivity and are particularly useful for detecting bacteria in both environmental and clinical samples.

In Vitro Culture

Samples from clinical, environmental, or animal sources are inoculated into selective enrichment broth and incubated at 30 °C for 72 h under a microaerophilic atmosphere. Aliquots are streaked onto *Arcobacter* selective agar plates and incubated for 2–7 days at 30 °C. Colonies appear small, round, white or grey, moist, and translucent with key identifying characters. Isolates provisionally identified as *Arcobacter* spp. require further differentiation from other Gram-negative oxidase-positive organisms (Table 48.4).

Serodiagnosis

Serodiagnosis is not widely applicable for diagnosing infections caused by *Arcobacter* spp. It involves detecting specific antibodies in a patient's serum, indicating prior exposure to the pathogen. However, serological assays for *Arcobacter* are still under development and are not routinely available in clinical practice. The limited availability of these tests is due to the

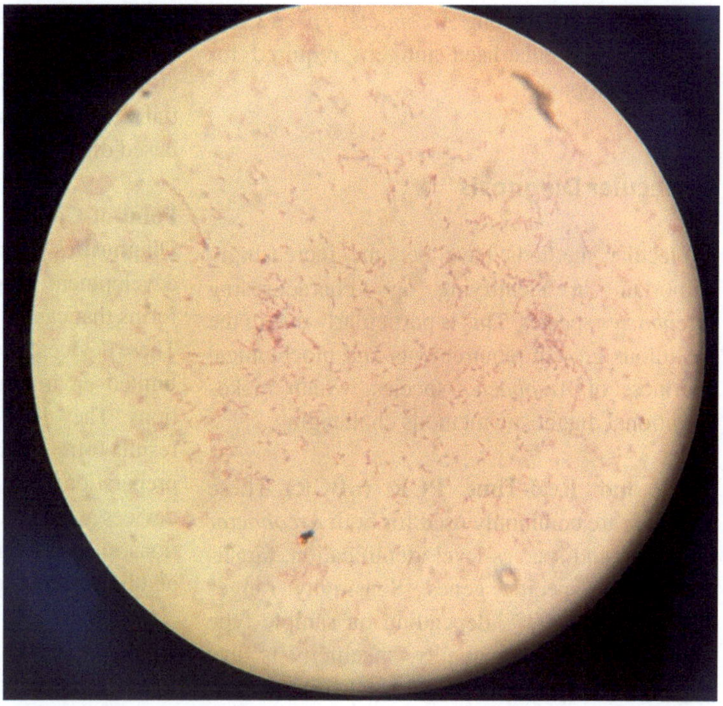

Fig. 48.4 Gram stain morphology of *Arcobacters* showing Gram-negative, small, slender, curved bacilli. 100× magnification. (Image source: Author's personal collection)

Table 48.4 Culture and biochemical characteristics for differentiation of Gram-negative, oxidase-positive bacterial pathogens isolated from stool

Features	*Arcobacter*	*Campylobacter*	*Aeromonas*	*Pseudomonas*	*Vibrio*
Gram stain	Gram-negative, curved rods	Gram-negative, curved rods	Gram--negative, straight rods	Gram-negative, straight rods	Gram-negative, curved rods
Motility	Darting, corkscrew--like motion	Darting, corkscrew--like motion	Single polar flagellum (darting)	Single or multiple polar flagella	Darting or shooting star motility
Colony morphology	Small, round, translucent	Small, moist, flat, greyish	Large, opaque, creamy	Large, flat, pigmented (greenish)	Smooth, shiny, yellow/greenish
Selective media	mCCDA, Arcobacter broth, Columbia agar	mCCDA, Skirrow or Campy blood agar	Ampicillin blood agar, CIN agar	Cetrimide agar, MacConkey agar	TCBS agar (green or yellow colonies)
Incubation temperature	30–37 °C	37–42 °C	30–37 °C	37 °C	35–37 °C
Atmosphere requirement	Microaerophilic or aerobic	Microaerophilic (5–10% O_2)	Aerobic or facultative anaerobe	Aerobic	Facultative anaerobe
Hippurate hydrolysis	negative	Positive (*Campylobacter jejuni*)	negative	negative	negative
Indole production	negative	negative	Positive	negative	negative
Growth in NaCl (6.5%)	No growth but grows at NaCl (3.5%)	No growth	Growth	Growth	Growth
H_2S production	negative	Positive (*Campylobacter lari*)	negative	negative	negative

genetic diversity among *Arcobacter species* and the lack of standardised antigens required for assay development.

Molecular Diagnosis

Molecular methods have become increasingly important in identifying and characterising *Arcobacter* species. This is particularly due to the fastidious growth requirements and biochemical inertness of *Arcobacter* species, which makes traditional diagnostic methods challenging.

PCR and Real-Time PCR (qPCR) These methods are commonly used for both *Arcobacter* detection and species-level identification, targeting species-specific genes. Sensitivity ranges from 95% to 100%, depending on sample type and DNA extraction quality. specificity is also high, generally >98%, especially when species-specific primers are used. These assays can detect as low as 10–100 CFU/mL in spiked samples.

Multiplex PCR Multiplex PCR assays allow the simultaneous detection of multiple *Arcobacter* species in a single reaction using species-specific primers. Commonly targeted species include *A. butzleri*, *A. cryaerophilus*, and *A. skirrowii*. This approach is efficient for epidemiological studies and routine screening.

16S rRNA Gene Sequencing This is considered the *gold standard* for confirming *Arcobacter* identification and differentiating between closely related species. It is highly accurate but more time-consuming and costlier compared to PCR. It also enables the identification of novel or atypical strains.

Loop-Mediated Isothermal Amplification (LAMP) LAMP is a newer method for rapid detection of *Arcobacter* DNA under isothermal conditions, eliminating the need for thermal cycling. It is fast and cost-effective and does not require sophisticated equipment, has a sensitivity comparable to PCR, and has detection limits around 10–100 CFU/mL. This test has the potential to be used as a point-of-care (PoC) or field-based diagnostic.

Point-of-Care (PoC) Molecular Diagnostics Recent advances have enabled the development of portable PCR and LAMP platforms that can be used in clinical or field settings. These devices are particularly valuable in resource-limited environments or for outbreak investigations. They have the advantage of giving rapid results (often within 30–60 min), minimal sample preparation and user-friendly, battery-operated devices available. However, there are a few limitations, such as the need for a cold chain for reagents and limited multiplexing capacity.

DNA Microarrays DNA microarrays are useful for rapid detection and preliminary typing of multiple species in the Campylobacteraceae family, including *Campylobater coli*, *C. jejuni*, and *A. butzleri*. The test is also used to classify isolates based on specific genomic regions like the LOS biosynthesis locus, contributing to diversity. The detection limit is around 10,000 cells for *C. jejuni*, similar to the infectious dose. Despite this moderate sensitivity, the method is highly specific due to the use of 70-mer oligonucleotide probes.

Whole-Genome Sequencing (WGS) WGS provides high-resolution typing using methods like multilocus sequence typing (MLST) and single-nucleotide polymorphism (SNP) analysis. It delivers detailed information on phylogeny, virulence, and resistance genes, making it ideal for comprehensive investigations. While WGS offers excellent specificity and strain-level identification, it requires high-quality DNA and is less sensitive for low-abundance organisms without enrichment. Its main drawbacks are high costs and the need for advanced sequencing and computational tools.

Emerging Technologies

Matrix-Assisted Laser Desorption/ Ionisation Time-of-Flight Mass Spectrometry (MALDI-TOF MS) This is a rapid and accurate method for identifying *Arcobacter* species by detecting species-specific protein profiles. It offers high sensitivity (>90%) and specificity (>95%) when supported by a reliable reference database. However, accurate results depend on good-quality culture isolates, as poor growth can affect protein expression. While the cost per test is low, the equipment is expensive and requires regular maintenance and reagents. Despite this, MALDI-TOF MS is widely used in advanced labs due to its speed, accuracy, and efficiency.

Raman Spectroscopy This is a promising technique for quickly identifying *Arcobacter* directly from clinical or environmental samples without the need for culture. It works by detecting the unique molecular vibrations of bacterial cells, acting like a chemical fingerprint. Sensitivity and specificity generally range from 85% to 95%, though results can vary depending on the sample and device. It delivers results within minutes but is currently limited by high equipment costs and the need for complex data analysis.

Table 48.5 summarises an overview of the different diagnostic methods. Moreveor, Summary of molecular methods for detection of *Arcobacter* species is presented in Table 48.6.

Treatment

Arcobacter infections are generally self-limiting, but severe or chronic cases may require antimicrobial therapy. There are no standard treatment guidelines due to limited clinical data. Most

Table 48.5 Diagnosis of *Arcobacter* infection

Diagnostic approaches	Methods	Targets	Remarks
Direct microscopy	Gram stain	Small, curved/spiral Gram-negative rods with tapered ends	Economical and rapid
	Modified Gram stain	Acid-fast bacilli coloured red	Improves visibility in clinical samples
	Fluorescent staining/FISH	Apple-green-coloured bacilli	Enhances detection from environmental and clinical samples
In vitro culture	Selective enrichment broth	Growth of bacilli visualised as turbidity	Allows maintaining the bacilli for further studies
	Arcobacter selective agar	Small, round, moist, translucent, white/grey colonies	Selective growth from stool/ environmental sample inhibiting commensal or non-pathogenic bacteria
	Catalase/ oxidase test	Positive reaction	A part of genus identification
	Key identification tests	Aero tolerance, growth on MacConkey agar, NaCl tolerance (2–3.5%), indoxyl acetate hydrolysis, nitrate reduction	Helps to differentiate from other genus and species
Serodiagnosis	Not routinely used	Targets *Arcobacter* species-specific antigens	The antibody-based tests are under development due to lack of standardised antigens and genetic diversity among *Arcobacter* species
Molecular diagnosis	PCR and real-time PCR	Accurate species identification after culture	Rapid and specific species detection
	16S rRNA sequencing	Gold standard for species identification and virulence/ resistance gene detection	Promising tool for direct detection from samples

Table 48.6 Summary of molecular methods for detection of *Arcobacter species*

Method	Target genes	Sensitivity	Specificity	Application
PCR/qPCR	*16S rRNA, gyrB, cadF, rpoB*	95–100%	>98%	Rapid detection and species identification
Multiplex PCR	*16S rRNA* + species-specific genes	90–95%	>95%	Simultaneous detection of multiple spp.
16S rRNA sequencing	*16S rRNA*	~100%	~100%	Gold standard for confirmation
LAMP	*16S rRNA, cadF*	95–100%	>98%	Rapid, PoC detection
PoC PCR devices	*cadF, gyrB,* etc.	~95%	~95%	On-site diagnosis, low-resource settings
DNA microarrays	-	Moderate	High	Strain typing
WGS (high cost)	-	High	High	Phylogeny, virulence, and resistance genes

strains are susceptible to nalidixic acid but resistant to cephalothin. Empirical therapy includes macrolides, fluoroquinolones, aminoglycosides, tetracyclines, and β-lactam/β-lactamase inhibitor combinations. However, resistance to fluoroquinolones, penicillins, macrolides, and tetracyclines is increasingly reported.

Prevention and Control

The widespread detection of *Arcobacter* in various food products and water sources carries significant public health implications. Contaminated food and water serve as the primary source of infection and pose a potential threat to human health. Person-to-person transmission appears to be infrequent and typically is associated with specific outbreak scenarios. Therefore, there is a critical and immediate need to develop and implement standardised detection methodologies to identify and quantify *Arcobacter* in food and water accurately. Furthermore, continuous surveillance programmes are essential to monitor its prevalence and identify emerging trends, enabling timely intervention. Finally, establishing effective control strategies across the food production chain and water treatment processes is paramount to mitigate the risks associated with *Arcobacter* contamination and safeguard public health.

Effective control measures are critical for reducing the risk of *Arcobacter* infections. Food safety is enhanced by thoroughly cooking meat and poultry. Cross-contamination should be prevented by using separate utensils and cutting boards for raw and cooked foods. Applying 1–2% food-grade citric or lactic acid has been shown to reduce *Arcobacter* contamination in food products. In addition, proper water treatment and purification processes are necessary to prevent transmission through contaminated water sources. Regular surveillance of food products, particularly those of animal origin, is important for detecting *Arcobacter* contamination. Furthermore, monitoring antimicrobial resistance (AMR) patterns in *Arcobacter* isolates is vital to guide appropriate treatment strategies and support public health interventions.

Case Study

A 10-year-old girl was admitted with a month-long history of watery diarrhoea, occasional mucus, stomach cramps, reduced appetite, and signs of mild malnutrition. She had previously experienced a febrile gastrointestinal illness, which resolved without a confirmed diagnosis. On examination, she was stable and afebrile, with normal vitals and no dehydration, but mild abdominal distension and malnutrition were noted. Stool microscopy was negative for parasites, and culture showed no growth on mCCDA at 42 °C. However, small translucent colonies grew at 37 °C by day 4. The isolate was Gram-negative, motile, oxidase- and catalase-positive, and biochemically non-fermentative, with *A. butzleri* identified by MALDI-TOF MS. She was treated with hydration, nutritional support, probiotics, and azithromycin, leading to rapid symptom resolution and discharge in good condition.

1. What additional diagnostic tests could have been performed to confirm the aetiology of prolonged diarrhoea in this child? Discuss the role of advanced molecular techniques and their advantages over traditional culture methods.
2. Considering the self-limiting nature of *Arcobacter* infections, was the use of azithromycin justified in this case? What alternative therapeutic strategies could have been considered?
3. Discuss the preventive measures that could be implemented to reduce the risk of *Arcobacter* infections in children. Include recommendations for food safety, hygiene practices, and public health interventions.

Research Questions

1. Do metagenomic approaches improve over the traditional PCR methods in identifying *Arcobacter* from complex samples?
2. What are the challenges in developing an effective oral vaccine for livestock against *Arcobacter* infection?
3. Can One Health approaches be implemented to monitor and control *Arcobacter* infections across animal, human, and environmental interfaces?

Declaration During the preparation of this chapter, we have used OpenAI's ChatGPT to assist in the preparation of this article. The assistance was limited to refining the English language for clarity and fluency, organising and structuring the content for improved readability, and generating illustrative images not available in the public domain. After using it, we reviewed and edited the content as needed and take full responsibility for the content of the publication.

Further Readings

Alegbeleye OO, Sant'Ana AS. Understanding the public health burden of unconventional produce-associated enteropathogens. Curr Opin Food Sci. 2020;32:37–44. https://doi.org/10.1016/j.cofs.2020.01.008.

Anandhi A, Blocksome CE. Developing adaptation strategies using an agroecosystem indicator: Variability in crop failure temperatures, Ecol Ind. 2017;76:30–41.

Atabay HI, Corry JEL. Evaluation of a new Arcobacter enrichment medium and comparison with two media developed forenrichment of Campylobacter species. Int J Food Microbiol. 1998;41(1):53–58. https://doi.org/10.1016/s0168-1605(98)00034-8.

Baztarrika I, Salazar-Sánchez A, Hernaez Crespo S, López Mirones JI, Canut A, Alonso R, Martínez-Ballesteros I, Martinez-Malaxetxebarria I. Virulence genotype and phenotype of two clinical isolates of *Arcobacter butzleri* obtained from patients with different pathologies. Arch Microbiol. 2023;205(12):369. https://doi.org/10.1007/s00203-023-03709-3.

Baztarrika I, Martinez-Malaxetxebarria I, Martínez-Ballesteros I, Wösten MM. Human Toll-like receptor activation by pathogenic *Arcobacter* species. Microb Pathog. 2025;198:107189. https://doi.org/10.1016/j.micpath.2024.107189.

Collado L, Figueras MJ. Taxonomy, epidemiology, and clinical relevance of the genus *Arcobacter*. Clin Microbiol Rev. 2011;24(1):174–92. https://doi.org/10.1128/CMR.00034-10.

De Smet S, De Zutter L, Houf K. Spatial distribution of the emerging foodborne pathogen Arcobacter in the gastrointestinal tract of pigs. Foodborne Pathog. Dis. 2012;9:1097–1103. https://doi.org/10.1089/fpd.2012.1184.

Douidah L, De Zutter L, Baré J, Houf K. Identification of five human and mammal associated *Arcobacter* species by a novel multiplex-PCR assay. J Microbiol Methods. 2010;80(3):281–6. https://doi.org/10.1016/j.mimet.2009.12.010.

Douidah L, De Zutter L, Bare´ J et al. Towards a typing strategy for Arcobacter species isolated from humans and animals and assessment of the in vitro genomic stability. Foodborne Pathog Dis. 2017;11:272–80.

Fanelli F, Di Pinto A, Mottola A, Mule G, Chieffi D, Baruzzi F, Tantillo G, Fusco V. Genomic characterization of *Arcobacter butzleri* isolated from shellfish: novel insight into antibiotic resistance and virulence determinants. Front Microbiol. 2019;10:670. https://doi.org/10.3389/fmicb.2019.00670.

Ferreira S, Júlio C, Queiroz JA, Domingues FC, Oleastro M. Molecular diagnosis of Arcobacter and Campylobacter in diarrhoeal samples among Portuguese patients. Diagn Microbiol Infect Dis. 2014;78:220–25.

Gölz G, Alter T, Bereswill S, Heimesaat MM. The immunopathogenic potential of *Arcobacter butzleri* – lessons from a meta-analysis of murine infection

studies. PLoS One. 2016;11(7):e0159685. https://doi.org/10.1371/journal.pone.0159685.
Levican A, Alkeskas A, Gunter C, Forsythe SJ, Figueras MJ. Adherence to and invasion of human intestinal cells by Arcobacter species and their virulence genotypes. Appl. Environ. Microbiol. 2013;79:4951–57. https://doi.org/10.1128/AEM.01073-13.
Mateus C, Martins R, Luís A, Oleastro M, Domingues F, Pereira L, Ferreira S. Prevalence of *Arcobacter*: from farm to retail – a systematic review and meta-analysis. Food Control. 2021;128:108177. https://doi.org/10.1016/j.foodcont.2021.108177.
Medina, G., Neves, P., Flores-Martin, S., Manosalva, C., Andaur,M., Otth, C., … Fernandez, H. Transcriptional analysis of flagellar and putative virulence genes of Arcobacter butzlerias anendocytobiont of Acanthamoeba castellanii. Archives of Microbiology. 2019;201:1075–1083. https://doi.org/10.1007/s00203-019-01678-0.
Novak A, Vuko-Tokić M, Žitko V, Tonkić M. Prolonged watery diarrhea and malnutrition caused by *Aliarcobacterbutzleri* (formerly *Arcobacter butzleri*): the first pediatric case in Croatia and a literature review. Infez Med. 2024;32(2):241–7. https://doi.org/10.53854/liim-3202-12.
Patyal A, Rathore RS, Mohan HV, Dhama K, Kumar A. Prevalence of *Arcobacter* spp. in humans, animals and foods of animal origin including seafood from India. Transbound Emerg Dis. 2011;58(5):402–10. https://doi.org/10.1111/j.1865-1682.2011.01221.x.
Ramees TP, Dhama K, Karthik K, Rathore RS, Kumar A, Saminathan M, Tiwari R, Malik YS, Singh RK. *Arcobacter*: an emerging foodborne zoonotic pathogen, its public health concerns and advances in diagnosis and control – a comprehensive review. Vet Q. 2017;37(1):137–61. https://doi.org/10.1080/01652176.2017.1323355.
Rajalakshmi S. Different types of PCR techniques and its applications, Int J Pharma Bio Sci. 2017;7:285–92.
Sharma R. Assessment of clinical profile and management of cases of diarrhoea in emergency ward. J Adv Med Dent Scie Res. 2019;7(8):187–90.
Uljanovas D, Gölz G, Fleischmann S, Kudirkiene E, Kasetiene N, Grineviciene A, et al. Genomic characterisation of *Arcobacter butzleri* strains isolated from various sources in Lithuania. Microorganisms. 2023;11(6):1425. https://doi.org/10.3390/microorganisms11061425.
Vandamme P, Falsen E, Rossau R, Hoste B, Segers P, Tytgat R, De Ley J. Revision of Campylobacter, Helicobacter, and Wolinella taxonomy: Emendation of generic descriptions and proposal of Arcobacter gen. nov. Int J Syst Bacteriol. (1991);41(1):88–103. https://doi.org/10.1099/00207713-41-1-88.
Wang K, Chen L, Ma X, Ma L, Chou KC, Cao Y, Khan IUH, Gölz G, Lu X. *Arcobacter* identification and species determination using Raman spectroscopy combined with neural networks. Appl Environ Microbiol. 2020;86(20):e00924-20. https://doi.org/10.1128/AEM.00924-20.
Zhang P, Liu Y, Fu M, Wang B, Ding S, Ma X, Zhang X, Shen Z. Meta-analysis on the global prevalence of *Arcobacter* in food-producing animals and humans. One Health Adv. 2024;2:13. https://doi.org/10.1186/s44280-024-00046-y.
Zitzler V, Šarkić D, Müller N, Häner A, Beckmann C. *Arcobacter butzleri* in stool samples from children with diarrhea. J Clin Microbiol. 2024;62(4):e01888-23. https://doi.org/10.1128/JCM.01888-23.

Bordetella Infections

49

Sanjay Bhattacharya, Sakhare Geetanjali Pramod, and Malabika Biswas

Learning Objectives

1. To have a broad understanding of zoonotic infections caused by *Bordetella* species, with special emphasis on *Bordetella bronchiseptica*
2. To inform about common clinical presentations of *B. bronchiseptica* infections and clinical management strategies, including diagnostic, therapeutic, and preventive approaches in animals and humans
3. To study the immunology and epidemiology of the *Bordetella* infections

Introduction

The genus *Bordetella* is a group of small, Gram-negative coccobacilli belonging to the family *Alcaligenaceae*. It encompasses a diverse array of species primarily known for causing respiratory tract infections in mammals and birds. *Bordetella* species of medical importance, also called classical *Bordetella*, include *Bordetella pertussis*, *Bordetella parapertussis*, and *Bordetella bronchiseptica*. Other less common nonclassical species are *Bordetella hinzii*, *Bordetella pseudoinzii*, *Bordetella holmesii*, *Bordetella avium*, *Bordetella trematum*, and *Bordetella petrii*.

These species exhibit diversity in their host range and the types of infections they cause. For instance, *B. pertussis* is primarily a human pathogen responsible for whooping cough. It is considered a strict human pathogen with no known animal reservoir. However, *B. bronchiseptica* has a broader host range, affecting various mammals and causing respiratory diseases like kennel cough in dogs and atrophic rhinitis in pigs. Some species, like *B. avium* and *B. hinzii*, are mainly associated with respiratory diseases in birds, and others, such as *B. petrii*, have been isolated from diverse environmental sources.

Among all these species, *B. bronchiseptica* is the most well-established zoonotic species within the genus. While human infections with *B. bronchiseptica* are considered relatively rare, the bacterium possesses the capacity to cross the species barrier and infect humans. The primary routes of transmission from infected animals to humans typically involve close contact with the animal, facilitating direct bacterial transfer, or the inhalation of aerosolized bacteria expelled by an infected animal through coughing or sneezing. This proximity, particularly in households with pets or in occupational settings involving animal handling, creates a potential pathway for human exposure and subsequent infection.

S. Bhattacharya (✉) · S. G. Pramod · M. Biswas
Department of Microbiology, Tata Medical Center, Kolkata, West Bengal, India
e-mail: sanjay.bhattacharya@tmckolkata.com

S. C. Parija, A. Poddar (eds.), *Textbook of Bacterial Zoonoses*, Microbial Zoonoses,
https://doi.org/10.1007/978-981-95-6615-0_49

History

Outbreaks of whooping cough (pertussis) were first described in the sixteenth century by Guillaume de Baillou. The key breakthrough came in 1906 when Belgian bacteriologist Jules Bordet and his brother-in-law Octave Gengou successfully isolated *B. pertussis* in pure culture. Bordet had observed the organism as early as 1900 while examining samples from his daughter, who had whooping cough. This isolation definitively identified the bacterium as the causative agent of whooping cough. Bordet and Gengou quickly realized that while symptoms of whooping cough lasted for weeks, the bacteria were only culturable very early in the disease. This led them to hypothesize the action of toxic compounds, qualifying pertussis as a toxin-mediated disease. The genus *Bordetella* was formally established later, named in honor of Jules Bordet.

It was in the year 1911 when Ferry first described a bacterium that he called *Bacillus bronchicanis*. This bacterium was later found to be related to *B. pertussis* and *B. parapertussis* and hence named *B. bronchiseptica.* In the year 1993, the KM22 strain of *B. bronchiseptica* was first isolated. The source was a swine herd with atrophic rhinitis in Hungary. These strains have been used extensively in laboratory settings to develop a reproducible swine respiratory disease model. This model helps researchers understand the progression of infections in pigs and study host-to-host transmission of the bacteria.

Research from the Sanger Institute in Cambridge, UK, has indicated that *B. pertussis* and *B. parapertussis* originated from *B. bronchiseptica*-like ancestors. Large-scale gene loss and inactivation of certain genes lead to the evolution and host specificity of *B. pertussis* and *B. parapertussis* from *B. bronchiseptica*-like ancestors.

Taxonomy

The genus *Bordetella* is classified under the phylum Pseudomonadota, class Betaproteobacteria, the order *Burkholderiales*, and family Alcaligenaceae. The genus currently comprises 17 species, with a few having clinical relevance. The phylogenetic relationship between the strains is shown in Fig. 49.1, indicating clear evolutionary divergence between classical and non-classical species.

Genomics and Proteomics

Genomics

The genome of *B. bronchiseptica*, containing a circular chromosome, is typically larger than those of its close relatives, *B. pertussis* and *B. parapertussis.* Genome sizes generally range from approximately 5.0–5.8 Mb. It is interesting that the G + C (guanine and cytosine) content of the DNA is maximum in the case of *B. bronchiseptica* among the important *Bordetella* species (68.1–69.5 mol%).

Comparative genomics studies have revealed significant genetic diversity among *B. bronchiseptica* strains isolated from different animal hosts and geographic locations. Multi-locus sequence typing (MLST) and whole-genome single nucleotide polymorphism (SNP) analysis have identified distinct lineages and clonal complexes within the species, some of which show host-specific associations. For example, certain lineages are more commonly found in pigs, while others are more prevalent in dogs or humans.

For instance, the complete genome of *B. bronchiseptica* KM22 consists of 5,205,646 bp with a G + C content of 68.2%, a total of 4827 predicted protein-coding sequences, 9 rRNA operons, 1 transfer-messenger RNA (tmRNA), 3 noncoding RNAs (ncRNAs), and 56 tRNAs. *B. bronchiseptica* has a G + C content in its DNA of about 68.9%. The melt temperature, or Tm, of its 16S rDNA is 57.4 °C. Based on multi-locus sequence type (MLST) analysis, *B. brochiseptica* KM22 is

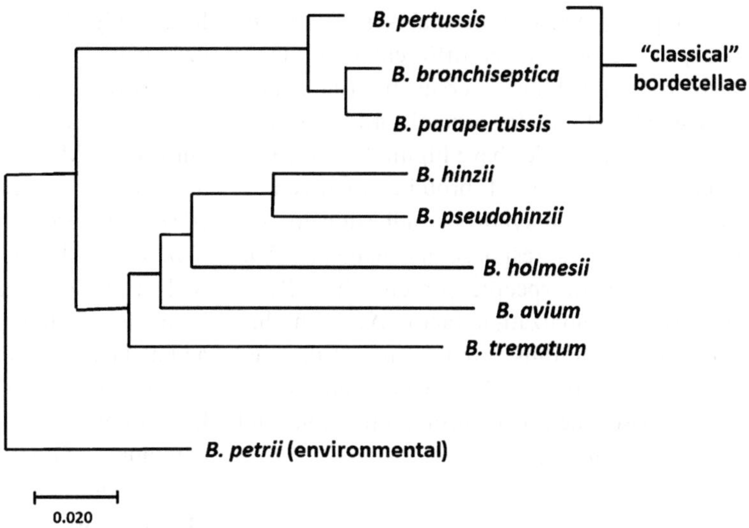

Fig. 49.1 Phylogenetic tree showing the "classical" and common nonclassical *Bordetella* species constructed using neighbor-joining method. (With permission from Eric T Harvill. Figure source: Rivera et al. (2020))

sequence type 7 (ST7), in clonal complex 1 of an MLST-based *Bordetella* phylogeny.

Proteomics

Proteomics is a powerful tool for gaining a deeper understanding of the biology of *B. bronchiseptica.* Information regarding the *B. bronchiseptica* proteome, encompassing protein sequences and their functional annotations, is accessible through several comprehensive protein databases. UniProt stands out as a key resource, offering extensive details on proteins, including those specific to *B. bronchiseptica*. Within UniProt, the proteome of *B. bronchiseptica* strain ATCC BAA-588/NCTC 13252/RB50 can be specifically accessed using the identifier UP000001027. Similarly, the NCBI Protein database, maintained by the National Center for Biotechnology Information, serves as another significant repository hosting a vast collection of protein sequences, including those identified in *B. bronchiseptica*. Furthermore, the Integrated Microbial Genomes (IMG) database provides valuable comparative genomic and proteomic information across a wide array of microbial genomes, including *B. bronchiseptica*, facilitating broader evolutionary and functional analyses.

One key area of insight is the direct identification and quantification of major virulence factors. Techniques like mass spectrometry have confirmed the expression levels of essential adhesins such as filamentous hemagglutinin (FHA), fimbriae (FIM), and pertactin (Prn), which mediate the initial attachment to host cells. Quantitative proteomics has further elucidated the regulatory control of these factors, often linked to the BvgAS two-component system, revealing their increased abundance under conditions promoting virulence.

Beyond well-established toxins, proteomic analyses of the *B. bronchiseptica* secretome have identified a broader repertoire of secreted proteins. This includes various enzymes, such as proteases and lipases, which are likely involved in tissue damage and nutrient acquisition during infection. The identification of these novel secreted proteins expands our understanding of the bacterium's arsenal for interacting with its host and presents potential new targets for therapeutic interventions aimed at disrupting these processes.

The expression and regulation of the classical toxins, adenylate cyclase toxin (ACT) and dermonecrotic toxin (DNT), have also been extensively studied using proteomic approaches. These studies have not only quantified the levels of these toxins but have also provided insights into the specific conditions that favor their production and secretion.

Comparative proteomic analyses of *B. bronchiseptica* grown under different conditions, such as in planktonic versus biofilm states, have identified proteins crucial for biofilm formation. These studies have highlighted the differential expression of proteins involved in exopolysaccharide synthesis, adhesion to surfaces, and stress response mechanisms. The identification of specific proteins like BcfA (*Bordetella* colonization factor A) as highly abundant in biofilms provides molecular targets for disrupting biofilm development, which is often associated with chronic infections and increased antibiotic resistance.

Morphology

B. bronchiseptica are small, coccoid-shaped, Gram-negative, motile bacteria with a size of about 0.2–0.5 μm in width and 0.5–2 μm in length (Fig. 49.2). These bacteria appear singly or in pairs under a microscope. It is motile and has peritrichous flagella.

In a study published by the Cornell University Medical College in 1967, Richter and Knees described the electron microscopic (EM) structures of a strain of *B. bronchiseptica*. For EM, the investigators used fixation with 1.5% glutaraldehyde, negative staining with phosphotungstic acid or uranyl acetate, and metal shadowing with platinum-palladium. It was found that the *Bordetella* cell membranes and cytoplasm were like those of other common Gram-negative bacilli (e.g., *Escherichia coli*). Peritrichous flagella 200 Å in width were observed. The cell wall of *Bordetella* was found to be multilayered and lobulated.

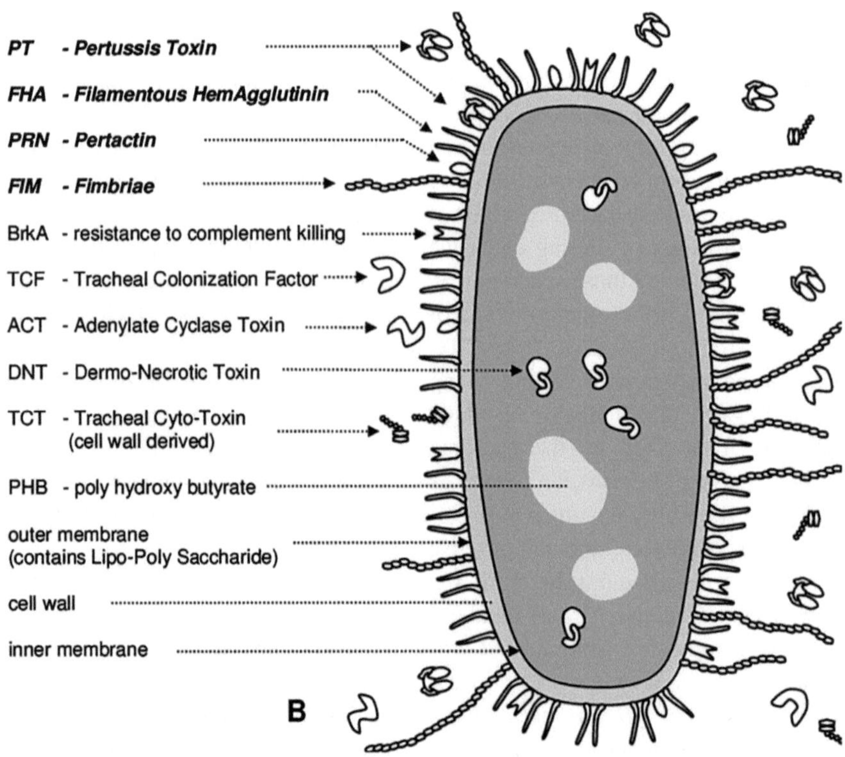

Fig. 49.2 Schematic representation of *Bordetella pertussis*, including the antigens and structures that play an important role in the pathogenesis. The antigens that are components of acellular whooping cough vaccines are indicated in bold italics. (With permission from: Marcel Thalen. Source: Thalen et al. (2006))

Cultivation

B. bronchiseptica is an obligate aerobe. The bacteria require a temperature of 37 °C for optimal growth. On culture, *B. bronchiseptica* colonies are visible in 1 day, as opposed to 1–2 days for *B. parapertussis* and 3 days for *B. pertussis*. However, *B. avium* and *B. trematum* colonies also grow within a day.

Peptone agar and MacConkey agar support the growth of all *Bordetella* species, including *B. bronchiseptica*, except *B. pertussis*. *B. bronchiseptica* produces non-lactose-fermenting colonies on MacConkey agar. Growth on Simmons citrate agar is always present in *B. bronchiseptica*, variable in *B. parapertussis*, and absent in *B. pertussis*. *B. pertussis* and *B. parapertussis* demonstrate beta-haemolysis on blood agar; however, in *B. bronchiseptica* and *B. avium*, beta-haemolysis is variable. Not all, but only some species of *Bordetella* are motile. Motile species of *Bordetella* include *B. bronchiseptica*, *B. avium*, *B. hinzii* and *B. trematum*. These motile species have peritrichous flagella.

Animal Models

Animal models play a crucial role in laboratory investigations of *B. bronchiseptica* infections. These provide valuable platforms to study pathogenesis, host immune responses, and the efficacy of potential interventions. Laboratory animals used to study *B. bronchiseptica* include guinea pigs, mice, rats, dogs, and pigs.

Mice, for instance, are frequently employed due to their affordability, ease of manipulation, and the availability of a wide array of genetically modified strains. This model enables the study of specific host factors and immune mechanisms during *B. bronchiseptica* colonization of the respiratory tract. These murine models are particularly useful for (i) examining the early stages of infection, (ii) elucidating the roles of individual virulence factors, and (iii) assessing the development of adaptive immunity, as well as for the preliminary evaluation of vaccine candidates.

Rats, being a natural host for *B. bronchiseptica*, offer an advantage in studying the cough reflex associated with infection. These models have been instrumental in investigating the underlying mechanisms of *Bordetella*-induced coughing and the contribution of specific bacterial factors to this clinical sign. Rabbits, natural hosts for certain *B. bronchiseptica* strains causing "snuffles," serve as valuable models for studying chronic infections and the development of long-term immunity against the bacterium.

B. bronchiseptica isolate KM22 has been used in experimental infections of swine as a model of clinical *B. bronchiseptica* infection and to study host-to-host transmission. As a natural host for *B. bronchiseptica*, swine develop the characteristic nasal turbinate hypoplasia associated with this condition, allowing for detailed studies of the pathogenesis, including the role of toxins like dermonecrotic toxin.

Transmission Cycle

The transmission of *B. bronchiseptica* mainly occurs through droplet spread from the respiratory tract of infected or colonized animals (Fig. 49.3). Other modes of transmission include direct contact with infected animals and through contaminated fomites. It has been reported that <100 colony-forming units (CFU) of *B. bronchiseptica* could infect mice and were able to displace cultivable host microbiota. In contrast, a much higher inoculum of 10,000 CFU of *B. pertussis* was required to colonize mice's nasal cavities and did not result in the displacement of host microbiota.

Pathogenesis and Pathology

The pathogenesis of *B. bronchiseptica* is initiated by its adhesion and colonization of the respiratory tract, mirroring mechanisms observed in other *Bordetella* species. This crucial first step is mediated by key adhesins, including fimbriae and filamentous hemagglutinin (FHA), which facilitate attachment to respiratory cilia and interac-

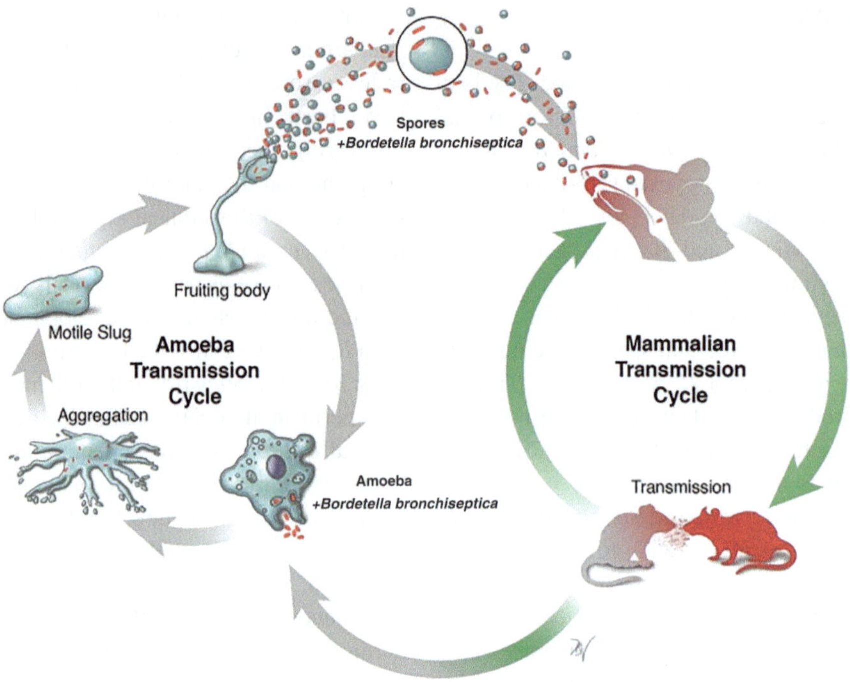

Fig. 49.3 Zoonotic transmission cycle of *Bordetella bronchiseptica*. (With permission from Eric T. Harvill. Figure source: Dewan and Harvill (2019))

tion with epithelial cells. Notably, anti-FHA antibodies are synthesized early during infection, highlighting their role in the host response. Furthermore, *B. bronchiseptica* induces ciliostasis, effectively hindering the host's natural clearance mechanisms and promoting efficient colonization. The persistence of *B. bronchiseptica* may also be attributed to its ability to form biofilms.

Following colonization, *B. bronchiseptica* employs a range of virulence factors (Table 49.1) to establish infection and modulate the host immune response, consistent with the pathogenic strategies of the *Bordetella* genus.

Virulence Factors

A significant factor is the *adenylate cyclase-hemolysin toxin*, which increases intracellular cAMP levels, leading to immune modulation. Interestingly, the host synthesizes anti-adenylate cyclase-haemolysin antibodies early in infection, and these antibodies are considered protective, helping in both cellular and humoral immune responses.

Another crucial virulence mechanism is the *Type III Secretion System* (*T3SS*). This plays an important role in both colonization and persistence by injecting Bte effector proteins into the host cell cytosol. While the precise role of T3SS in immunity and immunization remains uncertain, its contribution to pathogenesis is evident.

Lipopolysaccharide (*LPS*) also contributes to pathogenesis through colonization and immunomodulation by activating Toll-Like Receptor 4 (TLR4), leading to the production of bactericidal antibodies. *Pertactin*, another adhesin, is important for colonization and shedding, potentially facilitating transmission, and acts as a protective antigen in piglets. While *dermonecrotic toxin* production is variable among *B. bronchiseptica* isolates, its role in the overall virulence, particularly in the mouse model, may not be as significant.

Table 49.1 Virulence factors produced by *Bordetella bronchiseptica*

Virulence factors	Function in *Bordetella bronchiseptica*	Observation	Significance
Adenylate cyclase-hemolysin	Increases cAMP immuno-modulation	Anti-adenylate cyclase-haemolysin antibodies are synthesized very early during infection	Major protective antigen against *Bordetella bronchiseptica* infection. Responsible for cellular immune response and neutralizing antibody production in mice
Type III secretion system	Important role in colonization and persistence	Injects Bte protein into cytosol	Role in immunity and immunization uncertain
Lipopolysaccharide (LPS)	Colonization and Immuno-modulation	Toll-like receptor (TLR4) activation	Bactericidal antibodies produced in mice
Fimbriae and filamentous hemagglutinin	Adhesion, transmission, persistence	Anti-filamentous hemagglutinin antibodies are synthesized very early during infection	Anti-FIM antibodies are responsible for humoral immune response in rats
Pertactin	Colonization and shedding	Important for transmission	Protective antigen in piglets
Dermonecrotic toxin	Virulent isolate may not produce dermonecrotic toxin	Production of this toxin is variable	May not have an important role in the virulence of this bacteria in the mice model

Beyond direct damage, *B. bronchiseptica* actively modulates the host immune response. Adhesins contribute to immune response modulation and can even suppress inflammation. The interplay of these virulence factors allows *B. bronchiseptica* to evade host defenses, establish a persistent infection in the respiratory tract, and ultimately lead to the characteristic signs of respiratory disease in its diverse range of mammalian hosts.

Immunology

The immunology of *B. bronchiseptica* infections shares some commonalities with *B. pertussis*, although it presents unique aspects due to its broader host range and ability to cause both acute and chronic respiratory diseases in various mammals. The broad host range of *B. bronchiseptica* also implies variations in immune responses and outcomes across different animal species. Therefore, making a generalized immunological description remains challenging.

Innate Immune Response and Evasion

Upon encountering B. *bronchiseptica*, the host's innate immune system mounts an immediate response. Pattern recognition receptors (PRRs) on host cells, including Toll-like receptors (TLRs), recognize conserved bacterial components like LPS, triggering inflammatory pathways. Phagocytic cells, such as macrophages and neutrophils, are recruited to the site of infection to engulf and eliminate the bacteria.

B. bronchiseptica employs several strategies to evade or subvert these early defenses. The ACT directly impairs the function of phagocytes by elevating intracellular cAMP levels, reducing their ability to engulf and kill bacteria. While *B. bronchiseptica* generally does not produce pertussis toxin (PTx) (a key difference from *B. pertussis*), its other virulence factors effectively suppress or redirect early innate responses. The T3SS effectors play a crucial role in this subversion, actively modifying host cell signaling to reduce inflammation and promote bacterial survival. The net effect is often a delay in effective bacterial clearance, allowing *B. bronchiseptica* to establish persistent colonization in the respiratory tract.

Adaptive Immune Response and Chronic Infection

The adaptive immune response is essential for controlling *B. bronchiseptica* infection, but it often struggles to achieve complete clearance, leading to chronic or recurrent disease. Both humoral (antibody) and cellular (T-cell) immunity contribute to protection. Antibodies, particularly IgA in the mucosa and IgG in serum, neutralize toxins and block bacterial adherence, thereby reducing colonization and disease severity. However, antibodies alone are often insufficient to clear established infections. Cellular immunity, involving CD4+ and CD8+ T cells, is critical for mediating bacterial clearance. Th1 and Th17 responses are important for orchestrating effective antibacterial immunity and promoting macrophage activation and neutrophil recruitment, respectively. The host's immune response, while attempting to clear the infection, may also contribute to the pathology seen in chronic *B. bronchiseptica* diseases, such as the persistent coughing and tissue damage.

Infection in Humans

Human infections are relatively rare and mostly occur in individuals with weakened immune systems. The clinical presentation of *B. bronchiseptica* infection in humans varies considerably.

In some individuals, the infection may be asymptomatic, with no noticeable signs or symptoms. Others may develop a mild respiratory illness characterized by a cough that resembles the whooping cough caused by *B. pertussis*. However, in more susceptible individuals, particularly those with underlying health conditions or compromised immune systems, *B. bronchiseptica* can cause more severe and invasive infections. These serious manifestations include pneumonia, an inflammation of the lungs; endocarditis, peritonitis, and even meningitis.

Depending on the site of infection, stage of illness, comorbidities, and immunity of the host, *B. bronchiseptica* may manifest as coughing, sneezing, and nasal discharge. In immunocompetent individuals, laryngitis, persistent cough, whooping cough-like illness, broncho-pneumonia, and pneumonia have been reported. Pneumonia and sepsis are more common in individuals with comorbidities or those with immune deficiency bacteremia. If there is a history of exposure to animals, especially pets (dogs/cats), infection with *B. bronchiseptica* needs to be considered among the differential diagnoses.

Zoonotic transmission of live attenuated *Bordetella* vaccine strains from animals to a patient with impaired immunity has been reported. The patient received the tumor necrosis factor alpha (TNF-α) inhibitor Etanercept. Two weeks after intranasal vaccination of a dog with a live attenuated *B. bronchiseptica* vaccine (Nobivac KC, MSD Animal Health, USA), the patient started developing bronchitis with malaise and a mild fever.

No major human outbreaks have been reported to date due to *B. bronchiseptica*. Only sporadic cases have been reported in humans who are in close contact with pets.

Infection in Animals

The incubation period for *B. bronchiseptica* in animals, for example, in dogs, is generally between 2 and 14 days, with a mean incubation period of 6–7 days. The period of infectivity (infectious period) of the infected animals may vary. Prolonged shedding of the bacteria, even for 3–4 months, has been reported.

B. bronchiseptica is a respiratory tract pathogen of the rabbit and the dog. It causes kennel cough in puppies. A strong affinity of *B. bronchiseptica* for the canine respiratory tract was demonstrated experimentally by Bemis, Greisen, and Appel in 1977. In 1980, Matsuyama and Takino used scanning electron microscopy to describe large numbers of *B. bronchiseptica* bacteria becoming entangled with the cilia of the tracheal epithelial cells. It has been found in subsequent studies that *B. bronchiseptica* infects a variety of animals, often establishing chronic infections that range from asymptomatic carriage to lethal pneumonia, and is capable of surviving in the environment.

B. bronchiseptica is one of the pathogens associated with canine infectious respiratory disease (CIRD). This disease is a syndrome where many viral and bacterial pathogens may be involved. These include: *Mycoplasma cynos*, *Streptococcus equi*, canine adenovirus (CAV-2), canine parainfluenza virus (CPIV), canine respiratory coronavirus (CRCoV), and influenza A. These pathogens may act sequentially or in synergy to cause disease (e.g. chronic cough in dogs).

Epidemiology and Public Health

B. bronchiseptica resides primarily in the respiratory tract of various animal species, including dogs, cats, rabbits, and rodents. These animals act as the primary infection reservoir for human infection. Rarely the bacteria colonize the respiratory tract of humans.

The global burden or even the country-wise prevalence of *B. bronchiseptica* is not known with certainty. This is because of the absence of proper surveillance systems or case reporting mechanisms at local, regional, or national levels, especially in low-middle-income countries. Case identification would require formalization of case definitions based on clinical, laboratory, and epidemiologic criteria. Most approximate estimates of prevalence are therefore based on case reports, case series, or sero-prevalence studies within a restricted population, mainly animals.

A report from China documented the isolation rates of *B. bronchiseptica* among pigs with respiratory illness to be between 15.2% and 25.7%. A study from Turkey found 12% of dogs to be positive for *B. bronchiseptica* antibodies by enzyme-linked immunosorbent assay (ELISA). Among shelter-housed cats, the prevalence of *B. bronchiseptica* has been reported to vary between 3% and 14%. An outbreak of multidrug-resistant *B. bronchiseptica* infection among 16 shelter-housed cats was also reported in the same study from the USA. The study reported that 5/16 affected cats (31.3%) were euthanized or died due to pneumonia, including one kitten. The risk factors associated with animal infections include crowded environments, poor animal husbandry practices, comorbidities, and exposure to affected animals. Previously, Japan had reported an outbreak of *B. bronchiseptica* infection among 22 newborn puppies who did not receive colostrum.

In a review, 23 reports of human infections due to *B. bronchiseptica* were reported. In another review of *B. bronchiseptica* infection among immunocompromised populations, pneumonia was reported to be the commonest clinical presentation. Hematologic malignancy was the commonest underlying risk factor. Patient outcome was reported in only seven patients, out of whom four had died.

Diagnosis

Clinical specimens for the laboratory diagnosis of *B. bronchiseptica* in human and animal infections include nasopharyngeal swabs, nasopharyngeal secretions, and bronchoalveolar lavage.

Microscopy

On Gram stain, *B. bronchiseptica* appears as a small, Gram-negative coccobacilli with a size of about 0.2–0.5 μm by 0.5–2 μm. They are usually observed singly or in small clusters. The bacterium is motile, and this may be demonstrated either by hanging drop or in a motility test medium.

In Vitro Cultures

B. bronchiseptica grows under aerobic conditions. Growth can occur within a day of incubation at 37 °C. The bacteria grow not only on chocolate agar and blood agar but also on MacConkey agar, where non-lactose-fermenting colonies are formed. Brown pigment is formed only by *B. parapertussis* when grown on heart infusion agar with L-tyrosine.

The Bordet-Gengou agar and the Regan-Lowe agar support the growth of all common Bordetella species. However, blood agar, chocolate agar, and MacConkey agar support the growth of only *B. parapertussis* and *B. bronchiseptica* and not *B. pertussis*. The Regan-Lowe agar is composed of charcoal, agar, and defibrinated horse blood (as an enrichment agent). Charcoal plus starch neutralizes fatty acids and peroxides, which are toxic to *Bordetella*.

B. holmesii and *B. bronchiseptica* are inhibited by the antibiotic cephalexin, usually present in Bordet-Genou and Rean-Lowe medium. Hence, the isolation of these organisms will be missed if cephalexin-containing media are used for transport or culture. The use of methicillin or oxacillin in place of cephalexin would allow the growth of *B. holmesii* and *B. bronchiseptica*.

Bordetella species are catalase-positive. *B. bronchiseptica* is the only important species of *Bordetella* that forms nitrite. It is noteworthy that not all *Bordetella* species are oxidase positive. The oxidase-positive species include *B. pertussis*, *B. bronchiseptica*, *B. hinzii*, and *B. avium*. Urease is positive in *B. parapertussis* and *B. bronchiseptica*.

Colony morphologies of different *Bordetella* species in bacteriological media are presented in Table 49.2. Moreover, different biochemical properties of the members of the genus *Bordetella* are presented in Table 49.3.

Automated Biochemical Identification Systems

Phenotypic identification systems like API 20NE and VITEK2 (BioMerieux France) can be used for identification. VITEK 2 GN ID card can identify *B. bronchiseptica*, *B. hinzii*, and *B. trematum*. There are reports of misidentification of *B. bronchiseptica* by Vitek cards (Biomerieux). *B. bronchiseptica* may be misidentified as *Oligella ureolytica* or *Ralstonia paucula* by biochemical identification systems. However, another report suggested that identification of *B. bronchiseptica* by Vitek ID cards was excellent when compared to 16S rRNA sequencing. Therefore, caution is required while using automated biochemical identification cards for suspected *Bordetella*. It is advisable that biochemical identification be confirmed by other methods such as MALDI-TOF, Polymerase Chain Reaction (PCR), or DNA sequencing.

Antimicrobial Antibiotic Susceptibility Testing

CLSI documents VET01-A4 (2013) and VET01-S3 (2015), which provide performance standards for susceptibility tests for bacteria isolated from animals. CLSI document VET01-S provides clinical breakpoints for *B. bronchiseptica*. However, breakpoints are available only for ampicillin (which may be extrapolated to amoxicillin), florfenicol, tildipirosin, and tulathromycin. CLSI document VET01-S is the only antimicrobial susceptibility testing document that contains approved clinical breakpoints specific to *B. bronchiseptica*. Isolates can be tested by agar disk diffusion or by determining the minimal inhibitory concentration (MIC) by agar dilu-

Table 49.2 Colony morphologies of different *Bordetella* species in bacteriological media

Species	*Bordetella pertussis*	*Bordetella parapertussis*	*Bordetella bronchiseptica*
Colony characteristics (Bordet-Gengou agar, BG)	Smooth, shiny, highly domed (mercury droplets). Beta hemolysis in confluent growth areas	Gray or slightly brown in color. More beta-hemolytic	Large, flatter, and have a dull appearance
Colony characteristics (Regan-Lowe agar, RL)	Small, domed, and shiny, with a white "mother-of-pearl opalescence"	–	Growth may be inhibited by cephalexin in the media
Time for colony development	2–4 days (sooner on antibiotic-free media)	2–3 days	Within 24 h

Table 49.3 Common tests for the identification of *Bordetella* species (WHO, 2014)

Organism	Hanging drop or motility medium	Brown pigment	Catalase	Oxidase	Urease	Nitrate	Citrate
Bordetella pertussis	−	−	+	+	−	−	−
Bordetella parapertussis	−	+	+	−	+	−	−
Bordetella bronchiseptica	+	−	+	+	+	+	+

tion or by broth micro- or macro-dilution. CLSI-approved media are Mueller-Hinton agar for disk diffusion and agar dilution, as well as cation-adjusted Mueller-Hinton broth for broth dilution assays.

Serodiagnosis

Direct immune-fluorescence assay may be performed on nasopharyngeal specimens or colony emulsion (No.3 McFarland) from culture isolates. It is generally used for the laboratory diagnosis of *B. pertussis* and *B. parapertussis*. The antibodies used may be monoclonal or polyclonal in origin. Fluorescein isothiocyanate (FITC) conjugate is used. If polyclonal antibody reagents are used, they may cross-react with several non-*B. pertussis* bacteria, resulting in false-positive results.

Enzyme-linked immunosorbent assay (ELISA) and indirect immuno-fluorescent antibody tests have been used to detect antibodies against *B. bronchiseptica* in guinea pigs. In a study from Turkey, dogs positive for *B. bronchiseptica* antibodies by an in-house ELISA were also found to have elevated levels of serum amyloid A protein.

Common target antigens include bacterial lipopolysaccharide (LPS). Some assays use whole inactivated *B. bronchiseptica* cells as the antigen. This provides a broad antigenic target but may lead to cross-reactivity with other *Bordetella* species or even other Gram-negative bacteria. Additionally, specific virulence factors are also targeted. Antibodies against virulence factors like filamentous hemagglutinin (FHA) or adenylate cyclase toxin (ACT) can be measured. While these are more specific than whole-cell antigens, *B. bronchiseptica* does not produce pertussis toxin (PTx), which is the highly specific diagnostic antigen for *B. pertussis.*

Molecular Diagnosis

Molecular diagnosis has revolutionized the detection and identification of *B. bronchiseptica* due to its high sensitivity and specificity. The most employed molecular diagnostic methods include *Polymerase Chain Reaction (PCR)* and its variations. PCR is a highly sensitive in vitro DNA amplification technique that can detect even very small numbers of *B. bronchiseptica* DNA in clinical samples.

Target Genes for PCR

Several target genes have been used for the specific detection of *B. bronchiseptica* by PCR. Some are:

flaA: The gene encoding flagellin, a major component of the flagella.

IS481: An insertion sequence present in multiple copies in the *B. bronchiseptica* genome, increasing the sensitivity of detection. However, IS481 is also found in *B. pertussis* and *B. parapertussis*, so assays targeting this sequence alone may not be species-specific and require careful interpretation or multiplexing with other targets.

bpra: *Bordetella* pertussis regulator A homolog, a gene present in *B. bronchiseptica* and other *Bordetella* species but with sufficient sequence variation for species-specific primer design.

Specific Regions Within Virulence Factor Genes

Genes encoding FHA, pertactin, or other species-specific sequences can also be targeted.

Real-Time PCR (qPCR)

Real-time PCR allows for the simultaneous amplification and quantification of the target DNA. qPCR assays for *B. bronchiseptica* often target the same genes as conventional PCR but incorporate fluorescent detection methods for faster and precise quantification.

Multiplex PCR

Multiplex PCR assays can simultaneously detect multiple targets in a single reaction. This approach can be used to differentiate *B. bronchiseptica* from other *Bordetella* species like *B. pertussis* and *B. parapertussis* by including species-specific primers and probes for each in the same reaction. This is particularly useful for detection of infections caused by these related bacteria.

Matrix-Assisted Laser Desorption/Ionization Time-of-Flight Mass Spectrometry (MALDI-TOF)

In an article in 2019, it was reported that MALDI-TOF MS was unsuitable for the differentiation of *B. pertussis* and *B. bronchiseptica* (Hou et al. 2019). However, Zintgraff et al. reported from Argentina in 2018 that reliable identification of the classical *Bordetella* species might be possible by MALDI-TOF technology using the "10% rule" and the "5% breaking point." In the context of MALDI-TOF, the "10% rule" refers to the use of a strategy that excludes the identity of isolates that score less than 10% compared to the top-scoring match in the database, whereas in the "5% breaking point" criteria, isolated scoring ≥5% of the threshold is considered to have reliable identification, and those 5% below the threshold are considered less reliable.

Commercial Diagnostic Kits

Kits are available for the detection of *B. bronchiseptica* in animals like dogs, cats, and rabbits in nasopharyngeal/oropharyngeal swabs. Example includes the *B. bronchiseptica* end-point PCR kit from Norgen Biotek Corporation (Ontario, Canada). Commercially available kits in the market for the diagnosis of bordetellosis in humans mainly use IS481 for the detection of *B. pertussis* and IS1001 for the detection of *B. parapertussis*. IS481 may be present in a small proportion of isolates in *B. holmesii*. BactoReal® *Bordetella bronchiseptica*, based on the amplification of bfrA gene, is available for the detection of *B. bronchiseptica* by real-time PCR.

An overview of the diagnosis is presented here and summarized in Table 49.4.

Treatment

Antibiotic treatment options for the management of *B. bronchiseptica* include piperacillin-tazobactam, imipenem, meropenem, ciprofloxacin, levofloxacin, and amikacin. Other treatments reported include trimethoprim-sulfamethoxazole, rifampin, moxifloxacin, doxycycline, and minocycline. The duration of treatment and the specific choice of antimicrobial therapy would depend on the source and severity of infection, as well as the response to treatment. Pneumonia caused by *B. bronchiseptica* has been previously treated for 3 weeks. Consultation with infectious disease physicians or clinical microbiologists is recommended to optimize antimicrobial therapy.

A multicentric study from China involving 181 isolates (collected over 3 years from the lungs of dead pigs) tested *B. bronchiseptica* isolates for antimicrobial susceptibility using disk diffusion method. The study revealed that majority of the *B. bronchiseptica* isolates were resistant to ampicillin (83.98%). Resistance to cefotaxime (30.39%), chloramphenicol (12.71%), gentamicin (11.60%), tetracycline (8.84%), tobramycin (6.63%), and cefepime (0.55%) was also reported. There were no isolates demonstrating resistance to imipenem, meropenem, ciprofloxacin, and amikacin. The same study detected the following antimicrobial resistance genes: aac (3)-IV, aac6′-Ib (aminoglycoside modifying enzymes: aminoglycoside acetyl transferase), rmtA (encoding for 16S rRNA methyltransferase leading to aminoglycoside resistance), bla_{TEM}, bla_{SHV} (extended spectrum beta lactamases), oqxB (encoding an efflux pump protein that can lead to resistance to fluoroquinolone), and tetA (gene conferring resistance to tetracycline).

A rare case of human infection (pneumonia) in a patient with lung cancer (in Italy) reported *B. bronchiseptica* isolated from bronchoalveolar lavage (BAL) to be resistant to cefepime (minimal inhibitory concentration [MIC] ≥64), cefotaxime (MIC ≥64), and ceftazidime (MIC 16), but susceptible to amikacin (MIC 8), gentamicin (MIC 2), imipenem (MIC 0.5), meropenem (MIC ≤0.25), piperacillin/tazobactam (MIC ≤4), and ciprofloxacin (MIC 1). The patient was successfully treated with piperacillin/tazobactam.

Table 49.4 Laboratory diagnosis of *Bordetella bronchiseptica* infection

Diagnostic approaches	Method	Advantages	Disadvantages
Direct microscopy	Gram stain	Rapid; inexpensive	Low sensitivity; low specificity
	Fluorescence microscopy	Sensitive	Requires fluorescence microscopy and expertise to differentiate true positive from nonspecific fluorescence
In vitro cultures	Blood agar, chocolate agar, and MacConkey agar in aerobic incubation at 37 °C	Grows within a day; enables applications such as MALDI-TOF-based identification, antimicrobial susceptibility testing, and whole-genome sequencing	Relatively insensitive; requires an appropriate sample at an appropriate time; may be confused with other *Bordetella* species or commensal flora
Serodiagnosis	Indirect immunofluorescence antibody test	Useful for sero-epidemiological studies	Requires fluorescence microscopy; not useful for diagnosis of acute infections
	ELISA	Useful for sero-epidemiological studies	Requires fluorescence microscopy; not useful for diagnosis of acute infections
Molecular diagnosis	PCR (end-point PCR and real-time PCR)	Sensitive, specific, higher accuracy than other tests	Relatively expensive; availability and accessibility to tests are limited
	Mass spectrometry (MALDI-TOF)	Rapid; low consumable cost	May not reliably distinguish *Bordetella bronchiseptica* from other *Bordetella* species

Prevention and Control

Vaccines

Vaccines against *B. bronchiseptica* are currently available for dogs. These vaccines may have different routes of administration, such as intramuscular, intranasal, oral, or epicutaneous (topical).

B. bronchiseptica has three types of vaccines: live attenuated, whole cell vaccine, and cellular antigen extract vaccine. Unlike *B. pertussis*, there is no adjuvanted acellular vaccine for *B. bronchiseptica*. The live attenuated *B. bronchiseptica* vaccine is an oral/nasal mucosal vaccine, whereas the whole cell vaccine and the cellular antigen extract vaccines in *B. bronchiseptica* are injectable vaccines for dogs. The nasal and oral routes of vaccination, which are already licensed for *B. bronchiseptica*, induce mucosal immune responses and help to stop circulation of the pathogen.

Live Attenuated Intranasal Vaccines

These vaccines deliver live, weakened *B. bronchiseptica* directly to the nasal passages, stimulating a local immune response. Nobivac KC is an example, containing live *B. bronchiseptica* strain B-C2 and canine parainfluenza virus for comprehensive kennel cough protection. Vanguard B intranasal is another intranasal option.

Oral Live Attenuated Vaccines

These vaccines, such as Nobivac Intra Trac Oral BBPI and Recombitek® Oral *Bordetella*, utilize a similar principle of live, weakened bacteria administered orally. Recombitek® is notable for inducing protective IgA and IgG levels after a single dose, negating the need for a booster.

Inactivated/Subunit Injectable Vaccines

These vaccines contain killed *B. bronchiseptica* cells or purified antigenic extracts. Vanguard B injectable is an example of a non-adjuvanted antigenic extract vaccine. These typically stimulate a systemic immune response.

Case Study

A 50-year-old male cancer patient under treatment for Hodgkin's Lymphoma reported to the emergency room of a hospital with complaints of cough, shortness of breath, and fever for the last 3 days. He was diabetic and lived alone with a pet dog. During hospital admissions, the pet owner sheltered the dog in a local dog shelter. The dog has also been suffering from a chronic cough and nasal discharge since the last visit to the dog shelter.

On examination in the emergency room, the patient was found to have tachypnea, normal blood pressure, tachycardia, and a temperature of 101 °F. Chest X-ray showed unilateral consolidation of his right lower lobe. His oxygen saturation on room air was 92%.

1. What laboratory investigations should be suggested in view of the above case history, clinical presentation, and preliminary findings?
2. What is the provisional diagnosis? What are the differential diagnoses?
3. Which infections are transmissible from pet dogs to humans?

Research Questions

1. What is the estimated prevalence of *B. bronchiseptica* colonization and infection among pet animals?
2. What is the estimated burden of antibiotic-resistant *B. bronchiseptica* infections in humans and animals?
3. How can we investigate the molecular epidemiology and transmission dynamics of *B. bronchiseptica* infection, especially transmission between pet animals and farm animals and from animals to humans?
4. What is the estimated efficacy of *B. bronchiseptica* vaccines among pet animals?

Declaration The authors declare that they have not used artificial intelligence (AI) for the purpose of writing this chapter.

Further Readings

Akar K, Sanioğlu Gölen G, Ekin İH. Investigation of the correlation between ELISA and serum amyloid A in the diagnosis of *Bordetella bronchiseptica* in dogs. Vet Med Sci. 2025;11(3):e70323. https://doi.org/10.1002/vms3.70323.

Bappy MNI, Ahmed F, Lasker T, Sajib EH, Islam MS. Screening of novel drug targets and drug design for *Bordetella* pertussis: a subtractive proteomics approach. Curr Res Microb Sci. 2024;7:100291. https://doi.org/10.1016/j.crmicr.2024.100291.

Dewan KK, Harvill ET. Did new transmission cycles in anthropogenic, dense, host populations encourage the emergence and speciation of pathogenic *Bordetella*? PLoS Pathog. 2019;15(3):e1007600. https://doi.org/10.1371/journal.ppat.1007600.

Gueirard P, Guiso N. Virulence of *Bordetella bronchiseptica*: role of adenylate cyclase-hemolysin. Infect Immun. 1993;61(10):4072–8. https://doi.org/10.1128/iai.61.10.4072-4078.1993.

Hou TY, Chiang-Ni C, Teng SH. Current status of MALDI-TOF mass spectrometry in clinical microbiology. J Food Drug Anal. 2019;27(2):404–14. https://doi.org/10.1016/j.jfda.2019.01.001.

Kadlec K, Schwarz S. Antimicrobial resistance in *Bordetella bronchiseptica*. Microbiol Spectrum. 2018;6(4):10. https://doi.org/10.1128/microbiolspec.ARBA-0024-2017.

Kraai I, Knoester M, Bathoorn E, Bakker M, Nijland M. Zoonotic transmission of vaccine-derived *Bordetella bronchiseptica*. Open forum. Infect Dis. 2023;10(8):ofad421. https://doi.org/10.1093/ofid/ofad421.

Lanotte P, Plouzeau C, Burucoa C, Grélaud C, Guillot S, Guiso N, Garnier F. Evaluation of four commercial real-time PCR assays for detection of *Bordetella* spp. in nasopharyngeal aspirates. J Clin Microbiol. 2011;49(11):3943–6. https://doi.org/10.1128/JCM.00335-11.

Ma L, Linz B, Caulfield AD, Dewan KK, Rivera I, Harvill ET. Natural history and ecology of interactions between *Bordetella* species and amoeba. Front Cell Infect Microbiol. 2022;12:798317. https://doi.org/10.3389/fcimb.2022.798317.

Maboni G, Seguel M, Lorton A, Berghaus R, Sanchez S. Canine infectious respiratory disease: new insights into the etiology and epidemiology of associated pathogens. PLoS One. 2019;14(4):e0215817. https://doi.org/10.1371/journal.pone.0215817.

Matsuyama T, Takino T. Scanning electronmicroscopic studies of *Bordetella bronchiseptica* on the rabbit tracheal mucosa. J Med Microbiol. 1980;13(1):159–61. https://doi.org/10.1099/00222615-13-1-159.

Miguelena Chamorro B, De Luca K, Swaminathan G, Longet S, Mundt E, Paul S. *Bordetella bronchiseptica* and *Bordetella pertussis*: similarities and differences in infection, immuno-modulation, and vaccine considerations. Clin Microbiol Rev. 2023;36(3):e0016422. https://doi.org/10.1128/cmr.00164-22.

Rivera I, Linz B, Harvill ET. Evolution and conservation of *Bordetella* intracellular survival in eukaryotic host cells. Front Microbiol. 2020;11:557819. https://doi.org/10.3389/fmicb.2020.557819.

Thalen M, Venema M, Dekker A, Berwald L, van den IJssel J, Zomer B, Beuvery C, Martens D, Tramper J. Fed-batch cultivation of *Bordetella pertussis*: metabolism and pertussis toxin production. Biologicals. 2006;34(4):289–97. https://doi.org/10.1016/j.biologicals.2005.12.001.

Washington MA, Agee WA, Kajiura L, Hawley-Molloy JS, Staege CM, Barnhill JC. A case of *Bordetella brochiseptica* at a Military Medical Facility in Hawaii: phenotypic and molecular testing of an uncommon human pathogen. Hawaii J Med Public Health. 2015;74(7):230–3.

WHO. Laboratory manual for the diagnosis of whooping cough caused by Bordetella pertussis-Bordetella parapertussis. 2014. ISBN: WHO_IVB_14.03_eng. https://www.who.int/publications/i/item/laboratory-manual-for-the-diagnosis-of-whooping-cough-caused-bybordetella-pertussis-bordetella-parapertussis.

Part III

Future Perspectives in Bacterial Zoonoses

Bacteriome in Health and Diseases

50

Susanthika Jayachandiran, Roja Suresh,
Bharath Prithiviraj, and Dhamodharan Ramasamy

Learning Objectives

1. To know the types, composition, and diversity of the human bacteriome
2. To learn about the roles of bacteriome in health and zoonotic diseases
3. To understand the bacteriome-based therapeutics for zoonotic infection
3. To learn the methods for bacteriome analysis

Introduction

Microorganisms, including bacteria, viruses, fungi, and archaea, are present in various environments such as soil, fresh water, oceans, air, hot springs, polar ice, deserts, caves, plants, animals, insects, birds, and the human body, where they develop complex, dynamic ecosystems that are critical for maintaining health and ecological balance. The microorganisms that inhabit a particular environment are collectively referred to as the "*microbiota*," and the microorganisms, along with their genomes, are collectively referred to as the "*microbiome*." The microbiome is unique in different anatomic regions of the human body, including the gut, skin, oral cavity, lungs, and genital tract. Each of these regions has a distinct microbial community that has adapted to its environment.

The microbiome consists of several components. The *bacteriome***,** which is solely composed of bacteria, has been studied the most because of its functional importance, as it is the largest part of the microbiome in terms of mass and volume. The *virome***,** which encompasses viruses infecting human cells and bacteriophages that influence bacterial density and host immunity, is a relatively recent aspect of microbiome research. The *mycobiome*, which is composed of fungi, plays a role in nutrient metabolism and interacts with the host and bacterial communities. The *archaeome* comprises archea involved in methane metabolism and other unique biochemical functions, but this microbial component remains poorly understood.

Gut Bacteriome

It is among the most studied and essential contributor, acting as the interface of the digestive system, digestion, absorption, and even protection from pathogens. Another important role of the gut bacteriome is as a contact mediator for the immune system. This means that the gut bacteriome acts as a modulator of innate and adaptive

S. Jayachandiran · R. Suresh · D. Ramasamy (✉)
Mahatma Gandhi Medical Advanced Research Institute (MGMARI), Sri Balaji Vidyapeeth (Deemed to be University), Pondicherry, India

B. Prithiviraj
Reckitt Health US, Microbiome Science Platforms, Montvale, NJ, USA

S. C. Parija, A. Poddar (eds.), *Textbook of Bacterial Zoonoses*, Microbial Zoonoses,
https://doi.org/10.1007/978-981-95-6615-0_50

immunity, activating the Treg cells (regulatory T cells) and Th17 cells. Subsequently, they initiate immune signalling via the activation of Toll-like receptors and NOD-like receptors (TLR/NLR). Also, they maintain the intestinal barrier via the production of short-chain fatty acids (SCFA) to create an environment that mitigates inflammatory conditions. An example of a gut microbe that serves a beneficial role is *Faecalibacterium prausnitzii*. The abundance of *F. prausnitzii* induce the production of anti-inflammatory compounds and maintains homeostasis against ulcerative colitis, inflammatory bowel disease (IBD) and sepsis.

Other anatomical regions also harbor distinct and functionally significant microbial ecosystems.

The skin bacteriome serves as a frontline defence against external pathogens while also maintaining skin homeostasis. It assists in producing antimicrobial peptides such as cathelicidins and defensins, the education of Langerhans cells and T cells, and the induction of immune tolerance to innocuous stimuli.

The oral bacteriome is vital for dental and periodontal health, and it interacts with epithelial cells and salivary antimicrobial proteins to cause the release of cytokines and chemokines via pattern recognition receptors (PRRs). Disruption of this ecosystem has been linked to dental caries, periodontitis, and systemic inflammatory diseases.

*The pulmonary bacteriome***,** once considered sterile, is now placed in the perspective of respiratory health. Microorganisms enter the lungs via the oral and nasal cavities and are involved in immune regulation via TLR and NOD signalling, activity of alveolar macrophages, and potential crosstalk with the gut via the gut-lung axis.

The genital bacteriome is particularly relevant to women's health, providing some immunological protection against infections of the reproductive tract and promoting vaginal health. The genital bacteriome is primarily composed of *Lactobacillus* species. These species produce lactic acid that provides an acidic pH, modulates mucosal cytokine signalling, protects against sexually transmitted infections (STIs), and helps maintain immune tolerance during pregnancy.

Bacteriome Influences in Host Defence Mechanisms

The bacteriome influences host defence mechanisms in several ways.

Competitive exclusion occurs when beneficial or commensal microbes dominate and occupy ecological niches, thereby preventing the colonisation of harmful pathogens. In response to infections, several commensal microorganisms and host cells produce antimicrobial peptides (AMPs) that inhibit the growth and colonisation of pathogens by either inhibiting their growth or directly killing them.

Modulation of the host immune responses is influenced by the bacteriome, which activates both innate and adaptive immune pathways. These include the stimulation of pattern recognition receptors (e.g. Toll-like receptors), increasing mucosal immunity through immunoglobulin A (IgA) production, and regulating cytokine profiles.

Dysbiosis/imbalance in the host bacteriome may compromise the host's defence mechanism against the invading zoonotic pathogen, thereby facilitating the host's susceptibility to disease, disease onset, and pathogen colonisation. Dysbiosis entails a microbial imbalance and poses a significant threat by contributing to the development of healthcare-associated infections, autoimmune diseases, and metabolic disorders, such as type 2 diabetes, where a disturbed gut microbiota may impair immune responses to infections. Engineered probiotics are being increasingly recognised as a new therapeutic strategy for disease management, restoring a proper microbial balance by targeting or eliminating specific pathogens while enhancing beneficial microbial functions.

In recent years, advances in metagenomics and high-throughput sequencing technology have begun to provide insight into the diversity and function of microbial communities. This advances have catalysed the development of bacteriome-based applications in diagnostics, therapeutics, agriculture, and environmental science. One of the newly emerging areas of research focus on the relationship between the bacteriome and zoonotic infections, i.e. diseases transmitted from animals to humans. For instance, zoonotic infections may alter the human bacteriome by significantly changing its composition and modulating immune responses. In addition, pre-existing dysbiosis can result in the loss of colonisation resistance and immune homeostasis, thereby increasing susceptibility to more serious infections. While microbial dysbiosis is known to influence disease severity and immune response, the role of the host bacteriome in zoonotic infections remains largely unexplored.

Composition and Diversity of the Human Bacteriome

The human bacteriome exhibits remarkable variation from site to site, influenced by anatomical location, geography, and host-associated factors. Body habitat specialisation shapes the composition of the healthy human bacteriome, resulting in distinct microbial communities across different regions of the body. The communities in the gut, which are mainly made up of *Bacteroides* and *Firmicutes*, are completely dissimilar from those in the skin *(Corynebacterium)*, oral cavity *(Streptococcus)*, and vagina *(Lactobacillus)*. Collectively, these microbial communities contribute a significantly greater number of protein-coding genes than the human genome itself, encompassing functions essential for vitamin synthesis, immune modulation, and the production of short-chain fatty acids that support metabolic stability. Functional redundancy among these microbes enhances ecosystem resilience, with different species capable of performing similar roles. While some microbes undertake diverse metabolic tasks, core functions such as the degradation of complex carbohydrates are consistently maintained by the host's microflora, albeit through varying community compositions.

The human bacteriome has a different composition based on the body site. The gut and oral bacteriome exhibits very high *alpha diversity*, indicating a diverse and rich community of microbial species within a single individual or site. In contrast, the skin bacteriome exhibits high *beta diversity*, indicating substantial differences in microbial composition from person to person. A variety of factors can influence these differences, both within individuals and across sites. These include host genetics, diet, geographical location, and mode of birth in early life. In healthy individuals, 81%–99% of core microbial genera and metabolic functions are shared, indicating a common baseline of essential microbial roles across the population. Although microbial communities adjust to the pressures of their specific environments, they ultimately converge by carrying out similar roles, such as nutrient processing, pathogen protection, and immune regulation. Due to the functional similarities within the bacteriome, it is a stable and dependable contributor to human health, regardless of the species composition.

Gut Bacteriome

The highest bacteriome diversity is observed in the human gut microbiome. This is primarily composed of Firmicutes, Bacteroidetes, Actinobacteria, Proteobacteria, Verrucomicrobia, Fusobacteria, and Spirochaetes. Africa's largest microbiome study (2025) uncovered 1005 novel bacterial species and over 40,000 viral metagenomes, highlighting previously underrepresented diversity. The link between gut microbiota and inflammatory bowel syndrome gained momentum in 2024, with small intestine dysbiosis playing a potential role in symptom development. Current perspectives emphasise that a healthy gut bacteriome is a dynamic system influenced by diet, faecal microbial load, and gut physiology rather than a fixed composition.

Skin Bacteriome

The skin bacteriome varies from area to area. However, it is primarily dominated by the phyla Actinobacteria (51.8%), Firmicutes (24.4%), Proteobacteria (16.5%), and Bacteroidetes (6.3%), with *Cutibacterium* predominating in sebaceous areas such as the face. At the same time, sites like the dry forearms exhibit a richer bacterial flora. A recent investigation of Indian families reported that geographic location and family membership were strong influences on beta diversity, whereas diet and age had a limited influence. Advances such as co-culture systems now enabled adetailed view into microbial interactions in lipid-rich environments, mimicking the sebaceous niches. Dysbiosis in skin microbial communities is associated with disorders such as atopic dermatitis, while engineered *Staphylococcus epidermidis* strains show promising therapeutic potential through immune modulation.

Oral Bacteriome

The human oral microbiome is primarily dominated by the phyla Firmicutes, Bacteroidetes, Proteobacteria, Actinobacteria, and Fusobacteria. In the oral bacteriome, oral squamous cell carcinoma (OSCC) is associated with increased alpha diversity and a disruption of normal age- and gender-associated microbial patterns. In adolescents with untreated depression, alterations in oral microbiota, such as increased *Streptococcus* and reduced *Neisseria*, correlate with cognitive impairments, supporting their role as potential neuropsychiatric biomarkers. These observations underscore the expanding clinical relevance of oral microbes in both cancer and mental health contexts.

Upper Respiratory Bacteriome

The upper respiratory tract bacteriome represents a complex microbial ecosystem with significant implications for respiratory health and susceptibility to zoonotic pathogens. This dynamic community forms the first line of defence against air-borne bacterial agents, including those of zoonotic origin. The nasopharyngeal bacteriome typically harbours diverse commensals, predominantly *Corynebacterium*, *Dolosigranulum*, *Moraxella*, *Staphylococcus*, and *Streptococcus* species, which establish competitive colonisation resistance against pathogenic invaders.

Recent metagenomic (sequencing-based study of all microbial DNA to identify and understand microbial communities) analyses have revealed that the upper respiratory bacteriome composition may influences susceptibility to respiratory pathogens. Individuals with *Corynebacterium*-dominant nasopharyngeal profiles showed significantly reduced colonisation by *Streptococcus* species, suggesting bacteriome-mediated protective mechanisms. The respiratory bacteriome exhibits temporal stability in healthy individuals but demonstrates substantial perturbations during infections. These alterations may persist beyond pathogen clearance, potentially contributing to post-infectious complications. Contemporary research suggests that interspecies bacterial communication through quorum-sensing molecules has a significant impact on whether respiratory pathogens successfully establish infections in the human respiratory tract.

Vaginal Bacteriome

The vaginal bacteriome is primarily composed of Firmicutes, Actinobacteria, and Bacteroidetes phyla, with a high abundance of *Lactobacillus*. It displays ethnic variation, particularly in the abundance of *Lactobacillus crispatus*, which is lower in African and Latin American women compared to European and Asian populations. A 2023 global analysis reframed bacterial vaginosis as a context-dependent microbial imbalance rather than a strictly pathological condition. Additionally, emerging evidence suggests that beta-carboline compounds with anti-inflammatory properties, produced by vaginal *Lactobacillus* species, may offer promising therapeutic potential.

Urinary Tract Bacteriome (UroBiome)

The urinary tract, historically considered sterile in the absence of infection, is now recognised to harbour a complex bacterial (or microbial) community termed the "*UroBiome*." High-throughput sequencing technologies have revealed diverse microbial populations inhabiting the urinary tract, with significant implications for urogenital health and susceptibility to zoonotic infections. The healthy UroBiome typically contains facultative anaerobes from the *Lactobacillus*, *Streptococcus*, *Staphylococcus*, *Corynebacterium*, and *Gardnerella genera*, though considerable interpersonal variation exists.

Recent studies have demonstrated that the UroBiome influences susceptibility to zoonotic uropathogens, particularly *Leptospira interrogans* and uropathogenic *Escherichia coli* strains of animal origin. A 2024 comprehensive metagenomic analysis revealed that *Lactobacillus*-dominant urinary communities confer substantial protection against leptospiral colonisation through competitive exclusion and bacteriocin production. Additionally, specific *Lactobacillus* strains modulate urothelial immune responses, potentially limiting the establishment and dissemination of zoonotic pathogens. The UroBiome exhibits gender-specific differences, with female urinary tracts typically harbouring a greater abundance of *Lactobacillus*, potentially explaining gender-associated disparities in susceptibility to certain zoonotic uropathogens. Furthermore, age-related changes in the UroBiome composition may contribute to altered vulnerability to zoonotic infections across the lifespan, with implications for elderly populations.

Role of Bacteriome in Human Health

With trillions of bacteria inhabiting and adorning our bodies, the human bacteriome holds a pivotal role in health and disease prevention. Rather than working in isolation, these microbial communities function as deeply embedded partners in an array of physiological systems, ranging from metabolism and immune response to hormonal regulation and mental well-being. The bacteriome's ability to maintain immune balance is one of its most crucial functions. By constantly interacting with host cells, commensal organisms help educate the immune system in distinguishing between harmful pathogens and harmless antigens. This immune education helps mitigate the risk of autoimmune disorders and enables a full-fledged immune response from the body when needed. The microbiota also regulates inflammation, the underlying cause of many chronic diseases, through the action of biological mediators that modulate immune signalling pathways and maintain tissue homeostasis.

The bacteriome interacts with metabolic health by regulating energy balance, insulin sensitivity, and fat storage. Complex compounds that the human body cannot digest can be metabolized by bacterial enzymes, producing nutrients and bioactive compounds that modulate metabolic function. Changes in microbial community composition have been observed to be associated with obesity, diabetes, and cardiovascular disease, suggesting a relationship with high metabolic outcomes. Furthermore, the bacteriome essentially acts as a natural barrier against pathogenic invasion. Beneath the epithelium, commensal bacteria compete with the harmful pathogens to acquire available niches and nutrients and produce antimicrobial substances that reinforce the integrity of epithelial barriers, and hence serve as protection against pathogens. This is indeed beneficial for preventing infections and maintaining equilibrium within the internal ecosystems.

The health-promoting effects of the bacteriome extend to areas such as wound healing, hormonal balance, drug metabolism, and even cancer resistance. These diverse capabilities have reaffirmed that the bacteriome need not be regarded simply as a repository of microorganisms but rather as a living and vital dynamic component of human biology. Preserving its diversity and balance is therefore extremely important for preventive healthcare, new treatment strategies, and wellness in general.

Dysbiosis and Its Link to Diseases

The human bacteriome, which comprises the gut, oral cavity, skin, lung, and urogenital bacteriomes, is important to immune regulation, pathogen resistance, and overall health. Dysbiosis is increasing the susceptibility of the host to various infectious diseases and inflammatory diseases (Fig. 50.1). In the human gut, pathogens can disrupt the microbial balance, weakening the intestinal barrier. This increases susceptibility to infections such as lower respiratory tract infec-

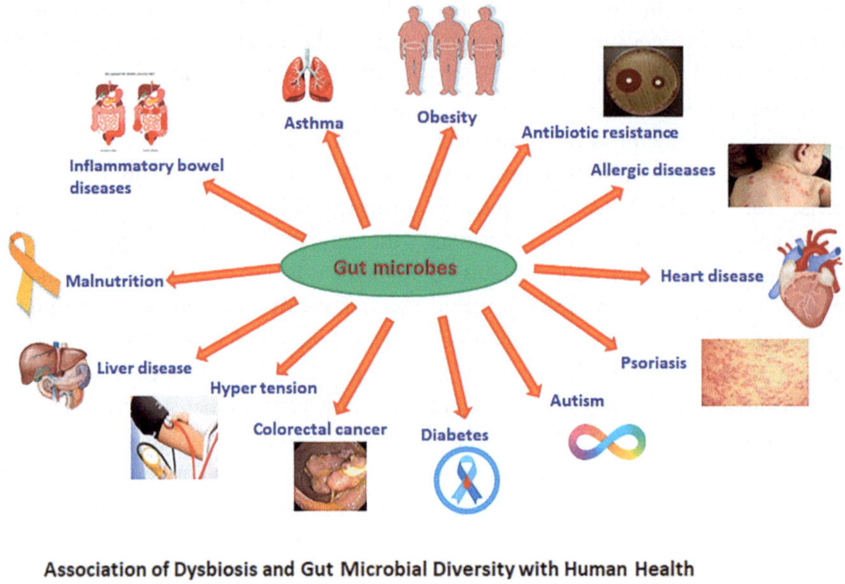

Fig. 50.1 Dysbiosis-associated diseases and clinical symptoms. (Image source: Authors' personal collection)

tions, *Clostridioides difficile*, sepsis, tuberculosis, and systemic viral diseases like COVID-19, HIV, and influenza. These interactions highlight the significance of the gut-lung axis in systemic immunity.

Immune responses mediated by intestinal microbiota are most important when respiratory infections challenges host defense and shifts in microbial diversity can influence the severity of gastrointestinal infections. *Streptococcus mutans* and *Porphyromonas gingivalis* are key bacteria capable of rapidly entering the bloodstream, contributing to systemic infections like infective endocarditis and aspiration pneumonia. Commensals such as *S. epidermidis* form protective barriers but, when disrupted, may allow pathogens like *S. aureus* to cause infections such as cellulitis, atopic dermatitis, and candidiasis. The vaginal microbiota, primarily composed of *Lactobacillus* species, maintains an acidic environment that prevents pathogenic overgrowth; its imbalance is associated with bacterial vaginosis, UTIs, and increased risk of HIV and HPV infections.

Microbiome-targeted treatments (probiotics, prebiotics, synbiotics, postbiotics, faecal microbiota transplantation (FMT), engineered microbial consortia, and dietary measures) have a lot of potential yet to be investigated. Ultimately, the complexity of microbial interactions and their effects on host immunity continues to be at the centre of microbiome-based medicine and infectious disease management.

Bacteriome and Zoonoses

Zoonoses are infectious diseases that are transmitted between animal reservoirs and human hosts (Fig. 50.2). In zoonotic disease, the microbiome of both animal and human hosts plays a crucial role in ecology and pathogenesis by influencing host susceptibility, enhancing the functioning of the immune system, and conferring colonisation resistance against invading zoonotic pathogens. Despite this evidence on microbial imbalance/dysbiosis, which can lead to increased susceptibility to zoonotic infection, this field of research remains underexplored.

A comprehensive list of zoonotic pathogens, their target cells and possible host microbiomes

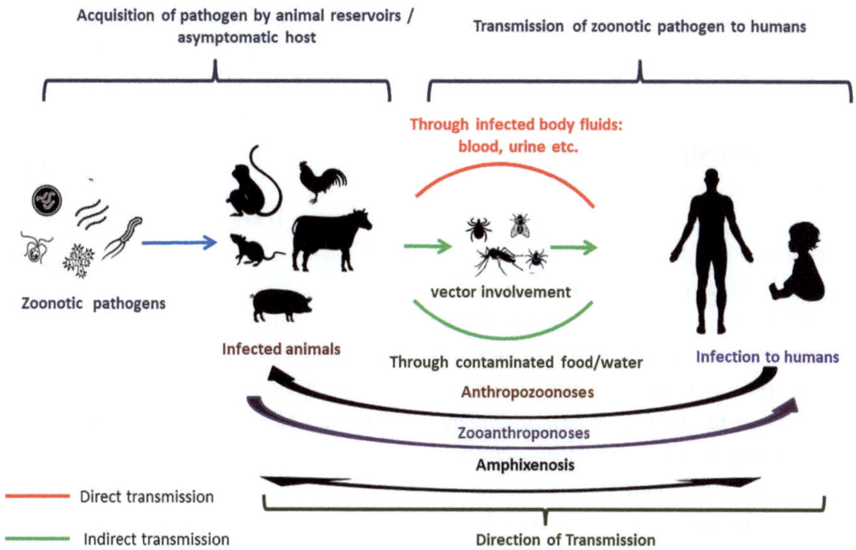

Fig. 50.2 Mode of transmission of zoonotic pathogens to humans. (Image source: Authors' personal collection)

Table 50.1 List of zoonotic pathogens, their target cells, and possible host microbiomes associated with disease susceptibility

Bacterial pathogens and associated diseases	Zoonotic source	Target cell/tissue in humans	Possibly associated microbiome
Bacillus anthracis (anthrax)	Livestock, wild herbivores	Macrophages, epithelial cells, especially in lungs, skin, gut	Disruption of lungs, skin, and gut microbiota
Brucella spp. (brucellosis)	Cattle, goats, sheep, pigs	Macrophages (replicate intracellularly)	Gut microbiome alterations
Francisella tularensis (tularaemia)	Rabbits, rodents, and ticks	Macrophages, dendritic cells	Lung microbiome shifts
Yersinia pestis (plague)	Rodents (via fleas)	Macrophages, neutrophils, and lymphatic tissue	Gut microbiome dysbiosis
Leptospira interrogans (leptospirosis)	Rodents (via urine--contaminated water)	Renal tubular cells, endothelial cells	Gut and urinary microbiome alterations
Coxiella burnetii (Q fever)	Livestock (sheep, goats, cattle)	Alveolar macrophages, monocytes	Lung microbiome changes
Campylobacter jejuni	Poultry, cattle, and dogs	Intestinal epithelial cells	Gut dysbiosis
Salmonella enterica (non-typhoidal)	Poultry, reptiles, cattle	Intestinal epithelial cells, macrophages	Gut microbiome disruption
Listeria monocytogenes	contaminated dairy/ meat from animals	Epithelial cells, hepatocytes, and macrophages	Gut microbiota shifts
Mycobacterium bovis	Cattle (raw milk or aerosol)	Alveolar macrophages, lymphoid tissue	Lung microbiome alterations
Pasteurella multocida	Cats, dogs (via bites/ scratches)	Subcutaneous tissue, epithelial, and connective tissue	Skin microbiome disruption
Bartonella henselae (cat scratch disease)	Cats	Endothelial cells, red blood cells, and macrophages	Possible interaction with skin microbiome
Chlamydia psittaci (psittacosis)	Birds, especially parrots and pigeons	Respiratory epithelial cells, macrophages	Lung microbiome shifts
Rickettsia rickettsii (Rocky mountain spotted fever)	Ticks (via rodents, dogs)	Endothelial cells	Skin microbiome alterations

associated with disease susceptibility is presented in Table 50.1.

Bacteriome Dysbiosis and Zoonotic Pathogens

Bacterial dysbiosis of the host bacteriome is often associated with a weakened immune system and a disrupted epithelial barrier, resulting in the invasion of zoonotic pathogens. Studies suggest that people residing near animal farms often harbour bacteria such as *Staphylococcus aureus*, *Clostridioides difficile*, *Streptococcus pneumoniae*, *Klebsiella pneumoniae*, *E. coli*, and other Enterobacteriaceae, which exhibit strong pro-inflammatory response properties. These bacteria may promote bacterial dysbiosis, thereby making the host more susceptible to infections.

Gut Bacteriome Dysbiosis Linked to Zoonotic Infection

Animal models have been widely used in zoonotic infection research, while human studies are limited, which provides insight into microbial alteration in response to infection with zoonotic pathogens. For instance, mice infected with zoonotic pathogens *Leptospira interrogans* and *Brucella* exhibited significant changes in gut bacteriome composition, resulting in more weight loss and higher pathogen loads in their organs compared to control mice. The infected mice group shows an increased abundance of certain bacterial species, such as *Prevotella* and *Staphylococcus*, and a reduced abundance of anaerobic bacteria compared to healthy mice. This indicates a potential role of the microbiome in disease severity and pathogenesis. Similarly, goats infected with *Brucella* showed increased levels of Firmicutes, Spirochaetota, Planctomycetota, and Proteobacteria. In contrast, healthy goats harboured more Bacteroidota. This finding further supports the notion that changes in the gut bacteriome can influence the host's ability to resist zoonotic infections. Nevertheless, the exact mechanism behind this function remains to be explored.

Gut-Lung Axis Dysbiosis in Zoonotic Infection

The gut bacteriome of the host is significantly associated with the progression of tuberculosis (TB) through the gut-lung axis. The patient with pulmonary TB caused by *Mycobacterium tuberculosis* shows reduced gut bacterial diversity and short-chain fatty acid (SCFA)-producing bacteria such as Firmicutes and Lachnospiraceae. This leads to a reduced level of butyrate production, which is involved in anti-inflammatory signalling and epithelial barrier production, and an altered level of SCFAs like acetate and propionate, making the host more susceptible to extrapulmonary TB.

Although many bacteriome-related TB studies focus on *M. tuberculosis*, the zoonotic TB pathogen *Mycobacterium bovis*, which causes tuberculosis, especially in cattle, and can be transmitted to humans, remains understudied in this context. Considering their close phylogenetic relationship and shared pathogenic mechanism between *M. tuberculosis* and *M. bovis*, it is plausible that *M. bovis* infection may also result in similar microbial disruptions.

Skin Bacteriome Dysbiosis in Zoonotic Infection

The skin is the largest organ in the human body. It acts as a primary site for the bites of arthropod vectors, where they co-inoculate both pathogens and salivary compounds during feeding, potentially playing a role in vector-borne disease transmission. The skin bacteriome harbours several commensal bacteria that produce AMP in response to certain infections or induce the host cells to produce AMP, which is involved in modulating the local immune response and maintaining microbial balance. During vector-borne bacterial zoonotic infections, such as Lyme borreliosis and spotted fever rickettsiosis, the cutaneous immune response may facilitate pathogen proliferation and even persistence before systemic dissemination. The bacteriome shift, or dysbiosis, has been reported in both human hosts and animal models infected with rickettsial pathogens.

Transcriptomic studies (the study of a complete set of RNA transcripts in an organism) of skin lesions from tick bites show significant overlap in bacterial genera found in both ticks and human skin. Common genera include *Pseudomonas*, *Acinetobacter*, *Corynebacterium*, and *Propionibacterium*.

Also, some of the pathogenic species, such as *C. burnetii* and *Rickettsia cervicalis*, were observed to be present at the site of the tick bite on the human skin. This suggests that the skin bacteriome plays a role in modulating immune responses at the bite site by influencing the inflammation and potentially affecting the host's tolerance to both the pathogen and the vectors. Additionally, the *Rickettsia*-infected rat model shows a higher abundance of *Chlamydia* spp. at the tick bite site compared to the control groups. All these observations further suggest a zoonotic pathogen-induced bacteriome shift in the host.

Overall, the evidence suggests that a bacterial shift or dysbiosis occurs in the host microbiome following the inoculation of the zoonotic pathogen, potentially disrupting the host-bacteriome interaction and contributing to disease progression.

Role of Animal Bacteriome in Zoonosis Transmission

The animal (reservoir host) microbiome plays a crucial role in the uptake and transmission of zoonotic pathogens to humans. The microbiome influences pathogen colonisation, immune response, and external dissemination of pathogens. Several bacterial communities residing in the gut, skin, and mucosal surfaces of animals can serve as reservoirs for zoonotic pathogens, such as *Salmonella*, *Campylobacter*, *E. coli*, and *M. bovis*. These zoonotic pathogens survive in animals by utilising the benefits from the commensal or beneficial bacteria residing within them. These can be shed into the environment, increasing the risk of transmission via direct or indirect contact and vectors. The microbiome modulates the host's immunity, with certain metabolites interfering with the pathogens' ability to establish infections.

The dysbiosis/imbalance in the animals' bacterial communities mainly occurs due to the overuse of antibiotics, stress, or dietary changes, associated with reduced colonisation resistance and enhanced pathogen multiplication and dissemination. In the case of vector-borne zoonotic disease transmission, the bacteriome of vectors, such as mosquitoes, ticks, and mites, can either promote or inhibit the transmission of vector-borne zoonoses. For example, *Wolbachia* in mosquitoes inhibits the replication of arboviruses, while the gut microbiota of ticks supports the colonisation of *Borrelia*, *Anaplasma*, and *Rickettsia* species. However, some of the symbiotic bacteria in the tick microbiome contribute to immune modulation, reducing the immune response and favouring the survival and replication of pathogens.

In essence, the bacteriome of the animal reservoir serves both as a protective and supportive factor for the spread of zoonotic pathogens. Hence, vector bacteriome manipulation through the introduction of symbiotic bacteria or microbiota-based vaccines is currently explored as a strategy to block pathogen transmission and reduce bacterial zoonoses.

Bacteriome-Based Therapeutics for Zoonotic Infection

Bacteriome-based therapeutics could serve as a promising methods for preventing several zoonotic infections by restoring the healthy bacterial balance, which helps promote the host's defence mechanism. The dysbiosis in the gut environment is restored through faecal microbiota transplantation (FMT). It involves transferring microorganisms from stool samples of a healthy donor to the recipient to reestablish a balanced microbiota. Through FMT, the depletion of gut microbiota was reversed in mice infected with the zoonotic pathogen. Similarly, probiotic supplements (e.g. *Lactobacillus plantarum*) are also implicated in restoring the imbalanced bacterial community caused by zoonotic infections in mice. Additionally, a reduced microbial ecology could impair the production of SCFA, which in turn disrupts immune modulation through the gut-lung axis. Along with probiotics, postbiotics, the non-viable microbial products, are being utilised to reverse microbial dysbiosis.

Collectively, the bacteriome-targeted treatments such as *probiotics* (live beneficial microbes that improve gut health), *prebiotics* (non-digestible fibres that feed good gut bacteria), *synbiotics* (combinations of probiotics and prebiotics for synergistic effect), *postbiotics* (metabolic byproducts or components from probiotics that confer health benefits), FMT, engineered microbial consortia, and dietary measures have a lot of potential. Ongoing research is therefore necessary to reconcile the observed efficacy of these approaches, optimise personalised treatment approaches, and develop new approaches to prevent communicable diseases through bacteriome homeostasis. Ultimately, the complexity of bacterial interactions and their effects on host immunity remains at the centre of bacteriome-based medicine and infectious disease management.

Methods for Bacteriome Analysis

An overview of the bacteriome analysis is presented in Fig. 50.3.

Sample Collection and Metagenomic DNA Extraction

The bacteriome analysis begins with the collection of biological samples such as skin swabs, faeces, saliva, sputum, or blood. Then, the samples are subjected to metagenomic DNA isolation, which involves targeting the total DNA present in the sample. The metagenomic DNA extraction procedure primarily involves a combination of mechanical disruption (e.g. bead beating) and chemical lysis of the samples. These rupture the microbial cell wall, followed by purification to remove proteins and other contaminants. The DNA quality can be checked by using a NanoDrop spectrophotometer. The instrument evaluates purity based on the absorbance ratio (A260/280) and fluorescence analysis, such as Qubit. DNA of good quality is essential for sequencing and downstream analysis.

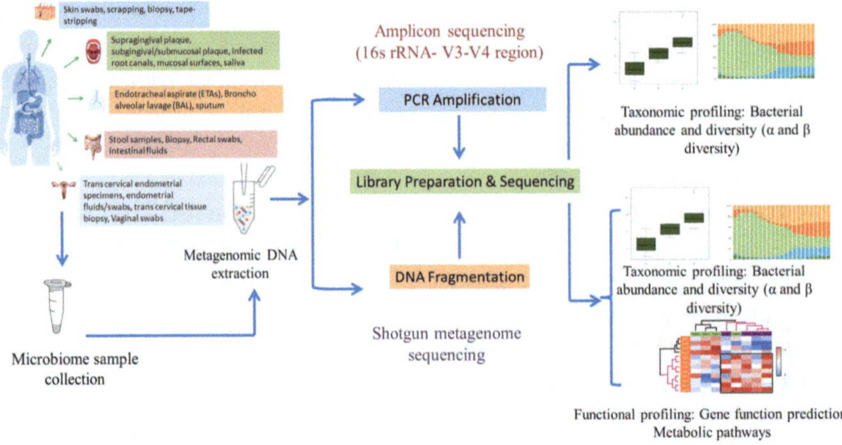

Fig. 50.3 The workflow showing the bacteriome profiling for taxonomic and functional analysis. (Image source: Authors' personal collection)

Amplicon Sequencing (16S rRNA Gene Sequencing)

Amplicon sequencing is a targeted sequencing approach that focuses on specific taxonomic marker genes, especially the 16S rRNA gene (a widely used target gene for bacteria). It targets the hypervariable regions (V3–V4) of the 16S rRNA gene, then amplifies them using PCR with barcoded primers to allow for multiplexed sequencing. This also enables taxonomic identification and classification based on phylogenetic relationships. It is also a cost-efficient method that utilises reference databases such as SILVA, RDP, and Greengenes. However, while it is cost-efficient and relatively easy to use, a major limitation is the resolution at the species level, and it does not give functional genomic information.

Shotgun Metagenomic Sequencing

Shotgun metagenomic sequencing enables the sequencing of all DNA fragments present in a sample, without focusing on specific genes. This method can characterise all the microorganisms, including bacteria, fungi, viruses, and archaea, in the samples. This approach enables analysis of both the phylogenetic classification and functional gene information of the microbial community. Shotgun metagenomic sequencing begins with fragmentation, followed by library preparation through end-repair, adapter ligation, and indexing, and then proceeds to sequencing on high-throughput sequencers like Illumina MiSeq, HiSeq, NovaSeq, and PacBio.

Illumina platforms offer short-read sequencing, usually 150–300 base pairs. It provides high accuracy and sequencing depth and is mainly used for taxonomic profiling or microbial abundance profiling. In contrast, long-read sequencing, over 1000 bp, is offered by platforms such as PacBio and Oxford Nanopore technologies. It gives a higher resolution of complex or repetitive genomic regions, enabling improved genome assembly. Long-read sequencing has a higher error rate than the short-read sequencing method. Sequencing is performed using a paired-end method, which allows sequencing of both ends of a DNA fragment to enhance alignment and improve accuracy. In contrast, the single-end method allows the sequencing of a DNA fragment at a single end. However, paired-end sequencing is typically slower and more expensive than single-end sequencing; however, single-end sequencing offers less resolution.

Bioinformatics and Data Analysis

Bioinformatics analysis involves interpreting raw sequence data in ecological and functional contexts. The first step is to cluster the sequences into operational taxonomic units (OTUs) for 16S sRNA amplicon sequencing data or to identify Amplicon Sequence Variants (ASVs) based on matching with available reference sequences to delineate microbial taxa. Taxonomic classification is performed using databases like SILVA or Greengenes. For shotgun metagenomic data, tools such as MetaPhlAn, HUMAnN, and MG-RAST are frequently employed for taxonomic profiling, functional gene prediction, and metabolic pathway reconstruction (Table 50.2).

To describe microbial community composition, it is most common to calculate species abundances based on read counts normalised for sequencing depth. Alpha diversity was calculated using indices such as Shannon, Simpson, or Chao1. These are used to measure the microbial diversity within a sample, reflecting both taxonomic richness (the number of taxa) and taxonomic evenness (the distribution of taxa). Beta diversity identifies the differences in microbial characteristics between samples using indices such as the Bray-Curtis dissimilarity index, Jaccard, or UniFrac. Often, these metrics are calculated and visualised through ordination procedures such as Principal Coordinates Analysis (PCoA) or Non-metric Multidimensional Scaling (NMDS). These procedures help compare microbial communities among groups, i.e., healthy versus diseased, and identify ecological patterns. In addition, several visualisation approaches, such as heatmaps, bar plots, or hierarchical clustering, are also used in the interpretation of species abundances and distribution patterns.

Table 50.2 Common databases and tools used for microbiome analysis

Analysis/method(s)	Tool(s)	Purpose	Application
Quality control	FastQC, Trimmomatic, Cutadapt	Filter and trim low-quality reads, dereplication	Removing adapters and low--quality sequences
Metagenome assembly	MG-RAST, MEGAHIT, metaSPAdes, IDBA-UD, MetaPhlAn2, HUMAnN2, SOAPdenovo	Assemble reads into contigs and scaffolds	De novo genome reconstruction
Reference database	Ribosomal Database Project (RDP), SILVA, Greengenes, EzBiocloud, NCBI RefSeq, GTDB	Chimera removal, classifying 16S sequences or proteins	Taxonomic/ functional annotation using reference database
Taxonomic profiling and diversity analysis	QIIME, Mothur, Kraken2, MetaPhlAn, DADA2, VEGAN, Phyloseq	Identifying microbial taxa	Identifying bacteria, archaea, and fungi present in samples
Gene prediction tools	Prodigal, Prokka, Genemark, Glimmer	Predicting coding and non-coding sequences	For functional diversity analysis
Functional profiling	HUMAnN3, PICRUSt2, EggNOG-mapper, KEGG, CARD, IslandViewer, VFDB, Patric, IMG, RAST, COG	Predict functional genes and their metabolic potential	Enzyme/ pathway prediction from gene content, virulence genes, antimicrobial genes, orthologous genes, and phages
Phylogenetic analysis	FastTree, RAxML, iTOL, MEGA	Build phylogenetic trees	Evolutionary relationships between microbial taxa
Statistical analysis	R packages, STAMP	Diversity analysis, differential abundance	Alpha/beta diversity, ordination plots
Visualisation	R, Python (matplotlib, seaborn), Krona, GraPhlAn	Plotting and interactive data visualisation	Bar plots, heatmaps, krona charts, PCA, PCoA

Integrative Analysis

To strengthen the biological relevance of microbiome studies, the microbial profile data are correlated with clinical, environmental, or demographic metadata. This approach enables the examination of relationships between microbial diversity and host characteristics, disease phenotypes, or environmental exposures. Multivariate statistical analyses, correlation networks, and machine learning approaches are usually utilised to uncover microbial features or biomarkers that enable reliable predictions. Ultimately, the incorporation of microbiome data with the host metadata provides insights into the understanding of host-microbe interaction that leads to the development of bacteriome-based diagnostics, therapeutics, and risk prediction models.

Case Study

A 32-year-old software engineer presented to the emergency department with severe abdominal cramps, watery diarrhoea, nausea, vomiting, and a low-grade fever. Her symptoms, which began 2 days prior, followed her consumption of undercooked chicken from a street food stall 3 days before. Notably, she had completed a 7-day course of broad-spectrum antibiotics 2 weeks prior to the onset of symptoms for a presumed sinus infection. She had participated in a wellness programme that included regular gut microbiome profiling. Her baseline faecal sample, collected 3 months prior, showed a healthy and diverse gut microbiome dominated by beneficial bacteria, including *Faecalibacterium prausnitzii* and other Firmicutes and Bacteroidetes, with very few Proteobacteria. However, a faecal sample taken 5 days after her antibiotic course revealed a significant disruption: a 40% decrease in overall diversity, a marked reduction in beneficial bacteria, and a notable increase in Proteobacteria, including *E. coli*. The faecal sample collected upon admission during her acute illness revealed an even more pronounced dysbiosis, characterised by a massive increase in Proteobacteria and the confirmed presence of *Salmonella enterica* serovar therapeutics and the culture confirmed *Salmonella enterica* serovar Typhimurium, and she recovered with supportive care, without requiring antibiotics. Her symptoms gradually resolved over 7 days. Follow-up faecal microbiome analysis revealed a slow recovery of her gut diversity, although some imbalances, particularly increased Proteobacteria levels, persisted for several weeks after her clinical recovery. She was advised on probiotic supplementation and dietary changes to aid her microbiome recovery.

1. How did her recent antibiotic course likely contribute to her susceptibility to *Salmonella* infection and the severity of her symptoms, considering the observed changes in her gut bacteriome?
2. What are the potential mechanisms by which gut bacteriome dysbiosis could influence the virulence or shedding dynamics of a zoonotic pathogen like *Salmonella* in humans?
3. What targeted interventions could be considered to reduce the risk or severity of food-borne zoonotic infections, particularly in individuals with compromised gut microbiota?

Research Questions

1. What are the processes and pathways linking microbiome dysbiosis to vulnerability and severity of zoonotic infections, such as brucellosis?
2. How are different zoonotic factors influencing and/or interacting with the host microbiome in different populations, geographical locations, and environments?
3. What is the efficacy and safety of specific targeted interventions on the microbiomes, such as probiotics, prebiotics, or symbiotics, for the prevention and treatment of bacterial zoonotic infections?
4. How can microbiome surveillance be integrated into One Health approaches to potentially predict, monitor, and control zoonotic outbreaks at the human-animal-environment interface?

Declaration During the preparation of this chapter, we have used [ChatGPT] to improve the readability of the text. After using, we reviewed and edited the content as needed and take full responsibility for the content of the publication.

Further Readings

Byrd AL, Belkaid Y, Segre JA. The human skin microbiome. Nat Rev Microbiol. 2018;16(3):143–55.

Cantas L, Suer K. Review: the important bacterial zoonoses in "one health" concept. Front Public Health. 2014;2:144.

Failloux AB, Moutailler S. Zoonotic aspects of vector-borne infections. Rev Sci Tech. 2015;34(1):175–83, 165–74.

Flandroy L, Poutahidis T, Berg G, Clarke G, Dao MC, Decaestecker E, Furman E, Haahtela T, Massart S, Plovier H, Sanz Y, Rook G. The impact of human activities and lifestyles on the interlinked microbiota and health of humans and of ecosystems. Sci Total Environ. 2018;627:1018–38.

Galloway-Peña J, Hanson B. Tools for analysis of the microbiome. Dig Dis Sci. 2020;65(3):674–85.

Gomaa EZ. Human gut microbiota/microbiome in health and diseases: a review. Antonie Van Leeuwenhoek. 2020;113(12):2019–40.

Kamel M, Aleya S, Alsubih M, Aleya L. Microbiome dynamics: a paradigm shift in combatting infectious diseases. J Pers Med. 2024;14(2):217.

Kunath BJ, De Rudder C, Laczny CC, Letellier E, Wilmes P. The oral-gut microbiome axis in health and disease. Nat Rev Microbiol. 2024;22(12):791–805.

Lloyd-Price J, Abu-Ali G, Huttenhower C. The healthy human microbiome. Genome Med. 2016;8(1):51.

Ma LC, Zhao HQ, Wu LB, Cheng ZL, Liu C. Impact of the microbiome on human, animal, and environmental health from a One Health perspective. Sci One Health. 2023;2:100037.

Manos J. The human microbiome in disease and pathology. APMIS. 2022;130(12):690–705.

Ojo-Okunola A, Nicol M, du Toit E. Human breast milk bacteriome in health and disease. Nutrients. 2018;10(11):1643.
Panthee B, Gyawali S, Panthee P, Techato K. Environmental and human microbiome for health. Life (Basel). 2022;12(3):456.
Rahman MT, Sobur MA, Islam MS, Ievy S, Hossain MJ, El Zowalaty ME, Rahman AT, Ashour HM. Zoonotic diseases: etiology, impact, and control. Microorganisms. 2020;8(9):1405.
Rubaker L, Putonti C, Dong Q, Wolfe AJ. The human urobiome. Mamm Genome. 2021;32(4):232–8.

Biosafety in Laboratory Settings

51

Sandhya Chouhan and Vikrant Singh Rajput

Learning Objectives

1. To understand the basic concepts and terminologies of laboratory biosafety
2. To study risk assessments and risk management to execute suitable biosafety measures
3. To learn the utilisation of personal protective equipment and manage biohazardous waste properly

Introduction

The rising prevalence of pathogenic organisms and biohazards presents substantial risks to world health, demanding strong biosafety standards in laboratory settings. Laboratories play an important role in detecting, studying, and managing zoonotic diseases. This necessitates a better understanding of biosafety standards to avoid laboratory-acquired illnesses and environmental contamination. The major goal of biosafety in laboratories is to safeguard staff, the community, and the environment against infectious organisms and dangerous biological materials.

The concept of biosafety has developed greatly since its origin, owing to historical events and the ongoing introduction of novel infections. The epidemic of the Ebola virus infection, which is zoonotic in origin (carried by monkeys, chimpanzees, gorillas, and apes), in Yambuku, Zaire (now the Democratic Republic of the Congo), in 1976, was one of the first and most significant occurrences to underscore the necessity for strong biosafety precautions. The virus's high fatality rate and quick dissemination highlighted the importance of effective containment and safety standards when dealing with highly deadly species.

Numerous zoonotic diseases, including anthrax, tuberculosis, brucellosis, bubonic plague, Lyme disease, bordetellosis, COVID-19, rabies, chikungunya, dengue, AIDS, monkeypox, trichinellosis, fascioliasis, aspergillosis, Q-fever, scrub typhus, chlamydiosis, trypanosomiasis, Chagas' disease, leishmaniasis, and many more, pose significant biosafety concerns in laboratories handling them. The global attention to vulnerabilities in biosafety and biosecurity policies has forced significant improvements in laboratory safety standards and laws.

The implementation of systematic methods and protocols meant to limit hazards associated with pathogenic organism handling and manipulation serves as the cornerstone of laboratory biosafety. These practices include a wide range of

S. Chouhan
Department of Biomedical Engineering, Central University of Rajasthan, Ajmer, Rajasthan, India

School of Medical Science and Technology, Indian Institute of Technology, Kharagpur, West Bengal, India

V. S. Rajput (✉)
Department of Biomedical Engineering, Central University of Rajasthan, Ajmer, Rajasthan, India
e-mail: vikrant.rajput@curaj.ac.in

S. C. Parija, A. Poddar (eds.), *Textbook of Bacterial Zoonoses*, Microbial Zoonoses,
https://doi.org/10.1007/978-981-95-6615-0_51

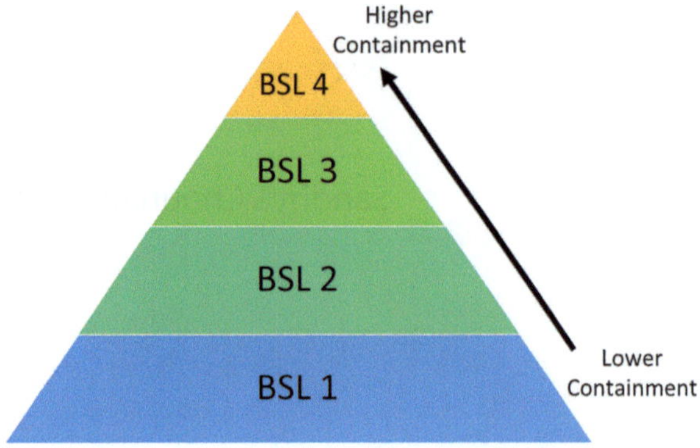

Fig. 51.1 Biosafety level containment hierarchy. (Image source: Authors' personal collection)

precautions, including engineering controls, personal protective equipment (PPE), administrative rules, and procedural safeguards. The effective deployment of biosafety measures is dependent on a rigorous risk assessment process that identifies possible dangers of pathogenicity (*Brucella* spp., *Bacillus anthracis*, etc.), transmission routes, infectious dose, and the availability of treatments and vaccines, and establishes the appropriately needed biosafety level (BSL) for safe operation. Biosafety levels, which range from BSL-1 to BSL-4, provide a tiered approach to containment, with each level offering greater levels of protection (Fig. 51.1). These levels are determined by specialised laboratory designs, safety equipment, and procedural requirements adapted to the risk posed by the microorganisms under investigation. In addition to physical confinement, good biosafety standards include correct management and disposal of biohazardous waste, which ensure that hazardous biological elements do not endanger people or the environment.

Principles of Laboratory Biosafety

Laboratory biosafety refers to the containment concepts, technology, and procedures used to avoid unintended exposure to infections and toxins or their accidental discharge. Its goal is to safeguard laboratory staff, the environment, and the public from possible risks connected with the handling of infectious organisms. Biosafety guidelines, such as those outlined in the *Biosafety in Microbiological and Biomedical Laboratories* (BMBL), are intended to control the hazards associated with biological research and diagnostic operations, ensuring that harmful microorganisms are handled safely and responsibly.

The principles of laboratory biosafety are based on risk assessment, containment, and operating requirements. They start with identifying and assessing the hazards presented by biological agents and then calculating the appropriate biosafety level (BSL) for safe handling. Primary containment entails direct protection using safety equipment and procedures. However, secondary containment is achieved by the laboratory's architecture and infrastructure to prevent disease escape. Personal protective equipment (PPE) acts as a barrier to exposure, and adhering to good microbiological procedures is critical for ensuring safety. Proper waste management, including sterilisation and disposal, is critical for avoiding environmental pollution. These safeguards jointly protect the staff, the public, and the environment in biological research and diagnostics.

Risk Assessment for Biosafety Preparedness

Risk assessment is critical for biosafety preparation since it involves the systematic identification, analysis, and control of dangers associated with infectious pathogens and hazardous biological materials. The periodic practice for risk assessment ensures that proper biosafety precautions are in place to safeguard laboratory staff, the environment, and the public.

Steps of Risk Assessment

Hazard Identification

Understanding pathogen features and analysing laboratory operations that may be hazardous, such as aerosol formation and high pathogen loads, are crucial. Advanced techniques, such as quantitative microbial risk assessment (QMRA), may model various exposure scenarios.

Evaluating Risks

Modelling parameters such as procedure frequency, pathogen concentration, and clinical outcomes is used to determine the chance of exposure and the severity of repercussions. Quantitative methods, such as fault tree analysis (FTA) and epidemiological data, increase the efficiency of risk assessment.

Risk Control Measures

Engineering controls such as high-efficiency particulate air (HEPA) filtration and negative--pressure systems in the laboratories, administrative controls such as SOPs, training and regular health check-ups and monitoring, and PPE customised to pathogen- and laboratory activity-specific hazards are all examples of risk control measures.

Monitoring and Review

Regular audits, incident reporting, and root cause analysis (RCA) are critical for the ongoing improvement of biosafety standards.

Risk Group of Microorganisms

Microorganisms are classified into risk groups (RGs) based on their pathogenicity, transmission risks, and underlying illness severity after infection (Table 51.1). These categories assist in determining the appropriate biosafety level for handling certain substances. The categorisation of risk categories is critical for understanding BSLs since each biosafety level is designed to address the dangers associated with these groups.

Types of Biosafety Level (BSL) Laboratories

Biosafety levels (BSLs) are extensive biocontainment measures that comprise not just procedural precautions but also specialised infrastructure, engineering controls, and safety equipment intended to isolate hazardous biological pathogens in a laboratory setting. These measures protect the laboratory staff, the surrounding environment, and the population from exposure to potentially pathogenic agents. BSLs vary from 1 to 4, with each level indicating a higher level of containment and safety precautions.

Each biosafety level comprises a set of laboratory procedures and techniques, safety equipment, and laboratory facilities appropriate for the activities and risks of the infectious organisms. BSL-1 is the lowest level of containment, whereas BSL-4 is the highest.

Table 51.1 Classification of risk groups of microorganisms

Risk group	Characteristics	Examples of zoonotic bacteria	Other important microorganisms
1	Agents that are not associated with illness in healthy adults	–	*Bacillus subtilis*, *Saccharomyces cerevisiae*, and *Escherichia coli K-12*
2	Agents linked to a human disease that is seldom serious and frequently treated with preventative or therapeutic measures	*Bordetella bronchiseptica*, *Campylobacter fetus*, *Borrelia burgdorferi*	Hepatitis B virus, HIV, *Salmonella* spp., *Toxoplasma gondii*
3	Agents linked to serious or fatal human diseases for which preventative or therapeutic strategies may be accessible (high individual risk, low population risk)	*Yersinia pestis*, *Bacillus anthracis*, *Brucella* spp.	*Mycobacterium tuberculosis*, *Bacillus anthracis*, West Nile virus
4	Agents capable of causing serious or fatal human illness for which no preventative or therapeutic measures are typically available (high individual and group risk)	–	Ebola virus, Marburg virus, Lassa fever virus

BSL-1: Basic Containment Laboratory

BSL-1 laboratories are basic containment facilities that can handle risk group 1 microorganisms, including the zoonotic bacteria that fall in this group. BSL-1 laboratories follow simple procedures and very limited safety standards. The overall layout of a BSL-1 laboratory is presented in Fig. 51.2.

Codes of Practice

The following codes of practice are essential for operation in a BSL-1 laboratory:

Access

The lab is typically accessible; however, access may be limited to avoid contamination. To keep people informed of ongoing activities and any risks, appropriate signs are displayed.

Personal Protection

Standard safety precautions include wearing lab coats, gloves, and eye protection. To preserve sterility and avoid cross-contamination, lab coats and gloves are secured and replaced frequently, particularly after handling biological materials.

Procedures

Laboratory activities adhere to conventional microbiological standards such as complete hand washing, appropriate waste disposal, sterilising work surfaces before and after tests, and disposing of sharps waste in puncture-resistant containers.

Laboratory Design and Construction

BSL-1 laboratories need to have elements that prioritise safety, functionality, and simplicity of maintenance. The work surfaces need to be chemically resistant, waterproof, and able to bear moderate heat and organic solvents, assuring durability and ease of cleaning. While specialised ventilation is not necessary, maintaining appropriate airflow via regular Heating, Ventilation and Air Conditioning unit (HVAC) and its maintenance is critical to preventing the development of hazardous vapours and providing a comfortable working environment. Emergency equipment, such as fire extinguishers, first aid kits, spill kits, eyewash stations, and safety showers, must be easily available and regularly maintained to guarantee safe operation during an emergency.

The laboratory usually has many important instruments such as microscopes, centrifuges, incubators, and autoclaves, which need to be inspected and calibrated regularly to ensure maximum performance. Autoclaves, which are essential for sterilising biological waste and laboratory equipment, are routinely maintained and certified. Refrigerators and freezers are vital for storing biological samples, and reagents are checked regularly for optimum temperature management. Personal protective equipment (PPE), such as lab coats, gloves, and eye protection, must be readily available and selected based on individual duties and potential dangers to ensure complete protection against biological agents. Decontamination facilities, such as autoclaves and chemical disin-

Fig. 51.2 Biosafety level 1 laboratory layout. (Image source: Authors' personal collection)

fectants, are to be readily available for sterilising reusable equipment before and after use.

Laboratory Equipment

Pipetting aids and disposable Pasteur pipettes, screw cap tubes and bottles, sterile loops, and spreaders are available. Equipment like an autoclave may also be available in a BSL-1 laboratory.

Laboratory Operation

Organisation is essential in a BSL-1 work environment to ensure safety. Workspaces should be well-organised and properly labelled, with all equipment and supplies kept at designated places. Hand washing stations with soap and disposable towels should be easily accessible and strategically located to reduce contamination hazards. Effective waste management techniques, such as autoclaving biological waste and storing chemical waste in appropriate containers, are required to comply with safety and environmental laws.

BSL-2: Enhanced Laboratory Safety

BSL-2 laboratories handle moderately hazardous substances (risk group 2) for both workers and the environment. These organisms can cause human sickness, which is seldom serious and frequently treatable. The BSL-2 procedures expand on BSL-1 standards, implementing tougher safety precautions to address the extra dangers. The overall layout of a BSL-2 laboratory is presented in Fig. 51.3.

Codes of Practice

The following codes of practice are essential for operation in a BSL-2 laboratory:

Access Control

Only authorised staff are permitted to enter, with stringent access controls and closed doors while not in use. A logbook is commonly kept for security and accountability purposes.

Personnel Training

Employees must get specialised training on how to handle pathogenic agents, use safety equipment, and follow emergency procedures. Regular refresher training is routine for everyone to be updated on safety procedures.

Medical Surveillance Systems

These may involve baseline health evaluations, continuous monitoring, and vaccination as needed to provide quick medical intervention in the event of exposure to infectious pathogens.

Laboratory Design and Construction

BSL-2 laboratories are physically separated from common spaces and ordinary building circulation to reduce the possibility of contamination spread. These laboratories have designated areas with limited access to efficiently control biohazards. Hand hygiene stations with hands-free controls, such as foot pedals or motion sensors, are strategically located near exits to prevent contamination from manual handling. Although HEPA filtering is not required in a BSL-2 laboratory, a directional airflow system is used to transport air

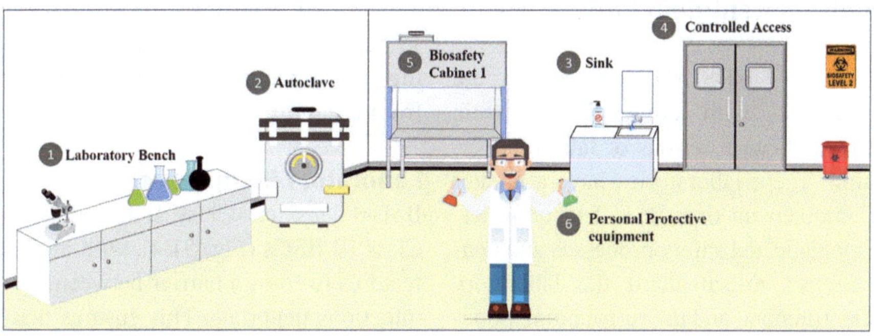

Fig. 51.3 Biosafety level 2 laboratory layout. (Image source: Authors' personal collection)

from clean to possibly contaminated locations, preventing pollutants from moving into non-laboratory places.

Laboratory Equipment

BSL-2 laboratories require specialised equipment and methods to properly control moderately dangerous compounds and pathogens. Biological safety cabinets (BSCs), such as Class II BSCs, are required for activities that produce aerosols or handle high-risk microbes since they provide a controlled environment to protect both the operator and the laboratory area. As shown in Fig. 51.3, autoclaves are required for sterilising trash and equipment, with simple access and sufficient training for all people to guarantee successful decontamination. Centrifuges must have sealed rotor heads or safety cups to contain aerosols. Frequent maintenance is required to prevent mechanical failures that might result in exposure to dangers.

Laboratory Operation

Operating methods are also critical in a BSL-2 facility. Manipulations of infectious agents that may create aerosols are carried out in biosafety cabinets, with techniques designed to reduce aerosol formation. Spill response protocols necessitate the availability of spill kits and qualified workers to manage spills quickly, cleanly, and safely. Regular cleaning of work surfaces and equipment with proper disinfectants is critical for ensuring a safe working environment.

BSL-3: High-Containment Laboratory

Biosafety level 3 (BSL-3) laboratories are intended for work with indigenous or exotic agents that might cause serious or fatal illnesses by inhalation. These laboratories have a greater degree of confinement than BSL-2 laboratories. They use sophisticated safety processes and containment devices to safeguard the laboratory staff, the environment, and the surrounding population. The overall layout of a BSL-3 laboratory is presented in Fig. 51.4.

Codes of Practice

The following codes of practice are essential for operation in a BSL-3 laboratory:

Access Control

A two-door entrance system (interlock or airlock) called an anteroom is necessary to restrict the entry of experienced workers and prevent air escape.

Medical Surveillance and Immunisation

Regular health checks, screenings, and vaccines are required, with quick medical attention accessible.

Personal Protection

To avoid contamination, personnel must wear respiratory protection (N95 masks or Powered Air-Purifying Respirators (PAPRs)), disposable gowns, shoe coverings, and double gloves, in addition to specialised lab gowns.

Laboratory Design and Construction

BSL-3 laboratories are built with strict safety precautions to handle high-risk disease pathogens safely. These facilities are physically segregated from common areas and normal traffic and are usually placed in discrete zones inside the facility. They have sealed windows, doors, walls, floors, and ceilings to provide airtight containment and prevent the spread of illnesses. The laboratory's ventilation system is carefully designed to move air from clean to contaminated areas. This is carried out by high-efficiency particulate air (HEPA) filters cleaning the air before it is exhausted and maintaining a negative pressure differential of 12.5 Pa (0.05 in.w.g.) relative to the surrounding environment to keep airborne contaminants from escaping.

Laboratory Equipment

BSL-3 labs use Class II biosafety cabinets or Class III BSCs (Fig. 51.4), with increased safety features to form a barrier between operators and infectious materials. This ensures that any aerosols created during operations remain within the cabinet. Secure storage is essential for storing

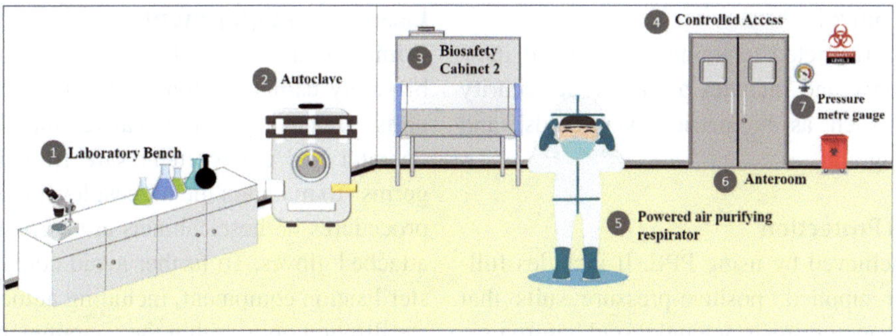

Fig. 51.4 Biosafety level 3 laboratory layout. (Image source: Authors' personal collection)

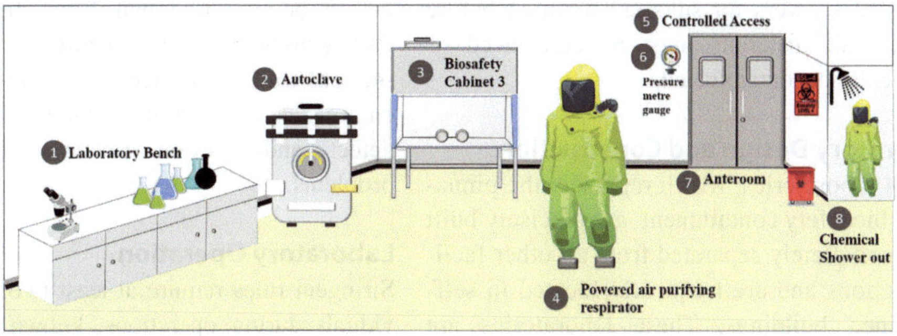

Fig. 51.5 Biosafety level 4 laboratory layout. (Image source: Authors' personal collection)

infectious agents and poisons in sealed, guarded locations with limited access and thorough inventory records within the laboratory. A two-way autoclave is present in BSL-3 laboratories (one face opening inside the laboratory to place biohazardous waste into the autoclave and the other face opening outside to collect sterilised trash).

Laboratory Operation

Operating methods in BSL-3 labs are intended to provide the utmost safety. To reduce the danger of exposure, all infectious agent handling must take place within a biosafety cabinet or other main containment equipment. Sharp items are used sparingly and with utmost caution. Aerosol management is very important owing to the inhalation danger associated with BSL-3 pathogens. Hence, all operations that may create aerosols are subjected to tight controls to limit aerosol generation. In the case of an infectious substance leak, urgent and comprehensive decontamination processes are implemented to guarantee rapid containment and cleaning while preserving a safe working environment.

BSL-4: Maximum Containment Laboratories

Biosafety level 4 (BSL-4) laboratories provide the greatest level of biosafety containment and are intended to work with the most hazardous and exotic diseases that represent a high risk of aerosol-transmitted illnesses and for which no vaccinations or therapies are available. These agents are often life-threatening and need the strictest containment and safety precautions. The overall layout of a BSL-4 laboratory is presented in Fig. 51.5.

Codes of Practice

The following codes of practice are essential for operation in a BSL-4 laboratory:

Access Control
Access is restricted to trained, authorised individuals only and includes a variety of security methods such as biometrics, key cards, and monitoring.

Personal Protection
This is achieved by using PPE. It includes full-body, air-supplied, positive-pressure suits that must be puncture-resistant and tested before use.

Decontamination Procedures
Chemical showers for suits are required before leaving; all materials must be autoclaved or chemically decontaminated.

Laboratory Design and Construction
BSL-4 laboratories, which represent the pinnacle of biosafety containment, are precisely built to be completely separated from all other facility sections and are frequently located in self-contained buildings. These laboratories are designed with airtight seals, which ensure that all penetrations, including ducts and pipes, are completely sealed to avoid hazardous substance leakage. The laboratory environment is kept under a strict negative pressure of 50–200 Pa compared to surrounding spaces, which forces inward airflow and eliminates the danger of microbial escape. This negative-pressure system is supplemented with an innovative air handling system in which exhaust air is rigorously filtered through multiple levels of HEPA filters, ensuring complete containment of infectious particles.

In addition to these containment elements, BSL-4 facilities are outfitted with redundant systems to ensure the continuous functioning of crucial safety measures. Redundant power supply, including backup generators and numerous air filtering units, is in place to reduce the possibility of system failures that may jeopardise containment integrity. These redundant systems are an important part of the laboratory's infrastructure. This is because they provide a fail-safe mechanism that ensures the strict biosafety regulations required for dealing with the deadliest diseases.

Laboratory Equipment
Primary containment devices, such as Class III biosafety cabinets, along with HEPA-ventilated units, create a gas-tight barrier that keeps the operator and environment safe from infectious germs. To maintain the highest level of safety, all procedures in these cabinets are carried out with attached gloves. To further avoid contamination, sterilisation equipment, including autoclaves and sterilisation units within the containment zone, is subjected to rigorous monitoring and testing to verify that all materials are properly disinfected before being withdrawn from the laboratory. Pathogens are housed in extremely secure, sealed sections with limited access, built to endure harsh circumstances. Continuous monitoring is used to detect and prevent potential breaches or problems.

Laboratory Operation
Stringent rules require at least two trained individuals during operations, known as the two-person rule, ensuring a quick emergency reaction. To reduce aerosol exposure concerns, activities that may produce aerosols are carried out within primary containment devices, with supplementary safeguards such as enclosed centrifuge rotors. There are well-defined incident response systems for dealing with exposures, spills, and containment breaches. These include immediate evacuation, decontamination, medical examination, and timely notification to health authorities. Regular training exercises are held to ensure that all workers are well-prepared to deal with situations effectively.

Safe Handling of Hazardous Biological Materials

The handling of dangerous biological agents in laboratory settings necessitates extreme rigour and adherence to painstakingly designed safety protocols. These materials, which include a wide range of infectious agents, powerful poisons, and other biologically active compounds, require an initial and thorough risk assessment to determine their possible manipulation risks. The implementation of safety measures, such as carefully selecting and using PPE, is crucial. The accuracy with which PPE is donned and doffed is critical since any lapse might result in unintentional exposure or contamination.

Containment methods are an important safeguard against the discharge of dangerous chemicals into the environment. The use of biosafety cabinets (BSCs), sealed centrifuge systems, and hermetically sealed workplaces ensures that potentially pathogenic or poisonous chemicals are contained, reducing the risk of laboratory-acquired infections and environmental contamination. More advanced containment solutions, such as glove boxes and negative-pressure laboratories, are required when dealing with high-risk substances. These systems are meant to form an impenetrable barrier to hazardous chemicals, protecting both humans and the surrounding environment from exposure.

In parallel, the regulations controlling the storage and labelling of hazardous biological items are distinguished to ensure security and accuracy. To prevent unauthorised handling, secure, access-controlled storage rooms are required. In addition, each container must be scrupulously labelled with detailed information, such as the nature of its contents, the related danger classification, and expiration dates. Secondary containment solutions, such as the use of hermetically sealed, leak-proof containers, provide an extra layer of protection against inadvertent leaks, preserving the storage environment.

The procedural side of handling these materials includes intensive cleaning of workplaces, equipment, and PPE, using powerful disinfectants selected particularly for the nature of the biological agents involved. Waste management, which is an essential component of biosafety, demands that all by-products of material handling be classified as hazardous or infectious. Before disposal, these wastes must undergo extensive decontamination techniques, such as autoclaving, chemical sterilisation, or incineration, to ensure that any remaining biological activity has been completely neutralised.

Furthermore, efficient handling of hazardous biological materials requires timely education and training of laboratory workers. A comprehensive and continuous training routine is required, including the most recent approaches in safe handling, emergency response, and biohazardous waste management. Concurrently, continuous monitoring systems, including periodical audits, are key tools for maintaining strict adherence to specified biosafety guidelines. These procedures help maintain a high level of safety, safeguarding both the laboratory staff and the larger environment from the potentially disastrous effects of mishandling dangerous biological materials.

Personal Protective Equipment

Personal protective equipment (PPE) protects the laboratory staff from harmful biological contaminants. Its success is dependent on careful selection, vigorous application, and attentive maintenance, which together form the foundation of laboratory biosafety.

Types of Personal Protective Equipment

Chemically resistant gloves are replaced on a frequent basis to prevent breaches and contamination. Lab jackets and gowns protect against spills and splashes and should be made of impermeable, easily decontaminated fabrics. Face shields and goggles are essential to protect the eyes and face from splashes and aerosols while also ensuring their safety and impact resistance (Fig 51.6). Respirators, which are designed to protect against airborne pollutants, are properly installed and maintained. These include both particle respirators and more modern air supply devices for enhanced protection. In high-risk situations, full-body suits are used to give comprehensive coverage.

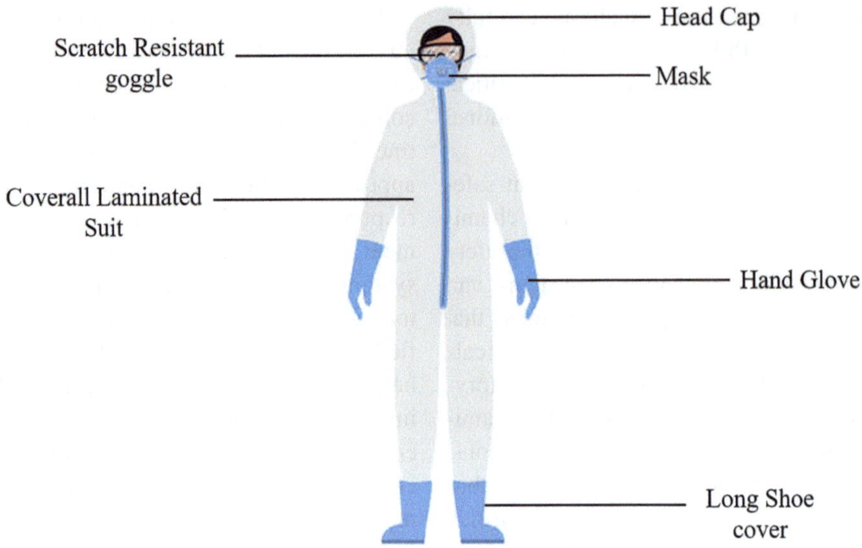

Fig. 51.6 Personal protective equipment, essential for laboratory safety, including gloves, gowns, masks, and face shields, ensuring protection from hazardous materials. (Image source: Authors' personal collection)

Decontamination and Disposal

Before removal, PPE is decontaminated with suitable disinfectants. To remove pollutants, reusable goods are thoroughly cleaned. To avoid environmental contamination, disposable PPE is treated as biohazardous waste and disposed of according to rigorous guidelines.

Biosafety and Biohazardous Waste Management

Effective biosafety measures and biohazardous waste management are critical for ensuring laboratory safety while also safeguarding public health and the environment. This entails establishing strict standards for handling, treating, and disposing of hazardous items because of their infectious, poisonous, or reactive qualities. Robust waste management techniques are critical for avoiding inadvertent exposure, environmental pollution, and the spread of pathogens.

Waste Classification and Segregation

Classification

Biohazardous waste is divided into three categories: biological, chemical, and sharps. Biological waste comprises pathogen-infected materials, such as wasted cultures and contaminated PPE. Chemical waste contains reactive or hazardous chemicals. Sharps include objects that might cause damage, such as needles, blades, and shattered glass.

Segregation

Segregating garbage at its source is crucial for effective waste management. Colour-coded bins are used to ease this segregation and guarantee that garbage is disposed of appropriately based on its kind and related dangers (Table 51.2).

To guarantee correct utilisation by all laboratory staff, the colour-coded trash container system should be properly labelled, and instructions should be provided. Regular training on trash segregation methods and respect for local rules is essential for sustaining a successful waste management system.

Table 51.2 Colour-coded bins for segregation

Container colour	Purpose	Content	Special notes
Red	For infectious waste	Used laboratory cultures, contaminated PPE, and other biologically hazardous materials	This indicates that the substance is and must be handled and disposed of with caution
Yellow	For chemical waste and certain biological waste	Chemical substances that are reactive, poisonous, or harmful; biological waste intended for combustion	Indicates the necessity for specific treatment owing to chemical characteristics
Black	For non-hazardous waste	General laboratory waste that is not contaminated with biohazardous materials or chemicals	Distinguishes non-hazardous waste from hazardous categories
Blue	For recyclable plastic	Clean, recyclable plastic items that have not been contaminated by biohazards	Separating from other garbage promotes recycling and resource-saving

Handling and Storage

To minimise danger, biohazardous waste is treated according to strict guidelines. Personnel wear suitable PPE and follow standardised procedures while transferring trash into approved containers, ensuring minimal aerosol formation and safe confinement. Waste is then kept in safe, labelled containers that avoid leaks or spills. Storage spaces are designed to withstand contamination and have limited access to authorised people only. Regular inspections and maintenance are critical to ensuring that the storage system functions properly.

Treatment

Biohazardous wastes are treated by autoclaving, incineration, or chemical disinfection. Autoclaving sterilises trash using high-pressure steam, and incineration destroys germs by burning garbage at high temperatures. Chemical disinfection uses powerful chemicals to neutralise biological risks.

Disposal

Following treatment, wastes are disposed of in compliance with regulatory requirements. Treated garbage that is no longer a hazard is disposed of using regular waste streams. Some materials may necessitate specialised disposal procedures, depending on residual dangers.

Waste Management

Reducing trash is critical for reducing environmental impact and alleviating the pressure on waste management systems. Laboratories achieve this by optimising procedures and reducing wasteful material use. Reusing laboratory supplies, such as glassware and containers, wherever feasible, saves resources and lowers waste volume. By developing successful recycling procedures, laboratories help promote sustainability and resource conservation. Integrating these strategies promotes a more environmentally conscious approach to trash management.

Effective biosafety measures and biohazardous waste management methods are critical for protecting laboratory staff, the public, and the environment. Laboratories may efficiently manage hazards related to biohazardous materials by following strict waste categorisation, handling, storage, treatment, and disposal processes and using the 3R (Replacement, Reduction, and Refinement) principles.

Accident Management and Reporting

Accident Management

Accident management in biosafety entails an organised method of dealing with occurrences involving hazardous chemicals and infectious materials. Immediate containment and control are essential. The first stage in responding to a spill, leak, or exposure is to secure the area and prevent hazardous or infectious material from spreading. Personnel adhere to the emergency protocols stated in the laboratory's biosafety plan. These generally involve the use of PPE, incident containment with approved means, and cleaning of impacted locations.

Emergency response procedures are crucial in handling laboratory incidents effectively. The initial response involves quickly assessing the situation and implementing emergency protocols to contain the threat. This includes notifying the biosafety officer, isolating the affected area, and utilising spill kits or other emergency response equipment to prevent further contamination. Once the immediate threat is contained, decontamination procedures are carried out meticulously. Contaminated surfaces and equipment are cleaned using appropriate decontamination chemicals, ensuring that all processes are strictly followed to eliminate any remaining pollutants.

Reporting Procedures

Accurate and timely accident reporting is critical for ensuring regulatory compliance and improving safety. Document the occurrence completely, including the nature of the mishap, the materials used, the reaction activities, and the consequences. To guarantee consistency, use a standardised incident report form that includes facts such as the date, time, persons involved, and measures followed.

Case Study

In a diagnostic laboratory, a trainee technician handled a blood sample from a febrile patient for culturing. The sample processing was carried out on an open bench without using the required PPE and a biosafety cabinet. Within a few days, several staff members of the laboratory began to exhibit flu-like symptoms. Investigations confirmed the causative agent to be *Brucella melitensis*. Following serological tests, multiple cases of brucellosis among the laboratory staff were confirmed, indicating a laboratory-acquired infection outbreak.

After controlling the incident and appropriate treatment of the infected staff members, the management of the laboratory undertook response measures like a mandatory refresher training on handling of probable infectious samples in the appropriate biosafety level laboratory and exposure response for all trainees and stakeholders and updated biosafety protocols.

1. What specific failures in biosafety procedures led to the laboratory exposure to *Brucella melitensis*, and what measures could have prevented them?
2. Determine the ways in which training can be enhanced and laboratory infrastructure upgraded to minimise the risk of zoonotic pathogen exposure in laboratories?
3. What steps do clinical and laboratory teams follow to promptly identify and respond to suspected cases of laboratory-acquired brucellosis?

Research Questions

1. How can laboratory biosafety processes be improved to avoid unintentional exposure to high-risk pathogens?
2. How can laboratories improve the design and performance of biosafety cabinets to increase containment and user safety?
3. What are the best waste management techniques in laboratories working with biohazardous materials?

4. How can real-time monitoring systems be used in laboratory settings to improve biosafety and accident prevention?

Declaration During the preparation of this chapter, AI tools such as QuillBot and ChatGPT (OpenAI) were used to improve the grammar and readability of the text. The chapter was reviewed, and the content was edited as needed; therefore, the authors take full responsibility for the content of the publication.

Further Readings

Beeckman DSA, Rüdelsheim P. Biosafety and biosecurity in containment: a regulatory overview. Front Bioeng Biotechnol. 2020;8:650.

BioRRAP. Regulations and guidelines on recombinant DNA research and biocontainment. Department of Biotechnology, Government of India; 2017.

Brewster DT, et al. Safety and perceptions of risk in the handling of laboratory chemicals in a biological research community. ACS Chem Health Saf. 2023;30(2):63–71.

Burnett LC, Lunn G, Coico R. Biosafety: guidelines for working with pathogenic and infectious microorganisms. Curr Protoc Microbiol. 2009;13(1):1A.

Division on Engineering, et al. Biosafety in the laboratory: prudent practices for handling and disposal of infectious materials. Washington, DC: National Academies Press; 1989.

Eggers B, Mackenzie R. The Cartagena protocol on biosafety. J Int Econ Law. 2000;3(3):525–43.

Hou M, et al. Quality management in a high-containment laboratory. J Biosaf Biosecur. 2019;1(1):34–8.

Janosko K, et al. Safety precautions and operating procedures in an (A) BSL-4 laboratory: 1. Biosafety level 4 suit laboratory suite entry and exit procedures. J Vis Exp. 2016;(116):52317.

Li N, et al. Biosafety laboratory risk assessment. J Biosaf Biosecur. 2019;1(2):90–2.

Meechan PJ, Potts J. Biosafety in microbiological and biomedical laboratories. National Institutes of Health; 2020.

Mourya DT, et al. Establishment of biosafety level-3 (BSL-3) laboratory: important criteria to consider while designing, constructing, commissioning & operating the facility in Indian setting. Indian J Med Res. 2014;140(2):171–83.

Tang Q, et al. Enhancing laboratory biosafety management: a comprehensive strategy from theory to practice. Front Public Health. 2024;12:1439051.

WHO. Laboratory biosafety manual. 4th ed. WHO; 2020.

Zaki AN. Biosafety and biosecurity measures: management of biosafety level 3 facilities. Int J Antimicrob Agents. 2010;36(Suppl 1):S70–4.

Geoclimatic Changes and Bacterial Zoonotic Diseases

52

Subhash Chandra Parija and Abhijit Poddar

Learning Objectives

1. To discuss the impact of geoclimatic changes on the distribution and transmission dynamics of bacterial zoonoses
2. To identify the direct and indirect effects of climate change on biodiversity and human health
3. To describe the impact of geoclimatic changes on plague, leptospirosis, and Lyme disease
4. To discuss the need for diagnostics and improved public health strategies for disease management

Introduction

Geoclimatic changes are characterised by an increase in global temperature and a rise in sea level. While they directly impact the environmental health, they have a cascading effect on zoonotic diseases. The relationship between geoclimatic changes and bacterial zoonoses is multifaceted and results in a shift in ecosystem services and functions. The distribution, behaviour, and survival of bacteria and their vectors are changing and reshaping the epidemiology of these diseases. For example, zoonoses previously not reported from a region are being reported, showing emerging hotspots. The disease transmission dynamics is altering, which is making management difficult. More concerning is the rise of antimicrobial resistance, where these zoonotic bacteria are gaining resistance against known antimicrobials and resulting in a large--scale public health risk.

Historically, some bacterial zoonoses, such as plague (*Yersinia pestis*), leptospirosis (*Leptospira* spp.), and Lyme disease (*Borrelia burgdorferi*), had a defined geographical distribution. For example, plague was common in Eurasia. Leptospirosis was common in the tropical and subtropical regions of the world. Lyme disease was primarily found in Europe, North America, and Eastern Asian areas. However, global warming has a noticeable effect on their distribution. This is catalysed by climate-influenced movement of arthropod vectors that are carrying the diseases across the corners of the globe. Parallelly, the chances of humans getting the infections are also rising because of habitat destruction and agricultural intensification. At the same time, floods, droughts, and hurricanes are acting as potential disruptors of ecosystems. All of this is promoting bacterial spillover.

As such, there is an urgent need to improve the understanding of the intricate relationship

S. C. Parija (✉)
Formerly at Sri Balaji Vidyapeeth (Deemed to be University), Pondicherry, India

A. Poddar
Mahatma Gandhi Medical Advanced Research Institute, Sri Balaji Vidyapeeth (Deemed to be University), Pondicherry, India

S. C. Parija, A. Poddar (eds.), *Textbook of Bacterial Zoonoses*, Microbial Zoonoses,
https://doi.org/10.1007/978-981-95-6615-0_52

between geoclimatic changes and bacterial zoonoses. This is essential for predicting outbreaks and developing adaptive strategies to protect global health in a rapidly changing world.

Overview

Geoclimatic change refers to significant alterations in the Earth's climate and geographic features over time. These are the results of natural and human-driven activities. Historically, even before human civilisations, geoclimatic changes had driven cycles of warming and cooling periods, such as the ice ages and interglacial periods.

However, modern geoclimatic change is mostly associated with unregulated anthropogenic activities that include but are not limited to fossil fuel combustion, deforestation, and industrial production. These are leading towards an increase of greenhouse gases (GHGs), which, after being trapped in the environment, prevent heat from escaping. The immediate and observable effects of geoclimatic changes can be categorised as direct and indirect impacts (Table 52.1).

Direct Impact

The direct impact of geoclimatic change is the rise in global temperature, a phenomenon popularly known as "*global warming*." This is characterised by the increase in Earth's average surface temperature due to higher greenhouse gas (GHG) emissions. Human activities such as industrialisation, deforestation, and the burning of fossil fuels play a major role in GHG emissions. GHGs trap heat in the atmosphere, intensifying the greenhouse effect. Therefore, reports suggest that the Earth's average surface temperature has risen by approximately 2 °F (1 °C) since the late nineteenth century. The past 40 years have been alarming, as the rise has become very much prominent.

This global warming also promotes the melting of polar ice caps and glaciers. With these melting, sea levels rise. This has a profound impact on coastal regions as they become more susceptible to flooding and erosion. Likewise, ecosystems such as coral reefs suffer from bleaching due to warmer ocean temperatures.

Indirect Impact

The direct impacts of climate change create a cascade of indirect effects that negatively impact global biodiversity. Many species struggle to adapt to rapidly changing habitats and face the risk of extinction. As ecosystems shift, some species may thrive, while others decline rapidly. This alters the food webs and ecosystem services.

These interconnected impacts influence human health and well-being in different ways (Fig. 52.1). Rising global temperatures lead to more frequent and intense heatwaves, resulting in increased cases of heatstroke, dehydration, and cardiovascular stress, particularly among vulnerable populations like the elderly and those with pre-existing health conditions. Changes in air quality, driven by increased wildfires, dust storms, and the production of ground-level ozone, intensify respiratory conditions like asthma and chronic obstructive pulmonary disease (COPD).

Shifting precipitation patterns and increased extreme weather events also disrupt agricultural production. It results in shortage of food and leads to malnutrition. The stress and trauma associated with extreme weather events, displacement, and the loss of livelihoods contribute to anxiety, depression, and post-traumatic stress disorder (PTSD).

Finally, climate change-induced sea level rise, desertification, and extreme weather events force populations to migrate, leading to increased vulnerability to disease, malnutrition, and social instability.

Geoclimatic Change and Zoonoses

The impacts of geoclimatic change on human health are many. However, among all, the risk of zoonotic disease transmission is very significant. The unprecedented changes in temperature and precipitation patterns are triggering a change in the distribution and population dynamics of

Table 52.1 Direct and indirect impact of geoclimatic changes

Impact type	Category	Impact	Consequences	Examples
Direct	Temperature changes	Global warming	Melting of polar ice caps and glaciers, increased ocean temperatures	Arctic Sea ice loss: The dramatic reduction in Arctic Sea ice extent observed over the past decades
	Sea level rise	Increased coastal flooding and erosion	Inundation of coastal cities, loss of coastal habitats	Miami, Florida: Increased frequency of "sunny day flooding" due to rising sea levels
	Weather pattern disruption	Increased frequency and intensity of extreme weather events	Hurricanes, droughts, heavy rainfall, heatwaves	Australian bush fires: The intense and prolonged bush fire seasons in Australia, exacerbated by prolonged drought and heatwaves
	Ecosystem impacts	Coral reef bleaching, habitat shifts, biodiversity loss	Decline of coral ecosystems, species migration or extinction, changes in food webs	Great barrier reef: Extensive coral bleaching events due to warmer ocean temperatures
	Geographical impacts	Changes to earth's geographic features	Changes to shorelines due to rising sea levels, changes in land use due to changing temperatures and rainfall patterns	Coastal erosion in Bangladesh: Increased erosion and land loss in the Ganges Delta due to rising sea levels and increased storm surges
Indirect	Biodiversity impacts	Species struggle to adapt, changes in food webs	Species extinction; some species thrive, while others decline; changes in ecosystem services like pollination and clean water	Amphibian decline: The rapid decline of amphibian populations worldwide, linked to habitat loss and climate-related diseases
	Zoonotic disease	Increased risk of disease transmission, expansion of vector-borne diseases	Changes in vector distribution, increased human–animal interactions, altered pathogen survival	Spread of Lyme disease: The expansion of tick populations due to warmer temperatures, leading to increased cases of Lyme disease in new regions. Also, increased spread of West Nile virus due to mosquito population increases
	Human health impacts	Increased diseases and heat-related illnesses, respiratory problems, food and water insecurity, mental health impacts, displacement and migration	Increased heat stroke, increased migration, etc.	Migration due to desertification in the Sahel region of Africa Increased cases of asthma due to wildfire smoke in the Western United States

disease-carrying vectors like mosquitoes and ticks. It is increasing the potential for disease spread.

At the same time, human–animal interactions are increasing. This is being triggered by rapid and consistent habitat disruption and changes in land use. As a result, the pathogens present in the wild are getting more chances to be in contact with humans. This is facilitating their transfer in a phenomenon called spillover. This is particularly concerning in regions where wildlife habitats are encroached for human civilisation.

Climate change also influences the life cycle of pathogens within these vectors. For example,

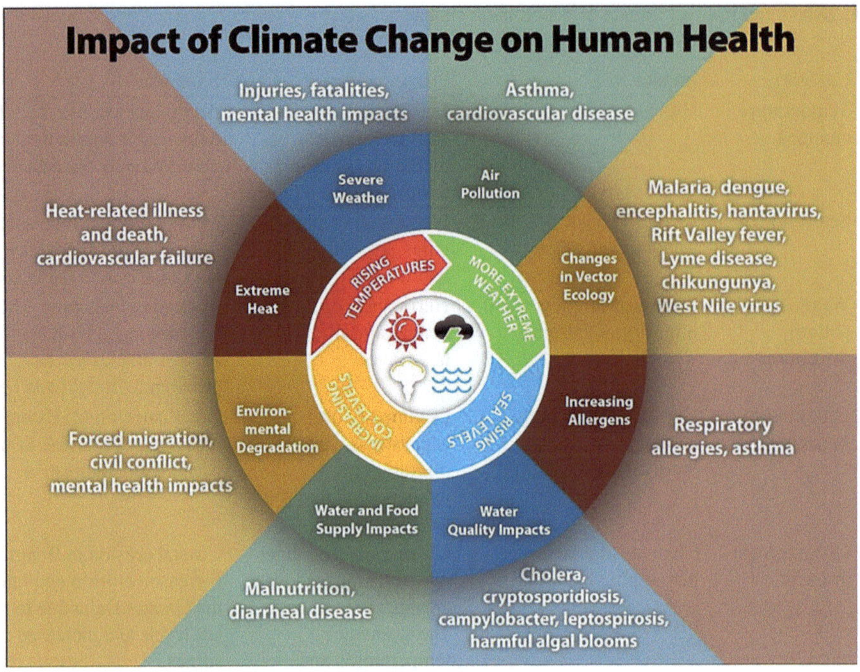

Fig. 52.1 Impact of geoclimatic change on human health. (Image source: US CDC)

warmer temperatures can shorten the *extrinsic incubation* period (i.e. the time it takes for a pathogen to develop within a vector), thus resulting in faster transmission cycles. Moreover, shifts in humidity and rainfall are affecting vector breeding sites. It is increasing their populations and thereby the risk of disease outbreaks.

The factors for the emerging zoonoses are the altered survival and reproduction rates of pathogens. Rising temperatures have made previously inhospitable areas hospitable for the pathogens. For example, rising temperatures are linked to the increased prevalence of bacterial pathogens such as *Vibrio* spp., which now thrive in warmer sea surface temperatures. These bacteria, responsible for diseases like cholera, have been observed to expand their range into previously cooler regions, such as Northern Europe. Similarly, the spread of foodborne pathogens is also influenced by climate change, as warmer temperatures enhance the persistence and dispersal of bacteria like *Salmonella* and *Escherichia coli* in the environment.

In addition, climate change also poses stress on animal populations. It weakens their immune systems, making them more susceptible to infections. It increases the likelihood of pathogen shedding (i.e. the release of infectious agents from a host) while amplifying the risks of zoonotic transmissions. However, the phenomenon of pathogen shedding exhibits considerable variation depending on the pathogen, host species, physiological state, and environmental factors. For instance, a study conducted on *Coxiella burnetii* exposed that no detectable pathogen shedding was found in sheep in Idaho. This absence of shedding suggests that factors such as natural immunity developed over time or that other biological mechanisms might play a role in reducing the risk of transmission in this population. In contrast, research focusing on dairy goats revealed a significant pattern of *C. burnetii* shedding in various body fluids, including vaginal mucus, faeces, and milk. Notably, this shedding was particularly pronounced following abortion events in goats. Beyond ruminants, wild rodents in agricultural landscapes can shed zoonotic parasites like *Cryptosporidium* and *Giardia*, as well as bacteria such as *Salmonella*, with prevalence influenced by ecological conditions and rodent biodiversity. Domestic animals like dogs, especially those in communal settings such as kennels and dog parks, are recognised shedders of

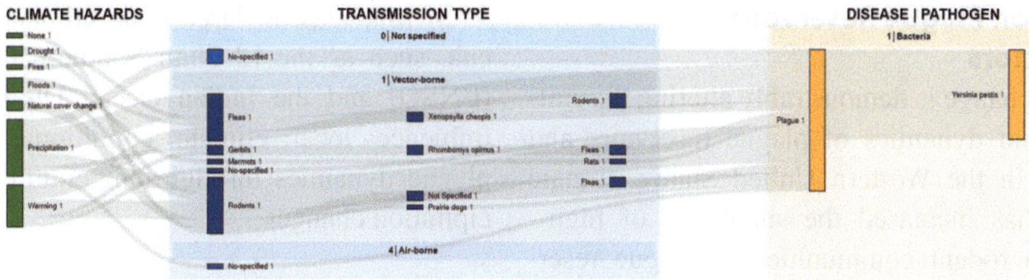

Fig. 52.2 Relationship between climate impact and plague infection. (Image source: The image was created using the web tool available at https://camilo-mora.github.io/Diseases/. Free to use. Article reference: Mora C, McKenzie T, Gaw IM, Dean JM, von Hammerstein H, Knudson TA, Setter RO, Smith CZ, Webster KM, Patz JA, Franklin EC (2022) Over half of known human pathogenic diseases can be aggravated by climate change. Nature Climate Change. https://doi.org/10.1038/s41558-022-01426-1)

Campylobacter spp. Furthermore, petting zoos have been identified as environments where antibiotic-resistant bacteria, including (Extended-Spectrum βbeta β-lactamases) ESBL/AmpC-producing Enterobacterales, can be shed by animals, posing a risk of zoonotic transmission due to close human–animal interaction and underscoring the importance of robust infection control practices.

Impact on Bacterial Zoonoses

Bacterial zoonoses are significantly influenced by geoclimatic changes. Warmer temperatures can extend the survival and replication rates of certain bacteria in the environment, increasing the risk of infection. Changes in precipitation patterns can lead to increased waterborne bacterial infections, particularly in regions experiencing flooding or drought. Additionally, the disruption of ecosystems also alters the distribution of animal reservoirs for bacterial pathogens, bringing them into closer contact with human populations. Some specific examples of climate-induced impacts of bacterial zoonoses are described here.

Plague

Climatic changes significantly influence plague caused by *Y. pestis* epidemiology by altering ecological and environmental factors (Fig. 52.2). These changes affect the distribution and behaviour of the pathogen's wildlife reservoirs and vectors, consequently increasing the risk of human outbreaks. The interplay between climate variables and plague dynamics is intricate, involving direct effects on the pathogen and its hosts, as well as indirect effects mediated through ecological and economic systems.

Historical correlations between climate shifts and plague pandemics, such as the Justinian Plague and the Black Death, stress the potential for ongoing climate change to sustain or escalate plague activity in vulnerable regions. Palaeoclimatic data from Central Asia over the past millennium further indicate that climate fluctuations have significantly shaped plague dynamics, with implications for future activity under global warming.

While direct climatic effects on plague are clear, broader ecological and socio-economic contexts are also critical transmission drivers. Climate change modifies the habitats and behaviours of plague reservoirs and vectors. Moreover, it impacts human societies, potentially worsening conditions that favour outbreaks. For instance, historical analyses in Europe suggest that climate-driven economic changes, rather than direct climatic effects, were influential in shaping plague dynamics. Fluctuations in temperature and precipitation affected economic systems, subsequently impacting the frequency of plague outbreaks.

Impact on Plague Reservoirs and Vectors

Climate change is demonstrably altering the suitability and dynamics of plague reservoirs and vectors. In the Western United States, climate change has increased the suitability of high--elevation rodent communities as plague reservoirs by up to 40%, heightening the risk of spillover to humans at mid-elevations. Likewise, in Central Asia, warmer springs and wetter summers have been linked to an increased prevalence of *Y. pestis* in great gerbils, a key reservoir species. Research indicates that a 1 °C increase in spring temperature can lead to a more than 50% increase in plague prevalence. Furthermore, the aridification of Central Asia and subsequent climatic shifts have historically contributed to the evolution of *Y. pestis* from its ancestral form, *Yersinia pseudotuberculosis* through changes in the behaviour of host species, such as the Tarbagan marmot and its fleas.

Seasonal and Regional Climate Effects

In Brazil, plague hotspots are associated with specific precipitation and vegetation patterns, with human cases correlating with seasonal rainfall and the positivity rates of rodents and fleas for *Y. pestis*. Historical data from British India show that moderate humidity levels (60%–80%) strongly correlate with plague outbreaks, with changes in humidity affecting the timing of these outbreaks.

In Madagascar, large-scale climate phenomena, such as the El Niño Southern Oscillation (ENSO) and the Indian Ocean Dipole (IOD), influence local climate conditions, affecting plague dynamics through temperature and precipitation changes.

Leptospirosis

Climate change has a profound impact on the incidence and distribution of leptospirosis, a zoonotic disease caused by *Leptospira* spp. (Fig. 52.3). These changes influence the environmental conditions that favour the survival and transmission of the bacteria, thereby affecting human and animal health. Climate change is predicted to alter the geographical distribution of leptospirosis-risk areas. For example, in Western Java, Indonesia, models indicate a shift in leptospirosis occurrence towards the northern and central parts of the islands, driven by changes in temperature and precipitation patterns. Similarly, in Argentina, regions with climatic conditions conducive to *Leptospira* survival, such as high humidity and frequent rainfall, are at an increased risk of outbreaks. Statistical models incorporating hydroclimatic variables have been used to predict leptospirosis cases, highlighting the importance of climate in disease forecasting.

While the impact of geoclimatic changes on leptospirosis is evident, it is important to con-

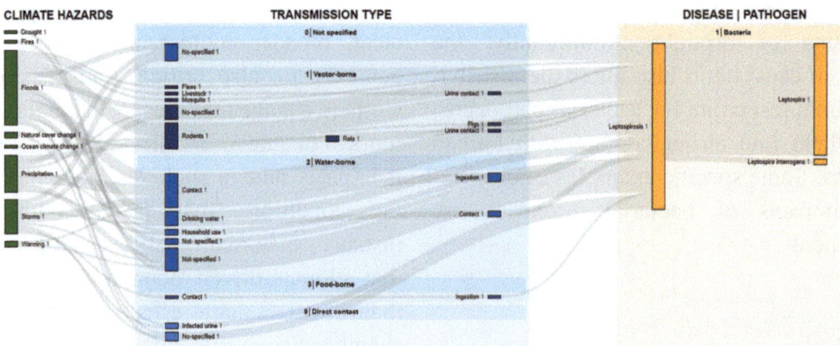

Fig. 52.3 Relationship between climate impact and leptospirosis infection. (Image source: The image was created using the web tool available at https://camilo-mora.github.io/Diseases/. Free to use. Article reference: Mora C, McKenzie T, Gaw IM, Dean JM, von Hammerstein H, Knudson TA, Setter RO, Smith CZ, Webster KM, Patz JA, Franklin EC (2022) Over half of known human pathogenic diseases can be aggravated by climate change. Nature Climate Change. https://doi.org/10.1038/s41558-022-01426-1)

sider the broader context of these changes. Climate change not only affects the environmental conditions conducive to *Leptospira* survival but also influences human and animal behaviours, which can either mitigate or exacerbate disease transmission.

Impact on Reservoirs and Vectors

Rodents are the most important reservoirs of leptospirosis, and their population dynamics are closely tied to climatic conditions. In New Caledonia, rodent abundance and *Leptospira* carriage were higher during hot and rainy periods, which also corresponded to increased human exposure to contaminated environments. In Indonesia, the distribution of leptospirosis-confirmed rats was significantly correlated with ecosystem variables such as trap location, altitude, and temperature, although not directly with weather variables like precipitation.

Wetlands provide a conducive environment for the survival and transmission of *Leptospira*, as they are home to many animal species that can act as reservoirs. The presence of *Leptospira* spp. in wetlands is primarily determined by water availability, which is influenced by precipitation and hydrometeorological conditions. This highlights the importance of understanding environmental factors in predicting leptospirosis outbreaks in areas with intensive livestock farming.

Seasonal and Regional Climate Effects

High temperatures and heavy rainfall, exacerbated by climate change, are significant drivers of leptospirosis outbreaks. In Greece, post-flooding events following Storm Daniel led to a notable increase in leptospirosis cases, emphasising the role of floodwaters in disease transmission. The El Niño Southern Oscillation (ENSO) significantly influences leptospirosis dynamics in Colombia, with increased cases during La Niña and decreased cases during El Niño, demonstrating the impact of rainfall variability on disease incidence.

Lyme Disease

The impact of geoclimatic change on Lyme disease caused by *B. burgdorferi* is multifaceted (Fig. 52.4). It involves shifts in tick distribution, changes in disease incidence, and alterations in the timing of disease onset. Climate change significantly influences the ecology of Lyme disease vectors, primarily *Ixodes* ticks, and consequently affects the transmission dynamics of the disease. This complex interaction between climate and Lyme disease is evident in various regions, with notable differences in how these changes manifest geographically.

While climate change is a significant driver, other factors such as land use changes and human activities also play crucial roles in the spread of Lyme disease. These factors can complicate the direct attribution of changes in disease patterns solely to climate change.

Impact on Reservoirs and Vectors

The changing climate affects ticks' life cycles, geographic distribution, and the prevalence of

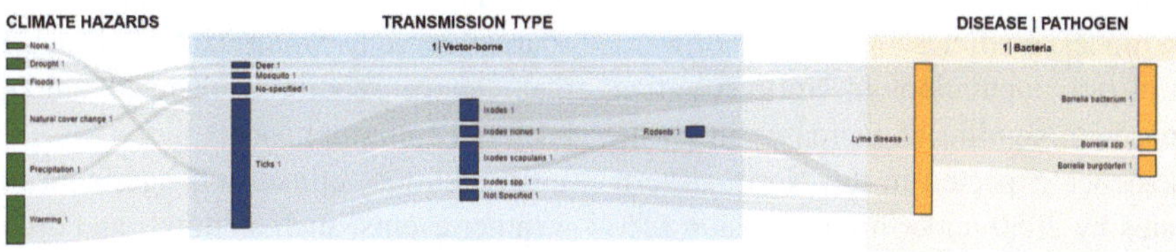

Fig. 52.4 Relationship between climate impact and Lyme disease. (Image source: The image was created using the web tool available at https://camilo-mora.github.io/Diseases/. Free to use. Article reference: Mora C, McKenzie T, Gaw IM, Dean JM, von Hammerstein H, Knudson TA, Setter RO, Smith CZ, Webster KM, Patz JA, Franklin EC (2022) Over half of known human pathogenic diseases can be aggravated by climate change. Nature Climate Change. https://doi.org/10.1038/s41558-022-01426-1)

Lyme disease. Warmer temperatures and altered precipitation patterns have been linked to increased tick activity, expanded geographic ranges, and higher disease incidence. Warmer temperatures have been shown to increase the reproductive capacity of ticks, extend their active seasons, and promote their questing behaviour, which is the process of seeking hosts for feeding. At the same time, rising temperatures and changes in humidity have been linked to increased tick survival and activity, which, in turn, elevate the incidence of Lyme disease. For example, a 2 °C increase in the annual average temperature could lead to a 20% rise in Lyme disease cases in the United States.

Climate change also results in shifts in the distribution of *Ixodes* ticks. In Europe, for instance, suitable habitats for *Ixodes ricinus* are projected to increase in regions like Latvia, while decreasing in Ukraine, reflecting broader patterns of vector redistribution driven by global warming. In North America, warming temperatures have facilitated the northward expansion of *Ixodes scapularis*, the primary vector in the United States, increasing the risk of Lyme disease in previously unaffected areas.

Seasonal and Regional Climate Effects

Climate change affects the seasonality of Lyme disease. In the United States, the onset of Lyme disease cases is projected to occur earlier in the year due to warmer winter and spring temperatures, with the most significant shifts expected in the mid-Atlantic states. Similarly, in Northern Europe, the peak of Lyme borreliosis cases has shifted earlier by 6 weeks over the past 25 years, indicating a significant change in the disease system's seasonal dynamics.

Climate models predict a continued northward expansion of tick populations in North America and Europe, with significant increases in Lyme disease incidence expected in the North-eastern United States by 2050 under high-emission scenarios. Similarly, in Europe, climate models suggest that Lyme disease will spread to higher latitudes and altitudes, with some areas becoming unsuitable for ticks due to excessive heat and dryness.

Climate Change and Antimicrobial Resistance

The convergence of climate change and antimicrobial resistance (AMR) in zoonotic bacteria constitutes an escalating and formidable threat to global public health security. Climate change acts as a critical multiplier of existing health vulnerabilities, and its influence on AMR within bacteria capable of zoonotic transmission is particularly alarming. Rising global temperatures result in the selective pressures on bacterial physiology, potentially accelerating mutation rates and enhancing the horizontal transfer of antibiotic resistance genes. For instance, studies have indicated a correlation between elevated temperatures and increased resistance in clinically significant pathogens such as *Escherichia coli*, *Klebsiella pneumoniae*, and *Staphylococcus aureus*, with resistance rates reportedly increasing by 4.2%, 2.2%, and 2.7%, respectively, for each 10 °C rise in ambient temperature.

Furthermore, the bacterial heat shock response, triggered by thermal stress, can upregulate mechanisms that facilitate horizontal gene transfer, thereby promoting the dissemination of resistance determinants across bacterial populations. Climate change-induced environmental stressors, including more frequent and intense freeze–thaw cycles, have also been demonstrated to promote horizontal gene transfer. This leads to the enrichment of antibiotic resistance genes (ARGs) within soil ecosystems, as evidenced by the increased abundance of ARGs following such treatments. Compounding this issue, climate change-driven water scarcity and pollution further contribute to the propagation of resistant bacteria. This is because contaminated water sources serve as significant reservoirs for ARGs that can be readily exchanged among diverse bacterial communities.

In addition, climate change-induced extreme weather events, such as floods and droughts, can disrupt sanitation systems and increase environmental contamination with both antibiotics and resistant bacteria from human and animal waste. Moreover, changes in agricultural practices in response to climate change (e.g., increased use of antimicrobials in livestock to prevent disease under stressful conditions) have been reported to exacerbate the problem of AMR in zoonotic bacteria.

Diagnostics

In the evolving landscape of geoclimatic changes, diagnostics are pivotal in preventing bacterial zoonoses influenced by these environmental shifts. Robust and rapid diagnostic tools are essential for the early detection of pathogens in animal reservoirs, vectors, and the environment. They enable the timely implementation of targeted interventions to interrupt transmission pathways before they reach human populations. Given the often non-specific nature of symptoms associated with many bacterial zoonotic infections, accurate and early diagnosis is paramount to avoid misdiagnosis and ensure prompt intervention. This is especially true for resource-poor settings where misdiagnosis leads to poor patient outcomes, as seen with diseases like leptospirosis and brucellosis, which are often mistaken for other febrile illnesses.

Enhanced surveillance, facilitated by advanced diagnostics, identifies emerging bacterial strains or the expansion of known pathogens into new geographical areas, impacted by climate change. Furthermore, diagnostics are crucial for understanding the prevalence and dynamics of antimicrobial resistance, which can be exacerbated by environmental stressors and impact the effectiveness of treatments for zoonotic infections. By providing critical data on pathogen presence, distribution, and characteristics, diagnostics underpin risk assessments, inform public health strategies, and guide the development of preventive measures. The latter includes targeted vaccination campaigns in animal populations or vector control programmes in areas rendered newly susceptible due to geoclimatic changes.

Molecular diagnostic techniques have revolutionised the detection of zoonotic pathogens. Polymerase chain reaction (PCR) and next-generation sequencing (NGS) are gaining prominence due to their high sensitivity and specificity. These methods are particularly useful for identifying fastidious or difficult-to-culture organisms, such as those belonging to the genera *Anaplasma*, *Bartonella*, and *Coxiella*. NGS has demonstrated significant potential in the accurate typing of complex bacterial groups like *Brucella* species, which pose challenges for traditional identification methods. The detailed genomic analysis afforded by NGS enhances diagnostic precision. It provides crucial insights for epidemiological tracking, enabling a better understanding of pathogen transmission dynamics and facilitating targeted control measures.

Imaging techniques such as thoracic, abdominal, and musculoskeletal imaging are valuable tools in the diagnostic armamentarium for bacterial zoonotic infections. Radiologists with expertise in recognising specific imaging patterns associated with these infections play a vital role in differentiating them from other clinical conditions. This thereby contributes to more accurate and timely diagnoses, which, in turn, guide appropriate treatment strategies. Furthermore, the utility of clinical diagnostics is significantly enhanced by a thorough understanding of the epidemiological context. This includes a patient's exposure history, occupational risks, and potential contact with animals or contaminated environments. Integrating this contextual information with clinical findings and diagnostic test results is crucial for guiding the diagnostic process effectively and ultimately improving patient outcomes in cases of suspected bacterial zoonoses.

Public Health Adaptations and Mitigation Strategies

The growing threat of climate-driven bacterial zoonoses demands a multifaceted approach combining One Health strategies, policy interventions, and community engagement.

One Health Strategies

One Health is a collaborative, multisectoral, and transdisciplinary approach that recognises the interconnectedness of the health of humans, animals (domestic and wild), plants, and their shared environment. It emphasises that the health of each of these domains is inextricably linked and that sustainable solutions to complex health challenges require a unified effort across various disciplines and sectors.

The One Health approach is crucial and multifaceted in addressing bacterial zoonoses. Fostering collaboration among human health professionals, veterinarians, environmental scientists, and other relevant disciplines benefits One Health in many ways: (i) One Health enables a more comprehensive understanding of the complex ecological dynamics that drive the emergence and spread of these pathogens; (ii) the integrated surveillance systems as part of One Health facilitate the early detection of bacterial zoonotic outbreaks, allowing for timely and coordinated interventions to prevent further transmission; and, furthermore, (iii) a One Health perspective is vital in understanding and mitigating the drivers of bacterial zoonoses, such as agricultural practices, wildlife interactions, antimicrobial resistance development in both human and animal populations, and environmental contamination.

One Health Approach for India

India is actively implementing the One Health approach through a multidisciplinary strategy, spearheaded by the National One Health Mission, operated under the Office of the Principal Scientific Adviser to the Government of India. The mission aims to establish an integrated disease control and pandemic preparedness system across human, animal, and environmental sectors while bringing all relevant stakeholders into one platform. This involves (i) robust inter-ministerial collaboration, particularly between health, animal husbandry, and environment ministries, alongside (ii) the establishment of dedicated institutions like the Centre for One Health at the National Centre for Disease Control (NCDC) and (iii) the implementation of national programmes such as the National One Health Programme for Prevention and Control of Zoonoses (NOHP-PCZ).

Key initiatives include (i) developing integrated disease surveillance systems, (ii) strengthening laboratory networks for high-risk pathogens, (iii) establishing joint outbreak investigation mechanisms, (iv) promoting targeted research and development, and (v) building human resource capacity in One Health principles. Furthermore, India is (i) undertaking specific projects focusing on zoonotic disease surveillance, (ii) implementing a One Health-integrated National Action Plan on Antimicrobial Resistance, and (iii) leveraging digital solutions like "Zoo-WIN" for improved data collection and response. In parallel, the Department of Biotechnology's "One Health" consortium and pilot projects in various states are focused on bringing about innovations in diagnostics and response mechanisms, demonstrating a comprehensive effort to address health challenges at the human–animal–environment interface.

Healthcare Capacity Building

Healthcare capacity building is a critical component of public health adaptations and mitigation strategies in the face of emerging and re-emerging bacterial zoonoses, particularly those influenced by geoclimatic changes. It has several components that involve (i) investing in human resources through training and education on zoonotic diseases, One Health principles, and (ii) climate change impacts on health. It also necessitates strengthening laboratory infrastructure and diagnostic capabilities, ensuring access to advanced molecular and imaging techniques for accurate and timely identification of bacterial pathogens.

Furthermore, it includes enhancing surveillance systems by integrating data across human, animal, and environmental health sectors and developing robust early warning systems to predict and prepare for potential outbreaks. Ultimately, healthcare capacity building aims to create a well-prepared and agile public health workforce and system. These can effectively address the complex challenges posed by bacterial zoonoses in a changing world, protecting both human and animal health.

Policy Interventions

Policy interventions are essential to address both direct and indirect climate impacts on disease transmission. In this, international frameworks, including the WHO-Office International des Epizooties (OIE)- United Nations Environment Proramme (UNEP) tripartite collaboration, are notable for standardising border surveillance and response protocols. At the national level, deforestation regulations and livestock vaccination programmes can mitigate spillover risks, particularly in agricultural frontiers where land use change drives human–wildlife contact. Economic measures, such as subsidies for climate-smart farming and outbreak insurance schemes, can also help reduce vulnerabilities by discouraging high-risk practices that amplify transmission.

Community Engagement

Community engagement is vital for sustainable adaptation and the reduction of impacts of zoonoses. Local education campaigns improve recognising of zoonotic symptoms and promote preventive measures, such as tick avoidance in Lyme-endemic zones. Participatory surveillance programmes empower farmers and indigenous communities to report animal illnesses, bridging gaps in formal health systems. When combined with modern science, traditional knowledge offers innovative solutions. For example, native plants can be used for natural vector control.

Future Direction

Climate change is becoming irreversible due to the continuing carbon footprint due to anthropogenic activities. Consequently, global adaptations to pathogens, including zoonotic bacteria, will be inevitable. Therefore, to move forwards, a concerted global effort must prioritise the scaling up of One Health platforms to foster seamless collaboration and information sharing across human, animal, and environmental health sectors. This expansion should involve establishing robust national and international networks, promoting joint research initiatives, and developing standardised protocols for surveillance, outbreak investigation, and risk assessment.

Crucially, prioritising equity in intervention strategies is of paramount importance because it ensures that vulnerable populations, often disproportionately affected by zoonotic diseases and climate change impacts, have equitable access to diagnostic tools, preventive measures, and healthcare services. This requires culturally sensitive and community-engaged approaches that address social determinants of health and empower local communities in disease prevention and control efforts.

Furthermore, significant investment in next-generation vaccines and therapeutics is essential to develop effective countermeasures against emerging and re-emerging bacterial zoonoses. This includes supporting research into (i) novel vaccine platforms, (ii) broad-spectrum antimicrobials, and (iii) innovative diagnostic technologies that rapidly and accurately identify pathogens.

Integrating cutting-edge scientific innovation with informed policy action is crucial for translating research findings into practical, scalable interventions. This necessitates strong political will, supportive regulatory frameworks, and dedicated funding mechanisms. Finally, fostering grassroot participation and public awareness is vital for building community resilience and promoting responsible behaviours that reduce the risk of zoonotic disease transmission.

Fostering robust collaboration between clinical practitioners and basic science researchers is paramount for driving innovation in preventing and managing bacterial zoonoses. Clinicians possess invaluable first-hand knowledge of disease presentation, patient responses, and the practical challenges encountered in diagnosis and treatment within real-world settings. This clinical insight effectively guides basic science research by identifying critical knowledge gaps and unmet needs.

Conversely, basic scientists bring expertise in fundamental biological processes, pathogen mechanisms, and cutting-edge technologies. All these can lead to the development of novel diagnostic tools, therapeutic targets, and preventive strategies. These are achieved by establishing synergistic partnerships. Clinical observations can inspire fundamental research questions, and basic science discoveries can be translated into tangible clinical applications. These, in turn, will accelerate the development of innovative solutions to combat the evolving threat of bacterial zoonoses. This iterative process of knowledge exchange and collaborative investigation is essential for advancing the understanding of these complex diseases and ultimately improving public health outcomes.

Case Study

One village has historically experienced seasonal spikes in leptospirosis cases, typically linked to the monsoon season and flooding in low-lying agricultural areas. However, local health officials have observed a concerning trend over the past decade. The traditional monsoon-related peak is becoming more intense and prolonged, and, alarmingly, cases are now being reported with increasing frequency outside the typical monsoon months, even in previously unaffected coastal fishing communities.

In the past year, several clusters of leptospirosis have emerged in coastal villages following unusually heavy pre-monsoon rainfall and localised flooding. These communities rely heavily on fishing and closely interact with brackish water environments. Public health workers investigating these clusters noted several factors: increased rodent populations in and around homes due to disrupted drainage systems after the floods, domestic animals (cattle and dogs) frequently roaming in flooded areas, and a lack of awareness among the fishing community regarding leptospirosis transmission beyond the traditional monsoon context.

One case involves a 45-year-old fisherman from a coastal village. He presented to the local primary health centre in early February with a sudden onset of high fever, severe muscle pain (especially in his calves), headache, and jaundice. He reported no recent travel outside his village. However, he mentioned frequent exposure to the backwaters and occasional wading through stagnant water near his home after the recent heavy rains. The initial diagnosis was suspected to be viral fever. However, when his condition worsened and liver function tests showed abnormalities, leptospirosis was suspected. A PCR test confirmed the presence of *Leptospira* spp. in his blood. Further investigation revealed that several other individuals in his neighbourhood presented with similar symptoms around the same time.

1. Based on the information provided, how are geoclimatic changes likely contributing to the increased incidence and geographical spread of leptospirosis in coastal areas?
2. Considering the One Health approach, what specific interdisciplinary collaborations and actions could be implemented in the patient's village and similar coastal communities in adjoining areas to address rising leptospirosis cases?
3. Beyond initial clinical suspicion, what diagnostic strategies would be most crucial for the timely and accurate identification of leptospirosis cases in resource-limited settings such as primary health centres in coastal areas?
4. What public health interventions and community engagement strategies would most effectively prevent future leptospirosis outbreaks in these newly affected coastal communities, considering their specific environmental exposures and livelihoods?

Research Questions

1. How do specific geoclimatic changes, such as increased temperatures and altered precipitation patterns, directly influence bacterial zoonotic diseases' emergence and transmission dynamics in human populations?
2. What role do habitat destruction and agricultural intensification play in facilitating closer interactions between wildlife and humans, thereby increasing the risk of zoonotic disease spillover?

3. How can public health strategies be adapted to mitigate the impacts of climate-induced shifts in the epidemiology of bacterial zoonoses, particularly in emerging hotspots?
4. How do extreme weather events, such as floods and droughts, correlate with increased incidence rates of bacterial zoonotic diseases in various ecosystems?
5. What are the long-term implications of rising sea levels and changing coastal ecosystems on the prevalence of zoonotic diseases, and how can communities best prepare for these challenges?

Further Readings

Cantas L, Suer K. The important bacterial zoonoses in "one health" concept. Front Public Health. 2014;2:144.

Carlson CJ, Bevins SN, Schmid BV. Plague risk in the western United States over seven decades of environmental change. Glob Change Biol. 2022;28(3):753–69.

Douchet L, Menkes C, Herbreteau V, Larrieu J, Bador M, Goarant C, Mangeas M. Climate-driven models of leptospirosis dynamics in tropical islands from three oceanic basins. PLoS Negl Trop Dis. 2024;18(4):e0011717.

Dumic I, Severnini E. "Ticking bomb": the impact of climate change on the incidence of Lyme disease. Can J Infect Dis Med Microbiol. 2018;2018(1):5719081.

García-Moreno J. Zoonoses in a changing world. Bioscience. 2023;73(10):711–20.

Grilo ML, Pereira A, Sousa-Santos C, Robalo JI, Oliveira M. Climatic alterations influence bacterial growth, biofilm production and antimicrobial resistance profiles in *Aeromonas* spp. Antibiotics (Basel). 2021;10(8):1008.

Parija SC, Bobhate PS. Mitigating the adverse impact of climate change: an Indian perspective. SBV J Basic Clin Appl Health Sci. 2022;5(4):87.

Zambrano MM. Interplay between antimicrobial resistance and global environmental change. Annu Rev Genet. 2023;57(1):275–96.

Economic Impact of Bacterial Zoonotic Infections

53

Sefinew Alemu Mekonnen and Yitayew Demessie

Learning Objectives

1. To identify direct and indirect costs associated with bacterial zoonotic infections
2. To know the health burden of bacterial zoonotic infections in humans
3. To understand the economic burden of bacterial zoonotic infections on various sectors such as livestock, animal health, tourism, and agriculture
4. To realise possible economic returns that could be obtained if zoonotic bacterial infections are controlled

Introduction

Zoonotic diseases spread naturally from vertebrate animals to humans. The persistent and unavoidable interaction between humans and animals poses a continual threat of zoonotic diseases to the health and survival of humans, livestock, pets, and wildlife. There are numerous pathogens affecting humans; a significant proportion of these pathogens are zoonotic. Out of the 1400 species of pathogenic organisms affecting humans identified by Cleveland and colleagues, 868 (61%) were classified as zoonotic.

Classification of Zoonotic Diseases

Based on their epidemiological characteristics and impact, zoonotic diseases are classified into three groups:

Endemic Zoonoses

Endemic zoonoses are widespread illnesses that affect both humans and animals. They are responsible for most human diseases (an estimated 99.9%) and life loss (an estimated 96%), besides causing a marked reduction in livestock productivity. Examples of bacterial endemic zoonoses include (i) brucellosis, caused by *Brucella* spp., which is often acquired from infected livestock or unpasteurised dairy; (ii) bovine tuberculosis, primarily caused by *Mycobacterium bovis* and transmissible via infected cattle or raw milk; (iii) leptospirosis, a widespread disease caused by *Leptospira* spp., linked to contaminated water or soil from animal urine; (iv) salmonellosis, often food-borne but also residing in various animal

S. A. Mekonnen (✉)
Department of Veterinary Epidemiology and Public Health, College of Veterinary Medicine and Animal Sciences, University of Gondar, Gondar, Ethiopia

Y. Demessie
Department of Veterinary Biomedical Sciences, College of Veterinary Medicine and Animal Sciences, University of Gondar, Gondar, Ethiopia

S. C. Parija, A. Poddar (eds.), *Textbook of Bacterial Zoonoses*, Microbial Zoonoses,
https://doi.org/10.1007/978-981-95-6615-0_53

guts; and (v) anthrax, caused by *Bacillus anthracis*, endemic in certain soil regions and primarily affecting herbivores, with human infection through animal contact.

Epidemic Zoonoses

Epidemic zoonoses are unpredictable in their temporal and spatial distribution, typically occurring as outbreaks. Bacterial examples include plague, caused by *Yersinia pestis,* which is historically responsible for devastating epidemics transmitted by fleas from rodents, with sporadic outbreaks still occurring where enzootic cycles are active. Q fever, caused by *Coxiella burnetii*, is frequently linked to exposure to infected livestock, particularly during birthing. Tularaemia, caused by *Francisella tularensis*, is responsible for localised outbreaks, often associated with the handling of infected animals, such as rabbits or rodents, or through tick bites.

Emerging Zoonoses

These are new infections in a population or have existed in the past. However, these spread rapidly in frequency or have increased geographical range. Many occur as outbreaks. Notable bacterial examples include Lyme disease, caused by *Borrelia burgdorferi* and transmitted by ticks, which has seen its geographical range expand due to environmental changes. Certain virulent strains of *Escherichia coli*, such as O157:H7, are also considered emerging concerns due to their association with cattle reservoirs and outbreaks resulting from contaminated food products. Additionally, the emergence of novel or re-emerging *Brucella* species or strains, such as a significant increase in human cases of *Brucella canis* infections from dogs in new regions, exemplifies an emerging zoonotic concern.

Impact of Zoonotic Diseases on Human Populations

Zoonotic diseases pose a dual impact on human populations. Firstly, they cause direct health risk and are most likely to result in indirect consequences on livelihoods and impact food security. Secondly, zoonotic diseases cause multifaceted economic burdens that consist of direct and indirect costs and costs associated with the death of animals. It is estimated that zoonotic infections account for more than 220$ billion in losses to growing economies globally. However, zoonotic diseases have a greater impact on the disadvantaged population in low- and middle-income countries.

There are several zoonotic bacterial infections, including anthrax, salmonellosis, bovine tuberculosis, leptospirosis, listeriosis, and brucellosis, which are among the common diseases affecting human health and the economy. However, unlike zoonotic diseases with pandemic potential, such as avian influenza, bacterial zoonoses often receive comparatively little attention. Furthermore, endemic countries often lack dependable data (both qualitative and quantitative) on their disease burden. The public health importance of bacterial endemic zoonotic diseases is usually overlooked because determining their actual incidence is challenging, and the extent of their health and socio-economic impacts is frequently uncertain.

Direct Costs of Bacterial Zoonosis

Direct costs of bacterial zoonotic infections refer to expenses incurred in response to any zoonotic bacterial disease. Direct costs are usually known as expenditures. These costs include medical expenses for human patients, veterinary interventions, and the costs of laboratory testing and diagnostic procedures. Direct costs entail the cost of all resources spent by the intervention, such as neutralisation owing to side effects, and their current or future implications.

In human health, direct costs are categorised into two main types. The first is direct healthcare expenditures, which include physician services, hospital care, and prescription medicines. The second is direct non-healthcare costs, which are incurred outside the formal healthcare system. These include informal care provided by family members and costs associated with the patient's transportation to a healthcare facility. When calculating direct healthcare costs, it may be difficult to use readily available hospital or physician charges.

In animal health, direct costs encompass losses resulting from ill health or mortality. Unlike in human health, direct costs are classified into visible losses and invisible losses. Visible losses include loss of milk production, loss of draught power, lower weight gains, and the death of animals. Invisible losses include lower fertility and changes in the herd and flock structure. The impact of brucellosis is presented in Fig. 53.1 as an example.

Indirect Costs of Bacterial Zoonosis

The indirect costs of zoonotic bacterial infections include treatment costs, medical expenses, containment efforts, reduced productivity, trade restrictions, and limited tourism. In human health, additional costs in this category include costs related to missing work due to absenteeism or early retirement, reduced efficiency, and nonfinancial burdens, such as lower quality leisure time and premature mortality. In animal health, indirect costs include the opportunity costs to the farmer for seeking diagnosis or treatment for their sick animal, as well as managing the sick animal.

In a broad sense, indirect costs encompass expenses related to disease risk, including those associated with controlling and managing the disease, as well as additional costs incurred to prevent or mitigate the disease's incidence. These costs include preventive and outbreak control expenses, such as vaccination, enforced quaran-

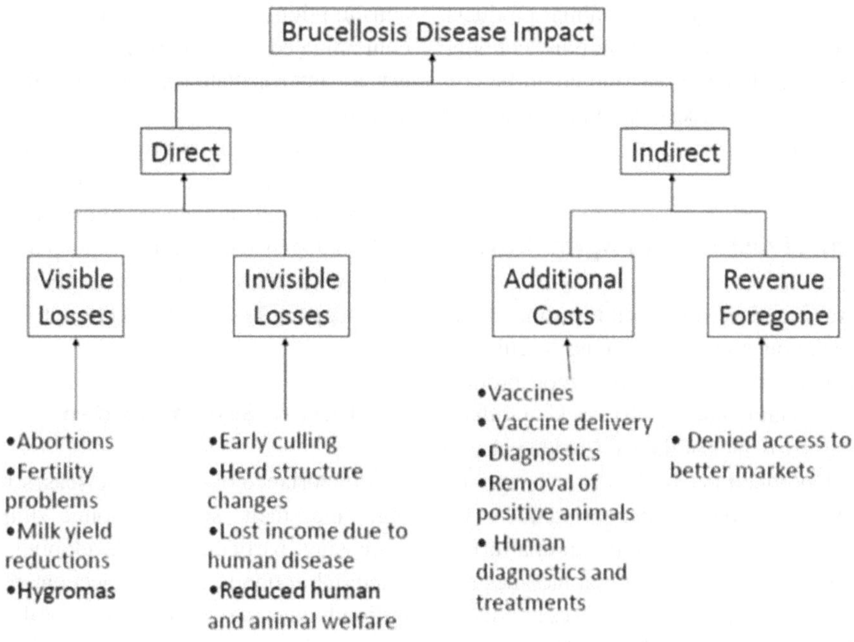

Fig. 53.1 Direct and indirect costs of brucellosis, a zoonotic disease caused by the genus *Brucella*. (Image source: More et al. (2017). Permitted for redistribution under CC BY-ND 4.0 license)

tine, movement restrictions that disrupt production, cleaning and disinfection of infected premises, surveillance, and compensation costs for affected farmers whose animals are culled due to outbreaks.

There are also costs due to trade restrictions that limit the movement of humans, animals, and their products within or outside a region or country. In animals, the presence of a disease restricts the development of commercial farming and limits access to better markets (e.g. foot and mouth disease).

Besides, there are costs related to *revenue foregone*. These are losses resulting from a limit in the adoption of improved breeds due to the presence of a disease. For instance, diseases restrict the use of high-productivity breeds, as they are more susceptible to disease, and the development of more intensive production systems based on these breeds.

There are also intangible costs, both in animals and humans, related to pain and suffering, as well as grief in humans and a drop in employee morale and costs due to customer disappointment with a decline in the quality of animal products. These costs are difficult to measure in economic terms. However, they are real and could be crucial to a business's success or failure in animal production.

Impact on Livestock and Agriculture

Agricultural practices refer to the various methods and techniques used in farming to cultivate crops and rear livestock, ensuring food production and sustainability. Livestock products account for nearly half of the worldwide value of agricultural production. Their share in agriculture is expected to continue increasing due to the sustained growth in demand for animal-origin food products, driven by rising real incomes and population growth. Infectious diseases are one of the few reasons authorised by the World Trade Organization for blocking imports of agricultural products. The best way to mitigate the propagation of bacterial zoonotic diseases is to impose movement control at all levels, which may lead to resource management failure and efficiency loss, which, in turn, will have consequences on agricultural production. For policymakers, two critical questions are paramount: which agricultural subsector, crops or livestock is more adversely affected during epidemics? And how are these subsectors affected by zoonotic disease? The answer to the first question identifies which subsector requires more support, and the answer to the second question determines the mechanism through which support can be provided.

Although the importance of both food animal and crop production in agriculture and their influences on One Health are recognised, attention should be given to livestock and poultry production, particularly on host–pathogen interactions from a One Health perspective. That means bacterial zoonotic diseases have consequences for farm commodities. At the same time, livestock is impacted more severely than crops.

Zoonotic diseases spread through several routes. Humans acquire the disease by direct or indirect contact with animals or their surroundings. Transmission can also occur through the consumption of food of animal origin or water contaminated by animals. In some cases, insects act as intermediate vectors and facilitate the spread.

For a pathogen to cross the species barrier, multiple conditions often need to align. Environmental influences, host susceptibility, and pathogen characteristics all play important roles. Once these factors converge, the pathogen may adapt and establish infection in humans.

Healthcare System Burden

The importance of zoonotic diseases is expanding, and their socio-economic impacts are increasing, particularly in developing countries. Bacterial zoonotic diseases account for about a billion human cases and millions of deaths annually. In developing countries, establishing and implementing effective measures to safeguard livestock and consumer health, particularly against newly emerging threats, have been a significant challenge. Consequently, endemic zoonotic diseases remain a burden on public health systems and challenge efforts to boost livestock production and trade.

Tourism and Travel Industry Effects

Tourism leads to the transformation of forests into hotels, restaurants, entertainment facilities, and residential areas. This is carried out to accommodate the growing population working in the tourism sector, which represents tourism-induced land changes. This impacts biodiversity and increases contact between hosts, their microbes, and humans. This results in the likelihood of zoonoses.

The tourism industry contributes to the emergence and distribution of zoonoses. Several human activities contribute to the spread of zoonotic diseases. These include land use changes driven by tourism, large-scale meat production from intensive animal farming, and the global movement of people, which increases close contact and exposure risks. For instance, small farmers in developing countries struggle to provide quality meat in sufficient quantities and at a competitive price for the increasing number of tourists, as hotels often adhere to standard international menus. This requires an intensive production system that suppresses the immune system, making it more susceptible to infectious diseases.

The intensification of agriculture and changes in environments are linked with a heightened risk of zoonotic disease emergence. This is influenced by the intensifying human migration and the consequences of human activities on the environment. A shortage leads to the movement of animals and the transportation of meat from place to place, which facilitate the emergence and spread of diseases. Tourism activities often bring people into close contact in confined spaces. This occurs during travel on cruise ships, aeroplanes, trains, or buses and in hotels, restaurants, and nightclubs. Such settings increase the chance of disease spread among humans. Tourism also alters the environment and promotes human–animal interactions through zoos and animal-focused attractions. These changes raise the frequency of direct contact between people and animals, which heightens the risk of zoonotic transmission and potential outbreaks.

Travel significantly contributes to the emergence and dissemination of diseases. Over the past 50 years, the number of international travellers has grown by more than 1300%. Moreover, there are over 80 million people, primarily from developing nations, who migrate legally or illegally. The impacts of travel affect not only the travellers themselves but also the populations they visit and the surrounding ecosystems. Travel and tourism facilitate the introduction of diseases to naive populations. In recent decades, increasing worldwide people movement and trade exchange, increasing human and livestock populations, and changing behaviour have been linked with a rise in disease emergence. The movement of people through tourism and immigration can contribute to the mixing of distinct genetic and ecological characteristics and social and cultural influences, which hold significant implications for zoonoses. Tourists who adopt the behavioural patterns prevalent in their destination country may be at an increased risk of infection due to different habits regarding food preparation, clothing, or accommodation arrangements.

During epidemics, governments often limit the movement of people to reduce disease spread. Travel restrictions may include denying visas to individuals from affected regions or completely closing national borders. In addition, governments issue travel advisories warning citizens against visiting countries experiencing outbreaks of infectious diseases.

Cost-Benefit Analysis of Control and Prevention Strategies

Several analytical tools are used to measure the financial effects of diseases, develop methodologies for optimising decisions, and determine the costs and benefits of disease control measures.

Cost-benefit analysis (CBA) is used to assess the costs of interventions against the benefits of these interventions, as well as the costs and benefits of control and prevention strategies. Both costs and benefits are typically expressed in the same units, usually standardised in monetary terms. Cost-benefit analysis allows for the evaluation of all costs and gains across time, encompassing those that extend beyond the intervention period. Its emphasis is on social welfare and policy, along with indirect (productivity) or intangible benefits; however, CBAs should be conducted from a societal perspective. Cost-benefit analysis is used to evaluate regulatory proposals that may involve substantial implementation costs but offer significant potential economic gains for society. By expressing all costs and outcomes in monetary terms, it enables an objective comparison of programmes based on their net benefits, allowing policymakers to allocate limited resources towards initiatives that provide the greatest societal return on investment. If the monetary value of benefits exceeds the cost, implementing the programme is important. The disadvantage of cost-benefit analysis is that assigning economic values to medical outcomes is challenging, and there is no universally accepted method for accomplishing this.

A key consideration before conducting CBA is the selection of the analytical perspective. A societal perspective is commonly recommended, as it accounts for both direct and indirect impacts on social welfare. In a CBA, all costs and benefits are expressed in monetary terms, allowing for objective comparisons between programmes. Benefits may be classified into direct (medical and non-medical cost savings), indirect (productivity gains), and intangible (such as psychological well-being) categories.

A benefit–cost ratio greater than one suggests that the benefits outweigh the costs. Despite its strengths, a notable challenge in CBA is the difficulty of allocating a monetary value to health outcomes, as there is no universally accepted method for doing so.

The monetary value of health is commonly estimated using two methods:

The *human capital approach* assumes that the economic value of an individual is equivalent to their income because the competitive market would not pay workers more than they are worth. Therefore, it is used for estimating the monetary value of health by evaluating the economic productivity lost due to illness, disability, or premature death. Income data can be derived from several sources, including self-reports or any other datasets that can provide income estimates. Work absence period due to illness can be gained through self-report. Determining a daily wage requires evaluating both annual earnings and the total number of workdays, considering weekends, holidays, and sick leave. The costs are estimated by multiplying the number of workdays lost due to illness, disability, or death by the individual's wage rate after discounting future earnings to their present value. There are limitations to this method; the first is that it undervalues individuals who have no formal wages, such as children, the elderly, and unemployed individuals. Another drawback of this method is that it assumes that wages reflect the total social value, thereby excluding non-market contributions, such as caregiving. Moreover, it overlooks individual preferences and quality of life, potentially underestimating the actual societal value of life and health.

The *willingness-to-pay approach* evaluates the value of health by examining the cost required to achieve a specific health outcome. It reflects individual preferences for health and safety improvements. In this method, data will be collected either by directly asking individuals how much they would pay to reduce health risks in surveys (contingent valuation) or by inferring from behaviour, e.g. wage premiums in risky jobs or the purchase of safety products. Unlike the human capital approach, this method reflects individual preferences and risk aversion, and it indirectly includes intangible aspects of health (e.g. pain and fear). However, it is subject to hypothetical bias (what people say may not match their real behaviour) and may vary with income levels, which may introduce inequality in valuation.

When direct monetary valuation is challenging, alternative approaches like cost-effectiveness analysis (CEA) and cost–utility analysis (CUA) are used. CEA compares interventions based on their cost per unit of health outcome (e.g. life years gained and cases prevented), without converting outcomes into monetary terms. It is particularly useful when evaluating interventions with a single measurable goal. However, it fails to assess the broader value of programmes or compare programmes with diverse outcomes. CUA extends CEA by incorporating both the length and quality of life into a single measure, typically quality-adjusted life years (QALYs), which facilitates comparisons across interventions with varied health effects.

In animal health, economic assessments are complicated by the diversity of species and the roles animals play in society. For livestock, many impacts are directly monetisable, such as reduced productivity, treatment costs, and premature death. Intangible costs, such as reputational loss, although difficult to quantify, are also acknowledged. The costs of control programmes include labour, veterinary services, supplies, and indirect costs, such as increased feed to support production.

Socio-economic Disparities and Vulnerability

Socio-economic disparities play a pivotal role in the distribution and impact of bacterial zoonotic infections. Vulnerable populations, including economically disadvantaged households, rural residents, and marginalised groups, are disproportionately affected. These populations may have limited access to healthcare, face barriers to diagnosis and treatment, and experience poor living conditions. The latter includes overcrowding, a lack of sanitation, and limited access to resources and preventive measures, which make them more susceptible to infection.

The socio-economic disparities in the impact of zoonotic diseases are further exacerbated by the fact that many of the affected populations rely on agriculture or animal husbandry for their livelihood. In this context, developing countries whose economies are largely dependent on agriculture are at a significant risk. In developing countries, individuals such as livestock owners/keepers, veterinarians, laboratory personnel, consumers, children, and women are among the most susceptible populations to zoonotic bacterial infections. Reducing these inequalities calls for targeted interventions that consider the social determinants of health and ensure fair access to essential resources.

Policy Implications and Economic Mitigation

The public health importance of bacterial endemic zoonotic diseases is often overlooked because it is difficult to determine their true incidence, and the severity of their health and socio-economic impacts often remains uncertain. Clinicians and policymakers frequently have limited awareness of these diseases in humans. They can be easily mistaken for more common causes of illnesses, such as febrile conditions, respiratory infections, or diarrheal diseases. For many endemic zoonotic diseases, clinical diagnosis is often restricted to referral hospitals and specialised reference laboratories.

Endemic specialised zoonotic diseases are considered "neglected" because they are inadequately tackled at either a national or an international level. These diseases are often neglected by health facility managers and policymakers in both the developed and developing worlds. However, as they affect the world's poorest population, control of neglected zoonotic diseases (NZDs) is done by protecting human health, saving lives, and securing livelihoods, offering an effective and affordable strategy for alleviating poverty, especially in remote rural settings and disadvantaged peri-urban communities.

Health investment and policies related to health require efficient allocation of available resources, best recognising that using labour, materials, and other resources for one purpose means that they cannot be used for alternative uses. Economic evaluation approaches, such as cost-effectiveness and benefit–cost analyses, are unified by the primary goal of assessing policy impacts to build a robust evidence base that supports informed decision-making. Benefit–cost analysis is distinct in its explicit monetisation of both health and non-health outcomes, enabling direct comparison. While it traditionally overlooks distributional effects within populations, it can be adapted to address them.

Based on the interconnections they have, the framework offers concrete managerial action recommendations for tourism businesses and policymakers to help prevent future disease outbreaks.

Animal diseases, whether in commercial production or subsistence settings, can negatively affect consumers, producers, and the broader economy. The rising global demand for animal-derived food underscores the need for safe and efficient production systems. Understanding how animal diseases impact and move through the value chain is crucial. This knowledge informs policies that support safe consumption, well-regulated markets, and more effective avenues for investment.

Intangible costs should be presented to the decision-maker, along with suitable descriptive information, allowing them to be weighed against the measured variables in the decision-making process. It may be possible to measure some costs and benefits in physical units but not to express them in monetary terms.

Research Gaps and Future Directions

Zoonotic diseases originate from animals, including both domestic species and wildlife. Neglected and endemic zoonoses, which continuously circulate between livestock and humans, have a significant impact on public health and the livelihoods of affected communities. The spread of zoonotic infections relies on intricate relationships among susceptibility, temporal patterns, and human activities, which are affected by a range of ecological, political, and socio-economic factors. Particularly, poverty and low socio-economic status are key factors in people's vulnerability to disease. These individuals have their livelihoods jeopardised by conflict or disasters and, hence, are seen to be at an elevated risk. However, the dynamics of zoonotic diseases and associated risks in displacement are poorly understood.

Although having animals may make zoonotic pathogens more prevalent, keeping livestock may enhance livelihoods, nutrition, and mental health while also potentially lowering a person's susceptibility to illness. Notwithstanding the existence of these disease-causing factors during forced migration, there is scant proof of widespread zoonotic disease outbreaks that are directly connected to displaced livestock. Future emergence or re-emergence rates of zoonotic diseases will be directly correlated with how the agriculture–environment relationship develops. However, the complexity and interdependence of the biological, social, economic, and environmental aspects of zoonotic pathogen emergence are not sufficiently addressed by the research that is currently available. All these severely impair our capacity to predict, prevent, and react to the emergence of zoonotic diseases.

There is a knowledge gap regarding the global burden of animal diseases and how these burdens are distributed along value chains. Additionally, there is a need to create an analytical framework grounded in established methods. Such a framework would guide future assessments of animal disease burden, offering clearer insights into the impacts on animal health.

One Health analyses examine the diverse economic and ecological determinants of zoonotic diseases. This frequently relates to risk maps that direct intervention, where geographical "hotspots" are noted. The technical "anti-politics" of One Health is created by such representations, which rarely consider the underlying political, social, and cultural elements that may have an impact on the emergence of zoonotic diseases. The One Health perspective acknowledges how various factors overlap and, therefore, advocates for close collaboration among veterinarians, doctors, and social scientists. Therefore, studies on zoonotic diseases should be addressed within the One Health approach.

Case Study

Anthrax imposes a substantial economic and health burden on both humans and animals, particularly in sub-Saharan Africa and South-east Asia. The World Health Organization (WHO) estimates that between 2000 and 20,000 human anthrax cases occur globally each year, with a higher concentration reported in Africa. Human cases often result from contact with infected animals or contaminated animal products. The economic impact in Europe is minor compared to endemic regions, largely due to better veterinary control systems, routine livestock vaccination, and effective surveillance mechanisms. However, rare outbreaks, such as the 2016 incident in Siberia, which led to one human death and over 2000 reindeer deaths, highlight the risk of re-emergence under changing environmental conditions. In comparison, the annual global economic burden of anthrax in animals is estimated to exceed USD 50 million, with sub-Saharan Africa and South-east Asia bearing the brunt.

1. What specific interventions or programmes could be most effective in mitigating economic losses and supporting the livelihoods of communities?
2. What public health communication strategies and community-level interventions would be most effective in reducing human exposure and infection rates in high-burden areas of South-east Asia?
3. What specific environmental factors or changes might contribute to the re-emergence or increased incidence of anthrax in currently low-burden regions?

Research Questions

1. How can the economic burden of bacterial zoonotic infections be more accurately measured in low-income countries and poor-resource settings?
2. What are the most resource-efficient strategies for preventing and controlling bacterial zoonotic diseases in rural populations?
3. How can One Health approaches be optimised to reduce the economic burden of zoonotic infections?

Declaration During the preparation of this chapter, we have used *ChatGPT* to improve the grammar and/ or readability of the text. After using it, we have reviewed and edited the content as needed, and we take full responsibility for the content of the publication.

Further Readings

Beyer W, Turnbull PC. Anthrax in animals. In: Anthrax. In: Springer; 2009. p. 61–96. https://doi.org/10.1007/978-0-387-09643-3_3.

Braam HD, Jephcott LF, Wood NLJ. Identifying the research gap of zoonotic disease in displacement: a systematic review. Glob Health Res Policy. 2021;6:25. https://doi.org/10.1186/s41256-021-00205-3.

Esposito MM, Turku S, Lehrfield L, Shoman A. The impact of human activities on zoonotic infection transmissions. Animals. 2023;13:1646. https://doi.org/10.3390/ani13101646.

Halliday JE, Allan KJ, Ekwem D, Cleaveland S, Kazwala RR, Crump JA. Endemic zoonoses in the tropics: a public health problem hiding in plain sight. Vet Rec. 2015;176(9):220–5.

Hotez PJ, Fenwick A, Savioli L, Molyneux DH. Rescuing the bottom billion through control of neglected tropical diseases. Lancet. 2009;373:1570–5.

Leach M, Scoones I. The social and political lives of zoonotic disease models: narratives, science, and policy. Soc Sci Med. 2013;88:10–7.

Lloyd-Smith JO, George D, Pepin KM, Pitzer VE, Pulliam JR, Dobson AP, et al. Epidemic dynamics at the human-animal interface. Science. 2009;326(5958):1362–7.

Molyneux D, Hallaj Z, Keusch GT, McManus DP, Ngowi H, Cleaveland S, Ramos-Jimenez P, Gotuzzo E, Kar K, Sanchez A, Garba A, Carabin H, Bassili A, Chaignat CL, Meslin F, Abushama HM, Willingham AL, Kioy D. Zoonoses and marginalised infectious diseases of poverty: where do we stand? Parasit Vectors. 2011;4:106.

Morand, et al. Domesticated animals and human infectious diseases of zoonotic origins: domestication time matters. Infect Genet Evol. 2014;24:76–81.

More SJ, Bøtner A, Butterworth A, Calistri P. Assessment of listing and categorisation of animal diseases within the framework of the Animal Health Law (Regulation (EU) No 2016/429): infection with *Brucella abortus, B. melitensis* and *B. suis*. EFSA J. 2017;15(7):4889. https://doi.org/10.2903/j.efsa.2017.4889.

Narrod, et al. A one health framework for estimating the economic costs of zoonotic diseases on society. EcoHealth. 2012;9:150–62.

Owczarczak-Garstecka S. Understanding risk in human–animal interactions. Forced Migr Rev. 2018;58:78–80.

Rahman MT, et al. Zoonotic diseases: etiology, impact, and control. Microorganisms. 2020;8:1405.

Scoones I, Jones K, Lo Iacono G, Redding DW, Wilkinson A, Wood JLN. Integrative modelling for one health: pattern, process, and participation. Philos Trans R Soc Lond Ser B Biol Sci. 2017;372(1725):1–11. https://doi.org/10.1098/rstb2016.0164.

Weld ED, Waitt C, Barnes K, Garcia BF. Twice neglected? Neglected diseases in neglected populations. Br J Clin Pharmacol. 2022;88(2):367–73. https://doi.org/10.1111/bcp.15148.

Zhang T, Nickerson R, Zhang W, Peng X, Shang Y, Zhou Y. The impacts of animal agriculture on one health—bacterial zoonosis, antimicrobial resistance, and beyond. One Health. 2024;18:100748. https://doi.org/10.1016/j.onehlt.2024.100748.

Index

S. C. Parija, A. Poddar (eds.), *Textbook of Bacterial Zoonoses*, Microbial Zoonoses,
https://doi.org/10.1007/978-981-95-6615-0

MIX
Papier aus verantwortungsvollen Quellen
Paper from responsible sources
FSC® C105338

If you have any concerns about our products,
you can contact us on
ProductSafety@springernature.com

In case Publisher is established outside the EU,
the EU authorized representative is:
Springer Nature Customer Service Center GmbH
Europaplatz 3, 69115 Heidelberg, Germany

Printed by Libri Plureos GmbH
in Hamburg, Germany